Nelson MindTap + **You** = Learning amplified

"I love that everything is interconnected, relevant and that there is a clear learning sequence. I have the tools to create a learning experience that meets the needs of all my students and can easily see how they're progressing."

— **Sarah,** Secondary School Teacher

nelson maths. 9

Rachel Theunissen
Rashmi Bhagwati
Klaas Bootsma
David Badger
Sarah Hamper
Series editor
Robert Yen

LEARNING DISCOVERY

RAINBOW COLOURS

When white light enters a transparent triangular prism, it splits into 7 colours: red, orange, yellow, green, blue, indigo and violet. That is because light is a wave and each colour has its own wavelength that it is refracted (bent) at a different angle when leaving the prism. Red has the longest wavelength and smallest angle of refraction. Rainbows are made this way when light is refracted in raindrops.

Australian Curriculum

Nelson Maths 9 for the Australian Curriculum WA
1st Edition
Rachel Theunissen
Rashmi Bhagwati
Judy Binns
Gaspare Carrozza
ISBN 9780170465618

Publisher: Robert Yen
Project editor: Alan Stewart
Series cover design: Leigh Ashforth (Watershed Art & Design) and Cengage Creative Studio
Series text design: Leigh Ashforth (Watershed Art & Design)
Series designer: Danielle Maccarone
Cover image: iStock.com/Andrei Naumenka
Permissions researcher: Catherine Kerstjens
Production controller: Karen Young
Proofreader / Answer checker: MPS Limited
Typesetter: KGL

For product information and technology assistance,
in Australia call **1300 790 853;**
in New Zealand call **0800 449 725**

For permission to use material from this text or product, please email **aust.permissions@cengage.com**

ISBN 978 0 17 046561 8

Cengage Learning Australia
Level 7, 80 Dorcas Street
South Melbourne, Victoria Australia 3205

Cengage Learning New Zealand
Unit 4B Rosedale Office Park
331 Rosedale Road, Albany, North Shore 0632, NZ

For learning solutions, visit **cengage.com.au**

Printed in China by 1010 Printing International Limited.
1 2 3 4 5 6 7 26 25 24 23 22

ACKNOWLEDGEMENT OF COUNTRY

Nelson acknowledges the Traditional Owners and Custodians of the lands of all First Nations Peoples of Australia. We pay respect to their elders past, present and emerging.

We recognise the continuing connection of First Nations Peoples to the land and waters, and thank them for protecting these land, waters and ecosystems since time immemorial.

Warning – First Nations Australians are advised that this book and associated learning materials may contain images, videos or voices of deceased persons.

Preface

Nelson Maths 7–10 has been designed for the 2020s classroom, focussing on core skills with explicit grading of exercise questions, 'flipped classroom' video tutorials, online interactivity, more applications and problem-solving questions, and worked solutions to every question.

This new series is carefully mapped to the Australian Curriculum and built on solid pedagogical foundations that integrate into every chapter practical classroom activities, engaging investigations, problem-solving, reasoning, communicating, reflecting, summarising, extension, revision, mental calculation, technology, numeracy and literacy.

This book, *Nelson Maths 9,* has been designed for Year 9 students, but includes Years 7–8 revision and Year 9 extension work. There is also available the *Nelson Maths 9 Advanced* book that covers the course at a faster pace, with less revision and more Years 9 and 10 content.

The *Nelson MindTap* online learning platform contains print and multimedia content: worksheets, video tutorials, topic tests, worked solutions and much more. We have provided an abundance of resources for teachers to plan and teach for a variety of pathways. *Nelson Maths* is clear, concise, fresh and smart. We have designed this series to be user-friendly and uncomplicated so that teachers and students everywhere can pick it up and use straight away. So let's get started.

About the authors

Rachel Theunissen is the Head of Mathematics at Lesmurdie Senior High School in Perth, where she also leads the Literacy and Numeracy Team. She is a Curriculum Support Teacher with the Department of Education and a volunteer of MAWA and the Australian Mathematics Trust (AMT). She has been teaching for more than 25 years and presents regularly at conferences on mathematics and problem solving.

Rashmi Bhagwati teaches at a high school in southeast Brisbane and has been a mathematics educator for over 20 years. She has been involved in external exam marking and assessment processes, and was a panellist for Mathematics A (senior General Mathematics). Rashmi wrote topic tests for the Nelson Senior Maths Essential Mathematics series for Queensland and WA.

Klaas Bootsma was head of mathematics at Ambarvale High School in Sydney and has been a mathematics author for over 25 years.

David Badger was principal of Toongabbie Christian School in Sydney and currently teaches at Wollondilly Anglican College.

Sarah Hamper is head of mathematics at Cheltenham Girls High School in Sydney. She has an interest in gifted and talented students, technology and girls education.

Series editor **Robert Yen** taught at Hurlstone Agricultural High School in Sydney and works for Cengage as the mathematics publisher.

Contributing authors

Kahlia Dreyer, Lachlan Grierson, Brook Johns, Tracey Johnson, Megan Boltze, Robert Yen and Deborah Smith wrote many of the worksheets and topic tests.

Rashmi Bhagwati, Jade Lori, Deborah Da Cruz, Monique Ellement, John Drake, Katie Jackson, Joanne Magner, Scott Smith and **Robert Yen** created the video tutorials.

With thanks to **Dr Caty Morris** on behalf of ATSIMA (Aboriginal and Torres Strait Islander Mathematics Alliance) for her content review and advice.

Table of contents

Curriculum grids

Australian Curriculum

Strand and substrand	*Nelson Maths 9* chapter		*Nelson Maths 10* chapter	
NUMBER AND ALGEBRA				
NUMBER	2	Pythagoras' theorem	3	Interest and depreciation
	3	Numeracy and calculation		
	8	Earning money		
ALGEBRA	1	Algebra	1	Graphing lines
	5	Indices	3	Algebra
	7	Equations and inequalities	6	Equations and inequalities
	11	Coordinate geometry and graphs	7	Graphing curves
			10	Simultaneous equations
MEASUREMENT	2	Pythagoras' theorem	2	Surface area and volume
	3	Numeracy and calculation	7	Trigonometry
	4	Trigonometry	12	Congruent and similar figures
	5	Indices		
	8	Earning money		
	10	Surface area and volume		
	13	Congruent and similar figures		
SPACE	2	Pythagoras' theorem	9	Networks
	4	Trigonometry	12	Congruent and similar figures
	6	Geometry and networks		
	13	Congruent and similar figures		
STATISTICS	9	Analysing data	5	Comparing data
PROBABILITY	12	Probability	11	Probability

Year 9 content descriptions

Australian Curriculum descriptions (© ACARA 2022).

Content description	*Nelson Maths 9* chapter	
NUMBER (N)		
AC9M9N01: recognise that the real number system includes the rational numbers and the irrational numbers, and solve problems involving real numbers using digital tools	2	Pythagoras' theorem
	3	Numeracy and calculation
ALGEBRA (A)		
AC9M9A01: apply the exponent laws to numerical expressions with integer exponents and extend to variables	5	Indices
AC9M9A02: simplify algebraic expressions, expand binomial products and factorise monic quadratic expressions	1	Algebra
AC9M9A03: find the gradient of a line segment, the midpoint of the line interval and the distance between 2 distinct points on the Cartesian plane	11	Coordinate geometry and graphs
AC9M9A04: identify and graph quadratic functions, solve quadratic equations graphically and numerically, and solve monic quadratic equations with integer roots algebraically, using graphing software and digital tools as appropriate	7	Equations and inequalities
	11	Coordinate geometry and graphs
AC9M9A05: use mathematical modelling to solve applied problems involving change including financial contexts; formulate problems, choosing to use either linear or quadratic functions; interpret solutions in terms of the situation; evaluate the model and report methods and findings	7	Equations and inequalities
	11	Coordinate geometry and graphs
AC9M9A06: experiment with the effects of the variation of parameters on graphs of related functions, using digital tools, making connections between graphical and algebraic representations, and generalising emerging patterns	11	Coordinate geometry and graphs
MEASUREMENT (M)		
AC9M9M01: solve problems involving the volume and surface area of right prisms and cylinders using appropriate units	10	Surface area and volume
AC9M9M02: solve problems involving very small and very large measurements, time scales and intervals expressed in scientific notation	5	Indices
	10	Surface area and volume
AC9M9M03: solve spatial problems, applying angle properties, scale, similarity, Pythagoras' theorem and trigonometry in right-angled triangles	2	Pythagoras' theorem
	4	Trigonometry
	13	Congruent and similar figures
AC9M9M04: calculate and interpret absolute, relative and percentage errors in measurements, recognising that all measurements are estimates	10	Surface area and volume
AC9M9M05: use mathematical modelling to solve practical problems involving direct proportion, rates, ratio and scale, including financial contexts; formulate the problems and interpret solutions in terms of the situation; evaluate the model and report methods and findings	3	Numeracy and calculation
	8	Earning money
	13	Congruent and similar figures

Content description		*Nelson Maths 9* chapter
SPACE (SP)		
AC9M9SP01: recognise the constancy of the sine, cosine and tangent ratios for a given angle in right-angled triangles using properties of similarity	4	Trigonometry
AC9M9SP02: apply the enlargement transformation to shapes and objects using dynamic geometric software as appropriate; identify and explain aspects that remain the same and those that change	13	Congruent and similar figures
AC9M9SP03: design, test and refine algorithms involving a sequence of steps and decisions based on geometric constructions and theorems; discuss and evaluate refinements	6	Geometry and networks
	13	Congruent and similar figures
	2	Pythagoras' theorem
STATISTICS (ST)		
AC9M9ST01: analyse reports of surveys in digital media and elsewhere for information on how data was obtained to estimate population means and medians	9	Analysing data
AC9M9ST02: analyse how different sampling methods can affect the results of surveys and how choice of representation can be used to support a particular point of view	9	Analysing data
AC9M9ST03: represent the distribution of multiple data sets for numerical variables using comparative representations; compare data distributions with consideration of centre, spread and shape, and the effect of outliers on these measures	9	Analysing data
AC9M9ST04: choose appropriate forms of display or visualisation for a given type of data; justify selections and interpret displays for a given context	9	Analysing data
PROBABILITY (P)		
AC9M9P01: list all outcomes for compound events both with and without replacement, using lists, tree diagrams, tables or arrays; assign probabilities to outcomes	12	Probability
AC9M9P02: calculate relative frequencies from given or collected data to estimate probabilities of events involving "and", inclusive "or" and exclusive "or"	12	Probability
AC9M9P03: design and conduct repeated chance experiments and simulations, using digital tools to compare probabilities of simple events to related compound events, and describe results	12	Probability

About this book

Coverage of the Australian Curriculum

- *Nelson Maths 9* covers the Australian curriculum, as shown by the table of contents and curriculum grid on the previous pages.
- This book contains Year 9 content. It also contains revision of some Years 7 and 8 content, and some extension work marked in **red** and by *.
- Each chapter begins with a **chapter outline** that includes the 4 curriculum proficiencies covered in each section.

U = UNDERSTANDING

Understanding is 'knowing and relating' maths. It is more than just learning facts. It's deep understanding, seeing how mathematical content is interconnected, knowing 'why' as well as 'how'.

F = FLUENCY

Fluency is 'applying' maths. It is being able to use mathematics competently and effectively. When you are fluent in a language, you have mastered it so that you can improvise and confidently use the correct word or phrase. Fluency in maths is choosing an appropriate skill, method or formula to use at the right place and time.

PS = PROBLEM SOLVING

Problem-solving is 'modelling and investigating' with maths. It involves interpreting a rich, elaborate problem, selecting an appropriate strategy or model, solving the problem, then evaluating, communicating and justifying the solution.

R = REASONING

Reasoning is 'generalising and proving' with maths, using higher-order thinking to connect specific facts to general principles, using algebra, logic, proof and justification.

To these proficiencies, we have added **C = COMMUNICATING.**

Communicating is 'describing and explaining' maths, representing mathematical theory and solutions in words, algebraic symbols, special notations, diagrams, graphs and tables.

Understanding and **Fluency** can be found in every exercise and activity, while **Problem solving**, **Reasoning** and **Communicating** are found in the **Investigations**, **Technology**, **Mental skills**, **Language of Maths** and **Topic summary** activities, and explicitly labelled in every exercise (see below).

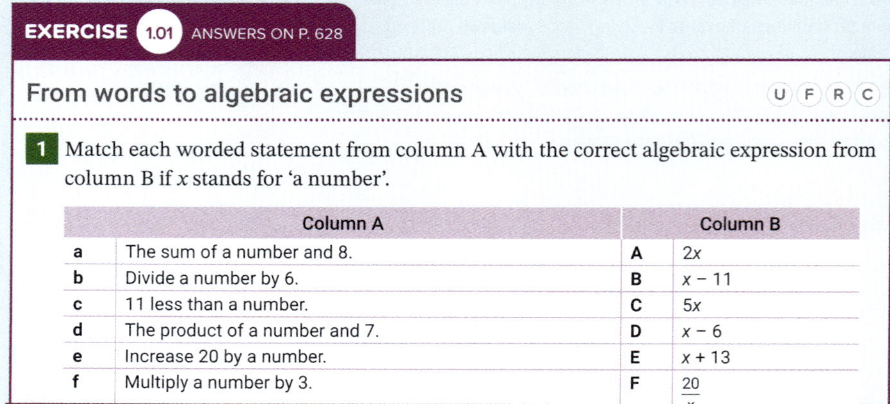

EXERCISE 1.01 ANSWERS ON P. 628

From words to algebraic expressions U F R C

1 Match each worded statement from column A with the correct algebraic expression from column B if x stands for 'a number'.

	Column A		Column B
a	The sum of a number and 8.	A	$2x$
b	Divide a number by 6.	B	$x - 11$
c	11 less than a number.	C	$5x$
d	The product of a number and 7.	D	$x - 6$
e	Increase 20 by a number.	E	$x + 13$
f	Multiply a number by 3.	F	$\frac{20}{x}$

ABOUT THIS BOOK

At the beginning of each chapter

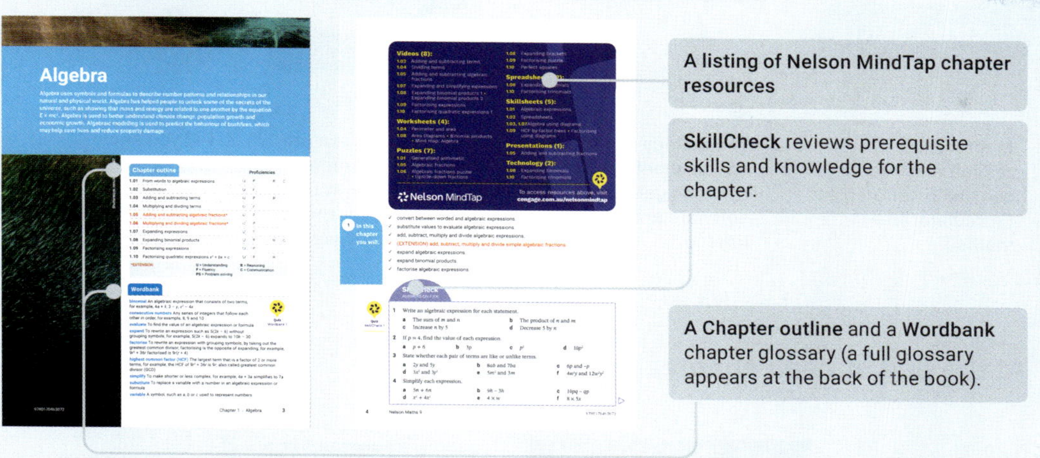

A listing of Nelson MindTap chapter resources

SkillCheck reviews prerequisite skills and knowledge for the chapter.

A **Chapter outline** and a **Wordbank** chapter glossary (a full glossary appears at the back of the book).

In each chapter

Important facts and formulas are highlighted in a shaded box.

Glossary terms are printed in **blue**.

Graded exercises are linked to worked examples and include multiple-choice questions, exam-style problems and realistic applications.

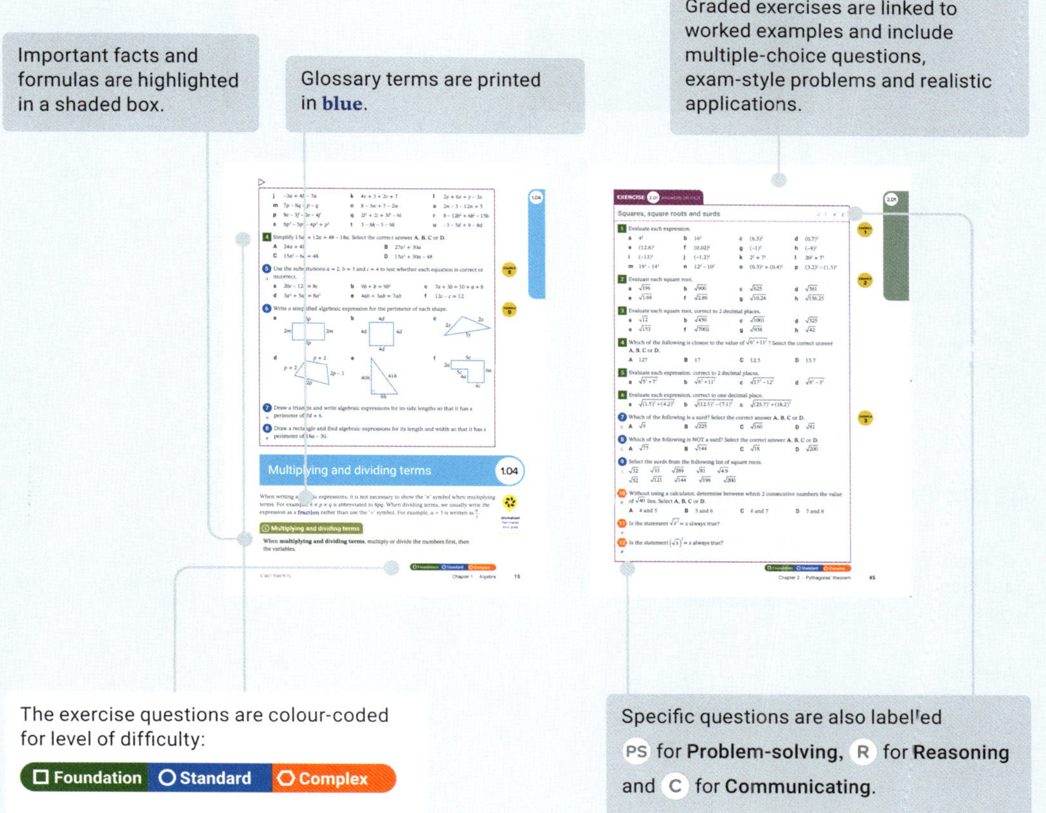

The exercise questions are colour-coded for level of difficulty:

Foundation | Standard | Complex

Specific questions are also labelled **PS** for **Problem-solving**, **R** for **Reasoning** and **C** for **Communicating**.

☆ MENTAL SKILLS

TECHNOLOGY

INVESTIGATION

DID YOU KNOW?

Mental skills reinforce mental calculation strategies ('calculator-free maths').

Technology includes spreadsheets, dynamic geometry software and the Internet.

Investigations explore the syllabus in more detail, through group work, discovery and modelling activities.

Did you know? contains interesting facts and applications of the mathematics learned in the chapter.

At the end of each chapter

Power plus is an extension/challenge exercise.

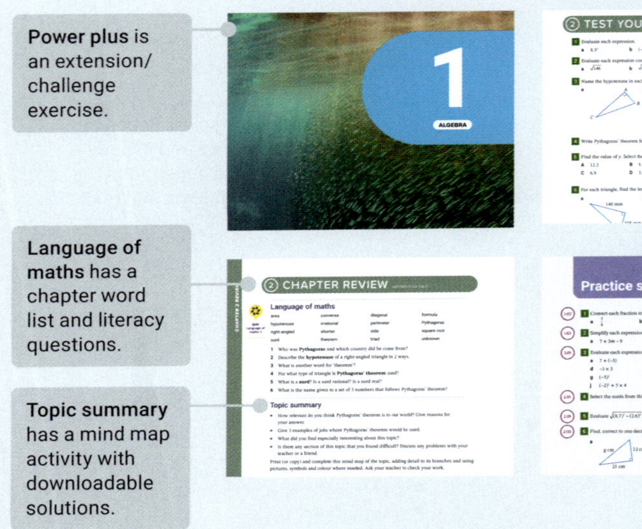

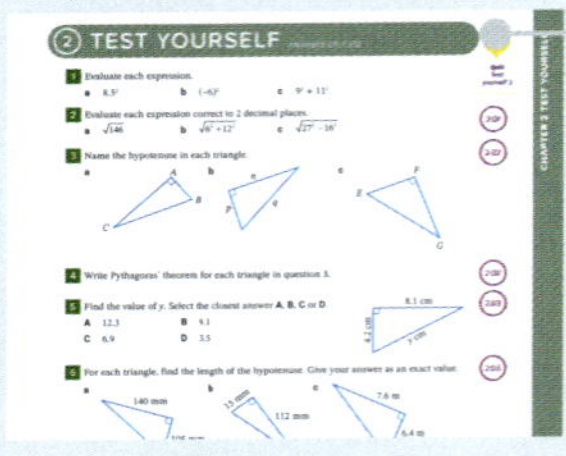

Test Yourself contains chapter revision linked to the relevant exercise set.

Language of maths has a chapter word list and literacy questions.

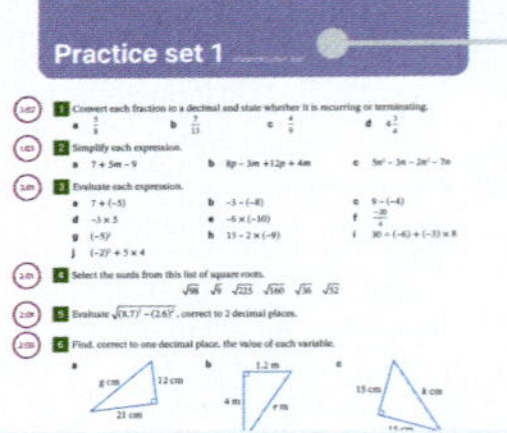

Practice sets after every 3 chapters.

Topic summary has a mind map activity with downloadable solutions.

At the end of the book

General practice exercise

Answers (worked solutions available to teachers.

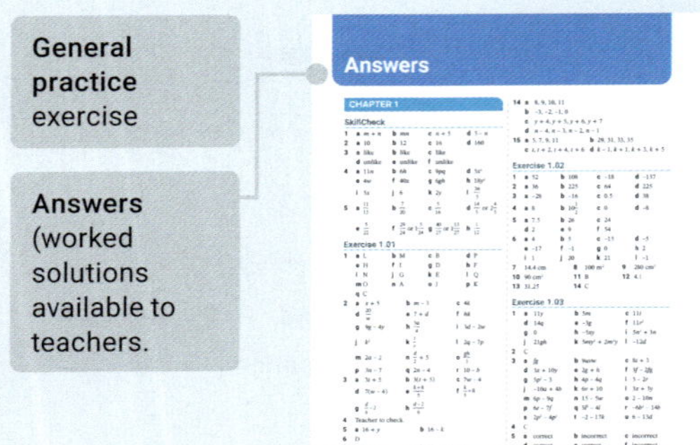

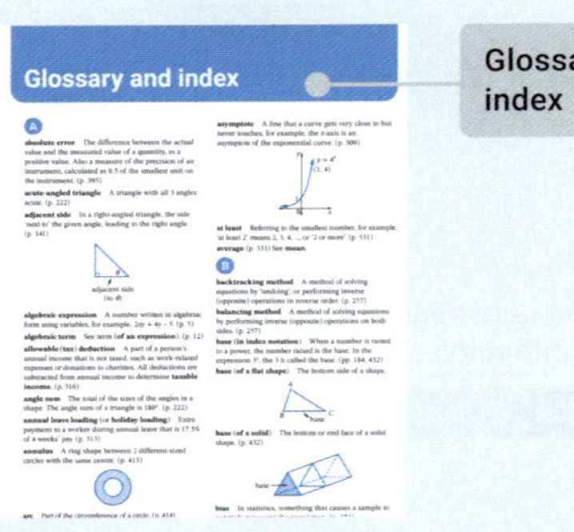

Glossary and index

Nelson MindTap

An online learning space that provides students with tailored learning experiences.

- Access tools and content that make learning simpler yet smarter to help you achieve maths mastery.
- Includes an eText with integrated interactives and online assessment.
- Margin links in the student book signpost multimedia student resources found on MindTap.

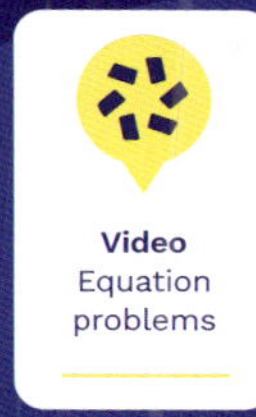

Video
Equation problems

For students:

- **Watch** video tutorials featuring expert teacher advice to unpack new concepts and develop your understanding.
- **Revise** using quizzes, worksheets and skillsheets to practise your skills and build your confidence.
- **Navigate** your own path, accessing the content, analytics and support as you need it.
- ***Twig* mini-documentaries** showing the background or context of mathematics
- ***PhET* interactives**: maths simulations

For teachers*:

- Tailor content to different learning needs – assign directly to the student, or the whole class.
- Monitor progress using assessment tools like Gradebook and Reports.
- Integrate content and assessment directly within your school's LMS for ease of access.
- Access topic tests, teaching plans and worked solutions to each exercise set.

* Complimentary access to these resources is only available to teachers who use this book as part of a class set, book hire or booklist. Contact your Cengage Education Consultant for information about access and conditions

Nelson Maths 7–10 series

Symbols and abbreviations

$=$	is equal to	$\parallel$	is parallel to
$\neq$	is not equal to	$\perp$	is perpendicular to
$\approx$	is approximately equal to	$\therefore$	therefore
$<$	is less than	x^2	x squared, $x \times x$
$>$	is greater than	x^3	x cubed, $x \times x \times x$
$\leq$	is less than or equal to	$\sqrt{\ }$	square root, radical sign
$\geq$	is greater than or equal to	$\sqrt[3]{\ }$	cube root
( )	parentheses, round brackets	$P(E)$	the probability of event E
[ ]	(square) brackets	LHS	left-hand side
{ }	braces	RHS	right-hand side
$\pm$	plus or minus	%	percentage
-3	negative 3	p.a.	per annum (per year)
$0.\dot{1}5\dot{2}$	the recurring decimal 0.152152 ...	$\bar{x}$	the mean (average)
$°$	degree	μ	micro-, mu
$\angle A$	angle A		
$\triangle ABC$	triangle ABC		

Mathematical verbs

A glossary of 'doing words' commonly found in mathematics problems

analyse: study in detail the parts of a situation

bisect: cut in half

calculate: *see* **evaluate**

classify, identify: state the type, category or feature of an item or situation

comment: express an observation or opinion about a result

complete: fill in detail to make a statement, diagram or table correct or finished

compare: show how 2 or more things are similar or different

construct: draw an accurate diagram

convert: change from one form to another, for example, from a fraction to a decimal, or from kilograms to grams

decrease: make smaller

describe: state the features of a situation

estimate: make an educated guess for a number, measurement or solution, to find roughly or approximately

evaluate, calculate: find the value of a numerical expression, for example, 3×8^2 or $4x + 1$ when $x = 5$

expand: remove brackets in an algebraic expression by multiplying, for example, expanding $3(2y + 1)$ gives $6y + 3$

explain: describe why or how

factorise: take out the highest common factor (HCF) of an expression and insert brackets, for example, factorising $5x - 20$ gives $5(x - 4)$. The opposite of **expand**.

give reasons: show the rules or thinking used when solving a problem. See also **justify**.

graph: display on a number line, number plane or statistical graph

hence find/prove: find an answer or prove a result using previous answers or information supplied

identify: *see* **classify**

increase: make larger

interpret: find meaning in an answer or result

justify: give reasons or evidence to support your argument or conclusion. See also **give reasons**

measure: use an instrument to find the size of something, for example, use a ruler to determine the length of a pen

prove that: *see* **show that**

recall: remember and state

reduce (a fraction) to its lowest terms: see **simplify (a fraction)**

round (a number): find the nearest approximation of a number. For example, 4.3 rounded to the nearest whole number is 4, \$12.9598 rounded to the nearest cent is \$12.96, 0.166 66 rounded to 3 decimal places is 0.167

show that, prove that: (in questions where the answer is given) use calculation, procedure or reasoning to prove that an answer or result is true

show working: show the steps you used to find the answer

simplify: give a result in its most basic, shortest, neatest form, for example, simplifying a ratio or algebraic expression

simplify (a fraction): reduce the numerator and denominator of a fraction by dividing by their highest common factor (HCF), for example, $\frac{16}{20}$ simplified is $\frac{4}{5}$

simplify (a ratio or rate): reduce the terms or units of a ratio or rate by dividing by their highest common factor (HCF), for example, 10 : 4 simplified is 5 : 2

sketch: draw a rough diagram that shows the general shape or idea, less accurate than **construct**

solve: find the value(s) of an unknown variable in an equation or inequality

state: *see* **write**.

substitute: replace a variable by a number and evaluate

verify: check that a solution or result is correct, usually by substituting back into the equation or referring back to the problem

write correct to: See **round (a number)**

write, state: give the answer, formula or result without showing any working or explanation (this usually means that the answer can be found mentally, or in one step)

1

ALGEBRA

Algebra

Algebra uses symbols and formulas to describe number patterns and relationships in our natural and physical world. Algebra has helped people to unlock some of the secrets of the universe, such as showing that mass and energy are related to one another by the equation $E = mc^2$. Algebra is used to better understand climate change, population growth and economic growth. Algebraic modelling is used to predict the behaviour of bushfires, which may help save lives and reduce property damage.

Chapter outline

		Proficiencies
1.01	From words to algebraic expressions	U F R C
1.02	Substitution	U F
1.03	Adding and subtracting terms	U F R
1.04	Multiplying and dividing terms	U F
1.05	Expanding expressions	U F
1.06	Expanding binomial products	U F R C
1.07	Factorising expressions	U F
1.08	Factorising quadratic expressions $x^2 + bx + c$	U F R

U = Understanding
F = Fluency
PS = Problem solving
R = Reasoning
C = Communication

Wordbank

binomial An algebraic expression that consists of 2 terms, for example, $4a + 9$, $3 - y$, $x^2 - 4x$

consecutive numbers Any series of integers that follow each other in order, for example, 8, 9 and 10

evaluate To find the value of an algebraic expression or formula

expand To rewrite an expression such as $5(2k - 6)$ without grouping symbols; for example, $5(2k - 6)$ expands to $10k - 30$

factorise To rewrite an expression with grouping symbols, by taking out the greatest common divisor; factorising is the opposite of expanding; for example, $9r^2 + 36r$ factorised is $9r(r + 4)$

highest common factor (HCF) The largest term that is a factor of 2 or more terms, for example, the HCF of $9r^2 + 36r$ is $9r$; also called greatest common divisor (GCD)

simplify To make shorter or less complex, for example, $4a + 3a$ simplifies to $7a$

substitute To replace a variable with a number in an algebraic expression or formula

variable A symbol, such as a, b or c used to represent numbers

Quiz
Wordbank 1

Videos (8):

- **1.03** Adding and subtracting terms
- **1.04** Multiplying terms • Dividing terms
- **1.05** Expanding and simplifying expressions
- **1.06** Expanding binomial products 1 • Expanding binomial products 2
- **1.07** Factorising expressions
- **1.08** Factorising quadratic expressions 1

Twig video (1):

- **1.02** Heptathlon

PhET interactives (2):

- **1.03** Expression Exchange
- **1.07** Area model Algebra

Quizzes (6):

- Wordbank 1
- SkillCheck 1
- Mental skills 1A
- Mental skills 1B
- Language of maths 1
- Test yourself 1

Skillsheets (5):

- **1.01** Algebraic expressions
- **1.02** Spreadsheets
- **1.03, 1.05** Algebra using diagrams
- **1.07** HCF by factor trees • Factorising using diagrams

Worksheets (4):

- **1.04** Perimeter and area
- **1.06** Area diagrams • Binomial products

Mind map: Algebra

Puzzles (4):

- **1.01** Generalised arithmetic
- **1.06** Expanding brackets
- **1.07** Factorising puzzle
- **1.08** Perfect squares

Technology (2):

- **1.06** Expanding binomials
- **1.08** Factorising trinomials

Spreadsheets (2):

- **1.06** Expanding binomials
- **1.08** Factorising trinomials

Nelson MindTap

To access resources above, visit **cengage.com.au/nelsonmindtap**

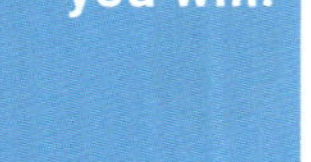

1 In this chapter you will:

- ✓ convert between worded and algebraic expressions.
- ✓ substitute values to evaluate algebraic expressions.
- ✓ add, subtract, multiply and divide algebraic expressions.
- ✓ expand algebraic expressions.
- ✓ expand binomial products.
- ✓ factorise algebraic expressions, including simple (monic) quadratic expressions of the form $x^2 + bx + c$.

SkillCheck

ANSWERS ON P. 628

Quiz SkillCheck 1

1 Write an algebraic expression for each statement.

a The sum of m and n **b** The product of n and m

c Increase n by 5 **d** Decrease 5 by n

2 If $p = 4$, find the value of each expression.

a $p + 6$ **b** $3p$ **c** p^2 **d** $10p^2$

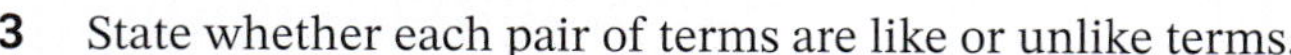

3 State whether each pair of terms are like or unlike terms.

a $2y$ and $5y$ b $8ab$ and $7ba$ c $6p$ and $-p$
d $3x^2$ and $3y^2$ e $5m^2$ and $3m$ f $4w^3y$ and $12w^2y^2$

4 Simplify each expression.

a $5n + 6n$ b $9h - 3h$ c $10pq - qp$ d $x^2 + 4x^2$
e $4 \times w$ f $8 \times 5x$ g $2g \times 3h$ h $3y \times 6y$
i $10x \div 2$ j $6b \div b$ k $\frac{16xy}{8x}$ l $\frac{15n^2}{25n}$

From words to algebraic expressions 1.01

In algebra, letters of the alphabet are used to represent numbers. Such letters are called **variables**. For example, a general way of stating 'any number plus 5' is the **algebraic expression** '$n + 5$' where the variable n stands for 'any number.'

- n is called a variable because its value can change (vary).
- n is also called a **pronumeral** because it stands for a number.

Variables are useful for writing general rules or formulas.

For example, the average of the 2 numbers 7 and 11 is $\frac{7+11}{2} = 9$. In general, we can say that the average of *any* 2 numbers, x and y, is the algebraic expression $\frac{x+y}{2}$.

Puzzle Generalised arithmetic

Skillsheet Algebraic expressions

Example 1

Convert each statement to an algebraic expression.

a The difference between 2 numbers m and n.

b The **product** of 4 and a number h.

SOLUTION

a $m - n$ Difference is the answer to a subtraction.

b $4 \times h = 4h$ Product is the answer to a multiplication.

Example 2

Write an algebraic expression for the result of each statement.

a Think of a number (n), multiply it by 3 and then subtract 5.

b Think of a number (n), subtract 5 and then multiply by 3.

c Think of a number (n), divide by 2 and then add 7.

d Think of a number (n), add 7 and then divide by 2.

SOLUTION

a $n \times 3 - 5 = 3n - 5$

b $(n - 5) \times 3 = 3(n - 5)$

c $n \div 2 + 7 = \frac{n}{2} + 7$

d $(n + 7) \div 2 = \frac{n+7}{2}$

Example 3

a Find the number of hours in:

i 180 minutes **ii** t minutes.

b A mobile phone plan costs \$24 per month plus \$10/GB for excess data use over 8 GB. Find the total cost of the plan for:

i 12 months and 10 GB of data used per month.

ii w months and 15 GB of data used per month.

SOLUTION

a i Number of hours $= 180 \div 60$
$= 3$

ii Number of hours $= t \div 60$
$= \frac{t}{60}$

b i Total cost $= 12 \times (\$24 + 2 \times \$10)$ 10 GB is 2 GB over 8 GB
$= \$528$

ii Total cost $= w \times (\$24 + 7 \times \$10)$ 15 GB is 7 GB over 8 GB
$= \$94w$

EXERCISE 1.01 ANSWERS ON P. 628

From words to algebraic expressions

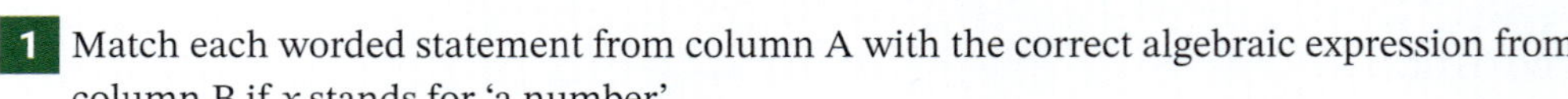

EXAMPLE 1

1 Match each worded statement from column A with the correct algebraic expression from column B if x stands for 'a number'.

	Column A		Column B
a	The sum of a number and 8.	A	$2x$
b	Divide a number by 6.	B	$x - 11$
c	11 less than a number.	C	$5x$
d	The product of a number and 7.	D	$x - 6$
e	Increase 20 by a number.	E	$x + 13$
f	Multiply a number by 3.	F	$\frac{20}{x}$
g	The difference between a number and 6.	G	$20 - x$
h	Divide 20 by a number.	H	$20 + x$
i	Decrease a number by 2.	I	$3x$
j	20 decreased by a number.	J	$30 - x$
k	13 more than a number.	K	$x - 30$
l	A number multiplied by itself.	L	$x + 8$

☐ Foundation ○ Standard ○ Complex

	Column A		Column B
m	Subtract 7 from a number.	**M**	$\frac{x}{6}$
n	Double a number.	**N**	$x - 2$
o	Remove a number from 30.	**O**	$x - 7$
p	30 less than a number.	**P**	$7x$
q	5 lots of a number.	**Q**	x^2

2 Write an algebraic expression for each statement.

C **a** 5 more than x. **b** 3 less than m.

c The product of 4 and k. **d** 20 divided by w.

e Increase 7 by d. **f** The product of h and k.

g The difference between $9g$ and $4y$. **h** $5a$ divided by 4.

i Decrease $3d$ by $2w$. **j** The square of k.

k How many times r goes into t. **l** Subtract $7p$ from $2q$.

m 2 less than double a. **n** 5 more than half d.

o One-third of the product of g and h. **p** 3 times n less 7.

q 4 less than twice n **r** 10 more than negative b.

3 Write an algebraic expression for the result of each statement.

C **a** Think of a number (t), multiply it by 3 and then add 5.

b Think of a number (t), add 5 and then multiply by 3.

c Think of a number (w), multiply it by 7 and then subtract 4.

d Think of a number (w), subtract 4 and then multiply by 7.

e Think of a number (k), add 8 and then divide by 5.

f Think of a number (k), divide by 5 and then add 8.

g Think of a number (d), divide by 6 and then subtract 2.

h Think of a number (d), subtract 2 and then divide by 6.

4 Write each algebraic expression in words.

C **a** $b + 5$ **b** $5b$ **c** $7 - b$ **d** $\frac{b}{3}$

e $2b + 3$ **f** $4 - 3b$ **g** $5+\frac{b}{4}$ **h** $2b^2$

5 If Rimsha is 16 years old, use an algebraic expression to write her age:

a in y years time **b** k years ago.

R C

6 The suggested cooking time for pork in a moderately hot oven is 40 minutes per kilogram of pork plus 15 minutes extra (for the crackling). Which expression gives the cooking time for a leg of pork of mass w kg? Select the correct answer **A**, **B**, **C** or **D**.

A $40(w + 15)$ **B** $\frac{40}{w}+15$ **C** $t + 55$ **D** $40w + 15$

7 The cost of a taxi trip is \$3.80 plus \$2.30/km. Find the total cost of a taxi trip for:

a 12 km **b** D km.

Foundation Standard Complex

8 The cost of an international phone call to parts of Europe is a connection fee of \$0.32 plus \$0.50 per minute. Which expression gives the cost of such a call lasting k minutes? Select **A**, **B**, **C** or **D**.

A $\$(0.32 + 50k)$ **B** $\$(0.32 + 0.5k)$ **C** $\$(32 + 0.5k)$ **D** $\$(0.5 + 32k)$

9 Write an algebraic expression for each statement.

C **a** The cost of p books at \$$Q$ each.

b The selling price after a GST of \$$t$ is added to the marked price of \$40.

c The number of metres in h km.

d The average of m, p, v and w.

e The number of dollars in y cents.

f The discount when a \$950 large screen TV sells for a price of \$$x$.

g The change in dollars from \$100 after buying w tickets at \$12 each.

h The amount of each share when \$450 is divided equally among n people.

i The perimeter of this rectangle.

b cm

a cm

j The area of this square.

s mm

s mm

10 **a** Two angles of a triangle are 130° and 25°. Find the size of the third angle.

b Two angles of a triangle have size $a°$ and $b°$. Find an algebraic expression for the size of the third angle.

11 **a** The unequal angle in an isosceles triangle is 70°. Find the size of one of the equal angles.

R C **b** The unequal angle in an isosceles triangle is $x°$. Find an algebraic expression for the size of one of the equal angles.

12 If d is an even number, what is the next:

a even number? **b** odd number?

R C

13 **Consecutive numbers** are integers that follow one another, such as 3, 4, 5 and 6. What is the next consecutive number after:

a 11? **b** −9? **c** x? **d** $n + 4$?

14 Write 4 consecutive numbers beginning with:

a 8 **b** −3 **c** $y + 4$ **d** $n - 4$

R C

15 Write 4 consecutive *odd* numbers beginning with:

a 5 **b** 29 **c** t **d** $k - 1$

R C

☐ Foundation ○ Standard ⬡ Complex

Substitution

1.02

When the variables in an algebraic expression are replaced with numbers, this is called **substitution**. When the value of the algebraic expression is then calculated, it is called **evaluation**.

E-*valu*-ate means 'find the value of'.

Video
Heptathlon

Example 4

If $f = 5$, $g = -4$ and $r = 2$, evaluate each of the following expressions.

a $2f + g$ **b** $8 + rf$ **c** $f^2 - r^2$

d $3g^2 - 5$ **e** $r(f - g)$ **f** $\frac{5r+f}{3}$

SOLUTION

a $2f + g = 2 \times 5 + (-4)$
$= 6$

b $8 + rf = 8 + 2 \times 5$
$= 18$

c $f^2 - r^2 = 5^2 - 2^2$
$= 21$

d $3g^2 - 5 = 3 \times (-4)^2 - 5$
$= 43$

e $r(f - g) = 2 \times [5 - (-4)]$
$= 18$

f $\frac{5r+f}{3} = \frac{5\times2+5}{3}$
$= 5$

Substituting into formulas

A **formula** is an algebraic rule that shows the relationship between variables. For example, if a rectangle has length l and width w, then:

- the formula $P = 2l + 2w$ gives the perimeter of the rectangle.
- the formula $A = lw$ gives the area of the rectangle.

Note: $3g^2 - 5$ is called an 'expression' but $P = 2l + 2w$ is called an 'equation' or 'formula' because it has an equals sign.

Example 5

A rectangle has length 3.4 m and width 2.5 m.

Use the formula $P = 2l + 2w$ to calculate its perimeter.

SOLUTION

$l = 3.4$, $w = 2.5$

$P = 2l + 2w$

$= 2 \times 3.4 + 2 \times 2.5$

$= 11.8$

The perimeter is 11.8 m.

EXERCISE 1.02 ANSWERS ON P. 628

Substitution

1 Find the value of $7a + 3$ if:

a $a = 7$ b $a = 15$ c $a = -3$ d $a = -20$

2 Find the value of w^2 if:

a $w = 6$ b $w = 15$ c $w = -8$ d $w = -15$

3 Evaluate $8 - 3x$ if:

a $x = 12$ b $x = 8$ c $x = 2.5$ d $x = -10$

4 Evaluate $\frac{h}{2} + 7$ if:

a $h = 2$ b $h = 7$ c $h = -14$ d $h = -30$

5 Evaluate each expression.

a $2a + b$, when $a = 3$ and $b = 1.5$

b $3k - \frac{d}{3}$, when $k = 10$ and $d = 12$

c $\frac{f}{5} + f$, when $f = 20$

d $3x + 2y$, when $x = -4$ and $y = 7$

e $\frac{m+n}{2}$, when $m = 7$ and $n = 11$

f $n^2 - 3n$, when $n = -6$

6 Evaluate each expression if $m = 2$, $n = -3$ and $p = 5$.

a $m + n + p$ b $m - n$ c np

d $m^2 - n^2$ e $np - m$ f $mn + p$

g $5n + 3p$ h $\sqrt{m+n+p}$ i $\frac{p+n}{m}$

j $5m^2$ k $m^2 + p - 4n$ l $\frac{p}{n-m}$

7 If the perimeter of a rectangle with length l and width w has the formula $P = 2l + 2w$, find the perimeter of a rectangle with length 4.7 cm and width 2.5 cm.

8 If the area of a rectangle has the formula $A = lw$, find the area of a rectangle with length 12.5 m and width 8 m.

9 If the volume of a rectangular pyramid with base area A, and height h is $V = \frac{1}{3}Ah$, find the volume of the rectangular prism with base area 28 cm^2 and height 30 cm.

10 The area of a triangle with base length b and perpendicular height h is $A = \frac{1}{2}bh$. Find the area of a triangle with base length 15 cm and perpendicular height 12 cm.

11 Given that $A = \frac{h(a+b)}{2}$, evaluate A when $h = 4.5$, $a = 6$ and $b = 8$. Select **A**, **B**, **C** or **D**.

A 17.5 **B** 31.5 **C** 63 **D** 108

12 If $M = \frac{x+y+z}{3}$, find M when $x = 3.7$, $y = 8.1$ and $z = 0.5$.

13 Given $m = \frac{5k}{8}$, find m when $k = 50$.

□ Foundation ○ Standard ○ Complex

14 Given that $P = -2d^2 - 5d + 8$, find P when $d = -4$. Select **A**, **B**, **C** or **D**.

A 60 **B** 4 **C** -4 **D** 50

DID YOU KNOW?

Einstein's formula

Alamy Stock Photo/GL Archive

In 1905, when the German scientist Albert Einstein was only 26, he proposed a new theory of physics for small particles of matter (such as atoms) moving at very high speeds. He proposed that matter (mass) could be converted into large amounts of energy, describing this in his Special Relativity formula $E = mc^2$, where E stands for energy, m stands for mass and c stands for the speed of light (3 billion metres per second). Einstein's theory revolutionised conventional laws of physics and led to the development of nuclear energy.

Find out why Einstein renounced his German citizenship and became an American in 1940.

TECHNOLOGY

Substitution

Skillsheet
Spreadsheets

This spreadsheet shows the values of 5 variables, x, y, z, M and N in cells B1 to B5. To show a fraction in B4, right-click and select **Format Cells**, **Number**, **Fraction** and **Up to 1 digit**.

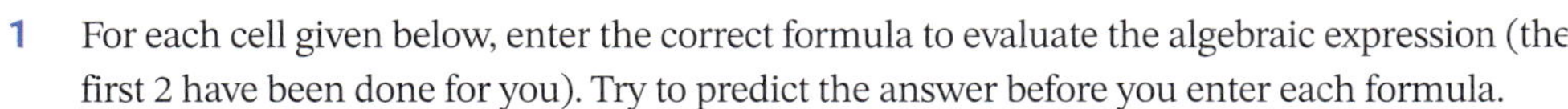

1 For each cell given below, enter the correct formula to evaluate the algebraic expression (the first 2 have been done for you). Try to predict the answer before you enter each formula.

a C1, $2x$ Enter: =2*B1 **b** C2, $\frac{y}{3}$ Enter: =B2/3 **c** C3, $x - y$

d C4, $5z + y$ **e** C5, $x + y + z$ **f** C6, zN

g C7, $\frac{2z}{5}$ **h** C8, xyz **i** C9, M^2

j C10, $N \div M$ **k** C11, $11 + \frac{z}{4}$ **l** C12, $yz + x$

m C13, $4(y - z)$ **n** C14, $z^3 + 2z$

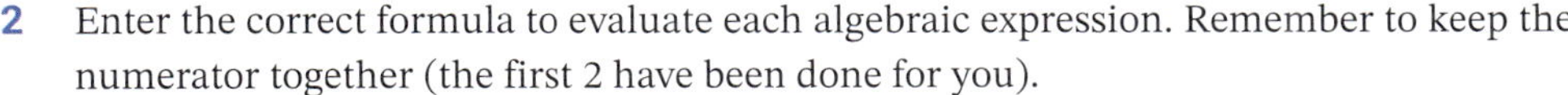

2 Enter the correct formula to evaluate each algebraic expression. Remember to keep the numerator together (the first 2 have been done for you).

a D1, $\frac{x+y}{5}$ Enter: =(B1+B2)/5 **b** D2, $\frac{3}{x+y}$ Enter: =3/(B1+B2)

c D3, $\frac{5y-6x}{7}$ **d** D4, $\frac{y}{zN}$

e D5, $\frac{xy}{MN}$ **f** D6, $\frac{y+x}{y-x}$

☐ Foundation ○ Standard ○ Complex

g D7, $\frac{xy}{x+y}$

h D8, $\frac{N(5+z)}{M}$

i D9, $\frac{yz}{2}+\frac{x}{N}$

j D10, $\sqrt{x^2+y^2}$

3 Does $2x^2 = (2x)^2$?

a In cell E1, enter **=2*B1^2** and in cell E2, enter **=(2*B1)^2**.

b Are the values in cells E1 and E2 equal?

4 Does $3z^2 = (3z)^2$?

a In cell E3, enter **=3*B3^2** and in cell E4, enter **=(3*B3)^2**.

b Are the values in cells E3 and E4 equal?

c Does $ax^2 = (ax)^2$ where a is a positive integer?

5 Does $-5y^2 = (-5y)^2$?

a In cell F1, enter **=-5*B2^2** and in cell F2, enter **=(-5*B2)^2**.

b Are the values in cells F1 and F2 equal?

6 Does $-6N^2 = (-6N)^2$?

a In cell F3, enter **=-6*B5^2** and in cell F4, enter **=(-6*B5)^2**.

b Are the values in cells F3 and F4 equal?

c Does $ax^2 = (ax)^2$, where a is a negative integer?

1.03 Adding and subtracting terms

Video Adding and subtracting terms

Skillsheet Algebra using diagrams

Interactive Expression exchange

An algebraic expression is made up of **algebraic terms**. For example, $3y - x + 5$ has 3 terms: $3y$, x and 5. **Like terms** are terms whose variables are exactly the same, for example, $3k$ and $7k$, $6ab$ and $2ba$, $9x^2$ and x^2. **Unlike terms** are terms whose variables are different, for example, $2d$ and $5m$, $4xy$ and $9yz$, $4x$ and $2x^2$.

We can add or subtract like terms because their variables represent the same number.

Example 6

Simplify each algebraic expression.

a $3a + 8a$ b $7r - r$ c $5xy - 4yx$

SOLUTION

a $3a + 8a = 11a$ — '3 lots of any number a' plus '8 lots of the same number a' will give '11 lots of the number a'.

b $7r - r = 6r$ — '7 lots of any number r' minus '1 lot of the same number r' will leave '6 lots of the number r'.

c $5xy - 4yx = 1xy$ — xy and yx are the same and can be subtracted.

$= xy$ — $xy = 1xy$.

Algebraic terms are usually written in alphabetical order.

ⓘ Adding and subtracting terms

Only **like terms** can be added or subtracted.

For example, $2x - 5y$ *cannot* be simplified because x and y represent different numbers.

Example 7

Simplify each expression.

a $2k + 3 + 8k$ **b** $5r + 4s - 7r - 2s$ **c** $11u^2 - 5u + 3u - 7u^2$

SOLUTION

a $2k + 3 + 8k = (2k + 8k) + 3$ Combine the like terms $2k$ and $8k$.

$= 10k + 3$ $10k$ and 3 cannot be simplified further.

b $5r + 4s - 7r - 2s = 5r - 7r + 4s - 2s$ Group the like terms.

$= -2r + 2s$ Rewrite with positive term first.

$= 2s - 2r$

c $11u^2 - 5u + 3u - 7u^2 = 11u^2 - 7u^2 - 5u + 3u$ Each + and − sign belongs to the term that follows it.

$= 4u^2 - 2u$

Example 8

Sonja wrote $7a + 3a = 10a^2$.

Use the substitution $a = 5$ to test whether her equation is correct or incorrect.

SOLUTION

LHS $= 7a + 3a$ LHS means 'left-hand side' of the equation.

$= 7 \times 5 + 3 \times 5$ Using $a = 5$.

$= 50$

RHS $= 10a^2$ RHS means 'right-hand side' of the equation.

$= 10 \times 5^2$ Using $a = 5$.

$= 250$

LHS ≠ RHS, so $7a + 3a = 10a^2$ is incorrect.

Example 9

Write a simplified algebraic expression for the perimeter of this rectangle.

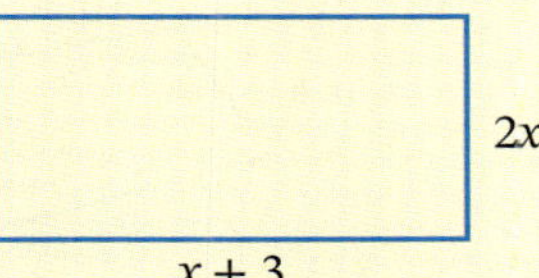

SOLUTION

Perimeter $= x + 3 + 2x + x + 3 + 2x$ Adding the lengths of the 4 sides.

$= x + 2x + x + 2x + 3 + 3$ Group the like terms.

$= 6x + 6$

EXERCISE 1.03 ANSWERS ON P. 628

Adding and subtracting terms

U F R

1 Simplify each algebraic expression.

a	$5y + 6y$	**b**	$12m - 7m$	**c**	$10t + t$	**d**	$15q - q$
e	$5g - 8g$	**f**	$4r^2 + 7r^2$	**g**	$6def - 6def$	**h**	$4xy - 9yx$
i	$5n^2 + 3n$	**j**	$9gh + 12hg$	**k**	$5my^2 + 2m^2y$	**l**	$d - 13d$

2 Simplify $15 + 8m - 6 + 3m$. Select the correct answer **A**, **B**, **C** or **D**.

A $21 + 11m$ **B** $9 + 5m$ **C** $9 + 11m$ **D** $9 + 5m^2$

3 Simplify each algebraic expression.

a	$2fg + 3fg - 4fg$	**b**	$6uvw - uvw + 4uvw$	**c**	$3t + 3 + 5t$
d	$3x + 2y + 8y$	**e**	$6g + h - 4g$	**f**	$2fg + 3f - 4fg$
g	$7p^2 - 3 - 2p^2$	**h**	$4p - q - 3q$	**i**	$-8r + 6r + 3$
j	$-3a + 4b - 7a$	**k**	$4v + 3 + 2v + 7$	**l**	$2y + 6x + y - 3x$
m	$7p - 8q - p - q$	**n**	$8 - 3w + 7 - 2w$	**o**	$2n - 3 - 12n + 5$
p	$8e - 3f - 2e - 4f$	**q**	$2l^2 + 2l + 3l^2 - 6l$	**r**	$b - 12b^2 + 6b^2 - 15b$
s	$6p^3 - 5p^2 - 4p^3 + p^2$	**t**	$3 - 8k - 5 - 9k$	**u**	$-3 - 5d + 9 - 8d$

4 Simplify $15a^2 + 12a + 48 - 18a$. Select the correct answer **A**, **B**, **C** or **D**.

A $24a + 48$ **B** $27a^2 + 30a$

C $15a^2 - 6a + 48$ **D** $15a^2 + 30a - 48$

5 (R) Use the substitutions $a = 2$, $b = 3$ and $c = 4$ to test whether each equation is correct or incorrect.

a	$20c - 12c = 8c$	**b**	$9b + b = 9b^2$	**c**	$7a + 3b = 10 + a + b$
d	$3a^2 + 5a^2 = 8a^2$	**e**	$4ab + 3ab = 7ab$	**f**	$12c - c = 12$

6 Write a simplified algebraic expression for the perimeter of each shape.

a

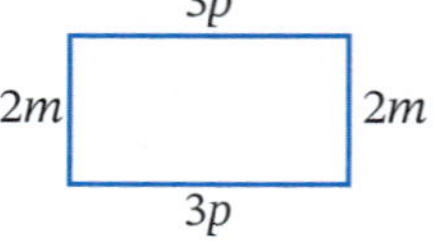

b

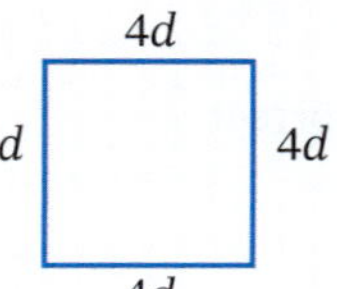

c

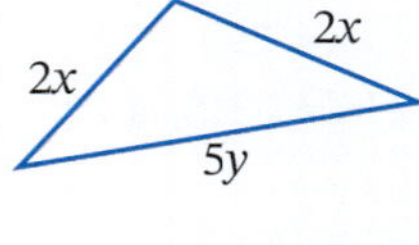

d

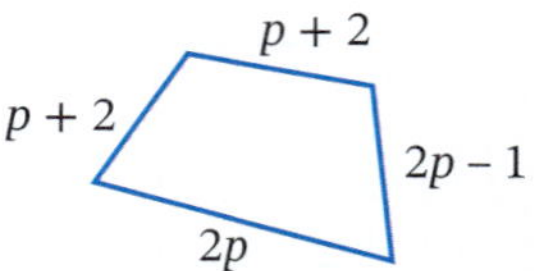

e

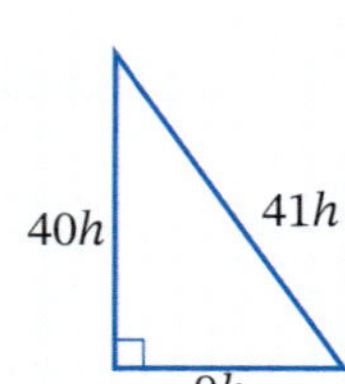

f

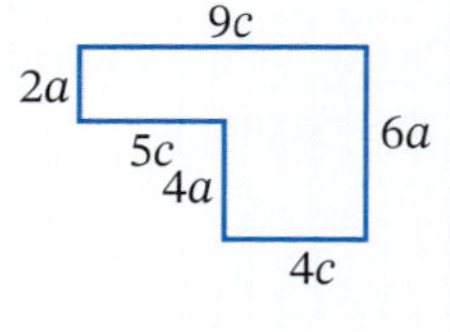

☐ Foundation ○ Standard ⬡ Complex

9780170465618

7 Draw a triangle and write algebraic expressions for its side lengths so that it has a perimeter of $7d + 6$.

8 Draw a rectangle and find algebraic expressions for its length and width so that it has a perimeter of $18a - 30$.

Multiplying and dividing terms 1.04

When writing algebraic expressions, it is not necessary to show the '×' symbol when multiplying terms. For example, $8 \times p \times q$ is abbreviated to $8pq$. When dividing terms, we usually write the expression as a **fraction** rather than use the '÷' symbol. For example, $a \div 3$ is written as $\frac{a}{3}$.

Video Multiplying terms

Worksheet Perimeter and area

Multiplying and dividing terms

When **multiplying and dividing terms**, multiply or divide the numbers first, then the variables.

Example 10

Simplify each expression.

a $2 \times 5n \times 4m$ **b** $3d \times 4d$ **c** $2g \times 3h \times (-5f)$

SOLUTION

a $2 \times 5n \times 4m = 2 \times 5 \times 4 \times n \times m$
$= 40mn$

Multiply the numbers and multiply the variables.

b $3d \times 4d = 3 \times 4 \times d \times d$
$= 12d^2$

$d \times d = d^2$

c $2g \times 3h \times (-5f) = 2 \times 3 \times (-5) \times g \times h \times f$
$= -30fgh$

Remember to write the variables in alphabetical order.

Example 11

Video Dividing terms

Simplify each expression.

a $\frac{15abc}{3ab}$ **b** $\frac{k^2m}{4km}$ **c** $27cf \div (-3fc)$ **d** $35p^2 \div 15pw$

SOLUTION

a $\frac{15abc}{3ab} = \frac{\cancel{15}^5\,\cancel{ab}^1 c}{\cancel{3}^1\,\cancel{ab}^1}$

We can cross out the ab in the **numerator** and **denominator** because $\frac{a}{a} = 1$ and $\frac{b}{b} = 1$

$= 5c$

Divide the numbers and divide the variables.

Foundation Standard Complex

b $\frac{k^2m}{4km} = \frac{kkm}{4km}$ — $k^2 = k \times k = kk$

$= \frac{1}{4}\ \frac{k\cancel{km}^1}{\cancel{km}^1}$ — Divide the numbers and divide the variables.

$= \frac{k}{4}$

c $27cf \div (-3fc) = \frac{27cf}{-3fc}$ — The variables divide by themselves to give 1.

$= \frac{27}{-3}\ \frac{\cancel{cf}^1}{\cancel{cf}^1}$

$= -9$

d $35p^2 \div 15pw = \frac{35p^2}{15pw}$ — 15 and 35 can both be divided by 5.

$= \frac{\cancel{35}^7}{\cancel{15}^3}\ \frac{\cancel{p}^1p}{\cancel{p}^1w}$

$= \frac{7p}{3w}$

Example 12

Maddison wrote $7a \times 3a = 21a^2$. Use the substitution $a = 5$ to test whether her equation is correct or incorrect.

SOLUTION

LHS $= 7a \times 3a$ — LHS means 'left-hand side' of the equation.

$= 7 \times 5 \times 3 \times 5$ — Using $a = 5$.

$= 525$

RHS $= 21a^2$ — RHS means 'right-hand side' of the equation.

$= 21 \times 5^2$ — $a = 5$

$= 525$

LHS = RHS, so $7a \times 3a = 21a^2$ is correct.

EXERCISE 1.04 ANSWERS ON P. 628

Multiplying and dividing terms

EXAMPLE 10

1 Simplify each expression.

a	$5 \times 8x$	**b**	$5n \times 3$	**c**	$b \times 5a$
d	$7m \times 6t$	**e**	$-4 \times 3w$	**f**	$2k \times (-3p)$
g	$4d \times 3e$	**h**	$6v \times 4w$	**i**	$11q \times 8q$
j	$-5a \times (-4a)$	**k**	$2y \times 3 \times 4b$	**l**	$-6w \times 3w \times 2$
m	$-9c \times (-4h) \times 2k$	**n**	$\frac{1}{3}a \times 24bc$	**o**	$-5d \times (-3e) \times (-2w)$

□ Foundation ○ Standard ⬡ Complex

p $(5n)^2$ **q** $(-3r)^2$ **r** $(at)^2$

s $-15d \times \frac{2}{3}ad$ **t** $-4k \times (-\frac{1}{4}ek)$ **u** $(-2xy)^2$

2 Simplify each expression. EXAMPLE 11

a $\frac{12x}{3}$ **b** $\frac{15y}{y}$ **c** $\frac{36a}{4a}$

d $18mn \div 6m$ **e** $8vw \div 48vw$ **f** $\frac{64abc}{16b}$

g $\frac{18g}{-3g}$ **h** $-25yz \div 5y$ **i** $\frac{9p}{-45pq}$

j $\frac{-75xyz}{-25x}$ **k** $10r^2 \div 40r$ **l** $-12hk^2 \div 2hk$

m $24ak^2 \div 8a^2$ **n** $16mwx \div 12mx$ **o** $7gh \div 21g$

3 Simplify $\frac{9yx^2}{24xy^2}$. Select the correct answer **A**, **B**, **C** or **D**.

A $\frac{3x^2}{8y^2}$ **B** $\frac{3x^3}{8y^3}$ **C** $\frac{3x}{8y}$ **D** $\frac{3y}{8x}$

4 Simplify each expression.

a $2a \times 3b \times 5c$ **b** $4p \times 3 \times 2q$ **c** $40n^2 \div 5n$

d $8b^2 \div 40b^2$ **e** $6h \times 2k \times (-3)$ **f** $-2 \times 5k \times (-4k)$

g $\frac{6d}{12}$ **h** $\frac{7r}{49r}$ **i** $(-33m) \div 3$

j $\frac{-9w}{-45w}$ **k** $8ab \div 2bc$ **l** $2ef \times (-3fe)$

m $18ac^2d \div 27acd$ **n** $-3pq \times (-2q) \times (-4)$ **o** $\frac{24v^2w}{18v}$

5 Simplify $-5k \times 4km \times (-3m)$. Select **A**, **B**, **C** or **D**.

A $-60k^2m$ **B** $60km$ **C** $-60km^2$ **D** $60k^2m^2$

6 Use the substitutions $p = 5$, $q = 4$ and $r = 3$ to test whether each equation is correct or incorrect. EXAMPLE 12

a $3 \times 4p = 12p$ **b** $q + q = q^2$

c $2r \times 3r = 6r^2$ **d** $3q^2 \times 5q^2 = 15q^2$

e $\frac{12p}{3p} = 4$ **f** $\frac{7qr}{qr} = 7$

g $3r \times 5q = 15qr$ **h** $p \times q \times q = 2pq$

i $\frac{14r^2}{7r} = 2r$ **j** $\frac{5pq}{10p} = \frac{q}{2}$

Foundation Standard Complex

7 Write a simplified algebraic expression for the area of each shape.

a

b

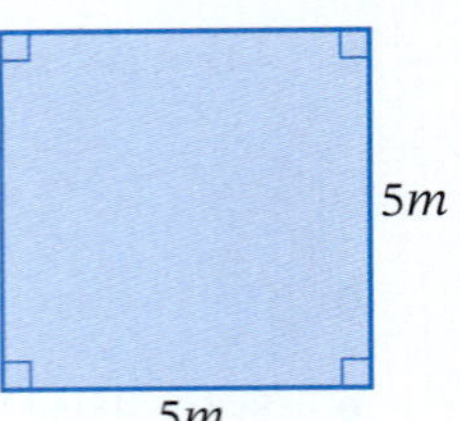

c

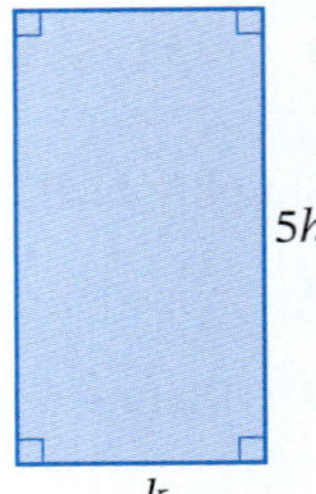

d

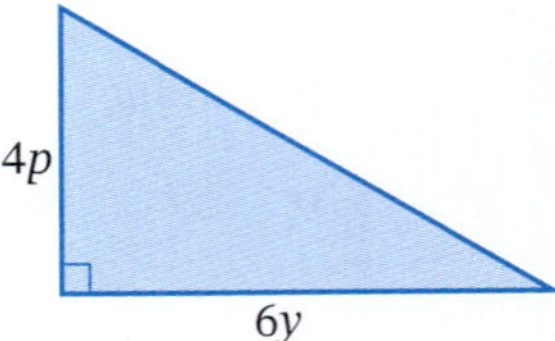

e

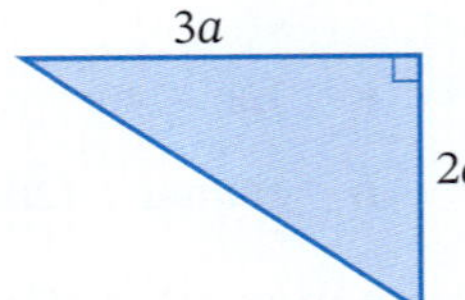

f

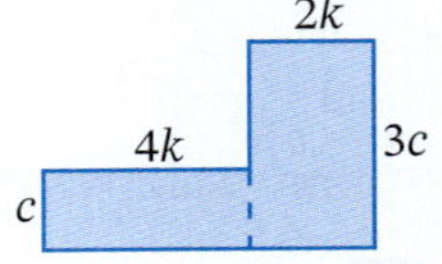

8 Find an algebraic expression for the volume of each rectangular prism.

a

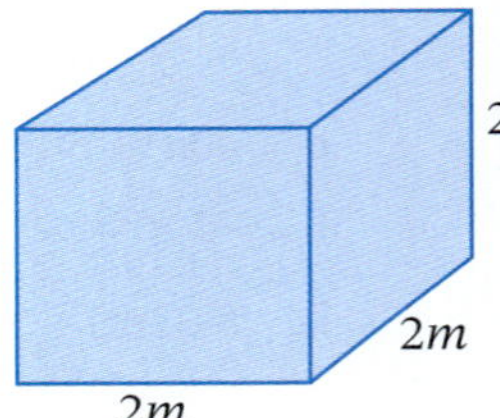

b

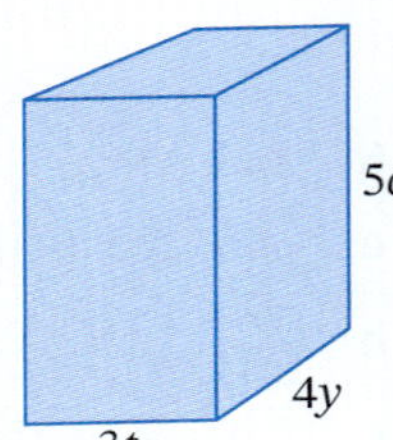

c

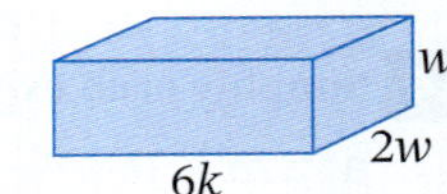

Quiz
Mental skills 1A

☆ MENTAL SKILLS 1A ANSWERS ON P. 629 **Maths without calculators**

Multiplying and dividing by 5, 15, 25 and 50

It is easier to multiply or divide a number by 10 than by 5. So whenever we multiply or divide a number by 5, we can double the 5 (to make 10) and then adjust the first number.

1 Study each example.

a **To multiply by 5**, halve the number, then multiply by 10.

$18 \times 5 = 18 \times \frac{1}{2} \times 10$ (or $9 \times 2 \times 10$)

$= 9 \times 10$

$= 90$

b **To multiply by 50**, halve the number, then multiply by 100.

$26 \times 50 = 26 \times \frac{1}{2} \times 100$ (or $13 \times 2 \times 100$)

$= 13 \times 100$

$= 1300$

☐ Foundation ◯ Standard ⬡ Complex

c **To multiply by 25**, quarter the number, then multiply by 100.

$44 \times 25 = 44 \times \frac{1}{4} \times 100$ (or $11 \times 4 \times 25$)

$= 11 \times 100$

$= 1100$

d **To multiply by 15**, halve the number, then multiply by 30.

$8 \times 15 = 8 \times \frac{1}{2} \times 30$ (or $4 \times 2 \times 15$)

$= 4 \times 30$

$= 120$

e **To divide by 5**, divide by 10 and double the answer. We do this because there are two 5s in every 10.

$140 \div 5 = 140 \div 10 \times 2$

$= 14 \times 2$

$= 28$

f **To divide by 50**, divide by 100 and double the answer. This is because there are two 50s in every 100.

$400 \div 50 = 400 \div 100 \times 2$

$= 4 \times 2$

$= 8$

g **To divide by 25**, divide by 100 and multiply the answer by 4. This is because there are four 25s in every 100.

$600 \div 25 = 600 \div 100 \times 4$

$= 6 \times 4$

$= 24$

h **To divide by 15**, divide by 30 and double the answer. This is because there are two 15s in every 30.

$240 \div 15 = 240 \div 30 \times 2$

$= 8 \times 2$

$= 16$

2 Now evaluate each expression.

a 32×5 **b** 14×5 **c** 48×5 **d** 18×50

e 52×50 **f** 36×25 **g** 28×5 **h** 12×25

i 12×15 **j** 22×35 **k** $90 \div 5$ **l** $170 \div 5$

m $230 \div 5$ **n** $1300 \div 50$ **o** $900 \div 50$ **p** $300 \div 25$

q $1000 \div 25$ **r** $360 \div 45$ **s** $210 \div 15$ **t** $360 \div 15$

1.05 Expanding expressions

Skillsheet Algebra using diagrams

Video Expanding and simplifying expressions

Interactive Area model Algebra

When an algebraic expression such as $5(x + 2)$ is simplified by removing the grouping symbols (brackets) by multiplying, we call it **expanding** the expression.

For example, $5(x + 2)$ means '5 times $(x + 2)$', so if we expand 'the long way':

$$\begin{aligned} 5(x + 2) &= (x + 2) + (x + 2) + (x + 2) + (x + 2) + (x + 2) \\ &= x + x + x + x + x + 2 + 2 + 2 + 2 + 2 \\ &= 5x + 10 \end{aligned}$$

We can expand $5(x + 2)$ more quickly by noticing that:

$$\begin{aligned} 5(x + 2) &= (5 \times x) + (5 \times 2) \\ &= 5x + 10 \end{aligned}$$

Each term inside the grouping symbols $(x + 2)$ is multiplied by the term outside the grouping symbols (5). This is called the **distributive law** because the 5 is distributed across the x and the 2 inside the grouping symbols. This can also be demonstrated using an area diagram.

Area of rectangle $= 5(x + 2) = 5x + 10$

We can expand $4(y - 1)$ in similar ways.

This can also be demonstrated using an area diagram.

$$\begin{aligned} 4(y - 1) &= (y - 1) + (y - 1) + (y - 1) + (y - 1) \\ &= y + y + y + y - 1 - 1 - 1 - 1 \\ &= 4y - 4 \end{aligned}$$

or more quickly:

$$\begin{aligned} 4(y - 1) &= (4 \times y) - (4 \times 1) \\ &= 4y - 4 \end{aligned}$$

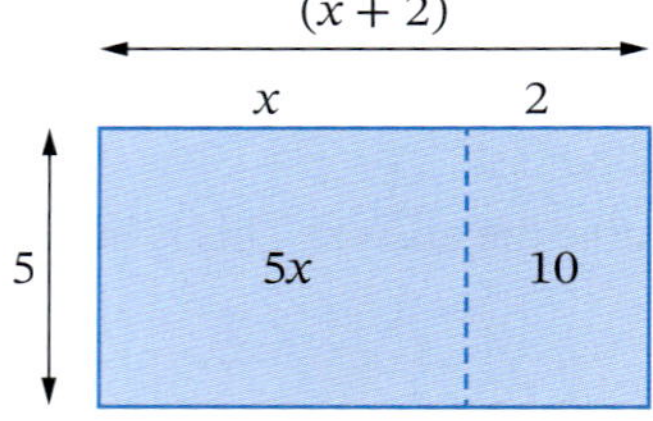

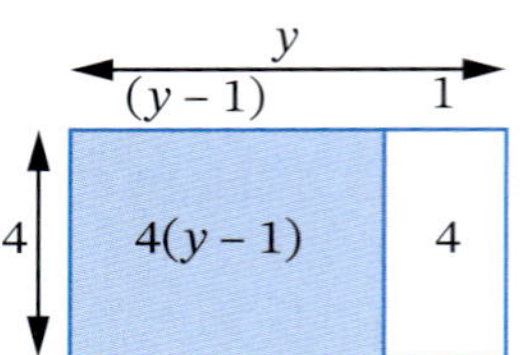

Area of left rectangle (shaded) $= 4(y - 1)$ (area of whole rectangle – area of right triangle)

$= 4y - 4$

ⓘ The distributive law for expanding an expression

Multiply each term inside the brackets by the term outside

$a(b + c) = ab + ac$

$a(b - c) = ab - ac$

1.05

Example 13

Expand each expression.

a $4(5 + d)$ **b** $2w(3w + 10)$

SOLUTION

a $4(5 + d) = 4 \times 5 + 4 \times d$ $a(b + c) = ab + ac$

$= 20 + 4d$ Multiply each term inside the brackets by the term outside.

b $2w(3w + 10) = 2w \times 3w + 2w \times 10$

$= 6w^2 + 20w$

Example 14

Expand each expression.

a $3g(h - 2)$ **b** $r(10 - 4r)$

SOLUTION

a $3g(h - 2) = 3g \times h - 3g \times 2$ $a(b - c) = ab - ac$

$= 3gh - 6g$

b $r(10 - 4r) = r \times 10 - r \times 4r$

$= 10r - 4r^2$

Example 15

Expand each expression.

a $-(y + 4)$ **b** $-3(2m - 7)$

SOLUTION

a $-(y + 4) = -1(y + 4)$

$= -y + (-4)$

$= -y - 4$

b $-3(2m - 7) = -3 \times 2m - (-3) \times 7$

$= -6m - (-21)$

$= -6m + 21$

Example 16

Expand and simplify by collecting like terms.

a $4(3p + 2) - 5p$ **b** $5(2e - 3) - 4(1 - 5e)$ **c** $7(n - 3) + n(n - 1)$

SOLUTION

a $4(3p + 2) - 5p = 4 \times 3p + 4 \times 2 - 5p$ Expanding

$= 12p + 8 - 5p$ Collecting like terms to simplify.

$= 12p - 5p + 8$

$= 7p + 8$

b $5(2e - 3) - 4(1 - 5e) = 5 \times 2e - 5 \times 3 - 4 \times 1 - (-4) \times 5e$ Expanding

$= 10e - 15 - 4 - (-20e)$ Collecting like terms to simplify.

$= 10e - 19 + 20e$

$= 30e - 19$

c $7(n - 3) + n(n - 1) = 7n - 21 + n^2 - n$ It's conventional to place n^2 first.

$= n^2 + 7n - n - 21$ Collecting like terms to simplify.

$= n^2 + 6n - 21$

EXERCISE 1.05 ANSWERS ON P. 629

Expanding expressions

U F

1 Use an area diagram to expand each expression.

a $3(x + 5)$ **b** $8(e + 9)$ **c** $2n(n + 3)$

EXAMPLE 13

2 Expand each expression.

a $4(h + 6)$ **b** $5(3 + t)$ **c** $3(2b + 7)$

d $5(8 + 3x)$ **e** $x(x + 9)$ **f** $p(5 + p)$

g $3r(2r + 1)$ **h** $4y(1 + 3y)$ **i** $10e(2e + 4f)$

3 Use an area diagram to expand each expression.

a $10(b - 2)$ **b** $6(2y - 3)$ **c** $4t(5t - 4)$

EXAMPLE 14

4 Expand each expression.

a $3(t - 2)$ **b** $7(5 - d)$ **c** $8(8g - 3)$

d $w(w - 7)$ **e** $a(a - 1)$ **f** $6h(3h - 1)$

g $4x(3x - 1)$ **h** $5x(2x - 4y)$ **i** $3a(4b - 7a)$

5 Expand $-5u(8 + 2u)$. Select the correct answer **A**, **B**, **C** or **D**.

A $3u - 3u^2$ **B** $-40u + 10u^2$ **C** $-40u - 10u^2$ **D** $-40u - 3u^2$

EXAMPLE 15

6 Use the substitution $x = 5$ to test whether each equation is correct or incorrect.

a $3(x + 3) = 3x + 6$ **b** $x(4 - 2x) = 4x - 2x^2$

7 Expand each expression.

a $-(a - 5)$ **b** $-(a + 5)$ **c** $-2(x + 6)$

d $-2(x - 6)$ **e** $-(11 + w)$ **f** $-(11 - w)$

g $-y(y + 9)$ **h** $-x(8 - z)$ **i** $-4t(t - 8)$

EXAMPLE 16

8 Expand $-k(3k - 7)$. Select the correct answer **A**, **B**, **C** or **D**.

A $-2k^2 - 8k$ **B** $-3k^2 + 7k$ **C** $-3k^2 - 7k$ **D** $-3k + 7k^2$

☐ Foundation ○ Standard ⬡ Complex

9780170465618

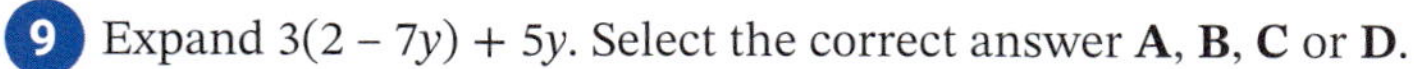

9 Expand $3(2 - 7y) + 5y$. Select the correct answer **A**, **B**, **C** or **D**.

A $6 - 26y$ **B** $6 + 16y$ **C** $6 - 5y$ **D** $6 - 16y$

10 Expand and simplify each expression.

a	$5(3m + 2) + 4m$	**b**	$3(1 - 5e) + 6e$	**c**	$4w - 2(5 + 2w)$
d	$8 - 5(2x - 7)$	**e**	$5(2a + 3) + 4(a + 7)$	**f**	$3(2d + 3) - 5(d + 4)$
g	$7(3g - 1) + 4(2g - 3)$	**h**	$4(3 - 4w) - 2(w - 5)$	**i**	$5(6c - 3) - 3(4 - 3c)$
j	$t(t + 4) + 3(t + 4)$	**k**	$4(3 + h) + h(7 - 2h)$	**l**	$6(2e - 1) - (5 - 3e)$
m	$3x(2x + 5) + 4(2x + 5)$	**n**	$v(2v + 3) - 6(v + 1)$	**o**	$3(1 - 2w) - w(2 - w)$
p	$2y(3y - 7) - 5(3y - 7)$				

11 Expand $3(1 - 4k) - 2(3k + 7)$. Select the correct answer **A**, **B**, **C** or **D**.

A $17 - 18k$ **B** $11 - 6k$ **C** $-11 - 18k$ **D** $-17 + 6k$

Expanding binomial products 1.06

$(x + 5)$ and $(x - 1)$ are called **binomial expressions** because they each have exactly 2 terms.
$(x + 5)(x - 1)$ is called a **binomial product** because it is a product of 2 binomial expressions.

binomial = '2 terms'

Worksheets
Area diagrams
Binomial products

Puzzle
Expanding brackets

Technology
Expanding binomials

Spreadsheet
Expanding binomials

Example 17

Expand each binomial product using an area diagram.

a $(a + 2)(a + 5)$ **b** $(n + 4)(n - 3)$

SOLUTION

a Draw an area diagram (rectangle) with length $(a + 2)$ and width $(a + 5)$ and divide the diagram into 4 smaller rectangles.

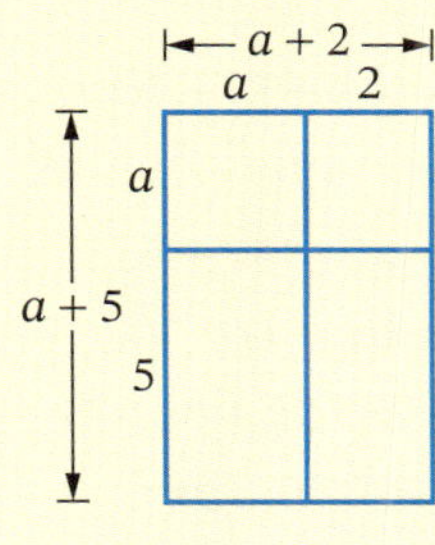

Find the area of each smaller rectangle.

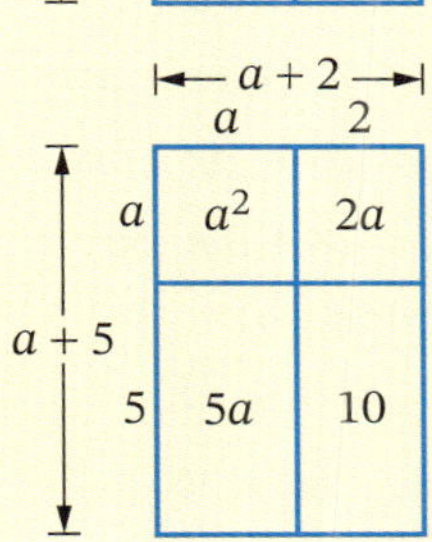

Foundation Standard Complex

Expand $(a + 2)(a + 5)$ by adding the areas of the 4 rectangles.

$(a + 2)(a + 5) = a^2 + 2a + 5a + 10$ Expanding

$= a^2 + 7a + 10$ Simplifying by collecting like terms.

b Draw an area diagram with length $(n + 4)$ and width $(n - 3)$, then increase the width to n by adding 3. This creates 4 rectangles.

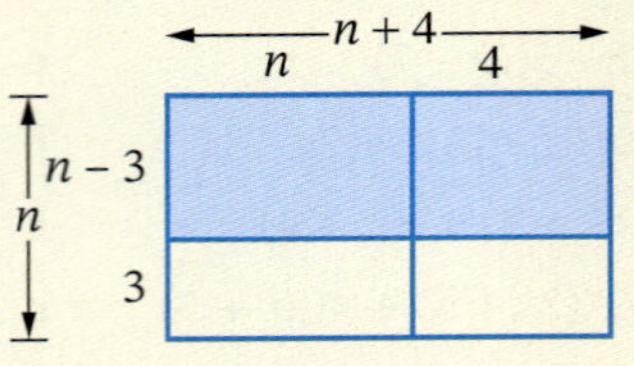

The 2 shaded upper rectangles together give the product $(n + 4)(n - 3)$. To find its area, we subtract the areas of the 2 lower rectangles from the area of the whole rectangle.

n + 4
n 4
n − 3 (n + 4)(n − 3)
n
3 3n 12

$(n + 4)(n - 3) = n(n + 4) - 3n - 12$ Whole rectangle – 2 lower rectangles

$= n^2 + 4n - 3n - 12$ Expanding

$= n^2 + n - 12$ Simplifying by collecting like terms

When expanding a binomial product algebraically, each term in the first binomial is multiplied by each term in the second binomial to give **4 terms**, which are collected and simplified.

Video
Expanding binomial products 1

Example 18

Expand each binomial product.

a $(x + 5)(x + 9)$ **b** $(k + 3)(k - 7)$

c $(7 - m)(4 + m)$ **d** $(a - 6)^2$

SOLUTION

a $(x + 5)(x + 9) = x(x + 9) + 5(x + 9)$ Each term in $(x + 5)$ is multiplied by $(x + 9)$.

$= x^2 + 9x + 5x + 45$ Expanding to make 4 terms.

$= x^2 + 14x + 45$ Adding $9x$ and $5x$.

b $(k + 3)(k - 7) = k(k - 7) + 3(k - 7)$ Each term in $(k + 3)$ is multiplied by $(k - 7)$.

$= k^2 - 7k + 3k - 21$ Expanding to make 4 terms.

$= k^2 - 4k - 21$ Adding $-7k$ and $3k$.

c $(7 - m)(4 + m) = 7(4 + m) - m(4 + m)$ Each term in $(7 - m)$ is multiplied by $(4 + m)$.

$= 28 + 7m - 4m - m^2$ Expanding to make 4 terms.

$= 28 + 3m - m^2$ Adding $7m$ and $-4m$.

d $(a - 6)^2 = (a - 6)(a - 6)$ Each term in $(a - 6)$ is multiplied by $(a - 6)$.

$= a(a - 6) - 6(a - 6)$ Expanding to make 4 terms.

$= a^2 - 6a - 6a + 36$ Adding $-6a$ and $-6a$.

$= a^2 - 12a + 36$

ⓘ The distributive law for expanding a binomial product

Multiply each term in the first binomial by each term in the second binomial.

$(a + b)(c + d) = ac + ad + bc + bd$

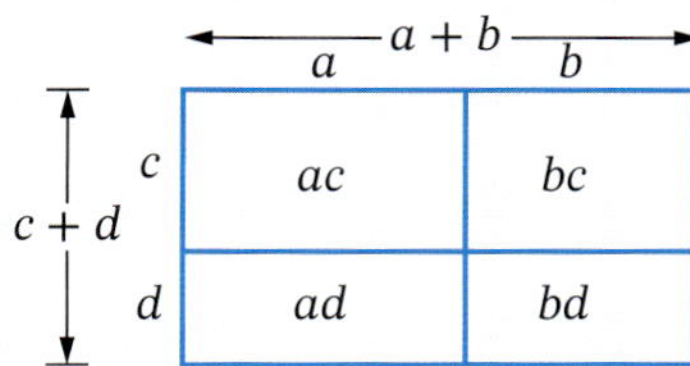

One way of remembering which pairs of terms to multiply together in a binomial product is called the **FOIL method**, as shown below.

$$(k + 3)(k - 7) = k^2 - 7k + 3k - 21$$
$$= k^2 - 4k - 21$$

- **F** means multiply the **first** terms: $k \times k = k^2$.
- **O** means multiply the **outside** terms: $k \times (-7) = -7k$.
- **I** means multiply the **inside** terms: $3 \times k = 3k$.
- **L** means multiply the **last** terms: $3 \times (-7) = -21$.

Example 19

Expand each binomial product.

a $(x - 6)(4x + 2)$ **b** $(3t - 1)(2t - 5)$ **c** $(2x + 5)(3y - 4)$

Videos
Expanding binomial products 1
Expanding binomial products 2

SOLUTION

a $(x - 6)(4x + 2) = x(4x + 2) - 6(4x + 2)$ Expanding
$= 4x^2 + 2x - 24x - 12$ Simplifying
$= 4x^2 - 22x - 12$

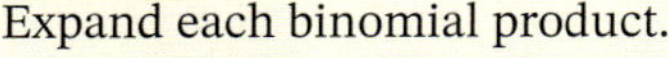

b $(3t - 1)(2t - 5) = 3t(2t - 5) - 1(2t - 5)$ Expanding
$= 6t^2 - 15t - 2t + 5$ Simplifying
$= 6t^2 - 17t + 5$

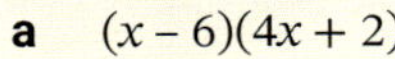

c $(2x + 5)(3y - 4) = 2x(3y - 4) + 5(3y - 4)$ Expanding
$= 6xy - 8x + 15y - 20$

EXERCISE 1.06 ANSWERS ON P. 629

Expanding binomial products

U F R C

EXAMPLE 17

1 Expand each binomial product by copying and completing the area diagram.

a $(x + 10)(x + 8)$

x 10
x --- ---
8 --- ---

b $(4h + 1)(3h + 7)$

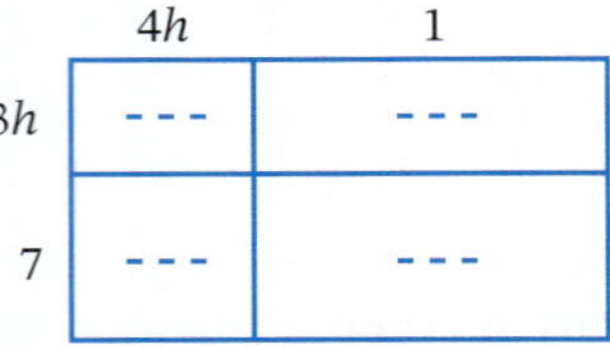

2 Expand each binomial product by copying and completing the area diagram. (Find the shaded area.)

a $(x + 9)(x - 6)$

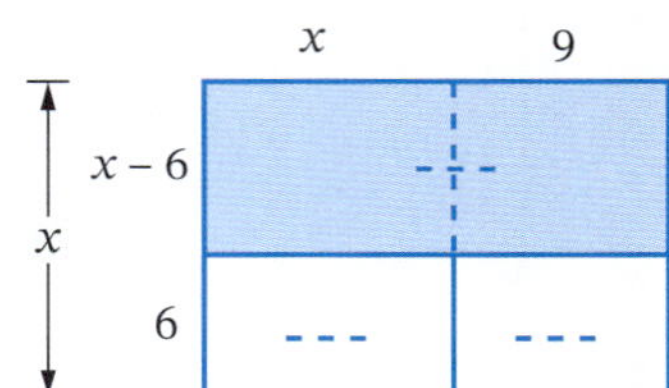

b $(n - 5)(n - 7)$

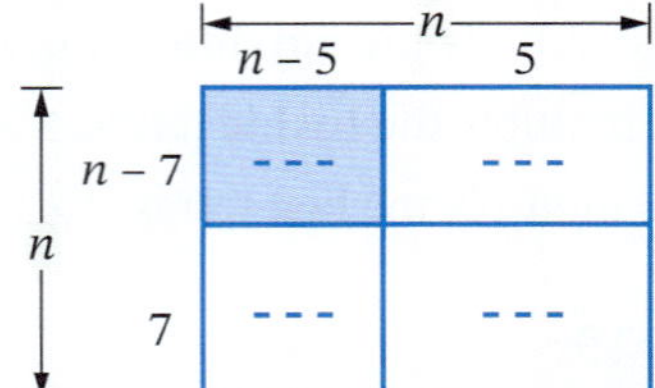

3 Draw an area diagram for each binomial product and use it to expand the product.

a $(2y + 5)(y + 3)$ **b** $(w - 5)(w + 3)$ **c** $(p - 3)(p - 5)$

4 Expand each binomial product algebraically.

a $(n + 6)(n + 3)$ **b** $(t + 6)(t + 5)$ **c** $(p + 10)(p - 10)$

d $(x - 8)(x + 3)$ **e** $(b - 2)(9 + b)$ **f** $(u - 8)(u - 7)$

g $(15 - r)(r + 1)$ **h** $(a - 10)(a - 9)$ **i** $(5 - c)(3 - c)$

j $(t - 1)(t + 2)$ **k** $(y - 4)(y + 10)$ **l** $(n - 9)(11 + n)$

m $(e - 2)^2$ **n** $(5 + w)^2$ **o** $(g + 11)^2$

5 Expand $(b + 7)^2$. Select the correct answer **A**, **B**, **C** or **D**.

A $b^2 + 49$ **B** $b^2 + 49b$

C $b^2 + 7b + 49$ **D** $b^2 + 14b + 49$

6 Expand $(x + 8)(3 - x)$. Select **A**, **B**, **C** or **D**.

A $24 - 5x - x^2$ **B** $-x^2 + 24$

C $x^2 + 5x - 24$ **D** $-x^2 + 5x + 24$

☐ Foundation ○ Standard ⬡ Complex

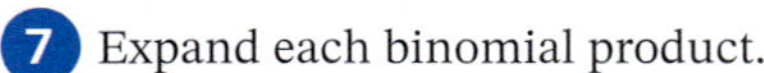

EXAMPLE 19

1.06

7 Expand each binomial product.

a	$(2x + 5)(x + 3)$	**b**	$(3e + 2)(4e + 5)$	**c**	$(10 + 3p)(p - 1)$
d	$(7d - 2)(7d - 2)$	**e**	$(2f - 2)(3f + 5)$	**f**	$(4m - 5)(5 + 3m)$
g	$(3 - 4h)(2 + 5h)$	**h**	$(4p - 5)(2p - 7)$	**i**	$(2m - 3)(4 - 5m)$
j	$(3t + 5)(2t - 1)$	**k**	$(5y - 5)(5y + 5)$	**l**	$(6 - 2a)(2a - 6)$
m	$(3k + 8)^2$	**n**	$(4c - 5)^2$	**o**	$(2d + 9)(3x - 1)$
p	$(5 - 2g)(h + 3)$	**q**	$(7 - 4w)^2$	**r**	$(3n - 5)(4n + 9)$

8 Expand $(3h + 8)(2h - 5)$. Select **A**, **B**, **C** or **D**.

A $6h^2 - 31h - 40$ **B** $6h^2 - h + 40$

C $6h^2 + 31h - 40$ **D** $6h^2 + h - 40$

9 Write a simplified algebraic expression for the area of each shape.

a

$x + 2$

$x - 3$

b

$3y - 7$

$3y - 7$

c

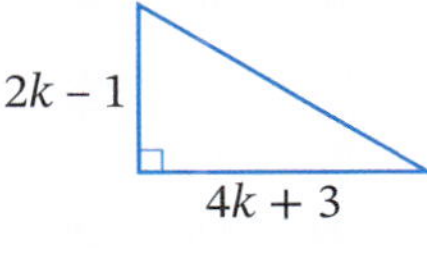

d

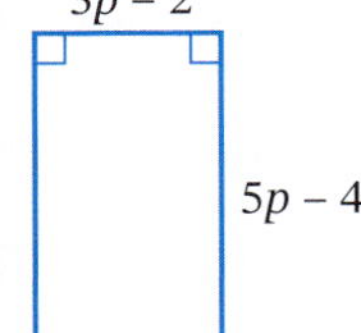

e

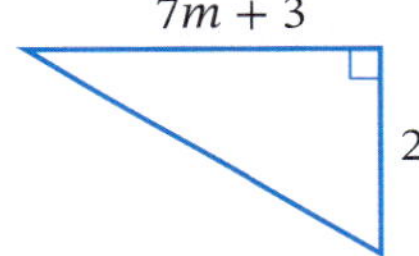

f

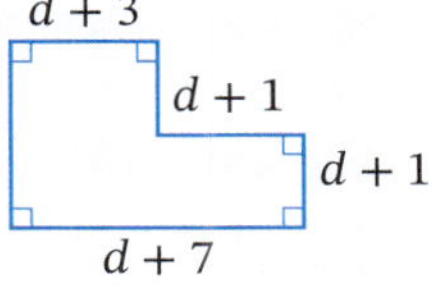

10 A rectangular barbecue plate has a length of 100 cm and a width of 75 cm. The length and width are both increased by x cm.

R C

a Write an expression for the new length of the plate in cm.

b Write an expression for the new width of the plate in cm.

c Hence find a simplified expression for the new area of the plate in cm^2.

d By how much has the area of the plate increased?

e If $x = 0.1$, find the increase in the area of the plate.

11 A family room in a house is to be extended. The room is 4 m long and 3 m wide. The length is to be increased by x m and the width by y m.

R C

a Write down expressions for the new length and width in metres.

b Write down a binomial expression for the new area of the room in square metres.

c Expand and simplify your expression for the area.

d Write the algebraic expression for how much the area of the room has increased.

Foundation Standard Complex

12 Prove that:

R **a** $(a - b)^2 = (b - a)^2$

C **b** $(a + b)(a - b) = a^2 - b^2$

c $(a - b)^2 = a^2 - 2ab + b^2$

1.07 Factorising expressions

Skillsheets HCF by factor trees

Factorising using diagrams

Puzzle Factorising puzzle

Factors

A **factor** (or **divisor**) is a number or algebraic expression that **divides evenly** into a larger number or algebraic expression. For example, 5 is a factor of 10 because $10 \div 5 = 2$, and $5x$ is a factor of $10xy$ because $10xy \div 5x = 2y$.

- The factors of 10 are 1, 2, 5, 10.
- *Some* of the factors of $10xy$ are 1, 5, $2x$, $5x$, $10x$, $10y$, xy, $2xy$ and $10xy$.

$3a$ is a factor of both $15a^2$ and $30a$ because it divides into both terms. We can say that $3a$ is a **common factor** of $15a^2$ and $30a$.

The highest common factor (HCF)

The **highest common factor (HCF)** or **greatest common divisor (GCD)** of 2 or more terms is the largest term that is a factor of all of the terms. To find the HCF of algebraic terms:

- find the HCF of the numbers.
- find the HCF of the variables.
- multiply them together.

Example 20

Find the highest common factor (HCF) of each pair of terms.

a $20x^2$ and $15xy$ **b** 16 and $12b$

SOLUTION

Find the HCF of the numbers and the HCF of the variables. Their product is the HCF of the expression.

a The HCF of 20 and 15 is 5.

The HCF of x^2 and xy is x, since it is the 'largest' common part of x^2 and xy.

$\therefore$ The HCF of $20x^2$ and $15xy = 5 \times x = 5x$

b The HCF of 16 and 12 is 4.

There is **no** HCF of the variables because b is in $12b$ but not in 16.

$\therefore$ The HCF of 16 and $12b$ is 4.

Foundation Standard Complex

9780170465618

Factorising algebraic expressions

1.07

When $4(2y + 5)$ is **expanded**, the answer is $8y + 20$.

Factorising is the reverse of expanding. Factorisation breaks an expression into factors. To factorise $8y + 20$, we take out the greatest common divisor and **insert** brackets. The result is $4(2y + 5)$. The factors are 4 and $(2y + 5)$.

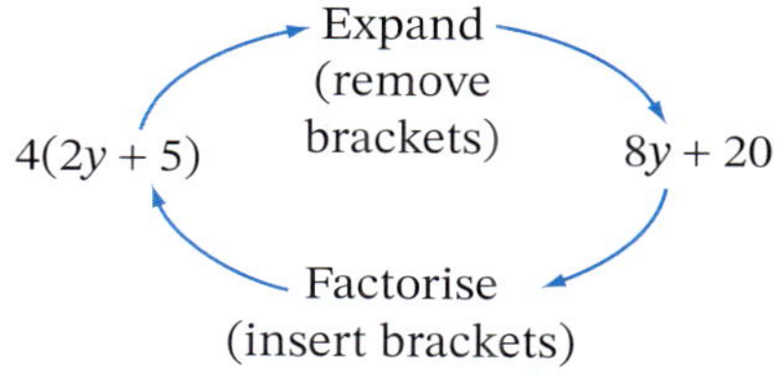

Example 21

Complete each factorisation by inserting the highest common factor.

a $5a + 15 = \_\_\_(a + 3)$

b $x^2 - 7x = \_\_\_(x - 7)$

c $y(y + 1) - 3(y + 1) = \_\_\_\_(y - 3)$

SOLUTION

a $5a + 15 = 5(a + 3)$ — The HCF is 5.

You can always check your answer by expanding it: $5(a + 3) = 5a + 15$.

b $x^2 - 7x = x(x - 7)$ — The HCF is x.

c $y(y + 1) - 3(y + 1) = (y + 1)(y - 3)$ — The HCF is $(y + 1)$.

ⓘ Factorising an expression

- Find the HCF of the terms and write it outside the brackets.
- Divide each term by the HCF and write the answers inside the brackets.

  $ab + ac = a(b + c)$

  $ab - ac = a(b - c)$
- To check that the factorised answer is correct, expand it.

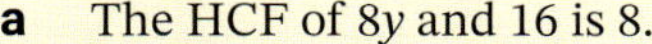

Example 22

Factorise each expression.

a $8y + 16$ **b** $25b^2 - 20ab$ **c** $v(4 + w) + 2(4 + w)$

Video
Factorising expressions

SOLUTION

a The HCF of $8y$ and 16 is 8.

$\therefore 8y + 16 = 8 \times y + 8 \times 2$

$= 8(y + 2)$

Rewrite the expression using the HCF 8.

Write the HCF at the front of the brackets.

b The HCF of $25b^2$ and $20ab$ is $5b$.

$\therefore 25b^2 - 20ab = 5b \times 5b - 5b \times 4a$

$= 5b(5b - 4a)$

Rewrite the expression using the HCF $5b$.

Write the HCF at the front of the brackets.

c The HCF of $v(4 + w) + 2(4 + w)$ is $(4 + w)$.

$\therefore v(4 + w) + 2(4 + w) = (4 + w) \times v + (4 + w) \times 2$

$= (4 + w)(v + 2)$

Factorising with negative terms

Example 23

Factorise each expression.

a $-x^2 + 3x$

b $-a - ab$

SOLUTION

When factorising expressions that begin with a **negative** term, we use the 'negative' HCF.

a The highest 'negative' common factor of $-x^2$ and $3x$ is $-x$. $(-x) \times (-3) = +3x$

$\therefore -x^2 + 3x = (-x) \times x + (-x) \times (-3)$

$= (-x)[x + (-3)]$

$= -x(x - 3)$

Check your answer by expanding.

b The highest 'negative' common factor of $-a$ and $-ab$ is $-a$. $(-a) \times b = -ab$

$\therefore -a - ab = -a \times 1 + (-a) \times b$

$= -a(1 + b)$

ⓘ Factorising with negative terms

- Find the 'negative' HCF of the terms and write it outside the brackets.
- Divide each term by the HCF and write the answers inside the brackets.

  $-ab - ac = -a(b + c)$

  $-ab + ac = -a(b - c)$
- To check that the factorised answer is correct, expand it.

EXERCISE 1.07 ANSWERS ON P. 629

Factorising expressions

1 Find the HCF of each pair of terms.

a 12, $6y$	**b** $8a^2$, 24	**c** $20b$, 15
d 6, $9p^2$	**e** $24mn$, m	**f** c, c^2
g $18pq$, $12p^2$	**h** $24w^2$, $16w$	**i** $3(x - 5)$, $x(x - 5)$

2 Copy and complete each factorisation. EXAMPLE 21

a $4a + 12 = \_\_\_(a + 3)$

b $a^2 + 10a = \_\_\_(a + 10)$

c $10b - 40 = \_\_\_(b - 4)$

d $18p^2 + 24 = 6(\_\_\_ + \_\_\_)$

e $9m^2 + 3m = 3m(\_\_\_ + \_\_\_)$

f $20d - 30d^2 = 10d(\_\_\_ - \_\_\_)$

g $6x^2y - 8xy = 2xy(\_\_\_ - \_\_\_)$

h $x(x - 2) + 5(x - 2) = (x - 2)(\_\_\_ + \_\_\_)$

i $n(3 + n) - 3(3 + n) = (3 + n)(\_\_\_ - \_\_\_)$

j $v(1 - w) + (1 - w) = (\_\_\_ + \_\_\_)(1 - w)$

3 Factorise each expression. EXAMPLE 22

a $3f + 6$ **b** $4m - 28$ **c** $9n + 27$

d $16 + 24t$ **e** $12q - 18$ **f** $24x + 30$

g $20g^2 - 64$ **h** $xy + y$ **i** $mn^2 - 3n$

j $2y - y^2$ **k** $16r^2 - 12r$ **l** $6t^2 + 27t$

4 Factorise each expression.

a $12x^2y - 16x$ **b** $18p^2 + 16pr$ **c** $4m^2n - 4mn^2$

d $14abc + 21bc$ **e** $28vw - 21v^2w^2$ **f** $45rt + 54r^2t$

5 Factorise completely $24xy^2 - 16xyw$. Select the correct answer **A**, **B**, **C** or **D**.

A $8xy(3y - 2w)$ **B** $2x(12y^2 - 8yw)$

C $4xy(6y - 4w)$ **D** $8xy^2(3 - 2w)$

6 Factorise each expression.

a $a(a - 3) + 6(a - 3)$

b $t(8 + t) - 3(8 + t)$

c $b(b + 5) - 2(b + 5)$

d $x(2 - y) - 6(2 - y)$

e $5(a - 7) + b(a - 7)$

f $s(a + 3) + 4(a + 3)$

g $3p(g + 5) - 2(g + 5)$

h $5(2 - 3m) + 4n(2 - 3m)$

i $r(8 + r) + (8 + r)$

j $(y - 6) - y(y - 6)$

7 Factorise each expression using the 'negative' HCF. EXAMPLE 23

a $-4q - 8$ **b** $-9u + 18$ **c** $-3g + 6$

d $-18a + 12$ **e** $-n - 1$ **f** $-n + 1$

g $-y^2 - 9y$ **h** $-6t + 10t^2$ **i** $-3a^2 - 6ab$

j $-ap + aq$ **k** $-20e^2 - 22e$ **l** $-9m + 3m^2$

8 Factorise $-10kr + 4rn$. Select **A**, **B**, **C** or **D**.

A $-2r(5k - 4n)$ **B** $-2r(5k - 2n)$

C $-5r(2k - 2n)$ **D** $-2r(5k + 2n)$

Foundation Standard Complex

Quiz
Mental skills 1B

☆ MENTAL SKILLS (1B) ANSWERS ON P.630 Maths without calculators

Multiplying by 9, 11, 99 and 101

We can use expansion when multiplying by a number near 10 or near 100.

1 Study each example.

a $25 \times 11 = 25 \times (10 + 1)$
$= 25 \times 10 + 25 \times 1$
$= 250 + 25$
$= 275$

b $14 \times 9 = 14 \times (10 - 1)$
$= 14 \times 10 - 14 \times 1$
$= 140 - 14$
$= 126$

c $32 \times 12 = 32 \times (10 + 2)$
$= 32 \times 10 + 32 \times 2$
$= 320 + 64$
$= 384$

d $7 \times 99 = 7 \times (100 - 1)$
$= 7 \times 100 - 7 \times 1$
$= 700 - 7$
$= 693$

e $27 \times 101 = 27 \times (100 + 1)$
$= 27 \times 100 + 27 \times 1$
$= 2700 + 27$
$= 2727$

f $18 \times 8 = 18 \times (10 - 2)$
$= 18 \times 10 - 18 \times 2$
$= 180 - 36$
$= 144$

2 Now evaluate each product.

a 16×11	**b** 33×11	**c** 29×9	**d** 45×9
e 62×11	**f** 7×101	**g** 18×101	**h** 36×99
i 19×8	**j** 45×12	**k** 21×102	**l** 6×98
m 32×9	**n** 7×99	**o** 39×101	**p** 71×12

INVESTIGATION

Factorising quadratic expressions

1 **a** Show that $(x + 3)(x + 5) = x^2 + 8x + 15$.

b The quadratic expression $x^2 + 8x + 15$ has 3 terms. The coefficient of x is 8, the number in front of the x. How are the 3 and 5 in $(x + 3)(x + 5)$ related to the 8?

c The constant term in $x^2 + 8x + 15$ is 15, the number with no x at the end. How are the 3 and 5 related to the 15?

2 **a** Expand $(x + 9)(x + 2)$.

b What is the coefficient of x? How are 9 and 2 related to it?

c What is the constant term? How are 9 and 2 related to it?

3 **a** Expand $(x + 8)(x - 3)$.

b What is the coefficient of x? How are 8 and -3 related to it?

c What is the negative constant term? How are 8 and -3 related to it?

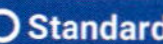 Foundation  Standard ⬡ Complex

9780170465618

4 a Expand $(x - 4)(x - 1)$.

b What is the negative coefficient of x? How are -4 and -1 related to it?

c What is the positive constant term? How are -4 and -1 related to it?

5 In the expansion of any binomial product, how are the coefficient of x and the constant term related to the numbers in the binomials?

6 Copy and complete:

a $(x + \_\_\_)(x + \_\_\_) = x^2 + 5x + 4$

b $(x + \_\_\_)(x + \_\_\_) = x^2 + 8x + 15$

c $(x + \_\_\_)(x + \_\_\_) = x^2 + 7x + 12$

d $(x + \_\_\_)(x - \_\_\_) = x^2 - 4x - 32$

e $(x + \_\_\_)(x - \_\_\_) = x^2 + 2x - 3$

f $(x - \_\_\_)(x - \_\_\_) = x^2 - 9x + 20$

Factorising quadratic expressions $x^2 + bx + c$ 1.08

A quadratic expression such as $x^2 - 5x + 7$ is called a **trinomial** because it has 3 terms. Other examples of quadratic trinomials are $x^2 + x - 15$, $2x^2 - 3x + 9$ and $-4x^2 + 9x + 20$.

The expansion of $(x + 2)(x + 4)$ is $x^2 + 6x + 8$.

$\therefore$ The factorisation of $x^2 + 6x + 8$ is $(x + 2)(x + 4)$.

When factorising a quadratic trinomial such as $x^2 + 6x + 8$:

- each factor must have an x term to give x^2

  $x^2 + 6x + 8 = (x + 2)(x + 4)$
- $2 + 4 = 6$, which is the **coefficient** of x, the number in front of the x

  $x^2 + 6x + 8 = (x + 2)(x + 4)$
- $2 \times 4 = 8$, which is the **constant term** with no x.

  $x^2 + 6x + 8 = (x + 2)(x + 4)$

Puzzle Perfect squares

Technology Factorising trinomials

Spreadsheet Factorising trinomials

Example 24

Factorise $a^2 + 7a + 12$.

SOLUTION

Find the 2 numbers that have a sum of 7 and a product of 12.

It is best to test numbers that have a product of 12 and then check if their sum equals 7:

Pair of numbers	Product	Sum
6, 2	6 × 2 = 12	6 + 2 = 8
3, 4	3 × 4 = 12	3 + 4 = 7

The correct numbers are 3 and 4.

$\therefore a^2 + 7a + 12 = (a + 3)(a + 4)$

Video Factorising quadratic expressions 1

Example 25

Video Factorising quadratic expressions 1

Factorise $x^2 + x - 6$.

SOLUTION

Find the 2 numbers with a sum of 1 and a product of –6.

Test numbers that have a product of –6 and check their sums:

Pair of numbers	Product	Sum
+6, −1	$+6 \times (-1) = -6$	$+6 + (-1) = 5$
+2, −3	$+2 \times (-3) = -6$	$2 + (-3) = -1$
+3, −2	$+3 \times (-2) = -6$	$+3 + (-2) = 1$

The correct numbers are 3 and –2.

$\therefore x^2 + x - 6 = (x + 3)(x - 2)$

ⓘ Factorising quadratic trinomials of the form $x^2 + bx + c$

- Find 2 numbers that have a sum of b and a product of c.
- Use these 2 numbers to write a binomial product of the form $(x$ ____ $)(x$ ____ $)$.

Example 26

Video Factorising quadratic expressions 1

Factorise each quadratic expression.

a $a^2 - 2a - 15$ **b** $y^2 - 6y + 8$ **c** $x^2 - 2x + 1$

SOLUTION

a $a^2 - 2a - 15$

Find 2 numbers that have a product of –15 and a sum of –2. Since the product is negative, one of the numbers must be negative. They are –5 and +3.

$\therefore a^2 - 2a - 15 = (a - 5)(a + 3)$

b $y^2 - 6y + 8$

Product = 8, sum = −6.

Since the sum is negative, one of the numbers must be negative, but since the product is positive, *both* of the numbers must be negative. They are −2 and −4.

$\therefore y^2 - 6y + 8 = (y - 2)(y - 4)$

c $x^2 - 2x + 1$

Product = 1, sum = −2. Since the sum is negative, one of the numbers must be negative, but since the product is positive, both of the numbers must be negative.

They are −1 and −1.

$\therefore x^2 - 2x + 1 = (x - 1)(x - 1)$

$= (x - 1)^2$ *Note:* $(x - 1)^2$ is a perfect square.

Example 27

Factorise each quadratic trinomial.

a $3g^2 + 12g - 36$ **b** $48 - 8p - p^2$

SOLUTION

a $3g^2 + 12g - 36 = 3(g^2 + 4g - 12)$ Taking out the HCF of 3 first

$= 3(g - 2)(g + 6)$ Product = −12, sum = 4

b $48 - 8p - p^2 = -p^2 - 8p + 48$ Rearranging the terms to make the p^2 term first.

$= -1(p^2 + 8p - 48)$ Taking out a common factor of −1

$= -(p + 12)(p - 4)$ Product = −48, sum = 8

EXERCISE 1.08 ANSWERS ON P. 630

Factorising quadratic expressions $x^2 + bx + c$

U F R

1 Find the 2 numbers whose:

R **a** product is 6 and their sum is −7 **b** product is −12 and their sum is 1

c product is −15 and their sum is −2 **d** product is 12 and their sum is 7

e product is 20 and their sum is −9 **f** product is −14 and their sum is 5

g product is −10 and their sum is 3 **h** product is −25 and their sum is 0

i product is −2 and their sum is −1 **j** product is −18 and their sum is −7

2 Factorise each quadratic expression.

EXAMPLE 24

R **a** $m^2 + 7m + 12$ **b** $x^2 + 12x + 35$ **c** $k^2 + 5k + 4$

d $p^2 + 7p + 10$ **e** $w^2 + 9w + 20$ **f** $b^2 + 6b + 5$

g $e^2 + 5e + 6$ **h** $r^2 + 4r + 4$ **i** $n^2 + 11n + 10$

j $a^2 + 11a + 30$ **k** $d^2 + 10d + 24$ **l** $y^2 + 15y + 44$

3 Factorise each quadratic expression.

EXAMPLE 25

R **a** $p^2 - 5p + 4$ **b** $v^2 - 11v + 30$ **c** $y^2 - 7y + 12$

d $x^2 - 12x + 20$ **e** $a^2 - 9a + 18$ **f** $e^2 - 8e + 12$

g $c^2 - 14c + 48$ **h** $w^2 - 5w + 6$ **i** $d^2 - 18d + 77$

j $u^2 - 3u + 2$ **k** $n^2 - 8n + 16$ **l** $b^2 - 15b + 56$

4 Factorise each quadratic expression.

EXAMPLE 26

R **a** $x^2 + 3x - 4$ **b** $g^2 + 5g - 24$ **c** $y^2 - 2y - 3$

d $r^2 - 5r - 14$ **e** $m^2 + 2m - 15$ **f** $a^2 + a - 2$

g $m^2 - 9m + 20$ **h** $h^2 - 3h - 4$ **i** $d^2 - 6d + 9$

j $w^2 + 4w - 12$ **k** $a^2 - 4a - 12$ **l** $k^2 + 5k - 14$

m $c^2 - 7c - 18$ **n** $e^2 - 6e - 27$ **o** $u^2 + 6u - 55$

p $p^2 + 9p - 36$ **q** $x^2 + 9x - 52$ **r** $q^2 - 9q - 70$

Foundation Standard Complex

5 Factorise each expression, given it is a perfect square.

R a $m^2 + 4m + 4$
b $p^2 + 20p + 100$
c $a^2 - 10a + 25$
d $y^2 - 14y + 49$
e $w^2 + 24w + 144$
f $a^2 - 3a + 2\frac{1}{4}$

6 Factorise each quadratic expression. Look for the highest common factor first.

R a $3m^2 + 9m + 6$
b $2y^2 + 2y - 4$
c $5t^2 - 10t - 400$
d $5e^4 + 25e^3 - 120e^2$
e $x^3 - x^2 - 110x$
f $4b^2 - 4b - 168$
g $4w^2 + 4w - 48$
h $3a^3 - 9a^2 - 12a$
i $2e^2 + 18e + 40$
j $24 - 5t - t^2$
k $42 + u - u^2$
l $28 + 3x - x^2$
m $12 - b - b^2$
n $7k - 12 - k^2$
o $12x - 35 - x^2$

7 Factorise each quadratic expression.

R a $h^2 - 3h - 18$
b $18 - 7t - t^2$
c $w^2 + 8w + 7$
d $k^2 - 12k - 45$
e $v^2 + 8v - 20$
f $3c^2 + 30c + 75$
g $q^2 - 6q + 5$
h $a^2 - 4a + 3$
i $6x^2 + 36x - 96$
j $x^2 - 16x + 64$
k $8u^2 - 24u - 32$
l $b^2 + 11b + 30$
m $y^2 - 12y + 36$
n $5r^2 - 5r - 10$
o $4l^2 - 8l - 32$
p $g^2 - 24g + 80$
q $108 - 6d - 2d^2$
r $26 + 11n - n^2$

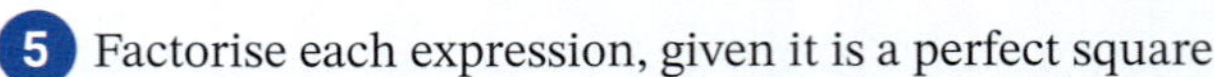

POWER PLUS ANSWERS ON P.630

1.08

1 If each expression represents an integer, write expressions for the next 3 consecutive integers.

a p b $w + 1$ c $2n + 1$ d $h - 3$

2 Evaluate each expression given $p = 2$, $q = 5$ and $r = -8$.

a $\frac{p+q+r}{3}$ b $\frac{pqr}{3}$ c $q^2 - 2q$

d $\frac{p-r}{q}$ e $\frac{qr}{p}$ f $\frac{p}{q}+\frac{r}{q}$

g $(p + q)^2 - 2pq$ h $p^3 - q^3 - r^3$ i $(p - q)^2 - r^2$

3 For this shape, write an expression for:

a the length of CD.

b the length of BC.

c its perimeter.

d its area.

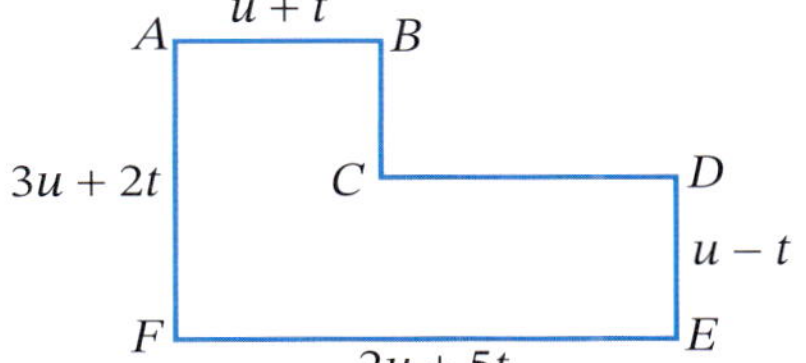

4 Use order of operations to simplify each expression.

a $5 \times 6a \div 3$ b $12r \div 2r \times 3$

c $8 \times (6nm \div 3n)$ d $10p \times (6pq \div 3q)$

e $15h - 9h \div 3$ f $3t + 5 \times 2t$

g $3 \times (x - 5x)$ h $(6c + 14c) \div 5$

i $3r + r \times 5 - 1$ j $(3r + r) \times (5 - 1)$

5 Expand and simplify each expression.

a $7 - (2m + 5)$ b $5x - 3x(2 - x)$

c $5q - 2(3 - 2q) + 5$ d $6 - (1 + 3m) - 2m$

e $(2a + 5)^2$ f $(y + 1)(y - 1) - (2y - 3)$

6 Find the average of $6r$, $2r + 8$, $r - 5$, $2r$, $r + 7$ and $3r + 8$.

7 A rectangular garden has its longer sides each 5 m longer than its shorter sides. If its longer sides each have a length of y m, then write a simplified expression for:

a the area of the garden b the perimeter of the garden.

8 How many hours does it take a car to travel a distance of $20t^2$ km at an average speed of $4t$ km/h?

9 Factorise each expression.

a $8a + 12b - 4$ b $20xy + 5y - 10x^2$

c $m^2t + mt + mt^2$ d $32 - 24y + 4y^2$

e $12v^2 + 4vw - 16w^2$ f $-12 - 9f - 6f^2$

g $\frac{1}{2}p+\frac{1}{2}q$ h $\frac{3}{4}k-\frac{1}{4}kt$

1 CHAPTER REVIEW

Quiz
Language of maths 1

Language of maths

binomial product	common factor	consecutive	difference
distributive law	evaluate	expand	expression
factor	factorise	formula	highest common factor (HCF)
like terms	product	quotient	simplify
substitute	sum	term	variable

1 Why is it possible to add or subtract terms like $6n$ and $4n$ but not $7x$ and $2y$?

2 What word means to rewrite an algebraic expression like $3(u - 5)$ without the grouping symbols?

3 What word means to replace a variable with a number in an algebraic expression?

4 Explain what **product** means.

5 In algebra, what is the opposite of **expand**?

6 Copy and complete: A binomial is an algebraic expression with ____ _________.

Worksheet
Mind map: Algebra

Topic summary

- Write down a list of the new things you have learnt, in your own words.
- What parts of this topic did you like?
- What parts of the topic did you find difficult or not understand?
- Make a list of the skills you have learnt in this topic, such as substituting and expanding.

Print (or copy) and complete this mind map of the topic, adding detail to its branches and using pictures, symbols and colour where needed. Ask your teacher to check your work.

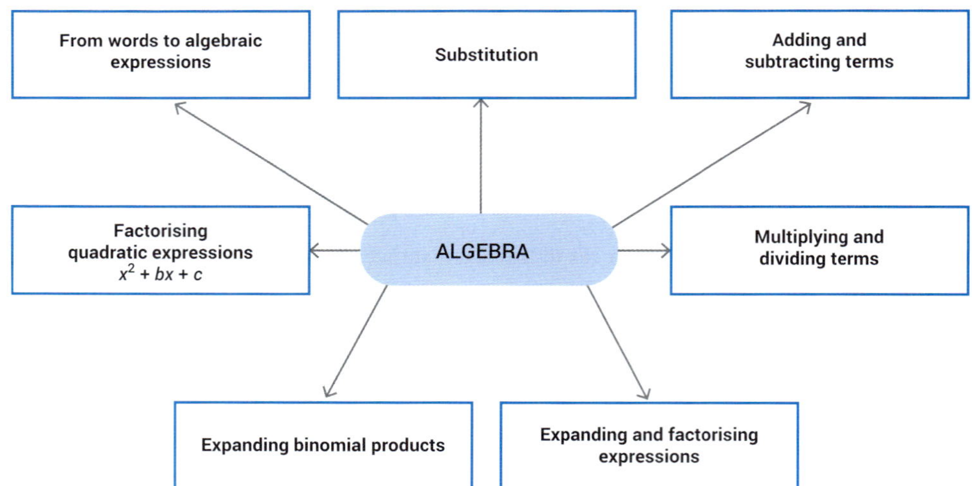

1 TEST YOURSELF

ANSWERS ON P. 630

1 Convert each statement to an algebraic expression. 1.01

Quiz
Test yourself 1

- **a** the cost of n pizzas at \$$P$ each
- **b** how many 3s divide into m
- **c** 5 rides at \$$A$ and 8 rides at \$$B$
- **d** decrease \$100 by \$$y$
- **e** the number of minutes in H hours
- **f** the number of litres in m mL
- **g** multiply P by 4 and add 5
- **h** add 5 to T, then multiply by 4
- **i** Q divided by 4, then minus 5
- **j** Q minus 5, then divided by 4

2 If $a = 3$, and $b = 2$, evaluate each expression. 1.02

- **a** $a + b$
- **b** ab
- **c** $3a + 5b$
- **d** $7b - a$
- **e** $8a \div b$
- **f** $\frac{a+b}{2}$
- **g** $a^2 - b^2$
- **h** $2a^2$

3 Simplify each expression. 1.03

- **a** $9y + 7y$
- **b** $4ab - 7ab$
- **c** $3x^2 + 5x^2$
- **d** $7p + 2q - 5p$
- **e** $4w - 5 + 2w$
- **f** $8x^2 - 13x - 5x^2$
- **g** $2m + 3n + 5m + 7n$
- **h** $5u + 7 - 2u - 3$
- **i** $12d^2 + 5d - 7d + 4$

4 Simplify each expression. 1.04

- **a** $9 \times 2n$
- **b** $g \times 3h$
- **c** $a \times d$
- **d** $(-3) \times 2r$
- **e** $12y^2 \div 3y$
- **f** $2xy \div y$
- **g** $36a \div 9a$
- **h** $15t^2 \div 5t^2$
- **i** $20pq \div (-5q)$

5 Expand and simplify each expression. 1.05

- **a** $5(r - 2)$
- **b** $a(a - 10)$
- **c** $3y(2x - 5y)$
- **d** $3p(4p - 7q)$
- **e** $-6n(n - 9)$
- **f** $-6t(t + 7)$
- **g** $b(b + 1) + 5(b + 1)$
- **h** $3x(2x - 7) - 5(2x - 7)$
- **i** $12 - 6(2x - 3)$

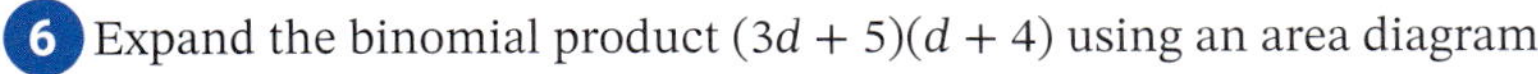

6 Expand the binomial product $(3d + 5)(d + 4)$ using an area diagram. 1.06

7 Expand each binomial product. 1.07

- **a** $(a + 2)(a + 3)$
- **b** $(y + 3)(y - 7)$
- **c** $(t - 3)(t + 8)$
- **d** $(h - 5)(h - 4)$
- **e** $(4 - g)(3 + g)$
- **f** $(3p + 1)(2p + 3)$
- **g** $(5x + 1)(4x - 3)$
- **h** $(4r - 5)(3r + 4)$
- **i** $(5 - 3q)(2q - 1)$
- **j** $(3m + 5)^2$
- **k** $(7 - 2y)^2$
- **l** $(3w - 5)(2 - w)$

8 Factorise each expression. 1.08

- **a** $4m - 12$
- **b** $xy + xz$
- **c** $-3g - 30$
- **d** $16w^2 + 24w$
- **e** $-8q + 16$
- **f** $15k - 10hk^2$
- **g** $3(y - 6) + y(y - 6)$
- **h** $3x(2x - 1) - 2(2x - 1)$
- **i** $p(p + 6) - 2(p + 6)$

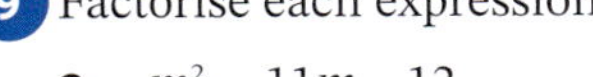

9 Factorise each expression. 1.08

- **a** $m^2 - 11m - 12$
- **b** $y^2 - 4y + 3$
- **c** $h^2 - 8h - 65$
- **d** $p^2 + 6p - 27$
- **e** $3n^2 + 9n + 6$
- **f** $4y^2 + 4y - 8$

2

MEASUREMENT, NUMBER

Pythagoras' theorem

Pythagoras was an ancient Greek mathematician who lived in the 5th century BCE. The theorem (or rule) which carries his name was well-known before his time, but Pythagoras may have been the first to prove it. Over 300 proofs for the theorem are known today. Pythagoras' theorem is the most famous mathematical formula, and it is still used today in architecture, engineering, surveying and astronomy.

iStock.com/kievith

Chapter outline

		Proficiencies				
2.01	Squares, square roots and surds	U	F		R	C
2.02	Pythagoras' theorem	U	F	PS	R	C
2.03	Finding the hypotenuse	U	F			
2.04	Finding a shorter side	U	F	PS	R	
2.05	Mixed problems	U	F	PS	R	
2.06	Testing for right-angled triangles	U	F		R	C
2.07	Pythagorean triads	U	F		R	C
2.08	Pythagoras' theorem problems	U	F	PS	R	

U = Understanding
F = Fluency
PS = Problem solving
R = Reasoning
C = Communication

Wordbank

converse A rule or statement turned back-to-front; the reverse statement

hypotenuse The longest side of a right-angled triangle; the side opposite the right angle

irrational number A number such as π or $\sqrt{2}$ that cannot be expressed as a fraction

Pythagoras An ancient Greek mathematician who discovered an important formula about the sides of a right-angled triangle

Pythagorean triad A set of 3 numbers that follow Pythagoras' theorem, such as {3, 4, 5}.

surd A square root (or other root) whose exact value cannot be found

theorem Another name for a formal rule or formula

Quiz
Wordbank 2

Videos (4):

2.03, 2.04 Pythagoras' theorem 1 • Pythagoras' theorem 2

2.06 Testing for right-angled triangles

2.08 Pythagoras' theorem problems

***Twig* videos (3):**

2.02 Irrational numbers: Pythagoras • Proving Pythagoras • Building the pyramids

***PhET* interactives (1):**

2.02 TBA

Quizzes (6):

- Wordbank 2
- SkillCheck 2
- Mental skills 2A
- Mental skills 2B
- Language of maths 2
- Test yourself 2

Skillsheets (1):

2.02 Pythagoras' theorem

Worksheets (6):

2.02 Pythagoras' discovery • Pythagoras 1

2.06 Testing for right-angled triangles

2.07 Pythagorean triads

2.08 Applications of Pythagoras' theorem

Mind map: Pythagoras' theorem

Puzzles (2):

2.05 Pythagoras 1 • Pythagoras 2

Technology (2):

2.03, 2.04 Pythagoras' theorem

2.07 Pythagorean triples

Spreadsheets (2):

2.03, 2.04 Pythagoras' theorem

2.07 Pythagorean triples

Nelson MindTap

To access resources above, visit **cengage.com.au/nelsonmindtap**

2 In this chapter you will:

✓ identify irrational numbers, surds and real numbers.

✓ solve problems involving Pythagoras' theorem, writing the answers in decimal or surd form.

✓ test whether a triangle is right-angled.

✓ investigate Pythagorean triads.

SkillCheck

ANSWERS ON P. 631

Quiz SkillCheck 2

1 Copy and complete:

a 50 cm = ____ mm **b** 200 cm = ____ m **c** 1.4 km = ____ m

d 62 mm = ____ cm **e** 8500 mm = ____ m **f** 1200 cm = ____ km

2 Evaluate each expression.

a 4^2 **b** 10.3^2 **c** $3^2 + 5^2$

d $8^2 - 6^2$ **e** $\sqrt{49}$ **f** $\sqrt{121}$

3 Round each number to one decimal place.

a 2.454 **b** 831.9342 **c** 372.658

4 Round each number to 2 decimal places.

a 28.777 **b** 7.9515 **c** 536.856

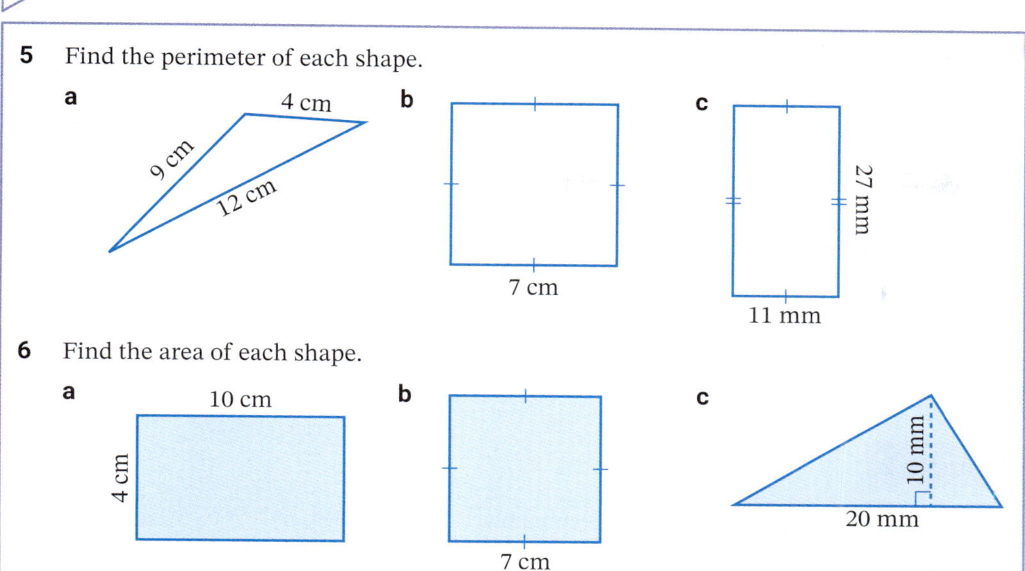

Squares, square roots and surds 2.01

- $\sqrt{25} = 5$ because $5^2 = 25$ 'the square root of 25'.
- $\sqrt{81} = 9$ because $9^2 = 81$ 'the square root of 81'.

Most square roots do not give exact answers like the ones above.
For example, $\sqrt{7} = 2.645751311\ldots \approx 2.6$. Such roots are called **surds**.

A surd is a square root ($\sqrt{\ }$), cube root ($\sqrt[3]{\ }$), or any type of root whose exact decimal or fraction value cannot be found. As a decimal, its digits run endlessly *without repeating* (like π), so they are neither terminating nor recurring decimals.

Rational numbers such as fractions, decimals and percentages, can be expressed in the form $\frac{a}{b}$, where a and b are integers (and $b \neq 0$), but surds are **irrational numbers** because they cannot be expressed in this form. The rational and irrational numbers together make up the real numbers.

Example 1

Evaluate each expression.

a 5.3^2 **b** $(-18)^2$ **c** $7^2 + 9^2$

SOLUTION

a $5.3^2 = 28.09$ On a calculator, enter: 5.3 [x^2] [=]

b $(-18)^2 = 324$ [(] [(−)] 18 [)] [x^2] [=]

c $7^2 + 9^2 = 130$ 7 [x^2] [+] 9 [x^2] [=]

Example 2

Evaluate each expression, correct to 2 decimal places where necessary.

a $\sqrt{1024}$ **b** $\sqrt{89}$ **c** $\sqrt{15^2 + 19^2}$

SOLUTION

a $\sqrt{1024} = 32$ [√] 1024 [=]

b $\sqrt{89} = 9.433981...$ [√] 89 [=] [*]

≈ 9.43

c $\sqrt{15^2 + 19^2} = 24.207436$ [√] [(] 15 [x^2] [+] 19 [x^2] [)] [=]

≈ 24.21

Example 3

Select the surds from this list of square roots: $\sqrt{72}$, $\sqrt{121}$, $\sqrt{64}$, $\sqrt{90}$, $\sqrt{28}$.

SOLUTION

$\sqrt{72} = 8.4852...$

$\sqrt{72}$ is called an '*exact* value' while 8.4852 is called its '*approximate* value'.

$\sqrt{121} = 11$

$\sqrt{64} = 8$

$\sqrt{90} = 9.4868...$

$\sqrt{28} = 5.2915...$

So the surds are $\sqrt{72}$, $\sqrt{90}$ and $\sqrt{28}$.

Squares, square roots and surds

U F R C

EXAMPLE 1

1 Evaluate each expression.

a 4^2 **b** 16^2 **c** $(6.3)^2$ **d** $(0.7)^2$

e $(12.8)^2$ **f** $(0.02)^2$ **g** $(-1)^2$ **h** $(-9)^2$

i $(-13)^2$ **j** $(-1.2)^2$ **k** $2^2 + 7^2$ **l** $20^2 + 7^2$

m $19^2 - 14^2$ **n** $12^2 - 10^2$ **o** $(0.3)^2 + (0.4)^2$ **p** $(3.2)^2 - (1.5)^2$

EXAMPLE 2

2 Evaluate each square root.

a $\sqrt{196}$ **b** $\sqrt{900}$ **c** $\sqrt{625}$ **d** $\sqrt{361}$

e $\sqrt{1.69}$ **f** $\sqrt{2.89}$ **g** $\sqrt{10.24}$ **h** $\sqrt{156.25}$

3 Evaluate each square root, correct to 2 decimal places.

a $\sqrt{12}$ **b** $\sqrt{450}$ **c** $\sqrt{1001}$ **d** $\sqrt{325}$

e $\sqrt{153}$ **f** $\sqrt{7002}$ **g** $\sqrt{938}$ **h** $\sqrt{42}$

4 Which of the following is closest to the value of $\sqrt{6^2+11^2}$? Select the correct answer **A**, **B**, **C** or **D**.

A 127 **B** 17 **C** 12.5 **D** 13.7

5 Evaluate each expression, correct to 2 decimal places.

a $\sqrt{5^2+7^2}$ **b** $\sqrt{6^2+11^2}$ **c** $\sqrt{17^2-12^2}$ **d** $\sqrt{8^2-3^2}$

6 Evaluate each expression, correct to one decimal place.

a $\sqrt{(1.5)^2+(4.2)^2}$ **b** $\sqrt{(12.5)^2-(7.1)^2}$ **c** $\sqrt{(25.7)^2+(18.2)^2}$

EXAMPLE 3

7 Which of the following is a surd? Select the correct answer **A**, **B**, **C** or **D**.

C **A** $\sqrt{9}$ **B** $\sqrt{225}$ **C** $\sqrt{160}$ **D** $\sqrt{81}$

8 Which of the following is NOT a surd? Select the correct answer **A**, **B**, **C** or **D**.

C **A** $\sqrt{77}$ **B** $\sqrt{144}$ **C** $\sqrt{18}$ **D** $\sqrt{200}$

9 Select the surds from the following list of square roots.

C $\sqrt{32}$ $\sqrt{33}$ $\sqrt{289}$ $\sqrt{81}$ $\sqrt{4.9}$

$\sqrt{52}$ $\sqrt{121}$ $\sqrt{144}$ $\sqrt{196}$ $\sqrt{200}$

10 Without using a calculator, determine between which 2 consecutive numbers the value of $\sqrt{40}$ lies. Select **A**, **B**, **C** or **D**.

R **A** 4 and 5 **B** 5 and 6 **C** 6 and 7 **D** 7 and 8

11 Is the statement $\sqrt{x^2} = x$ always true?

R

12 Is the statement $\left(\sqrt{x}\right)^2 = x$ always true?

R

Foundation Standard Complex

INVESTIGATION

Pythagoras' theorem

Pythagoras' theorem is a rule that describes the relationship between the longest side of a right-angled triangle and the other 2 (shorter) sides.

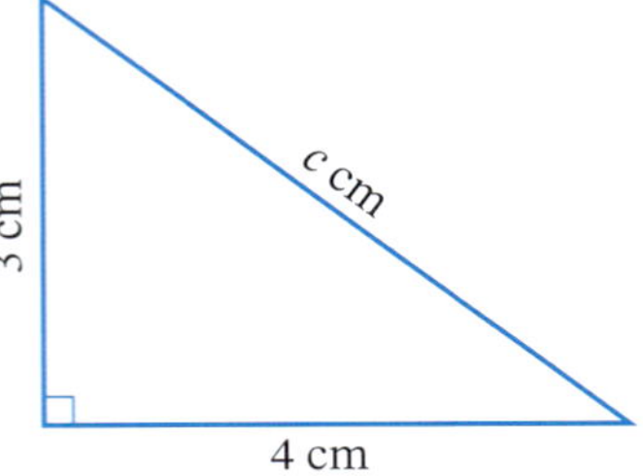

Use *dynamic geometry software* (or a pencil and ruler) to discover Pythagoras' theorem by following the steps below.

1 a Draw a right-angled triangle with the shorter sides of lengths 3 cm and 4 cm.

b Measure the length of the longest side, c cm.

c Copy the table below and enter the value for the longest side c when the 2 shorter sides are $a = 3$ and $b = 4$.

Shorter sides (cm)		Longest side (cm)				
a	b	c	a^2	b^2	$a^2 + b^2$	c^2
3	4					
6	2.5					
6	8					
5	12					

d Complete the row by evaluating a^2, b^2, $a^2 + b^2$ and c^2.

2 Complete the remaining rows by drawing right-angled triangles using the measurements given for the shorter sides a and b, finding the length of the longest side, c, and evaluating a^2, b^2, $a^2 + b^2$ and c^2.

3 What is the relationship between the longest side and the other 2 sides of a right-angled triangle?

2.02 Pythagoras' theorem

Worksheet
Pythagoras' discovery

Skillsheet
Pythagoras' theorem

Videos
Irrational numbers: Pythagoras

Proving Pythagoras

Building the pyramids

A **right-angled triangle** has one **right angle** (90°) and 2 acute angles. Its longest side is called the **hypotenuse**, which is always opposite the right angle.

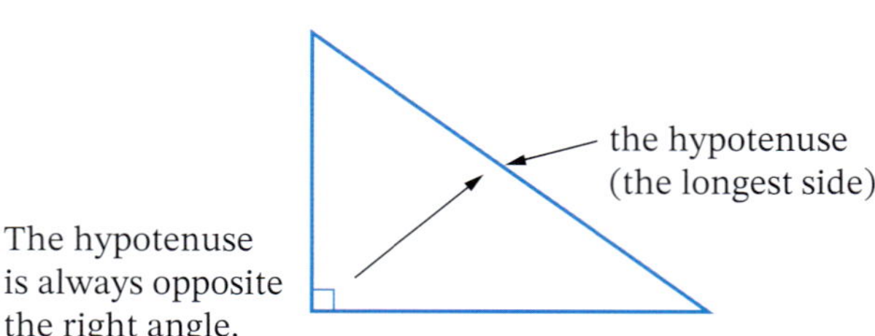

The hypotenuse in the below diagram is AB, opposite to the right angle $\angle C$.

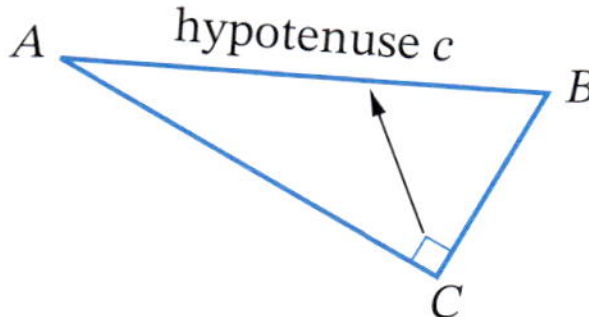

A **theorem** is a formal rule or formula. The ancient Greek mathematician Pythagoras discovered the following theorem.

Pythagoras' theorem

For any right-angled triangle, the square of the hypotenuse is equal to the sum of the squares of the other 2 sides.

If c is the length of the hypotenuse, and a and b are the lengths of the other 2 sides, then:

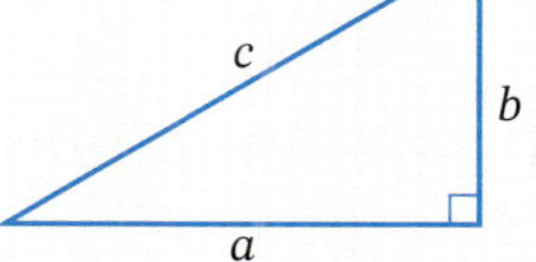

$$c^2 = a^2 + b^2$$

that is,

$$(\text{hypotenuse})^2 = (\text{shorter side 1})^2 + (\text{shorter side 2})^2$$

Example 4

Write **Pythagoras' theorem** for this triangle.

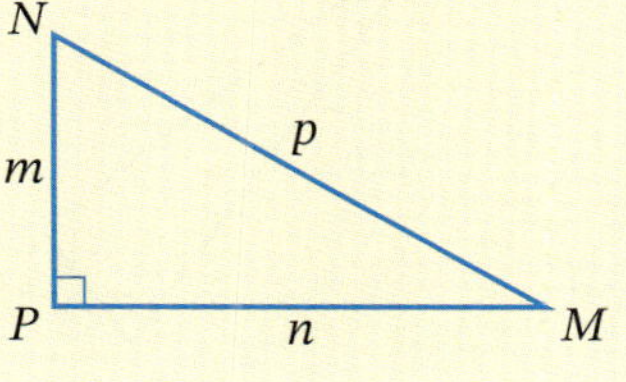

SOLUTION

p is the hypotenuse, so $p^2 = m^2 + n^2$.

OR

NM is the hypotenuse, so $NM^2 = NP^2 + PM^2$.

EXERCISE 2.02 ANSWERS ON P.631

Pythagoras' theorem

U F PS R C

1 In a right-angled triangle, what is the *hypotenuse*? Select the correct answer **A**, **B**, **C** or **D**.

A the 90° angle **B** an acute angle **C** the shortest side **D** the longest side

Foundation Standard Complex

2 For each right-angled triangle, name the hypotenuse.

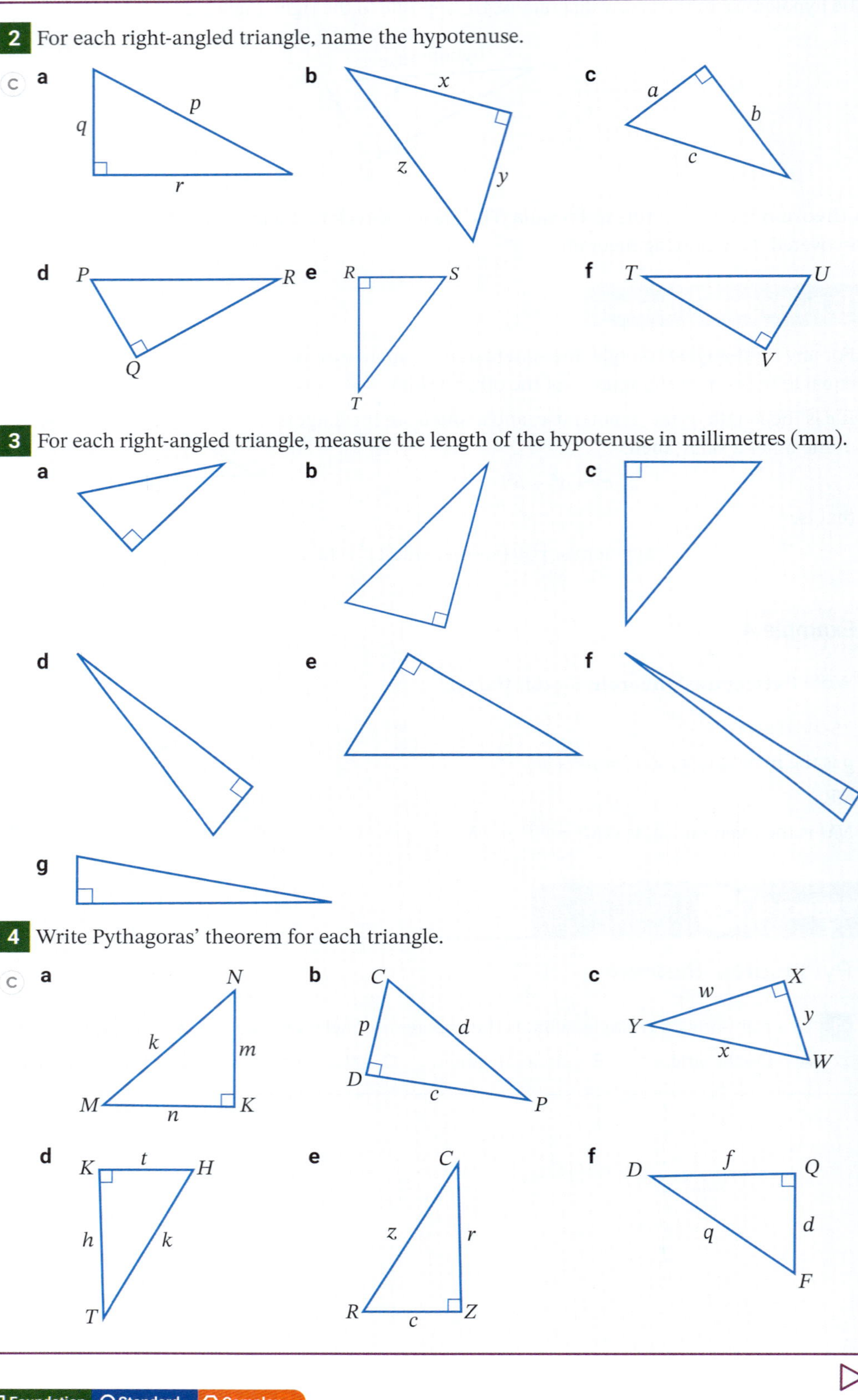

3 For each right-angled triangle, measure the length of the hypotenuse in millimetres (mm).

4 Write Pythagoras' theorem for each triangle.

☐ Foundation ○ Standard ⬡ Complex

5 For each triangle, select the correct statement of Pythagoras' theorem.

C **a**

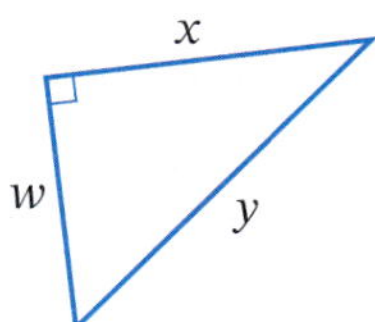

A $y^2 = x^2 + w^2$ **B** $x^2 = w^2 + y^2$

C $w^2 = x^2 + y^2$ **D** $y = x + w$

b

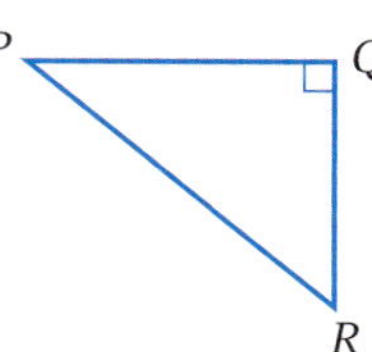

A $PR = PQ + QR$ **B** $QR^2 = PQ^2 + PR^2$

C $RP^2 = RQ^2 + PQ^2$ **D** $PQ^2 = PR^2 + QR^2$

c

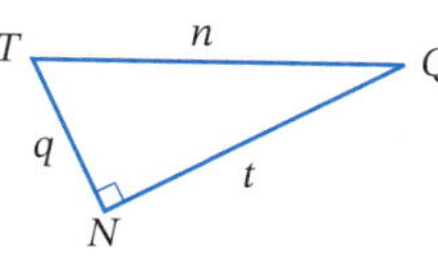

A $q^2 = n^2 + t^2$ **B** $TN^2 + TQ^2 = NQ^2$

C $n^2 = q^2 + t^2$ **D** $TN^2 = TQ^2 + NQ^2$

d

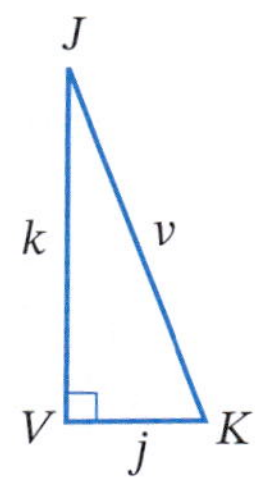

A $VK^2 + VJ^2 = JK^2$ **B** $j^2 = v^2 + k^2$

C $v^2 = j^2 + k^2$ **D** $VK^2 = JV^2 + JK^2$

6 Write Pythagoras' theorem for each triangle.

C **a**

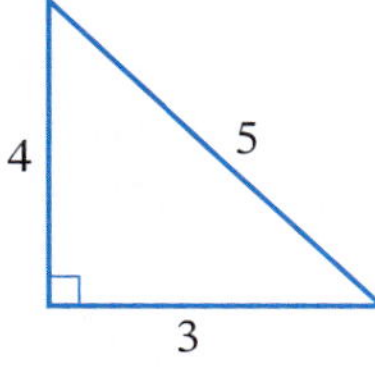

b

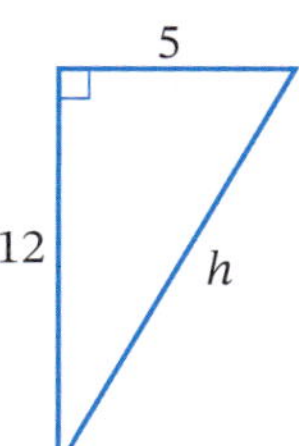

c

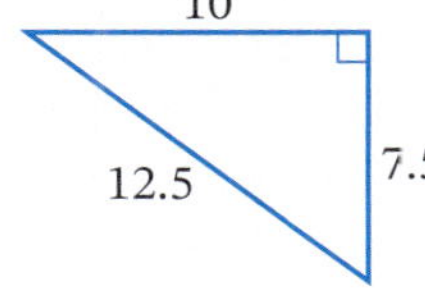

d

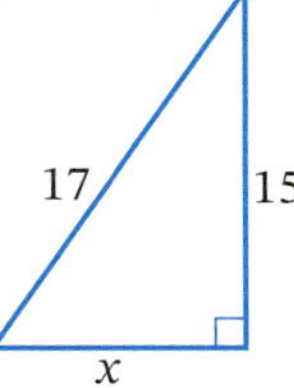

e

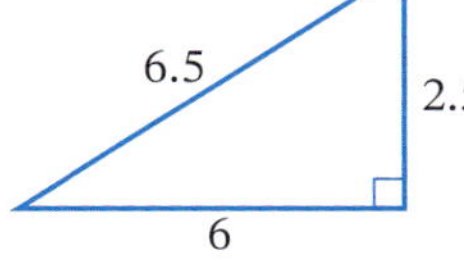

f

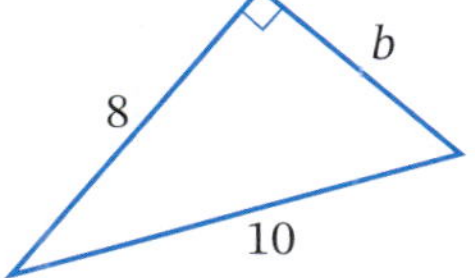

Foundation Standard Complex

Worksheet
Pythagoras 1

7 Pythagoras' theorem can be demonstrated visually by dissection (cutting up shapes).

PS R

a On a sheet of paper, construct a right-angled triangle and draw squares on all 3 sides. Alternatively, download the worksheet 'Pythagoras 1'.

b In the 2 smaller squares, draw lines perpendicular to the hypotenuse as shown, and in the bottom square, draw a line parallel to the hypotenuse as shown.

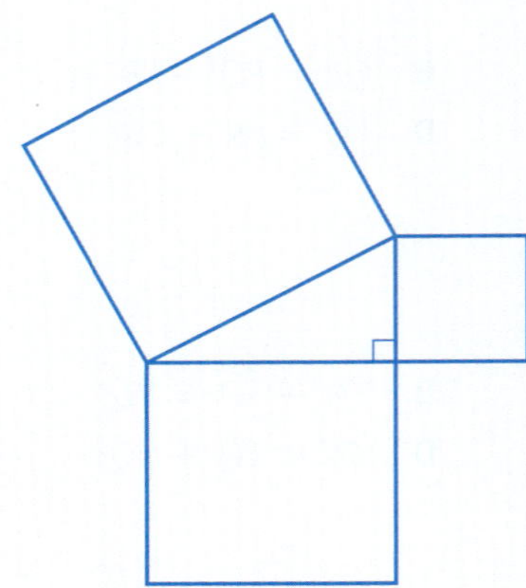

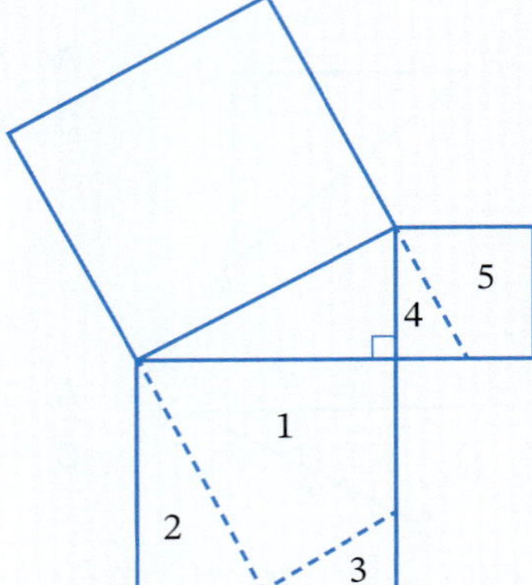

c Number the regions 1, 2, 3, 4, 5 as shown.

d Cut out the 5 pieces and rearrange them to form the square on the hypotenuse.

e Does the area of the 2 smaller squares equal the area of the larger square?

f How does this 'prove' Pythagoras' theorem?

g Repeat this question on another right-angled triangle.

DID YOU KNOW?

Egyptian rope stretchers

In ancient times, whenever the Nile River flooded, Egyptian farmers would have to rebuild fences around their fields afterwards. To form right angles for this, they used a secret 'magical' technique called rope stretching to create a 90° angle, making a length of rope with 13 knots tied in it at equal intervals.

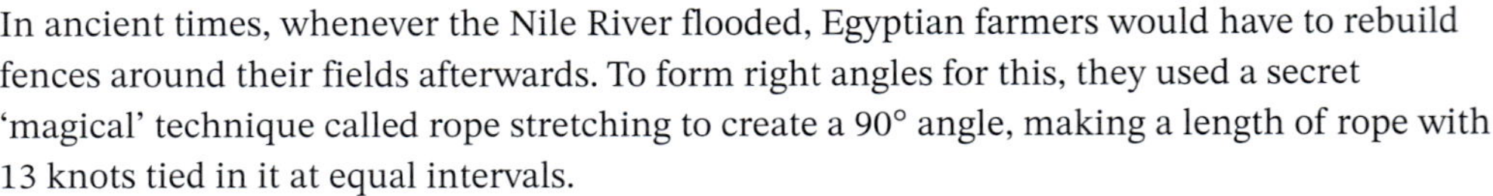

The rope was fixed to the ground at the 4th and 8th knots, and stretched around to create a triangle with side lengths 3, 4 and 5 units. The right angle is formed at the 4th knot.

☐ Foundation ○ Standard ○ Complex

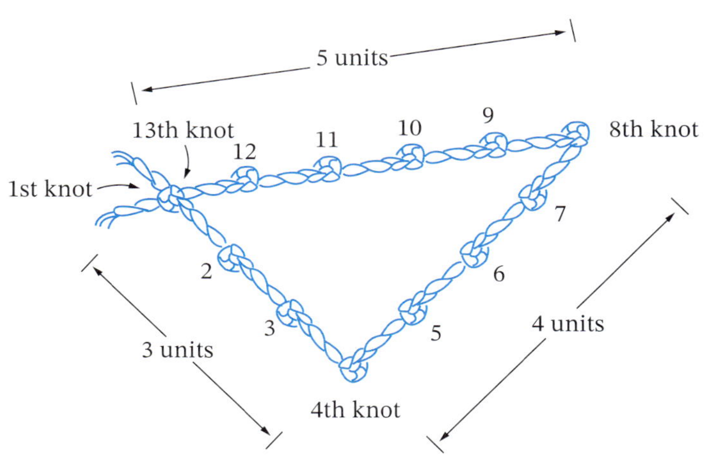

The builders of the great pyramids also used rope stretching, but could not understand why this 'trick' produced a right angle. Pythagoras studied with the rope stretchers and eventually worked it out.

Make and stretch a knotted rope to create a right angle using a 3-4-5 triangle.

Finding the hypotenuse 2.03

Pythagoras' theorem is used to find the length of one side of a right-angled triangle when the other 2 sides are known. In this section, we will look at finding the hypotenuse when the lengths of the 2 shorter sides are known.

Technology Pythagoras' theorem

Spreadsheet Pythagoras' theorem

Video Pythagoras' theorem 1

Example 5

Find the value of c in this triangle.

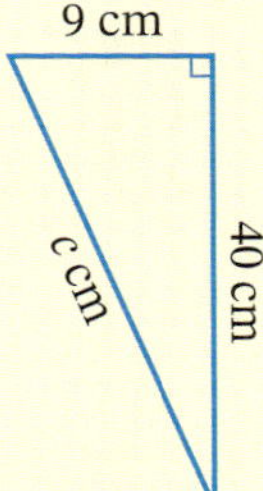

SOLUTION

We want to find the length of the hypotenuse.

Using Pythagoras' theorem:

$$c^2 = 9^2 + 40^2$$
$$= 1681$$
$$c = \sqrt{1681} \quad \text{Use the square root to find } c.$$
$$= 41$$

An answer of $c = 41$ looks reasonable because:

- the hypotenuse is the longest side.
- from the diagram, the hypotenuse looks a little longer than the side that is 40 cm.

Video
Pythagoras' theorem 2

Example 6

Find the length of the cable supporting this flagpole:

a as a surd.

b correct to one decimal place.

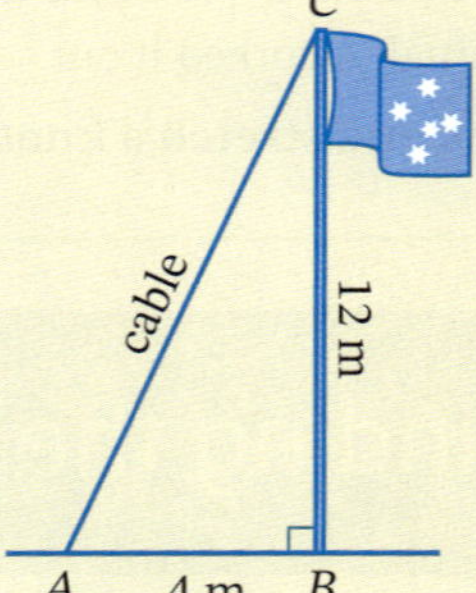

SOLUTION

a $AC^2 = 12^2 + 4^2$ — This is the answer as a surd (in $\sqrt{\ }$ form).
$= 160$
$AC = \sqrt{160}$ m

b $AC = \sqrt{160}$ — From part **a**.
$= 12.6491...$ — Rounded to one decimal place.
≈ 12.6 m

Example 7

In $\triangle PQR$, $\angle P = 90°$, $PQ = 25$ cm and $PR = 32$ cm. Sketch the triangle and find the length of the hypotenuse, correct to one decimal place.

9780170465618

SOLUTION

$p^2 = 32^2 + 25^2$

$= 1649$

$p = \sqrt{1649}$

$= 40.607\,881\,01...$

≈ 40.6

p is the side opposite to P and is also the hypotenuse.

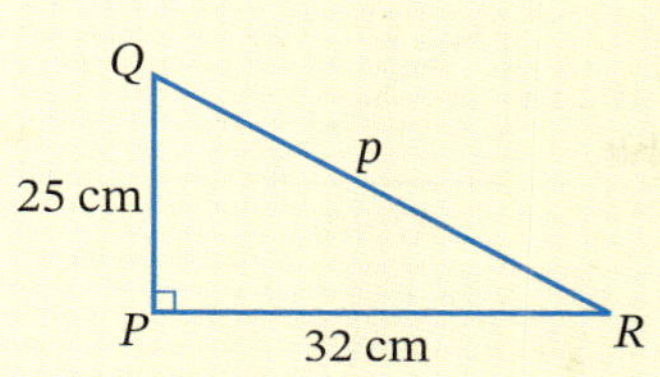

$\therefore$ The length of the hypotenuse is 40.6 cm.

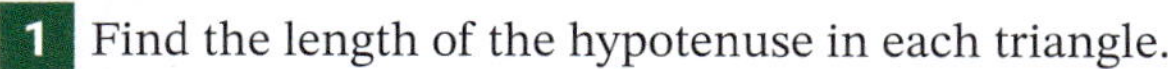

EXERCISE 2.03 ANSWERS ON P.631

ANSWERS ON P.631

Finding the hypotenuse

U F

EXAMPLE 5

1 Find the length of the hypotenuse in each triangle.

a

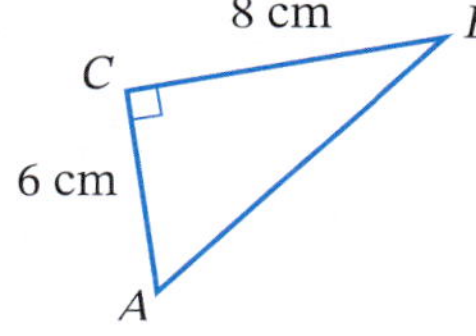

b

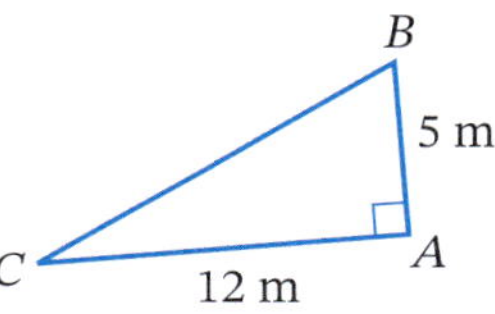

c

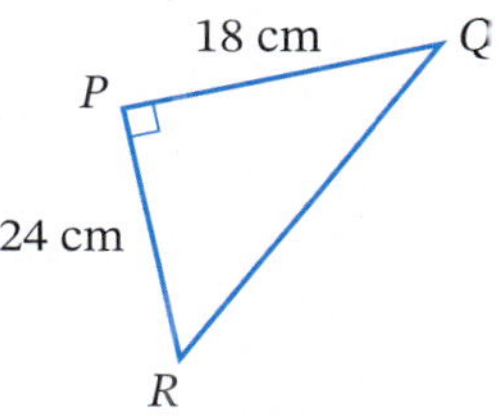

d

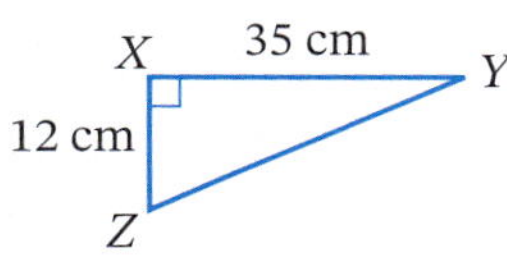

e

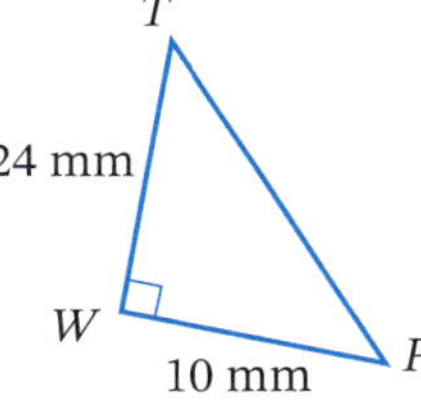

f

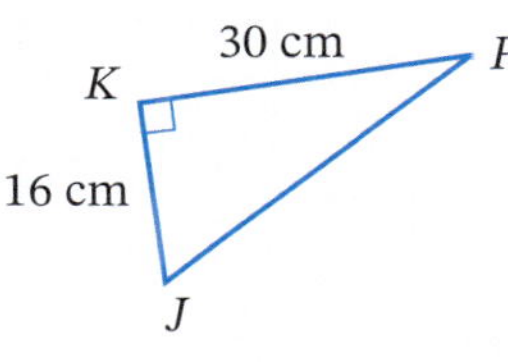

g

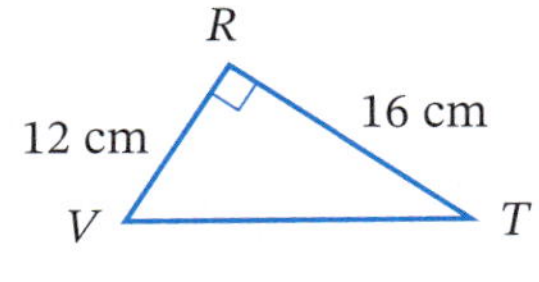

h

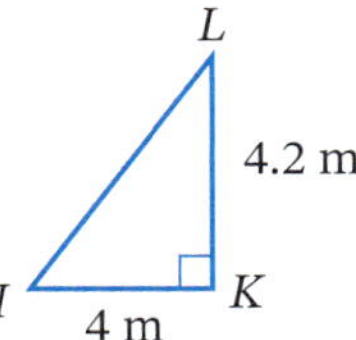

i

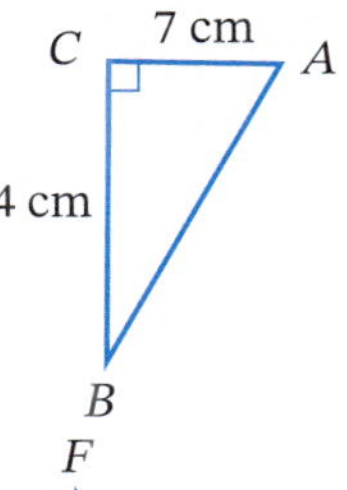

j

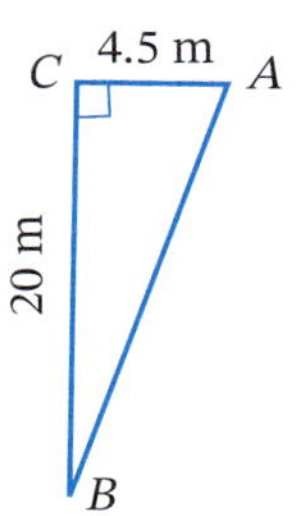

k

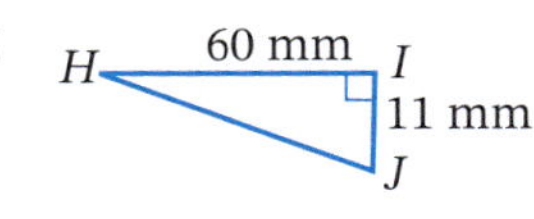

l

F

18 m

D

7.5 m

L

Foundation Standard Complex

EXAMPLE 6

2 Find the length of the hypotenuse in each triangle, as a surd.

a

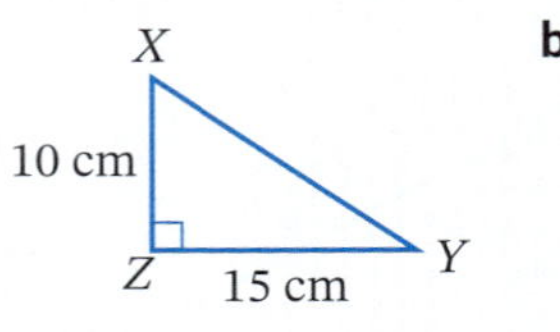

b

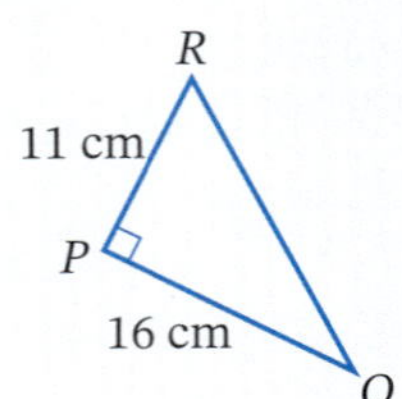

c

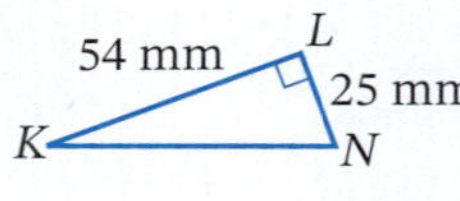

d

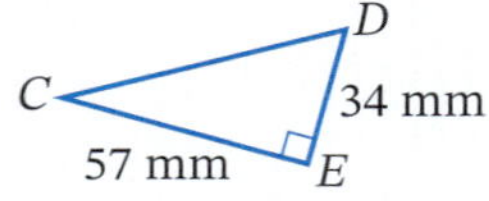

e

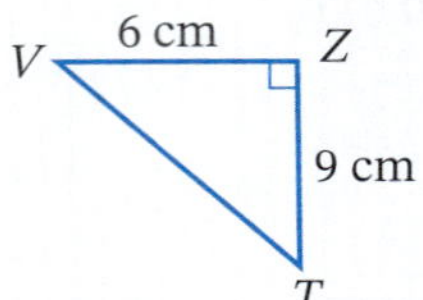

f

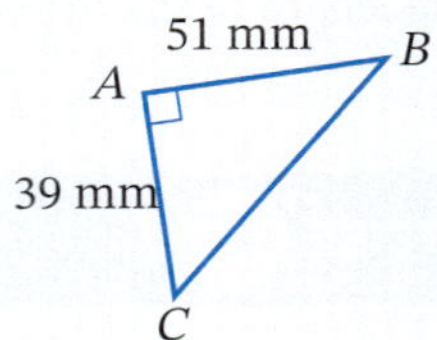

3 Find the length of the hypotenuse in each triangle, correct to one decimal place.

a

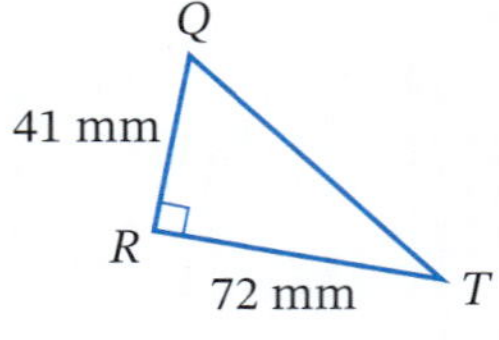

b

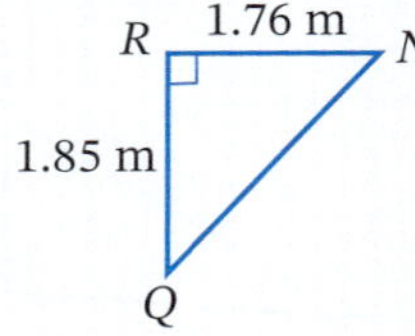

c

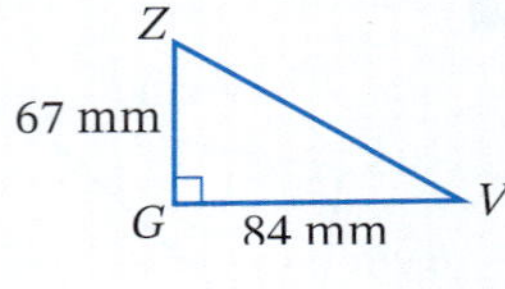

d

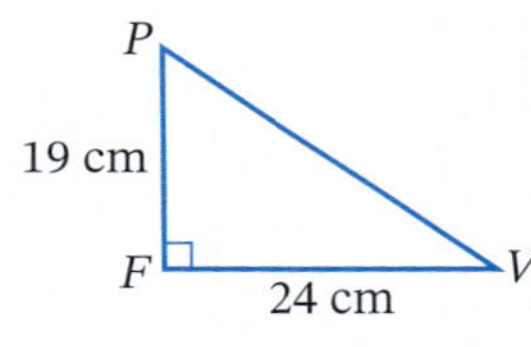

e

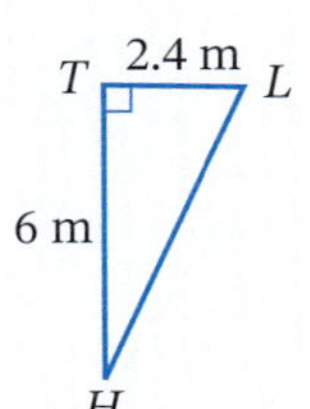

f

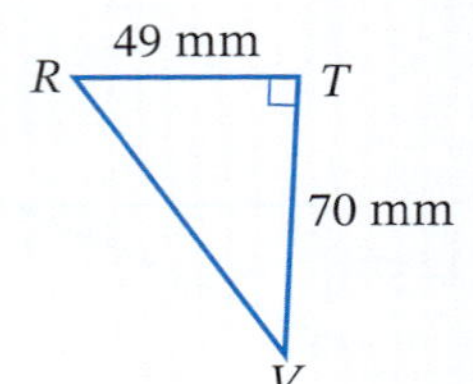

4 A rectangular field is 100 m long and 50 m wide. How far is it from one corner to the opposite corner, along the diagonal? Select the correct answer **A**, **B**, **C** or **D**.

A 150 m **B** 111.8 m

C 100.2 m **D** 98.3 m

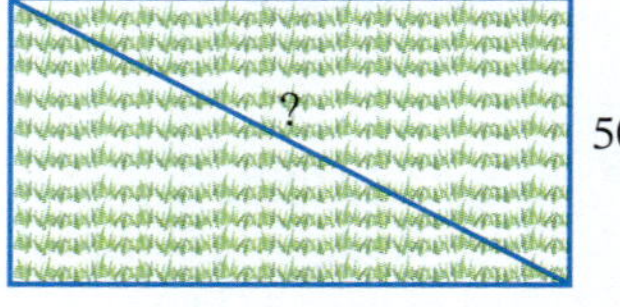

▷

Foundation Standard Complex

5 A firefighter places a ladder on a window sill 4.5 m above the ground. If the foot of the ladder is 1.6 m from the wall, how long is the ladder, correct to one decimal place?

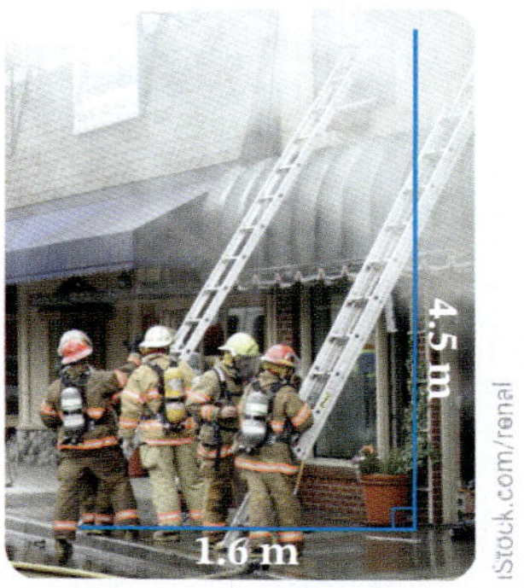

iStock.com/renal

6 The gable of a roof is 6 m long. If the roof is 1.6 m high, how long is the rafter?

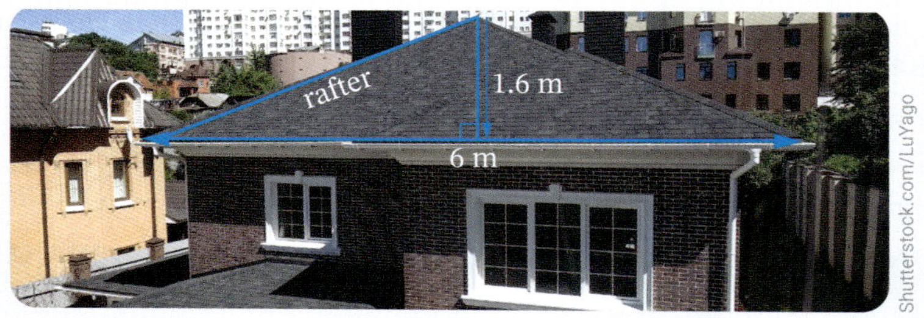

Shutterstock.com/LuYago

7 **a** In $\triangle ABC$, $\angle ABC = 90°$, $AB = 39$ cm and $BC = 57$ cm. Find AC correct to one decimal place.

b In $\triangle MPQ$, $\angle PQM = 90°$, $QM = 2.4$ m and $PQ = 3.7$ m. Find PM, as a surd.

c In $\triangle RVJ$, $\angle J = 90°$, $JV = 12.7$ cm and $JR = 42$ mm. Find RV, correct to the nearest millimetre.

d In $\triangle EGB$, $EG = EB = 127$ mm and $\angle GEB = 90°$. Find the length of BG, correct to one decimal place.

e In $\triangle VZX$, $\angle V = 90°$, $VX = 547$ cm and $VZ = 11.6$ m. Find ZX in metres, correct to the nearest 0.1 m.

f In $\triangle PQR$, $\angle RPQ = 90°$, $PQ = 2350$ mm and $PR = 5.8$ m. Find QR in metres, correct to 2 decimal places.

TECHNOLOGY

Finding the hypotenuse

Use a spreadsheet to calculate the length of the hypotenuse, c, of a right-angled triangle, given the lengths of the other 2 sides (a and b).

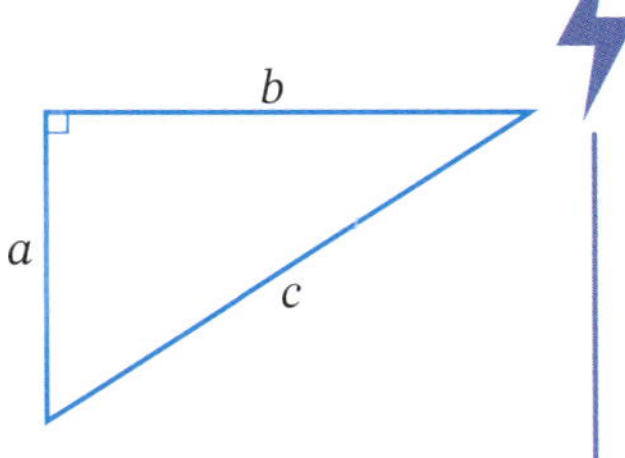

1 Enter the headings shown below into cells A1, B1, C1 and D1. Then enter the values shown into columns A and B.

☐ Foundation ○ Standard ⬡ Complex

	A	B	C	D	E
1	a	b	a^2+b^2	c	
2	3	4			
3	5	12			
4	6	8			
5	7	24			
6					
7					

2 Click on cell C2. Type the formula =**A2^2+B2^2**. Use **Fill Down** to copy this formula into cells C3 to C5.

3 To calculate each hypotenuse, we need to find the square root of the values in column C each time. So, in cell D2, enter =**sqrt(C2)**. Then click on cell D2 and use **Fill Down** to copy this formula into cells D3 to D5.

4 Use your spreadsheet to find the length of the hypotenuse, given the following pairs of values for the other 2 sides.

a $a = 12, b = 9$ **b** $a = 10, b = 10$ **c** $a = 7, b = 6$

d $a = 1, b = 8$ **e** $a = 7.5, b = 11.9$ **f** $a = 6.4, b = 2.3$

2.04 Finding a shorter side

Pythagoras' theorem can also be used to find the length of a shorter side of a right-angled triangle, if the hypotenuse and the other side are known.

Video Pythagoras' theorem 2

Technology Pythagoras' theorem

Spreadsheet Pythagoras' theorem

Example 8

Find the value of d in this triangle.

6 mm

d mm

10 mm

SOLUTION

We want to find the length of a shorter side.

Using Pythagoras' theorem:

$10^2 = d^2 + 6^2$

$100 = d^2 + 36$

$d^2 + 36 = 100$

$d^2 = 100 - 36$

$= 64$

$d = \sqrt{64}$

$= 8$

From the diagram, a length of 8 mm looks reasonable because it must be shorter than the hypotenuse, which is 10 mm.

Example 9

Find the value of y as a surd for this triangle.

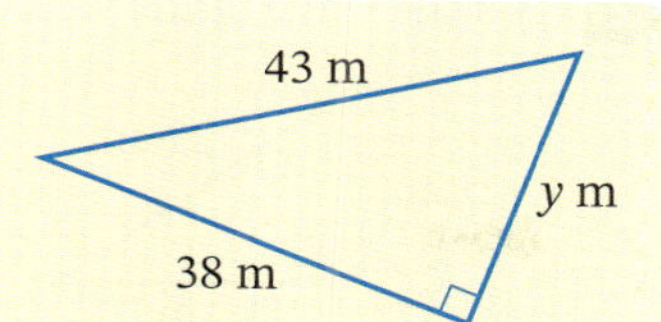

Video
Pythagoras' theorem 1

SOLUTION

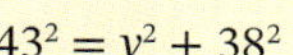

$43^2 = y^2 + 38^2$

$1849 = y^2 + 1444$

$y^2 + 1444 = 1849$

$y^2 = 1849 - 1444$

$\quad = 405$

$y = \sqrt{405}$ Leave the answer as a surd.

EXERCISE 2.04 ANSWERS ON P.631

Finding a shorter side

U F PS R

EXAMPLE 8

1 Find the value of the variable in each triangle.

a

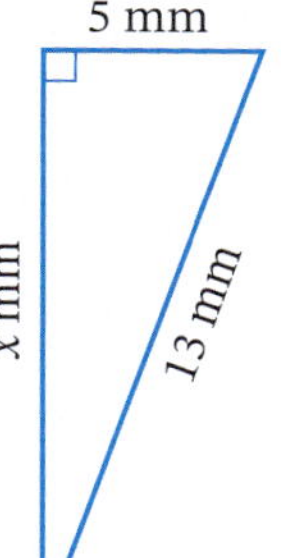

b

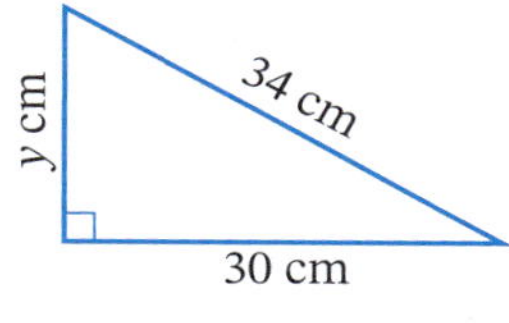

c

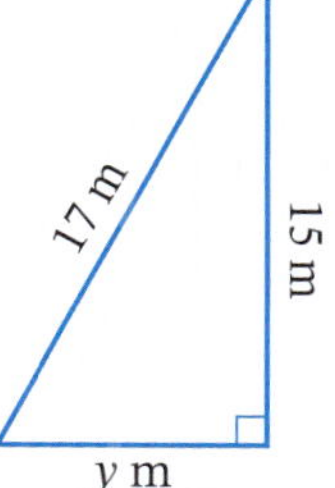

d

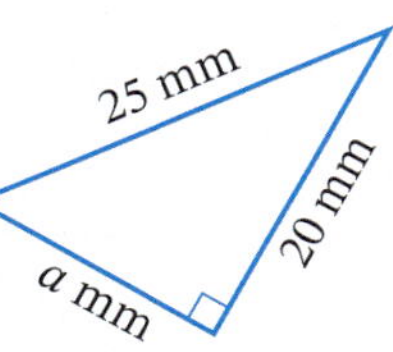

e

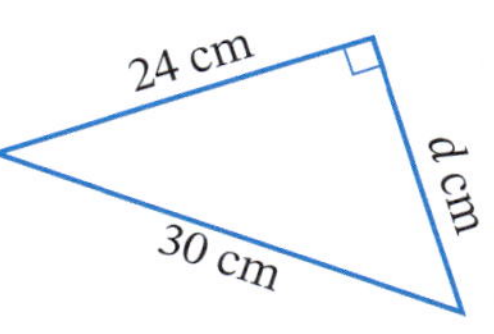

f

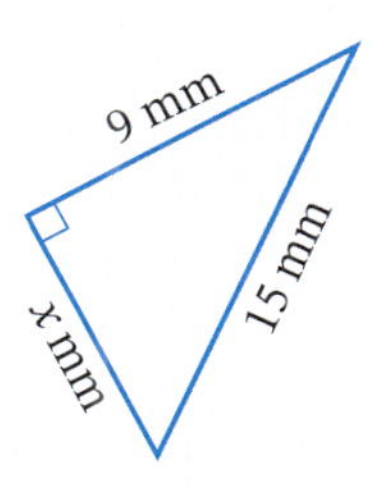

2 Find the value of the variable in each triangle, correct to one decimal place.

a

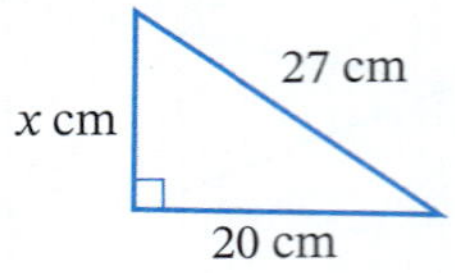
27 cm
x cm
20 cm

b

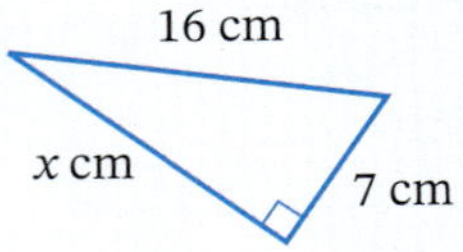
16 cm
x cm
7 cm

c

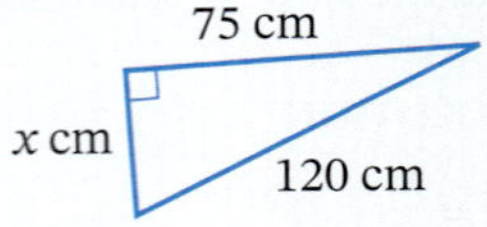
75 cm
x cm
120 cm

d

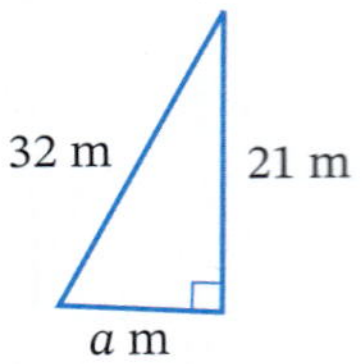
32 m
21 m
a m

e

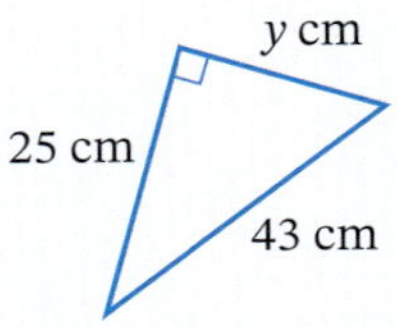
y cm
25 cm
43 cm

f

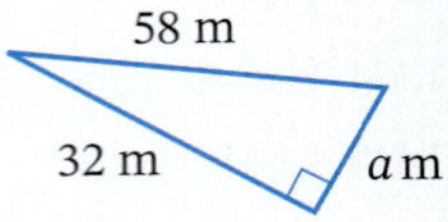
58 m
32 m
a m

EXAMPLE 9

3 Find the value of the variable in each triangle as a surd.

a

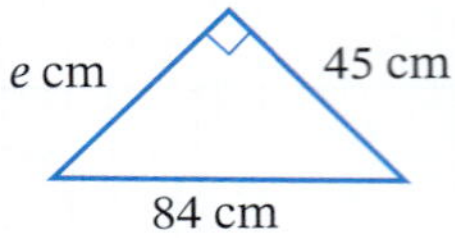
e cm
45 cm
84 cm

b

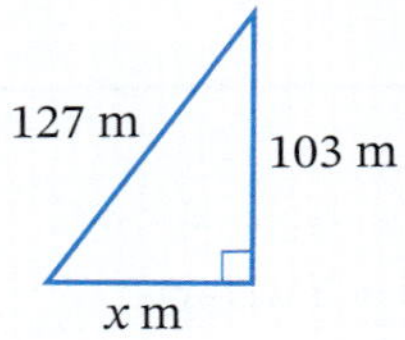
127 m
103 m
x m

c

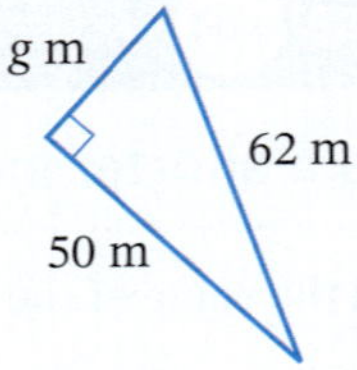
g m
62 m
50 m

d

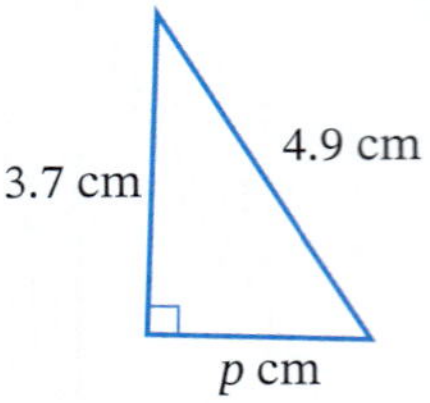
3.7 cm
4.9 cm
p cm

e

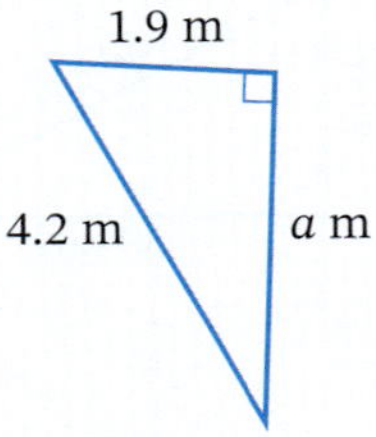
1.9 m
4.2 m
a m

f

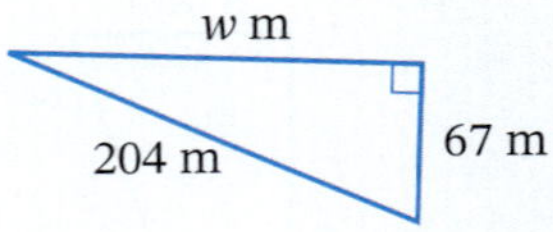
w m
204 m
67 m

4 Find the value of x in this rectangle.
Select the correct answer **A**, **B**, **C** or **D**.

A $\sqrt{20}$

B $\sqrt{80}$

C $\sqrt{136}$

D $\sqrt{208}$

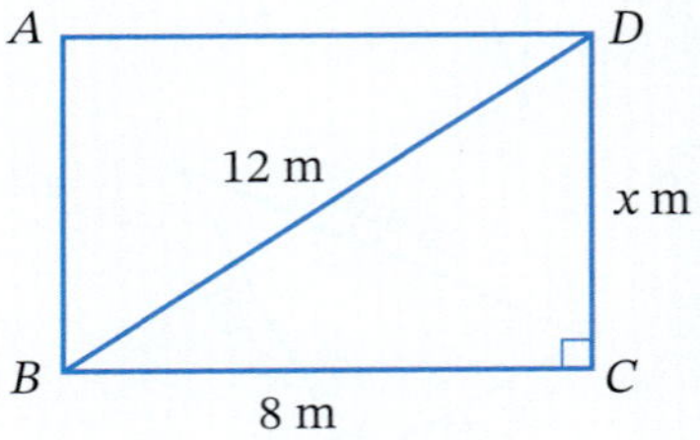
A
D
12 m
x m
B
C
8 m

5 Find the value of p in this rectangle.
Select **A**, **B**, **C** or **D**.

A 40

B 36

C 32

D 28

111 mm
p mm
105 mm

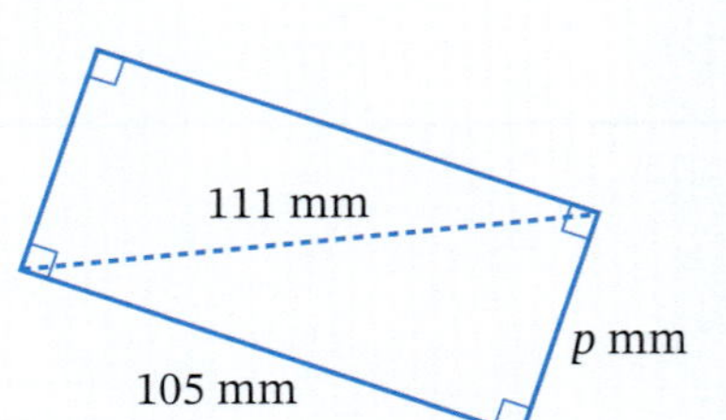

6 PS R In this diagram, O is the centre of a circle. A perpendicular line is drawn from O to B such that $OB \perp AC$ and $AB = BC$. Calculate the length of OB. Select **A**, **B**, **C** or **D**.

A 1.3 m

B 1.8 m

C 1.9 m

D 3.4 m

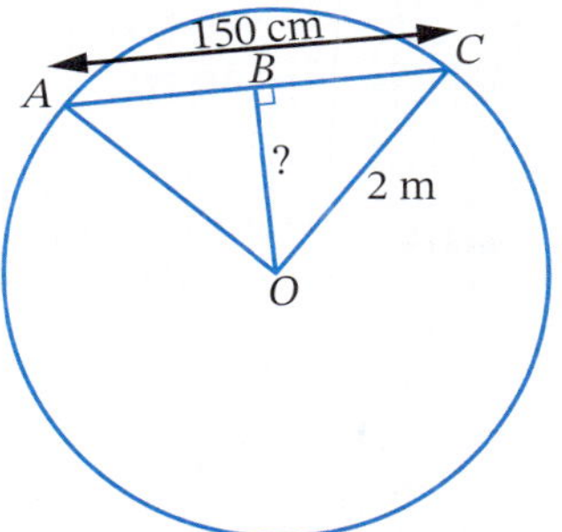

7 PS R An equilateral triangle has sides of length 12 cm. Find the perpendicular height, h cm, of the triangle, correct to 2 decimal places.

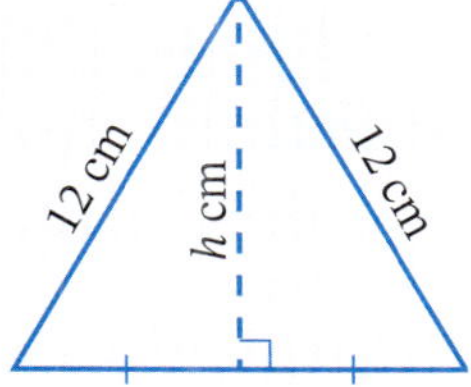

8 A square has a diagonal of length 30 cm. What is the length of each side of the square, correct to the nearest millimetre?

PS R

TECHNOLOGY

Finding a shorter side

Use a spreadsheet to calculate the length of a shorter side, b, of a right-angled triangle, given the length of the hypotenuse and the other side (c and a).

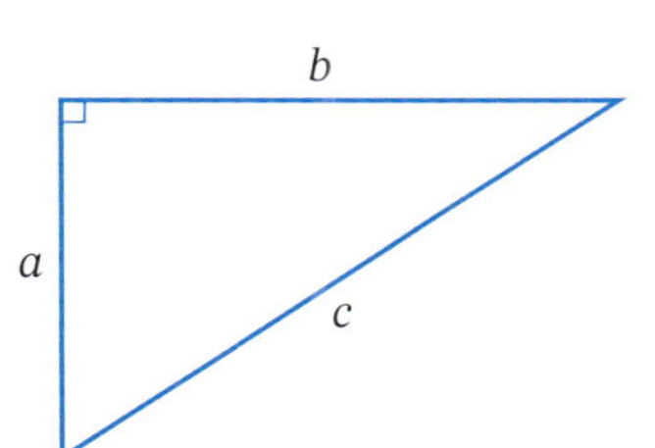

1 Enter the headings shown below into cells A1, B1, C1 and D1. Then enter the values shown into columns A and B.

	A	B	C	D
1	c	a	c^2 - a^2	b
2	25	20		
3	37	12		
4	18	16		
5	11	7		
6				

2 Click on cell C2. Type the formula =**A2^2-B2^2**. Use **Fill Down** to copy this formula into cells C3 to C5.

3 To calculate the values of b, we need to find the square root of each of the values in column C. So, in cell D2, enter = **sqrt(C2)**. Then click on cell D2 and use **Fill Down** to copy this formula into cells D3 to D5.

4 Use your spreadsheet to find the length of the unknown shorter side, given the following pairs of values for the hypotenuse and other side.

a $c = 61, a = 11$

b $c = 87, a = 60$

c $c = 35, a = 22$

d $c = 9.2, a = 2.7$

e $c = 158, a = 63$

f $c = 24.5, a = 12.8$

Quiz Mental skills 2A

☆ **MENTAL SKILLS** 2A ANSWERS ON P.631 **Maths without calculators**

Multiplying and dividing by a power of 10

Multiplying a number by 10, 100, 1000, etc. moves the decimal point to the right and makes the number bigger. We place zeros at the end of the number if necessary.

- When multiplying a number by **10**, move the decimal point **1** place to the right.
- When multiplying a number by **100**, move the decimal point **2** places to the right.
- When multiplying a number by **1000**, move the decimal point **3** places to the right.

The number of places the decimal is moved to the right matches the number of zeros in the 10, 100 or 1000 we are multiplying by.

1 Study each example.

a $26.32 \times 10 = 26.32 = 263.2$ The point moves 1 place to the right.

b $8.701 \times 100 = 8.701 = 870.1$ The point moves 2 places to the right.

c $6.01 \times 1000 = 6.010 = 6010$ The point moves 3 places to the right after a zero is placed at the end.

d $17 \times 100 = 17.00 = 1700$ The point moves 2 places to the right after 2 zeros are placed at the end.

2 Now evaluate each expression.

a 89.54×10

b 3.7×10

c 0.831×100

d 42×100

e 5.2716×1000

f 156.1×10

g 31.84×1000

h 64.3×100

i 0.0224×1000

j 4.894×10

k 7.389×1000

l 11.42×100

Dividing a number by 10, 100, 1000, etc. moves the decimal point to the left and makes the number smaller. We place zeros at the start of the decimal if necessary.

- When dividing a number by **10**, move the decimal point **1** place to the left.
- When dividing a number by **100**, move the decimal point **2** places to the left.
- When dividing a number by **1000**, move the decimal point **3** places to the left.

9780170465618

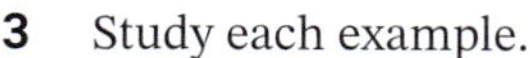

3 Study each example.

a $145.66 \div 10 = 145.66 = 14.566$ The point moves 1 place to the left.

b $2.357 \div 100 = 002.357 = 0.023\ 57$ The point moves 2 places to the left after 2 zeros are inserted at the start.

c $14.9 \div 1000 = 0014.9 = 0.0149$ The point moves 3 places to the left after 2 zeros are inserted at the start.

d $45 \div 100 = 045. = 0.45$ The point moves 2 places to the left after 1 zero is inserted at the start.

4 Now evaluate each expression.

a	$733.4 \div 10$	b	$9.4 \div 10$	c	$652 \div 100$
d	$10.4 \div 100$	e	$704 \div 1000$	f	$198.5 \div 100$
g	$2 \div 100$	h	$4159 \div 1000$	i	$123 \div 10$
j	$0.758 \div 100$	k	$8.49 \div 100$	l	$25.1 \div 1000$

Mixed problems 2.05

In this section, you will need to decide whether to find the hypotenuse or a shorter side.

Puzzles
Pythagoras 1
Pythagoras 2

Example 10

Find the value of y as a surd.

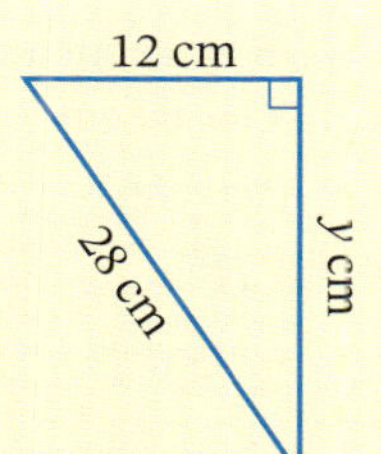

SOLUTION

y is the length of a shorter side.

$$28^2 = y^2 + 12^2$$
$$784 = y^2 + 144$$
$$y^2 + 144 = 784$$
$$y^2 = 784 - 144$$
$$= 640$$
$$y = \sqrt{640}$$

Example 11

A ship sails 80 nautical miles south and then 45 nautical miles east. How far is it from its starting point, correct to one decimal place?

SOLUTION

Let x be the distance the ship is from the starting point.

$x^2 = 80^2 + 45^2$

$= 8425$

$x = \sqrt{8425}$ — From the diagram, this looks like a reasonable answer

$= 91.7877...$

$\approx 91.8\text{M}$ — M is the symbol for nautical miles

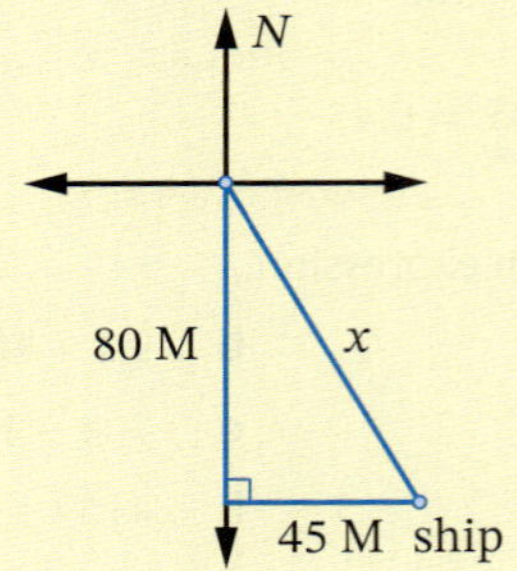

Example 12

Find the length of the unknown side in this trapezium, correct to one decimal place.

SOLUTION

The unknown side is PS. Form the right-angled $\triangle PSA$, as shown.

By comparing sides:

$PA = QR = 47$ cm and

$AR = PQ = 60$ cm.

$\therefore SA = 72 - 60 = 12$ cm

$\therefore PS^2 = 47^2 + 12^2$

$= 2353$

$PS = \sqrt{2353}$

$= 48.5077...$ — From the diagram, this looks like a reasonable answer.

≈ 48.5 cm

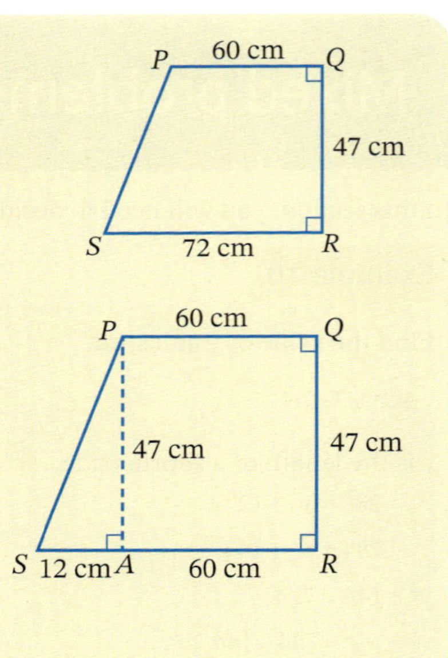

Example 13

Find the value of y, correct to 2 decimal places.

SOLUTION

We need to find BD first.

In $\triangle ABD$,

$BD^2 = 15^2 + 13^2$

$= 394$

$BD = \sqrt{394}$ Leave BD as a surd for further working.

In $\triangle BCD$,

$y^2 = \left(\sqrt{394}\right)^2 + 12^2$ $\left(\sqrt{394}\right)^2 = 394$

$= 538$

$y = \sqrt{538}$

$= 23.1948...$

≈ 23.19

From the diagram, this looks like a reasonable answer

EXERCISE 2.05 ANSWERS ON P.632

Mixed problems

U F PS R

1 What is the length of k? Select the correct answer **A**, **B**, **C** or **D**.

A 27 cm **B** 41.9 cm

C 46.2 cm **D** 49.8 cm

2 What is the length of YZ? Select **A**, **B**, **C** or **D**.

A 15.9 m **B** 13.4 m

C 11.7 m **D** 8.6 m

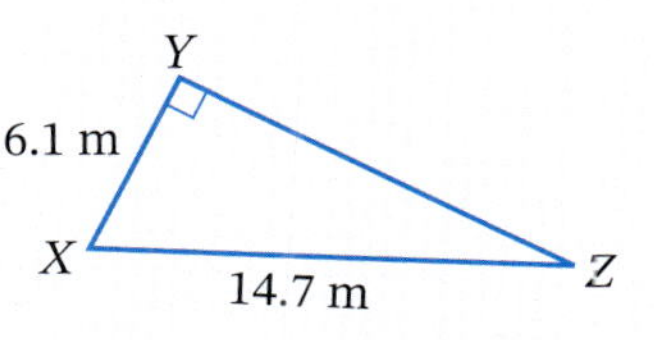

Foundation Standard Complex

3 Find the value of each variable, correct to one decimal place.

a

17 cm
m cm
27 cm

b

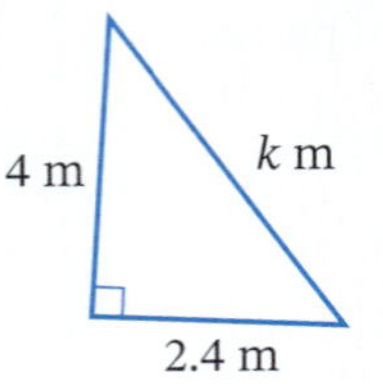

c

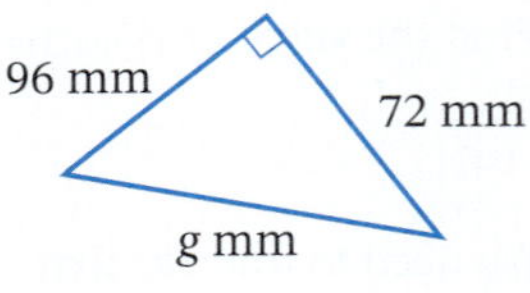

d

21.6 cm
p cm
18.0 cm

e

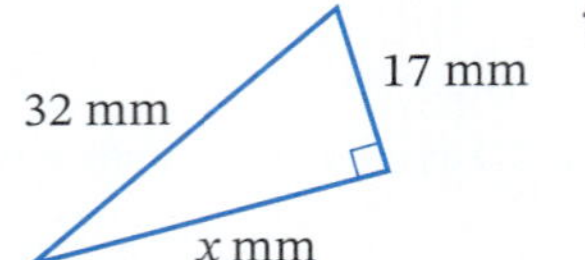

f

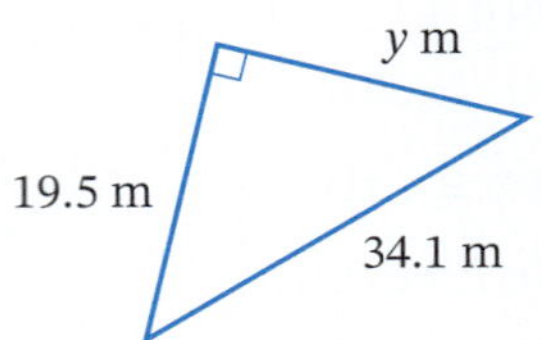

g

6.8 cm
5.1 cm
t cm

h

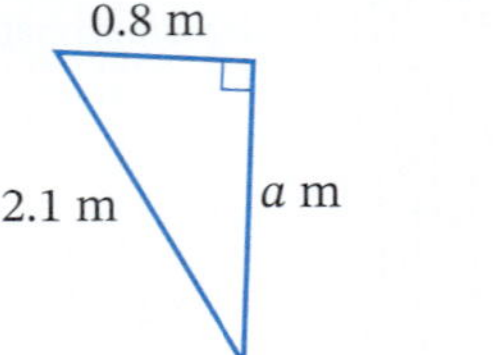

i

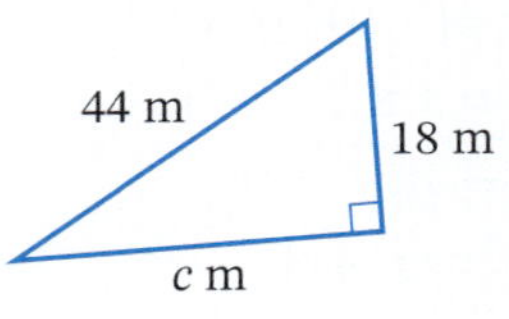

j

r m
1.9 m
3.5 m

k

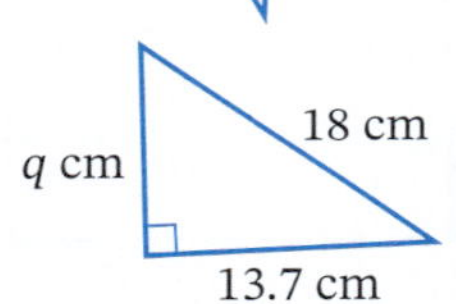

l

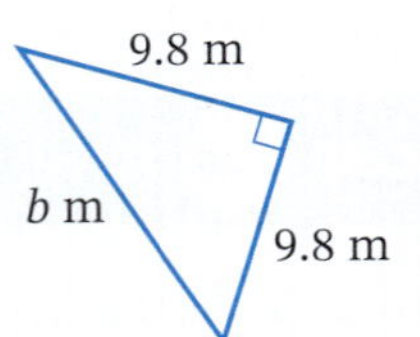

4 A ladder 5 m long is leaning against a wall. If the base of the ladder is 2 m from the bottom of the wall, how far does the ladder reach up the wall? (Answer correct to 2 decimal places.)

iStock.com/bmcent1

5 The size of a TV screen is described by the length of its diagonal. If a TV screen is 133 cm wide and 75 cm high, what is the size of the screen? Answer to the nearest centimetre.

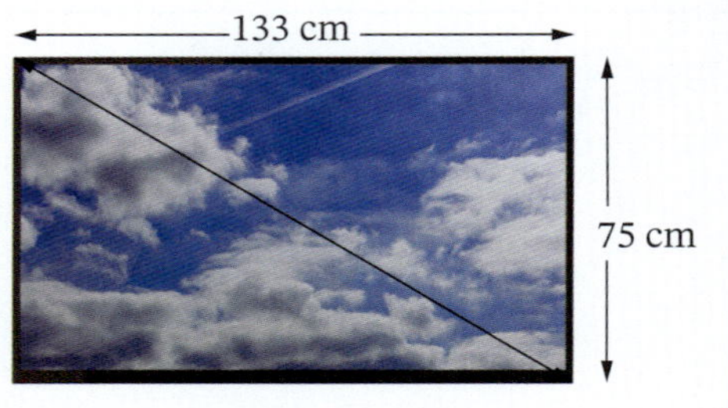

Shutterstock.com/rawf8

☐ Foundation ○ Standard ⬡ Complex

6 A ship sails 70 nautical miles west and then 60 nautical miles north. How far is it from its starting point, correct to one decimal place?

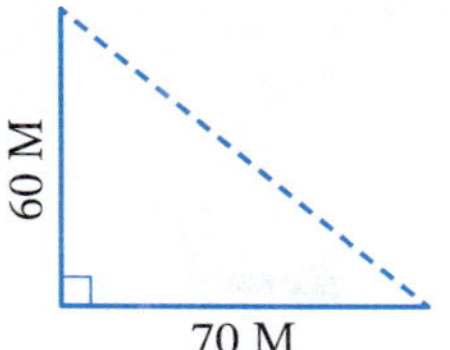

7 Ashleigh is standing on a tennis court at A and serves the ball to Ellen, who hits it back from B. Ashleigh then hits the ball from A to C. How far does Ellen have to run to hit a return shot? Answer to the nearest metre.

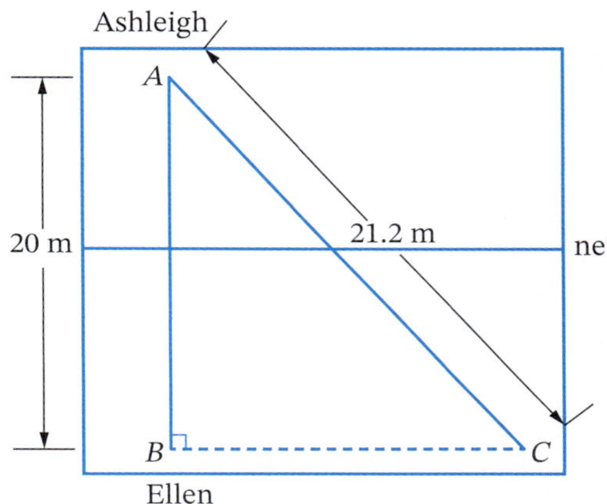

8 Use Pythagoras' theorem to calculate the distance between the 2 points on each number plane. In part **b**, write the answer as a surd.

a

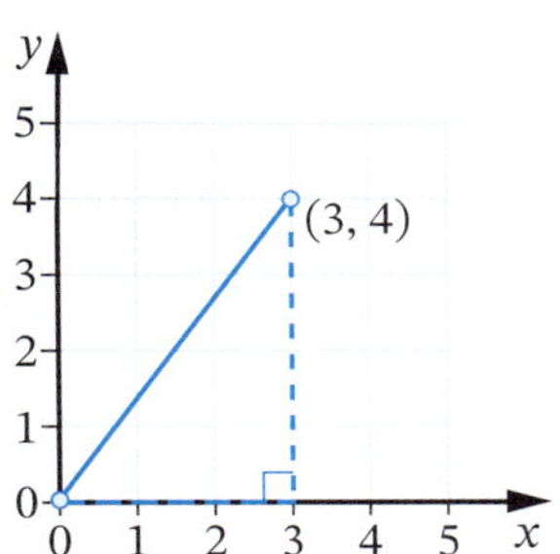

b

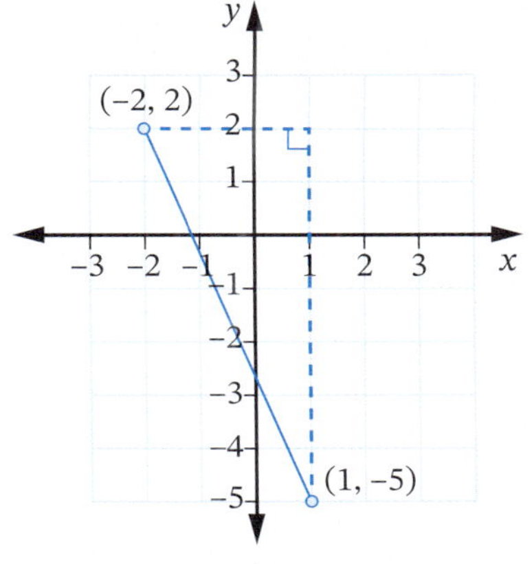

EXAMPLE 12

9 For each trapezium, find the length of the unknown side, correct to 2 decimal places where necessary.

R

a **b** **c**

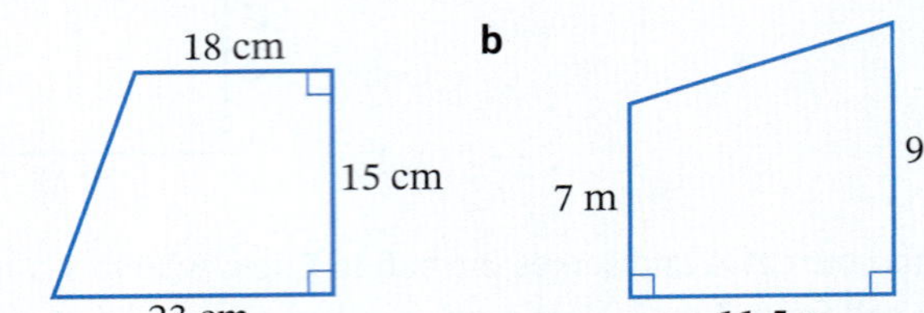

EXAMPLE 13

10 **a** Find the value of x, correct to one decimal place.

R

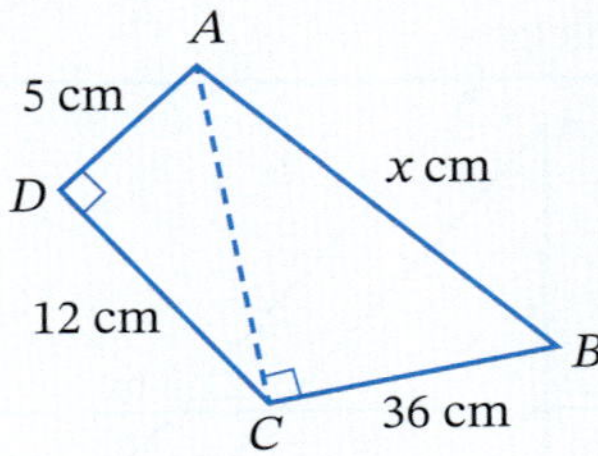

b Find the length of AD, correct to 2 decimal places.

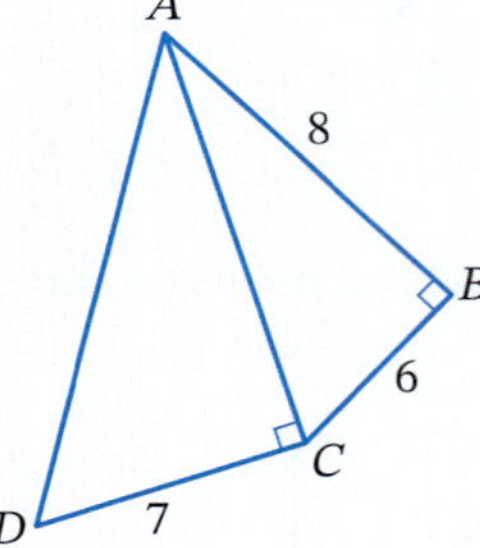

c Calculate the length of x as a surd.

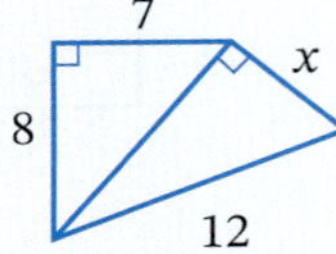

11 The diagonal of a rectangle is 200 mm in length. One of the side lengths of the rectangle is 167 mm. Find the length of the other side, correct to the nearest millimetre.

P

12 To keep a power pole vertically upright, it is supported by 2 wires as shown. Find, correct to one decimal place, the length of the second wire.

R

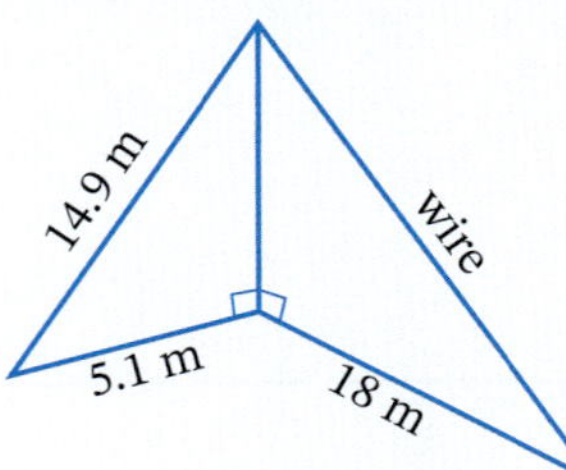

☐ Foundation ○ Standard ⬡ Complex

DID YOU KNOW?

The Pythagoreans

The Pythagoreans were a group of men who were Pythagoras' followers. Apparently, they were so upset about the discovery of surds that they tried to keep it a secret. Hippasus, one of the Pythagoreans, was drowned for revealing the secret to outsiders.

Find the special name for this symbol involving a star inside a pentagon, the symbol of the Pythagoreans.

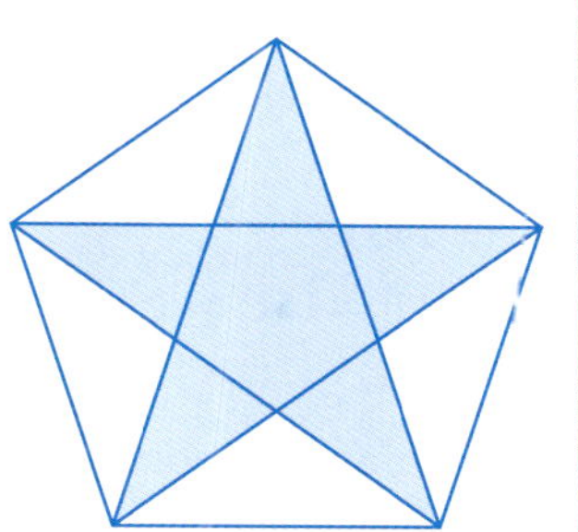

Testing for right-angled triangles 2.06

Pythagoras' theorem says that if a right-angled triangle has sides of length a, b, and c, then $c^2 = a^2 + b^2$. The reverse of this is also true:

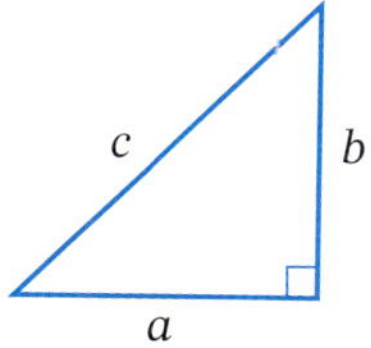

Worksheet Testing for right-angled triangles

Video Testing for right-angled triangles

ⓘ The converse of Pythagoras' theorem

If any triangle has sides of length a, b, and c that follow the formula $c^2 = a^2 + b^2$, then the triangle must be right-angled. The right angle is opposite the hypotenuse.

This is called the **converse** (or opposite) of Pythagoras' theorem, because it is the 'back-to-front' version of the theorem.

Example 14

Test whether $\triangle ABC$ is right-angled.

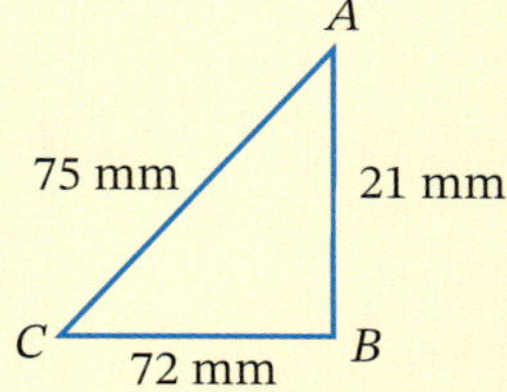

SOLUTION

$75^2 = 5625$	Square the longest side.
$21^2 + 72^2 = 5625$	Square the 2 shorter sides, then add.
$\therefore 75^2 = 21^2 + 72^2$	The sides of this triangle follow $c^2 = a^2 + b^2$.
$\therefore \triangle ABC$ is right-angled.	The right angle is $\angle B$.

Example 15

Rahul constructed a triangle with sides of length 37 cm, 12 cm and 40 cm. Show that these measurements do not form a right-angled triangle.

SOLUTION

$40^2 = 1600$	Square the longest side.
$12^2 + 37^2 = 1513 \neq 1600$	Square the 2 shorter sides, then add.
$\therefore 40^2 \neq 12^2 + 37^2$	The sides of this triangle do not follow $c^2 = a^2 + b^2$.
$\therefore$ The triangle is not right-angled.	

EXERCISE 2.06 ANSWERS ON P.632

Testing for right-angled triangles

U F R C

EXAMPLE 14

1 Test whether each triangle is right-angled.

C **a**

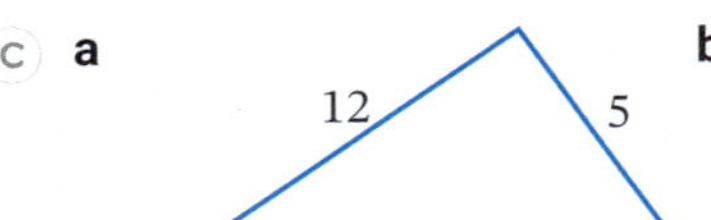

b

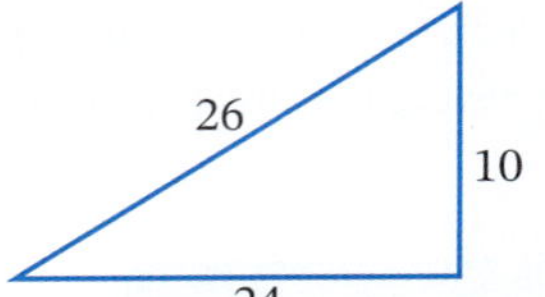

c

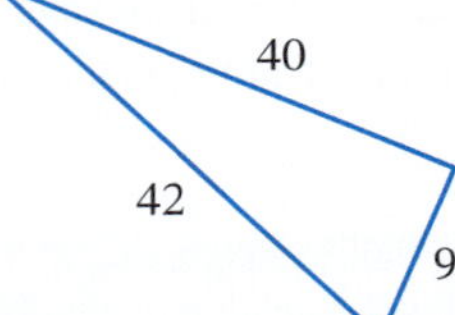

d

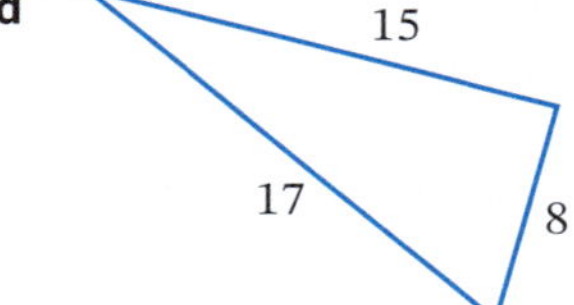

e

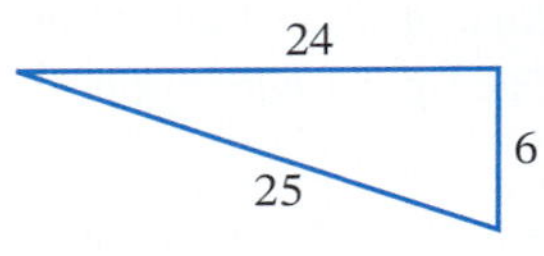

f

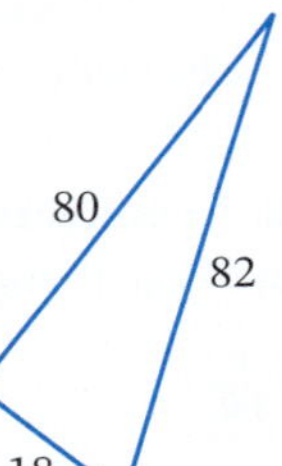

g

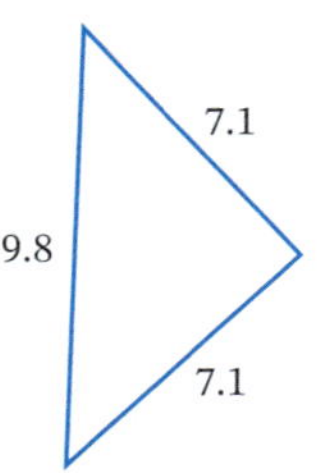

h

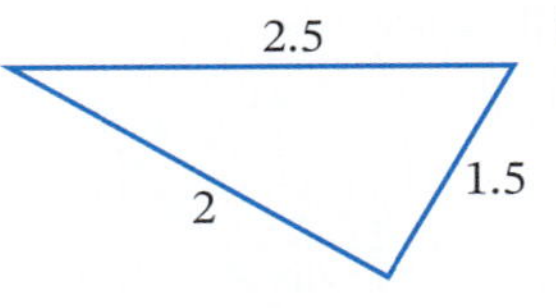

i

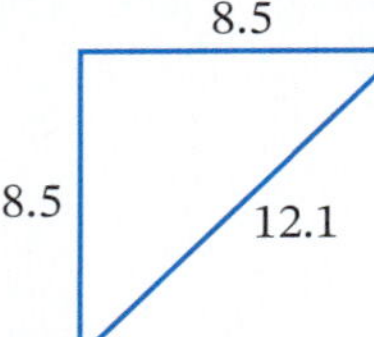

j

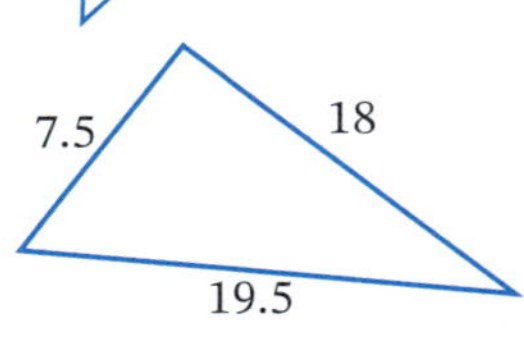

k

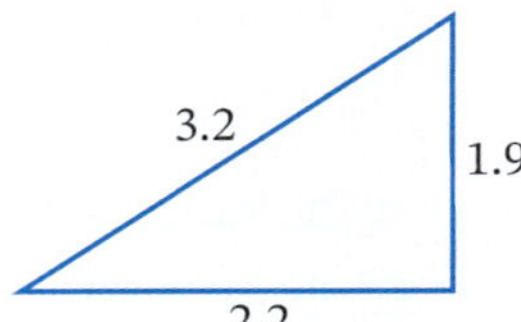

l

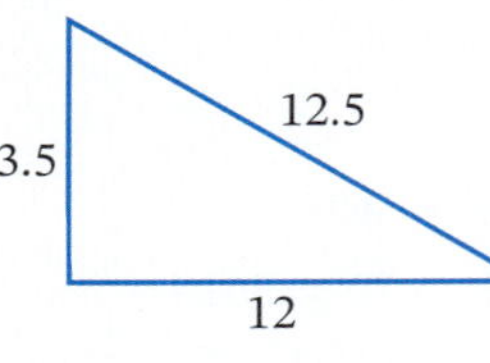

 Foundation  Standard Complex

2 Paba constructed a triangle with sides of length 17 cm, 21 cm and 30 cm.

C Show that these measurements do *not* form a right-angled triangle.

EXAMPLE 15

3 Which set of measurements would make a right-angled triangle?

C Select the correct answer **A**, **B**, **C** or **D**.

A 2 cm, 3 cm, 4 cm

B 5 mm, 10 mm, 15 mm

C 12 cm, 16 cm, 20 cm

D 7 m, 24 m, 31 m

4 Which one of these triangles is not right-angled? Select the correct answer **A**, **B**, **C** or **D**.

C **A**

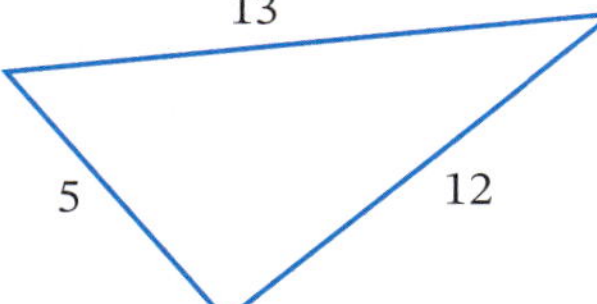

B

D

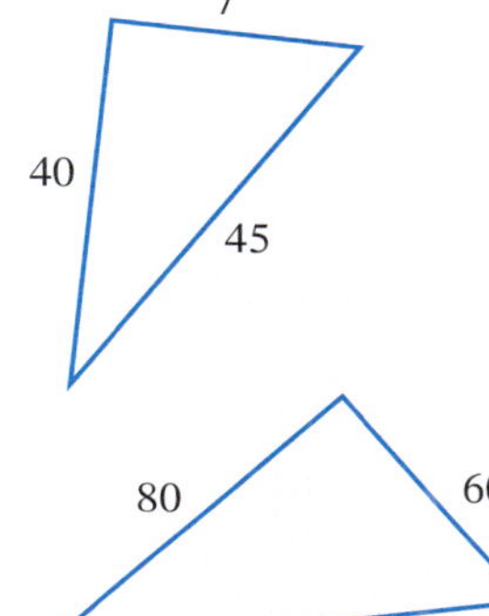

C

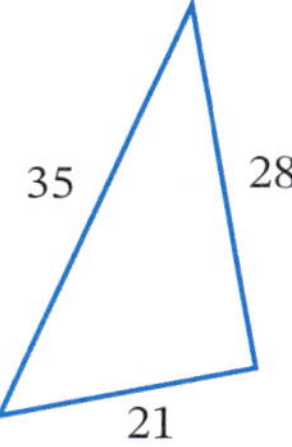

5 A pole 2.1 m tall is supported by a wire 2.9 m long. The other end of the wire is attached to the ground 2 m from the base of the pole. Is the pole standing vertically upright?

R C

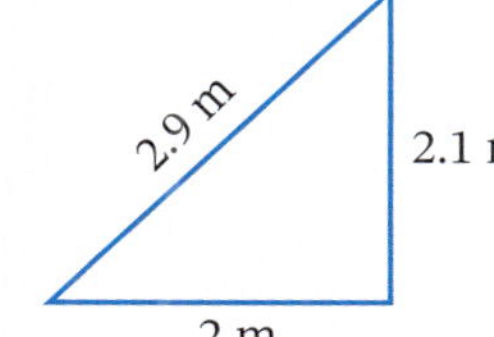

2.07 Pythagorean triads

A **Pythagorean triad** or **Pythagorean triple** is any set of 3 numbers that follow Pythagoras' theorem, for example, {3, 4, 5} or {2.5, 6, 6.5}. The word **triad** means a group of 3 related items ('tri-' means 3).

Worksheet Pythagorean triads

Technology Pythagorean triples

Spreadsheet Pythagorean triples

Pythagorean triad

$\{a, b, c\}$ is a Pythagorean triad if $c^2 = a^2 + b^2$.

Any multiple of $\{a, b, c\}$ is also a Pythagorean triad.

Example 16

Test whether {5, 12, 13} is a Pythagorean triad.

SOLUTION

$13^2 = 169$ — Squaring the largest number.

$5^2 + 12^2 = 169$ — Squaring the 2 smaller numbers and adding them.

$\therefore 13^2 = 5^2 + 12^2$ — These 3 numbers follow Pythagoras' theorem.

$\therefore$ {5, 12, 13} is a Pythagorean triad.

Example 17

{3, 4, 5} is a Pythagorean triad. Create other Pythagorean triads by multiplying (3, 4, 5) by:

a 2 **b** 9 **c** $\frac{1}{2}$

SOLUTION

a $2 \times \{3, 4, 5\} = \{6, 8, 10\}$

Checking: $10^2 = 100$

$6^2 + 8^2 = 100$

$\therefore 10^2 = 6^2 + 8^2$

$\therefore$ {6, 8, 10} is a Pythagorean triad.

b $9 \times \{3, 4, 5\} = \{27, 36, 45\}$

Checking: $45^2 = 2025$

$27^2 + 36^2 = 2025$

$\therefore 45^2 = 27^2 + 36^2$

$\therefore$ {27, 36, 45} is a Pythagorean triad.

c $\frac{1}{2} \times \{3, 4, 5\} = \{1.5, 2, 2.5\}$

Checking: $2.5^2 = 6.25$

$1.5^2 + 2^2 = 6.25$

$\therefore 2.5^2 = 1.5^2 + 2^2$

$\therefore$ {1.5, 2, 2.5} is a Pythagorean triad.

EXERCISE 2.07 ANSWERS ON P.632

Pythagorean triads

1 Test whether each triad is a Pythagorean triad.

a {8, 15, 17} **b** {10, 24, 26} **c** {30, 40, 50} **d** {5, 7, 9}
e {9, 40, 41} **f** {4, 5, 9} **g** {11, 60, 61} **h** {7, 24, 25}
i {15, 114, 115}

2 Which of the following is a Pythagorean triad? Select the correct answer **A**, **B**, **C** or **D**.

A {4, 6, 8} **B** {5, 10, 12} **C** {6, 7, 10} **D** {20, 48, 52}

3 Use the spreadsheet you created in **Technology: Finding the hypotenuse** on page 54 to check your answers to questions **1** and **2**.

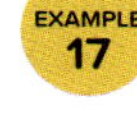

4 For each Pythagorean triad, create another Pythagorean triad by multiplying each number in the triad by:

i a whole number **ii** a fraction **iii** a decimal.

a {5, 12, 13} **b** {8, 15, 17} **c** {30, 40, 50} **d** {7, 24, 25}

Check that each answer follows Pythagoras' theorem.

5 Pythagoras developed formulas for finding Pythagorean triads $\{a, b, c\}$. If one number in the triad is a, the formulas for the other 2 numbers are $b = \frac{1}{2}(a^2 - 1)$ and $c = \frac{1}{2}(a^2 + 1)$.

a If $a = 5$, use the formulas to find the values of b and c.

b Show that $\{a, b, c\}$ is a Pythagorean triad.

6 Use the formulas to find Pythagorean triads for each value of a.

a $a = 7$ **b** $a = 11$
c $a = 15$ **d** $a = 4$
e $a = 9$ **f** $a = 19$
g $a = 10$ **h** $a = 51$

7 Prove that if $b = \frac{1}{2}(a^2 - 1)$ and $c = \frac{1}{2}(a^2 + 1)$, then $c^2 = a^2 + b^2$ is true.

R C

8 There are many other formulas for creating Pythagorean triads. Use the Internet to research some of them.

Foundation Standard Complex

2.08 Pythagoras' theorem problems

Worksheet
Applications of Pythagoras' theorem

When using Pythagoras' theorem to solve problems, it is useful to follow these steps.

- Read the problem carefully.
- Draw a diagram involving a right-angled triangle and label any given information.
- Choose a variable to represent the length or distance you want to find.
- Use Pythagoras' theorem to find the value of the variable.
- Answer the question.

Video
Pythagoras' theorem problems

Example 18

A tower is supported by a wire that is 20 m long and attached to the ground 10 m from the base of the tower. How far up the tower does the wire reach? Answer to the nearest 0.1 m.

SOLUTION

First draw a diagram. Let h represent how far up the tower the wire reaches in metres.

The diagram

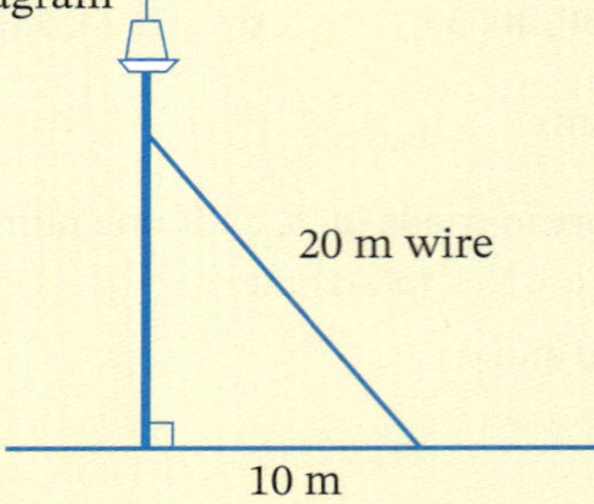

The triangle

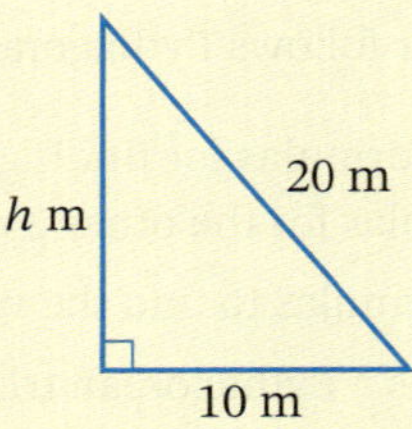

$20^2 = h^2 + 10^2$

$400 = h^2 + 100$

$h^2 + 100 = 400$

$h^2 = 400 - 100$

$= 300$

$h = \sqrt{300}$

$= 17.3205...$

≈ 17.3

$\therefore$ The wire reaches 17.3 m up the tower.

Video
Pythagoras' theorem problems

Example 19

Find the perimeter of this triangle, correct to one decimal place.

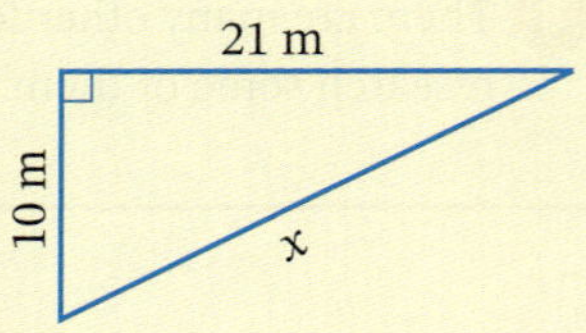

SOLUTION

Let x be the length of the hypotenuse.

$x^2 = 10^2 + 21^2$

$= 541$

$x = \sqrt{541}$

$\approx 23.3\text{m}$

Perimeter $\approx 10 + 21 + 23.3$

$= 54.3$ m

EXERCISE 2.08 ANSWERS ON P.632

Pythagoras' theorem problems

U F PS R

EXAMPLE 18

1 A flagpole is supported by a piece of wire that reaches 4 m up the flagpole. The wire is attached to the ground 2.4 m from the base of the flagpole. Find the length of the wire, correct to one decimal place.

PS R

2 Mia is walking home from the library. She can walk around Hamper Park, entering the park at gate A, through B (stopping at the playground) and then out of the park at gate C, to her home nearby. Alternatively, Mia can take a shortcut across the park from gate A to gate C.

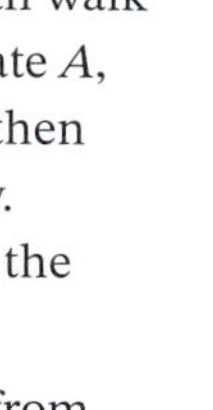

a Find the total distance Mia would walk from A to C via B.

b How far is it if Mia walks from A directly to C? Answer to the nearest metre.

c How much further would Mia walk if she takes the longer route home?

3 Cooper wanted to find the length XY of the lake shown. He placed a marker at Z so that $\angle YXZ = 90°$. He measured XZ to be 450 m long and ZY to be 780 m long. What is the length of the lake? Answer correct to one decimal place.

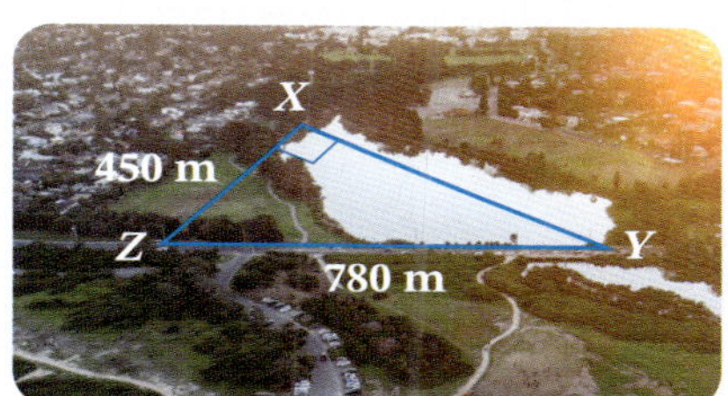

Shutterstock.com/Jamen Percy

4 The swim course of a triathlon race has the shape of a right-angled triangle joined by a 2-m link to the beach as shown. Calculate the total distance covered in this course (starting and finishing on the beach).

PS

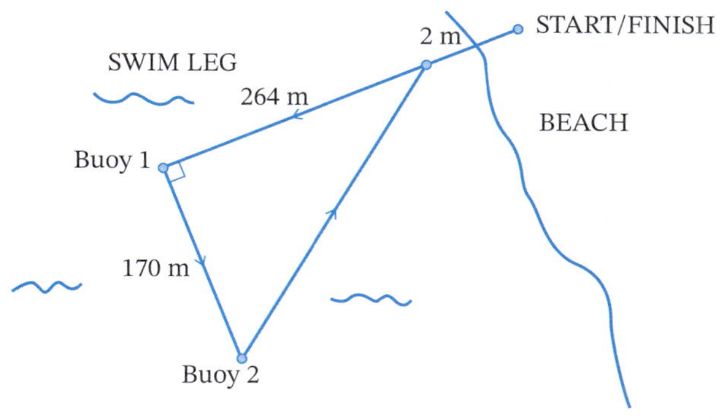

☐ Foundation ○ Standard ○ Complex

5 A kite is attached to a 24 m piece of string. The rope is held 1.2 m above the ground and covers a horizontal distance of 10 m.

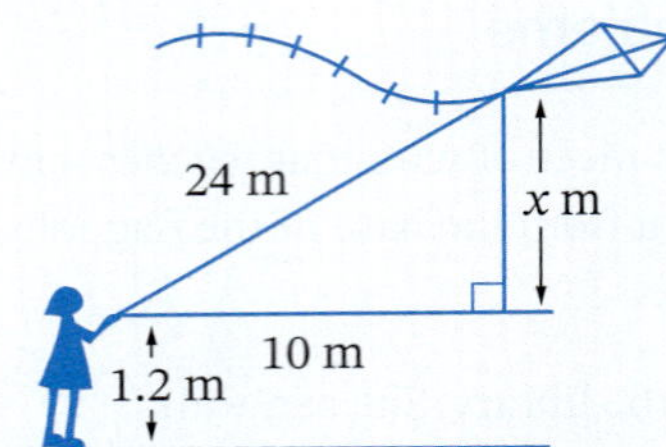

a Find the value of x, correct to one decimal place.

b Find the height of the kite above the ground, correct to the nearest metre.

6 Find the length of the electricity line from the top of the pole to the house, correct to one decimal place.

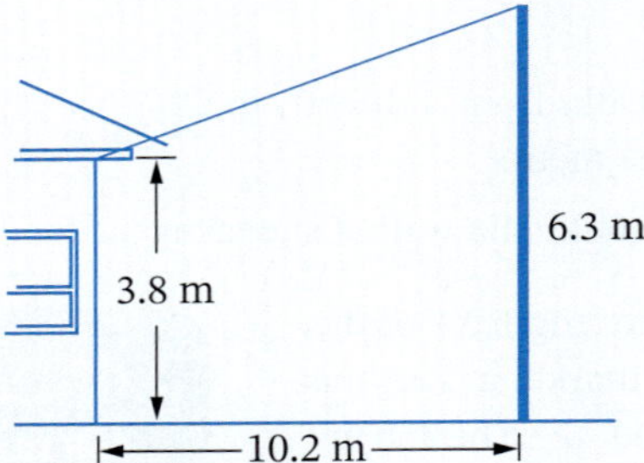

7 A screen door is reinforced by 2 metal braces that diagonally cross the door. Find the total length of the 2 braces in metres, correct to 2 decimal places.

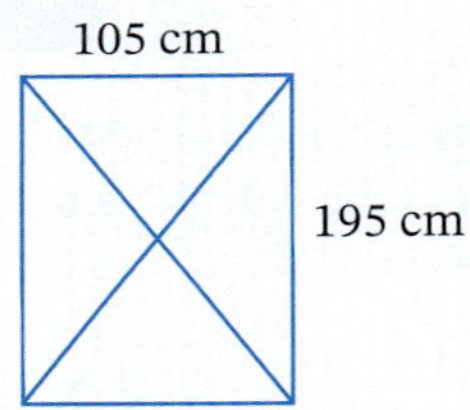

8 For the below shown playground slide, find correct to 3 decimal places:

a the height (h) of the slide

b the length (l) of the slide.

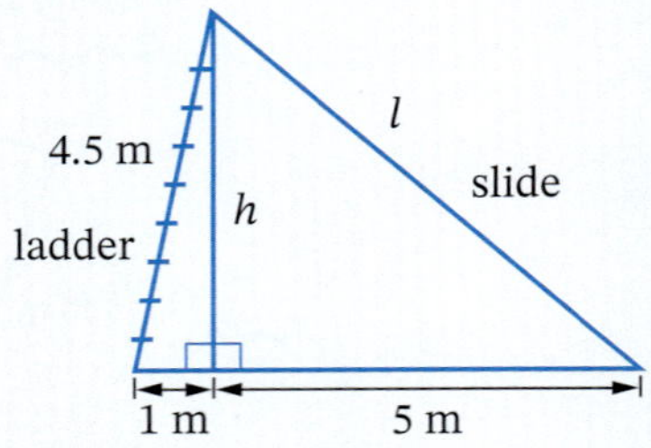

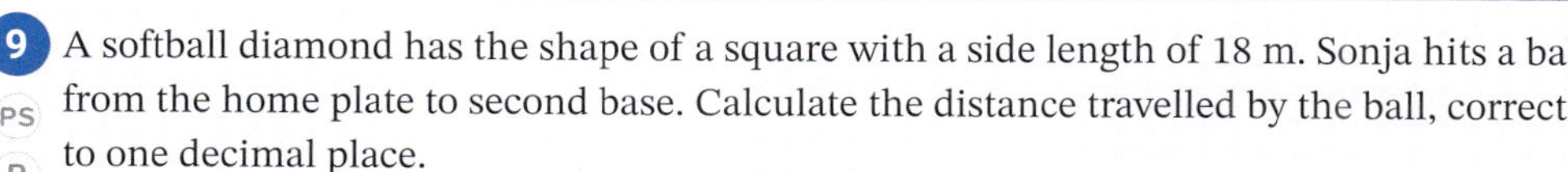

9 A softball diamond has the shape of a square with a side length of 18 m. Sonja hits a ball from the home plate to second base. Calculate the distance travelled by the ball, correct to one decimal place.

PS R

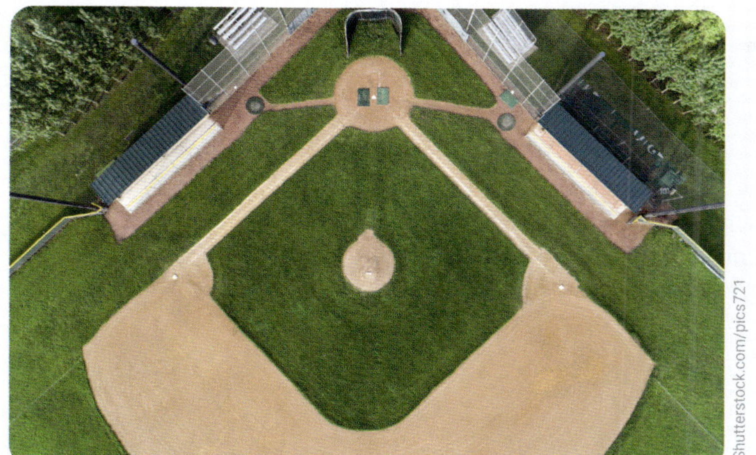

Shutterstock.com/pics721

10 Calculate the perimeter of each shape, correct to one decimal place where necessary.

EXAMPLE 19

a

b

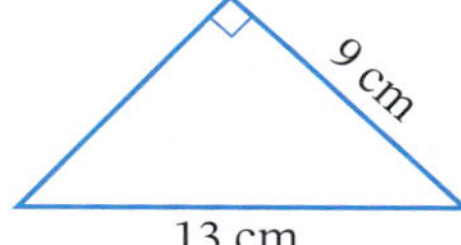

c

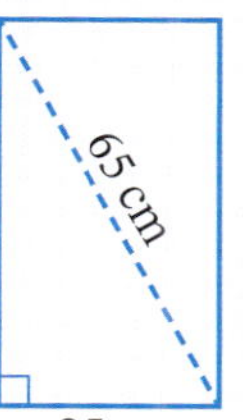

d

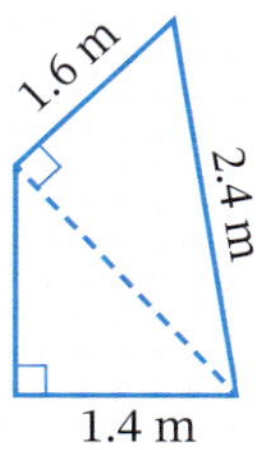

e

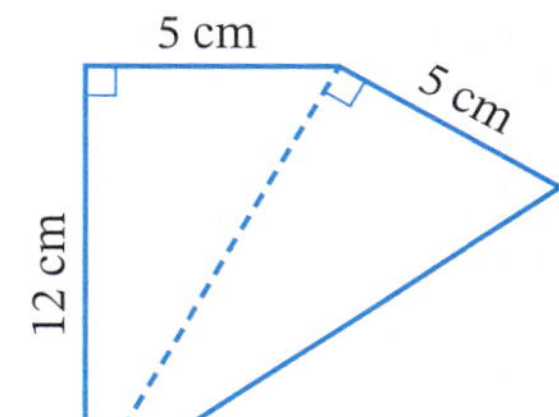

f

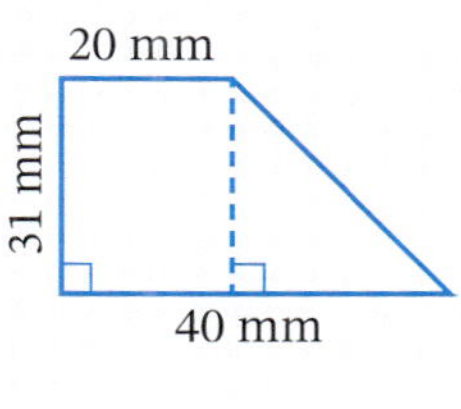

11 Calculate the area of each shape, correct to one decimal place where necessary.

a

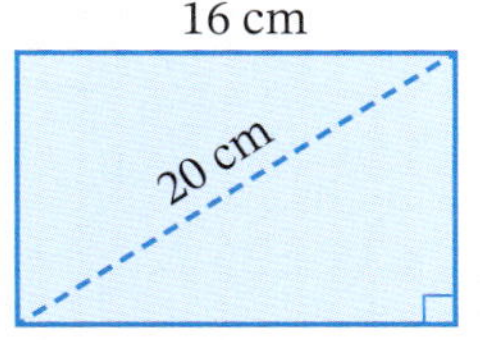

b

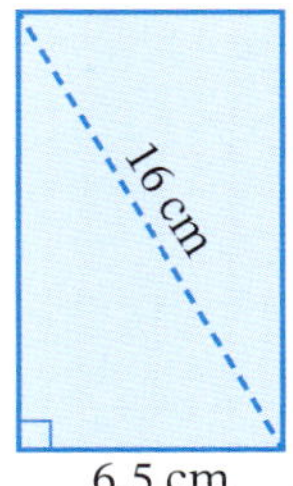

c

d

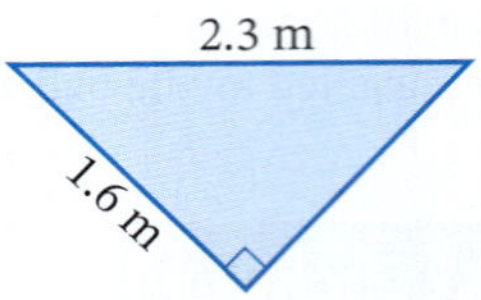

e

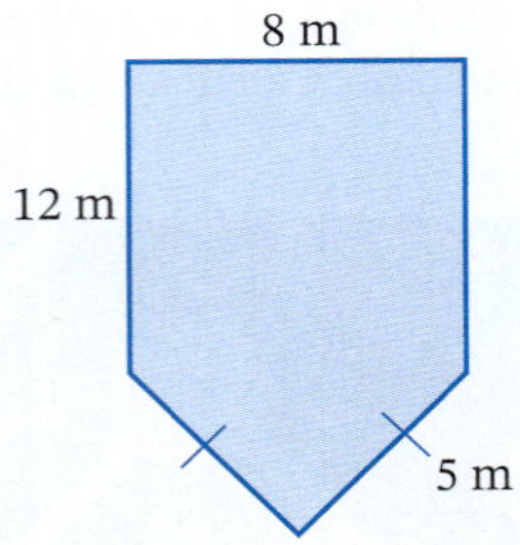

f

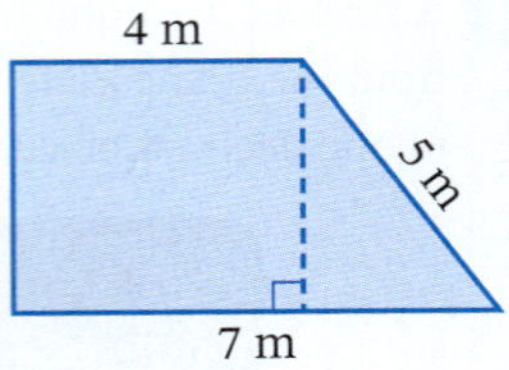

12 A cone has a vertical height of 14 cm and a slant height of 27 cm. Find r cm, the radius of its circular base, correct to 2 decimal places.

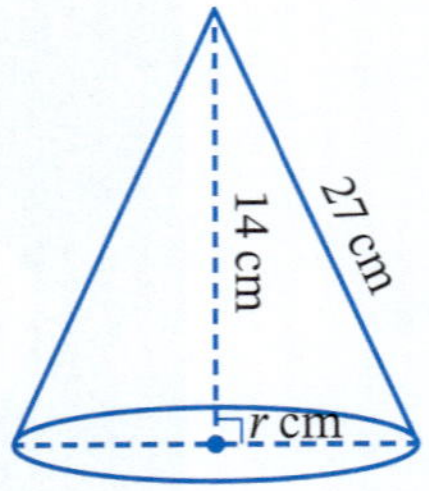

13 PS R *DCBA* is a rhombus with diagonals $AC = 18$ cm and $DB = 12$ cm. Find the perimeter of the rhombus, correct to one decimal place.

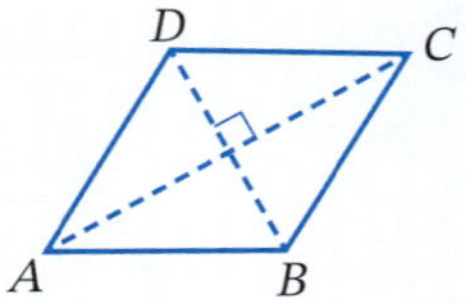

14 PS R A set of 24 steps is to be replaced by a ramp as shown. Each step has a tread of 45 cm and a riser of 28 cm. Find the length of the ramp in metres.

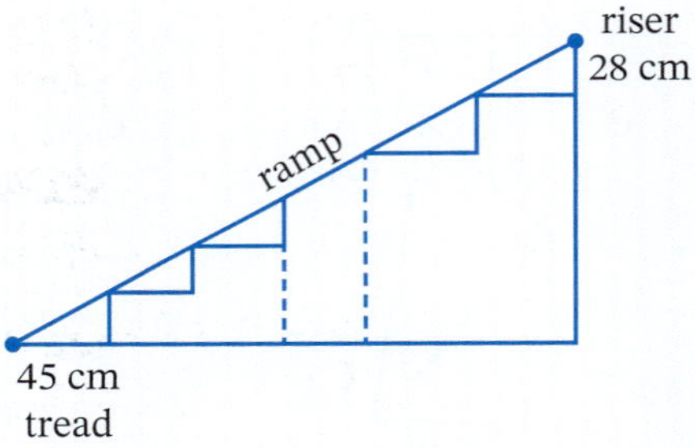

☐ Foundation ○ Standard ⬡ Complex

9780170465618

☆ **MENTAL SKILLS 2B** ANSWERS ON P.632 Maths without calculators

2.08

Quiz Mental skills 2B

Multiplying and dividing by a multiple of 10

1 Consider each example.

a $4 \times 700 = 4 \times 7 \times 100 = 28 \times 100 = 2800$

b $5 \times 60 = 5 \times 6 \times 10 = 30 \times 10 = 300$

c $12 \times 40 = 12 \times 4 \times 10 = 48 \times 10 = 480$

d $3.2 \times 30 = 3.2 \times 3 \times 10 = 9.6 \times 10 = 96$ (by estimation, $3 \times 30 = 90 \approx 96$)

e $4.6 \times 50 = 4.6 \times 5 \times 10 = 23 \times 10 = 230$ (by estimation, $5 \times 50 = 250 \approx 230$)

f $9.4 \times 200 = 9.4 \times 2 \times 100 = 18.8 \times 100 = 1880$ (by estimation, $9 \times 200 = 1800 \approx 1880$)

2 Now evaluate each product.

a 8×2000 **b** 3×70 **c** 11×900 **d** 2×300

e 4×4000 **f** 5×80 **g** 7×70 **h** 1.3×40

i 2.5×600 **j** 5.8×200 **k** 3.6×50 **l** 4.4×3000

3 Consider each example.

a $8000 \div 400 = 8000 \div 100 \div 4 = 80 \div 4 = 20$

b $200 \div 50 = 200 \div 10 \div 5 = 20 \div 5 = 4$

c $6000 \div 20 = 6000 \div 10 \div 2 = 600 \div 2 = 300$

d $282 \div 30 = 282 \div 10 \div 3 = 28.2 \div 3 = 9.4$

e $3520 \div 40 = 3520 \div 10 \div 4 = 352 \div 4 = 88$

f $8940 \div 200 = 8940 \div 100 \div 2 = 89.4 \div 2 = 44.7$

4 Now evaluate each quotient.

a $560 \div 70$ **b** $2500 \div 500$ **c** $3200 \div 400$ **d** $440 \div 20$

e $160 \div 40$ **f** $1500 \div 30$ **g** $450 \div 50$ **h** $744 \div 80$

i $2550 \div 300$ **j** $846 \div 200$ **k** $576 \div 60$ **l** $2160 \div 90$

DID YOU KNOW?

Pythagoras and President Garfield

According to *The Guinness Book of Records*, Pythagoras' theorem has been proved the greatest number of times, with a book published in 1940 containing 370 different proofs, including one by US president James Garfield (1831–1881). Before becoming involved in politics, Garfield studied mathematics and taught in a public school. He published his proof in 1876 and was the 20th president of the United States, governing for only 150 days in 1881.

Alamy Stock Photo/Prisma Archivo

Find the main mathematical feature of President Garfield's proof.

POWER PLUS ANSWERS ON P.632

1 Name the 5 hypotenuses on this diagram.

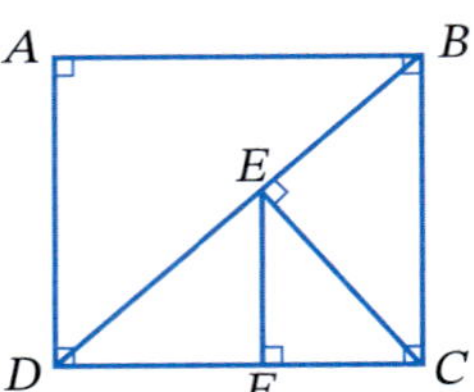

2 The rectangle $ABDE$ in the diagram has an area of 160 m^2. Find the value of x as a surd if $\triangle BDC$ is an isosceles triangle.

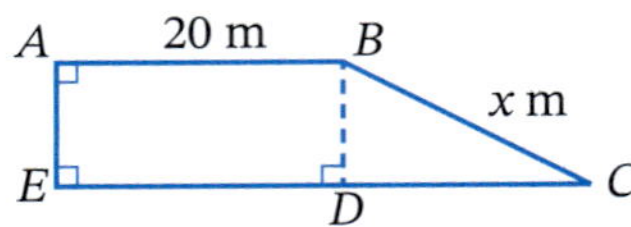

3 Find the value of x in each triangle, correct to one decimal place.

a

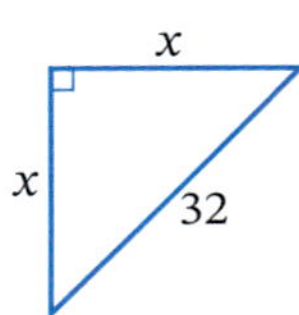

b

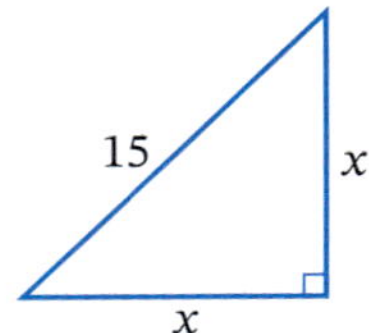

4 For this rectangular prism, find, correct to one decimal place, the length of diagonal:

a HD **b** DE

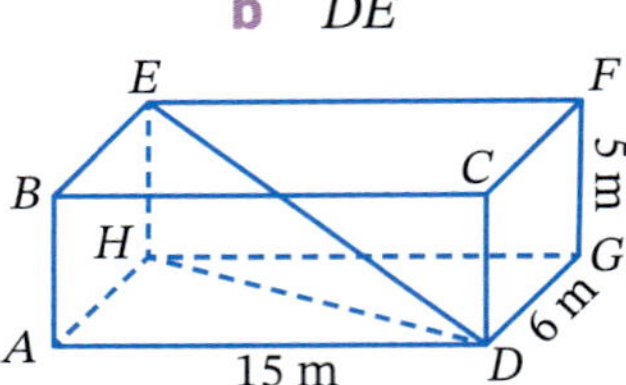

5 Find, correct to one decimal place, the length of the interval joining points $A(-2, 1)$ and $B(3, -2)$ on the number plane.

6 For this cube, find, correct to 2 decimal places, the length of diagonal:

a QS **b** QT

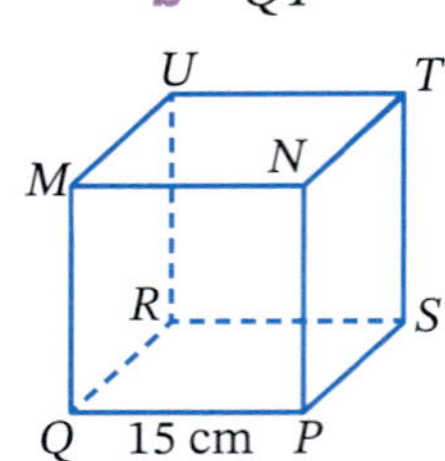

7 For this square pyramid, find, as a surd, the slant height EF.

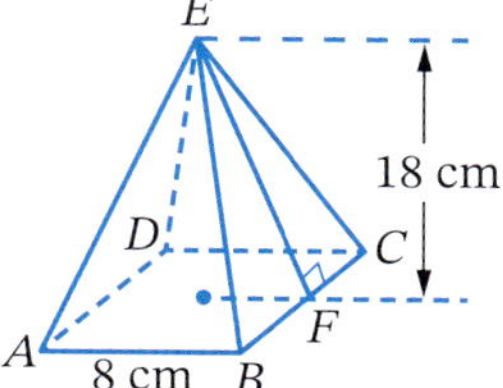

8 Use Pythagoras' theorem to prove that for a right-angled triangle, the area of the semicircle constructed on the hypotenuse is equal to the areas of the semicircles constructed on the other 2 sides.

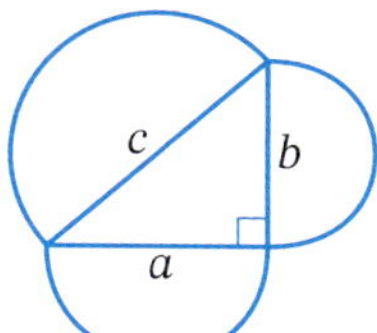

2 CHAPTER REVIEW

ANSWERS ON P.632

Language of maths

Quiz Language of maths 2

area	converse	diagonal	formula
hypotenuse	irrational	perimeter	Pythagoras
right-angled	shorter	side	square root
surd	theorem	triad	unknown

1 Who was **Pythagoras** and which country did he come from?

2 Describe the **hypotenuse** of a right-angled triangle in 2 ways.

3 What is another word for 'theorem'?

4 For what type of triangle is **Pythagoras' theorem** used?

5 What is a **surd**? Is a surd rational? Is a surd real?

6 What is the name given to a set of 3 numbers that follows Pythagoras' theorem?

Topic summary

- How relevant do you think Pythagoras' theorem is to our world? Give reasons for your answer.
- Give 3 examples of jobs where Pythagoras' theorem would be used.
- What did you find especially interesting about this topic?
- Is there any section of this topic that you found difficult? Discuss any problems with your teacher or a friend.

Print (or copy) and complete this mind map of the topic, adding detail to its branches and using pictures, symbols and colour where needed. Ask your teacher to check your work.

Worksheet Mind map: Pythagoras' theorem

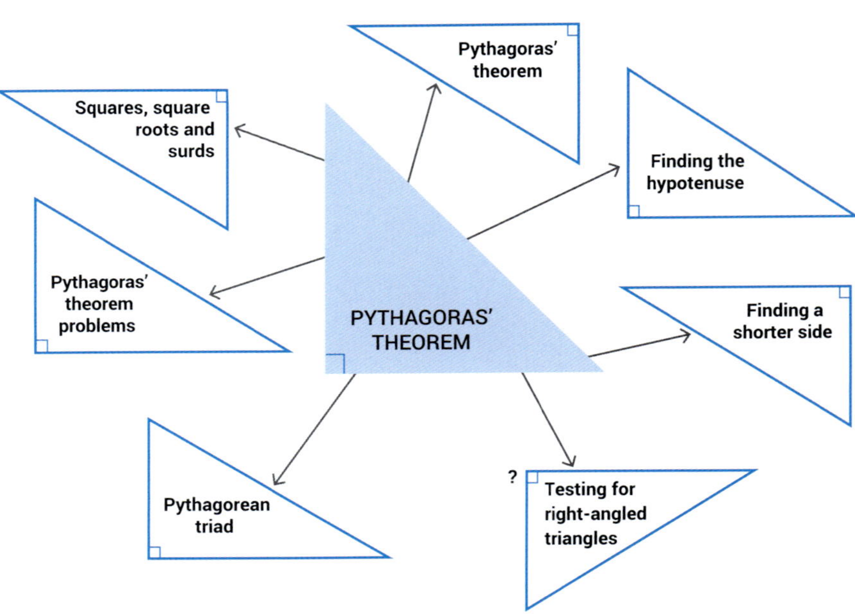

② TEST YOURSELF

ANSWERS ON P.632

Quiz
Test yourself 2

1 Evaluate each expression.

a 8.5^2 **b** $(-6)^2$ **c** $9^2 + 11^2$

2 Evaluate each expression correct to 2 decimal places.

a $\sqrt{146}$ **b** $\sqrt{6^2 + 12^2}$ **c** $\sqrt{27^2 - 16^2}$

3 Name the hypotenuse in each triangle.

a

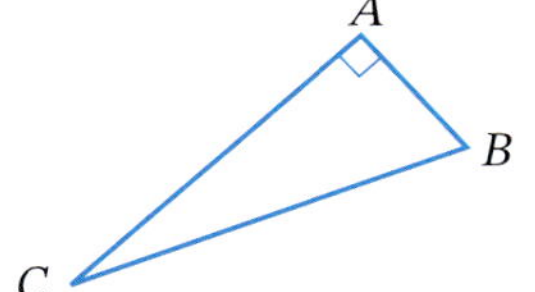

b

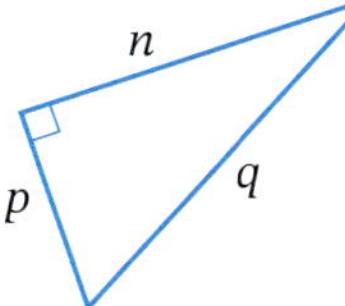

c

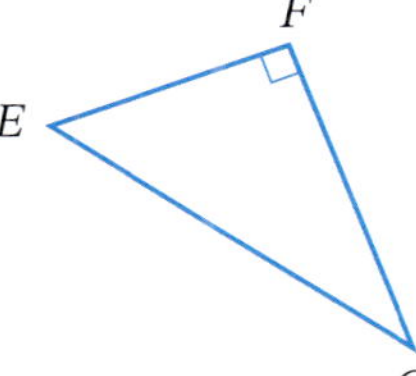

4 Write Pythagoras' theorem for each triangle in question **3**.

5 Find the value of y. Select the closest answer **A**, **B**, **C** or **D**.

A 12.3 **B** 9.1

C 6.9 **D** 3.5

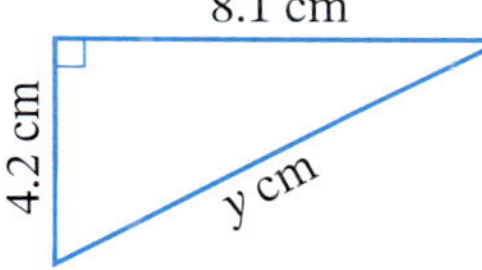

2.03

6 For each triangle, find the length of the hypotenuse. Give your answer as an exact value.

a

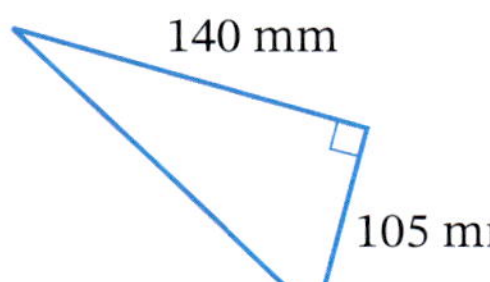

b

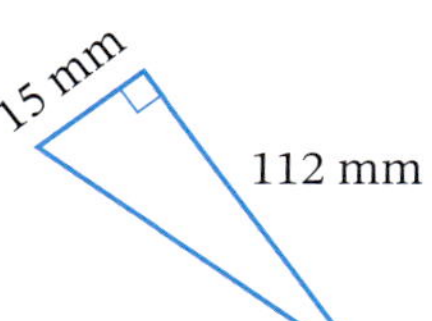

c

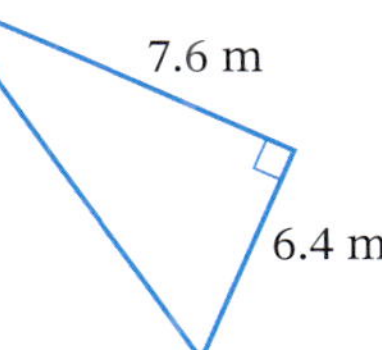

7 Find the value of d. Select the correct answer **A**, **B**, **C** or **D**.

A 12 m **B** 15 m

C 16.8 m **D** 18.9 m

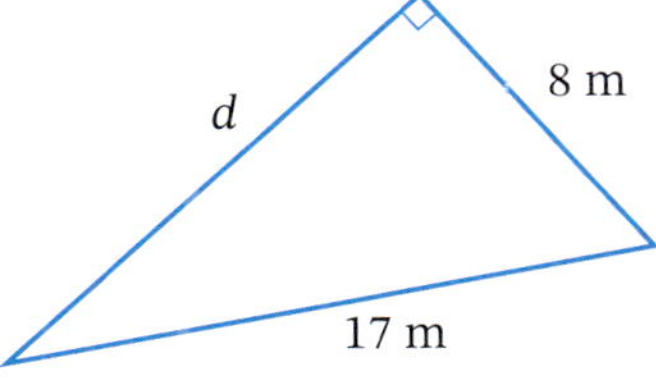

☐ Foundation ○ Standard ○ Complex

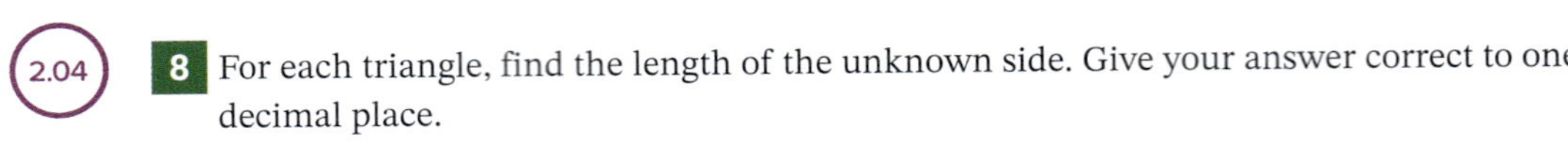

2.04 **8** For each triangle, find the length of the unknown side. Give your answer correct to one decimal place.

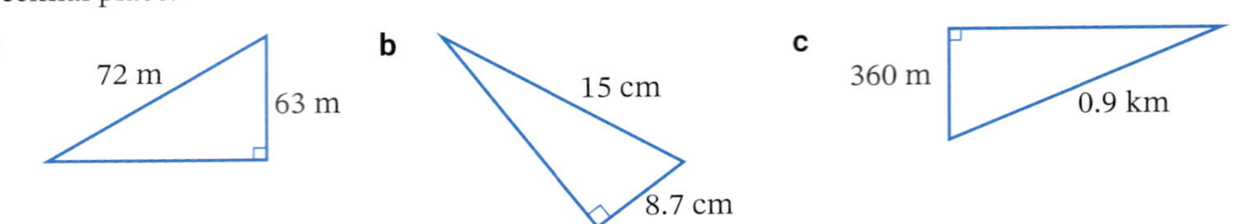

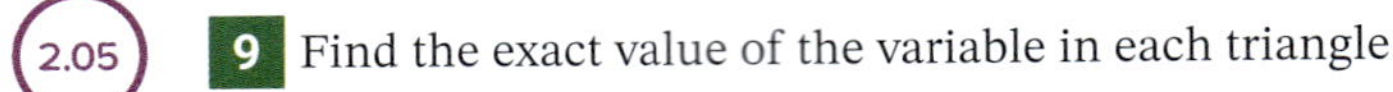

2.05 **9** Find the exact value of the variable in each triangle.

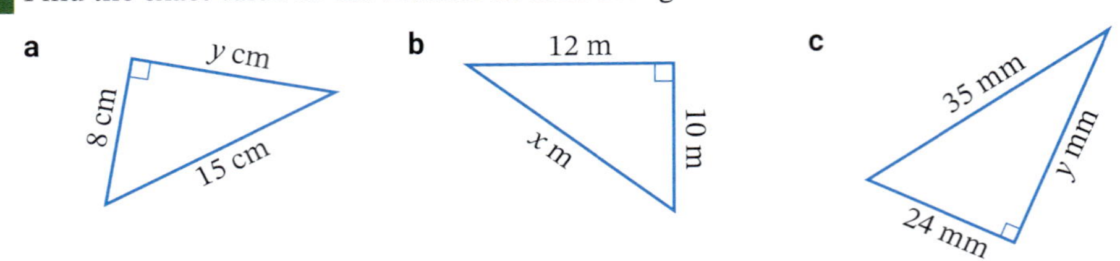

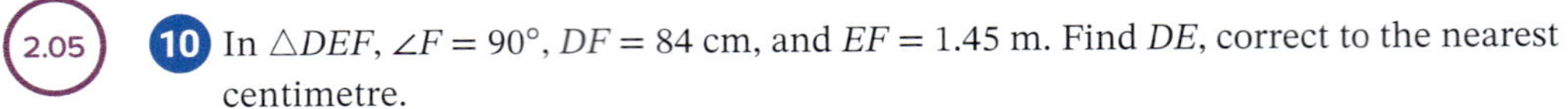

2.05 **10** In $\triangle DEF$, $\angle F = 90°$, $DF = 84$ cm, and $EF = 1.45$ m. Find DE, correct to the nearest centimetre.

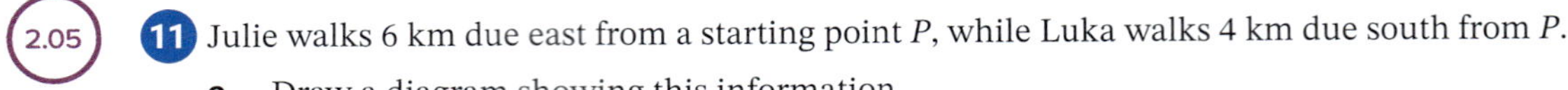

2.05 **11** Julie walks 6 km due east from a starting point P, while Luka walks 4 km due south from P.

- **a** Draw a diagram showing this information.
- **b** How far are Julie and Luka apart? Give your answer correct to one decimal place.

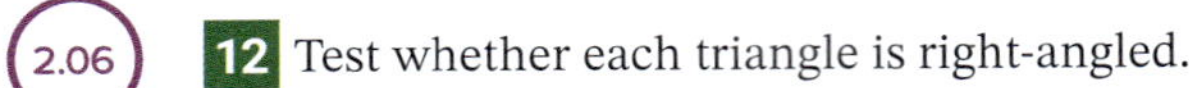

2.06 **12** Test whether each triangle is right-angled.

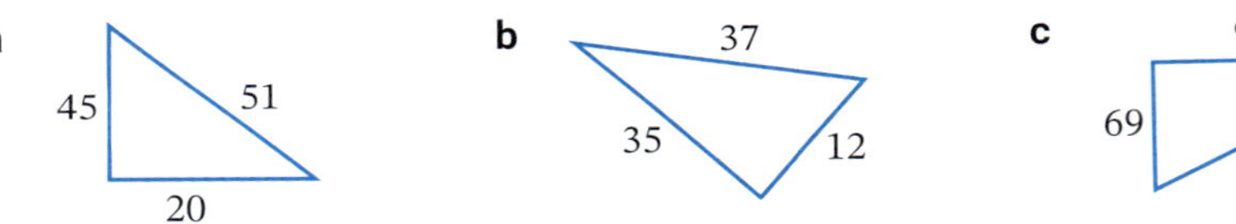

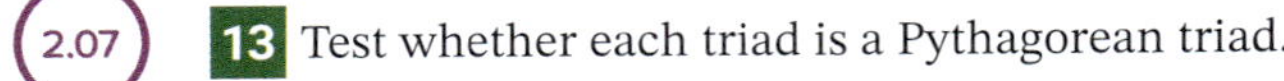

2.07 **13** Test whether each triad is a Pythagorean triad.

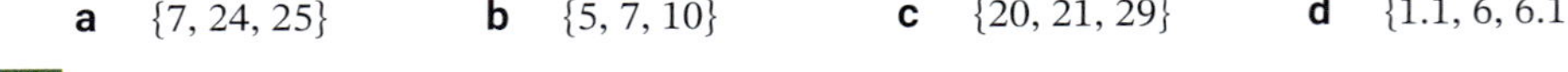

a $\{7, 24, 25\}$ **b** $\{5, 7, 10\}$ **c** $\{20, 21, 29\}$ **d** $\{1.1, 6, 6.1\}$

2.08 **14** On a cross-country ski course, checkpoints C and A are 540 m apart and checkpoints C and B are 324 m apart. Find the distance:

- **a** between checkpoints A and B.
- **b** skied in one lap of the course in kilometres.

B
324 m
C
540 m
A

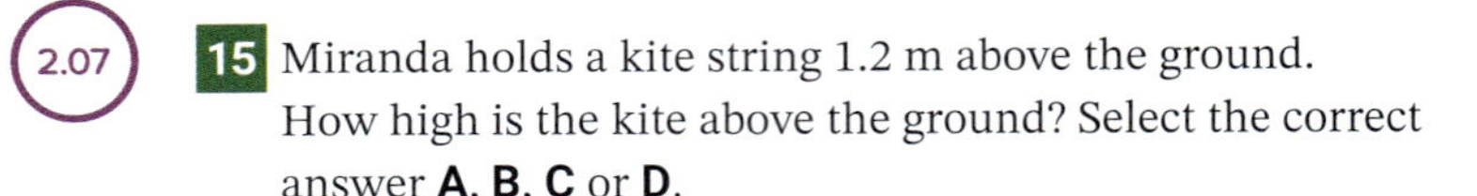

2.07 **15** Miranda holds a kite string 1.2 m above the ground. How high is the kite above the ground? Select the correct answer **A**, **B**, **C** or **D**.

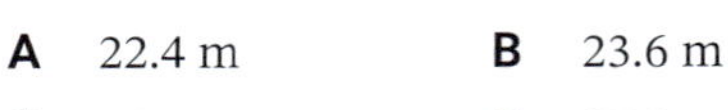

A 22.4 m **B** 23.6 m

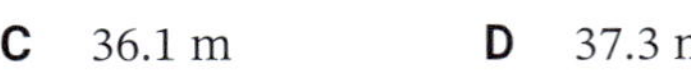

C 36.1 m **D** 37.3 m

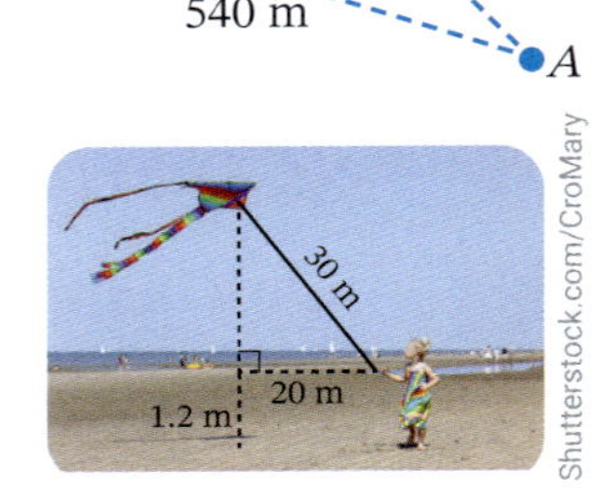

Shutterstock.com/CroMary

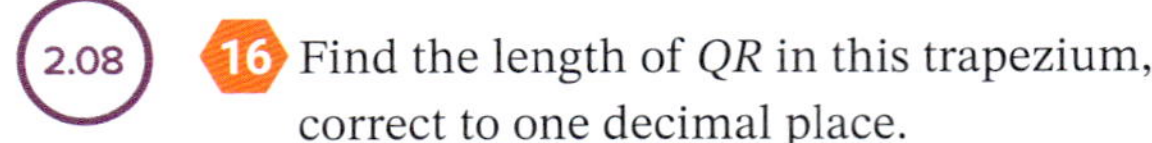

2.08 **16** Find the length of QR in this trapezium, correct to one decimal place.

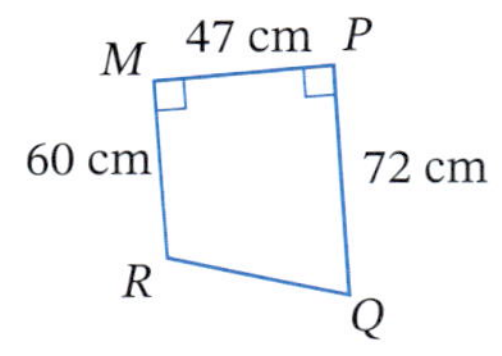

□ Foundation ○ Standard ○ Complex

3

NUMBER, MEASUREMENT

Numeracy and calculation

Numbers are an important part of our lives. We work with numbers when we shop for the best deal and when we borrow or invest money. Calculations are made when using cooking recipes and in the design and construction of houses. Whenever numbers are used to help us solve problems in our lives, calculations need to be made.

iStock.com/spukkato

Chapter outline

U = Understanding
F = Fluency
PS = Problem solving
R = Reasoning
C = Communication

Wordbank

profit The amount made when selling an item at a higher price

loss The amount lost when selling an item at a lower price

cost price The price an item costs the retailer

GST Goods and services tax, a 10% tax added to the original price of an item or service

recurring decimal A decimal with one or more digits that repeat endlessly, for example, $0.1\dot{6} = 0.1666...$

principal An amount of money invested or borrowed, on which interest is given or charged

per annum (p.a.) Per year

unitary method A method of finding a quantity by finding one part or 1% first

Quiz
Wordbank 3

Videos (9):

3.05 Percentage increase and decrease • The unitary method
3.06 Discounts and GST • GST
3.07 Simple interest
3.08 Ratio problems • Dividing a quantity in a given ratio • Rate problems
3.10 Time differences

Twig video (1):

SkillCheck Chinese development of maths

PhET interactives (3):

3.08 Ratio and proportion • Proportion playground • Unit rates

Quizzes (6):

- Wordbank 3
- SkillCheck 3
- Mental skills 3A
- Mental skills 3B
- Language of maths 3
- Test yourself 3

Skillsheets (11):

3.01 Integers • Integers using diagrams
3.02 Decimals • Multiplying by 10, 100, 1000
3.03 Improper fractions and mixed numerals • Equivalent fractions • Simplifying fractions
3.04 Fractions, decimals and percentages • Mental percentages
3.07 Simple interest
3.08 Ratios

Worksheets (6):

3.02 Where's the point? • Fraction families
3.05 Working with percentages
3.06 Profit and loss
3.08 Rates exercise
Mind map: Numeracy and calculation

Puzzles (4):

3.04 Percentages to fractions • Fractions to percentages • Percentages of an amount
3.06 Percentage profit and loss

Technology (2):

3.04 Percentages, fractions and decimals
3.07 Profit and loss

Spreadsheet (1):

3.04 Percentages, fractions and decimals

Nelson MindTap

To access resources above, visit **cengage.com.au/nelsonmindtap**

3 In this chapter you will:

- ✓ add, subtract, multiply and divide integers, decimals and fractions.
- ✓ convert between fractions and decimals.
- ✓ solve problems involving percentages, ratios, rates and time.
- ✓ perform calculations mentally, with pen-and-paper and using technology.
- ✓ calculate time differences, including with 24-hour time.
- ✓ solve problems involving profit, loss, discounts, GST and simple interest.
- ✓ convert between units for rates, for example, kilometres per hour to metres per second.

SkillCheck

ANSWERS ON P. 632

Quiz SkillCheck 3

Video Chinese development of maths

1 Evaluate each expression.

a	8×7	**b**	$42 \div 6$	**c**	$49 \div 7$	**d**	$14 \div 2$
e	6×9	**f**	$8 \div 8$	**g**	9×6	**h**	7×4
i	4×12	**j**	$81 \div 9$	**k**	3×12	**l**	$0 \div 8$
m	$20 \div 5$	**n**	$56 \div 8$	**o**	5×7	**p**	$54 \div 6$

2 Simplify each fraction.

a $\frac{12}{24}$ **b** $\frac{45}{50}$ **c** $\frac{36}{80}$ **d** $\frac{21}{15}$

3 Evaluate each expression.

a $6 + 5 \times 2$
b $15 \div 5 + 4 \times 7$
c $28 \div 7 - 8 \div 2$
d $80 \div 4 \div 5$
e $36 \div 9 \times 5$
f $42 - 4 \times 8$

4 Express each number as a percentage.

a $\frac{1}{4}$
b 0.3
c 0.07
d $\frac{3}{5}$

Integers 3.01

Integers are the positive and negative whole numbers and zero.

A **rational number** can be expressed in the form $\frac{a}{b}$, where a and b are integers and $b \neq 0$.

Because all integers can be written as a fraction with a denominator of 1, for example, $3 = \frac{3}{1}$, $-12 = -\frac{12}{1}$, an integer is also a **rational number** and therefore also a **real number**.

Skillsheets
Integers
Integers using diagrams

Integers can be represented on a number line as shown below.

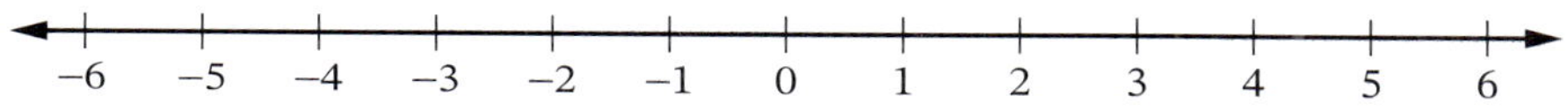

ⓘ Adding and subtracting a negative number

When **adding** a negative number, **subtract** its opposite.

When **subtracting** a negative number, **add** its opposite.

Example 1

Evaluate each expression.

a $12 + (-7)$
b $-3 + (-9)$
c $9 - (-8)$
d $(-12) - (-7)$

SOLUTION

a $12 + (-7) = 12 - 7$
$= 5$
Adding (−7) is the same as subtracting 7.

b $-3 + (-9) = -3 - 9$
$= -12$
$+ (-9) = - 9$

c $9 - (-8) = 9 + 8$
$= 17$
Subtracting (−8) is the same as adding 8.

d $-12 - (-7) = -12 + 7$
$= -5$
$-(-7) = + 7$

ⓘ Multiplying and dividing positive and negative numbers

- Multiplying or dividing 2 positive numbers gives a positive answer.
- Multiplying or dividing a positive number by a negative number gives a negative answer.
- Multiplying or dividing 2 negative numbers gives a positive answer.

Example 2

Evaluate each expression.

a $-5 \times (-8)$ **b** -4×11 **c** $7 \times (-9)$

d $-72 \div (-8)$ **e** $-42 \div 6$ **f** $15 \div (-3)$

SOLUTION

a $-5 \times (-8) = 40$ — 2 negatives

b $-4 \times 11 = -44$ — a negative and a positive

c $7 \times (-9) = -63$ — a positive and a negative

d $-72 \div (-8) = 9$ — 2 negatives

e $-42 \div 6 = -7$ — a negative and a positive

f $15 \div (-3) = -5$ — a positive and a negative

EXERCISE 3.01 ANSWERS ON P. 633

Integers

1 For each pair of integers, write > or < to make a true statement.

(C) **a** -4 ____ 7 **b** 9 ____ 12 **c** 8 ____ -3 **d** -5 ____ -8

2 Write these integers in **ascending order** (smallest to largest): 8, −13, −5, 2, −1, 0.

3 Write these integers in **descending order**: −7, −10, 2, −15, 6, −2.

4 Evaluate each expression.

a $8 + (-5)$ **b** $-7 + 9$ **c** $-10 + 7 + (-5)$ **d** $-5 + (-7) + 8$

e $6 - 10$ **f** $3 - (-2)$ **g** $-7 - 8$ **h** $7 - 5 - 6$

i $-9 - 11$ **j** $-8 - 5 + 4$ **k** $-10 - 3 - 5$ **l** $-12 - (-12)$

m $3 - (-2) - (-3)$ **n** $-1 - (-3) + 5$ **o** $5 - 6 - 7$ **p** $-7 + (-5) - 6$

5 (PS) The temperature at Thredbo, NSW, at 6 a.m. was −12 °C. By 1 p.m., the temperature had risen by 7°, but by 6 p.m., it had fallen by 5°. What is the temperature at 6:00 p.m.?

6 The highest temperature recorded in Australia was 51 °C at Oodnadatta, South Australia, in 1960 and at Onslow, Western Australia, in 2022. The lowest temperature recorded in Australia was −23 °C at Charlotte Pass, NSW, in 1994. The lowest temperature recorded in WA was −7 °C at Eyre in 2008. What was the difference between the highest and lowest recorded temperatures in:

a Australia? **b** WA?

□ Foundation ○ Standard ○ Complex

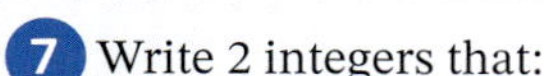

7 Write 2 integers that:

a have a sum of –2. b have a difference of –3.

8 Sejuti was asked to find 2 integers less than 10 that have a difference of 15. What integers did Sejuti choose? Is there more than one answer? Give reasons for your answer.

PS R

9 Evaluate each expression.

EXAMPLE 2

a -3×5 b $8 \times (-1)$ c -4×20 d $(-6) \times (-10)$

e $(-9)^2$ f $12 \times (-20)$ g -8^2 h -50×9

i $45 \div (-9)$ j $-80 \div (-5)$ k $-24 \div 3$ l $\frac{36}{-6}$

m $\frac{-40}{-10}$ n $120 \div (-6)$ o $\frac{-18}{-3}$ p $52 \div (-4)$

10 Evaluate each expression.

a $7 \times (-8)$ b $-120 \div (-6)$ c $7 - (-15)$

d $-10 \times 5 \times (-2)$ e $\frac{27}{-3}$ f $20 - 5 \times 8$

g $(-3)^2 + (-2)^2$ h $(-10 + 7) \times 4$ i $-24 \div 6 \times (-5)$

j $\frac{-100}{-5} \times 4$ k $-8 - (10 - 15)$ l $(-4)^3$

m $6 - (-30) \div 6$ n $40 + 8 \times (-9)$ o $(-5)^2 + 4 \times (-7)$

p $24 \div (-6) + (-5) \times 3$ q $12 - (4 \times 8)$ r $64 \div (-4)^2$

11 Write 2 integers that:

a have a product of –20. b have a quotient of –4.

Decimals 3.02

Decimals are based on powers of 10. The decimal point separates the whole number part from the fractional part. Every decimal can be converted to a fraction, so decimals are **rational numbers**, and therefore **real numbers**.

Skillsheet Decimals

Worksheet Where's the point?

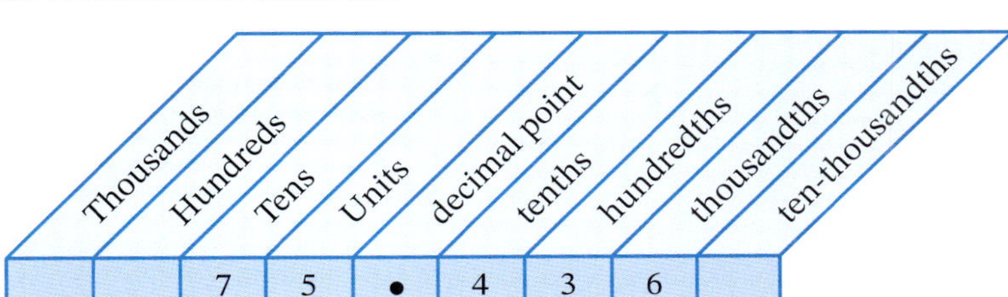

$$75.436 = 7 \times 10 + 5 \times 1 + 4 \times \frac{1}{10} + 3 \times \frac{1}{100} + 6 \times \frac{1}{1000}$$

Foundation Standard Complex

Example 3

Write each fraction as a decimal.

a $\frac{3}{10}$ **b** $\frac{8}{100}$ **c** $\frac{97}{1000}$

SOLUTION

a $\frac{3}{10} = 0.3$ **b** $\frac{8}{100} = 0.08$ **c** $\frac{97}{1000} = 0.097$

Ordering decimals

To order decimals, write them so that they all have the same number of decimal places, putting in zeros where necessary, then compare the numbers.

Example 4

Arrange 35.2, 35.08 and 35.107 in ascending order.

SOLUTION

35.107 has the most decimal places (3), so write all decimals with 3 decimal places.

35.200, 35.080, 35.107.

Then arrange the numbers in ascending order.

35.080, 35.107, 35.200.

So the correct answer is 35.08, 35.107, 35.2.

Adding and subtracting decimals

When adding or subtracting decimals, keep the decimal points under one another.

Example 5

Evaluate each expression.

a 37.6 + 42.85 **b** 327 – 156.25

SOLUTION

a

$$\begin{array}{r} 37.60 \\ +\,42.85 \\ \hline 80.45 \\ \hline \end{array}$$

b

$$\begin{array}{r} 327.00 \\ -\,156.25 \\ \hline 170.75 \\ \hline \end{array}$$

Adding zeros to give the same number of decimal places.

Multiplying and dividing by powers of 10

Skillsheet Multiplying by 10, 100, 1000

3.02

- When multiplying decimals by 10, 100, 1000, ... move the decimal point 1, 2, 3, ... places to the right (the same number of places as the number of zeros in the multiplier).
- When dividing decimals by 10, 100, 1000, ... move the decimal point 1, 2, 3, ... places to the left (the same number of places as the number of zeros in the divisor).

Example 6

Evaluate each product.

a 46.718×100 **b** $15.82 \div 1000$

SOLUTION

a $46.718 \times 100 = 4671.8$ For $\times$ 100, move the point 2 places to the right.

b $015.82 \div 1000 = 0.015\ 82$ For $\div$ 1000, move the point 3 places to the left.

Multiplying decimals

When multiplying decimals, the number of decimal places in the answer must be the same as the total number of decimal places in the question.

Example 7

Evaluate each product.

a 7.62×9 **b** 5.23×0.002

SOLUTION

a $762 \times 9 = 6858$ Multiply without decimal points first.

$7.62 \times 9 = 68.58$ The question has 2 decimal places so the answer must have 2 decimal places.

b $523 \times 2 = 1046$ Multiply without decimal points first.

$5.23 \times 0.002 = 0.010\ 46$ The question has a total of 5 decimal places so the answer must have 5 decimal places.

Dividing decimals

- When dividing by a whole number, keep the decimal points under one another.
- When dividing by another decimal, first move the decimal point in both decimals the same number of places to the right so that we are dividing by a whole number.

Example 8

Evaluate each quotient.

a 72.31 ÷ 4 **b** 8.46 ÷ 0.005

SOLUTION

a $$4\overline{)7\,{}^{3}2.31\,{}^{3}0\,{}^{2}0} = 18.0775$$

72.31 ÷ 4 = 18.0775

b Change the second decimal (0.005) to a whole number (5) by moving the decimal point 3 places to the right.

Then move the decimal point in the first number (8.46) 3 places to the right.

8.460 ÷ 0.005 = 84 60 ÷ 5 Decimal point moves 3 places to the right.

$$5\overline{)8460} = 1692$$

8.46 ÷ 0.005 = 1692

Rounding decimals

Example 9

Round 56.1835:

a to 2 decimal places. **b** to 3 decimal places.

SOLUTION

a 56.1835 ≈ 56.18 The 8 has not changed because the figure after it is 3, which is less than 5.

b 56.1835 ≈ 56.184 The 3 has been rounded up to 4 because the figure after it is 5.

Worksheet Fraction families

Recurring decimals

Recurring decimals have a pattern of numbers that repeat or recur. We use dots to indicate the digits that repeat. Here are some examples:

$0.4444... = 0.\dot{4}$

$0.588\ 88... = 0.5\dot{8}$

$0.636\ 363... = 0.\dot{6}\dot{3}$

$0.415\ 415\ 415... = 0.\dot{4}1\dot{5}$

Terminating decimals do not repeat (terminating means 'stopping'). Here are some examples:

0.6 0.875 0.057

Some decimals are neither recurring nor terminating. They have an infinite number of digits, but there is no pattern. These are **irrational numbers**. Here are some examples:

$\pi = 3.141\ 592\ 653...$

$\sqrt{3} = 1.732\ 050\ 808...$

To convert a fraction to a decimal, divide its **numerator** by its **denominator**.

Example 10

Convert each fraction to a decimal and state whether it is recurring or terminating.

a $\frac{5}{6}$ **b** $\frac{11}{40}$

SOLUTION

a $\frac{5}{6} = 5 \div 6$

$= 0.8333\ ...$

$= 0.8$ (a recurring decimal)

b $\frac{11}{40} = 11 \div 40$

$= 0.275$ (a terminating decimal).

EXERCISE 3.02 ANSWERS ON P. 633

Decimals

EXAMPLE 3

1 Write each fraction as a decimal.

a $\frac{8}{10}$ **b** $\frac{31}{100}$ **c** $\frac{456}{1000}$ **d** $\frac{57}{100}$

e $\frac{3}{100}$ **f** $\frac{92}{10\,000}$ **g** $\frac{6}{100}$ **h** $\frac{124}{10\,000}$

2 Write each decimal as a fraction.

a 0.35 **b** 0.008 **c** 0.4 **d** 0.0017

EXAMPLE 4

3 Write each set of decimals in ascending order.

a 0.7, 0.707, 7.007, 0.7007 **b** 0.202, 0.22, 0.022 02, 0.002

4 Write each set of decimals in descending order.

a 6.03, 6.1, 6.33, 0.6003, 6.3033 **b** 3.0667, 3.6, 3.66, 3.606

5 Write 3 decimals between 5.3 and 5.4 in ascending order. R C

6 Evaluate each expression.

a 2.34 + 5.62 **b** 8.04 + 9.761 **c** 3.458 – 2.237

d 69.24 – 41.1 **e** 34.74 – 18 **f** 24.932 + 12.05

g 15.3 – 12.582 **h** 45.804 – 27.26 **i** 231.56 + 18.407

j 6.56 + 12 **k** 32 – 16.479 **l** 45.6 – 0.007

Foundation Standard Complex

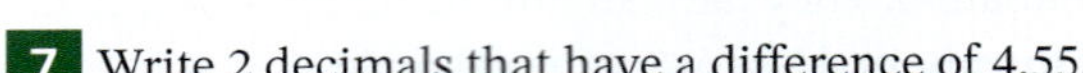

7 Write 2 decimals that have a difference of 4.55.

8 Declan receives the following scores from 7 judges for diving:

PS

7.2 7.8 7.1 8.3 7.9 8.1 7.6

The highest and lowest scores are crossed out, and the remaining 5 scores are added to make Declan's final score.

a Which 2 scores are crossed out?

b What is Declan's final score?

9 **a** A piece of timber 4 m long is cut into 2 pieces. If one piece is 1.85 m, what is the length of the other piece?

b Cate's time for swimming 100 m was 61.35 s. After training for 6 months, she reduced her time by 2.76 s. What is Cate's new time for swimming 100 m?

c Mack ran 400 m in 59.54 s. After training for 12 months, he reduced his time to 55.6 s. By how much had Mack improved his time?

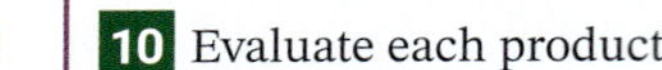

EXAMPLE 6

10 Evaluate each product.

a 7.23×10 **b** 15.012×100 **c** 0.0532×1000

d 8.92×1000 **e** $35.67 \div 100$ **f** $9.7 \div 10$

g $0.987 \div 10$ **h** $234.5 \div 1000$ **i** $0.054 \div 100$

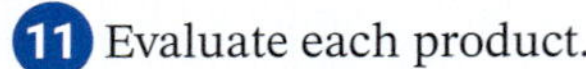

EXAMPLE 7

11 Evaluate each product.

a 2.3×4 **b** 3.6×0.2 **c** 5.23×0.4

d 0.06×0.4 **e** 1.11×0.03 **f** $(0.5)^2$

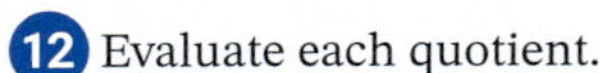

EXAMPLE 8

12 Evaluate each quotient.

a $34.5 \div 5$ **b** $12.08 \div 2$ **c** $1.56 \div 3$

d $12 \div 8$ **e** $4.8 \div 0.3$ **f** $8.0004 \div 0.04$

13 Evaluate each expression.

a $0.45 \times 0.2 \times 0.1$ **b** $12.6 + 0.4 \times 7$ **c** $36.9 \div 3 \times 0.02$

14 Robyn had to calculate $36.024 \div 0.08$ without using a calculator. Which of these calculations will give her the same answer? Select the correct answer **A**, **B**, **C** or **D**.

A $36\,024 \div 8$ **B** $36.024 \div 8$ **C** $3602.4 \div 8$ **D** $0.360\,24 \div 8$

EXAMPLE 9

15 Round each decimal to the number of decimal places indicated in the brackets.

a 55.347 (2) **b** 325.75 (1) **c** 4.99 (1)

d 0.0695 (3) **e** 200.71 (0) **f** 23.8547 (3)

g 19.9 (0) **h** 46.053 (1) **i** 7.5989 (2)

j 245.67 (1) **k** 35.995 (2) **l** 301.9835 (3)

☐ Foundation ○ Standard ⬡ Complex

16 Which one of these decimals can be rounded to 6.75? Select **A**, **B**, **C** or **D**.

A 6.745 **B** 6.736 **C** 6.744 **D** 6.7349

17 Write each recurring decimal using dot notation.

C **a** 0.166 666... **b** 0.273 273 27... **c** 0.888 88... **d** 0.209 99...

18 Show the repeating pattern for each recurring decimal.

C **a** $0.\dot{1}09\dot{8}$ **b** $0.\dot{1}\dot{7}$ **c** $0.2\dot{5}$ **d** $8.\dot{3}1\dot{8}$

19 Convert each fraction to a decimal and state whether it is recurring or terminating.

C **a** $\frac{7}{8}$ **b** $\frac{3}{11}$ **c** $4\frac{8}{9}$ **d** $\frac{13}{5}$

e $\frac{19}{25}$ **f** $\frac{2}{3}$ **g** $\frac{13}{20}$ **h** $1\frac{7}{12}$

i $8\frac{13}{40}$ **j** $\frac{17}{24}$

20 **a** Convert $\frac{1}{3}$ to a recurring decimal.

C **b** Use your answer to part **a** to determine the value of:

R **i** $0.\dot{6}$ **ii** $0.\dot{9}$

21 A decimal with 3 decimal places has been rounded to 8.53.

PS **a** What is the smallest decimal that can be rounded to 8.53?

R **b** What is the largest decimal that can be rounded to 8.53?

22 Aditi and James were asked to multiply 7.28 by 5.56 and express their answers to one decimal place. Their methods are shown below.

C PS R

Aditi	James
7.28 × 5.56 ≈ 7.3 × 5.6 = 40.88 ≈ 40.9	7.28 × 5.56 = 40.4768 ≈ 40.5

Discuss the methods used by both students and decide which is the more accurate answer. Give reasons.

23 Without using a calculator, evaluate:

a $(0.1)^6$ **b** $(0.2)^3$ **c** $(0.3)^4$

Foundation Standard Complex

TECHNOLOGY

Terminating and recurring decimals

Create the spreadsheet below that shows the numerators and denominators of different fractions. For example, row 2 shows $\frac{1}{2}$ and row 3 shows $\frac{3}{4}$.

J2 ▾ fx

	A	B	C
1	**Numerator**	**Denominator**	**Decimal**
2	1	2	
3	3	4	
4	7	11	
5	3	10	
6	1	3	
7	5	6	
8	8	9	
9	12	15	
10	23	40	
11	67	99	
12	5	8	
13			

1. To calculate the equivalent decimal in column C, enter **=A2/B2** in cell C2, then **Fill Down** to cell C12.
2. In column D, classify each decimal as either T (terminating) or R (recurring).
3. Use **Format cells** for each cell in column C that has a recurring decimal and set the number of decimal places to 8.
4. Compare your answers with other students in your class.
5. Try other fractions by entering different numerators and denominators, and determine whether they convert to recurring or terminating decimals.
6. Investigate which denominators give recurring decimals by testing the unit fractions $\frac{1}{2}, \frac{1}{3}, \frac{1}{4}, \ldots, \frac{1}{20}$. Create the spreadsheet below and **Fill Down** cells A3, B3 and C2 to row 21.

	A	B	C
1	Numerator	Denominator	Decimal
2	1	2	=A2/B2
3	1	=B2+1	
4			
5			
6			

Fractions

3.03

All **fractions** are **rational numbers** and therefore **real numbers**.

Skillsheet Improper fractions and mixed numerals

Improper fractions and mixed numerals

Example 11

a Convert $5\frac{3}{8}$ to an **improper fraction**.

b Convert $\frac{13}{4}$ to a mixed numeral.

c What whole number could be placed in the box so that $\frac{\square}{7}$ has a value between 4 and 5?

SOLUTION

a $5\frac{3}{8} = \frac{8 \times 5 + 3}{8}$

$= \frac{48}{8}$

b $\frac{13}{4} = 13 \div 4$

$= 3\frac{1}{4}$

c $4 = \frac{4 \times 7}{7} = \frac{28}{7}$

$5 = \frac{5 \times 7}{7} = \frac{35}{7}$

∴ The number in the box must be between 28 and 35.

So the number in the box could be 29, 30, 31, 32, 33 or 34.

Ordering fractions

Skillsheet
Equivalent fractions

We can compare the size of fractions if they have the same denominator. A common denominator can be found by multiplying the denominators of the fractions together.

Example 12

a Which fraction is larger: $\frac{3}{5}$ or $\frac{2}{3}$?

b Arrange the fractions $\frac{5}{8}, \frac{2}{3}, \frac{5}{7}$ in descending order.

SOLUTION

a A common denominator is $5 \times 3 = 15$.

Rewrite the fractions with the common denominator.

$$\frac{3}{5} = \frac{3 \times 3}{5 \times 3} = \frac{9}{15}$$

$$\frac{2}{3} = \frac{2 \times 5}{3 \times 5} = \frac{10}{15}$$

$10 > 9$ so $\frac{10}{15} > \frac{9}{15}$

Compare the numerators.

$\therefore \frac{2}{3}$ is the larger fraction.

b A common denominator is $8 \times 3 \times 7 = 168$.

$$\frac{5}{8} = \frac{105}{168}, \frac{2}{3} = \frac{112}{168} \text{ and } \frac{5}{7} = \frac{120}{168}$$

(× 21, × 56, × 24)

$120 > 112 > 105$

So the fractions in descending order are $\frac{5}{7}, \frac{2}{3}, \frac{5}{8}$.

Simplifying fractions

Skillsheet
Simplifying fractions

Example 13

Simplify $\frac{27}{36}$.

SOLUTION

$$\frac{27}{36} = \frac{27 \div 9}{36 \div 9}$$

$$= \frac{3}{4}$$

Divide the numerator and denominator by a large common factor, preferably the highest common factor (HCF).

Adding and subtracting fractions

- When **adding or subtracting fractions**, convert them (if needed) so that they have the same denominator.
- When **adding or subtracting mixed numerals**, add or subtract the whole numbers and fractions separately.

Example 14

Evaluate each expression.

a $\frac{3}{5}+\frac{4}{7}$ **b** $\frac{5}{6}-\frac{3}{8}$ **c** $3\frac{1}{2}+2\frac{4}{5}$ **d** $4\frac{1}{5}-2\frac{5}{8}$

SOLUTION

a $\frac{3}{5}+\frac{4}{7}=\frac{3\times7}{35}+\frac{4\times5}{35}$

$=\frac{21}{35}+\frac{20}{35}$

$=\frac{41}{35}$

$=1\frac{6}{35}$

b $\frac{5}{6}-\frac{3}{8}=\frac{5\times8}{48}-\frac{3\times6}{48}$

$=\frac{40}{48}-\frac{18}{48}$

$=\frac{22\div2}{48\div2}$

$=\frac{11}{24}$

c $3\frac{1}{2}+2\frac{4}{5}=5+\frac{1}{2}+\frac{4}{5}$ Adding the whole numbers.

$=5+\frac{5}{10}+\frac{8}{10}$ Common denominator is $2\times5=10$.

$=5+\frac{13}{10}$

$=6\frac{3}{10}$ Converting the improper fraction to a mixed numeral.

d $4\frac{1}{5}-2\frac{5}{8}=2+\frac{1}{5}-\frac{5}{8}$ Subtracting the whole numbers.

$=2+\frac{8}{40}-\frac{25}{40}$ Common denominator is $5\times8=40$.

$=2-\frac{17}{40}$

$=1\frac{40}{40}-\frac{17}{40}$ $2=1+1=1+\frac{40}{40}$

$=1\frac{23}{40}$

Multiplying and dividing fractions

- When **multiplying fractions**, multiply the numerators and multiply the denominators, then simplify if possible (sometimes, it is easier to simplify the fractions first).
- To **divide by a fraction** $\frac{a}{b}$, multiply by its **reciprocal** $\frac{b}{a}$.
- When **multiplying or dividing mixed numerals**, first convert them to improper fractions.

Example 15

Find:

a $\frac{2}{3}\times\frac{4}{7}$ **b** $2\frac{1}{4}\times1\frac{2}{3}$ **c** $2\times\frac{3}{5}$ **d** $\frac{3}{5}\div\frac{7}{11}$ **e** $1\frac{3}{7}\div1\frac{3}{5}$

SOLUTION

a $\frac{2}{3}\times\frac{4}{7}=\frac{2\times4}{3\times7}$ Multiplying numerators and multiplying denominators.

$=\frac{8}{21}$

b $2\frac{1}{4}\times1\frac{2}{3}=\frac{9}{4}\times\frac{5}{3}$

$=\frac{45}{12}$

$=3\frac{9}{12}$

$=3\frac{3}{4}$

c $2\times\frac{3}{5}=\frac{2}{1}\times\frac{3}{5}$

$=\frac{2\times3}{1\times5}$

$=\frac{6}{5}$

$=1\frac{1}{5}$

d $\frac{3}{5}\div\frac{7}{11}=\frac{3}{5}\times\frac{11}{7}$ Multiply by the reciprocal of $\frac{7}{11}$.

$=\frac{33}{35}$

e $1\frac{3}{7}\div1\frac{3}{5}=\frac{10}{7}\div\frac{8}{5}$ Convert to improper fractions.

$=\frac{10}{7}\times\frac{5}{8}$

$=\frac{50}{56}$

$=\frac{25}{28}$

EXERCISE 3.03 ANSWERS ON P. 633

Fractions

U F R C

1 Convert each mixed numeral to an improper fraction. EXAMPLE 11

a $1\frac{2}{3}$ b $4\frac{1}{2}$ c $2\frac{3}{5}$ d $5\frac{11}{20}$

2 Convert each improper fraction to a mixed numeral. EXAMPLE 12

a $\frac{7}{3}$ b $\frac{11}{7}$ c $\frac{16}{5}$ d $\frac{37}{10}$

3 For each pair of fractions, determine which is larger.

a $\frac{5}{6}, \frac{7}{8}$ b $\frac{7}{10}, \frac{3}{4}$ c $\frac{7}{8}, \frac{9}{11}$

4 Write each set of fractions in descending order.

a $\frac{3}{5}, \frac{3}{4}, \frac{6}{7}$ b $\frac{5}{11}, \frac{1}{2}, \frac{2}{3}$ c $\frac{4}{5}, \frac{7}{8}, \frac{17}{20}$

5 Copy the number line below and clearly mark the positions of the fractions $\frac{1}{3}, \frac{4}{5}, \frac{3}{4}, \frac{1}{2}$.

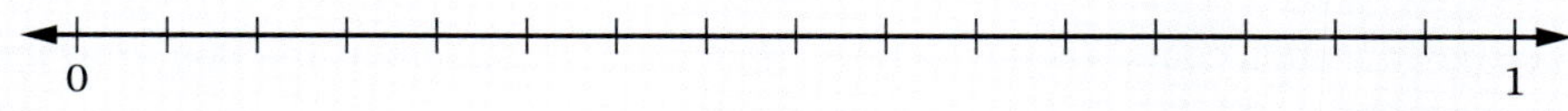

6 Simplify each fraction. EXAMPLE 13

a $\frac{18}{27}$ b $\frac{48}{64}$ c $\frac{15}{33}$ d $\frac{28}{35}$

7 Evaluate each expression. Answer in simplest form. EXAMPLE 14

a $\frac{2}{3}+\frac{4}{7}$ b $\frac{3}{5}+\frac{1}{2}$ c $\frac{5}{9}-\frac{1}{4}$ d $\frac{9}{10}-\frac{2}{3}$

e $\frac{3}{8}+\frac{7}{10}$ f $\frac{5}{6}-\frac{3}{4}$ g $\frac{5}{11}-\frac{1}{3}$ h $\frac{17}{25}-\frac{1}{4}$

i $2\frac{3}{4}+1\frac{2}{5}$ j $3\frac{7}{10}+2\frac{1}{3}$ k $2\frac{4}{5}-1\frac{2}{3}$ l $4\frac{3}{7}-1\frac{4}{5}$

8 Evaluate each product. EXAMPLE 15

a $\frac{5}{7}\times\frac{3}{4}$ b $\frac{1}{2}\times\frac{3}{5}$ c $6\times\frac{3}{4}$ d $\frac{2}{3}\times 7$

e $\frac{7}{8}\times\frac{2}{3}$ f $\frac{7}{8}\times 24$ g $\frac{1}{4}\times\frac{7}{9}$ h $\frac{4}{5}\times 8$

i $\frac{4}{5}$ of 10 m j $\frac{3}{8}$ of 40 L k $\frac{7}{10}$ of \$500 l $\frac{3}{4}$ of 76 kg

m $1\frac{1}{2}\times 3\frac{5}{6}$ n $5\frac{2}{3}\times 2\frac{3}{5}$ o $\frac{5}{6}\times 1\frac{1}{5}$ p $\left(2\frac{1}{2}\right)^2$

9 Evaluate each quotient.

a $\frac{3}{4}\div\frac{1}{2}$ b $\frac{2}{3}\div\frac{4}{5}$ c $\frac{3}{7}\div 5$ d $\frac{6}{7}\div\frac{2}{3}$

e $2\frac{1}{2}\div 4$ f $1\frac{7}{8}\div 2\frac{2}{3}$ g $3\frac{3}{4}\div 1\frac{1}{5}$ h $5\frac{1}{2}\div 2\frac{1}{4}$

10 Write a fraction between $\frac{3}{5}$ and $\frac{2}{3}$.

☐ Foundation ○ Standard ⬡ Complex

EXAMPLE 11

11 For each improper fraction below, select the correct number **A**, **B**, **C** or **D** that should go in the box.

R

a $\frac{\square}{5}$ has a value between 7 and 8.

A 40 **B** 37 **C** 35 **D** 33

b $\frac{\square}{9}$ has a value between 4 and 5.

A 30 **B** 35 **C** 40 **D** 45

c $\frac{35}{\square}$ has a value between 3 and 4.

A 6 **B** 8 **C** 10 **D** 12

12 The sum of 2 numbers is $5\frac{1}{2}$. If one of the numbers is $2\frac{3}{5}$, what is the other number?

13 At a football match, 14 players are given three-quarters of an orange to eat. How many oranges are eaten?

PS

14 Taylor worked $5\frac{1}{2}$ h yesterday and $\frac{1}{2}$ of that today. How long did Taylor work today?

15 A driver education course requires at least 20 hours of classroom instruction. How many lessons are required if each lesson lasts $1\frac{1}{2}$ hours?

PS

16 Cassie spent $\frac{1}{3}$ of the day working, $\frac{5}{12}$ of the day sleeping, $\frac{1}{8}$ of the day travelling and the rest of the day relaxing. What **fraction** of the day did she spend relaxing?

17 Find 2 mixed numerals that have:

R

a a sum of $10\frac{1}{2}$

b a product of $10\frac{1}{2}$

18 At Nelson College's athletics carnival, $\frac{7}{9}$ of the students attended. There are 1188 students enrolled at Nelson College. How many students attended the carnival?

19 Chloe ran $4\frac{1}{2}$ laps of an oval in 20 minutes. How long did it take her to run one lap?

DID YOU KNOW?

Diophantus of Alexandria

Diophantus of Alexandria is known as the 'Father of Higher Arithmetic'. He was an outstanding Greek mathematician who, along with Pappus, dominated Greek mathematics from 250 CE to 350 CE. Diophantus introduced some algebraic symbolism into mathematics, and collected and catalogued what the Greeks had achieved in algebra.

Not much is known about the life of Diophantus. However, part of what is known comes from a problem that was inscribed on his tombstone, as shown.

What was Diophantus' age when he died?

DIOPHANTUS PASSED $\frac{1}{6}$ OF HIS LIFE IN CHILDHOOD, $\frac{1}{12}$ IN YOUTH, AND $\frac{1}{7}$ MORE AS A BACHELOR. FIVE YEARS AFTER HIS MARRIAGE WAS BORN A SON WHO DIED FOUR YEARS BEFORE HIS FATHER AT HALF THE AGE AT WHICH HIS FATHER DIED.

□ Foundation ○ Standard ⬡ Complex

☆ MENTAL SKILLS 3A ANSWERS ON P. 634 Maths without calculators

Quiz Mental skills 3A

3.03

Squaring a number ending in 5, 1 or 9

The square of a number ending in 5 always ends in 25.

For example, $35^2 = 1225$ and $105^2 = 11\ 025$.

A mental calculation trick requires three easy steps:

- delete the 5 from the number.
- multiply the remaining number by the next consecutive number.
- write '25' at the end of the product.

1 Study each example.

a To calculate 35^2:

- deleting the 5 from 35 leaves 3.
- multiply 3 by the next consecutive number: $3 \times 4 = 12$.
- write '25'at the end: 1225.

$35^2 = 1225$

b To calculate 105^2:

- deleting the 5 from 105 leaves 10.
- multiply 10 by the next consecutive number: $10 \times 11 = 110$.
- write 25 at the end: 11 025.

$105^2 = 11\ 025$

2 Now calculate each square number.

a 25^2	**b** 55^2	**c** 45^2	**d** 65^2
e 75^2	**f** 2.5^2	**g** 4.5^2	**h** 205^2
i 6.5^2	**j** 7.5^2	**k** 275^2	**l** 345^2

The square of a number ending in 1 always ends in 1.

For example, $41^2 = 1681$ and $71^2 = 5041$.

A mental calculation trick requires 3 steps.

- Round the number to the nearest 10 (by subtracting 1) to make a new number.
- Square the new number.
- Add to your answer the new number and its next consecutive number.

3 Study each example.

a To calculate 41^2:

- Round 41 down to 40.
- Square 40: $40^2 = 1600$.
- $1600 + 40 + 41 = 1681$.

$41^2 = 1681$

b To calculate 71^2:

- $70^2 = 4900$ Rounding down and squaring
- $4900 + 70 + 71 = 5041$

$71^2 = 5041$

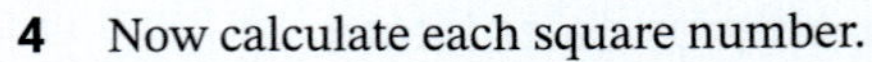

4 Now calculate each square number.

a 21^2	**b** 101^2	**c** 31^2	**d** 91^2	**e** 5.1^2
f 81^2	**g** 61^2	**h** 201^2	**i** 2.1^2	**j** 9.1^2

The square of a number ending in 9 also ends in 1.

For example, $29^2 = 841$ and $99^2 = 9801$.

A mental calculation trick requires 3 steps:

- round the number to the nearest 10 (by adding 1) to make a new number.
- square the new number.
- subtract from your answer the new number and its *previous* consecutive number.

5 Study each example.

a To calculate 29^2:

- round up to 30.
- square 30: $30^2 = 900$.
- subtract 30 and 29: $900 - 30 - 29 = 841$.

$29^2 = 841$

b To calculate 99^2:

- round up to 100.
- square 100: $100^2 = 10\,000$.
- $10\,000 - 100 - 99 = 9801$.

$99^2 = 9801$

6 Now calculate each square number.

a 59^2	**b** 69^2	**c** 89^2	**d** 19^2	**e** 109^2
f 5.9^2	**g** 6.9^2	**h** 119^2	**i** 39^2	**j** 8.9^2

Percentages

3.04

A **percentage** is a fraction 'out of 100' (denominator of 100). For example, 15% means $\frac{15}{100}$ (which can be simplified to $\frac{3}{20}$). A percentage is a **rational number** and **real number**.

Skillsheets
Fractions, decimals and percentages

Mental percentages

Puzzles
Percentages to fractions

Fractions to percentages

Technology
Percentages, fractions and decimals

Spreadsheet
Percentages, fractions and decimals

ⓘ Converting from percentages

- To **convert a percentage to a fraction**, write it as a fraction with denominator 100 and simplify.
- To **convert a percentage to a decimal**, divide by 100.

Example 16

Convert each percentage to a simplified fraction.

a 24% **b** $5\frac{1}{2}\%$ **c** 8.2%

SOLUTION

a
$$24\% = \frac{24 \div 4}{100 \div 4} = \frac{6}{25}$$

b
$$5\frac{1}{2}\% = \frac{5\frac{1}{2}}{100} = \frac{5\frac{1}{2} \times 2}{100 \times 2} = \frac{11}{200}$$

c
$$8.2\% = \frac{8.2}{100} = \frac{8.2 \times 10}{100 \times 10} = \frac{82}{1000} = \frac{41}{500}$$

Example 17

Convert each percentage to a decimal.

a 18% **b** 3.5% **c** 215%

SOLUTION

a $18\% = 18 \div 100 = 0.18$

b $3.5\% = 3.5 \div 100 = 0.035$

c $215\% = 215 \div 100 = 2.15$

ⓘ Converting to percentages

To **convert a fraction or a decimal to a percentage**, multiply by 100%.

Example 18

Convert each fraction or decimal to a percentage.

a $\frac{5}{8}$ **b** 0.085 **c** 0.3 **d** $\frac{5}{6}$

SOLUTION

a $\frac{5}{8} = \frac{5}{8} \times 100\%$
$= 62\frac{1}{2}\%$ (or 62.5%)

b $0.085 = 0.085 \times 100$
$= 8.5\%$

c $0.3 = 0.30 \times 100\%$
$= 30\%$

d $\frac{5}{6} = \frac{5}{6} \times 100\%$
$= 83\frac{1}{3}\%$ (or 83.3%)

Ordering fractions, decimals and percentages

Example 19

Arrange $\frac{3}{5}$, 5%, 0.5 and $\frac{5}{9}$ in ascending order.

SOLUTION

Convert each number to a percentage or decimal.

Method 1: as percentages

$\frac{3}{5} = 60\%$, $5\% = 5\%$, $0.5 = 50\%$, $\frac{5}{9} = 55.\dot{5}\%$

In ascending order:

5%, 50%, $55.\dot{5}\%$, 60%

The ascending order is: 5%, 0.5, $\frac{5}{9}$, $\frac{3}{5}$.

Method 2: as decimals

$\frac{3}{5} = 0.6$, $5\% = 0.05$, $0.5 = 0.5$, $\frac{5}{9} = 0.\dot{5}$

In ascending order:

0.05, 0.50, $0.\dot{5}$, 0.60

The ascending order is: 5%, 0.5, $\frac{5}{9}$, $\frac{3}{5}$.

Percentage of a quantity

Puzzle
Percentages of an amount

Example 20

Find:

a 12% of \$90 **b** $5\frac{1}{2}\%$ of \$540

SOLUTION

a $12\% \times \$90 = \frac{12}{100} \times \$90$ or $0.12 \times \$90$
$= \$10.80$

Entering the decimal 0.12 for 12% on a calculator is faster

b $5\frac{1}{2}\%$ of $\$540 = 5\frac{1}{2}\% \div 100 \times \$540$ or $0.055 \times \$540$
$= \$29.70$

EXERCISE 3.04 ANSWERS ON P. 634

3.04

Percentages

1 Convert each percentage to a simplified fraction. EXAMPLE 16

a 6% **b** 45% **c** 84% **d** 2%
e 0.5% **f** 150% **g** 51% **h** $4\frac{1}{2}\%$
i $8\frac{3}{4}\%$ **j** $33\frac{1}{3}\%$ **k** $12\frac{1}{2}\%$ **l** 4.75%
m 6.25% **n** 5.5% **o** 30% **p** $16\frac{2}{3}\%$

2 Convert each percentage to a decimal. EXAMPLE 17

a 85% **b** 5.1% **c** 10% **d** 8.45%
e $6\frac{1}{2}\%$ **f** 12.25% **g** 0.4% **h** 450%

3 Convert each fraction or decimal to a percentage. EXAMPLE 18

a $\frac{7}{100}$ **b** $2\frac{3}{8}$ **c** 0.48 **d** 0.055
e $\frac{36}{50}$ **f** 0.016 **g** $\frac{2}{3}$ **h** $\frac{3}{4}$
i $\frac{11}{25}$ **j** 0.73 **k** $\frac{3}{15}$ **l** $\frac{7}{11}$
m 1.015 **n** $\frac{1}{6}$ **o** $\frac{18}{20}$ **p** 0.9

4 State whether each statement is true (T) or false (F). EXAMPLE 19

a $55\% > 0.58$ **b** $\frac{6}{11} < 55\%$ **c** $24.5\% > \frac{2}{9}$ **d** $0.87 < \frac{7}{8}$

5 Write each set of numbers in ascending order:

a $46\%, \frac{3}{7}, 0.427, \frac{2}{5}$ **b** $\frac{29}{40}, 70\%, 0.73, \frac{8}{11}$

6 Write each set of numbers in descending order:

a $0.333, \frac{4}{11}, 33.33\%, \frac{1}{3}$ **b** $\frac{4}{7}, 55\%, 0.58, \frac{1}{2}$

7 Find: EXAMPLE 20

a 15% of \$85 **b** 120% of 60 kg **c** 45% of 16 m
d 2.4% of \$350 **e** $5\frac{1}{4}\%$ of \$80 **f** 12% of 95 kg
g $2\frac{1}{2}\%$ of \$870 **h** 150% of 16 L **i** 3.75% of \$2850
j $33\frac{1}{3}\%$ of \$6540 **k** 6.25% of 25 t **l** 0.4% of \$3720

8 A total of 84 600 people attended the State of Origin match between NSW and Queensland in Sydney. If 71% of the crowd supported NSW:

a what percentage of the crowd supported Queensland?
b how many people in the crowd supported Queensland?

Foundation Standard Complex

9 980 students attend St Klaas' College. If 19% of them are in Year 9, calculate the number of Year 9 students.

10 Buzz Coffee advertises an extra 20% of coffee for the same price. If the normal amount is 700 g, how much extra coffee will you get for the same price?

11 Jakub's weight is now 125% of what it was a year ago. If his weight then was 85 kg, what is his weight now?

DID YOU KNOW?

Be positive

There are 4 main blood types, and each type can be positive or negative. Your blood type depends on your genes. The distribution of blood types in Australia are shown in the table.

O		A		B		AB	
+	−	+	−	+	−	+	−
40%	9%	31%	7%	8%	2%	2%	1%

1 **If Australia's population is 26 million, how many people would be in each blood type?**

2 **Do you know what your blood type is? How would you find out?**

TECHNOLOGY

Converting fractions to percentages

In this activity, we will use a spreadsheet to convert 12 fractions to percentages and sort them in descending order.

	A	B	C
1	Numerator	Denominator	Percentage
2	1	4	
3	3	5	
4	9	10	
5	7	8	
6	2	3	
7	28	50	
8	45	75	
9	5	6	
10	27	60	
11	76	80	
12	7	9	
13	375	1000	
14			

1 Create a spreadsheet like the one at right that shows the numerators and denominators of different fractions. For example, row 2 shows $\frac{1}{4}$ and row 3 shows $\frac{3}{5}$.

2 Enter **= A2/B2*100** in cell C2 to convert $\frac{1}{4}$ to a **decimal** first.

3 **Fill Down** from cell C2 to C13 to convert each fraction to a decimal.

4 Use **Format cells** to set column C in **percentage** format to 5 decimal places: this converts all the decimals to percentages.

5 Highlight rows 2 to 13 and click **Sort** column C by **descending (largest to smallest)**.

6 Try converting different fractions to percentages and sorting them.

☐ Foundation ○ Standard ⬡ Complex

Applying percentages

3.05

Video Percentage increase and decrease

Worksheet Working with percentages

Example 21

a Increase \$200 by 7%.

b Decrease \$150 by 12%.

SOLUTION

a Increase = 7% × \$200
= \$14
New amount = \$200 + \$14
= \$214

100% + 7% = 107%

or New amount = 107% of \$200
= 1.07 × \$200
= \$214

b Decrease = 12% × \$150
= \$18
New amount = \$150 − \$18
= \$132
or New amount = 88% of \$150

100% − 12% = 88%

= 0.88 × \$150
= \$132

ⓘ Expressing quantities as percentages

To **express one quantity as a percentage of another**, write the appropriate fraction and then change it to a percentage:

$$\text{Percentage} = \frac{\text{amount}}{\text{whole amount}} \times 100\%$$

or Percentage = amount ÷ whole amount × 100%

Example 22

a Jamila scores 43 out of 60 in a History test. What is her percentage mark to the nearest whole number?

b What percentage (correct to one decimal place) is 35 s of 5 min?

SOLUTION

a Percentage mark $= \frac{43}{60} \times 100\%$
$= 71.666...\%$
$\approx 72\%$

Use $\text{Percentage} = \frac{\text{amount}}{\text{whole amount}} \times 100\%$

b Percentage $= \frac{35\text{ s}}{5\text{ min}} \times 100\%$
$= \frac{35\text{ s}}{300\text{ s}} \times 100\%$
$= 11.666...\%$
$\approx 11.7\%$

Change minutes to seconds so that the same units are being used.
5 min = 5 × 60 s = 300 s

Video
The unitary method

The **unitary method** can be used to find the whole amount (100%) when part of the amount is known. 'Unit' means 'one' and the unitary method involves finding 1% first.

ⓘ The unitary method

Given a percentage of an amount, to find the whole amount:

- find 1% of the amount by dividing by the known percentage.
- multiply by 100 to find the whole amount (100%).

Example 23

This week, Julio received a bonus of \$86, which was 9% of his weekly wage. Calculate Julio's weekly wage, correct to the nearest dollar.

SOLUTION

9% of the wage = \$86

∴ 1% of the wage = \$86 ÷ 9
= \$9.555...

∴ Julio's wage = \$9.55... × 100 — 100% of the wage
= \$955.555...
≈ \$956

EXERCISE 3.05 ANSWERS ON P. 634

Applying percentages

U F PS R C

EXAMPLE 21

1 Increase:

a \$455 by 10% **b** \$80 by 15.5% **c** \$680 by 200%
d 80 kg by 2.5% **e** 35 m by 30% **f** 120 L by 140%

2 Decrease:

a \$830 by 35% **b** 120 min by 5% **c** 90 L by 12%
d 6.8 m by 25% **e** 130 kg by 18% **f** \$96.50 by 10%

EXAMPLE 22

3 Matt buys a bike for \$860 and sells it, making a profit of 5%.

a How much profit did Matt make?
b For what price did Matt sell the bike?

4 T-shirts priced at \$38 are reduced by 45% at an end-of-summer sale. What is the sale price of the T-shirts?

5 Due to severe storms in summer, the price of fresh fruit and vegetables increased by 34%. How much will broccoli cost if its original price was \$4.99/kg?

6 Lucy's weekly wage of \$895 increased by 5.5%. Find her new weekly wage.

☐ Foundation ○ Standard ⬡ Complex

9780170465618

7 A car dealer is offering New Year sales. What will you pay for a used car marked at \$16 500 if a **discount** of 12% is given?

8 Express each proportion as a percentage, correct to one decimal place if necessary.

C
a 55 out of 80 **b** 80 out of 120 **c** 45 out of 70
d 15 out of 18 **e** \$8 out of \$12 **f** \$9 out of \$60
g 35 s out of 4 min **h** 15 min out of 2 h **i** 340 mm out of 2 m
j 3 h out of 1 day **k** 85 cm out of 5 m **l** 140c out of \$10

9 Selina achieved the following marks in 2 tests:
PS English: 32 out of 40
R Mathematics: 47 out of 60
C Convert each mark to a percentage to decide which is the better mark.

10 Travis earns \$840 per week, of which \$280 goes to rent. Find what percentage of Travis' wage goes to rent, correct to one decimal place.

11 Bozena earns \$480 per week while she is completing Year 12 at school. If she spends \$95 of it on petrol, what percentage of her wage is this, correct to one decimal place?

12 The nutritional information on a food label is shown.
C

	Per 100 g
Energy	2105 kJ
Total Fat (of which is Saturated fat)	25 g 16 g
Carbohydrate (of which is sugars)	62 g 38 g
Protein	5 g
Salt	1 g

a What percentage of a 100 g serve is:
  i protein? **ii** carbohydrate?

b For a 100 g serve, what percentage of the total fat is saturated fat?

13 In the Netball World Cup final, Australia scored 15 goals in the second quarter. If this
PS represented 29.4% of the goals Australia scored in the match, how many goals did
R Australia score in total?

14 A jeweller charges \$105 for valuing a diamond ring. If this charge represents 1.25% of the value of the ring, find this value correct to the nearest dollar.

15 A can of energy food drink contains 20% more for the same price as the standard can. If the extra amounts to 150 g, how much does the standard can contain?

Foundation Standard Complex

16 Last year, Ben paid 28% of his salary in tax. If he paid $23 423 in tax, what was Ben's salary? Answer to the nearest dollar.

17 An amount is increased by 10% and then decreased by 10%. What is the overall percentage change?

PS R

18 An amount is decreased by 10% and then increased by 10%. What is the overall percentage change?

PS R

TECHNOLOGY

Converting marks to percentages

In this activity, you will convert students' test marks from a score out of 60 to a percentage.

1 Create the spreadsheet shown below and enter the formula shown in C3. Use it to convert each mark out of 60 to a percentage.

	A	B	C
1	**Student**	**Mark**	**Percentage**
2		60	%
3	Sam	48	=B3/B2*100
4	Tranh	37	
5	Mithra	59	
6	Anka	47	
7	Serg	51	
8	Rebecca	39	
9	Suriya	50	
10			

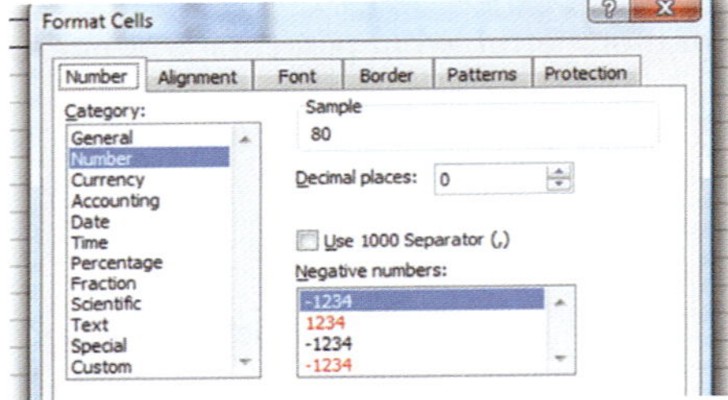

2 Highlight column C and choose **Number** and set the decimal places to 0.

3 Use your spreadsheet to answer these questions. Write your answers in the cells indicated in brackets [ ].

a Who scored the highest mark? [D1]

b What mark out of 60 is equivalent to 80%? [D2]

c How many students scored between 70% and 90%? [D3]

d How many students scored less than 70%? [D4]

e What was the difference between the highest and lowest percentage marks? [D5]

4 Select rows 3 to 9 and sort the marks in descending order.

☐ Foundation ○ Standard ⬡ Complex

Profit, discounts and GST

3.06

Worksheet Profit and loss

Puzzle Percentage profit or loss

Profit and loss

- When an item is sold for **more** than it costs, a **profit** is made.
- When an item is sold for **less** than it costs, a **loss** is made.
- The percentage profit and loss are usually calculated as a percentage of the **cost price**.

Example 24

MindTech online store buys notebook computers for \$550 each and sells them for \$680. Find:

a the profit on each computer.

b the percentage profit on the cost price, correct to one decimal place.

SOLUTION

a Profit = \$680 – \$550

= \$130

b Percentage profit $= \dfrac{\text{profit}}{\text{cost price}} \times 100\%$

$= \dfrac{130}{550} \times 100\%$

$= 23.6363...$

$\approx 23.6\%$

Discounts

Video Discounts and GST

Example 25

At *Jeans Minus*' end-of-season sale, jeans are discounted by 25%. What is the discount price of a pair of jeans marked at \$148?

SOLUTION

Method 1

Discount = 25% of \$148

= \$37

Discount price = \$148 – \$37

= \$111

Method 2

Discount price = 75% of \$148

= \$111

Video
Discounts and GST

Example 26

A home cinema costs \$865 after a 24% discount. What was its original price, rounded to the nearest dollar?

SOLUTION

Sale price = \$865, discount = 24%

$\therefore$ 76% of the original price = \$865 — 100% − 24% = 76%

$1\% = \frac{\$865}{76}$ — Using the unitary method.

$= \$11.381\,578...$

$\therefore$ Original price $= \$11.381\,578\ldots \times 100$ — To find 100%.

$= \$1138.157\ldots$

$\approx \$1138$

Videos
Discounts and GST
GST

GST

Goods and Services Tax (GST) is a tax paid to the government on most goods (items) and services that we purchase. In Australia, GST is charged at 10% of the original price and is generally included in the marked price of the good or service.

Example 27

The **selling price** of a lounge suite is \$2695 with GST included. How much of this price is the GST?

SOLUTION

Selling price = 100% + 10% — Adding 10% GST.

= 110%

110% of the original price = \$2695

10% GST = \$2695 ÷ 11

= \$245

EXERCISE 3.06 ANSWERS ON P. 634

Profits, discounts and GST

U F PS R

In this exercise, express all percentages correct to one decimal place and all prices to the nearest cent.

EXAMPLE 24

1 A home theatre system with a cost price of \$650 is sold for \$1250. Find:

a the profit

b the percentage profit on the cost price.

2 A motorbike originally priced at \$16 000 is sold for \$14 500. Find:

a the loss

b the percentage loss.

☐ Foundation ○ Standard ○ Complex

3 Alla bought an old car for \$3000, spent another \$1500 on repairs and then sold it for \$6200. Find:

a the total amount that Alla spent on the car.

b the profit made when she sold the car.

c her percentage profit.

4 A house was bought for \$850 000. A year later, it was sold for \$720 000.

Calculate:

a the loss

b the percentage loss.

5 Towels with a cost price of \$25 each are sold for \$35. Calculate the percentage profit.

6 A car costing \$12 000 is sold for \$8000. Calculate the percentage loss.

7 *Home Necessities* store buys cutlery sets for \$120 that are then sold for \$230. Calculate the percentage profit.

8 GIA supermarket reduces the price of some items by 35%. Find the discount price of the following items given their original prices.

a 250 g butter, \$2.40

b 2 L milk, \$3

c 1 kg cheese, \$11

d 24 cans of soft drink, \$15.99

9 Calculate the selling price of a computer marked at \$760, now discounted by 15%. Select the correct answer **A**, **B**, **C** or **D**.

A \$114 **B** \$660 **C** \$646 **D** \$636

10 The price of a punnet of strawberries marked at \$4.95 was reduced by 25%. Calculate its discounted price. Select the correct answer **A**, **B**, **C** or **D**.

A \$4.74 **B** \$3.74 **C** \$3.71 **D** \$1.25

11 At a book sale, all books are discounted by 35%. What is the discount price of a book marked at \$25.50?

12 A factory-seconds washing machine has its price reduced from \$980 to \$750. What is the percentage discount?

13 A TV priced at \$1599 is sold for \$1250. Find the discount as a percentage of the marked price.

14 Calculate the percentage discount for each item.

a

Mens Shirts
Usually **\$80**
THIS WEEK ONLY
\$55

b

15 A jacket discounted by 40% after Christmas sells for \$84. What was the original price of the jacket?

16 A bedroom suite reduced by 15% is sold for \$2801. What was its original price? Answer to the nearest dollar.

17 Running shoes are sold to make a profit of 85%. If they sold for \$164.65, what was their cost price? Select the correct answer **A**, **B**, **C** or **D**.

A \$140 **B** \$89 **C** \$76 **D** \$65

18 A plane fare from Brisbane to Sydney has been discounted by 45%. If Amy paid \$159 for a discounted ticket, what was the full price to the nearest dollar?

19 For each item, calculate the GST payable and the final price.

a a car priced at \$20 900
b a home cinema priced at \$1810
c a theatre ticket priced at \$145
d plumber's fees of \$180

20 Given the selling price of each item with GST included, calculate the GST and the original price, correct to the nearest cent.

PS R

a haircut: \$46.20
b car repairs: \$739.20
c photo frame: \$31.35
d pair of jeans: \$65.89
e refrigerator: \$924
f piano lessons: \$198 per term

21 A discount warehouse has a 30% discount on all computers with a further 5% off the discount price if you pay in cash. Carlos wants to buy a computer priced at \$1899.

PS R

a What is the discount price of the computer?
b Carlos pays by cash. How much will he need to pay now, to the nearest 5 cents?
c What is the total discount Carlos received?
d Calculate his total percentage discount (on the original cost).

22 Justine buys 20 boxes of paper at \$29.95 a box.

PS R

a What is the cost of the 20 boxes of paper?
b Justine receives a business discount of 15%. How much will she need to pay?
c Justine pays by cash and receives a further discount of 5%. How much will she now pay, to the nearest 5 cents?
d Find the amount Justine would pay if she received a single discount of 20% on the cost of the paper.
e Is the single discount of 20% the same as the successive discounts of 15% and 5%? Give reasons for your answer.

Foundation Standard Complex

TECHNOLOGY

Profit or loss

Technology
Profit and loss

The spreadsheet below shows items for sale in *Nina's Furniture Warehouse*. Each item is shown with its cost price and selling price. Devesh and Sarita want to buy these items for their new family home.

1 Create a new spreadsheet as shown.

	A	B	C	D	E
1	Item	Cost price	Selling price	Profit or Loss	% Profit or Loss
2	lounge	599	799		
3	bookshelf	189	169		
4	childrens table	110	95		
5	child's cot	350	475		
6	armchair	239	310		
7	rug	695	490		
8	mirror	45	79		
9	dining table	865	1060		
10					

2 Highlight columns B, C and D and choose **Format cells** and **Currency**.

3 In cell D2, enter the formula **=C3 - B2** to calculate the profit or loss for the lounge (A loss will be shown as a negative value).

4 Use **Fill down** in column D to calculate the profit or loss for each item.

5 In cell E2, enter **=D2/B2*100** to calculate the percentage profit or loss for the lounge.

6 Use **Fill down** in column E.

7 Highlight column E and choose **Format cells** and **Number** and set the number of decimal places to 0.

8 How much money did Devesh and Sarita spend altogether on furniture? In cell C10, enter **=sum(C2:C9)**.

9 Calculate the total profit or loss made by the store on these items. In cell D10, enter **=sum(D2:D9)**.

Simple interest 3.07

Skillsheet
Simple interest

Banks, credit unions and other financial institutions reward investors by paying them interest on their savings or investments. Conversely, they charge borrowers by making them pay interest on their loans.

- The original amount of money invested or borrowed is called the **principal**.
- **Interest** is calculated as a percentage of the principal.
- This percentage is called the **interest rate**, usually written as a rate **per annum** ('per year'), abbreviated 'p.a.'
- **Simple interest** (or **flat rate interest**) is interest calculated simply on the original principal.

Example 28

Find the simple interest earned on an investment of \$18 000 at 3.5% p.a. for 5 years.

SOLUTION

Principal = \$18 000, interest rate = 3.5% p.a., term = 5 years.

Term means the amount of time the investment or loan has occurred.

Interest = 3.5% × \$18 000 × 5 — Over 5 years.

= \$3150

ⓘ The simple interest formula

$$I = Prn,$$

where:

I is the simple interest.

P is the principal.

r is the interest rate per year, expressed as a decimal.

n is the number of years.

The simple interest formula can also be written as $I = Pin$ or $I = Prt$.

Applying this to the above example:

$P = \$18\ 000$, $r = 3.5\% = 0.035$, $n = 5$ years

$I = Prn$

$= \$18\ 000 \times 0.035 \times 5$

$= \$3150$

Example 29

Video Simple interest

Find the simple interest on:

a \$8620 at 2.4% p.a. for 7 months

b \$5600 at 6.25% p.a. for 220 days

SOLUTION

a $P = \$8620$, $r = 2.4\% = 0.024$, $n = 7$ months $= \frac{7}{12}$ years — n must be in years.

$I = Prn$

$= \$8620 \times 0.024 \times \frac{7}{12}$

$= \$120.68$

b $P = \$5600$, $r = 6.25\% = 0.0625$, $n = 220$ days $= \frac{220}{365}$ years — Rounded to the nearest cent.

$I = Prn$

$= \$5600 \times 0.0625 \times \frac{220}{365}$

$= \$210.9589...$

$\approx \$210.96$

Example 30

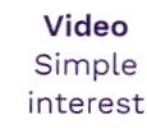
Video Simple interest

3.07

After 2 years, an investment of \$1560 has earned \$87.36 in simple interest.
What is the annual interest rate?

SOLUTION

$I = \$87.36, P = \$1560, n = 2$ years

$I = Prn$

$\$87.36 = \$1560 \times r \times 2$

$\$87.36 = \$3120r$

$r = \frac{\$87.36}{\$3120}$

$= 0.028$

$= 2.8\%$

$\therefore$ Annual interest rate $= 2.8\%$

Example 31

For how long will \$12 000 need to be invested to earn \$1350 if the flat rate of interest is 4.5% p.a.?

SOLUTION

$I = \$1350, P = \$12\,000, n = 4.5\% = 0.045$

$I = Prn$

$1350 = 12\,000 \times 0.045 \times n$

$1350 = 540n$

$n = \frac{1350}{540}$

$= 2.5$ years

EXERCISE 3.07 ANSWERS ON P. 634

Simple interest

U F PS R

In this exercise, give all money answers correct to the nearest cent.

EXAMPLE 28

1 Calculate the simple interest earned on each investment.

- **a** \$18 000 at 7% p.a. for 5 years
- **b** \$5280 at 2.85% p.a. for 2 years
- **c** \$3000 at 5.5% p.a. for 4 years
- **d** \$9450 at $2\frac{1}{2}\%$ p.a. for 3 years
- **e** \$6800 at 6.2% p.a. for 6 years
- **f** \$12 500 at 7.3% p.a. for 2 years

2 Steffi invests \$17 000 at 6.45% p.a. simple interest for 4 years. To what final value will her investment grow?

EXAMPLE 29

3 Calculate the simple interest for each situation.

- **a** \$2100 invested at 6% p.a. for 9 months
- **b** \$3450 invested for 5 months at 3.25% p.a.
- **c** \$10 000 invested at 6.2% p.a. for 150 days

Foundation Standard Complex

d $4000 borrowed for 14 months at 6.55% p.a.

e $280 000 borrowed at 8.3% p.a. for 30 weeks

f $18 400 borrowed for 160 days at 14.7% p.a.

4 Find the annual interest rate if:

a an investment of $3000 earns $756 simple interest after 4 years.

b an investment of $17 630 earns $2715.02 simple interest after 2 years.

c a flat rate loan of $9000 is charged $2115 after 5 years.

5 For how long will $12 350 need to be invested to earn $1778.40 if the flat rate of interest is 2.4% p.a.?

PS R

6 How many months will it take for an investment of $22 000 to earn $12 650 in interest if the interest rate is 6.9% p.a.?

7 How many days will it take for $15 600 to earn $436.97 in interest if the flat rate of interest is 7.1%?

8 William took 2 years to pay off a flat rate loan of $7000. His total loan repayments amounted to $8400. Calculate:

a the interest charged.

b the interest rate.

9 Simone and Jeff borrowed $4000 to build a garden in their backyard. They repaid the loan at $128 per month for 4 years.

PS

a Calculate the total amount they paid for the garden.

b How much interest did they pay on the loan?

c Calculate the flat rate of interest charged.

Quiz Mental skills 3B

☆ MENTAL SKILLS 3B ANSWERS ON P. 634 **Maths without calculators**

Estimating square roots

We can use the square numbers to estimate square roots.

n	1	2	3	4	5	6	7	8	9	10	11	12
n^2	1	4	9	16	25	36	49	64	81	100	121	144

1 Study each example.

a Estimate $\sqrt{7}$.

7 is between the square numbers 4 and 9 (2^2 and 3^2 respectively).

Because 7 is closer to 9, $\sqrt{7}$ is closer to 3.

An estimate is $\sqrt{7}$ ≈ 2.6. (Actual answer = 2.645...)

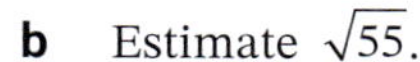

b Estimate $\sqrt{55}$.

55 is between the square numbers 49 and 64 (7^2 and 8^2 respectively).

Since 55 is closer to 49, $\sqrt{55}$ is closer to 7.

An estimate is $\sqrt{55} \approx 7.4$. (Actual answer = 7.416...)

2 Now estimate each square root correct to one decimal place.

a	$\sqrt{8}$	**b**	$\sqrt{18}$	**c**	$\sqrt{28}$	**d**	$\sqrt{111}$
e	$\sqrt{80}$	**f**	$\sqrt{31}$	**g**	$\sqrt{12}$	**h**	$\sqrt{65}$
i	$\sqrt{75}$	**j**	$\sqrt{29}$	**k**	$\sqrt{40}$	**l**	$\sqrt{126}$

Ratios and rates 3.08

A **ratio** compares quantities of the same type measured in the same units.

It is written in the form '$a : b$' (pronounced 'a to b'), where a and b are numbers.

Skillsheet Ratios

Video Ratio problems

Interactives Ratio and proportion

Proportion playground

Unit rates

Simplifying ratios

To **simplify a ratio**, keep dividing each term by the same number, preferably a large number such as their highest common factor (HCF), until each term is as small as possible.

Example 32

Simplify each ratio.

a 36 : 48 : 18 **b** 425 mL to 5 L

SOLUTION

a $36 : 48 : 18 = \frac{36}{6} = \frac{48}{6} = \frac{18}{6}$ Dividing all terms by 6, their HCF.

$= 6 : 8 : 3$

b 425 mL : 5 L = 425 mL : 5000 mL Expressing both ratios in the same units.

$= 425 : 5000$

$= \frac{425}{25} : \frac{5000}{25}$ Dividing both terms by 25, their HCF.

$= 17 : 200$

Ratio problems can be solved using **equivalent ratios** or the **unitary method**. The unitary method requires finding one part first.

Example 33

During summer, the ratio of jeans to shorts sold at a beachside store was 3 : 10. If 240 pairs of jeans were sold, how many pairs of shorts were sold?

SOLUTION

Method 1: Equivalent ratios

Jeans : shorts = 3 : 10 = 240 : ______

Since $3 \times 80 = 240$,

Number of shorts $= 10 \times 80$

$= 800$

Method 2: Unitary method

Jeans : shorts = 3 : 10

Since 240 jeans were sold, 3 parts = 240

$\therefore$ 1 part $= 240 \div 3$ — Finding 1 part first.

$= 80$

$\therefore$ Number of shorts $= 80 \times 10$ — Finding 10 parts.

$= 800$

ⓘ Dividing a quantity in a given ratio

- Find the total number of parts.
- If using the **unitary method**, find the size of one part, then multiply to calculate each share.
- If using the **fraction method**, calculate fractions of the quantity to find each share.
- Check that the shares add up to the original quantity.

Example 34

Video
Dividing a quantity in a given ratio

Divide a cash prize of \$420 between Elyse and Henry in the ratio 3 : 4.

SOLUTION

Method 1: Unitary method

Total number of parts $= 3 + 4 = 7$

$\therefore$ 7 parts = \$420

1 part $= \$420 \div 7 = \$60$

$\therefore$ Elyse's share $= 3 \times \$60 = \$180$

Henry's share $= 4 \times \$60 = \$240$

Checking: $\$180 + \$240 = \$420$

Method 2: Fraction method

Total number of parts $= 3 + 4 = 7$

$\therefore$ Elyse's share $= \frac{3}{7} \times \$420 = \$180$

Henry's share $= \frac{4}{7} \times \$420 = \$240$

Checking: $\$180 + \$240 = \$420$

Simplifying rates

3.08

A **rate** compares 2 quantities of different types measured in different units.

It measures how one quantity changes with another quantity.

It is written in the form 'a/b' (pronounced 'a per b'), where a and b are units of measurement.

Example 35

Write each statement as a rate.

a Travelling 600 km in 12 h.

b A phone call costs $6.45 for 15 min.

SOLUTION

a $600 \text{ km in } 12 \text{ h} = 600 \div 12$

$= 50 \text{ km/h}$

b $\$6.45 \text{ for } 15 \text{ min} = \$6.45 \div 15$

$= \$0.43/\text{min}$

Rate problems

Video
Rate problems

Worksheet
Rates exercise

Example 36

A car travels at 92 km/h.

a How far will it travel in $4\frac{1}{2}$ h?

b How long will it take to travel 800 km? Answer correct to one decimal place.

a Write a pair of equivalent rates in fraction form. On the left, the given rate: the car travels 92 km in every hour. On the right, what we're trying to find (the number of km in 4.5 h):

$$\frac{92 \text{ km}}{1 \text{ h}} = \frac{? \text{ km}}{4.5 \text{ h}}$$

$\text{Distance} = 92 \times 4.5$

$= 424 \text{ km}$

OR

Write the units of the rate as a fraction: $\text{km/h} = \frac{\text{km}}{\text{h}}$

To find the distance (km) in the numerator, multiply by the rate.

$\text{Distance} = 4\frac{1}{2} \times 92$

$= 414 \text{ km}$

b $$\frac{92 \text{ km}}{1 \text{ h}} = \frac{800 \text{ km}}{? \text{ h}}$$

The car travels 92 km per hour. To find the time taken to travel 800 km, calculate how many times 92 km goes into 800 km.

$\text{Time} = 800 \div 92$

$= 8.6956...$

$\approx 8.7 \text{ h}$

OR

To find the time (h) in the denominator, divide by the rate.

$\text{Time} = 800 \div 92$

$= 8.6956...$

$\approx 8.7 \text{ h}$

Problems involving rates can be solved usually by multiplying or dividing.

- Write the units of the rate x/y as a fraction: $\frac{x}{y}$
- To find the quantity in the numerator, x, **multiply** by the rate.
- To find the quantity in the denominator, y, **divide** by the rate.

EXERCISE 3.08 ANSWERS ON P. 635

Ratios and rates

1 Simplify each ratio.

a	48 : 28	**b**	15 : 36	**c**	125 : 60	**d**	84 : 56
e	124 : 62	**f**	$\frac{3}{5}:\frac{1}{4}$	**g**	0.68 : 0.24	**h**	1.25 : 4.5
i	$5:\frac{4}{5}$	**j**	$\frac{5}{8}:\frac{2}{3}$	**k**	4.8 : 0.64	**l**	15 : 18 : 6
m	27 : 36 : 9	**n**	120 : 20 : 80	**o**	$\frac{5}{6}:3$	**p**	$2\frac{1}{2}:5$

2 Simplify each ratio.

a	75c : $4	**b**	5 kg : 800 g	**c**	2 h : 45 min
d	600 mm : 2 m	**e**	8 h : 1 day	**f**	8 mm : 4 cm
g	300 mL : 2 L	**h**	$3.50 : 25c	**i**	6 min : 45 s

3 The ratio of Ken's age to his daughter Jess' age is 10 : 3. If Ken is 30, how old is Jess?

4 Bella and Suri invest in a business in the ratio 3 : 5. If the larger investment is $520 000, find the amount of the smaller investment.

5 A survey of car buyers found that they purchased cars with colours in the ratio of white : grey : blue : other colour = 8 : 5 : 11 : 10.

A car dealer ordered 20 cars in colours other than white, grey and blue. How many white cars should be ordered?

6 A concrete mixture of gravel to sand to cement of 4 : 3 : 1 is needed for strong foundations. How much of each is needed to make 40 cubic metres of concrete?

PS R

7 At a busy intersection involving a red-light camera, the ratio of cars running a red light to cars stopping was 1 : 79. If 90 000 cars passed through the intersection last week, how many ran the red light?

8 Shahid and Bridie won $60 000 in a lottery. If they share the prize in the ratio 23 : 27, how much does each person receive?

9 An alloy consists of nickel and copper in the ratio 4 : 7. If the alloy weighs 3.41 kg, how much copper was used? Select the correct answer **A**, **B**, **C** or **D**.

A 0.21 kg **B** 0.85 kg **C** 1.24 kg **D** 2.17 kg

Foundation Standard Complex

9780170465618

EXAMPLE 35

3.08

10 Write each statement as a rate using the units in brackets. Round answers to one decimal place if needed.

- **a** 540 km in 8 h (km/h)
- **b** 5 kg for \$42 (\$/kg)
- **c** 2800 words in 50 min (words/min)
- **d** 14 L for 170 km (km/L)
- **e** 200 m in 26 s (m/s)
- **f** 5 kg of seed for 120 m^2 (g/m^2)
- **g** 40 days to walk 900 km (km/day)
- **h** \$215 for 3 hours (\$/h)

EXAMPLE 36

11 A truck maintains an average speed of 65 km/h. Calculate how far it will travel:

- **a** in $2\frac{1}{2}$ h
- **b** from 6:30 a.m. to 3:00 p.m.

12 Walid earns \$220 for an 8-hour day.

PS R

- **a** Express this pay as an hourly rate (\$/h).
- **b** How much would he earn in a 40-h week?
- **c** How many days will it take him to earn \$1760?

13 5 kg of lamb loin chops cost \$79.95.

- **a** Express this as a rate (\$/kg).
- **b** How much would 3 kg cost?
- **c** How many kilograms (correct to one decimal place) can be bought for \$40?

14 James' heart beats at 78 beats/min.

- **a** How many times will it beat in 1 h?
- **b** How long will it take to beat 1000 times? Answer to the nearest second.

15 Gold jewellery is classified according to its gold content. The ratio of gold to other metals is given in carats. Pure gold is 24 carats, so 10-carat gold has gold mixed with other metals in the ratio 10 : 14.

PS R

- **a** Write the ratio of gold to other metals in:
  - **i** a 9-carat bracelet
  - **ii** an 18-carat ring
- **b** Nami purchased a 14-carat gold necklace with a mass of 50 g. How much gold is in the necklace, correct to one decimal place?

16 A utility van has a fuel consumption of 11.6 L/100 km. Calculate how much fuel it will use for a trip of:

- **a** 640 km
- **b** 58 km
- **c** 140 km

17 Calculate the fuel consumption, in L/100 km, of a Toyota Prius hybrid that uses 30.6 L to travel 900 km.

18 Birth rates are given as births/1000 people. A city's birth rate is 16 births/1000 people. If there were 520 births in the city last year, what was its population?

19 David claims that 5 kg of Greenie grass seed will plant an area of 1200 m^2. How much grass seed is needed for an area of 60 m^2? Select the correct answer **A**, **B**, **C** or **D**.

- **A** 0.25 kg
- **B** 0.4 kg
- **C** 1.44 kg
- **D** 4 kg

□ Foundation ○ Standard ○ Complex

20 **a** Indonesia has a population of 278 million and an area of 1 919 440 km². Calculate its population density in persons/km², correct to one decimal place.

b Australia has a much smaller population of 26.0 million but a much larger area of 7 692 000 km². Calculate its population density, correct to one decimal place.

c If Australia was as densely populated as Indonesia, what would its population be? Answer to the nearest hundred.

DID YOU KNOW?

The beat of life

The heart of an unfit person works harder whether at rest, during activity or even while recovering from activity. A fit person's heart copes with activity better and its beat returns to normal sooner.

	Heartbeats per minute		
	Fit	Unfit	Difference
Before activity	65	80	15
During activity	95	135	40
3 minutes after	80	115	35
6 minutes after	70	100	30
9 minutes after	65	95	30

What patterns do you notice in the differences in heart rates between fit and unfit persons?

3.09 Converting rates

It is often necessary to convert rates from one set of units to another, for example, from kilometres/hour to metres/second, or from millilitres/minute to litres/day.

Example 37

Convert:

a 80 L/kg to L/g **b** 5 m/s to km/h **c** 2 mL/min to L/day

SOLUTION

a $80 \text{ L/kg} = \frac{80}{1000} \text{ L/g}$ $1 \text{ kg} = 1000 \text{ g}$

$= 0.08 \text{ L/g}$

☐ Foundation ○ Standard ⬡ Complex

b $5\text{ m/s} = 5 \times 3600\text{ m/h}$ — $1\text{ h} = 3600\text{ s}$

$= 18\,000\text{ m/h}$ — $1\text{ km} = 1000\text{ m}$

$= \frac{18\,000}{1000}\text{ km/h}$

$= 18\text{ km/h}$

c $2\text{ mL/min} = 2 \times 60\text{ mL/h}$ — $1\text{ h} = 60\text{ min}$

$= 120\text{ mL/h}$ — $1\text{ day} = 24\text{ h}$

$= 120 \times 24\text{ mL/day}$ — $1\text{ L} = 1000\text{ mL}$

$= 2880\text{ mL/day}$

$= \frac{2880}{1000}\text{ L/day}$

$= 2.88\text{ L/day}$

EXERCISE 3.09 ANSWERS ON P. 635

Converting rates

U F PS R

EXAMPLE 37

1 Convert:

a 5 m/s to m/h
b 30 m/s to m/h
c \$10/kg to \$/g
d 20 mL/min to mL/h
e 40 mL/min to mL/h
f 5 mL/s to mL/h
g 5 kg/m^2 to kg/ha
h 2.5 tonnes/day to kg/day
i \$750/week to \$/year
j 0.5 km/min to km/h
k 15 sheep/h to sheep/day
l \$60/day to \$/week

2 Convert (round to one decimal place if needed):

a 5 m/s to km/h
b 35 cm/s to m/h
c 10 cents/g to \$/kg
d 40 mL/min to L/h
e 40 L/min to mL/s
f 5 mL/s to L/day
g 25 g/m^2 to kg/ha
h 2.5 tonnes/h to kg/day
i 110 km/h to m/s
j 5 kg/m to g/cm
k 8 km/L to m/mL
l 70c/min to \$/h

3 Convert each speed into kilometres per hour. Round your answers to 2 decimal places.

a An African cheetah was measured running at 27 m/s.
b A German peregrine falcon dived at 97 m/s.
c A Tanzanian snake can travel at a speed of 3.3 m/s.
d A sailfish off the coast of Florida was estimated to swim 30 m/s.
e A racing cyclist rode at 23 m/s.

4 If Nate's reaction time is 0.9 s, how far will his car travel in this time if its speed is:

a 60 km/h?
b 80 km/h?
c 110 km/h?

PS R

Foundation Standard Complex

3.10 Time differences

Video
Time differences

Example 38

Calculate the time difference between:

a 9:25 a.m. and 4:40 p.m. **b** 06:35 and 15:20

SOLUTION

a The time from 9:25 a.m. to 4:40 p.m. can be broken up as follows.

9:25 a.m. to 10:00 a.m. = 35 min

10:00 a.m. to 4:00 p.m. = 6 h

4:00 p.m. to 4:40 p.m. = 40 min

Time difference = 6 h + 35 min + 40 min

= 6 h 75 min

= 6 h + 1 h + 15 min

= 7 h 15 min

OR Convert to 24-hour time first, then use the calculator's ° ′ ″ or DMS key to enter hours and minutes, and subtract the times.

b The time from 06:35 to 15:20 can be broken up as follows.

06:35 to 07:00 = 25 min

07:00 to 15:00 = 8 h

15:00 to 15:20 = 20 min

Time difference = 8 h + 25 min + 20 min

= 8 h 45 min

OR Use the calculator's ° ′ ″ or DMS key to enter hours and minutes, and subtract the times.

EXERCISE 3.10 ANSWERS ON P. 635

Time differences

1 Find the time difference between:

a 6:00 a.m. and 1:00 p.m. **b** 9:00 a.m. and 4:00 p.m.

c 4:30 p.m. and 8:00 p.m. **d** 11:50 a.m. and 4:30 p.m.

e 8:17 a.m. and 5:00 p.m. **f** 10:35 a.m. and 4:55 p.m.

g 1:20 a.m. and 7:50 p.m. **h** 6:22 a.m. and 11:09 a.m.

i 12:47 p.m. and 11:43 p.m.

☐ Foundation ○ Standard ○ Complex

2 Change each of these to 24-hour time.

C

a	3:40 a.m.	**b**	5:00 p.m.	**c**	12:35 a.m.
d	7:18 a.m.	**e**	9:51 p.m.	**f**	12:24 p.m.

3 Change each of these to 12-hour (a.m./p.m.) time.

a	22:15	**b**	08:30	**c**	13:44
d	18:52	**e**	02:05	**f**	19:27

4 Find the time difference between:

a	05:00 and 13:00	**b**	13:00 and 19:30	**c**	11:40 and 17:55
d	03:25 and 14:47	**e**	09:12 and 23:20	**f**	10:05 and 11:20
g	09:44 and 12:50	**h**	18:16 and 22:40	**i**	14:32 and 20:09

5 What is the time 2h 50 min:

a	after 10:20 a.m.?	**b**	before 12:40 p.m.?	**c**	before 6:23 a.m.?
d	after 13:00?	**e**	after 19:45?	**f**	before 01:05?

6 Phoebe catches the 7:05 a.m. train from Melbourne. At what time will she arrive in Albury if the trip takes 4 h 21 min?

7 The school day starts at 8:35 a.m. and finishes at 2:55 p.m. How many hours and minutes is this? Select the correct answer **A**, **B**, **C** or **D**.

A 5 h 20 min **B** 5 h 30 min **C** 6 h 20 min **D** 6 h 30 min

8 Find the time difference between:

a 2:00 p.m. today and 2:00 a.m. tomorrow
b 5:00 p.m. today and 3:30 a.m. tomorrow
c 8:00 a.m. today and 9:00 p.m. tomorrow
d 7:15 p.m. today and 9:45 a.m. tomorrow

9 What is the time:

a	115 min after 8:40 p.m.?	**b**	4 h 35 min before 2:20 p.m.?
c	3 h 50 min before 1:00 p.m.?	**d**	154 min after 11:06 a.m.?

10 The length of a movie is 112 min.

a Express 112 min in hours and minutes.
b At what time will the movie finish if it started at 8:30 p.m.?
c At what time did the movie start if it finished at 1:15 p.m.?

☐ Foundation ○ Standard ⬡ Complex

POWER PLUS ANSWERS ON P. 635

1 What fractions do ***A***, ***B*** and ***C*** represent on this number line if the intervals between the fractions are equal?

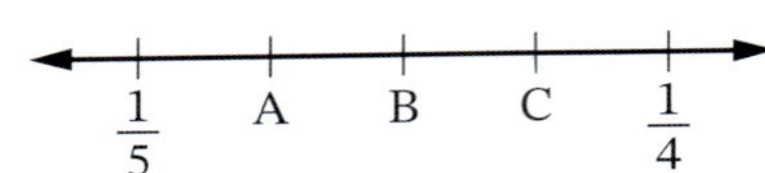

2 Deepti sold a car for \$14 700, making a loss of 30% on its original price. Find the original price.

3 A store marks up the price of a lounge suite by 80%. What percentage discount would bring the price of the lounge suite back to its original price?

4 Ineke went to a supermarket and bought 5 kg of potatoes. If 3 kg of potatoes cost \$4.99, how much change would Ineke receive from \$20?

5 2 L of water is to be shared equally between 3 people.

a Calculate how much water each person would receive (correct to 3 decimal places).

b Explain why 3 decimal places would be too accurate for practical purposes.

6 Melinda bought a pair of shoes that were reduced by $\frac{1}{3}$. If the shoes sold for \$72, what was their original price?

7 After training for 6 months, Owen improved his time for running 100 m by 8%. If his new time is 10.2 s, what was his old time? Answer correct to one decimal place.

8 At a local football game, the ratio of men to children was 5 : 8 and the ratio of women to children was 7 : 10. What is the ratio of men to women?

9 Hannah, Nicola and Olivia share an amount of money in the ratio 2 : 3 : 5. If Nicola gives \$35 to Hannah, both of them will have an equal amount of money. How much money will Olivia receive?

10 James and Jordan each drove their cars to the beach. James drove his car at a speed of 40 km/h for half the distance and at 60 km/h for the other half. Jordan drove at 50 km/h for the entire distance. Who arrived first, James or Jordan? Give reasons for your answer.

11 a How long will it take for an investment of \$5000 to double if it is invested at 5% p.a. simple interest?

b How long will it take for an investment of \$10 000 to double at the same interest rate?

9780170465618

3 CHAPTER REVIEW

Language of maths

24-h time	ascending	decimal	decimal places
descending	discount	fraction	GST
integer	interest rate	irrational	loss
per annum (p.a.)	percentage	principal	profit
rate	ratio	rational	reciprocal
recurring	simple interest	terminating	unitary method

Quiz
Language of maths 3

1 What is another name for **flat rate interest**?

2 What is the product of a fraction and its reciprocal?

3 Find a non-mathematical meaning for:

a reciprocal   b recurring   c terminating.

4 What is the difference between the selling price and the cost price called when the selling price is higher?

5 What does 'annum' mean, as in 'per annum'?

6 Which method involves finding the value of one part when given the value of the whole amount or several parts?

Topic summary

- Write 10 questions (with solutions) that could be used in a test for this chapter. Include some questions that you have found difficult to answer.

Print (or copy) and complete this mind map of the topic, adding detail to its branches and using pictures, symbols and colour where needed. Ask your teacher to check your work.

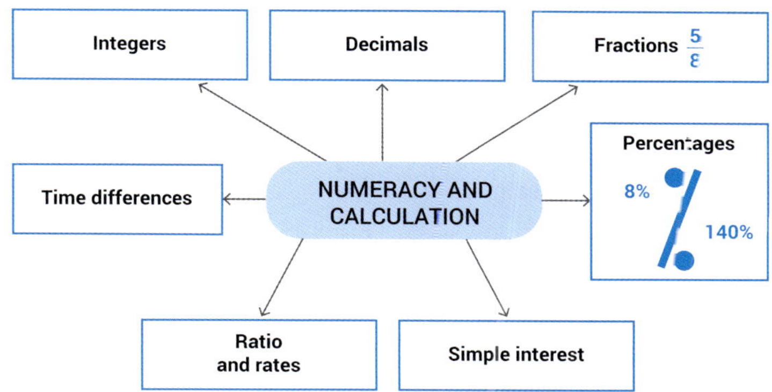

Worksheet
Mind map: Numeracy and calculation

CHAPTER 3 REVIEW

3 TEST YOURSELF

ANSWERS ON P. 635

Quiz
Test yourself 3

1 Evaluate each expression.

a $5 \times (-6) \div (-10)$ b $8 - (9 + 4)$ c $-3 \times 4 - 9$

d $5 \times 3 - 8 \times 5$ e $(-5)^2 - 7 \times 3$ f $\frac{4 \times -8 + (-10)}{3 \times (-2)}$

2 Write these decimals in descending order: 0.401, 0.0483, 4.001, 0.05635.

3 Evaluate each expression.

a $12 + 0.56$ b $14.5 - 6.782$ c 4.2×5

d 3.45×0.03 e $5.026 \div 4$ f $80.45 \div 0.05$

g 7.432×1000 h $345.7 \div 100$ i $0.605 \div 10$

j $34.1 \times 10\,000$ k 856.03×1000 l $84.6 \div 100\,000$

4 Which expression is NOT the same as $948 \div 4$? Select the correct answer **A**, **B**, **C** or **D**.

A $9.48 \div 0.04$ **B** $94\,800 \div 400$ **C** $0.948 \div 0.004$ **D** $0.948 \div 0.4$

5 Round 4.9948 to 3 decimal places.

6 Convert $\frac{4}{11}$ to a recurring decimal.

7 Which fraction is smaller, $\frac{2}{5}$ or $\frac{5}{12}$?

8 Evaluate each expression.

a $\frac{4}{5} + \frac{3}{8}$ b $\frac{5}{11} - \frac{2}{3}$ c $2\frac{4}{7} - 1\frac{2}{5}$

d $\frac{1}{4} \div \frac{4}{5}$ e $\frac{4}{5}$ of \$163 f $\frac{1}{2} \times 2\frac{2}{3}$

g $5 \div \frac{3}{4}$ h $2\frac{4}{5} \div 1\frac{1}{4}$ i $\frac{4}{5} \div \frac{3}{4} - \frac{1}{2}$

9 Express each percentage as a simplified fraction.

a 74% b 0.5% c $66\frac{2}{3}\%$ d 14.8%

10 Express each percentage as a decimal.

a 21.4% b 45% c $8\frac{1}{3}\%$ d 0.2%

11 Express each decimal and fraction as a percentage.

a 0.245 b $\frac{7}{8}$ c 1.275 d $\frac{19}{25}$

12 Find:

a 42% of \$580 b $8\frac{1}{4}\%$ of 90 kg c 2.45% of \$358

□ Foundation ○ Standard ⬡ Complex

9780170465618

13 **a** Increase 120 L by 35%. (3.05)

b Decrease $25 by 12.5%

14 What percentage is (correct to one decimal place): (3.05)

a 5 min of 1 h?

b 75c out of $4.50?

15 In a recent test cricket match, Steve made 286 runs in both innings, which represented 37% of the team total. How many runs did the team make in the match? (3.05)

16 Calculate the selling price of a smartphone marked at $485 and discounted by 15%. Select the correct answer **A**, **B**, **C** or **D**. (3.06)

A $72.75 **B** $470 **C** $412.20 **D** $412.25

17 A block of land was purchased for $250 000. Four years later, it was sold for $630 000. Calculate: (3.06)

a the profit.

b the profit as a percentage of cost price.

18 At a sale, Garrett and Kay bought a cutlery set, saving $150 on the original price. If this discount represented 55% of the price, what was the price of the cutlery? (3.06)

19 Calculate, to the nearest cent, the simple interest on: (3.07)

a $8050 invested for 5 years at 4.8% p.a.

b $3890 borrowed for 7 months at 2.4% p.a.

20 Calculate the annual simple interest rate if an investment of $7000 earns $1820 in interest over 2 years. (3.07)

21 Simplify each ratio. (3.08)

a 200 : 450 **b** 35 mm : 2 m **c** 5 h : 45 min

22 Bill and Ben win $800 000 in Lotto. If they agree to share the prize in the ratio of 25 : 17, how much will each person receive? (3.08)

23 A truck travels 581 km in 7 h. (3.08)

a Calculate its average speed in km/h.

b How far does the truck travel in 4 h?

c How long will it take to travel 1000 km? Answer correct to one decimal place.

24 Convert: (3.09)

a 0.8 mL/s to mL/h **b** 140 g/m^2 to kg/m^2

c 30 km/h to m/s **d** 345.6 m/day to cm/min

25 Calculate the time difference between: (3.10)

a 8:45 a.m. and 6:55 p.m. **b** 09:19 and 21:06

Foundation Standard Complex

Practice set 1 ANSWERS ON P. 640

1 Convert each fraction to a decimal and state whether it is recurring or terminating.

a $\frac{5}{8}$ **b** $\frac{7}{13}$ **c** $\frac{4}{9}$ **d** $4\frac{3}{4}$

2 Simplify each expression.

a $7 + 5m - 9$ **b** $8p - 3m + 12p + 4m$ **c** $5n^2 - 3n - 2n^2 - 7n$

3 Evaluate each expression.

a $7 + (-5)$ **b** $-3 - (-8)$ **c** $9 - (-4)$

d -3×5 **e** $-6 \times (-10)$ **f** $\frac{-20}{4}$

g $(-5)^2$ **h** $15 - 2 \times (-9)$ **i** $30 \div (-6) + (-3) \times 8$

j $(-2)^3 + 5 \times 4$

4 Select the surds from this list of square roots.

$\sqrt{98}$ $\sqrt{9}$ $\sqrt{225}$ $\sqrt{160}$ $\sqrt{36}$ $\sqrt{52}$

5 Evaluate $\sqrt{(8.7)^2 - (2.6)^2}$, correct to 2 decimal places.

2.03 **6** Find, correct to one decimal place, the value of each variable.

a

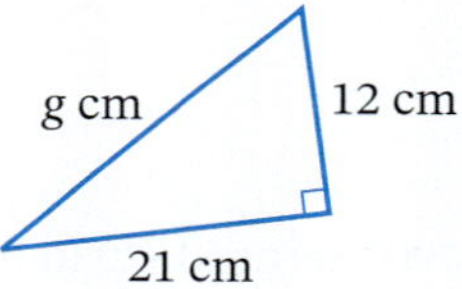

b

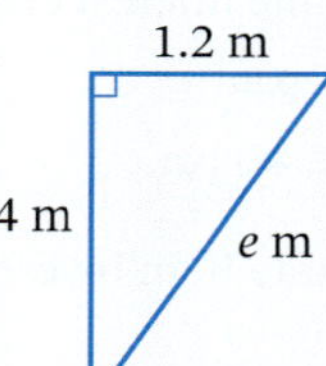

c

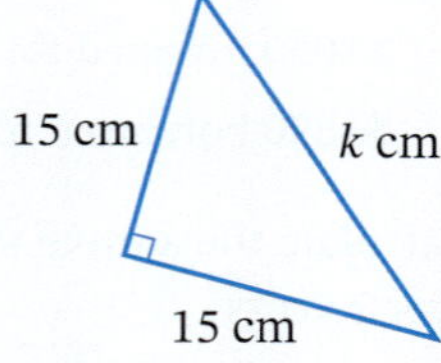

3.03 **7** Convert $\frac{35}{12}$ to a mixed numeral.

3.04 **8** Which of these numbers is larger than $\frac{1}{3}$? Select the correct answer **A**, **B**, **C** or **D**.

A 0.3 **B** $\frac{4}{9}$ **C** $\frac{2}{8}$ **D** 0.3333

3.02 **9** Evaluate each expression.

a 93.45×10 **b** $17.8 \div 100$ **c** 9.3×1000 **d** $234.7 \div 1000$

e $9.76 + 13.8$ **f** $15.86 + 4$ **g** $27.9 - 0.86$ **h** $12 - 8.4$

i 9.2×4 **j** 13.4×0.3 **k** $5.65 \div 0.05$ **l** $9 \div 4$

3.02 **10** Write these decimals in descending order: 5.85, 0.5867, 0.59, 5.095.

☐ Foundation ○ Standard ⬡ Complex

9780170465618

11 Find the value of each variable. (2.04)

a

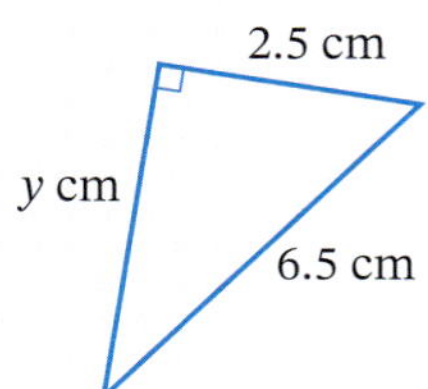

b

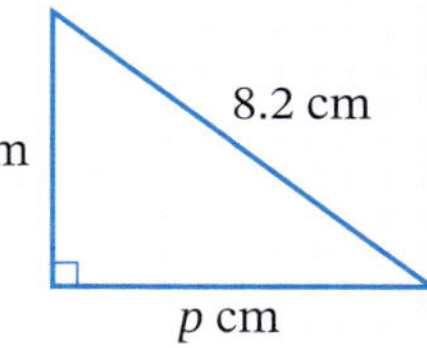

c

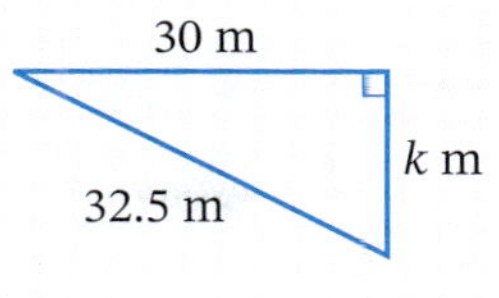

12 Write an algebraic expression for each statement. 

a r increased by 7 **b** The difference between T and 15

c Multiply k by 4 **d** 10 divided by z

13 Evaluate each expression. 

a $\frac{1}{4}+\frac{3}{5}$ **b** $\frac{7}{10}-\frac{2}{3}$ **c** $3\frac{1}{5}+1\frac{2}{3}$ **d** $4\frac{1}{2}-2\frac{4}{5}$

e $\frac{4}{7}\times\frac{3}{5}$ **f** $\frac{5}{8}\div\frac{2}{3}$ **g** $\frac{4}{5}\div 3$ **h** $1\frac{2}{7}\times 2\frac{1}{2}$

14 Arrange the fractions $\frac{2}{3}, \frac{3}{5}$ and $\frac{5}{8}$ in ascending order. 

15 Adeeb calculated $5.64 \div 0.2$ without using a calculator. Which of these calculations will give him the correct answer? Select the correct answer **A**, **B**, **C** or **D**. 

A $564 \div 2$ **B** $56.4 \div 0.02$ **C** $56.4 \div 20$ **D** $56.4 \div 2$

16 Convert each number to a percentage.  

a $\frac{27}{40}$ **b** 0.27 **c** $\frac{3}{11}$ **d** 0.085

17 Calculate: 

a 170% of \$60 **b** 14% of \$250 **c** $3\frac{1}{2}$% of \$670

18 Test whether each triangle is right-angled. 

a

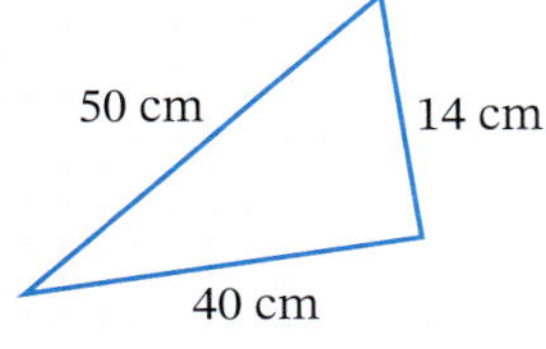

b

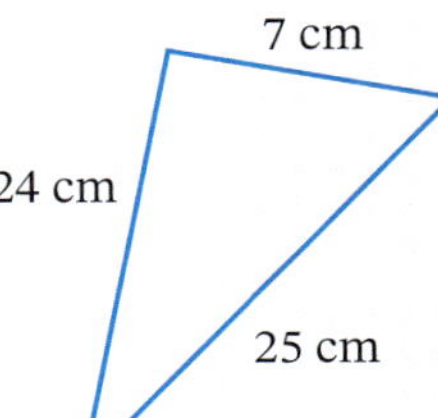

c

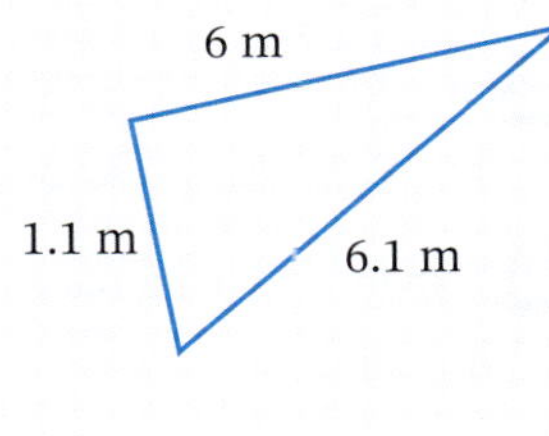

19 Paba bought 55 L of petrol at 189.9 cents per litre. (3.08)

a How much did Paba pay (to the nearest cent)?

b How many litres can be bought for \$60? Answer to one decimal place.

☐ Foundation ○ Standard ⬡ Complex

PRACTICE SET 1

20 At a sale, the price of a TV was reduced by 30%. Find the original price of the TV if its sale price is $339.50.

21 If $a = 7$, $k = -2$, and $w = 5$, evaluate each expression.

a $a + k + w$ **b** $5kw$ **c** $\frac{a+k}{w}$

d $a^2 - w^2$ **e** $\frac{w-k}{a}$ **f** $3w + 4k$

22 Simplify each expression.

a $2a^2 \times 4a$ **b** $5 \times 4p \times 2pd$ **c** $9b^2d \times 3bd$

23 Simplify each expression.

a $\frac{21mn}{7n}$ **b** $36x \div 12y$ **c** $\frac{w^2v^3}{2vw}$

24 Convert:

a 35% to a fraction in simplest form.

b 0.214% to a decimal.

25 A dishwasher costing $525 is sold for $699.95. Calculate the percentage profit on the cost price, correct to one decimal place.

26 Simplify:

a 160 : 96 **b** 3 h : 45 min

c 1275 words in 15 minutes (as words/min).

27 The foot of a ladder is placed 0.8 m from the base of a wall. If the ladder reaches 2.2 m up the wall, calculate, correct to one decimal place, the length of the ladder.

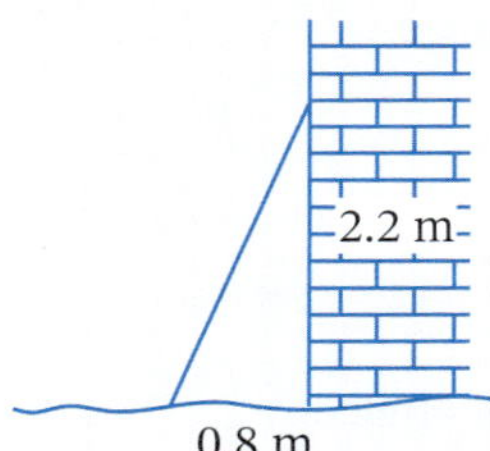

3.05 **28** **a** Increase 85 kg by 30%.

b Decrease $6.40 by 18%.

1.05 **29** Expand each expression.

a $8(5 - x)$ **b** $3h(2 - 7h)$ **c** $9c(2d - 3c)$

1.05 **30** Expand and simplify each expression.

a $2(5w - 3) + 4(1 + 3w)$ **b** $3(1 - 4k) - 2(2k + 1)$ **c** $4(3t - 4) - 7(1 - 2t)$

3.08 **31** Divide $450 between Klaas, David and Sarah in the ratio 3 : 5 : 7.

3.09 **32** The speed of a spacecraft leaving the Earth's gravitational pull must reach 11.2 km/s. Express this speed in km/h.

□ Foundation ○ Standard ⬡ Complex

33 Find the time difference between 10:44 a.m. and 3:17 p.m. 

34 A rectangle has side lengths of 5 cm and 7 cm. Find the length of a diagonal as a surd. 

35 A discount of \$75 represents 4% of the marked price. What is the marked price? 

36 Expand each expression. 

a $(p + 4)(p - 8)$ **b** $(2n + 1)(3n + 5)$ **c** $(7 - y)(7 + y)$ 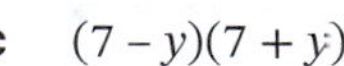

d $(4g - 3)^2$ **e** $(2e + 5)^2$ **f** $(3 - 2x)(x + 5)$

37 Which expression is $16mn + 24m^2$ completely factorised? Select **A**, **B**, **C** or **D**. 

A $4(4n + 6m)$ **B** $8m(2n + 3m)$

C $4m(4n + 6m)$ **D** $m(16n + 24m)$

38 Factorise each expression. 

a $3p - 15$ **b** $5(m + 4) - t(m + 4)$ **c** $-8x^2y - 12xy$

39 Sarita spent $\frac{1}{4}$ of the day working, $\frac{1}{3}$ of the day sleeping and $\frac{1}{12}$ of the day travelling to and from work. The rest of the time was spent relaxing. 

a What fraction of the day was spent relaxing?

b How many hours did Sarita spend travelling to and from work and at work?

40 Patrick bought a new computer and received a 20% discount. He paid by cash and received a further 8% discount off the discount price. What is the total percentage discount (on the original cost) that Patrick received? Select the correct answer **A**, **B**, **C** or **D**. 

A 8% **B** 28% **C** 23.6% **D** 26.4%

41 Calculate the length of steel required to make the frame of the gate to a paddock shown in the diagram (correct to one decimal place). 

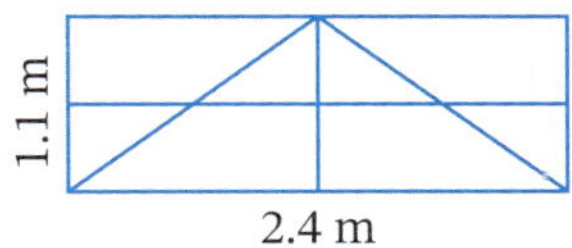

4 MEASUREMENT

Trigonometry

In the second century BCE, the Greek astronomer Hipparchus could calculate distances to the moon and the Sun and he was the first scientist to chart the positions of over 850 stars. How was he able to achieve this over 2100 years ago? Hipparchus started a branch of mathematics called **trigonometry**, meaning 'triangle measure', which uses angles, triangles and circles to calculate lengths and distances that cannot be physically measured. Trigonometry is used widely today in engineering, surveying, navigation, astronomy, electronics and construction.

Shutterstock.com/Alex Stemmers

Chapter outline

		Proficiencies			
4.01	The sides of a right-angled triangle	U	F		C
4.02	The trigonometric ratios	U	F	R	C
4.03	Similar right-angled triangles	U	F	R	C
4.04	Trigonometry on a calculator	U	F		
4.05	Finding an unknown side	U	F PS		C
4.06	Finding more unknown sides	U	F PS		
4.07	Finding an unknown angle	U	F PS		

U = Understanding
F = Fluency
PS = Problem solving
R = Reasoning
C = Communication

Note: Bearings and angles of elevation and depression will be covered in Year 10 (or Year 9 Advanced).

Wordbank

Quiz
Wordbank 4

adjacent side In a right-angled triangle, the side that is next to a given angle and pointing to the right angle

hypotenuse The longest side of a right-angled triangle, the side opposite the right angle

minute (′) A unit for measuring angle size, $\frac{1}{60}$ of a degree

opposite side In a right-angled triangle, the side that is facing a given angle and not one of its arms

theta (θ) A letter of the Greek alphabet used as a variable for angles

trigonometric ratio The ratio of 2 sides in a right-angled triangle, for example, sine is the ratio of the opposite side to the hypotenuse

Videos (9):

4.02 The trigonometric ratios
4.04 Trigonometry on a calculator
4.05 Finding an unknown side • Trigonometry
4.06 Finding the hypotenuse
4.07 Finding an unknown angle • Trigonometry

Twig videos (3):

SkillCheck Measuring the Earth • The tunnel of Samos • Distance to the Sun and Moon

Quizzes (5):

- Wordbank 4
- SkillCheck 4
- Mental skills 4
- Language of Maths 4
- Test Yourself 4

Worksheets (3):

4.03 Investigating the tangent ratio
4.07 Trigonometry problems

Mind map: Trigonometry

Puzzles (5):

4.02 Trigonometry match-up
4.04 Trigonometry squaresaw
4.06 Trigonometry equations 1
4.07 Trigonometry squaresaw • Trigonometry: Finding angles • Trigonometric equations 2

Technology (1):

4.02 Trigonometry values

Spreadsheet (1):

4.02 Trigonometry values

Presentation (1):

4.07 Trigonometry

Nelson MindTap

To access resources above, visit **cengage.com.au/nelsonmindtap**

4 In this chapter you will:

✓ investigate the sine, cosine and tangent ratios in similar right-angled triangles.
✓ use trigonometry to solve problems involving right-angled triangles.
✓ find unknown sides and angles in right-angled triangles, where the angle is measured in degrees.
✓ find unknown sides and angles in right-angled triangles, where the angle is measured in degrees and minutes.

SkillCheck

ANSWERS ON P. 636

Quiz SkillCheck 4

Video Measuring the Earth

The tunnel of Samos

Distance to the Sun and Moon

1 Simplify each fraction.

a $\frac{15}{25}$ **b** $\frac{9}{12}$ **c** $\frac{4}{10}$

2 Convert each fraction to a decimal, correct to 3 decimal places.

a $\frac{3}{8}$ **b** $\frac{5}{7}$ **c** $\frac{2}{9}$

3 For each triangle, name the hypotenuse.

a

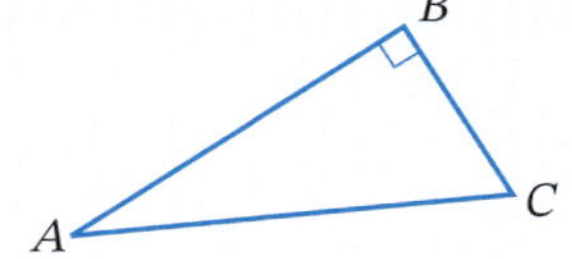

b

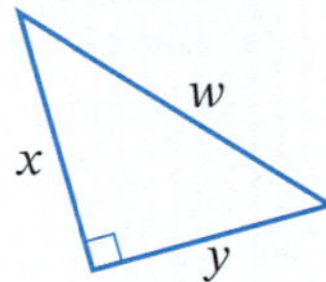

c

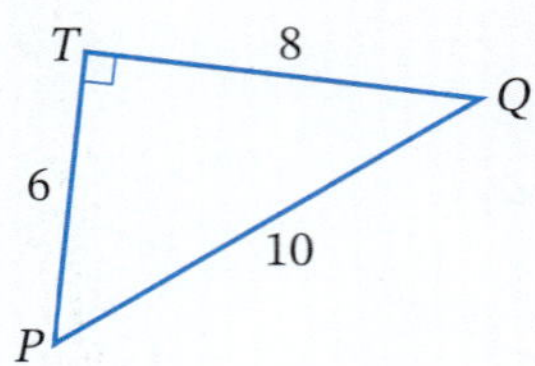

4 Solve each equation.

a $\frac{x}{5} = 7$ b $\frac{h}{4} = 8.3$ c $\frac{45}{y} = 9$

5 For each triangle, find the value of n.

a
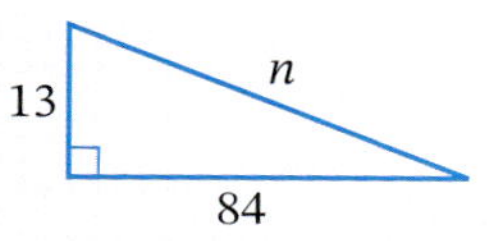

b
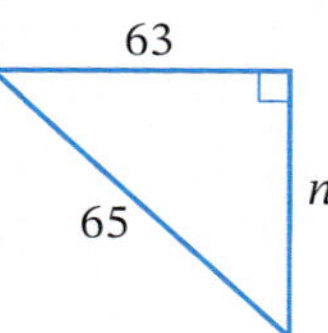

c
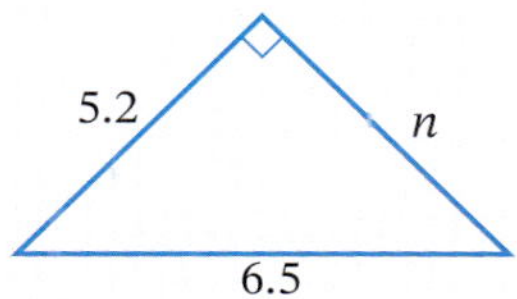

6 Round each time to the nearest hour.

a 8 h 18 min b 3 h 45 min c 1 h 30 min

7 Convert each time to hours and minutes.

a 4.7 h b 2.25 h c 6.85 h

The sides of a right-angled triangle 4.01

The 3 sides of a **right-angled triangle** have special names. These names depend on the position of a given angle in the triangle.

- The **hypotenuse** is the longest side and is always opposite the right angle.

We have already learnt about the hypotenuse in Pythagoras' theorem.

- The **opposite side** directly faces the given angle.
- The **adjacent side** runs from the given angle to the right angle.

Adjacent means 'next to'.

In this right-angled triangle, the marked angle $\angle W$ has also been labelled with the Greek letter, θ, ('theta').

The hypotenuse is WY.

The opposite side is XY.

The adjacent side is WX.

In this right-angled triangle, the marked angle $\angle Y$ has also been labelled with the Greek letter α ('alpha'), so the opposite and adjacent sides are different.

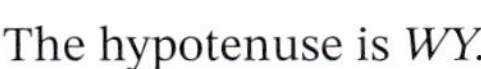

The hypotenuse is WY.

The opposite side is WX.

The adjacent side is XY.

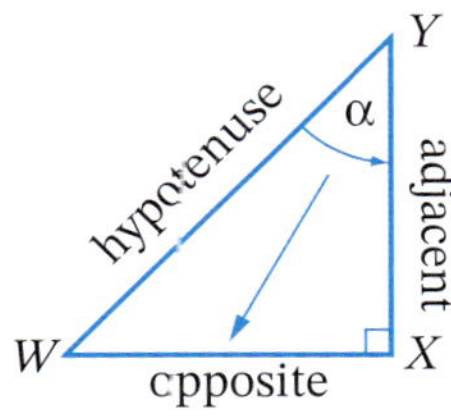

Example 1

For each triangle, name the hypotenuse, opposite and adjacent sides for angle θ.

a

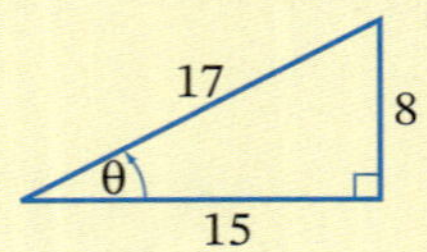

b

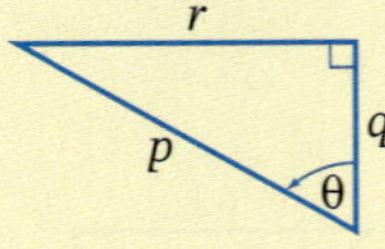

c

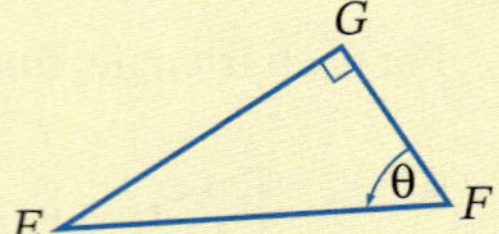

SOLUTION

a Hypotenuse is 17.
Opposite side is 8.
Adjacent side is 15.

b Hypotenuse is p.
Opposite side is r.
Adjacent side is q.

c Hypotenuse is EF.
Opposite side is EG.
Adjacent side is FG.

Example 2

For angles α and β, name the adjacent side.

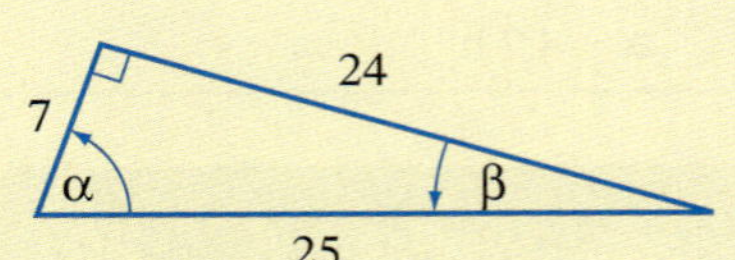

SOLUTION

For α, the adjacent side is 7.

For β, the adjacent side is 24.

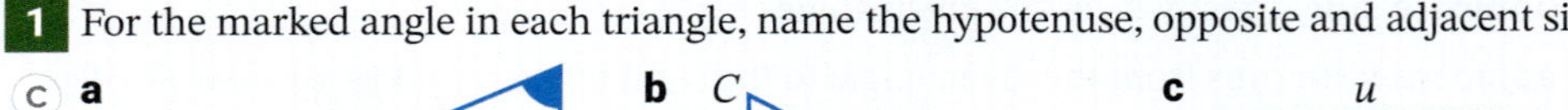

EXERCISE 4.01 ANSWERS ON P. 636

The sides of a right-angled triangle

U F C

1 For the marked angle in each triangle, name the hypotenuse, opposite and adjacent sides.

C **a**

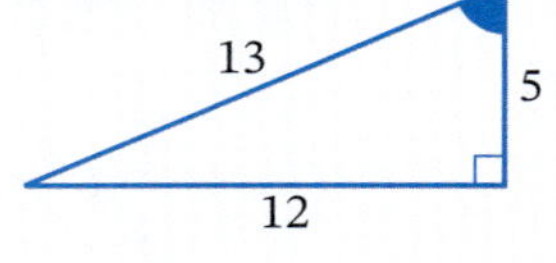

b

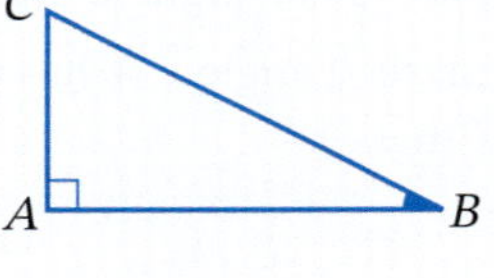

c

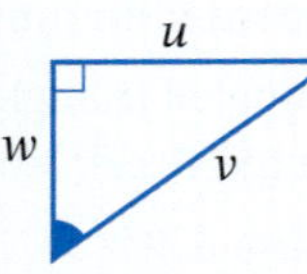

d

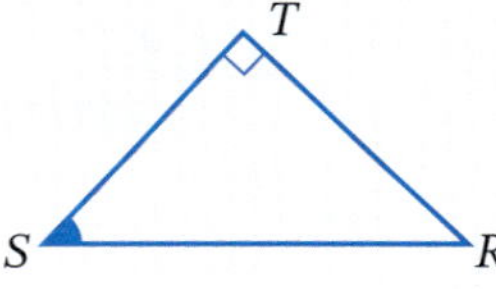

e

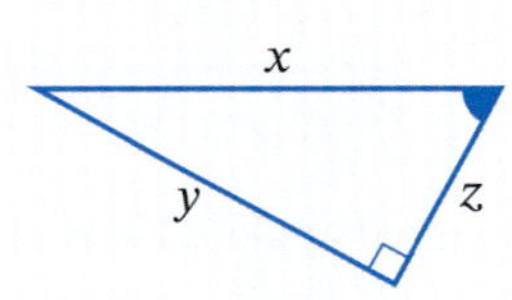

f

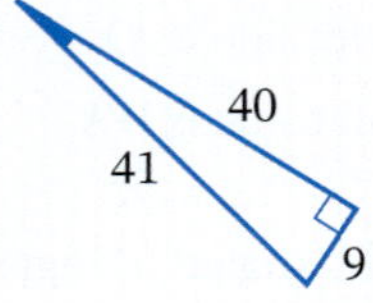

2 For △LKM, name the angle:

a opposite the hypotenuse.

b opposite side KM.

c opposite side LK.

d adjacent to side KM.

e adjacent to side LK.

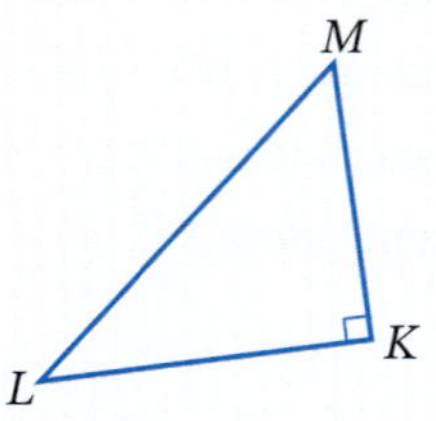

3 Which one of these statements is **false** about $\triangle TSU$?

Select the correct answer **A**, **B**, **C** or **D**.

A The adjacent side to $\angle S$ is US.

B The adjacent side to $\angle T$ is TU.

C The hypotenuse is UT.

D The opposite side to $\angle T$ is US.

4 For each triangle, find the opposite sides for angles θ ('theta') and ϕ ('phi').

a

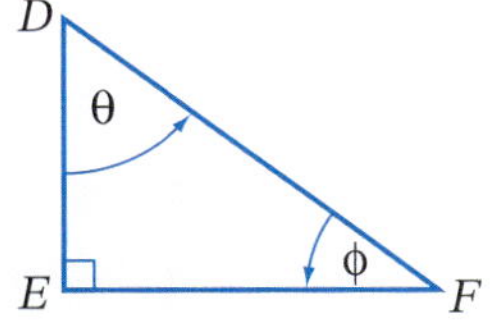

b

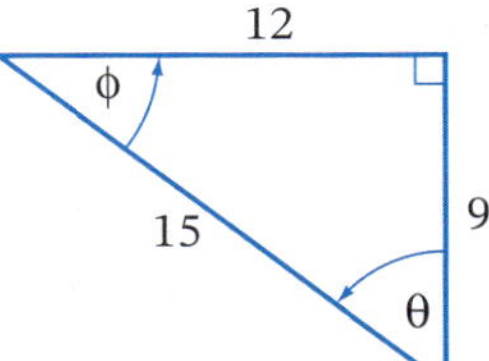

c

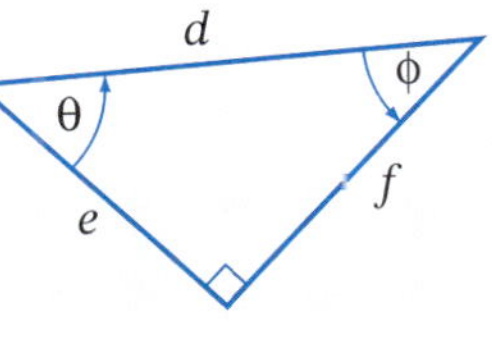

5 For each triangle, find the adjacent sides for angles θ and ϕ.

a

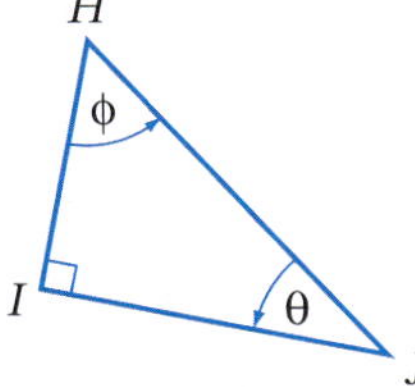

b

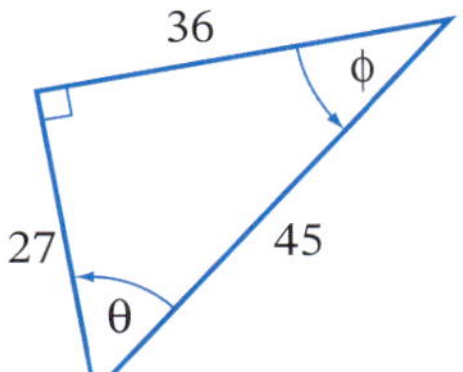

c

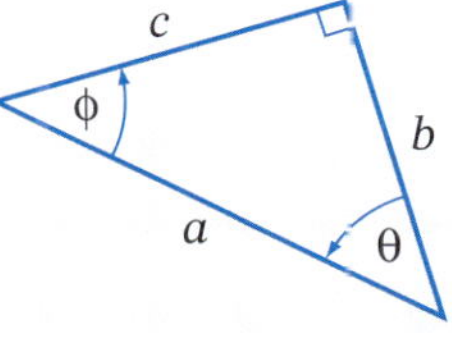

6 Which side of a right-angled triangle is fixed and does not depend on a given angle? Select the correct answer **A**, **B**, **C** or **D**.

A adjacent **B** hypotenuse **C** opposite **D** shortest

7 Given each description of a right-angled triangle, sketch the triangle with the correctly labelled vertices and angle.

a $\triangle ABC$ has hypotenuse AB and side AC opposite angle θ.

b $\triangle XYZ$ has hypotenuse YZ and side XZ adjacent to angle α.

c $\triangle PRQ$ has side RQ opposite $\angle P$ and adjacent to $\angle R$.

d $\triangle DEF$ is right-angled at E, with the opposite and adjacent sides of $\angle D$ equal.

Foundation Standard Complex

DID YOU KNOW?

The Greek alpha-bet

Here are 8 letters (in lower-case and capitals) from the Greek alphabet:

α, A alpha	β, B beta	γ, Γ gamma	δ, Δ delta
θ, Θ theta	π, Π pi	σ, Σ sigma	ω, Ω omega

Greek letters are commonly used as variables, particularly in geometry and trigonometry.

1 **Find how many letters there are in the Greek alphabet, and name each one.**

2 **Compare the Greek alphabet with our Roman alphabet.**

3 **Can you see where the word 'alphabet' comes from? Explain how it originated.**

4.02 The trigonometric ratios

There are 3 special fractions called **trigonometric ratios** that relate the lengths of 2 sides of a right-angled triangle: **sine**, **cosine** and **tangent**.

Puzzle Trigonometry match-up

Technology Trigonometry values

Spreadsheet Trigonometry values

The trigonometric ratios

Ratio	Abbreviation	Meaning
sine	sin	$\sin\theta = \dfrac{\text{opposite}}{\text{hypotenuse}}$
cosine	cos	$\cos\theta = \dfrac{\text{adjacent}}{\text{hypotenuse}}$
tangent	tan	$\tan\theta = \dfrac{\text{opposite}}{\text{adjacent}}$

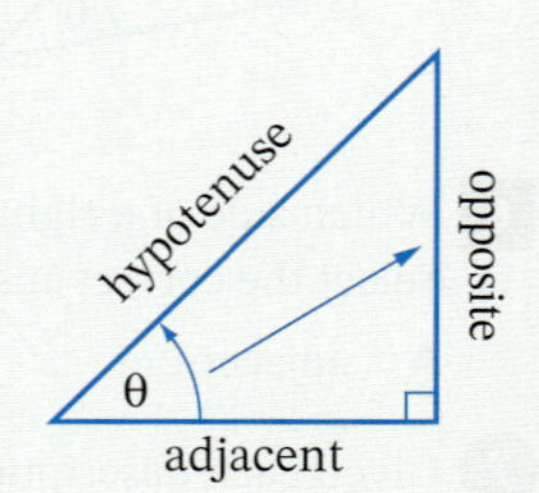

Example 3

In $\triangle AXP$, find sin θ, cos θ and tan θ.

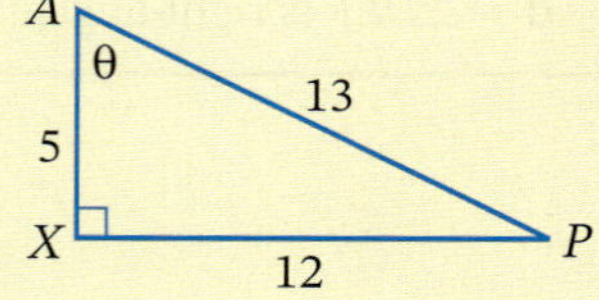

SOLUTION

For angle θ, opposite = 12, adjacent = 5, hypotenuse = 13.

$\sin\theta = \dfrac{\text{opposite}}{\text{hypotenuse}} = \dfrac{12}{13}$ $\qquad \cos\theta = \dfrac{\text{adjacent}}{\text{hypotenuse}} = \dfrac{5}{13}$

$\tan\theta = \dfrac{\text{opposite}}{\text{adjacent}} = \dfrac{12}{5}$

A useful **mnemonic** (memory aid) for remembering the 3 ratios is to look at the initials of the words in the ratios:

$$\sin = \frac{\text{opposite}}{\text{hypotenuse}} = \frac{O}{H} \quad \text{'SOH'}$$

$$\cos = \frac{\text{adjacent}}{\text{hypotenuse}} = \frac{A}{H} \quad \text{'CAH'}$$

$$\tan = \frac{\text{opposite}}{\text{adjacent}} = \frac{O}{A} \quad \text{'TOA'}$$

If you remember **SOH-CAH-TOA** (pronounced 'so-car-toe-ah'), then you can remember the ratios for sin, cos and tan. Some students also learn a phrase where the first letter of each word follows the SOH-CAH-TOA sequence, for example, '**S**un **O**ver **H**ead **C**aused **A** **H**uge **T**an **O**n **A**rms'. Find your own mnemonic for the 3 ratios.

Example 4

For the triangle shown, find:

a $\sin A$

b $\cos B$

c $\tan A$

d $\sin B$

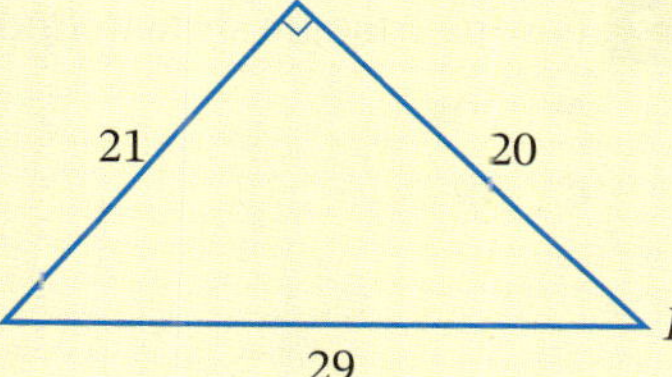

Video
The trigonometric ratios

SOLUTION

For angle A, opposite = 20, adjacent = 21, hypotenuse = 29

For angle B, opposite = 21, adjacent = 20, hypotenuse = 29

a $\sin A = \frac{\text{opposite}}{\text{hypotenuse}} = \frac{20}{29}$

b $\cos B = \frac{\text{adjacent}}{\text{hypotenuse}} = \frac{20}{29}$

c $\tan A = \frac{\text{opposite}}{\text{adjacent}} = \frac{20}{21}$

d $\sin B = \frac{\text{opposite}}{\text{hypotenuse}} = \frac{21}{29}$

Example 5

If $\tan R = \frac{8}{15}$, find the value of $\sin R$ and $\cos R$.

SOLUTION

Since $\tan R = \frac{\text{opposite}}{\text{adjacent}} = \frac{8}{15}$, draw a right-angled triangle that has an angle R with opposite side 8 and adjacent side 15. Let x be the length of the hypotenuse. Find x using Pythagoras' theorem.

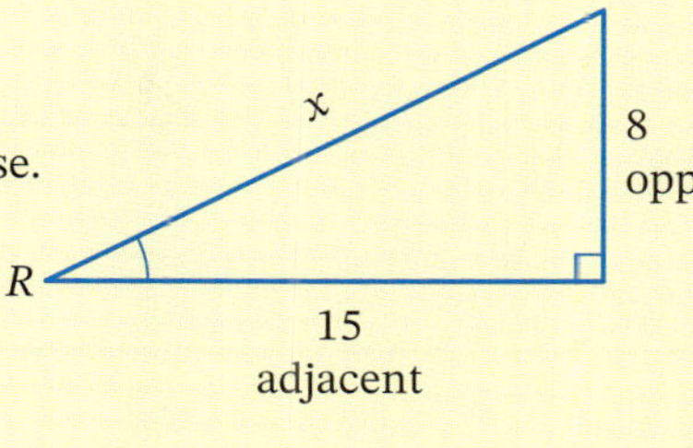

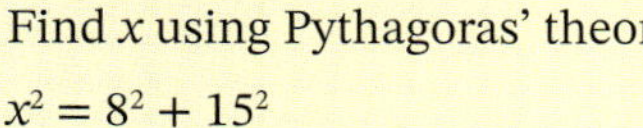

$$x^2 = 8^2 + 15^2$$
$$= 289$$
$$x = \sqrt{289}$$
$$= 17$$

$\therefore \sin R = \frac{\text{opposite}}{\text{hypotenuse}} = \frac{8}{17}$ $\qquad \cos R = \frac{\text{adjacent}}{\text{hypotenuse}} = \frac{15}{17}$

EXERCISE 4.02 ANSWERS ON P. 637

The trigonometric ratios

U F R C

EXAMPLE 3

1 For each marked angle, find the sine, cosine and tangent ratios.

C **a**

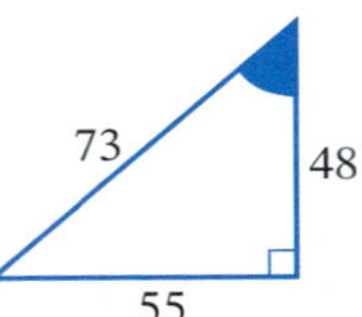

b

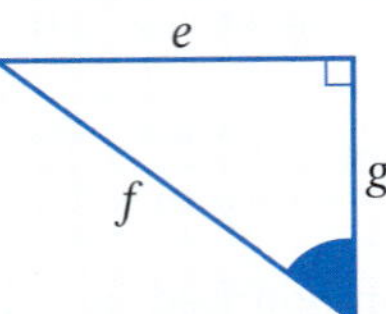

c

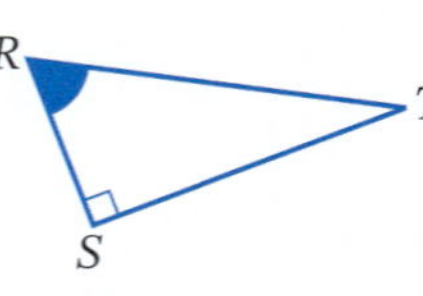

d

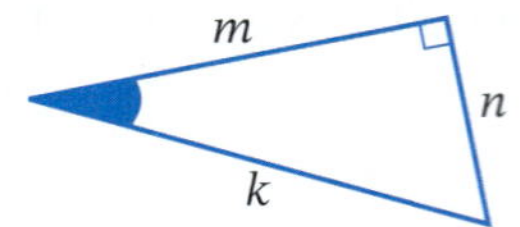

e

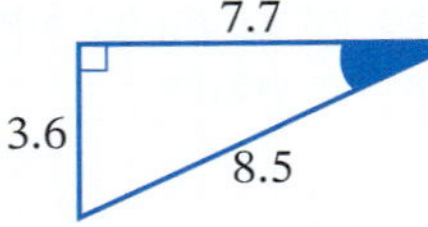

f

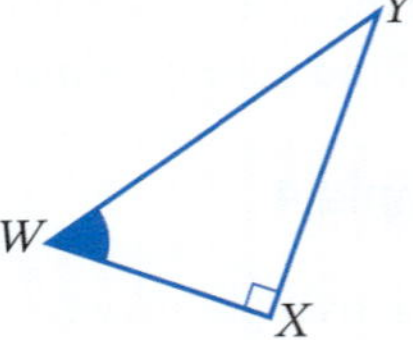

EXAMPLE 4

2 For the triangle shown, find:

a $\cos X$ **b** $\tan Y$

c $\sin X$ **d** $\sin Y$

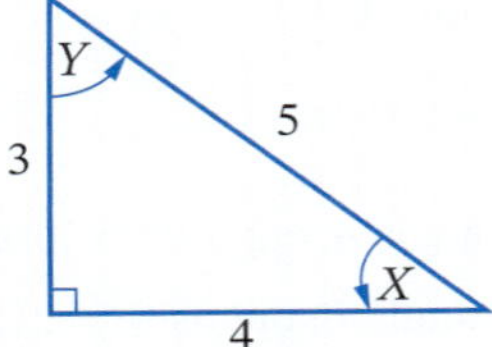

3 Complete each statement below with the correct angle (α or β).

a $\sin\square = \frac{IJ}{HJ}$ **b** $\sin\square = \frac{HI}{HJ}$ **c** $\cos\square = \frac{IJ}{HJ}$

d $\cos\square = \frac{HI}{HJ}$ **e** $\tan\square = \frac{IJ}{HI}$ **f** $\tan\square = \frac{HI}{IJ}$

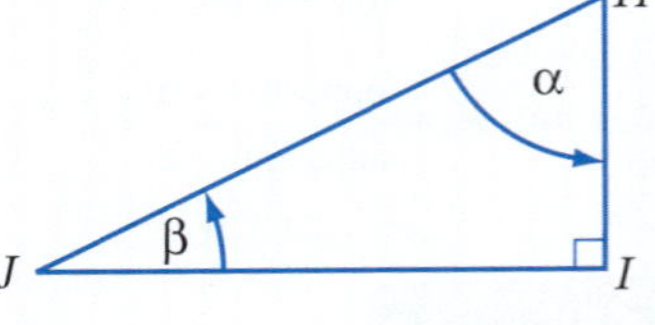

4 For each triangle below, find:

i $\tan\theta$ **ii** $\cos\theta$ **iii** $\cos\phi$ **iv** $\tan\phi$

ϕ = 'phi'.

a

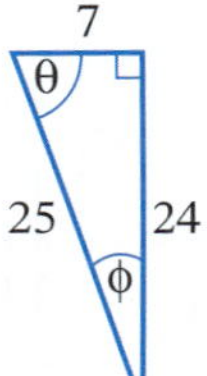

b

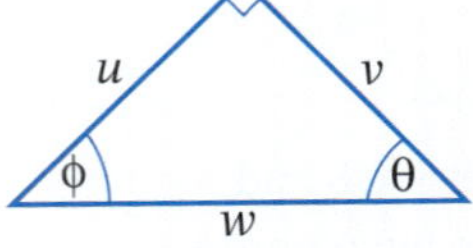

c

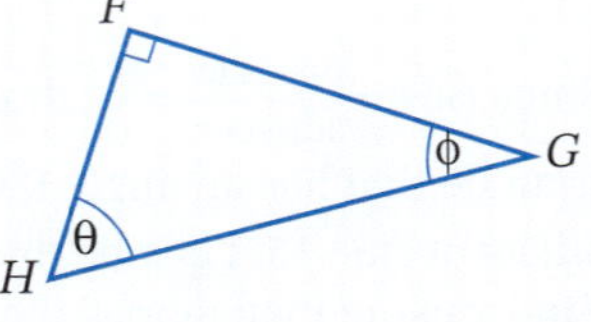

☐ Foundation ○ Standard ⬡ Complex

d

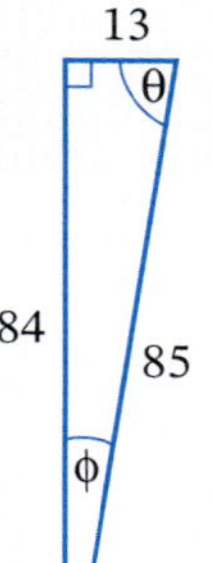

e

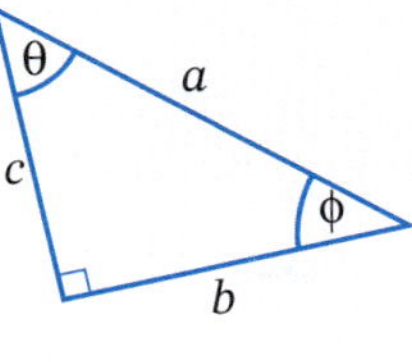

f

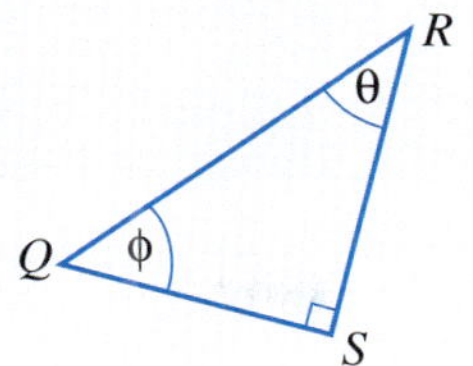

5 For each fraction, write a correct trigonometric ratio involving angle X or Y in the triangle.

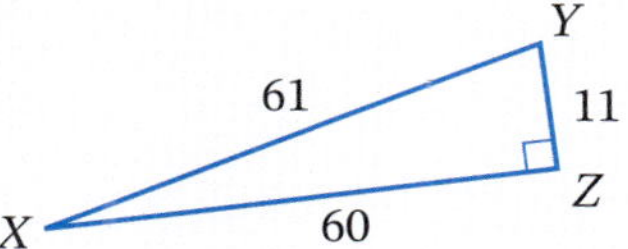

a 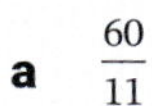 $\frac{60}{11}$ **b** $\frac{11}{60}$

c $\frac{11}{61}$ **d** $\frac{60}{61}$

6 Which ratio is equal to $\frac{p}{m}$? Select the correct answer **A**, **B**, **C** or **D**.

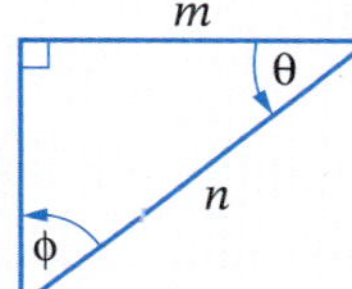

A 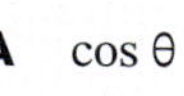 $\cos\theta$ **B** $\cos\phi$

C $\tan\theta$ **D** $\tan\phi$

7 Which statement is true for this triangle? Select **A**, **B**, **C** or **D**.

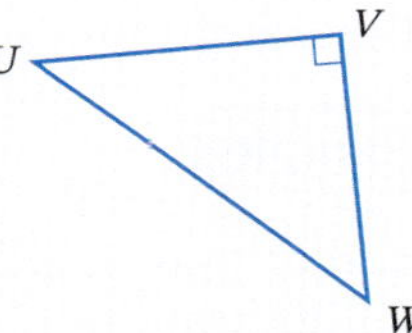

A 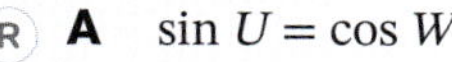 $\sin U = \cos W$ **B** $\tan U = \sin W$

C $\cos U = \tan W$ **D** $\tan U = \tan W$

8 Sketch a right-angled triangle for each trigonometric ratio, then use Pythagoras' theorem to find the length of the unknown side and the other 2 trigonometric ratios for the same angle.

a 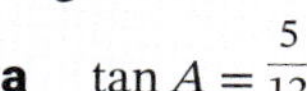 $\tan A = \frac{5}{12}$ **b** $\sin B = \frac{3}{5}$ **c** $\cos X = \frac{9}{41}$ **d** $\sin Y = \frac{7}{25}$

4.03 Similar right-angled triangles

Worksheet
Investigating the tangent ratio

In each right-angled triangle below, $\angle A = \mathbf{35°}$.

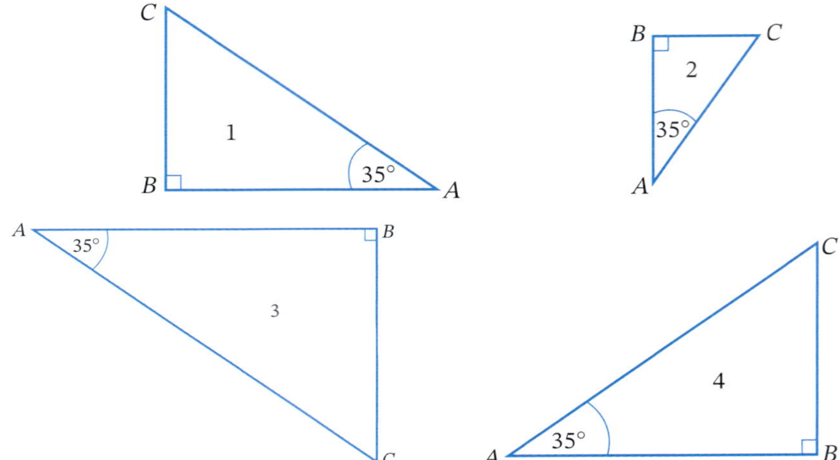

Furthermore, because $\angle B = 90°$, $\angle C = 180° - 90° - 35° = 55°$ because of the angle sum of a triangle.

These four triangles are called **similar triangles** because they have the same shape but are not the same size. In fact, they are enlargements or reductions of one another.

We learned about similar triangles in Year 8.

Example 6

For each triangle above, measure the length of each side (correct to the nearest mm) and then calculate sin A, cos A and tan A as decimals (correct to 2 decimal places). Write your results in the table below.

	Side length (mm)			Trigonometric ratio		
	BC (opp)	AB (adj)	AC (hyp)	$\sin A = \frac{BC}{AC}$	$\cos A = \frac{AB}{AC}$	$\tan A = \frac{BC}{AB}$
1						
2						
3						
4						

SOLUTION

	Side length (mm)			Trigonometric ratio		
	BC (opp)	AB (adj)	AC (hyp)	$\sin A = \frac{BC}{AC}$	$\cos A = \frac{AB}{AC}$	$\tan A = \frac{BC}{AB}$
1	25	36	44	$25 \div 44 \approx 0.57$	$36 \div 44 \approx 0.82$	$25 \div 36 \approx 0.69$
2	14	20	24	0.58	0.83	0.70
3	39	57	69	0.57	0.83	0.68
4	29	41	50	0.58	0.82	0.71

Note: See **Technology: Similar right-angled triangles** after Exercise 4.03 for a dynamic geometry activity based on this example.

Similar triangles have matching equal angles. We will learn more about them in Chapter 13, *Congruent and similar figures*.

4.03

EXERCISE 4.03 ANSWERS ON P. 637

Similar right-angled triangles

U F R C

EXAMPLE 6

1 R C **a** For each similar right-angled triangle below, measure the length of each side (correct to the nearest mm) and then calculate sin Y, cos Y and tan Y as decimals (correct to 2 decimal places). Copy and complete the table below.

Triangle 1: Y 60°, Z (right angle), X

Triangle 2: X, Y 60°, Z (right angle)

Triangle 3: X, Z (right angle), Y 60°

Triangle 4: X, Z (right angle), Y 60°

	Side length (mm)			Trigonometric ratio		
	XZ (opp)	ZY (adj)	XY (hyp)	$\sin Y = \frac{XZ}{XY}$	$\cos Y = \frac{ZY}{XY}$	$\tan Y = \frac{XZ}{ZY}$
1						
2						
3						
4						

b What do you notice about the value of sin Y for all four similar right-angled triangles?

c Use your calculator to evaluate sin 60° by pressing sin 60 =. What do you notice about your answer?

□ Foundation ○ Standard ⬡ Complex

d What do you notice about the value of cos Y for all four similar triangles?

e Use your calculator to evaluate cos 60°. What do you notice about your answer?

f What do you notice about the value of tan Y for all four similar triangles?

g Use your calculator to evaluate tan 60°. What do you notice about your answer?

2 R C **a** Draw four similar right-angled triangles that have an angle of 48°, measure the length of each side (correct to the nearest mm) and then calculate sin 48°, cos 48° and tan 48° as decimals (correct to 2 decimal places). Copy and complete the table.

	Side length (mm)			Trigonometric ratio		
	Opp	Adj	Hyp	sin 48° = $\frac{\text{opp}}{\text{hyp}}$	cos 48° = $\frac{\text{adj}}{\text{hyp}}$	tan 48° = $\frac{\text{opp}}{\text{adj}}$
1						
2						
3						
4						

b Examine the value of sin 48° for all four similar triangles, then evaluate sin 48° on a calculator. What do you notice?

c Examine the values of cos 48°, then evaluate cos 48° on a calculator.

d Examine the values of tan 48°, then evaluate tan 48° on a calculator.

3 R C For each trigonometric ratio, draw a large right-angled triangle with the given angle, then by measurement and calculation, find the value of the ratio, correct to 3 decimal places. Compare your answer to the calculator's answer.

a tan 55° **b** cos 39° **c** sin 67° **d** cos 21°

TECHNOLOGY

Similar right-angled triangles

In this activity, use dynamic geometry software to measure and calculate trigonometric ratios.

1 **a** Before you start, set the measuring tool to measure angles to degrees.

b By constructing intervals, draw 2 similar right-angled triangles. Then use the measuring tool to measure the right angle and another angle (as shown right).

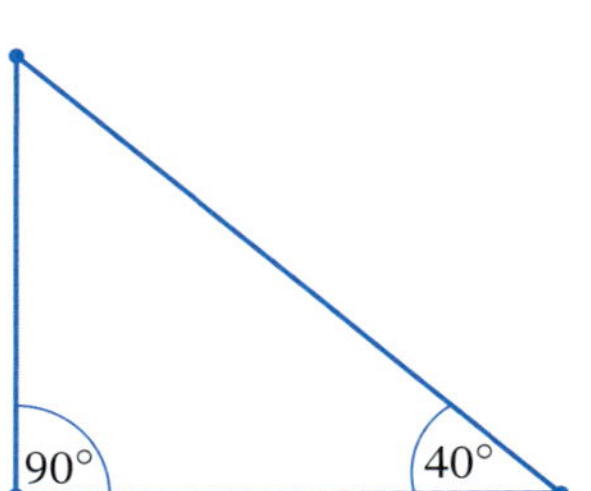

c Now keeping the triangle right-angled, move any of the other 2 vertices so that one angle is 40°.

2 Draw 2 more similar right-angled triangles with a 40° angle.

3 Label the vertices of each triangle as A, B and C.

Make sure $\angle A = 40°$ and $\angle B = 90°$.

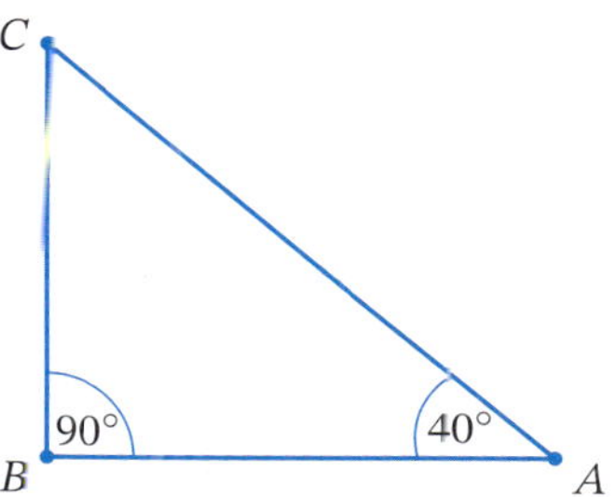

4 a Copy this table.

	Side length (mm)			Trigonometric ratio		
	BC (opp)	*AB* (adj)	*AC* (hyp)	$\sin A = \frac{BC}{AC}$	$\cos A = \frac{AB}{AC}$	$\tan A = \frac{BC}{AB}$
1						
2						
3						

b Use the measurement tool to measure the sides of each triangle and write them in the table.

c Then, for each triangle, calculate the trigonometric ratios, correct to 2 decimal places.

d What do you notice about the values of each ratio for all of the triangles?

Trigonometry on a calculator 4.04

In the previous section, we discovered that for any particular angle, the sine, cosine and tangent ratios stay constant (the same) for all right-angled triangles with that angle. For example, $\sin 32° \approx 0.53$ always, no matter what size the similar right-angled triangle.

ⓘ The trigonometric ratios of an angle size

For any given angle, the values of the sine, cosine and tangent ratios are constant.

This means that the value of a trigonometric ratio can be easily found on a calculator rather than through constructing and measuring triangles.

Puzzle
Trigonometry squaresaw

Angles are measured in **degrees**, but one degree can be subdivided into 60 **minutes**.

One minute can be further subdivided into 60 **seconds**. The abbreviations for minutes and seconds are shown below.

ⓘ Degrees and minutes

$1° = 60'$ (1 degree = 60 minutes)

$1' = 60''$ (1 minute = 60 seconds)

For example, an angle size of 48° 35′ 56′′ is 48 degrees, 35 minutes and 56 seconds, about halfway between 48° and 49°.

To enter degrees, minutes and seconds into a scientific calculator, use the °′′′ or DMS (Degrees-Minutes-Seconds) key.

Video Trigonometry on a calculator

Example 7

Evaluate each expression correct to 2 decimal places.

a sin 46° **b** tan 57.4° **c** 4 cos 20°

d 68.3 sin 38°25′ **e** $\frac{23}{\cos 18°50'}$

SOLUTION

Make sure that your calculator is in the degrees mode (D or DEG) or your answer will be incorrect.

a sin 46° = 0.71933...
≈ 0.72

On calculator: sin 46 =

b tan 57.4° = 1.56365...
≈ 1.56

On calculator: tan 57.4 =

This angle is 57.4° (a decimal), not 57°4′.

This means 4 × cos 20°.

c 4 cos 20° = 3.75877...
≈ 3.76

On calculator: 4 cos 20 =

d 68.3 sin 38°25′ = 42.43996...
≈ 42.44

On calculator: 68.3 sin 38 °′′′ 25 °′′′ =

e $\frac{23}{\cos 18°50'}$ = 24.30103...
≈ 24.30

On calculator: 23 ÷ cos 18 °′′′ 50 °′′′ =

Example 8

Round each angle correct to the nearest degree.

a 73° 27′ **b** 9° 41′ **c** 42° 30′ **d** 128° 29′ 47″

SOLUTION

a 73°27′ ≈ 73° 27′ < 30′, so round down

b 9°41′ ≈ 10° 41′ ≥ 30′, so round up

c 42° 30′ ≈ 43° 30′ ≥ 30′, so round up

d 128° 29′ 47″ ≈ 128° 29′ < 30′, so round down

Rounding angle sizes

When rounding an angle to the nearest degree or minute, use 30 as the halfway mark.

Example 9

Round each angle correct to the nearest minute.

a 33°53′30″ **b** 44°15′40″

SOLUTION

a $33°53'\,30'' \approx 33°54'$ $30'' \geq 30''$, so round up

b $44°15'40'' \approx 44°16''$ $40'' \geq 30''$, so round up

Example 10

Convert each angle size to degrees and minutes, correct to the nearest minute.

a 82.5° **b** 60.81°

SOLUTION

a $82.5° = 82°\ 30'$ On a calculator: 82.5 [=] [°'"]

b $60.81° = 60°\ 48'\ 36''$ On a calculator: 60.81 [=] [°'"]

$\approx 60°\ 49'$

EXERCISE 4.04 ANSWERS ON P. 637

Trigonometry on a calculator

1 Evaluate each expression correct to 2 decimal places. EXAMPLE 7

a	tan 84°	**b**	cos 15°	**c**	tan 47°
d	sin 33°	**e**	sin 77°	**f**	cos 60.1°
g	tan 39.55°	**h**	cos 18°	**i**	8 tan 75°
j	14 sin 56°	**k**	12 ÷ tan 20°	**l**	$\frac{7}{\sin 43°}$
m	50 × sin 70° 34′	**n**	66.2 cos 81° 42′	**o**	18.53 sin 11° 8′
p	$\frac{27}{\cos 35°22'}$	**q**	$\frac{44.5}{\tan 65°58'}$	**r**	$\frac{200}{\sin 54°47'}$
s	24.1 ÷ tan 63° 2′	**t**	$\frac{15.7}{\cos 21°8'}$		

2 Round each angle size correct to the nearest degree.

a	27° 54′	**b**	40° 30′	**c**	19° 18′	**d**	33° 7′ 25″
e	33° 41′ 5″	**f**	56.4°	**g**	29.75°	**h**	44° 18′

Foundation Standard Complex

3 Round each angle size correct to the nearest minute.

a	$68° 39' 42''$	**b**	$54° 22' 21''$	**c**	$24° 46' 30''$	**d**	$18° 30' 27''$
e	$9° 10' 55''$	**f**	$47° 59' 9.5''$	**g**	$3° 45' 35''$	**h**	$57° 10' 29''$

4 Convert each angle size to degrees and minutes, correct to the nearest minute.

a	55.5°	**b**	14.15°	**c**	72.38°	**d**	33.77°
e	66.41°	**f**	7.875°	**g**	28.123°	**h**	31.046°
i	34.45°	**j**	71.087°	**k**	5.4829°	**l**	69.4545°

5 By guess-and-checking with your calculator, find the angle size, θ (to the nearest degree), that gives each value.

a	$\sin \theta = 0.7880$	**b**	$\tan \theta = 0.2493$	**c**	$\tan \theta = 1.2799$
d	$\cos \theta = 0.5$	**e**	$\sin \theta = 0.5446$	**f**	$\cos \theta = 0.8829$
g	$\tan \theta = 0.7265$	**h**	$\sin \theta = 0.9998$		

Quiz
Mental skills 4

☆ MENTAL SKILLS 4 ANSWERS ON P. 638

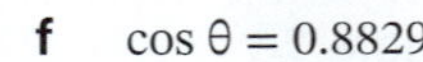

Maths without calculators

Estimating answers

A quick way of estimating an answer is to round each number in the calculation.

1 Study each example.

a $631 + 280 + 51 + 43 + 96 \approx 600 + 300 + 50 + 40 + 100$
$= (600 + 300 + 100) + (50 + 40)$
$= 1000 + 90$
$= 1090$ (Actual answer = 1101)

b $55 + 132 - 34 + 17 - 78 \approx 60 + 130 - 30 + 20 - 80$
$= (60 + 20 - 80) + (130 - 30)$
$= 0 + 100$
$= 100$ (Actual answer = 92)

c $78 \times 7 \approx 80 \times 7$
$= 560$ (Actual answer = 546)

d $510 \div 24 \approx 500 \div 20$
$= 50 \div 2$
$= 25$ (Actual answer = 21.25)

2 Now estimate each answer.

a	$27 + 11 + 87 + 142 + 64$	**b**	$55 + 34 - 22 - 46 + 136$
c	$684 + 903$	**d**	$35 + 81 + 110 + 22 + 7$
e	$517 - 96$	**f**	$210 - 38 - 71 + 151 - 49$
g	$766 - 353$	**h**	367×2

☐ Foundation ○ Standard ○ Complex

i 83×81

j 984×16

k $828 \div 3$

l $507 \div 7$

3 Study each example involving decimals.

a $20.91 - 11.3 + 2.5 \approx 21 - 11 + 3$
$= 13$ (Exact answer = 12.11)

b $4.78 \times 19.2 \approx 5 \times 20$
$= 100$ (Exact answer = 91.776)

c $75.13 \div 8.4 \approx 75 \div 8$
$< 80 \div 8$
< 10
≈ 9 (Exact answer = 8.944...)

d $\frac{37.6+9.3}{41.2-12.7} \approx \frac{38+9}{40-13}$
$= \frac{47}{27}$ (Exact answer = 1.645...)
$\approx \frac{50}{30}$
≈ 1.6

4 Now estimate each answer.

a $3.75 + 9.381 + 4.6 + 10.5$

b $14.807 + 6.6 - 7.22$

c 18.47×9.61

d 4.27×97.6

e $\frac{11.07+18.4}{12.2}$

f $\frac{38.18}{17.2-9.6}$

g $54.75 - 18.6 - 14.4$

h $\frac{18.46 \times 4.9}{39.72-15.2}$

i $62.13 \div 10.7$

j $(4.89)^2$

DID YOU KNOW?

Degrees, minutes and seconds

We use base 10 systems in number, measurement and currency: there are 100 centimetres in a metre, 1000 grams in a kilogram and 100 cents in a dollar. So why are there 360° in a revolution, 60 minutes in a degree and 60 seconds in a minute?

In 2000 BCE, the Babylonians used a base 60 or sexagesimal system of numeration, because 60 is a rounder, more convenient number than 10. This is because 60 has more factors and is divisible by 3, 4 and 6. Furthermore, $6 \times 60 = 360$, which was the Babylonian approximation for the number of days in a year, so that each day the Earth would travel 1° around the Sun.

As measuring devices and calculations required greater precision, each degree was subdivided into 60 equal parts called minutes, and these were further divided into 60 parts called seconds. This level of accuracy is essential in navigation and mapping.

1 **A minute is a 'small' part of a degree. Investigate how an alternative meaning (and pronunciation) of 'minute' is 'tiny'.**

2 **A second is the 'second' subdivision of a degree. Explain how there are two different meanings of 'second'.**

4.05 Finding an unknown side

Videos Finding an unknown side

Trigonometry

Since the trigonometric ratio of any angle is a constant number, we can use it to calculate the length of an unknown side in a right-angled triangle if one other side is known. We need to select the correct ratio that links the given angle to the unknown side and known side.

Example 11

Find the value of each variable, correct to 2 decimal places.

a

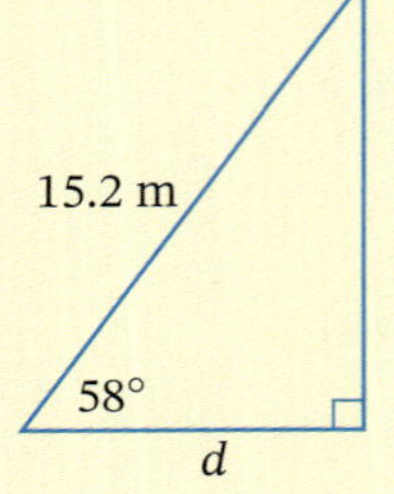

b

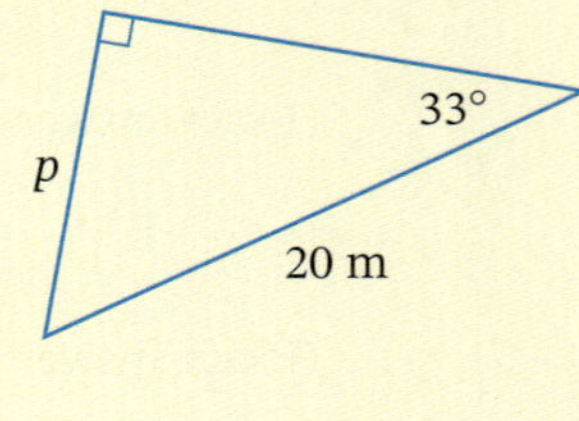

c

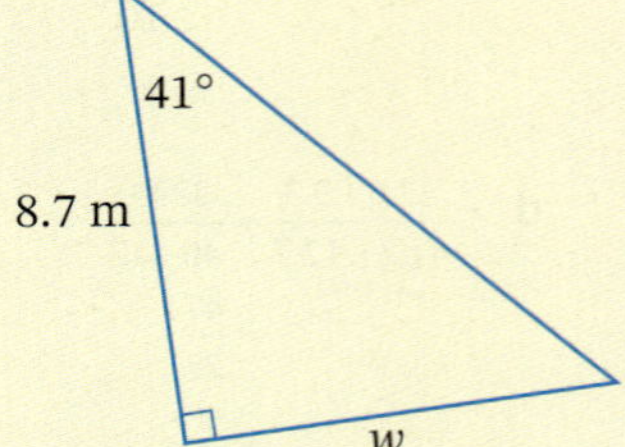

SOLUTION

a SOH, CAH or TOA?

The marked sides related to 58° are the adjacent side (A) and the hypotenuse (H), so use cos.

15.2 m hypotenuse, 58°, d adjacent

$$\cos 58° = \frac{\text{adjacent}}{\text{hypotenuse}}$$

$$= \frac{d}{15.2}$$

$$15.2 \times \cos 58° = 15.2 \times \frac{d}{15.2}$$

Multiplying both sides by 15.2.

$$15.2 \cos 58° = d$$

$$d = 15.2 \cos 58°$$

$$= 8.05477...$$

$$= 8.05$$

From the diagram, a length of 8.05 m looks reasonable.

b SOH, CAH or TOA?

The marked sides related to 33° are the opposite side (O) and the hypotenuse (H), so use sin.

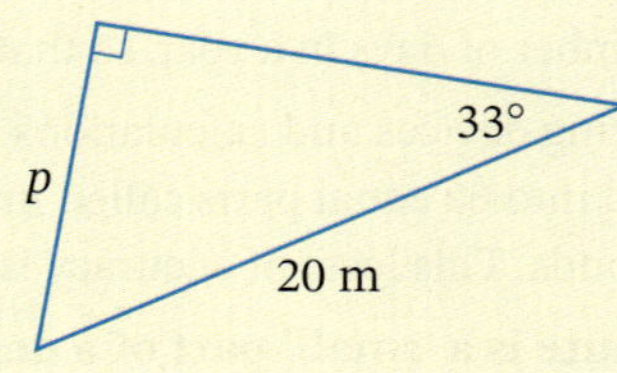

$$\sin 33° = \frac{\text{opposite}}{\text{hypotenuse}}$$

$$= \frac{p}{20}$$

$20 \times \sin 33° = 20 \times \frac{p}{20}$

$20 \sin 33° = d$

$d = 20 \sin 33°$

$= 10.8927...$

$= 10.89$

From the diagram, a length of 10.89 m looks reasonable.

c SOH, CAH or TOA?

The marked sides related to 41° are the opposite side (O) and the adjacent side (A), so use tan.

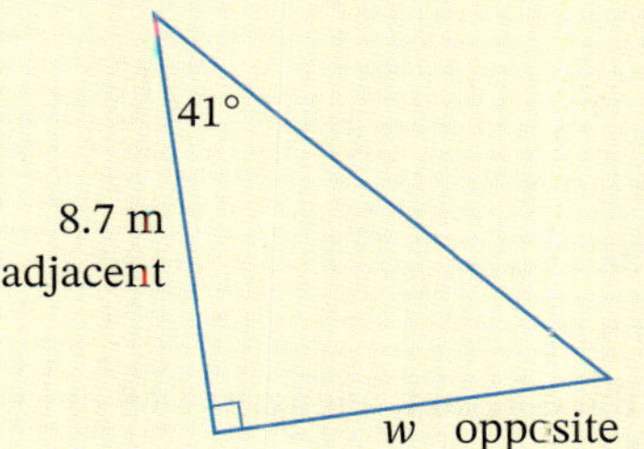

$\tan 41° = \frac{\text{opposite}}{\text{adjacent}}$

$= \frac{w}{8.7}$

$w = 8.7 \tan 41°$

$= 7.5627...$

$= 7.56$

From the diagram, a length of 7.56 m looks reasonable.

ⓘ Finding an unknown side in a right-angled triangle

- Identify the 2 labelled sides and decide whether to use sin, cos or tan.
- Write an equation using the ratio, the given angle and the variable.
- Solve the equation to find the value of the variable.

Example 12

Find the value of q, correct to the nearest centimetre.

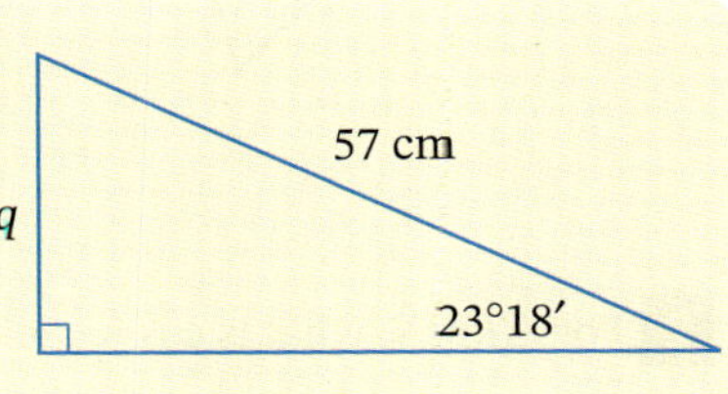

SOLUTION

q is opposite 23° 18′, 57 cm is the hypotenuse, so use sin.

$\sin 23° 18' = \frac{\text{opposite}}{\text{hypotenuse}}$

$\sin 23°18' = \frac{q}{57}$

$q = 57 \sin 23° 18'$

$= 22.5460...$

$= 23$ cm

From the diagram, a length of 23 cm looks reasonable.

Example 13

$\triangle JKL$ is right-angled at K, $JK = 35$ m and $\angle J = 63°$. Find the length of LK, correct to the nearest metre.

SOLUTION

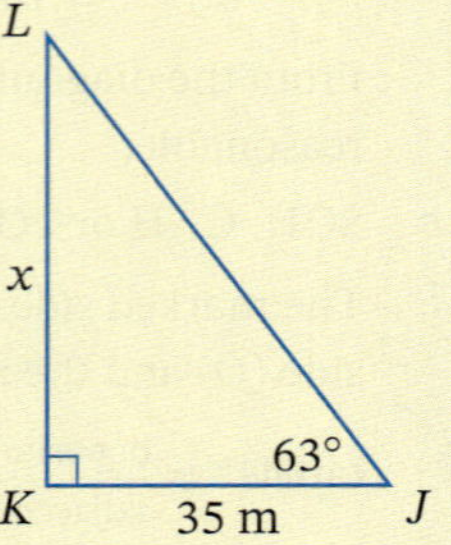

Draw a diagram.

Let the length of LK be x.

x is opposite, 35 m is adjacent, so use tan.

$\tan 63° = \frac{x}{35}$

$x = 35 \tan 63°$

$\quad = 68.6913...$

$LK \approx 69$ m

From the diagram, a length of 69 m looks reasonable.

EXERCISE 4.05 ANSWERS ON P. 638

Finding an unknown side

1 For each triangle, which trigonometric ratio ($\sin \theta$, $\cos \theta$ or $\tan \theta$) is equal to $\frac{a}{b}$?

a

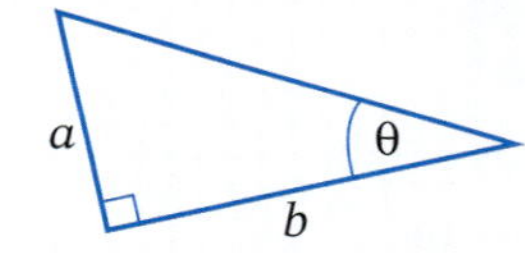

b

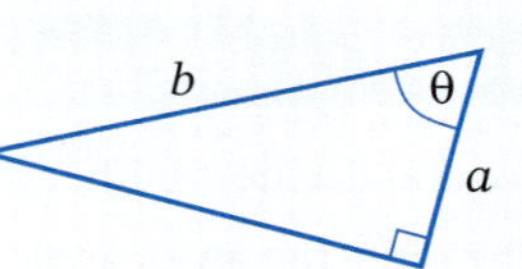

c

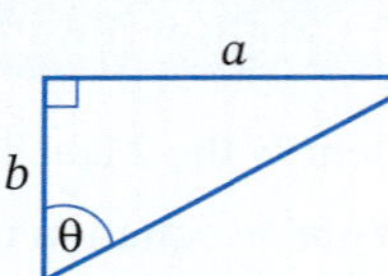

d

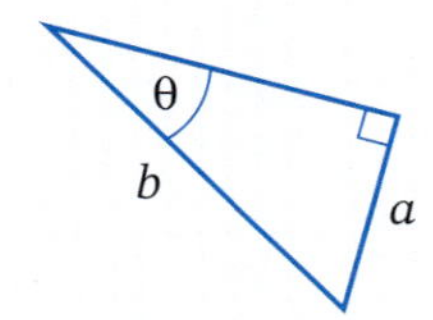

e

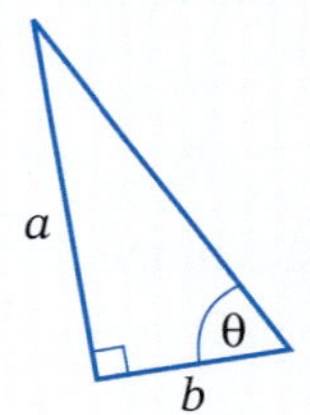

f

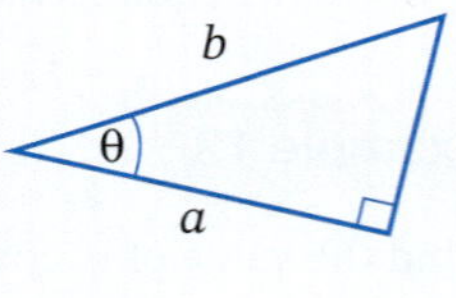

2 For $\triangle KMN$, what is $\cos K$? Select the correct answer **A**, **B**, **C** or **D**.

A $\frac{k}{m}$ **B** $\frac{m}{n}$

C $\frac{n}{m}$ **D** $\frac{k}{n}$

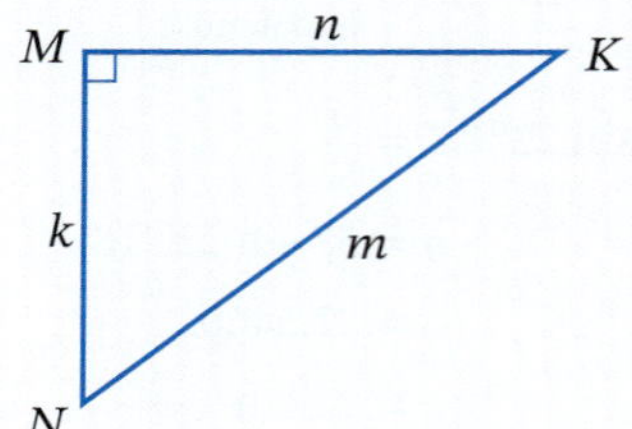

☐ Foundation ○ Standard ⬡ Complex

9780170465618

3 Find the value of the variable in each triangle, correct to 2 decimal places.

a

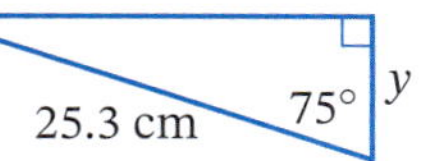

b

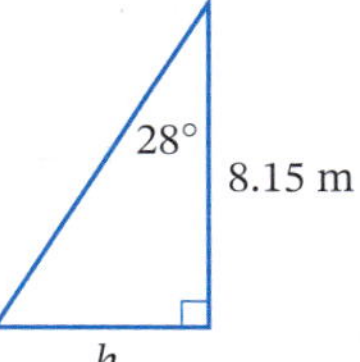

c

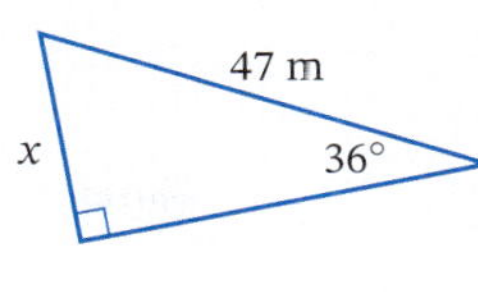

d

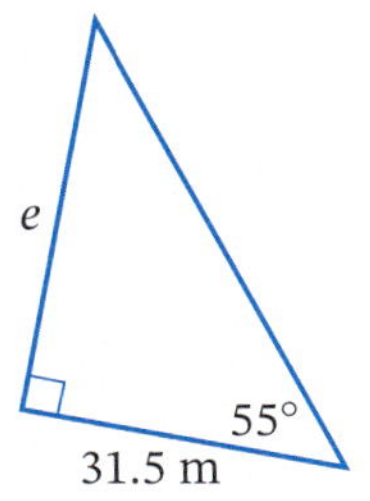

e

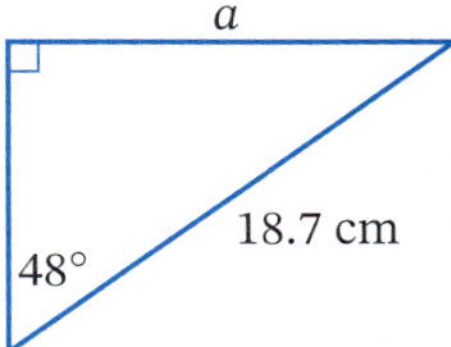

f

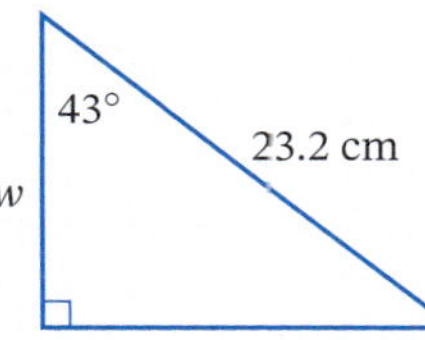

g

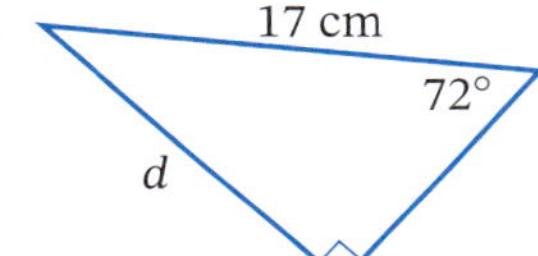

h

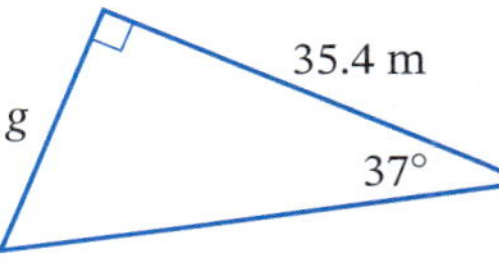

i

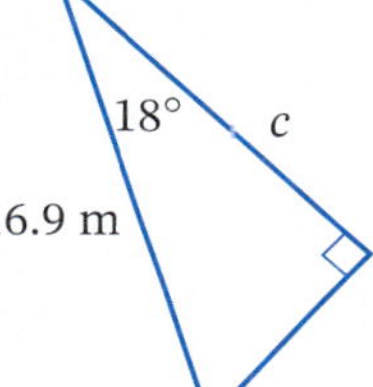

4 Which of the following is the correct expression for y in this triangle? Select **A**, **B**, **C** or **D**.

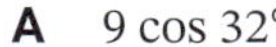

A 9 cos 32°

B 9 tan 32°

C 9 sin 32°

D 9 tan 58°

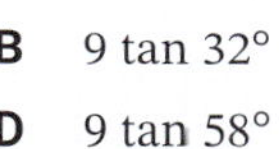

32°
y
9

5 What is the length of d in this triangle?

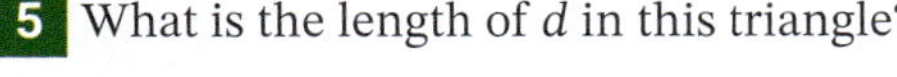

A 10.67 cm

B 10.68 cm

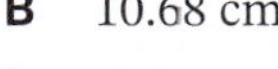

C 16.44 cm

D 16.43 cm

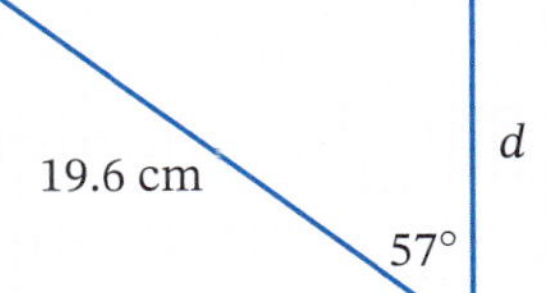

EXAMPLE 12

6 Find the value of the variable in each triangle, correct to 2 decimal places.

a 45.87 cm, 57°33′, *b* cm

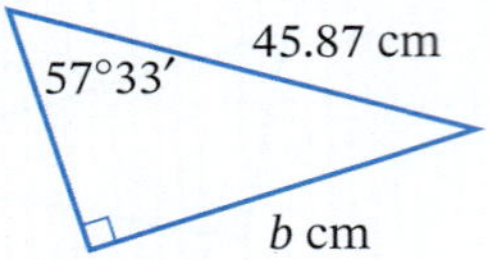

b 7.3 m, 45°18′, *n* m

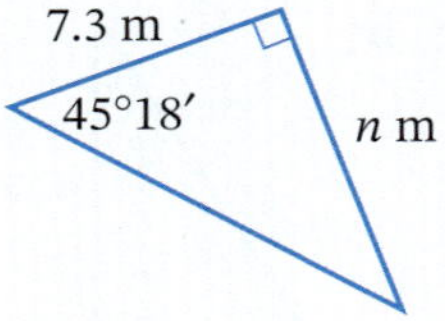

c 20.7 cm, 33°45′, *g* cm

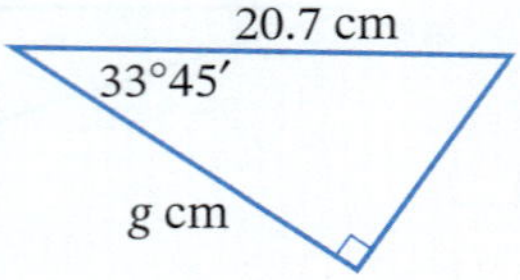

d 150 m, 20°7′, *c* m

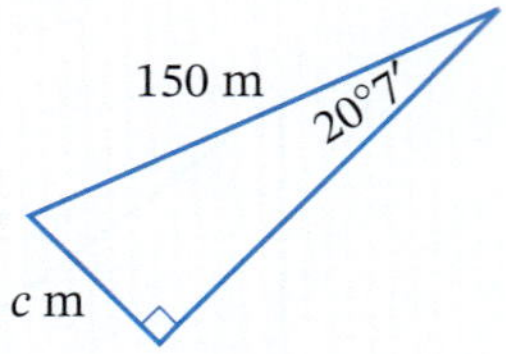

e 38.25 m, 50°11′, *s* m

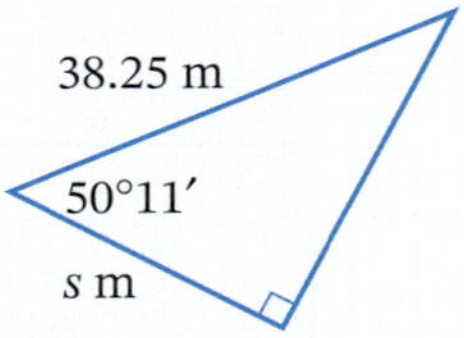

f *p* cm, 42°50′, 143 cm

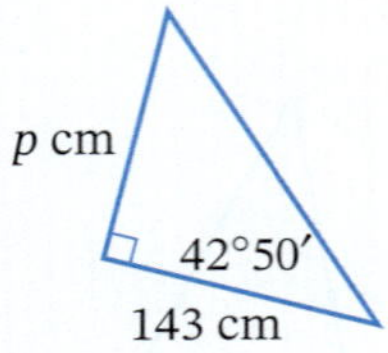

g *x* cm, 37°29′, 43.8 cm

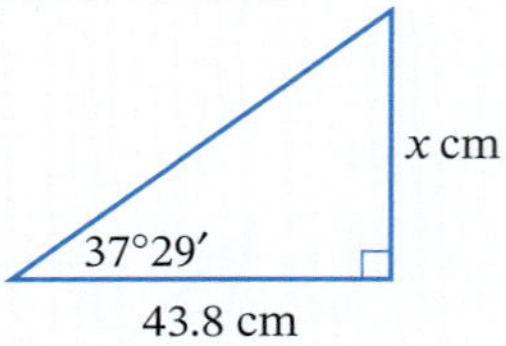

h 78°40′, *r* m, 12.3 m

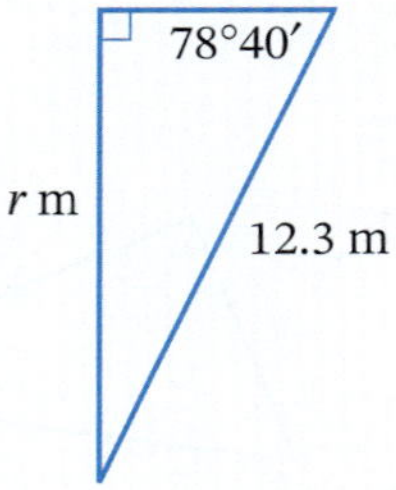

i 84.7 cm, 64°48′, *q* cm

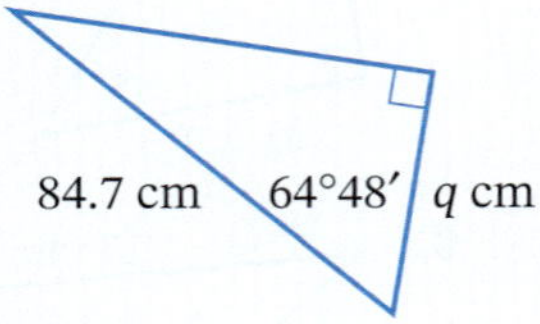

7 What is the value of *h*?

A 7.52 **B** 23.32

C 45.78 **D** 76.00

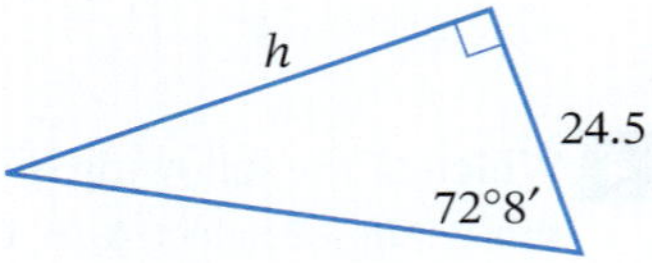

h, 24.5, 72°8′

8 What is the height of this tree?

A 2.68 m **B** 3.27 m

C 3.83 m **D** 6.68 m

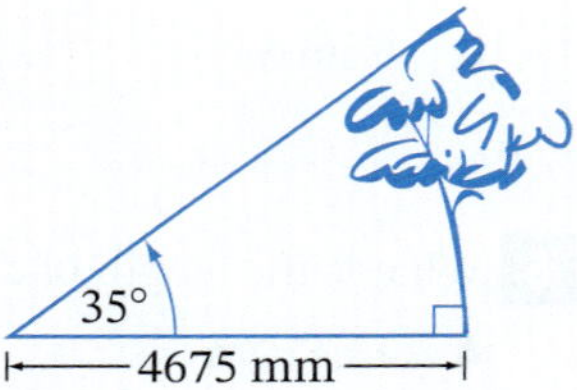

35°, 4675 mm

9 Find each length or distance, correct to one decimal place.

PS **a** How high the ladder reaches up the wall.

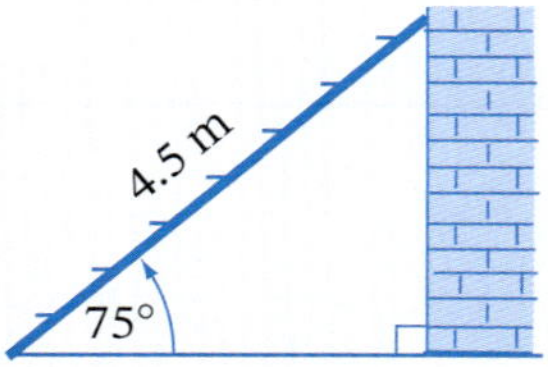

4.5 m, 75°

b The distance between the boat and the start.

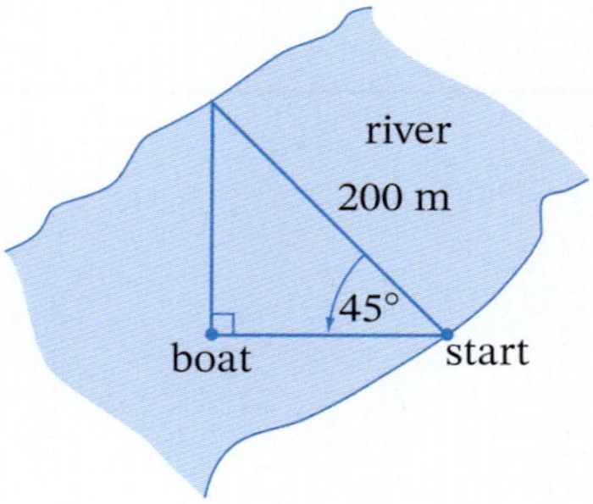

river, 200 m, 45°, boat, start

c The distance from the observer to the base of the building.

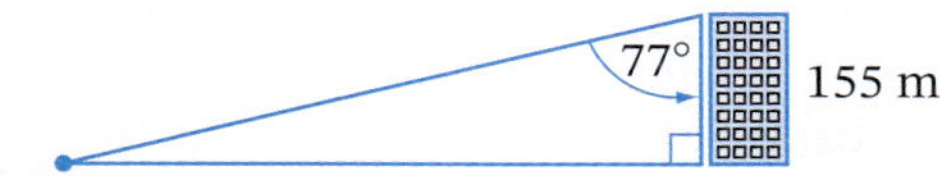

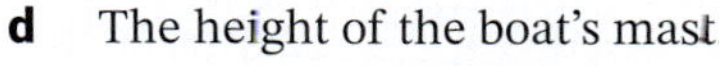

d The height of the boat's mast.

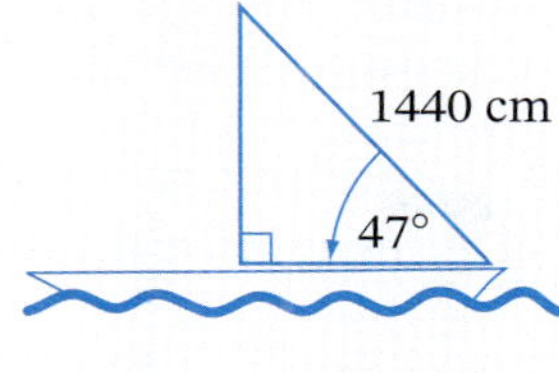

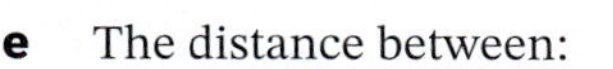

e The distance between:

i checkpoints 1 and 2

ii checkpoint 1 and the start.

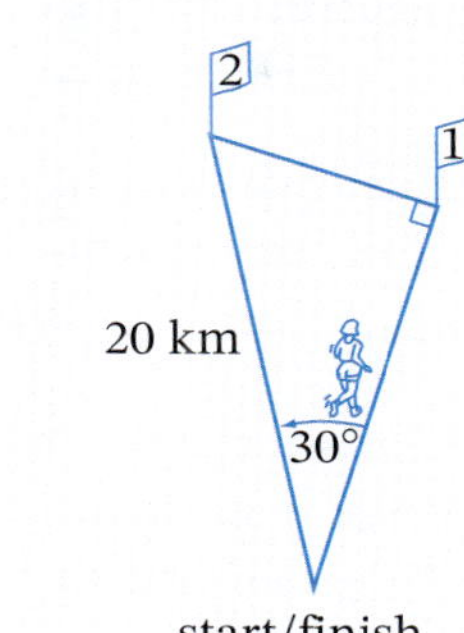

10 In $\triangle PQR$, $\angle P$ is right-angled, $\angle Q = 32°$, $PR = 8$ cm and $QR = 15$ cm. Which diagram fits this description?

C

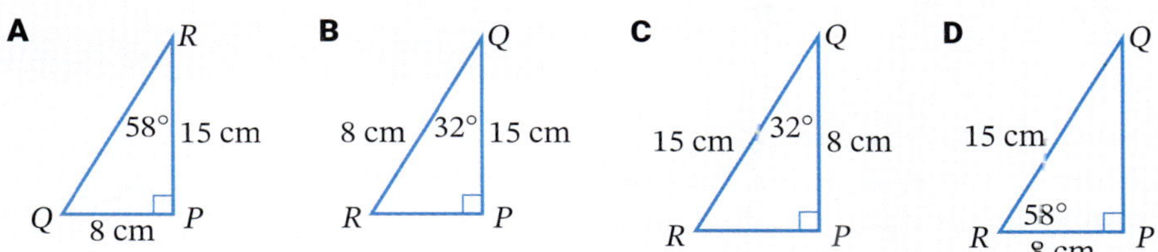

11 $\triangle ABC$ is right-angled at B, $AC = 14.8$ m and $\angle C = 56°$. Find the length of side AB, correct to one decimal place.

12 $\triangle MNR$ is right-angled at M, $MN = 19$ cm and $\angle N = 27°$. Find the length of MR, correct to the nearest centimetre.

13 In $\triangle XYW$, $\angle X = 90°$, $\angle Y = 43.7°$ and $WY = 8.34$ m. Find the length of XW, correct to 2 decimal places.

14 $\triangle AHK$ is right-angled at K, $\angle H = 76°$ and $AH = 13.9$ m. Find the length of HK, correct to one decimal place.

15 A tree casts a shadow 20 m long. If the Sun's rays meet the ground at 25°, find the height of the tree, correct to the nearest cm.

PS

16 A 6-m ladder is placed against a pole. If the ladder makes an angle of 17° with the pole, how far up the pole does the ladder reach? Answer to the nearest mm.

PS

17 A golfer is 180 m in a straight line from the 8th hole. The ball is hit 15° to the right of the hole but still ends up level with the hole. How far is the ball from the hole? Answer to the nearest metre.

PS

☐ Foundation ◯ Standard ◯ Complex

18 PS A park is in the shape of a rectangle. A path 450 m in length crosses the park diagonally. If the path makes an angle of 36° with the longer side, find the dimensions of the park. Answer to the nearest metre.

19 PS A wheelchair ramp is 6 m long and makes an angle of 4.5° with the ground. How high is the top of the ramp above the ground (correct to 2 decimal places)?

20 PS A boat is anchored by a rope 5.5 m long. If the anchor rope makes an angle of 23° with the vertical, calculate the depth of the water (correct to one decimal place).

21 PS A rectangular gate has a diagonal brace that makes an angle of 60° with the bottom of the gate. If the length of the diagonal brace is 1860 mm, calculate the height of the gate.

A 2148 mm **B** 930 mm **C** 1611 mm **D** 3221 mm

22 PS Tayyab is flying a kite that is attached to a string 155 m long. The string makes an angle of 35° to the horizontal. How high, to the nearest metre, is the kite above Tayyab's hand?

A 78 m **B** 89 m
C 109 m **D** 127 m

Shutterstock.com/pirita

23 PS A wind turbine is supported by 2 wires that are attached to the pole 6 m above the ground. The wires make an angle of 28° with the pole. Calculate the distance from the bottom of the wire to the base of the pole, correct to the nearest centimetre.

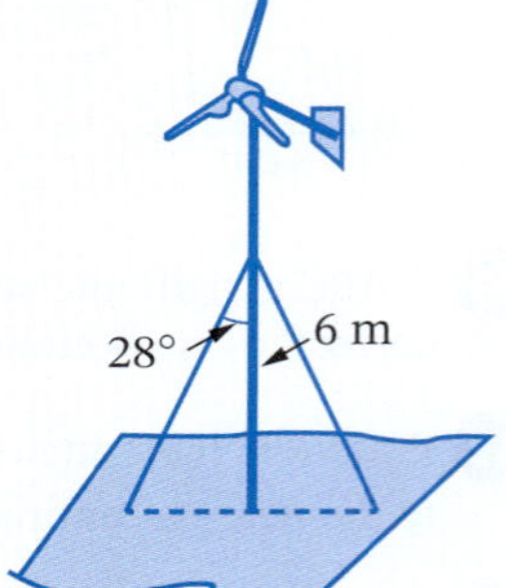

☐ Foundation ◯ Standard ⬡ Complex

INVESTIGATION

Calculating the height of an object

You will need: tape measure or trundle wheel, a clinometer (or protractor) to measure the angle.

Trigonometry can be used to find the heights of buildings, flagpoles and trees without actually measuring them. This can be done by measuring the distance along the ground from the base of the object to a person. The person then measures the angle to the top of the object. For example, the height of a flagpole can be calculated using the set-up shown in the diagram.

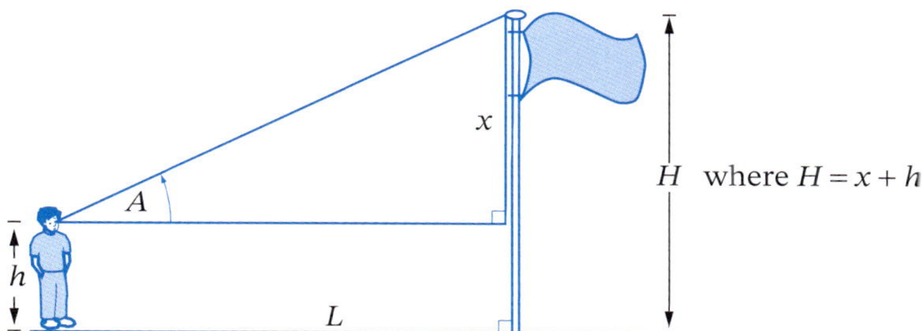

- h is the eye height of the person who measures the angle, A, to the top of the flagpole.
- L is the distance the person is from the base of the flagpole.
- x is the height of the flagpole above the person's eye height.
- $H = x + h$ is the height of the flagpole above the ground.

1 Select a tall object outside to measure.

2 Work with a partner to measure (in cm) the distance, L, along the ground, the height, h, of the person, and the angle (in degrees) to the top of the object. Copy the table below and record your information in the first row.

Distance, L (cm)	Angle, $A°$	Height of person, h (cm)	Calculated height, x cm	Height of flagpole, H cm

3 Use the tan ratio to calculate the value of x, correct to the nearest whole number.

4 Hence find H, the height of the object, correct to the nearest centimetre. Write your answers in the table.

5 Repeat the measurements and calculations 3 more times from different positions, with different persons measuring the angle. This will help to improve the accuracy of your results and minimise errors. Write your results in the table.

6 Did you find similar values for H? Do they seem reasonable for the height of the object?

7 Calculate the average value for H.

4.06 Finding more unknown sides

Puzzle Trigonometry equations 1

Video Finding the hypotenuse

In the following examples, the unknown appears in the **denominator** of the equation.

Using sin or cos to find the hypotenuse

Example 14

Find the value of x, correct to one decimal place.

x m
7 m
32°

SOLUTION

7 m is the opposite side, x m is the hypotenuse, so use sin.

$\sin 32° = \frac{7}{x}$ Note that the variable x appears in the denominator of the equation.

Multiply both sides by x.

$\sin 32° \times x = \frac{7}{x} \times x$

$x \sin 32° = 7$

$\frac{x \sin 32°}{\sin 32°} = \frac{7}{\sin 32°}$ Divide both sides by sin 32°.

$x = \frac{7}{\sin 32°}$

$= 13.2095...$

≈ 13.2

Note that when the unknown appears in the denominator of an equation, it can swap positions with the trigonometric ratio, so that $\sin 32° = \frac{7}{x}$ becomes $x = \frac{7}{\sin 32°}$.

Example 15

$\triangle PQR$ is right-angled at Q, $QR = 41$ m and $R = 25°$. Find RP, correct to the nearest metre.

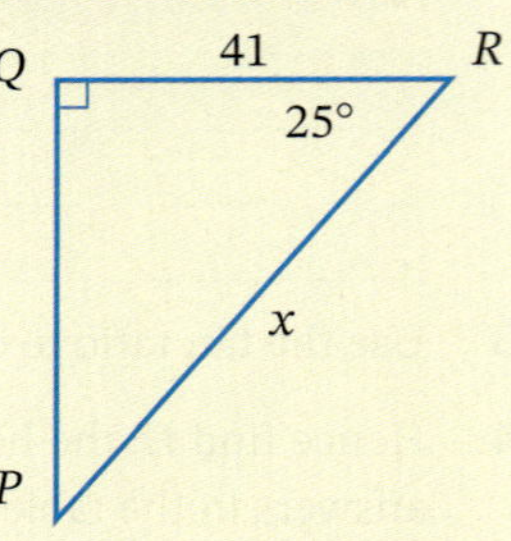

SOLUTION

Let $x = RP$.

41 m is the adjacent side, x is the hypotenuse, so use cos.

$\cos 25° = \frac{41}{x}$ Note that the variable x appears in the denominator of the equation.

$x = \frac{41}{\cos 25°}$ Swap the position of x with cos 25°.

$= 45.2384...$

≈ 45 m

Using tan to find the adjacent side

Example 16

Find the length of x, correct to 2 decimal places.

SOLUTION

18 cm is the opposite side, x is the adjacent side, so use tan.

$\tan 34° = \frac{18}{x}$ — x appears in the denominator.

$x = \frac{18}{\tan 34°}$ — Swap the position of x with tan 34°.

$= 26.6860...$

≈ 26.69 cm

Alternative method

To avoid having x in the denominator, we could use tan with the 3rd angle of the triangle.

3rd angle $= 180° - 90° - 34° = 56°$

$\tan 56° = \frac{x}{18}$

$x = 18 \tan 56°$

$= 26.6860...$

≈ 26.69 cm

EXERCISE 4.06 ANSWERS ON P. 638

Finding more unknown sides

U F PS

1 Find the value of each variable, correct to one decimal place. EXAMPLE 14

a

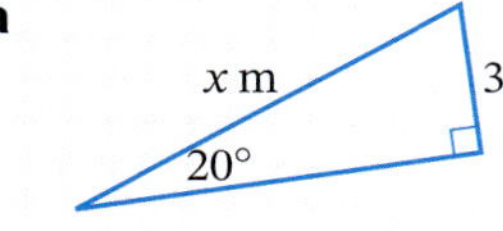

b

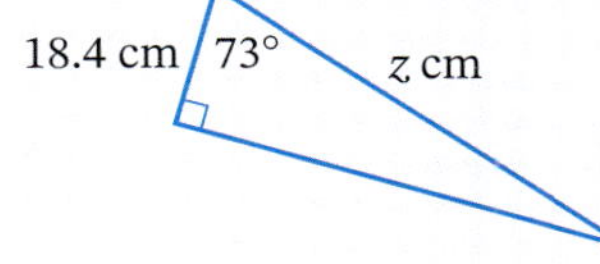

c

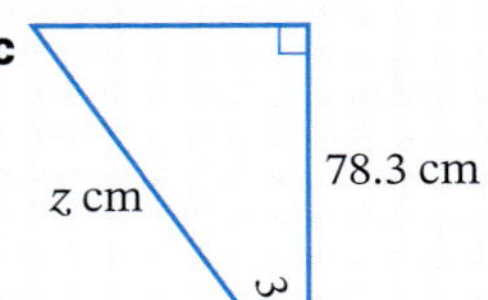

d

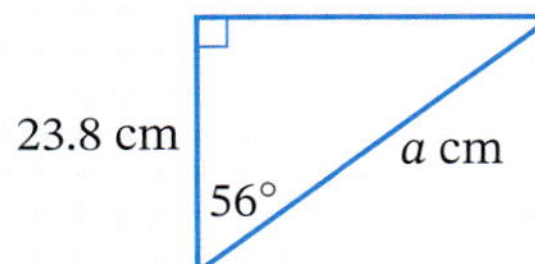

e

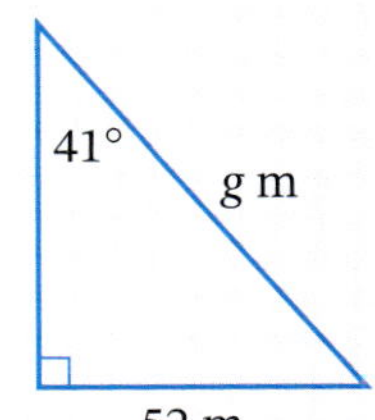

f

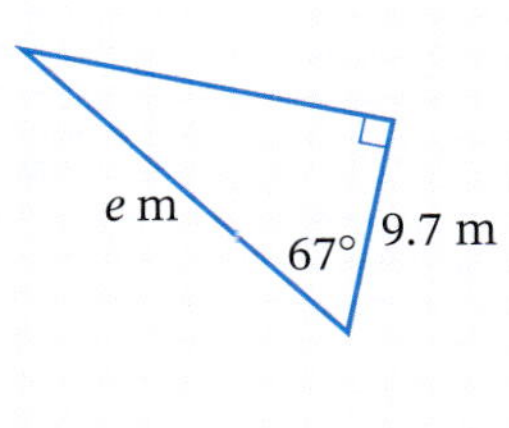

Foundation Standard Complex

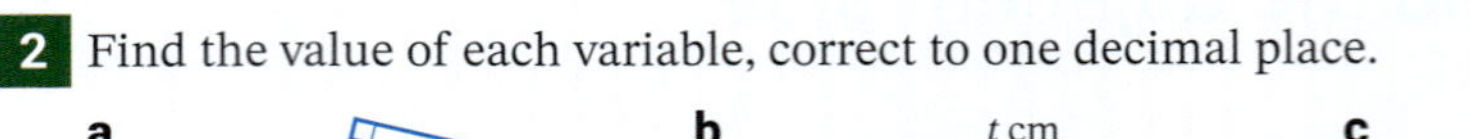

2 Find the value of each variable, correct to one decimal place.

a

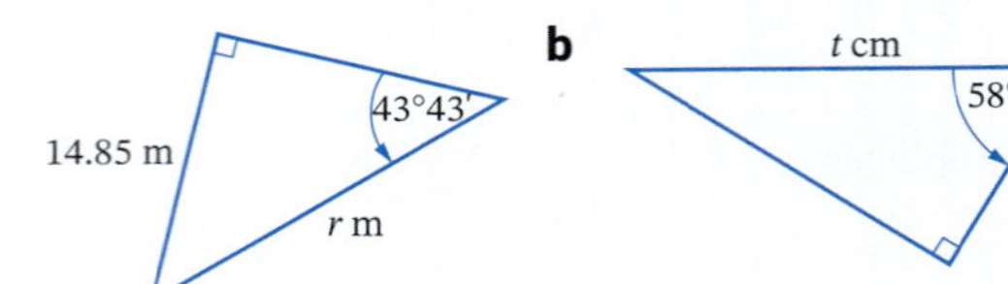

c

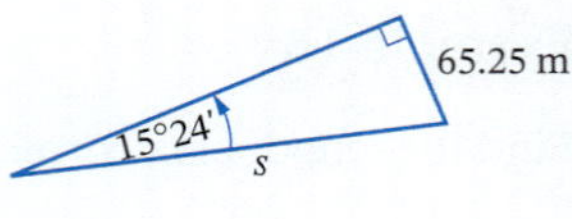

3 △XYZ is right-angled at Z, $ZY = 230$ mm and $\angle Y = 45°$. Find the length of XY, correct to the nearest millimetre.

4 In △KLW, $\angle L = 90°$, $KL = 12$ m and $\angle W = 75.2°$. Find KW, to one decimal place.

5 Find the value of each variable, correct to 2 decimal places.

a

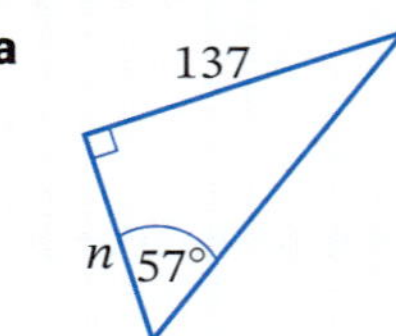

b

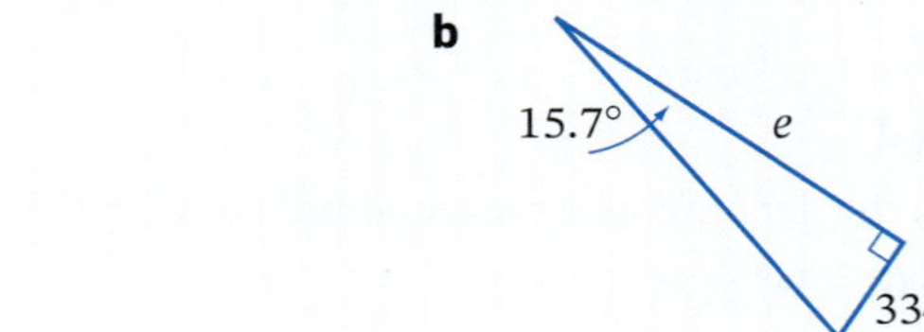

c

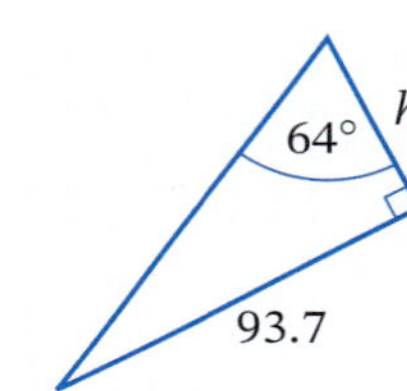

d

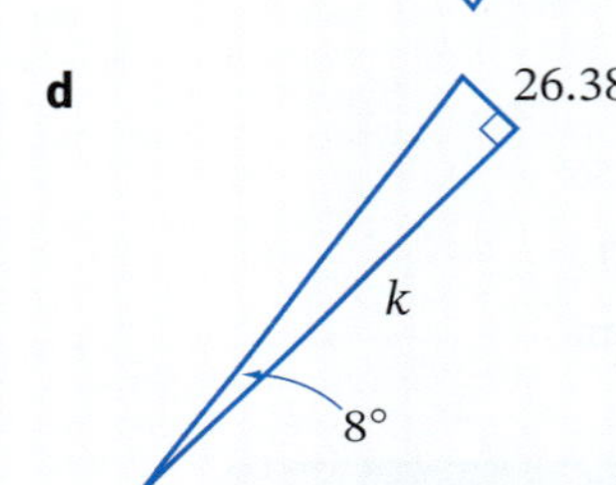

6 △CDE is right-angled at D, $\angle E = 36°$ and $CD = 5$ m. Find the length of side DE, correct to 2 decimal places.

7 What is the correct expression for p in this triangle? Select the correct answer **A**, **B**, **C** or **D**.

A $\dfrac{46}{\cos 23°}$ **B** $\dfrac{46}{\sin 23°}$

C $46 \cos 23°$ **D** $46 \sin 23°$

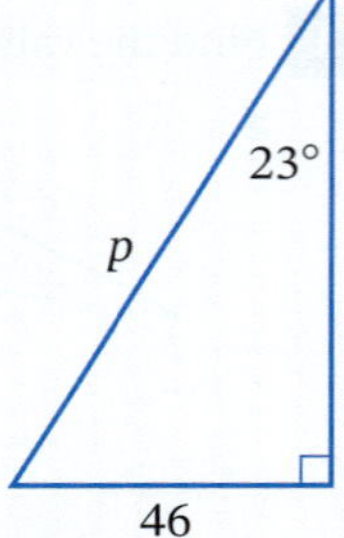

☐ Foundation ◯ Standard ⬡ Complex

9780170465618

8 What is the value of v in this triangle?

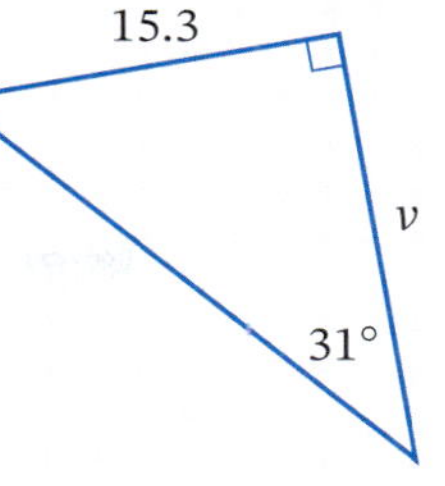

A 25.46 **B** 9.19

C 29.71 **D** 17.85

9 In $\triangle HMT$, $\angle T = 90°$, $\angle M = 19° 47'$ and side $HT = 18.4$ cm. Find the length of side HM, correct to one decimal place.

10 Find the value of each variable, correct to one decimal place.

a

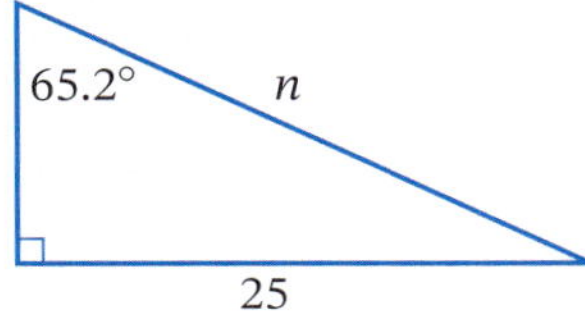

b

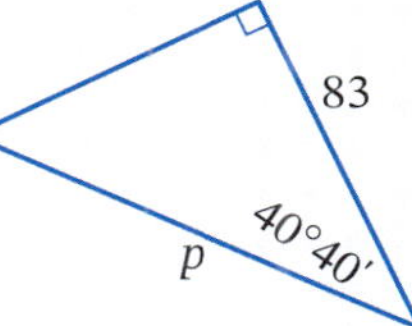

c

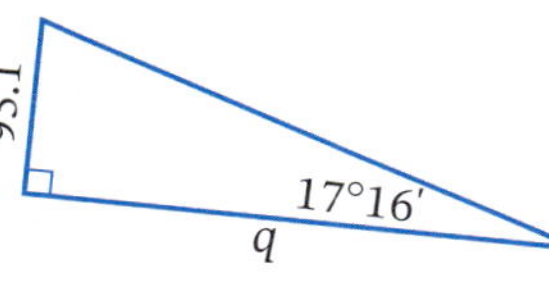

d

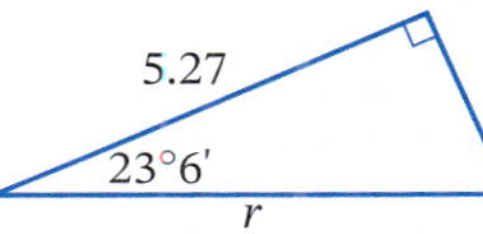

e

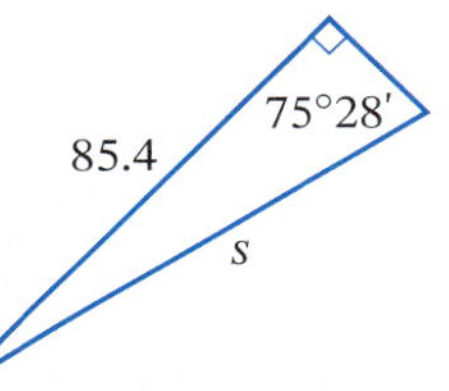

f

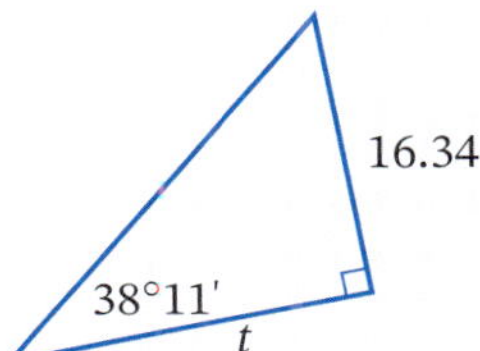

11 Find the length of this ladder.

A 159 cm **B** 171 cm

C 243 cm **D** 638 cm

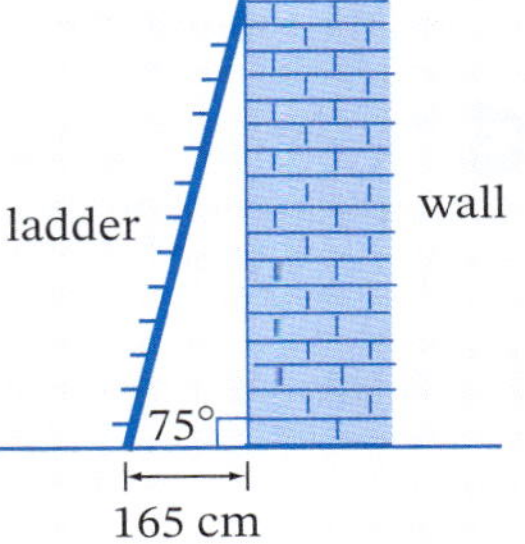

12 Find each length or distance, to one decimal place.

PS **a** How far the person is from being directly under the birds.

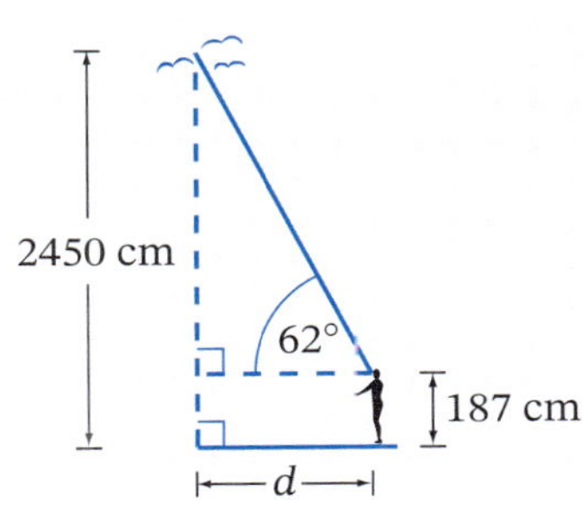

Foundation Standard Complex

b The length of the ramp.

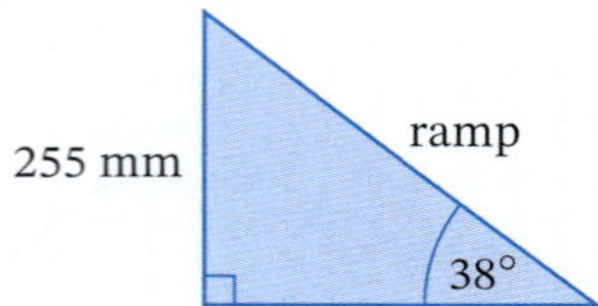

c The length of the support wire.

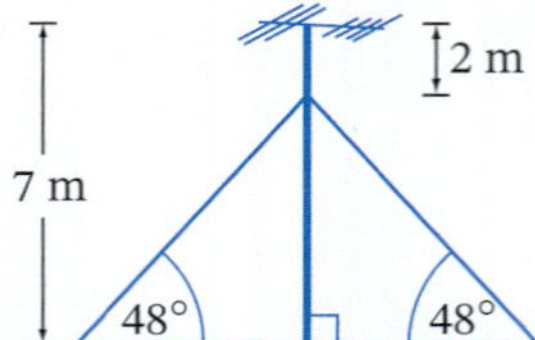

d The length of:

i the shortest road.

ii the longest road.

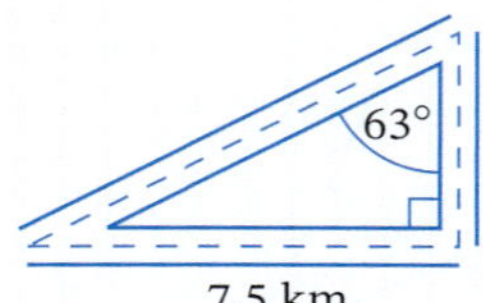

e The slant height of the roof.

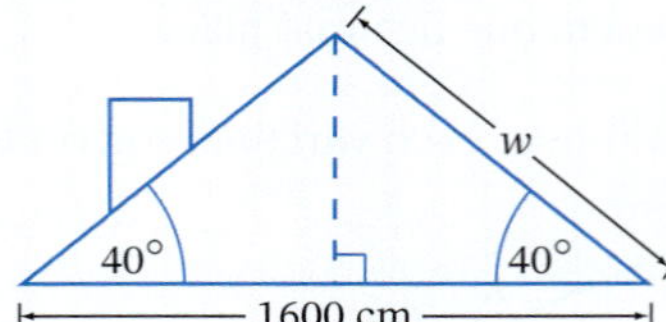

13 $\triangle FGW$ is right-angled at F, $\angle W = 84°$ and $WF = 42.1$ m. Find the length of WG, correct to one decimal place.

14 PS A ladder rests against a wall. The foot of the ladder is 355 cm from the wall and makes an angle of 63° with the ground. How long (to the nearest cm) is the ladder?

15 PS A supporting wire is attached to the top of a flagpole. The wire meets the ground at an angle of 51° and the flagpole is 15 m high. How far, to the nearest 0.1 m, from the base of the flagpole is the wire anchored to the ground?

16 PS A glider is flying at an altitude (height) of 1.5 km. To land, it descends at an angle of 18° to the ground. How far, to the nearest 0.1 km, must the glider travel before landing?

17 PS The entrance to the school library is 60 cm above ground level. A wheelchair ramp is built to the entrance at an angle of 5° with the ground. How long (to the nearest 0.01 m) is the ramp?

18 PS A shooter aims directly at a target, but just before firing, the rifle is lifted 1° off target. The shot misses the target by 67 mm. How far is the shooter standing from the target?

A 1169 mm **B** 3838 mm **C** 3839 mm **D** 6701 mm

19 PS A hot air balloon is anchored to the ground by a rope. When it drifts 20 m sideways, it makes an angle of 75° with the ground. How long is the rope (to one decimal place)?

☐ Foundation ○ Standard ⬡ Complex

20 For $\triangle ABC$, $\angle A = 40°$, $\angle C = 65°$, $AB = 28$ cm and BX is perpendicular to AC. Calculate, correct to one decimal place, the length of:

a BX

b CX

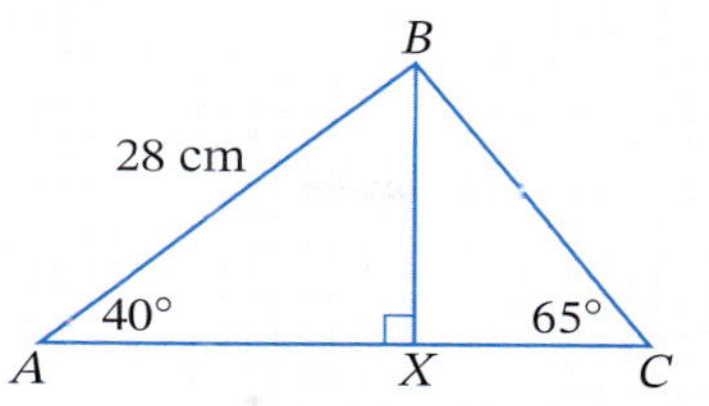

4.06

INVESTIGATION

Finding an angle, given a trigonometric ratio

You will need: a ruler, compasses and a calculator.

1 Copy and complete this table, calculating each ratio as a decimal, correct to 3 decimal places.

θ	sin θ	cos θ	tan θ
0°			
15°			
30°			
45°			
60°			
75°			
90°			undefined

2 Can you work out why there is no answer for tan 90°?

3 What are the minimum and maximum values of sin θ?

4 What are the minimum and maximum values of cos θ?

5 Is there a pattern between the values of sin θ and cos θ?

6 Check the value of sin 30° by constructing a right-angled triangle with one angle that is 30°, measuring the opposite side and hypotenuse and dividing them.

7 Check the value of tan 45° by constructing a right-angled triangle with one angle that is 45°, measuring the opposite and adjacent sides and dividing them.

8 Use the table to estimate each trigonometric ratio and check your estimate using a calculator.

a sin 80° **b** cos 34° **c** tan 55°

9 If $\sin \theta = \frac{3}{8}$, find the value of the unknown angle θ, correct to the nearest degree:

a by using the table and estimating (change $\frac{3}{8}$ to a decimal first).

b using a calculator to guess-and-check.

c constructing a right-angled triangle with one angle θ, opposite side 3 cm and hypotenuse 8 cm, then measuring the size of θ.

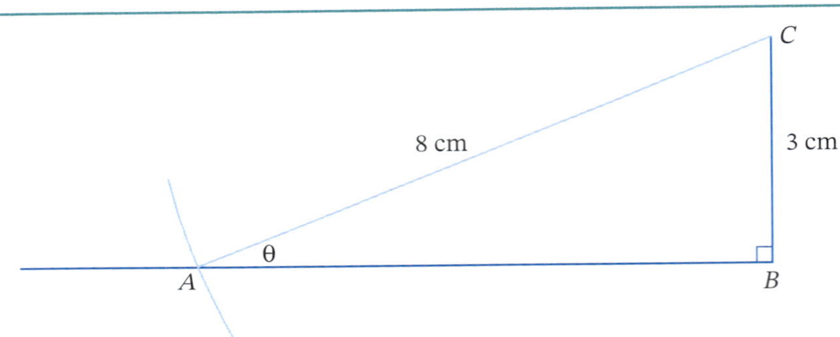

10 If $\cos \theta = \frac{2}{5}$, find the value of the unknown angle θ, correct to the nearest degree:

- **a** by using the table and estimating (change $\frac{2}{5}$ to a decimal first).
- **b** using a calculator to guess-and-check.
- **c** constructing a right-angled triangle with one angle θ, adjacent side 2 cm and hypotenuse 5 cm, then measuring the size of θ.

11 If $\tan \theta = \frac{7}{10}$, find the value of the unknown angle θ, correct to the nearest degree:

- **a** by using the table and estimating (change $\frac{7}{10}$ to a decimal first).
- **b** using a calculator to guess-and-check.
- **c** constructing a right-angled triangle with one angle θ, opposite side 7 cm and adjacent side 10 cm, then measuring the size of θ.

4.07 Finding an unknown angle

Worksheet Trigonometry problems

Puzzles Trigonometry squaresaw

Trigonometry: Finding angles

Trigonometric equations 2

Presentation Trigonometry

A scientific calculator can be used to evaluate a trigonometric ratio such as sin 38°, but it can also be used to find an unknown angle, θ, if the trigonometric ratio of the angle is known, for example, if $\sin \theta = 0.9063$.

An unknown angle can be found using the $\boxed{\sin^{-1}}$ $\boxed{\cos^{-1}}$ $\boxed{\tan^{-1}}$ keys on the calculator. These are called the **inverse sin**, **inverse cos** and **inverse tan** functions, found by pressing the SHIFT or 2nd F key before the sin, cos or tan key.

Example 17

a If $\sin \theta = 0.9063$, find angle θ, correct to the nearest degree.

b If $\tan X = 3.754$, find angle X, correct to the nearest minute.

c If $\cos \alpha = \frac{4}{7}$, find angle α, correct to the nearest degree.

SOLUTION

a $\sin \theta = 0.9063$

$\theta = 64.9989...°$ On a calculator: SHIFT sin 0.9063 =

$\approx 65°$

b $\tan X = 3.754$

$X = 75.0837\ldots°$ SHIFT tan 3.754 =

$= 75° \ 5' \ 1.62''$ °''' or DMS

$\approx 75° \ 5''$

c $\cos \alpha = \frac{4}{7}$

$\alpha = 55.1500\ldots°$ SHIFT cos 4 ⊟ 7 =

≈ 55

Note: ⊟ or $a^{b/c}$ is the fraction key.

Example 18

Find the size of angle θ, correct to the nearest degree.

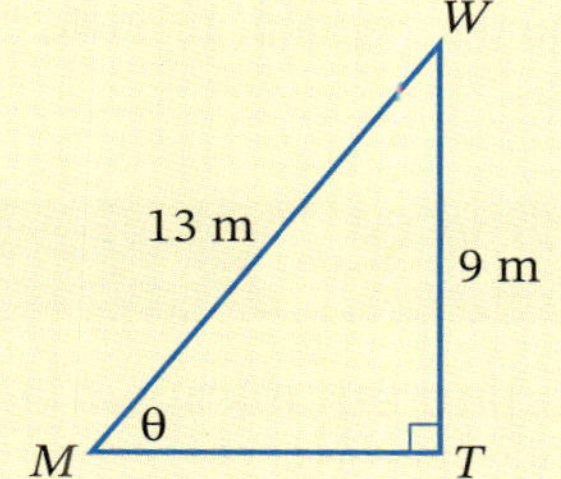

SOLUTION

SOH, CAH or TOA?

The known sides for θ are the opposite (*O*) side and the hypotenuse (*H*), so use sin.

$\sin\theta = \frac{9}{13}$

$\theta = 43.8130\ldots°$

$\approx 44°$

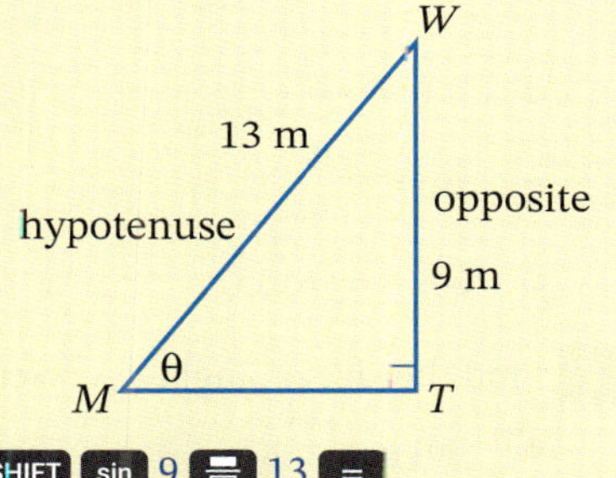

From the diagram, an angle size of 44° looks reasonable. SHIFT sin 9 ⊟ 13 =

ⓘ Finding an unknown angle in a right-angled trianglen

- Identify the 2 known sides and decide whether to use the sin, cos or tan ratic.
- Write an equation using the ratio, the angle and the 2 sides as a fraction.
- Use the calculator's inverse trigonometric function to find the size of the angle.

Example 19

$\triangle XYZ$ is right-angled at Y, with $XY = 35$ cm and $YZ = 47$ cm. Find $\angle Z$, correct to the nearest degree.

SOLUTION

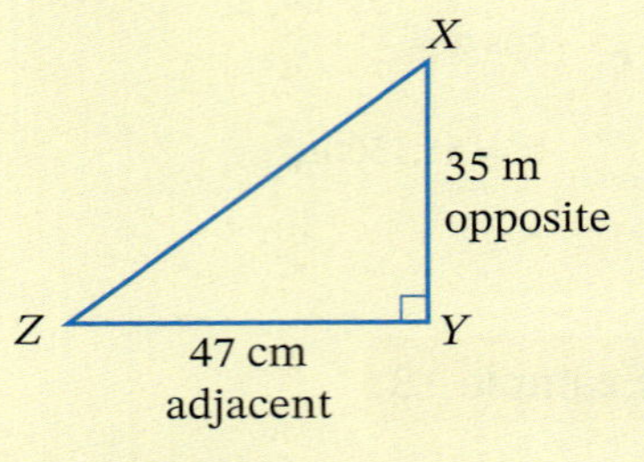

Sketch a diagram.

SOH, CAH or TOA?

The known sides to $\angle Z$ are the opposite (O) and the adjacent (A), so use tan.

$\tan \mathbf{Z} = \frac{35}{47}$.

$Z = 36.6743...°$

$\approx 37°$

From the diagram, an angle size of 37° looks reasonable.

SHIFT tan 35 ÷ 47 =

EXERCISE 4.07 ANSWERS ON P. 638

Finding an unknown angle

U F PS

1 Find the size of angle θ, correct to the nearest degree.

a	$\cos\theta = 0.76$	**b**	$\tan\theta = 2.0532$	**c**	$\sin\theta = \frac{\sqrt{3}}{2}$
d	$\tan\theta = 6$	**e**	$\sin\theta = \frac{7}{8}$	**f**	$\cos\theta = \frac{13}{15}$
g	$\sin\theta = \frac{1}{10}$	**h**	$\cos\theta = \frac{1}{\sqrt{2}}$	**i**	$\tan\theta = \sqrt{3}$
j	$\cos\theta = 0.1352$	**k**	$\tan\theta = 8.836$	**l**	$\sin\theta = \frac{1}{4}$

2 Find the size of angle A, correct to the nearest minute.

a	$\tan A = \frac{15}{7}$	**b**	$\sin A = 0.815$	**c**	$\cos A = \frac{4}{5}$
d	$\cos A = 0.9387$	**e**	$\tan A = \frac{19}{20}$	**f**	$\cos A = \frac{3}{10}$
g	$\sin A = \frac{5}{11}$	**h**	$\sin A = 0.88$	**i**	$\tan A = 15.07$
j	$\cos A = \frac{1}{7}$	**k**	$\tan A = \frac{\sqrt{2}}{2}$	**l**	$\sin A = \frac{7}{9}$

EXAMPLE 18

3 Find θ, correct to the nearest degree.

a

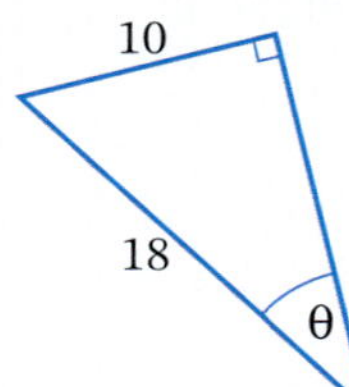

b

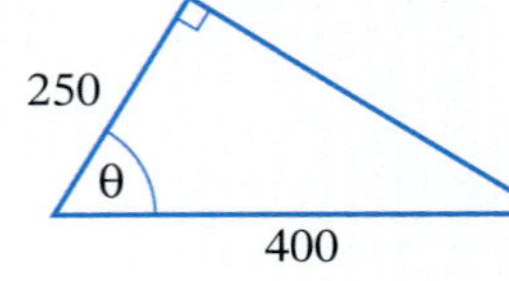

c

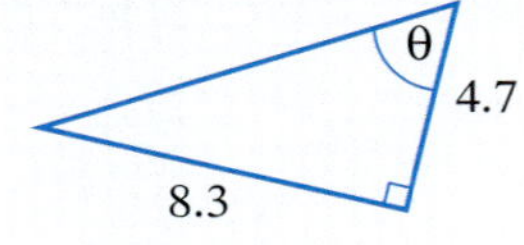

☐ Foundation ○ Standard ⬡ Complex

d

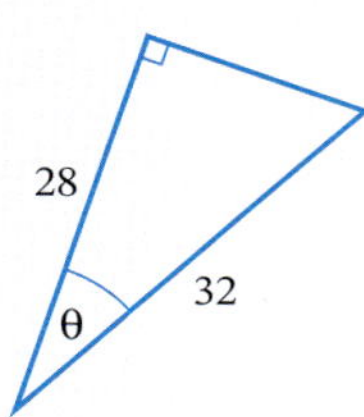

e

f

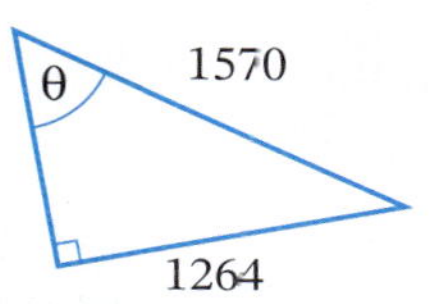

4 Find the size of angle α, correct to the nearest minute. Select the correct answer **A**, **B**, **C** or **D**.

A 30° 52′ **B** 30° 53′

C 36° 42′ **D** 36° 43′

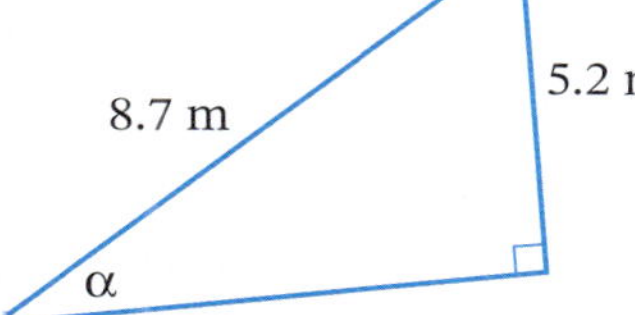

5 Find the size of angle α, correct to the nearest minute.

a

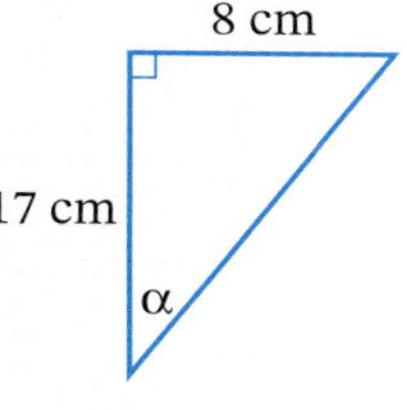

b

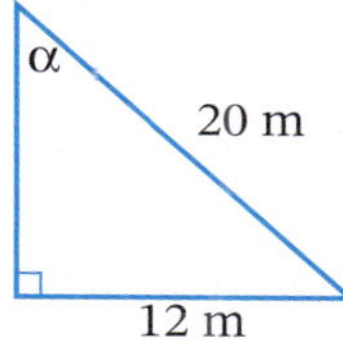

c

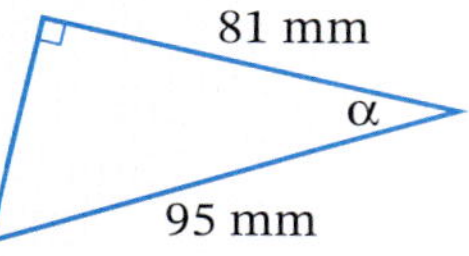

d

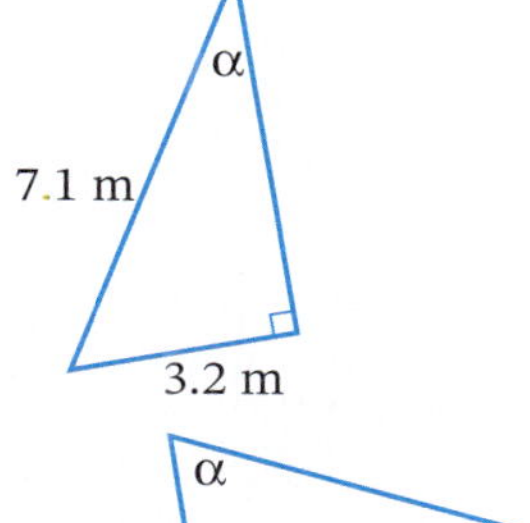

e

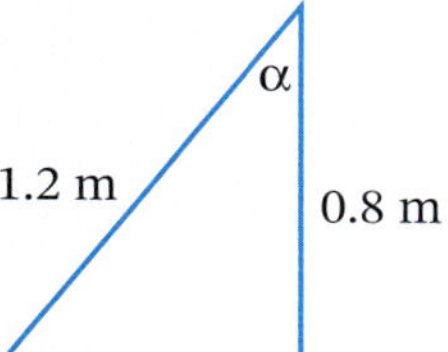

f

Foundation Standard Complex

6 Find the size of angle θ, correct to the nearest degree.

a

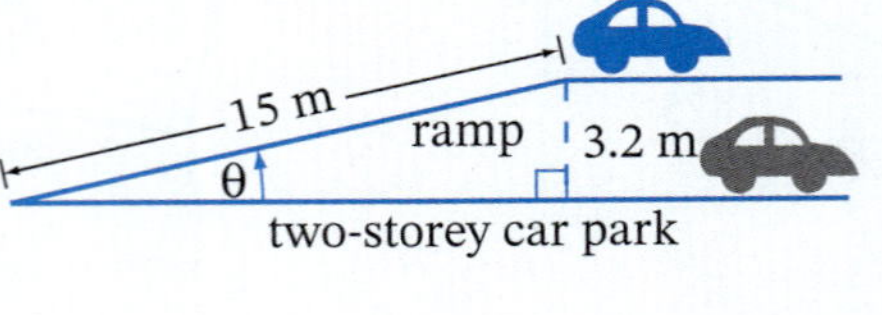

b

80 m
string
θ
30 m

c

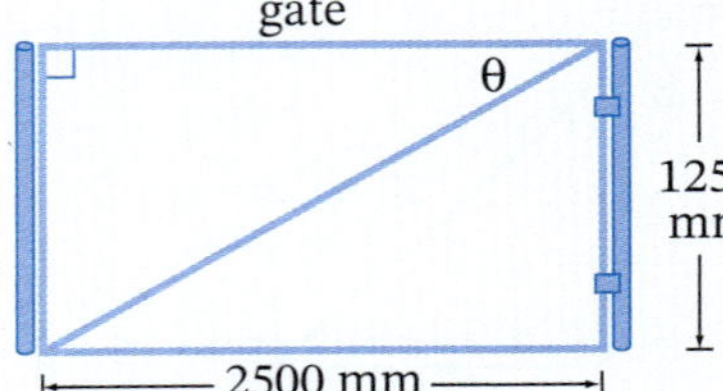

d

175 cm
shadow
θ
340 cm

e

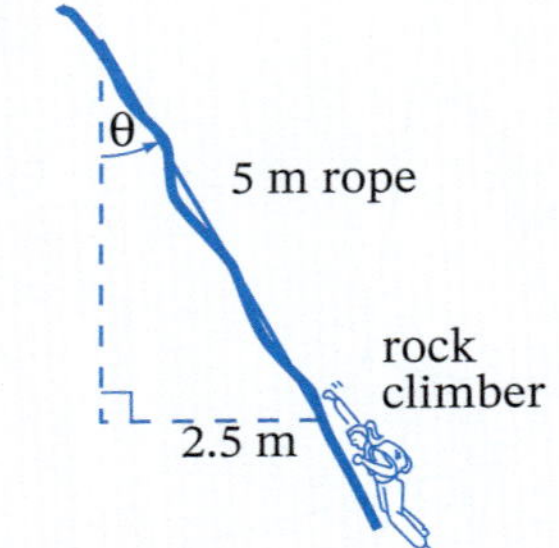

f

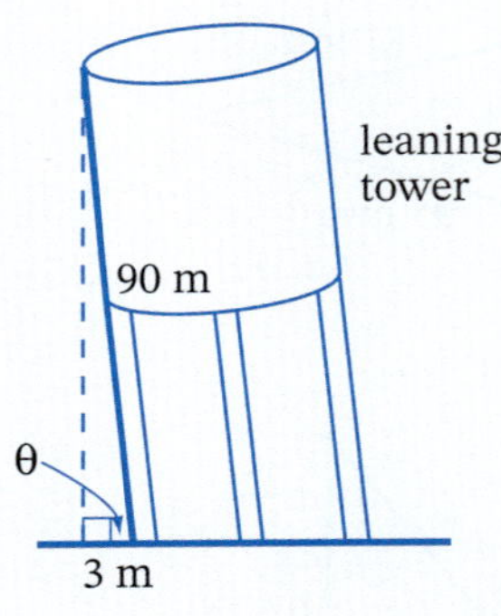

7 In $\triangle XYW$, $\angle X = 90°$, $XY = 8$ cm and $XW = 10$ cm. Find $\angle W$, correct to the nearest degree.

8 $\triangle HTM$ is right-angled at T, $HM = 45$ m and $MT = 35$ m. Find $\angle M$, correct to one decimal place.

9 In $\triangle FGH$, $\angle G = 90°$, $GH = 3.7$ m and $FH = 19.5$ m. Find the size of angle F, correct to the nearest minute.

10 In $\triangle CDE$, $\angle C = 90°$, $CD = 54$ mm and $CE = 42$ mm. Find $\angle E$, to the nearest minute.

□ Foundation ○ Standard ⬡ Complex

For questions 11 to 17, write answers correct to the nearest degree.

11 PS A stretch of freeway rises 55 m for every 300 m travelled along the road. Find the angle at which the road is inclined to the horizontal.

12 PS A ladder 20 m long is placed against a building. If the ladder reaches 16 m up the building, find the ladder's angle of inclination to the building.

13 PS An aircraft is descending in a straight line to an airport. At a height of 1270 m, it is 1500 m horizontally from the airport. Find its angle of descent to the horizontal, to the nearest degree. Select the correct answer **A**, **B**, **C** or **D**.

A 32° **B** 40° **C** 50° **D** 58°

14 PS A tree 8.5 m high casts a shadow 3 m long. What is the angle of the Sun from the ground?

15 PS At a resort, an artificial beach slopes down at a steady angle. After walking 8.5 m down the slope from the water's edge, the water has a depth of 1.6 m. At what angle is the beach inclined to the horizontal?

16 PS A pile of wheat is in the shape of a cone that has a diameter of 35 m and measures 27 m up the slope to the apex. Calculate the angle of repose of the wheat (the angle the sloping side makes with the horizontal base).

17 PS A ship is anchored in water 40 m deep by a 65 m anchor chain. Find the angle at which the chain is inclined to the sea floor.

18 PS The arms of this fold out ladder are 3.5 m long. When fully spread, the feet of the ladder are 1.8 m apart.

Calculate, correct to the nearest minute, the angle between the arms of the ladder when fully spread.

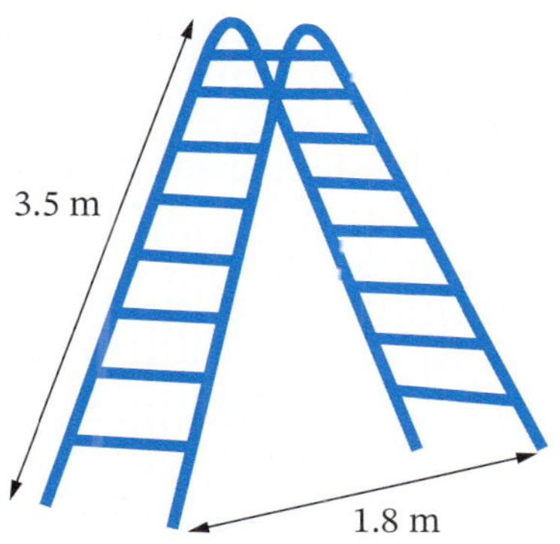

POWER PLUS ANSWERS ON P. 638

1 a Copy and complete each pair of trigonometric ratios, correct to 3 decimal places.

i $\sin 20° =$ ______, $\cos 70° =$ ______ ii $\sin 47° =$ ______, $\cos 43° =$ ______

iii $\sin 55° =$ ______, $\cos 35° =$ ______ iv $\sin 85° =$ ______, $\cos 5° =$ ______

b What do you notice about each pair of answers in part **a**?

c What do you notice about each pair of angles in part **a**?

d If $\cos 30° \approx 0.8660$ and $\sin \theta \approx 0.8660$, what is the value of θ?

e Copy and complete each equation.

i $\sin 75° = \cos$ ____ ii ____ $80° = \cos 10°$

iii $\cos$ ____ $= \sin 72°$ iv $\sin 30° =$ ____ $60°$

v $\cos 65° = \sin$ ____ vi $\sin$ ____ $= \cos 58°$

f Copy and complete this general rule: $\sin x = \cos$ (__________).

g Use a right-angled triangle with one angle x and sides a, b and c to prove that the above rule is true.

2 A plane is flying at an angle of 15° inclined to the horizontal.

a How far will the plane have to travel along its line of flight to increase its altitude (height) by 500 m?

b At what angle must the plane climb to achieve an increase in altitude of 500 m in half the distance needed at an angle of 15°?

3 If $\sin 30° = \frac{1}{2}$, find, as a surd, the value of:

a $\cos 30°$ b $\tan 30°$

4 Find angle θ, correct to the nearest second.

a

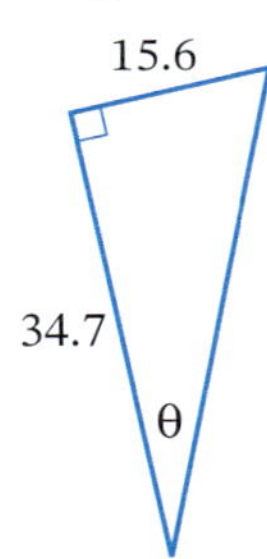

b

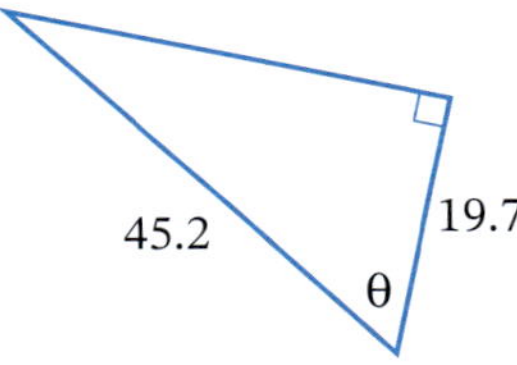

5 By drawing an appropriate triangle, prove that:

a $\tan 45° = 1$ b $\sin 45° = \frac{1}{\sqrt{2}}$ c $\cos 45° = \frac{1}{\sqrt{2}}$

6 Sketch a right-angled triangle with side lengths a, b and c, where c is the hypotenuse. Let θ be one of the acute angles. Prove that $(\sin \theta)^2 + (\cos \theta)^2 = 1$.

4 CHAPTER REVIEW

Language of maths

Quiz Language of maths 4

adjacent	alpha (α)	clinometer	cosine (cos)
degree (°)	denominator	horizontal	hypotenuse
inverse ($^{-1}$)	minute (′)	mnemonic	opposite
phi (φ)	Pythagoras' theorem	right-angled	second (″)
similar triangles	sine (sin)	tangent (tan)	theta (θ)
trigonometry	trigonometric ratio	unknown	vertical

1 When measuring angle size, what is a second and what is its symbol?

2 What word means 'next to'?

3 Which side of a right-angled triangle is fixed and does not depend on the position of an angle?

4 What are the first two letters of the Greek alphabet?

5 The word 'minute' has an alternative pronunciation and meaning. What is its alternative meaning?

6 What does 'inverse' mean and how is it used in trigonometry?

Topic summary

Worksheet Mind map: Trigonometry

Print (or copy) and complete this mind map of the topic, adding detail to its branches and using pictures, symbols and colour where needed. Ask your teacher to check your work.

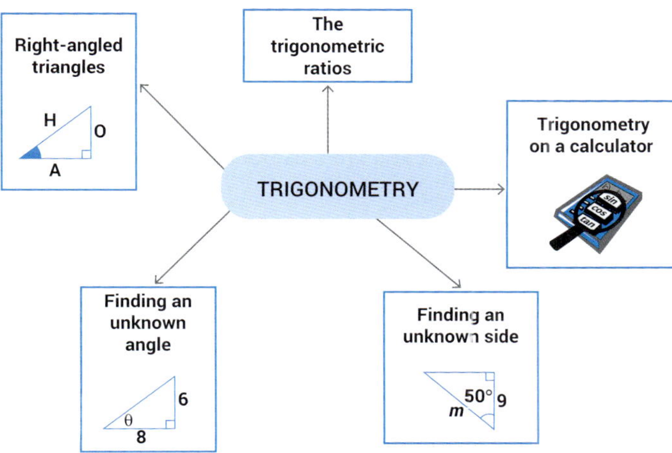

4 TEST YOURSELF

ANSWERS ON P. 638

1 For angle U, name the opposite and adjacent sides and the hypotenuse.

Quiz
Test yourself 4

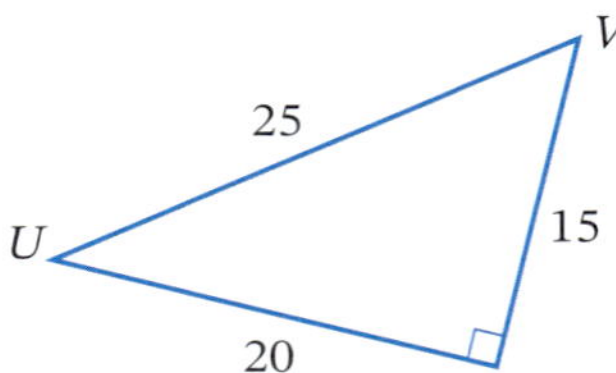

2 For angle V, name the opposite and adjacent sides and the hypotenuse.

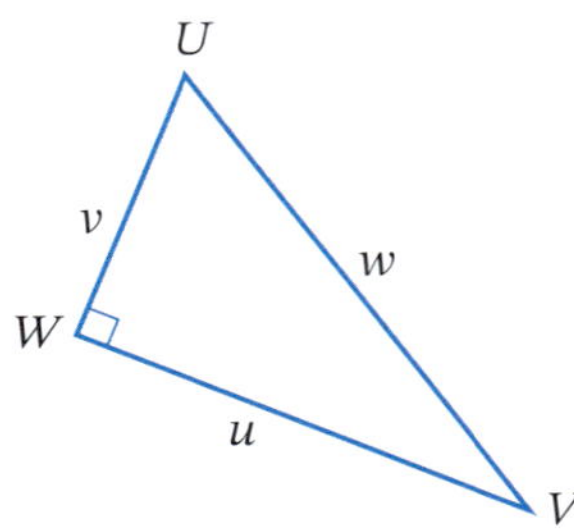

4.02

3 For this triangle, write as a fraction:

a $\sin Y$ **b** $\tan Y$ **c** $\sin X$ **d** $\cos X$

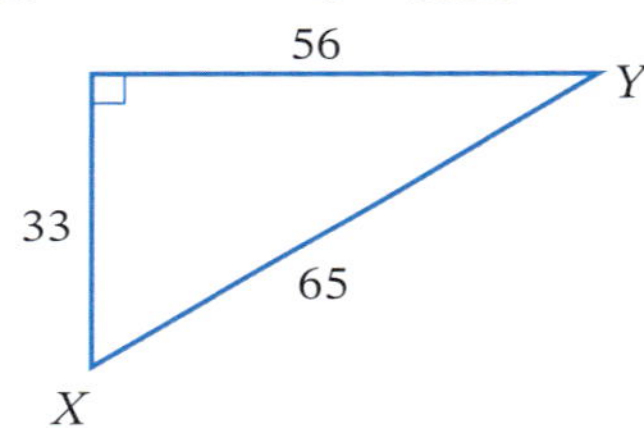

4.02

4 If $\sin \alpha = \frac{36}{85}$, find the values of $\cos \alpha$ and $\tan \alpha$ as fractions. (Draw a diagram.)

4.03

5 Construct a large right-angled triangle with an angle of 42°, then by measurement and calculation, find the value of each trigonometric ratio, correct to 3 decimal places.

a tan 42° **b** cos 42° **c** sin 42°

Compare your answers to the calculator's answers.

4.04

6 Round each angle to the nearest degree.

a 64° 27′ **b** 25° 43′ **c** 12° 8′ 50″

4.04

7 Round each angle to the nearest minute.

a 50° 19′ 26″ **b** 31° 55′ 55″ **c** 64° 18′ 30″

☐ Foundation ○ Standard ○ Complex

8 Evaluate each expression, correct to four decimal places. 

a $\cos 32°$ **b** $\sin 50° 9'$ **c** $\tan 8° 45'$

d $200 \tan 18°$ **e** $14 \sin 87° 40'$ **f** $\dfrac{13}{\cos 18° 27'}$

9 Convert each angle size to degrees and minutes, correct to the nearest minute. 

a $45.8°$ **b** $33.175°$ **c** $5.346°$

10 Find the value of each variable, correct to 2 decimal places. 4.05

a

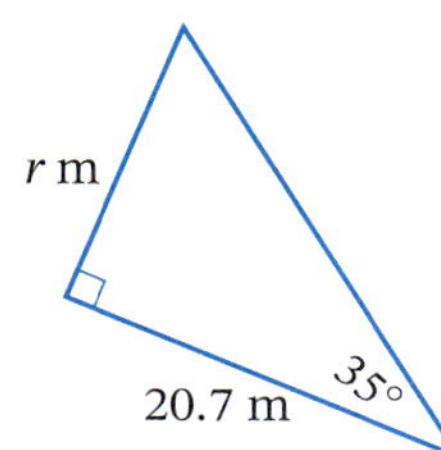

b

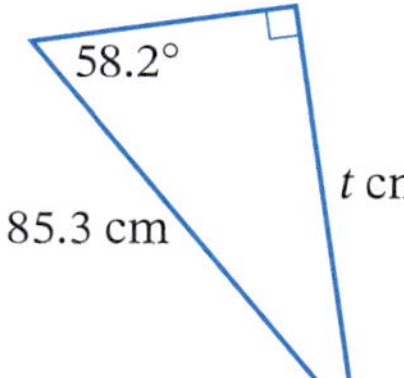

c

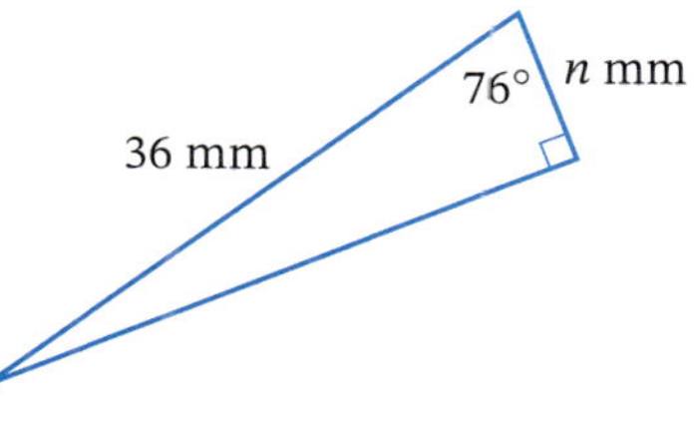

d

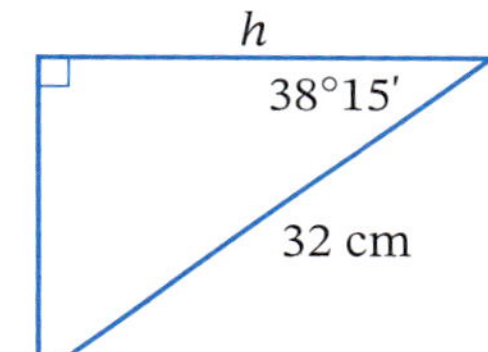

e

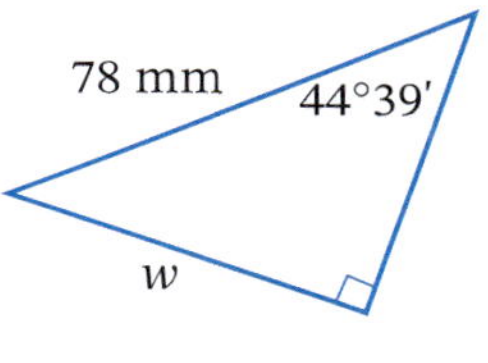

f

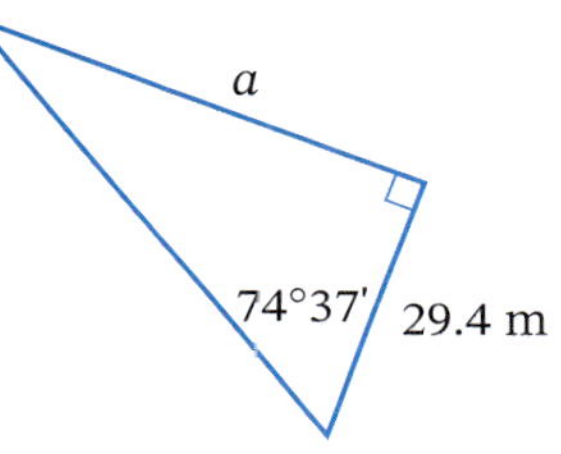

11 For each triangle, find the length of side AC, correct to one decimal place. 

a

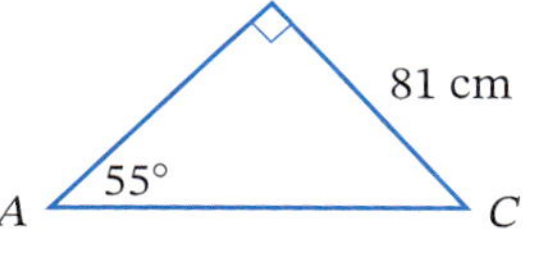

b

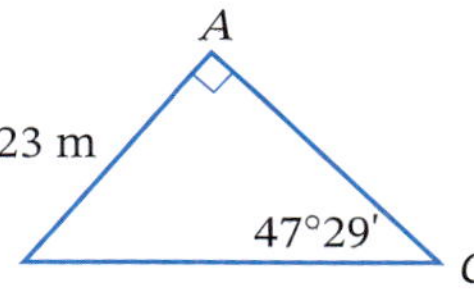

c

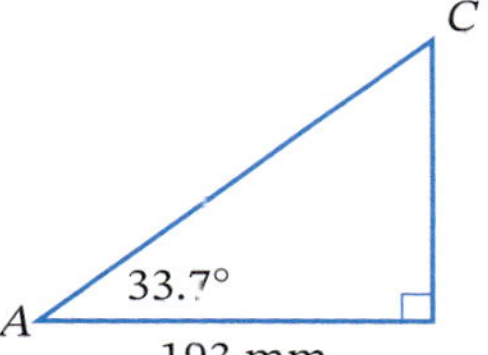

12 Find θ, correct to the nearest degree. 

a $\tan \theta = 2.57$ **b** $\cos \theta = \dfrac{4}{7}$ **c** $\sin \theta = \dfrac{1.5}{1.6}$ 

13 Find the size of angle α, correct to the nearest degree. 

a

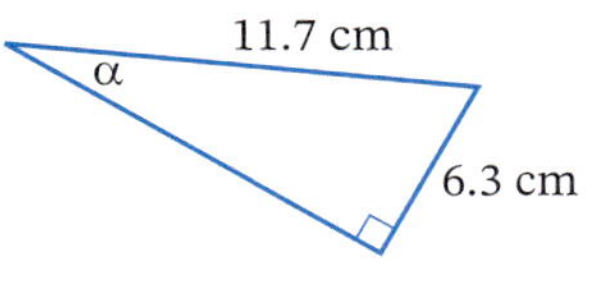

b

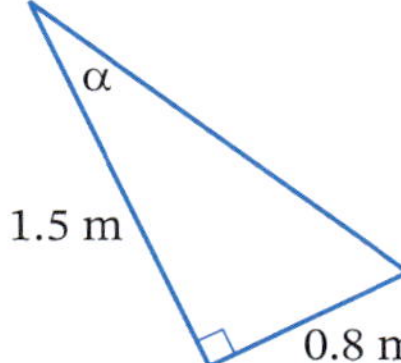

c

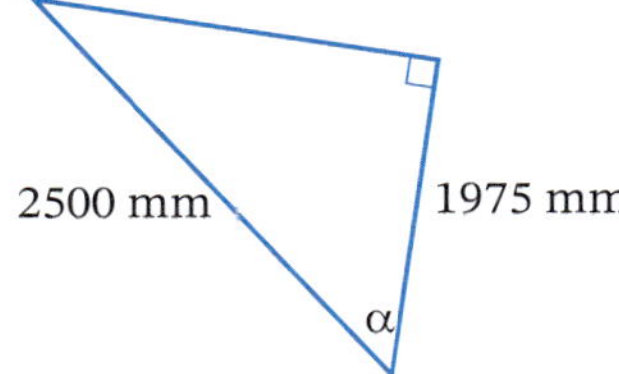

14 In $\triangle AEC$, $\angle C = 90°$, $CE = 3.9$ m and $AE = 4.2$ m. Find $\angle A$, correct to the nearest minute. 4.07

☐ Foundation ○ Standard ⬡ Complex

5

ALGEBRA, MEASUREMENT

Indices

The speed of light is about 300 000 000 m/s. In one year, light travels approximately 9 460 000 000 000 km and this distance is called a light year. Light from the stars travels for many years before it is seen on Earth. The Milky Way contains over 100 billion stars and the approximate diameter of our solar system is 10 000 000 000 km. On a smaller scale, the mass of a human cell is about 0.000 000 000 001 kg. Powers or indices provide a way to work easily with very large and very small numbers.

iStock.com/coffeekai

Chapter outline

		Proficiencies				
5.01	Multiplying and dividing terms with the same base	U	F		R	
5.02	Power of a power	U	F		R	
5.03	Powers of products and quotients	U	F		R	
5.04	The zero index	U	F		R	
5.05	Negative indices	U	F		R	C
5.06	Summary of the index laws	U	F		R	C
5.07	Significant figures*	U	F		R	C
5.08	Scientific notation	U	F		R	C
5.09	Scientific notation on a calculator	U	F	PS	R	C

*EXTENSION

U = Understanding
F = Fluency
PS = Problem solving
R = Reasoning
C = Communication

Wordbank

base A number that is raised to a power, meaning it is multiplied by itself repeatedly, for example, in 2^5, the base is 2.

index laws Rules for simplifying algebraic expressions involving powers of the same base, for example, $a^m \div a^n = a^{m-n}$.

index notation A way of writing repeated multiplication using indices (powers), in the form a^n, for example, $2 \times 2 \times 2 \times 2 \times 2$ in index notation is 2^5.

negative power A power that is a negative number, as in the term 3^{-2}.

power (or index or exponent) The number of times a base appears in a repeated multiplication, for example, in 2^5, the power is 5.

scientific notation A shorter way of writing very large or very small numbers using powers of 10. For example, 9 460 000 000 000 in scientific notation is 9.46×10^{12}.

significant figures Meaningful digits in a numeral that tell 'how many'. For example, 28 000 000 has 2 significant figures: 2 and 8.

Quiz
Wordbank 5

Videos (6):

SkillCheck Powers, indices and exponents
5.01, 5.02, 5.03 Simplifying with the index laws
5.04 Index laws
5.05 Negative indices
5.07 Significant figures
5.08 Scientific notation

Twig videos (3):

5.07 Rounding: Snails vs rockets
5.08 What does the Internet weigh?
5.09 Binary: the computer language

Quizzes (5):

- Wordbank 5
- SkillCheck 5
- Mental skills 5
- Language of maths 5
- Test yourself 5

Skillsheet (1):

SkillCheck Indices

Worksheets (8):

SkillCheck Powers review
5.05 Power calculations
5.07 Significant figures
5.08 Scientific notation 1 • Scientific notation 2
5.09 Scientific notation problems • Binary number system
Mind map: Indices

Puzzles (3):

5.02 Indices puzzle
5.06 Indices squaresaw
5.09 Scientific notation: accomplishing great things

Technology (1):

5.08 Scientific notation

Spreadsheet (1):

5.08 Scientific notation

To access resources above, visit **cengage.com.au/nelsonmindtap**

5 In this chapter you will:

- ✓ apply index laws to numerical expressions with integer indices.
- ✓ apply index laws to algebraic expressions with integer indices.
- ✓ interpret and use zero and negative integer indices.
- ✓ (EXTENSION) round numbers to significant figures.
- ✓ interpret, write and order numbers in scientific notation.
- ✓ interpret and use scientific notation on a calculator.
- ✓ solve problems involving scientific notation.

Quiz SkillCheck

Video Powers, indices and exponents

Worksheet Powers review

Skillsheet Indices

SkillCheck

ANSWERS ON P. 639

1 For each term:

i state the base. **ii** state the index. **iii** write the expression in words.

a 3^5 **b** 4^8 **c** h^5 **d** 7^y

2 Evaluate each product.

a 5×5 **b** $7 \times 7 \times 7$ **c** $10 \times 10 \times 10 \times 10$

d $2 \times 2 \times 2 \times 2 \times 2 \times 2$ **e** $\frac{2}{3} \times \frac{2}{3} \times \frac{2}{3}$ **f** 9^2

g 3^4 **h** 6^3 **i** 4^4

j $\left(\frac{1}{2}\right)^3$ **k** $\left(\frac{3}{4}\right)^2$ **l** 5^3

3 Express each repeated multiplication in index notation.

a $2 \times 2 \times 2 \times 2 \times 2$
b $3 \times 3 \times 3 \times 3 \times 7 \times 7 \times 7$
c $5 \times 5 \times 5 \times 5 \times 5 \times 5 \times 8 \times 8$
d $\frac{3}{7} \times \frac{3}{7} \times \frac{3}{7} \times \frac{3}{7}$
e $10 \times x \times x \times x \times x \times x \times x$
f $6 \times 6 \times 6 \times k \times k$
g $x \times y \times x \times y \times x$
h $a \times b \times b \times b \times a$
i $5 \times n \times 5 \times n \times n$
j $q \times p \times q \times p \times q \times q$

4 Write each term in expanded form.

a 9^3 b 7^2 c d^5 d k^2 e $\left(\frac{4}{5}\right)^3$

5 Evaluate each expression.

a $4^2 \times 4^3$ b $10^6 \div 10^2$ c $(3^3)^2$ d 6^0
e 9^1 f $5^5 \times 5$ g $2^4 \div 2$ h $(-8)^2$

6 For each equation, find the missing power.

a $8 = 2^{\square}$ b $81 = 3^{\square}$ c $216 = 6^{\square}$ d $144 = 12^{\square}$
e $4096 = 4^{\square}$ f $2401 = 7^{\square}$ g $64 = 2^{\square}$ h $625 = 5^{\square}$

INVESTIGATION

Multiplying and dividing terms with powers

1 Write each expression in expanded form, then evaluate it.

a i $2^2 \times 2^3$ ii 2^5
b i $3^4 \times 3^3$ ii 3^7
c i $4^3 \times 4^3$ ii 4^6
d i $5^5 \times 5^3$ ii 5^8

2 What do you notice about each pair of answers in question **1**?

3 Is it true that $2^4 \times 2^6 = 2^{10}$? Give a reason for your answer.

4 Determine whether each equation is true (T) or false (F). Justify your answer.

a $2^5 \times 2^5 = 2^{10}$
b $6^3 \times 6^7 = 6^{21}$
c $4^3 \times 4^9 = 4^{27}$
d $3^5 \times 3^{10} = 3^{15}$

5 Write in words and as a formula the rule for multiplying a^m and a^n, two terms with the same base.

6 Use the rule to copy and complete each equation.

a $5^4 \times 5^2 = 5...$ b $4^5 \times 4^3 = 4...$ c $10^5 \times 10^7 = ...$
d $9^3 \times 9^2 = ...$ e $n^3 \times n^8 = ...$ f $p^3 \times p^7 = ...$

7 Use a calculator to evaluate each expression.

a i $3^6 \div 3^3$ ii 3^3
b i $2^8 \div 2^6$ ii 2^2
c i $5^8 \div 5^3$ ii 5^5
d i $10^8 \div 10^4$ ii 10^4

8 What do you notice about each pair of answers in question **7**?

9 Is it true that $4^8 \div 4^6 = 4^2$? Give a reason for your answer.

10 Use a calculator to determine whether each equation is true (T) or false (F).

a $3^{10} \div 3^6 = 3^4$
b $4^8 \div 4^2 = 4^4$
c $2^{12} \div 2^3 = 2^4$
d $6^{10} \div 6^5 = 6^5$

11 Write in words and as a formula the rule for dividing a^m and a^n, two terms with the same base.

12 Use the rule to copy and complete each equation.

a $2^6 \div 2^3 = 2...$ b $10^8 \div 10^6 = 10...$ c $3^7 \div 3^2 =$

d $4^{11} \div 4^6 = ...$ e $x^8 \div x^5 = ...$ f $g^{12} \div g^{10} = ...$

5.01 Multiplying and dividing terms with the same base

Video Simplifying with the index laws

Consider $5^4 \times 5^3 = (5 \times 5 \times 5 \times 5) \times (5 \times 5 \times 5)$

$= 5 \times 5 \times 5 \times 5 \times 5 \times 5 \times 5$

$= 5^7$

$\therefore 5^4 \times 5^3 = 5^{4+3}$

$= 5^7$

ⓘ Multiplying terms with the same base

When multiplying terms with the same base, add the powers:

$$a^m \times a^n = a^{m+n}$$

The rule above is called an **index law**. **Index** is another name for **power.** The plural of index is **indices** (pronounced 'in-de-sees').

Proof:

$$a^m \times a^n = \underbrace{a \times a \times \cdots \times a}_{m \text{ factors}} \times \underbrace{a \times a \times \cdots \times a}_{n \text{ factors}}$$

$$= \underbrace{a \times a \times \cdots \times a}_{(m+n) \text{ factors}}$$

$$= a^{m+n}$$

Example 1

Simplify each expression, writing the answer in **index notation**.

a $8^4 \times 8^5$ b 10×10^3 c $d^3 \times d^5$

d $4m^2 \times 3m^6$ e $3r^2t \times 6rt^3$

SOLUTION

a $8^4 \times 8^5 = 8^{4+5}$
$= 8^9$

b $10 \times 10^3 = 10^1 \times 10^3$
$= 10^{1+3}$
$= 10^4$

c $d^3 \times d^5 = d^{3+5}$
$= d^8$

5.01

d $4m^2 \times 3m^6 = (4 \times 3) \times (m^2 \times m^6)$
$= 12m^{2+6}$
$= 12m^8$

Multiply the numbers and add the indices.

e $3r^2t \times 6rt^3 = 3r^2t^1 \times 6r^1t^3$
$= (3 \times 6) \times (r^2 \times r^1) \times (t^1 \times t^3)$
$= 18r^{2+1}t^{1+3}$
$= 18r^3t^4$

$3r^2t = 3r^2t^1$ and $6rt^3 = 6r^1t^3$

Multiply the numbers and add the indices.

Consider $5^6 \div 5^4 = \frac{5^6}{5^4}$

$$= \frac{5\times5\times5\times5\times5\times5}{5\times5\times5\times5}$$
$$= 5\times5$$
$$= 5^2$$
$$\therefore 5^6 \div 5^4 = 5^{6-4}$$
$$= 5^2$$

ⓘ Dividing terms with the same base

When dividing terms with the same base, subtract the powers:

$$a^m \div a^n = \frac{a^m}{a^n} = a^{m-n}$$

This is another **index law.**

Proof:

$$a^m \div a^n = \frac{a^m}{a^n}$$
$$= \frac{a\times a\times a\times a\times a\times\cdots\times a}{a\times a\times a\times\cdots\times a} \quad \left(\frac{m \text{ factors}}{n \text{ factors}}\right)$$
$$= a\times a\times\ldots\times a \quad (m-n) \text{ factors}$$
$$= a^{m-n}$$

Example 2

Simplify each expression, writing the answer in index notation.

a $8^5 \div 8^3$ **b** $\frac{10^8}{10}$ **c** $d^{20} \div d^4$ **d** $20w^{10} \div 5w^2$ **e** $\frac{8x^3y^7}{24x^2y}$

SOLUTION

a $8^5 \div 8^3 = 8^{5-3}$
$= 8^2$

b $\frac{10^8}{10} = \frac{10^8}{10^1}$
$= 10^{8-1}$
$= 10^7$

c $d^{20} \div d^4 = d^{20-4}$
$= d^{16}$

d $20w^{10} \div 5w^2 = \frac{\cancel{20}^{4} w^{10}}{\cancel{5}^{1} w^2}$

$= 4w^{10-2}$

$= 4w^8$

Divide the numbers and subtract the indices.

e $\frac{8x^3y^7}{24x^2y^1} = \frac{{}^{1}\cancel{8}x^{3-2}y^{7-1}}{{}_{3}\cancel{24}}$

$= \frac{xy^6}{3}$

Divide the numbers and subtract the indices.

Example 3

Simplify each expression, writing the answer in index notation.

a $5^3 \times 5^7 \times 5^4$ **b** $2^7 \times 2^3 \div 2^5$ **c** $m^7 \times m^6 \div m^4$

d $3e^5 \times 2e^4 \times 5e^2$ **e** $a^{12} \div a^4 \div a^5$ **d** $12y^4 \times 4y^3 \div 20y^5$

SOLUTION

a $5^3 \times 5^7 \times 5^4 = 5^{3+7+4}$

$= 5^{14}$

b $2^7 \times 2^3 \div 2^5 = 2^{10} \div 2^5$

$= 2^5$

c $m^7 \times m^6 \div m^4 = m^{13} \div m^4$

$= m^9$

d $3e^5 \times 2e^4 \times 5e^2 = (3 \times 2 \times 5) \times (e^5 \times e^4 \times e^2)$

$= 30e^{11}$

e $a^{12} \div a^4 \div a^5 = a^{12-4-5}$

$= a^3$

f $12y^4 \times 4y^3 \div 20y^5 = \frac{12y^4 \times 4y^3}{20y^5} = \frac{{}^{12}\cancel{48}y^{7-5}}{{}_{5}\cancel{20}}$

$= \frac{12y^2}{5}$

Divide the numbers and subtract indices.

EXERCISE 5.01 ANSWERS ON P. 639

Multiplying and dividing terms with the same base

1 Which expression is equal to $5^{12} \times 5^3$? Select the correct answer **A**, **B**, **C** or **D**.

A 5^9 **B** 5^{15} **C** 25^{15} **D** 25^{36}

2 Simplify each expression, writing the answer in index notation.

(R) **a** $10^3 \times 10^2$ **b** 2×2^4 **c** $3^2 \times 3^5$

d $7^4 \times 7$ **e** $8 \times 8^3 \times 8^4$ **f** $5^4 \times 5 \times 5^4$

g $6 \times 6^2 \times 6^3 \times 6^4$ **h** $4^4 \times 4^4 \times 4^4$ **i** $3^4 \times 3^0 \times 3^7$

j $x \times x^4$ **k** $g^4 \times g^4$ **l** $w^7 \times w$

m $b^3 \times b^{10}$ **n** $p^{10} \times p^{10}$ **o** $r \times r$

p $y^5 \times y^7$ **q** $m^{15} \times m^4$ **r** $n^8 \times n^2$

Foundation Standard Complex

3 Which expression is equal to $10^4 \times 10$? Select **A**, **B**, **C** or **D**.

A 100^5 **B** 100^4 **C** 10^4 **D** 10^5

EXAMPLE 2

4 Which expression is equal to $5^{12} \div 5^3$?

A 5^4 **B** 5^9 **C** 1^4 **D** 1^9

5 Simplify each expression, writing the answer in index notation.

(R) **a** $10^7 \div 10^5$ **b** $8^5 \div 8$ **c** $20^{15} \div 20^5$

d $\frac{5^8}{5^2}$ **e** $\frac{9^{12}}{9^3}$ **f** $\frac{2^{27}}{2^3}$

g $7^4 \div 7^3$ **h** $\frac{2^{20}}{2}$ **i** $11^4 \div 11^4$

j $p^{15} \div p^{10}$ **k** $n^7 \div n$ **l** $w^{24} \div w^6$

m $\frac{h^{20}}{h^4}$ **n** $\frac{y^8}{y^2}$ **o** $\frac{a^{12}}{a^4}$

p $b^{16} \div b^{15}$ **q** $\frac{w^{25}}{w}$ **r** $m^{16} \div m^{16}$

6 Which expression is equal to $10^4 \div 10$?

A 10^4 **B** 1^4 **C** 1^3 **D** 10^3

EXAMPLE 3

7 Which expression is equal to $6^7 \times 6^2 \times 6^4$?

A 216^{13} **B** 6^{56} **C** 6^{13} **D** 216^{56}

8 Which expression is equal to $2^7 \div 2^2 \times 2^4$?

A 2^9 **B** 2^{13} **C** 8^{13} **D** 4^9

9 Simplify each expression.

(R) **a** $3p^2 \times 2p^5$ **b** $4y^{10} \times 3y^2$ **c** $6m \times 3m^8$

d $h^3 \times 5h^8$ **e** $3q \times 8q^8$ **f** $2a^2 \times 5a^5$

g $5n^8t \times 6n^8t^4$ **h** $2ab^3 \times 15ab$ **i** $3e^4g^3 \times e^6g^2$

j $8p^4m^5 \times 4p^3m^5$ **k** $16qr^8 \times 3q^7$ **l** $9u^3v \times 6uv^2w^8$

10 Simplify each expression.

(R) **a** $10y^{15} \div 5y^3$ **b** $20w^9 \div 4w^3$ **c** $24r^8 \div 3r$

d $\frac{30x^4}{x^3}$ **e** $\frac{10m^{10}}{2m}$ **f** $\frac{4g^{12}}{8g^6}$

g $14d^4h^{10} \div 7hd^2$ **h** $15x^6y^8 \div 15xy^4$ **i** $6e^{25}d^{40} \div 18e^5d^4$

j $\frac{12q^5t^4}{16q^4t^3}$ **k** $\frac{45a^{10}b^8}{5a^5}$ **l** $\frac{36pq^3r^5}{24qr}$

11 Which expression is equal to $5d^7 \times 6d^2 \div 15d^4$?

A $2d^{13}$ **B** $2d^5$ **C** $15d^5$ **D** $5d^5$

Foundation Standard Complex

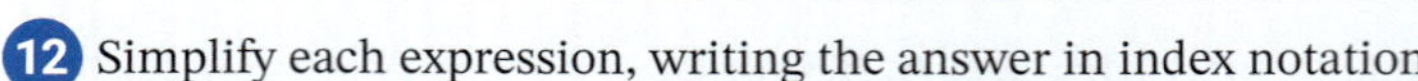

12 Simplify each expression, writing the answer in index notation.

a	$3^3 \times 3^5 \times 3^2$	**b**	$10^{15} \div 10^7 \div 10^2$	**c**	$7^6 \times 7^8 \div 7^{10}$
d	$8^9 \div 8^8 \times 8^4$	**e**	$a^5 \times a^5 \times a^4$	**f**	$m^3 \times m^8 \div m^{10}$
g	$w^9 \div w^6 \times w^{10}$	**h**	$x \times x^7 \div x^4$	**i**	$p^{12} \div p \div p^6$
j	$2c^5 \times 5c^8 \times 3c$	**k**	$6y^4 \times 4y^8 \div 8y^4$	**l**	$5d^2 \times 12d^4 \times 2d^5$
m	$8q^7 \times 3q^3 \div 12q^8$	**n**	$24g^{11} \div 8g^8 \div g$	**o**	$6h^{10} \div h^8 \times 5h^{10}$
p	$3v^5 \times 6v^4 \div 12v^6$	**q**	$8b^6 \times 2b \div 10b^5$	**r**	$24k^5 \div 6k^2 \div 2k^2$

13 Which expression is equal to $4p^3y^5 \times 5p^7y^3 \div 8p^6y^6$?

A $160p^4y^2$ **B** $\frac{20p^4y}{8}$ **C** $\frac{5p^4y}{2}$ **D** $\frac{5p^4y^2}{2}$

14 Simplify each expression, writing the answer in index notation.

a	$a^{-3} \times a^3$	**b**	$y \times y^{-2}$	**c**	$n^{\frac{1}{2}} \times n^{\frac{1}{2}}$
d	$h^{-1} \times h^{-4}$	**e**	$b^4 \div b^4$	**f**	$\frac{r^3}{r^5}$

INVESTIGATION

Powers of powers

1 Write each expression in expanded form, then evaluate it.

a	i	$(2^3)^2$	ii	2^6
b	i	$(3^4)^3$	ii	3^{12}
c	i	$(5^2)^3$	ii	5^6
d	i	$(2^5)^4$	ii	2^{20}

2 What do you notice about each pair of answers in question **1**?

3 Is it true that: $(2^7)^3 = 2^{21}$? Give a reason for your answer.

4 Determine whether each equation is true (T) or false (F). Justify your answer.

a	$(3^5)^3 = 3^{15}$	b	$(2^3)^2 = 2^5$	c	$(2^{10})^4 = 2^{14}$
d	$(4^2)^5 = 4^{10}$	e	$(3^3)^6 = 3^{18}$	f	$(5^2)^4 = 5^6$

5 Write in words and as a formula the rule for raising a^m to a power of n, that is, $(a^m)^n$.

6 Use the rule to copy and complete each equation.

a	$(3^7)^2 = 3...$	b	$(5^2)^6 = 5...$	c	$(4^5)^2 = 4...$
d	$(a^3)^4 = a...$	e	$(8^3)^7 = ...$	f	$(k^4)^6 = ...$

☐ Foundation ○ Standard ⬡ Complex

9780170465618

Power of a power

5.02

Video Simplifying with the index laws

Puzzle Indices puzzle

Consider $(5^3)^4 = 5^3 \times 5^3 \times 5^3 \times 5^3$

$= (5 \times 5 \times 5) \times (5 \times 5 \times 5) \times (5 \times 5 \times 5) \times (5 \times 5 \times 5)$

$= 5 \times 5 \times 5 \times 5 \times 5 \times 5 \times 5 \times 5 \times 5 \times 5 \times 5 \times 5$

$= 5^{12}$

$\therefore (5^3)^4 = 5^{3\times4}$

$= 5^{12}$

ⓘ Power of a power

When raising a term with a power to another power, multiply the powers:

$$(a^m)^n = a^{m\times n}$$

Proof:

$$(a^m)^n = \underbrace{a^m \times a^m \times \cdots \times a^m}_{n \text{ factors}}$$

$$= \underbrace{\underbrace{a \times a \times \cdots \times a}_{m \text{ factors}} \times \underbrace{a \times a \times \cdots \times a}_{m \text{ factors}} \times \ldots \times \underbrace{a \times a \times \cdots \times a}_{m \text{ factors}}}_{n \text{ lots of } m \text{ factors}}$$

$$= a^{m\times n}$$

Example 4

Simplify each expression, writing the answer in index notation.

a $(8^5)^2$ **b** $(d^3)^5$ **c** $(2g)^4$

d $(5v^4)^3$ **e** $(-n)^6$ **f** $(-3t^4)^3$

SOLUTION

a $(8^5)^2 = 8^{5\times2}$
$= 8^{10}$

b $(d^3)^5 = d^{3\times5}$
$= d^{15}$

c $(2g)^4 = 2^4 \times g^4$
$= 16g^4$

d $(5v^4)^3 = 5^3 \times (v^4)^3$
$= 125 \times v^{4\times3}$
$= 125v^{12}$

e $(-n)^6 = (-1)^6 \times n^6$
$= 1 \times n^6$
$= n^6$

f $(-3t^4)^3 = (-3)^3 \times (t^4)^3$
$= -27 \times t^{4\times3}$
$= -27t^{12}$

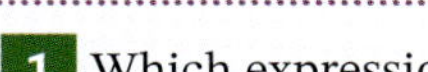

EXERCISE 5.02 ANSWERS ON P. 639

Power of a power

1 Which expression is equal to $(10^3)^3$? Select the correct answer **A**, **B**, **C** or **D**.

A 30^3 **B** 1000^9 **C** 10^9 **D** 10^6

□ Foundation ○ Standard ○ Complex

2 Simplify each expression, writing the answer in index notation.

R
a $(4^3)^2$ **b** $(5^2)^8$ **c** $(3^3)^4$ **d** $(2^7)^4$
e $(2^1)^2$ **f** $(9)^3$ **g** $(10^0)^2$ **h** $(6^4)^5$
i $(5^3)^5$ **j** $(e^2)^4$ **k** $(t^5)^5$ **l** $(y^3)^7$
m $(c^1)^5$ **n** $(m^7)^5$ **o** $(y^4)^4$ **p** $(h^0)^6$
q $(q^6)^3$ **r** $(w^4)^1$ **s** $(2x)^{10}$ **t** $(5n^3)^8$
u $(4d^3)^3$ **v** $(-k^5)^9$ **w** $(-d^3)^4$ **x** $(2a^8)^8$

3 Which expression is equal to $(-3)^5$? Select **A**, **B**, **C** or **D**.

A -3^6 **B** -3^5 **C** 3^5 **D** -15

4 Simplify each expression.

R
a $(2d^3)^4$ **b** $(5m^3)^2$ **c** $(4y^5)^2$ **d** $(3x^2)^4$
e $(5u^6)^5$ **f** $(2w^5)^3$ **g** $(10d^5)^4$ **h** $(3e)^3$
i $(2b^4)^1$ **j** $(6d^6)^2$ **k** $(3f^4)^5$ **l** $(2c^3)^{10}$
m $(-2r)^4$ **n** $(-5t)^3$ **o** $(-3m^3)^2$ **p** $(-y^3)^{12}$
q $(-x)^3$ **r** $(-m^3)^{10}$ **s** $(-4w^5)^4$ **t** $(-3f)^5$
u $(-3p^2)^3$ **v** $(-3h^5)^4$ **w** $(-10k)^2$ **x** $(-8y^3)^1$

5 Which expression is equal to $(-4m^3)^3$? Select **A**, **B**, **C** or **D**.

A $-64m^9$ **B** $-64m^6$ **C** $64m^6$ **D** $64m^9$

6 Simplify each expression.

R
a $\left(d^{-1}\right)^3$ **b** $\left(x^2\right)^{-2}$ **c** $\left(w^{\frac{1}{2}}\right)^{\frac{1}{2}}$
d $\left(z^{-2}\right)^{-1}$ **e** $\left(u^{\frac{1}{2}}\right)^2$ **f** $\left(p^{-1}\right)^{-1}$

5.03 Powers of products and quotients

Video
Simplifying with the index laws

Consider $(2 \times 5)^3 = (2 \times 5) \times (2 \times 5) \times (2 \times 5)$

$= 2 \times 2 \times 2 \times 5 \times 5 \times 5$

$= 2^3 \times 5^3$

$\therefore (2 \times 5)^3 = 2^3 \times 5^3$

ⓘ Powers of products

When raising a product of terms to a power, raise each term to that power:

$$(ab)^n = a^n b^n$$

☐ Foundation ○ Standard ⬡ Complex

Proof:

$$(ab)^n = \underbrace{ab \times ab \times \cdots \times ab}_{n \text{ factors}}$$
$$= \underbrace{a \times a \times \cdots \times a}_{n \text{ factors}} \times \underbrace{b \times b \times \cdots \times b}_{n \text{ factors}}$$
$$= a^n b^n$$

Example 5

Simplify each expression.

a $(-2gh^2)^5$ **b** $(p^3q^4)^2$ **c** $(a^5x^2y)^3$

SOLUTION

a $(-2gh^2)^5 = (-2)^5 \times g^5 \times (h^2)^5$
$= -32 \times g^5 \times h^{2\times 5}$
$= -32g^5h^{10}$

b $(p^3q^4)^2 = (p^3)^2 \times (q^4)^2$
$= p^{3\times 2} \times q^{4\times 2}$
$= p^6q^8$

c $(a^5x^2y)^3 = (a^5)^3 \times (x^2)^3 \times y^3$
$= a^{5\times 3} \times x^{2\times 3} \times y^3$
$= a^{15}x^6y^3$

Consider $\left(\frac{5}{8}\right)^6 = \frac{5}{8} \times \frac{5}{8} \times \frac{5}{8} \times \frac{5}{8} \times \frac{5}{8} \times \frac{5}{8}$

$$= \frac{5\times 5\times 5\times 5\times 5\times 5}{8\times 8\times 8\times 8\times 8\times 8}$$
$$= \frac{5^6}{8^6}$$
$$\therefore \left(\frac{5}{8}\right)^6 = \frac{5^6}{8^6}$$

ⓘ Powers of quotients

When raising a quotient of terms to a power, raise each term to that power:

$$\left(\frac{a}{b}\right)^n = \frac{a^n}{b^n}$$

Proof:

$$\left(\frac{a}{b}\right)^n = \underbrace{\frac{a}{b} \times \frac{a}{b} \times \cdots \times \frac{a}{b}}_{n \text{ factors}}$$
$$= \frac{a \times a \times \cdots \times a \ (n \text{ factors})}{b \times b \times \cdots \times b \ (n \text{ factors})}$$
$$= \frac{a^n}{b^n}$$

Example 6

Simplify each expression.

a $\left(\frac{7c}{d}\right)^2$ **b** $\left(\frac{4k^2}{5}\right)^3$ **c** $\left(\frac{6m^5p^3}{m^2p}\right)^4$ **d** $\frac{2x^4y^8}{(xy^2)^3}$

SOLUTION

a $\left(\frac{7c}{d}\right)^2 = \frac{(7c)^2}{d^2}$

$= \frac{7^2c^2}{d^2}$

$= \frac{49c^2}{d^2}$

b $\left(\frac{4k^2}{5}\right)^3 = \frac{(4k^2)^3}{5^3}$

$= \frac{4^3(k^2)^3}{125}$

$= \frac{64k^6}{125}$

c $\left(\frac{6m^5p^3}{m^2p}\right)^4 = \frac{(6m^5p^3)^4}{(m^2p)^4}$

$= \frac{6^4m^{5\times4}p^{3\times4}}{m^{2\times4}p^4}$

$= \frac{1296m^{20}p^{12}}{m^8p^4}$

$= 1296m^{20-8}p^{12-4}$

$= 1296m^{12}p^8$

OR $\left(\frac{6m^5p^3}{m^2p}\right)^4 = (6m^{5-2}p^{3-1})^4$

$= (6m^3p^2)^4$

$= 6^4m^{3\times4}p^{2\times4}$

$= 1296m^{12}p^8$

d $\frac{2x^4y^8}{(xy^2)^3} = \frac{2x^4y^8}{x^3(y^2)^3}$

$= \frac{2x^4y^8}{x^3y^6}$

$= 2x^{4-3}y^{8-6}$

$= 2xy^2$

EXERCISE 5.03 ANSWERS ON P. 639

Powers of products and quotients

U F R

EXAMPLE 5

1 Which expression is equal to $(-4\times5)^2$? Select the correct answer **A**, **B**, **C** or **D**.

A 16×25 **B** -16×25 **C** -8×10 **D** 8×10

2 Simplify each expression.

R **a** $(ab)^3$ **b** $(x^2y)^5$ **c** $(l^3m^5)^6$ **d** $(6dp^2)^4$

e $(-8k^4y^5)^2$ **f** $(3m^2n)^5$ **g** $(ek^3)^3$ **h** $(-w^3x^4)^7$

i $(-8d^3y^5)^2$ **j** $(4b^2c^3)^4$ **k** $(-3a^3d)^3$ **l** $(2p^2q^3)^4$

Foundation Standard Complex

3 Express each of the following in simplest form.

(R) **a** $(x^2yw^3)^2$ **b** $(h^2y^3m^5)^4$ **c** $(3d^7k^2p^3)^2$

d $(5a^3q^3r^4)^3$ **e** $(2d^4ew^3)^5$ **f** $(km^3p^2x^6)^4$

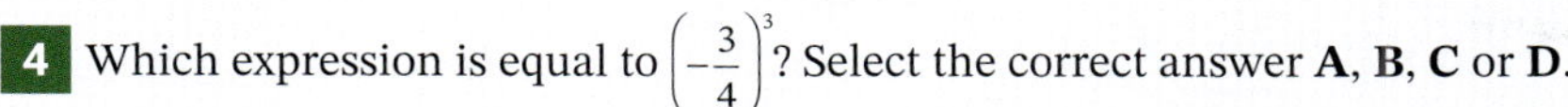

4 Which expression is equal to $\left(-\frac{3}{4}\right)^3$? Select the correct answer **A**, **B**, **C** or **D**.

EXAMPLE 6

A $-\frac{9}{12}$ **B** $\frac{9}{12}$ **C** $\frac{27}{64}$ **D** $-\frac{27}{64}$

5 Simplify each expression.

(R) **a** $\left(\frac{6}{7}\right)^2$ **b** $\left(\frac{m}{2}\right)^3$ **c** $\left(\frac{5}{x}\right)^4$ **d** $\left(\frac{2n^5}{p}\right)^8$

e $\left(\frac{w^2}{t^3}\right)^5$ **f** $\left(\frac{m^2}{4n}\right)^4$ **g** $\left(-\frac{2}{3}\right)^4$ **h** $\left(-\frac{5h}{6}\right)^3$

i $\left(\frac{7k^4}{10}\right)^2$ **j** $\left(\frac{3r}{t^2}\right)^2$ **k** $\left(\frac{a^2b}{d^5}\right)^4$ **l** $\left(-\frac{2}{3c^2}\right)^5$

6 Simplify each expression.

(R) **a** $(2x^{10}y^{15})^3 \times 5x^2y^3$ **b** $(2x^{10}y^{15} \times 5x^2y^3)^3$ **c** $18q^5r^8 \div (3qr^2)^2$

d $(18q^5r^8 \div 3qr^2)^2$ **e** $\left(\frac{3a^5x^6}{ax}\right)^3$ **f** $\frac{3a^5x^6}{(ax)^4}$

g $(4p^3h^{10})^2 \times 2p^2h^9$ **h** $(4p^3h^{10})^2 \div 2p^2h^9$ **i** $(4p^3h^{10} \times 2p^2h^9)^2$

j $\left(\frac{2a^4d^5}{a^3d^2}\right)^3$ **k** $\left(\frac{7e^4y^2}{e^2y}\right)^2$ **l** $\frac{5x^7y^{10}}{(x^2y^3)^3}$

m $\frac{6h^9k^{12}}{(2h^2k^3)^3}$ **n** $(3m^5w^3)^4 \div 6m^9w^5$ **o** $\left(\frac{3c^5c^2}{c^2d}\right)^4$

INVESTIGATION

The power of 0

What is the value of a number raised to a power of 0, for example, 2^0?

Copy and complete each table of decreasing powers. Notice the pattern in your answers.

a

Power of 2	Number
2^5	32
2^4	16
2^3	
2^2	
2^1	
2^0	

b

Power of 3	Number
3^5	243
3^4	
3^3	
3^2	
3^1	
3^0	

9780170465618

2 Simplify each expression in index notation.

a $3^4 \times 3^0$ b $5^2 \times 5^0$ c $2^0 \times 2^7$

d $7^0 \times 7^3$ e $4^5 \times 4^0$ f $5^0 \times 5^7$

g $2^5 \div 2^0$ h $3^5 \div 3^0$ i $4^2 \div 4^0$

j $9^3 \div 9^0$ k $5^6 \div 5^0$ l $8^4 \div 8^0$

3 Any number will remain unchanged when multiplied by what?

4 Any number will remain unchanged when divided by what?

5 What is the answer when any number is raised to the power of 0, that is, a^0? Justify your answer.

5.04 The zero index

Video
Index laws

Consider $5^3 \div 5^3 = \frac{5^3}{5^3}$

$= 1$ Any number divided by itself equals 1.

But also $5^3 \div 5^3 = 5^{3-3}$

$= 5^0$

$\therefore 5^0 = 1.$

Power of 0

Any number raised to the power of 0 is equal to 1.

$$a^0 = 1$$

Proof:

$a^m \div a^m = 1$ Any number divided by itself equals 1.

But also $a^m \div a^m = a^{m-m}$

$= a^0$

$\therefore a^0 = 1.$

Example 7

Simplify each expression.

a 11^0 **b** $(-8)^0$ **c** g^0

d $(3r)^0$ **e** $3r^0$ **f** -8^0

SOLUTION

a $11^0 = 1$

b $(-8)^0 = 1$

c $g^0 = 1$

d $(3r)^0 = 1$

e $3r^0 = 3 \times r^0$
$= 3 \times 1$
$= 3$

f $-8^0 = -1 \times 8^0$
$= -1 \times 1$
$= -1$

EXERCISE 5.04 ANSWERS ON P. 639

5.04

The zero index

U F R

EXAMPLE 7

1 Simplify each each expression.

R **a** 2^0 **b** $(-2)^0$ **c** -2^0 **d** $(-m)^0$

e $-m^0$ **f** $(4a)^0$ **g** $\left(\frac{2}{3}\right)^0$ **h** $7x^0$

i -1000^0 **j** $(p+3)^0$ **k** $\left(\frac{p}{3}\right)^0$ **l** $-2b^0$

m $(9k)^0$ **n** $(x^2y)^0$ **o** $(xyw)^0$ **p** $(-ab)^0$

q $(6r)^0$ **r** $-(6r)^0$ **s** $-6r^0$ **t** $6(-r)^0$

u $(cd)^0$ **v** $-(7x^2)^0$ **w** $-3(a^2b^3)^0$ **x** $(-5v^5w^4)^0$

2 Simplify each expression.

R **a** $7^0 + 2^0$ **b** $12^0 - 5^0$ **c** $2m^0 + (2m)^0$

d $5w^0 - (5w)^0$ **e** $(3a)^0 + 6a^0$ **f** $(8a)^0 - 6x^0$

g $(5y)^0 - 4$ **h** $(5y)^0 - 4^0$ **i** $3^0 \times 5^0$

j $3^2 \times 5^0$ **k** $\left(\frac{3}{4}\right)^0 + \frac{3}{4}y^0$ **l** $\left(\frac{1}{3}\right)^0 + \left(\frac{1}{3}d\right)^0$

m $2w^0 \times 3p^0$ **n** $12u^0 \div 3$ **o** $(5d^0)^3$

p $8b^0 - (3b^0)^2$ **q** $\frac{12p^0}{(2p)^0}$ **r** $6n^3 \div 2n^3$

s $\frac{12q^5}{36q^5}$ **t** $(3x^3)^3 \div x^9$ **u** $(5h^4x^3)^0 - 4(h^3x)^0$

v $4m^0 + 7$ **w** $9c^0 + 2d^0$ **x** $3 - 2(8h)^0$

☆ MENTAL SKILLS 5 ANSWERS ON P. 640 Maths without calculators

Quiz Mental skills 5

Adding or multiplying in any order

Numbers can be added or multiplied in any order. We can use this property to make our calculations simpler.

1 Study each example.

a $19 + 5 + 5 + 1 = (19 + 1) + (5 + 5)$

$= 20 + 10$

$= 30$

b $13 + 8 + 20 + 27 + 80 = (13 + 27) + (20 + 80) + 8$

$= 40 + 100 + 8$

$= 148$

☐ Foundation ○ Standard ⬡ Complex

c $2 \times 36 \times 5 = (2 \times 5) \times 36$
$= 10 \times 36$
$= 360$

d $25 \times 11 \times 4 \times 7 = (25 \times 4) \times (11 \times 7)$
$= 100 \times 77$
$= 7700$

2 Now evaluate each sum.

a $45 + 16 + 45 + 4 + 7$
b $38 + 600 + 50 + 12 + 40$
c $18 + 91 + 9 + 20$
d $75 + 33 + 7 + 25$
e $24 + 16 + 80 + 44 + 10$
f $56 + 5 + 20 + 15 + 4$
g $100 + 36 + 200 + 10 + 90$
h $54 + 27 + 9 + 16 + 3$
i $70 + 50 + 30 + 25 + 25$
j $32 + 120 + 40 + 80 + 40$

3 Now evaluate each product.

a $8 \times 4 \times 5$
b $50 \times 7 \times 2$
c $3 \times 5 \times 6$
d $5 \times 11 \times 40$
e $12 \times 2 \times 3$
f $2 \times 4 \times 25 \times 8$
g $3 \times 20 \times 7 \times 5$
h $6 \times 8 \times 5 \times 2$
i $2 \times 3 \times 2 \times 11$

INVESTIGATION

Negative powers

What is the value of a number raised to a negative power, for example, 2^{-1} or 2^{-2}?

1 Copy and complete each table showing decreasing powers. Notice the pattern in your answers.

a

Power of 2	Number
2^3	8
2^2	4
2^1	
2^0	
2^{-1}	
2^{-2}	
2^{-3}	

b

Power of 10	Number
10^3	1000
10^2	
10^1	
10^0	
10^{-1}	
10^{-2}	
10^{-3}	

Copy and complete this table showing decreasing powers in expanded form. Notice the pattern in your answers.

a

Power of 5	Expanded form
3^3	$3 \times 3 \times 3$
3^2	3×3
3^1	3
3^0	1
3^{-1}	$\frac{1}{3}$
3^{-2}	$\frac{1}{3 \times 3} = \frac{1}{3^2}$
3^{-3}	$\frac{1}{3 \times 3 \times 3} = \frac{1}{3^3}$
3^{-4}	
3^{-5}	

b

Power of 5	Expanded form
5^3	$5 \times 5 \times 5$
5^2	
5^1	
5^0	
5^{-1}	
5^{-2}	
5^{-3}	
5^{-4}	
5^{-5}	

9780170465618

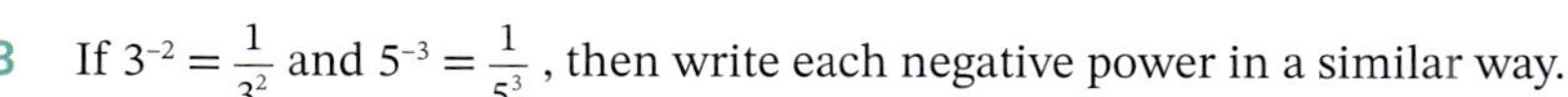

3 If $3^{-2} = \frac{1}{3^2}$ and $5^{-3} = \frac{1}{5^3}$, then write each negative power in a similar way.

a 4^{-1} **b** 7^{-4} **c** 2^{-6}

4 Simplify each expression in index notation.

a $10^4 \div 10^7$ **b** $2^3 \div 2^8$ **c** $3^4 \div 3^5$

d $5^2 \div 5^8$ **e** $a^4 \div a^6$ **f** $a \div a^4$

5 Consider $\frac{10^4}{10^7} = \frac{\cancel{10} \times \cancel{10} \times \cancel{10} \times \cancel{10}}{\cancel{10} \times \cancel{10} \times \cancel{10} \times \cancel{10} \times 10 \times 10 \times 10}$

$= \frac{1}{10 \times 10 \times 10}$

$= \frac{1}{10^3}$

But also $\frac{10^4}{10^7} = 10^{4-7}$

$= 10^{-3}$

$\therefore 10^{-3} = \frac{1}{10^3}$

Use the method above to show that:

a $\frac{2^3}{2^8} = 2^{-5} = \frac{1}{2^5}$ **b** $\frac{3^4}{3^5} = 3^{-1} = \frac{1}{3}$

c $\frac{5^2}{5^8} = 5^{-2} = \frac{1}{5^6}$ **d** $\frac{a^4}{a^6} = a^{-2} = \frac{1}{a^2}$

6 Write in words and as a formula the rule for raising a to a negative power $-n$, that is, a^{-n}.

TECHNOLOGY

Negative powers

In this activity, we will discover the pattern for negative powers. We will consider base values from 2 to 10 shown in column A and indices (powers) −1, −2 and −3 shown in row 1. On a spreadsheet, the symbol for power is ^ (called a **carat,** press **SHIFT 6**) . For example, 3^{-1} is entered as **3^−1.**

1 Create a spreadsheet as shown.

	A	B	C	D
1	**Index**	-1	-2	-3
2				
3	**Base**			
4	2			
5	3			
6	4			
7	5			
8	6			
9	7			
10	8			
11	9			
12	10			
13				

2 We will first examine the power of –1. In cell B4, enter **=A4^B1** to calculate 2^{-1}. B1 is an **absolute cell reference**, which ensures that the cell does not change when a formula is copied. This means that in column B, the power will always refer to cell B1 (–1) only. **Fill Down** from cell B4 to B12.

3 Use **Format cells** to set column B decimals to **Fraction** and **Up to three digits**.

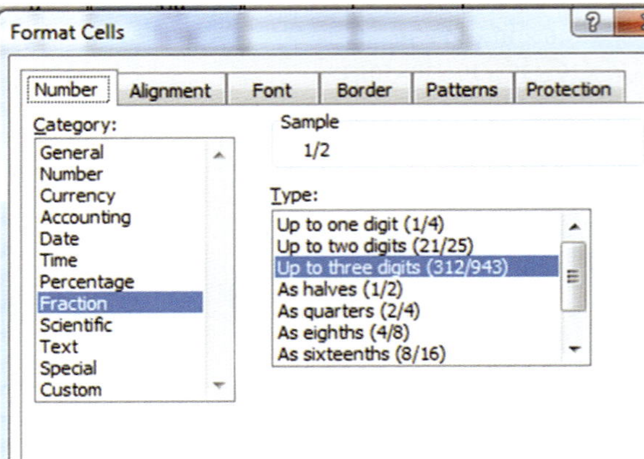

4 Compare your answers in column B with the original values in column A. Can you describe the pattern when a base is raised to a power of –1?

5 Now consider powers of –2. Adapt steps from 1 to 3 for **column C**. Use **Fill Down** from cell C4 to C12.

	A	B	C	D
1	**Index**	-1	-2	-3
2				
3	**Base**			
4	2	1/2	=A4^C1	
5	3	1/3		
6	4	1/4		
7	5	1/5		

6 Compare your answers in column C with the original values in column A. Can you describe the pattern when a base is raised to a power of –2?

7 Now consider powers of –3. Adapt steps for **column D**. In cell D4, enter the formula **=A4^D1**.

Note: D12's fraction is missing as it has 4 digits in the denominator, which the spreadsheet doesn't allow for. Can you figure out what the fraction should be?

8 Compare your answers in column D with the original values in column A. Can you describe the pattern when a base is raised to a power of –3?

9 Write a rule for negative powers, given the answers you have found in this activity. Discuss with other students in your class.

Negative indices

5.05

Worksheet
Power calculations

Consider $2^0 \div 2^3 = \frac{2^0}{2^3}$

$= \frac{1}{2^3}$

But also $2^0 \div 2^3 = 2^{0-3}$

$= 2^{-3}$

$\therefore 2^{-3} = \frac{1}{2^3}$

ⓘ Negative powers

A number raised to a negative power gives a fraction (with a numerator of 1):

$$a^{-n} = \frac{1}{a^n}$$

Proof:

$a^0 \div a^n = \frac{a^0}{a^n}$

$= \frac{1}{a^n}$

But also $a^0 \div a^n = a^{0-n}$

$= a^{-n}$

$\therefore a^{-n} = \frac{1}{a^n}$

Example 8

Simplify each expression using a positive index (power).

a 2^{-1} **b** 5^{-3} **c** g^{-5}

Video
Negative indices

SOLUTION

a $2^{-1} = \frac{1}{2^1} = \frac{1}{2}$ **b** $5^{-3} = \frac{1}{5^3}$ **c** $g^{-5} = \frac{1}{g^5}$

Example 9

Evaluate 2^{-6}, giving your answer in fraction form.

SOLUTION

$2^{-6} = \frac{1}{2^6} = \frac{1}{64}$

Example 10

Write each expression with a negative index.

a $\frac{1}{c^3}$ **b** $\frac{5}{k^2}$

SOLUTION

a $\frac{1}{c^3} = c^{-3}$ **b** $\frac{5}{k^2} = 5k^{-2}$

Example 11

Video Negative indices

Simplify each expression using a positive index.

a $3n^{-2}$ **b** $(3n)^{-2}$ **c** $p^{-2}q^3$ **d** ab^{-6}

SOLUTION

a
$$\begin{aligned}3n^{-2} &= 3 \times n^{-2}\\ &= \frac{3}{1} \times \frac{1}{n^2}\\ &= \frac{3}{n^2}\end{aligned}$$

b
$$\begin{aligned}(3n)^{-2} &= \frac{1}{(3n)^2}\\ &= \frac{1}{9n^2}\end{aligned}$$

c
$$\begin{aligned}p^{-2}q^3 &= \frac{1}{p^2} \times q^3\\ &= \frac{q^3}{p^2}\end{aligned}$$

d
$$\begin{aligned}ab^{-6} &= a \times \frac{1}{b^6}\\ &= \frac{a}{b^6}\end{aligned}$$

Power of −1

Consider $9^{-1} = \frac{1}{9}$

$\frac{1}{9}$ is the **reciprocal** of 9.

Consider
$$\begin{aligned}\left(\frac{2}{3}\right)^{-1} &= \frac{1}{\left(\frac{2}{3}\right)}\\ &= 1 \div \frac{2}{3}\\ &= 1 \times \frac{3}{2}\\ &= \frac{3}{2}\\ &= 1\frac{1}{2}\end{aligned}$$

ⓘ Power of -1

A number raised to a power of −1 gives its **reciprocal.**

$$a^{-1} = \frac{1}{a}$$

$$\left(\frac{a}{b}\right)^{-1} = \frac{b}{a}$$

Example 12

Simplify each expression.

a $\left(\frac{4}{3}\right)^{-1}$ **b** $\left(\frac{y}{5}\right)^{-1}$ **c** $\left(\frac{2}{3m}\right)^{-1}$

SOLUTION

a $\left(\frac{4}{3}\right)^{-1} = \frac{3}{4}$ **b** $\left(\frac{y}{5}\right)^{-1} = \frac{5}{y}$ **c** $\left(\frac{2}{3m}\right)^{-1} = \frac{3m}{2}$

EXERCISE 5.05 ANSWERS ON P. 640

Negative indices

1 Simplify each expression using a positive index. EXAMPLE 8

R C

a 6^{-2} **b** 5^{-7} **c** 3^{-1} **d** 10^{-2}
e g^{-5} **f** z^{-1} **g** n^{-3} **h** t^{-2}
i a^{-4} **j** 5^{-3} **k** y^{-d} **l** r^{-m}

2 Evaluate each expression, giving your answers in fraction form. EXAMPLE 9

R C

a 3^{-2} **b** 5^{-4} **c** 6^{-1} **d** 7^{-2}
e 25^{-1} **f** 2^{-7} **g** 4^{-3} **h** 10^{-6}
i 2^{-10} **j** 3^{-3} **k** 6^{-2} **l** 9^{-4}

3 Write each expression using a negative index. EXAMPLE 10

R C

a $\frac{1}{n^2}$ **b** $\frac{1}{n}$ **c** $\frac{1}{8^3}$ **d** $\frac{1}{8}$
e $\frac{1}{10^5}$ **f** $\frac{2}{a^4}$ **g** $\frac{1}{3}$ **h** $\frac{-1}{b}$
i $\frac{6}{a}$ **j** $\frac{4}{t^2}$ **k** $\frac{-2}{w^5}$ **l** $\frac{5}{d^3}$

4 Simplify each expression using positive indices. EXAMPLE 11

R C

a $5h^{-1}$ **b** $2b^{-5}$ **c** $3e^{-3}$ **d** $4n^{-2}$
e pb^{-2} **f** r^2s^{-4} **g** $w^{-2}y$ **h** $d^{-3}y^3$
i $(2m)^{-1}$ **j** $(xy)^{-1}$ **k** $(4h)^{-2}$ **l** $(5k)^{-3}$
m $3m^3p^{-2}$ **n** $15k^{-1}w^{-4}$ **o** $12x^{-2}y^{-3}$ **p** $12x^{-2}y^3$
q $(3h)^{-2}$ **r** $(4k)^{-3}$ **s** $(2c)^{-4}$ **t** $(8y)^{-1}$
u $4pq^{-3}$ **v** $4p^{-1}q^{-3}$ **w** vm^{-2} **x** $v^{-1}m^{-2}$

Foundation Standard Complex

EXAMPLE 12

5 Simplify each expression.

R C

a $\left(\frac{2}{7}\right)^{-1}$ **b** $\left(\frac{8}{5}\right)^{-1}$ **c** $\left(\frac{9}{10}\right)^{-1}$ **d** $\left(\frac{3}{2}\right)^{-1}$

e $\left(-\frac{3}{4}\right)^{-1}$ **f** $\left(\frac{5}{2}\right)^{-1}$ **g** $\left(\frac{x}{3}\right)^{-1}$ **h** $\left(\frac{5}{a}\right)^{-1}$

i $\left(-\frac{m}{2}\right)^{-1}$ **j** $\left(\frac{5r}{4}\right)^{-1}$ **k** $\left(\frac{2}{3z}\right)^{-1}$ **l** $\left(\frac{1}{v}\right)^{-1}$

5.06 Summary of the index laws

Puzzle Indices squaresaw

ⓘ The index laws

$a^m \times a^n = a^{m+n}$	$a^m \div a^n = \frac{a^m}{a^n} = a^{m-n}$
$(a^m)^n = a^{m\times n}$	$(ab)^n = a^n b^n$
$\left(\frac{a}{b}\right)^n = \frac{a^n}{b^n}$	$a^0 = 1$
$a^{-1} = \frac{1}{a}$	$a^{-n} = \frac{1}{a^n}$
$\left(\frac{a}{b}\right)^{-1} = \frac{b}{a}$	

EXERCISE 5.06 ANSWERS ON P. 640

Summary of the index laws

U F R C

1 Simplify each expression, writing the answer in index notation.

R C

a $3^4 \times 3^7$ **b** $2^{10} \div 2^7$ **c** $(4^4)^2$ **d** $9^8 \div 9$

e $(8^3)^4$ **f** $5^8 \times 5$ **g** $6^5 \times 6^4 \times 6^3$ **h** $3^8 \div 3^5 \times 3$

i $(2^3)^5 \times 2^4$ **j** $\frac{4^{20}}{(4^5)^2}$ **k** $(8^2)^3 \div 8$ **l** $\frac{(10^3)^5}{(10^2)^7}$

2 Simplify each expression.

R

a $a^4 \times a^3$ **b** $t^8 \times t$ **c** $n^8 \div n^2$ **d** $p^3 \div p$

e $(w^2)^4$ **f** $(g^3)^6$ **g** $2b^2 \times 3b^5$ **h** $4d^7 \times 5d^6$

i $\frac{30c^{12}}{5c^8}$ **j** $(5b^4)^4$ **k** $24m^6 \div 8m^4$ **l** $(3a)^2$

☐ Foundation ○ Standard ⬡ Complex

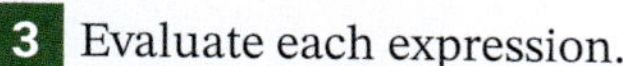

3 Evaluate each expression.

R

a 4^0 **b** $(-4)^0$ **c** 7×2^0 **d** $(7 \times 2)^0$

e $(-2)^3$ **f** $(-3)^2$ **g** $(5^2)^2$ **h** $2^4 \times 2^3$

i $(7^2)^0$ **j** $4^5 \div 4^2$ **k** $4^2 \div 4^5$ **l** $10^3 \div 10^3$

m $5^2 \div 5^0$ **n** $10^{-2} \div 10^2$ **o** $\left(\frac{1}{2}\right)^0$ **p** $10^{-2} \times 10^2$

4 Evaluate each expression, giving your answers in fraction form.

R C

a 5^{-2} **b** 2^{-5} **c** 20^{-1} **d** 10^{-3}

5 Simplify each expression.

a $(3mn^3)^2$ **b** $8a^2w^2 \times 5a^3w^7$ **c** $(4a^2b^5)^4$ **d** $\frac{20p^3q^8}{5p^2q^6}$

e $\left(\frac{4}{5}\right)^3$ **f** $6c^2d^0$ **g** $\frac{48u^5v^4}{16uv}$ **h** $\left(\frac{3x}{10}\right)^2$

i $(-4n^2t)^3$ **j** $\left(-\frac{2}{3}\right)^4$ **k** $\frac{7x^2y^6}{35x^5y^3}$ **l** $\left(\frac{p^5}{9y}\right)^2$

m $(2p^3q^2)^5$ **n** $\left(\frac{1}{7n}\right)^3$ **o** $-2n^0$ **p** $\frac{(a^2b)^4 \times a^3}{b^5}$

6 Simplify each expression using a positive index.

R C

a 8^{-7} **b** 3^{-5} **c** y^{-1} **d** x^{-3}

e $(5b)^{-2}$ **f** $5b^{-2}$ **g** $(ab)^{-1}$ **h** ab^{-1}

i $4t^{-8}$ **j** $(11t)^{-3}$ **k** p^3q^{-5} **l** mw^{-3}

m $8u^{-3}v^{-4}$ **n** $-2r^6y^{-5}$ **o** $10e^{-1}f^3$ **p** $\frac{1}{2}k^{-4}n^7$

7 Simplify each expression.

R

a $\left(\frac{7}{4}\right)^{-1}$ **b** $\left(\frac{5}{2}\right)^{-1}$ **c** $\left(\frac{2}{3}\right)^{-1}$ **d** $\left(\frac{1}{7}\right)^{-1}$

e $\left(\frac{r}{8}\right)^{-1}$ **f** $\left(\frac{1}{10p}\right)^{-1}$ **g** $\left(\frac{6y}{z}\right)^{-1}$ **h** $\left(\frac{2}{5a}\right)^{-1}$

8 Write each expression using a negative index.

R C

a $\frac{1}{4^3}$ **b** $\frac{1}{2}$ **c** $\frac{1}{10^4}$ **d** $\frac{1}{9^2}$

e $\frac{1}{k}$ **f** $\frac{9}{k^4}$ **g** $\frac{-1}{x^7}$ **h** $\frac{5}{p^3}$

9 Simplify each expression.

R

a $q^5 \times q^{-2}$ **b** $d^{-3} \times d^7$ **c** $m^{-6} \div m^5$ **d** $t \div t^{-1}$

e $5g^3 \times 6g^{-1}$ **f** $8a^{-2} \times 3a^3$ **g** $7x^{-2} \times 4x$ **h** $\frac{64p^{-1}}{16p^2}$

i $48q \div 3q^{-2}$ **j** $\frac{5t^3}{10t^{-1}}$ **k** $2(b^{-1})^4$ **l** $(3h)^{-2}$

Foundation Standard Complex

5.07 Extension: Significant figures

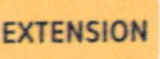

Worksheet
Significant figures

Video
Rounding: Snails vs rockets

A way of rounding a number is to give the most relevant or important digits of the number. For example, a crowd of 47 321 people can be written as 47 000, which is rounded to the nearest thousand, or to 2 **significant figures**.

The first significant figure in a number is the first **non-zero** digit.

Number	First significant digit	Number of significant digits
47 321	4	5
47 000	4	2
0.000 159 2	1	4
0.000 2	2	1

- When rounding to significant figures, start counting from the first digit that is not 0.
- If it is a large number, you may need to insert zeros at the end as placeholders.
- Zeros at the end of a whole number or at the beginning of a decimal are **not significant**: they are necessary placeholders.
- Zeros *between* significant figures or at the end of a decimal *are* significant. For example, the significant figures are shown in bold in this table.

Number	First significant digit	Number of significant digits
809 000	8	3
0.0**20 70**	2	4

Example 13

State the number of significant figures in each number.

a 63.70 **b** 0.003 05 **c** 7600

SOLUTION

a The 0 after 7 is significant.

$\therefore$ 63.70 has 4 significant figures.

b The first significant figure is 3, and the 0 between 3 and 5 is significant.

$\therefore$ 0.003 05 has 3 significant figures.

c The 0s after 6 are not significant.

$\therefore$ 7600 has 2 significant figures.

9780170465618

Example 14

Round each number to 3 significant figures.

a 56.357 **b** 9.249 **c** 548 307

SOLUTION

a **56.3**57 ≈ 56.4 **b** **9.24**9 ≈ 9.25 **c** **548** 307 ≈ 548 000

The zeros in part c are not significant, but they are placeholders that are necessary for showing the place values of the 5, 4 and 8.

Video
Significant figures

Example 15

Write each number correct to one significant figure.

a 0.007 39 **b** 0.025 **c** 0.963

SOLUTION

a 0.00**7** 39 ≈ 0.007 **b** 0.02**5** ≈ 0.03 **c** 0.**9**63 ≈ 1

The zeros at the beginning of a decimal are not significant: they are placeholders.

Video
Significant figures

EXERCISE 5.07 ANSWERS ON P. 640

Significant figures

U F R C

EXAMPLE 13

1 State the number of significant figures in each number.

a	457	**b**	0.23	**c**	15 000
d	4.0004	**e**	0.0005	**f**	5000
g	0.002 07	**h**	89 072	**i**	0.040
j	76 000 000	**k**	0.000 328	**l**	169.320

EXAMPLE 14

2 Round each number to 3 significant figures.

a	37.609	**b**	9435	**c**	168.39
d	2.813	**e**	15.99	**f**	60 522
g	1 769 000	**h**	385 764	**i**	10.2717

EXAMPLE 15

3 Write each number correct to 2 significant figures.

a	0.0637	**b**	0.903	**c**	0.084 55
d	0.000 158	**e**	0.007 625	**f**	0.038 71
g	0.2795	**h**	0.018 944	**i**	0.3145

Foundation Standard Complex

EXTENSION

4 What is 45 067 853 rounded to 3 significant figures?
Select the correct answer **A**, **B**, **C** or **D**.

A 45 167 853 **B** 45 100 000 **C** 45 067 900 **D** 45 070 000

5 What is 0.005 605 0 rounded to 2 significant figures? Select **A**, **B**, **C** or **D**.

A 0.01 **B** 0.010 000 0 **C** 0.0056 **D** 0.005 600 0

6 Round each number to one significant figure.

a 9.478 **b** 57.12 **c** 0.0367
d 0.007 66 **e** 0.5067 **f** 10 675
g 1856.78 **h** 0.000 28 **i** 56 239 400

7 A company makes a profit of $35 754 125.

C **a** Round the profit to the nearest million and state the number of significant figures in the answer.

b Round the profit to the nearest 10 million and state the number of significant figures in the answer.

8 Australia's population in 2019 was 25 498 000. To how many significant figures has this number been written? C

9 A total of 21 558 people attended a football match. Express this number to 3 significant figures. C

10 The radius of a hydrogen atom is 0.000 000 000 104 m. To how many significant figures has this number been written? Select the correct answer **A**, **B**, **C** or **D**.

A 12 **B** 9 **C** 3 **D** 2

11 Evaluate each expression, correct to the number of significant figures shown in the brackets.

a $45.6 \times 8.7 - 2.75 \times 78.32$ [2] **b** $15.5 - 9.87 \div 0.24 + 8.43 \times 2.4$ [1]

c $(63.73 - 27.89) \div 5.82$ [3] **d** $\frac{63.25+76.03}{55.89-89.24}$ [4]

e $\frac{9.732+2.765}{12.27\times15.8}$ [1] **f** $78.91 \div (23.6 + 94.7)$ [2]

g $\frac{1}{0.941}+\frac{253}{0.0076}$ [3] **h** $\sqrt{84.3}\times0.0715$ [4]

☐ Foundation ○ Standard ⬡ Complex

DID YOU KNOW?

Big numbers

This table lists the names of some big numbers and their meanings.

In 2009, due to hyperinflation, Zimbabwe printed the $100 000 000 000 000 note (10^{14} dollars), which was actually worth only $30 US.

What is the special name for the number:

a 10^{100}?

b $10^{(10^{100})}$?

Name	Numeral
million	$10^6 = 1\ 000\ 000$
billion	$10^9 = 1\ 000\ 000\ 000$
trillion	10^{12}
quadrillion	10^{15}
quintillion	10^{18}
sextillion	10^{21}
septillion	10^{24}
octillion	10^{27}
nonillion	10^{30}
decillion	10^{33}
centillion	10^{303}

Scientific notation 5.08

Scientific notation is a short way of writing very large or very small numbers using powers of 10. It was invented in the early 20th century when scientists needed to describe very large values, such as astronomical distances, and very small values, such as the masses of atoms.

Worksheet Scientific notation 1

Scientific notation 2

Technology Scientific notation

Spreadsheet Scientific notation

Videos Scientific notation

What does the Internet weigh?

ⓘ Scientific notation

Numbers written in **scientific notation** are expressed in the form

$$m \times 10^n$$

where m is a number between 1 and 10 and n is an integer.

Example 16

Express each number in scientific notation.

a 764 000 000 000 **b** 6000 **c** 0.0008 **d** 0.000 000 472

SOLUTION

a Use the significant figures in the number to write a value between 1 and 10: **7.64**

Count how many places the decimal point moves to the right to make 764 000 000 000.

764 000 000 000 11 places

$\therefore$ 764 000 000 000 = 7.64×10^{11}

or count the number of places after the first significant figure, 7.

b Use the significant figures in the number to write a value between 1 and 10: **6**

Count how many places the decimal point moves to the right to make 6000.

6000 3 places

$\therefore 6000 = 6 \times 10^3$

or count the number of places after the first significant figure, 6.

c Use the significant figures in the number to write a value between 1 and 10: **8**

Count how many places the decimal point moves to the left to make 0.0008.

0.0008 4 places

$\therefore 0.0008 = 8 \times 10^{-4}$

or count the number of decimal places to the first significant figure, 8

Note that small numbers are written with *negative* powers of 10.

d Use the significant figures in the number to write a value between 1 and 10: **4.72**

Count the number of places the decimal point moves to the left to make 0.000 000 472.

0.000 000 472 7 places

$\therefore 0.000\,000\,472 = 4.72 \times 10^{-7}$

or count the number of decimal places to the first significant figure, 4

Example 17

Express each number in decimal form.

a 2.7×10^4 **b** 3.56×10^{-2}

SOLUTION

a $2.7 \times 10^4 = 2.7000$ Move the decimal point 4 places to the right.

$= 27\,000$

b $3.56 \times 10^{-2} = 0.0356$ Move the decimal point 2 places to the left.

$= 0.0356$

Example 18

a Which number is the larger: 3.65×10^{12} or 8.1×10^{12}?

b Write these numbers in ascending order: 4.3×10^6, 2.8×10^7, 1.9×10^7

SOLUTION

To compare numbers in scientific notation, first compare the powers of 10.

If the powers of 10 are the same, then compare the decimal parts.

a The powers of 10 are the same. Compare the decimal parts: $8.1 > 3.65$.

$\therefore$ The larger number is 8.1×10^{12}.

b Compare the powers of 10: $10^6 < 10^7$.

Then compare the 2 numbers with 10^7: $1.9 < 2.8$.

$\therefore$ The numbers in ascending order are 4.3×10^6, 1.9×10^7, 2.8×10^7.

EXERCISE 5.08 ANSWERS ON P. 641

Scientific notation

U F R C

EXAMPLE 16

1 Express each number in scientific notation.

R C

a	2400	**b**	786 000	**c**	55 000 000	**d**	95
e	7.8	**f**	348 000 000	**g**	59 670	**h**	15
i	3 000 000 000	**j**	80	**k**	763	**l**	10
m	0.035	**n**	0.000 076	**o**	0.8	**p**	0.0713
q	0.000 003	**r**	0.913	**s**	0.000 007 146	**t**	0.009
u	0.000 000 1	**v**	0.000 89	**w**	0.000 000 078	**x**	0.1

2 Express the value shown in each measurement in scientific notation.

C

- **a** The world's largest mammal is the blue whale, which can weigh up to 130 000 kg.
- **b** The diameter of an oxygen molecule is 0.000 000 29 cm.
- **c** The thickness of a human hair is 0.000 08 m.
- **d** Light travels at a speed of 300 000 000 m/s.
- **e** The nearest star to Earth, excluding the Sun, is Alpha Centauri, which is 40 000 000 000 000 km away.
- **f** The thickness of a typical piece of paper is 0.000 12 m.
- **g** The small intestine of an adult is approximately 610 cm long.
- **h** The diameter of a hydrogen atom is 0.000 000 000 106 m.
- **i** The diameter of our galaxy, the Milky Way, is 770 000 000 000 000 000 000 m.
- **j** A microsecond means 0.000 001 s.
- **k** The Andromeda Galaxy is the most remote body visible to the naked eye, at a distance of 2 200 000 light years away.

EXAMPLE 17

3 Express each number in decimal form.

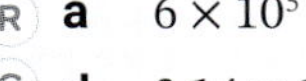

a	6×10^5	**b**	7.1×10^3	**c**	3.02×10^8
d	3.14×10^0	**e**	6×10^{-5}	**f**	7.1×10^{-3}
g	3.02×10^{-8}	**h**	5.9×10^{-10}	**i**	1.1×10^{12}
j	4×10^{-4}	**k**	5×10^3	**l**	4.76×10^{-4}
m	8.03×10^{-1}	**n**	6.32×10^4	**o**	1.6×10^{-2}
p	2.2×10^{-7}	**q**	9.0×10^6	**r**	1.11×10^{-1}

Foundation Standard Complex

EXAMPLE 18

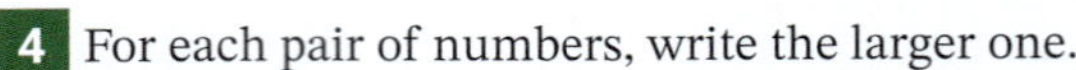

4 For each pair of numbers, write the larger one.

R **a** 3×10^5 or 4×10^5 **b** 8.4×10^5 or 2.7×10^6

c 8.4×10^0 or 1.3×10^7 **d** 3.6×10^{-7} or 6.3×10^{-7}

e 9.3×10^9 or 7.6×10^9 **f** 3.5×10^{-6} or 9.3×10^2

g 3.04×10^0 or 3.04×10^{-4} **h** 4.5×10^{-5} or 3.7×10^{-7}

i 2×10^{-15} or 2×10^{-17} **j** 6.23×10^{-5} or 9.7×10^{-5}

5 Write each set of numbers in ascending order.

R **a** 3.8×10^9, 7.3×10^9, 5.5×10^9

b 2.2×10^{-4}, 5.8×10^{-6}, 7×10^{-4}

c 3.5×10^0, 5.3×10^2, 4.9×10^2

6 Write each set of numbers in descending order.

R **a** 6×10^5, 2.9×10^2, 8×10^2

b 1.2×10^{-9}, 6.3×10^2, 8.1×10^{-4}

c 4.1×10^{-1}, 9.5×10^{-1}, 6.4×10^{-3}

5.09 Scientific notation on a calculator

Worksheet Scientific notation problems

Puzzle Scientific notation: accomplishing great things

To enter a number in scientific notation on a calculator, use the [×10ˣ] or [EXP] key.

Example 19

Evaluate each expression using scientific notation.

a $(4.25 \times 10^7) \times (8.2 \times 10^6)$

b $(1.08 \times 10^{-15}) \div (3 \times 10^{11})$

c $(4.9 \times 10^7)^2$

SOLUTION

a Enter 4.25 [×10ˣ] 7 [×] 8.2 [×10ˣ] 6 [=]

$(4.25 \times 10^7) \times (8.2 \times 10^6) = 3.485 \times 10^{14}$

Note that with scientific notation on a calculator, there is no need to enter brackets [(] [)] around the numbers.

b Enter 1.08 [×10ˣ] [−] 15 [÷] 3 [×10ˣ] 11 [=]

$(1.08 \times 10^{-15}) \div (3 \times 10^{11}) = 3.6 \times 10^{-27}$

c Enter 4.9 [×10ˣ] 7 [x^2] [=]

$(4.9 \times 10^7)^2 = 2.401 \times 10^{15}$

5.09

Example 20

Estimate the value of each expression in scientific notation, then evaluate it correct to 3 significant figures.

a $\dfrac{9.2\times10^9}{2.7\times10^5}$ **b** $(8.5\times10^4)\times(6.3\times10^7)$ **c** $(6.08\times10^5)^3$

SOLUTION

Estimate

a

$$\frac{9.2\times10^9}{2.7\times10^5}\approx\frac{9\times10^9}{3\times10^5}$$
$$=\frac{9}{3}\times\frac{10^9}{10^5}$$
$$=3\times10^4$$

Calculated answer

$$\frac{9.2\times10^9}{2.7\times10^5}=34\,074.074\,07$$
$$\approx 34\,000$$
$$=3.40\times10^4$$

With scientific notation, all digits in the decimal are significant, so 3 significant figures means 2 decimal places.

b

Estimate:

$$(8.5\times10^4)\times(6.3\times10^7)\approx(9\times10^4)\times(6\times10^7)$$
$$=(9\times6)\times(10^4\times10^7)$$
$$=54\times10^{11}$$
$$=5.4\times10\times10^{11}$$
$$=5.4\times10^{12}$$

Calculated answer:

$$(8.5\times10^4)\times(6.3\times10^7)=5.355\times10^{12}$$
$$\approx 5.36\times10^{12}$$

c

Estimate:

$$(6.08\times10^5)^3\approx(6\times10^5)^3$$
$$=6^3\times(10^5)^3$$
$$=216\times10^{15}$$
$$=2.16\times10^2\times10^{15}$$
$$=2.16\times10^{17}$$

Calculated answer:

$$(6.08\times10^5)^3=2.24755\ldots\times10^{17}$$
$$\approx 2.25\times10^{17}$$

EXERCISE 5.09 ANSWERS ON P. 641

Scientific notation on a calculator

U F PS R C

EXAMPLE 19

1 Evaluate each expression using scientific notation.

a $(2 \times 10^3) \times (3 \times 10^5)$ **b** $(8 \times 10^7) \div (4 \times 10^2)$ **c** $(2 \times 10^5)^3$

d $\sqrt{9 \times 10^{12}}$ **e** $(4 \times 10^7) \times (6 \times 10^8)$ **f** $(1 \times 10^8) \div (2 \times 10^3)$

g $(4 \times 10^3)^5$ **h** $24.08 \div (8 \times 10^6)$ **i** $\sqrt{3.969 \times 10^{19}}$

j $(2 \times 10^5)^{-2}$ **k** $\sqrt[3]{8 \times 10^{-9}}$ **l** $\dfrac{7.62 \times 10^9}{2 \times 10^{-4}}$

EXAMPLE 20

2 Estimate the value of each expression in scientific notation, then evaluate correct to 3 significant figures.

R

a $(5.7 \times 10^3) \times (2.3 \times 10^5)$ **b** $(8 \times 10^5) \times (3.7 \times 10^7)$

c $(9.1 \times 10^{20}) \div (3.2 \times 10^5)$ **d** $(1.2 \times 10^8)^2$

e $(7.13 \times 10^{10}) \times (9.8 \times 10^8)$ **f** $(1.9 \times 10^{11}) \div (2.1 \times 10^7)$

g $(5.85 \times 10^4)^3$ **h** $(6 \times 10^{12}) \div (2.8 \times 10^3)$

3 The human body consists of approximately 6×10^9 cells, and each cell consists of 6.3×10^9 atoms. Roughly how many atoms are there in a human body?

PS R

4 *The Complete Sherlock Holmes* is a book that is 4.5 cm thick and contains 2000 pages. Find the thickness, in *millimetres*, of one page in scientific notation.

PS R

5 Evaluate each expression in scientific notation, correct to 2 significant figures.

a $(7.4 \times 10^{30}) - (3.59 \times 10^{29})$ **b** $(1.076 \times 10^{17}) + (2.3 \times 10^{16})$

c $\sqrt{6.6 \times 10^{27}}$ **d** $(7.5 \times 10^{23}) \div (3.3 \times 10^{-13})$

e $(8.17 \times 10^{16})^3$ **f** $\sqrt{2.69 \times 10^{45}}$

g $(7.05 \times 10^3) \div (3.9 \times 10^7)$ **h** $\dfrac{(5.6 \times 10^4) \times (3.9 \times 10^5)}{(2.3 \times 10^7)}$

i 1595×1959 **j** 5^{20}

k 801^{-1} **l** 3^{-10}

m 9^9 **n** $(0.7)^{-5}$

6 The Earth is 1.50×10^8 km from the Sun and the speed of light is 3×10^5 km/s. How long does it take for light to travel from the Sun to Earth? Express your answer in:

PS R C

a seconds

b minutes and seconds.

Foundation Standard Complex

7 The Sun burns 6 million tonnes of hydrogen a second. Calculate to 2 significant figures how many tonnes of hydrogen it burns in a year (that is, 365.25 days).

PS R

8 Sound travels at approximately 330 m/s. If Mach 1 is the speed of sound, how fast is Mach 5, which is 5 times the speed of sound? Convert your answer to kilometres per second to 2 significant figures.

PS R C

9 The distance that light travels in one year is called a light year. If the speed of light is approximately 3×10^5 km per second, how far to 2 significant figures does light travel in a leap year?

PS R

10 A thunderstorm is occurring 30 km from where you are standing. Use the speed of light (3×10^5 km per second) and the speed of sound (330 m/s) to calculate in seconds how long it takes:

PS R C

a the light from the lightning to reach you.

b the sound from the thunder to reach you.

11 **a** What is the largest number that can be displayed on your calculator?

b What is the smallest number that can be displayed?

PS R C

INVESTIGATION

A lifetime of heartbeats

How many times does your heart beat in an average lifetime of 80 years?

1 Work in pairs and copy this table.

Name	Trial 1	Trial 2	Average beats per minute

2 Use 2 fingers to measure your pulse. Have your partner time you for a minute. Do this twice, record your results in the table and find the average.

3 Repeat Step **2** for your partner.

4 Calculate how many times your heart (and your partner's heart) beats in the following periods. Write your answers in scientific notation, correct to 2 significant figures.

a an hour
b a day
c a week
d a year (use 365.25 days)
e an average lifetime of 80 years

DID YOU KNOW?

Hairy numbers

Straight hair
• round follicle

Wavy hair
• oval follicle

Curly hair
▬ flat follicle

Left to right: Shutterstock.com/Tracy Whiteside, Alamy Stock Photo/Stephen Flint, Shutterstock.com/Cookie Studio

There are about 110 000 hairs on your head. Each hair grows at the rate of about 1.3×10^{-3} cm per hour. A single hair lasts about 6 years. Every day you lose between 30 and 60 hairs. Each hair grows from a small depression in the skin called a follicle (a gland). After the hair falls out, the follicle rests for about 3 to 4 months before the next hair starts growing. Hair follicles are either oval, flat or round in shape. How straight, wavy or curly your hair is depends on the shape of your hair follicles.

How many hairs are on all the heads in China if its population is approximately 1.448×10^9? Answer in both scientific notation and decimal notation.

POWER PLUS ANSWERS ON P. 641

Worksheet Binary number system

Video Binary: the computer language

1 If $u = 2$, $v = 4$ and $w = 5$, evaluate each expression.

a $5w^{-2}$
b $-u^{-3}$
c vu^3
d $(wu)^3$
e $\left(\frac{2v}{w^2}\right)^{-1}$
f $\left(\frac{w}{u}\right)^{-2}$
g $u^0(vw)^{-2}$
h $(uvw)^{-2}$

2 Simplify each expression.

a $5^a \times 5^a$
b $4^x \div 4^x$
c $2^{y-1} \times 2^{y+1}$
d $10^{b+3} \div 10^b$
e $(3^p)^3$
f $(3^3)^p$
g $8^n \times (8^n)^2$
h $(6^t)^3 \div 6^t$
i $p \times p + p \times p$
j $n \times n + n \times n + n \times n$
k $\frac{q \times q}{q + q}$

3 Write each number in scientific notation.

a 438.2×10^9
b 0.52×10^{-7}
c 0.0004×10^{12}
d 2013×10^{-3}
e 57.8 thousand
f 57.8 thousandths
g 6.7 millionths
h 3.2 billion
i 3.2 billionths

4 For how many values of a and b does $a^b = b^a$? What if a and b are different?

5 The terms in the pattern 3, 5, 17, 257, 65 537, ... can all be generated by a simple method, using only the numbers 1 and 2.

a What is this method?
b What is the next number in the sequence?

5 CHAPTER REVIEW

Quiz
Language of maths 5

Language of maths

ascending	base	descending	estimate
expanded form	exponent	index	index laws
index notation	indices	negative power	placeholder
product	power	quotient	reciprocal
significant figures	scientific notation	term	zero power

1 What does a power of 0 mean?

2 Which 2 words from the list mean 'power'?

3 What is the $\times 10^x$ or EXP key on a calculator used for?

4 What is the index law for dividing terms with the same base?

5 Which digits in 0.006 701 are **significant figures?**

6 What power is associated with the **reciprocal** of a term or number?

7 What type of numbers when written in **scientific notation** have **negative** powers of 10?

Topic summary

- What was this topic about? What was the main theme?
- What content was new and what was revision?
- What are the index laws?
- Write 10 questions (with solutions) that could be used in a test for this chapter.
- Include some questions that you have found difficult to answer.
- List the sections of work in this chapter that you did not understand. Follow up this work with a friend or your teacher.

Print (or copy) and complete this mind map of the topic, adding detail to its branches and using pictures, symbols and colour where needed. Ask your teacher to check your work.

Worksheet
Mind map: Indices

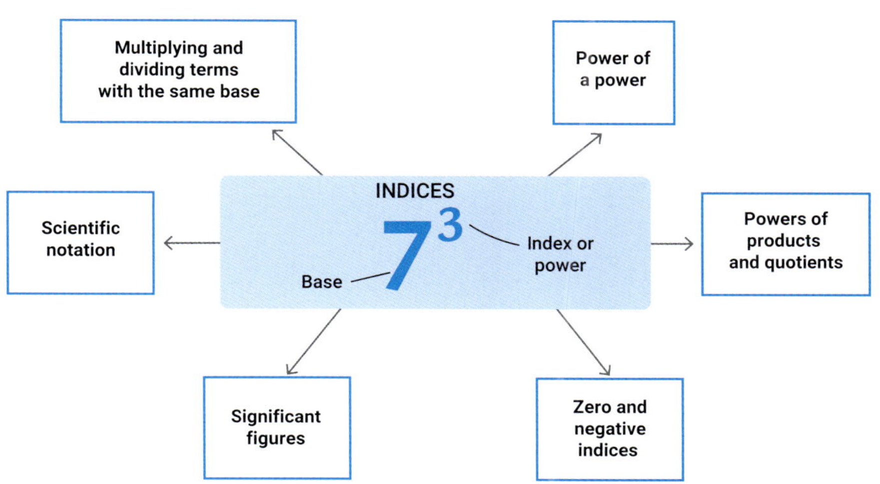

5 TEST YOURSELF

ANSWERS ON P. 641

5.01

1 Simplify each expression, writing the answer in index notation.

Quiz
Test yourself 5

a $10^3 \times 10^7$ b $4^{20} \div 4^4$ c $a^{12} \div a^2$

d $h^8 \times h^2$ e $3n^3 \times 4n$ f $10d^{15} \div 5d^3$

g $20m^9 \div 4m$ h $3v^4w^2 \times 2v^3w^5$ i $5x^5y^2 \times 3xy$

j $\frac{24t^8h^8}{3t^4h^2}$ k $\frac{p^6q^{10}}{p^2q^2}$ l $\frac{100a^2b^4}{5ab^2}$

5.02

2 Simplify each expression, writing the answer in index notation.

a $(2^2)^3$ b $(k^5)^5$ c $(x)^4$

d $(2y^3)^{10}$ e $(5t^2)^2$ f $(10g)^3$

g $(-2)^5$ h $(-2k)^5$ i $(-5m^3)^2$

5.03

3 Simplify each expression.

a $(ab^2)^4$ b $(5x^3y^2)^2$ c $(4t^2)^3$

d $(-4h^2g)^3$ e $\left(\frac{a}{7}\right)^4$ f $(2pqr)^5$

g $\left(\frac{3m}{2}\right)^5$ h $(-3np^2)^4$ i $\left(\frac{2a^7}{b}\right)^4$

j $(4t^4u^5)^3 \times 8t^2u$ k $\left(\frac{b^8y^6}{8b^2y}\right)^3$ l $45c^6d^8 \div (3cd^2)^2$

5.04

4 Simplify each expression.

a 7^0 b $(-7)^0$ c e^0

d $(-e)^0$ e $-e^0$ f g^0h

g $(gh)^0$ h $\left(\frac{2p}{3}\right)^0$ i $\frac{2p^0}{3}$

5.05

5 Simplify each expression using a positive index.

a 8^{-3} b 19^{-2} c x^{-1} d p^{-5}

5.05

6 Simplify each expression using a positive index.

a $(4m)^{-1}$ b $(4m)^{-2}$ c $(5b)^{-1}$ d $5b^{-1}$

e $-2x^{-4}$ f $\left(\frac{3}{5a}\right)^{-1}$ g $c^{-4}d^2$ h $\left(\frac{100}{9}\right)^{-1}$

5.05

7 Write each expression using a negative index.

a $\frac{1}{10^3}$ b $\frac{1}{r^5}$ c $\frac{1}{r}$ d $\frac{3}{b}$

5.06

8 Simplify each expression.

a $96u^5t^8 \div 24ut^4$ b $(5ab)^2$ c $4hd^3 \times 5h^3d^6$

d $e^5 \times e \times (e^2)^3$ e $(r^3)^4 \div r^2$ f $(pn^2)^3 \div (pn^3)^2$

g $\frac{a^8b^3c}{a^4b^3c^2}$ h $\frac{60n^4m^3}{12n^8m}$ i $\frac{144t^4u^3v}{16u^9v^4}$

Foundation Standard Complex

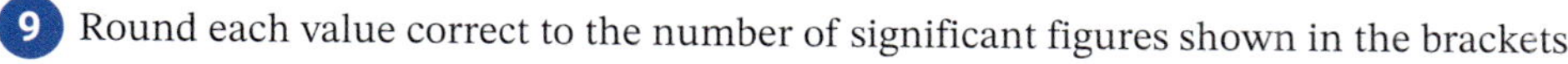

9 Round each value correct to the number of significant figures shown in the brackets.

EXTENSION 5.07

a 8.5678 [2] b 15 712 [3] c 476 [1]
d 0.007 126 6 [4] e 0.9041 [3] f 301 378 [2]
g 4805.28 [3] h 0.000 87 [1] i 67 000 000 [1]

10 Express each number in scientific notation. 5.08

a 37 000 b 0.61 c 250 000
d 0.000 49 e 13 f 0.000 000 000 08

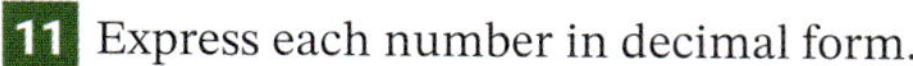

11 Express each number in decimal form. 5.08

a 8.1×10^3 b 6×10^7 c 3.075×10^0
d 8.1×10^{-3} e 6×10^{-7} f 3.075×10^{-2}

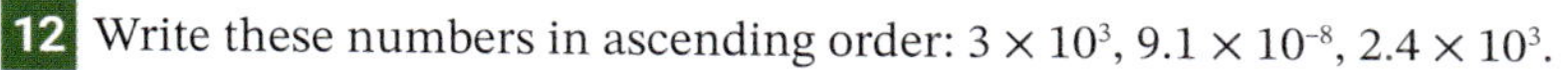

12 Write these numbers in ascending order: 3×10^3, 9.1×10^{-8}, 2.4×10^3. 5.08

13 Evaluate each expression using scientific notation. 5.09

a $(3.65 \times 10^{22}) \times (7.4 \times 10^8)$ b $(1.44 \times 10^{10}) \div (3.6 \times 10^4)$
c $(5 \times 10^5)^3$ d $\sqrt{6.25 \times 10^{-8}}$

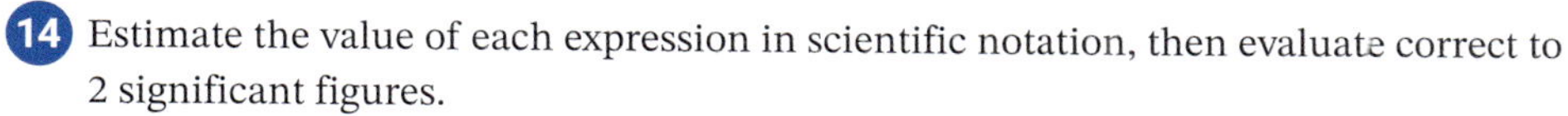

14 Estimate the value of each expression in scientific notation, then evaluate correct to 2 significant figures. 5.09

a $(8.9 \times 10^9) \times (1.1 \times 10^7)$ b $(9.3 \times 10^{15}) \times (4 \times 10^2)$ c $(3.1 \times 10^4)^2$

CHAPTER 5 TEST YOURSELF

☐ Foundation ○ Standard ⬡ Complex

6

SPACE

Geometry

'Geometry' comes from the Greek word *geometria*, which means 'land measuring'. The principles and ideas of geometry are evident everywhere – in road signs, buildings, bridges and patterns for tiles and wallpaper. Many examples of geometric patterns can be seen in architecture, such as in this ground view of an apartment building in Berlin, Germany.

Alamy Stock Photo/Stefan Sollfors

Chapter outline

		Proficiencies
6.01	Triangles	U F R C
6.02	Quadrilaterals	U F R C
6.03	Polygons	U F R C
6.04	Networks*	U F R C
6.05	Polyhedra*	U F R C

*YEAR 10 EXTENSION

U = Understanding
F = Fluency
PS = Problem solving
R = Reasoning
C = Communication

Wordbank

angle sum The total of the sizes of all the angles in a shape, such as a triangle

bisect To cut in half

convex polygon A polygon whose vertices all point outwards

diagonal An interval joining two non-adjacent vertices of a shape

edge (of a network) A line that connects vertices (points) together

network An arrangement of edges (lines) that connect a set of vertices (points) connected by edges

regular polygon A polygon with all angles equal and all sides equal, such as a square

Quiz
Wordbank 6

Videos (3):

6.01 Angle sum of a triangle • Exterior angle of a triangle 1 • Exterior angle of a triangle 2

Twig videos (3):

6.04 The seven bridges of Konigsberg • Networks: Labyrinths and mazes

6.05 Polyhedra: Platonic solids

Quizzes (5):

- Wordbank 6
- SkillCheck 6
- Mental skills 6
- Language of maths 6
- Test yourself 6

Skillsheets (2):

SkillCheck Types of angles

6.02 Naming shapes

Worksheets (10):

6.02 Classifying quadrilaterals • Properties of triangles and quadrilaterals • Deductive geometry • Diagonal properties of quadrilaterals • Shapes and angles review

6.03 Angle sum of a polygon • Equal angles

6.05 Nets of Platonic solids

Geometry summary poster

Mind map: Geometry

Puzzles (3):

6.02 Mixed angle problems • Parallel lines • Triangles and quadrilaterals

Technology (1):

6.03 Angle sum of a polygon

Presentation (1):

6.02 Angles and shapes

Nelson MindTap

To access resources above, visit **cengage.com.au/nelsonmindtap**

6 In this chapter you will:

- ✓ classify triangles, quadrilaterals and polygons.
- ✓ solve geometrical problems involving the angle sum of a triangle and quadrilateral and the exterior angle of a triangle.
- ✓ solve geometrical problems involving the angle sum of a polygon and the exterior angle sum of a convex polygon.
- ✓ identify the faces, vertices and edges of a network.
- ✓ identify the faces, vertices and edges of a polyhedron.
- ✓ investigate the properties of polyhedra including the Platonic solids.
- ✓ investigate Euler's rule for networks and polyhedra.

Quiz SkillCheck 6

Skillsheet Types of angles

SkillCheck

ANSWERS ON P. 642

1 Classify each angle as acute, obtuse, reflex, right, straight or a revolution.

a

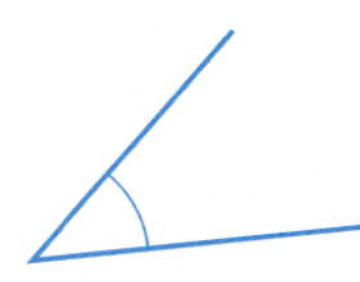

b

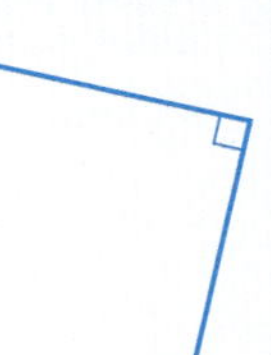

c

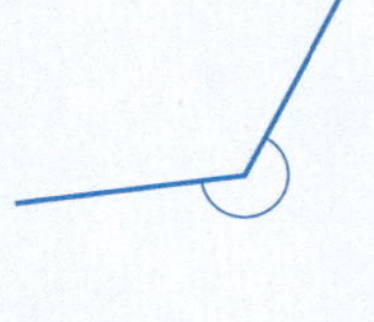

d

e

f

g

h

2 What type of angle is 95°? Select the correct answer **A**, **B**, **C** or **D**.

A acute
B a revolution
C reflex
D obtuse

3 Which of the following is a reflex angle? Select the correct answer **A**, **B**, **C** or **D**.

A 70°
B 270°
C 115°
D 360°

4 An angle of 270° is which of the following? Select the correct answer **A**, **B**, **C** or **D**.

A an obtuse angle
B a right angle
C equal to a complete turn
D equal to three right angles

Triangles 6.01

Triangles can be classified in two ways: by their sides or by their angles.

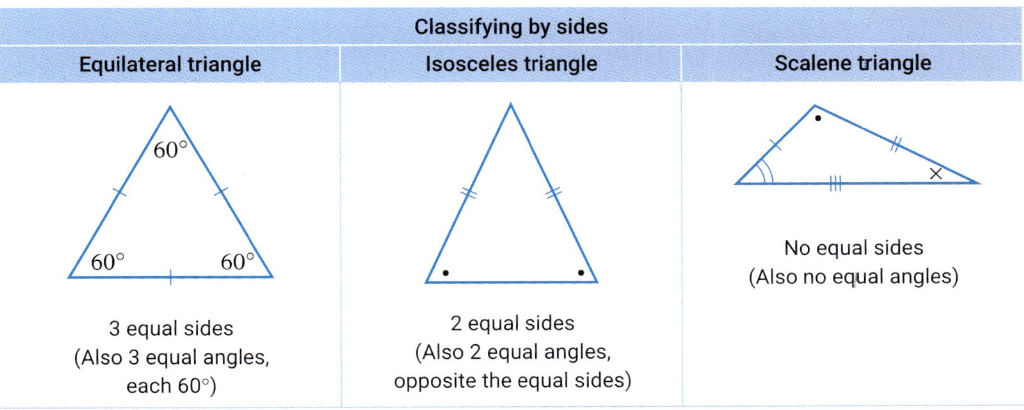

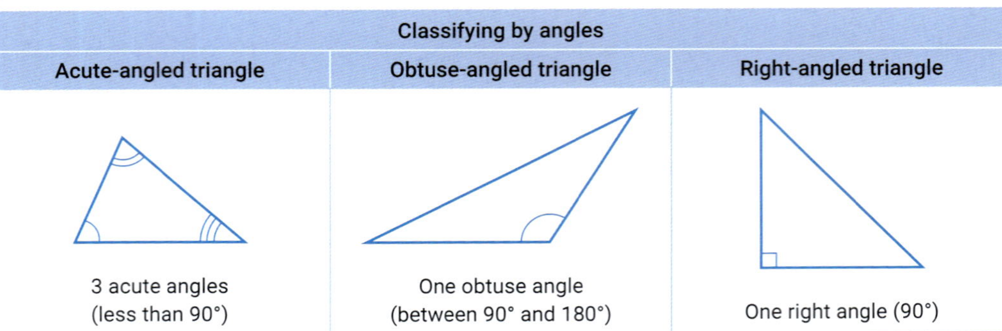

Classifying by angles		
Acute-angled triangle	Obtuse-angled triangle	Right-angled triangle
3 acute angles (less than 90°)	One obtuse angle (between 90° and 180°)	One right angle (90°)

Video Angle sum of a triangle

ⓘ Angle sum of a triangle

The **angle sum of a triangle** is 180°.

$$a + b + c = 180$$

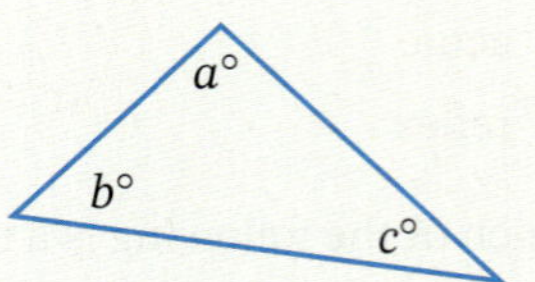

Proof:

Consider any triangle PQR with angles $a°$, $b°$ and $c°$.

Construct a line through P parallel to QR.

$\therefore \angle UPQ = b$ (alternate angles, $UV \parallel QR$)

and $\angle VPR = c$ (alternate angles, $UV \parallel QR$)

But $\angle UPQ + \angle QPR + \angle VPR = 180$ (angles on a straight line)

$\therefore a° + b° + c° = 180°$

$\therefore$ The sum of the angles of a triangle is 180°.

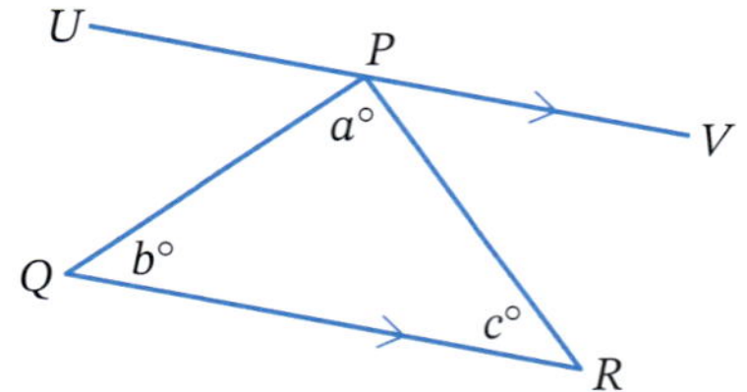

Example 1

Find the value of each variable, giving reasons.

a

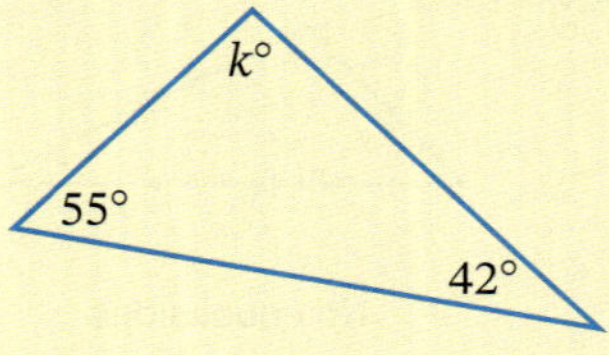

b

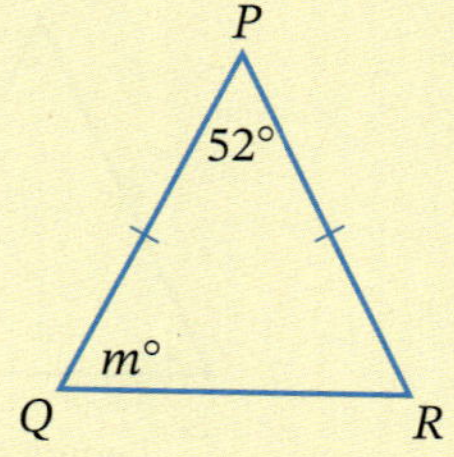

SOLUTION

a $k + 55 + 42 = 180$ (angle sum of a triangle)

$k = 180 - 55 - 42$

$= 83$

b $\angle R = \angle Q = m°$ ($\triangle PQR$ is isosceles)

$m + m + 52 = 180$ (angle sum of a triangle)

$2m + 52 = 180$

$2m = 128$

$m = 64$

ⓘ The exterior angle of a triangle

Video Exterior angle of a triangle 1

The **exterior angle of a triangle** is equal to the sum of the two interior opposite angles.

$$z = x + y$$

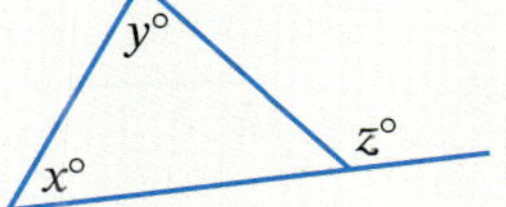

Example 2

Video Exterior angle of a triangle 2

Find the value of k in each diagram.

a

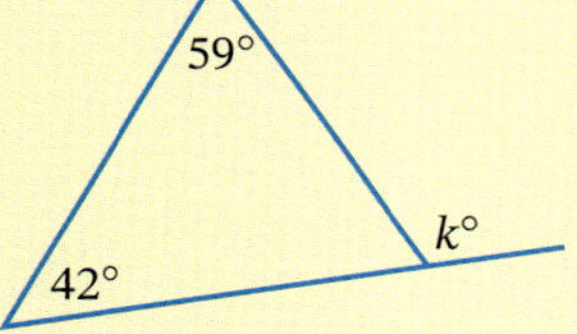

b

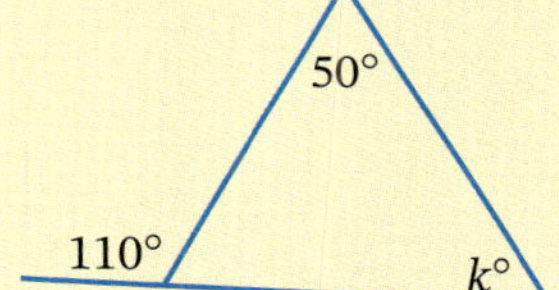

SOLUTION

a $k = 42 + 59$ (exterior angle of triangle)

$= 101$

b $k + 50 = 110$ (exterior angle of triangle)

$k = 60$

Example 3

Video Exterior angle of a triangle 2

Find the size of the marked angle $\angle BDF$.

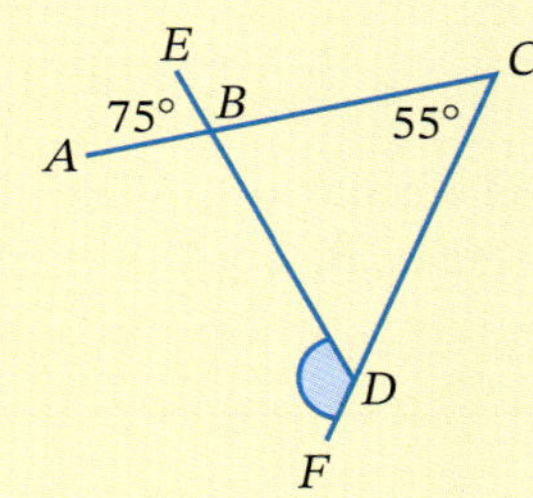

SOLUTION

$\angle DBC = \angle ABE = 75°$ (vertically opposite angles)

$\angle BDF = 75° + 55°$ (exterior angle of $\triangle BCD$)

$= 130°$

EXERCISE 6.01 ANSWERS ON P. 642

Triangles

1 Find the value of each variable, giving reasons.

a

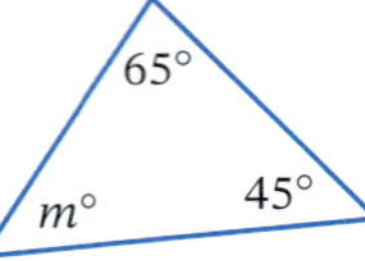

b

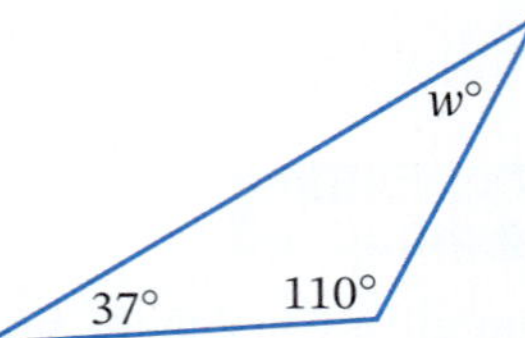

c

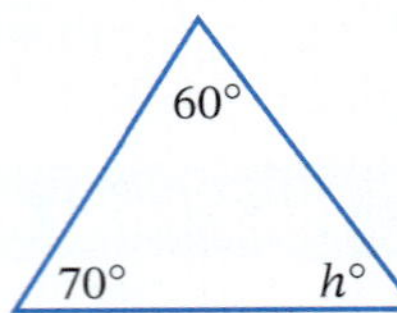

d

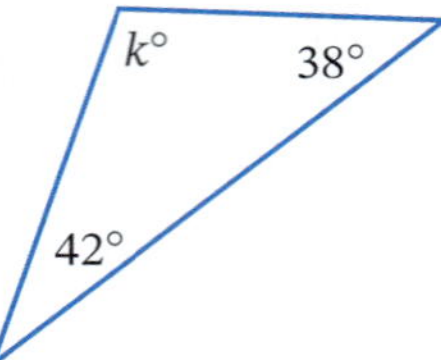

e

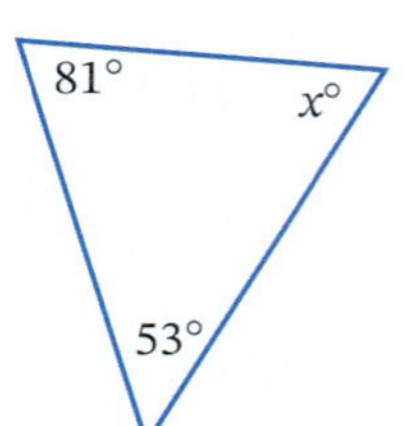

f

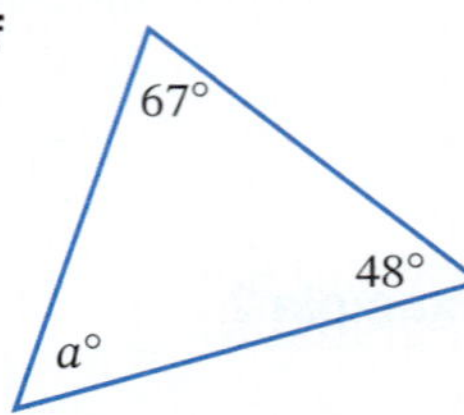

g

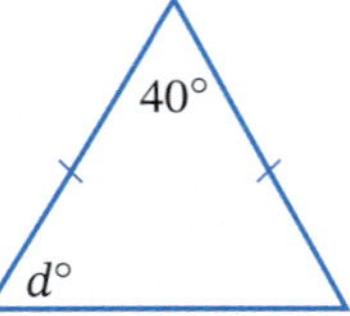

h

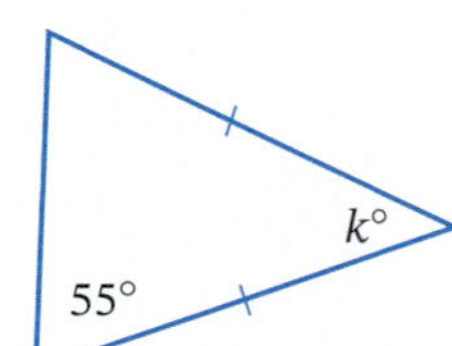

i

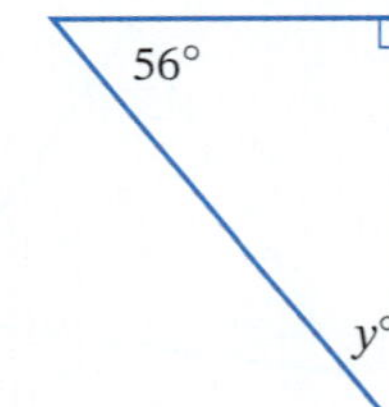

j

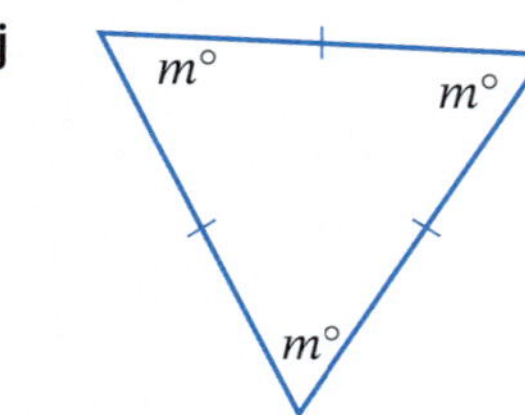

k

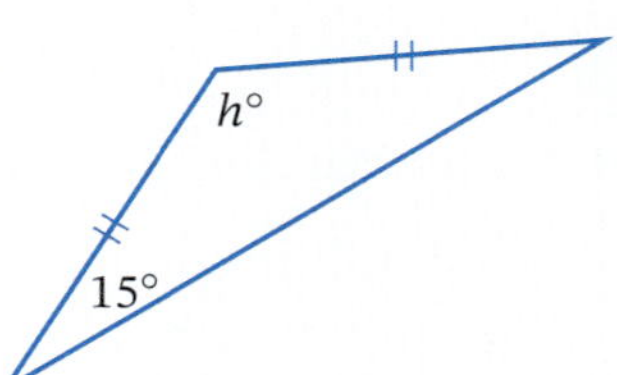

l

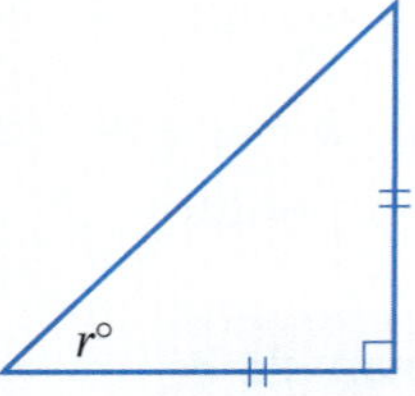

2 What is the size of each angle in an equilateral triangle? Select the correct answer **A**, **B**, **C** or **D**.

A 30 **B** 45 **C** 60 **D** 90

3 One angle of an isosceles triangle is equal to 80. Which 2 of the following could be the sizes of the other angles? Select 2 of **A**, **B**, **C** or **D**.

A 80 and 20 **B** 40 and 60 **C** 40 and 40 **D** 50 and 50

☐ Foundation ○ Standard ⬡ Complex

4 Find the value of m.

R C **a**

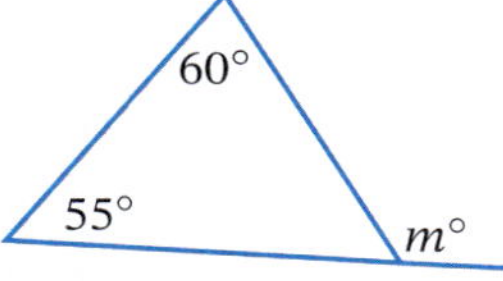

b

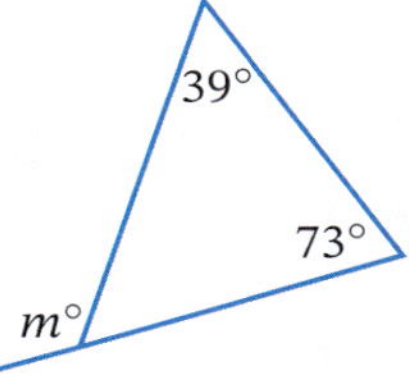

c

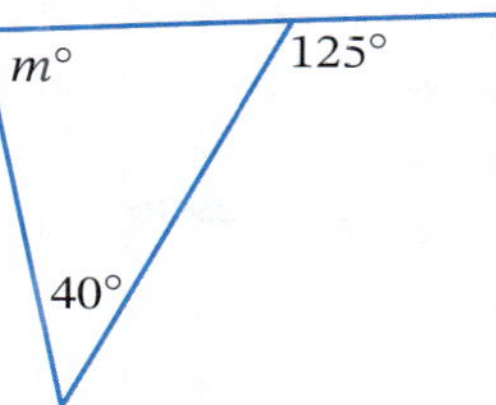

d

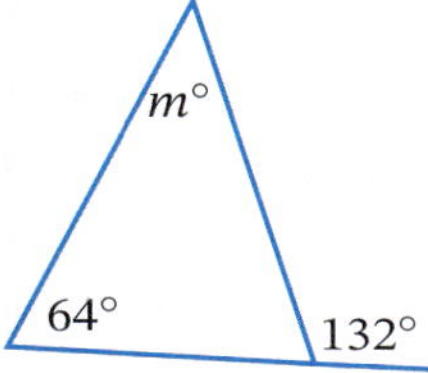

e

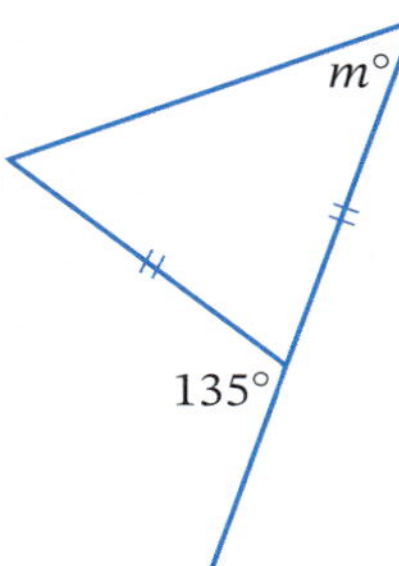

f

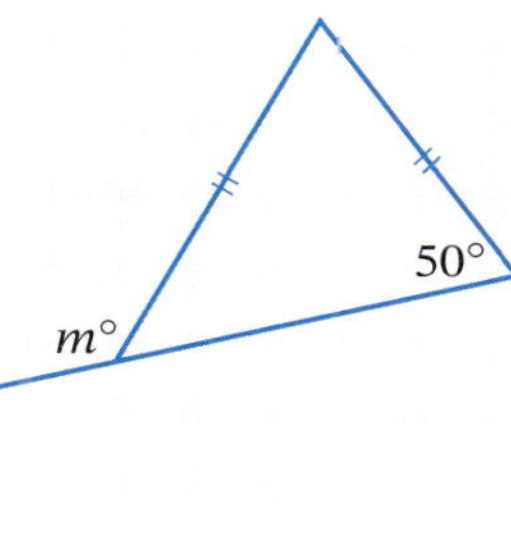

5 What is the size of the unknown angles in this triangle?
R Select the correct answer **A**, **B**, **C** or **D**.

A 30° **B** 45°

C 60° **D** 90°

6 One angle of an obtuse-angled triangle is 50°. Which of the following could be the sizes
R of the other angles? Select the correct answer **A**, **B**, **C** or **D**.

A 80° and 50° **B** 100° and 30° **C** 65° and 65° **D** 45° and 95°

EXAMPLE 3

7 Find the size of $\angle CBF$ in each diagram.

R C **a**

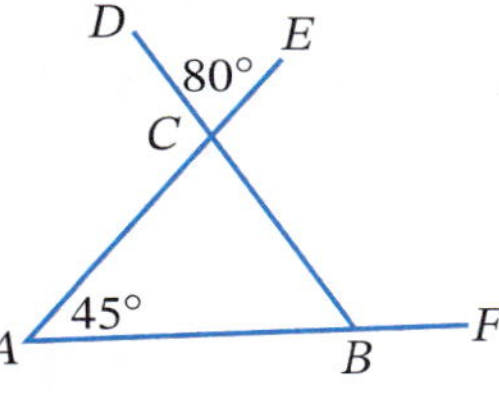

b

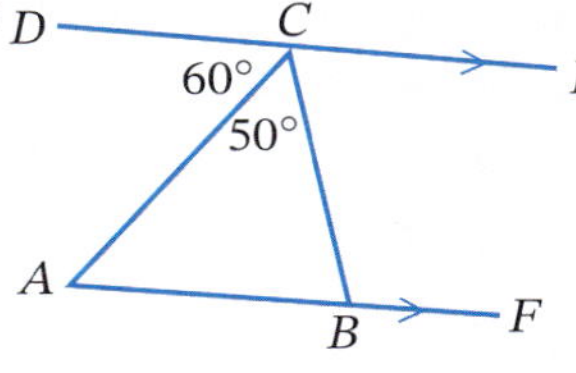

c

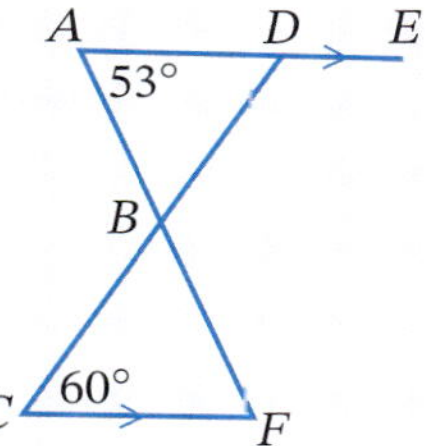

d

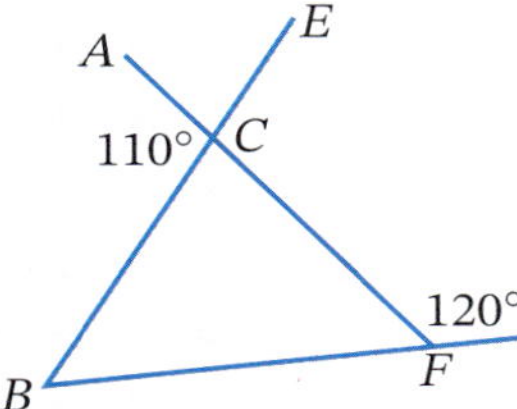

e

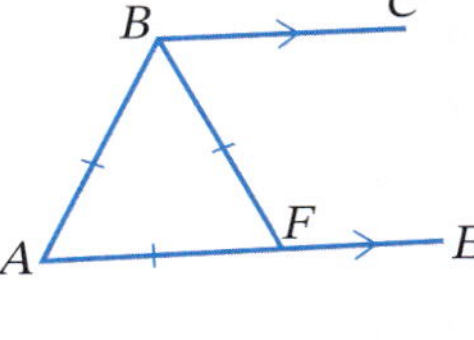

f

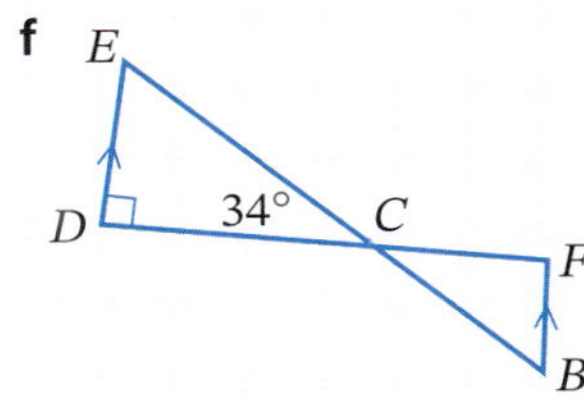

EXAMPLE 2

6.01

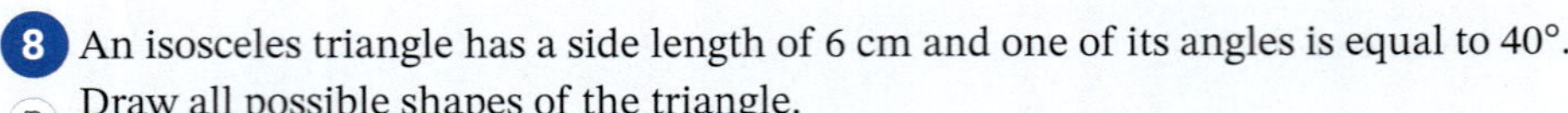

8 An isosceles triangle has a side length of 6 cm and one of its angles is equal to 40°.
R Draw all possible shapes of the triangle.

9 The diagram shows the shape of a roof truss. Find the value of each variable.
R

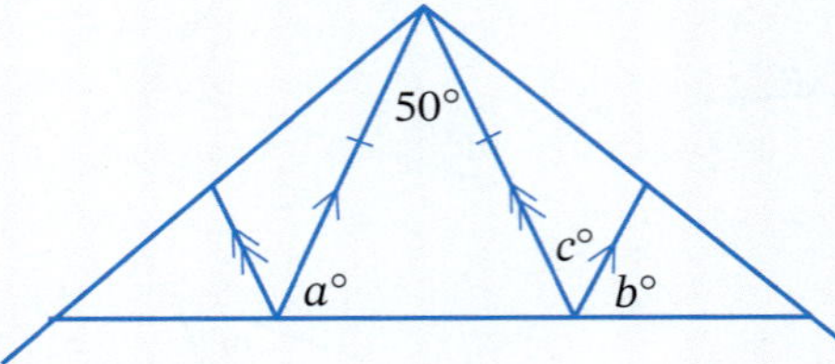

10 Copy this diagram and mark all angles equal to the angle marked d.
R

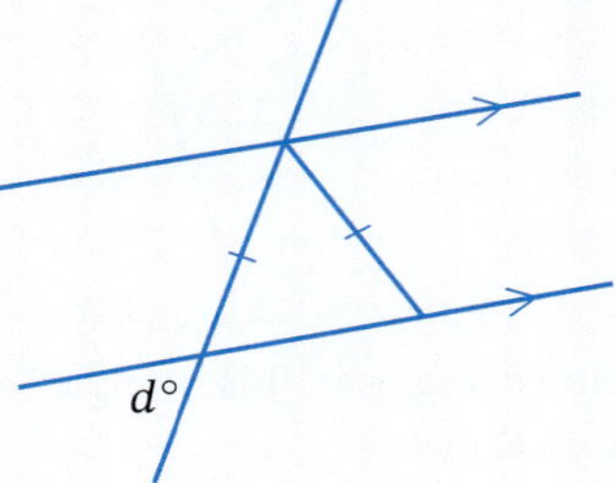

11 Find the value of each variable.
R C

a

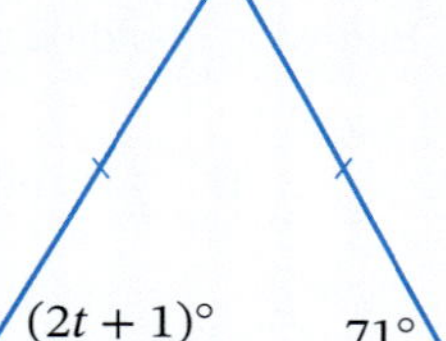

b

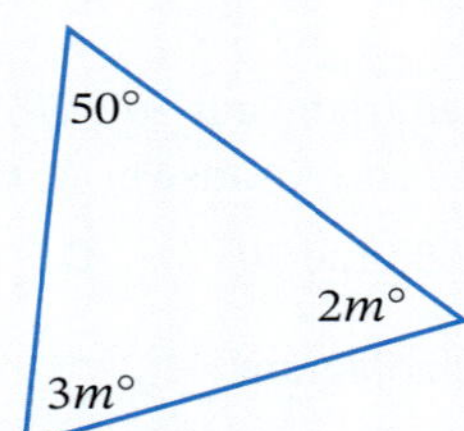

c

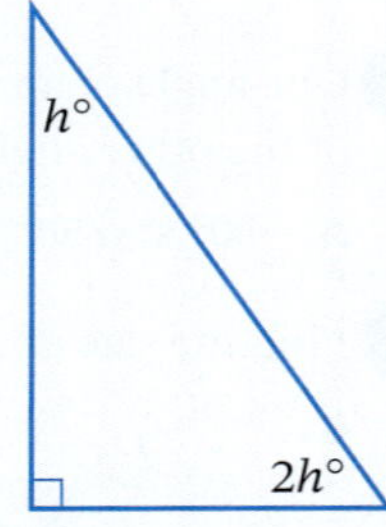

d

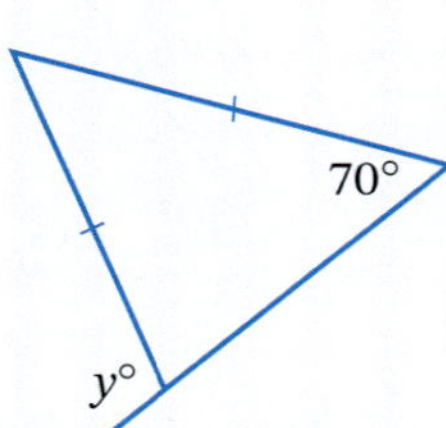

e

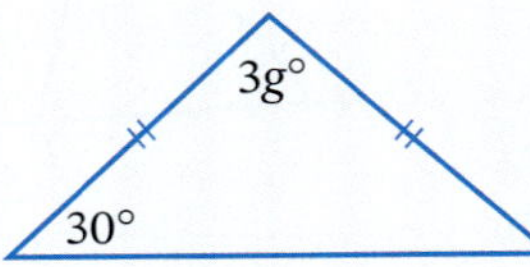

f

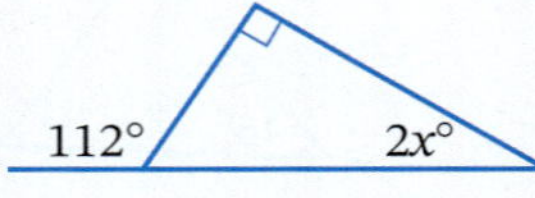

TECHNOLOGY

Sketching parallelograms and rectangles

In this activity, you will use dynamic geometry software to draw parallelograms and rectangles.

1 **Use** the **line tool to construct quadrilateral** *ABCD*. Now manipulate the vertices to change the opposite sides so that they are parallel.

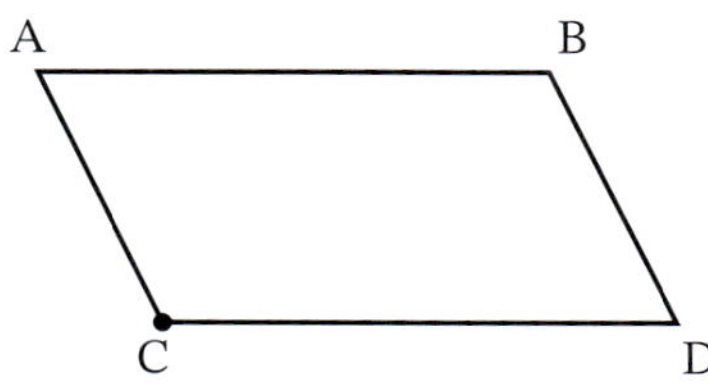

OR

2 Select **distance or length** to measure the sides. In both parallelograms and rectangles, opposite sides are of same length. Also select **angle** to measure the angle sizes. In rectangles, all angles are 90°.

3 Manipulate the quadrilateral you have constructed to create:

a a parallelogram. b a rectangle.

4 Construct other quadrilaterals using the **polygon** tool and take their **angle** and **length** measurements. Swap their dimensions with other students in your class and try to reproduce their quadrilaterals using geometry software.

Quadrilaterals 6.02

A **quadrilateral** is any shape with four straight sides. A quadrilateral may be either **convex** or **non-convex** (**concave**).

Worksheets
Classifying quadrilaterals
Properties of triangles and quadrilaterals
Deductive geometry
Diagonal properties of quadrilaterals
Shapes and angles review

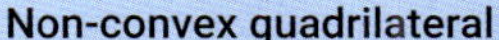

Convex quadrilateral	Non-convex quadrilateral
	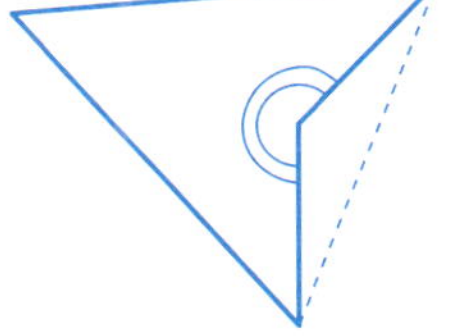
• All **vertices** (corners) point outwards. • All **diagonals** lie within the shape. • All angles are less than 180°.	• One vertex points inwards. • One diagonal lies outside the shape. • One angle is more than 180° (reflex angle).

Skillsheet
Naming shapes

Puzzles
Mixed angle problems

Parallel lines

Triangles and quadrilaterals

Presentation
Angles and shapes

There are 6 special types of quadrilaterals.

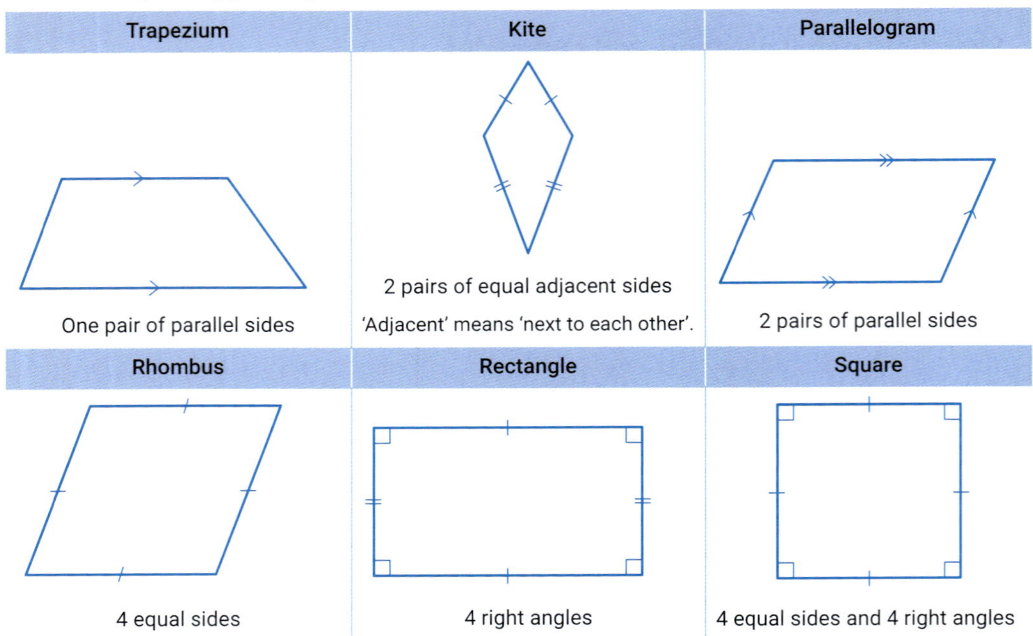

ⓘ Angle sum of a quadrilateral

The **angle sum of a quadrilateral** is 360°.

$$a + b + c + d = 360$$

This property is true for both convex and non-convex quadrilaterals.

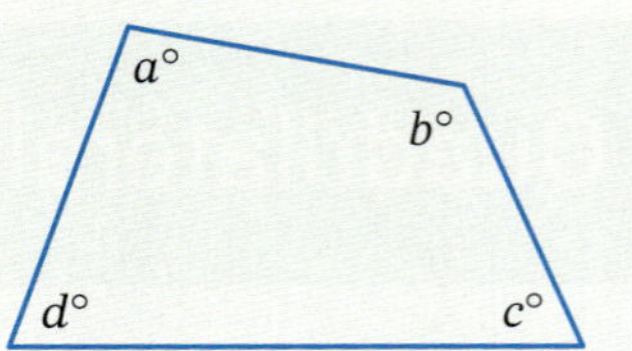

ⓘ Properties of quadrilaterals

Trapezium

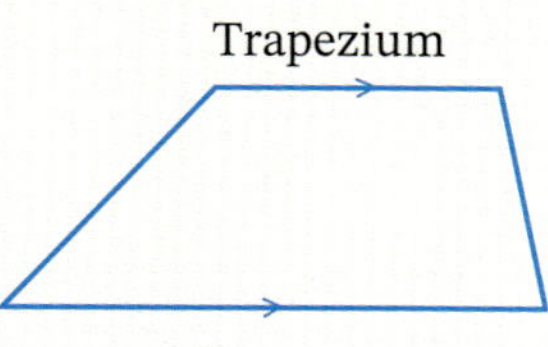

- One pair of opposite sides parallel
- No axes of symmetry

Kite

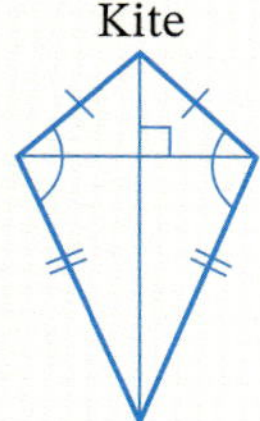

- 2 pairs of equal adjacent sides
- One pair of opposite angles equal
- One axis of symmetry
- Diagonals intersect at right angles

Parallelogram

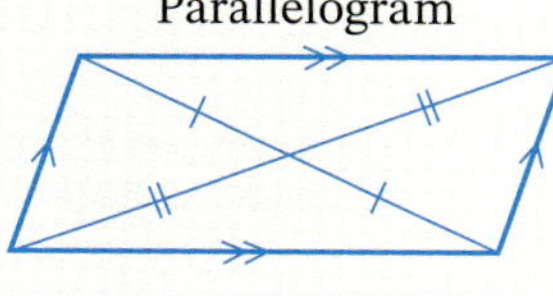

- Opposite sides are parallel and equal
- Opposite angles are equal
- No axes of symmetry
- Diagonals bisect each other

Rhombus

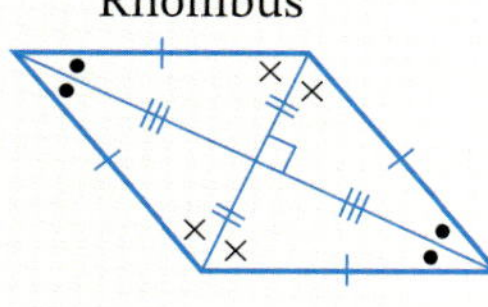

- All sides are equal
- 2 axes of symmetry
- A special type of parallelogram
- Diagonals bisect each other at right angles
- Diagonals bisect the angles of the rhombus

Rectangle

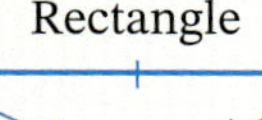

- All angles are 90° (right angles)
- 2 axes of symmetry
- A special type of parallelogram
- Diagonals are equal (in length)
- Diagonals bisect each other

Square

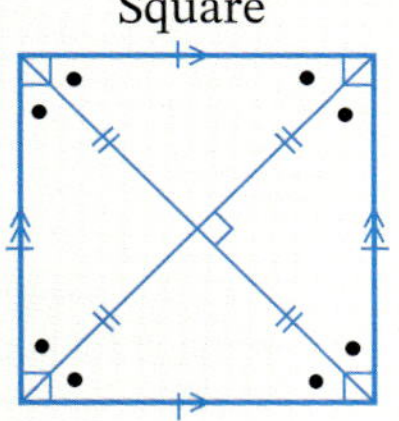

- All sides are equal
- All angles are 90° (right angles)
- 4 axes of symmetry
- A special type of rhombus and rectangle
- Diagonals are equal
- Diagonals bisect each other at right angles
- Diagonals bisect the angles of the square

Example 4

Find the value of each variable.

a

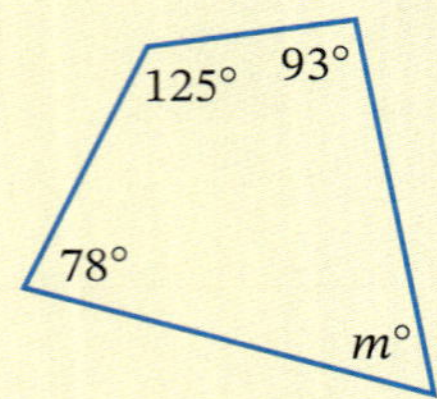

b

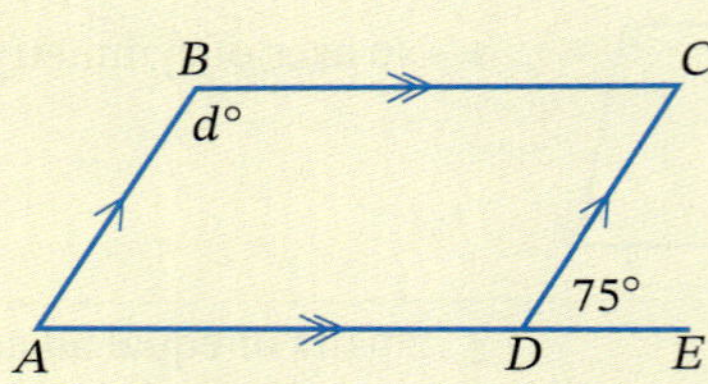

SOLUTION

a $m + 78 + 125 + 93 = 360$ (angle sum of a quadrilateral)

$m = 360 - 78 - 125 - 93$

$= 64$

b $\angle CDA = 180° - 75°$ (angles on a straight line)

$= 105°$

$d = 105$ (opposite angles of a parallelogram)

Can you see another method for finding *d*?

Example 5

Find the size of $\angle BED$.

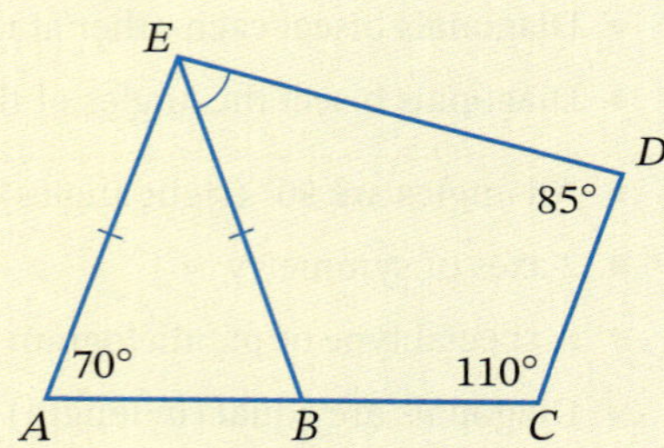

SOLUTION

$\angle ABE = 70°$ (equal angles in isosceles $\triangle ABE$)

$\angle EBC = 180° - 70°$ (angles on a straight line)

$= 110°$

$\angle BED = 360° - 85° - 110° - 110°$ (angle sum of quadrilateral *BCDE*)

$= 55°$

Can you see another method for finding $\angle BED$?

EXERCISE 6.02 ANSWERS ON P. 642

Quadrilaterals

U F R C

EXAMPLE 4

1 Find the value of m in each diagram.

R

a

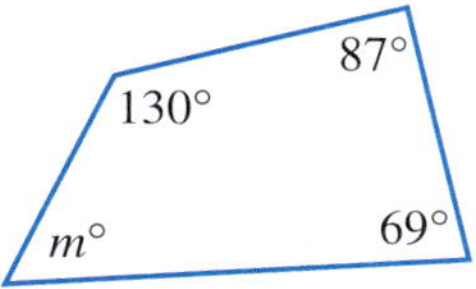

b

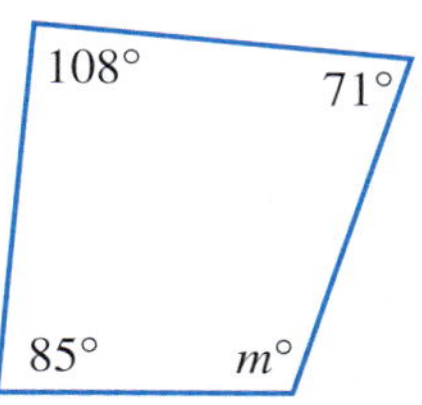

c

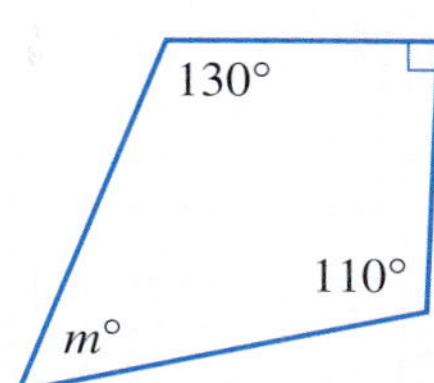

d

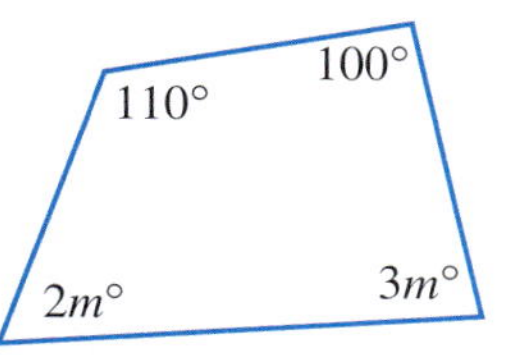

e

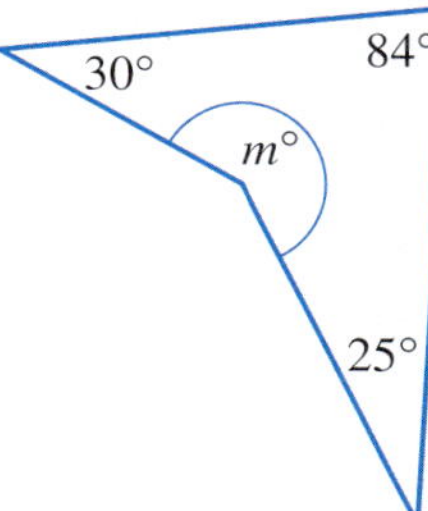

f

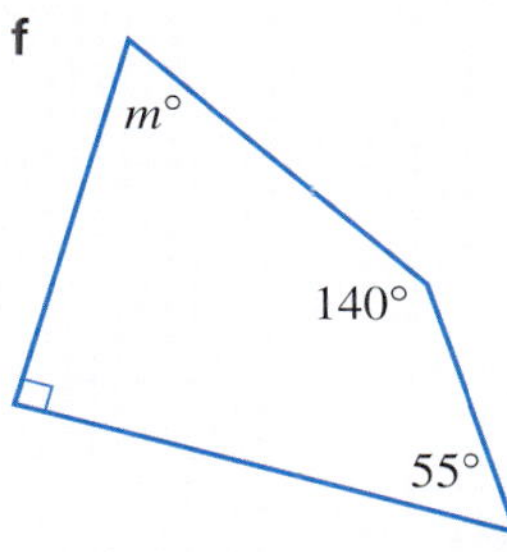

2 Copy and complete this table.

R C

Property	Trapezium	Kite	Parallelogram	Rhombus	Rectangle	Square
One pair of opposite sides parallel						
Opposite sides parallel						
Opposite sides equal			✓			
All sides equal						
Two pairs of adjacent sides equal						
Diagonals equal					✓	
Diagonals bisect each other						
Diagonals meet at right angles						
Diagonals bisect the angles of the shape						
Opposite angles equal						
One pair of opposite angles equal						
All angles 90°						
Axes of symmetry	0					

☐ Foundation ○ Standard ○ Complex

3 Refer to the table you completed in question **2**, and name all quadrilaterals that have:

R **a** no axes of symmetry

b one pair of parallel sides

C **c** four equal sides

d equal diagonals

e opposite sides equal

f four axes of symmetry

g adjacent sides equal

h one axis of symmetry

i opposite sides parallel

j all angles measuring 90°

k two axes of symmetry

l diagonals which bisect each other

m opposite angles equal

n diagonals meeting at 90°

4 Each triangle in this diagram is an equilateral triangle.

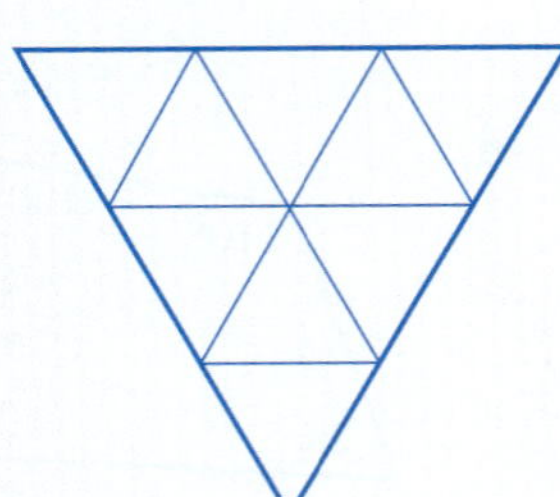

R **a** Name the different types of quadrilaterals that you can find in the diagram.

b How many of each type of quadrilateral are there in the diagram?

c How many triangles are there? (It may help to copy the diagram so that you can draw on it using coloured pencils.)

5 Which statements are *always* true?

R **A** A rhombus is a parallelogram.

B The diagonals of a parallelogram meet at right angles.

C A square is a rhombus.

D A parallelogram is a quadrilateral with a pair of opposite sides equal and parallel.

E A square is a rectangle.

F The diagonals of an isosceles trapezium bisect each other.

G The opposite angles of a rhombus are equal.

H The diagonals of a rhombus are equal and bisect each other at right angles.

I A rectangle is a parallelogram.

6 Name each quadrilateral using the properties marked on it.

R **a**

b

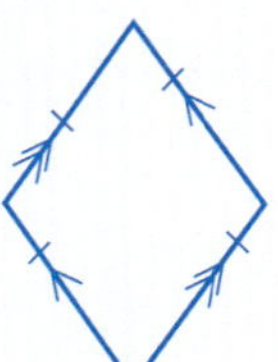

c

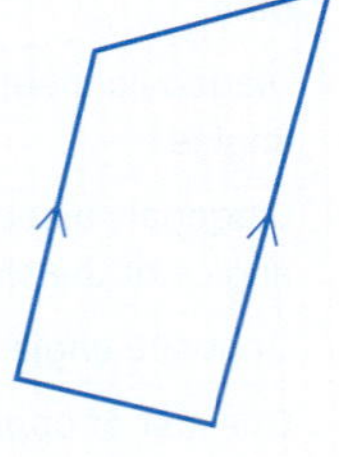

☐ Foundation ○ Standard ⬡ Complex

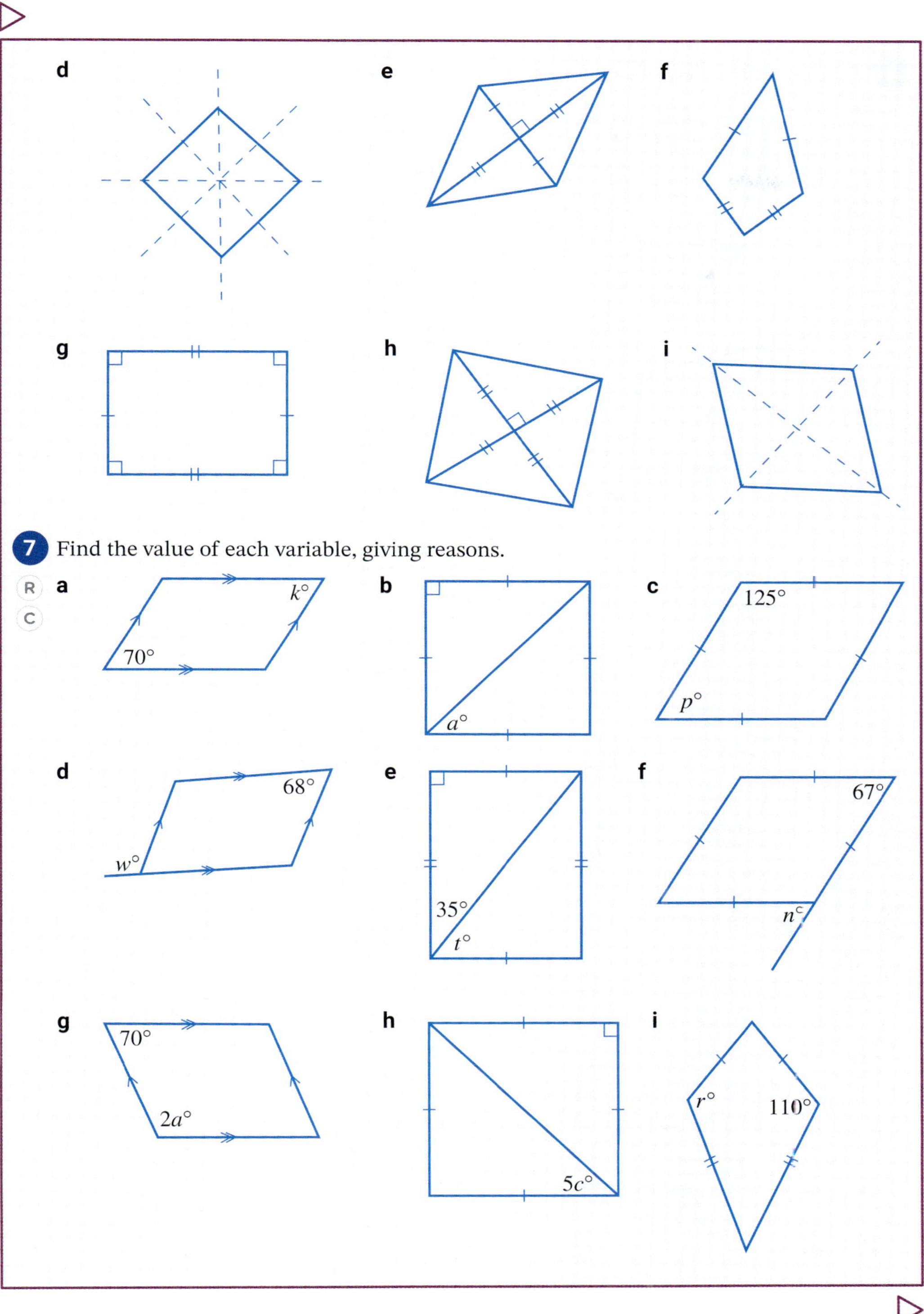

Foundation Standard Complex

EXAMPLE 5

8 Find the size of $\angle PQR$ in each diagram, giving reasons.

R C

a

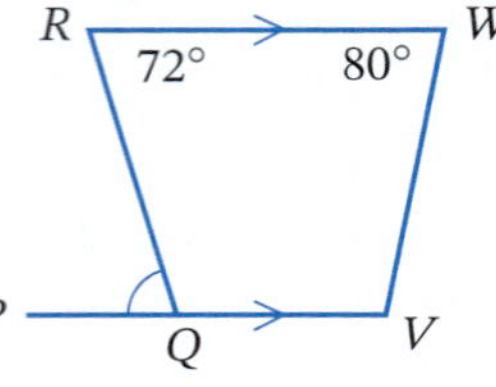

b

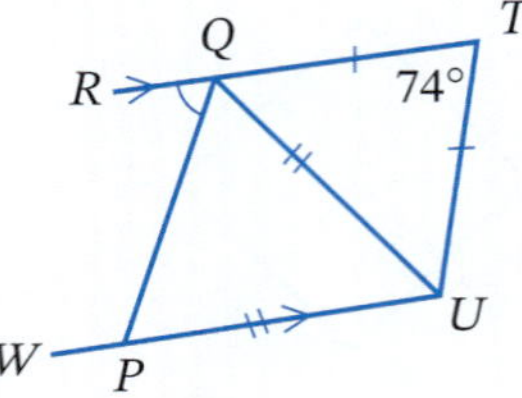

c

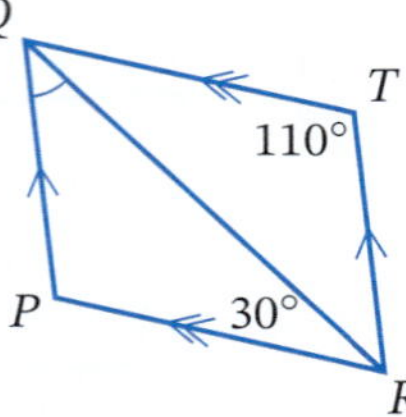

d

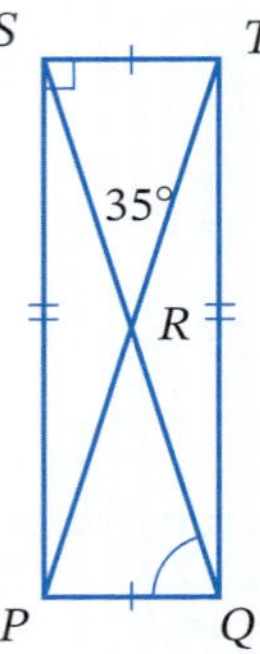

e

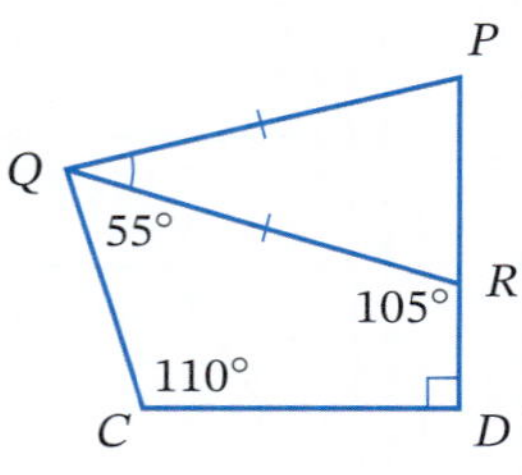

f

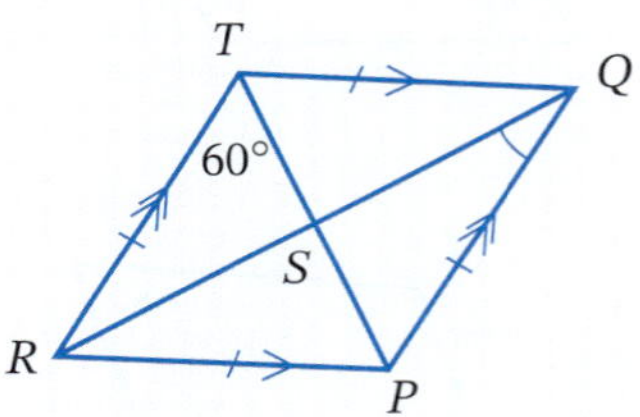

9 Use geometrical instruments or dynamic geometry software to construct each quadrilateral.

a Parallelogram $TUVW$ with sides $TU = 30$ mm, $UV = 60$ mm and $\angle TUV = 113°$.

b Rhombus with side length 5 cm.

c Convex quadrilateral $WXYZ$ with $WX = 65$ mm, $WZ = 40$ mm, $\angle ZWX = 54°$, $\angle WZY = 114°$ and $XY = 24$ mm.

d Non-convex quadrilateral with one side 4.5 cm and one angle 200°.

e $DEFG$ as shown in the diagram.

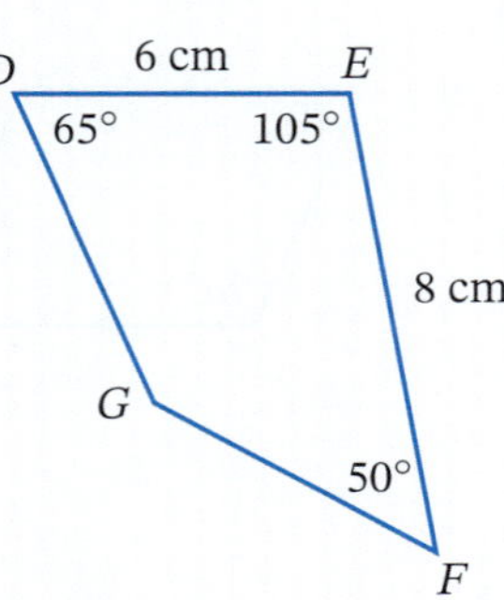

Foundation Standard Complex

DID YOU KNOW?

Geometry in art

Geometry has been used in artwork for centuries.

Tessellations can be used to create artwork, such as the regular hexagons used in the tessellation shown.

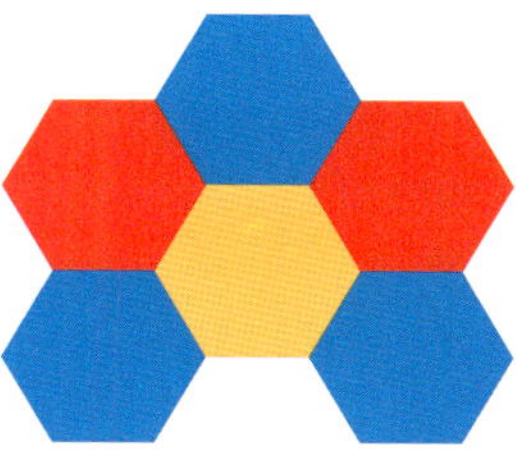

Indian and Islamic artworks show incredible intricacy, detail and colour within the geometric images used.

iStock.com/sharrocks

Fractals use geometric formulas to create infinitely occurring images, so they are usually created using computer software. A famous fractal is the Mandelbrot set, shown here.

1 Design a tessellation or find examples of geometry in art.

2 Investigate fractals such the Mandelbrot set and the Koch snowflake, and their history.

Alamy Stock Photo/Visharo

INVESTIGATION

Angle sum of a polygon

To find the angle sum of a pentagon *ABCDE* (5 sides), follow this reasoning.

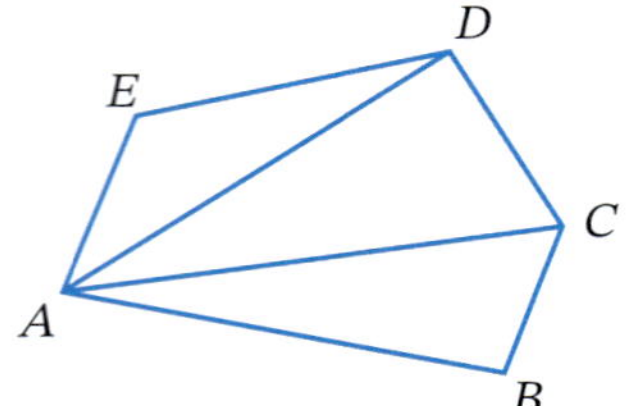

From one vertex of the pentagon, draw all the diagonals. The diagonals from vertex *A* have divided the pentagon into three triangles.

∴ Angle sum of a pentagon = angle sum of the three triangles

$= 3 \times 180°$

$= 540°$

1 **a** Draw a hexagon (six sides) and from one vertex draw all the diagonals.

b How many diagonals are there?

c How many triangles did you form?

d Hence find the angle sum of a hexagon.

2 Repeat the procedure to find the angle sum of:

a an octagon (8 sides) **b** a decagon (10 sides)

3 Copy and complete this table.

Polygon	Number of sides	Number of triangles	Angle sum
Triangle	3	1	180°
quadrilateral	4	2	
Pentagon	5		
Hexagon			
Octagon			
Decagon			

4 Copy and complete this pattern: The number of triangles formed is always 2 ________ than the number of __________ of the polygon.

5 **a** Using your own words, describe the rule for finding the angle sum of a polygon.

b What is the angle sum of a polygon with 20 sides?

c For a polygon with *n* sides, write a formula for the sum of its angles. Discuss your result with other students.

6 **a** Draw a non-convex octagon.

b Divide the polygon into triangles as shown.

c How many triangles have been formed?

d Find the angle sum of this non-convex octagon.

e Does your rule for the angle sum of a polygon also apply to the non-convex polygon?

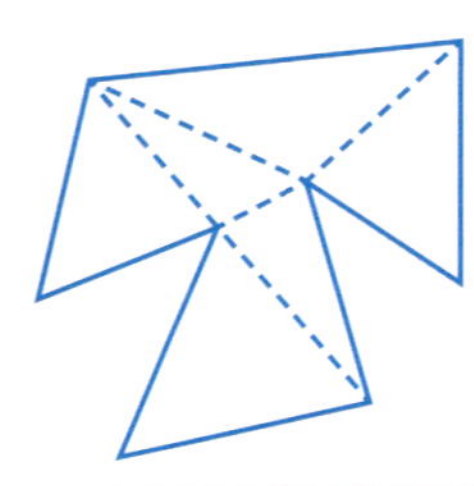

Polygons

A **polygon** is a general name for any shape with straight sides. The word is derived from the Greek, meaning 'many angles'. Shapes with curved sides, such as circles, ellipses and semicircles, are *not* polygons.

A polygon may be either convex or **non-convex** (**concave**).

Worksheet Angle sum of a polygon

Technology Angle sum of a polygon

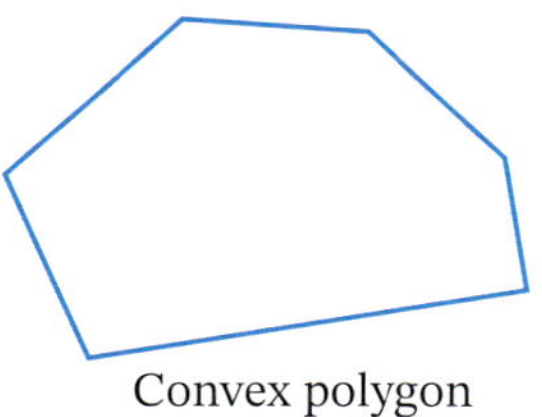

Convex polygon

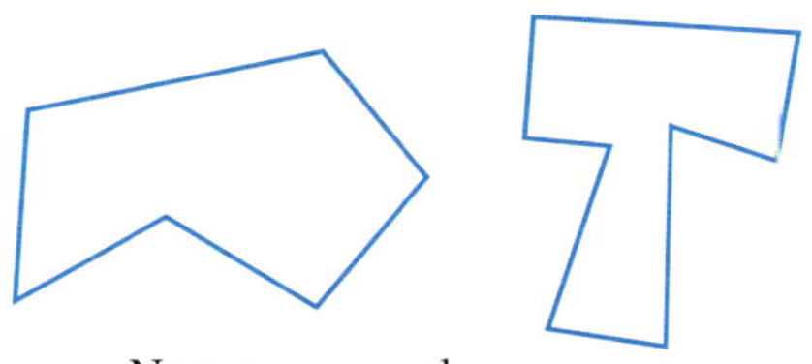

Non-convex polygons

In a convex polygon, all vertices point outwards, all diagonals lie within the shape and all angles are less than 180°.

In a non-convex polygon, some vertices point inwards, some diagonals lie outside the shape and some angles are more than 180° (reflex angles).

A polygon's name is determined by its number of sides.

Name	Number of sides
Pentagon	5
Hexagon	6
Heptagon	7
Octagon	8
Nonagon	9
Decagon	10
Undecagon	11
Dodecagon	12

Shutterstock.com/goturk_06

iStock.com/TimZillion

Shutterstock.com/Stephen Finn

The images above show the Pentagon building in the USA, a 50 p coin from the UK, and a stop sign in Hong Kong, China.

ⓘ Angle sum of a polygon

The **angle sum of a polygon** with n sides is given by the formula $A = 180(n - 2)°$.

This formula applies to both convex and non-convex polygons.

Example 6

Find the angle sum of a nonagon (9 sides).

SOLUTION

Substitute $n = 9$ into $A = 180(n - 2)$

$$\begin{aligned}\text{Angle sum} &= 180(9 - 2)\\ &= (180 \times 7)\\ &= 1260°\end{aligned}$$

Example 7

Find the number of sides of a polygon that has an angle sum of 720°.

SOLUTION

$$180(n - 2) = 720$$

$A = 720$

$$\begin{aligned}180n - 360 &= 720\\ 180n &= 1080\\ n &= \frac{1080}{180}\\ &= 6\end{aligned}$$

∴ The polygon has six sides (hexagon).

Regular polygons

Worksheet Equal angles

A **regular polygon** has all angles equal and all sides equal. For example, a regular pentagon has five equal sides and five equal angles. A square is a regular polygon but a rhombus is not.

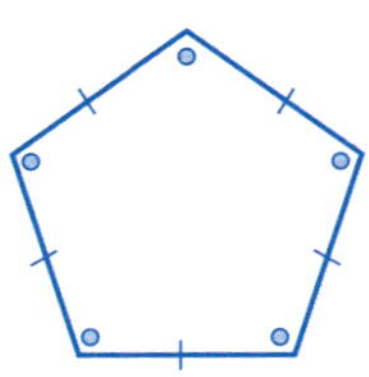

ⓘ Angle in a regular polygon

The size of each angle in a **regular polygon** with n sides is

$$\frac{\text{Angle sum}}{\text{No. of sides}} = \frac{180(n - 2)°}{n}$$

Example 8

Find the size of one angle in a regular hexagon.

SOLUTION

A hexagon has six sides ($n = 6$).

$$\text{Size of one angle} = \frac{180(6-2)^\circ}{6}$$

$$= \frac{(180 \times 4)^\circ}{6}$$

$$= 120^\circ$$

Each angle in a regular hexagon is 120°.

EXERCISE 6.03 ANSWERS ON P. 643

Polygons

U F R C

1 How many sides has:

- **a** a hexagon?
- **b** a quadrilateral?
- **c** a nonagon?
- **d** a decagon?
- **e** a heptagon?
- **f** a pentagon?
- **g** a dodecagon?
- **h** an octagon?
- **i** an undecagon?

2 Name each polygon.

R C

a

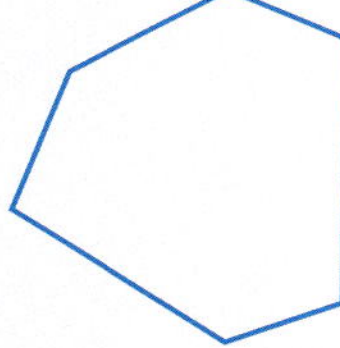

b

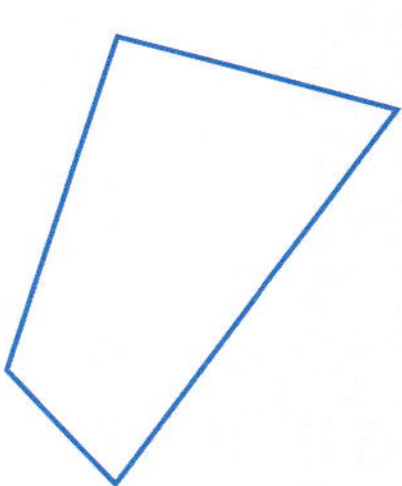

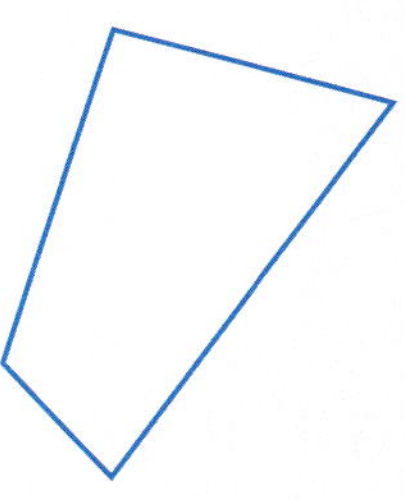

c

d

e

f

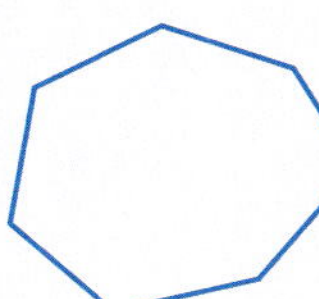

g

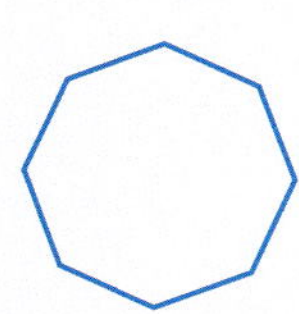

h

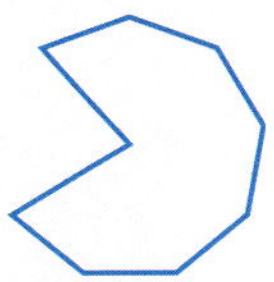

3 Which polygons from question **2** are:

R C

a convex? **b** regular? **c** concave?

4 Which shape is a regular octagon? Select the correct answer **A**, **B**, **C** or **D**.

A

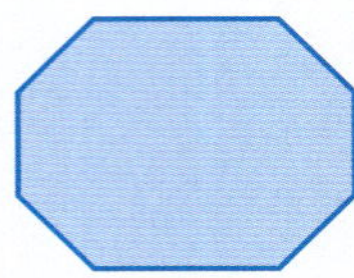

B

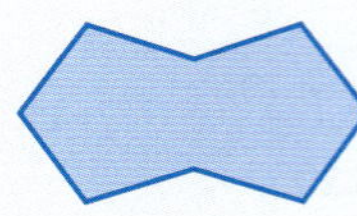

C

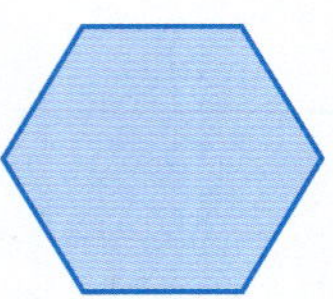

D

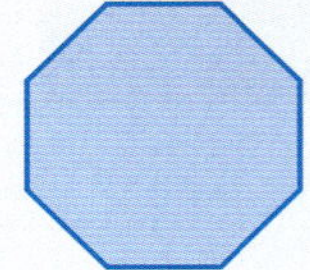

5 What is the more common name for:

R C

a a regular triangle? **b** a regular quadrilateral?

EXAMPLE 6

6 Find the angle sum of a polygon with:

a 5 sides **b** 8 sides **c** 15 sides **d** 12 sides

e 7 sides **f** 10 sides **g** 20 sides **h** 11 sides

7 Find the value of each variable.

R

a

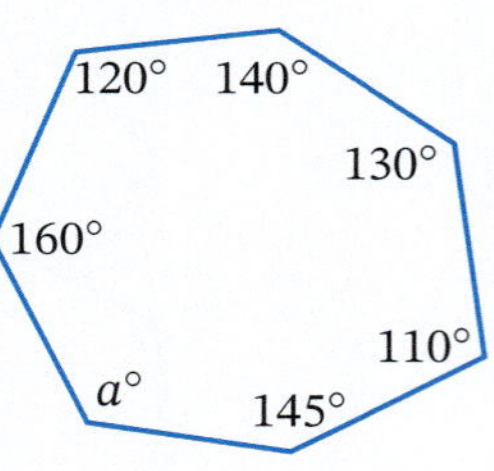

b

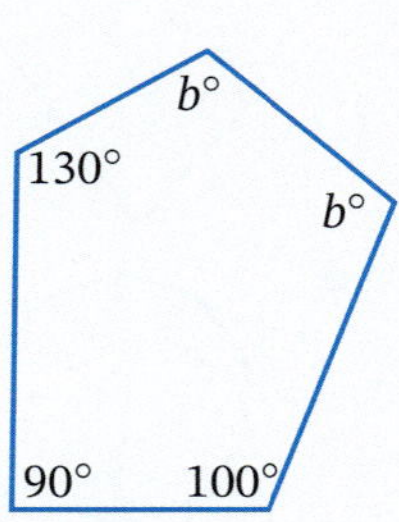

c

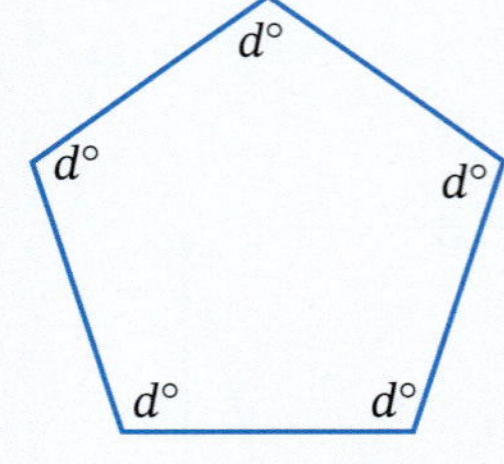

d

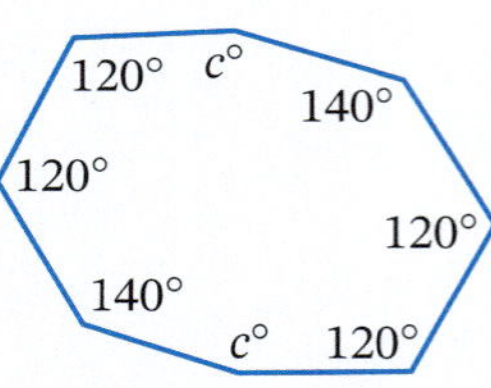

e

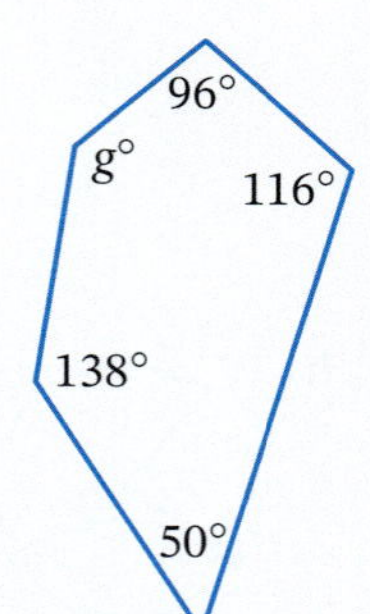

f

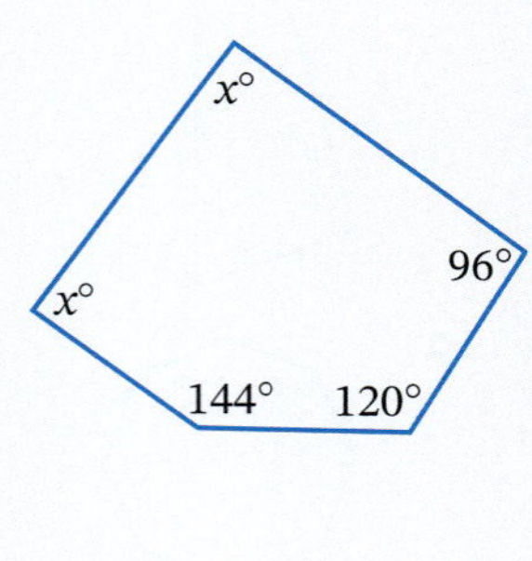

☐ Foundation ○ Standard ⬡ Complex

g

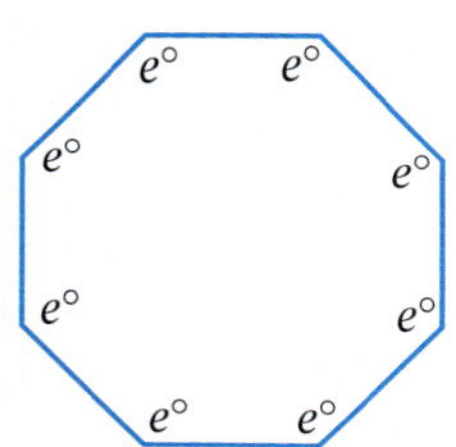

h

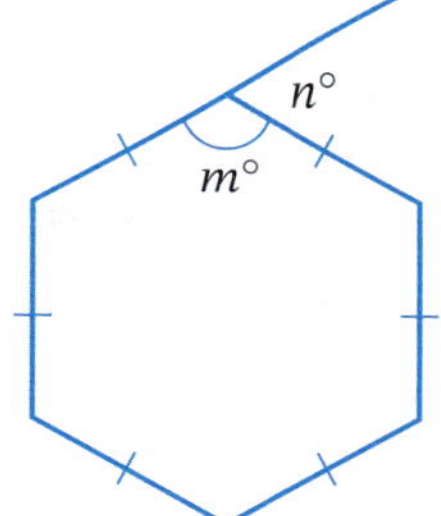

i

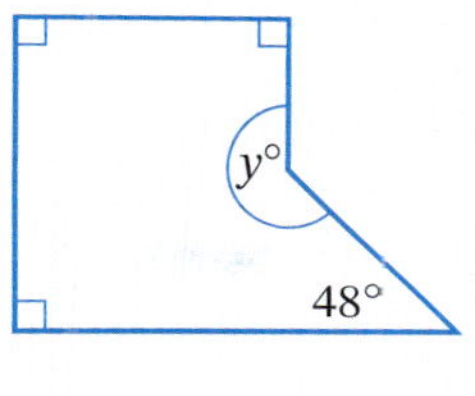

EXAMPLE 7

8 Find the number of sides of a polygon that has an angle sum of:

R **a** 2160° **b** 1620° **c** 3960° **d** 2700°

9 Which polygon has an angle sum of 1440°? Select the correct answer **A**, **B**, **C** or **D**.

R **A** pentagon **B** decagon **C** nonagon **D** octagon

EXAMPLE 8

10 Find the size of one angle in a regular:

R **a** octagon **b** decagon **c** dodecagon **d** hexagon

11 The angle sum of a regular polygon is 6840.

R **a** How many sides does the polygon have?

b Find the size of each angle.

12 How many sides does a regular polygon have if each of its angles is:

R **a** 165°? **b** 170°? **c** 144°?

☆ MENTAL SKILLS 6 ANSWERS ON P. 644 Maths without calculators

Quiz
Mental skills 6

Dividing decimals

To divide one decimal by another, first move the decimal points in both decimals the same number of places to the right so that the second decimal is a whole number.

1 Study each example.

a $0.24 \div 0.06 = 24 \div 6 = 4$ **b** $0.45 \div 0.5 = 4.5 \div 5 = 0.9$

c $0.006 \div 0.3 = 0.06 \div 3 = 0.02$ **d** $27 \div 0.9 = 270 \div 9 = 30$

e $1.6 \div 0.4 = 16 \div 4 = 4$ **f** $5.6 \div 0.07 = 560 \div 7 = 80$

2 Now evaluate each quotient.

a $0.25 \div 0.5$ **b** $63 \div 0.7$ **c** $3.2 \div 0.4$ **d** $0.18 \div 0.2$

e $2.7 \div 0.03$ **f** $0.042 \div 0.06$ **g** $4 \div 0.5$ **h** $1.2 \div 0.04$

i $0.072 \div 0.9$ **j** $0.35 \div 0.1$ **k** $0.28 \div 0.07$ **l** $0.033 \div 0.11$

Foundation Standard Complex

3 Study each example.

Given that $112 \div 14 = 8$, evaluate each expression.

a $112 \div 1.4 = 112.0 \div 1.4$
$= 1120 \div 14$
$= 112 \times 10 \div 14$
$= 112 \div 14 \times 10$
$= 8 \times 10$
$= 80$

Estimate: $112 \div 1.4 \approx 112 \div 1 = 112$

b $0.112 \div 0.14 = 0.112 \div 0.14$
$= 11.2 \div 14$
$= 112 \div 10 \div 14$
$= 112 \div 14 \div 10$
$= 8 \div 10$
$= 0.8$

Estimate: $0.112 \div 0.14 \approx 0.1 \div 0.1 = 1$

c $1120 \div 1.4 = 11\,200 \div 14$
$= 112 \times 100 \div 14$
$= 112 \div 14 \times 100$
$= 8 \times 100$
$= 800$

Estimate: $1120 \div 1.4 \approx 1120 \div 1 = 1120$

d $1.12 \div 14 = 112 \div 100 \div 14$
$= 112 \div 14 \div 100$
$= 8 \div 100$
$= 0.08$

Estimate: $1.12 \div 14 \approx 1.12 \div 10 = 0.112$

4 Now given that $368 \div 23 = 16$, evaluate each quotient.

a $36.8 \div 2.3$ **b** $368 \div 2.3$ **c** $3.68 \div 2.3$ **d** $0.368 \div 0.23$
e $36.8 \div 23$ **f** $3.68 \div 0.23$ **g** $36.8 \div 0.23$ **h** $0.368 \div 2.3$
i $0.368 \div 23$ **j** $3.68 \div 0.023$ **k** $3.68 \div 23$ **l** $0.368 \div 230$

6.04 Extension: Networks

YEAR 10 EXTENSION

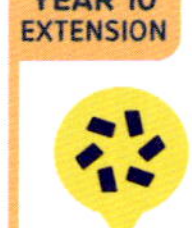

Videos
The seven bridges of Konigsberg
Networks: Labyrinths and mazes

A **network** is a set of points called **vertices** connected together by lines called **edges**. It represents a collection of objects connected to each other in some way, for example, computer network, electrical wiring inside a building, roads connecting towns, and friendships between a large group of people (such as on social media). A network is also called a **graph**, and if its edges don't cross each other (overlap), then it is called a **planar graph** (planar meaning 2D, flat, of a plane).

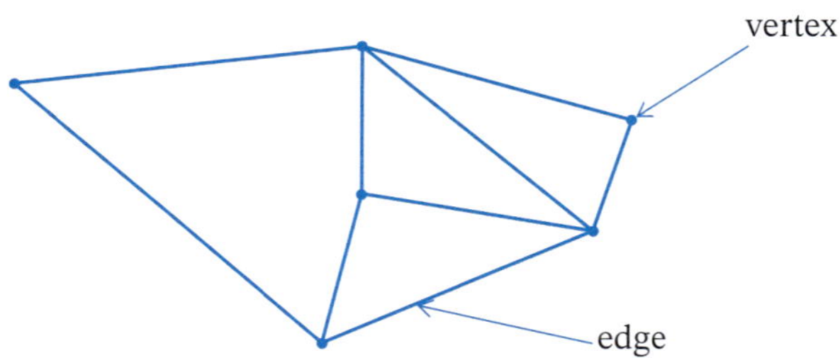

Degree of a vertex

The **degree** (or **order**) of a vertex is the number of edges connected to it. An **even vertex** has an even number of edges connected to it. An **odd vertex** has an odd number of edges.

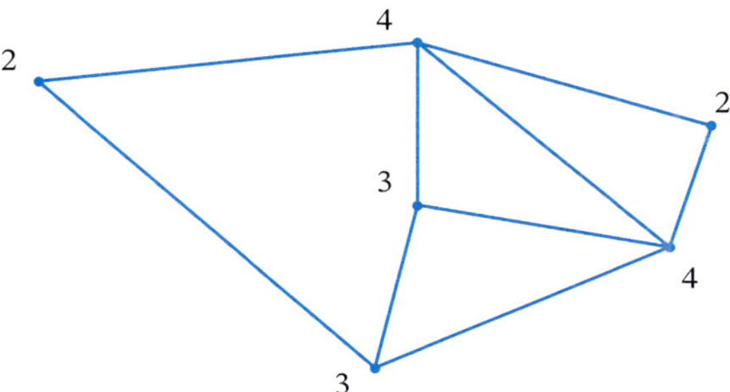

Faces

The **faces** or **regions** of a network are the areas bounded by its edges, plus 1 for the outside region.

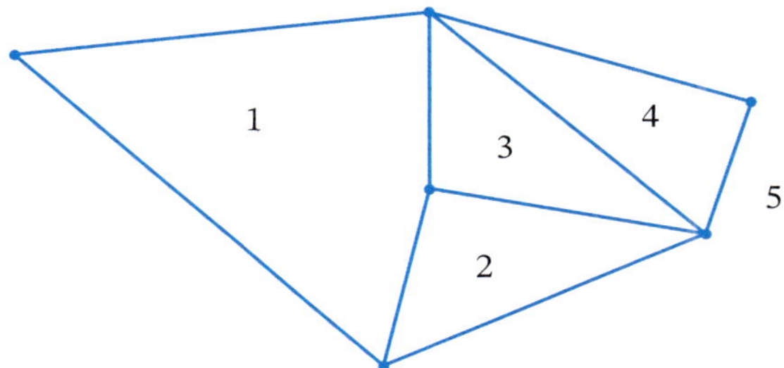

This network has 5 faces.

Traversable networks

If you can start at one vertex and trace over every edge of the network once, and end up back at the starting vertex, then the network is called **traversable**.

If you can trace over every edge once but do not start and finish at the same vertex then the network is called **semi-traversable**.

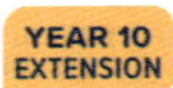

EXERCISE 6.04 ANSWERS ON P. 644

Networks

U F R C

1 Copy each network and write down the degree of each vertex next to it.

a

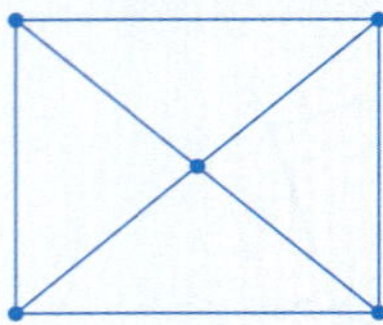

b

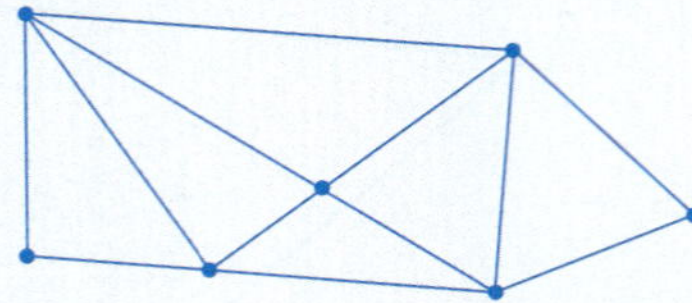

2 Copy each network and write odd or even next to each vertex based on its degree.

a

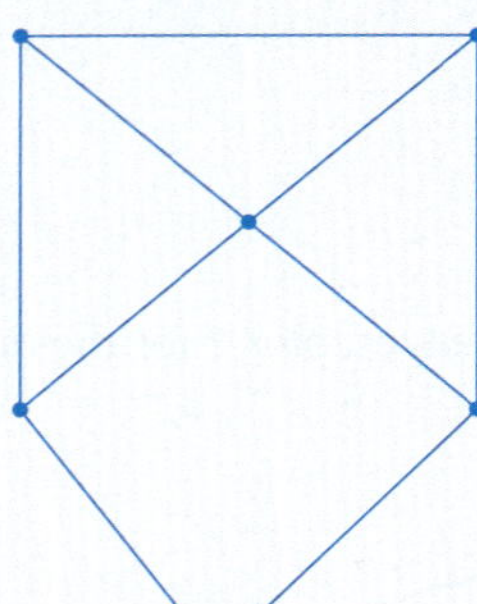

b

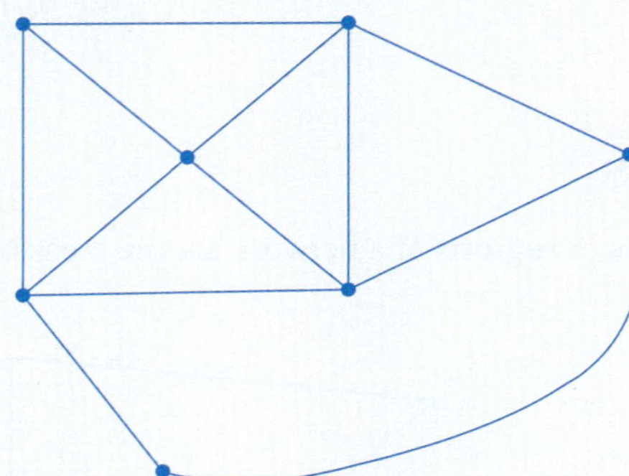

3 Copy each network and count its number of faces.

a

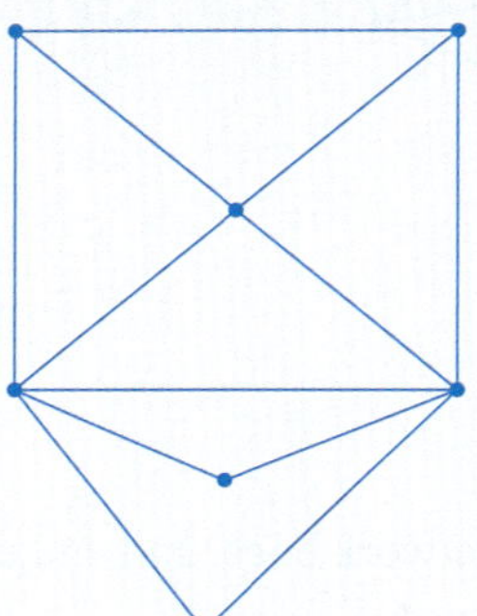

b

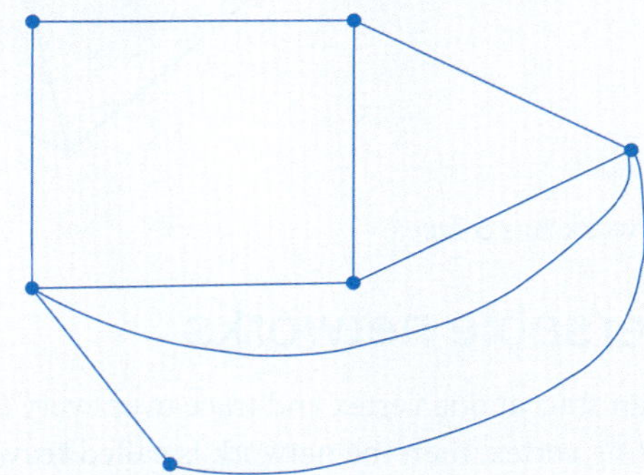

4 **a** Copy and complete this table, using the networks from questions **1** to **3**.

Network	Number of faces (F)	Number of vertices (V)	$F + V$	Number of edges (E)
1 a				
1 b				
2 a				
2 b				
3 a				
3 b				

Foundation Standard Complex

b Look at the values in the above table. The Swiss mathematician Leonhard Euler discovered a relationship between $F + V$ and E for all networks that are planar graphs (no overlapping edges). What is Euler's rule?

5 **a** For each network you drew from Questions **1** to **3**, see if you can trace over each edge of the network once and return to the starting vertex. If you can, then the network is traversable. Copy and complete the table.

Network	Traversable?	Number of odd vertices
1 a		
1 b		
2 a		
2 b		
3 a		
3 b		

6 In the ancient Prussian city of Königsberg, there were seven bridges crossing the river, as shown below.

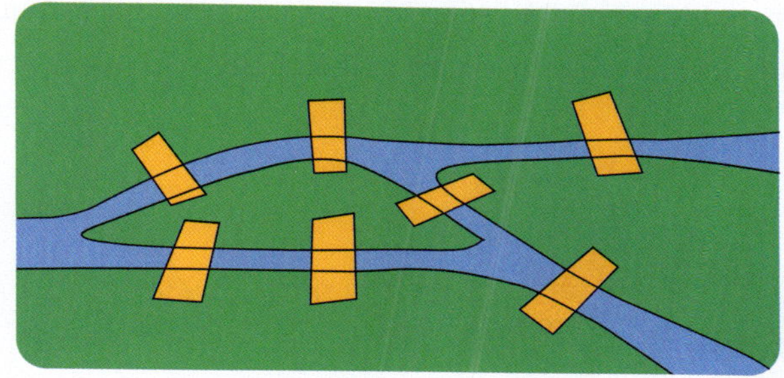

If you were planning a tour of the city, could you travel over each bridge only once and end up back at your starting point? Explain why or why not.

7 Draw a network that has 3 vertices, and 8 edges and state how many faces it has.

8 Is the network you drew in question 7 traversable? If not, add edges to make it traversable.

6.05 Extension: Polyhedra

YEAR 10 EXTENSION

Video
Polyhedra: Platonic solids

A **polyhedron** is a three-dimensional solid with flat faces made up of polygons. The plural of polyhedron is **polyhedra** or **polyhedrons**. A **polyhedron** has **faces**, **vertices** (corners) and **edges** (lines around the faces).

This cube has 6 faces, 8 vertices and 12 edges.

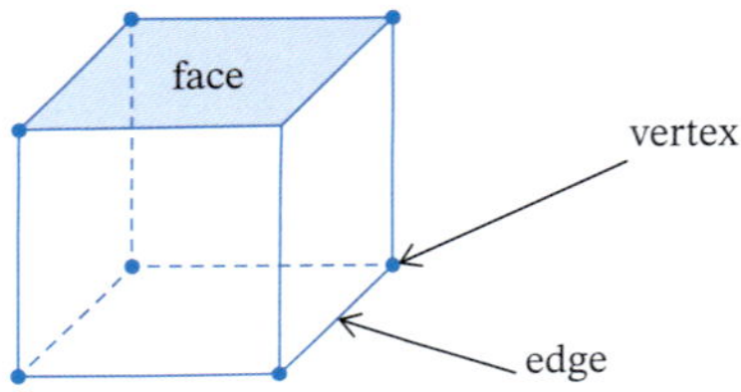

Prisms and **pyramids** are examples of polyhedra. A cylinder is not a polyhedron because it has a curved face. A **prism** has the same **cross-section** along its length, and each cross-section (including the base) is a polygon. A **pyramid** has a polygon for the base and each vertex of the polygon joins to one vertex at the **apex** of the pyramid. A pyramid's cross-sections have the same shape as the base but are of different sizes. The side faces of a pyramid are all triangles.

Prisms and pyramids are named by the shape of their base.

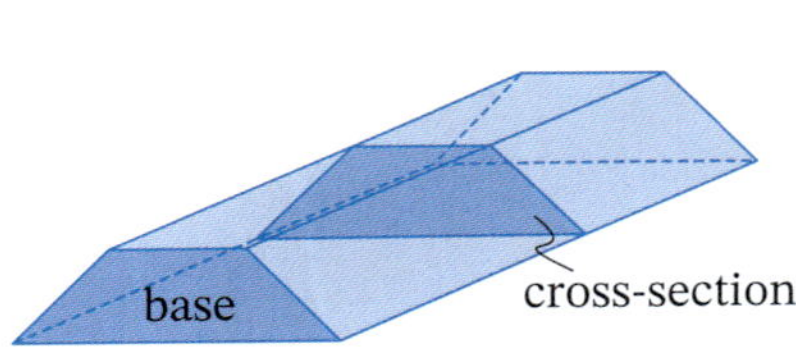

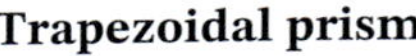

Trapezoidal prism

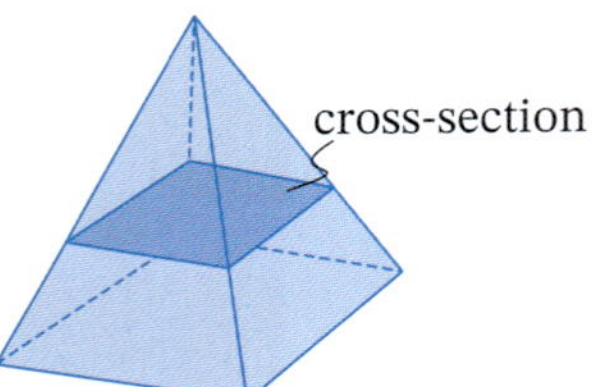

Square pyramid

A **regular polyhedron** or **Platonic solid** is a polyhedron whose faces are the same **regular polygon**. A **cube** is one example of a Platonic solid because all its faces are identical squares. There are only five Platonic solids, each named by the number of faces, as shown below.

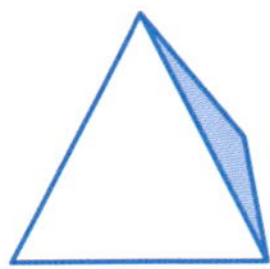

tetrahedron

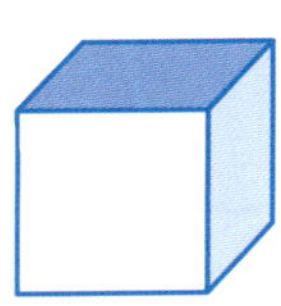

hexahedron (cube)

octahedron

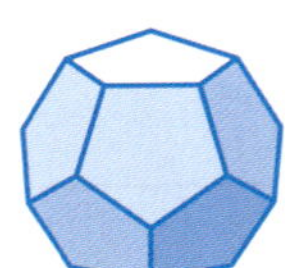

dodecahedron

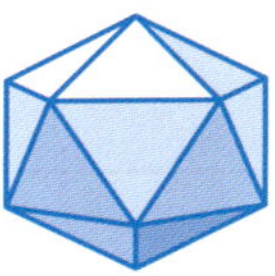

icosahedron

9780170465618

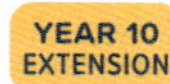

6.05

Platonic solid	Number of faces	Shape of each face
Regular tetrahedron (Equilateral triangular pyramid)	tetra = 4	Equilateral triangle
Regular hexahedron (Cube)	hexa = 6	Square
Regular octahedron	octa = 8	Equilateral triangle
Regular dodecahedron	dodeca =16	Regular pentagon
Regular icosahedron	icosa = 20	Equilateral triangle

Polyhedra

U F R C

Worksheet Nets of Platonic solids

1 The nets of the Platonic solids are shown below and can also be downloaded as the worksheet 'Nets of Platonic solids'. Make a copy of each net (on cardboard for best results), cut them out and build the solids.

Tetrahedron

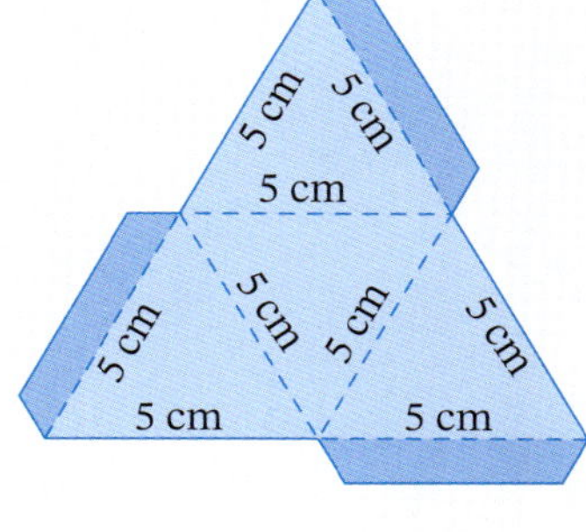

Cube (hexahedron)

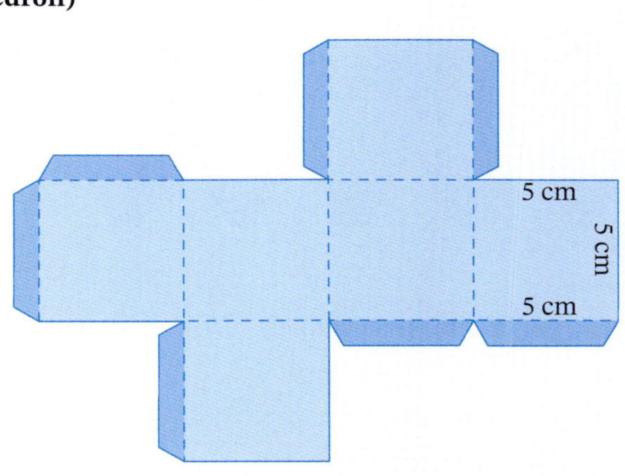

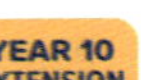

Octahedron

5 cm
5 cm
5 cm

Dodecahedron

3 cm
3 cm
3 cm
3 cm
3 cm

Icosahedron

4 cm
4 cm
4 cm

2 Which Platonic solid:

a has 12 faces?
b is a pyramid?
c has 12 edges?
d has 12 vertices?
e is a prism?
f has pentagonal faces?
g has a flattened 'soccer ball' pattern?
h is a composite shape of 2 square pyramids?

YEAR 10 EXTENSION

6.05

3 **a** Copy and complete this table below for each Platonic solid.

Platonic solid	Number of faces (*F*)	Number of vertices (*V*)	*F* + *V*	Number of edges (*E*)
Tetrahedron				
Cube				
Octahedron				
Dodecahedron				
Icosahedron				

b Euler's formula also applies to the number of faces, vertices and edges of a polyhedron. What is the relationship between $F + V$ and E for each polyhedron?

4 For each polyhedron, count the number of faces, vertices and edges. Copy and complete this table.

	Polyhedron	Number of faces (*F*)	Number of vertices (*V*)	*F* + *V*	Number of edges (*E*)
a	Triangular prism				
b	Hexagonal prism				
c	Square pyramid				
d	Pentagonal prism				
e	Trapezoidal prism				

a

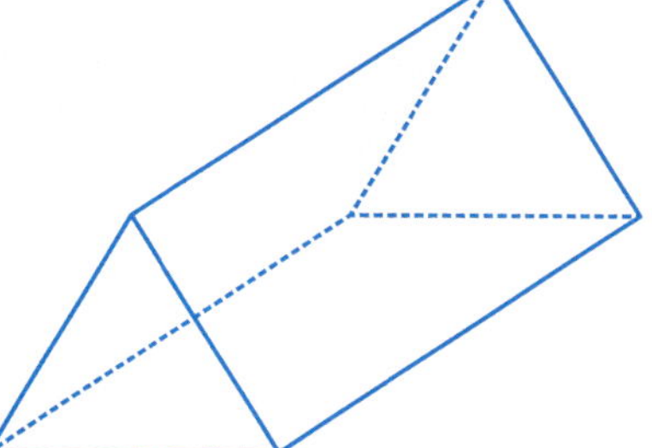

b

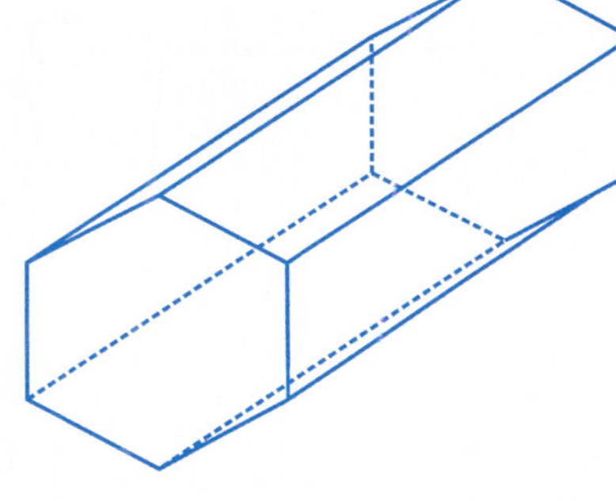

c

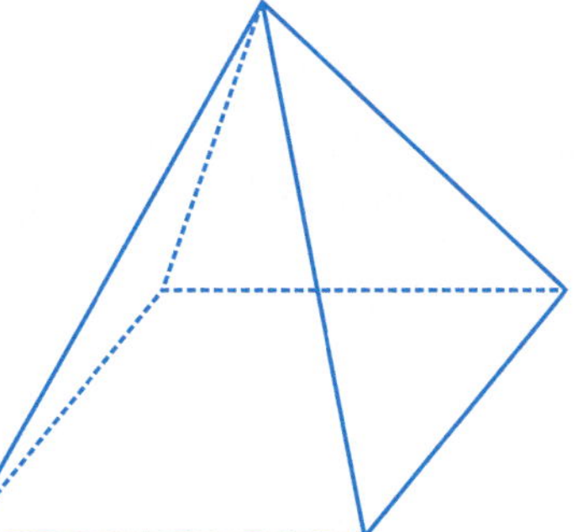

d

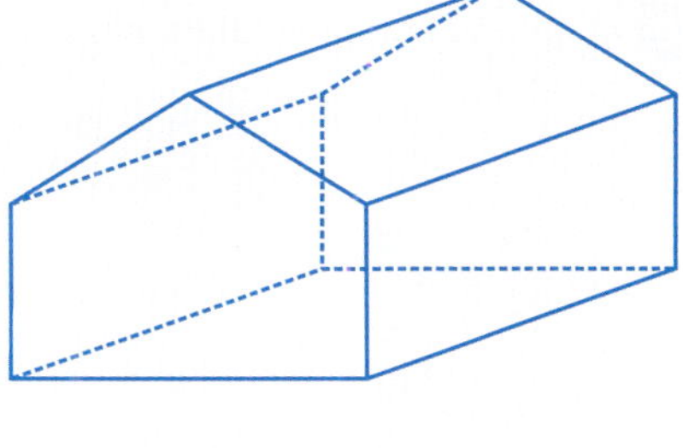

☐ Foundation ○ Standard ⬡ Complex

e

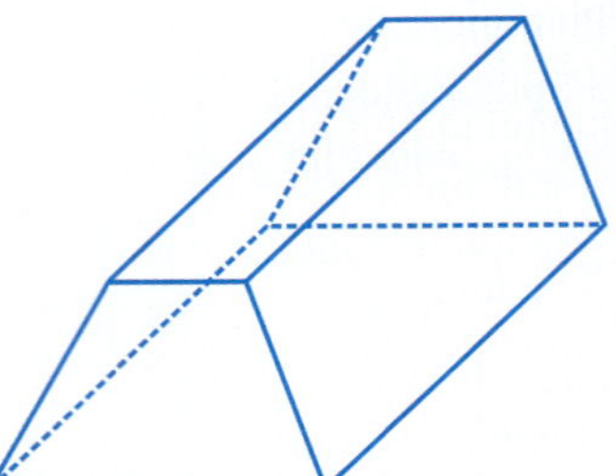

5 Euler's formula states that $F + V = E + 2$. For each shape below consider the number of faces, vertices and edges and determine whether Euler's formula is true.

a

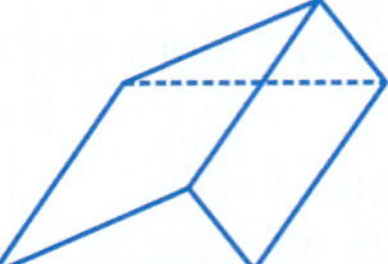

b

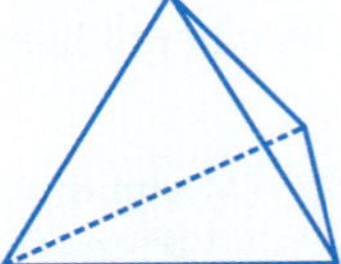

c

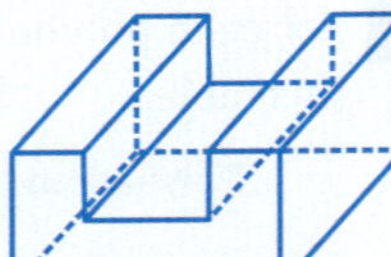

d

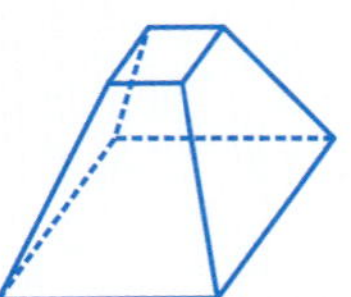

e

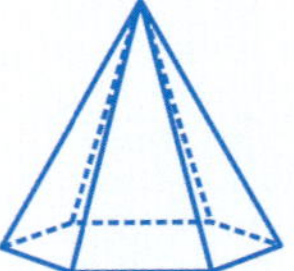

Polyhedron	Number of faces (F)	Number of vertices (V)	$F + V$	Number of edges (E)	Euler's formula true?
a					
b					
c					
d					
e					

6 Copy and complete this table using Euler's formula.

	Number of faces (F)	Number of vertices (V)	$F + V$	Number of edges (E)
a	6	6		
b		4		6
c	6			12
d		10		15
e	8			12

7 Draw a polyhedron that has 8 faces and 18 edges.

☐ Foundation ○ Standard ⬡ Complex

POWER PLUS ANSWERS ON P. 645

1 How many diagonals has:

a a quadrilateral? b an octagon? c a dodecagon?

2 Name all quadrilaterals whose diagonals:

a bisect each other at right angles

b bisect each other

c intersect at right angles

d have equal length

e bisect the angles of the quadrilateral

f are equal and bisect each other

3 Find the value of each variable, giving reasons.

a

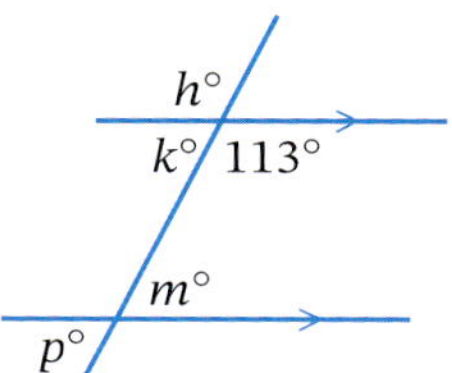

b

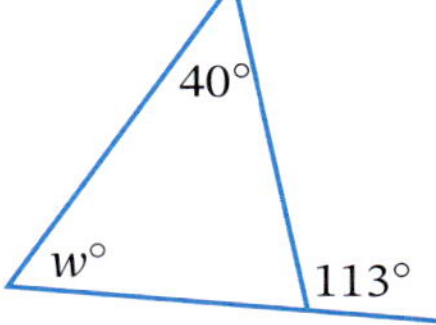

c

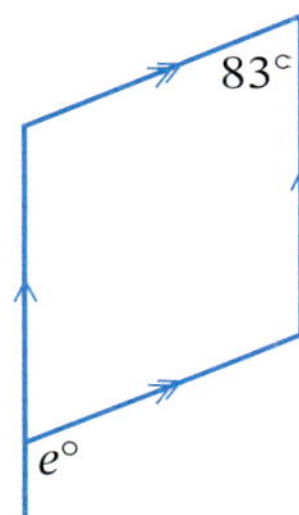

d

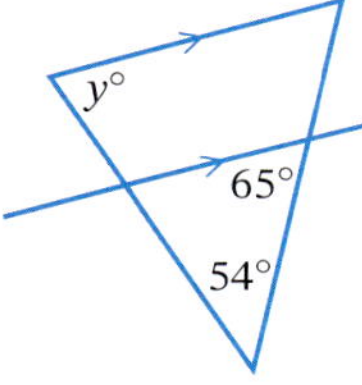

e

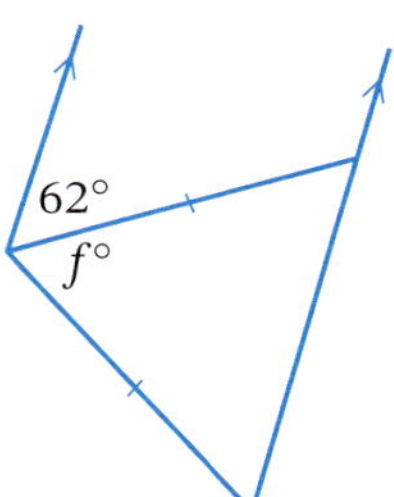

f

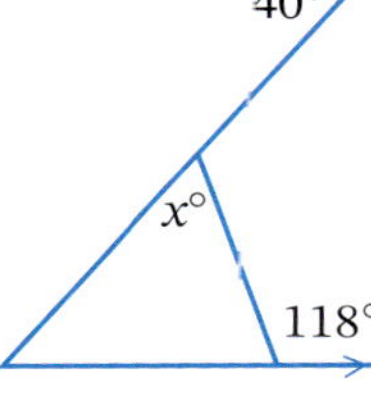

g

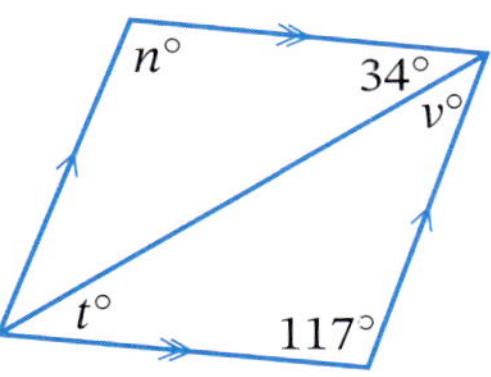

h

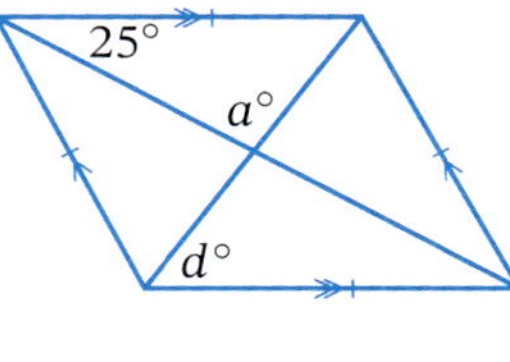

i

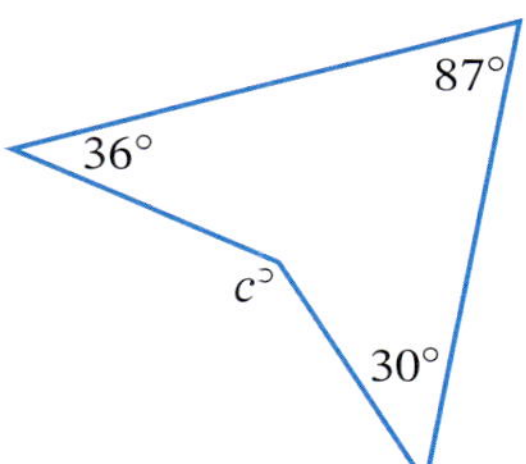

j

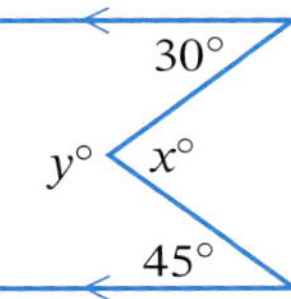

k

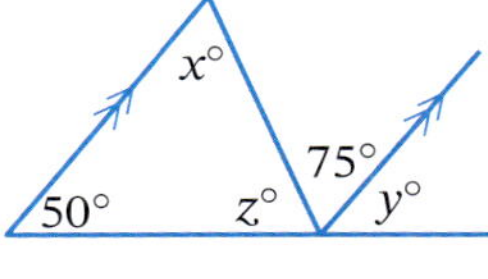

l

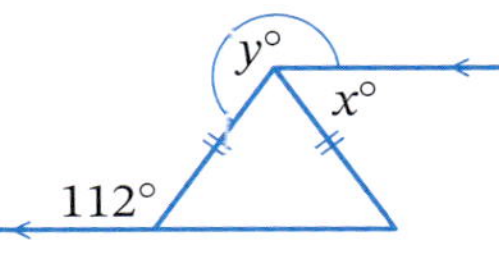

6 CHAPTER REVIEW

Quiz
Language of maths 6

Language of maths

angle sum	bisect	convex	diagonal
equilateral	exterior angle	interior	isosceles
kite	parallelogram	polygon	quadrilateral
rectangle	regular	rhombus	right angle
scalene	square	trapezium	vertex

1 What shape has 2 pairs of equal adjacent sides?

2 Name one property of a convex quadrilateral.

3 What type of angle is created when one side of a triangle is extended?

4 What is the sum of the angles of an octagon?

5 What is a **regular polygon?** Are all regular polygons also **convex**?

6 Copy and complete: The exterior angle of a triangle is equal to the ______ of the ________ __________ ___________ angles.

Worksheets
Geometry summary poster

Mind map: Geometry

Topic summary

- Write 3 ideas from this topic that were new to you.
- Summarise what you know about the angle sum of a triangle, quadrilateral and polygon.
- Name the types of triangles, choose one and list all of its properties.
- Name the 6 special quadrilaterals, choose one and list all of its properties.

Print (or copy) and complete this mind map of the topic, adding detail to its branches and using pictures, symbols and colour where needed. Ask your teacher to check your work.

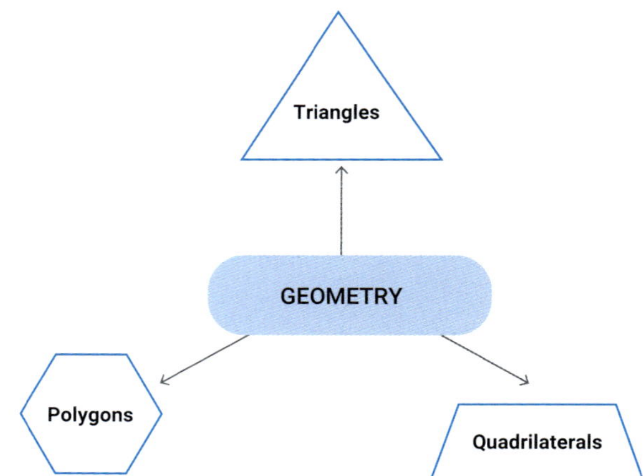

6 TEST YOURSELF

ANSWERS ON P. 645

1 Find the value of each variable, giving reasons. 6.01

Quiz
Test yourself 6

a

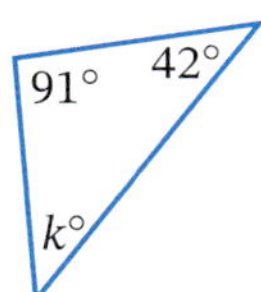

b

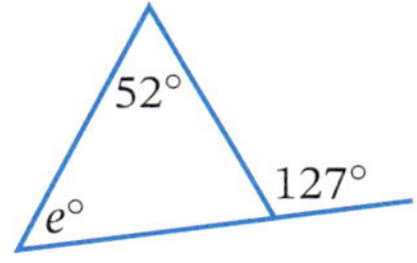

c

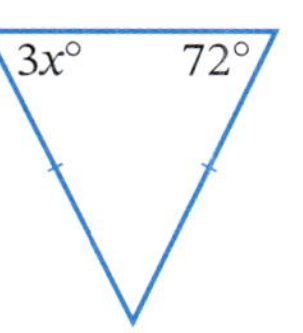

d

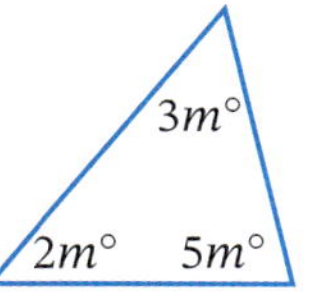

e

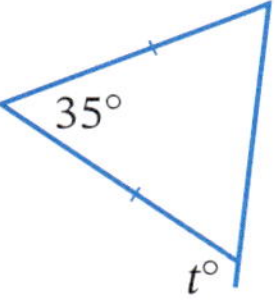

f

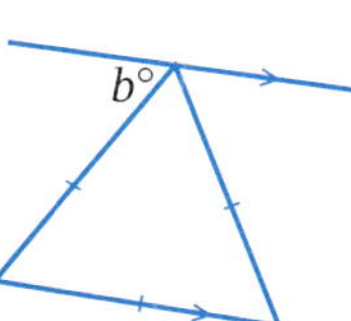

g

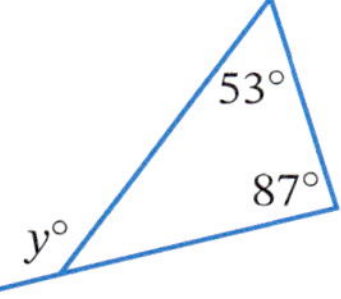

h

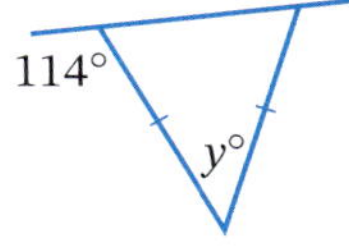

i

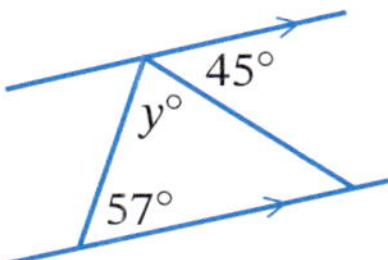

2 What are the sizes of the angles in: 6.01

a an equilateral triangle? **b** a right-angled isosceles triangle?

3 Find the value of each variable, giving reasons. 6.02

a

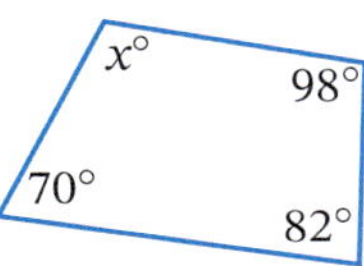

b

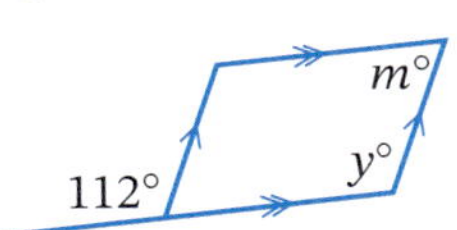

c

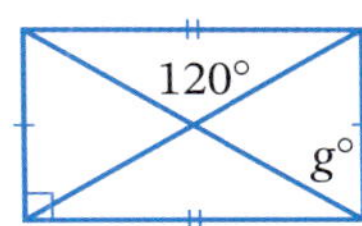

d

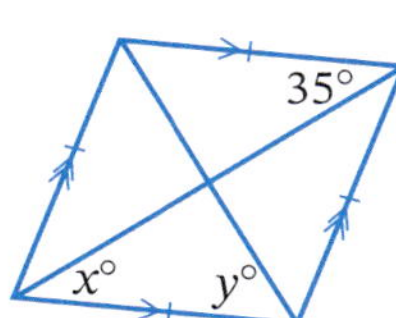

e

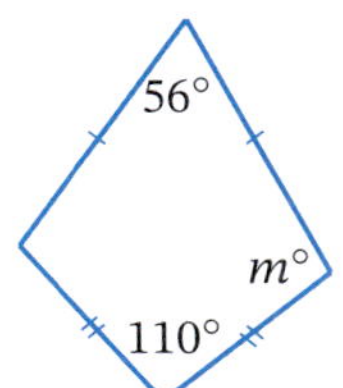

f

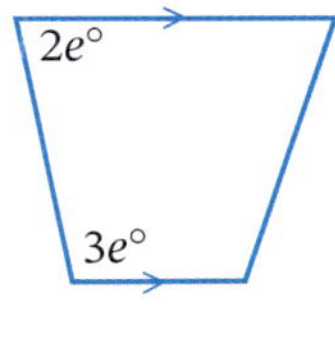

4 Name all quadrilaterals that have: 6.02

a both pairs of opposite sides parallel **b** 2 equal diagonals

c all sides equal **d** diagonals that bisect each other.

5 Draw each pentagon. 6.03

a regular pentagon **b** non-regular pentagon **c** concave pentagon

6 **a** Show that the angle sum of a decagon is 1440°. 6.03

b Find the size of one angle in a regular nonagon.

☐ Foundation ○ Standard ○ Complex

7

ALGEBRA

Equations and inequalities

Equation-solving has been recorded as far back as 1500 BCE. It was first used in ancient Babylon and Egypt and was brought to Europe from India by the Arabs during the 9th century. The word 'algebra' comes from the Arabic word *al-jabr*, meaning restoration, the process of performing the same operation on both sides of an equation to solve the equation.

Getty Images/iStock/WildImages

Chapter outline

U = Understanding
F = Fluency
PS = Problem solving
R = Reasoning
C = Communication

Wordbank

equation A mathematical statement that two quantities are equal, involving algebraic expressions and an equals sign (=)

formula A rule written as an algebraic equation, using variables

inequality A mathematical statement that two quantities are unequal, involving algebraic expressions and an inequality sign (>, <, ≥ or ≤)

linear equation An equation involving a variable that is not raised to a power, such as $2x + 9 = 17$.

quadratic equation An equation involving a variable squared (power of 2), such as $3x^2 - 6 = 69$.

solution The answer to an equation or problem, the correct value(s) of the variable that makes an equation true

solve (an equation) To find the value of an unknown variable in an equation

Quiz
Wordbank 7

Videos (7):

7.01 Two-step equations

7.02 Equations with variables on both sides

7.03 Solving equations • Equations with brackets

7.04 Equation problems

7.06 Quadratic equations by factorising

7.08 Graphing inequalities on the number line

Twig videos (2):

7.01 The Arabic science of balancing

7.08 Inequalities: The most populous country

PhET interactives (2):

7.01 Equality explorer: Basics • Equality explorer

Quizzes (6):

- Wordbank 7
- SkillCheck 7
- Mental skills 7A
- Mental skills 7B
- Language of maths 7
- Test yourself 7

Skillsheets (3):

7.01 Solving equations by balancing • Solving equations by backtracking • Solving equations using diagrams

Worksheets (5):

7.03 Equations 2 • Checking solutions

7.04 Angle problems with algebra

7.09 Inequalities review

Mind map: Equations and inequalities

Puzzles (6):

7.01 Solving equations • Backtracking

7.02 Equations with unknowns on both sides

7.03 Equations • Equations order activity

7.04 Writing and solving equations

Presentation (1):

7.04 Applying linear equations

To access resources above, visit **cengage.com.au/nelsonmindtap**

7 In this chapter you will:

- ✓ solve linear equations and problems involving equations.
- ✓ solve quadratic equations of the form $ax^2 = c$ and $x^2 + bx + c = 0$.
- ✓ use formulas to solve problems.
- ✓ graph inequalities of a number line.
- ✓ solve linear inequalities.

SkillCheck

ANSWERS ON P. 646

Quiz SkillCheck 7

1 Use $x = -4$ to evaluate each expression.

a $5x$ **b** $x + 10$ **c** $x - 1$ **d** x^2

e $15 - x$ **f** $3(x - 1)$ **g** $7(x + 5)$ **h** $\frac{x}{2}+3$

2 What operation is the opposite to:

a adding? **b** dividing? **c** multiplying? **d** subtracting?

3 Simplify each expression.

a $6n - 2n$ **b** $8x + 3x$ **c** $-r - 4r$ **d** $-8t + 2t$

4 Expand each expression.

a $4(2k + 7)$ b $5(3d - 1)$ c $-3(1 + 2y)$
d $-4(b - 8)$ e $5(3d + 2) + 8$ f $2(a + 5) + 5(a + 2)$
g $4(1 + d) - 3(1 - 5d)$

5 Solve each equation. Use substitution to check your solutions.

a $m + 10 = 16$ b $5a = 45$ c $y - 6 = 2$
d $\frac{w}{5} = 10$ e $-3y = 27$ f $27 = m + 4$
g $6a = -42$ h $\frac{a}{3} = 4$ i $\frac{w}{-5} = 6$

6 Using n to represent 'the number', write an expression for each of these statements.

a The product of the number and 7.
b The square of the number.
c 5 times the sum of the number and 8.
d The number decreased by 20.
e The product of 6 and the number, decreased by nine.
f If the number is even, the *next* even number.

7 Factorise each quadratic expression.

a $x^2 + 8x + 15$ b $m^2 + 3m - 4$ c $b^2 + 11b + 28$
d $t^2 - 4t - 12$ e $u^2 - 7u + 10$ f $h^2 + 17h + 72$

Two-step equations 7.01

An **equation** is a statement involving a variable (such as x), numbers and an equals (=) sign. If the variable is not raised to a power, then the equation is a **linear equation**. For example, $2x - 5 = 11$ is a linear equation, but $x^2 + 8 = 24$ is not because the x is squared (raised to the power of 2).

The value of the variable that makes the equation true is called the **solution** of the equation.

There are two algebraic methods for solving equations.

The **balancing method** involves representing both sides of an equation as balance scales, and solving the equation by performing the same operation on both sides to keep it 'balanced'.

The **backtracking method** involves using a flowchart to show what operations are performed on the variable (say, x) to create the equation, then using a reverse flowchart to undo each operation by performing the inverse (opposite) operation in reverse order.

To **solve an equation**, aim to have the variable (such as x) on one side of the equation and a number on the other side, in the form: $x =$ _____

Check your solution by substituting it back into the equation.

Two-step equations require 2 steps or operations to solve.

Puzzles
Solving equations

Backtracking

Skillsheets
Solving equations by balancing

Solving equations by backtracking

Solving equations using diagrams

Interactives
Equality explorer: Basics

Equality explorer

Videos
The Arabic science of balancing

Two-step equations

Shutterstock.com/Olesia Bilkei

Example 1

Solve each two-step equation.

a $5x + 1 = 11$ **b** $\frac{5m}{2} = -3$ **c** $\frac{w}{3} - 6 = 5$

SOLUTION

a **Method 1: The balancing method**

$5x + 1 = 11$

$5x + 1 - 1 = 11 - 1$ *Step 1:* Subtracting 1 from both sides

$5x = 10$

$\frac{5x}{5} = \frac{10}{5}$ *Step 2:* Dividing both sides by 5

$x = 2$ Check: $5 \times 2 + 1 = 11$

Method 2: The backtracking method

First consider how we get from x to $5x + 1$ using a flowchart.

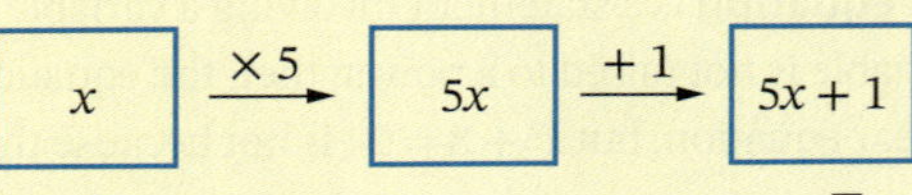

= =

2 ← ÷ 5 ← 10 ← − 1 ← 11

In this equation, $5x + 1 = 11$. To get back to x, use the 'reverse flowchart' to undo the operations 'multiply by 5' and 'add 1' in the **reverse order**.

Subtract 1 and divide by 5 to find the solution $x = 2$.

Using algebra, we have:

$5x + 1 = 11$

$5x = 11 - 1$ *Step 1:* Undo '+ 1' by subtracting 1

$5x = 10$

$x = \frac{10}{5}$ *Step 2:* Undo '× 5' by dividing by 5

$x = 2$ Check: $5 \times 2 + 1 = 11$

7.01

b **Method 1: The balancing method**

$$\frac{5m}{2} = -3$$

$$\frac{5m}{2} \times 2 = -3 \times 2$$ *Step 1:* Multiply both sides by 2

$$5m = -6$$

$$\frac{5m}{5} = \frac{-6}{5}$$ *Step 2:* Divide both sides by 5

$$m = -1\frac{1}{5}$$ Check: $\frac{5 \times \left(-1\frac{1}{5}\right)}{2} = \frac{-6}{2} = -3$

Method 2: The backtracking method

$$\frac{5m}{2} = -3$$

$$5m = -3 \times 2$$ *Step 1:* Undo '÷ 2' by multiplying by 2

$$5m = -6$$

$$m = \frac{-6}{5}$$ *Step 2:* Undo '× 5' by dividing by 5

$$m = -1\frac{1}{5}$$ Check: $\frac{5 \times \left(-1\frac{1}{5}\right)}{2} = \frac{-6}{2} = -3$

c **Method 1: The balancing method**

$$\frac{w}{3} - 6 = 5$$

$$\frac{w}{3} - 6 + 6 = 5 + 6$$ *Step 1:* Add 6 to both sides

$$\frac{w}{3} = 11$$

$$\frac{w}{3} \times 3 = 11 \times 3$$ *Step 2:* Multiply both sides by 3

$$w = 33$$ Check: $\frac{33}{3} - 6 = 11 - 6 = 5$

Method 2: The backtracking method

$$\frac{w}{3} - 6 = 5$$

$$\frac{w}{3} = 5 + 6$$ *Step 1:* Undo '– 6' by adding 6

$$\frac{w}{3} = 11$$

$$w = 11 \times 3$$ *Step 2:* Undo '÷ 3' by multiplying by 3

$$w = 33$$ Check: $\frac{33}{3} - 6 = 11 - 6 = 5$

EXERCISE 7.01 ANSWERS ON P. 646

Two-step equations

1 Solve each equation. Use substitution to check your solutions.

R **a** $3x + 2 = 8$ **b** $4p - 3 = 13$ **c** $2w + 7 = 16$

d $9k + 3 = 12k$ **e** $8x = x - 3$ **f** $10 + 5m = 20m$

g $4 - 2y = 12$ **h** $3 - 7x = 10$ **i** $22 + 7h = 36$

j $1.2 - 5q = 7.2$ **k** $-12 - 8w = -2w$ **l** $-6 - 4d = -20$

2 Solve $4x + 6 = 26$. Select the correct answer **A**, **B**, **C** or **D**.

A $x = 3$ **B** $x = 5$ **C** $x = 6$ **D** $x = 8$

3 Solve each equation. Use substitution to check your solutions.

R **a** $\frac{2n}{3} = 4$ **b** $\frac{5x}{2} = 7$ **c** $\frac{6a}{7} = 8$

d $\frac{-2y}{9} = -8$ **e** $\frac{-5n}{3} = -9$ **f** $\frac{-6p}{11} = -4$

g $\frac{x-1}{3} = 7$ **h** $\frac{n+6}{2} = 8$ **i** $2 = \frac{d+9}{5}$

j $\frac{a-3}{4} = -3$ **k** $\frac{n+3}{-5} = -1$ **l** $\frac{3-x}{-4} = 2$

4 Copy and complete this flowchart to solve $3k + 5 = 20$ using the backtracking method.

R

k $\xrightarrow{\times 3}$ [ ] $\xrightarrow{+5}$ $3k + 5$ This is what has been done to k.

= =

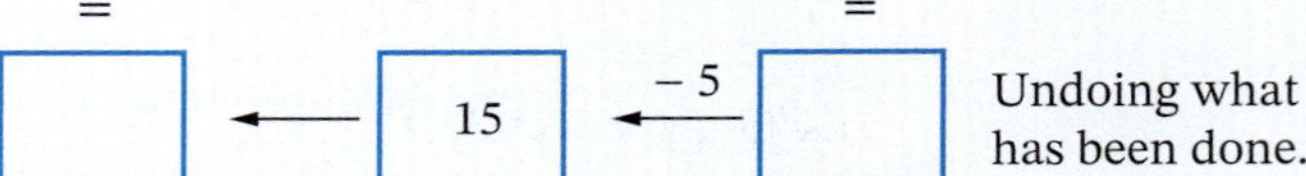

[ ] ← 15 $\xleftarrow{-5}$ [ ] Undoing what has been done.

5 Copy and complete each flowchart to solve the equation using the backtracking method.

R **a** $3c + 6 = -24$

c $\xrightarrow{\times 3}$ [ ] $\xrightarrow{+6}$ [ ]

= =

[ ] ← [ ] ← -24

b $\frac{5n}{3} = 10$

n $\xrightarrow{\times 5}$ ☐ $\xrightarrow{\div 3}$ ☐

= =

☐ ⟵ ☐ ⟵ 10

c $\frac{m}{5} - 12 = 3$

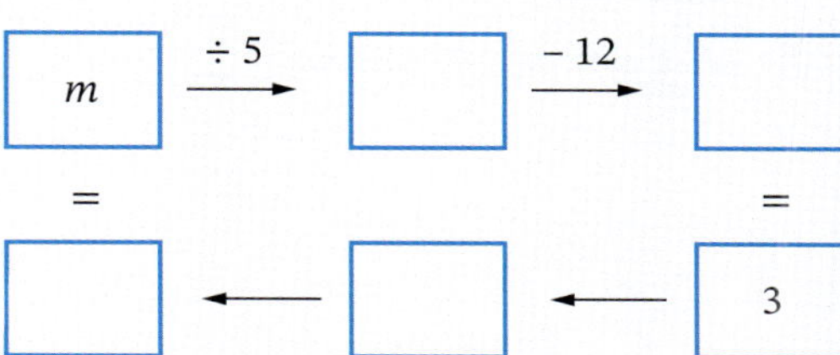

6 Solve each equation.

R **a** $\frac{w}{2} + 3 = 4$ **b** $\frac{a}{6} - 1 = 7$ **c** $\frac{m}{3} - 12 = 8$

d $2 - \frac{a}{3} = 3$ **e** $1 - \frac{x}{8} = -1$ **f** $4 - \frac{a}{2} = -3$

7 Which one of these equations has the solution $x = -3$? Select **A**, **B**, **C** or **D**.

A $\frac{3x}{2} = 3$ **B** $\frac{x-1}{2} = -1$ **C** $\frac{x}{3} - 1 = 0$ **D** $2x + 1 = -5$

8 Solve each equation.

R **a** $\frac{x+8}{6} = 3$ **b** $\frac{w}{7} + 10 = 10$ **c** $\frac{2y}{11} = -6$

d $13 + \frac{k}{4} = 10$ **e** $\frac{x-3}{-4} = 12$ **f** $-6 - 14d = -20$

9 What is the solution to $\frac{k}{10} - 5 = 3$? Select **A**, **B**, **C** or **D**.

A $k = 80$ **B** $k = \frac{4}{5}$ **C** $k = -12$ **D** $k = -20$

7.02 Equations with variables on both sides

Puzzle Equations with unknowns on both sides

Video Equations with variables on both sides

For equations with variables on both sides, such as $3x + 4 = 2x + 7$, we can only use the **balancing method**, not the backtracking method.

ⓘ Equations with variables on both sides

For **equations with variables on both sides**, perform operations on both sides to move:

- all the variables onto one side of the equation.
- all the numbers onto the other side of the equation.

Example 2

Solve each equation.

a $7x + 7 = 2x + 2$ **b** $9 - 6y = 10 - 2y$

SOLUTION

a

$7x + 7 = 2x + 2$

$7x - 2x + 7 = 2x - 2x + 2$ Subtracting $2x$ from both sides.

$5x + 7 = 2$

$5x + 7 - 7 = 2 - 7$ Subtracting 7 from both sides.

$5x = -5$

$\frac{5x}{5} = \frac{-5}{5}$ Dividing both sides by 5.

$x = -1$

Check:

LHS $= 7 \times (-1) + 7 = 0$

RHS $= 2 \times (-1) + 2 = 0$

LHS = RHS. LHS = left-hand side, RHS = right-hand side.

b

$9 - 6y = 10 - 2y$

$9 - 6y + 2y = 10 - 2y + 2y$ Adding 2y to both sides.

$9 - 4y = 10$

$9 - 4y - 9 = 10 - 9$ Subtracting 9 from both sides.

$-4y = 1$

$\frac{-4y}{-4} = \frac{1}{-4}$ Dividing both sides by (−4).

$y = -\frac{1}{4}$

Check:

$$\text{LHS} = 9 - 6 \times \left(-\frac{1}{4}\right) = 10\frac{1}{2}$$

$$\text{RHS} = 10 - 2 \times \left(-\frac{1}{4}\right) = 10\frac{1}{2}$$

LHS = RHS.

EXERCISE 7.02 ANSWERS ON P. 646

Equations with variables on both sides

U F R

1 Solve each equation and check your solutions. EXAMPLE 2

R **a** $5w + 3 = 2w + 21$ **b** $2q - 10 = q - 4$ **c** $13x + 1 = 8x + 26$

d $12n + 3 = 5n - 11$ **e** $8y - 10 = 10y - 30$ **f** $3m - 2 = 10 - 3m$

g $9 - 2a = a - 9$ **h** $9 - 2x = 18 + 7x$ **i** $12y + 6 = 6 + 9y$

j $-12 - 10u = -20 - 18u$ **k** $15 - 7x = 22 - 3x$ **l** $10 - 6x = -15 - 11x$

2 For each equation, select the correct solution **A**, **B**, **C** or **D**.

a $6x - 1 = 2x + 11$

A $x = 12$ **B** $x = 3$ **C** $x = 0$ **D** $x = 2.5$

b $11 - 4p = 2p + 2$

A $p = 6.5$ **B** $p = 2$ **C** $p = 1.5$ **D** $p = 3$

3 Solve each equation.

R **a** $7w + 15 = w + 3$ **b** $10 - 3t = 16 + t$ **c** $4a + 2 = 10 - 4a$

d $50 + 7y = 20 - 3y$ **e** $8y - 2 = 10y + 1$ **f** $9y + 3 = 9 - y$

g $9 - t = 7t - 2$ **h** $5y + 2 = 17 - y$ **i** $25 - 12k = 15 - 6k$

4 Solve $-3n - 8 = -7n - 12$. Select **A**, **B**, **C** or **D**.

A $n = 5$ **B** $n = 2$ **C** $n = -1$ **D** $n = -0.4$

5 Each diagram below shows a balanced scale that represents an equation.

R Let each blank box be x. For each scale:

i write the equation that the scale represents and solve the equation.

ii check your answer by substitution to see if the scale remains balanced.

a

b

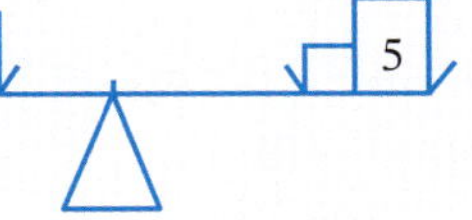

c

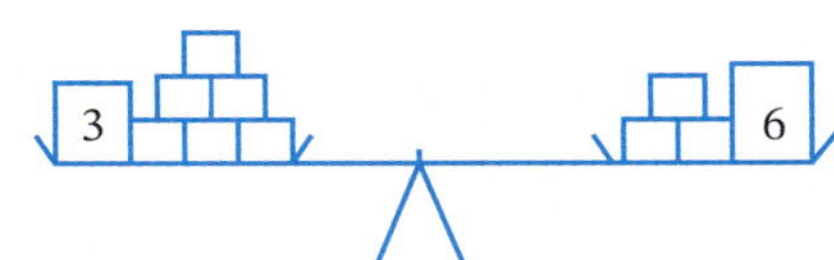

d

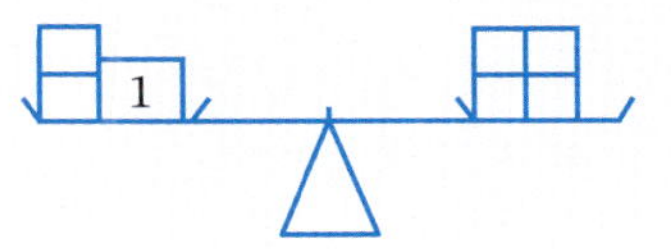

Foundation Standard Complex

7.03 Equations with brackets

ⓘ Equations with brackets

For **equations with brackets (grouping symbols)**, expand the expressions and then solve as usual.

Example 3

Videos
Solving equations
Equations with brackets

Worksheets
Equations 2
Checking solutions

Puzzles
Equations
Equations order activity

Solve each equation.

a $3(a + 7) = 6$ **b** $9(m - 5) = 7(m + 1)$ **c** $10y - 3(2y - 5) = 6(8 - 3y)$

SOLUTION

a

$3(a + 7) = 6$

$3a + 21 = 6$ Expanding the expression to make it a two-step equation.

Can you think of another way to solve this equation without expanding?

$3a + 21 - 21 = 6 - 21$ Subtract 21 from both sides.

$3a = -15$

$\frac{3a}{3} = \frac{-15}{3}$ Divide both sides by 3.

$a = -5$ Check: $3(-5 + 7) = 3 \times 2 = 6$.

b

$9(m - 5) = 7(m + 1)$

$9m - 45 = 7m + 7$ Expanding brackets on both sides.

$9m - 7m - 45 = 7m - 7m + 7$ Subtracting $7m$ from both sides.

$2m - 45 = 7$

$2m - 45 + 45 = 7 + 45$ Adding 45 to both sides.

$2m = 52$

$\frac{2m}{2} = \frac{52}{2}$ Dividing both sides by 2.

$m = 26$

Check:

$\text{LHS} = 9 \times (26 - 5) = 9 \times 21 = 189$

$\text{RHS} = 7 \times (26 + 1) = 7 \times 27 = 189$

LHS = RHS.

c

$10y - 3(2y - 5) = 6(8 - 3y)$

$10y - 6y + 15 = 48 - 18y$ Expanding brackets on both sides.

$4y + 15 = 48 - 18y$ Collecting like terms.

9780170465618

$$4y + 18y + 15 = 48 - 18y + 18y$$ Adding $18y$ to both sides.

$$22y + 15 = 48$$

$$22y + 15 - 15 = 48 - 15$$ Subtracting 15 from both sides.

$$22y = 33$$

$$\frac{22y}{22} = \frac{33}{22}$$ Dividing both sides by 22.

$$y = 1\frac{1}{2}$$

EXERCISE 7.03 ANSWERS ON P. 646

Equations with brackets

U F R C

EXAMPLE 3

1 Solve each equation.

R

a	$2(m + 3) = 8$	**b**	$3(x + 1) = 9$	**c**	$5(y - 2) = 15y$
d	$35 = 7(k + 1)$	**e**	$4(3 - a) = 16$	**f**	$11 = 9(1 + 2p)$
g	$3h = 4(h + 6)$	**h**	$6(m - 10) = -6$	**i**	$8u = 11(u - 3)$
j	$27 = 7(2y + 1)$	**k**	$5(2 + 3p) = -8$	**l**	$22x = 9(4x - 3)$

2 In which line has an error been made in solving $5(x - 3) = 25$?
Select the correct answer **A**, **B**, **C** or **D**.

$5(x - 3) = 25$

$5x - 8 = 25$ Line 1

$5x - 8 + 8 = 25 + 8$ Line 2

$5x = 33$ Line 3

$x = \frac{33}{5}$ Line 4

$= 6\frac{3}{5}$

A Line 1 **B** Line 2 **C** Line 3 **D** Line 4

3 Show that $k = 5$ is the solution to $12(k - 1) = 48$.

R C

4 Show that $a = 6$ is the solution to $10 + a = 2(2 + a)$.

R C

5 Solve each equation.

R

a	$8(m + 2) = 5(m + 5)$	**b**	$2(y - 3) = 4(y - 5)$	**c**	$3(2 + x) = 4(1 + x)$
d	$5(p + 2) = 3(6 + p)$	**e**	$5n + 6 = 2(2n + 1)$	**f**	$2(4 - 3x) = 4(7 - 3x)$
g	$4(3w - 1) = 5(4 + 3w)$	**h**	$-2(x + 1) = 16 - 5x$	**i**	$-8y - 5 = 5(2y - 3)$

☐ Foundation ○ Standard ⬡ Complex

6 Show that the solution to $5(2m - 2) = 6(m + 1)$ is $m = 4$.

R C

7 Solve each equation.

R

a $5(m + 6) + 10 = 3(m + 2) + 20$

b $3(y + 2) - 10 = 2(y - 1) + 5$

c $7y + 2(y + 5) = 4(y - 10)$

d $3x + 4(5 + x) = 6(2 + x) + 20$

e $5y + 2(y - 3) = 4y + 2(2y + 10)$

f $11 - 2(5 + y) = 4(3 + y) - 1$

g $8 - 3(1 - m) = 5(m + 3) + 4$

h $12 - 7(2y - 5) = 6 - 15(2 - 5y)$

INVESTIGATION

Make your own equation

Here are 2 equations that have the same solution, $x = 6$:

$5x - 1 = 23 + x$ and $\frac{3x+12}{10} = 3$.

1 For each solution below, make up three equations that have that solution.

a $x = 6$

b $x = \frac{1}{4}$

c $x = 12$

d $x = 1\frac{1}{2}$

e $x = 0$

f $x = -2$

2 Compare your answers with those of other students. Check that each equation is correct.

7.04 Equation problems

Worksheet Angle problems with algebra

Puzzle Writing and solving equations

ⓘ Steps in solving word problems using equations

1. Read the problem carefully and determine what needs to be found: 'What is the question?'
2. Use a variable to represent the unknown quantity.
3. Write the problem as an equation.
4. Solve the equation.
5. Answer the problem.

☐ Foundation ○ Standard ⬡ Complex

7.04

Example 4

When three-quarters of a number is decreased by 8, the result is 46. What is the number?

SOLUTION

Let the number be x.

$\frac{3x}{4} - 8 = 46$	Translating from words to algebra.
$\frac{3x}{4} - 8 + 8 = 46 + 8$	Adding 8 to both sides.
$\frac{3x}{4} = 54$	
$\frac{3x}{4} \times 4 = 54 \times 4$	Multiplying both sides by 4.
$3x = 216$	
$\frac{3x}{3} = \frac{216}{3}$	Dividing both sides by 3.
$x = 72$	Check: $\frac{3 \times 72}{4} - 8 = 54 - 8 = 46$

The number is 72.

Example 5

A rectangle is 3 times as long as it is wide. If its perimeter is 60 cm, find its dimensions.

SOLUTION

Let the width of the rectangle be w cm. Then the length is $3w$ cm.

The perimeter is $w + 3w + w + 3w$ and this is given as 60.

3w cm

w cm

$$w + 3w + w + 3w = 60$$
$$8w = 60$$
$$w = 7.5$$

$\therefore$ The width of the rectangle is 7.5 cm and the length is $3 \times 7.5 = 22.5$ cm.

Check: The perimeter of a rectangle with dimensions 7.5 cm and 22.5 cm is $7.5 + 22.5 + 7.5 + 22.5 = 60$ cm.

Example 6

The sum of 3 consecutive numbers is 150.
Find the numbers.

Consecutive numbers follow each other in order, such as 7, 8, 9.

SOLUTION

Let the first number be x.

The next number is $x + 1$ and the third number is $x + 2$.

Their sum is $x + (x + 1) + (x + 2)$ and this is given as 150.

$$x + x + 1 + x + 2 = 150$$
$$3x + 3 = 150$$
$$3x = 147$$
$$x = \frac{147}{3}$$
$$= 49$$

$\therefore$ The consecutive numbers are 49, 50 and 51.

Check: $49 + 50 + 51 = 150$.

Example 7

Video Equation problems

Presentation Applying linear equations

Ryan is 6 years older than his sister Kelsey. Their mother is 3 times Ryan's age.

a If the sum of the 3 ages is 59, write a simplified equation to find Ryan's age.

b Solve the equation and find each person's age.

Shutterstock.com/Syda Productions

SOLUTION

a Let x = Ryan's age.

Kelsey's age is $x - 6$. — Kelsey is 6 years younger than Ryan.

The mother's age is $3x$.

$x + (x - 6) + 3x = 59$

$5x - 6 = 59$ — Simplify the equation.

b $5x - 6 = 59$

$5x = 65$

$x = 13$

Ryan is 13 years old.

Kelsey is $13 - 6 = 7$ years old.

Their mother is $3 \times 13 = 39$ years old.

Check: $13 + 7 + 39 = 59$

EXERCISE 7.04 ANSWERS ON P. 646

Equation problems

1 When 7 is subtracted from 4 times a certain number, the answer is 37. What is the number?

PS R C

EXAMPLE 4

2 If 15 more than a number is 3 more than double the number, what is the number?

3 Two-thirds of a number is 16. What is the number?

4 When two-fifths of a number is increased by 15, the result is 27. What is the number?

5 A rectangle is 4 times as long as it is wide. The perimeter of the rectangle is 100 cm. Find the dimensions of the rectangle.

PS R C

EXAMPLE 5

6 The length of a rectangle is 7 cm longer than its width.

a Let w be the width of the rectangle. Write an equation for w if the perimeter of the rectangle is 94 cm.

b Solve the equation and find the dimensions of the rectangle.

7 Find the value of x in this triangle.

88°

$2(x + 11)°$

8 **a** Find the value of y.

b What is the size of each alternate angle?

$(7y + 19)°$

$5(y + 9)°$

9 Calculate the size of each marked angle.

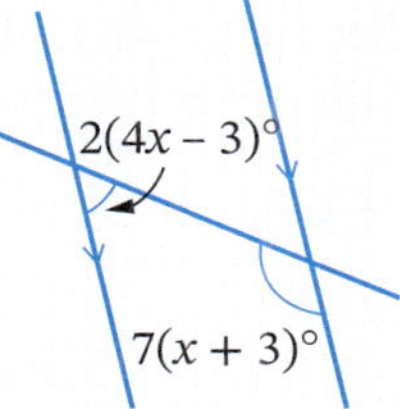

Foundation Standard Complex

10 The sum of 2 consecutive numbers is 87. Find the numbers.

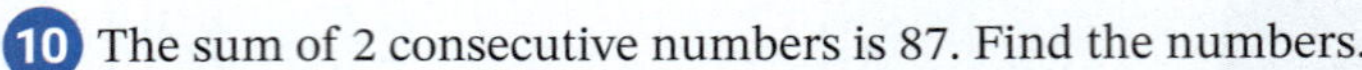

PS R C

11 The sum of 3 consecutive numbers is 87. Find the numbers.

12 The sum of 3 consecutive *even* integers is 168. What are the 3 integers?

13 The sum of 3 consecutive *odd* integers is 75. Find the integers.

14 Dean's father, Franco, is 5 times Dean's age. Dean is 8 years older than his sister, Helen. The sum of all their ages is 62 years. How old is each person? (*Hint:* Let Dean's age be x.)

PS R C

15 Bill is 3 times as old as his daughter, Rebekah, who is 6 years younger than her brother, Tyrone. How old is Rebekah if the sum of their 3 ages is 76 years?

16 In my money box I have only $1 and $2 coins. I have 240 coins in total, worth $318. How many $2 coins do I have in the money box?

17 Aerin bought 4 ice creams and received $2.80 change from his $20 note. How much did each ice cream cost?

18 Ashvin is twice as tall as his little sister, Sureti, and 30 cm shorter than his father. Their combined height is 3.8 m. Find the height of each person.

19 The length of a rectangle is 9.5 cm longer than its width. The perimeter of the rectangle is 87 cm. Find the dimensions of the rectangle.

20 Janine is 6 years younger than her husband, Vila, who is 3 times the age of their son Brett. Brett is 5 years older than his sister Amanda. The sum of all their ages is 125 years. How old is each person?

21 A house is constructed in the shape shown. If the house has a floor area of 54m^2, what is the value of d?

PS R C

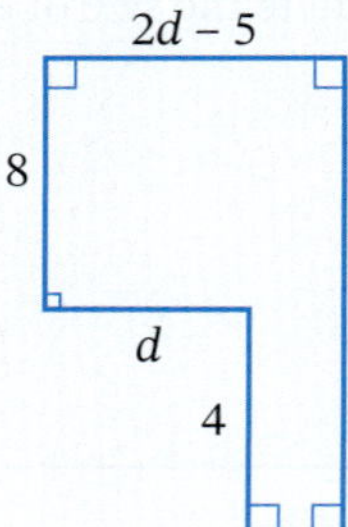

22 Ashleigh only collects 20-cent and 50-cent coins. Last week she counted 420 coins to the value of $156. How many 50-cent coins did Ashleigh have?

Foundation Standard Complex

☆ MENTAL SKILLS 7A ANSWERS ON P. 647

Maths without calculators

Quiz Mental skills 7A

Fraction of a quantity

Learn these commonly-used fractions and their decimal equivalents.

Fraction	$\frac{1}{2}$	$\frac{1}{4}$	$\frac{1}{8}$	$\frac{3}{4}$	$\frac{1}{5}$	$\frac{1}{10}$	$\frac{1}{20}$	$\frac{2}{5}$
Decimal	0.5	0.25	0.125	0.75	0.2	0.1	0.05	0.4

Now we will use them to find a fraction or decimal of a quantity.

1 Study each example.

a $\frac{1}{4} \times 72 = 72 \div 4$
$= 18$

b $\frac{2}{5} \times 40 = \left(\frac{1}{5} \times 40\right) \times 2$
$= 8 \times 2$
$= 16$

c $\frac{3}{4} \times 32 = \left(\frac{1}{4} \times 32\right) \times 3$
$= 8 \times 3$
$= 24$

d $0.5 \times 66 = \frac{1}{2} \times 66$
$= 33$

e $0.05 \times 80 = \frac{1}{20} \times 80$
$= 4$

f $0.125 \times 56 = \frac{1}{8} \times 56$
$= 7$

2 Now simplify each expression.

a $\frac{1}{2} \times 28$ **b** $\frac{1}{4} \times 36$ **c** $\frac{1}{10} \times 70$ **d** $\frac{1}{8} \times 64$

e $\frac{1}{5} \times 15$ **f** $\frac{1}{10} \times 80$ **g** $\frac{2}{5} \times 25$ **h** $\frac{1}{20} \times 100$

i $\frac{3}{4} \times 44$ **j** $\frac{1}{8} \times 40$ **k** 0.25×60 **l** 0.4×45

m 0.1×260 **n** 0.125×48 **o** 0.75×48 **p** 0.05×120

q 0.2×70 **r** 0.5×320 **s** 0.25×56 **t** 0.125×16

INVESTIGATION

Solving $x^2 = c$

1 $x^2 = 25$ has 2 solutions. What are they?

2 What are the possible solutions for each of the following?

a $x^2 = 9$ **b** $x^2 = 49$ **c** $x^2 = 100$

3 What is the inverse operation of 'squaring'?

4 Study this example:

$x^2 = 81$

$x = \pm\sqrt{81}$ which means $\sqrt{81}$ or $-\sqrt{81}$

$x = \pm 9$ which means 9 or -9

Check: When $x = 9$, $x^2 = 9^2 = 81$

When $x = -9$, $x^2 = (-9)^2 = 81$

Now use the same method to solve each equation and check your answers.

a $m^2 = 1$ b $k^2 = 64$

5 How many solutions does each quadratic equation have?

a $m^2 = 1$ b $k^2 = 64$ c $x^2 = 81$

6 Do the following quadratic equations have solutions? (Give reasons for your answers.)

a $w^2 = -1$ b $y^2 = -64$ c $h^2 = -81$

7 Write an example of a quadratic equation that has only *one* solution.

7.05 Simple quadratic equations $ax^2 = c$

An equation in which the highest power of the variable is 2 is called a **quadratic equation**, for example, $x^2 = 5$, $3m^2 + 7 = 10$, $d^2 - 4 = 0$ and $4t^2 - 3t = 8$.

The quadratic equation $x^2 = c$

The quadratic equation $x^2 = c$ (where c is a positive number) has two solutions:

$x = \pm\sqrt{c}$ (which means $x = \sqrt{c}$ and $x = -\sqrt{c}$).

Example 8

Solve each quadratic equation.

a $y^2 = 16$ b $p^2 = 65$ c $5a^2 = 245$

SOLUTION

a $y^2 = 16$

$y = \pm\sqrt{16}$ Finding the square root of both sides.

$= \pm 4$

b $p^2 = 65$

$p = \pm\sqrt{65}$ Finding the square root of both sides.

65 is not a square number, so leave the answer as a surd.

c $5a^2 = 245$

$a^2 = \frac{245}{5}$ Dividing both sides by 5.

$a^2 = 49$

$a = \pm\sqrt{49}$

$= \pm 7$

Example 9

Solve each quadratic equation, writing the solution correct to one decimal place.

a $4x^2 = 500$　　**b** $\frac{3h^2}{5} = 26$

SOLUTION

a

$$4x^2 = 500$$
$$x^2 = \frac{500}{4}$$
$$x^2 = 125$$
$$x^2 = \pm\sqrt{125}$$
$$= \pm 11.1803...$$
$$\approx \pm 11.2$$

b

$$\frac{3h^2}{5} = 26$$
$$3h^2 = 26 \times 5$$
$$3h^2 = 130$$
$$h^2 = \frac{130}{3}$$
$$h^2 = 43\frac{1}{3}$$
$$h = \pm\sqrt{43\frac{1}{3}}$$
$$= \pm 6.5828...$$
$$\approx \pm 6.6$$

EXERCISE 7.05 ANSWERS ON P. 647

Simple quadratic equations $ax^2 = c$

U F R C

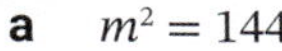

EXAMPLE 8

1 Solve each quadratic equation, writing the solutions as surds if necessary.

a $m^2 = 144$　**b** $x^2 = 400$　**c** $y^2 = 225$　**d** $k^2 = 59$　**e** $y^2 = 10$

f $w^2 = 36$　**g** $8x^2 = 200$　**h** $9t^2 = 81$　**i** $\frac{a^2}{2} = 8$　**j** $5k^2 = 40$

k $3w^2 = 30$　**l** $2d^2 = 288$　**m** $\frac{k^2}{2} = 8$　**n** $\frac{w^2}{10} = 7$　**o** $4x^2 = 1$

p $\frac{m^2}{3} = 27$　**q** $8y^2 = 40$　**r** $2p^2 + 3 = 21$　**s** $3k^2 = 48$　**t** $\frac{y^2}{5} - 2 = 9$

u $6x^2 = 42$

EXAMPLE 9

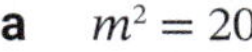

2 Solve each equation, writing the solution correct to one decimal place.

a $m^2 = 20$　**b** $b^2 = 17$　**c** $v^2 = 6$　**d** $2p^2 = 35$

e $9k^2 = 63$　**f** $\frac{x^2}{5} = 8$　**g** $\frac{k^2}{16} = 6$　**h** $\frac{7u^2}{10} = 2$

i $6y^2 = 84$　**j** $\frac{3w^2}{4} = 20$　**k** $a^2 + 11 = 28$　**l** $2y^2 - 14 = 63$

3 Explain why the quadratic equation $k^2 + 25 = 0$ has no solutions.

R C

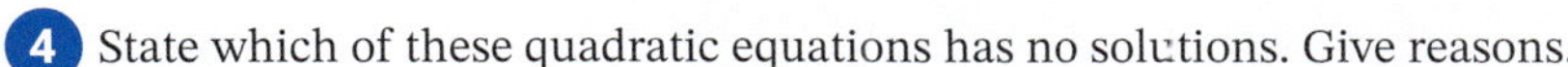

4 State which of these quadratic equations has no solutions. Give reasons.

R C

a $x^2 = -9$　**b** $2k^2 + 5 = 9$　**c** $3m^2 + 8 = 4$

d $\frac{9w^2}{2} - 1 = 1$　**e** $4 + \frac{d^2}{3} = 8$　**f** $\frac{5a^2}{2} + 3 = 2$

Foundation　Standard　Complex

7.06 Quadratic equations $x^2 + bx + c = 0$

ⓘ The quadratic equation $x^2 + bx + c = 0$

To solve quadratic equations of the form $x^2 + bx + c = 0$, factorise $x^2 + bx + c$ into a binomial product of the form $(x + d)(x + e)$.

Example 10

Video Quadratic equations by factorising

Solve $x^2 + 5x + 6 = 0$.

SOLUTION

To factorise $x^2 + 5x + 6$, find 2 numbers that have a sum of 5 and a product of 6.

The correct numbers are 2 and 3.

$$x^2 + 5x + 6 = 0$$
$$(x + 2)(x + 3) = 0$$

The LHS has been factorised into 2 binomials, $(x + 2)$ and $(x + 3)$, whose product is 0.

If 2 numbers have a product of 0, then one of the numbers *must* be 0.

$x + 2 = 0$ or $x + 3 = 0$

$x = -2$ or $x = -3$

∴ The solution to $x^2 + 5x + 6 = 0$ is

$x = -2$ or $x = -3$.

Check:

When $x = -2$,
$\text{LHS} = (-2)^2 + 5 \times (-2) + 6 = 0$
$\text{RHS} = 0$
LHS = RHS.

When $x = -3$,
$\text{LHS} = (-3)^2 + 5 \times (-3) + 6 = 0$
$\text{RHS} = 0$
LHS = RHS.

Example 11

Video Quadratic equations by factorising

Solve each quadratic equation.

a $x^2 - x - 2 = 0$

b $u^2 + 3u - 28 = 0$

c $3m^2 = 6m$

d $w^2 - 10w + 25 = 0$

SOLUTION

a $x^2 - x - 2 = 0$

Find 2 numbers that have a sum of -1 and a product of -2.

They are -2 and 1.

$(x - 2)(x + 1) = 0$

7.06

$x - 2 = 0$ or $x + 1 = 0$

$x = 2$ or $x = -1$

$\therefore$ The solution to $x^2 - x - 2 = 0$ is

$x = 2$ or $x = -1$.

Check:

When $x = 2$,
LHS $= 2^2 - 2 - 2 = 0$
RHS $= 0$
LHS = RHS.

When $x = -1$,
LHS $= (-1)^2 - (-1) - 2 = 0$
RHS $= 0$
LHS = RHS.

b $u^2 + 3u - 28 = 0$

$(u + 7)(u - 4) = 0$

$u + 7 = 0$ or $u - 4 = 0$

$u = -7$ or $u = 4$

$\therefore$ The solution to $u^2 + 3u - 28 = 0$ is $u = -7$ or $u = 4$.

c $3m^2 = 6m$

$3m^2 - 6m = 0$

You cannot divide both sides by *m* because if *m* = 0, then you are dividing by 0.

This requires a simpler factorisation as there are only 2 terms, both involving *m*.

$3m(m - 2) = 0$

$3m = 0$ or $m - 2 = 0$

$m = 0$ or $m = 2$

$\therefore$ The solution to $3m^2 - 6m = 0$ is $m = 0$ or $m = 2$.

d $w^2 - 10w + 25 = 0$

$(w - 5)(w - 5) = 0$

$w - 5 = 0$ or $w - 5 = 0$

$w = 5$ or $w = 5$

$\therefore$ The solution to $w^2 - 10w + 25 = 0$ is $w = 5$ (only one solution).

Note: Quadratic equations of the form $ax^2 + bx + c = 0$ can be found in Chapter 13, *Quadratic equations and the parabola*, of *Nelson Maths 10 Advanced*.

EXERCISE 7.06 ANSWERS ON P. 647

Quadratic equations $x^2 + bx + c = 0$

U F

1 Solve each equation.

a $x^2 + 3x + 2 = 0$ **b** $y^2 + 5y + 4 = 0$ **c** $y^2 + 16y + 48 = 0$

d $x^2 + x - 12 = 0$ **e** $x^2 + 2x - 3 = 0$ **f** $x^2 + 3x - 40 = 0$

2 Solve each equation.

a $x^2 - x - 30 = 0$ **b** $x^2 - 8x + 16 = 0$ **c** $x^2 - 5x - 66 = 0$

d $d^2 - 2d = 0$ **e** $x^2 - 3x - 10 = 0$ **f** $n^2 + 4n = 0$

g $k^2 - 7k = 0$ **h** $y^2 = 5y$ **i** $v^2 = 12v$

3 Solve each quadratic equation. Select **A**, **B**, **C** or **D**.

a $x^2 + 4x - 60 = 0$

A $x = -10, 6$ **B** $x = 12, -5$ **C** $x = 10, -6$ **D** $x = -12, 5$

b $q^2 + 3q = 0$

A $q = 3, -3$ **B** $q = 6, -3$ **C** $q = 0, -3$ **D** $q = 0, 3$

7.07 Equations and formulas

A **formula** is an algebraic equation that shows a relationship between variables. For example, the formula for the area of a circle is $A = \pi r^2$, where A is the area and r is the radius of the circle (π is a constant). Because the formula is for the area, A is called the **subject** of the formula and it is the variable on its own on the left side of the '=' sign.

Example 12

The formula for the perimeter (P) of a rectangle of length l and width w is given by $P = 2(l + w)$.

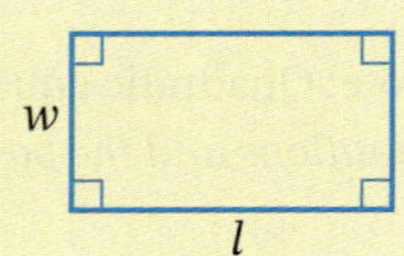

Use the formula to find:

a the perimeter of a rectangle with length 20 cm and width 9 cm.

b the width of a rectangle if its length is 12 m and its perimeter is 70 m.

c the length of a rectangle if its width is 42 cm and its perimeter is 1.8 m.

☐ Foundation ○ Standard ⬡ Complex

7.07

SOLUTION

a $l = 20, w = 9$:

$P = 2(l + w)$

$= 2(20 + 9)$

$= 2 \times 29$

$= 58$

$\therefore$ The perimeter is 58 cm.

b $l = 12, P = 70$:

$P = 2(l + w)$

$70 = 2(12 + w)$

$70 = 24 + 2w$

$46 = 2w$

$w = \frac{46}{2}$

$= 23$

$\therefore$ The width is 23 cm.

c $w = 42, P = 1.8 \text{ m} = 180 \text{ cm}$ since w is given in cm.

$P = 2(l + w)$

$180 = 2(l + 42)$

$180 = 2l + 84$

$96 = 2l$

$l = \frac{96}{2}$

$= 48$

$\therefore$ The length is 48 cm.

Example 13

The cost of hiring a sound system for a party is \$250 plus \$75 per hour. The cost can be represented by the formula $C = 250 + 75h$, where C is the total cost (in dollars), and h is the number of hours.

a Find the cost of hiring the sound system for 4 h.

b A family is willing to spend \$750 for hiring the sound system. What is the maximum number of *whole* rental hours that the family can afford?

SOLUTION

a $h = 4$:

$C = 250 + 75h$

$= 250 + 75 \times 4$

$= 550$

$\therefore$ The cost is \$550.

b $C = 750$:

$C = 250 + 75h$

$750 = 250 + 75h$

$500 = 75h$

$h = \frac{500}{75}$

$= 6\frac{2}{3}$

$\therefore$ The maximum number of whole hours is 6.

EXERCISE 7.07 ANSWERS ON P. 647

Equations and formulas

U F PS R

1 Given the formula $y = 5x + c$, find:

- **a** y if $x = 5$ and $c = 3$
- **b** y if $x = -1$ and $c = 16$
- **c** c if $y = 40$ and $x = 3$
- **d** c if $y = -6$ and $x = -1$
- **e** x if $y = 27$ and $c = 12$
- **f** x if $y = 64$ and $c = -16$

2 A temperature in degrees Celsius (°C) can be converted to degrees Fahrenheit (°F) using the formula $F = \frac{9C}{5} + 32$. Convert each temperature to °F.

- **a** 35 °C
- **b** −10 °C
- **c** 16 °C

3 Use the formula in question **2** to convert each temperature to °C, correct to one decimal place.

PS R

- **a** 100 °F
- **b** −45 °F
- **c** 78 °F

4 The formula $A = \frac{1}{2}(a + b)h$ is used to find the area of a trapezium, where A is the area, a and b are the lengths of the parallel sides, and h is the perpendicular height between them.

Use the formula to find:

- **a** the area of a trapezium with height 6 cm and parallel sides of length 9 cm and 15 cm.
- **b** the height of a trapezium if its area is 420 cm² and it has parallel sides of length 22 cm and 20 cm.
- **c** the length of one side of a trapezium if its parallel side is 20.5 m, its area is 318 m² and its height is 12 m.

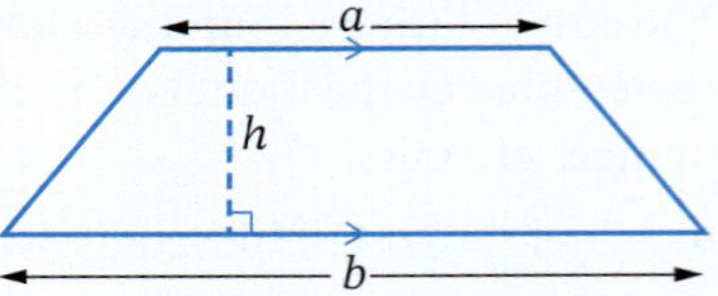

5 Find the value of S in the formula $M = \frac{kS}{5}$ if $M = 12.6$ and $k = 3.15$. Select the correct answer **A**, **B**, **C** or **D**.

- **A** 60
- **B** 7.938
- **C** 20
- **D** 0.8

6 The cost C (in dollars) of hiring a limousine is given by the formula $C = 2500 + 150h$, where h is the number of hours of hire. Find:

PS R

- **a** the cost of hiring a limousine for 4 h.
- **b** the cost of hiring a limousine for 2 days.
- **c** the number of hours for which you could hire a limousine for $4000.
- **d** the maximum number of whole hours a limousine could be hired for $5250.

7 The profit, $$P$, made by an online store selling games is given by $P = 35x - 750$, where x represents the number of games sold. Find:

PS R

- **a** the profit made when 230 games are sold.
- **b** the number of games sold if the profit is $18 010.

▷

□ Foundation ○ Standard ⬡ Complex

9780170465618

8 A catering company charges $\$C$ for a function with P people using the formula $C = 1500 + 75.5P$.

PS R

a How much does the company charge for a function with 10 guests?

b Find the cost of catering for a group of 60 people.

c Diane has \$10 000 to spend on catering for her wedding reception. What is the maximum number of people she can invite?

9 The perimeter of a rectangle is given by the formula $P = 2(l + w)$, where l is the rectangle's length and w is its width. What is the length of a rectangle with a width of 4 and a perimeter of 52? Select the correct answer **A**, **B**, **C** or **D**.

A 9 **B** 12 **C** 18 **D** 22

10 The temperature T (in °C) of a hot liquid as it cools is given by the formula $T = 100 - 17.5h$, where h is the number of hours it has been cooling. Find:

PS R

a the temperature of the liquid after 2 h.

b the temperature of the liquid after 30 min.

c the number of hours it takes for the temperature of the liquid to reach 30 °C.

11 Archeologists use the formula $H = 2.52t + 75.8$ to estimate the height, H cm, of a man when the tibia (shin bone) length, t cm, is measurable.

PS R

a A male tibia 42 cm long was found intact. Estimate the height of the male, correct to the nearest centimetre.

b Estimate, to the nearest centimetre, the length of the tibia of a male whose height is 174 cm.

Graphing inequalities on a number line 7.08

An **inequality** looks like an equation except that the equals sign (=) is replaced by an inequality symbol >, ≥, < or ≤.

$2x - 7 = 15$ is an **equation**. There is only one value of x that makes it true.

$2x - 7 \leq 15$ is an **inequality**. There is a range of values of x that make it true.

Video
Inequalities: The world's most populous country

ⓘ Inequality symbols

>	'is greater than'	≥	'is greater than or equal to'
<	'is less than'	≤	'is less than or equal to'

□ Foundation ○ Standard ⬡ Complex

For example, '$x \geq 3$' is read 'x is greater than or equal to 3'. It includes 3 and all the numbers above 3, such as 3.01, 4, 10, 20 000. '$x > 3$' is read 'x is greater than 3' and means all the numbers above 3, but **not** 3.

Inequality	In words	Meaning
$x > 3$	x is greater than 3	Values above 3
$x < 3$	x is less than 3	Values below 3
$x \geq 3$	x is greater than or equal to 3	Values above and including 3
$x \leq 3$	x is less than or equal to 3	Values below and including 3

An inequality can be graphed on a **number line**.

Example 14

Video Graphing inequalities on the number line

Graph each inequality on a number line.

a $x \geq 1$ **b** $x < 5$ **c** $x > -3$

SOLUTION

a $x \geq 1$ means that x can be any number greater than 1 or equal to 1.

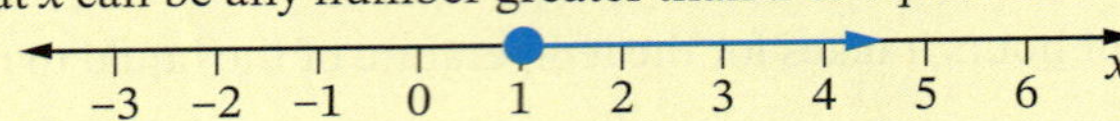

The filled circle at 1 means we include 1.

b $x < 5$ means that x can be any number less than 5, but not including 5.

The open circle on 5 means that 5 is not included.

c $x > -3$ means that x can be any number greater than −3, but not including −3.

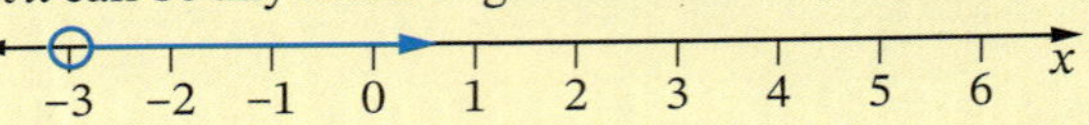

EXERCISE 7.08 ANSWERS ON P. 647

Graphing inequalities on a number line

EXAMPLE 14

1 Graph each inequality on a separate number line.

C **a** $x \geq 2$ **b** $x < -3$ **c** $x \leq 1$ **d** $x > 7$

e $x \leq 4$ **f** $x > 0$ **g** $x \geq -2$ **h** $x < 10$

2 Write the inequality shown on each number line.

C **a** −2 0 2 4 6 x

b 0 1 2 3 4 5 6 x

c −10 −8 −6 −4 −2 0 x

d −10 −8 −6 −4 −2 0 2 x

Foundation Standard Complex

9780170465618

3 Which inequality is graphed below? Select the correct answer **A**, **B**, **C** or **D**. C

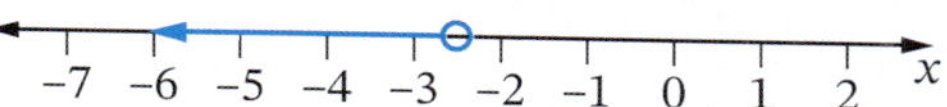

A $x > -2.5$ **B** $x < -2.5$ **C** $x < -3.5$ **D** $x > -3.5$

4 Write the inequality shown on each number line.

a

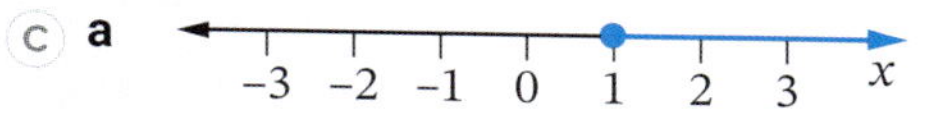

b −1 0 1 2 3 4 5 x

c 0 2 4 6 8 10 12 x

d −3 −2 −1 0 1 2 3 x

e −9 −6 −3 0 3 6 9 x

f −3 −2 −1 0 1 2 3 x

g −10 −8 −6 −4 −2 0 2 x

h −5 0 5 10 15 20 25 x

i −1 0 1 2 3 4 5 x

INVESTIGATION

The language of inequalities

Work in pairs to complete this activity.

Use inequality symbols to write each statement algebraically.

a The minimum height (H) for rides at an amusement park is 1.3 m.

b The speed limit in a school zone is 40 km/h.

c To be eligible to vote, you must be at least 18 years old (A = age).

d The overseas tour is only for people whose age (A) is from 18 to 35.

e The cost (A) of a tennis racquet will be at least \$95, but no more than \$360.

f A new flute (F) costs at least \$475.

g The price of units (U) in a new block start at \$240 000.

INVESTIGATION

Solving inequalities

We have solved equations by doing the same thing to both sides (keeping the equation 'balanced'). Will this method work with inequalities, such as $x + 4 > 10$ or $6x < 13$?

1 Start with an inequality that is true, such as $7 > 4$.

2 Add 5 (or any number you choose) to both sides of the inequality; for example, $7 > 4$ becomes $12 > 9$. Is the new inequality true or false?

3 Subtract 9 (or any number you choose) from each side of the original inequality; for example, $7 > 4$ becomes $-2 > -5$. Is the new inequality true or false?

Foundation Standard Complex

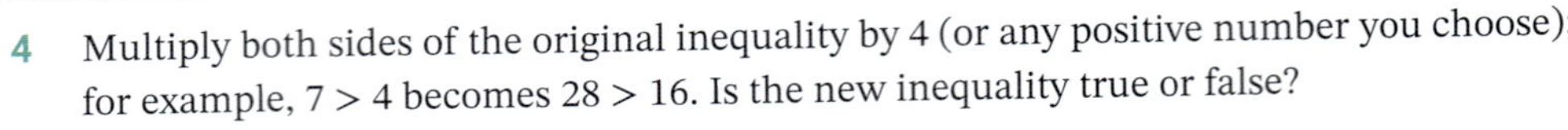

4 Multiply both sides of the original inequality by 4 (or any positive number you choose); for example, 7 > 4 becomes 28 > 16. Is the new inequality true or false?

5 Divide both sides of the original inequality by 2 (or any positive number you choose); for example, 7 > 4 becomes $3\frac{1}{2} > 2$. Is the new inequality true or false?

6 Multiply both sides of the original inequality by −3 (or any negative number you choose); for example, 7 > 4 becomes −21 > −12. Is the new inequality true or false?

7 Divide both sides of the original inequality by −4 (or any negative number you choose), for example, 7 > 4 becomes $-1\frac{3}{4} > -1$. Is the new inequality true or false?

8 Which of the 6 operations used in questions **2** to **7** can be used on inequalities to give a true result?

9 Which of the 6 operations used in questions **2** to **7** cannot be used with inequalities because they give a false result?

10 Copy and complete each set of inequality statements.

a $6 < 8$

$6 \times 3 < 8 \times$ ___ (multiplying both sides by 3)

$\therefore 18$ ___ 24

b $10 > -4$

$10 \div 2$ ___ $-4 \div$ ___ (dividing both sides by 2)

$\therefore$ __________

Does the inequality sign (< or >) stay the same when multiplying or dividing by a positive number?

11 a Is it true that 5 < 8?

b Multiply both sides by −2. Is it true that −10 < −16?

c What needs to be reversed to change −10 < −16 into a true inequality statement?

d Copy and complete to make a true statement: −10 ______ −16.

12 a Is it true that 18 > −6?

b Divide both sides by −3. Is it true that −6 > 2?

c What needs to be reversed to change −6 > 2 into a true inequality statement?

d Copy and complete to make a true inequality statement: −6 _____ 2.

13 Copy and complete: When multiplying or d__________ both sides of an inequality by a n__________ number, the inequality sign must be r__________.

9780170465618

DID YOU KNOW?

Film and game classification

In Australia, films and computer games are rated by the Classification Board. Films and videos are rated G, PG, M, MA15+ or R18+, with each category containing a list of guidelines related to the film's use of violence, coarse language, adult themes, sex and nudity.

General (G) means suitable for all ages. Children can watch films classified G without adult supervision.

Parental guidance (PG) means that parental guidance is recommended for persons under 15 years of age. These films contain mild content that may be confusing or upsetting to children, but not harmful or disturbing. Parents should watch the film with their children or preview it to check elements such as language used or inappropriate themes.

Mature (M) means recommended for mature audiences, 15 years and over. The film or computer game may contain moderate content that is harmful or disturbing to children, but the impact is not so strong as to require restriction.

Mature accompanied (MA15+) means persons under 15 are not allowed to see MA films unless accompanied by a parent or guardian, because they contain strong content that is likely to be harmful or disturbing to them.

Restricted (R18+) means legally restricted to adults, 18 years and over. It applies to films that deal with issues and scenes that require an adult perspective, and so are unsuitable for persons under 18. A person will be asked for proof of age before buying, hiring or viewing films or computer games in this category.

1 **Each of the classifications is represented by a logo (as shown) with the letter inside a particular shape. What shape is each logo?**

2 **Write each classification category as an inequality.**

7.09 Solving inequalities

Example 15

Worksheet Inequalities review

Solve each inequality and graph its solution on a number line.

a $2x - 10 \leq 16$ **b** $2(y - 1) \geq 12$ **c** $\frac{w+3}{2} > -1$

SOLUTION

a

$$2x - 10 \leq 16$$
$$2x - 10 + 10 \leq 16 + 10$$
$$2x \leq 26$$
$$\frac{2x}{2} \leq \frac{26}{2}$$
$$x \leq 13$$

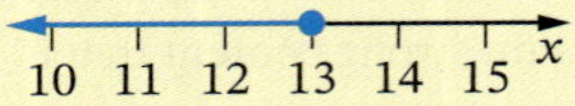

Check: Test a value less than 13, let $x = 11$.

LHS $= 2 \times 11 - 10 = 12$

RHS $= 16$

$12 \leq 16$

LHS $\leq$ RHS

b

$$2(y - 1) \geq 12$$
$$2y - 2 \geq 12$$
$$2y - 2 + 2 \geq 12 + 2$$
$$2y \geq 14$$
$$\frac{2y}{2} \geq \frac{14}{2}$$
$$y \geq 7$$

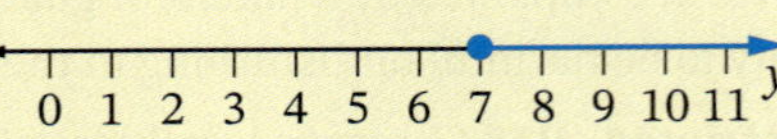

Check: Test a value greater than 7, let $y = 8$.

LHS $= 2(8 - 1) = 14$

RHS $= 12$

$14 \geq 12$

LHS $\geq$ RHS

c

$$\frac{w+3}{2} > -1$$
$$\frac{w+3}{2} \times 2 > -1 \times 2$$
$$w + 3 > -2$$
$$w + 3 - 3 > -2 - 3$$
$$w > -5$$

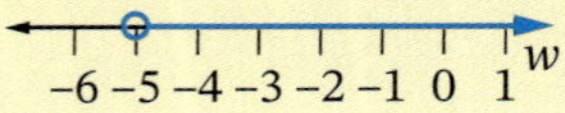

Check: Test a value greater than −5, let $w = -1$.

LHS $= \frac{-1+3}{2} = 1$

RHS $= -1$

$1 > -1$

LHS $>$ RHS

ⓘ Solving inequalities

- Inequalities can be solved algebraically in the same way as equations, using inverse operations.
- However, when multiplying or dividing both sides of an inequality by a negative number, you must reverse the inequality sign.

Example 16

Solve each inequality.

a $1 - 2x \geq -11$ **b** $4 - r < 7$ **c** $\frac{a+5}{-3} > 4$

SOLUTION

a

$$1 - 2x \geq -11$$
$$1 - 2x - 1 \geq -11 - 1$$
$$-2x \geq -12$$
$$\frac{-2x}{-2} \leq \frac{-12}{-2}$$
$$x \leq 6$$

Dividing both sides by a negative number reverses the inequality sign.

Check: Test a value less than 6, let $x = 5$.

LHS $= 1 - 2 \times 5 = -9$

RHS $= -11$

$-9 \geq -11$

LHS $\geq$ RHS

b

$$4 - r < 7$$
$$4 - r - 4 < 7 - 4$$
$$-r < 3$$
$$\frac{-r}{-1} > \frac{3}{-1}$$
$$r > -3$$

Dividing both sides by a negative number reverses the inequality sign.

Check: Test a value greater than -3, let $r = 0$.

LHS $= 4 - 0 = 4$

RHS $= 7$

$4 < 7$

LHS $<$ RHS

c

$$\frac{a+5}{-3} > 4$$
$$\frac{a+5}{-3} \times (-3) < 4 \times (-3)$$
$$a + 5 < -12$$
$$a + 5 - 5 < -12 - 5$$
$$a < -17$$

Multiplying both sides by a negative number reverses the inequality sign.

Check: Test a value less than -17, let $a = -20$.

LHS$=\frac{-20+5}{-3}=\frac{-15}{-3}=5$

RHS $= 4$

$5 > 4$

LHS $>$ RHS

EXERCISE 7.09 ANSWERS ON P. 647

Solving inequalities

U F R C

EXAMPLE 15

1 Solve each inequality and graph its solution on a number line.

R C

a $x - 1 > 6$ **b** $3y \geq 12$ **c** $m + 4 \leq 2$

d $\frac{x}{5} \geq -20$ **e** $12x < 60$ **f** $5y > -20$

g $4a \geq 2$ **h** $3w \leq -30$ **i** $8a + 5 \geq 45$

j $3a + 1 \leq 10$ **k** $6a + 4 \geq -2$ **l** $3w - 3 < -12$

m $5a + 3 \leq -27$ **n** $5y + 1 \leq 16$ **o** $4a + 5 < 15$

2 Solve $3a - 3 > -18$. Select the correct answer **A**, **B**, **C** or **D**.

A $a > -5$ **B** $a > -7$ **C** $a > 5$ **D** $a > 9$

3 Solve each inequality.

a $3(x + 2) \geq 9$ **b** $5(m - 4) \leq 10$ **c** $2(y + 5) \leq -6$

d $3(w - 2) > -6$ **e** $5(2w + 3) \leq 15$ **f** $4(2m - 5) \geq 8$

g $\frac{m+5}{3} \geq 1$ **h** $\frac{x-1}{2} \leq 2$ **i** $\frac{w-2}{5} > -1$

j $\frac{2a+1}{3} < 3$ **k** $\frac{5a+2}{4} \geq 8$ **l** $\frac{2(m+1)}{3} \leq 3$

m $\frac{5(m-1)}{4} > 3$ **n** $\frac{4(m-2)}{3} \geq -6$ **o** $3 + \frac{x}{5} < 10$

4 Solve $\frac{x-2}{5} \geq -1$. Select **A**, **B**, **C** or **D**.

A $x \geq -7$ **B** $x \leq -3$ **C** $x \leq 10$ **D** $x \geq -3$

EXAMPLE 16

5 Solve each inequality and graph its solution on a number line.

R C

a $5 - x \leq 2$ **b** $15 > 7 - y$ **c** $1 - k < 12$

d $7 - m \geq 7$ **e** $2 - p > 8$ **f** $-t + 6 \geq 10$

6 Solve each inequality.

a $-2x < 6$ **b** $\frac{k}{-3} \geq 4$ **c** $-5t > 12$

d $\frac{-x}{3} \leq -4$ **e** $4 - 3w > 7$ **f** $-4y + 3 \leq 11$

g $3 - 2x \geq -5$ **h** $8 - 5a < 3$ **i** $-2d - 3 > 8$

j $\frac{5+w}{-3} > 2$ **k** $\frac{x-4}{-4} \geq 3$ **l** $\frac{-p+2}{-3} < -2$

☐ Foundation ○ Standard ○ Complex

9780170465618

☆ MENTAL SKILLS 7B ANSWERS ON P. 648

Maths without calculators

Quiz Mental skills 7B

Percentage of a quantity

Learn these commonly-used percentages and their fraction equivalents.

Percentage	50%	25%	12.5%	75%	20%	10%	$33\frac{1}{3}\%$	$66\frac{2}{3}\%$
Fraction	$\frac{1}{2}$	$\frac{1}{4}$	$\frac{1}{8}$	$\frac{3}{4}$	$\frac{1}{5}$	$\frac{1}{10}$	$\frac{1}{3}$	$\frac{2}{3}$

Now we will use them to find a percentage of a quantity.

1 Study each example.

a $20\% \times 25 = \frac{1}{5} \times 25$
$= 5$

b $50\% \times 120 = \frac{1}{2} \times 120$
$= 60$

c $12.5\% \times 32 = \frac{1}{8} \times 32$
$= 4$

d $75\% \times 56 = \frac{3}{4} \times 60$
$= \left(\frac{1}{4} \times 60\right) \times 3$
$= 15 \times 3$
$= 45$

e $33\frac{1}{3}\% \times 27 = \frac{1}{3} \times 27$
$= 9$

f $66\frac{2}{3}\% \times 60 = \frac{2}{3} \times 60$
$= \left(\frac{1}{3} \times 60\right) \times 2$
$= 20 \times 2$
$= 40$

2 Now simplify each expression.

a $25\% \times 44$ **b** $33\frac{1}{3}\% \times 120$ **c** $20\% \times 35$ **d** $66\frac{2}{3}\% \times 36$

e $10\% \times 230$ **f** $12\frac{1}{2}\% \times 48$ **g** $50\% \times 86$ **h** $20\% \times 400$

i $75\% \times 24$ **j** $33\frac{1}{3}\% \times 45$ **k** $25\% \times 160$ **l** $10\% \times 650$

m $12.5\% \times 88$ **n** $66\frac{2}{3}\% \times 21$ **o** $20\% \times 60$ **p** $75\% \times 180$

POWER PLUS ANSWERS ON P. 648

1 Solve each equation.

a $11y = 16 - 3(10 - 4y)$

b $5(2s - 10) + 3(2s + 1) = 2(5 - s)$

c $17 - 5(3t - 2) = 12 - 10(9 - t)$

d $6(4 - x) - 2(x - 1) = 3(x + 10) - 5(x - 4)$

e $\frac{-3(2m+5)}{4} = 6$

f $\frac{d-3}{4} + \frac{2}{5} = 4$

g $\frac{5x}{2} - \frac{3x-1}{4} = 9$

h $\frac{2y-1}{4} = 5 + \frac{y+2}{3}$

i $\frac{5}{2y} = 14$

j $\frac{7}{r} = \frac{1}{r-3}$

k $\frac{y^2+8}{12} = 4\frac{2}{3}$

l $4m^2 - 3 = m^2 + 72$

2 Given $W = \frac{X-Y}{X+Y}$, find:

a W when $X = 15$ and $Y = 10$

b W when $X = 6$ and $Y = -12$

c X when $W = 25$ and $Y = 6$

d Y when $W = 5$ and $X = 1$

3 Karlie is a farmer who raises pigs and chickens. From a total of 42 animals, she can count 116 legs. Write an equation and solve it to find how many chickens she has.

4 Joe is twice as old as his daughter, Wendi. 10 years ago, he was 3 times as old as her. Write an equation and solve it to calculate how old Wendi is now.

5 Find the value of B in the balance scale below.

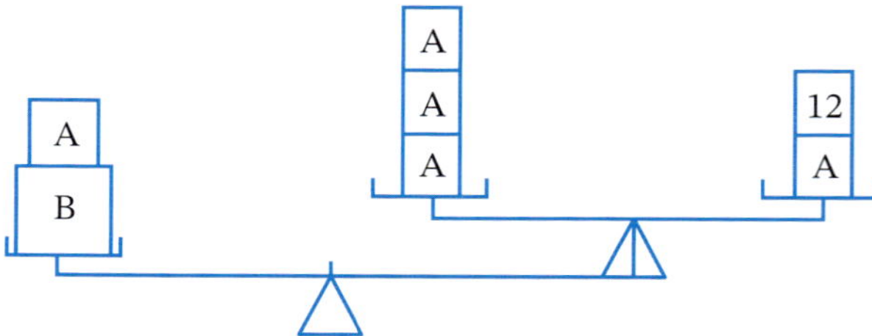

6 Graph each inequality on a number line.

a $1 \leq x \leq 4$

b $-2 \leq x \leq 3$

c $-12 < 4x \leq 4$

9780170465618

7 CHAPTER REVIEW

Language of maths

backtracking method	balancing method	brackets	consecutive
equation	expand	flowchart	formula
inequality	inverse operation	LHS	linear equation
number line	quadratic equation	RHS	solution
solve	square root	subject	substitute
surd	undoing	unknown	variable

Quiz
Language of maths 7

1 Which method of solving equations involves using inverse operations on both sides of the equation?

2 What is the **subject** of the formula $A = \frac{1}{2}(a + b)h$?

3 What name is given to numbers that follow each other in order, such as 9, 10, 11?

4 Write an example of:

a a quadratic equation **b** a linear equation

5 How many solutions does a linear equation have?

6 What does **LHS** stand for?

Topic summary

Print (or copy) and complete this mind map of the topic, adding detail to its branches and using pictures, symbols and colour where needed. Ask your teacher to check your work.

Worksheet
Mind map: Equations and inequalities

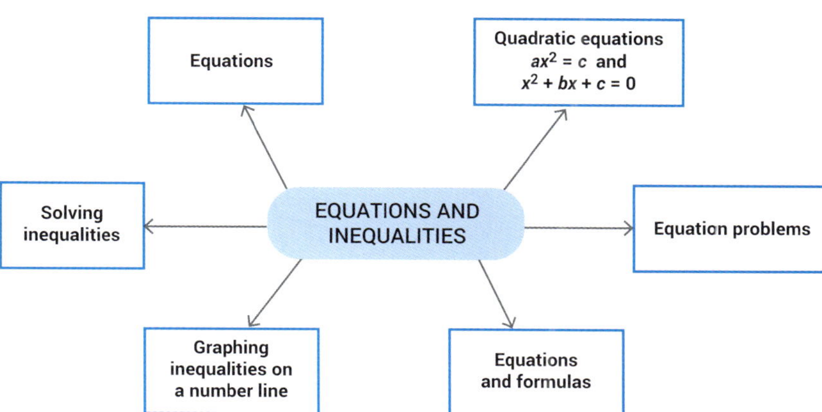

7 TEST YOURSELF

ANSWERS ON P.XX

Quiz
Test yourself 7

1 Solve each equation. Check your solutions.

a $3x - 10 = 5$ **b** $4 + 2y = 21$ **c** $12y + 5 = 23$

d $\frac{m}{4} - 6 = 2$ **e** $\frac{x-5}{4} = 7$ **f** $\frac{5x}{3} = 10$

g $\frac{-2x}{5} = 3$ **h** $11 - 2a = 17$ **i** $20 + 4d = -6$

2 Solve each equation. Check your solutions.

a $2x + 5 = 3x + 4$ **b** $7 + 4x = x - 8$ **c** $5 - 6w = 3w - 7$

d $3x + 4 = 2x + 7$ **e** $5n - 3 = 2n - 15$ **f** $2d - 8 = -5d - 71$

g $4t = 12 - 4t$ **h** $8j - 17 = 10j$ **i** $6 - 3q = 8 - q$

3 Write an equation with x on both sides that has the solution $x = 3$.

4 Solve each equation.

a $2(w - 5) = 4$ **b** $3(1 + 4n) = 15$ **c** $5(1 - 3p) = 20$

d $2(3 + x) = 5(x + 1)$ **e** $3(1 - y) = 4(2 - y)$ **f** $2(3 - 4x) = -(2x + 3)$

5 Solve each equation.

a $18 - 3a = 5a - 2(a - 6)$ **b** $11g + 2(g + 3) = 5(2g + 10)$

6 The length of a rectangle is 6 cm longer than it is wide. The perimeter of the rectangle is 76 cm. Find the dimensions of the rectangle.

7 Find the value of each variable.

a

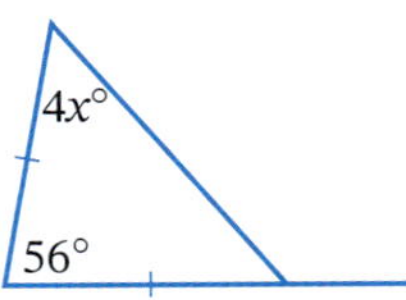

b

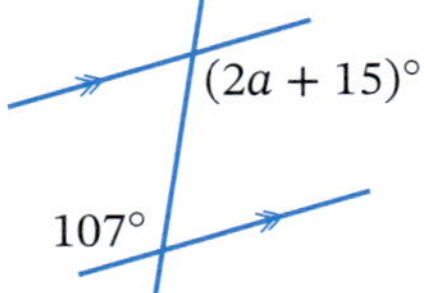

8 If 6 more than a number equals 5 more than double the number, what is the number?

7.06

9 Solve each quadratic equation.

a $d^2 = 64$ **b** $8p^2 = 288$ **c** $3z^2 = 105$

d $x^2 + 3x - 18 = 0$ **e** $m^2 - 10m + 16 = 0$ **f** $b^2 - 5b - 36 = 0$

g $t^2 - 14t + 49 = 0$ **h** $u^2 + 11u + 24 = 0$ **i** $h^2 - 7h - 8 = 0$

□ Foundation ○ Standard ○ Complex

9780170465618

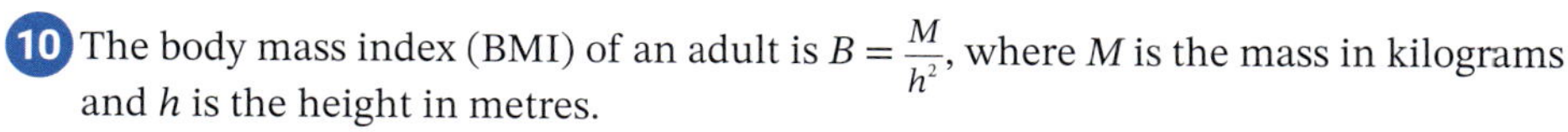

10 The body mass index (BMI) of an adult is $B = \frac{M}{h^2}$, where M is the mass in kilograms and h is the height in metres. 

a Find the BMI of Ned who is 1.85 m tall and has a mass of 72 kg.

b Find the mass of a person with a BMI of 24, who is 2.1 m tall.

11 The cost, C, in dollars, of hiring a taxi is $C = 5 + 2.4d$, where d is the distance travelled in kilometres. Find: 

a the cost of a taxi trip if the distance travelled is 15 km.

b the distance travelled if the cost of a taxi trip was \$78.20.

12 Graph each inequality on a number line. 

a $x \geq 0$ **b** $x < 3$ **c** $x \leq -2$ **d** $x > -5$

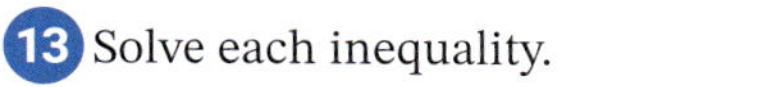

13 Solve each inequality. 7.09

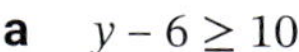

a $y - 6 \geq 10$ **b** $2y \leq -15$ **c** $3a + 10 > -5$

d $10 - 6x < 28$ **e** $\frac{a+2}{-4} > \frac{7}{2}$ **f** $\frac{3-5x}{2} \geq 9$

CHAPTER 7 TEST YOURSELF

☐ Foundation ○ Standard ⬡ Complex

Practice set 2 ANSWERS ON P.XX

1 Simplify each expression.

a $y^9 \times y$ **b** $3g^7 \times 4g^8$ **c** $2p^3q \times 9p^2q^2$

d $\frac{k^{18}}{k^6}$ **e** $\frac{x^{11}}{x^7}$ **f** $h^{10} \div h$

g $54k^7 \div 6k^3$ **h** $\frac{h^3y^7}{hy^4}$ **i** $12m^5n^8 \div 18mn^7$

2 Simplify each expression.

a $6y^0$ **b** $4^0 \times 3^0$ **c** $(4k)^0$

d $x \times (3y)^0$ **e** $2q^0 - (2q)^0$

3 Evaluate each expression, correct to one decimal place.

a $16 \cos 55°$ **b** $8.7 \tan 13°$ **c** $\frac{24.5}{\sin 37°}$ **d** $\frac{40}{\cos 76.3°}$

4 **a** Calculate the size of one angle in a regular hexagon.

b Name the polygon that has an angle sum of 1440°.

YEAR 10 EXTENSION

5 Copy this network and find:

a the number of faces, vertices and edges.

b the order of each vertex and whether it is odd or even.

c whether the network is traversable.

6 Evaluate each expression, giving your answers in fraction form.

a 2^{-1} **b** 5^{-3} **c** 10^{-2}

7 Simplify each expression.

a $\left(\frac{1}{2}\right)^{-1}$ **b** $\left(\frac{8}{3}\right)^{-1}$ **c** $\left(\frac{4d}{5}\right)^{-1}$

4.05

8 Calculate the value of each variable, correct to 2 decimal places.

a

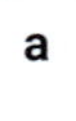

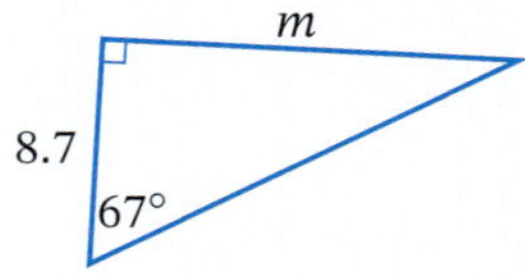

b

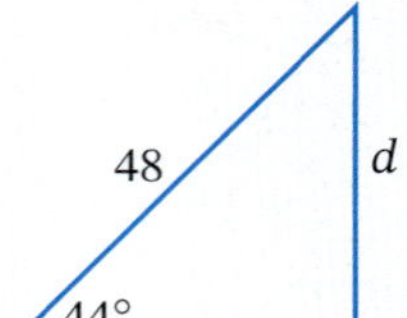

c

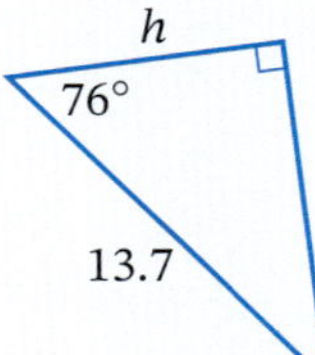

Foundation Standard Complex

9 Simplify each expression, giving the answer in index notation. 

a $(2^3)^5$ **b** $(4^2)^6$ **c** $(3p^7)^2$ **d** $(2w^5)^3$

10 Express each number in decimal form. 

a 1.54×10^7 **b** 3.2×10^{-3} **c** 9.837×10^{-1} **d** 2.4×10^4

11 Find A, correct to the nearest degree, if: 

a $\sin A = \frac{4}{5}$ **b** $\tan A = 4.713$ **c** $\cos A = 0.5148$

12 Find θ, correct to the nearest minute, if: 4.07

a $\cos \theta = 0.21$ **b** $\sin \theta = \frac{9}{10}$ **c** $\tan \theta = 1.5$

13 Write each expression using a negative index. 5.05

a $\frac{1}{n}$ **b** $\frac{1}{2^3}$ **c** $\frac{1}{k^5}$ **d** $\frac{7}{y^2}$

14 Write 7 280 000 in scientific notation. 5.08

15 Evaluate each expression, giving your answers in scientific notation, correct to 2 significant figures. 5.09

a $(7.08 \times 10^7) \times (3.8 \times 10^8)$ **b** $(4.65 \times 10^{-3}) \div (3.9 \times 10^7)$

16 Find θ, correct to the nearest degree. 4.07

a

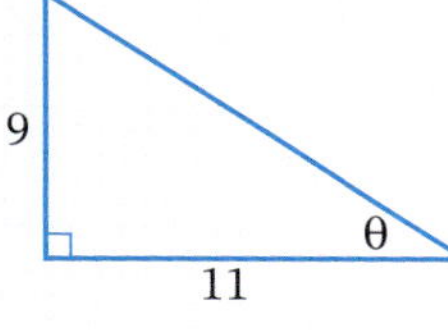

b

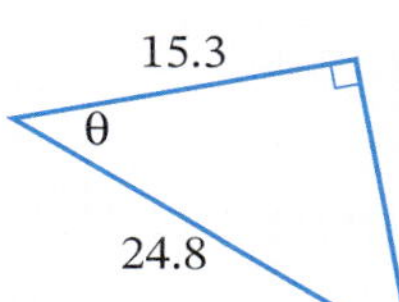

c

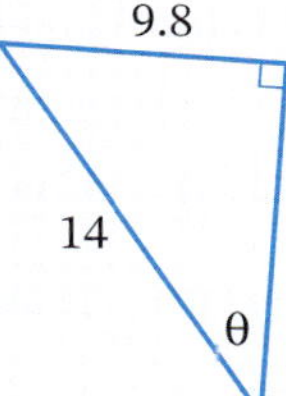

17 Solve each equation. 7.01

a $5x - 9 = 11$ **b** $1 - 4x = 10$ **c** $-3 = 7 + 8x$

d $\frac{y+2}{3} = 5$ **e** $1 + \frac{a}{3} = 7$ **f** $8 = \frac{2x-1}{7}$

18 Find the value of each variable, giving reasons. 6.01

a

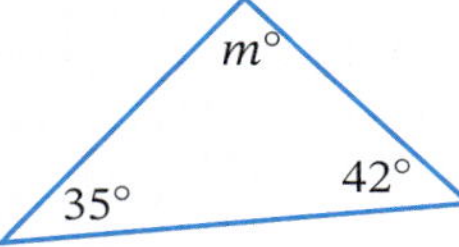

b

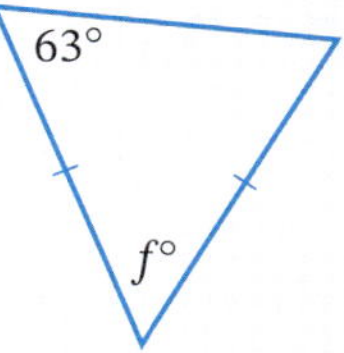

c

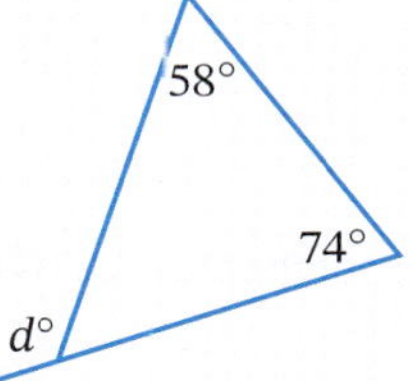

19 Solve $12 - 5k = 2k - 9$. Select the correct answer **A**, **B**, **C** or **D**. 

A $k = 3$ **B** $k = 1$ **C** $k = \frac{3}{7}$ **D** $k = 7$

☐ Foundation ○ Standard ○ Complex

5.08 **20** Select the larger number for each pair of numbers.

a 5.3×10^9 or 8.6×10^8 **b** 3.73×10^5 or 3.4×10^5

4.06 **21** Find the value of each variable, correct to one decimal place.

a

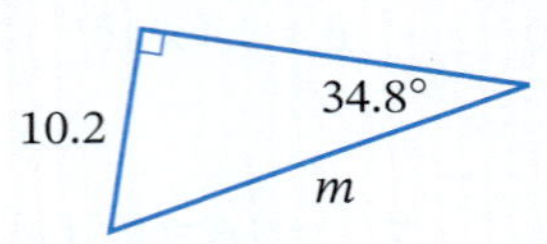

b

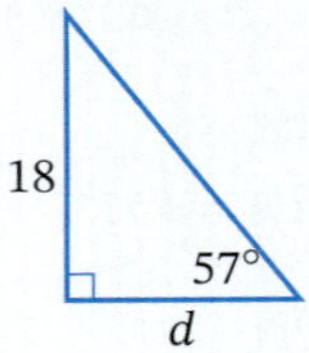

c

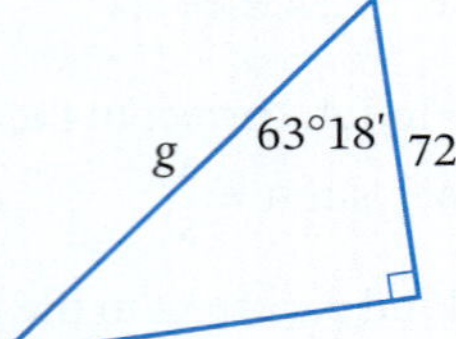

6.02 **22** Find the value of w in each quadrilateral.

a

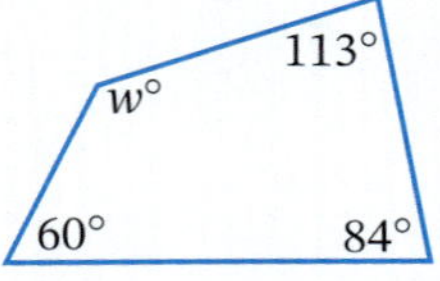

b

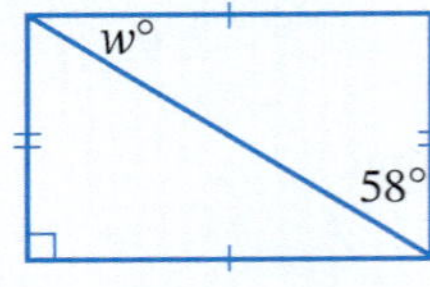

c

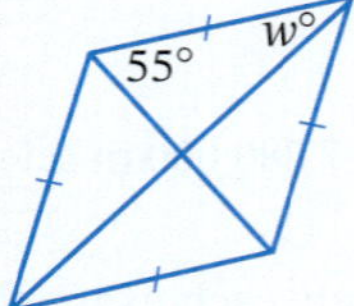

7.08 **23** The volume of useable timber (in cubic metres) in a tree is calculated using the formula:

$$v = 0.5hd^2 + 10$$

where d = the diameter of the tree trunk

h = the height to the first branch

Calculate the volume of useable timber in this tree.

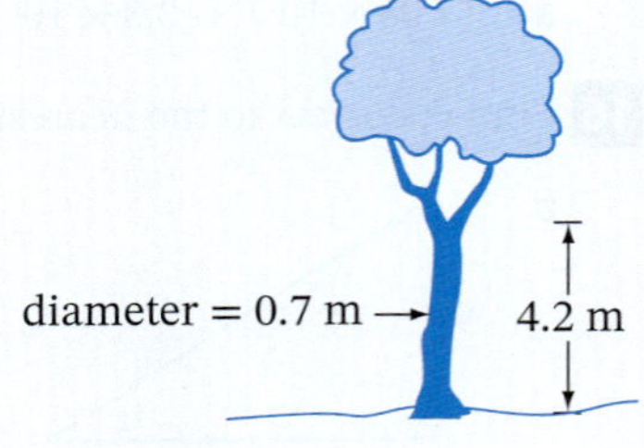

Select the correct answer **A**, **B**, **C** or **D**.

A 16.2 m^3 **B** 11.5 m^3 **C** 11.0 m^3 **D** 18.8 m^3

4.06 **24** How long is the diagonal brace of this gate, correct to the nearest centimetre?

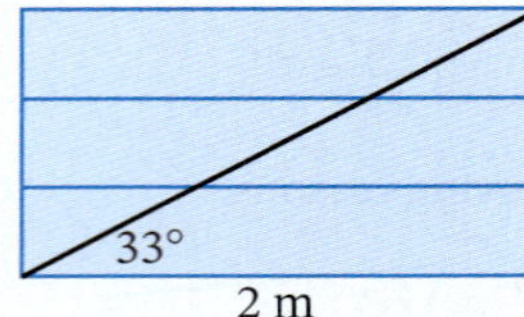

YEAR 10 EXTENSION

6.05 **25** A polyhedron has 8 faces and 6 vertices. How many edges does it have?

5.03 **26** Simplify each expression.

a $(2m^3n^2)^4$ **b** $\left(-\frac{2h}{5}\right)^2$ **c** $(7rt^2)^3$

7.07 **27** Solve each inequality.

a $x + 21 < 42$ **b** $3x - 5 \geq 1$

☐ Foundation ○ Standard ○ Complex

28 Find the size of angle A, correct to the nearest degree. 4.07

a

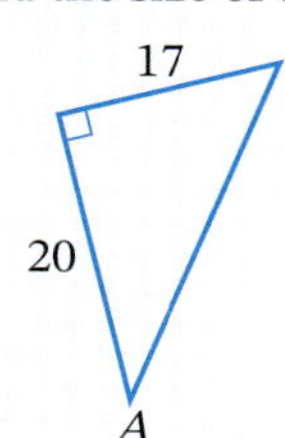

b

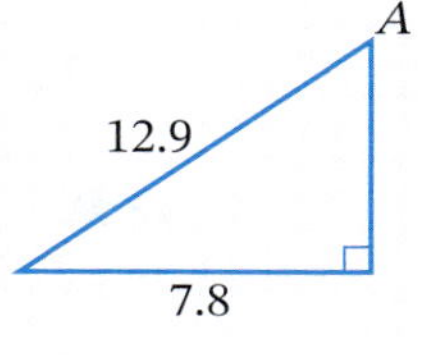

c

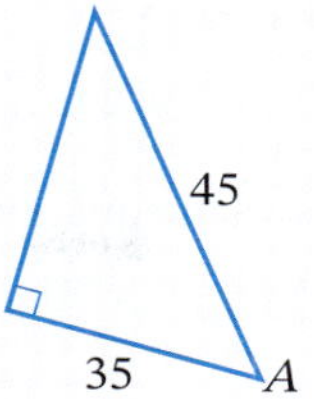

29 A formula for cooking roast beef is $T = 40M + 25$, where T is the cooking time in minutes and M is the mass of beef in kilograms.

a How long will it take to cook a 5 kg piece of beef?

b What mass of beef has a cooking time of $2\frac{1}{2}$ hours?

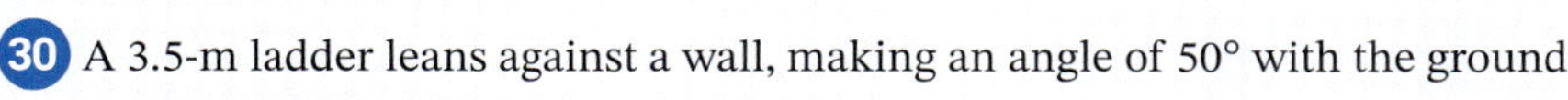

30 A 3.5-m ladder leans against a wall, making an angle of 50° with the ground.

a Draw a diagram showing this information.

b Calculate, correct to the nearest 0.1 m:

- **i** how far up the wall the ladder reaches.
- **ii** how far the foot of the ladder is from the base of the wall.

31 What type of quadrilateral is shown here?

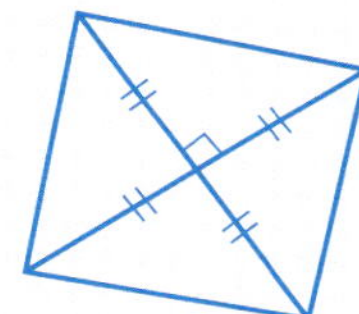

32 Solve each equation.

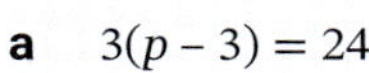

a $3(p - 3) = 24$ **b** $3(4y - 1) + 5 = 2(9 - 2y)$

33 **a** The sum of 4 consecutive numbers is 82. Find the 4 numbers.

b Sean collected \$1 and \$2 coins until he had 275 coins, having a total value of \$402. How many \$2 coins did Sean have?

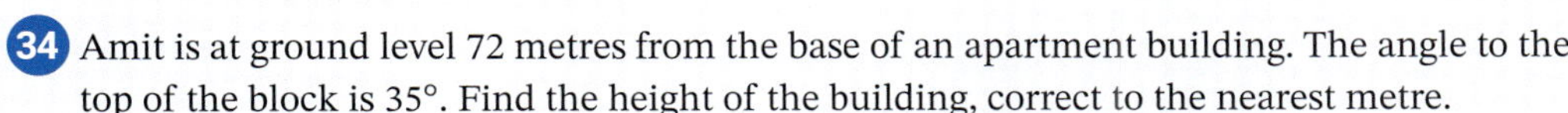

34 Amit is at ground level 72 metres from the base of an apartment building. The angle to the top of the block is 35°. Find the height of the building, correct to the nearest metre.

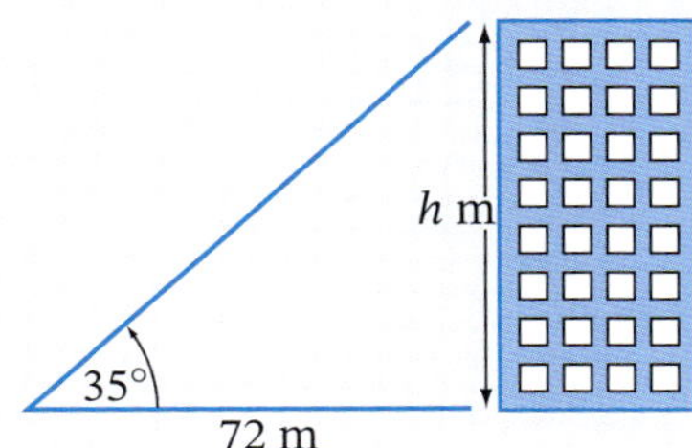

☐ Foundation ○ Standard ⬡ Complex

7.02

35 $a = 4$ is the solution to *more than one* of the following equations. Select all correct answers **A**, **B**, **C** or **D**.

A $3a - 5 = 2a - 1$ **B** $2a + 3 = 11 - a$ **C** $\frac{a}{2} - 2 = 0$ **D** $16 - a = 3a$

7.05

36 Solve each equation.

a $5m^2 + 1 = 81$ **b** $\frac{x^2}{3} = 12$

7.08

37 The formula $A = \frac{1}{2}(a + b)h$ is used to find the area of the trapezium. Find the height if the area is 920 cm² and the 2 parallel sides are 25 cm and 21 cm.

4.07

38 Solve $\tan\alpha = \frac{9}{13}$ for angle α, correct to the nearest minute. Select **A**, **B**, **C** or **D**.

A 34° 69′ **B** 35° 10′ **C** 34° 41′ **D** 34° 42′

Foundation Standard Complex

8

NUMBER

Earning money

An Australian person working full-time works an average of 35.7 hours per week. In 2020, the median weekly pay was $1712 for individuals and $2349 for households. How much does your household earn each week?

iStock.com/AzmanL

Chapter outline

		Proficiencies
8.01	Wages and salaries	U F PS R
8.02	Overtime pay	U F PS R
8.03	Commission, piecework and leave loading	U F PS
8.04	Income tax	U F PS
8.05	PAYG tax and net pay	U F PS R C

U = Understanding
F = Fluency
PS = Problem solving
R = Reasoning
C = Communication

Note: This topic is not strictly part of the Australian Curriculum but can be taught as an optional Financial Mathematics topic as an application of number.

Wordbank

Quiz
Wordbank 8

allowable deduction A part of a person's yearly income that is not taxed, such as work-related expenses and donations to charities

annual leave loading Extra payment to a worker based on 17.5% of 4 weeks annual leave

income tax A tax paid to the government based on the size of a person's income

net pay Pay received after deductions from gross pay; 'take-home' pay

overtime Time worked beyond normal working hours, such as at night or on weekends, at a higher rate of pay

time-and-a-half Overtime pay that is calculated at 1.5 times the normal pay rate

salary A fixed yearly amount of money that is paid weekly, fortnightly or monthly, not dependent on the number of hours worked

wage An amount of money paid to people for work, calculated on the number of hours worked

Videos (2):

8.02 Overtime

8.04 Income tax

***Twig* video (1):**

8.04 Percentages: Tax breaks

Quizzes (5):

- Wordbank 8
- SkillCheck 8
- Mental skills 8
- Language of maths 8
- Test yourself 8

Skillsheet (1):

8.03 Percentage calculations

Worksheets (6):

8.03 Earning money

8.04 Income tax tables • Tax return form • Income tax and Medicare levy

8.05 Pay day

Mind map: Earning money

Nelson MindTap

To access resources above, visit **cengage.com.au/nelsonmindtap**

8 In this chapter you will:

- ✓ calculate wages, salaries, overtime, commission and piecework.
- ✓ calculate weekly, fortnightly, monthly and yearly incomes.
- ✓ calculate annual leave loading.
- ✓ use tables to calculate income tax and PAYG instalments.
- ✓ calculate net pay from gross pay.

SkillCheck

ANSWERS ON P. 649

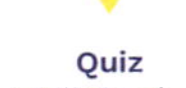

Quiz
SkillCheck 8

1 Write 8.2% as a decimal.

2 Find 15% of \$15.

3 How many:

a	days in 1 year?	**b**	weeks in 1 year?
c	days in 4 weeks?	**d**	months in 3 years?
e	days in a fortnight?	**f**	weeks in a quarter of a year?
g	months in 5 years?	**h**	fortnights in 1 year?

4 Round each amount to the nearest cent.

a	\$7.6831	**b**	\$21.4209	**c**	\$524.7373
d	\$92.3557	**e**	\$3085.6382	**f**	\$106.3446

5 Evaluate (correct to the nearest cent where necessary):

a	$38 \times \$12.87 \times 1.5$	**b**	$\$75\ 430 \div 12$	**c**	$0.42 \times \$7500$
d	$0.13 \times \$9217$	**e**	$\$4758 \times 0.073 \times 2$	**f**	$0.47 \times \$31\ 726$
g	$\$69\ 870 \div 52.18$	**h**	$\$2150 \times 1.09$	**i**	$\$3276 \times 0.175 \times 4$

6 Evaluate $\$3572 + 0.325 \times (\$43\ 210 - \$37\ 000)$.

Wages and salaries 8.01

People who work usually earn a wage or a salary.

A **wage** is calculated by the number of hours worked and is usually paid weekly. People such as supermarket cashiers, electricians and gardeners earn a wage. Wage earners can earn more income by working extra hours (overtime).

A **salary** is a fixed annual amount that is divided evenly into payments paid weekly, fortnightly or monthly. People such as architects, software designers and the police earn a salary. Salary earners do not earn overtime pay, but can receive benefits such as a company car, expense account, shares in the company or paid medical expenses.

Example 1

Peter's first job pays him \$16.75 per hour. If he works 40 hours each week, calculate his weekly income.

SOLUTION

Weekly income = 40 × \$16.75
= \$670

ⓘ Units of time

1 year = 12 months

1 fortnight = 2 weeks

1 year = 52 weeks or 26 fortnights for wage earners

1 year = 52.18 weeks for salary earners

To convert a salary to a weekly amount, divide by 52.18 because

1 year = $365\frac{1}{4}$ days = $365\frac{1}{4} \div 7 \approx 52.18$ weeks.

Example 2

Greta earns a salary of \$72 600 p.a. 'per annum' (p.a.) = per year'

How much (correct to the nearest cent) does she earn:

a each week? **b** each fortnight? **c** each month?

SOLUTION

a Weekly income = \$72 600 ÷ 52.18
= \$1391.3376...
≈ \$1391.34

b Fortnightly income = 2 × \$1391.34 — 1 fortnight = 2 weeks
= \$2782.68

c Monthly income = \$72 600 ÷ 12 — 1 year = 12 months
= \$6050.00

Example 3

Sarah earns a salary of $62 500 and works 38 hours per week. Anh earns $1087 per week for working 35 hours. Who has the higher hourly rate of pay?

SOLUTION

Sarah's hourly pay rate = $62500 ÷ 52.18 ÷ 38
= $31.5204....
≈ $31.52

Anh's hourly pay rate = $1087 ÷ 35
= $31.0571...
≈ $31.06

Sarah has the higher hourly rate of pay.

EXERCISE 8.01 ANSWERS ON P. 649

Wages and salaries

U F PS R

Express all answers correct to the nearest cent where necessary.

1 Find the weekly wage for each person.

- **a** Alex, who earns $26.24 per hour and works for 40 hours.
- **b** Bella, who earns $34.60 per hour and works for 38.5 hours.
- **c** Christy, who works 8 hours a day, Monday to Friday, and is paid $22.38 per hour.
- **d** Dilshan, who works Monday to Wednesday, 6 hours per day, and is paid $17.85 per hour.

2 What is Kim's annual salary if she earns $1980 per week?

3 If Sunil earns $6875 per month, what is his annual salary? Select the correct answer **A**, **B**, **C** or **D**.

A $357 500 **B** $178 750 **C** $89 375 **D** $82 500

4 Rhys earns $3038.47 per fortnight. How much does he earn in a year?

5 Kelly earns a salary of $158 340 p.a. How much does she earn:

a each week? **b** each fortnight? **c** each month?

6 Mitch earns $12.40 per hour before midnight and $18.60 per hour after midnight for each shift. Calculate his income for working each of these shifts:

- **a** 7:00 p.m. to 3:00 a.m.
- **b** 9:00 p.m. to 5:30 a.m.
- **c** 10:30 p.m. to 6:00 a.m.

7 PS R Danica earns $18.47 an hour and works for 38 hours each week. Trent earns $19.16 per hour for his 35 hours of work each week. Who earns more per week and by how much?

Foundation Standard Complex

8 Abdul considers 2 jobs, one with a salary of $97 640 p.a. and the other with a fortnightly income of $3865. Calculate the weekly income for each job to determine which pays more.

9 Who earns the higher hourly rate of pay for each pair of workers, and by how much?

PS R

a Mick worked 40 hours for $2584 pay.
Marisol worked 38 hours for $2501.60 pay.

b Phil worked 30 hours for $1441 pay.
Paula worked 27 hours for $1384.75 pay.

c Faith worked Monday to Friday, 8.5 hours per day, for $1092.50 pay.
Fabian worked Monday to Thursday, 9 hours per day, for $1015.60 pay.

d Ranjan earned $92 760 p.a., working 9:00 a.m. to 5:30 p.m., 5 days per week.
Rhonda earned $98 620 p.a., working 6:00 a.m. to 6:00 p.m., Monday to Thursday.

10 A printer earns $64 per hour, working from 8:00 a.m. to 4:30 p.m., Monday to Friday. Find how much he earns:

a each day **b** each week **c** each month.

11 Ziad earns $10 640 per month. Calculate his:

a annual salary **b** weekly income.

12 Priya earns a salary of $88 720 p.a. and works 42 hours per week. Find how much Priya earns:

a per month **b** per week **c** per hour.

13 Liam works 9 hours per day from Tuesday to Friday. He is paid $27.84 per hour. He also works on Saturday for 6 hours at a special rate of $41.76 per hour. How much did Liam earn for the week?

PS

14 Charlie works from 7:30 a.m. to 3 p.m. Monday to Wednesday and from 8 a.m. to 4 p.m. Thursday and Friday. His hourly pay rate is $27.38. Calculate Charlie's weekly wage.

15 Radka packs shelves at a supermarket and is paid $28.50 per hour on weekdays and $42.75 per hour on Saturdays. She worked the following hours last week.

PS R

Day	Hours worked
Monday	6:30 a.m. – 2:00 p.m.
Tuesday	8:00 a.m. – 4:00 p.m.
Wednesday	11:00 a.m. – 4:30 p.m.
Friday	9:30 a.m. – 3:30 p.m.
Saturday	10:00 a.m. – 4:00 p.m.

What was Radka's total income for the week?

16 Jack is a café manager and works the following hours in a week:

Day	Hours worked
Monday	6:30 a.m. – 2:30 pm.
Tuesday	7:00 a.m. – 4:00 p.m.
Thursday	11:30 a.m. – 8:30 p.m.
Friday	7:30 a.m. – 3:30 p.m.
Saturday	10:30 a.m. – 5:00 p.m.

He is paid at the following rates:

Day	Rate of pay
Monday to Friday	$24.22 per hour
Thursday after 4:00 p.m. and all day Saturday	$36.33 per hour

PS R What is Jack's total income for the week?

17 Andrew works from 8:00 a.m. to 4:00 p.m. from Monday to Wednesday and from 7:30 a.m. to 5:30 p.m. on Thursday and Friday. If his weekly wage is $1274, find his hourly rate of pay.

PS R

18 Lidija's monthly income is $7600. Convert this amount to a fortnightly income.

Foundation Standard Complex

INVESTIGATION

Positions vacant

1 a Find 5 job advertisements from a newspaper or employment website.

b For each advertisement, write:

- a brief job description, including whether the job is full-time or part-time, permanent or casual.
- whether a salary or wage is paid, and how much.
- whether there are any other payments, incentives, or benefits that may come with the job.

2 This job advertisement is for a courier.

COURIERS

Full-time Permanent Positions

Green Hills, Lidcombe, Pymble, Southlands, Dural and Hornsby

Part-time Permanent Positions

Thomson – 21 hours per week; shifts start at approximately 6:30am
Wetherill Park – 20 hours per week; shifts start at approximately 7:30am

Fixed-term Positions

PART-TIME

Regents Vale – until 14 June 2022; 20 hours per week; shifts start at approximately 6:00am
Thomson – until 10 June 2022; 24 hours per week; shifts start at approximately 6:30am

FULL-TIME

Regents Vale – until 10 August 2022
North Bank – until 19 June 2022
Camperdown
Blacktown – until 22 October 2022
Nelson Cove – until 31 May 2022

All full-time positions are from Monday to Friday, 36.75 hours per week. Shifts start at approximately 6:00am.

Essential Conditions/Requirements:

- Must hold a current unrestricted motorcycle licence

a How many hours would a full-time permanent employee be expected to work in a week?

b Explain what a 'fixed-term' position is.

c The advertisement does not give wage rates. Can you give a reason for this?

3 According to this advertisement, how much would a glasscutter earn in 1 year if he is expected to work 35 hours per week?

Glasscutters

$27 p/h

Exp. Cutters req. in Gladstone for immediate start. Must be able to cut up to 10 mm, be familiar with cutting machinery and have a minimum 2 years exp.

8.02 Overtime pay

Wage earners receive **overtime** pay when they work more than their standard number of hours.

Overtime pay

The 2 most common rates of overtime pay are:

- **time-and-a-half** = 1.5 × normal hourly rate
- **double time** = 2 × normal hourly rate

Video
Overtime

Example 4

Siki works 35 hours at \$28.94 per hour and 4 hours overtime at time-and-a-half. Calculate his total earnings.

SOLUTION

Normal pay = 35 × \$28.94 = \$1012.90

Overtime pay = 4 × \$28.94 × 1.5 = \$173.64

Total earnings = \$1012.90+ \$173.64 = \$1186.54

Alternative method:

Equivalent number of normal hours = 35 + (4 × 1.5) = 41

Total earnings = 41 × \$28.94

= \$1186.54

Example 5

Sinead earns time-and-a-half on Saturdays and double time on Sundays. She works 38 hours from Monday to Friday, 6 hours on Saturday and 4 hours on Sunday. Calculate Sinead's total earnings if her normal rate of pay is \$24.50 per hour.

SOLUTION

Normal pay (Monday to Friday) = 38 × \$24.50

= \$931

Saturday pay = 6 × \$24.50 × 1.5

= \$220.50

Sunday pay = 4 × \$24.50 × 2

= \$196

Total pay = \$931 + \$220.50 + \$196

= \$1347.50

Example 6

Carrie and Vinh are cadet hamburger chefs. Last week, Carrie worked for 40 hours and Vinh worked for 43 hours. Calculate their respective total earnings based on the following award.

Award for cadet hamburger chefs	
Normal rate = $15.80 per hour	
Normal rate	For first 38 hours worked
Time-and-a-half	For the next 3 hours
Double time	For each additional hour after that

SOLUTION

For Carrie (40 hours):

$$\begin{aligned}\text{Normal pay} &= 38 \times \$15.80 && \text{First 38 hours}\\ &= \$600.40\\ \text{Time-and-a-half pay} &= 2 \times \$15.80 \times 1.5 && \text{Remaining 2 hours}\\ &= \$47.40\\ \text{Total earnings} &= \$600.40 + \$47.40\\ &= \$647.80\end{aligned}$$

For Vinh (43 hours):

$$\begin{aligned}\text{Normal pay} &= \$600.40 && \text{First 38 hours}\\ \text{Time-and-a-half pay} &= 3 \times \$15.80 \times 1.5 && \text{Next 3 hours}\\ &= \$71.10\\ \text{Double time pay} &= 2 \times \$15.80 \times 2 && \text{Remaining 2 hours}\\ &= \$63.20\\ \text{Total earnings} &= \$600.40 + \$71.10 + \$63.20\\ &= \$734.70\end{aligned}$$

Example 7

Last week Rani worked her normal 38 hours, then 4 hours at time-and-a-half and 5 hours at double time. She was paid $1009.90 for the week. Find her hourly rate of pay.

SOLUTION

$$\begin{aligned}\text{Equivalent number of normal hours} &= 38 + (4 \times 1.5) + (5 \times 2)\\ &= 54\\ \text{Hourly rate of pay} &= \$1009.90 \div 54\\ &= \$18.70\end{aligned}$$

EXERCISE 8.02 ANSWERS ON P. 650

Overtime

U F PS C

1 The hourly rate of pay for workers in a factory is \$28.65. Calculate the total earnings for each employee.

- **a** Soo-jin works 28 hours at normal rates.
- **b** Jemima works 33.5 hours at normal rates.
- **c** Ethan works 38 hours at normal rates and 3 hours at time-and-a-half.
- **d** Stavros works 35 hours at normal rates and 5.5 hours at time-and-a-half.
- **e** Lana works 37 hours at normal rates, 6 hours at time-and-a-half and 4 hours at double-time.
- **f** Marco works 40 hours at normal rates, 5 hours at time-and-a-half and 4 hours at double-time.

2 Pavel works 20 hours at normal rates with an hourly pay rate of \$25.46 and 4 hours at time-and-a-half. Which calculation could be used to find his weekly wage? Select the correct answer **A**, **B**, **C** or **D**.

- **A** $\$25.46 \times 24$
- **B** $\$25.46 \times 20 + \$12.73 \times 24 \times 1.5$
- **C** $\$25.46 \times 20 + \$25.46 \times 4 \times 1.5$
- **D** $\$25.46 \times 24 \times 1.5$

3 Josie works 38 hours at \$26.87 per hour and 3 hours overtime at time-and-a-half. Calculate her total earnings.

4 Wayne is a store assistant manager who works 40 hours at \$37.32 per hour, 5 hours overtime at time-and-a-half and 3.5 hours double-time. Calculate his total earnings.

5 PS Pancake House pays employees normal rates for Monday to Friday, time-and-a-half for Saturday and double-time for Sunday. Mia works 36 hours Monday to Friday, 5.5 hours on Saturday and 4 hours on Sunday. Calculate her total earnings if her normal rate of pay is \$21.40 per hour.

6 Anu works at a mobile phone store. She is paid normal rates for shifts worked on Monday to Friday, time-and-a-half for Thursday nights from 5 p.m. onwards, and double-time on Saturdays. In a single week, Anu worked 37 normal hours and on Thursday night she worked from 5 p.m. to 8 p.m. On Saturday, she worked from 9:30 a.m. until 2:30 p.m. If her normal pay rate is \$23.74 per hour, what is Anu's total wage for the week?

7 Holly works from 9:00 a.m. until 4:00 p.m. each day on Monday, Tuesday and Wednesday at a normal rate of \$19.60 per hour. She works Thursday nights from 4:30 p.m. to 9:30 p.m. at time-and-a-half and Sundays from 8:00 a.m. until 4:00 p.m. at double-time. Which expression is used to calculate her weekly wage? Select **A**, **B**, **C** or **D**.

  Foundation Standard Complex

9780170465618

A $\$19.60 \times 34 + \$19.60 \times 1.5 \times 4 + \$19.60 \times 12 \times 2$

B $\$19.60 \times 21 + \$19.60 \times 1.5 \times 5 + \$19.60 \times 8 \times 2$

C $\$19.60 \times 7 + \$19.60 \times 1.5 \times 6 + \$19.60 \times 8 \times 2$

D $\$19.60 \times (13 + 1.5 \times 11 + 2 \times 12)$

EXAMPLE 6

8 *Elegant Jewellery* pays its staff at the following rates:

PS R Calculate the weekly wages for each employee.

Weekly Award Normal rate = $23.70/h	
Normal	For the first 35 hours
Time-and-a-half	For the next 5 hours
Double-time	For each hour after that

a Grant, who works 33 hours

b Tashi, who works 36 hours

c Kylie, who works 40 hours

d Mannix, who works 44.5 hours

e Noah, who works 38.5 hours

f Jennifer, who works 42 hours

9

PS R

Nuts and bolts machinists' enterprise agreement	
Normal rate	Applies Monday to Friday
Time-and-a-half	For any hours worked on Saturday
Double time	For any hours worked on Sunday

Calculate the total earnings for each employee.

			Hours worked		
	Name	Hourly rate	Mon–Fri	Sat	Sun
a	Alan Douglas	$19.60	36	4	8
b	Ava Hawkes	$27.30	38	0	6
c	Seetha Mehta	$28.60	38	4	2.5
d	Tarek Mifsud	$25.00	40	8	0
e	Ivan Vitsic	$38.00	34	4	6
f	John Hackett	$26.40	36	4.5	6

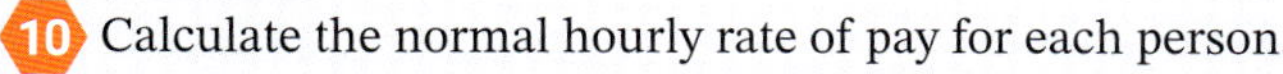

EXAMPLE 7

10 Calculate the normal hourly rate of pay for each person.

PS R

a Sarah earns $1262.40 by working 36 hours at normal rates and 4 hours at double time.

b Fiona earns $1469.60 by working 37 hours at normal rates, 3 hours at time-and-a-half and 4 hours at double time.

c Thierry earns $340 by working 6 hours at normal rates and 4 hours at double time.

d Sana earns $263.80 by working 4 hours at time-and-a-half and 3 hours at double time.

e Phillip earns $1127.60 by working 30 hours at normal rates, 5 hours at time-and-a-half and 3 hours at double time.

f Tara earns $126.50 by working 7 hours at normal rates and 1.5 hours at time-and-a-half.

☐ Foundation ◯ Standard ⬡ Complex

TECHNOLOGY

Calculating incomes

In this activity, we will use a spreadsheet to calculate the incomes of employees at J-mart department store. The award for all workers is as follows.

- Normal rate for up to 36 hours worked.
- Time-and-a-half for the next 4 hours worked.
- Double-time for any hours worked after that.

1 Create a spreadsheet as shown below, then for each employee enter the correct number of hours for columns D, E and F.

Name	Hourly Pay	Hours Worked	Normal	Time-and-a-half	Double time	TOTAL PAY
R Jones	\$42.85	38				
A Watson	\$39.40	34				
Q Tran	\$35.67	40				
L Djokovic	\$23.90	42				
C Lam	\$22.96	39				
J Hampen	\$20.76	41.5				
L Ngo	\$20.28	43				
K Barnett	\$19.22	45.5				

2 In cell G1, enter the heading **Total Pay.**

3 To calculate R Jones' total pay, in cell G2 enter the formula **=B2*D2+B2*1.5*E2+B2*2*F2**

4 **Fill Down** the formula from G2 to G9. Use **Format Cells** to change the values in column G to **Currency.**

DID YOU KNOW?

Labour Day

Each year, Australians have a public holiday on Monday called Labour Day or Eight-hours Day. This day celebrates the significant improvement of wages and working conditions in the 19th century.

In the early 1800s, the Australian worker was working up to 14 hours per day, 6 days per week, often under difficult conditions. Sick leave or holidays did not exist, and employees could be sacked immediately without reason.

In 1856, Victorian workers won the right to limit their working time to 8 hours per day. This was a world first, allowing workers more time in the day for recreation, reading, education and family, dividing the day into '8 hours labour, 8 hours recreation, 8 hours rest'. The Eight-hour Day was adopted in all Australian states by the 1920s. There is an Eight Hour monument that stands in central Melbourne with the numbers '888' featured at the top.

Labour Day is celebrated on a different day for each state in Australia.

Find out when these are.

Commission, piecework and leave loading

Some workers are not paid by the amount of time they work, but by the number of items they sell, make or process.

Worksheet Earning money

Skillsheet Percentage calculations

Commission

Commission is earned by salespeople or agents of a business or company. It is calculated as a percentage of the value of items sold or income made. A fixed amount called a **retainer** may also be paid. Real-estate agents and actors' agents earn a commission.

Example 8

Nathan sells cars and is paid a weekly retainer of \$750 plus 7.5% of the value of cars he sells. Calculate his income if he sells cars worth a total value of \$44 000 this week.

SOLUTION

Nathan's income = $\$750 + 7.5\% \times \$44\,000$

$= \$4050$

Example 9

Corrina is a real-estate agent and is paid according to the following commission rates.

- 2% on the first \$100 000 of property sold.
- 1.5% on any value thereafter.

Calculate Corrina's commission for selling an apartment for \$580 000.

SOLUTION

Commission on 1st \$100 000 = $2\% \times \$100\,000$

$= \$2000$

Remaining amount = $\$580\,000 - \$100\,000$

$= \$480\,000$

Commission on remaining amount = $1.5\% \times 480\,000$

$= \$7200$

Corrina's commission = $\$2000 + \$7200$

$= \$9200$

Piecework

Piecework is earned according to the number of items made or tasks completed. Fruit pickers and dressmakers earn money this way.

Example 10

Sue earns \$18 750 for each garden she designs. How many gardens must Sue design to earn over \$100 000?

SOLUTION

Number of gardens = \$100 000 ÷ 18 750

= 5.333...

≈ 6 Round up

Sue must design at least 6 gardens to earn over \$100 000.

Bonus

A **bonus** is extra money paid to employees who produce work of high quality or quantity, or for meeting an important goal or deadline. Some businesses reward their employees with a Christmas bonus for a productive year's work.

Example 11

After a successful year, a software company gave each employee a 2.5% bonus on their annual income. Find:

a the bonus received by Shaina, who earns \$5324.50 per week.

b the annual income of Kahil, whose bonus amounted to \$4175.

SOLUTION

a Shaina's annual income = 52 × \$5324.50

= \$276 874

Shaina's bonus = 2.5% of \$276 874

= \$6921.85

b 2.5% of Kahil's income = \$4175

1% of Kahil's income = \$4175 ÷ 2.5 Using the unitary method

= \$1670

∴ Kahil's annual income = \$1670 × 100 100%

= \$167 000

ⓘ Annual leave loading

Annual leave loading or **holiday loading** is extra pay given during annual leave (holidays). It is paid at the rate of 17.5% of 4 weeks' normal pay.

Example 12

For her Christmas holidays, Katrina received 4 weeks normal pay plus 17.5% annual leave loading for the 4 weeks. If Katrina's annual salary is $91 485, find:

a her normal weekly pay.

b her leave loading.

c her total pay for the 4-week holiday.

SOLUTION

a Weekly income $= \$91\,485 \div 52.18$
$= \$1753.2579...$
$\approx \$1753.26$

b Leave loading $= 17.5\% \times \$1753.26 \times 4$
$= \$1227.282$
$\approx \$1227.28$

c Total holiday pay $= 4 \times \$1753.26 + \$1227.28$
$= \$8240.32$

EXERCISE 8.03 ANSWERS ON P. 650

Commission, piecework and leave loading

U F PS

EXAMPLE 8

1 Anita is paid a commission of 3% on the value of kitchenware she sells. How much will Anita earn if she sells goods to the value of $24 560 in one week?

2 Dan is paid a commission of 5.5% of the value of the book sales he makes in excess of $5000. In 1 week, Dan sells books to the value of $37 427. Calculate his income for that week.

'in excess of' means to go over or above.

3 Karen is a real-estate agent who is paid a monthly retainer of $800 and a commission of 1.5% of the value of properties sold. Her sales in January were $1 780 400. Calculate Karen's income for January.

4 Consider these 2 salespeople and their method of payment.

- Perry: $1250 retainer and 2.5% of the value of holidays sold.
- Amy: no retainer and 7.5% of the value of holidays sold.

Calculate the income earned by Perry and Amy if they each sell holidays worth $54 000.

5 Lara earns 2.5% commission on the sale of thoroughbred horses. Last year, she sold horses worth $2 750 000. What is Lara's commission?
Select the correct answer **A**, **B**, **C** or **D**.

A $6 875 000 **B** $1 100 000 **C** $68 750 **D** $11 000

☐ Foundation ○ Standard ⬡ Complex

EXAMPLE 9

6 A real-estate agency pays the following rates of commission.

- 5% on the first $500 000.
- 2.5% on the next $250 000.
- 1.25% on any value thereafter.

Calculate the commission earned by selling property valued at:

a $500 000 b $750 000 c $1 350 000

7 Xanthe is paid 5% of the value of used cars she sells and 2.5% of the value of new cars. What commission is earned by selling new cars worth $108 400 and used cars worth $36 200?

EXAMPLE 10

8 Calculate the income for each amount of piecework.

a assembling 450 boxes at 75 cents per box

b delivering 2300 pamphlets at 10 cents per pamphlet

c assembling 95 prams at $5.85 per pram

d polishing 16 cars at $19.75 per car

e mowing nine lawns at $45 per lawn

f hemming 320 pairs of trousers at $1.50 per pair

g polishing 18 shoes at $2.50 per shoe

h laying 30.75 m^2 of paving at $24 per m^2

9 Fiona earns 85 cents for each toy she assembles.

PS a How much does she earn for assembling:

i 146 toys? ii 203 toys?

b Last week, Fiona earned $459. How many toys did she assemble?

c Find the number of toys that Fiona must assemble to earn (at least) the following amounts:

i $200 ii $620

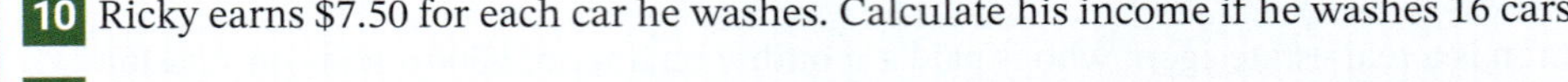

10 Ricky earns $7.50 for each car he washes. Calculate his income if he washes 16 cars.

EXAMPLE 11

11 A publishing house gives all employees a 4.75% annual bonus for a profitable year. Calculate the annual bonus for each income.

a $92 400 p.a. b $165 600 p.a. c $1795 per week

d $994 per week e $8750 per month f $4156 per fortnight

EXAMPLE 12

12 Frank works 40 hours a week and is paid $27.40 per hour. Calculate:

a Frank's weekly income.

b the annual leave loading of 17.5% on 4 weeks' pay.

c Frank's total pay for the 4-week holiday.

13 Astra is the manager of a computer megastore. Her salary is $252 000 p.a. Calculate:

a Astra's fortnightly pay.

b the holiday loading Astra would receive when she takes her 4 weeks' annual leave.

c her total holiday pay.

14 A process worker is paid $1.15 per article assembled, up to 500 articles, then $1.65 per article assembled after that. Calculate the total income earned for assembling each number of articles.

a 346 **b** 506 **c** 1406

15 Kevin works at the SuperGlo Hand Car Wash. He earns $7.50 per car for a wash only and $12.50 per car for a wash and polish. His job sheet for last week is shown.

Day	Wash only (number of cars)	Wash and polish (number of cars)
Monday	11	4
Tuesday	12	8
Wednesday	7	10
Thursday	16	0
Friday	18	13
Saturday	10	12

Calculate Kevin's:

a total earnings for the week.

b average daily earnings.

c average hourly rate of pay if he worked for 8.5 hours each day.

16 PS Zlatko, a construction supervisor, has signed a contract worth $970 000 to ensure that a factory is built by June 30. He will receive a bonus of 0.5% of his contract for each day the job is completed ahead of time. How much will Zlatko earn if the factory is completed on June 22?

17 Valentine, a storeman, works from 8:00 a.m. to 4:00 p.m., Monday to Friday. Valentine's rate of pay is $25.80 per hour. Calculate:

a his weekly wage.

b his holiday pay for 4 weeks (including leave loading).

Foundation Standard Complex

8.04 Income tax

Worksheets
Income tax tables

Tax return form

Income tax and Medicare levy

Videos
Income tax

Percentages: Tax breaks

Anyone who earns more than $18 200 in a **financial year** (1st July to 30th June the following year) must pay **income tax** to the federal government. Income includes money earned from a job or interest earned from investments. The government uses the money paid in tax to fund public services such as schools, hospitals, roads and the defence forces.

Not all of a person's income is taxed. If we use some of our income for work-related expenses or to donate money to charities, these amounts are called **allowable deductions** (or **tax deductions**) and are not taxed. Other examples of allowable deductions are tools of trade, uniforms, car-related expenses, subscriptions to professional organisations and journals.

ⓘ Taxable income

Income tax is calculated on a person's **taxable income,** which is the **gross income** (total earnings) less all allowable deductions, **rounded down to the nearest dollar**.

Taxable income = gross income – allowable deductions

The more a person earns, the higher the rate of tax to be paid. The table below shows how income tax is calculated.

Tax rates for Australian residents	
Taxable income	**Tax on this income**
0 – $18 200	Nil
$18 201 – $45 000	19c for each $1 over $18 200
$45 001 – $120 000	$5092 plus 32.5c for each $1 over $45 000
$120 001 – $180 000	$29 467 plus 37c for each $1 over $120 000
$180 001 and over	$51 667 plus 45c for each $1 over $180 000

Source: © Commonwealth of Australia

Example 13

Harry earned $65 660 last financial year and has allowable deductions of $727 in work-related car expenses and $259 in donations to charities.

a Calculate his taxable income.

b Use the tax table to calculate the income tax Harry must pay.

SOLUTION

a Taxable income = $65 660 – $727 – $259
= $64 674

b According to the table, a taxable income of $64 674 is in the $45 001 – $120 000 tax bracket.
Income tax = $5092 + 0.325 × ($64 674 – $45 000)
= $11 486.05

'32.5c for each $1' means 32.5% or 0.325'

Example 14

Alice is an accountant who earned $168 642 one financial year. She also earned $732.51 in interest from her bank savings. Alice has allowable deductions of $387.20 for office supplies and $75.88 in books. Calculate:

a Alice's taxable income.

b the amount of income tax payable.

SOLUTION

a Taxable income = $168 642 + $732.51 − $387.20 − $75.88

= $168 911.43

≈ $168 911   Rounded down to the nearest dollar

b According to the table, a taxable income of $168 911 is in the $120 001 – $180 000 tax bracket.

Tax payable = $29 467 + 0.37 × ($168 911 − $120 000)

= $47 564.07

EXERCISE  8.04 ANSWERS ON P. 650

Income tax

U F PS

In this exercise, use the tax table on the previous page to calculate income tax when needed.

1 Jake works as a plumber and earned $93 219 this year. He is entitled to these deductions: tools $1527.40, training courses $735.26 and office supplies $321.55. What is Jake's taxable income, rounded down to the nearest dollar?

2 Raina earns $75 660 in a year and has allowable deductions of $986.40.  EXAMPLE 13

a Calculate her taxable income, rounded down to the nearest dollar.

b Calculate the income tax Raina must pay.

3 Ali had a yearly income of $125 157 and earned $3240 in dividends from his shares. He also had allowable deductions of $540 for car expenses and $332 in the depreciation of his photocopier. Calculate:  EXAMPLE 14

a Ali's taxable income   **b** the amount of tax payable.

4 Kadin earns $2145 per fortnight and has allowable annual deductions of $337. Calculate his income tax for the financial year. Select the correct answer **A**, **B**, **C** or **D**.

A $3412.40   **B** $6824.60   **C** $8482.73   **D** $18 015.73

Foundation | Standard | Complex

5 Joemon is a dive instructor and earns $67 300 per year. He has allowable deductions of $1126 for diving equipment and $912 for a wetsuit. Calculate:

a his taxable income. **b** his income tax.

6 Cassie earns $1956.55 per week and claims deductions of $655.30 in travel expenses, $276 depreciation of computer equipment and $155 membership to a professional association. She also donated $32 each month to a charity. Calculate Cassie's:

PS

a taxable income **b** income tax payable.

7 Jamie is a swimming pool manager with a yearly income of $85 958. He also earned $327.58 in interest from his bank savings account. Jamie also had allowable deductions of $134 and $55 in donations to charity. Calculate Jamie's:

PS

a gross annual income

b taxable income

c the amount of tax payable.

8 Addison is a casual dance teacher who earns $32.50 per hour. She works 12 hours a week for 35 weeks of the year. How much should she pay in income tax? Select the correct answer **A**, **B**, **C** or **D**.

A $4717.95 **B** $3572 **C** $669.94 **D** no tax paid

9 Sefu earns $295 160 p.a. His allowable deductions are donations to charity of $3045, uniforms costing $595 and car-related work expenses worth $890. Calculate the income tax payable.

PS

10 Eerin earns $2846.87 per fortnight and claims deductions of $1955.30 in travel expenses, $572 depreciation of camera equipment, $180.20 membership to a professional association and $150 donated to a charity. Calculate Eerin's income tax payable.

PS

TECHNOLOGY

Online income tax calculator

The Australian Taxation Office (ATO) website www.ato.gov.au has an online calculator for income tax. Visit the website and search 'Simple Tax Calculator' to find the income tax calculator for individuals.

1 Enter the taxable income $63 000 as '63000' (no spaces).

2 Select the current financial year.

3 Select 'Resident for full year' and click 'Next'.

4 The estimated tax payable will be shown on a new screen.

Repeat for at least 2 more taxable incomes. Compare the amount of tax payable with the results of other students in your class.

☆ MENTAL SKILLS 8 ANSWERS ON P. 650

Maths without calculators

Quiz
Mental skills 8

Percentage increase and decrease

The fraction equivalents of commonly used percentages can help us when we need to increase or decrease a number by a percentage.

Percentage	1%	5%	10%	$33\frac{1}{3}$	20%	25%	%	50%
Fraction	$\frac{1}{100}$	$\frac{1}{20}$	$\frac{1}{10}$	$\frac{1}{8}$	$\frac{1}{4}$	$\frac{1}{4}$	$\frac{1}{3}$	$\frac{1}{2}$

1 Study each example.

a Increase 360 by 25%.

$$\begin{aligned} 25\% \text{ of } 360 &= \frac{1}{4} \times 360 \\ &= 360 \div 4 \\ &= 90 \end{aligned}$$

$$360 + 90 = 450$$

b Increase \$80 by 5%.

$$\begin{aligned} 5\% \text{ of } \$80 &= \frac{1}{20} \times \$80 \\ &= \$80 \div 20 \\ &= \$4 \end{aligned}$$

$$\$80 + \$4 = \$84$$

or

$$\begin{aligned} 10\% \text{ of } \$80 &= \$8 \\ \therefore 5\% \text{ of } \$80 &= \$8 \div 2 \\ &= \$4 \end{aligned}$$

2 Now increase:

a \$340 by 20% **b** 66 by 50% **c** 150 by $33\frac{1}{3}$ % **d** \$400 by 1%

e 640 by 5% **f** \$72 by $12\frac{1}{2}$ % **g** \$470 by 10% **h** 180 by 25%

i 420 by $33\frac{1}{3}$ % **j** \$80 by 5% **k** \$280 by 25% **l** 70 by 20%

3 Study each example.

a Decrease \$225 by $33\frac{1}{3}$ %.

$$\begin{aligned} 33\frac{1}{3}\% \text{ of } \$225 &= \frac{1}{3} \times \$225 \\ &= \$225 \div 3 \\ &= \$75 \end{aligned}$$

$$\$225 - \$75 = \$150$$

b Decrease \$70 by 15%.

$$10\% \text{ of } \$70 = \frac{1}{10}\$70 = \$7$$

$$\therefore 5\% \text{ of } \$70 = \frac{1}{2} \times \$7 = \$3.50$$

$$\begin{aligned} \therefore 15\% \text{ of } \$70 &= (10\% \times \$70) + (5\% \times \$70) \\ &= \$7 + \$3.50 \\ &= \$10.50 \end{aligned}$$

$$\$70 - \$10.50 = \$59.50$$

4 Now decrease:

a 440 by 25% **b** \$300 by 20% **c** 2400 by $33\frac{1}{3}$ % **d** \$500 by 15%

e \$250 by 10% **f** \$120 by 50% **g** \$72 by $12\frac{1}{2}$ % **h** 80 by 5%

i \$85 by 20% **j** \$3800 by 1% **k** \$440 by 15% **l** \$150 by $33\frac{1}{3}$%

8.05 PAYG tax and net pay

Worksheet
Pay day

To avoid paying income tax as a huge sum at the end of the financial year, your employer takes out an estimate of the tax from your pay every payday. This is called **PAYG (Pay As You Go) tax**.

The total amount of PAYG tax paid over the year is usually more than the actual income tax payable, so at the end of the financial year you will receive the difference as a **tax refund**. However, if the PAYG tax paid is less than the income tax payable, you will have a **tax debt** and have to pay more.

Example 15

Employers use a PAYG tax table to determine how much tax to deduct from their employees' pays.

Weekly earnings ($)	PAYG tax withheld ($)
182–185	40
186–189	41
190–194	42
195–198	43
199–202	44

While he was at university, Li worked casually in a factory outlet store and earned $196 per week for 8 weeks.

a Calculate Li's gross income for the 8 weeks.

b How much PAYG tax was deducted from his pay over the 8 weeks?

SOLUTION

a Gross income = $196 × 8
= $1568

b In the table, $196 falls in the $195–$198 range.
Weekly PAYG tax = $43
Total PAYG tax = $43 × 8
= $344

Gross and net pay

Gross pay is the total amount a person earns or receives each payday.

Most income earners have a variety of deductions made against their gross pay before they receive it, including PAYG tax, superannuation contributions, union fees and health fund payments. The amount of income left after the deductions is called **net pay**.

Net pay

Net pay = gross pay – tax – other deductions

8.05

Example 16

Paula earns a gross pay of $700.70 per fortnight. Her deductions are for PAYG tax, $45.60 for private health insurance and $35 for superannuation.

Fortnightly earnings ($)	PAYG tax withheld ($)
672–678	152
680–686	154
688–694	156
696–704	158
706–712	160

a Use the PAYG tax table to find Paula's PAYG tax per fortnight.

b Calculate Paula's net pay.

c What are Paula's total deductions as a percentage of her gross income (correct to one decimal place)?

SOLUTION

a In the table, $700.70 falls in the $696–$704 range.

Fortnightly PAYG tax = $158

b Net pay $= \$700.70 - \$158 - \$45.60 - \$35$

$= \$462.10$

c Total deductions $= \$158 + \$45.60 + \$35$

$= \$238.60$

$$\text{Deductions percentage} = \frac{\$238.60}{\$700.70} \times 100\%$$

$$= 34.0516...\%$$

$$\approx 34.1\%$$

EXERCISE 8.05 ANSWERS ON P. 650

PAYG tax and net pay

U F PS R C

Use the table from Example **15** on the previous page when calculating PAYG tax in questions **1** to **6.**

1 Xavier earned $200 per week for 16 weeks in a casual job as a party clown. EXAMPLE 15

a Calculate his gross income.

b How much PAYG tax was deducted from his pay in total?

2 Rose is a part-time dental nurse. She is paid $192 to work half a day per week.

a Calculate Rose's gross income over 42 weeks.

b Find the total amount she paid in PAYG tax.

3 Pushkara earned $184 per week over the summer holidays working as a casual at the local plant nursery. How much did he pay in tax for the 6 weeks that he worked?

4 Sonja is a part-time music teacher who earns $195 per week.

PS **a** How much tax will she pay per week?

b What percentage of her gross pay will she pay in tax? Answer correct to one decimal place.

c Calculate her net pay per week.

5 In the holidays, Phoebe earns $201 per week, while Jemima earns $191 per week. If they both work for 12 weeks in these jobs, who pays more in tax and by how much?

6 Huan wants to work during his school holidays over Christmas. He has been offered 2 different jobs for 6 weeks:

PS R C

- Cinema: Earn $368 per fortnight (fortnightly tax of $80)
- Electronics store: Earn $190 per week

Over 6 weeks, which job offers the higher net pay? Show all working to justify your answer.

7 Goran earns $698 per fortnight. He also pays $19.20 in union fees and $84.90 in superannuation per fortnight.

a Use the table from Example **16** on the previous page to calculate how much PAYG tax he pays per fortnight.

b Calculate Goran's fortnightly net pay.

c What percentage (correct to one decimal place) of his gross pay is made up of deductions?

8 Caleb paid $154 in tax for one fortnight's work at the local cinema. Use the fortnightly tax table to determine how much he could have possibly earned. Select the correct answer **A**, **B**, **C** or **D**.

A $672 **B** $679 **C** $685 **D** $690

9 Cooper has a gross income of $1125.86 per week. His deductions are $219.48 tax, $64.35 for private health insurance and $30 for superannuation. Calculate Cooper's:

a net income.

b total deductions as a percentage of his gross income (correct to one decimal place).

10 Stefan earns a gross income of $864.25 per week. His deductions are $141.94 tax and $51.33 for private health insurance. Calculate:

a Stefan's net income.

b the percentage (correct to one decimal place) of Stefan's gross income that is paid in tax.

☐ Foundation ○ Standard ○ Complex

11 Copy and complete each pay slip.

PS R

a

Employee: Janice Hall		Hourly pay rate: $28.80	
Hours worked		**Deductions**	
Normal	38	Tax: $215	Other: $124.60
Time-and-a-half	0	Gross weekly income	
Double time	0	Total deductions	
		Net weekly income	

b

Employee: Amber Menzies		Hourly pay rate: $32.50	
Hours worked		**Deductions**	
Normal	39	Tax: $275	Other: $229.10
Time-and-a-half	4	Gross weekly income	
Double time	1	Total deductions	
		Net weekly income	

12 Jodie earns a gross fortnightly pay of $1027.60. Apart from tax, she has the following deductions: superannuation ($26.80), health insurance ($21.60), union fees ($5.20) and savings plan ($65).

a If Jodie's net fortnightly pay is $657.30, find how much tax is taken out.

b Express the tax she pays as a percentage of her gross income (correct to one decimal place).

13 Lina earns a salary of $76 800 p.a. Each fortnight she has deductions of $186.50 for family health insurance and $65.12 for superannuation taken from her gross income. She also pays tax at the rate of 32.5 cents in the dollar. Calculate:

PS R

a Lina's fortnightly gross income.

b the amount of tax Lina pays per fortnight.

c Lina's total deductions each fortnight.

d Lina's net income for the fortnight.

14 Balun earns a salary of $128 900 p.a. Each month he has deductions of $43.26 for union fees and he pays $341.05 per month for health insurance taken from his gross income. He also pays $36 432 tax for the year. Calculate:

a Balun's annual union fees.

b his annual health insurance payments.

c Balun's annual net income.

d the percentage of Balun's gross income that is paid in tax (correct to one decimal place).

Foundation | Standard | Complex

TECHNOLOGY

Online PAYG tax calculator

Visit the Australian Tax Office website **www.ato.gov.au/taxtables** to use the online PAYG tax calculator or to download the PAYG tax tables. It contains the PAYG tax amounts for weekly, fortnightly and monthly incomes. Find the PAYG tax payable and net pay for a gross pay of:

a $909 weekly.

b $2639 fortnightly.

c $4914 monthly.

POWER PLUS ANSWERS ON P. 650

1 In one year, tennis player Roger Federer earned $136.7 million. How much is this per minute? Assume 365 days in a year.

2 Danielle is paid $2500 per month. On average, how much does she earn in 4 hours, if she works 9 hours per day, 5 days per week, for 48 weeks a year?

3 A fitness instructor conducts a class with 12 participants who pay $18 each. The class lasts for 45 minutes. What is the instructor's average hourly income?

4 Felicity and Rose have the same net pay. Felicity has $187.60 deducted from her fortnightly gross pay of $514. Rose, who is paid weekly, has $64.20 deducted from her gross pay each week. Calculate the difference between Felicity's and Rose's gross fortnightly pays.

5 Mirko earns $63 for working from 7:30 a.m. to 9:15 a.m. At this rate, how much will he earn for a 38-hour week?

6 Anya earned $903 for working 36 hours at normal rates and 4 hours at time-and-a-half. Find her hourly rate of pay.

7 When Bhavin works 6 hours overtime at time-and-a-half, his gross income increases by $202.50. Find his weekly income for working 38 hours at normal rates.

8 One year, Robert received a holiday loading of $589.05, which was 17.5% of 4 weeks' pay. What was his normal weekly pay?

9 If Ingrid paid income tax of $34 561 last financial year, what was her taxable income?

10 Maddy earns $2420 for a 40-hour week. How long, in hours and minutes, will Maddy need to work overtime at time-and-a-half so her overtime is equivalent to her annual leave loading?

CHAPTER REVIEW

Language of maths

Quiz
Language of maths 8

allowable deduction	annual leave loading	bonus	commission
deduction	double time	financial year	fortnight
gross pay	income tax	net pay	overtime
PAYG tax	per annum	piecework	retainer
salary	taxable income	time-and-a-half	wage

1 What is the difference between a **wage** and a **salary?**

2 What is an **allowable deduction?** Give an example of one.

3 What earnings are calculated by 17.5% of 4 weeks of normal pay?

4 What earnings are calculated as a percentage of the value of items sold?

5 What does **PAYG** stand for?

6 What is a **financial year?**

Topic summary

Worksheet
Mind map: Earning money

- Write 10 questions (with solutions) that could be used in a test for this chapter.
- Swap your questions with another student and check their solutions against yours.
- Write down any part of this topic that you did not understand and discuss it with your teacher.

Print (or copy) and complete this mind map of the topic, adding detail to its branches and using pictures, symbols and colour where needed. Ask your teacher to check your work.

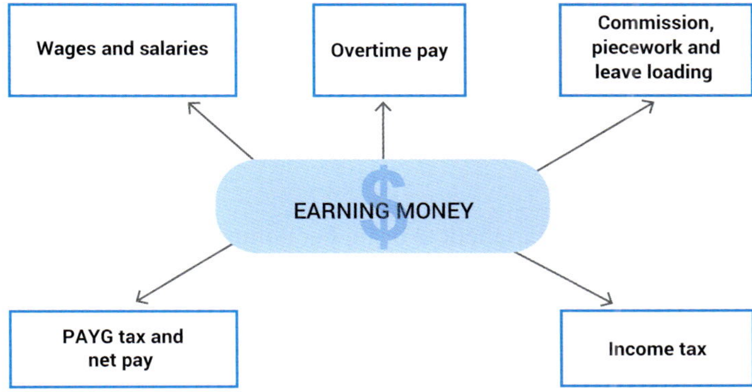

8 TEST YOURSELF

ANSWERS ON P. 650

Quiz
Test yourself 8

1 Milan has an annual salary of $87 410. How much would he be paid:

a per week? **b** per fortnight? **c** per month?

8.01 **2** Diane is paid $23.65 per hour and works a 37-hour week. Find her weekly wage.

8.02 **3** Riley works 38 hours at $37.38 per hour and 5 hours overtime at time-and-a-half. Calculate his total earnings.

8.02 **4** Liam works at a warehouse under the following weekly award:

Award schedule: Normal rate is $24.75 per hour	
Normal rate	For 0 to 38 hours worked
Time-and-a-half	For the next 3 hours worked
Double time	For each hour worked after that

Calculate his wage for working:

a 38 hours **b** 44 hours.

5 Nikky is a bookseller who earns 4% commission on the value of books she sells. She is also paid a monthly retainer of $900. If Nicole's sales for March were $82 500, find her total income for March.

8.03 **6** Sam is paid commission on a sliding scale of 3% on the first $50 000 of sales and 1.5% on all sales after that. Calculate Sam's income if he sells goods worth $80 000.

7 Calculate the piecework pay for:

a delivering 1500 pamphlets at 50 cents per pamphlet.

b washing 45 cars at $27.50 per car.

c laying 35.5 m^2 of tiles at $35 per m^2.

8.03 **8** A jockey earns $500 per race and 4.5% of the value of prize money his horses win. Last week, Dan had 12 rides and his horses won prize money worth $120 000. How much did Dan earn?

9 Zoe earns $1786 per week and is about to begin her annual holidays. Along with her 4 weeks' pay, she also receives a 17.5% annual holiday loading on the 4 weeks. Find Zoe's total holiday pay.

10 Emily is a department store manager who earned a gross salary last financial year of $187 542 and had the following allowable deductions: $378 uniform cleaning costs, $218 training course costs and $1492 travel expenses. Calculate:

a Emily's taxable income.

b the amount of income tax Emily should pay using the table on page 300.

☐ Foundation ○ Standard ⬡ Complex

9780170465618

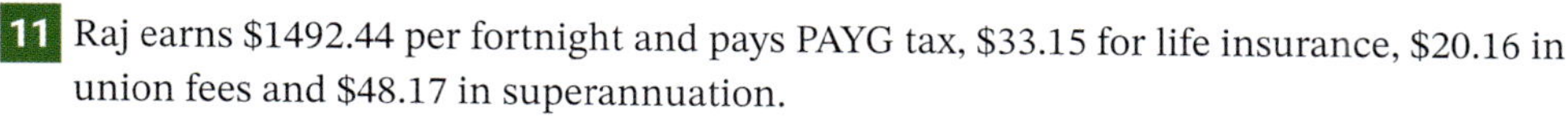

11 Raj earns $1492.44 per fortnight and pays PAYG tax, $33.15 for life insurance, $20.16 in union fees and $48.17 in superannuation. 

a Use the PAYG table to find Raj's PAYG tax paid per fortnight.

Fortnightly earnings ($)	PAYG tax withheld ($)
1470–1475	174
1476–1481	176
1482–1485	178
1486–1491	180
1492–1497	182

b Calculate Raj's fortnightly net pay.

c Calculate, correct to one decimal place, the percentage of Raj's gross pay that was paid in tax.

12 For the pay slip below, calculate the: 

a gross income **b** total deductions **c** net income.

Employee: Shang Chi Hours worked		Hourly pay rate: $22.80 Deductions	
Normal	37	Tax: $275.10	Other: $84.50
Time-and-a-half	4	Gross weekly income	
Double time	2	Total deductions	
		Net weekly income	

CHAPTER 8 TEST YOURSELF

☐ Foundation ○ Standard ⬡ Complex

9

STATISTICS

Analysing data

Statistical data are used by governments and businesses to make decisions about health and welfare, the population, production, tourism and the environment. Data collected from a census or through market research such as online surveys, questionnaires, street polls can lead to advertising campaigns for health issues and road safety, improved customer service and new products. However, before any action can be taken, the data must be carefully analysed in order to avoid misinterpretation and to check it for relevance and reliability.

iStock.com/BlackSalmon

Chapter outline

		Proficiencies				
9.01	The mean, median, mode and range	U	F		R	C
9.02	Frequency histograms and polygons	U	F		R	C
9.03	Stem-and-leaf plots	U	F		R	C
9.04	Choosing the best graph	U	F		R	C
9.05	The shape of a distribution	U	F		R	C
9.06	Comparing data sets	U	F		R	C
9.07	Sampling	U	F		R	C
9.08	Bias and questionnaires	U	F	PS	R	C

U = Understanding
F = Fluency
PS = Problem solving
R = Reasoning
C = Communication

Wordbank

bias Something unwanted that causes a sample to not truly represent the population

cluster Values in a data set that are close or 'bunched' together

frequency histogram A special column graph that shows the frequencies of numerical data

outlier An extreme value that is very different from the other values in a set of data

population All of the items being studied, the entire group

random sample A sample in which every item in the population has an equal chance of being selected

skewed distribution A distribution in which most of the values are clustered at one end, creating a 'tail' at the other end

symmetrical distribution A distribution in which all values are distributed equally on both sides of the centre, its shape having line symmetry

Quiz
Wordbank 9

Videos (5):

SkillCheck Statistics

9.01 The mode, median and mean • Analysing dot plots • Statistics from a frequency table

9.03 Back-to-back stem-and-leaf plots

Twig videos (6):

9.01 Cumulative frequency: You're fired?

9.02 Histograms: snapshot

9.07 Counting crowds • Can fish oil make you smarter? • Can you trust your IQ?

9.08 The wrong guy won

PhET interactive (1):

9.01 Centre and variability

Quizzes (5):

- Wordbank 9
- SkillCheck 9
- Mental skills 9
- Language of maths 9
- Test yourself 9

Skillsheet (1):

9.01 Statistical measures

Worksheets (7):

9.01 Collecting and describing data • The mean and fx tables

9.03 Stem-and-leaf plots

9.02 Analysing data

9.06 Comparing word lengths • Investigating young drivers

Mind map: Analysing data

Puzzles (4):

9.01 Mode, median and mean

9.02 Highway elephants

9.04 Data match-up

9.05 Statistical match-up

Technology (3):

9.01 Statistical measures calculator

9.06 Population polygons • World climate

Spreadsheets (2):

9.01 Statistical measures calculator

9.06 World climate

Nelson MindTap

To access resources above, visit **cengage.com.au/nelsonmindtap**

9 In this chapter, you will:

- ✓ calculate and interpret the mean, median, mode and range for a set of data.
- ✓ investigate the effect of an outlier on the mean and median.
- ✓ construct histograms and stem-and-leaf plots.
- ✓ group data into class intervals.
- ✓ identify the different types of data: categorical, numerical, discrete, continuous.
- ✓ choose the most appropriate graph to represent a set of data.
- ✓ describe the shape of a statistical distribution by identifying skewness, symmetry and modality.
- ✓ compare 2 or more sets of data by analysing their graphs and statistics.
- ✓ investigate sampling methods.
- ✓ collect and analyse data involving numerical and categorical variables.
- ✓ evaluate statistics found in the media and detect bias in reports of surveys.

SkillCheck
ANSWERS ON P. 651

Quiz SkillCheck 9

Video Statistics

1 This dot plot shows the number of apps purchased by students last month.

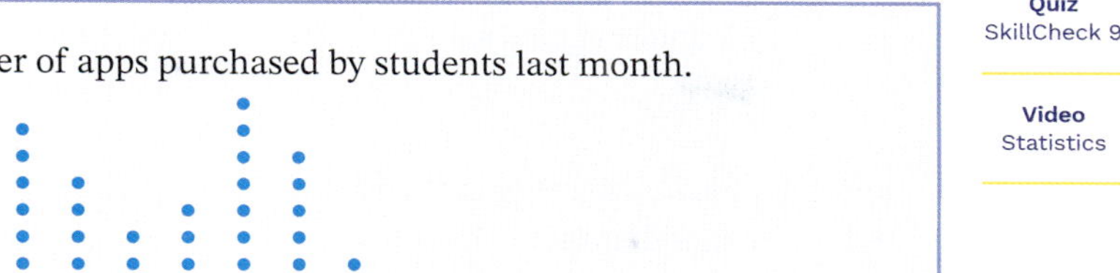

a How many students were surveyed?

b What was the most frequent number of apps purchased?

c What percentage (correct to one decimal place) of students purchased over 5 apps?

2 This stem-and-leaf plot shows the ages of people at a café during 8:00–8:30 a.m. on a Saturday morning.

a How many people were in the café?

b How many of them were aged in their 20s?

c What was the most common age?

d What was the youngest age?

e What percentage (correct to one decimal place) of people were 45 or older?

Stem	Leaf
1	4 6 7 9 9
2	0 4 5 7 8 8
3	1 5 5 6 6 6 7 8 9
4	2 3 3 5 7
5	0 0 1 4 5 6 8 8
6	3 5 6 7 7

3 The length of a queue at a supermarket checkout was recorded every 5 minutes for 3 hours.

4	3	6	5	1	4	7	5	4	1	4	4
4	3	6	3	1	3	4	2	4	5	1	3
5	4	2	3	5	2	5	2	6	5	1	3

a Arrange the data in a frequency table.

b Draw a frequency histogram and a frequency polygon.

c What was the most common number of people in the queue?

d How many times was the length of the queue longer than 3 people?

The mean, median, mode and range 9.01

The **mean**, **median** and **mode** are called **measures of central tendency** because they indicate a central value of a set of **data**. So they are also called **averages** or **measures of centre**.

The **range** is called a measure of spread because it indicates how widely the values are spread in a set of data.

Worksheet Collecting and describing data

Skillsheet Statistical measures

Video The mode, median and mean

Interactive Centre and variability

ⓘ Measures of central tendency and spread

Mean, $\bar{x} = \dfrac{\text{sum of data values}}{\text{number of data values}}$

$= \dfrac{\sum x}{n}$

where $\sum x$ means 'the sum of the values' and n is the total number of values.

Σ is the Greek letter **sigma**, and stands for 'the sum of'.

The **mode** is the value (or values) that occurs most often.

When the data values are arranged in order, the **median** is:

- the middle value, for an odd number of scores
- the average of the 2 middle values, for an even number of values.

Range = highest value − lowest value

An **outlier** (pronounced 'out-ly-er') is an extremely high or low value that is separated from the other values in the data set.

Puzzle Mode, median and mean

Technology Statistical measures calculator

Spreadsheet Statistical measures calculator

Comparing the mean, median and mode

The **mean** is usually the best measure of central tendency because it takes into account every data value.

If there are outliers in the set of data, then the mean may be affected by these extreme values and will not accurately represent all of the values. In this case, the **median** is the best measure because it is not affected by outliers.

The **mode** is useful when the most common data value is important, or when the data is categorical (such as make of car or hair colour). With categorical data, it is not possible to have a mean or median.

Example 1

The maximum daily temperatures (in °C) in Sydney for a fortnight in March were:

34	22	32	26	23	24	24
26	31	26	27	28	28	27

Find:

a the mean **b** the median

c the mode **d** the range

SOLUTION

a $\text{Mean} = \dfrac{34+22+32+\cdots+28+27}{14}$

$= \dfrac{378}{14}$

$= 27$

b Arranging the values from lowest to highest:

22 23 24 24 26 26 26 27 27
28 28 31 32 34

There is an even number of values (14), so we have 2 middle values, the 7th and 8th values 26 and 27.

$$\text{Median} = \frac{26+27}{2}$$
$$= 26.5$$

c Mode = 26   The most common value, occurring 3 times

d Range = 34 – 22   highest value – lowest value

$= 12$

Example 2

Video
Analysing dot plots

This **dot plot** shows the ages of the children at Soshanna's birthday party.

0 1 2 3 4 5 6 7 8 9 10
Ages

Find:

a the mean (correct to one decimal place)
b the median
c the mode
d the range
e the outlier

SOLUTION

a
$$\text{mean} = \frac{1+4+5+5+\cdots+10}{22}$$
$$= 6.636363...$$
$$\approx 6.6$$

b There are 22 values, so the median is the average of the 11th and 12th values. Counting dots from the left and upwards, these are 7 and 7.

$$\text{Median} = \frac{7+7}{2} = 7$$

c Mode = 7   The value with the most dots.

d Range = 10 – 1 = 9

e Outlier = 1   This value is much different from the rest.

The mean from a calculator's statistics mode

Scientific and graphics calculators have a statistics mode (SD or STAT). Follow the instructions in the table below to calculate the mean of the scores from Example **1** using your calculator's statistics mode.

Operation	Casio	Sharp
Start statistics mode.	MODE STAT 1-VAR	MODE STAT =
Clear the statistical memory.	SHIFT 1 Edit, Del-A	2nd F DEL
Enter data.	SHIFT 1 Data to get table 34 = 22 =, etc. to enter in column AC to leave table	34 M+ 22 M+, etc.
Calculate the mean. ($\bar{x}$ = 27:0714)	SHIFT 1 Var $\bar{x}$ =	RCL $\bar{x}$
Check the number of values. (n = 14)	SHIFT 1 Var n =	RCL n
Return to normal (COMP) mode.	MODE COMP	MODE 0

The mean from a frequency table

Worksheet The mean and fx tables

For a large set of data, we can use a **frequency distribution table** to organise the data. By adding an fx column, the mean can be easily found.

Score (x)	Frequency (f)
3	2
4	5
5	10
6	36
7	25
8	13
9	4
10	3

Video Statistics from a frequency table

Example 3

The scores of Year 9 students giving a speech in English (out of 10) are shown in the frequency table above.

a How many students were there?

b Use an fx column to calculate the mean score.

c Find the mode.

d What is the range?

SOLUTION

a The total sum of the Frequency column is 98.

There were 98 students.

b

Mark (x)	Frequency (f)	fx
3	2	6
4	5	20
5	10	50
6	36	216
7	25	175
8	13	104
9	4	36
10	3	30
	$\sum f = 98$	$\sum fx = 637$

$\sum fx$ means $f \times x$, which multiplies each score by its frequency.

$\sum f$ is the sum of the frequency column (that is, the total number of scores).

9.01

$$\bar{x} = \frac{\sum fx}{\sum f}$$

$$\textbf{Mean} = \frac{637}{98} = 6.5$$

Sum of all scores divided by the number of scores.

c Mode = 6

Score with the highest frequency.

d Range = 10 − 3 = 7

ⓘ Mean, mode and range for data in a frequency table

- **Mean**, $\bar{x} = \frac{\text{sum of } fx}{\text{sum of } f} = \frac{\sum fx}{\sum f}$
- The **mode** is the value(s) with the highest frequency, f.
- The **range** is the difference between the last and first values in the score (x) column.

This table shows calculator instructions for finding the mean of data presented in a frequency table. It uses the table of data from Example **3**.

Operation	Casio	Sharp
Start statistics mode.	MODE STAT 1-VAR SHIFT MODE scroll down to STAT Frequency? ON	MODE STAT =
Clear the statistical memory.	SHIFT 1 Edit, Del-A	2nd F DEL
Enter data.	SHIFT 1 Data to get table 3 = 4 =, etc. to enter in x column 2 = 5 =, etc. to enter in FREQ column AC to leave table	3 2nd F STO 2 M+ 4 2nd F STO 5 M+, etc.
Calculate the mean. ($\bar{x}$ = 6.5)	SHIFT 1 Var $\bar{x}$ =	RCL $\bar{x}$
Check the number of values. (n = 98)	SHIFT 1 Var n =	RCL n
Return to normal (COMP) mode.	MODE COMP	MODE 0

Cumulative frequency and the median

To find the median of data presented in a frequency table, add a **cumulative frequency** column, which is a progressive total of the frequency column.

Video Cumulative frequency: You're fired?

Example 4

Find the median of the Year 9 student speech scores from Example **3**.

SOLUTION

The cumulative frequency column is a running total of the frequencies.

$2 + 5 = 7$

$7 + 10 = 17$, etc.

There are 98 scores, so the 2 middle values are the 49th and 50th scores. Using the cumulative frequency column, the 17th score is a 5 and the 53rd score is 6, so the 49th and 50th scores must both be 6s.

Median $= \frac{6+6}{2} = 6$

Mark (x)	Frequency (f)	Cumulative frequency (cf)
3	2	2
4	5	7
5	10	17
6	36	53
7	25	78
8	13	91
9	4	95
10	3	98
	$\Sigma f = 98$	

EXERCISE 9.01 ANSWERS ON P. 651

The mean, median, mode and range

U F R C

Round all means to one decimal place where necessary, unless otherwise specified.

EXAMPLE 1

1 For each set of data, find:

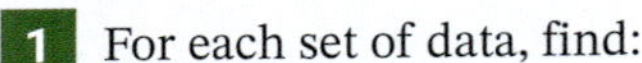

i the mean **ii** the median

iii the mode **iv** the range

a 23 20 12 15 28 19 15 14 18

b 5 3 8 7 1 6 7 7 5 15 12

c 32 45 12 45 38 32

d 8.8 8.2 8.8 8.3 8.5 8.9 9.2 8.8

e −5 0 −2 4 6 −2 5 −1 −2 0

2 The maximum and minimum daily temperatures (in °C) over a week at Point Perpendicular are listed.

Maximum	23	24	22	23	22	23	23
Minimum	17	17	15	16	15	15	14

a Calculate the mean daily maximum and mean daily minimum temperature over the week.

b What was the difference between the 2 means?

c Find the mode of daily maximum temperatures and the mode of daily minimum temperatures.

d What was the difference between the 2 modes?

▷

☐ Foundation ○ Standard ⬡ Complex

3 The monthly rainfall (in mm) in Cairns is shown in the table.

Month	J	F	M	A	M	J	J	A	S	O	N	D
Rainfall	413	435	442	191	94	49	28	27	36	38	90	175

a What is the range?

b Find the mean (correct to the nearest mm).

c Find the median.

4 A number of homes were surveyed on how many digital devices they had, with the results shown in the dot plot. Find:

EXAMPLE 2

a the mean **b** the median

c the mode **d** the range

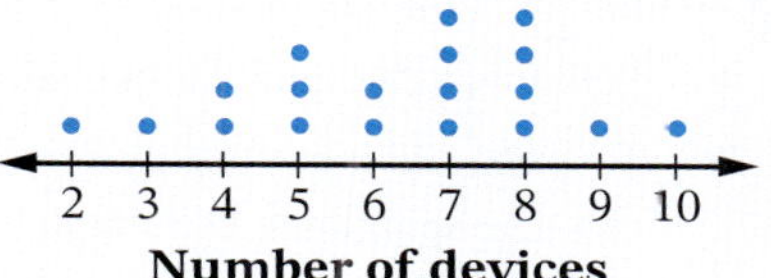

5 This dot plot shows the results of a maths quick quiz (out of 10).

a Are there any outliers? (Give reasons)

b What is the mode?

c Find the mean.

d Find the median.

e What percentage of results (correct to one decimal place) were above:

i the mean? **ii** the median?

0 1 2 3 4 5 6 7 8 9 10

Quiz results

6 This dot plot shows the number of goals scored by a soccer team during the season.

a Are there any outliers?

b Find the mean and median:

i without the outlier.

ii with the outlier.

c Which measure (mean or median) is affected by the outlier?

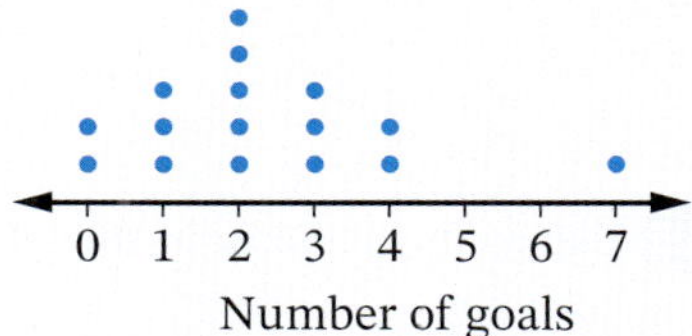

7 The salaries of 10 people working at a software company are:

R C

$80 000	$105 000	$70 000	$93 000	$78 500
$180 000	$90 000	$72 000	$93 000	$101 000

a Which salary is an outlier?

b Find the mean and median:

i without the outlier **ii** with the outlier

c Which measure, the mean or the median, represents the salaries better?

d Find the mode and range:

i without the outlier **ii** with the outlier

e Which measure, the mode or the range, is affected by the outlier?

Foundation Standard Complex

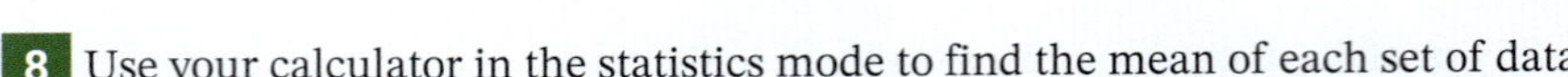

8 Use your calculator in the statistics mode to find the mean of each set of data.

a 5 7 8 3 6 8 5 4 8 3 8 10 6 8 4 9

b 13.5 14.2 11.9 15.7 18.4 17.6 12.0

c 56 72 84 57 49 65 59 79 80 76 57 66 68

9 Jay played in a golf tournament. His average score for the 4 rounds was 71. If the scores in his first 3 rounds were 72, 66 and 70, what was his score in the 4th round?

R

10 Tasnuba scored a mean of 74% for 5 maths tests that she completed.

R

a What was the total number of marks Tasnuba scored for the 5 tests?

b Tasnuba did a 6th test and her mean test mark increased to 77%. What is the total that she scored for the 6 tests?

c What mark did she achieve in the last test?

11 Alice's average mark for 4 Science exams is 68%. Can Alice score enough marks in a fifth exam so that her average mark increases to 75%? Give reasons.

12 Copy and complete each frequency table, then calculate the mean, correct to 2 decimal places.

a

Score, X	Frequency, f	fx
2	5	
3	3	
4	7	
5	6	
6	4	
	$\Sigma f =$	$\Sigma fx =$

b

X	f	fx
22	3	
23	8	
24	7	
25	9	
26	5	
27	4	
	$\Sigma f =$	$\Sigma fx =$

13 Use your calculator in the statistics mode to find the mean of each frequency table.

a

Score, X	Frequency, f
32	5
33	8
34	12
35	9
36	7
37	4

b

X	f
10	12
11	18
12	27
13	37
14	23
15	15
16	9

14 For each frequency table, add a cumulative frequency column and find:

i the median **ii** the mode **iii** the range.

☐ Foundation ○ Standard ⬡ Complex

a

x	f
2	2
3	5
4	11
5	7
6	10
7	9
8	6

b

x	f
17	3
18	8
19	20
20	15
21	11
22	7

15 The number of seedlings that have sprouted in each punnet at a nursery are shown:

10	5	7	7	9	8	8	11	7	8	5	8	4	6	4
12	11	9	4	6	6	10	7	6	9	10	9	7	8	8
8	7	4	7	7	11	10	9	8	9	9	8	8	8	9

a Complete a frequency table and find the median.

b What is the mode?

c Find the range.

d Calculate the mean.

INVESTIGATION

The effect of outliers

1 Find the mean, median, mode and range of each data set.

a	1	2	2	2	3	4	5	5	5
b	1	2	2	2	3	4	5	5	6
c	1	2	2	2	3	4	5	5	8
d	1	2	2	2	3	4	5	5	17

2 The 4 sets of scores in question **1** are the same, apart from the last score in each set. If one score is changed, what is the effect on:

a the mean?　　**b** the median?

c the mode?　　**d** the range?

3 Consider the 2 sets of data:

A:	1	10	11	14	15	12	13	11		
B:	3	4	2	6	3	14	6	3	5	4

a Find the mean, median and range of each set of data.

b Identify the outlier for each set.

c Find the mean, median and range of each set of data, excluding the outlier.

d What effect does the size of the outlier have on:

i the mean?　　**ii** the median?　　**iii** the range?

e Why is the median not affected by the size of the outlier?

TECHNOLOGY

Most valuable player

A team of basketball players would like to compare the number of points each player has scored per game over 10 games.

1 Enter the following data into a spreadsheet.

	A	B	C	D	E	F	G	H	I	J	K	L	M	N	O	P
1	NAME	Game 1	Game 2	Game 3	Game 4	Game 5	Game 6	Game 7	Game 8	Game 9	Game 10	Total	Mean	Mode	Median	Range
2	Sirisha	5	8	2	10	6	8	4	4	6	10					
3	Colin	4	8	6	7	7	9	10	6	5	8					
4	Eloise	12	8	8	16	4	5	7	10	11	8					
5	Diana	4	3	5	2	6	8	4	3	3	6					
6	Tran	10	14	7	8	12	16	10	10	8	6					
7	Michelle	12	4	6	10	9	9	5	16	8	12					
8	Anish	4	7	6	8	7	5	3	8	9	2					
9	Jamar	14	12	8	7	10	9	5	12	15	8					
10	Total															
11	Mean															

2 In B10, enter **=sum(B2:B9)** to calculate the **total score** for Game 1. **Fill Right** to repeat this step for cells C10 to K10.

3 In L2, enter **=sum(B2:K2)** to calculate the **total score** for Sirisha for the 10 games. **Fill Down** to repeat this step for cells L3 to L9.

4 In B11, enter **=average(B2:B9)** to calculate the **mean** score per player for Game 1. **Fill Right** from cell B11 across to K11.

5 In M2, enter **=average(B2:K2)** to calculate the **mean** score for Sirisha for the 10 games. **Fill Down** into cells M3 to M9.

6 In N2, enter **=mode(B2:K2)** to calculate the **mode** score for Sirisha. **Fill Down** to cell N9.

7 In column O, use the formula **=median(B2:K2)** and **Fill Down**.

8 There is no formula for the **range**, but we can create one using the formulas for the **maximum** and **minimum** values. In P2, enter **=max(B2:K2)–min(B2:K2)** for Sirisha's range, then **Fill Down**.

9 Now answer these questions, writing each answer in your spreadsheet in the given cell [ ].

a Which player had the best mean score? [A13]

b Which player was the most inconsistent? Why? [A14]

c Which of the statistics is the most useful for describing the players' overall contribution to team score? [A15]

d Who do you think was the MVP (Most Valuable Player)? Why? [A16]

Frequency histograms and polygons

A **frequency histogram** is a column graph of numerical data where the columns stand together without gaps between them.

A **frequency polygon** is a line graph that is constructed by joining the midpoints of the tops of the columns of a frequency histogram, starting and ending on the horizontal axis. It is called a 'polygon' because the graph and the horizontal axis together make a shape with straight sides.

Video Histograms: snapshot

Worksheet Analysing data

Puzzle Highway elephants

Example 5

45 students were surveyed on the number of hours of homework they did in the past week. The results were as follows:

3	7	5	8	8	7	7	4	8	7	9	8	8	5	7
4	2	5	3	8	6	6	7	9	6	5	4	9	8	4
2	3	6	7	8	2	8	8	4	5	6	8	7	2	5

a Arrange the data in a frequency table.

b Draw a frequency histogram and polygon for this data.

SOLUTION

a

No. of hours, x	Tally	Frequency, f
2	IIII	4
3	III	3
4	𝍸	5
5	𝍸 I	6
6	𝍸	5
7	𝍸 III	8
8	𝍸 𝍸 I	11
9	III	3
	Total	45

b

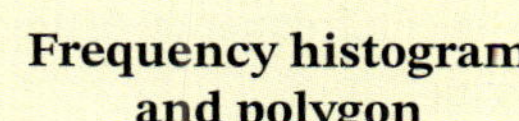

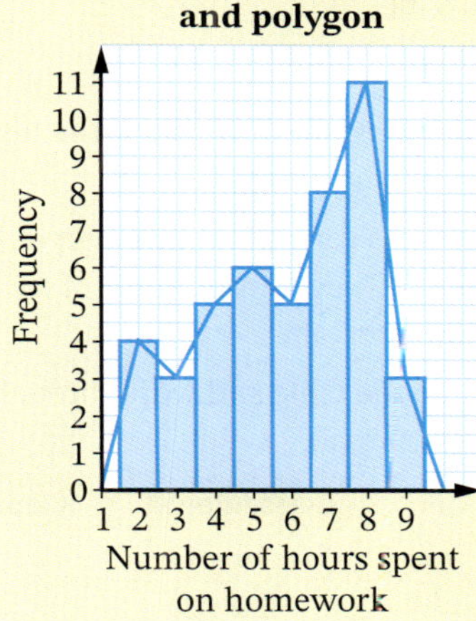

Note that each column of the histogram is centred on each value marked on the horizontal axis, so there is a half-column gap at the start between the Frequency axis and the first column.

Class intervals for grouped data

Collected data can sometimes have a large range depending on the variable being collected. For example, the heights of 28 students in a class could all be different heights making the range of results quite large. Constructing a frequency table and a histogram for every single value would not provide much useful information as the frequency of each value would be 1 or close to 1. To overcome this problem, the heights of students could be grouped together into **class intervals.**

When graphing this grouped data on a frequency histogram or polygon, we use the centres of the class intervals on the horizontal axis.

Example 6

The percentage marks that 50 students scored for a history test is shown:

65	56	26	34	67	70	34	59	72	89
37	75	48	64	72	61	53	48	57	68
19	34	56	66	84	90	71	66	37	26
48	29	78	63	59	54	68	74	81	69
45	63	67	74	68	54	44	38	57	60

a Sort this data into a frequency table using class intervals (0–9, 10–19, 20–29, ….) and identify the class centres.

b Construct a frequency histogram and a frequency polygon on the same set of axes for this data.

SOLUTION

a The lowest mark is 19 and the highest is 90 so the class intervals should start with 10 – 19 and end at 90 – 99.

The class centre for each interval should also be included into the frequency table. The half-way 'centre' value of the group 10 – 19 is $\frac{10+19}{2} = 14.5$. Do the same for the other intervals.

Set up the table and work through the data set to tally the individual values into the class interval.

Class Interval	Class Centre	Tally	Frequency
10 – 19	14.5	|	1
20 – 29	24.5	|||	3
30 – 39	34.5	||||| |	6
40 – 49	44.5	|||||	5
50 – 59	54.5	||||| ||||	9
60 – 69	64.5	||||| ||||| ||||	14
70 - 79	74.5	||||| |||	8
80 – 89	84.5	|||	3
90 - 99	94.5	|	1

b Use the class centres to construct both a histogram and frequency polygon on the same set of axes.

You should include the centre *before* the first group and the centre *after* the last group with frequencies of 0 so that the polygon can be brought down to connect with the horizontal axis.

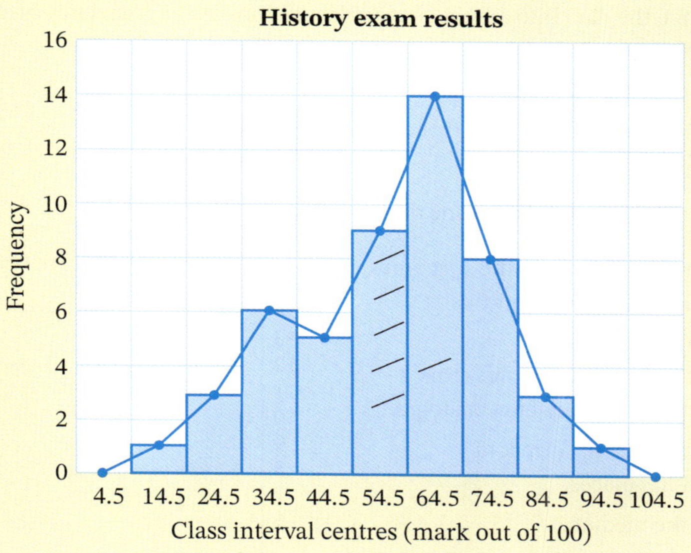

EXERCISE 9.02 ANSWERS ON P. 651

Frequency histograms and polygons

U F R C

EXAMPLE 5

1 The results of an Art project (out of 10) for 30 students are shown.

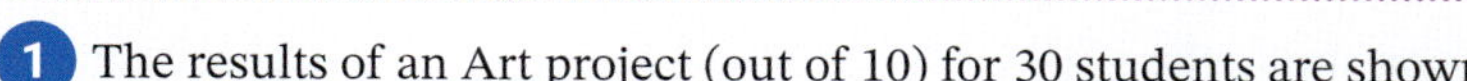

C

7	9	3	5	6	6	7	5	2	8
7	8	6	5	5	4	3	6	6	7
8	10	7	5	8	9	8	5	4	5

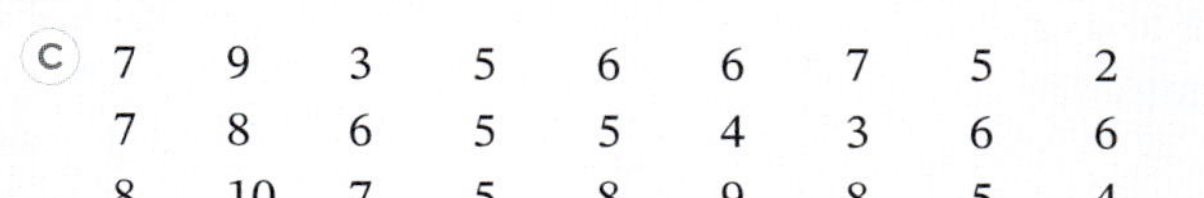

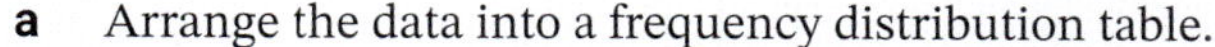

a Arrange the data into a frequency distribution table.

b Draw a frequency histogram and polygon.

c Find the mean, correct to one decimal place.

d Find the median.

e Find the mode.

f Find the range.

g Are there any outliers and is there any clustering?

Foundation Standard Complex

9780170465618

2 A sample of people were surveyed on the number of children in their families.

1	2	6	5	3	3	2	1	7	4	4	4	7	2
3	5	3	1	2	1	1	2	7	4	2	2	3	6
3	6	5	3	5	7	9	8	3	2	3	4		

a Arrange the data into a frequency table and construct a frequency histogram and polygon.

b Find the mode and the range.

c Are there any outliers and is there any clustering?

d Find the median.

e Find the mean, correct to one decimal place.

3 The numbers of letters in the surnames of 55 students are shown in this frequency histogram.

a What is the mode? How is this shown on the frequency histogram?

b Find the mean, correct to 2 decimal places.

c Find the median.

d Find the range.

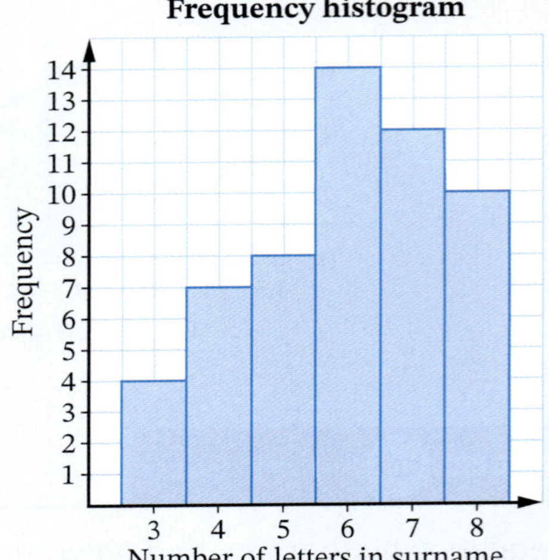

4 R C This frequency polygon shows the number of calls made per minute to a customer support line over a 30-minute period.

a How many calls were made altogether?

b How many times were there 3 or more calls received per minute?

c What is the mode? How is this shown on the frequency polygon?

d Find the range.

e Find the median.

f Find the mean, correct to 2 decimal places.

g Is there any clustering of data?

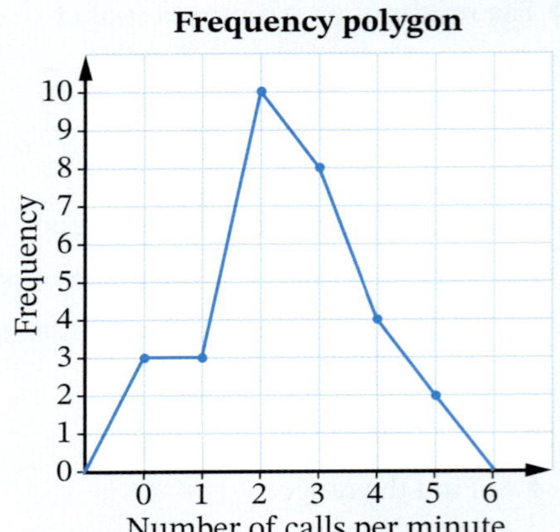

Foundation Standard Complex

EXAMPLE 6

5 Find the class centre of each class interval.

a 15–23 **b** 21–30 **c** 45–50

6 The pollution level of a large Australian city was tested over a period of 30 days. The levels of sulphur dioxide (in parts per million) were as follows:

0.07	0.17	0.18	0.11	0.10	0.09	0.08	0.08	0.04	0.15
0.12	0.06	0.12	0.13	0.18	0.14	0.07	0.16	0.19	0.23
0.22	0.26	0.25	0.19	0.05	0.13	0.05	0.03	0.3	0.11

a Sort this data into a frequency table using class intervals (0.00 – 0.04, 0.05 – 0.09, 0.10 – 0.14, 0.15 – 0.19, etc.) and identify the class centres.

b Construct a frequency histogram and a frequency polygon on the same set of axes for this data.

c Identify the modal class for this data set.

7 The following results are the marks obtained by a class in a maths test.

71	65	95	70	52	33	87	72	69	76
52	73	62	47	56	63	73	53	41	48
47	68	58	87	68	71	76	70	60	67

a What is the range of results for this data set?

b Sort this data into a frequency table using class intervals (30 – 39, 40 – 49, 50 – 59, etc.) and identify the class centres.

c Construct a frequency histogram and a frequency polygon on the same set of axes for this data.

8 This histogram shows the heights of students at a high school.

a How many students had their height measured?

b What are the class intervals?

c Identify the median class of the students.

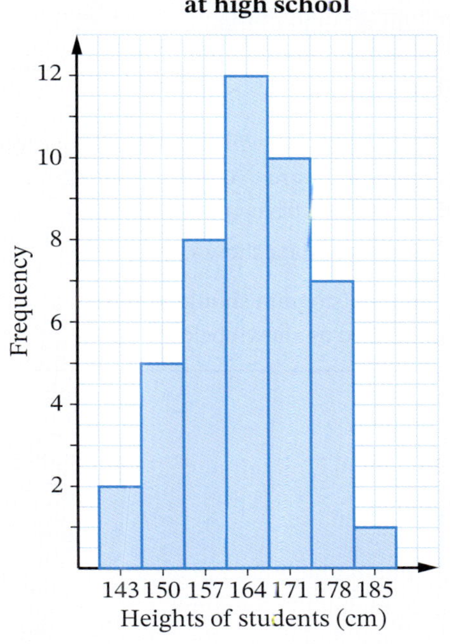

Foundation Standard Complex

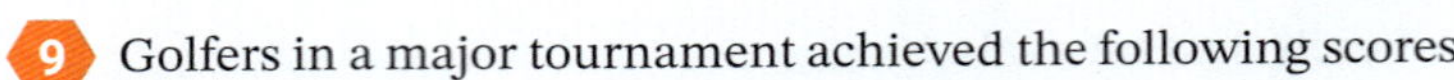

9 Golfers in a major tournament achieved the following scores.

71	72	70	73	67	66	78	65	79	78	80
74	75	75	69	68	66	68	67	70	70	71
64	69	65	66	76	69	70	73	77	76	81

a Sort this data into a frequency table using class intervals (60 – 64, 65 – 69, 70 – 74, etc.) and identify their class centres.

b Construct a frequency histogram and a frequency polygon on the same set of axes for this data.

c Identify the modal class for this data set.

d Identify the median class for the data set.

10 The heights (in centimetres) of 50 students are given below.

139	163	142	155	173	138	174	168	155	147
147	181	177	164	168	176	184	180	171	163
163	147	158	150	146	159	170	163	166	154
168	152	158	164	175	163	177	170	150	156
183	174	163	157	159	162	172	167	178	161

a Sort this data into a frequency table using class intervals (131 – 140, 141 – 150, 151 – 160, etc.) and identify the class centres.

b Construct a frequency histogram and a frequency polygon on the same set of axes for this data.

c Identify the modal class for this data set.

d What percentage of students were taller than 170 cm?

TECHNOLOGY

Histograms

In this activity, we will use statistical, spreadsheet or dynamic geometry software to construct a histogram of players' scores (out of 10) in a trivia quiz.

1 Enter the data shown.

2 Your histogram should also have a heading and the axes labelled as shown below.

Score	Frequency
5	1
6	2
7	6
8	4
9	3
10	2

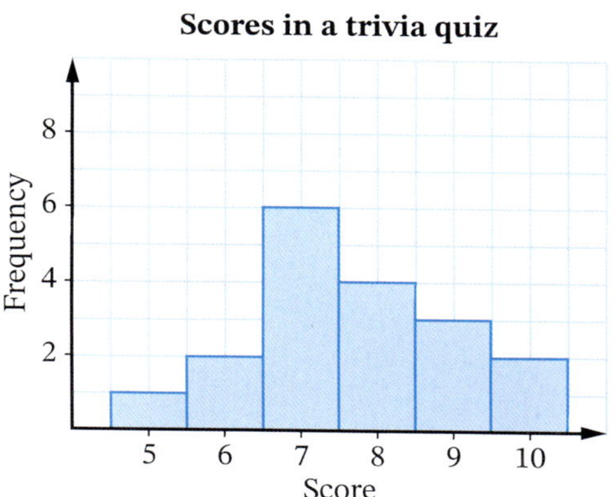

3 Remember, the data represents one-variable statistics. In the menu, select **1VAR** (or similar) to show calculations such as mean, median and mode.

Stem-and-leaf plots

A **stem-and-leaf plot** lists the values of a data set, usually in order. It shows:

- any **clusters** where values are grouped or bunched together.
- any **outliers.**
- how the values are spread out.

Worksheet Stem-and-leaf plots

Example 7

This stem-and-leaf plot shows the pulse rate (in heartbeats per minute) of 30 people in a shopping centre.

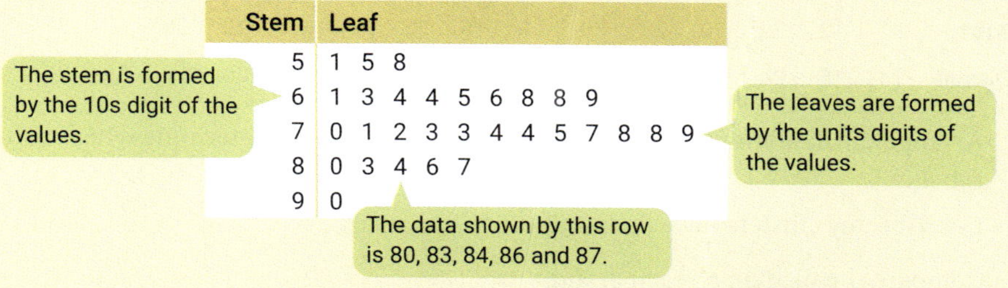

Stem	Leaf
5	1 5 8
6	1 3 4 4 5 6 8 8 9
7	0 1 2 3 3 4 4 5 7 8 8 9
8	0 3 4 6 7
9	0

a Are there any outliers?

b Is there any clustering of data?

c Find the mean, correct to one decimal place.

d What is the median?

e What is the mode?

f What is the range?

SOLUTION

a There are no outliers as no values are 'separate' from the main body of data.

b The scores are bunched in the 60s and 70s, so we say that there is clustering in the 60s and 70s.

c $\text{Mean} = \frac{51+55+58+\cdots+87+90}{30}$

$= \frac{2156}{30}$

$= 71.8666...$

≈ 71.9

d The 30 data values in the stem-and-leaf plot are in order, so the median is the average of the 15th and 16th values, 72 and 73 (counting from top to bottom, left to right).

$\text{Median} = \frac{72+73}{2}$

$= 72.5$

e Modes = 64, 68, 73, 74, 78 Each occurs 2 times.

f The lowest value is 51 and the highest is 90.

range = 90 − 51

= 39

Back-to-back stem-and-leaf plots

Video Back-to-back stem-and-leaf plots

When 2 related sets of data are compared, we can use **back-to-back stem-and-leaf plots.**

Example 8

Buzz batteries and *Eternal* batteries are to be compared by testing 15 batteries of each brand until they fail. The lengths of the battery lives are recorded to the nearest hour as shown.

Buzz	8	12	22	18	23	20	11	15	18	8	9	19	26	30	31
Eternal	14	16	24	32	35	36	37	28	31	30	24	26	19	35	32

a Arrange the data into a back-to-back stem-and-leaf plot to determine which battery is better.

b Describe any clusters in each distribution.

c Compare the median of each brand.

d Compare the mean of each brand.

e Which brand of battery is better? Justify your answer.

SOLUTION

a

Buzz		Eternal
9 8 8	0	
9 8 8 5 2 1	1	4 6 9
6 3 2 0	2	4 4 6 8
1 0	3	0 1 2 2 5 5 6 7

b Buzz is clustered in the 10–20 hours' life, Eternal is clustered in the 30s hours of life.

c Median for Buzz = 18 — The 8th value, counting from top to bottom, right to left.

Median for Eternal = 30 — The 8th value, counting from top to bottom, left to right.

Eternal has the higher median of 30 hours.

d Mean for Buzz = 18

Mean for Eternal = 27.933...

Eternal has the higher mean of approximately 27.9 hours.

e From looking at the statistics and the stem-and-leaf plot, Eternal is better as its median and mean are higher.

EXERCISE 9.03 ANSWERS ON P. 653

Stem-and-leaf plots

EXAMPLE 7

1 This unordered stem-and-leaf plot shows the marks of a class of students in a Science test.

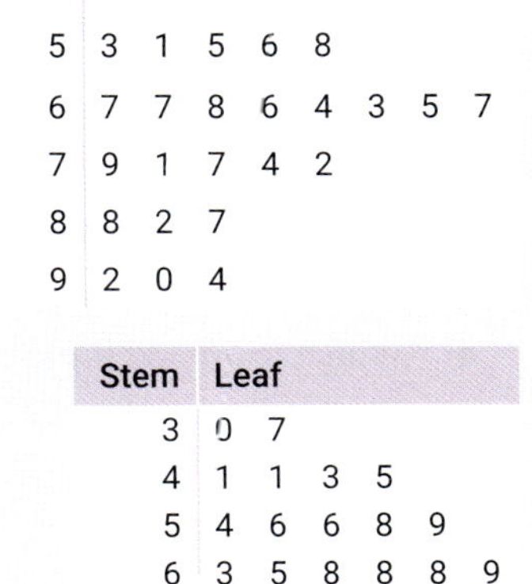

Stem	Leaf
4	0 1 5
5	3 1 5 6 8
6	7 7 8 6 4 3 5 7
7	9 1 7 4 2
8	8 2 7
9	2 0 4

a Construct an ordered stem-and-leaf plot.
b Find the range.
c Are there any outliers or clusters?
d How many students are there in the class?
e Find the mean, correct to 2 decimal places.
f Find the median.

2 This stem-and-leaf plot shows the points scored by a basketball team in different matches.

Stem	Leaf
3	0 7
4	1 1 3 5
5	4 6 6 8 9
6	3 5 8 8 8 9
7	2 4 5 5
8	0 2

a Find the range.
b What is the mode?
c Find the mean, correct to one decimal place.
d Find the median.

Foundation Standard Complex

9780170465618

3 When 30 students ran 100 m, their times (in seconds) were recorded.

13.5	15.0	12.1	12.5	17.6	13.7	15.7	18.7	13.1	11.8
12.6	15.7	14.4	15.4	14.9	15.0	18.3	16.3	15.4	15.1
11.9	16.1	14.3	14.5	16.3	14.6	14.9	15.0	16.1	13.5

Stem	Leaf
11	8 9
12	
13	
14	
15	
16	
17	
18	

a Copy and complete this stem-and-leaf plot.

b Are there any:

i outliers? **ii** clusters?

c How many students ran the 100 m in less than 13.7 seconds?

d What percentage of students took more than 14 seconds to run the 100 m?

e What percentage (to one decimal place) of students ran times that were below:

i the mean **ii** the median?

4 This ordered stem-and-leaf plot shows the marks obtained by students in a History exam.

Stem	Leaf
2	□
3	0 2 3 4
4	1 5 7 8
5	2 4 □ 6 9 9
6	0 2 3 3 6 8 8 7
7	2 5 6 7 7
8	2 5

R **a** One of the marks in the 50s is missing. What could this missing mark be?

b The entry for the lowest mark is missing. If the range of the marks is 63, find the lowest mark.

c Find the median.

5 This back-to-back stem-and-leaf plot shows the number of goals scored per match by 2 netball teams during a season.

R C

Rockets		Blues
7 5 4 3	1	7
8 7 6 5 4 2	2	0 5 6 8
7 6 4 3	3	4 5 7 7 8 9
5 4 0	4	2 3 7 8
2	5	1 3 7

a How many times did each team score more than 30 goals?

b Find the highest number of goals scored by each team.

c Which team had the greater range?

d Which team performed better? Give reasons.

e Which team had the greater median?

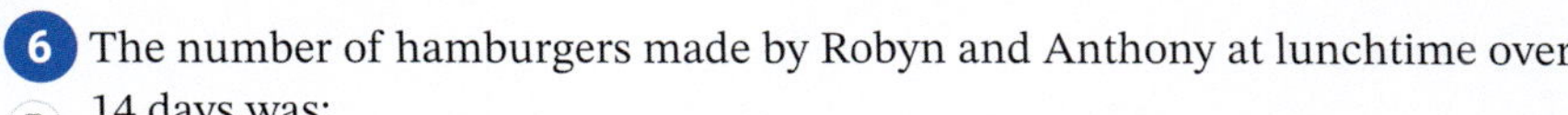

6 The number of hamburgers made by Robyn and Anthony at lunchtime over 14 days was:

R

Robyn	37	23	33	35	17	42	37	54	45	42	37	27	36	41
Anthony	35	28	29	36	43	37	55	42	47	51	36	34	35	39

a Construct an ordered back-to-back stem-and-leaf plot.

b Who made more hamburgers?

c On how many days did Robyn make more hamburgers than Anthony?

d Are there any outliers in the figures for either Robyn or Anthony?

e Is there any clustering of figures for either person?

f Which person had the higher mean?

7 A sample of commuters were surveyed on how far (in kilometres) they live from the closest railway station.

R C

1.5	4.8	6.5	1.1	3.3	3.4	3.5	4.5	4.5	12.5	3.1	3.6
3.8	4.0	4.2	2.3	2.5	1.9	4.7	5.5	2.7	3.2	5.2	

a Arrange the data in a stemplot.

b Are there any outliers? Give reasons.

c Are there any clusters?

d Find the mean, correct to 2 decimal places.

e Find the median.

f Which is the better measure of centre, the mean or median? Give reasons.

8 The results for a writing task for 20 girls and 20 boys are shown.

Girls:	18	48	56	59	67	73	76	94	76	78
	82	39	45	59	60	85	80	74	51	36
Boys:	50	88	76	72	10	28	40	95	74	63
	67	55	32	29	16	75	26	82	57	50

a Construct a back-to-back stem-and-leaf plot for the data.

b What was the highest mark? Was it scored by a girl or boy?

c What is the median for:

i girls **ii** boys?

d What percentage of boys scored more than 60?

e What percentage of girls scored more than 60?

f Find the mean mark for:

i girls **ii** boys.

g Is there a there a significant difference between the median mark and the mean mark for:

i girls? **ii** boys?

9780170465618

9.04 Choosing the best graph

Puzzle
Data match-up

Types of data

There are 2 types of data: categorical and numerical.

Categorical data are usually words and can be grouped into categories, such as a person's hair colour or cultural background.

Numerical data (or quantitative data) are numbers describing things that can be counted or measured, like the number of goals scored or a person's height. Numerical data is either discrete or continuous.

- **Discrete data** are counted or measured and can only take on separate, distinct values, with 'gaps' or 'jumps', for example, the number of girls in a family is discrete because it could be 0, 1, 2, etc., but not 2.3 or $1\frac{2}{5}$.
- **Continuous data** are measured on a smooth scale with no 'gaps' or 'jumps' between values, e.g., temperature during the day is continuous because it can be 28 °C, 29 °C, 29.5 °C, 30.1 °C, etc.

Data can be classified using the flowchart below:

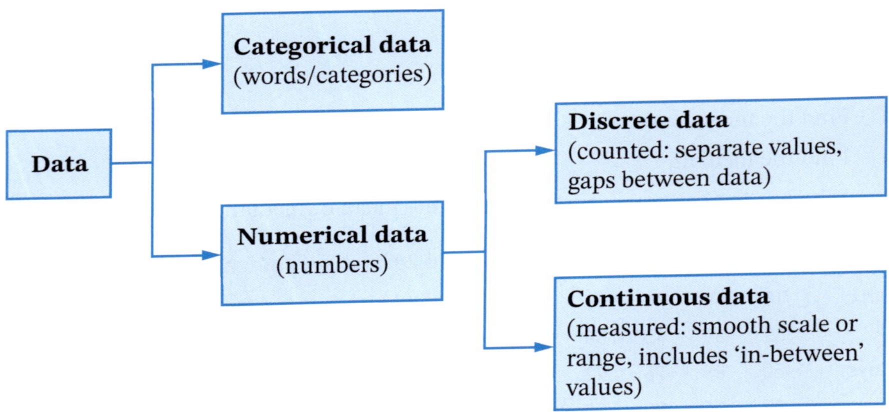

Example 9

Classify whether each type of data is categorical or numerical. If numerical, then classify whether it is discrete or continuous.

a School report grades (for example, A, B, C).

b The noise level at a concert.

c People's shoe sizes.

d Marital status (for example, single, married).

SOLUTION

a School report grades are categorical data. — Involves words or categories.

b Noise level is continuous numerical data. — Takes on a smooth scale of values.

c People's shoe sizes are discrete numerical data. — Takes on separate values.

d Marital status is categorical data. — Involves words or categories.

Selecting appropriate displays

The best type of statistical graph that should be used to represent a data set depends on the type of data being analysed. Here are some examples of the most appropriate graphs to use depending on the type of data you are working with.

Graph	Appropriate for	Description
Picture graph 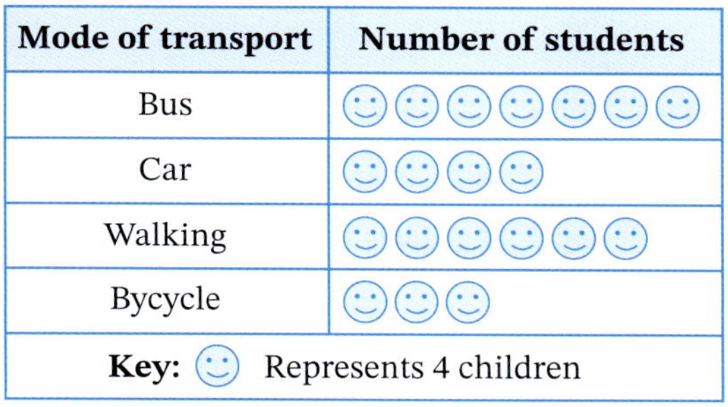	Categorical data	• Uses pictures or symbols to represent frequencies. • Has a key to show what one picture represents.
Column graph (or bar chart) 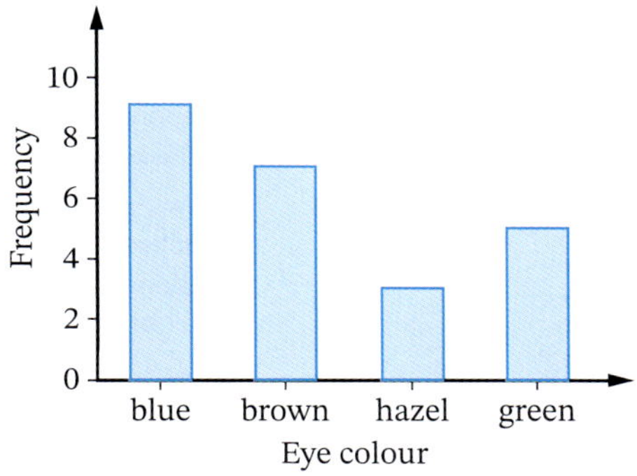	Categorical data Discrete numerical data	• Uses columns to show frequency, gaps between columns. • Frequency shown on the vertical axis.
Clustered column graph 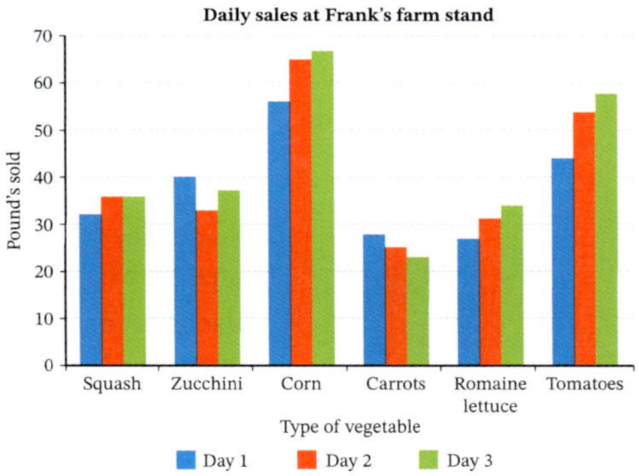	Categorical data Discrete numerical data	• Allows comparisons to be made between 2 or more data sets. • A key must be included to identify what the different colour bars represent.

Mode of transport	Number of students
Bus	☺ ☺ ☺ ☺ ☺ ☺ ☺
Car	☺ ☺ ☺ ☺
Walking	☺ ☺ ☺ ☺ ☺ ☺
Bycycle	☺ ☺ ☺
Key: ☺ Represents 4 children	

Sector graph (or pie chart) **Preferred drink** Milk, Cola, Juice, Lemonade	Categorical data Discrete numerical data	• Shows parts of a whole in circle form. • Uses sectors of a circle to show sizes of parts.
Divided bar graph **Favourite colour** red, green, blue, yellow, white	Categorical data Discrete numerical data	• Shows parts of a whole in rectangular form. • Uses sections of a rectangle to show sizes of parts.
Line graph **Temperature** Degrees Centigrade: 10, 8, 6, 4, 2, 0, −2, −4, −6 Time: 00.00, 04.00, 08.00, 12.00, 16.00, 20.00, 00.00	Discrete numerical data Continuous numerical data	• Dependent variable on the y-axis, independent variable on the x-axis. • Allows you to visualise the relationship between a pair of data sets by using connected line segments.
Frequency histogram **Tree Heights** Trees: 0, 10, 20, 30, 40, 50 Height (cm): 100, 150, 200, 250, 300, 350	Continuous numerical data	• No gaps between columns, to indicate continuous data. • A space at the beginning and end of the graph. • Frequency is shown on the y-axis. • Useful for data grouped into class intervals.

<table>
<tr><td>Frequency polygon 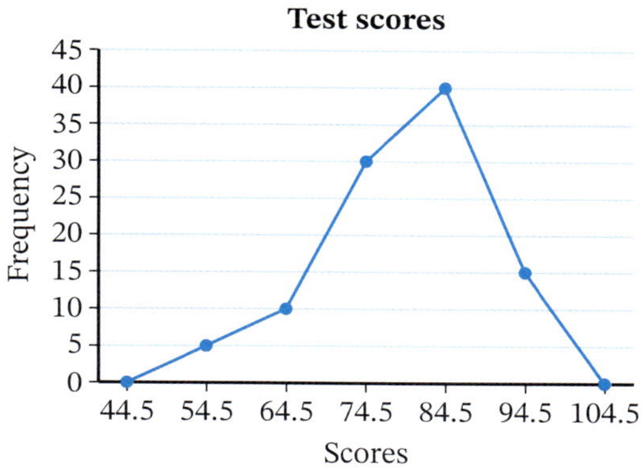

Test scores
Frequency: 0, 5, 10, 15, 20, 25, 30, 35, 40, 45
Scores: 44.5, 54.5, 64.5, 74.5, 84.5, 94.5, 104.5</td><td>Continuous numerical data</td><td>• Frequency is shown on the y-axis .
• Graph starts and finishes on the horizontal axis
• Useful for data grouped into class intervals.</td></tr>
<tr><td>Dot plot 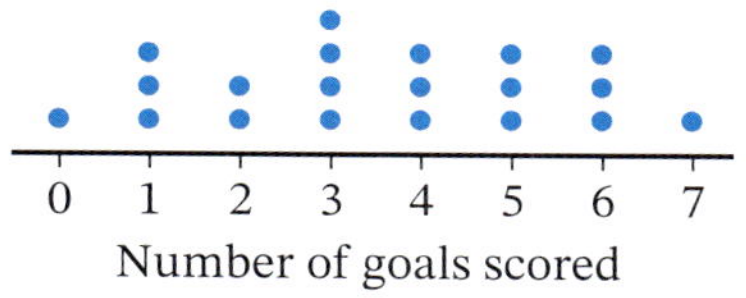

0, 1, 2, 3, 4, 5, 6, 7
Number of goals scored</td><td>Discrete numerical data</td><td>• Only useful for small data sets with small frequencies.</td></tr>
<tr><td>Stem-and-leaf plot

<table><tr><th>Stem</th><th>Leaf</th></tr><tr><td>0</td><td>1 1 2 2 3 4 4 4 4 5 8</td></tr><tr><td>1</td><td>0 0 0 1 1 3 7 9</td></tr><tr><td>2</td><td>5 5 7 7 8 8 9 9</td></tr><tr><td>3</td><td>0 1 1 1 2 2 2 4 5</td></tr><tr><td>4</td><td>0 4 8 9</td></tr><tr><td>5</td><td>1 6 7 7 8</td></tr><tr><td>6</td><td>3 6</td></tr></table>key: 6|3 = 63 years old</td><td>Discrete numerical data

Continuous numerical data</td><td>• Useful for large data sets.
• Lists every value in the data set.</td></tr>
</table>

Example 10

Identify what type of graphical display is appropriate to represent each data set described and explain why.

a The number of roast chickens sold each day from a supermarket over a week.

b The temperature of a chemical reaction taken every minute.

SOLUTION

a Column graph
Days of week is categorical data and number of roast chickens per day will be frequency. A column graph would easily show the differences in sales over the days with the heights of the bars.

b Line graph
Need to identify a relationship between time and temperature. Will therefore need to plot time on x-axis and temperature on y-axis producing a line graph.

EXERCISE 9.04 ANSWERS ON P. 653

Choosing the best graph

1 Classify each sample of data as either categorical or numerical.

- **a** The number of people absent from school today.
- **b** The distance travelled by a car on a trip.
- **c** The amount of petrol bought when filling up the car.
- **d** The different types of pies available at the supermarket.
- **e** The blood pressure of a patient in hospital.
- **f** The amount of time spent on devices during the week.
- **g** The time for swimming 100 m.
- **h** The country of birth of people living in Australia.
- **i** The computer games played by students.
- **j** The age you turn this year.

2 Students are asked to rate a movie on a scale of 1 to 5 stars. What type of data is this? Select the correct answer **A**, **B**, **C** or **D**.

a continuous **B** categorical **C** discrete **D** random

3 Classify whether each type of numerical data is discrete or continuous.

- **a** The number of people watching a cricket match.
- **b** The temperature in Perth over a 24-hour period.
- **c** The amount of time spent doing homework.
- **d** The speed of a car passing through an intersection.
- **e** The age you turn this year.
- **f** The results in a maths test.
- **g** The amount of rainfall in a month.
- **h** The number of dogs at the RSPCA.

4 A survey was conducted about the colour of people's cars. The results are shown in the table below:

Colour	Red	Blue	White	Green	Silver	Black
Frequency	42	16	53	5	45	32

- **a** What type of statistical graph would you use to represent the results of this survey?
- **b** Why is this an appropriate graph to use?
- **c** Draw the graph you chose in part **a**.
- **d** Is there a different type of graph that can be used to represent the same data set? If there is, construct it.

5 The pulse rate of a student was measured over a 10-minute period to track their fitness and the results are shown in the table below

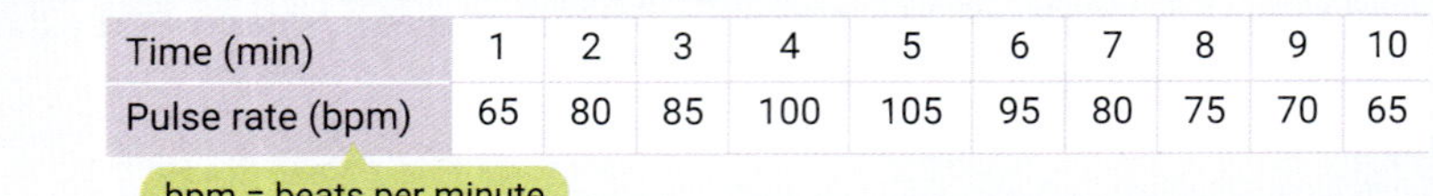

Time (min)	1	2	3	4	5	6	7	8	9	10
Pulse rate (bpm)	65	80	85	100	105	95	80	75	70	65

bpm = beats per minute

Foundation Standard Complex

9780170465618

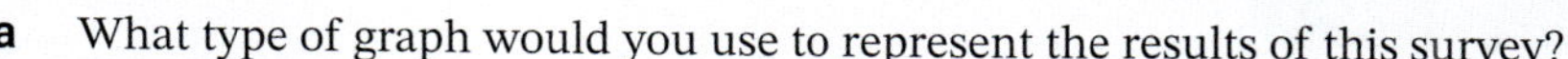

a What type of graph would you use to represent the results of this survey?

b Why is this an appropriate display to use?

c Draw the graph you chose in part **a**.

6 The average number of hours of homework completed by students in year 9 each week is shown in this table.

R C

Hours of homework	6	7	8	9	10
Frequency	4	8	5	3	2

a What statistical display would you use to represent the results of this survey?

b Why is this an appropriate display to use?

c Construct the display that you chose in part **a**.

7 The sports played by children living on Nelson Avenue are shown in this table.

R C

Sport	Netball	Hockey	Rugby	Soccer
Frequency	15	10	26	9

a What display would you use to represent the results of this survey?

b Why is this an appropriate display to use?

8 The ages of guests at the screening of a new movie are shown in this table.

Age (years)	20–29	30–39	40–49	50–59	60–69	70–79	80–39	90–99
Frequency	22	41	45	58	37	12	0	1

A column graph was constructed to represent this data. Was this the best way to display this data visually? Explain your reasoning.

9 This table below shows the number of different coloured lollies in a packet.

R C

Colour	Red	Purple	Yellow	Blue	Green	Pink	Brown	Orange
Frequency	8	5	4	2	7	2	4	9

a What display would you use to represent the results of this survey?

b Why is this an appropriate display to use?

10 The table below shows the sale of fruit (in kilograms) from a fruit shop over 2 days.

R C

Fruit	Monday	Tuesday
Apples	10	15
Bananas	32.5	22.5
Mangos	17.5	25
Oranges	27.5	35
Strawberries	15	20

a What display would you use to represent the results of this survey?

b Why is this an appropriate display to use?

Foundation Standard Complex

DID YOU KNOW?

Infographics

An **infographic** is a visual representation of information or data that allows readers to easily understand complicated concepts. This form of graphical display provides a colourful alternative to standard text and numbers.

This is an infographic about Aboriginal and Torres Strait Islander people published by the Australian Human Rights Commission. Read through it and notice the way it presents information in a clear and colourful way.

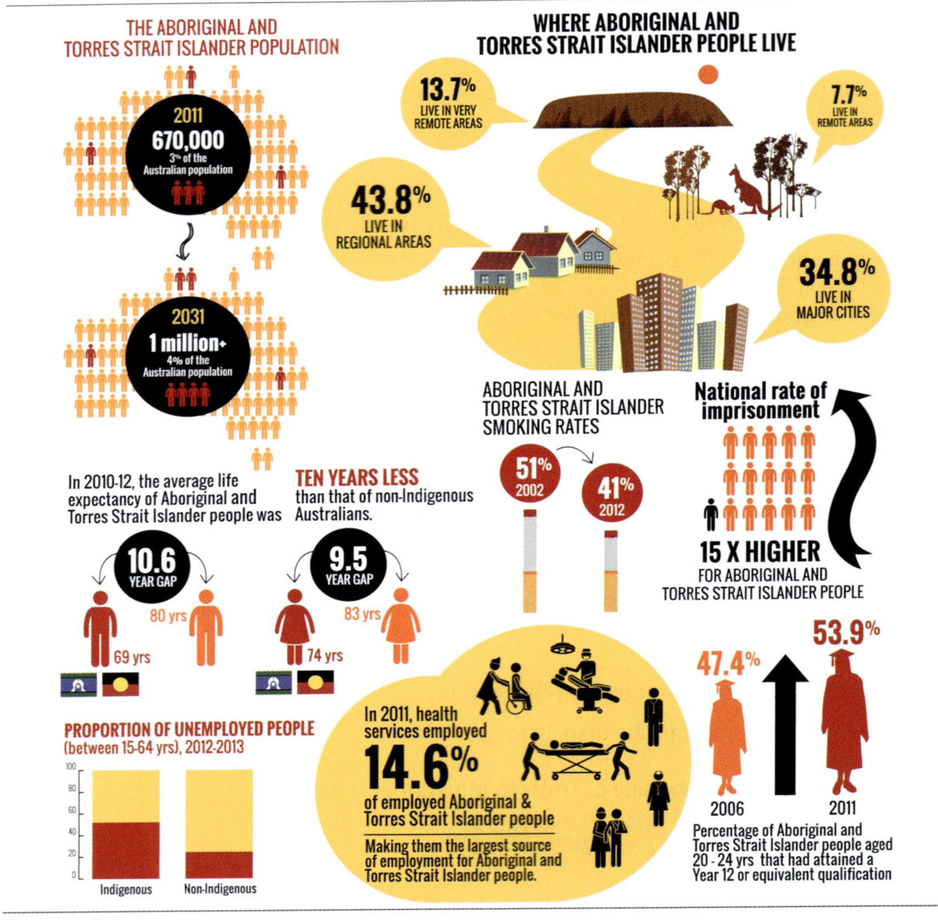

2014 Face the Facts www.humanrights.gov.au/face-facts

The rock art paintings found all over traditional sites in Australia can also be considered infographics. These paintings can illustrate aspects of ceremonies, show the indigenous people's bond with the land and record significant events in the people's lives, like successful hunts.

Can you remember a time when an infographic helped you to understand information and data?

Design an infographic about the students who make up your maths class.

The shape of a distribution

9.05

A **statistical distribution** is the way the values of a data set are arranged, especially when graphed. When looking at histograms, dot plots and stem-and-leaf plots, an overall pattern can be seen from the **shape of a distribution**.

Puzzle
Statistical match-up

The shape of a statistical distribution shows how the data is spread and can be seen by drawing a curve around the graph or display.

A distribution is **symmetrical** if the data is evenly spread or balanced about the centre. For example, the frequency histogram and stem-and-leaf plot shown below are both symmetrical in shape.

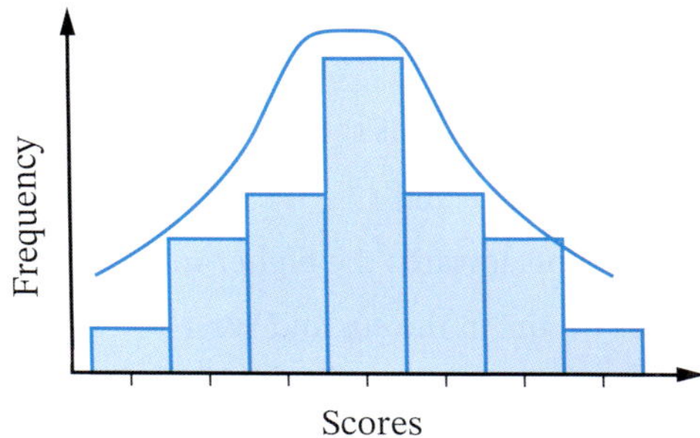

Stem	Leaf
2	6
3	3 7 9
4	5 8
5	0 2 3 3 7 8
6	0 4
7	6 7 8
8	9

A distribution is **skewed** if most of the data is bunched or clustered at one 'end' of the distribution and the other end has a 'tail', as shown on this dot plot and frequency histogram.

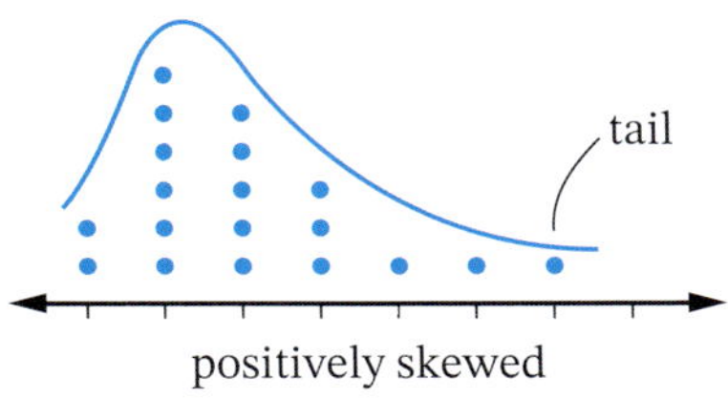

positively skewed

A distribution is **positively skewed** if its tail points to the right.

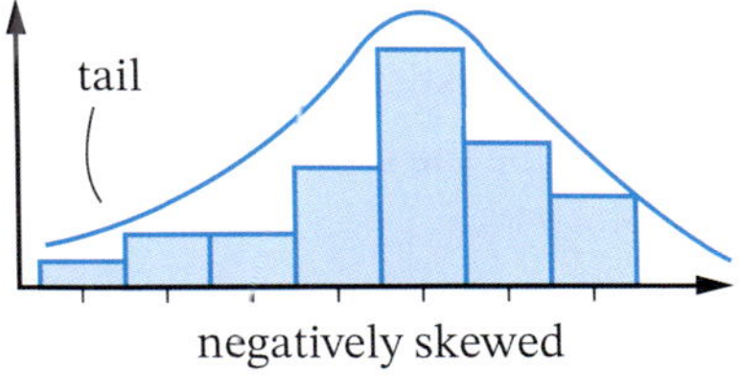

negatively skewed

A distribution is **negatively skewed** if its tail points to the left.

A distribution is **bimodal** if it has 2 peaks. The higher peak is the mode, while the other peak indicates another data value that has a high frequency.

For example, this dot plot has peaks at 2 and 6 so it is bimodal. The mode, however, is 6.

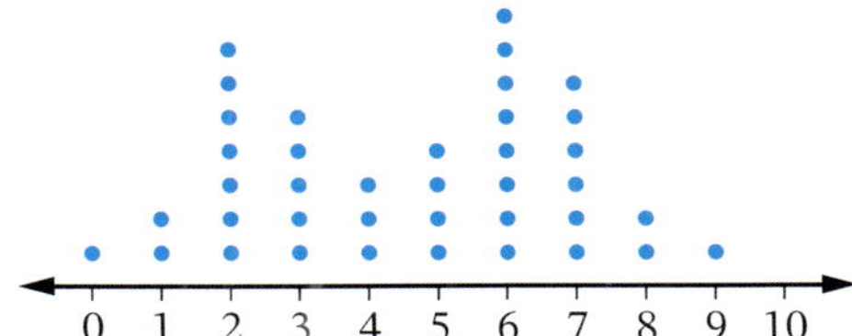

Example 11

For each statistical distribution:

i describe the shape **ii** identify any outliers and clusters.

a

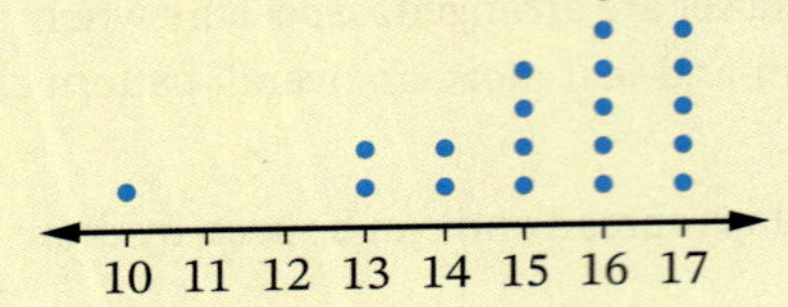

b

Stem	Leaf
3	0 1 2
4	1 3 4 4 5 6
5	0 4 5 7 8
6	3 7 8
7	0 1
8	4
9	8

SOLUTION

a **i** The shape is negatively skewed (tail points towards the lower values).

ii 10 is an outlier and clustering occurs at 15, 16 and 17.

b **i** The shape is positively skewed (tail points towards the higher values).

ii 84 and 98 are outliers and clustering occurs in the 40s and 50s.

EXERCISE 9.05 ANSWERS ON P. 654

The shape of a distribution

U F R C

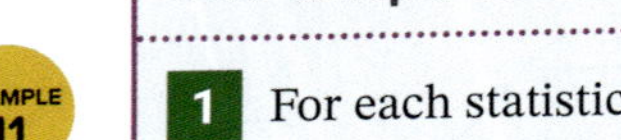

EXAMPLE 11

1 For each statistical distribution:

i describe the shape **ii** identify any outliers and clusters.

a

Frequency

5 6 7 8 9 10 11

Score

b

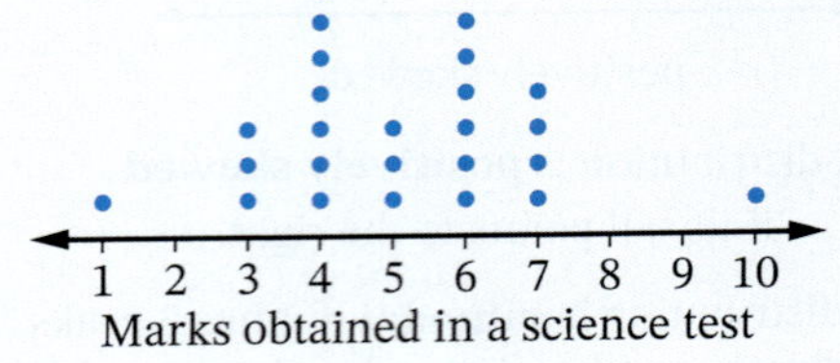

Foundation Standard Complex

c

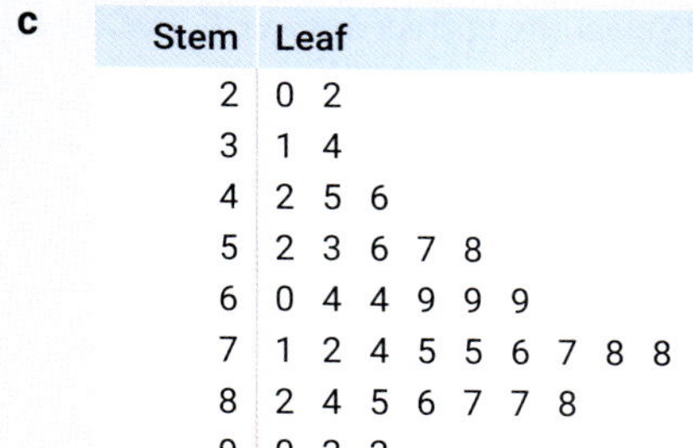

Stem	Leaf
2	0 2
3	1 4
4	2 5 6
5	2 3 6 7 8
6	0 4 4 9 9 9
7	1 2 4 5 5 6 7 8 8
8	2 4 5 6 7 7 8
9	0 2 2

d

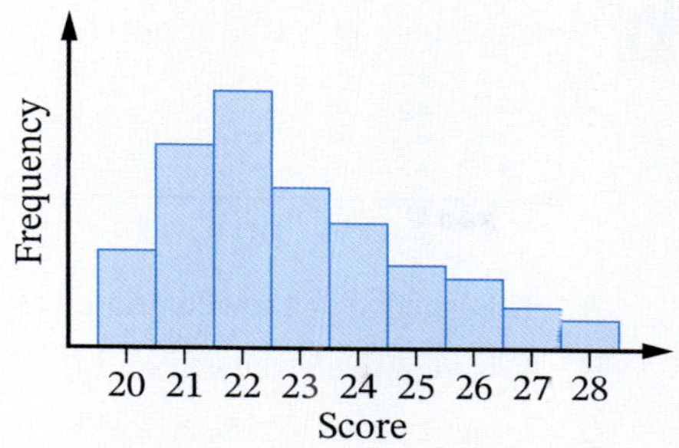

e

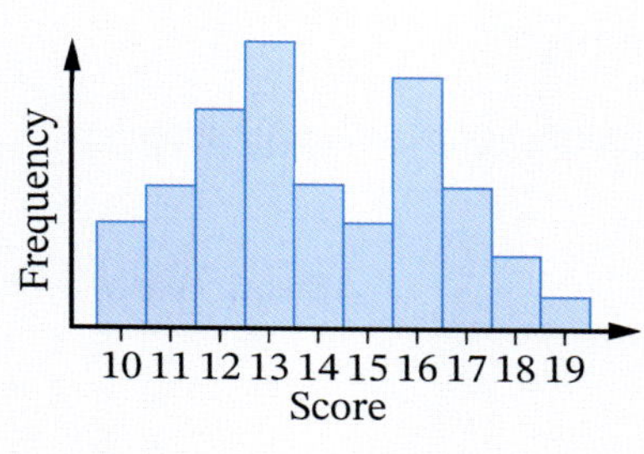

f

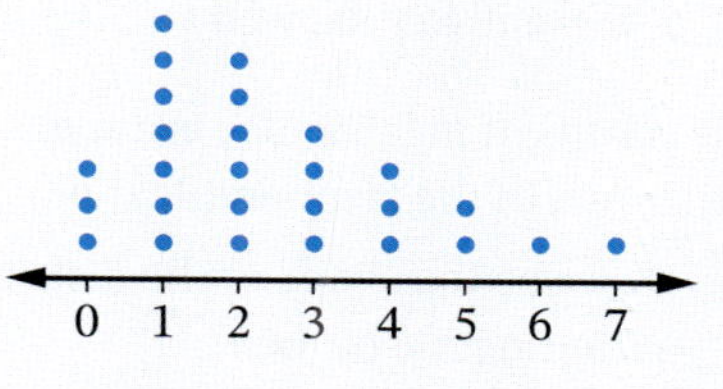

g

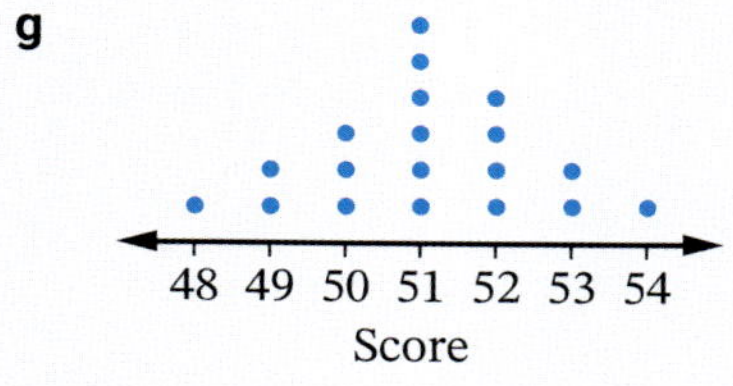

h

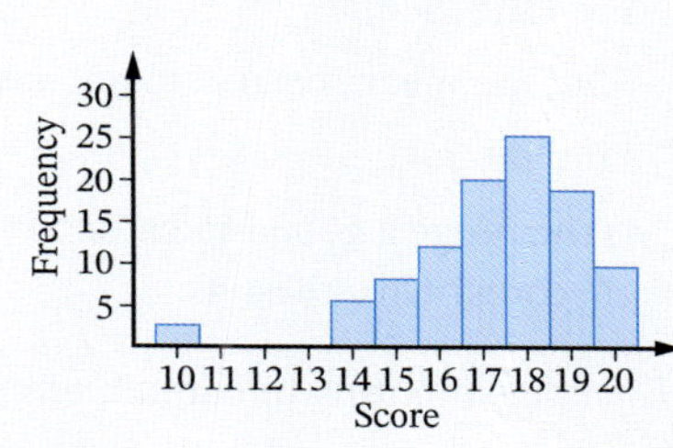

2 The results of a Mathematics test for a Year 9 class are shown in the stem-and -leaf plot.

Stem	Leaf
1	8
2	3 5 8 8
3	2 4 6 7 8 8 9
4	0 4 5 5 7 7
5	3 6 7 8
6	0 4 4
7	2 5
8	1 2
9	2

R C

a How many students are in the class?

b Where does the clustering occur?

c Describe the shape of the data.

d Give a possible reason for the shape of the distribution.

e Find the mean, median and mode.

3 The number of children in the families of students at Eastleigh High School are shown below.

4	4	7	3	3	6	8	6	3	3	4	2	9	5	1	2	3	3	4
7	2	7	2	2	3	5	3	1	4	1	4	1	2	5	6	3	5	2

a Arrange the data into a frequency table and construct a frequency histogram.

b Are there any outliers?

c Describe the shape of the distribution.

d Where does clustering occur?

e Find the mode, the mean (to one decimal place) and the median and show their position on the histogram.

Foundation | Standard | Complex

4 Which statement is true about the 2 sets of data? Select the correct answer **A**, **B**, **C** or **D**.

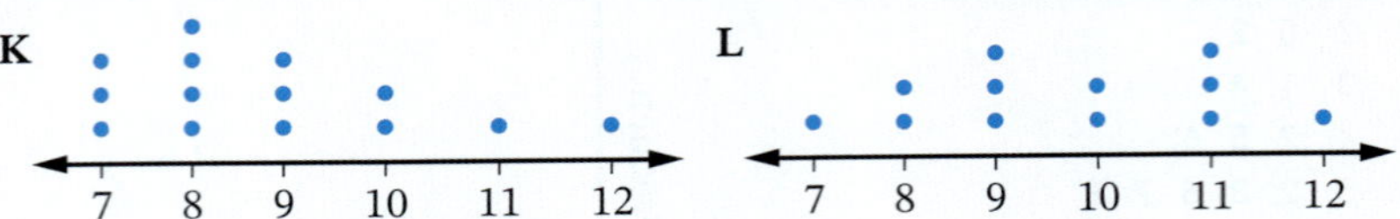

A K is positively skewed and L is symmetrical.

B K is negatively skewed and L is bimodal.

C L has 2 peaks and K is positively skewed.

D K is positively skewed and the median of L is 9.

5 The results of an entrance exam to select apprentices for an energy supplier are shown in the stem-and-leaf plot.

R C

a Describe the shape of the distribution, excluding the outlier.

b What is the mode?

c Find the median.

d Find the mean, correct to one decimal place.

e Find the range.

f Is the range a good indicator of the spread of scores? Give reasons.

Stem	Leaf
9	2 3 5 6 7 8
8	3 3 6 7
7	2 4 8
6	0 3 4 4
5	3 5 6 6 8 9
4	
3	
2	6

6 The average monthly maximum temperatures for Sydney and Melbourne are shown.

R C

Month	J	F	M	A	M	J	J	A	S	O	N	D
Sydney	26	26	25	22	19	17	16	18	20	22	24	26
Melbourne	26	26	24	20	17	14	13	15	17	20	22	24

a **i** Display the data for Sydney and Melbourne on dot plots.

ii For each city, describe the shape of the dot plot.

b Find the mean (to 2 decimal places), median, mode and range for each city.

c Which city is generally warmer? Give reasons for your answer.

☐ Foundation ○ Standard ⬡ Complex

Comparing data sets

9.06

Example 12

This back-to-back stem-and-leaf plot shows the results of a Year 9 class in a Health exam.

Girls		Boys
2	2	3 4 8
8 3 1	3	0 3 5 9 9
6 2	4	1 3 7 8 8 8 9 9
7 3 0	5	0 2 3 4 5 5 8 9 9
8 7 7 6 2 1	6	0 1 1 4 7 8
9 8 8 7 5 4 4 3	7	0 2 6 7
9 9 8 7 7 5 3 2 2 0	8	1 3 8
8 5 4 0	9	1 6

a Find the mean (to one decimal place) for:

i the girls **ii** the boys **iii** the whole group.

b Find the median for:

i the girls **ii** the boys

c Where are the scores clustered for:

i the girls? **ii** the boys?

d Compare the shapes of the two data sets.

e Which group, girls or boys, performed better? Justify your answer.

SOLUTION

a **i** Mean for girls $= \frac{2600}{37} \approx 70.3$

ii Mean for boys $= \frac{2234}{40} \approx 55.9$

iii Mean for group $= \frac{4834}{77} \approx 62.8$

b **i** Median for girls = 75 — 19th score

ii Median for boys $= \frac{54+55}{2} = 54.5$ — Average of the 20th and 21st values.

c **i** The girls' scores are clustered at 70s and 80s.

ii The boys' scores are clustered at 40s and 50s.

d The scores for the girls are negatively skewed, while the scores for the boys are positively skewed. The scores for the girls are generally higher than the scores for the boys.

e The girls performed better, which can be seen from their mean of 70.3 being greater than the mean of 55.8 for the boys. Also, the median for the girls was 75, which was higher than the median of 54.5 for the boys.

Worksheets
Comparing word lengths
Investigating young drivers

Technologies
Population polygons'
World climate

Spreadsheet
World climate

EXERCISE 9.06 ANSWERS ON P. 654

Comparing data sets

U F R C

1 The times (in seconds) taken for swimming 100 m by Brooke and Ryan are shown.

R C

Brooke	59.71	59.20	60.21	60.13	59.69	59.50	60.41	59.33	58.95	58.70
Ryan	55.60	56.20	57.50	56.70	57.15	57.65	58.10	59.70	60.15	56.65

a Find the mean for each swimmer.

b Find the median for each swimmer.

c Find the range for each swimmer.

d Which swimmer is faster? Give reasons.

e Which swimmer is more consistent? Give reasons.

2 The maximum daily temperatures in Sydney and Adelaide for a 15-day period in March are shown.

R C

Sydney	34.3	21.9	32.3	30.6	21.2	22.0	25.1	29.2
	31.5	29.2	25.9	30.6	32.1	27.3	24.7	
Adelaide	22.0	22.4	24.3	24.3	26.2	34.6	34.9	23.2
	22.1	21.6	26.8	32.0	28.3	27.1	29.0	

a Find the range for each city.

b Find correct to 2 decimal places the mean for each city.

c Find the median for each city.

d On how many days was Sydney warmer than Adelaide?

e Which city was warmer? Give reasons.

EXAMPLE 12

3 A class was given a topic test on trigonometry. The results were disappointing and, after some more work on the topic, the class was given a second test on trigonometry. The results of the 2 tests are shown in this back-to-back stem-and-leaf plot.

R C

Test 1		Test 2
1	9	0 9
5 2	8	4 8 8
4 3 2	7	0 1 5 5 8 8
7 4 4 1 0	6	0 3 4 4 7 7 7 8
9 9 9 6 5 3 2 2	5	3 4 4 9 9
9 8 7 4 3 0	4	5 7 8
4 2	3	

a Find the median for each test.

b Find correct to one decimal place the mean for each test.

c Find the range for each test.

d Describe the shape of each test.

e Has significant improvement been made from Test 1 to Test 2? Give reasons.

4 The dot plots show the temperatures (in °C) at 3 p.m. for Logan City, Queensland and Wollongong, NSW for 16 days in Autumn.

R C

Logan City

20 21 22 23 24 25 26 27 28 29 30 31 32 33

Wollongong

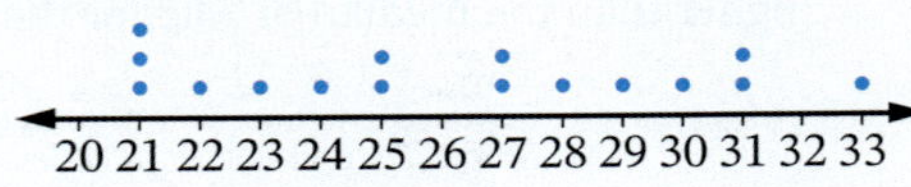

a Which city's temperatures are more skewed?

b Find, correct to 2 decimal places, the mean for each city.

c Find the range for each city.

d Which city had a more consistent temperature? Give reasons.

e Find the median for each city.

f Which city had cooler temperatures? Give reasons.

5 The reaction times (in seconds) of a group of men and women were measured.

R C

Men:	38	29	37	28	30	32	31	53	31	26
	49	47	33	55	29	44	32	48	43	29
Women:	51	34	40	35	45	40	36	48	57	39
	42	35	51	54	36	37	37	39	43	45

a Arrange the 2 sets of data in a back-to-back stem-and-leaf plot.

b Would dot plots be suitable for displaying the 2 sets of data? Give reasons.

c Are the mean and median the same or different for:

i men? ii women?

d Find the mode for each group.

e Describe the shape of each distribution. Are there any outliers or clusters?

f Which group has the faster reaction times?

g Which group has the greater spread of reaction times?

6 Use the Internet to find the gold medal winning times for the 100-m race for men and women for each Olympic Games from 1960 to 2021 (postponed from 2020 due to the pandemic).

R C

a Copy and complete this table by entering the winning times in seconds.

	1960	1964	1968	1972	1976	1980	1984	1988	1992	1996	2000	2004	2008	2012	2016	2021
Men																
Women																

b Calculate the mean winning time (correct to 2 decimal places) for:

i men? ii women?

c Find the median for:

i men? ii women?

d Where are the data values mainly clustered for:

i men? ii women?

e Identify any outliers for:

i men? ii women?

f Compare the shape of the 2 data sets.

g Did either group show a significant improvement in the winning time for the period 1960–2021? Identify any contributing factors.

Foundation Standard Complex

TECHNOLOGY

Comparing relative humidities

Relative humidity is the amount of moisture in the atmosphere, measured as a percentage of water vapour per area at a certain temperature. It is reported more often in summer. When presented together with a high temperature, it shows that a day has been particularly warm and 'muggy'. In this activity, we will compare the relative humidity in a town at 9 a.m. and again at 3 p.m. each day for the first fortnight in September 2018 and 2019.

1 Enter the data below into a spreadsheet (as shown).

	A	B	C	D	E	F	G
1	Sept 2018	Relative Humidity 9am Adelaide 2018 (%)	Relative Humidity 3pm Adelaide 2018 (%)		Sept 2019	Relative Humidity 9am Adelaide 2019 (%)	Relative Humidity 3pm Adelaide 2019 (%)
2	1	92	55		1	79	44
3	2	62	62		2	39	26
4	3	57	35		3	97	57
5	4	38	27		4	92	71
6	5	23	22		5	82	57
7	6	76	53		6	70	62
8	7	70	40		7	74	55
9	8	54	38		8	78	69
10	9	54	53		9	53	49
11	10	63	20		10	53	38
12	11	17	56		11	33	22
13	12	62	52		12	47	60
14	13	65	39		13	91	65
15	14	35	31		14	58	35

2 Select **Scatter graph with smooth lines and markers** to graph this data on one set of axes. Give the graph a title and label the axes.

3 In B16, C16, F16 and G16, enter formulas to calculate the **average (mean)** relative humidity at 9 a.m. and 3 p.m. for each fortnight.

4 In B17, C17, F17 and G17, enter formulas to calculate the **maximum** relative humidity.

5 In B18, C18, F18 and G18, enter formulas to calculate the **minimum** relative humidity.

6 In B19, C19, F19 and G19, enter formulas to calculate the **range**.

7 In B20, C20, F20 and G20, enter formulas to calculate the **median**.

8 Compare your graphs as well as your statistical values. What do they suggest about the relative humidity each year? Is there a year and time of day showing a more consistent relative humidity? Justify your answer.

9 Visit the Bureau of Meteorology website **www.bom.gov.au** and select **Daily Weather Observations** to find the relative humidity for the past fortnight in your area. Compare the data with this time last year.

   a Can you see any patterns?

   b Is the relative humidity consistent?

   c Are there any outliers?

   d Analyse the data and describe any similarities or differences you can see.

☆ MENTAL SKILLS 9 ANSWERS ON P. 655

Maths without calculators

Quiz Mental skills 9

Finding a percentage of a multiple of 10

We can use the unitary method to calculate a percentage of a multiple of 10. To do this, we first find 1% of the number by dividing it by 100. This is easily done by moving the decimal point 2 places to the left.

1 Study each example.

a $1\% \times 300 = 3.00$
$= 3$

b $4\% \times \$500$
$1\% \times \$500 = \$5$
$\therefore 4\% \times \$500 = 4 \times \$5$
$= \$20$

c $9\% \times 40$
$1\% \times 40 = 0.4$
$\therefore 9\% \times 40 = 9 \times 0.4$
$= 3.6$

d $12\% \times \$700$
$1\% \times \$700 = \$7$
$\therefore 12\% \times \$700 = 12 \times \$7$
$= \$84$

2 Now evaluate each expression.

a	$1\% \times 800$	**b**	$1\% \times 30$	**c**	$3\% \times \$400$	**d**	$4\% \times 20$
e	$8\% \times \$600$	**f**	$3\% \times 40$	**g**	$7\% \times 80$	**h**	$2\% \times \$900$
i	$6\% \times \$50$	**j**	$8\% \times \$90$	**k**	$11\% \times 700$	**l**	$13\% \times 200$

3 Study each example.

a $4\% \times 320$
$1\% \times 320 = 3.2$
$\therefore 4\% \times 320 = 4 \times 3.2$
$= 12.8$

b $1.5\% \times \$200$
$1\% \times \$200 = \$2$
$\therefore 1.5\% \times \$200 = 1.5 \times \$2$
$= \$3$

c $3\% \times 250$
$1\% \times 250 = 2.5$
$\therefore 3\% \times 250 = 3 \times 2.5$
$= 7.5$

d $4.5\% \times \$800$
$1\% \times \$800 = \$8$
$\therefore 4.5\% \times \$800 = 4.5 \times \$8$
$= \$36$

4 Now evaluate each expression.

a	$5\% \times 250$	**b**	$1.5\% \times \$300$	**c**	$2\% \times \$540$	**d**	$9\% \times 110$
e	$2.5\% \times \$600$	**f**	$3\% \times 160$	**g**	$4\% \times \$750$	**h**	$4.5\% \times \$2000$
i	$3.5\% \times \$400$	**j**	$6\% \times \$120$	**k**	$3\% \times 190$	**l**	$1.5\% \times \$700$

9.07 Sampling

Videos
Counting crowds

Can fish oil make you smarter?

Can you trust your IQ?

When the marketing department of a large company wishes to launch a new product, it collects information about consumer tastes. Governments collect information about the population so that they can provide for transport, housing, health and education needs. A school may need to collect information regarding school uniforms or food to be sold in the school canteen.

Sample vs census

In statistics, the **population** refers to all the items under investigation or being studied. For example, if the life of AAA batteries is being tested, then the population is all the different brands or types of AAA batteries.

A **census** is a survey of the entire population. If the population is too large or difficult to survey, then a **sample** of the population is selected instead. The sample must be representative of the population, by being:

- **random:** every item in the population should have an equal chance of being selected for the sample.
- **unbiased:** there should not be any factors that may influence the sample and prevent it from being a true indicator of the population.
- **large:** the bigger the sample, the more representative it will be.

Example 13

Decide whether a census or a sample should be used for each investigation.

a Finding the number of primary school children in Victoria.

b Determining people's opinions on public transport.

c Finding the most streamed TV series in Australia.

SOLUTION

a Census, because accurate figures are needed and these figures would be recorded.

b Sample, because it would be too expensive and time-consuming to survey all people.

c Census, because data can be collected about the number of accounts with a provider that streamed a certain TV series.

Types of samples

Samples can be selected using a variety of different methods, but it is important to ensure the sample is fair and representative of the population.

1 Random sample

Every item has an equal chance of being selected. Methods of random sampling include 'drawing' names out of a box or using a random number generator using technology.

2 Systematic sample

Items are selected at regular intervals from a list, for example, every 10th person or test every 50th car on a production line. The list should be prepared randomly and the starting point on the list should also be selected randomly.

3 Stratified sample

A sample where all identifiable groups within a population are represented in the sample in the same proportion as they appear in the population.

Example 14

Use the random number generator on your calculator (or an online random number generator) to select a random sample of 5 whole numbers between 120 and 180.

SOLUTION

Press RAN# or [RANDOM] and then = on your calculator to generate a random decimal from 0 to 1, correct to 3 decimal places. You may need to press SHIFT · or 2nd F 7 key to get this.

For more random numbers, keep pressing =. For numbers between 120 and 180, read the first 3 digits after the decimal point and only record the ones between 120 and 180.

For example, you could randomly generate the values 177, 143, 150, 151 and 168.

Example 15

A sample of 6 students must be selected from a class of 24 to participate in a 'Have your Say' day. The class list is alphabetical and includes student enrolment numbers from 50 to 74. Use systematic sampling to select your sample of 6 students.

SOLUTION

There are 24 students and we need a sample of 6.

$24 \div 6 = 4$

Select every 4th student.

Select a random starting point for the enrolment numbers. Close eyes and drop point of pencil onto starting point in student enrolment number chart.

50	51	52	53	54	55	56	57	58	59
60	61	62	**63**	64	65	66	67	68	69
70	71	72	73	74					

Start at student 63.

Pick every 4th student, starting at number **63**, until you have a sample of 6. If you come to the end of the list, go back to the start and continue counting from there.

50	51	52	53	**54**	55	56	57	**58**	59
60	61	62	**63**	64	65	66	**67**	68	69
70	**71**	72	73	74					

List the selected students.

Students 50, 54, 58, 63, 67 and 71 are selected for the sample.

Example 16

A restaurant has 2 managers, 5 kitchen staff and 18 waiting staff. A sample of 6 staff need to be surveyed regarding working conditions. For a stratified sample, how many people from each group of workers should be selected?

SOLUTION

$$\begin{aligned}\text{Total number of employees} &= 2 + 5 + 18\\ &= 25\end{aligned}$$

Use fractions to calculate number from each group required in sample

$$\begin{aligned}\text{Number of managers} &= \frac{2}{25} \times 6\\ &= 0.48\\ &\approx 1\end{aligned}$$

$$\begin{aligned}\text{Number of kitchen staff} &= \frac{5}{25} \times 6\\ &= 1.2\\ &\approx 1\end{aligned}$$

$$\begin{aligned}\text{Number of waiting staff} &= \frac{18}{25} \times 6\\ &= 4.32\\ &\approx 4\end{aligned}$$

The sample should include 1 manager, 1 kitchen staff member and 4 waiting staff.

EXERCISE 9.07 ANSWERS ON P. 655

Sampling

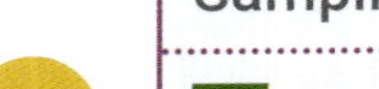

EXAMPLE 13

1 Determine whether a census or a sample should be used for each investigation.

R **a** Finding the number of people in Adelaide who were born overseas.

b Testing the quality of running shoes.

c Determining whether Australia needs a new flag.

d Determining the number of people in Darwin who earn \$60 000–\$70 000 per year.

e Determining the average family size in Australia.

2 In Australia, a national census was held in 2011, 2016 and 2021.

a How often is a census held in Australia?

b When will the next census be held?

Foundation Standard Complex

3 The table shows the number of students in each year at Southwest High.

Year	7	8	9	10	11	12
Students	148	151	160	153	132	128

a What is the school population?

b What percentage (to one decimal place) of the school population is formed by:

i Year 8? **ii** Year 11?

c A sample of 100 students is to be surveyed on the amount of time that they spend on the Internet. How many students should be selected from:

i Year 8? **ii** Year 11?

4 R C Students are surveyed about how they travel to school. The first 30 students arriving at school were selected for surveying. Is this a representative sample to use? Give reasons.

5 R C People were asked how often they eat take-away food. A sample of 100 people was surveyed at a shopping centre. Is this a representative sample to use? Give reasons.

6 Use the random number generator on your calculator or online to select:

a a random sample of 8 numbers between 20 and 98.

b a random sample of 10 values between the numbers 155 and 315.

c a random sample of 20 values between the numbers 1070 and 1460.

7 Use systematic sampling to select:

a 6 coffee mugs from a production run of 300.

b 10 participants for a competition from a population of 500 people that applied.

c select 15 students from a primary school, where the student ID numbers go from 4930 to 5725.

8 A sample of 15 local community members are required to help organise this year's multicultural festival. The local community is made up of 40 Australians, 25 Indians, 20 Europeans and 16 Asians. Use a stratified random sample to determine how many members from each group should be included in the sample.

9 A sample of 10 biscuit packets need to be tested for quality assurance purposes from a daily production run. The business has made 20 Gingernut Snap packets, 15 Scotch Finger packets, 30 Milk Arrowroot packets, 18 Monte Carlo packets and 25 Chocolate Wafer packets today. Use a stratified random sample to determine how many packets of each type of biscuit should be tested in the sample.

10 A sample of 30 students are needed to participate in a school opinion survey. There are 150 year 7 students, 120 year 8 students, 100 year 9 students, 110 year 10 students, 80 year 11 students and 70 year 12 students who attend the school. Use a stratified random sample to determine how many students from each year level should be included in the sample.

☐ Foundation ○ Standard ○ Complex

DID YOU KNOW?

Australia's first census

The first census in Australia was conducted in November 1828. Before this, statistical information was collected by musters on the same day each year. Worried about the accuracy of the figures, Governor Darling and the Legislative Council passed an Act requiring a census to be held at regular intervals. Each citizen needed to disclose his or her name, age, marital status, number of children, the name of the ship on which they arrived, the amount of land owned and the number of cattle and other stock owned. Collectors were employed to gather this information by personal interview.

The first census showed that the total non-Aboriginal population was 36 598 and nearly 30% of that population lived in Sydney.

How many national censuses have there been in Australia?

TECHNOLOGY

Random numbers

Random numbers can be used to select people or items for a sample. A calculator or spreadsheet can be set up to generate a random number from 1 to 100, for example, and that number can be matched to the name of a person or item from a list. For a scientific calculator:

Casio	Sharp
Enter the formula RanInt#(1,100) as shown: ALPHA . (1 SHIFT) 100) =	2nd F 7 (RANDOM), select 3 R-INT, + 1 =

This generates a random number from 1 to 100. Press = more times to generate more random numbers.

Alternatively, on a **spreadsheet**, type **=INT(RAND()*100+1)** into a cell and press the ENTER key, then the F9 key repeatedly for more random numbers.

1. Use your calculator or spreadsheet to generate 12 random numbers from 1 to 100.
2. How many students are there in your class?
3. Use your calculator or spreadsheet to generate 12 random numbers from 1 to the number of students in your class.
4. Use a class roll to number the students in your class from 1 onwards. Suppose you need to choose a random sample of 10 students from your class for a survey on homework. Use your calculator or spreadsheet to generate 10 random numbers to select the 10 students for your survey.
5. Choose 4 students from each Year 9 maths class to answer some questions. Devise several different methods to ensure that your selection is random. Prepare a short class talk about your methods.

INVESTIGATION

9.07

Australian Bureau of Statistics

The Australian Bureau of Statistics (ABS) is the official organisation in charge of collecting data for government departments. The data collected covers many areas, from the height of children in Australia, the alcohol consumption in the country to the number of Internet subscribers and Internet providers. The ABS conducts a census of the entire Australian population every 5 years.

Visit the ABS website www.abs.gov.au to answer the following questions.

1 a What is a population clock?
  b What is the current predicted population of Australia?
  c On average, another baby is born in Australia after how many minutes?
  d On average, another person dies in Australia after how many minutes?
  e A new migrant arrives in Australia after how many minutes?
  f Australia's population increases by one after how many minutes?

2 Go to Australian Demographic Statistics and find the populations of the states and territories.
  a What percentage of Australia's population lives in:
    i QLD?
    ii the ACT?
  b Which state had the greatest percentage increase over the previous year?

3 Find the population pyramid for Australia's population.
  a What is a population pyramid?
  b What percentage of Australian males are aged 15–19?
  c What is the modal age?
  d How many people are aged 100 years and over?
  e Which group, males or females, have the higher life expectancy? Justify your answer.

4 a Explain why Australia's population is said to be ageing.
  b What is the median age of Australia's population?
  c Find the median ages of the people for all states and territories and display them in a table. Which has the:
    i highest median age?
    ii lowest median age?

5 Use the ABS website to find the number of people that live in your local area.

9.08 Bias and questionnaires

Bias in sampling

Video The wrong guy won

To find a random sample of students to survey about school uniform, we could place students' names in a box, mix them up, and then choose 100 names without looking. Alternatively, we could use technology to generate random numbers to select students.

A sample that is not random is called a **biased sample**, and is not truly representative of the population. For example, if you survey the first 100 students to arrive at school today, or the first 100 students on the school roll, then not all students in the school have an equal chance of being selected. Your sample would not represent students who arrive at school later or whose names begin with letters at the end of the alphabet.

Example 17

A random sample of students is to be selected for their views on a new school uniform. Will each of the following sampling methods provide a sample that is representative or biased? If biased, explain why.

a Randomly selecting one student from each class.

b Selecting 20 students from your mathematics class.

c Selecting the students from the school's netball teams.

d Randomly selecting students from the school roll by rolling a die and counting down the roll.

SOLUTION

a Representative. Every student in the school has an equal chance.

b Biased. Because your mathematics class does not represent the other Year groups (or even the other Year 9 classes) at your school.

c Biased. Because the netball teams do not represent the whole school; it may also have mainly female students.

d Representative. Every student in the school has an equal chance.

Designing a questionnaire

Questionnaires are often used to collect information. They may ask for details about a person's opinions and attitudes, their lifestyle and behaviours, such as their recreational activities, TV viewing habits and types of foods eaten.

When designing a questionnaire, it is important that the questions asked are clear, fair and not biased.

- Decide what questions you want answered before you begin.
- Write the questions in plain English so that they are easily understood: make sure they are not vague, too open-ended or have a 'double meaning'.

9.08

- Ask for only one piece of information per question.
- Do not allow for too many different answers, provide choices if possible:
  - a tick or a cross beside each response
  - 'Yes', 'No', 'Don't know' or 'Other' responses
  - a rating on a scale
  - a one-word response
  - a sentence
- Determine how you are going to present and use the data you collect.
- Do not try to influence a particular answer unfairly (this is also called bias).
- Do not ask for information that may be too personal, such as phone numbers.

Example 18

Explain what is wrong with each survey question and suggest an alternative.

a 'How old are you?'

b 'How much pocket money do you receive each week?'

SOLUTION

a The question is not precise enough. Possible answers could be:

15 15 years, 4 months 14 years 10 months and 3 days $16\frac{1}{2}$

A better question might be:

'What age do you turn this year?'

14 ☐ 15 ☐ 16 ☐

or

'What is your date of birth?'

....../......./........ (dd/mm/yy)

b The question is too general and responses could be:

Nothing $5/day $40/week Whatever I need

A better question could be:

'Circle the amount of pocket money you receive in a normal week:'

$5 or less $6 to $10 $11 to $15 more than $15

EXERCISE 9.08 ANSWERS ON P. 655

Bias and questionnaires

U F PS R C

EXAMPLE 17

1 When collecting data, it is important to have an unbiased sample. Write, in your own words, what is meant by an 'unbiased sample'.
C

2 List the possible areas of bias in each sampling method.
R C
- **a** A survey about the amount of time spent doing homework, taken at a football match.
- **b** A survey about violence in sport, taken at a sporting event.
- **c** A survey about the use of alternative fuels, taken at a car show.
- **d** A survey about types of music listened to, taken at a rock concert.

3 A market research company uses a telephone survey to determine the most popular news program on TV. List some of the advantages and disadvantages of using this type of survey to collect data.
C

4 In order to determine what type of music should be played on the school radio station, 20 students were given a questionnaire.
PS R C
- **a** Is the sample suitable for obtaining the required information? Give reasons.
- **b** List some of the advantages of using a questionnaire to obtain the data.
- **c** List some of the disadvantages of using this method.
- **d** What other methods could be used to find out the type of music that should be played?

5 You need to determine how many students at your school have part-time jobs. Which of the following methods would be most suitable? Give reasons.
- **a** You ask 15 students in your class at school.
- **b** You randomly select 15 students from each Year level at your school.
- **c** You ask all students at your school to fill in a questionnaire and return them to you.

EXAMPLE 18

6 For each survey question, explain what is wrong with the question and suggest a better question.
R C
- **a** How much time do you spend studying?
- **b** What is your favourite sport?
- **c** What is your income?
- **d** What do you have for lunch each day?
- **e** What do you do in the holidays?
- **f** How often do you exercise?
- **g** How much time do you spend on social media websites?
- **h** How many text messages do you send in one day?

7 Select one topic for a questionnaire.
PS R C
- Amount of TV watched at home.
- How students in Year 9 spend their time on a Sunday.
- The amount of time spent on homework during the week.

Foundation Standard Complex

9780170465618

- The amount of time spent on the internet.
- Another topic of your choice.

a Design the questionnaire.
b Select a sample of students to answer it.
c Collect and present the information.
d Did the questions obtain the required data?
e Do you need to change any of the questions?
f Are there any questions you forgot to ask?
g Are there any questions you would omit next time?

8 Describe how you would collect data about each topic.

PS C

a The most popular fast food eaten by students at your school.
b The average daily amount of money spent by students at your school canteen.
c The most popular TV series watched by students at your school.
d The most popular Australian artist listened to by students at your school.
e The average number of girls in a classroom in NSW.
f The most commonly used brand of toothpaste.

INVESTIGATION

Media reports of surveys

We live in a world of 24-hour news, often quoting the results of surveys. To detect bias in news reports, we need to always consider where the news item comes from and what types of samples the statistics are based on. Is the information supplied by a reporter, the police, a company executive, a government official or an opinionated blogger? Are the statistics based on a small sample, large sample, unrepresentative sample, or a phone or Internet poll of volunteers? Each may have a particular bias that may influence how the story is reported.

Find a report of a statistical survey from a newspaper, magazine or the Internet and investigate the following questions.

a Where did the article come from?
b Who wrote the article and what claims does it make about the population?
c Does it show any bias?
d What type of sampling was used? Was it representative?
e How many people were questioned?
f When was the survey conducted?
g Was there any bias in the sample?
h Was there any bias in the survey questions?

INVESTIGATION

Year 9 student survey

The survey below is designed to collect data about Year 9 students. Questions can be changed or more can be added, but remember that the data is only useful if good questions are asked in a controlled way.

1 What is your gender? ☐ Male ☐ Female ☐ ____________

2 What is your age (as of last birthday)? ____________ years

3 How tall are you? ____________ cm

4 How many people are in your household? ____________ people

5 What is your position in your household (1 = oldest)? ____________w

6 How many pets do you have? ____________ pets

7 How far from school do you live (to nearest 0.1 km)? ____________ km

8 How did you travel to school today?

☐ Car ☐ Bus ☐ Walk ☐ Train

9 How long did the journey take? ____________ minutes

10 On a scale of 0–10, rate your enjoyment of school (10 = most). ____________

11 Which subject do you like best?

☐ Maths ☐ Science ☐ English ☐ Art ☐ PE ☐ History ☐ Other

12 Which subject do you study most?

☐ Maths ☐ Science ☐ English ☐ Art ☐ PE ☐ History ☐ Other

13 How much home study do you do each week? ____________ hours

14 How many hours do you study maths each week? ____________ hours

15 On a scale of 0–10, rate your enjoyment of maths (10 = most) ____________

16 How many hours each week do you play sport out of school? ____________ hours

17 How many sports do you play out of school? ____________ sports

18 Which sport do you prefer to watch?

☐ Football ☐ Soccer ☐ Netball ☐ Tennis ☐ Surfing ☐ Other

19 What is your favourite TV station? ____________

20 What is your favourite radio station? ____________

▷

9780170465618

Steps in conducting the survey

1 Finalise the questions and print the survey sheets.

2 Decide how many students you want to respond to the survey.

3 Decide how you will select the sample of students.

4 Decide whether the survey will be handed to selected students or if an interviewer will record responses.

5 Decide whether you will interpret the questions/responses or not.

6 Ask the selected students to complete the survey.

7 Collect all the sheets and label the first sheet 1, the second sheet 2 and so on.

8 Decide on the code you will use for each response, then code the responses and enter the results in a table, either on a spreadsheet or in your workbook.

9 Decide what to do if a question is not answered or the answer appears to be a mistake.

POWER PLUS ANSWERS ON P. 656

1 The number of yoghurts sold daily in a school canteen were:

25 28 22 35 26 36 33 48 42 37 25 25 37 44 19 40 23
38 42 32 28 21 36 22 39 46 28 45 28 41 43 26 17 28

a Complete a frequency table by grouping the data into class intervals like this:

Class interval	Frequency
16–21	
22–27	
28–33	
34–39	
40–45	
46–51	

b Draw a frequency histogram and describe the shape, identifying any outliers or clusters.

c What is the modal class?

d Add a cumulative frequency column to your frequency table and find the median class.

2 When would it be impractical to use a stem-and-leaf plot?

3 Seeds were sown in 40 large boxes. The number of seedlings that sprouted per box were:

14 23 18 18 17 28 28 33 4 24
17 27 11 19 13 18 18 7 25 16
14 3 31 4 5 12 23 26 11 17
19 20 24 7 30 16 25 8 19 20

a Organise the data into a stem-and-leaf plot using intervals of:

i 10 for the stem ii 5 for the stem (00 for 0–4, 05 for 5–9, 10 for 10–14, ...)

b Describe the shape of both plots, identifying outliers and clusters.

c Which plot gives a better indication of how the seedlings have grown? Explain your answer.

9 CHAPTER REVIEW

Language of maths

Quiz
Language of maths 9

bias	categorical data	census	cluster
continuous	discrete	dot plot	histogram
mean	measure of central tendency	median	mode
negatively skewed	numerical data	outlier	polygon
population	positively skewed	random	sample
stem-and-leaf plot	stratified	symmetrical	systematic

1 What is the collective name for the mean, median and mode?

2 Is a person's arm length an example of discrete data or continuous data?

3 What is the meaning of 'skewed'?

4 Explain the difference between a *population* and a *sample*.

5 What name is given to an extreme value that is very different from the other values in a data set?

6 Name one way of avoiding bias when selecting a sample.

Topic summary

Worksheet
Mind map: Analysing data

Print (or copy) and complete this mind map of the topic, adding detail to its branches and using pictures, symbols and colour where needed. Ask your teacher to check your work.

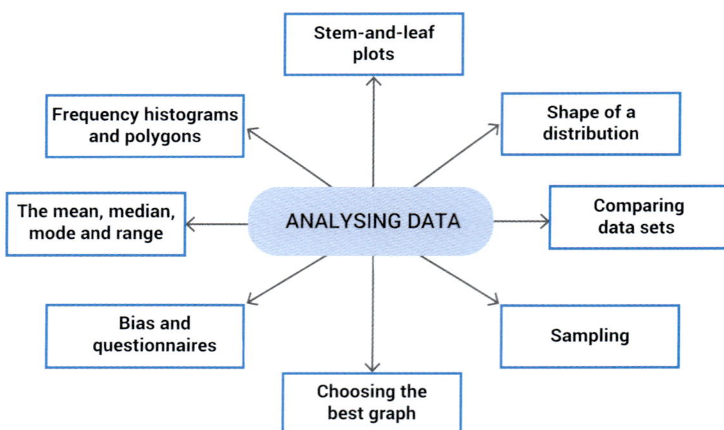

9 TEST YOURSELF

ANSWERS ON P. 656

1 Find the mean, median, mode and range of this set of data.

8 6 9 4 10 8 5 5 8

2 A real-estate agent sold 10 properties during one week. The prices of these properties are:

\$420 000	\$485 000	\$379 000	\$295 000	\$1 750 000
\$515 500	\$315 500	\$408 000	\$368 000	\$315 000

Quiz
Test yourself 9

a Find the mean price of the properties.

b Find the median price of the properties.

c Compare the mean price with the median price. Which is the better measure of centre? Give reasons.

d When data on sales of properties are given, the median price is used, not the mean. Explain why.

3 These percentage marks were scored by 30 students in a Maths test. 9.03

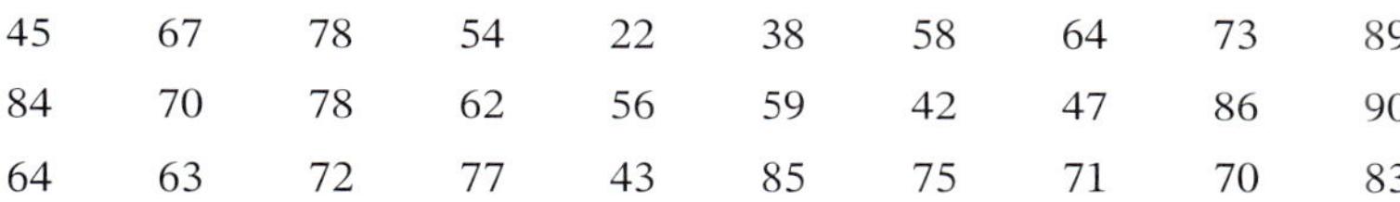

45	67	78	54	22	38	58	64	73	89
84	70	78	62	56	59	42	47	86	90
64	63	72	77	43	85	75	71	70	83

a Construct an ordered stem-and-leaf plot for the data.

b Find the mean, median, mode and range of the marks.

c Are there any outliers?

d Compare the mean and median. Which is more accurate? Give reasons.

4 9.02

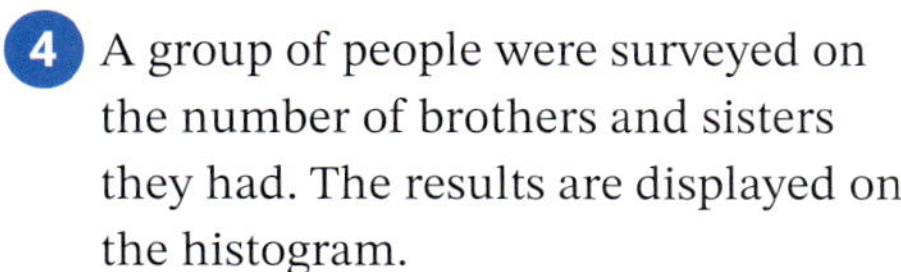

A group of people were surveyed on the number of brothers and sisters they had. The results are displayed on the histogram.

a What is the range of this data?

b What is the mode?

c Find the mean correct to 2 decimal places.

d What is the median?

e Which is the best measure of centre?

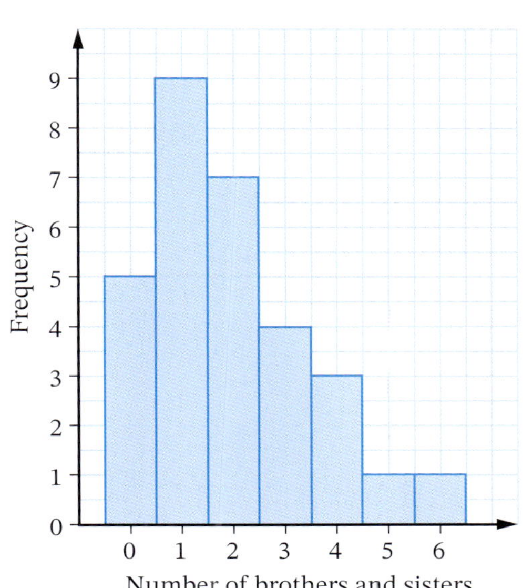

5 24 patients were chosen at random from 2 medical centres.
The waiting times (in minutes) for each patient are listed below.

Northbridge:	7	16	28	27	30	30	47	9	9	15	27	11
	38	35	42	53	23	22	45	32	37	37	18	16
Westvale:	44	13	7	9	25	26	37	35	8	18	18	15
	22	24	7	5	24	24	33	37	24	15	36	43

a Display the information in a back-to-back stem-and-leaf plot.

b Find the mean, median and range of both sets of data.

c Identify any outliers or clusters.

d Are there significant differences between the 2 sets of data? Explain your answer.

6 Classify each type of data as categorical or numerical. If numerical, then classify it as discrete or continuous.

a The temperature in Alice Springs.

b The number of views of a YouTube video.

c Suburb or town of home.

d The make of car passing through a motorway toll collection point.

e The probability of rain today.

7 The table below shows the results of a survey on how students travel to school

Travel method	School A	School B	School C
Walking	40	55	15
Bus	45	25	35

a What type of graph can be used to represent this data?

b Why is this an appropriate display to use?

c Construct the graph.

8 For each distribution, describe the shape and identify any outliers or clusters.

a

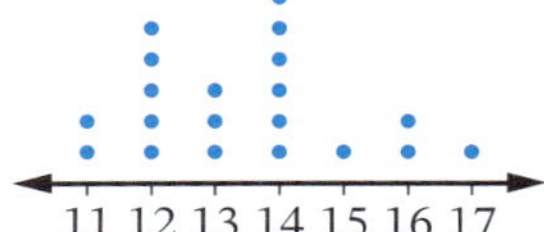

b

Stem	Leaf
8	7 8
9	1 3 4 4
10	0 2 7 8 8 9
11	1 2 2
12	0 7

c

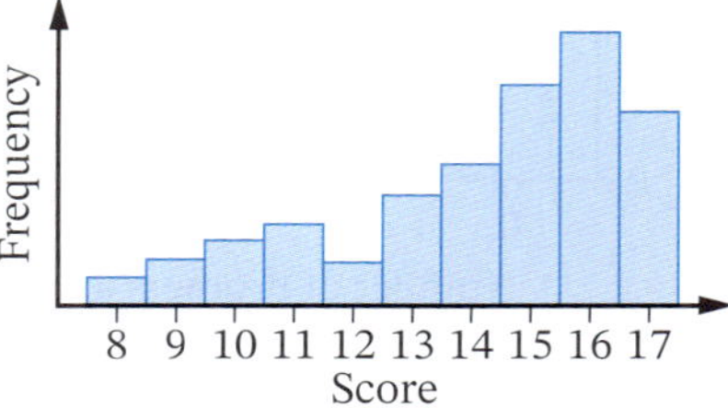

9 The heights (in centimetres) of students in a Year 9 class are:

Girls:	160	153	155	156	142	163	162
	148	150	150	131	138	147	145
Boys:	150	165	164	178	170	171	164
	148	154	176	134	180	176	175

a Arrange the data in a back-to-back stem-and-leaf plot.

b Describe the shape of the display for:

i the girls **ii** the boys.

c Are there any outliers or clusters for either group?

d Find the mean (correct to 2 decimal places) and median for both groups.

e Which measure is the better indicator of the measure of central tendency?

f Which group is shorter?

g Which group has the greater spread of heights?

10 Should a sample or a census be used to investigate each topic? 9.07

a The most popular car colour.

b Whether energy-saving light globes really use less electricity.

c The number of retired people living on the Gold Coast.

d People's views on which nation will win the soccer World Cup.

11 A sample of 45 employees is taken for a survey about work conditions at a city hospital. The hospital employs 73 doctors, 267 nurses, 184 administrative staff and 220 kitchen staff. For a stratified random sample, calculate how many employees from each department should be selected.

12 Explain why each sample is biased. 9.08

a A sample of online gamers were surveyed to investigate teenagers' views on road safety.

b Customers in a café can give feedback about their meal by filling in a survey and posting it to the café.

c At a factory, a sample of executives were surveyed about safety in the workplace.

d A sample of audience members at a movie premiere were surveyed about their views on the film.

e Classic Gold FM radio listeners were asked to vote for the best song of the year.

13 Explain what is wrong with each survey question and suggest an alternative.

a 'Are you happy?'

b 'How did you enjoy your meal at our restaurant?'

c 'Australian soldiers have fought under our country's flag. Do you think that it should be changed?'

☐ Foundation ○ Standard ⬡ Complex

10

MEASUREMENT

Surface area and volume

The Sydney Opera House was designed by Danish architect Jørn Utzon, with construction taking place between 1959 and 1973 at a cost of $102 million. Its shell design represents a ship under full sail, with each concrete shell being segments of a sphere of radius 75.2 metres. The shells are covered by 1 056 006 white and cream 120-mm square tiles. The Sydney Opera House stands on 1.3 hectares of land, is 183 metres long and 120 metres wide at its widest point.

Shutterstock.com/Ingus Kruklitis

Chapter outline

	Proficiencies				
10.01 Metric units	U	F	PS	R	R
10.02 Absolute error and percentage error	U	F	PS	R	C
10.03 Perimeters and areas of composite shapes	U	F	PS	R	
10.04 Areas of quadrilaterals	U	F	PS	R	
10.05 Circumferences and areas of circular shapes	U	F	PS	R	
10.06 Surface area of a prism	U	F	PS	R	
10.07 Surface area of a cylinder	U	F	PS	R	
10.08 Volumes of prisms and cylinders	U	F	PS	R	

U = Understanding
F = Fluency
PS = Problem solving
R = Reasoning
C = Communication

Wordbank

absolute error The difference between the actual value and the measured value of a quantity

capacity The amount of fluid (liquid or gas) in a container

cross-section A 'slice' of a solid, taken across the solid rather than along it

limits of accuracy How accurate a measured value using a measuring instrument can be, equal to half of a unit on the instrument's scale

prism A solid shape with identical cross-sections and ends with straight sides

sector A region of a circle cut off by 2 radii, shaped like a slice of pizza

surface area The total area of all the faces of a solid shape

volume The amount of space occupied by an object

Quiz
Wordbank 10

Videos (5):

10.05 Perimeter and area of a sector

10.06 Surface area of a prism • Surface area of prisms

10.07 Surface area of a cylinder

10.08 Volumes of prisms and cylinders

Twig videos (3):

10.02 Decimal places: Photofinish

10.05 Beating the U-boats

10.06 Cubist art

Quizzes (6):

- Wordbank 10
- SkillCheck 10
- Mental skills 10A
- Mental skills 10B
- Language of maths 10
- Test yourself 10

Skillsheets (3):

10.03 What is area?

10.06 Solid shapes

10.08 What is volume?

Worksheets (13):

10.02 Accuracy in measurement

10.05 A page of circular shapes • Applications of area • Back-to-front problems • Area and perimeter investigations • Nets of solids

10.06 Surface area 1 • Nets of solids • Solid shapes

10.07 Surface area 2 • Applications of area 3

10.08 Volume and capacity

Mind map: Surface area and volume

Puzzles (3):

10.05 Carpet talk

10.07 Car song

10.08 Formula matching game

Technology (3):

10.06, 10.07 Surface area calculator

10.08 Measuring pyramids • Approximating the volume of a cone

Spreadsheet (1):

10.06, 10.07 Surface area calculator

Presentation (1):

10.06 Surface area

Nelson MindTap

To access resources above, visit **cengage.com.au/nelsonmindtap**

10 **In this chapter you will:**

- ✓ learn the metric prefixes for very small and very large units of measurement.
- ✓ convert between units of digital memory, such as megabytes and gigabytes.
- ✓ calculate the limits of accuracy of measuring instruments.
- ✓ calculate the absolute error and percentage error of a measurement.
- ✓ calculate the perimeters and areas of quadrilaterals, circles, sectors and composite shapes.
- ✓ calculate the surface areas of rectangular and triangular prisms.
- ✓ calculate the surface areas of prisms and cylinders.
- ✓ calculate the volumes and capacities of prisms and cylinders

SkillCheck

ANSWERS ON P. 657

Quiz SkillCheck 10

1 The measurements shown on the shapes below are in centimetres. For each shape, find (correct to one decimal place if necessary):

i its perimeter **ii** its area

a 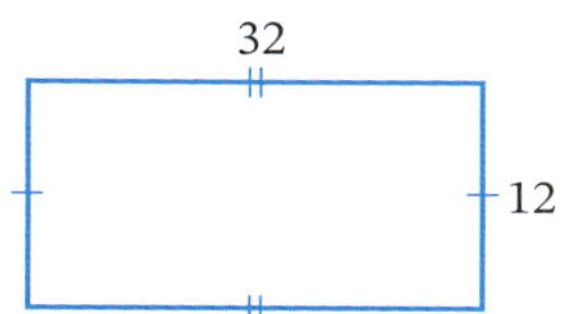

b 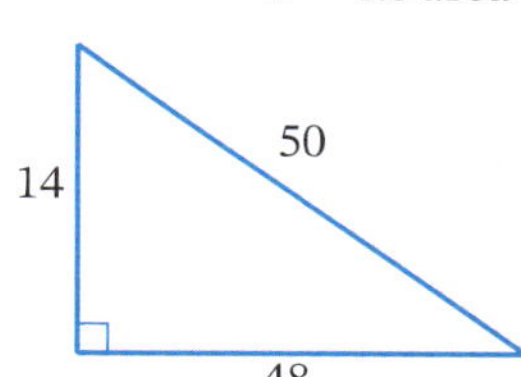

c 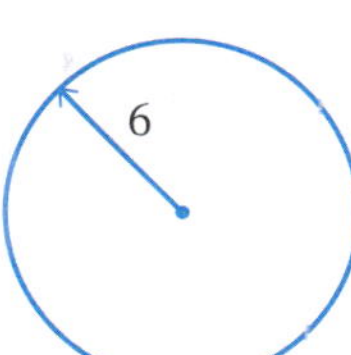

d 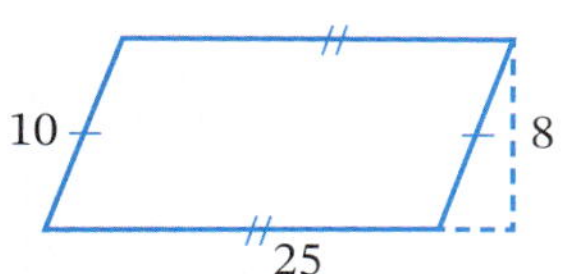

e 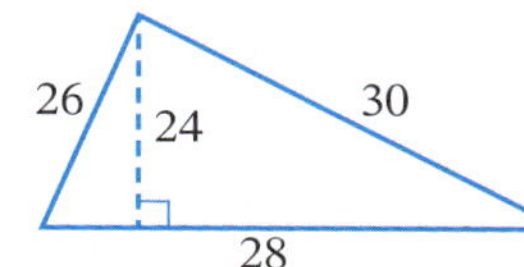

f 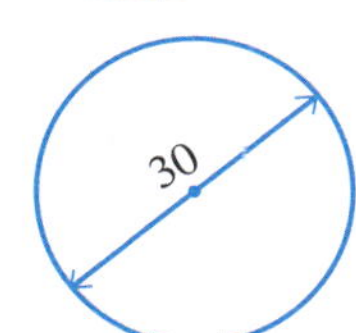

2 Convert:

a 34 m to mm **b** 925 cm to m **c** 1500 m to km

d 3750 mL to L **e** 38 cm to mm **f** 7.5 L to mL

3 Convert each fraction to a percentage, correct to one decimal place where required.

a 60 out of 120 **b** 34 out of 220 **c** 1200 out of 14 500

d 93 out of 4100 **e** 7 out of 18 **f** 375 out of 2500

4 What fraction of each circle is shaded?

a 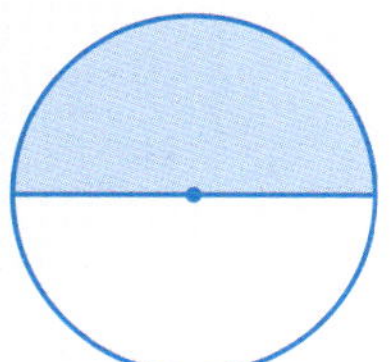

b 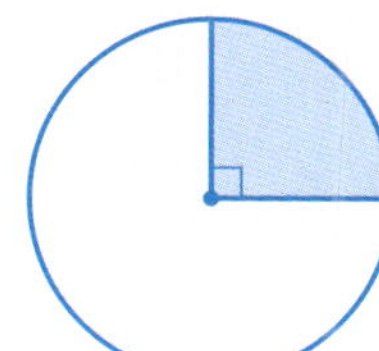

c 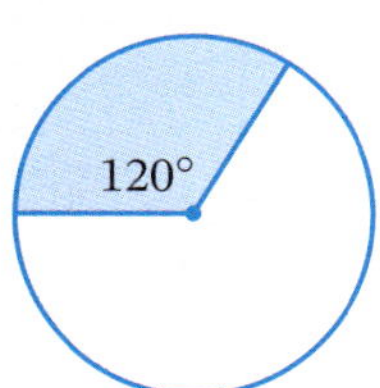

d 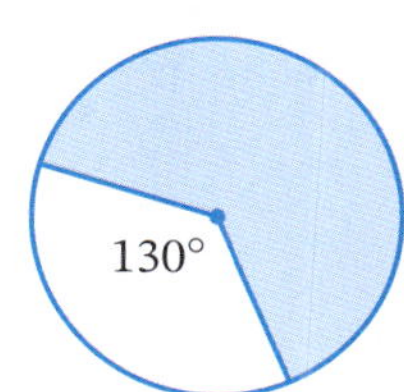

5 Calculate the volume of each prism (all units are in cm).

a 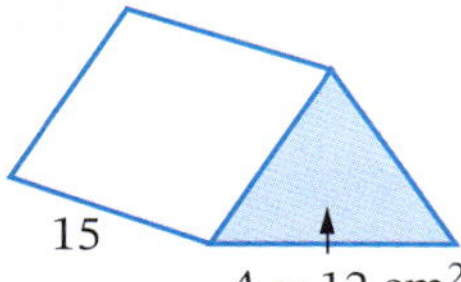

b 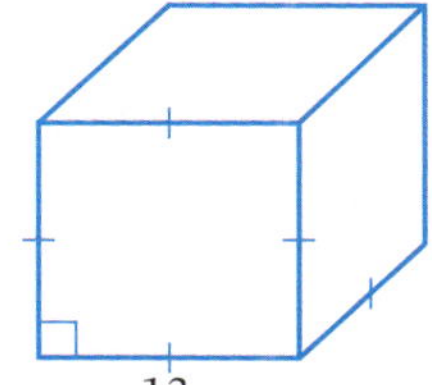

c

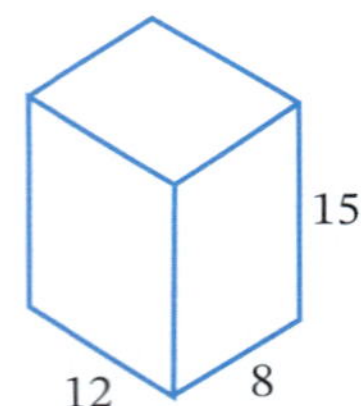

d

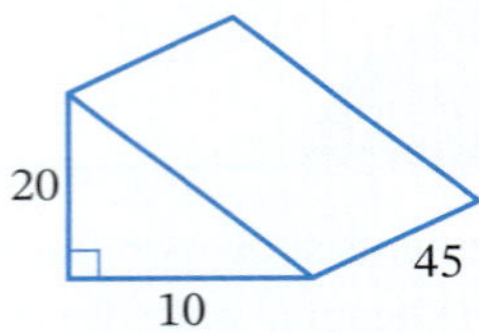

6 For each prism:

  i count the number of faces.

  ii name the shapes of the different faces.

  iii name the prism.

a

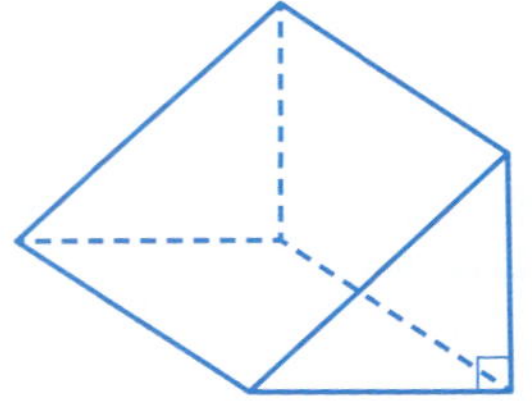

b

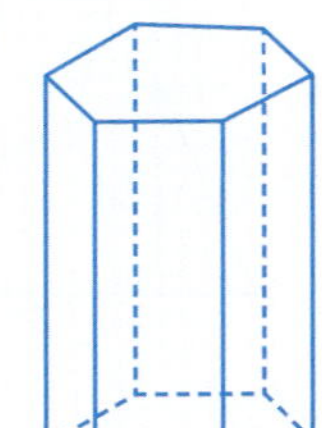

c

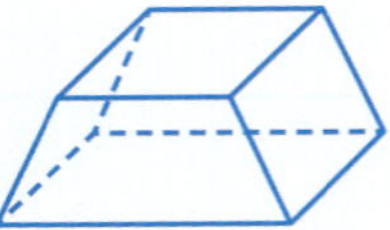

7 Find the value of each variable.

a

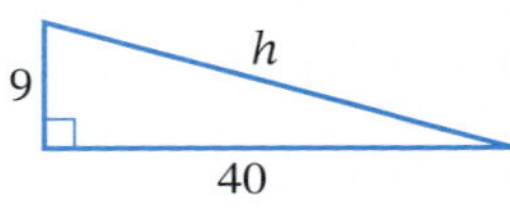

b

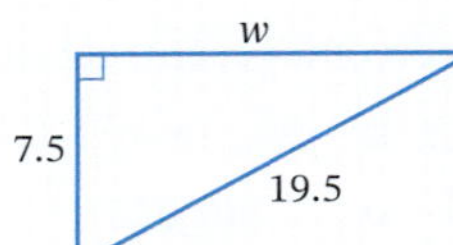

c

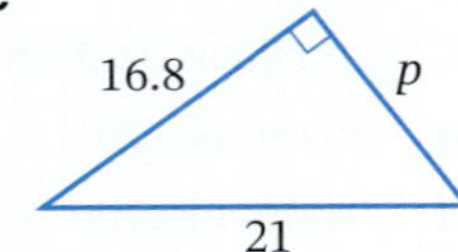

10.01 Metric units

Length

1 cm = 10 mm

1 m = 100 cm = 1000 mm

1 km = 1000 m

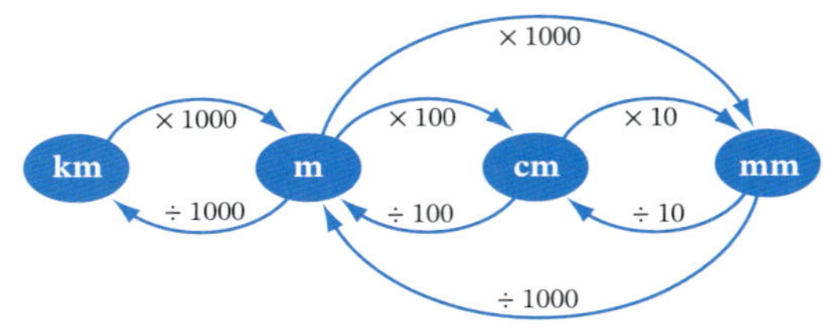

Capacity

1 L = 1000 mL

1 kL = 1000 L

1 ML = 1000 kL = 1000 000 L

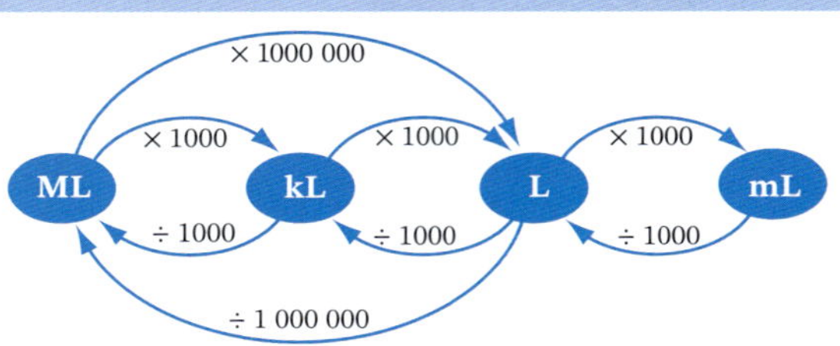

Mass

1 g = 1000 mg

1 kg = 1000 g

1 t = 1000 kg

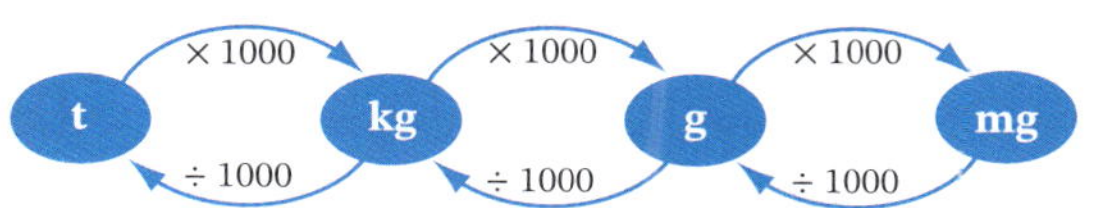

Time

1 min = 60 s

1 h = 60 min = 3600 s

1 day = 24 h

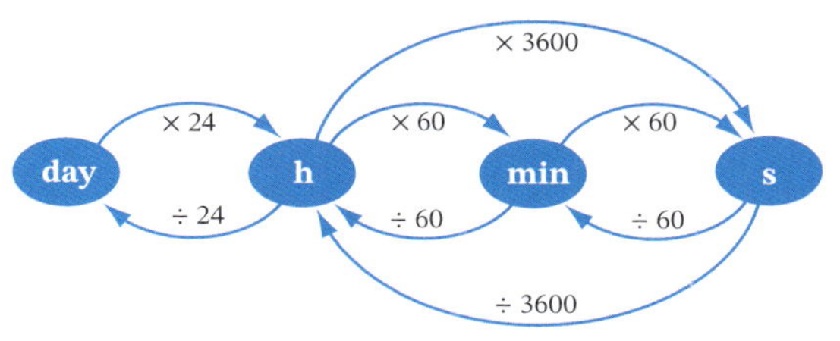

Other than for time, units of the metric system have prefixes based on powers of 10.

- **kilo-** means 1000 (thousand) or 10^3 — 1 kilogram = 1000 grams
- **mega-** means 1 000 000 (million) or 10^6 — 1 megalitre = 1 000 000 litres
- **centi-** means $\frac{1}{100}$ (hundredth) or 10^{-2} — 1 centimetre = $\frac{1}{100}$ metre
- **milli-** means $\frac{1}{1000}$ (thousandth) or 10^{-3} — 1 millilitre = $\frac{1}{1000}$ litre

Example 1

Convert:

a 5.8 t to kg **b** 16 km to cm **c** 23 700 mL to L **d** 1500 min to h

SOLUTION

a 5.8 t = 5.8 × 1000 kg
= 5800 kg

To convert from large to small units, multiply (×).

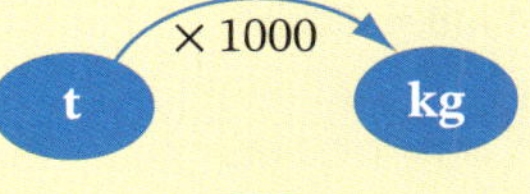

b 16 km = 16 × 1000 × 100 cm
= 1 600 000 cm

c 23 700 mL = 23 700 ÷ 1000 L
= 23.7 L

To convert from small to large units, divide (÷).

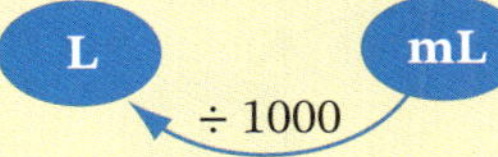

d 1500 min = 1500 ÷ 60 h
= 25 h

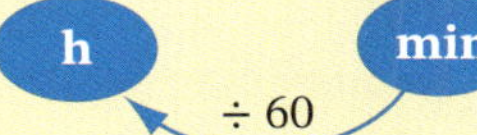

Metric prefixes

This table shows a more detailed list of metric prefixes.

Prefix (abbreviation)	Meaning	Example
pico (p)	10^{-12}	
nano (n)	$10^{-9} = \frac{1}{1000000000}$	nanosecond (ns) = billionth of a second
micro (μ)	$10^{-6} = \frac{1}{1000000}$	microgram (μg) = millionth of a gram
milli (m)	$10^{-3} = \frac{1}{1000}$	millibar (mbar) = thousandth of a bar (air pressure)
centi (c)	$10^{-2} = \frac{1}{100}$	centimetre (cm) = hundredth of a metre
deci (d)	$10^{-1} = \frac{1}{10}$	decibel (dB) = tenth of a bel (sound level)
deca (da)	$10^{1} = 10$	decametre (dam) = ten metres
hecto (h)	$10^{2} = 100$	hectopascal (hPa) = hundred pascals (air pressure)
kilo (k)	$10^{3} = 1000$	kilojoule (kJ) = thousand joules (energy)
mega (M)	$10^{6} = 1\ 000\ 000$	megahertz (MHz) = million hertz (frequency)
giga (G)	10^{9}	gigalitre (GL) = billion litres (water)
tera (T)	10^{12}	terawatt (TW) = trillion watts (power)

Digital memory

Metric prefixes are used in units of digital (computer) memory. A **byte**, abbreviated B, is the amount of memory used to store one character, such as @, g, #, 7 or ?.

Unit	Relationship	Size
kilobyte (kB)	1 kB = 1000 B	About half a page of text.
megabyte (MB)	1 MB = 1000 kB	A million bytes, about the size of a novel.
gigabyte (GB)	1 GB = 1000 MB	A billion bytes, about 500 photos or 7.5 hours of video.
terabyte (TB)	1 TB = 1000 GB	A trillion bytes, about the size of all of the books in a large library.

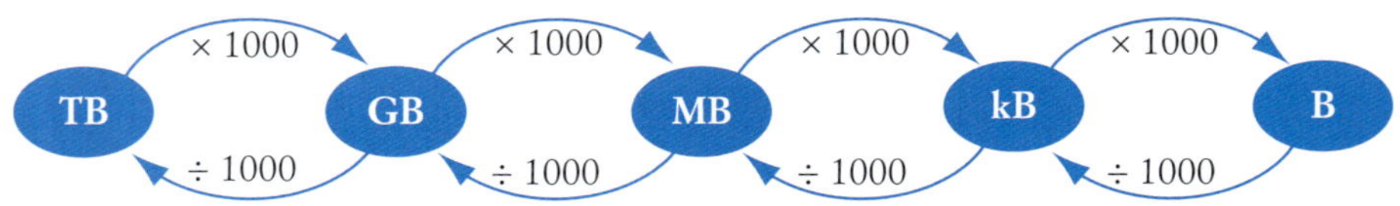

Some examples of digital memory sizes:	
E-mail (without attachment): 75 kB	Photo: 2 MB
Music file: 4 MB	USB drive: 8 to 128 GB
Streaming a video: 790 MB per hour	External hard drive: 1 to 5 TB
Blu-ray disc: 25 GB	Online game: 40 to 300 MB per hour

Example 2

Convert:

a 3.2 MB to kB **b** 147 000 MB to GB **c** 2 TB to GB

SOLUTION

a 3.2 MB = 3.2 × 1000 kB
= 3200 kB

To convert from large to small units, multiply (×).

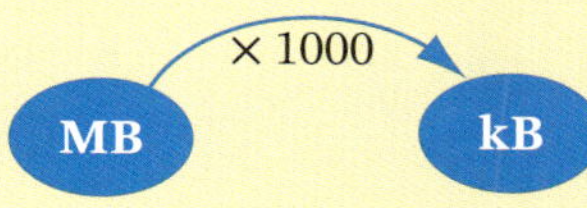

b 147 000 MB = 147 000 ÷ 1000 GB
= 147 GB

To convert from small to large units, divide (÷).

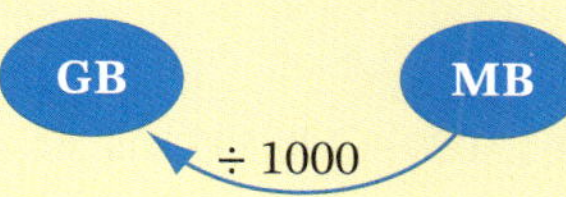

c 2.6 TB = 2.6 × 1000 GB
= 2600 GB

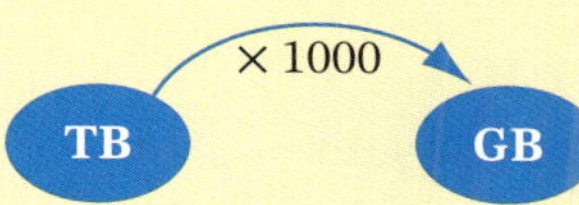

Very small and very large units

Some very small and very large units of length, distance and time also use metric prefixes.

Length and distance

- One **nanometre** (nm) = $\frac{1}{1\,000\,000\,000}$ m: your fingernails grown 1 nm every second, the diameter of a helium atom is 0.1 nm.
- One **micrometre** (μm) = $\frac{1}{1\,000\,000}$ m: used to measure diameters of bacteria.
- One **astronomical unit** (AU) = 149 597 871 km: the average distance between the Earth and the Sun, used to measure distances between planets in our solar system.
- One **light year** (ly) = 9.46×10^{12} km: the distance light travels in a year, for measuring distances between stars.
- One **parsec** (pc) = 3.0857×10^{13} km or 3.26 ly: for measuring distances between stars, the nearest star *Proxima Centauri* is 1.30 parsecs away.

Time

- One **nanosecond** (ns) = $\frac{1}{1\,000\,000\,000}$ s: the time it takes electricity to travel along a 30-cm wire.
- One **microsecond** (μs) = $\frac{1}{1\,000\,000}$ s.
- One **millisecond** (ms) = $\frac{1}{1\,000}$ s: the duration of the flash on a camera.
- One **decade** = 10 years.
- One **century** = 100 years.
- One **millennium** = 1000 years.

- One **mega-annum** (Ma) = 1 000 000 years: used in geology and paleontology, the *Tyrannosaurus Rex* dinosaur lived 65 Ma ago.
- One **giga-annum** (Ga) = 1 000 000 000 years: the Earth formed 4.57 Ga ago.

Example 3

Convert:

a 0.73 s to μs

b 25 610 000 km to AU

c 4.57 Ma to years

d 1 920 000 nm to m

SOLUTION

a $0.73\ \text{s} = 0.73 \times 1\ 000\ 000\ \mu\text{s}$
$= 730\ 000\ \mu\text{s}$

To convert from large to small units, multiply (×).

b $25\ 610\ 000\ \text{km} = 25\ 610\ 000 \div 149\ 597\ 871\ \text{AU}$
$= 0.17119...$
$\approx 0.17\ \text{AU}$

To convert from small to large units, divide (÷).

c $4.57\ \text{Ma} = 4.57 \times 1\ 000\ 000\ \text{years}$
$= 4\ 570\ 000\ \text{years}$

d $1\ 920\ 000\ \text{nm} = 1\ 920\ 000 \div 1\ 000\ 000\ 000\ \text{m}$
$= 1.92 \times 10^{-3}\ \text{m}$
$= 0.001\ 92\ \text{m}$

EXERCISE 10.01 ANSWERS ON P. 657

Metric units

U F PS R C

1 What metric unit would you use for each measurement?

R **a** Width of a computer screen

b Mass of a plane

C **c** Distance between 2 towns

d Capacity of a car's fuel tank

e Amount of water in a pool

f Time taken to fly overseas

EXAMPLE 1

2 Convert:

R

a 750 g to mg	**b** 36 kL to L	**c** 67 kg to g
d 1.5 L to mL	**e** 6.7 cm to mm	**f** 3.5 km to m
g 9.4 t to kg	**h** 3.75 g to mg	**i** 4.25 m to cm
j 0.2 kg to g	**k** 26.2 L to mL	**l** 0.3 km to m

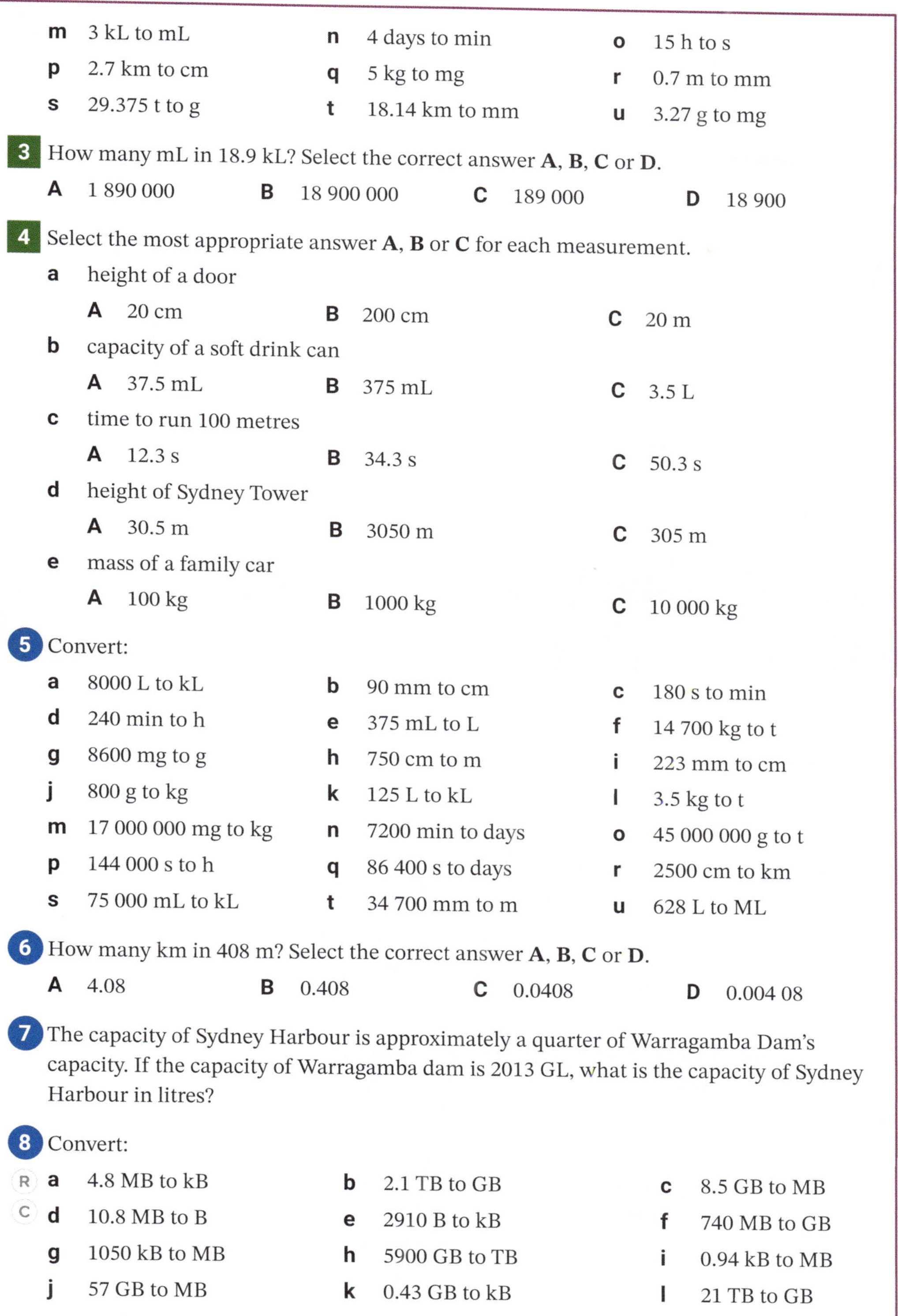

m 3 kL to mL **n** 4 days to min **o** 15 h to s
p 2.7 km to cm **q** 5 kg to mg **r** 0.7 m to mm
s 29.375 t to g **t** 18.14 km to mm **u** 3.27 g to mg

3 How many mL in 18.9 kL? Select the correct answer **A**, **B**, **C** or **D**.

A 1 890 000 **B** 18 900 000 **C** 189 000 **D** 18 900

4 Select the most appropriate answer **A**, **B** or **C** for each measurement.

a height of a door
A 20 cm **B** 200 cm **C** 20 m

b capacity of a soft drink can
A 37.5 mL **B** 375 mL **C** 3.5 L

c time to run 100 metres
A 12.3 s **B** 34.3 s **C** 50.3 s

d height of Sydney Tower
A 30.5 m **B** 3050 m **C** 305 m

e mass of a family car
A 100 kg **B** 1000 kg **C** 10 000 kg

5 Convert:

a 8000 L to kL **b** 90 mm to cm **c** 180 s to min
d 240 min to h **e** 375 mL to L **f** 14 700 kg to t
g 8600 mg to g **h** 750 cm to m **i** 223 mm to cm
j 800 g to kg **k** 125 L to kL **l** 3.5 kg to t
m 17 000 000 mg to kg **n** 7200 min to days **o** 45 000 000 g to t
p 144 000 s to h **q** 86 400 s to days **r** 2500 cm to km
s 75 000 mL to kL **t** 34 700 mm to m **u** 628 L to ML

6 How many km in 408 m? Select the correct answer **A**, **B**, **C** or **D**.

A 4.08 **B** 0.408 **C** 0.0408 **D** 0.004 08

7 The capacity of Sydney Harbour is approximately a quarter of Warragamba Dam's capacity. If the capacity of Warragamba dam is 2013 GL, what is the capacity of Sydney Harbour in litres?

8 Convert:

EXAMPLE 2

R **a** 4.8 MB to kB **b** 2.1 TB to GB **c** 8.5 GB to MB
C **d** 10.8 MB to B **e** 2910 B to kB **f** 740 MB to GB
g 1050 kB to MB **h** 5900 GB to TB **i** 0.94 kB to MB
j 57 GB to MB **k** 0.43 GB to kB **l** 21 TB to GB
m 7.8 kB to B **n** 2180 MB to GB **o** 42 000 B to MB
p 940 MB to kB **q** 0.64 TB to kB **r** 278 MB to GB

Foundation Standard Complex

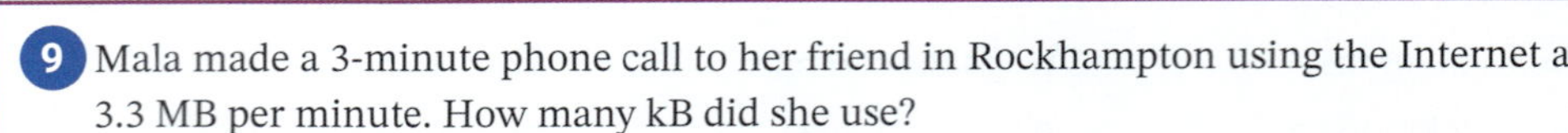

9 Mala made a 3-minute phone call to her friend in Rockhampton using the Internet at 3.3 MB per minute. How many kB did she use?

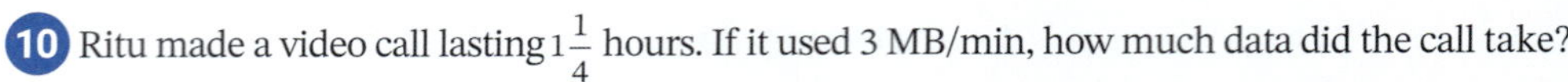

10 Ritu made a video call lasting $1\frac{1}{4}$ hours. If it used 3 MB/min, how much data did the call take?

11 Calculate the number of MP3 files of average size 5 MB that can be stored on a 64 GB MP3 player.

PS C

EXAMPLE 3

12 Convert, correct to 2 significant figures:

R C

a	3.5 AU to km	**b**	6000 ms to s	**c**	7.5 centuries to years
d	12 000 000 nm to m	**e**	0.64 Gm to m	**f**	2.1 ly to km
g	0.07 s to ms	**h**	0.25 Ma to years	**i**	4 800 000 μs to s
j	5 290 000 km to Mm	**k**	0.2 pc to km	**l**	6 mm to μm
m	0.94 Ma to millennia	**n**	849 years to centuries	**o**	8×10^{14} km to ly
p	0.000 008 cm to nm	**q**	20 ly to pc	**r**	156 millennia to Ma

10.02 Absolute error and percentage error

Video
Decimal places: Photofinish

All measurements are only approximations. No measurement is ever **exact**. This is because of two reasons:

1 Human error in taking the measurement.
2 The limitations of the measuring instrument.

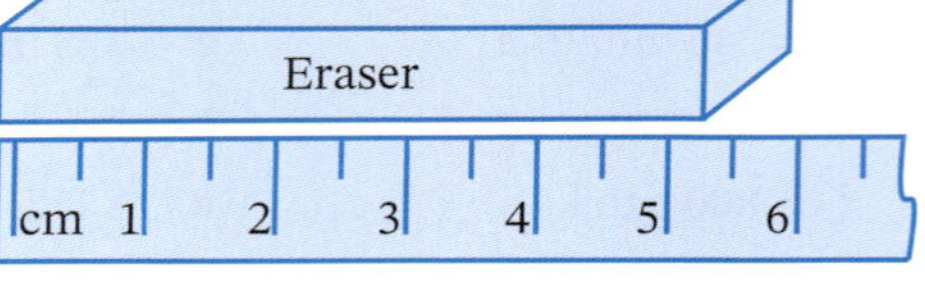

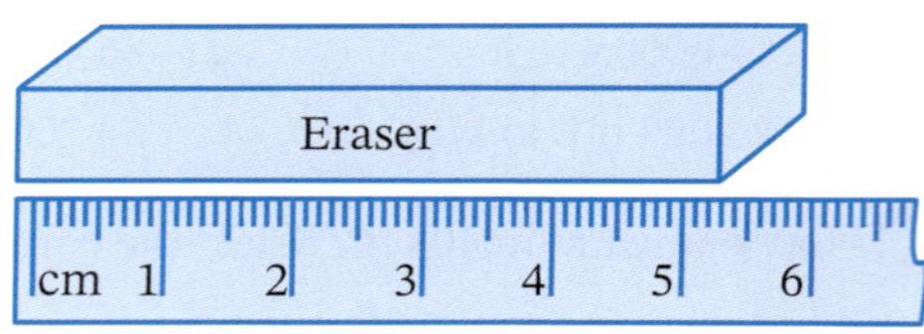

A measuring instrument has **limits of accuracy** due to its level of **precision**. 'Accuracy' means how close a measured value is to the true value, while 'precision' means how fine the scale is on the measuring instrument.

For example, we might measure this eraser to be **5.5 cm** long, because the scale on the ruler measures to the nearest 0.5 cm.

If we used a ruler with a more precise scale, such as 0.1 cm markings, then we may find the length to be **5.4 cm**.

If we used an instrument such as a micrometer that measured to the nearest 0.01 cm, we may find that the length is **5.41 cm**.

Notice that we can always find a more accurate measurement using a more precise measuring instrument. So all measurements are approximations.

☐ Foundation ○ Standard ⬡ Complex

Absolute error

The difference between a **measured value** (or approximate value) and the **actual value** (or exact value) is called the **absolute error**. It is usually expressed as a **positive value.**

ⓘ Absolute error

Absolute error = measured value – actual value (if measured value > actual value)

Absolute error = actual value – measured value (if measured value < actual value)

Example 4

Find the absolute error for each measurement.

a Actual value = 327, measured value = 330

b

Actual mass = 72.6 g

SOLUTION

a The measured value (330) is higher than the actual value (327).

Absolute error = 330 – 327

= 3

b The measured mass from the scale is 70 g.

It is smaller than the actual mass (72.6 g).

Absolute error = 72.6 g – 70 g

= 2.6 g

Relative error and percentage error

An absolute error of 1 metre may not be much when measuring the length of a bridge, but it is huge when measuring the length of a bathroom. To understand how big the errors are when something is measured and be able to compare errors in measurement, we can calculate **relative error** and **percentage error**.

Relative error is the absolute error as a fraction of the actual value.

Percentage error is the absolute error as a percentage of the actual value, or the relative error expressed as a percentage.

ⓘ Relative error and percentage error

$$\text{Relative error} = \frac{\text{absolute error}}{\text{actual value}}$$

$$\text{Percentage error} = \frac{\text{absolute error}}{\text{actual value}} \times 100\%$$

The smaller the relative or percentage error, the more accurate the measurement, the closer it is to the actual value.

Example 5

A cable that is measured to be 5 m long is actually 5.08 m long. Calculate the relative error correct to 3 decimal places.

SOLUTION

Measured value = 5 m, Actual value = 5.08 m

$$\begin{aligned}\text{Absolute error} &= 5.08 - 5\\ &= 0.08\\ \text{Relative error} &= \frac{\text{absolute error}}{\text{actual value}}\\ &= \frac{0.08}{5.08}\\ &= 0.0157\ldots\\ &\approx 0.016\end{aligned}$$

Example 6

Fan Xuan estimates a magazine article to contain 620 words. If there are actually 540 words in the article, calculate the percentage error of her estimation correct to one decimal place.

SOLUTION

Measured value = 620 words, Actual value = 543 words

$$\begin{aligned}\text{Absolute error} &= 620 - 543\\ &= 77\\ \text{Percentage error} &= \frac{\text{absolute error}}{\text{actual value}} \times 100\%\\ &= \frac{77}{543} \times 100\%\\ &= 14.1804\ldots\%\\ &\approx 14.2\%\end{aligned}$$

Absolute error of a measuring instrument

If the actual measurement value is not known, then we can examine the precision of the measuring instrument.

For example, this ruler is marked in centimetres, so any length measured with it can only be given to the nearest centimetre.

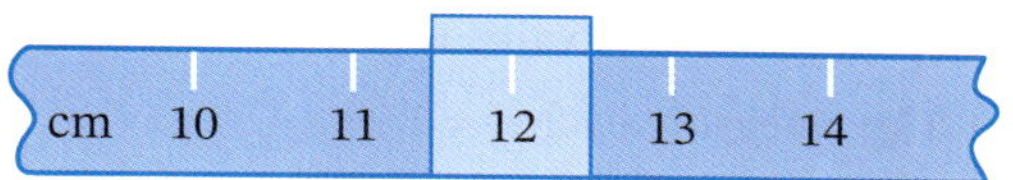

Worksheet
Accuracy in measurement

10.02

Any measurement in the shaded region should be recorded as 12 cm. The measured length is 12 cm, but the actual length is 12 ± 0.5 cm, meaning '12 centimetres, give or take 0.5 centimetres' or '11.5 to 12.5 cm'.

The **limits of accuracy** of this measurement is **11.5 to 12.5 cm.**

The **absolute error** of this ruler is **0.5 cm.**

Absolute error of measuring instruments

The **absolute error** of a measuring instrument is **0.5** of the unit shown on the instrument's scale.

Example 7

Find the absolute error for each measuring scale.

a

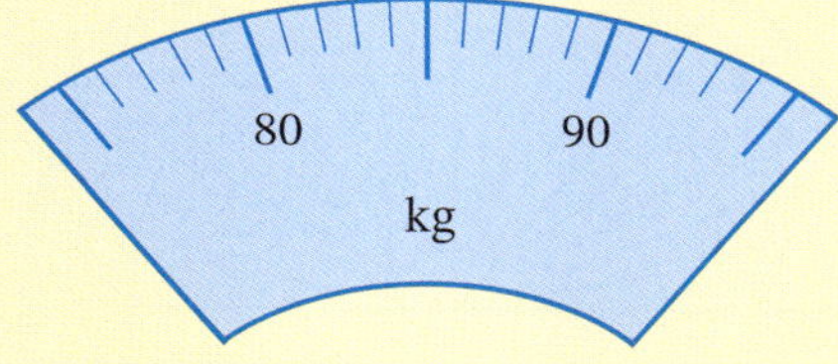

b

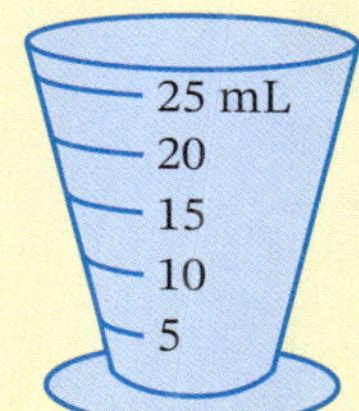

SOLUTION

a The size of one unit on the scale is 1 kg.

The absolute error is 0.5×1 kg = 0.5 kg.

b The size of one unit on the scale is 5 mL.

The absolute error is 0.5×5 mL = 2.5 mL.

Absolute error and percentage error

1 Calculate the absolute error for each pair of values.

- **a** Actual value = 120 cm ; measured value = 115 cm
- **b** Actual value = 1.8 kg ; measured value = 1.9 kg
- **c** Actual value = 382 mL ; measured value = 375 mL
- **d** Actual value = 146 minutes ; measured value = 155 minutes

2 Write the measured value on each scale shown.

a

30
20
10

b

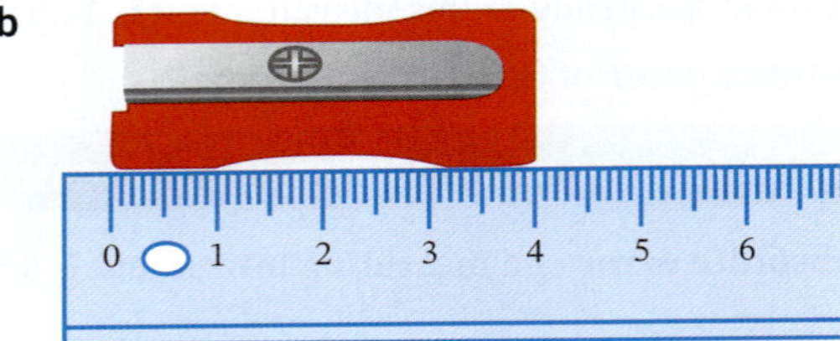

c

d

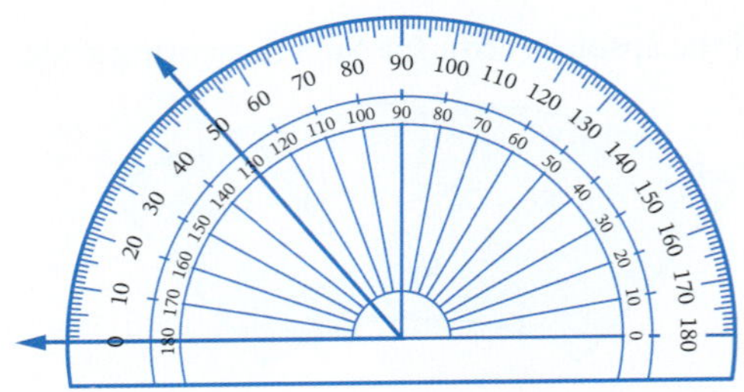

e

f

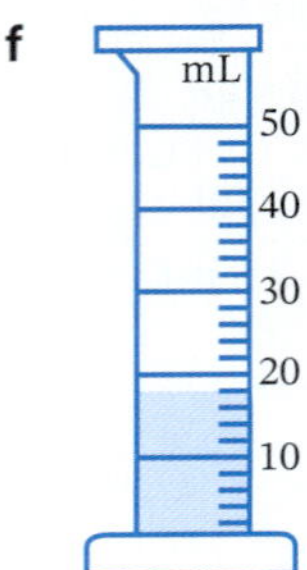

g

0 mm 1 2 3 4 5 6

h

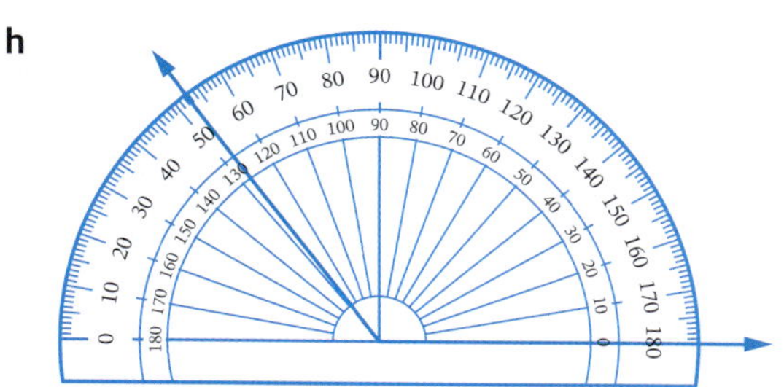

3 Use your measured values from question **2** above and the actual values listed below to calculate the absolute error for each measurement.

a 23.6 mL

b 3.92 cm

c 4.09 kg

d 49.5°

e 592 g

f 17.2 mL

g 2.98 cm

h 127.5°

4 Use the actual values and absolute errors from question **3** above to calculate the relative error for each measurement correct to 3 decimal places.

5 By comparing the relative error answers in question **4**, determine which measurement was:

a the most accurate

b the least accurate.

6 A cricket fan estimates there to be approximately 50 000 seats at the Adelaide Oval. If the actual number of seats is 53 500, calculate the percentage error of this estimation.

7 Miriam estimates that there are 233 coins in a piggy bank. If the actual number of coins in the piggy bank is 245, calculate the percentage error of Miriam's estimation.

8 R C The actual weight of a bag of lollies is 0.75 kg. Lisa measures the weight of the lollies to be 0.73 kg while Aryan measures the weight to be 785 g. Calculate the percentage error of each measurement correct to 2 decimal places to determine whose measurement is more accurate.

9 R C At an auction, Melissa valued the price of a vase to be \$3200 and Sandeep valued the price to be \$3400. The vase sold for \$3312. Calculate the percentage error of each valuation correct to 2 decimal places to determine whose valuation was more accurate.

Foundation Standard Complex

EXAMPLE 6

10 For each measuring instrument, state:

i the size of one unit on the scale **ii** the absolute error.

a

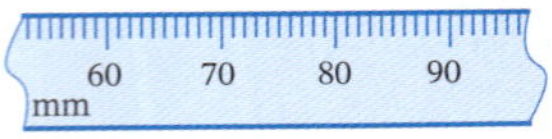

b

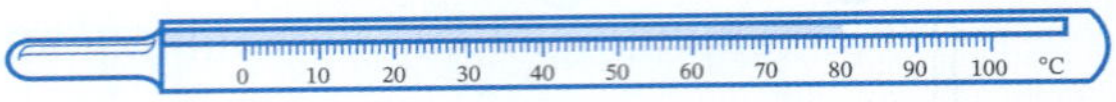

c

d

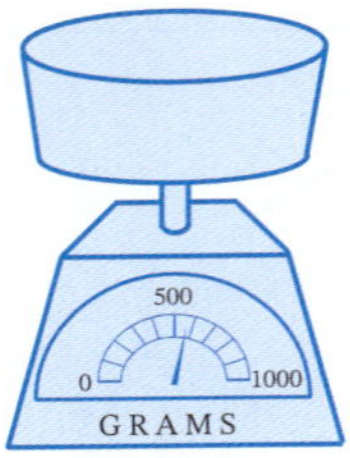

e

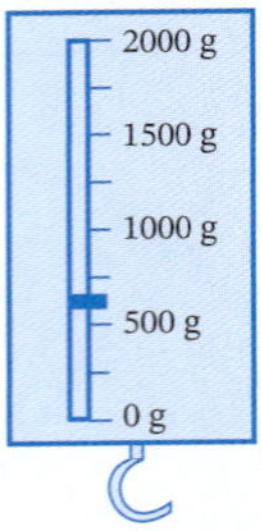

f

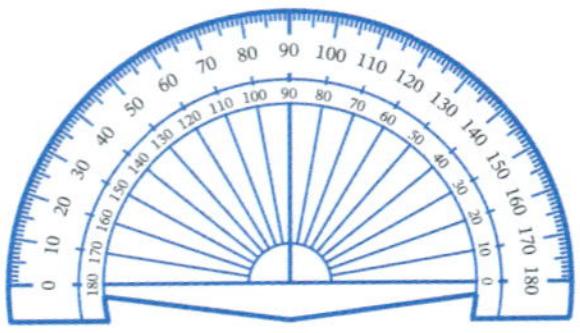

g

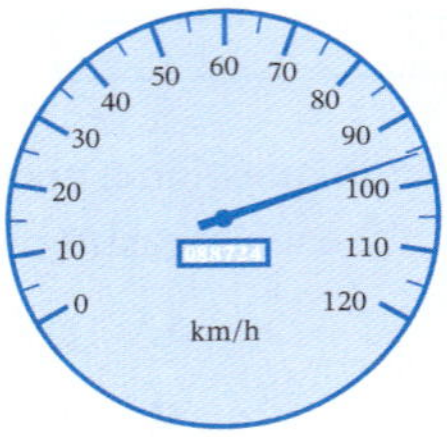

h

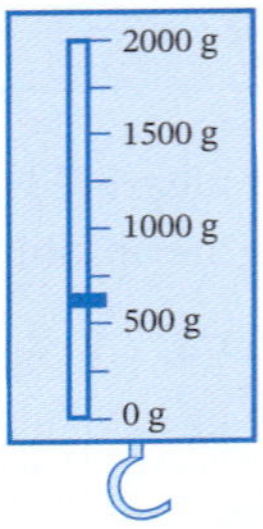

i

j

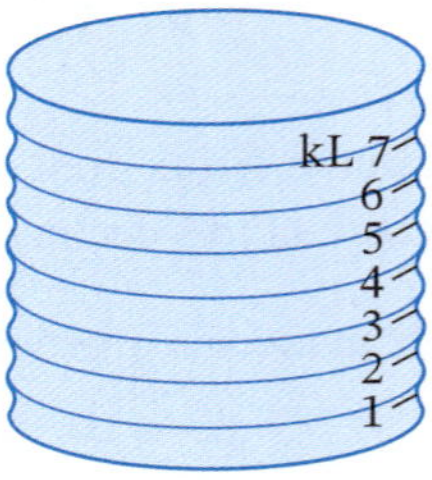

k

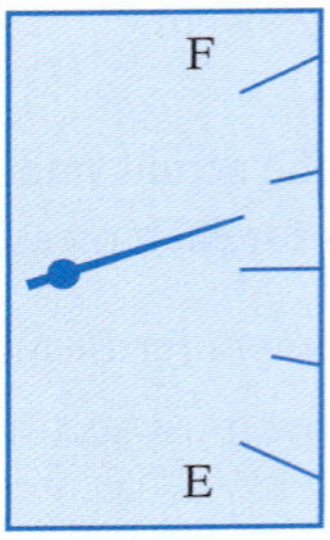

11 What is the absolute error for this measuring cylinder?
Select the correct answer **A**, **B**, **C** or **D**.

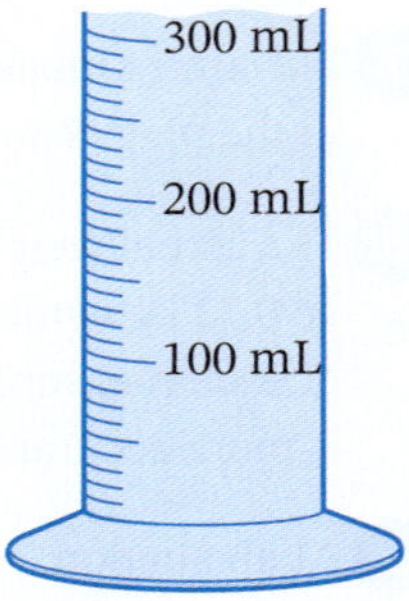

A 0.5 mL

B 10 mL

C 20 mL

D 5 mL

12 Measuring with a ruler, Gemma correctly gives a measurement as 26.5 to 27.5 cm.

a What is the size of a unit on the ruler?

b What is the absolute error of this measurement?

c Using a ruler marked in millimetres, Emily gives the same measurement as 26.8 cm. What is the absolute error of this measurement?

d What are the limits of accuracy between which this measurement must lie?

Foundation Standard Complex

If the actual value of a measurement is not known, we can calculate the percentage error as a percentage of the *measured value*:

$$\text{Percentage error} = \frac{\text{absolute error}}{\text{measured value}} \times 100\%.$$

Use this formula for questions **13** and **14**.

13 The weight of a box of laundry detergent is 500 g, correct to the nearest 10 g. Calculate the percentage error for a box of laundry detergent.

14 The height of a tree is approximately 16 m, correct to the nearest 0.2 m. Calculate the percentage error for this height.

15 R C An expert is hired to estimate the price of an antique oil painting. If the absolute error in the expert's estimate is $3785 and the percentage error in the estimate is 19.46%, can the actual value of the painting be greater than $20 000? Explain your answer.

16 Bassam measures the weight of 1500 identical paperclips to be 900 g. The percentage error of this measurement is 6.25%. If the actual weight of one paperclip is x g, find the possible range of x.

DID YOU KNOW?

Is 1 MB = 1000 or 1024 kB?

Computers represent all information as a series of 0s and 1s. These binary digits, called bits (b), give their name to the smallest unit of computer memory. Computer memory is measured in bytes, where 1 byte is equal to 8 bits. A byte (B) is the amount of memory needed to store one character such as #, d, D, 8 or $. For example, the letter 'd' is represented by the byte 01100100, while 'D' is the byte 01000100.

On most storage devices such as USB drives, hard drives and flash drives, 1 MB = 1000 kB. Internet and phone companies use the same. Originally, however, computer programmers used 1 MB = 2^{10} kB = 1024 kB because that was actually how computer hardware stored data. This table shows the original conversions.

Unit	Meaning	Number of bytes
byte (B)	1 B = 8 b	1 B
kilobyte (kB)	1 kB = 1024 B	1024 B or 2^{10} B (approx. 1000)
megabyte (MB)	1 MB = 1024 kB	1024 × 1024 = 1 048 576 B or 2^{20} B (approx. one million)
gigabyte (GB)	1 GB = 1024 MB	1024^3 = 1 073 741 824 B or 2^{30} B (approx. one billion)
terabyte (TB)	1 TB = 1024 GB	1024^4 = 1 099 511 627 776 B or 2^{40} B (approx. one trillion)

Find the byte (code) for '8' and '$'.

9780170465618

10.03 Perimeters and areas of composite shapes

Skillsheet
What is area?

The **perimeter** of a shape is the distance around the shape, the sum of the lengths of the sides of the shape.

The **area** of a shape is the amount of surface covered by the shape, measured in **square units**.

ⓘ Areas of rectangles and triangles

Rectangle

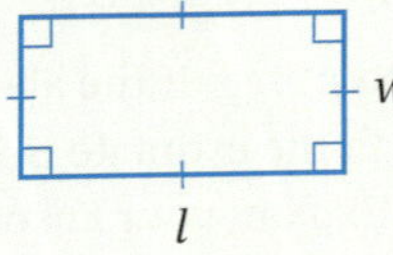

$A = \text{length} \times \text{width}$

$A = lw$

Triangle

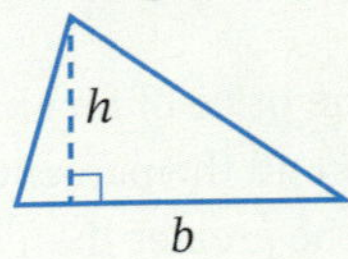

$A = \frac{1}{2} \times \text{base} \times \text{perpendicular height}$

$A = \frac{1}{2}bh$

Example 8

For each **composite shape**, find:

i its perimeter **ii** its area.

a

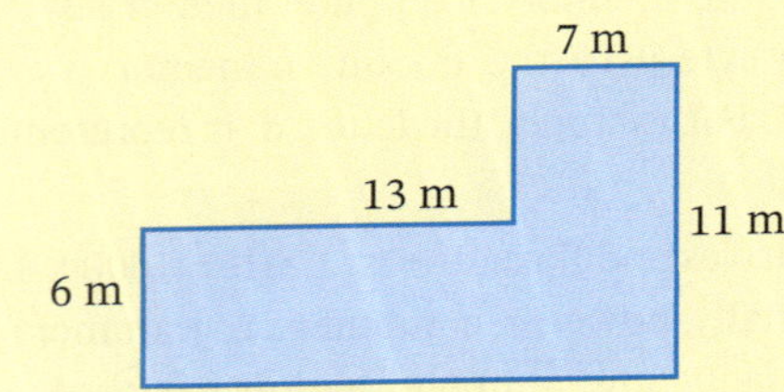

b

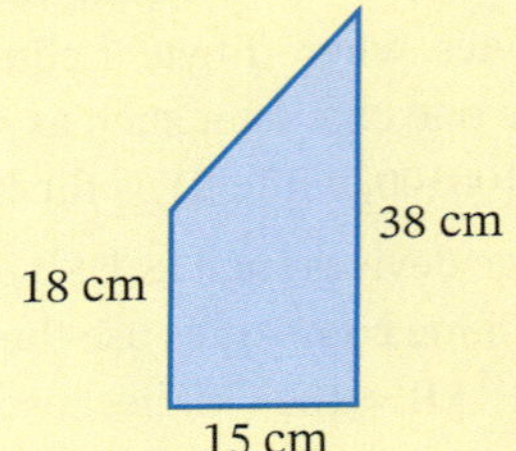

SOLUTION

a **i** Find any unknown lengths on the shape.

Perimeter $= 6 + 13 + 5 + 7 + 11 + 20$

$= 62$ m

ii Area $= (7 \times 5) + (20 \times 6)$

$= 155 \text{ m}^2$

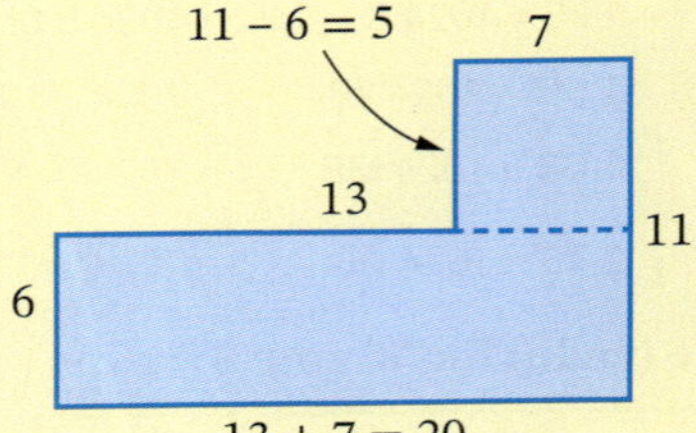

b **i** Divide the shape into a rectangle and a triangle, and find the unknown length, h, using Pythagoras' theorem.

$h^2 = 15^2 + 20^2$

$= 625$

$h = \sqrt{625}$

$= 25cm$

Perimeter = 18 + 25 + 38 + 15

= 96 cm

ii Area = $(\frac{1}{2} \times 15 \times 20) + (15 \times 18)$

Area of triangle + Area of rectangle

= 420 cm^2

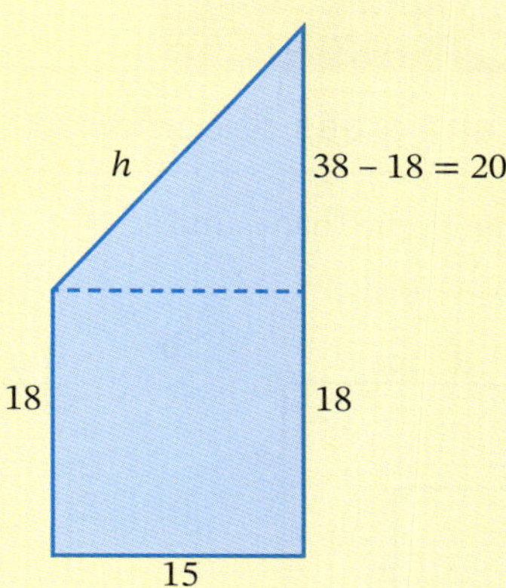

Example 9

This is the plan of a farm up for sale.

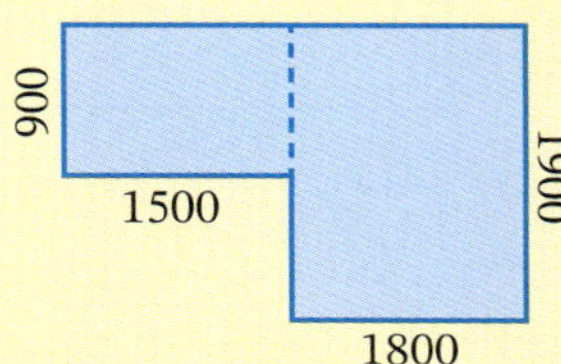

All measurements are in metres.

a Find the area of the farm.

b Express the area in square kilometres.

c Find the area of the property in **hectares**, where 1 ha = 10 000 m^2.

SOLUTION

a Area = 900 × 1500 + 1900 × 1800

= 4 770 000 m^2

b Since 1 km = 1000 m, 1 km^2 = 1000^2 m^2 = 1 000 000 m^2

∴ Area = 4 770 000 ÷ 1 000 000

= 4.77 km^2

c 1 ha = 10 000 m^2

∴ Area = 4 770 000 ÷ 10 000

= 477 ha

EXERCISE 10.03 ANSWERS ON P. 658

Perimeters and areas of composite shapes

U F PS R

EXAMPLE 8

1 For each composite shape, find:

PS R

i its perimeter **ii** its area.

a 33 cm, 15 cm, 28 cm, 15 cm

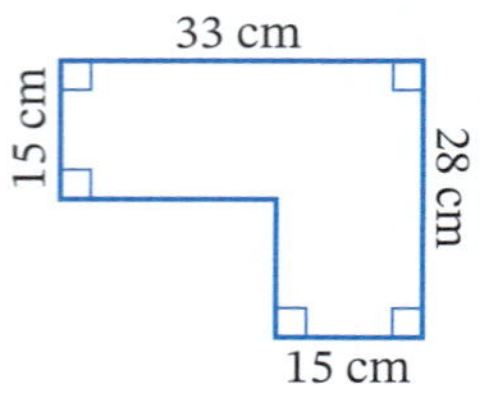

b 7 m, 16 m, 6 m, 11 m, 5 m, 22 m

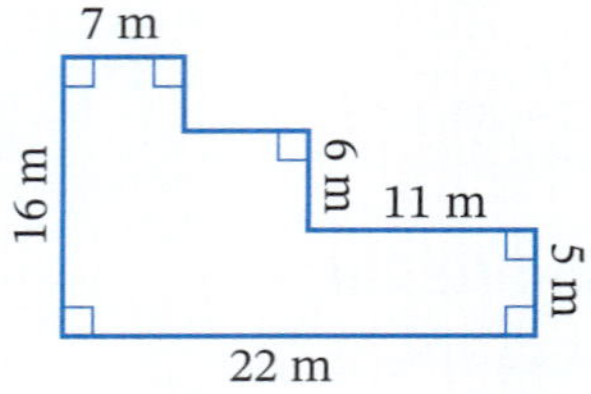

c 25 mm, 57 mm, 48 mm, 32 mm

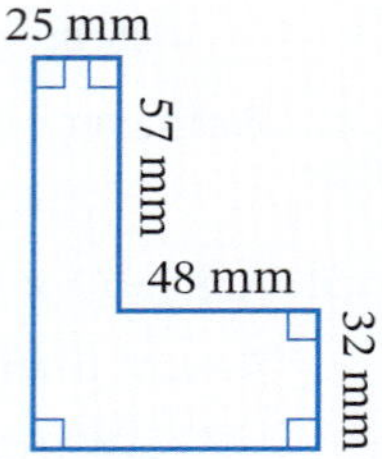

d 8 cm, 9 cm, 20 cm

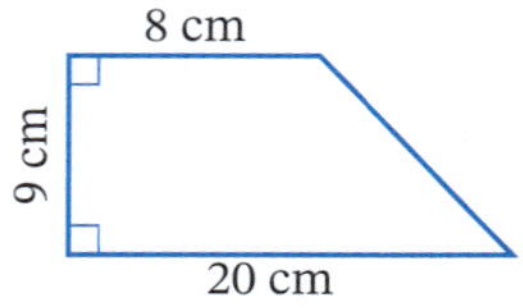

e 24 cm, 10 cm, 30 cm

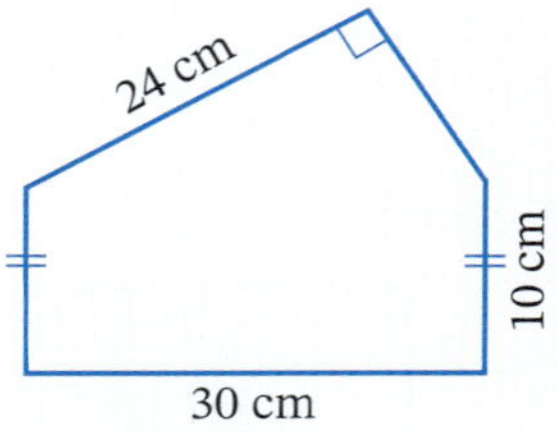

f 52 cm, 20 cm, 40 cm

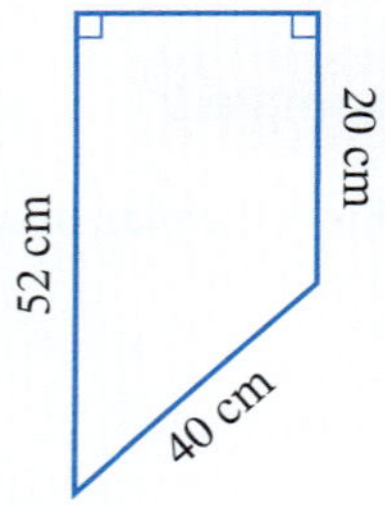

2 The length of a rectangular block of land is 112 m and its perimeter is 334 m.

R How wide is the block? Select the correct answer **A**, **B**, **C** or **D**.

A 222 m **B** 110 m **C** 55 m **D** 224 m

3 Find the (shaded) area of each composite shape.

PS R

a 9 cm, 15 cm, 8 cm, 24 cm

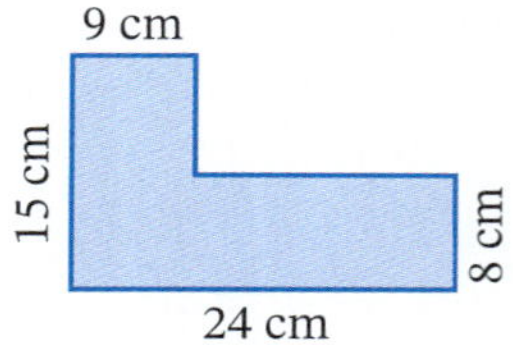

b 12 cm, 17 cm, 15 cm

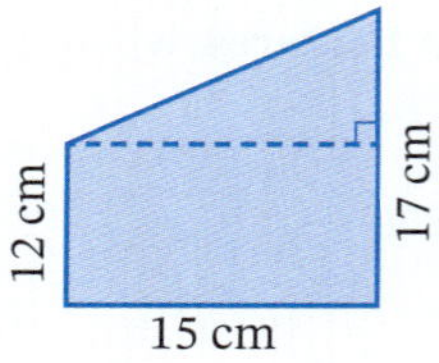

c 11 cm, 4 cm, 4 cm, 4 cm, 11 cm, 11 cm, 4 cm, 4 cm, 11 cm

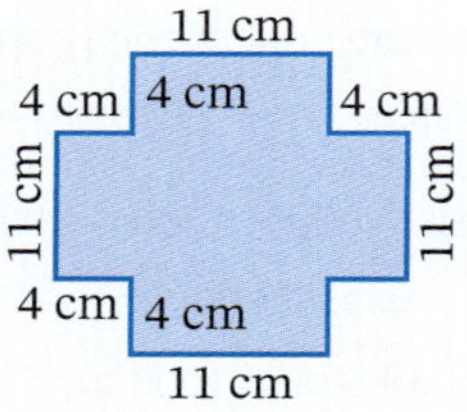

d 9 cm, 16 cm, 20 cm

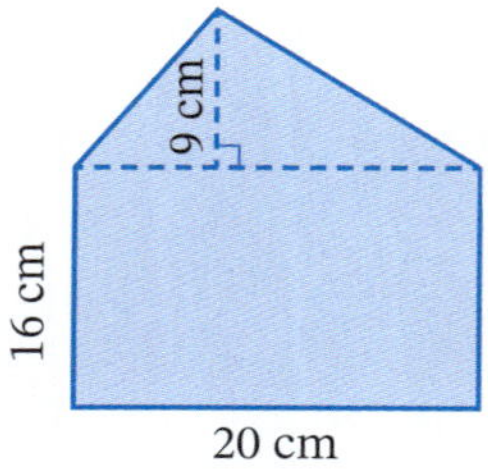

e 10 cm, 9 cm, 8 cm, 5 cm, 9 cm

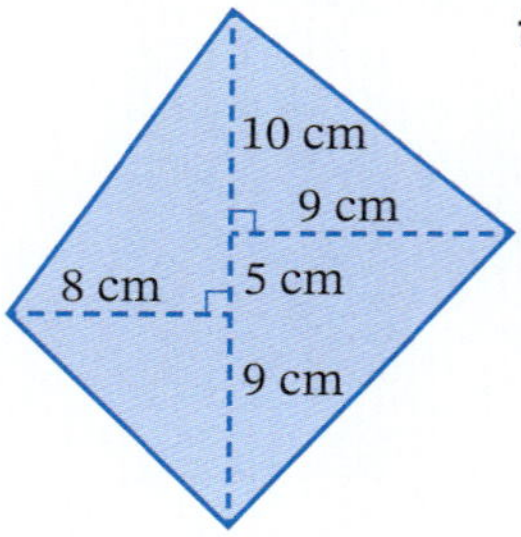

f 17 cm, 11 cm, 15 cm, 11 cm

▷

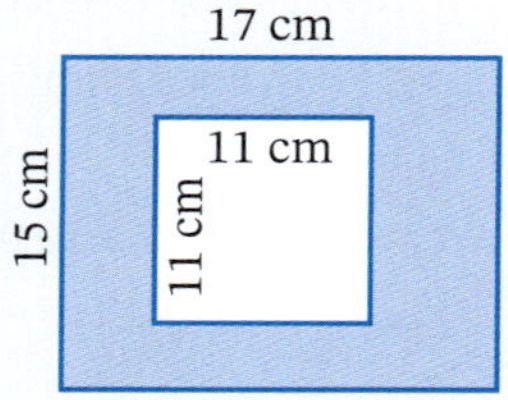

□ Foundation ○ Standard ⬡ Complex

9780170465618

g

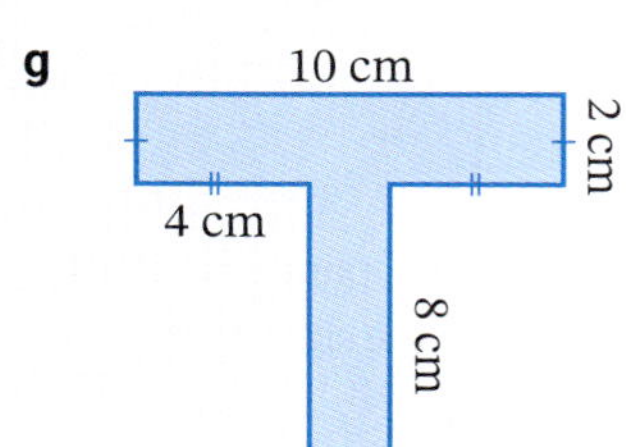

h

i

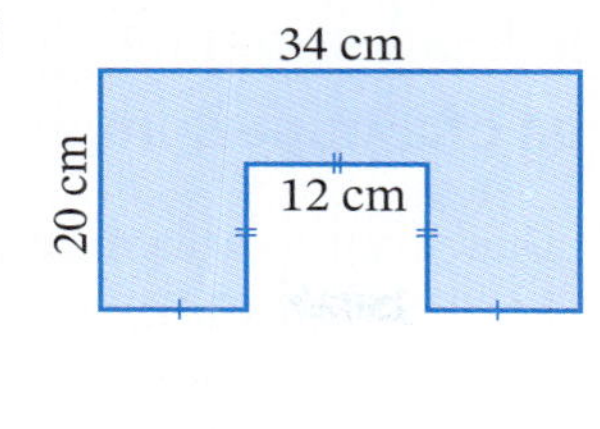

j

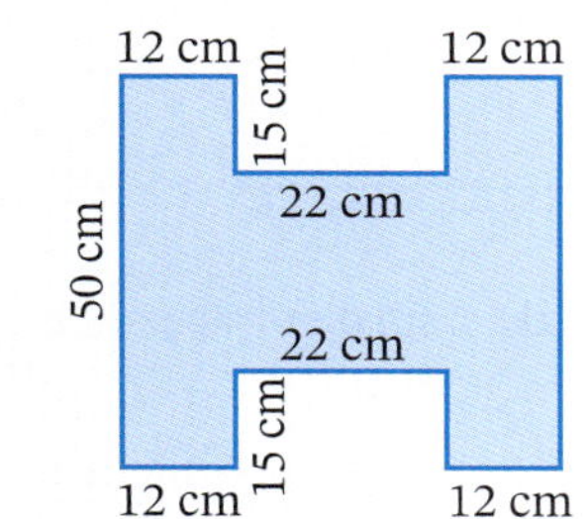

k

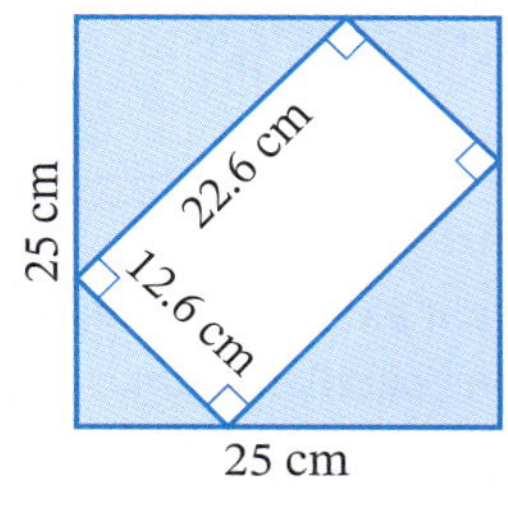

l

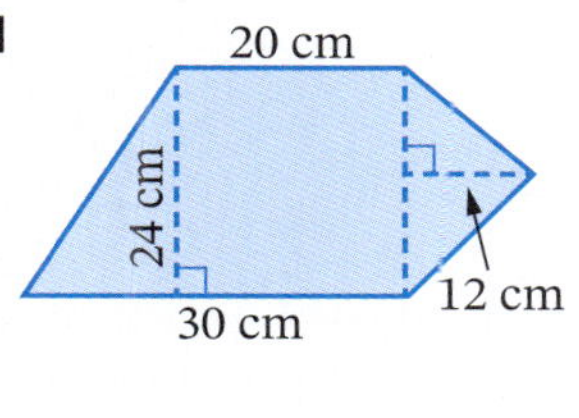

4 What is the area of the shaded arrow? Select the correct answer **A**, **B**, **C** or **D**.

A 14 cm² **B** 8 cm² **C** 9 cm² **D** 12 cm²

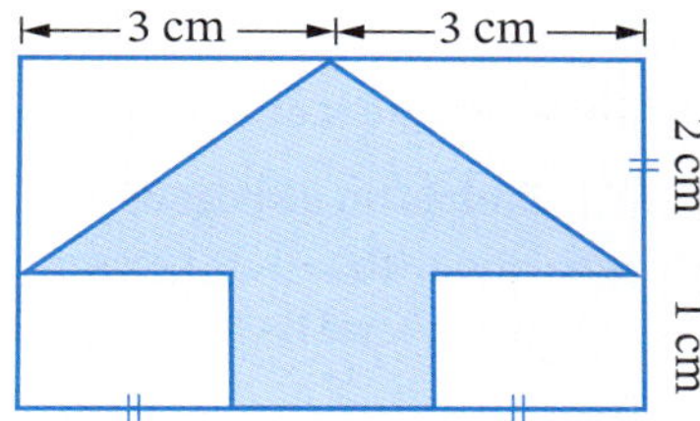

5 Find each shaded area.

PS R **a**

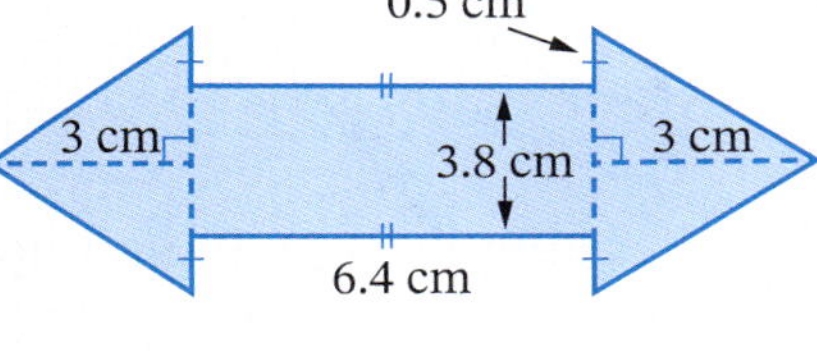

b

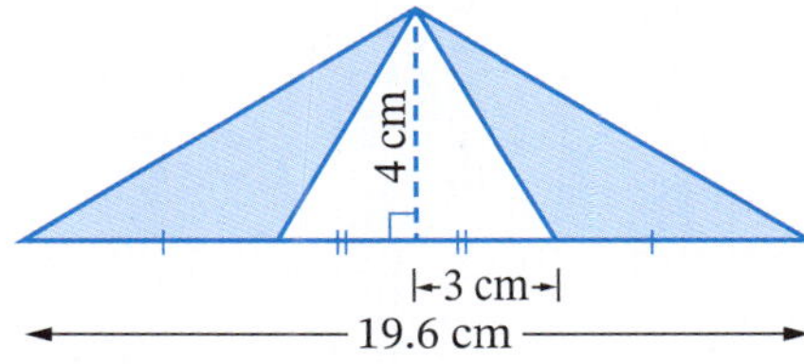

6 **a** Find the area of the large block of land shown in the diagram (in m²). EXAMPLE 9

b Express the area of the land in:

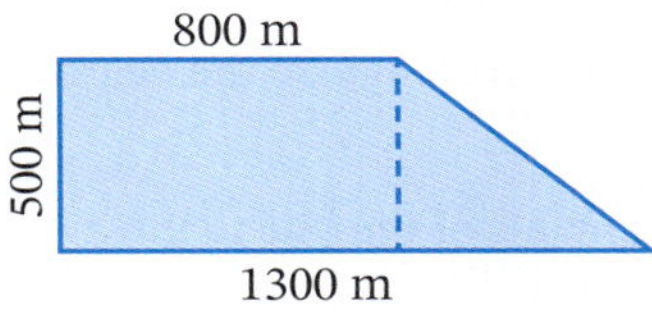

i km² **ii** hectares.

7 The area of a large rectangular property is 8.6 ha. Calculate the width if its length is 400 m. Select the correct answer **A**, **B**, **C** or **D**.

A 21.5 m **B** 215 m **C** 2150 m **D** 0.0215 km

INVESTIGATION

Area of a trapezium, kite and rhombus

1 Draw 2 copies of this trapezium onto a sheet of paper. Draw a dotted line, parallel to the sides labelled a and b, halfway down the height of each trapezium and label the sides as shown.

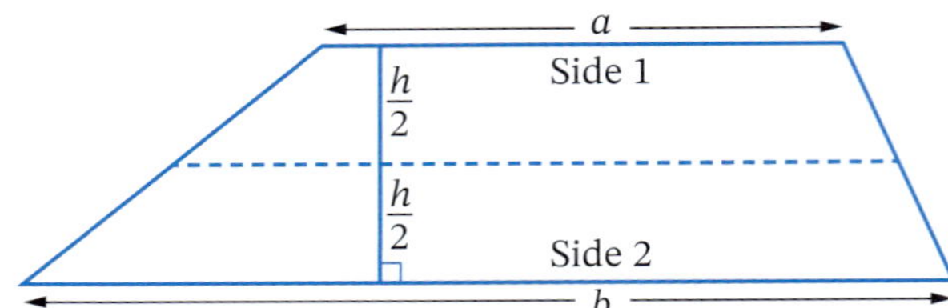

2 Cut out one trapezium and cut along the dotted line to make 2 smaller trapeziums. Join the pieces to make a long parallelogram.

3 By measuring the height and base of the parallelogram to the nearest millimetre, find its area.

4 By comparing the measurements of the parallelogram to the measurements of the original trapezium, suggest a general formula for finding the area of any trapezium. Check your answer with your teacher.

5 Cut out the original trapezium and paste it and the parallelogram into your book.

6 Draw 2 copies of the kite and rhombus on a sheet of paper. Draw diagonals of length x and y and label them as shown. Notice that they cross at right angles. For the kite, one diagonal (x) is bisected by the other (y), while for the rhombus, *both* diagonals bisect each other.

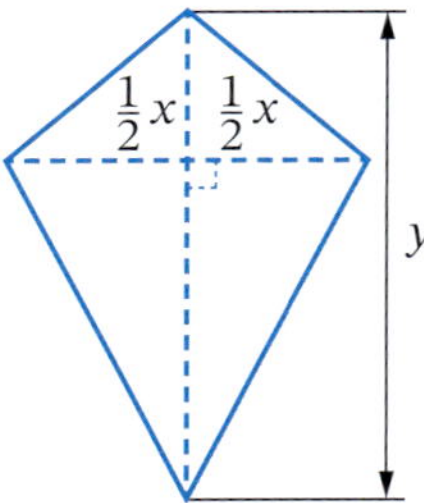

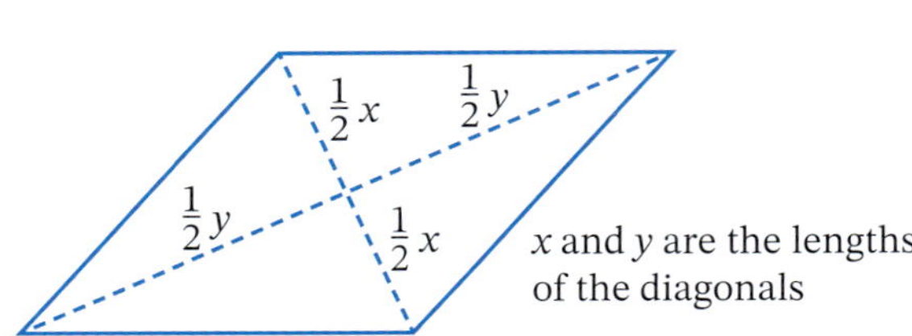

x and y are the lengths of the diagonals

7 Cut out one of your kites and rhombuses and cut along their diagonals to make 4 triangles each. Join each set of triangles to make a rectangle.

8 By measuring the sides of each rectangle to the nearest millimetre, find the area of each one.

9 By comparing the measurements of the kite to the measurements of the rectangle formed from it, suggest a general formula for finding the area of any kite. Check your answer with your teacher.

10 Is a rhombus a special type of kite?

11 Suggest a general formula for finding the area of any rhombus. Check your answer with your teacher.

12 Cut out the original kite and rhombus and paste all shapes into your book.

Areas of quadrilaterals 10.04

ⓘ Areas of special quadrilaterals

Parallelogram	Trapezium
Area = base × perpendicular height $A = bh$	Area = $\frac{1}{2}$ × (sum of parallel sides) × perpendicular height $A = \frac{1}{2}(a + b)h$

Rhombus	Kite
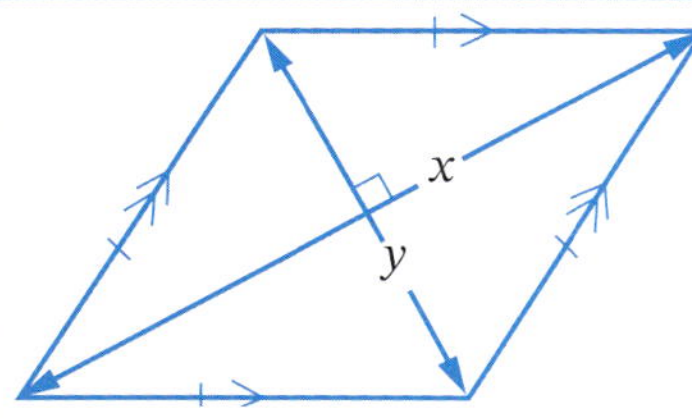 	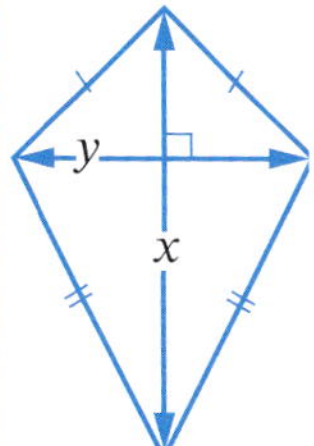
Area = $\frac{1}{2}$ × (product of diagonals) $A = \frac{1}{2}xy$	Area = $\frac{1}{2}$ × (product of diagonals) $A = \frac{1}{2}xy$

Example 10

Find the area of each quadrilateral.

a

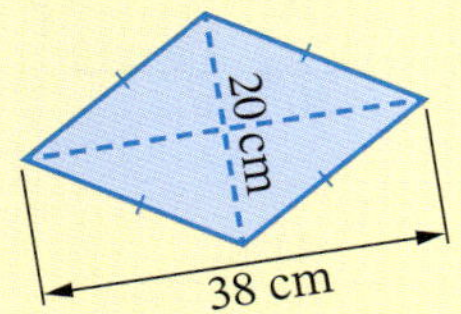

b

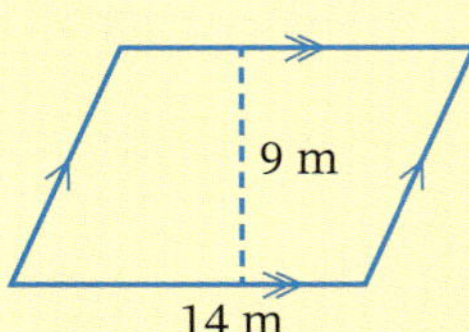

c

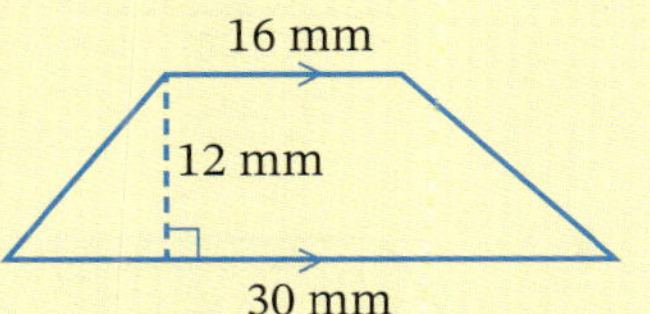

d

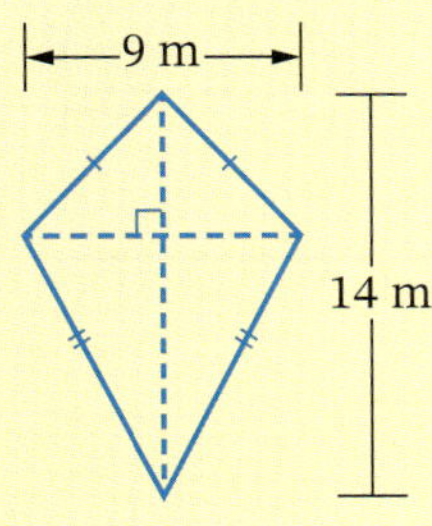

e

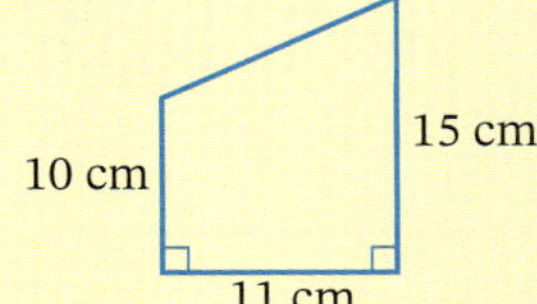

SOLUTION

a Area of rhombus $= \frac{1}{2} \times 38 \times 20$ $\qquad A = \frac{1}{2}xy$

$= 380 \text{ cm}^2$

b Area of parallelogram $= 14 \times 9$ $\qquad A = bh$

$= 126 \text{ m}^2$

c Area of trapezium $= \frac{1}{2} \times (16 + 30) \times 12$ $\qquad A = \frac{1}{2}(a + b)h$

$= 276 \text{ mm}^2$

d Area of kite $= \frac{1}{2} \times 9 \times 14$ $\qquad A = \frac{1}{2}xy$

$= 63 \text{ m}^2$

e Area of trapezium $= \frac{1}{2} \times (10 + 15) \times 11$

$= 137.5 \text{ cm}^2$

EXERCISE 10.04 ANSWERS ON P. 658

Areas of quadrilaterals

U F PS R

EXAMPLE 10

1 Find the shaded area of each shape.

a

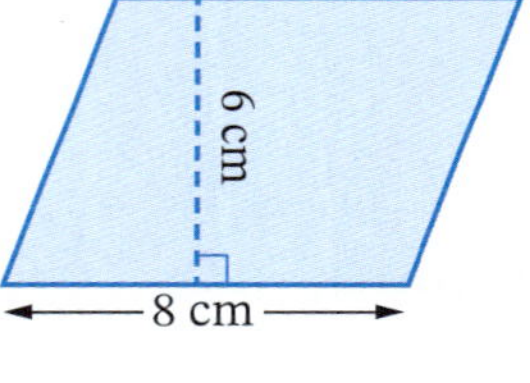

b

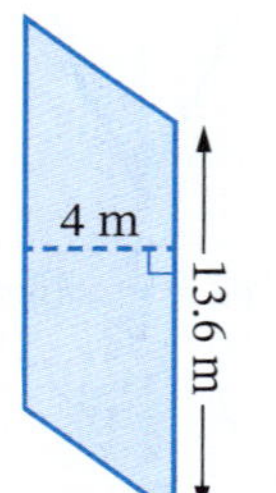

c

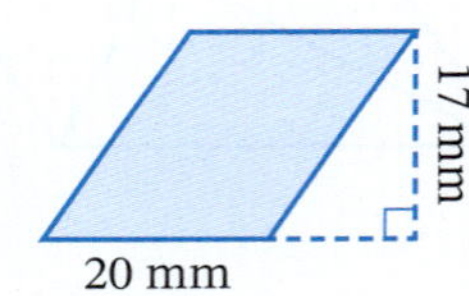

d

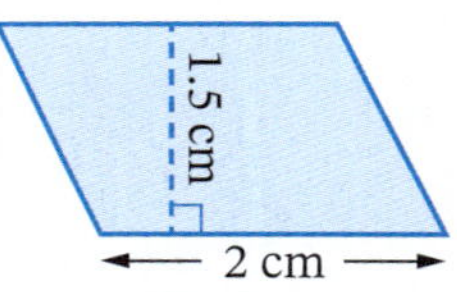

e

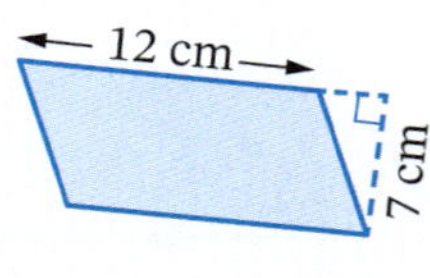

f

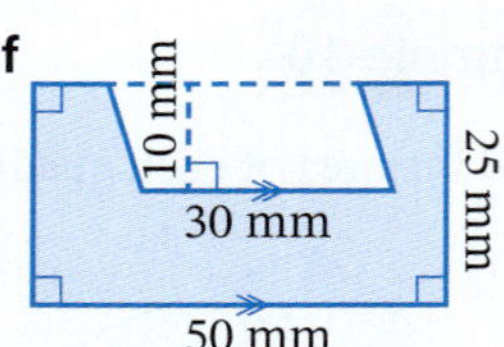

g

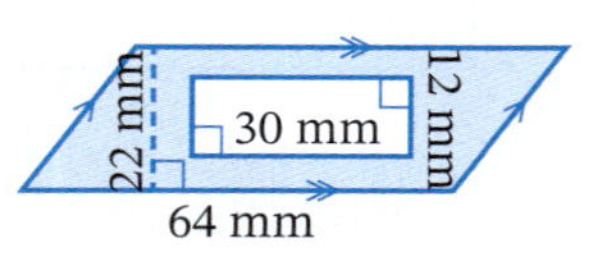

h

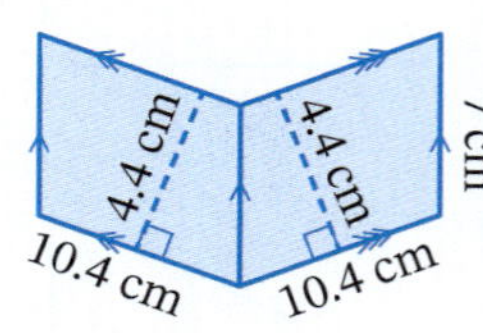

i

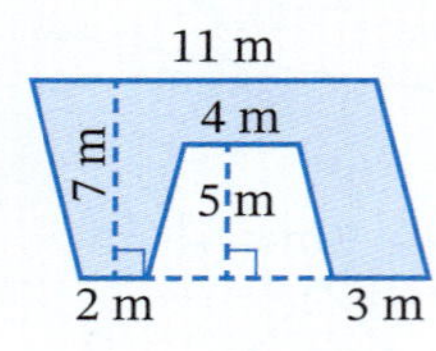

Foundation Standard Complex

9780170465618

2 Find the area of each shape.

a

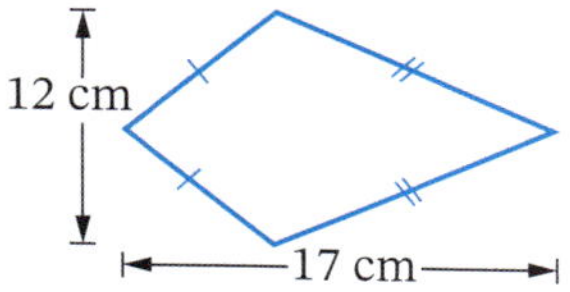

12 cm
17 cm

b

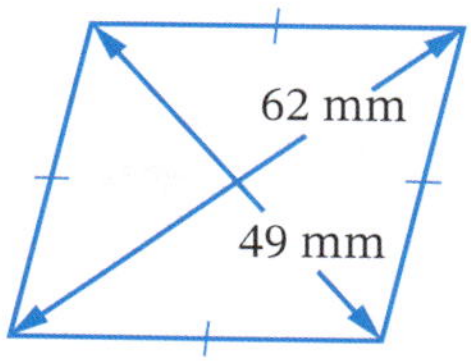

62 mm
49 mm

c

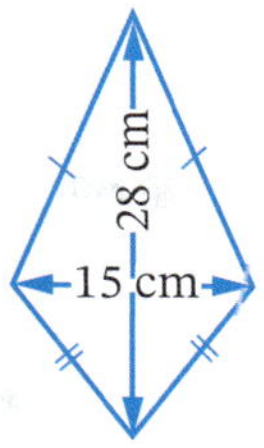

28 cm
15 cm

d

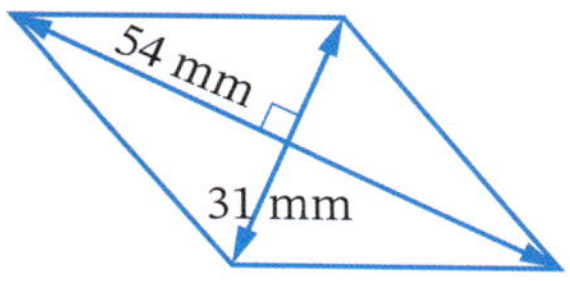

54 mm
31 mm

e

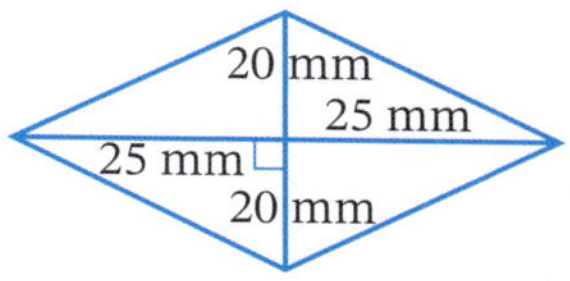

20 mm
25 mm
25 mm
20 mm

f

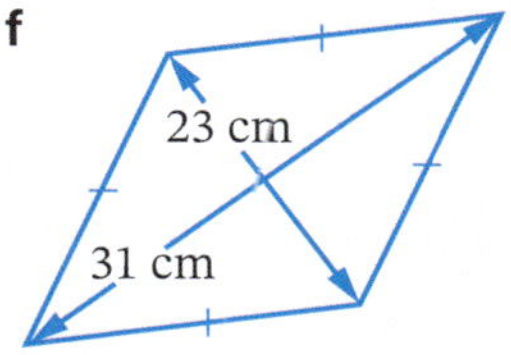

23 cm
31 cm

g

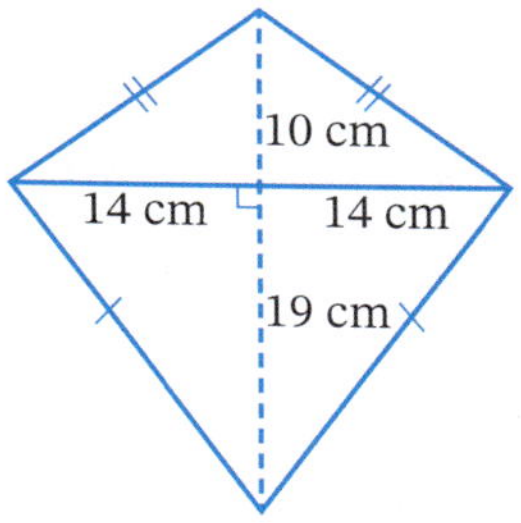

10 cm
14 cm
14 cm
19 cm

h

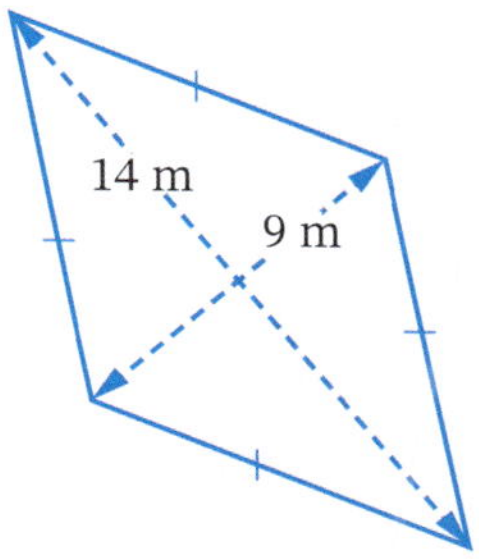

14 m
9 m

i

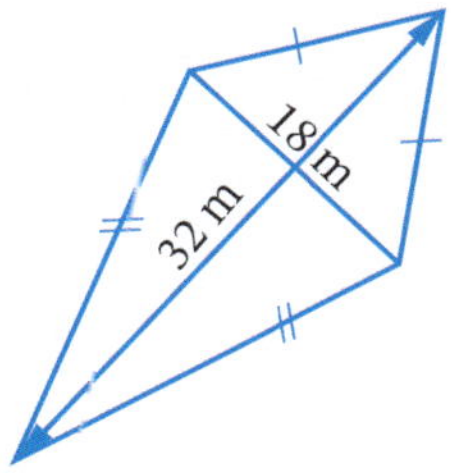

18 m
32 m

3 Find the area of each shape.

a

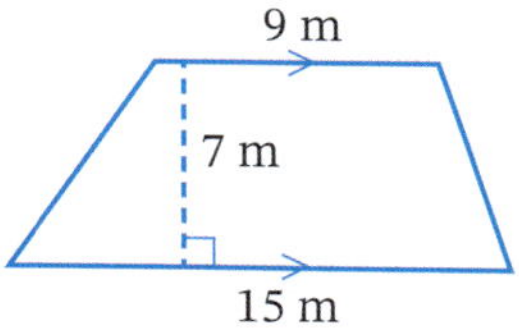

9 m
7 m
15 m

b

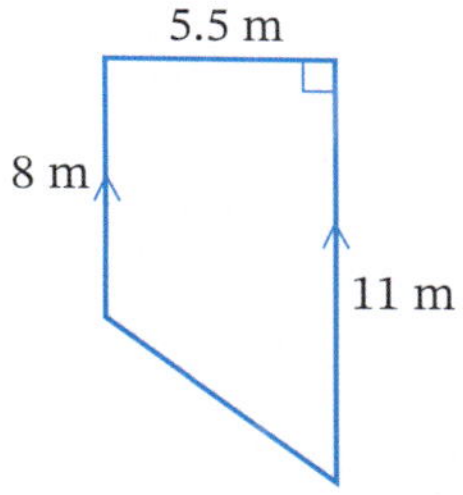

5.5 m
8 m
11 m

c

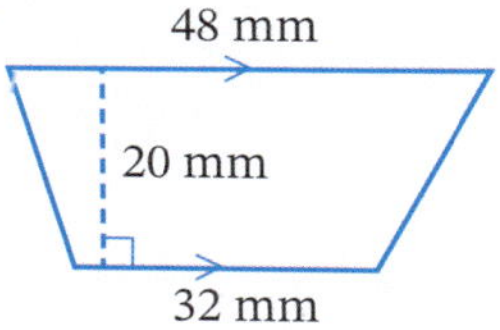

48 mm
20 mm
32 mm

d

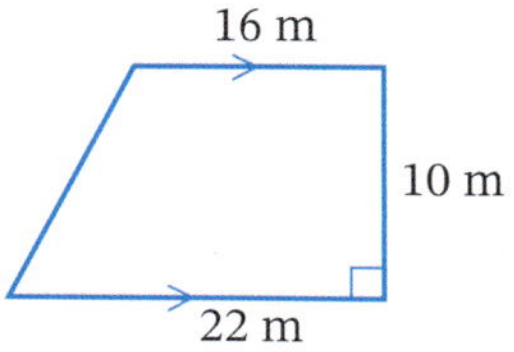

16 m
10 m
22 m

e

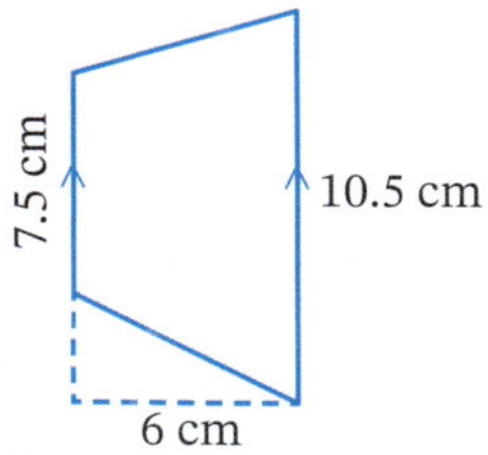

7.5 cm
10.5 cm
6 cm

f

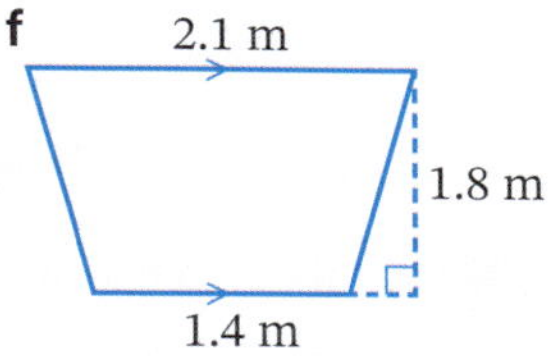

2.1 m
1.8 m
1.4 m

g

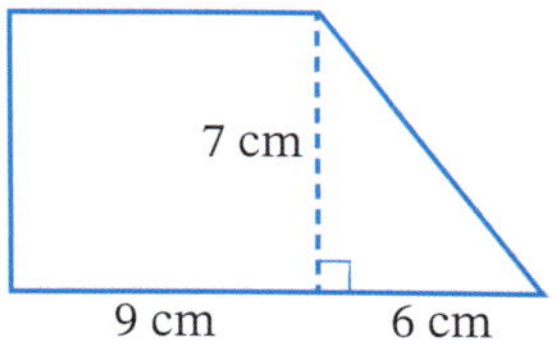

7 cm
9 cm
6 cm

h

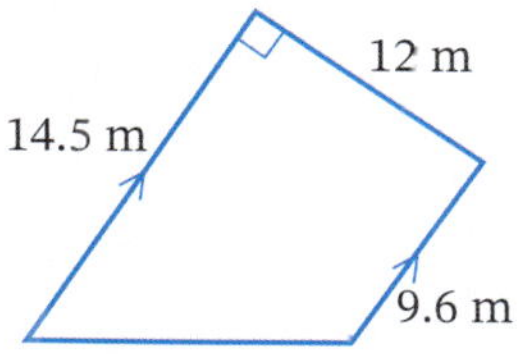

12 m
14.5 m
9.6 m

i

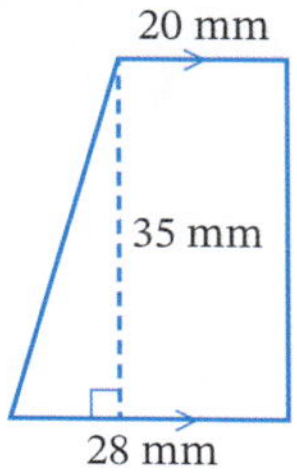

20 mm
35 mm
28 mm

Foundation Standard Complex

9780170465618

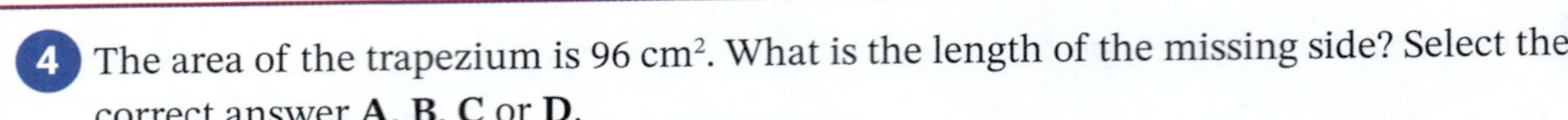

4 The area of the trapezium is 96 cm². What is the length of the missing side? Select the correct answer **A**, **B**, **C** or **D**.

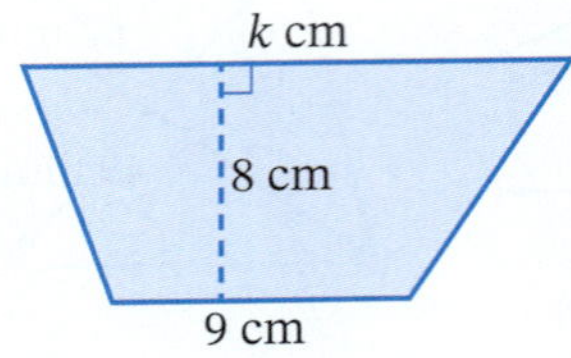

A 10 cm　　**B** 15 cm

C 17 cm　　**D** 12 cm

5 Find the area of each shape.

a

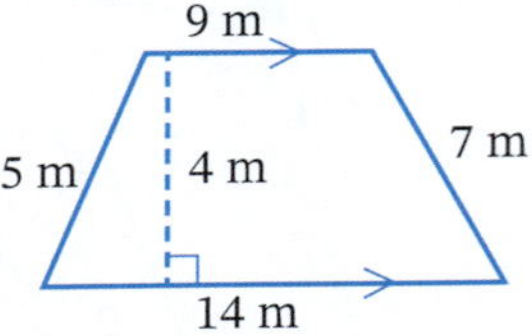

b

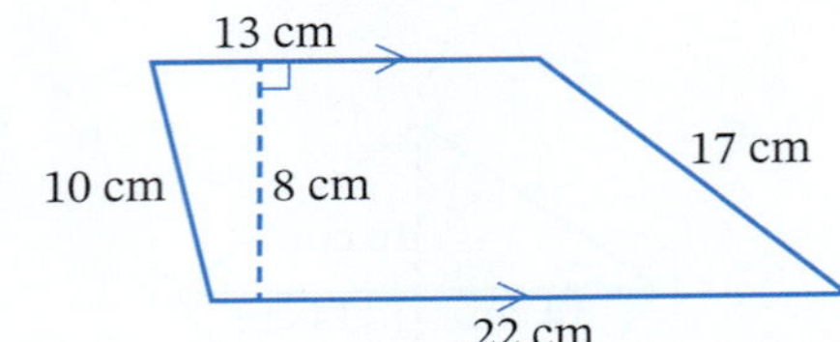

c

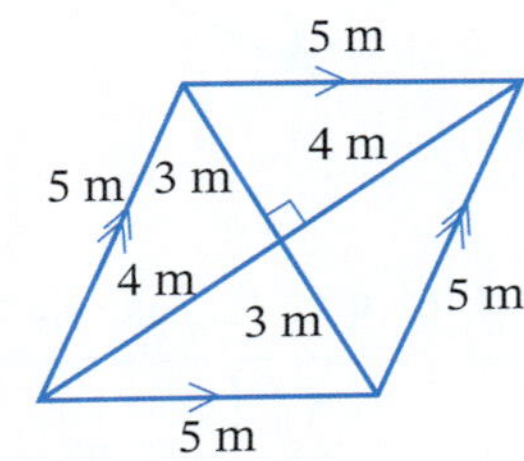

d

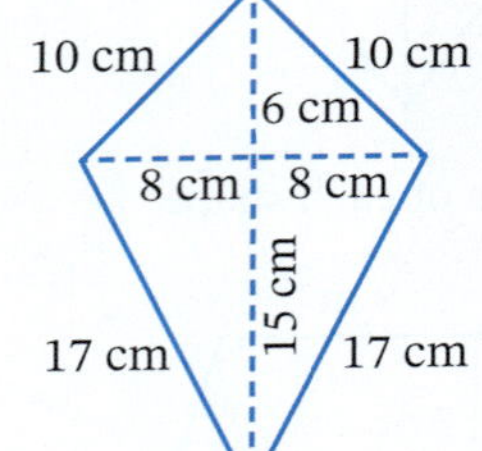

e

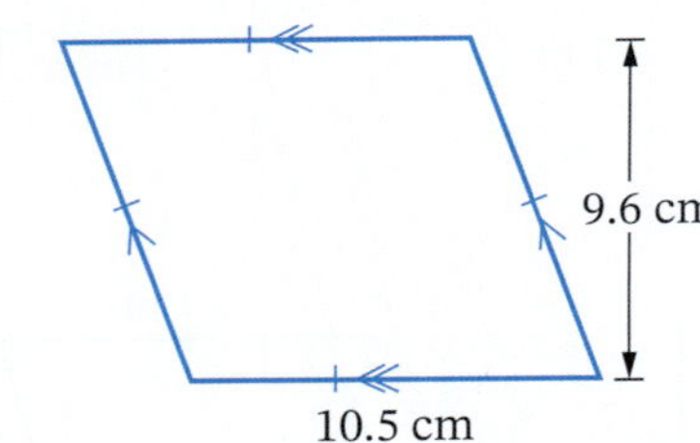

f

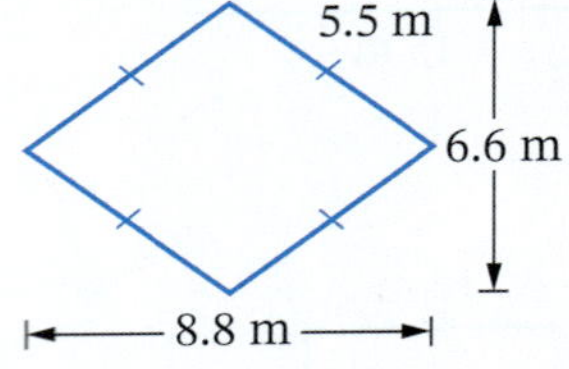

6 Calculate each shaded area (all measurements are in metres).

P R

a

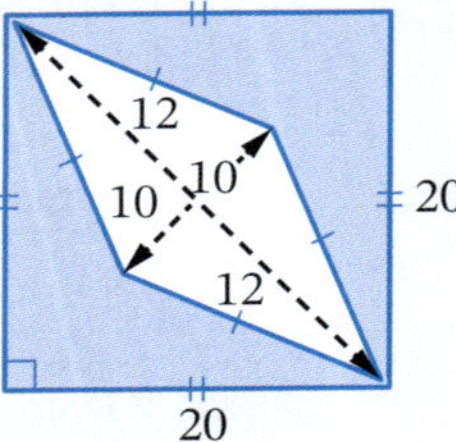

b

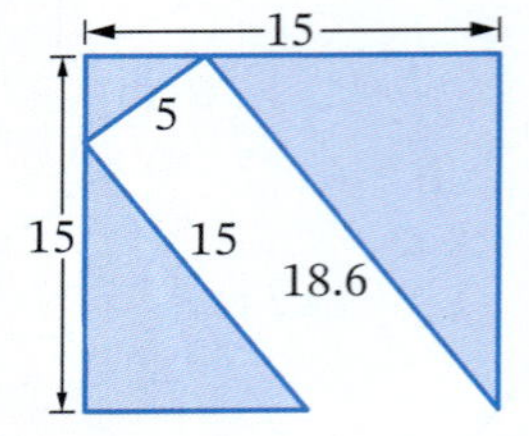

c

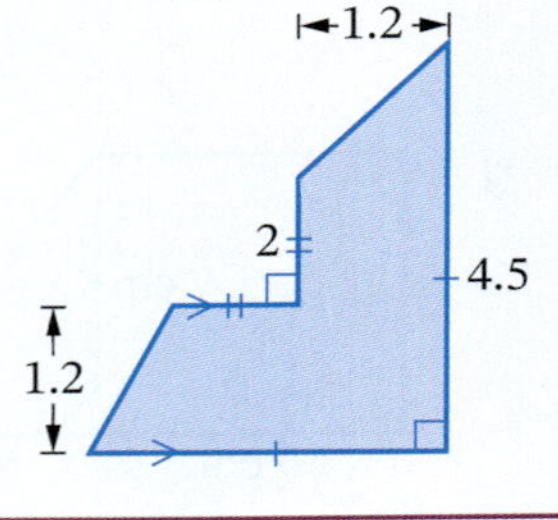

Foundation　Standard　Complex

Maths without calculators

Quiz Mental skills 10A

Finding 10%, 20% and 5%

To find 10% or $\frac{1}{10}$ of a number, simply divide the number by 10 by moving the decimal point one place to the left.

1 Study each example.

a $10\% \times 150 = 15\,0. = 15$

b $10\% \times \$1256.80 = \$1\,25\,6.8 = \$125.68$

c $10\% \times 4917 = 4\,91\,7. = 491.7$

d $10\% \times \$48.55 = \$\,4\,8.\,55 = \$4.885$

2 Now find 10% of each amount.

a	190	**b**	\$75	**c**	875	**d**	\$202
e	\$37.60	**f**	400	**g**	\$9.25	**h**	896
i	\$2700	**j**	\$3.80	**k**	\$1527.60	**l**	\$72.50
m	3154	**n**	\$10.70	**o**	426	**p**	\$24 317.60

20% is 10% doubled, so to find 20% of a number, first find 10% then double it.

3 Study each example.

a $20\% \times 700$

$10\% \times 700 = 70$

$\therefore 20\% \times 700 = 70 \times 2$

$= 140$

b $20\% \times \$876$

$10\% \times \$876 = \$87.60$

$\therefore 20\% \times \$876 = \$87.60 \times 2$

$= \$175.20$

c $20\% \times 325$

$10\% \times 325 = 32.5$

$\therefore 20\% \times 325 = 32.5 \times 2$

$= 65$

d $20\% \times \$38.50$

$10\% \times \$38.50 = \$3.85$

$\therefore 20\% \times \$38.50 = \$3.85 \times 2$

$= \$7.70$

4 Now find 20% of each amount.

a	50	**b**	620	**c**	\$2450	**d**	\$8.60
e	72	**f**	\$12 700	**g**	390	**h**	\$5.80
i	\$45	**j**	\$84	**k**	\$4600	**l**	320

5% is half of 10%, so to find 5% of a number, first find 10% then divide it by 2.

5 Study each example.

a $5\% \times 180$

$10\% \times 180 = 18$

$\therefore 5\% \times 180 = 18 \div 2$

$= 9$

b $5\% \times \$76$

$10\% \times \$76 = \$7.60$

$\therefore 5\% \times \$76 = \$7.60 \div 2$

$= \$3.80$

c $5\% \times 120$

$10\% \times 120 = 12$

$\therefore 5\% \times 120 = 12 \div 2$

$= 6$

d $5\% \times \$142.20$

$10\% \times \$142.20 = \$14.22$

$\therefore 5\% \times \$142.20 = \$14.22 \div 2$

$= \$7.11$

6 Now find 5% of each amount.

a	2000	**b**	\$12	**c**	50	**d**	\$27
e	\$36.80	**f**	\$84	**g**	800	**h**	130
i	\$9.60	**j**	\$138	**k**	\$72	**l**	840

10.05 Circumferences and areas of circular shapes

Worksheets
A page of circular shapes
Applications of area
Back-to-front problems
Area and perimeter investigations

Puzzle
Carpet talk

Video
Beating the U-boats

The **circumference** of a circle is found by multiplying its **diameter** by a special number called **pi** (pronounced 'pie'), represented by the Greek letter π.

The **area** of a circle is found by multiplying π by the **radius squared.**

ⓘ Circumference of a circle

The **circumference (perimeter) of a circle** is:

$C = \pi \times \text{diameter}$ or $C = 2 \times \pi \times \text{radius}$

$C = \pi d$ $\quad$ $C = 2\pi r$

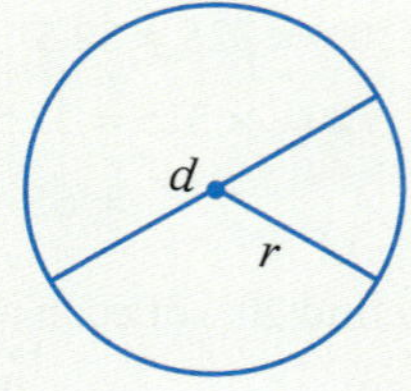

ⓘ Area of a circle

The **area of a circle** is:

$$A = \pi \times (\text{radius})^2$$

$$A = \pi r^2$$

$\pi = 3.141\ 592\ 653\ 897\ 93...$ and can be found by pressing the π key on your calculator. It has no exact decimal or fraction value, so like a surd such as $\sqrt{2}$, it is an **irrational number**.

10.05

Example 11

For each circle, calculate correct to 2 decimal places:

i its circumference **ii** its area.

a

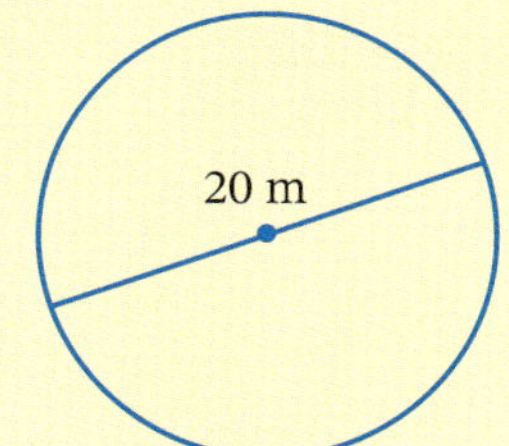

b

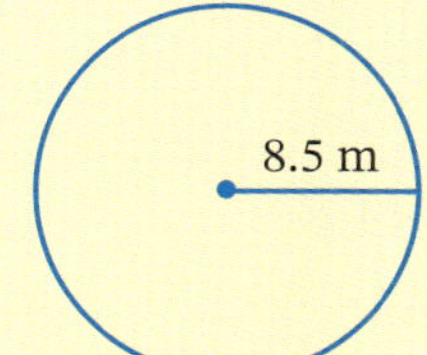

SOLUTION

a **i** $C = \pi \times 20$ $\quad C = \pi d$

$= 62.8318...$

≈ 62.83 m

ii $A = \pi \times 10^2$ $\quad A = \pi r^2$

$= 314.1592...$

≈ 314.16 m^2

b **i** $C = 2 \times \pi \times 8.5$ $\quad C = 2\pi r$

$= 53.4070...$

≈ 53.41 m

ii $A = \pi \times 8.5^2$ $\quad A = \pi r^2$

$= 226.9800 ...$

≈ 226.98 m^2

Example 12

For each sector, calculate correct to one decimal place:

i its perimeter **ii** its area.

A **sector** is a fraction of a circle 'cut' along 2 radii, like a pizza sl ce.

Video Perimeter and area of a sector

a

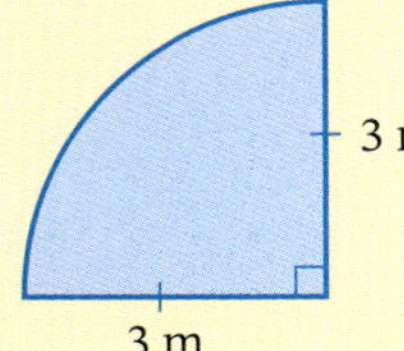

b

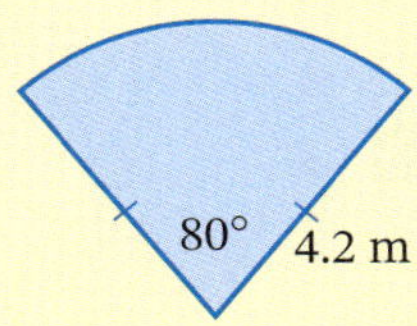

c

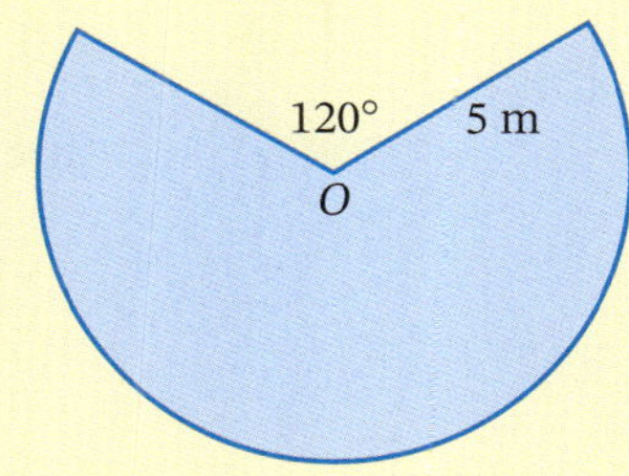

SOLUTION

a **i** This sector is a **quadrant**, a quarter of a circle. $\quad \frac{1}{4} \times$ circumference + radius + radius

Perimeter $= \frac{1}{4} \times 2 \times \pi \times 3 + 3 + 3$

$= 10.71238...$

$= 10.7$ m

ii Area $= \frac{1}{4} \times \pi \times 3^2$ — $\frac{1}{4} \times$ area of circle

$= 7.06858...$

$= 7.1 \text{ m}^2$

b i Perimeter $= \frac{80}{360} \times 2 \times \pi \times 4.2 + 4.2 + 4.2$ — $\frac{80}{360} \times$ circumference + radius + radius

$= 14.26430...$

$= 14.3 \text{ m}$

There are 360° in a circle, but a sector is a fraction of a circle.

ii Area $= \frac{80}{360} \times \pi \times 4.2^2$ — $\frac{80}{360} \times$ area of circle

$= 12.31504...$

$= 12.3 \text{ m}^2$

c i Sector angle $= 360° - 120°$

$= 240°$

Perimeter $= \frac{240}{360} \times 2 \times \pi \times 5 + 5 + 5$

$= 30.94395...$

$= 30.9 \text{ m}$

ⓘ Arc length, perimeter and area of a sector

For a sector with angle θ:

- Arc length $= \frac{\theta}{360} \times 2\pi r$
- Perimeter of the sector $= \frac{\theta}{360} \times 2\pi r + r + r$ (arc length + radius + radius)
- Area of the sector $= \frac{\theta}{360} \times \pi r^2$

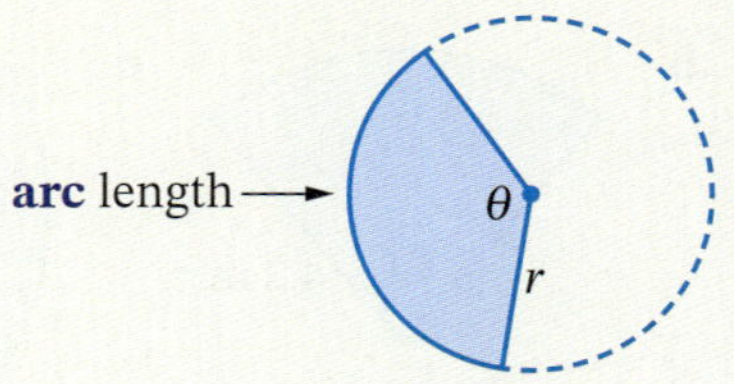

Example 13

Find, correct to one decimal place, the area of each shape.

a

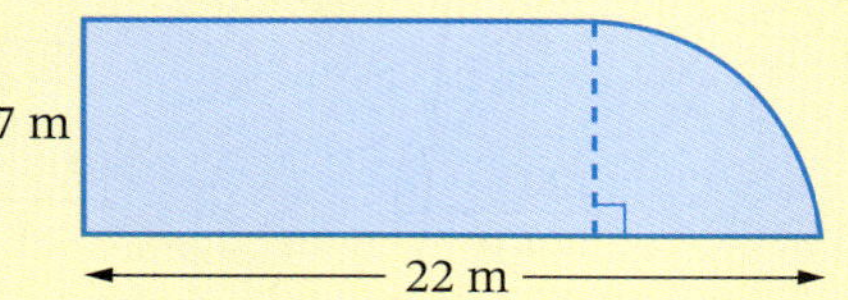

b

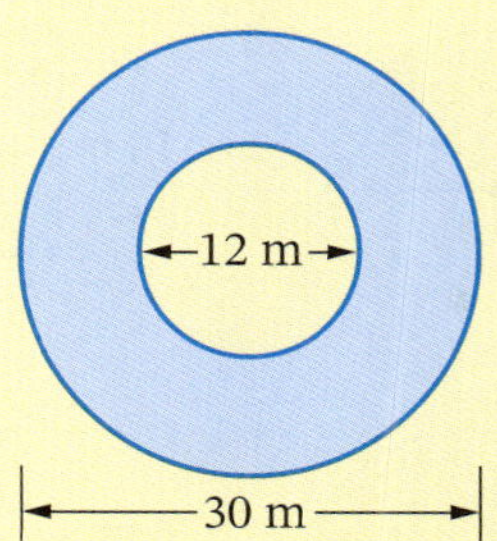

SOLUTION

a The shape is made up of a rectangle and a quadrant.

Radius of quadrant = 7 m

Length of rectangle = 22 − 7

= 15 m

Area of shape = area of rectangle + quadrant

$= 15 \times 7 + \frac{1}{4} \times \pi \times 7^2$

$= 143.4845...$

$\approx 143.5 \text{ m}^2$

b This ring shape is called an **annulus**, it is the area enclosed by 2 circles with the same centre.

Radius of large circle $= \frac{1}{2} \times 30$ m

= 15 m

Radius of small circle $= \frac{1}{2} \times 12$ m

= 6 m

Shaded area = large circle − small circle

$= \pi \times 15^2 - \pi \times 6^2$

$= 593.7610...$

$\approx 593.8 \text{ m}^2$

EXERCISE 10.05 ANSWERS ON P. 659

Circumferences and areas of circular shapes

U F PS R

EXAMPLE 11

1 For each circle, calculate correct to 2 decimal places:

i its circumference **ii** its area.

a

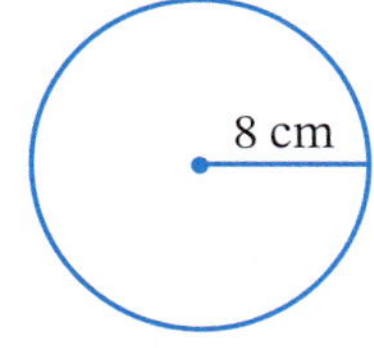

b

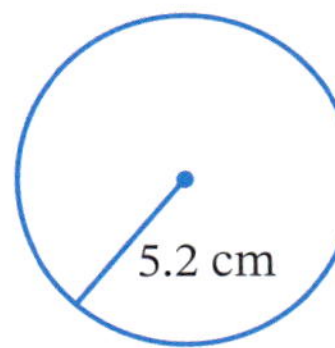

c

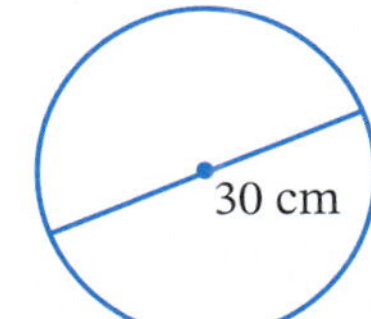

d

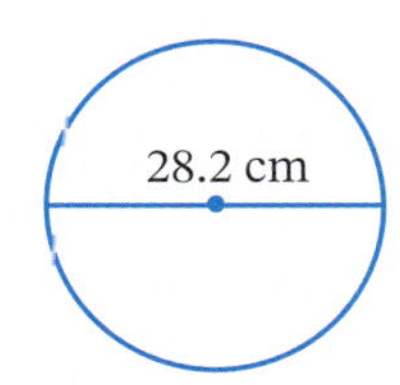

EXAMPLE 12

2 For each sector, calculate correct to one decimal place:

i its perimeter **ii** its area.

a 36 m

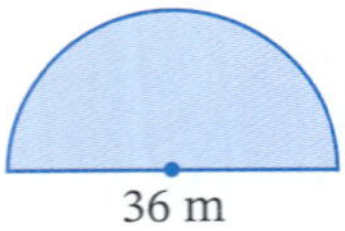

b 10 m, 10 m

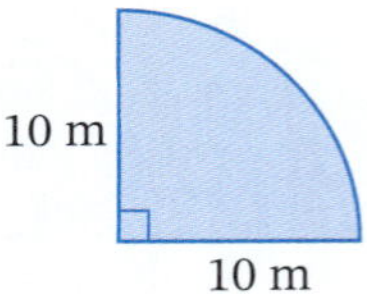

c 120°, 1 m

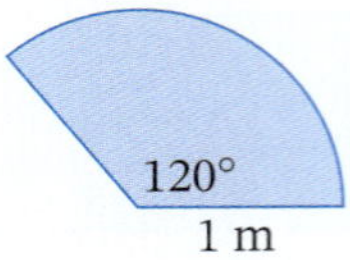

d 210°, 24 m

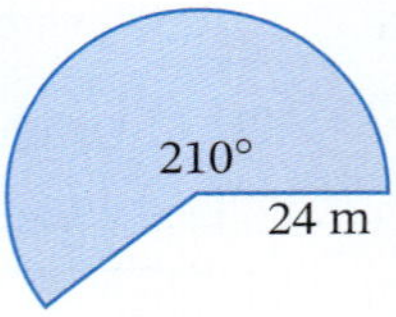

e 33 m

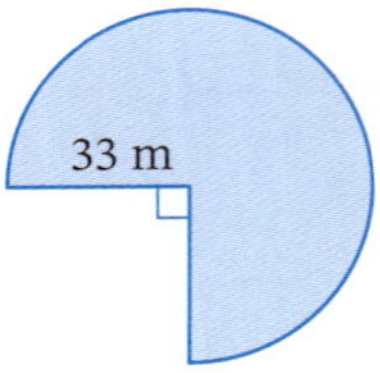

f 6 m, 30°, 6 m

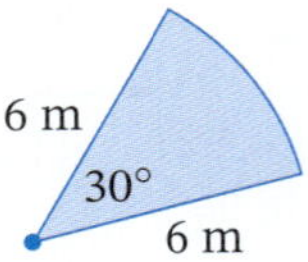

g 8 m, 10°, 8 m

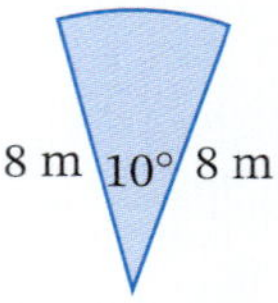

h 42 m, 60°

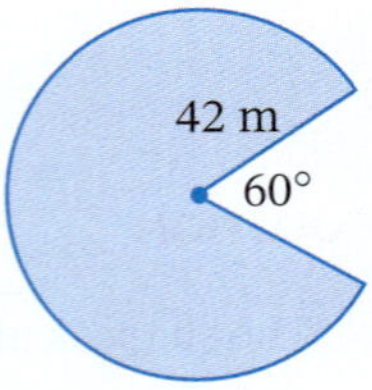

3 Find the perimeter of this sector. Select the correct answer **A**, **B**, **C** or **D**.

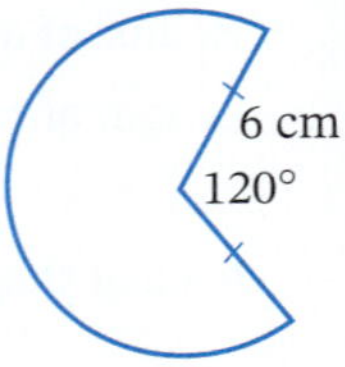

A 12.6 cm **B** 18.3 cm

C 24.6 cm **D** 37.1 cm

6 cm, 120°

4 Find the area of this sector. Select the correct answer **A**, **B**, **C** or **D**.

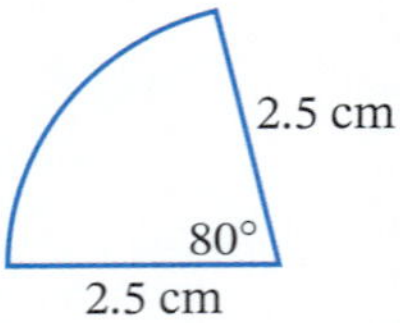

A 19.6 cm^2 **B** 15.3 cm^2

C 8.5 cm^2 **D** 4.4 cm^2

2.5 cm, 80°, 2.5 cm

5 Calculate the perimeter of each shape, correct to 2 decimal places.

a 40 cm, 40 cm

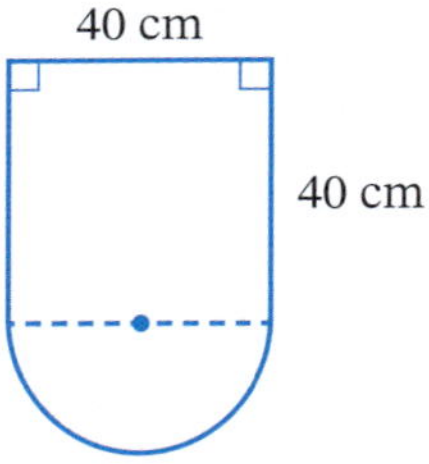

b 15 cm, 12 cm

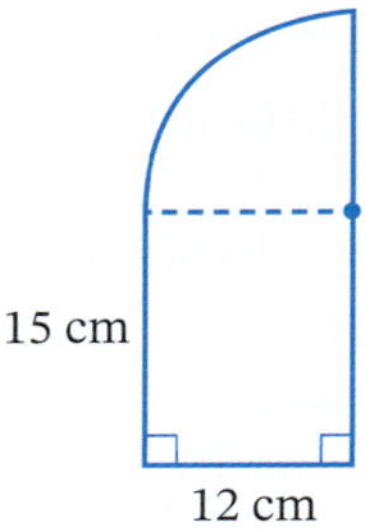

c 26 mm, 22 mm

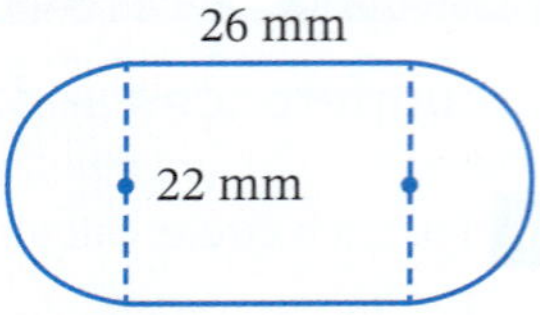

d 11 m, 11 m, 30 m

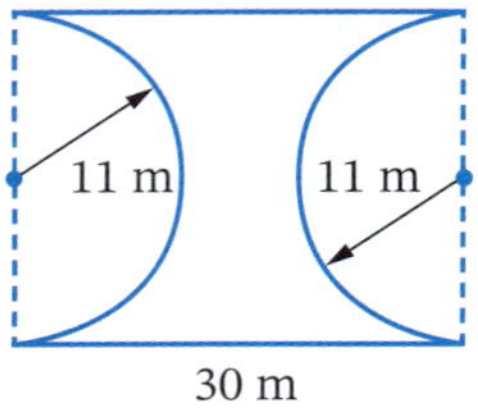

e 150 cm, 90 cm, 65 cm

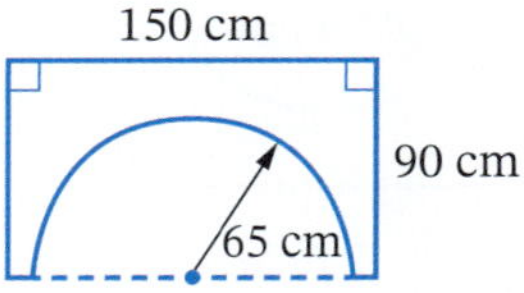

f 248 m, 248 m

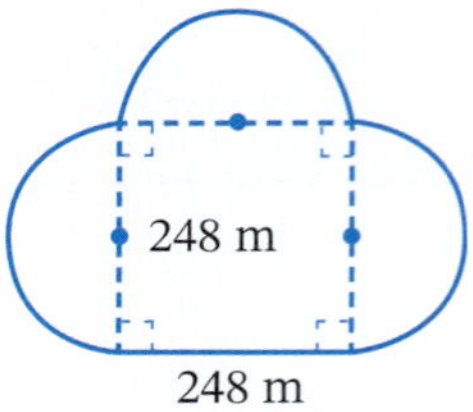

Foundation Standard Complex

g

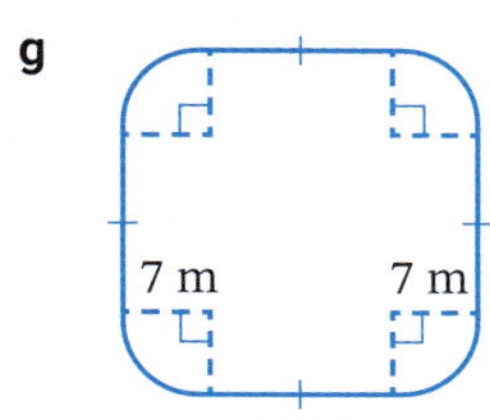

h

i

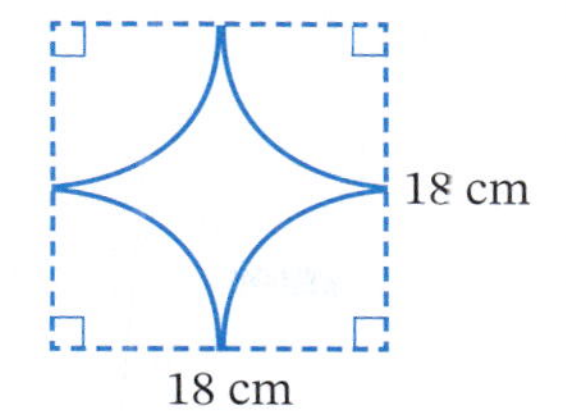

10.05

6 Calculate the area of the shape. Select the correct answer **A**, **B**, **C** or **D**.

EXAMPLE 13

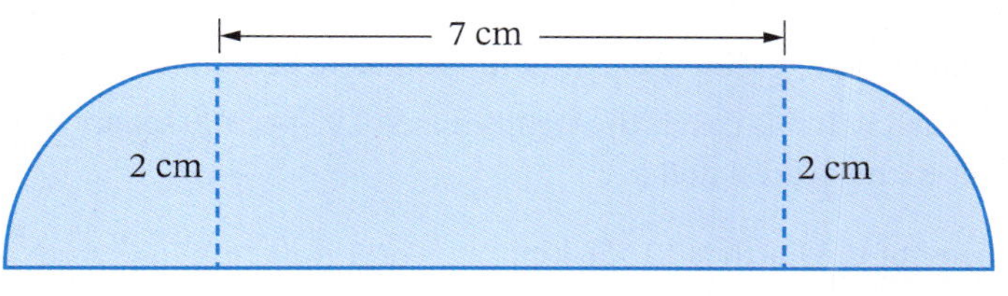

A 26.57 cm² **B** 20.28 cm² **C** 24.28 cm² **D** 28.28 cm²

7 Calculate the (shaded) area of each shape, correct to the nearest square metre. All measurements are in metres.

a

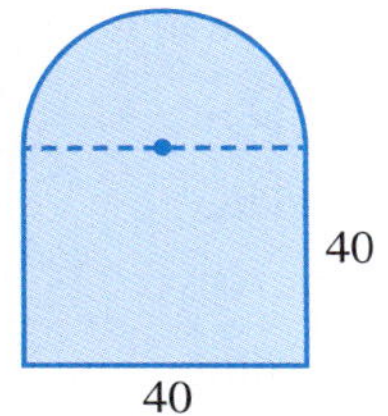

b

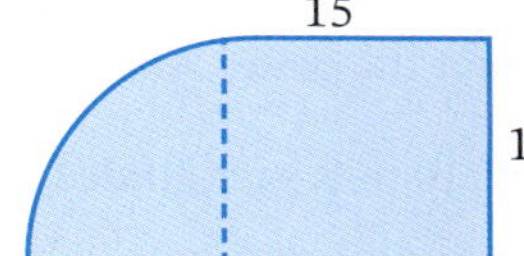

c

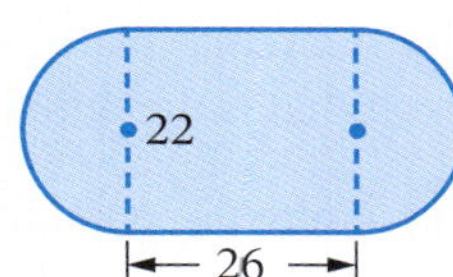

d

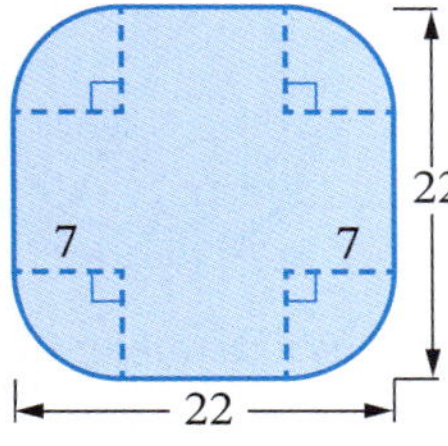

e

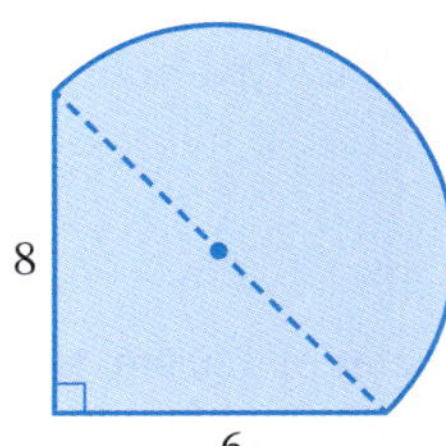

f

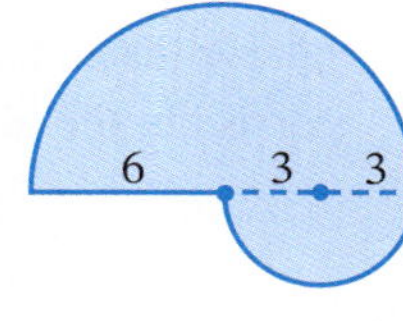

g

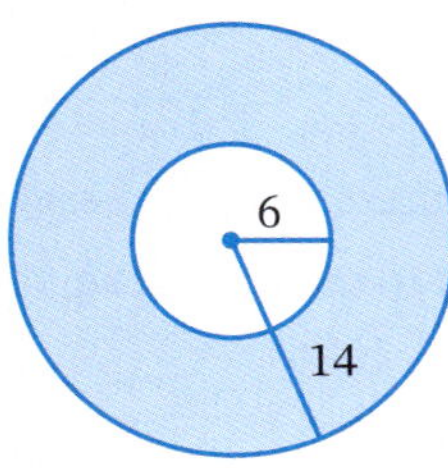

h

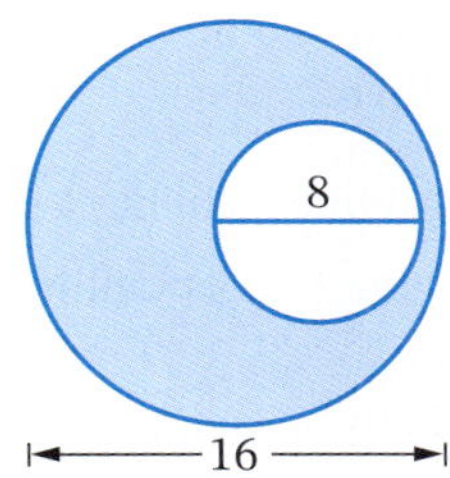

i

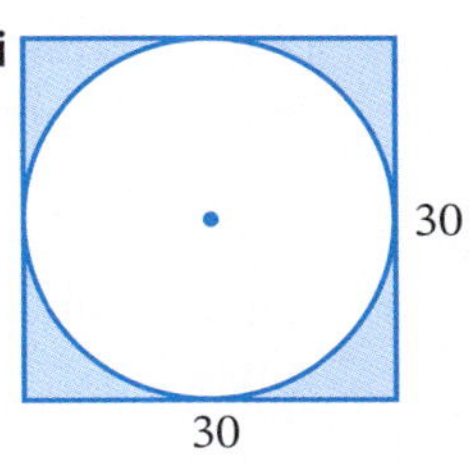

Foundation Standard Complex

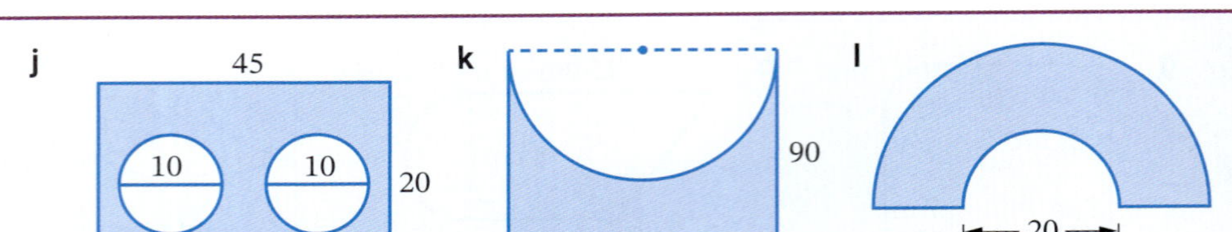

8 A circular playing field has a radius of 80 m. A rectangular cricket pitch measuring 25 m by 2 m is placed in the middle. The field, excluding the pitch, is to be fertilised.

PS R

a Calculate to the nearest square metre the area to be fertilised.

b How much will this cost if the fertiliser is $19.95 per 100 square metres? Give your answer to the nearest dollar.

9 The diameter of the Earth is 12 756 km.

a Find the circumference of the Earth to the nearest kilometre.

b The International Space Station orbits the Earth at an average height of 400 km above the Earth's surface. Find the distance it travels in one orbit of the Earth.

10 A circular plate of diameter 2 m has 250 holes of diameter 10 cm drilled in it. What is the remaining area of the plate? Answer to the nearest 0.1 m^2.

11 A circular pond of diameter 10 m is surrounded by a path 1 metre wide.

PS R

a Calculate the area of the path, correct to 2 decimal places.

b If pavers are $165 per square metre laid, what is the cost of the path?

12 A circular patch of grass has a diameter of 8.5 m. How much further does Arya walk if she walks around the border instead of directly across it? Give your answer correct to the nearest 0.1 m.

PS R

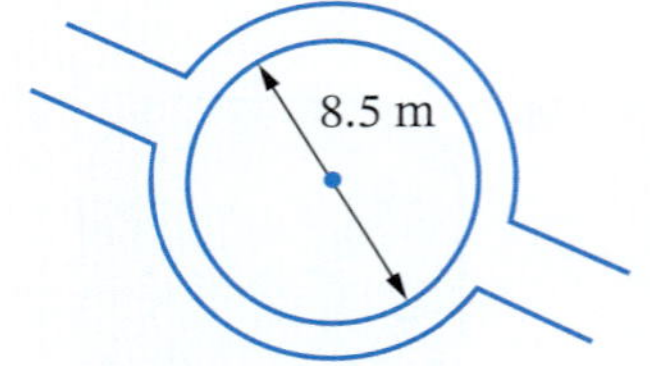

13 A new tractor tyre has a diameter of 120 cm, while a worn tyre has a diameter of 115 cm.

a Calculate the difference in circumference between a new and a worn tyre, correct to 3 decimal places.

b Over 1000 revolutions, how much further (to the nearest metre) will a new tyre travel compared to a worn tyre?

14 A square courtyard measuring 5 m by 5 m has a semi-circular area added to each side.

PS R

a Calculate the area of the semi-circular additions, correct to the nearest square metre.

b By what percentage has the area of the courtyard increased?

(This can be calculated as $\frac{\text{increase in area}}{\text{original area}} \times 100\%$)

☐ Foundation ○ Standard ○ Complex

9780170465618

15 The measurements of a running track are shown below. Each lane is 1 m wide and the athletes run along the insides of the lanes.

PS R

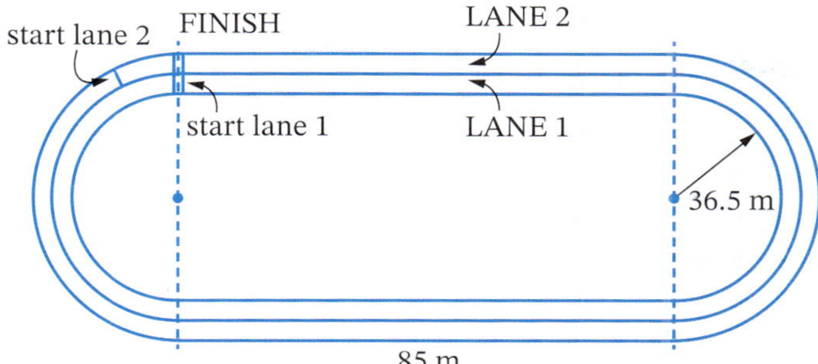

a Izabella runs one lap in lane 1. What distance does she cover, correct to one decimal place?

b Noah runs one lap in lane 2. What distance does he cover, correct to one decimal place?

c By how much should the start line be staggered in lane 2 (from lane 1) so that Izabella and Noah run the same distance in one lap?

16 The wheels of a bicycle have a diameter of 66 cm.

a How far (correct to 2 decimal places) will the bicycle move after one rotation of the wheel?

b How many rotations of the wheel are needed to cover a distance of 5 km?

DID YOU KNOW?

A piece of pi

$\pi = 3.141\ 592\ 653\ 589\ 793\ 238\ 462\ 643\ 383\ 279\ 502\ 884\ 197\ 169\ 399\ 375\ 105\ 820...$

Because π is an irrational number, its decimal digits run endlessly without repeating. Over history, mathematicians and scientists (and now computer scientists) have tried to calculate more accurate values of π, using more sophisticated formulas and calculation techniques. The ancient Greeks, Romans and Chinese first estimated π as 3. This value is also mentioned in the Bible. Since the first computer, the ENIAC, was invented in 1949, much progress has been made. Supercomputers have been used to calculate more decimal places of π and, in 2019, Emma Haruka Iwao (Japan), working for Google, calculated π to over 31.4 trillion places. In 2015, Suresh Sharma (India) memorised π to 70 030 decimal places. It took him 17 hours 14 minutes to recite it.

Research the modern history of π and find out about recent calculations of its value.

Worksheet
Nets of solids

INVESTIGATION

Nets of prisms

You will need: 1 cm grid paper.

If a solid is cut along some of its edges, the faces can all be folded down flat. The flat shape obtained is called the **net** of the solid.

Net of a cube

How many different nets of a cube can you make (or draw)? Do not count reflections or rotations because they are considered to be the same. For example, these are 2 versions of the same net.

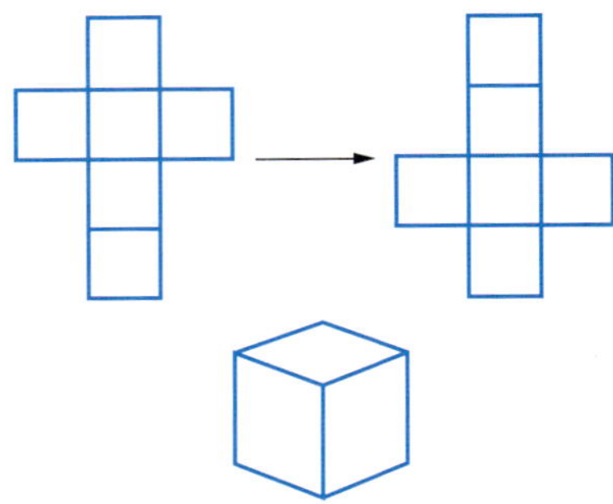

1 Draw the possible nets of a cube on 1 cm grid paper. Cut out the nets and fold them to see if they each form a cube, then paste them into your book.

2 If a cube has side length 1 cm, what is the area of one of its faces?

3 The surface area of a cube is the sum of the areas of all of its faces. What is the surface area of a cube with side length 1 cm?

Net of a rectangular prism

1 Draw each of the following nets of rectangular prisms on 1 cm grid paper. Cut them out and fold them to make rectangular prisms, then paste them into your book.

a

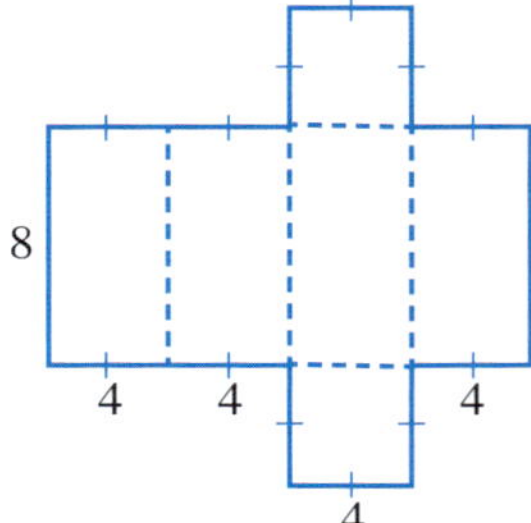

b

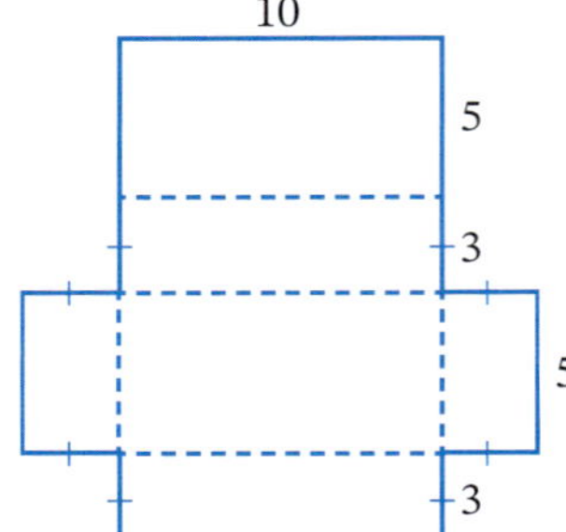

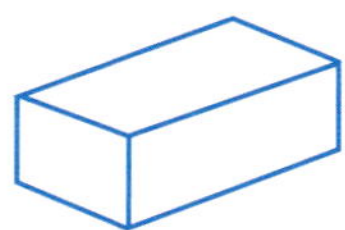

2 Calculate the surface area of each rectangular prism by adding the areas of all of its faces.

Net of a triangular prism

1 Draw each of the following nets on 1 cm grid paper. Measurements are in centimetres. Cut them out and fold them to make triangular prisms, then paste them into your book.

a

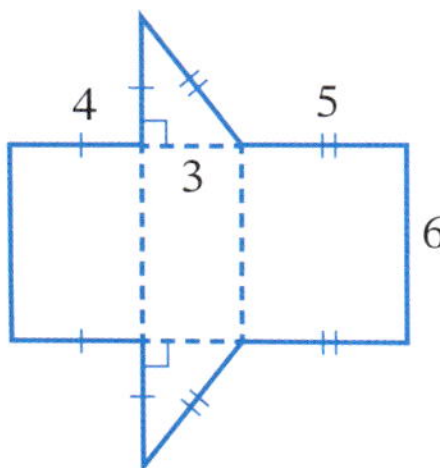

b

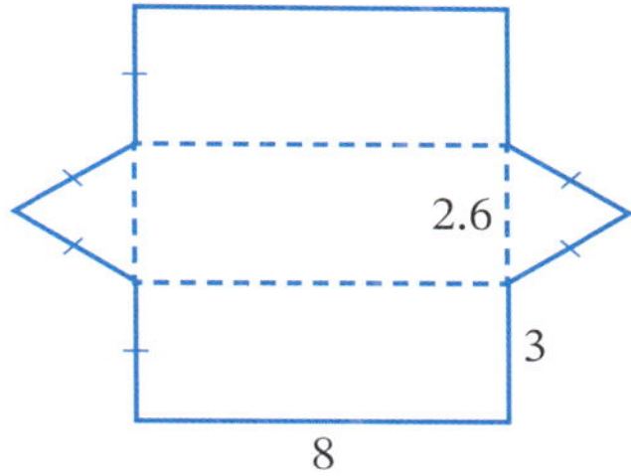

2 Calculate the surface area of each triangular prism.

Surface area of a prism 10.06

The **surface area** of a solid is the total area of all the faces of the solid. To calculate the surface area of a solid, find the area of each face and add the areas together.

Skillsheet
Solid shapes

Worksheets
Surface area 1
Nets of solids
Solid shapes

Presentation
Surface area

Technology
Surface area calculator

Spreadsheet
Surface area calculator

Video
Cubist art

Example 14

Calculate the surface area of each prism.

a cube

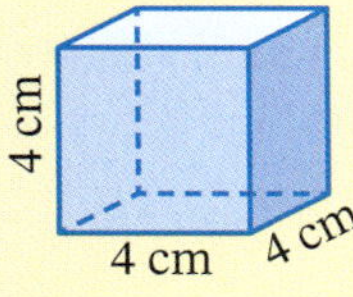

b rectangular prism

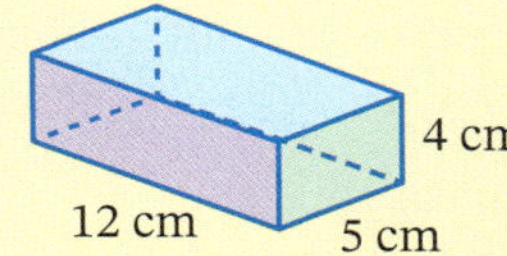

c open rectangular prism.

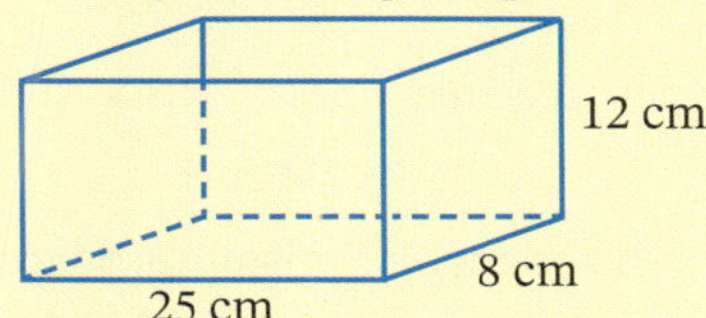

SOLUTION

a A cube has 6 identical faces in the shape of a square.

Surface area $= 6 \times 4^2$ — 6 × area of a square

$= 96$ cm^2

b A rectangular prism has 6 faces that are not all the same. However, opposite faces, such as the top and bottom, are the same.

Surface area = 2 top and bottom faces + 2 end faces + 2 side faces

$= (2 \times 5 \times 12) + (2 \times 5 \times 4) + (2 \times 12 \times 4)$

$= 256$ cm^2

c The open rectangular prism has no top so it has 5 faces that are not all the same.

Surface area = Bottom face + 2 front and back faces + 2 side faces

$= (25 \times 8) + (2 \times 25 \times 12) + (2 \times 8 \times 12)$

$= 992\text{ cm}^2$

For the surface area of an open solid, we only count the *external* surfaces, not the internal ones (so that we don't count each surface twice).

Videos
Surface area of a prism

Surface areas of prisms

Example 15

Calculate the surface area of each triangular prism.

a

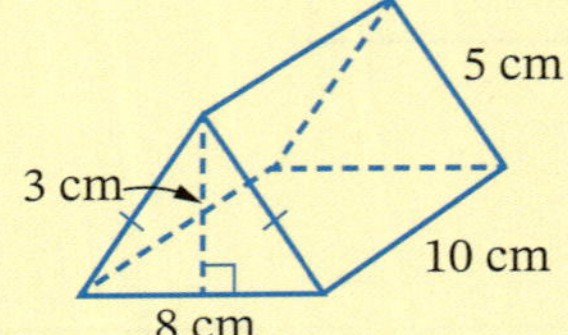

b

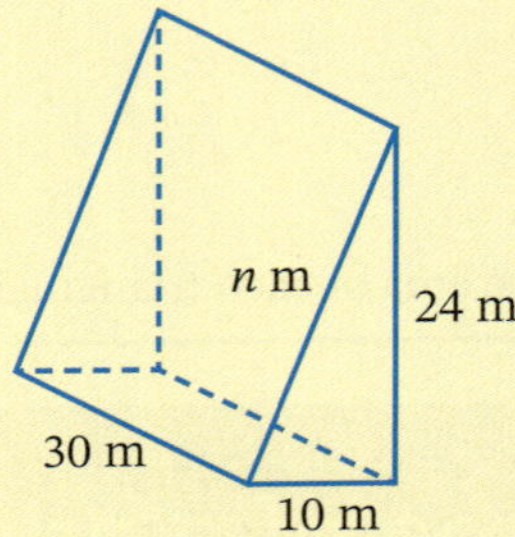

SOLUTION

a This triangular prism has 5 faces: 2 identical triangles (front and back), 2 identical rectangles (sides) and another rectangle (bottom). Drawing the net of the prism may help you see this.

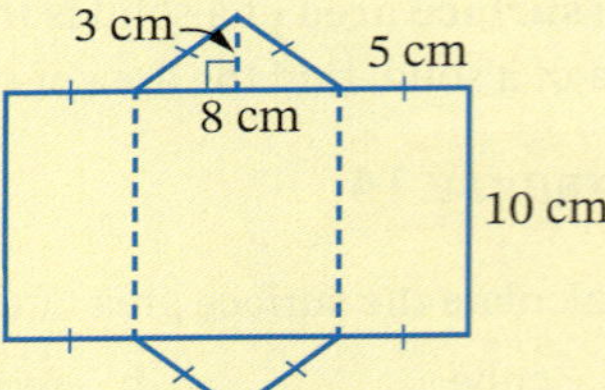

Surface area $= (2 \times \frac{1}{2} \times 8 \times 3) + (2 \times 10 \times 5) + (8 \times 10)$

$= 204\text{ cm}^2$

Front and back faces + side faces + bottom face

b This triangular prism has 5 faces: 2 identical triangles (front and back), 3 different rectangles.

To calculate the area of the left face, we need to know the value of n, which can be found using Pythagoras' theorem.

$$n^2 = 10^2 + 24^2$$
$$= 676$$
$$n = \sqrt{676}$$
$$= 26$$

Surface area $= (2 \times \frac{1}{2} \times 10 \times 24) + (30 \times 26) + (30 \times 10) + (30 \times 24)$

$= 2040\text{ m}^2$

Front and back + left + bottom + right

10.06

Example 16

Calculate the surface area of this trapezoidal prism.

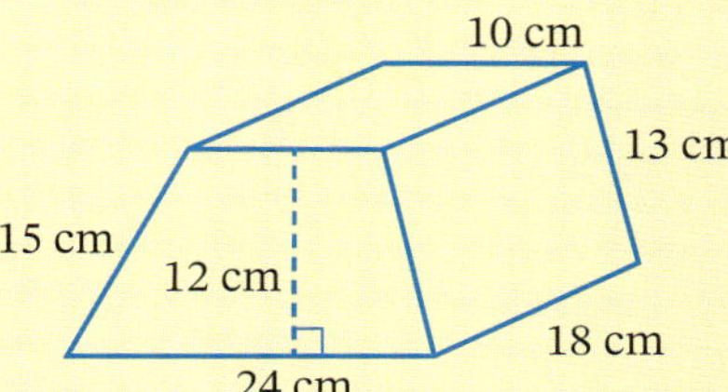

SOLUTION

This trapezoidal prism has 6 faces: 2 identical trapeziums (front and back), 4 different rectangles.

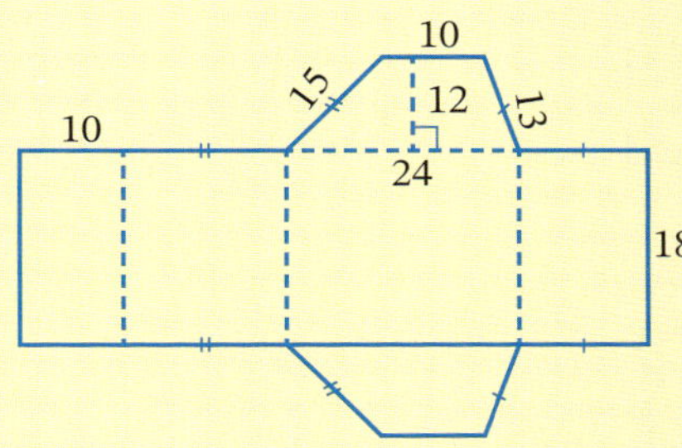

Area of each trapezium $= \frac{1}{2} \times (10 + 24) \times 12$

$= 204 \text{ cm}^2$

Surface area $= (2 \times 204) + (18 \times 10) + (18 \times 15) + (18 \times 24) + (18 \times 13)$ 2 trapeziums + 4 rectangles

$= 1524 \text{ cm}^2$

EXERCISE 10.06 ANSWERS ON P. 659

Surface area of a prism

U F PS R

1 What is the surface area of this rectangular prism? Select the correct answer **A**, **B**, **C** or **D**.

A 320 cm^2 **B** 352 cm^2

C 512 cm^2 **D** 768 cm^2

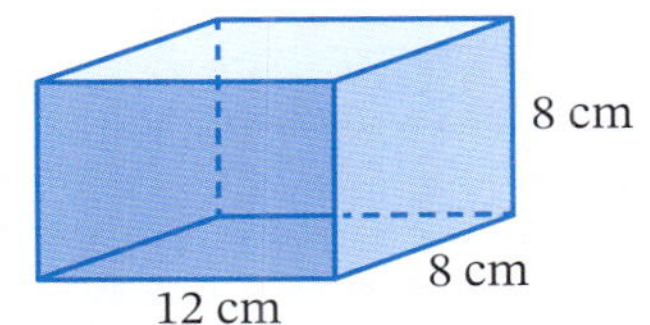

2 Draw a net of each solid.

a cube   **b** rectangular prism   **c** triangular prism   **d** cylinder

▷

Foundation Standard Complex

3 Match each net of a solid to the name of the solid.

triangular prism	cube	rectangular prism
square pyramid	cylinder	trapezoidal prism

a

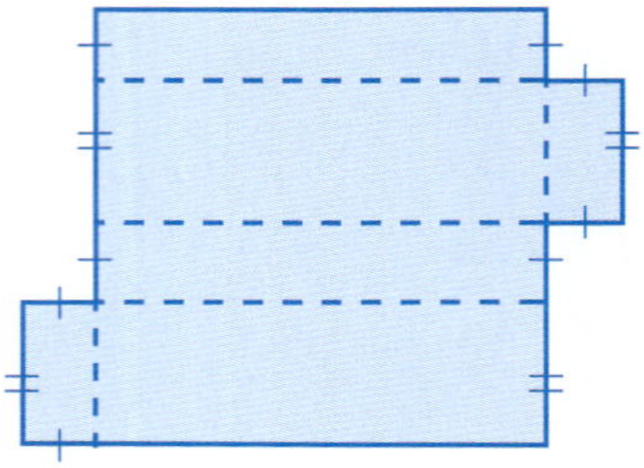

b

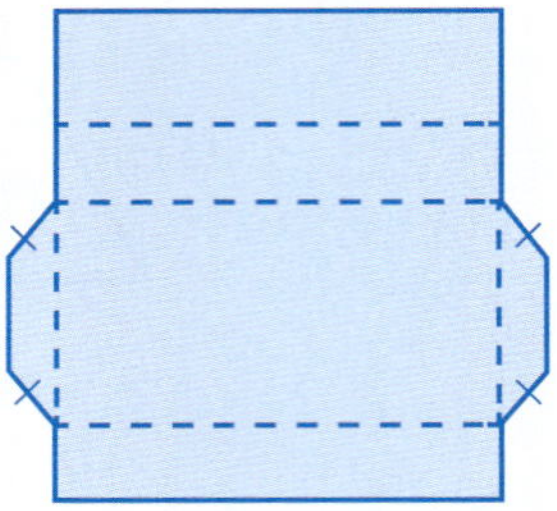

c

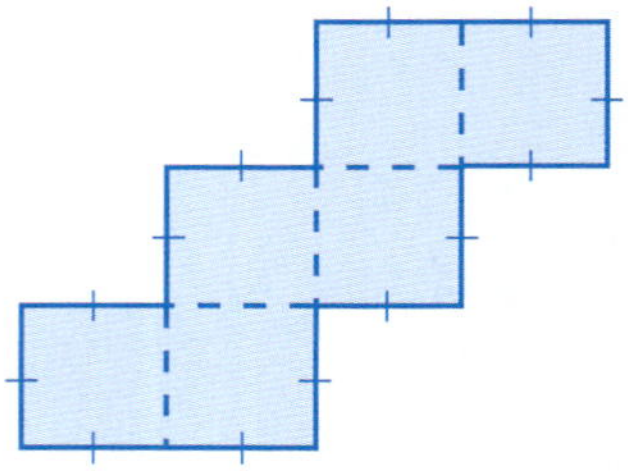

d

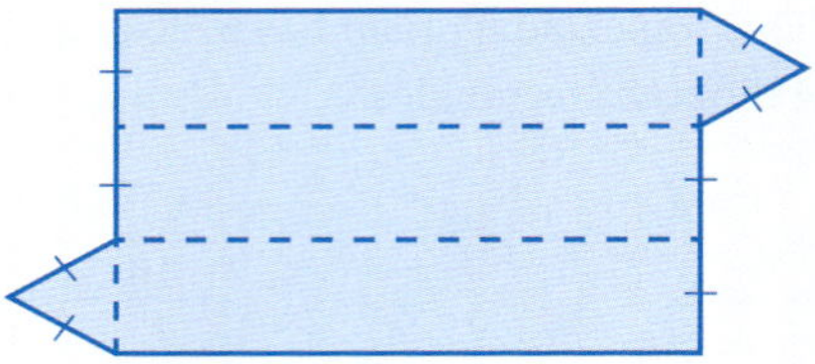

e

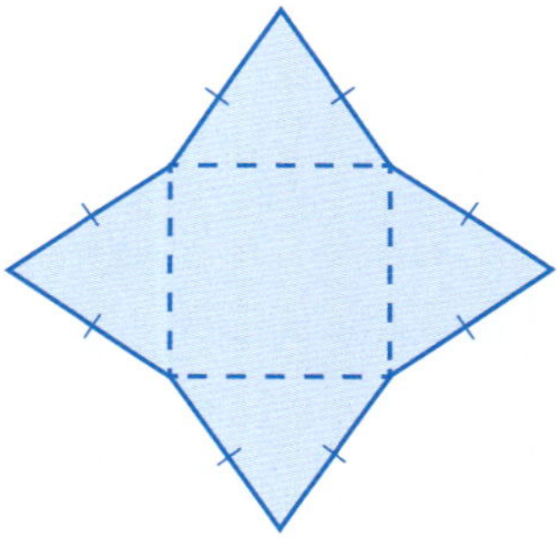

f

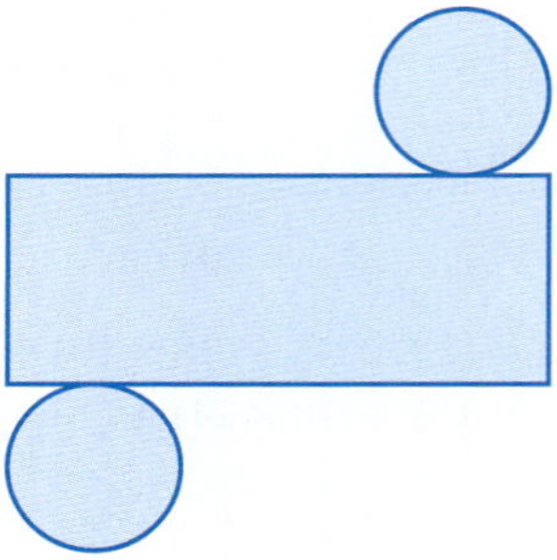

g

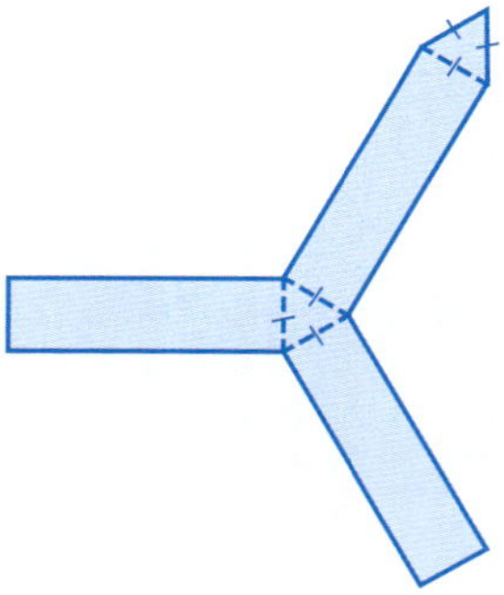

h

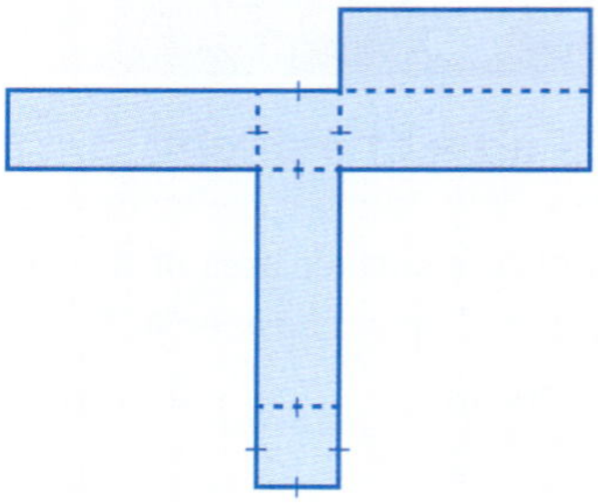

☐ Foundation ○ Standard ⬡ Complex

EXAMPLE 12

10.06

4 Name the prism that each net represents, then calculate the surface area of the prism. All lengths are in metres.

a

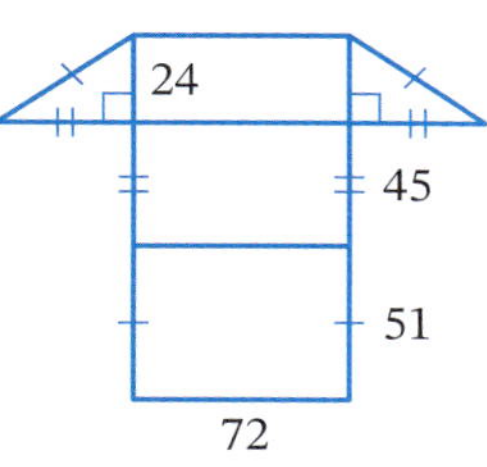

b

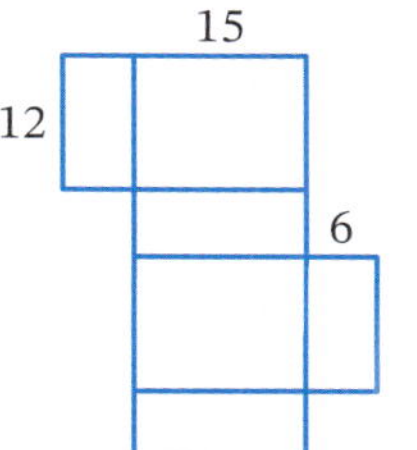

c

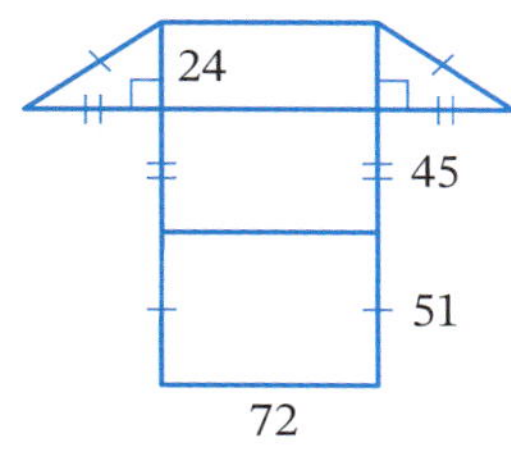

d

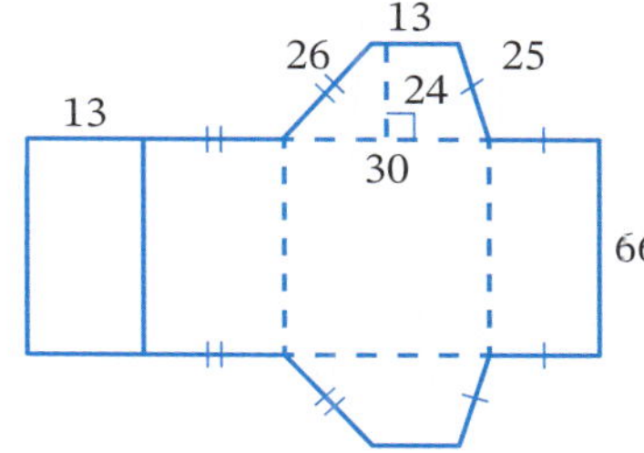

5 Calculate the surface area of each prism.

PS R

a cube

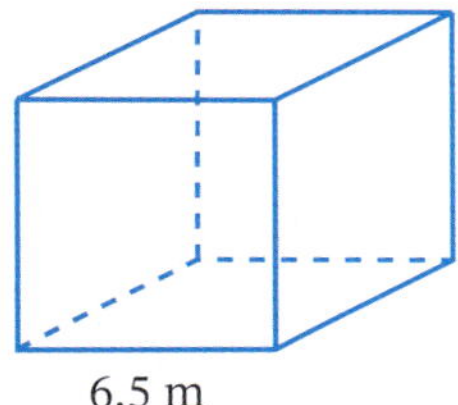

b rectangular prism

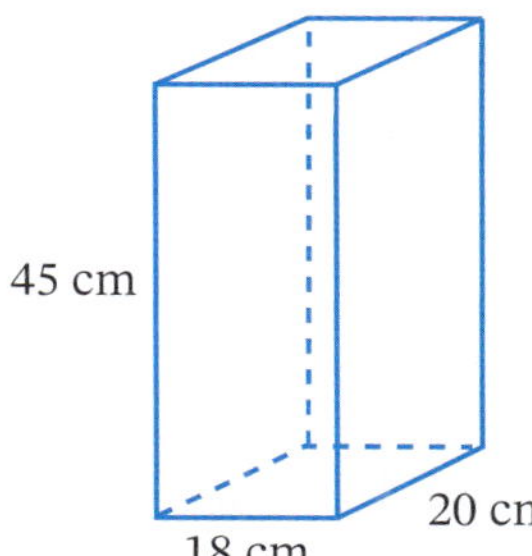

c open cube (no top)

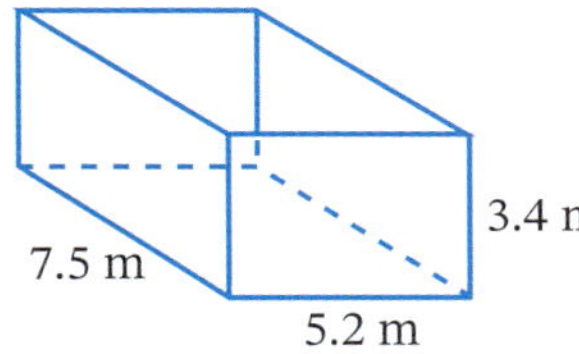

d open rectangular prism

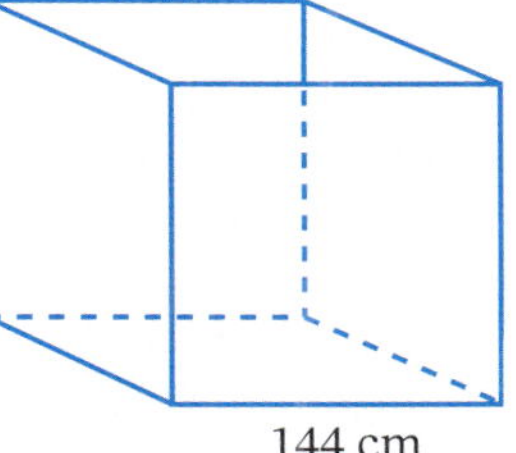

e rectangular prism

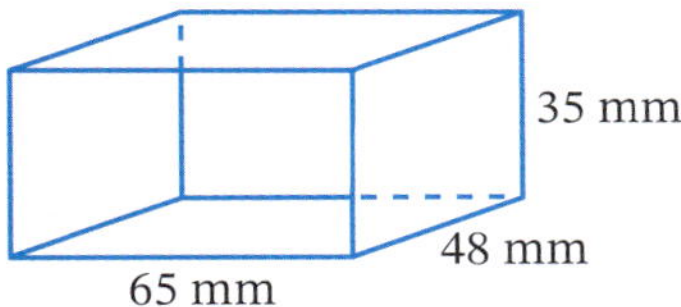

f cube, open one end open front and back

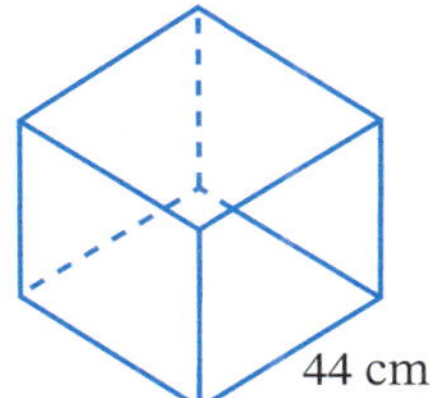

☐ Foundation ○ Standard ⬡ Complex

6 Find the surface area of each triangular prism.

PS R

a

8 m, 10 m, 15 m, 6 m

b

10 m, 6 m, 7 m, 16 m

c

29 cm, h, 40 cm, 15 cm

d

8 cm, 3.5 cm, r, 3.7 cm

e

12 m, 10 m, 5 m

f

9 m, 15 m, 12 m

7 Yasmin made this tent.

PS R

a Find the surface area of the tent, including the floor.

b Yasmin bought material at \$12 per square metre. How much did it cost her for the material to make the tent?

197 cm, 180 cm, 210 cm, 160 cm

8 5 cubes of side length 1 cm are joined to make this solid. What is the surface area of the solid? Select the correct answer **A**, **B**, **C** or **D**.

A 17 cm² **B** 18 cm²

C 5 cm² **D** 22 cm²

Foundation Standard Complex

9 PS R An art gallery has a width of 16 metres, a length of 30 metres and a slant height of 12 metres, as shown in the diagram.

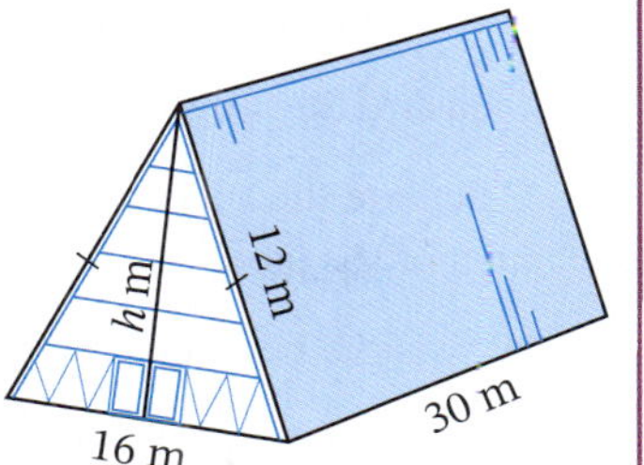

a Find the perpendicular height, h, of the building, correct to one decimal place.

b Find the area of the front wall.

c Find the surface area of the art gallery, without the floor.

10 PS R Calculate the surface area of each prism. Measurements are in centimetres.

EXAMPLE 16

a

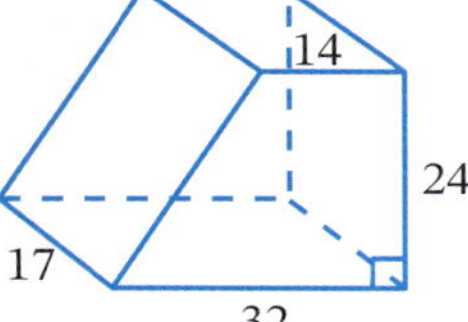

b

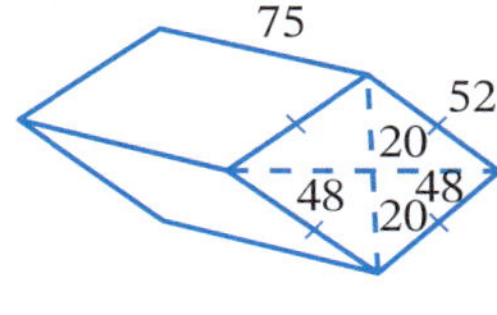

c

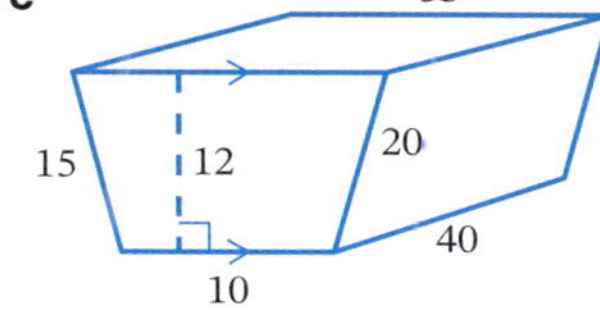

11 PS R This swimming pool is 15 m long and 10 m wide. The depth of the water ranges from 1 m to 3 m.

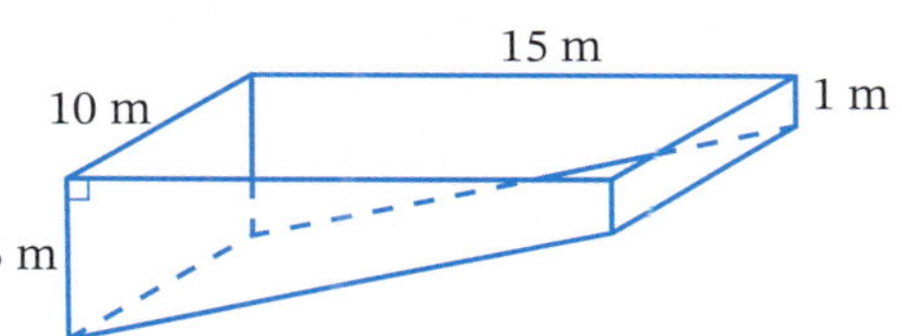

Calculate, correct to 2 decimal places:

a the area of the floor of the pool.

b the total surface area of the pool.

INVESTIGATION

Surface area of a cylinder

Collect 5 different cylindrical cans with paper labels.

1 Copy this table with rows for Cans 1 to 5 and complete it as you perform the following measurements and calculations.

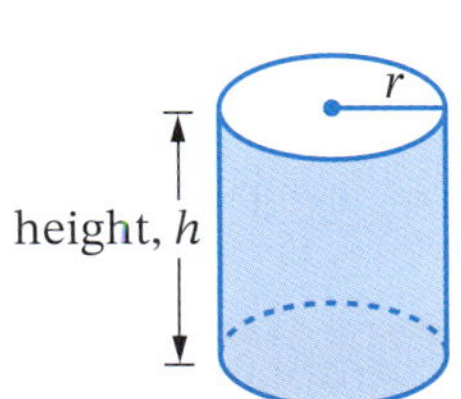

		Circular ends				Curved surface			Surface area
	Height	Diameter	Radius	Area	Circumference	Length	Width	Area	
Can 1									
Can 2									

2 Measure the height and diameter of the first can.

3 Calculate the radius, area and circumference of one circular end, correct to 2 decimal places.

4 Cut the label off the first can and lay it out flat. What shape is this curved surface?

5 Measure the length, width and area of the curved surface (label).

6 Calculate the surface area of the can by adding the areas of the 2 circular ends and the area of the label.

7 Why is the height of each can and the width of its label the same?

8 Why is the circumference of each circular end and the length of the label the same?

9 Repeat Steps **2** to **6** for the other 4 cans.

10.07 Surface area of a cylinder

Worksheets
Surface area 2

Applications of area 3

Puzzle
Car song

Technology
Surface area calculator

Spreadsheet
Surface area calculator

Video
Surface area of a cylinder

A closed **cylinder** has 3 faces made up of 2 circles (the circular ends) and a rectangle (the curved surface). The length of the rectangle is the circumference of the circular end, while the width of the rectangle is the height of the cylinder.

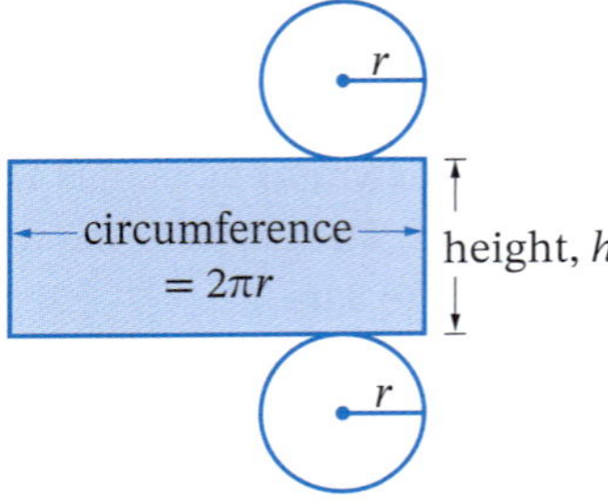

Surface area of a cylinder = area of 2 circles + area of rectangle

$$SA = 2 \times \pi r^2 + 2\pi r \times h$$
$$= 2\pi r^2 + 2\pi rh$$

ⓘ Surface area of a closed cylinder

$$SA = 2\pi r^2 + 2\pi rh$$

where r = radius of circular base

h = perpendicular height

The area of the 2 circular ends = $2\pi r^2$ and the area of the curved surface = $2\pi rh$.

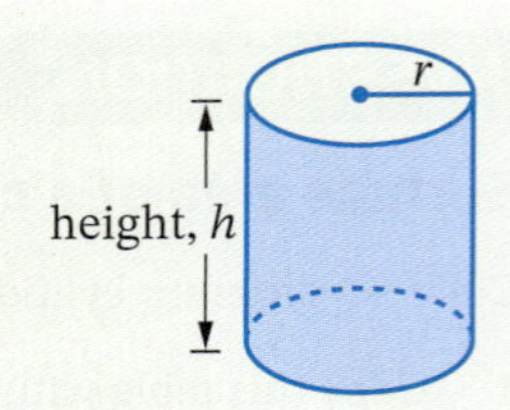

Example 17

Find, correct to one decimal place, the surface area of a cylinder with diameter 12 cm and height 20 cm.

SOLUTION

Radius = $\frac{1}{2} \times 12$ cm $\frac{1}{2}$ of diameter

= 6 cm

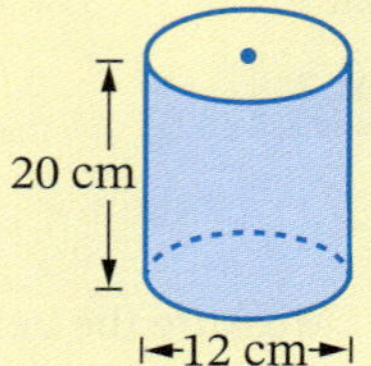

Surface area = area of 2 ends + area of the curved surface

$$SA = 2\pi r^2 + 2\pi rh$$
$$= 2\times\pi\times 6^2 + 2\times\pi\times 6\times 20$$
$$= 980.1769...$$
$$\approx 980.2 \text{ cm}^2$$

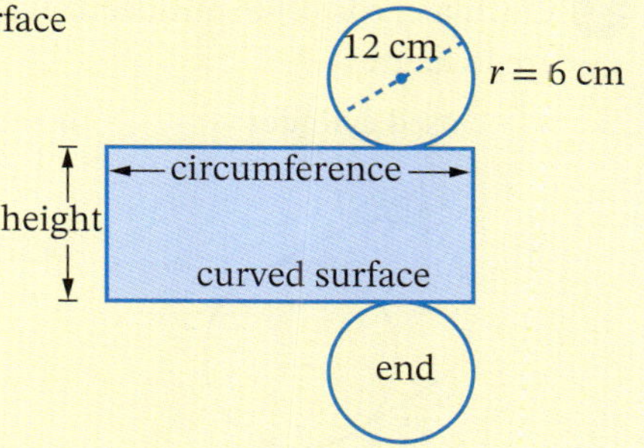

Example 18

Find, correct to 2 decimal places, the surface area of this open half-cylinder with radius 0.5 m and height 3 m.

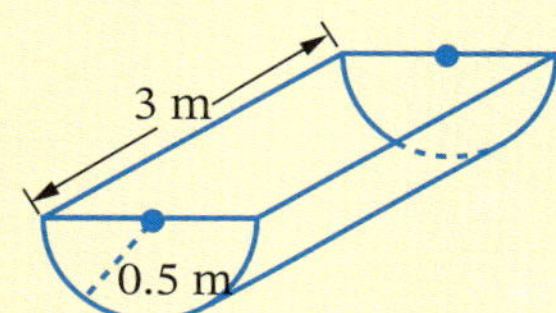

Note: For the surface area of an open solid, we only count the external surfaces, not the internal ones (so that we don't count each surface twice).

SOLUTION

Surface area = 2 semicircle ends + $\frac{1}{2}$ × curved surface

$$SA = 2\times(\tfrac{1}{2}\times\pi\times 0.5^2) + \tfrac{1}{2}\times(2\times\pi\times 0.5\times 3)$$
$$= 5.49778...$$
$$\approx 5.50 \text{ cm}^2$$

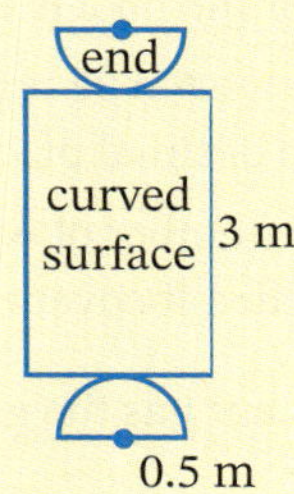

EXERCISE 10.07 ANSWERS ON P. 659

Surface area of a cylinder

U F PS R

1 Calculate, correct to 2 decimal places, the surface area of a cylinder with:

- **a** radius 7 m, height 10 m
- **b** diameter 28 cm, height 15 cm
- **c** diameter 6.2 m, height 7.5 m
- **d** radius 0.8 m, height 2.35 m

2 Find, correct to one decimal place, the curved surface area of a cylinder with:

- **a** diameter 9 cm, height 32 cm
- **b** radius 85 mm, height 16 mm

3 PS R Calculate, correct to one decimal place, the surface area of each solid. All lengths shown are in metres.

a closed cylinder

b cylinder with

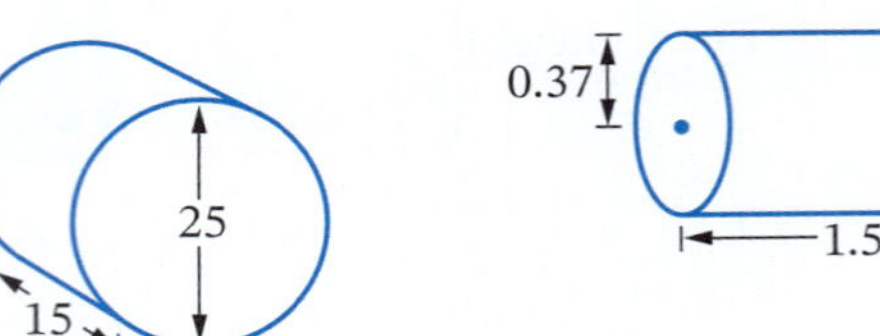

c closed half-cylinder one open end

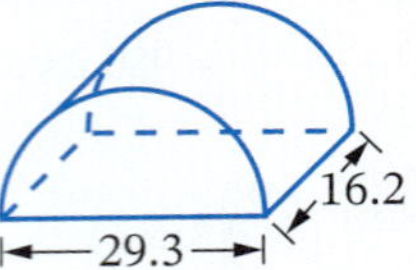

d half-cylinder open top

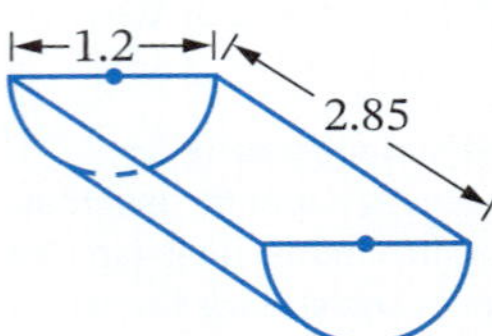

e half-cylinder with open top, one end open

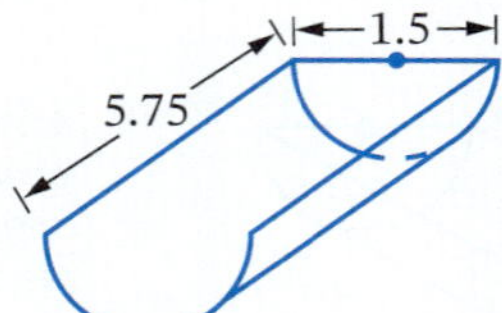

f cylinder open with at both ends

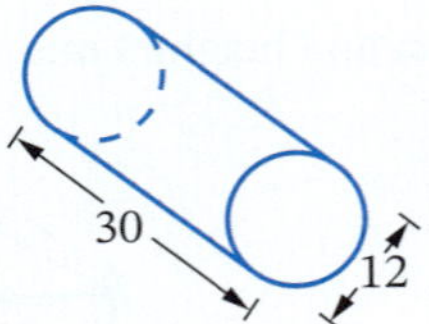

4 PS A swimming pool is in the shape of a cylinder 1.5 m deep and 6 m in diameter. The inside of the pool is to be repainted, including the floor. Find:

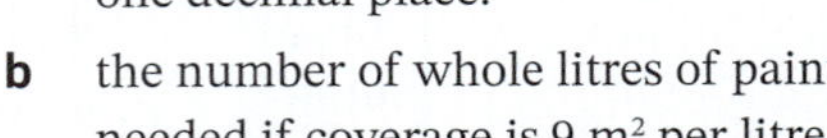

a the area to be repainted, correct to one decimal place.

b the number of whole litres of paint needed if coverage is 9 m^2 per litre.

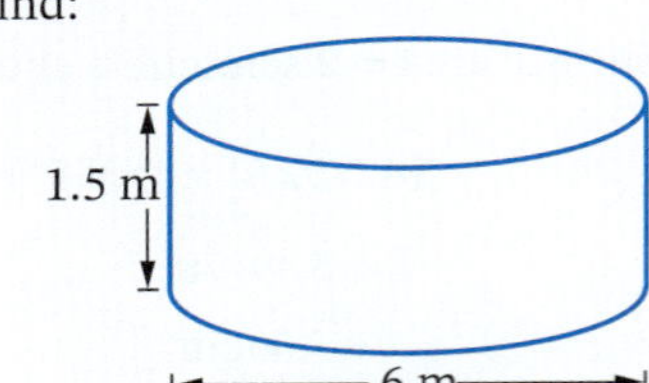

5 Which tent has the greater surface area?

a

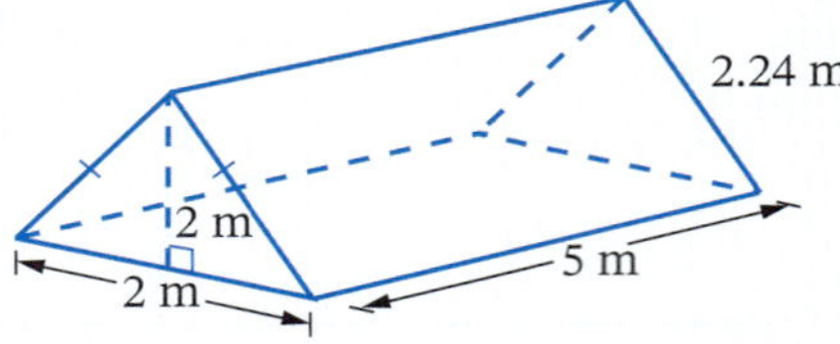

b

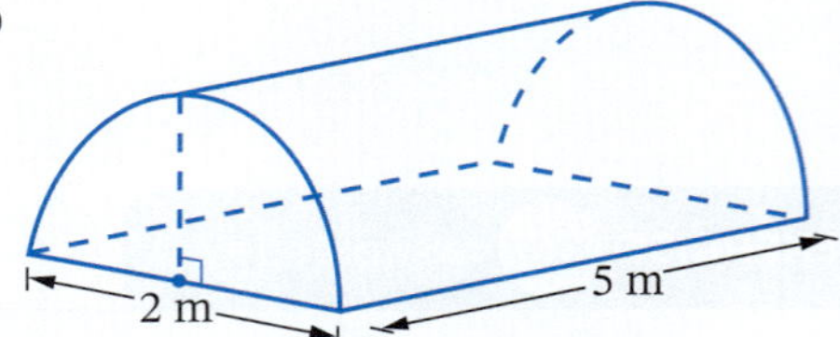

(*Note*: the floor is included for both tents)

Foundation Standard Complex

6 PS This lampshade is to be covered. Find the area of material needed, correct to one decimal place, if 10% extra is allowed for seams and overlaps. Note: The lampshade is a cylinder open at both ends.

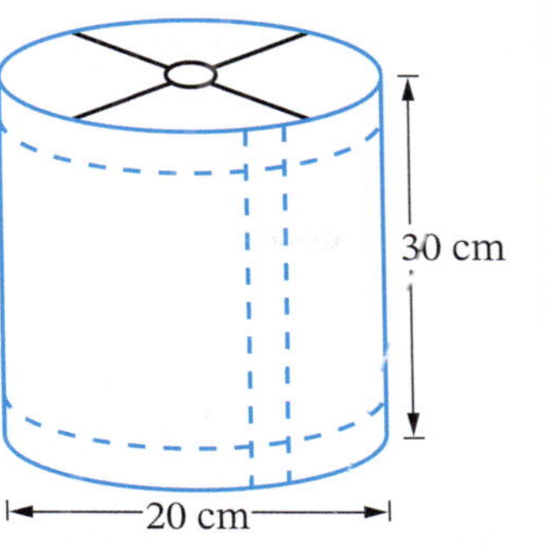

Volumes of prisms and cylinders 10.08

The **volume** of a solid is the amount of space it takes up. Volume is measured in **cubic units**, for example, cubic metres (m^3) or cubic centimetres (cm^3).

The **capacity** of a container is the amount of fluid (liquid or gas) it holds, measured in millilitres (mL), litres (L), kilolitres (kL) and megalitres (ML).

Worksheet Volume and capacity

Skillsheet What is volume?

Puzzle Formula matching game

Volume and capacity

1 cm^3 contains 1 mL.

1 m^3 contains 1000 L or 1 kL.

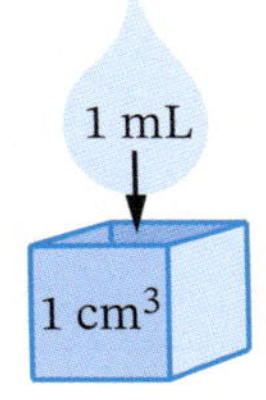

× 1 000 000 =

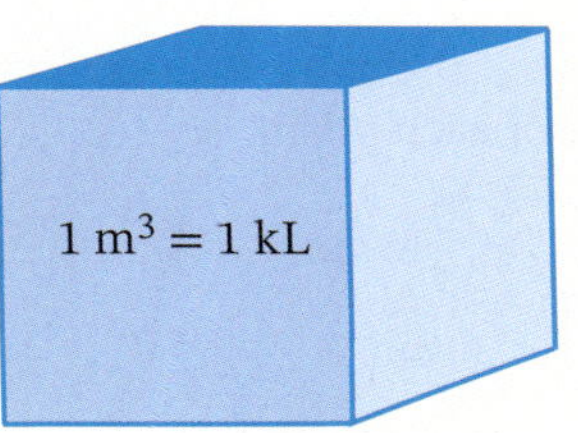

Volume of a prism

A **cross-section** of a solid is a 'slice' of the solid cut *across* it, parallel to its end faces, rather than along it. A **prism** has the same (uniform) cross-section along its length, and each cross-section is a **polygon** (with straight sides).

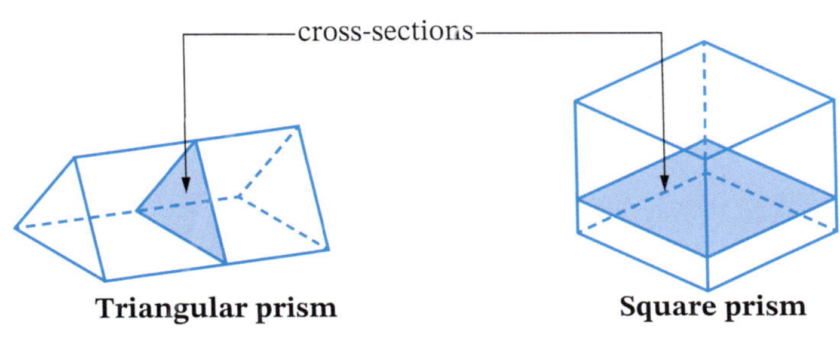

An end face of a prism is called its **base**. Prisms take their names from their base and cross-section. For example, this prism is a **trapezoidal prism** because its base and cross-sections are trapeziums.

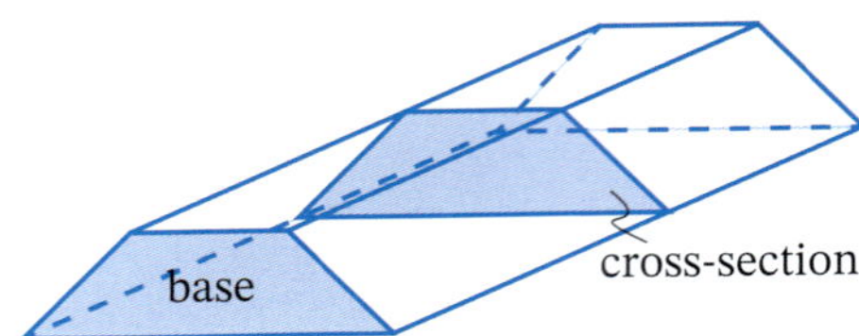

Trapezoidal prism

Because a prism is made up of identical cross-sections, its volume can be calculated by multiplying the **area of its base** by its **perpendicular height** (the length or depth of the prism).

ⓘ Volume of a prism

$V = Ah$

where A = area of base and

h = perpendicular height

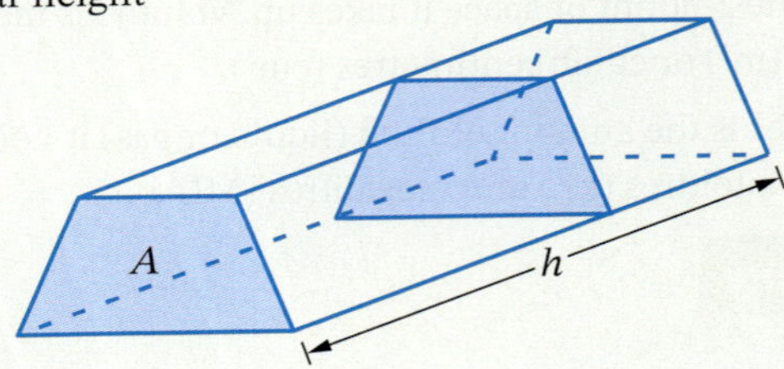

Example 19

Video Volumes of prisms and cylinders

Find the volume of each prism.

a

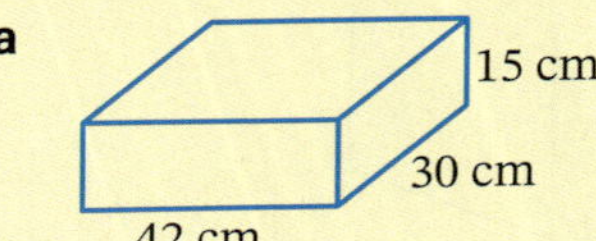

b

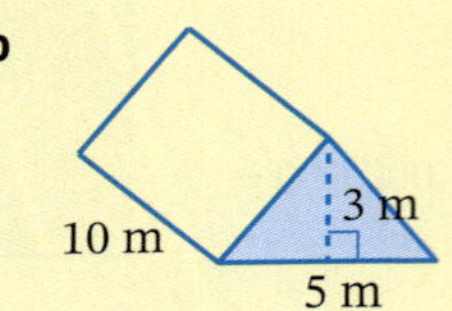

c

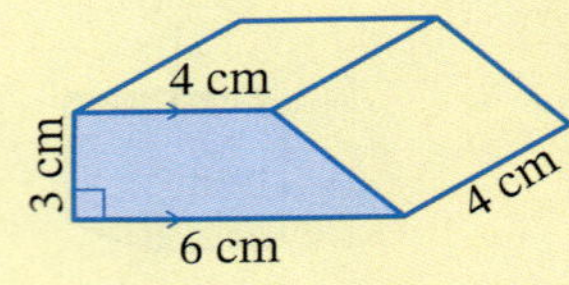

SOLUTION

a $V = 42 \times 30 \times 15$

$= 18\,900\text{ cm}^3$

For a rectangular prism, volume = length × width × height ($V = lwh$).

b $A = \frac{1}{2} \times 5 \times 3$

$= 7.5$

Area of a triangle.

$V = 7.5 \times 10$

$= 75\text{ m}^3$

$V = Ah$, where height $h = 10$

c $A = \frac{1}{2} \times (4 + 6) \times 3$

$= 15\text{ cm}^2$

Area of a trapezium.

$V = 15 \times 4$

$= 60\text{ cm}^3$

$V = Ah$, where height $h = 4$

9780170465618

Example 20

A storage box in the shape of a rectangular prism is 110 cm long, 65 cm wide and 70 cm high.

a Find the volume of the box in cm^3.

b What is the volume in m^3?

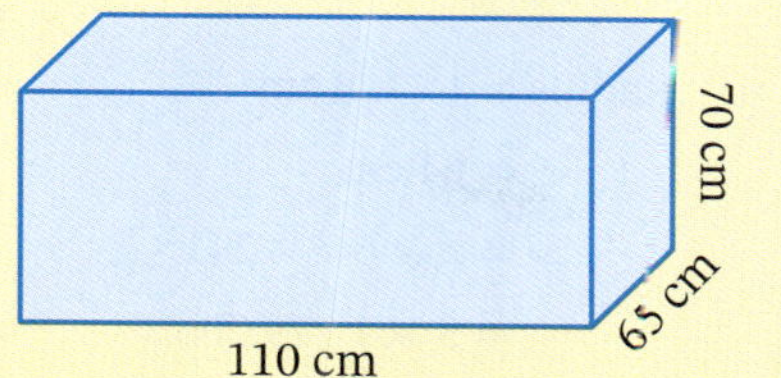

SOLUTION

a $V = 110 \times 65 \times 70$

$= 500\ 500\ cm^3$

Alternatively, you could convert 110 cm, 65 cm and 70 cm to metres, then multiply them together for the volume.

b $1\ m = 100\ cm$, so $1\ m^3 = 100^3\ cm^3 = 1\ 000\ 000\ cm^3$

$\therefore V = 500\ 500 \div 1\ 000\ 000$

$= 0.5005\ m^3$

Volume of a cylinder

A cylinder is like a 'circular prism' because its cross-sections are identical circles. Because of this, we can also use $V = Ah$ to find the volume of a cylinder. But for a circle, $A = \pi r^2$, so:

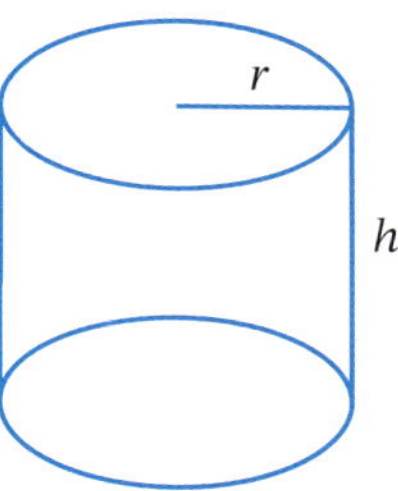

Volume $= Ah$

$V = \pi r^2 \times h$

$= \pi r^2 h$

ⓘ Volume of a cylinder

$V = \pi r^2 h$

where r = radius of circular base

h = perpendicular height

Example 21

For this cylinder, calculate:

a its volume, correct to the nearest cm^3.

b its capacity in kL, correct to one decimal place.

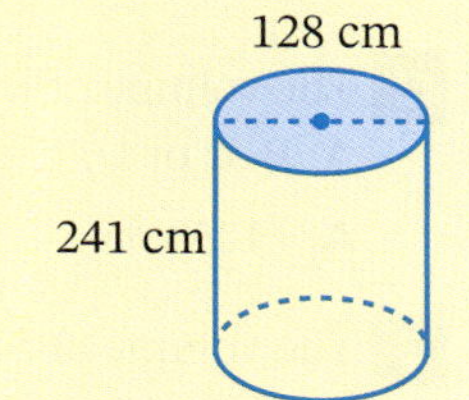

Video
Volumes of prisms and cylinders

SOLUTION

a Radius $= \frac{1}{2} \times 128$ cm 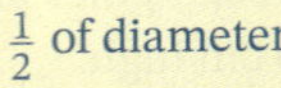 $\frac{1}{2}$ of diameter

$= 64$ cm $\quad V = \pi r^2 h$

$V = \pi \times 64^2 \times 241$

$= 3\,101\,179.206\ldots$

$\approx 3\,101\,179 \text{ cm}^3$

b Capacity $= 3\,101\,179$ mL $\quad 1 \text{ cm}^3 = 1$ mL

$= (3\,101\,179 \div 1000 \div 1000)$ kL

$= 3.101\,179$ kL

≈ 3.1 kL

kL ← ÷ 1000 ← L ← ÷ 1000 ← mL

EXERCISE 10.08 ANSWERS ON P. 659

Volumes of prisms and cylinders

1 Find the volume of each rectangular prism using the formula $V = lwh$.

a

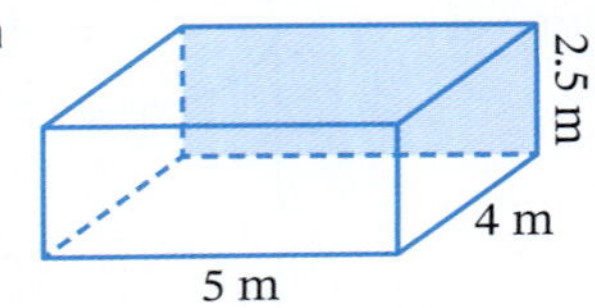

b

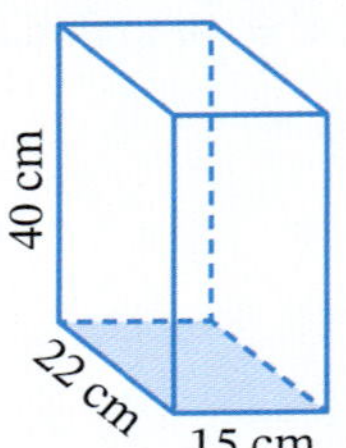

c

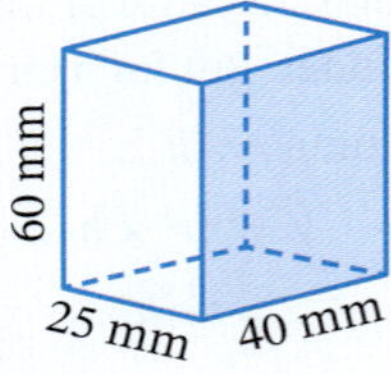

d

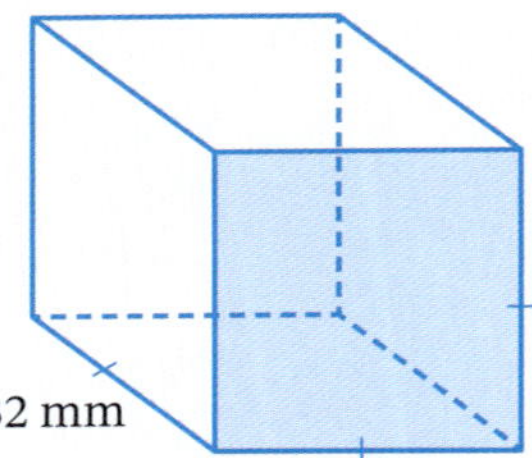

e

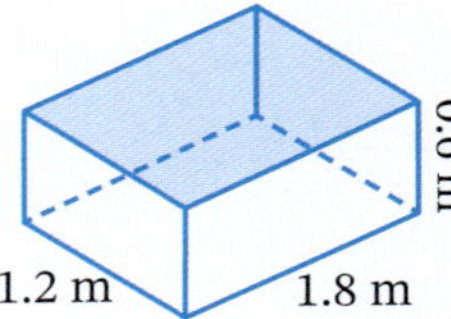

f

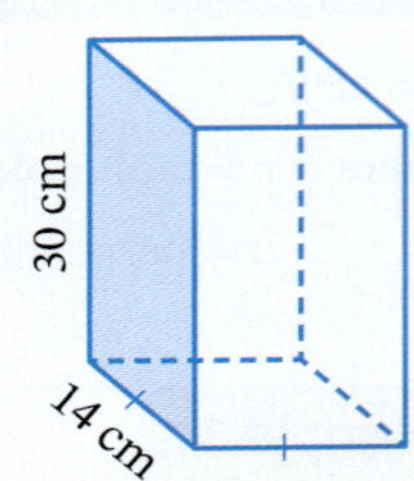

2 The volume of a cube is 125 000 cm^3. Convert this to m^3. Select the correct answer **A**, **B**, **C** or **D**.

A 1250 m^3 **B** 12.5 m^3 **C** 1.25 m^3 **D** 0.125 m^3

3 The volume of a small jewellery box is 160 000 mm^3. Which expression gives its volume in cm^3? Select the correct answer **A**, **B**, **C** or **D**.

A $160\,000 \div 10^3$ **B** $160\,000 \div 100$

C $160\,000 \div 100^2$ **D** $160\,000 \div 10^2$

▷

☐ Foundation ○ Standard ⬡ Complex

4 Find the volume of this toolbox in cubic metres by:

PS **a** calculating the volume in cm^3, then converting to m^3.

R

C **b** converting each length to metres first, then calculating the volume.

Which method is easier?

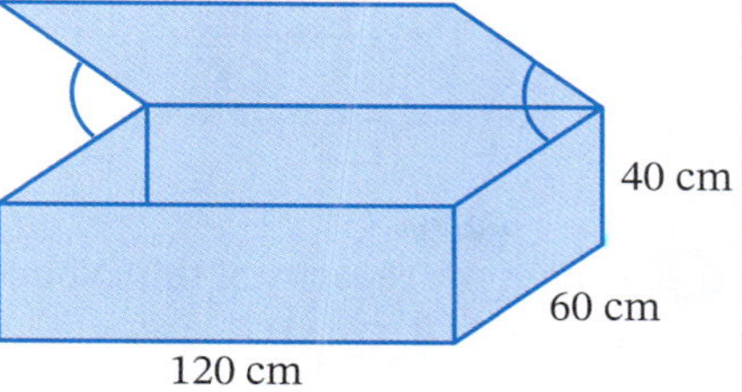

5 Find the volume of each solid.

a

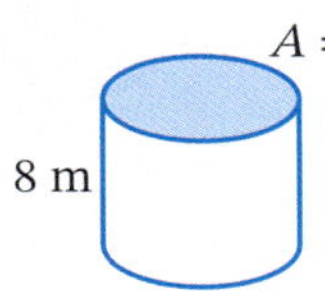

b

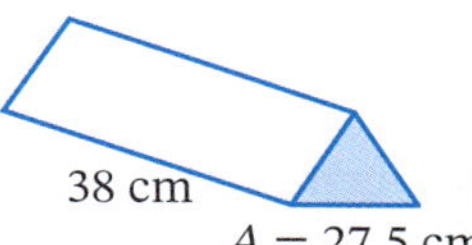

c

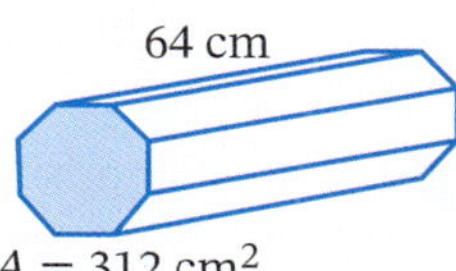

EXAMPLE 21

6 Find, correct to nearest whole unit, the volume of each cylinder.

a

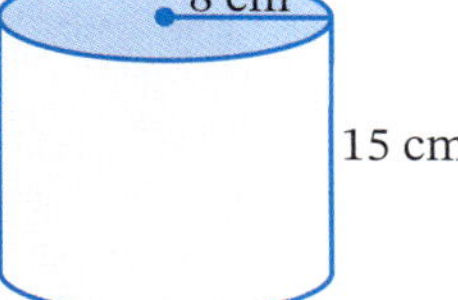

b

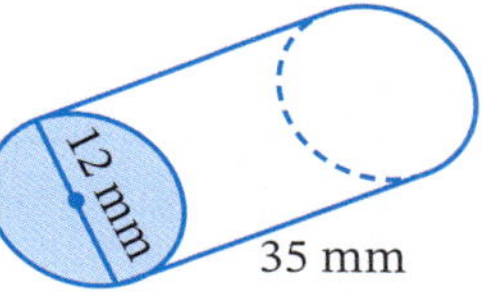

c

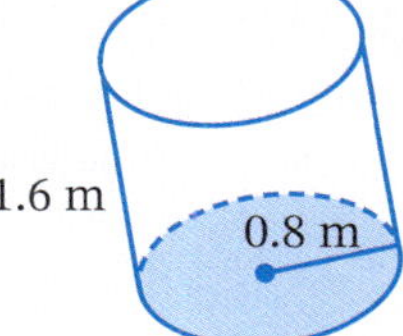

d

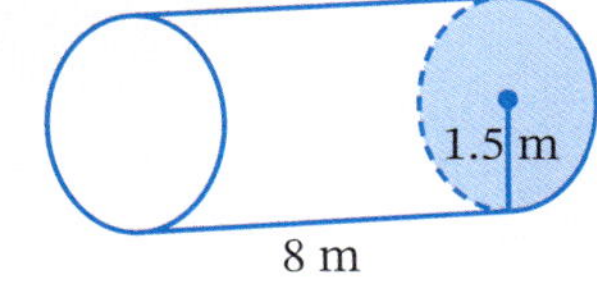

e

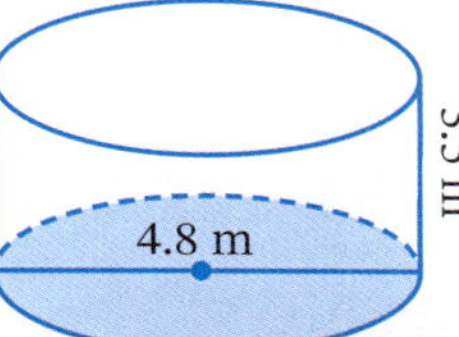

f

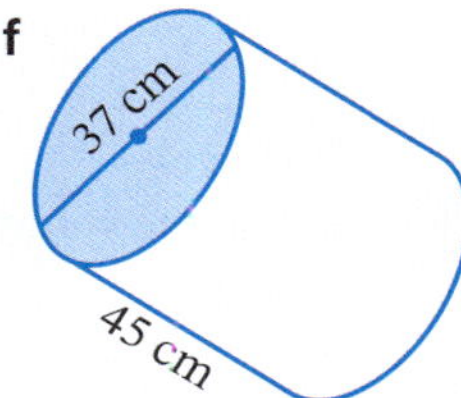

7 Calculate, correct to one decimal place, the volume of each solid. All lengths are in metres.

a

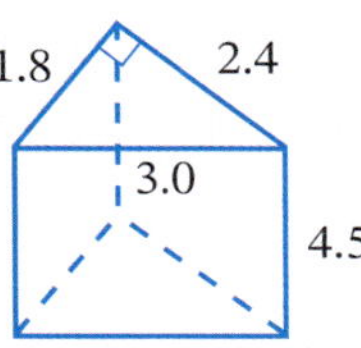

b

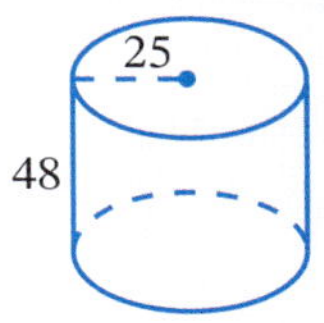

c

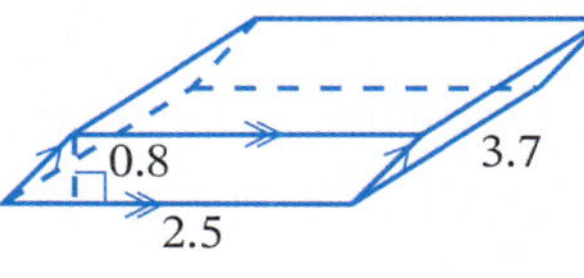

d

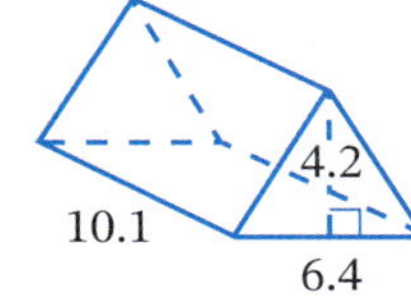

e

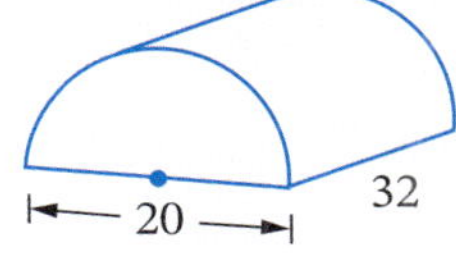

f

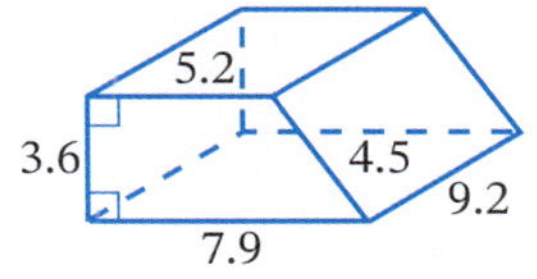

Foundation Standard Complex

g

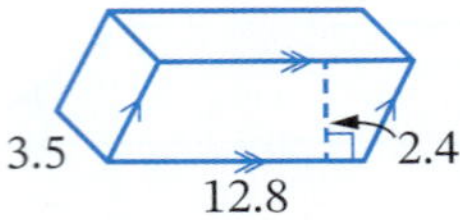

h

i

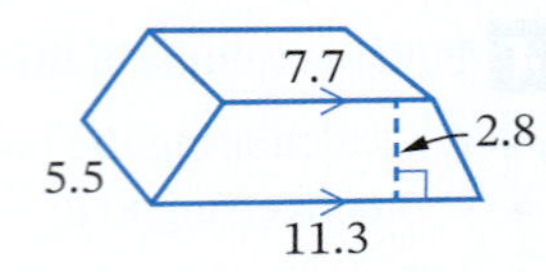

8 What is the capacity of this cylinder in litres? Select the closest answer **A**, **B**, **C** or **D**.

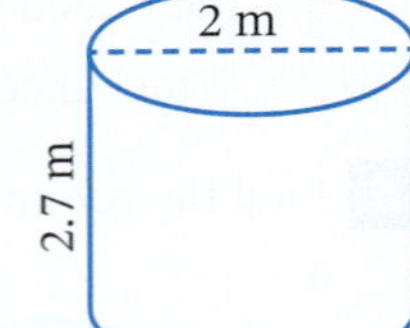

A 33 900 L

B 8482 L

C 6280 L

D 16 964 L

9 A triangular prism has base length 15 cm, height 8 cm and volume 630 cm^3. What is its length?

R

10 A fish tank that is 55 cm long, 24 cm wide and 22 cm high is filled to 4 cm below the top. Calculate the amount of water in the tank in litres.

11 A wedding cake with 3 tiers rests on a table. Each tier is 6 cm high. The layers have radii of 20 cm, 15 cm and 10 cm respectively. Find the total volume of the cake, correct to the nearest cm^3.

Shutterstock.com/John Wollwerth

12 Calculate the volume of the skip b in, correct to the nearest m^3.

Shutterstock.com/Stephen Rees

13 This swimming pool is 15 m long and 10 m wide. The depth of the water ranges from 1 m to 3 m. Calculate the capacity of this pool in kilolitres.

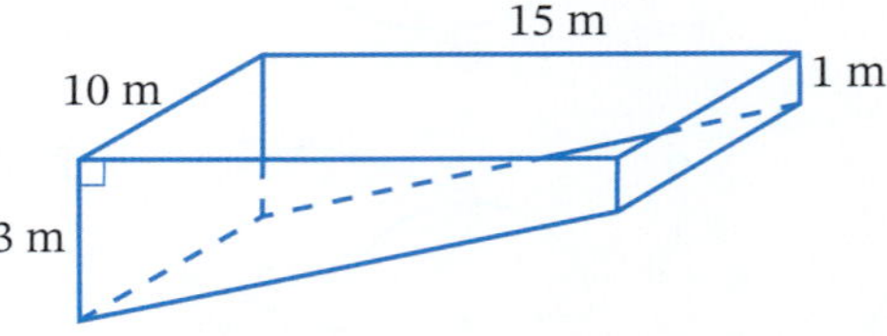

Foundation Standard Complex

10.08

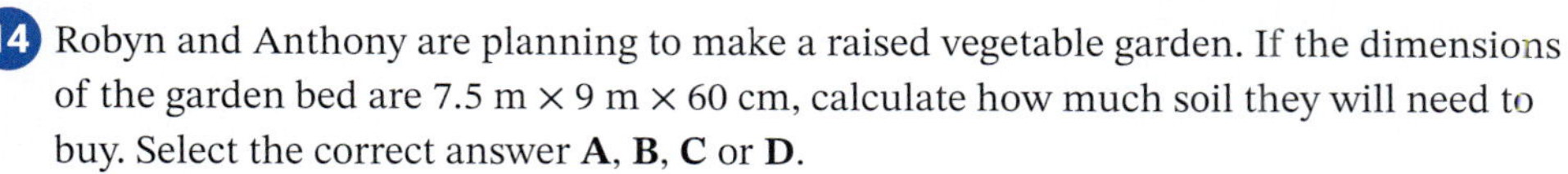

14 Robyn and Anthony are planning to make a raised vegetable garden. If the dimensions of the garden bed are 7.5 m × 9 m × 60 cm, calculate how much soil they will need to buy. Select the correct answer **A**, **B**, **C** or **D**.

A 405 m^3 **B** 40.5 m^3 **C** 4050 cm^3 **D** 4050 m^3

15 PS R This metal pipe has an inner diameter of 8.5 cm and an outer diameter of 9.5 cm. Calculate, correct to 2 decimal places, the volume of metal needed to make the pipe.

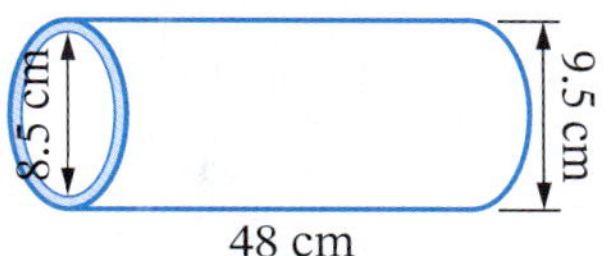

16 A cube has a volume of 2150 mm^3. What is its side length, to the nearest millimetre?

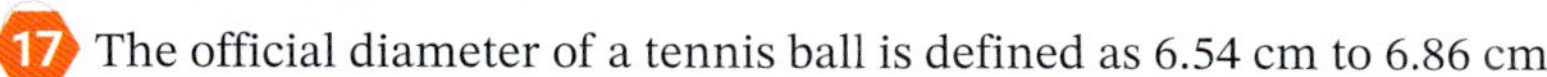

17 PS R The official diameter of a tennis ball is defined as 6.54 cm to 6.86 cm.

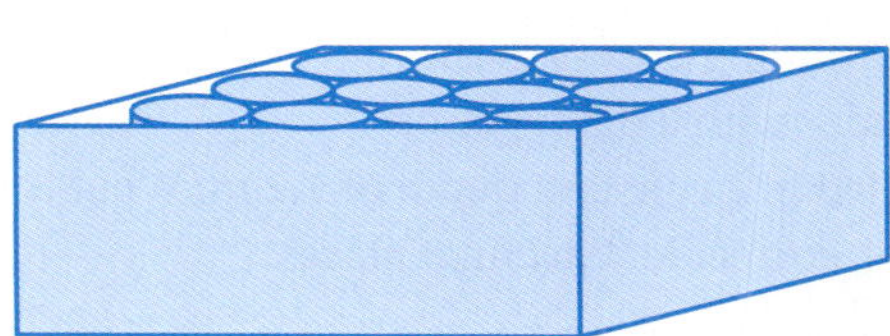

a Find the average diameter of a tennis ball.

b The tennis balls are packed in cylinders as shown. Using the larger diameter, calculate the radius and height of the cylinders.

c **i** To find the volume of the cylinder, its radius and height are rounded *up* to one decimal place. Explain why, and write the radius and height rounded up.

ii Hence, calculate the volume of a cylinder, correct to one decimal place.

d For packaging, 12 cylinders are put in a cardboard box, as shown above.

i What are the dimensions of the box?

ii Calculate the volume of the box.

e Calculate the volume of the box not taken up by the cylinders.

f What percentage (correct to one decimal place) of the box is space not taken up by the cylinders?

☐ Foundation ○ Standard ○ Complex

INVESTIGATION

Drinking the swimming pool dry

Charlotte's mum told her that it was healthy to drink 6 full glasses of water each day. Assume that Charlotte uses a cylindrical glass with a height of 9 cm and a diameter of 7 cm.

Shutterstock.com/Ground Picture

1 Calculate the volume of water Charlotte drinks each day, in millilitres, correct to 4 decimal places.

2 How much water will Charlotte drink in a year? Answer in litres, correct to 4 decimal places.

3 The family's swimming pool is in the shape of a cylinder with a diameter of 8.3 m, and is filled to a constant depth of 1.2 m. What is the amount of water in the pool? Answer in litres, correct to 4 decimal places.

4 How long would it take Charlotte to drink the equivalent of the swimming pool? Select the closest answer **A**, **B**, **C** or **D**.

A 6 years B 26 years C 56 years D 86 years

INVESTIGATION

Volume vs surface area

You will need: At least 20 centicubes

1 Copy the following table.

Length	Width	Height	Volume	Surface area
			8	
			8	
			⋮	

2 Build as many rectangular prisms as you can with a volume of 8 cubes, but with different dimensions. *Note*: $1 \times 1 \times 8$ is the same as $1 \times 8 \times 1$ and $8 \times 1 \times 1$.

3 Record your dimensions in the table and calculate the surface area of each prism.

4 What are the dimensions of the solid that has the smallest surface area?

5 What name do we give to this solid?

6 Repeat the experiment for a volume of:

a 12 cubes b 20 cubes.

☆ MENTAL SKILLS 10B ANSWERS ON P. 660

Maths without calculators

Quiz
Mental skills 10B

Finding 15%, $2\frac{1}{2}$%, 25% and $12\frac{1}{2}$%

- To find 10% or $\frac{1}{10}$ of a number, divide by 10.
- To find 5% of a number, find 10% first, then halve it (since 5% is half of 10%).
- So to find 15% of a number, find 10% and 5% of the number separately, then add the answers together.

1 Study each example.

a $15\% \times 80 = (10\% \times 80) + (5\% \times 80)$
$= 8 + 4$
$= 12$

b $15\% \times \$170 = (10\% \times \$170) + (5\% \times \$170)$
$= \$17 + \$8.50$
$= \$25.50$

c $15\% \times 3600 = (10\% \times 3600) + (5\% \times 3600)$
$= 360 + 180$
$= 540$

d $15\% \times \$28 = (10\% \times \$28) + (5\% \times \$28)$
$= \$2.80 + \$1.40$
$= \$4.20$

2 Now find 15% of each amount.

a 120 **b** \$840 **c** 260 **d** \$202
e \$50 **f** 72 **g** \$180 **h** 400
i \$1600 **j** \$22 **k** 6000 **l** \$350

To find $2\frac{1}{2}$% of a number, first find 5%, then halve it.

3 Study each example.

a $2\frac{1}{2}\% \times 600$
$10\% \times 600 = 60$
$5\% \times 600 = \frac{1}{2} \times 60 = 30$
$2\frac{1}{2}\% \times 600 = \frac{1}{2} \times 30 = 15$

b $2\frac{1}{2}\% \times \$820$
$10\% \times \$820 = \$82$
$5\% \times \$820 = \frac{1}{2} \times 82 = \$41$
$2\frac{1}{2}\% \times \$820 = \frac{1}{2} \times \$41 = \$20.50$

4 Now find $2\frac{1}{2}$% of each amount.

a 400 **b** 6640 **c** \$2000 **d** \$880
e 1500 **f** \$232 **g** 5400 **h** \$904

To find 25% of a number, halve the number twice as $25\% = \frac{1}{4}$.

5 Study each example.

a $25\% \times 700$

$50\% \times 700 = \frac{1}{2} \times 700 = 350$

$25\% \times 700 = \frac{1}{2} \times 350 = 175$

b $25\% \times \$86$

$50\% \times \$86 = \frac{1}{2} \times \$86 = \$43$

$\therefore 25\% \times \$86 = \frac{1}{2} \times \$43 = \$21.50$

6 Now find 25% of each of each amount.

a 2000 **b** \$80 **c** 18 **d** \$25
e \$324 **f** \$140 **g** 66 **h** 298
i \$780 **j** \$1700 **k** \$126 **l** 1160

To find $12\frac{1}{2}\%$ of a number, find 25% first, then halve it. In other words, halve 3 times because $12\frac{1}{2}\% = \frac{1}{8}$.

7 Study each example.

a $12\frac{1}{2}\% \times 400$

$50\% \times 400 = \frac{1}{2} \times 400 = 200$

$25\% \times 400 = \frac{1}{2} \times 200 = 100$

$12\frac{1}{2}\% \times 400 = \frac{1}{2} \times 100 = 50$

b $12\frac{1}{2}\% \times \$144$

$50\% \times \$144 = \frac{1}{2} \times \$144 = \$72$

$25\% \times \$144 = \frac{1}{2} \times \$72 = \$36$

$12\frac{1}{2}\% \times \$144 = \frac{1}{2} \times \$36 = \$18$

8 Now find $12\frac{1}{2}\%$ of each amount.

a 1280 **b** \$12 **c** 60 **d** \$260
e \$540 **f** \$250 **g** 304 **h** 1360

POWER PLUS ANSWERS ON P. 660

1 A spiral is formed from 4 semicircles as shown. The diameter of the smallest semicircle is 10 cm, and the semicircles are 10 cm apart. Find the total length of the spiral:

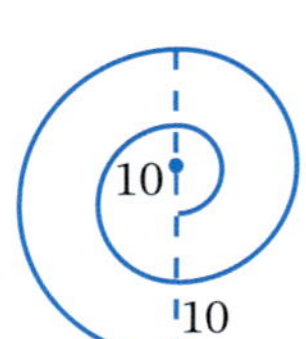

a correct to 2 decimal places

b in terms of π.

2 In this pattern, all lengths are in metres. Calculate, correct to 2 decimal places:

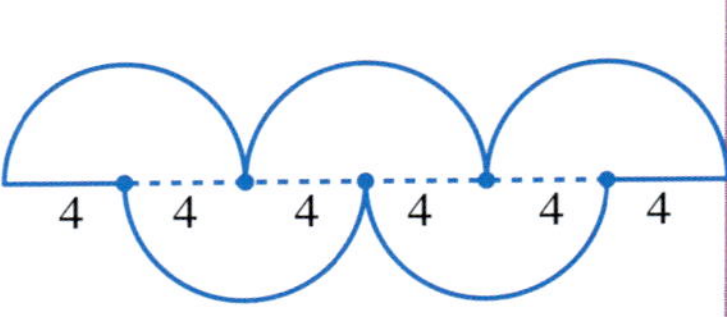

a its perimeter

b its area

3 A rhombus has diagonals of length 24 cm and 18 cm. Find its perimeter.

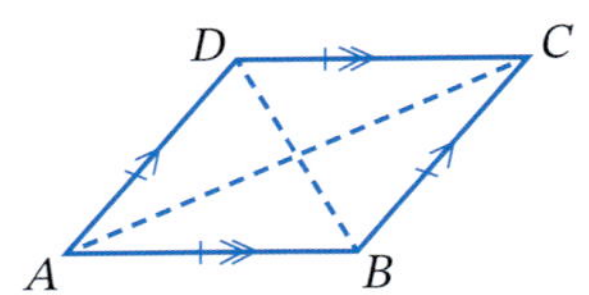

9780170465618

4 To qualify for the next round of the discus trials, Bronte must throw the discus beyond the 40 m arc. Find, correct to the nearest square metre, the shaded disqualifying area of the sector, given that the small circle has a radius of 1 m.

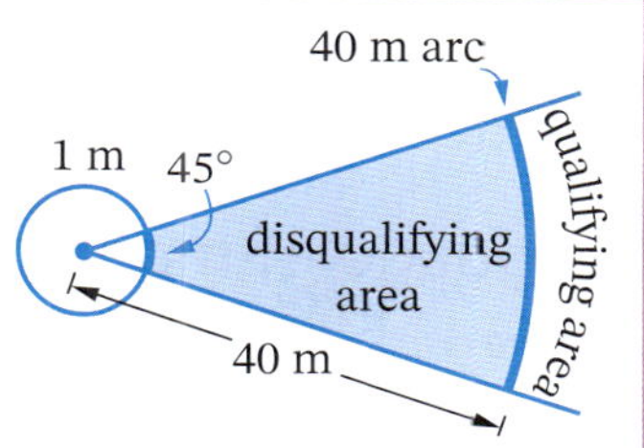

5 For each composite shape calculate, correct to one decimal place:

i its area

ii its perimeter.

All measurements are in centimetres.

a
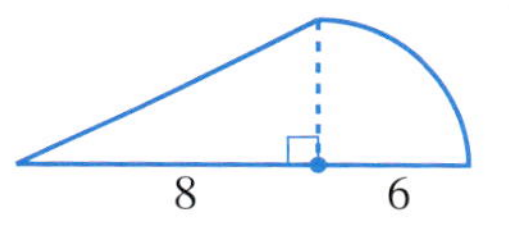

b
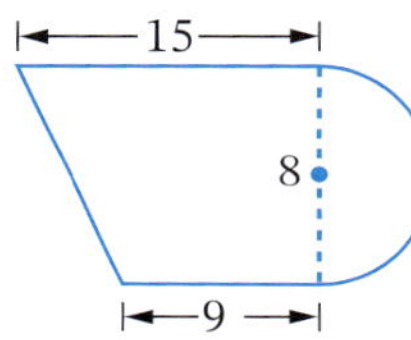

c
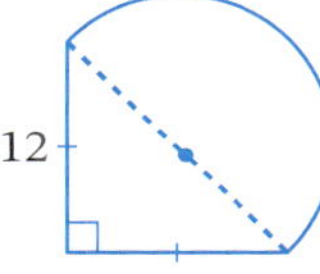

6 The 3 faces of a rectangular prism have areas as shown. Calculate the volume of the prism.

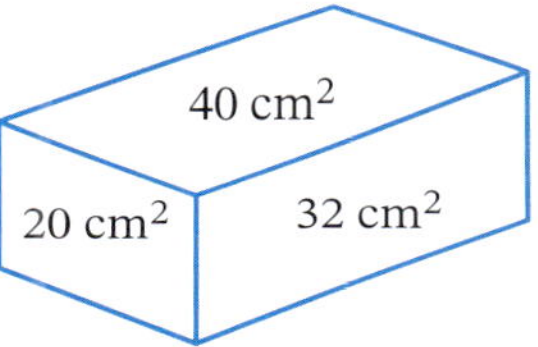

7 A cube opened at one end has an external surface area of 1125 cm^2. Find its volume.

8 A 10-m flat square roof drains into a cylindrical rainwater tank with a diameter of 4 m. If 5 mm of rain falls on the roof, by how much (to the nearest mm) does the level of water in the tank rise?

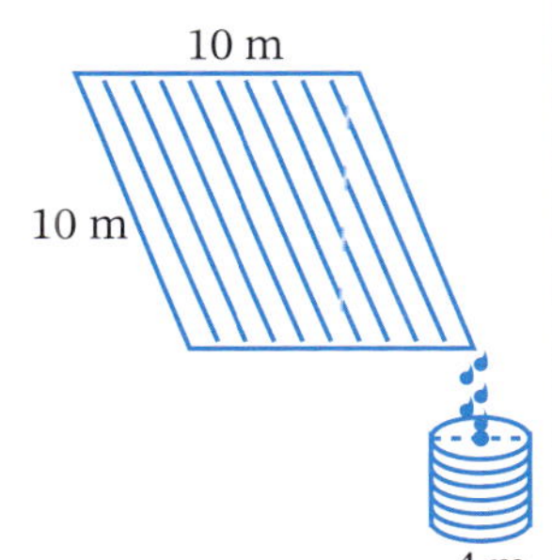

9 Calculate correct to one decimal place the volume of a tetrahedron with a side length of 24 cm.

10 CHAPTER REVIEW

Language of maths

Quiz Language of maths 10

absolute error	arc length	base	capacity
circumference	cross-section	curved surface	diameter
giga-	kilo-	mega-	micro- (μ)
net	percentage error	perimeter	perpendicular height
pi (π)	radius/radii	rectangular prism	relative error
sector	surface area	triangular prism	volume

1 What is the difference between a measured value and the actual value called?

2 Draw a circle and label on it:

a a sector **b** a diameter **c** an arc length

3 Which metric prefix means 'one-millionth'?

4 Write the definition of a **prism** using the word **cross-section**.

5 How do you find the **surface area** of a solid?

6 Name a solid that has a **curved surface**.

Topic summary

Worksheet Mind map: Surface area and volume

- Write 10 questions (with solutions) that could be used in a test for this chapter. Include some questions that you have found difficult to answer.
- Swap your questions with another student and check their solutions against yours.
- List the sections of work in this chapter that you did not understand. Follow up this work with your friend or teacher.

Print (or copy) and complete this mind map of the topic, adding detail to its branches and using pictures, symbols and colour where needed. Ask your teacher to check your work.

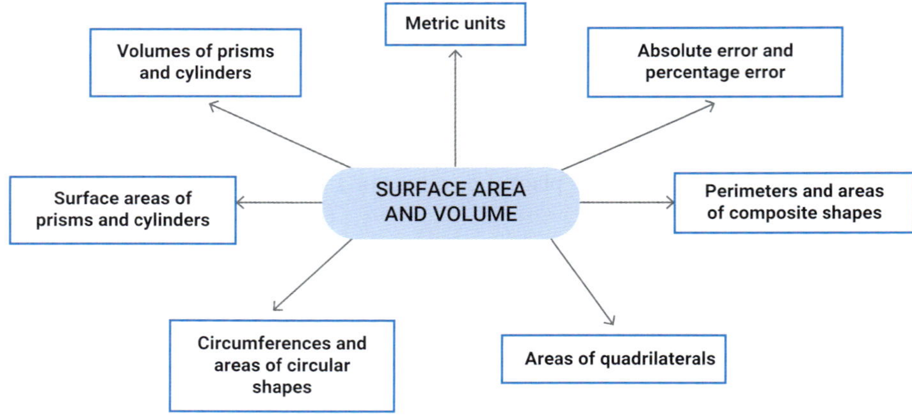

10 TEST YOURSELF

ANSWERS ON P. 664

1 Convert: 10.01

a	8 km to m	**b**	15 min to s	**c**	0.7 m to mm
d	250 000 g to kg	**e**	8.4 GB to MB	**f**	9.5 t to kg
g	300 min to h	**h**	34 000 B to kB	**i**	125 kL to L
j	2500 cm to km	**k**	17 000 000 kg to t	**l**	8900 GB to TB
m	10 500 ms to s	**n**	9.1 millennia to years	**o**	2.6 Mm to m
p	3 ly to km	**q**	0.000 000 4 s to μs	**r**	75 000 000 μm to m

Quiz
Test yourself 10

2 Yazmin estimates there are 550 jellybeans in a jar at the school fete. If there were actually 615 jellybeans in the jar, what is the absolute error in Yazmin's prediction? 10.02

3 A vet measures the weight of a puppy to be 5 kg. Its actual weight is 5.37 kg. Calculate the percentage error of the vet's measurement. 10.02

4 For each measuring instrument, state: 10.02

i the size of one unit on the scale

ii the absolute error.

a

b

c

Shutterstock.com/Sergey Melnikov

d

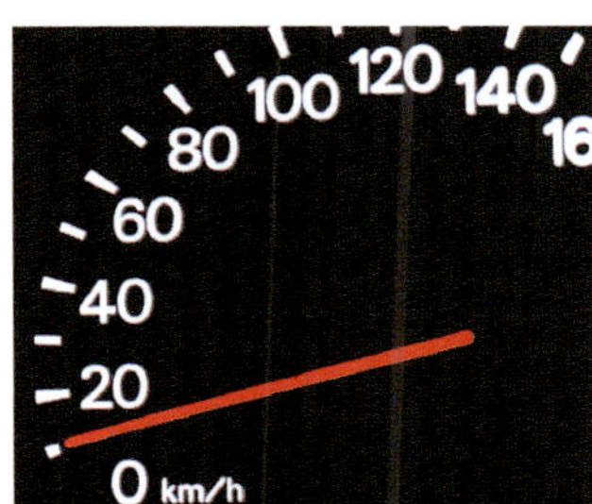

5 Calculate the perimeter of each shape. 10.03

a

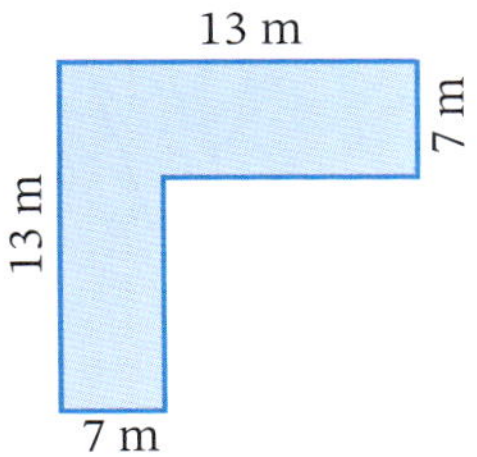

b

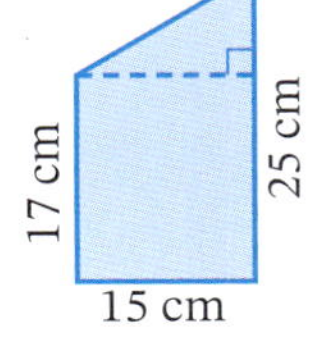

c
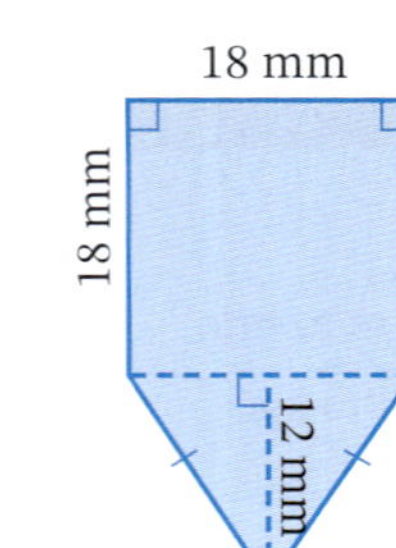

d
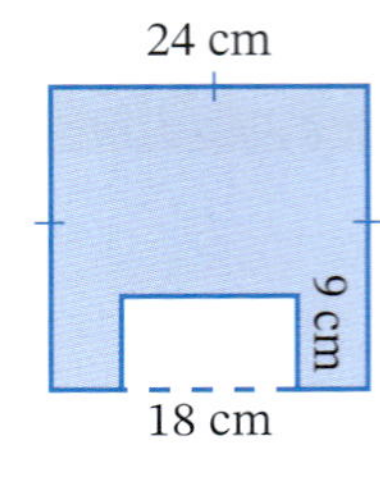

10.03

6 Calculate the shaded area of each shape.

a
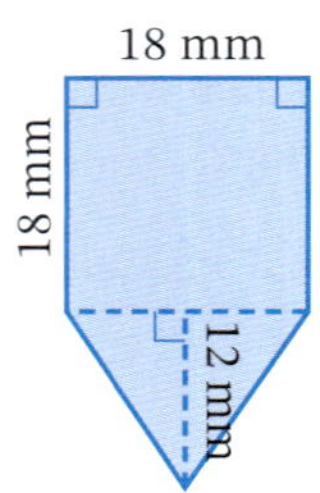

b
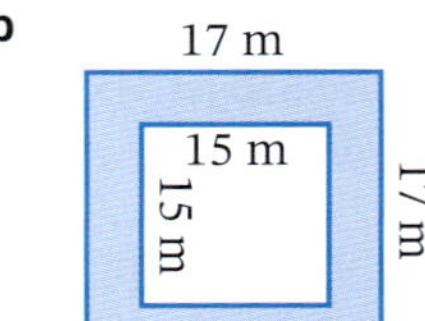

c
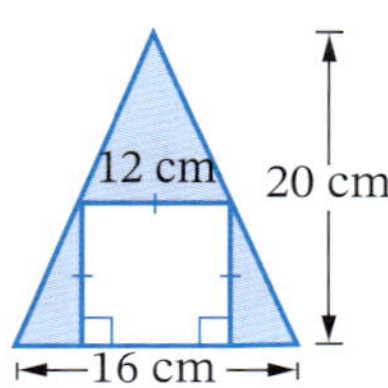

d
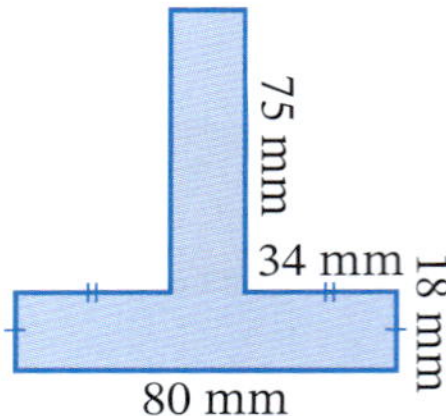

10.04

7 Calculate the area of each quadrilateral.

a
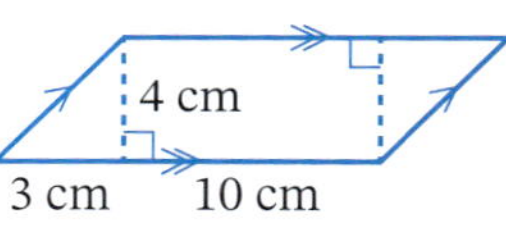

b
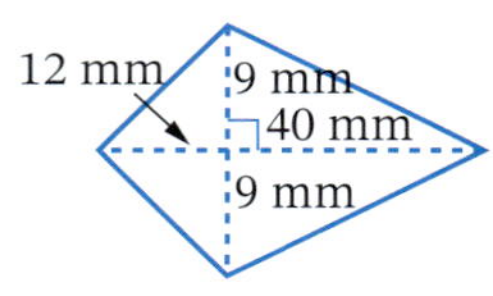

c
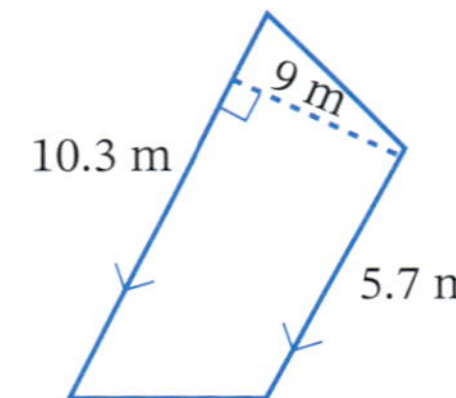

d
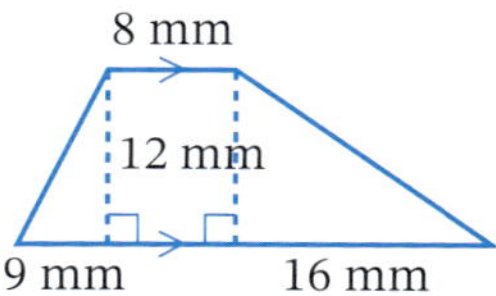

e
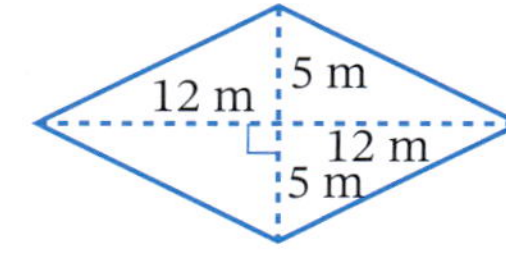

f
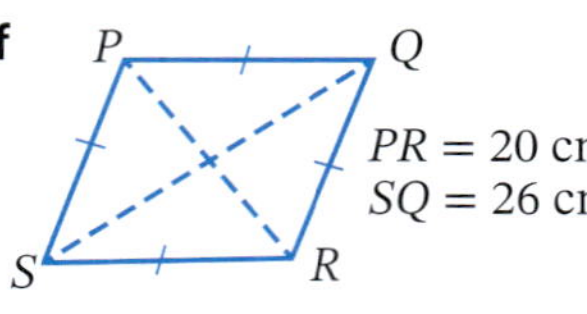

Foundation Standard Complex

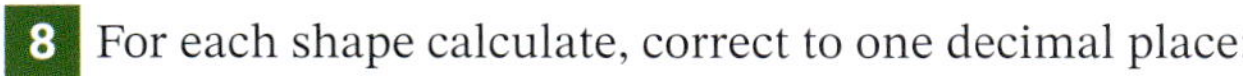

8 For each shape calculate, correct to one decimal place:

i its perimeter **ii** its area.

a

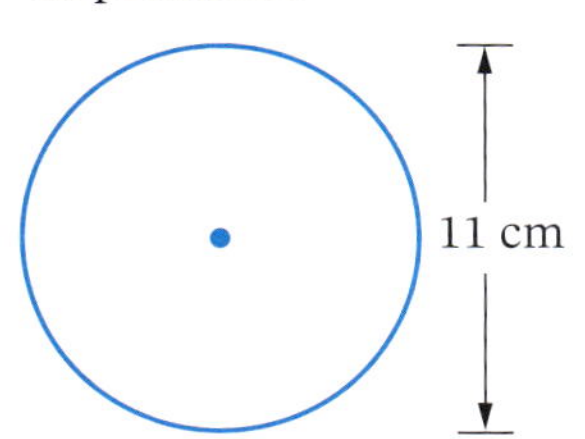

b

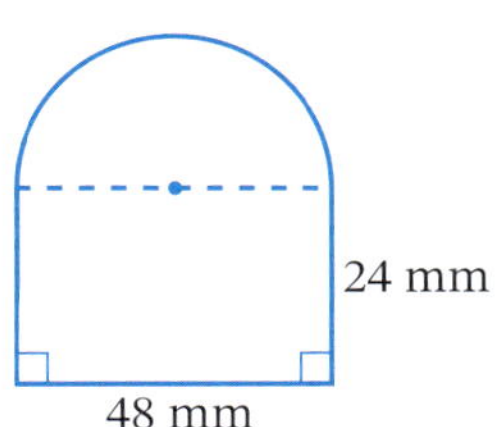

c

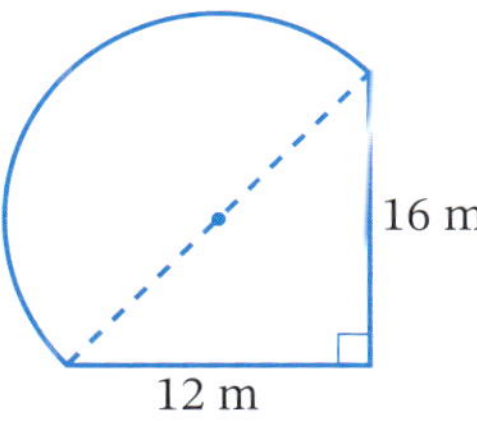

d

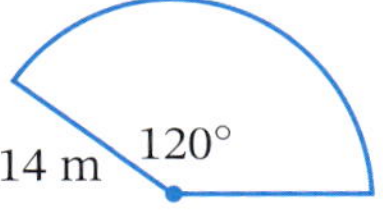

e

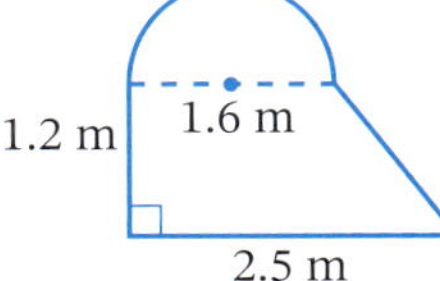

f

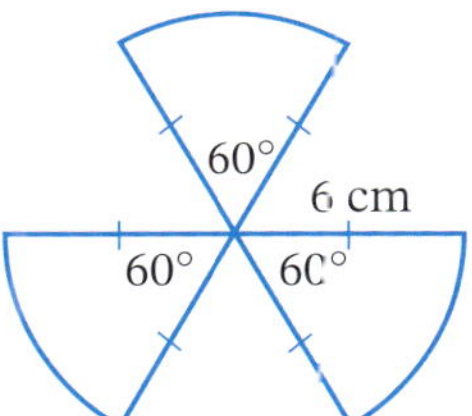

9 Calculate, correct to 2 decimal places, the area of each shape. 10.05

a

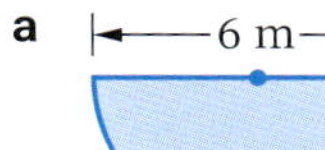

b

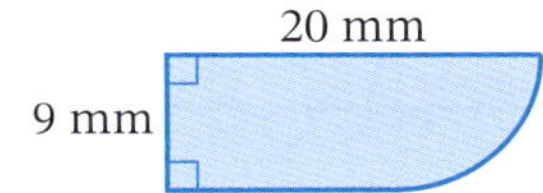

c

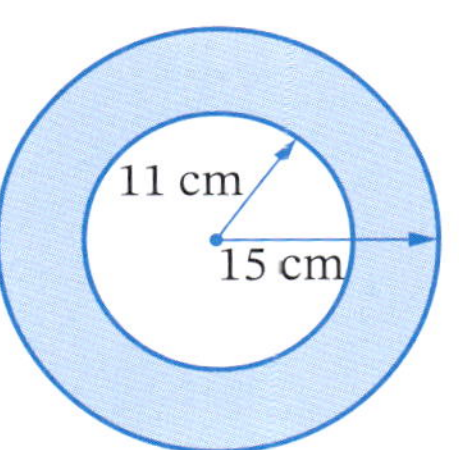

10 Calculate the surface area of each prism. 10.06

a closed cube

b open rectangular prism

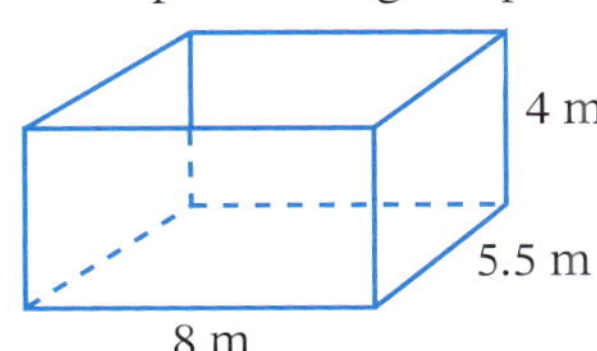

Cube: 3.2 m

c

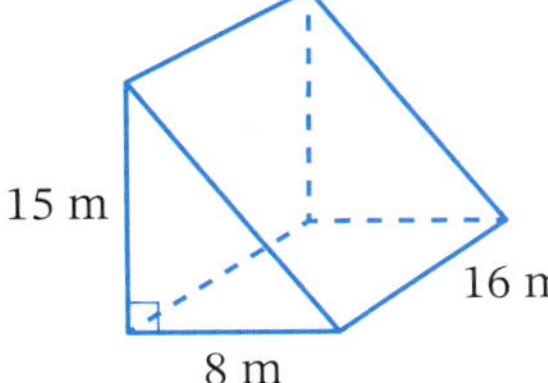

d

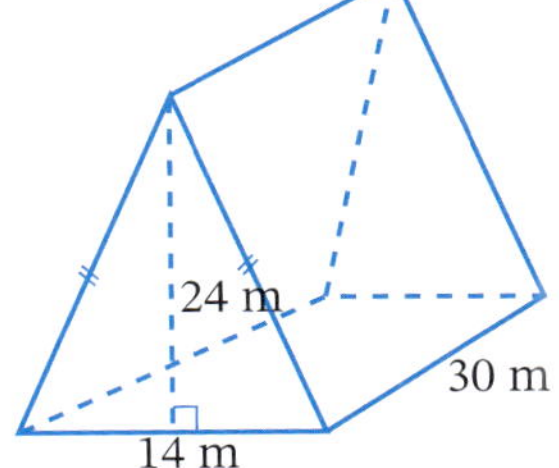

11 Calculate the volume of each prism in question **10**.

12 Calculate, correct to 2 decimal places, the surface area of each cylinder.

a

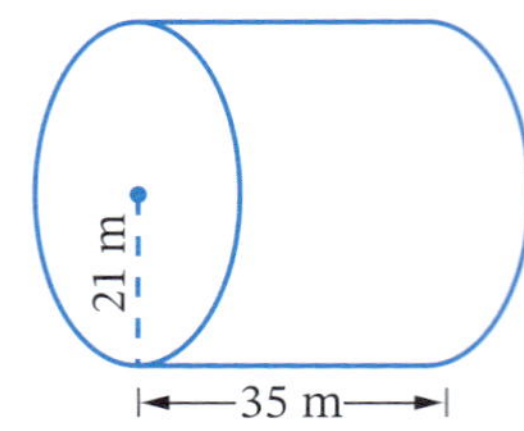

b

15 mm

23 mm

c

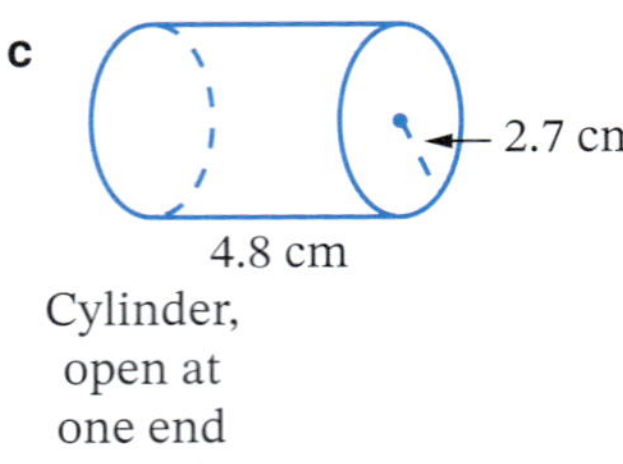

Cylinder, open at one end

13 Calculate, correct to 2 decimal places, the volume of each cylinder in question **12**.

14 A rectangular fish tank that is 75 cm long by 55 cm wide by 35 cm deep is filled to 5 cm from the top. How many litres of water is in the tank?

15 Calculate the volume of this lunchbox.

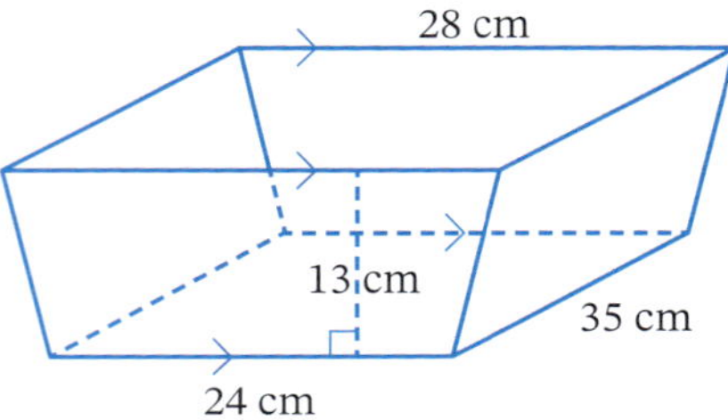

9780170465618

Practice set 3 ANSWERS ON P. 664

1 Copy and complete: 10.01

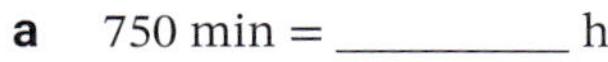

a 750 min = ________ h   **b** 1350 mL = ________ L

c 512 MB = ________ GB   **d** 3.005 kg = ________ g

e 2.7 km = ________ m   **f** 0.25 MB = ________ kB

2 Mason is paid $3.94 for each toy he assembles. 8.03

a How much does he earn for assembling 140 toys?

b How many toys did Mason assemble if he earned $1260.80?

3 Katrina's annual salary is $85 000 and she earns another $4800 from investments. Katrina's work-related tax deductions amount to $2380. Calculate Katrina's: 8.04

a taxable income

b income tax payable using the table below.

Taxable income	Tax on this income
0–$18 200	Nil
$18 201–$45 000	19% for each $1 over $18 200
$45 001–$120 000	$5092 plus 32.5 cent for each $1 over $45 000
$120 001–$180 000	$29 467 plus 37 cent for each $1 over $120 000
$180 01 and over	$51 667 plus 45 cent for each $1 over $180 000

Source: ATO

4 The shoe sizes of 30 Year 9 students are: 9.02

6	10	7	6	8	6	7	7	6	5
5	6	5	8	10	11	9	8	9	5
6	7	9	8	7	6	5	6	7	8

a Arrange the data in a frequency table.

b Draw a frequency histogram and polygon of this data.

c What percentage (correct to one decimal place) of students had shoe sizes less than 8?

d Find the mode and the range.

e Find the median.

f Calculate the mean, correct to one decimal place.

5 The height of a building is approximately 68 m, correct to the nearest 2 m. Calculate the percentage error for this height, correct to one decimal place. 10.02

6 For a salary of $89 760 p.a., calculate, correct to the nearest cent, the: 8.01

a weekly pay   **b** fortnightly pay   **c** monthly pay

Foundation | Standard | Complex

7 Bhavin works 38 hours a week and is paid \$43.50 per hour. Calculate:

a Bhavin's weekly income.

b the holiday loading of $17\frac{1}{2}\%$ on 4 weeks' pay.

c Bhavin's fortnightly pay when he receives his holiday loading.

8 Toby measured the mass of a pumpkin to be 1.9 kg. If the actual mass of the pumpkin was 1980 g, determine the absolute error in Toby's measurement.

9.01

9 For each set of data, find:

i the mode **ii** the mean (correct to one decimal place)

iii the median **iv** the range

a 47 46 50 51 48 48 52 46 49

b

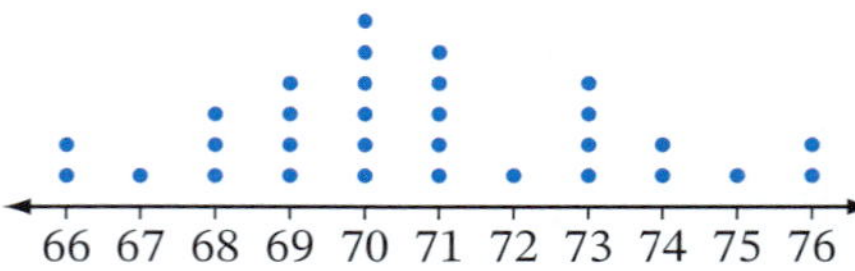

9.04

10 Classify whether each type of data is categorical or numerical. If numerical, then classify whether it is discrete or continuous.

a the age Year 9 students turn this year **b** colour of cars in a car park

c travelling time between home and work **d** amount of rainfall

e shoe size **f** film classification (e.g. PG, M)

11 For each composite shape, find:

i its perimeter **ii** its area

a

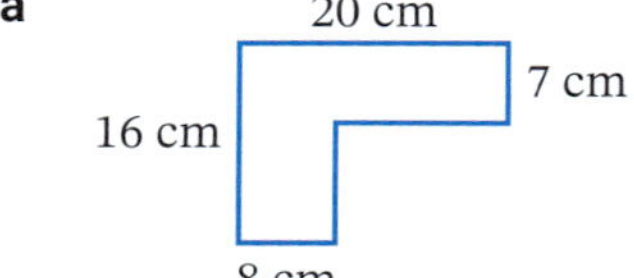

b

15 cm
15 cm
3 cm
19 cm

12 Nadia's gross pay for a fortnight is \$3038. She pays \$724.50 in tax and her other deductions amount to \$285.

a Calculate Nadia's net pay.

b Express her net pay as a percentage (correct to one decimal place) of her gross pay.

9.03

13 For the data in this stem-and-leaf plot, which statement is true? Select the correct answer **A**, **B**, **C** or **D**.

A The mean is 82.

B The mode is 6.

C The range is 46.

D The median is 81.

Stem	Leaf
6	0 1 5
7	3 4 4 8 9
8	0 1 2 2 5 6 6 6
9	3 4 7
10	6

☐ Foundation ○ Standard ○ Complex

14 Sefa is paid $27.40 per hour. Calculate his earnings for working 38 hours at normal time, 8 hours at time-and-a-half and 4 hours at double time. 8.02

15 What is the absolute error for these measuring scales?

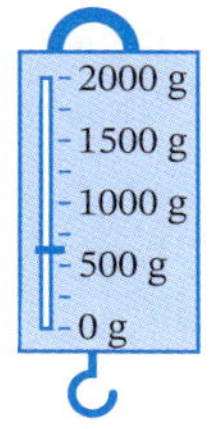

16 A survey was conducted about the colour of student's eyes. The results are shown below. 9.04

Colour	Blue	Brown	Hazel	Green
Frequency	37	52	21	16

a What is an appropriate statistical graph to represent the results of this survey?

b Why is this an appropriate display to use?

c Construct a graph of this data.

17 Find the area of each quadrilateral. 10.04

a

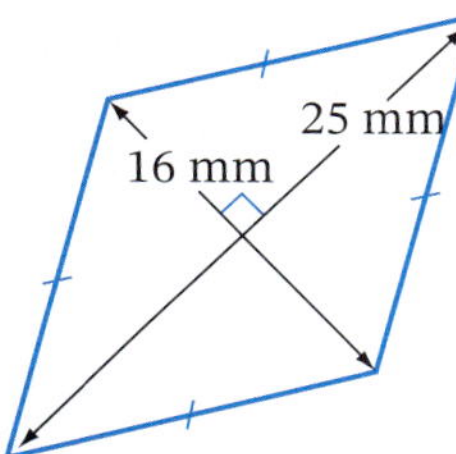

b

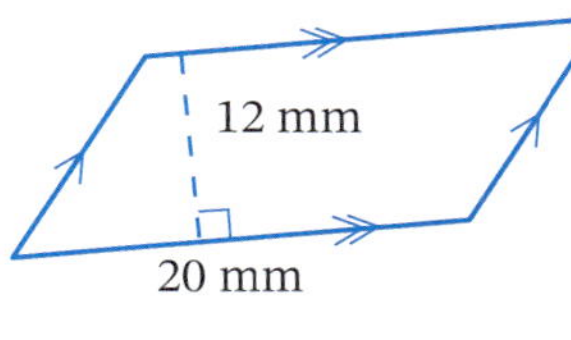

c

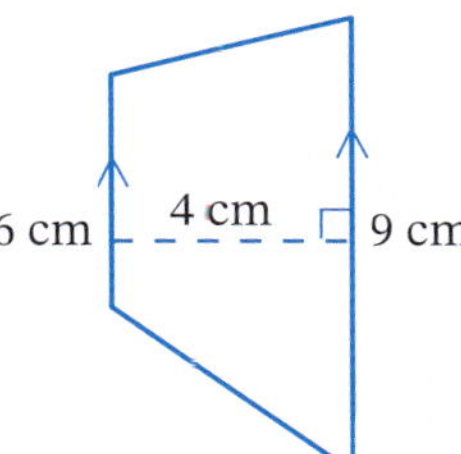

18 Determine whether a census or a sample should be used for each investigation.

a Finding the number of people in Brisbane who were born interstate.

b Testing the quality of tennis racquets.

c Determining people's opinions on mandatory vaccinations.

19 Thiago earns a commission of 2.75% on every property he sells. Calculate his commission for selling an apartment for $635 200.

20 For each circular shape, calculate, correct to one decimal place: 10.05

i its perimeter

ii its area.

a

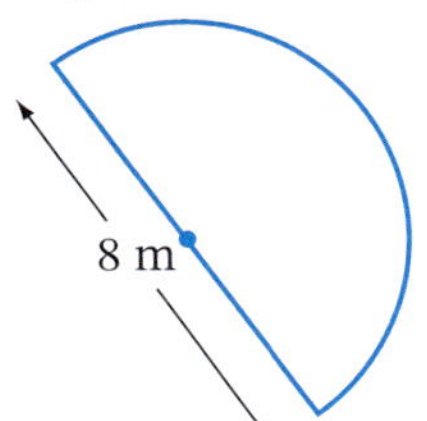

b

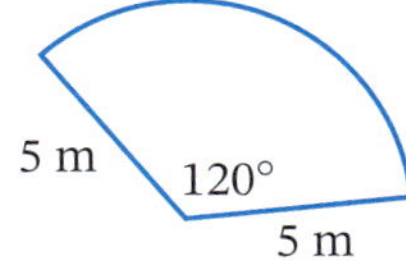

c

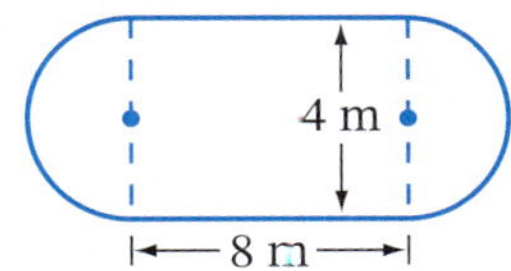

☐ Foundation ○ Standard ⬡ Complex

21 A sample of students was surveyed on the number of children in their families, and the results were graphed on a frequency histogram and polygon.

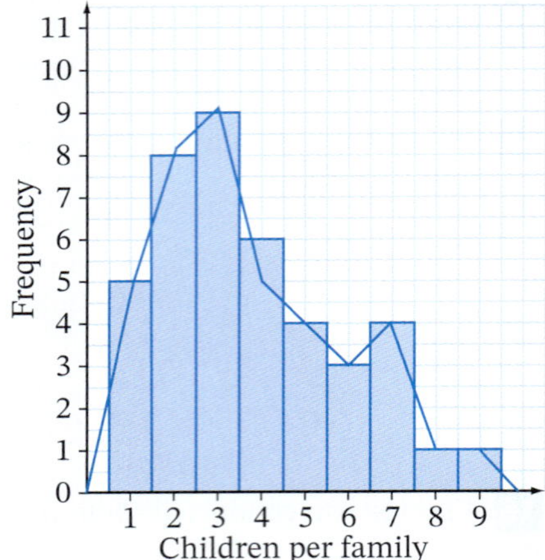

a Describe the shape of the distribution.

b Where does clustering occur?

c Find the mode and range.

10.06 **22** Find the surface area of each prism.

a

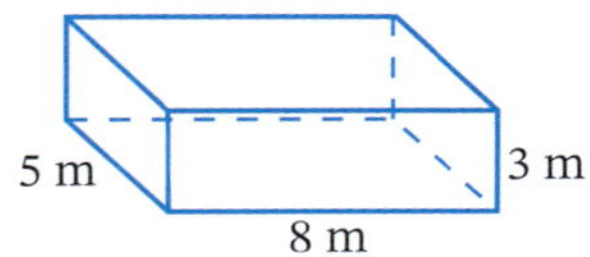

b

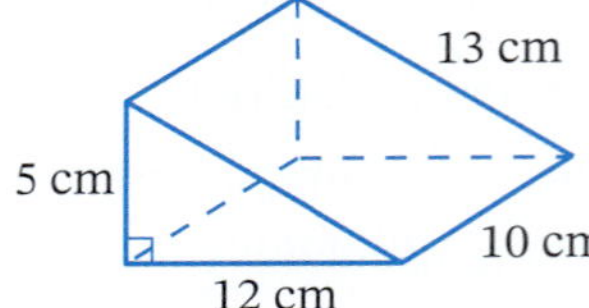

c

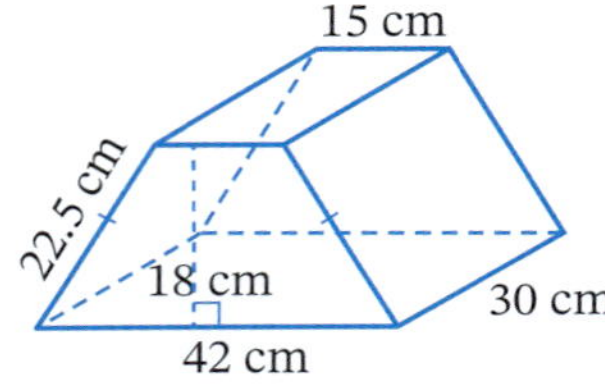

10.08 **23** Find the volume of each prism above.

9.08 **24** In a survey, a sample of people at a park were asked 'Are you employed?'

a Why might this sample of people be biased?

b What is wrong with the survey question?

c Suggest a better-worded question.

10.08 **25** Find correct to the nearest 10 mm^3 the volume of this cylinder.

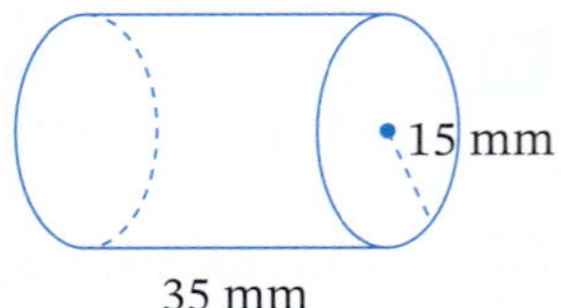

10.07 **26** Find correct to the nearest 10 mm^2 the surface area of the above cylinder if it is open at one end.

☐ Foundation ○ Standard ⬡ Complex

9780170465618

27 These dot plots show the ages of students in two Taekwondo classes.  9.06

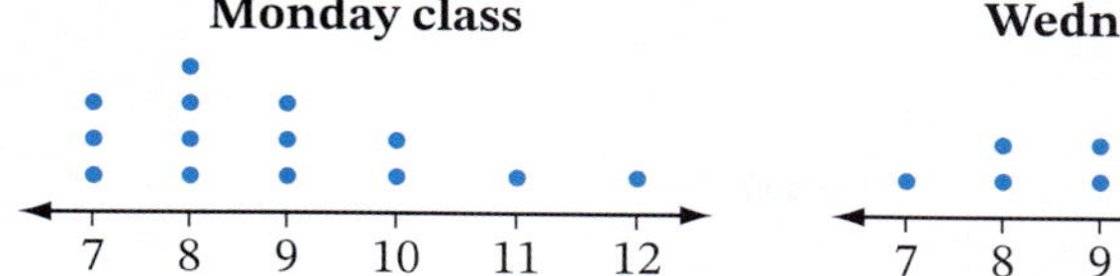

a Describe the shape of the age distribution for each class.

b Find the mean for each class.

c Find the median for each class.

d Which class had more consistent ages?

e Which class had more older students?

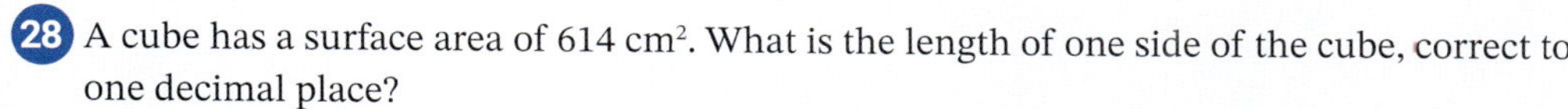

28 A cube has a surface area of 614 cm^2. What is the length of one side of the cube, correct to one decimal place?  10.06

29 A local business employs 3 managers, 22 sales assistants and 7 cleaning staff. A sample of 10 employees need to be surveyed regarding work conditions. Use a stratified random sample to determine how many employees from each group should be included in the sample.  9.07

30 Zahara works a 38-hour week and her hourly pay rate is \$28.70. Calculate Zahara's fortnightly pay.  8.01

31 The hourly pay rate for people working at a catering and party hire company are:   9.01

\$25.80	\$34.00	\$27.60	\$27.60	\$32.50	\$48.90
\$32.50	\$34.70	\$25.80	\$35.40	\$38.10	\$32.50

a Find the mean and median pay rate.

b Which pay rate is the outlier?

c Find the mean and median without the outlier.

d How does the outlier affect the mean and median?

PRACTICE SET 3

Foundation Standard Complex

9780170465618

11

ALGEBRA, MEASUREMENT

Coordinate geometry and graphs

The use of a coordinate system allows us to monitor the location of objects on Earth or in space. Whether reading street maps, using global positioning systems or managing road traffic flow, coordinates provide an accurate method for indicating position. Coordinates have important applications in many areas, including astronomy, architecture, engineering and art design.

Shutterstock.com/DedMityay

Chapter outline

		Proficiencies				
11.01	The length of an interval	U	F		R	C
11.02	The midpoint of an interval	U	F		R	C
11.03	The gradient of a line	U	F		R	C
11.04	Graphing linear functions	U	F		R	C
11.05	The gradient–intercept equation $y = mx + c$	U	F			
11.06	Finding the equation of a line	U	F			
11.07	Solving linear equations graphically	U	F			
11.08	Direct proportion	U	F	PS	R	
11.09	Applying linear functions	U	F	PS	R	C
11.10	Graphing quadratic functions $y = ax^2 + c$	U	F		R	C
11.11	Graphing quadratic functions $y = ax^2 + bx + c$	U	F		R	C
11.12	Applying quadratic functions	U	F	PS	R	C
11.13	Graphing exponential functions $y = a^x$	U	F		R	C

U = Understanding
F = Fluency
PS = Problem solving
R = Reasoning
C = Communication

Wordbank

asymptote A line that a curve gets very close to but never touches, for example, the *x*-axis is an asymptote of the exponential curve

direct proportion A relationship between 2 variables in which one variable is a constant multiple of the other, for example, if $y = 8.5x$, then y is directly proportional to x

exponential function A function involving a variable as a power, such as $y = 3^x$, whose graph is an exponential curve

gradient The steepness of a line or interval, measured by the fraction $\frac{\text{rise}}{\text{run}}$

linear function A function whose graph is a straight line

midpoint The point in the middle of an interval or halfway between 2 given points

parabola A U-shaped curve that is the graph of a quadratic function

quadratic function A function involving a variable squared (power of 2), such as $y = 3x^2 - 6$, whose graph is a curve called a parabola

Quiz
Wordbank 11

Videos (11):

11.01 The length of an interval

11.03 The gradient of a line • Gradient (slope) of a line

11.04 Graphing linear equations • Testing if a point lies on a line

11.05 Graphing $y = mx + c$

11.06 Finding the equation of a line

11.08 Direct proportion

11.11 Quadratic functions

11.13 The exponential curve • Graphing exponentials

Twig videos (2):

SkillCheck Coordinate geometry: Descartes

11.13 The emperor's chess board

PhET interactives (5):

11.04 Graphing lines • Function builder: Basics • Function builder

11.05 Graphing slope-intercept

11.10, 11.11 Graphing quadratics

Quizzes (5):

- Wordbank 11
- SkillCheck 11
- Mental skills 11
- Language of maths 11
- Test yourself 11

Skillsheets (1):

SkillCheck The number plane

Worksheets (15):

SkillCheck Tables of values

11.01 Length of an interval

11.03 A page of lines • Gradient between two points

11.04 Graphing linear equations • A page of lines • A page of number planes • Number plane grid paper • Straight-line equations

11.06 Finding the equation of a line • A page of lines

11.10 Graphing parabolas

11.11 Graphing quadratic functions • Graphing quadratics • Features of a parabola

Mind map: Coordinate geometry and graphs

Puzzles (4):

11.01 Length of a line segment

11.03 Find the gradient 1

11.05 Equation of a line

11.06 Straight-line equations

Technology (2):

11.02 Finding midpoints

11.10, 11.11 Curve sketcher

Spreadsheets (2):

11.02 Finding midpoints

11.10, 11.11 Curve sketcher

Nelson MindTap

To access resources above, visit **cengage.com.au/nelsonmindtap**

11 In this chapter you will:

- ✓ find the distance between 2 points on a number plane.
- ✓ find the midpoint and gradient of an interval on a number plane.
- ✓ graph a line on a number plane given the coordinates of 2 points on the line.
- ✓ graph and solve problems involving direct proportion.
- ✓ test whether a point lies on a line.
- ✓ use the gradient–intercept equation of a straight line $y = mx + c$.
- ✓ find the equation of a line from its graph.
- ✓ solve linear equations graphically.
- ✓ graph quadratic functions of the form $y = ax^2 + c$ and $y = ax^2 + bx + c$.
- ✓ examine applications of linear and quadratic functions.
- ✓ graph exponential functions of the form $y = a^x$.

SkillCheck

ANSWERS ON P. 661

Quiz
SkillCheck 11

Worksheet
Tables of values

Skillsheet
The number plane

Video
Coordinate geometry: Descartes

1 Write the coordinates of points Z, D, Q, W and P, shown on this number plane.

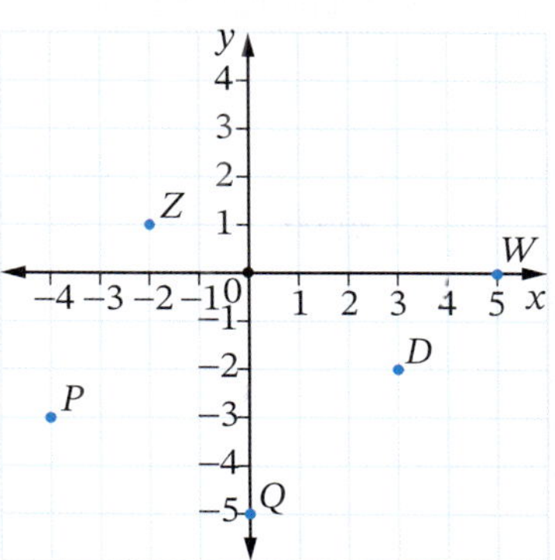

2 Complete each table of values using the given equation.

a $y = -2x + 1$

x	−2	−1	0	1	2
y					

b $y = \frac{x}{2} + 1$

x	−2	−1	0	1	2
y					

3 Find, correct to one decimal place, the length of the hypotenuse for each right-angled triangle.

a

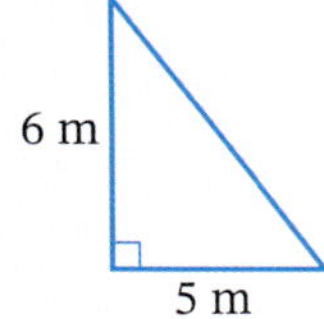

b

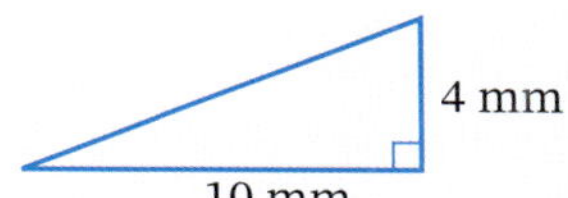

c

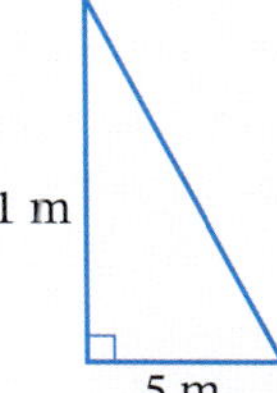

4 For each triangle in question **3**, write $\frac{\text{height}}{\text{base length}}$ as a simple fraction.

5 Calculate the average of:

a 1 and 3 **b** −2 and 6 **c** 5 and 12

The length of an interval 11.01

Worksheet
Length of an interval

Puzzle
Length of a line segment

A line is usually drawn with arrows at both ends to show that it runs infinitely in both directions. However, an **interval** or line segment is part of a line, with endpoints on both sides and a definite length. The interval joining points P and Q is called the interval PQ. We can calculate the length of an interval (or the distance between 2 points) on a **number plane** if we know the coordinates of its endpoints.

Example 1

Find the distance between the points:

a $P(-4, -2)$ and $Q(2, 6)$ **b** $A(-2, 5)$ and $B(2, -2)$

SOLUTION

a With P and Q plotted on a number plane, construct a right-angled triangle PTQ such that PQ is the hypotenuse.

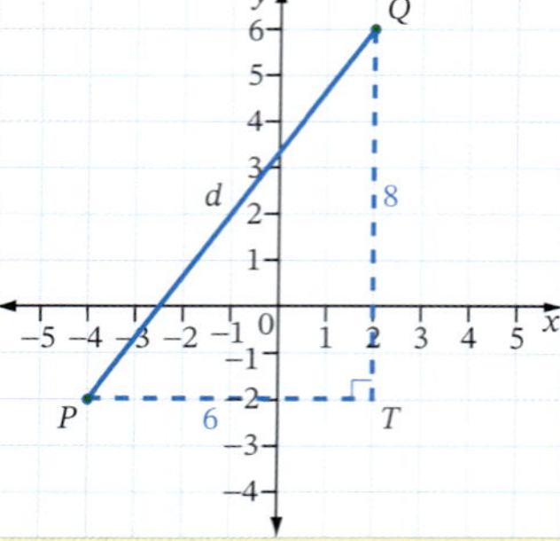

$PT = 6$ units, $TQ = 8$ units.

Let the length of PQ be d, so by Pythagoras' theorem:

$d^2 = 6^2 + 8^2$

$= 100$

$d = \sqrt{100} = 10$

The distance between the points P and Q is 10 units.

b By plotting A and B and constructing $\triangle ABC$, we have:

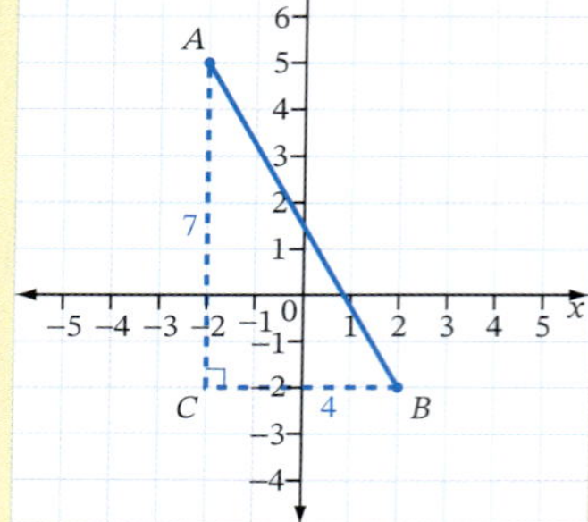

$AB^2 = 7^2 + 4^2$

$= 65$

$AB = \sqrt{65}$

$= 8.0622...$

≈ 8.06 units

We can write the distance AB in **surd** form, $\sqrt{65}$ units as an exact answer, or we can approximate (round) it as a decimal, 8.06 units.

Example 2

a Plot the points $A(0, 6)$, $B(5, 6)$, $C(5, 2)$ and $D(-4, 2)$ on a number plane and join them to make the quadrilateral $ABCD$.

b What type of quadrilateral is $ABCD$?

c Find the exact length of AD.

d Hence find the perimeter of $ABCD$, correct to 2 decimal places.

SOLUTION

a

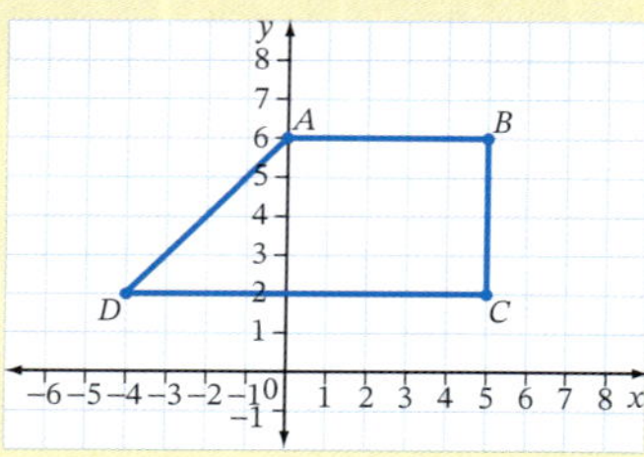

b *ABCD* is a trapezium.

c $AD^2 = 4^2 + 4^2$ By Pythagoras' theorem

$= 32$

$AD = \sqrt{32}$ units In exact surd form

d By counting grid squares, $AB = 5$, $BC = 4$, $CD = 9$.

Perimeter of $ABCD = 5 + 4 + 9 + \sqrt{32}$

$= 23.6568...$

≈ 23.66 units

11.01

INVESTIGATION

The distance formula

The method for finding the length of an interval can be summarised by a formula.

It is called the **distance formula** because it is used to calculate the distance (d) between any 2 points $P(x_1, y_1)$ and $Q(x_2, y_2)$, in other words, the length of the interval PQ.

Length of $PT = x_2 - x_1$.

Length of $TQ = y_2 - y_1$.

By Pythagoras' theorem: $d^2 = (x_2 - x_1)^2 + (y_2 - y_1)^2$

$\therefore d = \sqrt{(x_2 - x_1)^2 + (y_2 - y_1)^2}$

Use the distance formula to show that the distance between:

a $A(2, 6)$ and $B(4, 7)$ is approximately 2.2 units.

b $P(-3, 5)$ and $Q(9, -11)$ is 20 units.

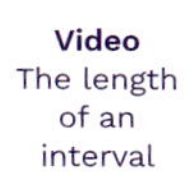

Video The length of an interval

ⓘ The distance formula

The length, d, of the interval joining 2 points (x_1, y_1) and (x_2, y_2) is:

$$d = \sqrt{(x_2 - x_1)^2 + (y_2 - y_1)^2}$$

EXERCISE 11.01 ANSWERS ON P. 661

The length of an interval

U F R C

EXAMPLE 1

1 Calculate the length of each interval shown on this number plane. Express your answers correct to one decimal place, where necessary.

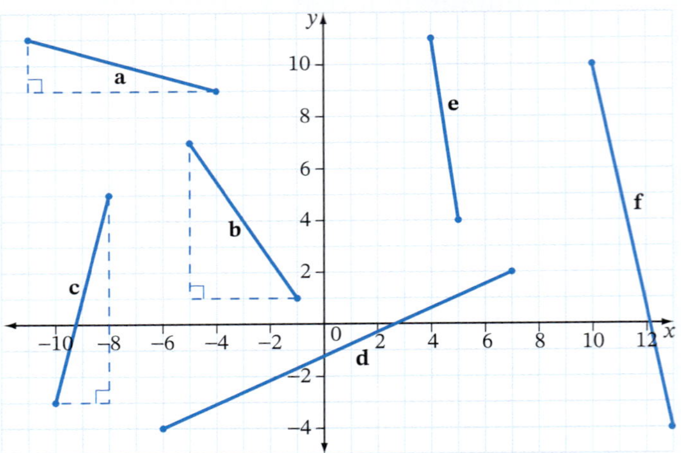

2 Plot each pair of points on a number plane and construct a right-angled triangle to find the distance between the 2 points. Express your answer:

i as a surd **ii** correct to one decimal place.

a $A(7, 6)$ and $B(1, 9)$ **b** $P(0, -5)$ and $Q(6, 4)$

c $G(9, -2)$ and $H(1, 6)$ **d** $W(-5, -4)$ and $X(-3, 4)$

3 **a** What are the lengths of a and b in the diagram? Select the correct answer **A**, **B**, **C** or **D**.

A $a = 3, b = 3$ **B** $a = 3, b = 2$

C $a = 5, b = 3$ **D** $a = 5, b = 2$

b What is the length of CD?

A 2 units **B** 5.8 units

C 3.2 units **D** 8 units

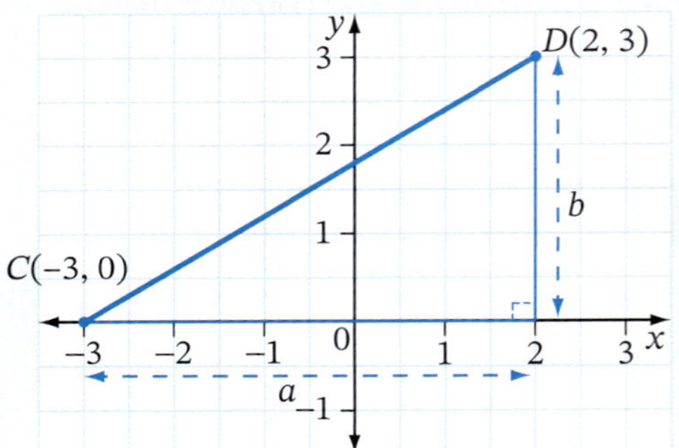

4 A triangle is formed by joining the points $A(5, 2)$, $B(-2, 3)$ and $C(1, 7)$. Find, correct to 3 decimal places:

a the length AB **b** the length BC

c the length AC **d** the perimeter of $\triangle ABC$

5 What is the exact distance between the points $Y(6, -3)$ and $Z(-2, 4)$?

A $\sqrt{113}$ units **B** $\sqrt{65}$ units **C** $\sqrt{57}$ units **D** $\sqrt{15}$ units

▷

☐ Foundation ○ Standard ○ Complex

9780170465618

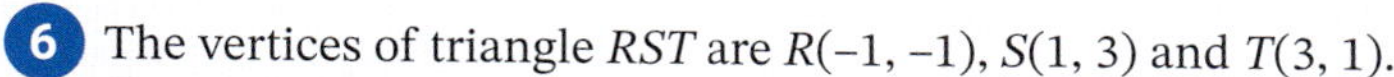

6 The vertices of triangle RST are $R(-1, -1)$, $S(1, 3)$ and $T(3, 1)$.

R **a** Find the exact length of each side of the triangle.

C **b** What type of triangle is $\triangle ABC$?

7 A quadrilateral is formed by joining the points $M(-3, 5)$, $N(5, 5)$, $P(8, 1)$ and $T(-3, 1)$.

R **a** What type of quadrilateral is $MNPT$?

C **b** Find the perimeter of $MNPT$.

c Find the area of $MNPT$.

8 The vertices of a quadrilateral are $H(1, 6)$, $I(7, 2)$, $J(3, -4)$ and $K(-3, 0)$.

R **a** What type of quadrilateral is $HIJK$?

C **b** Find the exact length of each side of the quadrilateral.

c Find its perimeter, correct to one decimal place.

d Calculate its area.

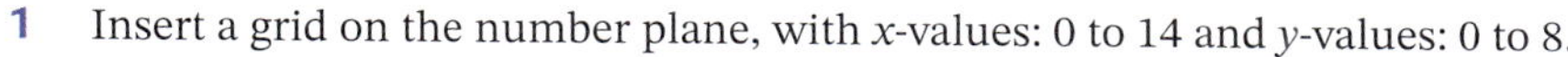

TECHNOLOGY

The length of an interval

In this activity, you will use dynamic geometry software to apply Pythagoras' theorem to find the length of an interval.

1. Insert a grid on the number plane, with x-values: 0 to 14 and y-values: 0 to 8.
2. Plot the points $A(2, 1)$, $B(6, 4)$ and $C(6, 1)$ on the grid. Join AB, BC and AC.
3. Find the lengths of AC and BC.
4. Calculate $(AC)^2 + (BC)^2$ and $\sqrt{(AC)^2 + (BC)^2}$.
5. Now find the length of AB.
6. Compare answer **5** with answer **4**. What do you notice?
7. Move the number plane until the axes are centred in the middle of the screen. [Approx: x: –10 to 10 and y: –10 to 10]

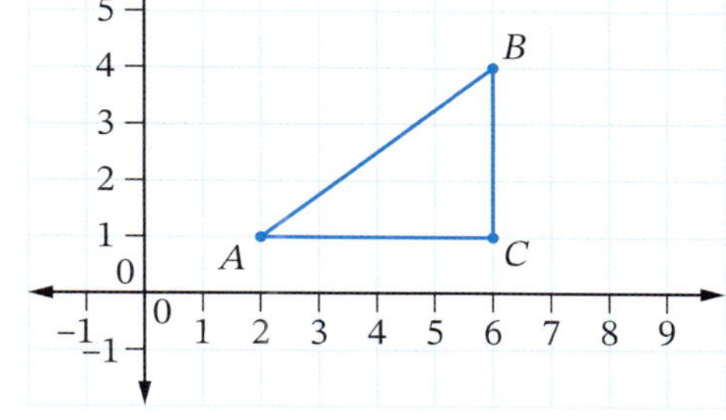

8. Plot each pair of points and calculate the distance between them by repeating all the steps above.

a $X(-4, 2)$ and $Y(4, -4)$ **b** $P(-7, 5)$ and $Q(-2, -7)$

DID YOU KNOW?

René Descartes

The number plane is also called the Cartesian plane, named after its inventor, the French philosopher and mathematician René Descartes (1596–1650). In an age when Shakespeare and Galileo had made contributions to drama and science respectively, Descartes showed that many relationships and formulas could be represented by simple graphs. In 1637, he made a great contribution to mathematics when he published a book linking algebra and geometry for the first time.

As a philosopher, the big question that Descartes spent many years thinking about was whether he, or the world, really existed. He finally concluded that he did exist, using a brilliant argument summarised by his famous quote.

Complete his famous quote: "I __________, therefore ____________"

INVESTIGATION

The midpoint of an interval

1 **a** What number is halfway between −2 and 5 on this number line?

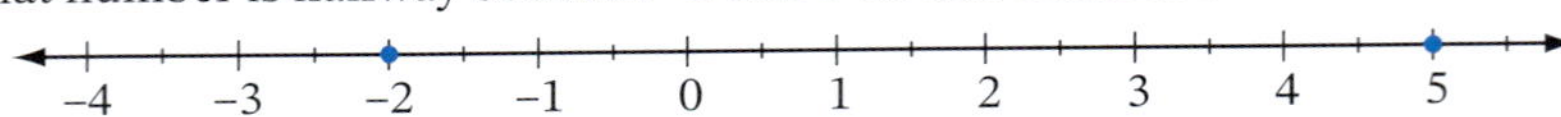

b What is the average of −2 and 5?

2 **a** What are the coordinates of A and B on this number plane?

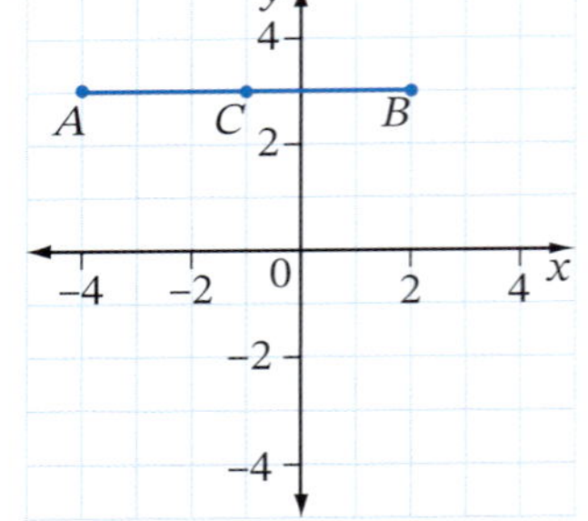

b What is the average of the x-values of A and B?

c What is the average of the y-values of A and B?

d What are the coordinates of C, the midpoint of AB?

e What do you notice about your answers for parts **b**, **c** and **d**?

3 **a** What are the coordinates of P and Q ?

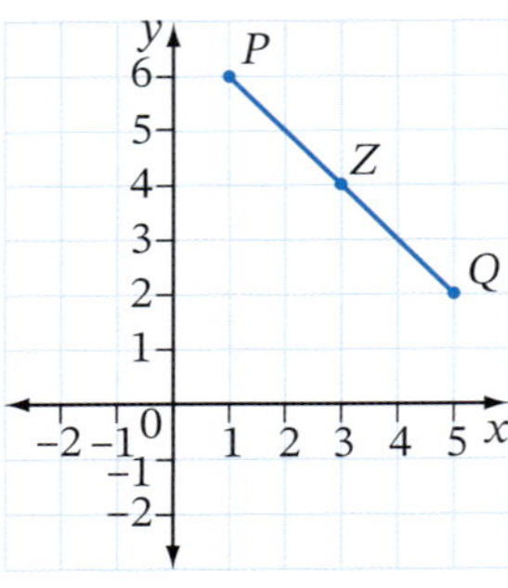

b What is the average of the x-values of P and Q?

c What is the average of the y-values of P and Q?

d What are the coordinates of Z, the midpoint of PQ?

e How do the coordinates of Z relate to your answers for parts **b** and **c**?

4 Describe how to find the midpoint of any interval on a number plane given the coordinates of its endpoints.

11.01

TECHNOLOGY

The midpoint of an interval

1 Use dynamic geometry software to create a number plane with x-values: 0 to 14 and y-values: 0 to 8.

2 Plot the points $A(1, 2)$, $B(5, 6)$ and $C(5, 2)$ and join AB, BC and AC.

3 What is the average x-value on AC?

4 What is the average y-value on BC?

5 Label the midpoint of AB, M (as shown below). What are its coordinates?

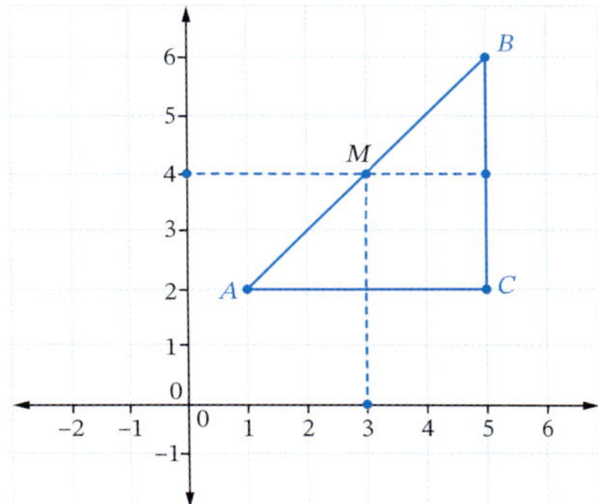

6 Redraw interval AB. Select the midpoint and label it M. Does it have the same coordinates as you found in step **5**?

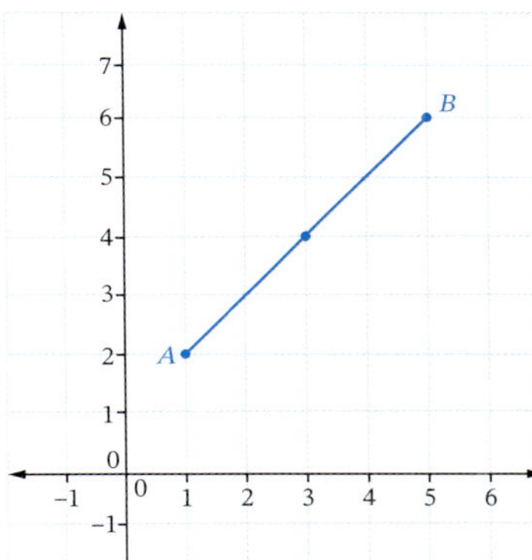

11.02 The midpoint of an interval

Technology
Finding midpoints

Spreadsheet
Finding midpoints

The **midpoint of an interval** is the point in the middle of an interval or halfway between 2 given points.

The midpoint of an interval

To find the midpoint of an interval joining 2 points A and B:

- calculate the average of the x-coordinates of A and B.
- calculate the average of the y-coordinates of A and B.

These 2 values give the coordinates of the midpoint.

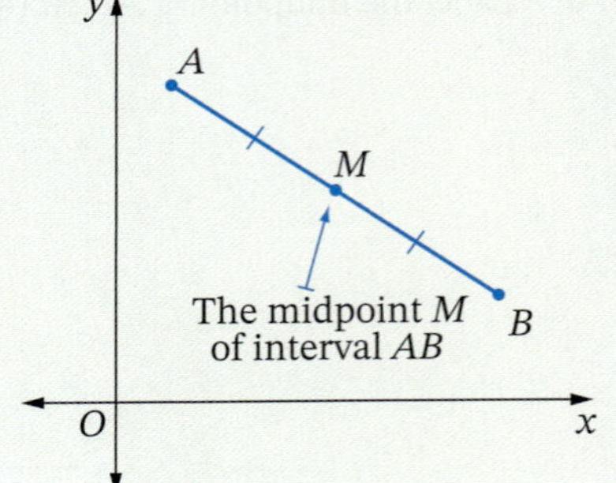

Example 3

Find the midpoint of the interval joining:

a $A(5, -2)$ and $B(1, 6)$ **b** $P(-2, 7)$ and $Q(4, 0)$

SOLUTION

a The x-coordinates are 5 and 1.

The average of 5 and 1 $= \frac{5+1}{2} = 3$.

The y-coordinates are −2 and 6.

The average of −2 and 6 $= \frac{-2+6}{2} = 2$

∴ The midpoint is the point (3, 2).

The diagram shows that this is true.

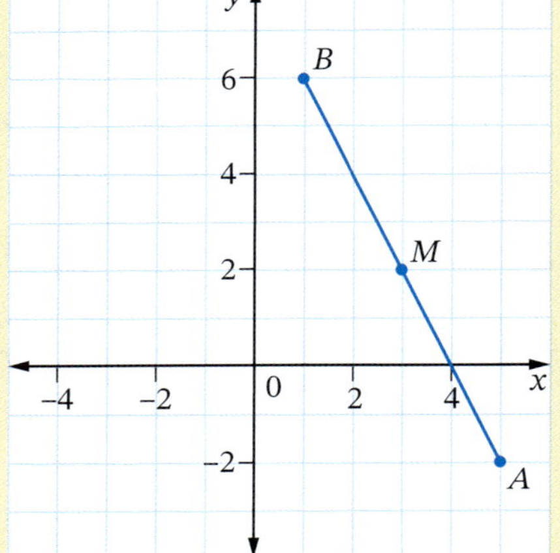

b The x-coordinates are −2 and 4.

The average of −2 and 4 $= \frac{-2+4}{2} = 1$.

The y-coordinates are 7 and 0.

The average of 7 and 0 $= \frac{7+0}{2} = 3\frac{1}{2}$.

∴ The midpoint is the point $(1, 3\frac{1}{2})$.

The method for finding the midpoint of an interval can be summarised by a formula.

The midpoint formula

The midpoint, M, of the interval joining 2 points (x_1, y_1) and (x_2, y_2) has coordinates:

$$M(x, y) = \left(\frac{x_1 + x_2}{2}, \frac{y_1 + y_2}{2}\right)$$

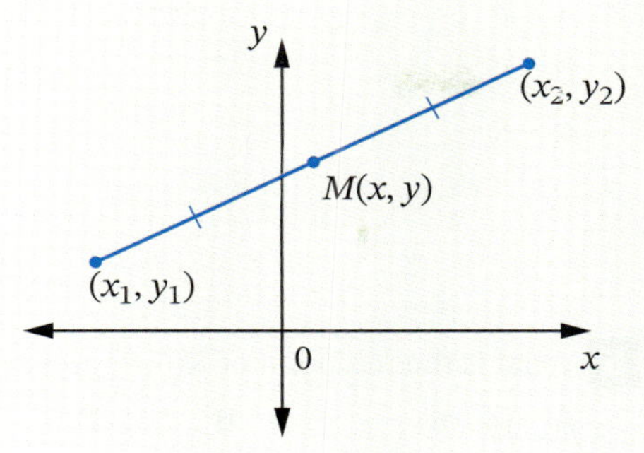

EXERCISE 11.02 ANSWERS ON P. 661

The midpoint of an interval

U F R C

1 What is the midpoint of interval XY? Select the correct answer **A**, **B**, **C** or **D**.

A (2, 3)

B (2, 4)

C (3, 2)

D (3, 3)

2 On a number plane, plot and join each pair of points, then find and plot the midpoint of each interval.

a (1, 3) and (7, 5)

b (0, 2) and (10, 4)

c (−3, 3) and (5, −1)

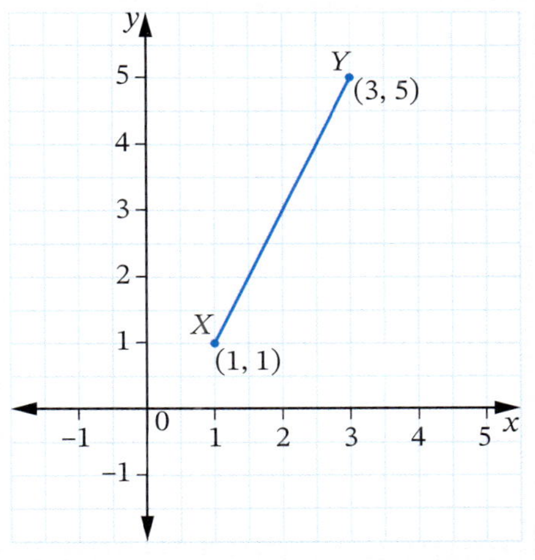

3 Find the midpoint of the interval joining:

EXAMPLE 3

a (10, 10) and (6, 4)
b (2, 7) and (12, 9)
c (0, 5) and (9, 3)
d (2, 2) and (2, 4)
e (−2, 6) and (10, 3)
f (9, −1) and (5, 2)
g (−8, 11) and (9, −4)
h (−6, −1) and (0, 0)
i (3, −3) and (−5, −10)
j (6, −1) and (−8, −1)
k (5, −6) and (9, −10)
l (−8, −7) and (−3, −5)

☐ Foundation ○ Standard ○ Complex

4 The interval AB has been subdivided into 4 equal parts. Find the coordinates of:

a P

b Q

c R

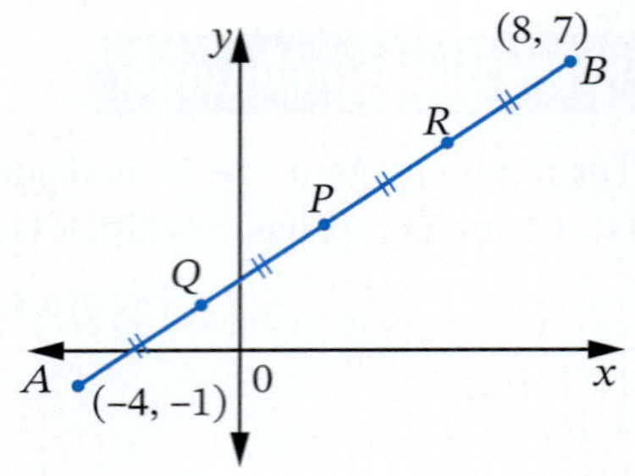

5 What is the midpoint of $K(6, 9)$ and $L(-2, 7)$?

A (4, 16) **B** (−4, 2) **C** (−2, 1) **D** (2, 8)

6 $WXYZ$ is a rectangle.

R C **a** Find the midpoint of:

i diagonal WY.

ii diagonal XZ.

b What do you notice about the 2 midpoints in part **a**? Which property for the diagonals of a rectangle does this show?

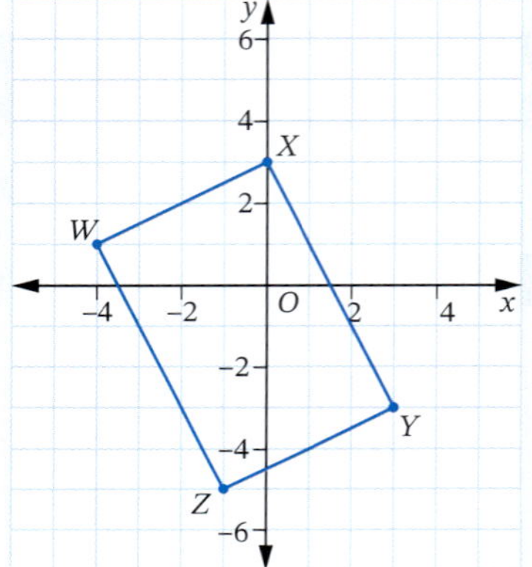

7 The diagram shows an isosceles triangle, PQR.

a Find M, the midpoint of PQ.

b Find, as a surd, the length of:

i PQ **ii** MR **iii** QR

c Calculate, correct to one decimal place

i the area of $\triangle PQR$.

ii the perimeter of $\triangle PQR$.

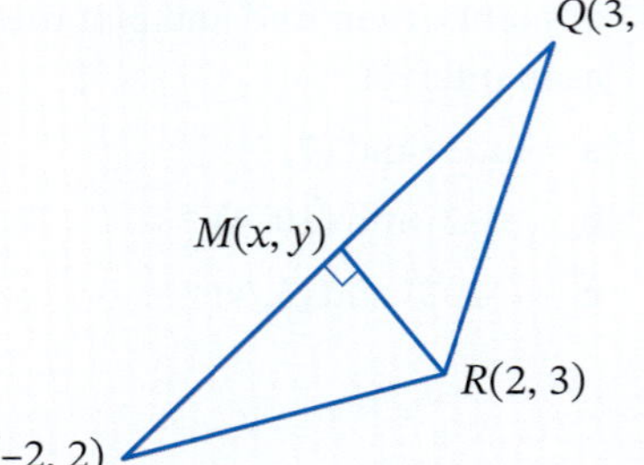

☐ Foundation ○ Standard ⬡ Complex

The gradient of a line

11.03

The **gradient** of a line (or interval) is a measure of how steep the line is. It is represented by the variable m and is the ratio (fraction) of the vertical **rise** to the horizontal **run** between any 2 points on a line.

Worksheets
A page of lines
Gradient between two points

Puzzle
Find the gradient 1

ⓘ The gradient of a line or interval

$$m = \frac{\text{vertical rise}}{\text{horizontal run}} = \frac{\text{rise}}{\text{run}}$$

- A line **sloping upwards** has a **positive rise** and a **positive gradient**.
- A line **sloping downwards** has a **negative rise** and a **negative gradient**.
- **The** run **is always** positive.

sloping upwards (positive gradient)
vertical rise
horizontal run

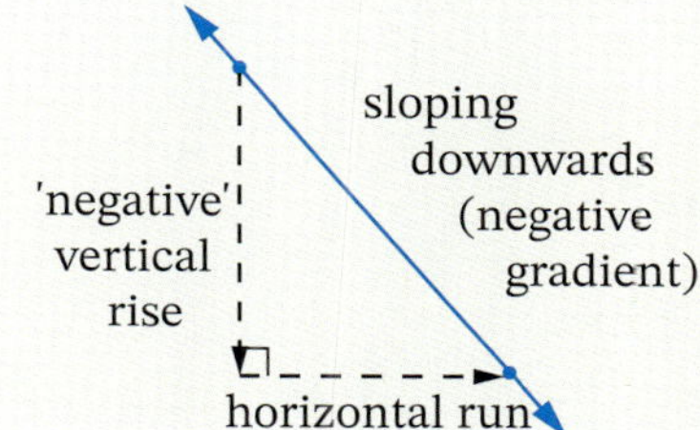

Example 4

Calculate the gradient of each line.

a

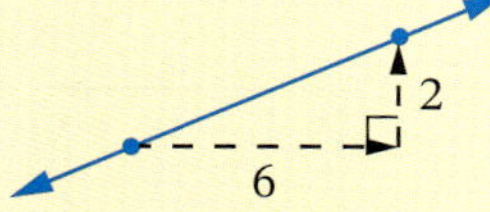

b

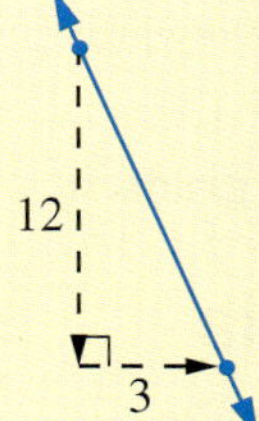

SOLUTION

a The line has a positive rise (↑).

$$m = \frac{2}{6}$$
$$= \frac{1}{3}$$

$m = \frac{\text{rise}}{\text{run}}$

rise = 2
run = 6

b The line has a negative rise or a drop (↓).

$$m = \frac{-12}{3}$$
$$= -4$$

$m = \frac{\text{rise}}{\text{run}}$

rise = −12
run = 3

Videos
The gradient of a line
Gradient (slope) of a line

Example 5

Calculate the gradient of the line going through:

a $A(-4, 2)$ and $B(4, 6)$

b $C(-2, 4)$ and $D(2, -4)$.

SOLUTION

Form a right-angled triangle with the interval as the hypotenuse, as shown.

a

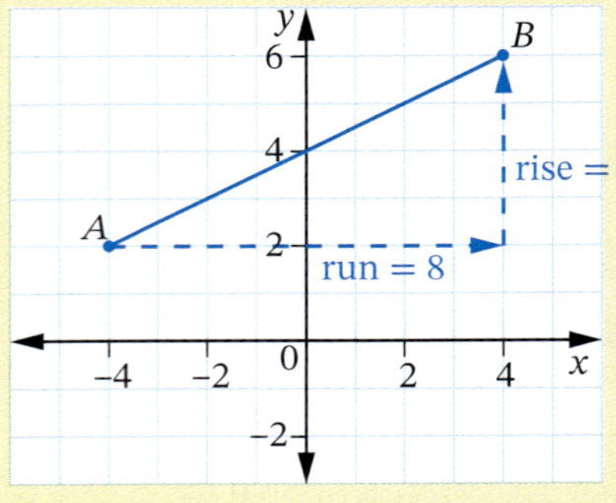

$$m = \frac{4}{8} = \frac{1}{2}$$

b

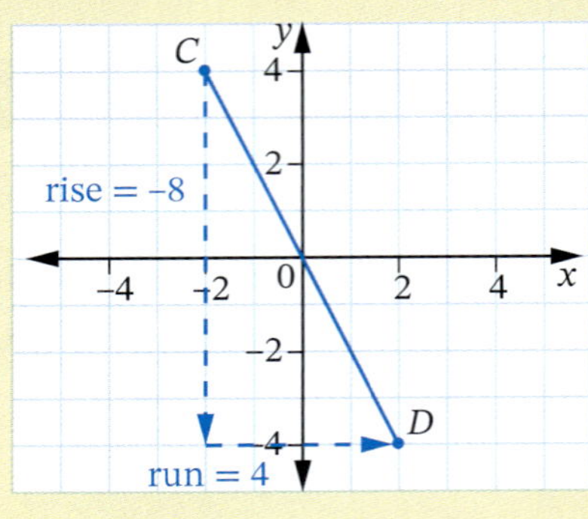

$$m = \frac{-8}{4} = -2$$

The size and sign of the gradient

The **size** of the gradient shows how quickly the line is going up or down: the higher the number, the steeper the line. The **sign** of the gradient shows whether the line is sloping upwards (positive) or downwards (negative) as you look from left to right.

A **horizontal (flat) line** has **zero gradient**.

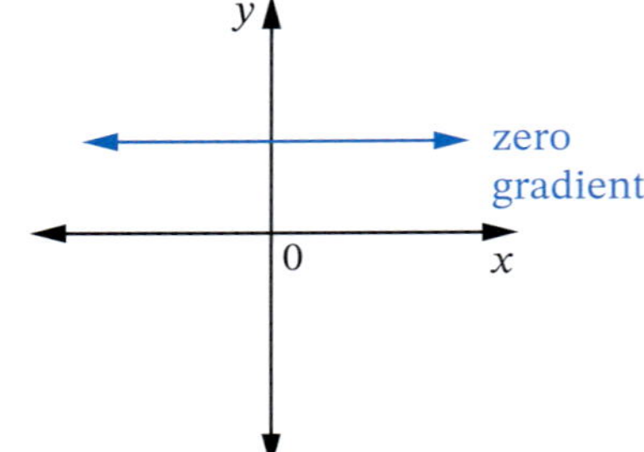

The gradient formula

The method for finding the gradient of an interval can be summarised by a formula.

Consider the line passing through the points (x_1, y_1) and (x_2, y_2).

Vertical rise = difference in y-coordinates = $y_2 - y_1$.

Horizontal run = difference in x-coordinates = $x_2 - x_1$.

ⓘ The gradient formula

The gradient of a line passing through the points (x_1, y_1) and (x_2, y_2) is:

$$\text{Gradient, } m = \frac{\text{difference in } y}{\text{difference in } x} = \frac{y_2 - y_1}{x_2 - x_1}$$

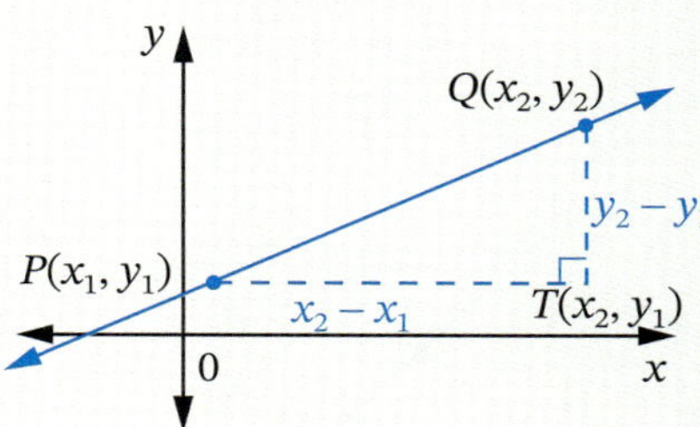

So applying the formula to Example **5a** on the previous page, the gradient of the line passing through $A(-4, 2)$ and $B(4, 6)$ is:

$$m = \frac{y_2 - y_1}{x_2 - x_1} \qquad x_1 = -4, x_2 = 4, y_1 = 2, y_2 = 6$$

$$= \frac{6-2}{4-(-4)}$$

$$= \frac{4}{8}$$

$$= \frac{1}{2}$$

EXERCISE 11.03 ANSWERS ON P. 662

The gradient of a line

1 State whether the gradient of each line is positive (+) or negative (–).

a

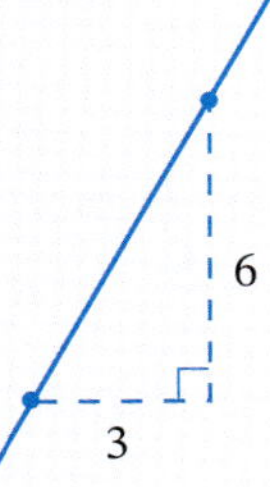

b

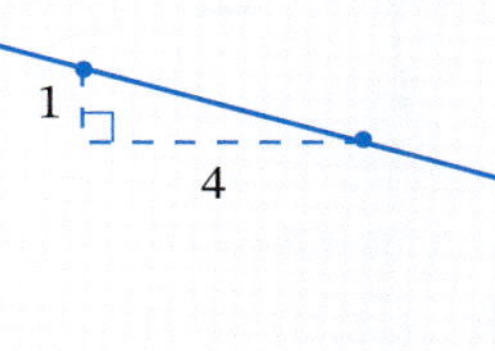

c

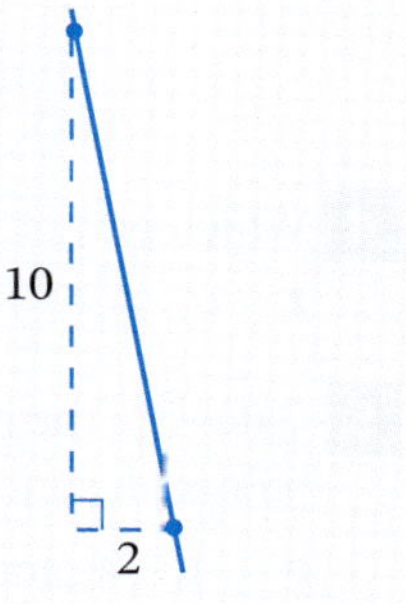

d

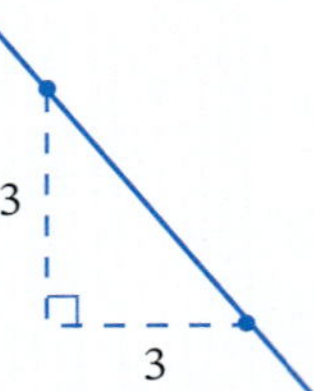

e

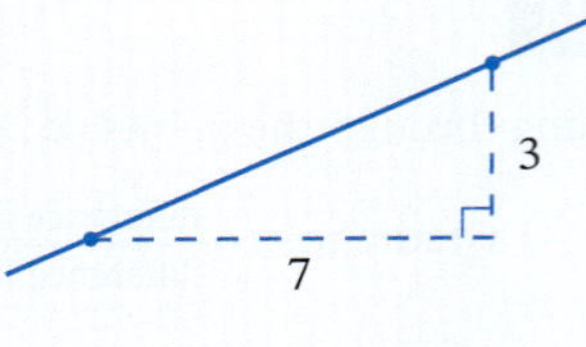

f

3
7

g

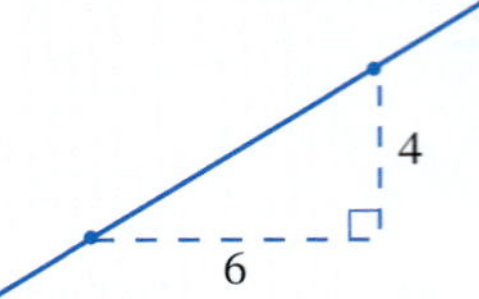

h

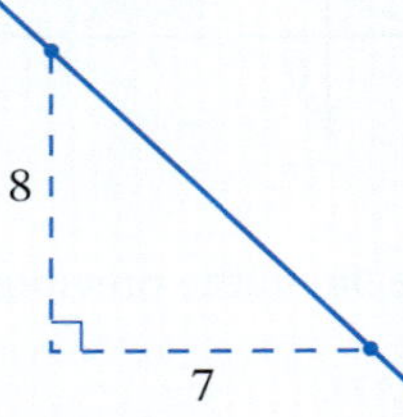

i

2
5

EXAMPLE 4

2 Calculate the gradient of each line in question **1**.

EXAMPLE 5

3 Calculate the gradient of each interval.

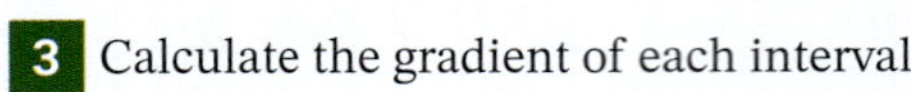

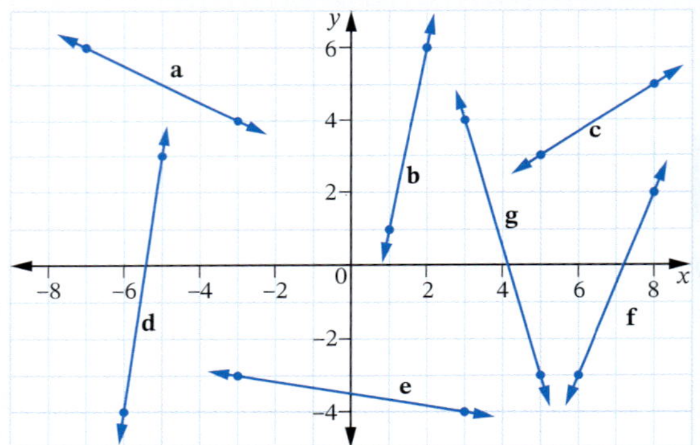

4 Which pair of points form an interval that has a gradient of 2? Select **A**, **B**, **C** or **D**.

A (0, 2), (2, 0) **B** (−5, 2), (5, 7) **C** (−2, −2), (1, 4) **D** (−6, −8), (−2, −6)

5 Calculate the gradient of the interval joining each pair of points, either by plotting the points on a number plane or by using the gradient formula.

a $A(1, 4)$ and $B(3, 6)$ **b** $Z(4, 7)$ and $T(1, 6)$ **c** $M(4, -2)$ and $D(3, 3)$

d $A(-2, -1)$ and $P(4, -4)$ **e** $D(0, -1)$ and $F(3, -1)$ **f** $H(9, -2)$ and $L(-1, 3)$

g $K(4, 9)$ and $M(-4, -5)$ **h** $Z(-8, -6)$ and $T(4, 0)$ **i** $F(0, 4)$ and $B(3, 4)$

6 Which pair of points form an interval that has a gradient of −1? Select **A**, **B**, **C** or **D**.

A (1, 2), (2, 1) **B** (−5, 6), (6, 7) **C** (2, 2), (3, −1) **D** (−5, −10), (4, −11)

☐ Foundation ○ Standard ⬡ Complex

7 Find the gradient of the lines labelled k and l.

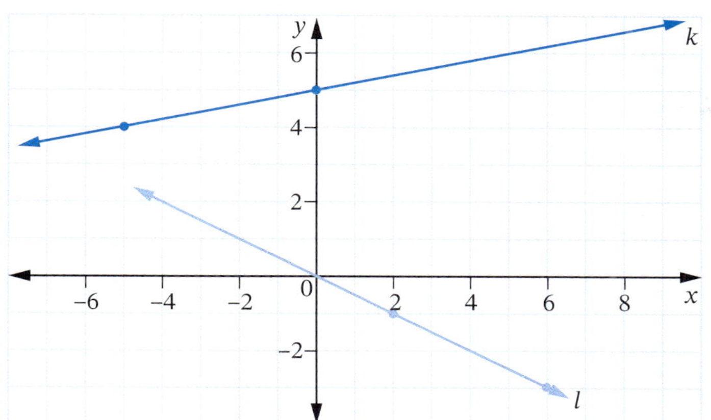

8 Which pair of points lie on a line with a gradient of 0?

A (3, 3), (−3, −3) **B** (−5, 7), (5, 7) **C** (1, −6), (1, 4) **D** (0, 9), (9, 0)

9 $PQRS$ is a quadrilateral.

R C

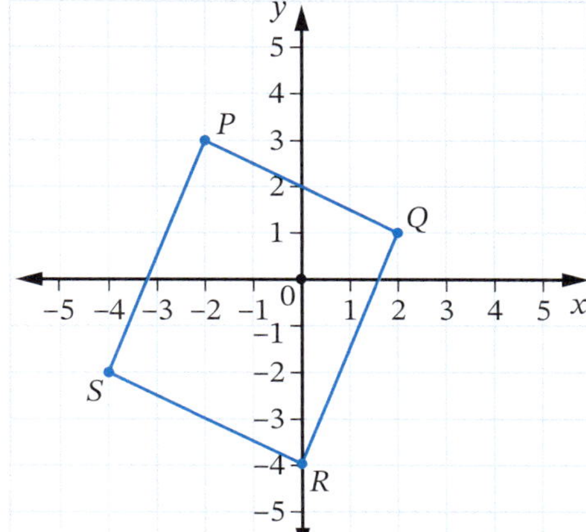

a Find the gradient of each side of $PQRS$.

b Which sides have the same gradient? What does that mean about those sides?

c What type of quadrilateral is $PQRS$?

☐ Foundation ○ Standard ○ Complex

DID YOU KNOW?

Stairs, ramps and hills

- Some stairs are easier to climb than others: the gradient of stairs should not be more than 5 in 6, that is, $\frac{5}{6}$.
- A ramp for walking should not have a gradient higher than $\frac{3}{10}$ and the gradient of a wheelchair ramp should be between $\frac{1}{14}$ and $\frac{1}{8}$.
- The maximum slope of a hill for parking your car safely is $\frac{2}{9}$.
- The steepest street in the world is Baldwin Street in Dunedin, New Zealand, with one section having a gradient of 35% or $\frac{7}{20}$.

Would it be safe to park your car on Gower St, Toowong, in Brisbane, which has a slope of $\frac{5}{16}$, or Mount St, Perth, which has a slope of $\frac{7}{58}$?

Shutterstock.com/Deyan Denchev

The steepest street in the world, in New Zealand.

TECHNOLOGY

The gradient of collinear points

In this activity, you will use dynamic geometry software to investigate collinear points.

1. Create a number plane with x-values: 0 to 30 and y-values: 0 to 16.
2. Plot the points $A(5, 2)$, $B(9, 5)$ and $C(17, 11)$. Create intervals AB and BC.
3. To check if the 3 points are collinear, select **Slope** for interval AB, and again for interval BC. Is the slope (gradient) the same for AB and BC? If so, these points are called **collinear points**. What do you think **collinear points** are? Discuss this with other students and explain it in your own words.
4. Now test whether each set of points are collinear.

   a (0, 3), (2, 4) and (6, 6)
   b (8, 0), (−8, 0) and (−6, −10)
   c (1, 8), (2, 9) and (4, 10)
   d (−2, 8), (−1, 6) and (0, 1)
   e (2, 7), (4, 10), (5, 13) and (6, 17)
   f (−14, 11), (−8, 9), (1, 6) and (16, 1)

Graphing linear functions 11.04

A relationship between 2 variables such as x and y whose graph is a **straight line** is called a **linear relationship**. The expression of that relationship as an algebraic formula, such as $y = x + 2$, is called a **linear function**.

Video
Graphing linear equations

Worksheets
Graphing linear equations
A page of lines
A page of number planes
Number plane grid paper

Interactives
Graphing lines
Function builder: Basics
Function builder

Example 6

Graph $y = 2x + 1$ on a number plane.

SOLUTION

Complete a table of values. Choose x-values close to 0 for easy calculation and graphing.

x	−1	0	1
y	−1	1	3

Graph the table of values on a number plane. Rule a straight line through the points, place arrows at each end, and label the line with its equation.

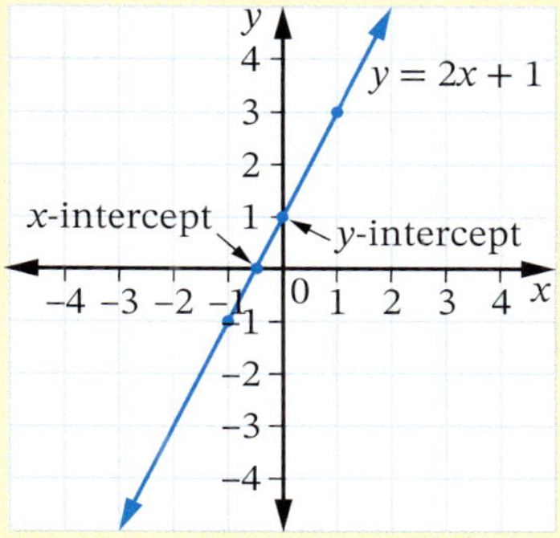

Note:

- the ***x*-intercept** of the line is $-\frac{1}{2}$: this is the x-value where the line cuts the ***x*-axis.**
- the ***y*-intercept** of the line is 1: this is the y-value where the line cuts the ***y*-axis.**
- every point on the line follows the linear equation $y = 2x + 1$: for example, (−4, −7), (−2, −3) and (2, 5) lie on the line and also follow the rule $y = 2x + 1$.
- there are an infinite number of points that follow the rule: arrows on both ends of the line indicate that it has infinite length.

ⓘ Testing if a point lies on a line

A point lies on a line if its (x, y) coordinates satisfy the equation of the line.

- Separate the equation into LHS (left-hand side) and RHS (right-hand side).
- Substitute the coordinates of the point into both sides.
- If LHS = RHS, the point satisfies the equation and lies on the line.
- If LHS ≠ RHS, the point does not lie on the line.

Worksheet
Straight-line equations

Video
Testing if a point lies on a line

Example 7

Test whether each point lies on the line $y = 2x + 3$.

a $(3, 9)$ **b** $(8, -4)$

SOLUTION

a Substitute $x = 3$ and $y = 9$ into $y = 2x + 3$.

$\text{LHS} = y$
$= 9$

$\text{RHS} = 2x + 3$
$= 2 \times 3 + 3$
$= 9$

LHS = RHS, so $(3, 9)$ lies on the line.

b Substitute $x = 8$ and $y = -4$ into $y = 2x + 3$.

$\text{LHS} = y$
$= -4$

$\text{RHS} = 2x + 3$
$= 2 \times 8 + 3$
$= 19$

$\text{LHS} \neq \text{RHS}$, so $(8, -4)$ does not lie on the line.

Horizontal and vertical lines

A **horizontal line** is flat and runs across (sideways), parallel to the x-axis.

Every point on the horizontal line has y-coordinate -2, and the y-intercept is -2.

x	−1	0	1	2
y	−2	−2	−2	−2

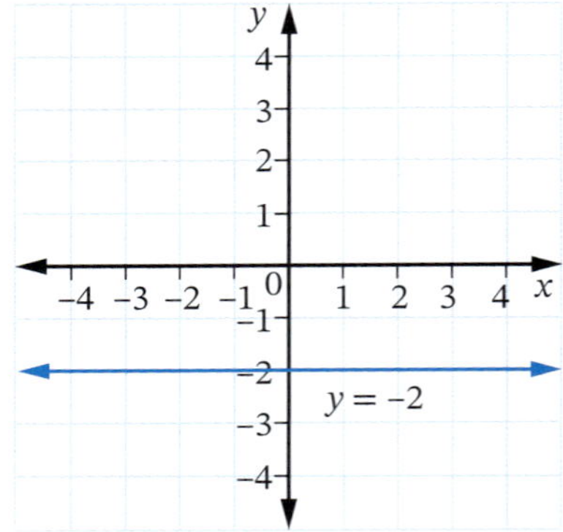

y is always equal to -2, so the equation of this horizontal line is $y = -2$.

A **vertical line** runs up and down, parallel to the y-axis and at right angles to a horizontal line.

Every point on the vertical line has x-coordinate 1, and the x-intercept is 1.

x	1	1	1	1
y	0	1	2	3

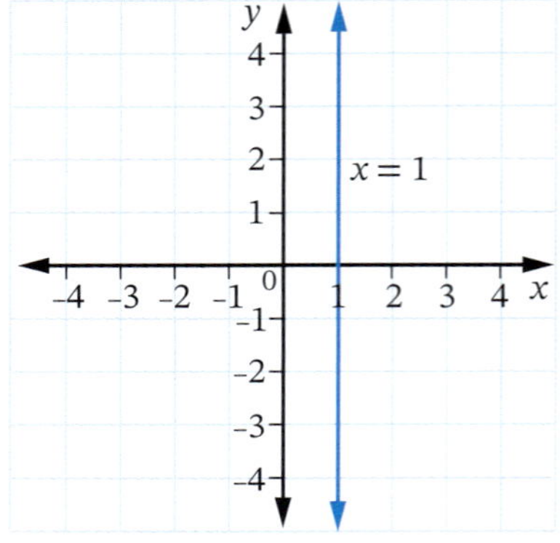

x is always equal to 1, so the equation of this vertical line is $x = 1$.

ⓘ Horizontal and vertical lines

The **equation of a horizontal line** is of the form $\boldsymbol{y = c}$ (where c is a constant number).

The **equation of a vertical line** is of the form $\boldsymbol{x = c}$ (where c is a constant number).

Example 8

Graph each line on the same number plane.

a $x = 4$ **b** $y = -2$ **c** $x = -\frac{1}{2}$ **d** $y = 3\frac{1}{2}$

SOLUTION

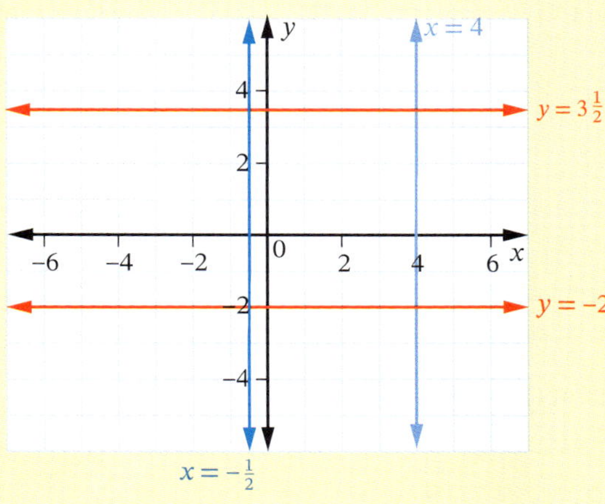

Example 9

Find the equations of lines A, B, C and D.

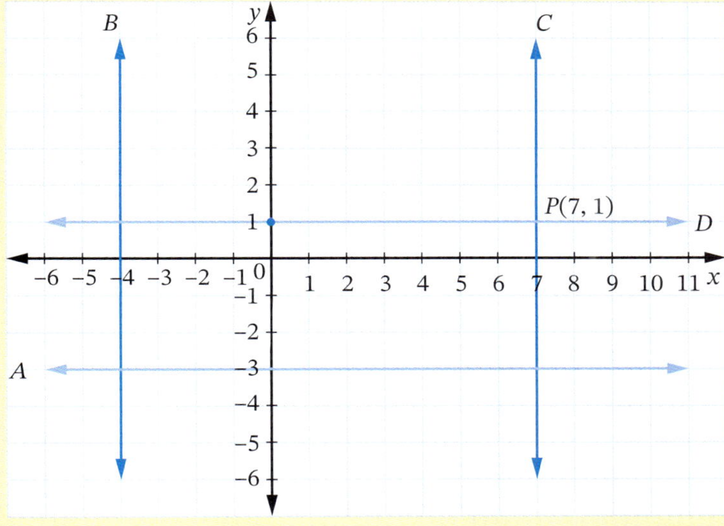

SOLUTION

- Line A is horizontal and has a y-intercept of −3, so its equation is $y = -3$.
- Line B is vertical and has an x-intercept of −4, so its equation is $x = -4$.
- Line C is vertical and goes through (7, 1), so its equation is $x = 7$.
- Line D is horizontal and goes through (7, 1), so its equation is $y = 1$.

EXERCISE 11.04 ANSWERS ON P. 662

Graphing linear functions

1 Graph each linear function on a number plane, and write:

i its x-intercept **ii** its y-intercept.

a $y = x - 3$ **b** $y = 2x + 2$ **c** $y = -x + 5$

d $y = 3x$ **e** $y = -x$ **f** $y = 10 - 4x$

g $y = \frac{x+2}{2}$ **h** $y = 2x + 1$ **i** $y = \frac{x}{3} - 2$

j $y = \frac{3x}{4}$ **k** $x + y = 4$ **l** $x - y = 2$

2 What is the x-intercept of the graph of $y = 6 - 3x$? Select the correct answer **A**, **B**, **C** or **D**.

A -3 **B** 2 **C** 3 **D** 6

EXAMPLE 7

3 Test whether the point $(3, -1)$ lies on each line:

a $y = 2x - 7$ **b** $y = -2x + 4$ **c** $x - y = 4$

d $x + 2y = 1$ **e** $y = -3x + 8$ **f** $x - 3y = 0$

4 Which of these points lies on the line $x - y = -2$? Select the correct answer **A**, **B**, **C** or **D**.

A $(1, 1)$ **B** $(-1, -1)$ **C** $(0, 2)$ **D** $(-2, 4)$

EXAMPLE 8

5 Graph each set of lines on a number plane.

a $x = 2, x = -3$ **b** $y = 6, y = -2$

c $x = 2\frac{1}{2}, x = 0, y = -5$ **d** $y = 1.5, y = 3, x = -2$

6 Write down the equation of each line shown below.

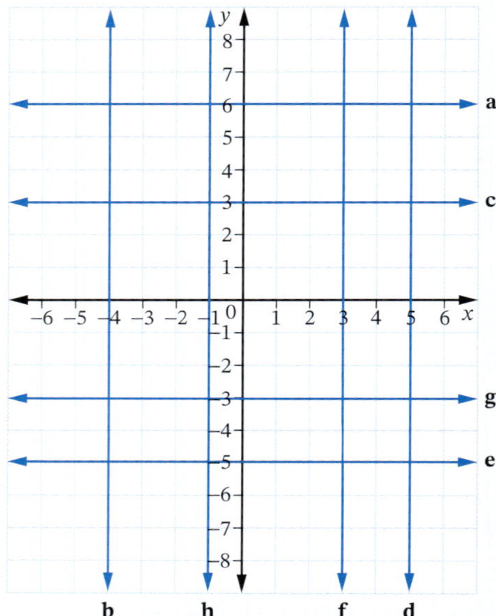

7 **a** Show that the point (3, 2) lies on both the lines $x = 2y - 1$ and $x + y = 5$.

b So at what point do the lines $x = 2y - 1$ and $x + y = 5$ cross?

R C

8 Show that the lines $y = -3x + 6$ and $y = -2x$ intersect at the point $(6, -12)$.

R C

9 Find the equation of each line described.

R **a** A horizontal line that passes through the y-axis at 6.

C **b** A vertical line that goes through $(-5, 2)$.

c Each point on the line is 2 units above the x-axis.

d Each point on the line is 5 units to the left of the y-axis.

e The line is parallel to the x-axis and passes through the point $(0, -2)$.

f A horizontal line that goes through $(4, -1)$.

g The line goes through the points $(-1, 8)$ and $(-1, 2)$.

10 What is the equation of:

a the y-axis? **b** the x-axis?

R C

TECHNOLOGY

Graphing linear functions

In this activity, we will use dynamic geometry software to graph a linear function.

1 Create a number plane with axes: x: −10 to 10 and y: −10 to 10.

2 Enter the following linear functions in column A of the spreadsheet.

A1: $y = 3x + 1$ A2: $y = 3x$ A3: $y = 2x + 4$

A4: $y = -2x - 5$ A5: $y = x + 2$ A6: $y = -x - 1$

3 Read the gradient of each line by selecting the **Slope tool**. Select a line and you will see the right-angled triangle showing $\frac{\text{rise}}{\text{run}}$.

4 Enter each linear function and find the gradient of each line.

a $y = x + 1$ **b** $y = -2x - 4$ **c** $y = 6x + 3$

d $y = 3x$ **e** $y = -8x$ **f** $y = -4x - 6$

Quiz
Mental skills 11

☆ MENTAL SKILLS 11 ANSWERS ON P. 664 Maths without calculators

Divisibility tests

How can you tell if a number is divisible by 2? Look at its last digit. If that digit is 2, 4, 6, 8 or 0, then the number is divisible by 2 (that is, it is even).

How can you tell if a number is divisible by 5? If its last digit is 0 or 5, then the number is divisible by 5.

These are examples of **divisibility tests**—rules for checking whether a number is divisible by a certain number. The table below shows some common divisibility tests.

A number is divisible by:	if:
2	its last digit is 2, 4, 6, 8 or 0
3	the sum of its digits is divisible by 3
4	its last 2 digits form a number divisible by 4
5	its last digit is 0 or 5
6	it is even and the sum of its digits is divisible by 3
9	the sum of its digits is divisible by 9
10	its last digit is 0

1 Study each example.

a Test whether 748 is divisible by 2, 3 or 4.

- Last digit is 8 (even). ∴ 748 is divisible by 2.
- Sum of digits $= 7 + 4 + 8 = 19$, which is not divisible by 3. ∴ 748 is not divisible by 3.
- 48 is divisible by 4. ∴ 748 is divisible by 4 ($748 \div 4 = 187$).

b Test whether 261 is divisible by 5 or 9.

- Last digit is 1, not 0 or 5. ∴ 261 is not divisible by 5.
- $2 + 6 + 1 = 9$, which is divisible by 9. ∴ 261 is divisible by 9 ($261 \div 9 = 29$).

c Test whether 570 is divisible by 4, 6 or 10.

- 70 is not divisible by 4. ∴ 570 is not divisible by 4.
- 570 is even and $5 + 7 + 0 = 12$, which is divisible by 3. ∴ 570 is divisible by 6 ($570 \div 6 = 95$).
- Last digit is 0. ∴ 570 is divisible by 10 ($570 \div 10 = 57$).

2 Test whether each number is divisible by 2, 3, 5 or 6.

a 250 **b** 189 **c** 78 **d** 465
e 1024 **f** 840 **g** 715 **h** 627

3 Test whether each number is divisible by 4, 9 or 10.

a 144 **b** 280 **c** 522 **d** 4170
e 936 **f** 726 **g** 342 **h** 5580

9780170465618

INVESTIGATION

Comparing gradients and y-intercepts

1 This number plane shows the graphs of 4 linear functions.

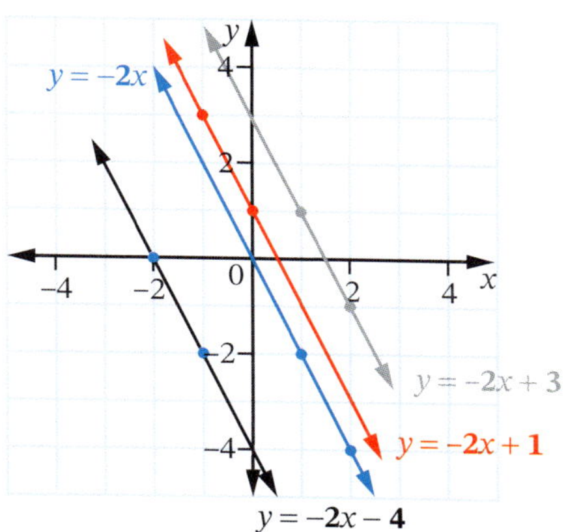

a Copy and complete this table by finding the gradient and y-intercept of each line.

Equation	Gradient	y-intercept
$y = -2x - 4$	-2	-4
$y = -2x$		
$y = -2x + 1$		
$y = -2x + 3$		

b How are the 4 lines:

i similar?

ii different?

2 This number plane shows the graphs of another 4 linear functions.

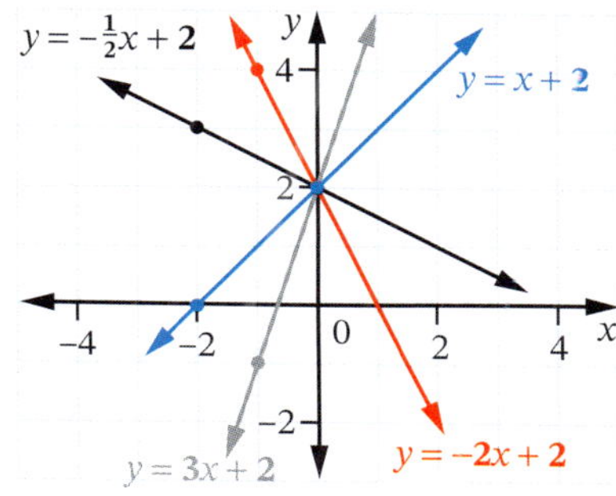

a Copy and complete this table.

Equation	Gradient	y-intercept
$y = -2x + 2$		
$y = -\frac{1}{2}x + 2$		
$y = x + 2$		
$y = 3x + 2$		

b How are the 4 lines:

i similar? **ii** different?

3 **a** Copy and complete this table for the 3 lines graphed below.

Equation	Gradient	*y*-intercept
$y = -2x + 3$		
$y = \frac{1}{2}x - 2$		
$y = x - 6$		

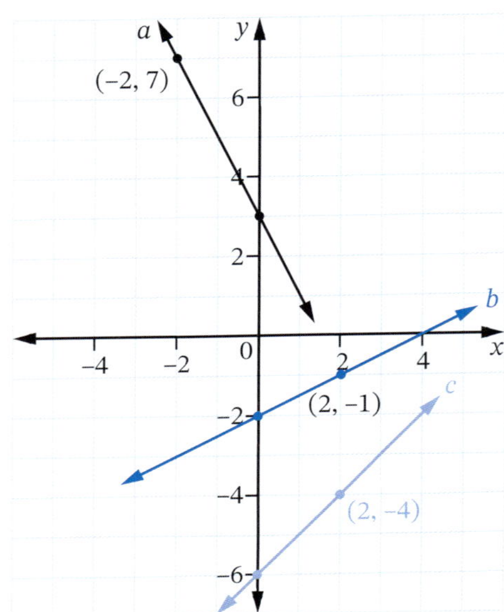

b Explain how the values for the gradient and *y*-intercept of each line relate to its equation.

c What is the equation of a line that has a gradient of 4 and a *y*-intercept of −1?

d What is the gradient and *y*-intercept of the line with equation $y = -x + 7$?

9780170465618

The gradient–intercept equation $y = mx + c$

11.05

ⓘ The gradient–intercept form of a linear equation

The equation of a straight line is $y = mx + c$, where m is the gradient and c is the y-intercept. For this reason, $y = mx + c$ is also called the **gradient–intercept form** of a linear equation.

Puzzle Equation of a line

Interactive Graphing slope-intercept

Example 10

Find the gradient and y-intercept of the line with equation:

a $y = 4x - 7$ **b** $y = \frac{1}{2}x + 3$ **c** $y = 3 - 5x$ **d** $y = \frac{3x}{4} - 1$

SOLUTION

a $y = 4x - 7$ is in the form $y = mx + c$.

$\therefore$ Gradient $m = 4$ and y-intercept $c = -7$.

b For $y = \frac{1}{2}x + 3$, gradient $m = \frac{1}{2}$ and y-intercept $c = 3$.

c $y = 3 - 5x$ can be rewritten as $y = -5x + 3$.

$\therefore$ Gradient $m = -5$ and y-intercept $c = 3$.

d For $y = \frac{3x}{4} - 1$, gradient $m = \frac{3}{4}$ and y-intercept $c = -1$.

Example 11

Write the equation of the line that has a gradient of –4 and a y-intercept of –6.

SOLUTION

$m = -4, c = -6$

$\therefore$ The equation of the line is $y = -4x - 6$. Using $y = mx + c$.

Video
Graphing
y = mx + c

Example 12

Graph each linear function by finding the gradient and y-intercept first.

a $y = 2x - 1$ **b** $y = -2x + 5$ **c** $y = \frac{3}{4}x - 2$

SOLUTION

a $y = 2x - 1$ has a gradient of 2 and a y-intercept of -1.

- Plot the y-intercept -1 on the y-axis.
- Make a gradient of 2 by moving **across** 1 unit (**run**) and **up** 2 units (**rise**) and marking the point at (1, 1).
- Rule a line through this point and the y-intercept.
- Label the line with the equation $y = 2x - 1$.

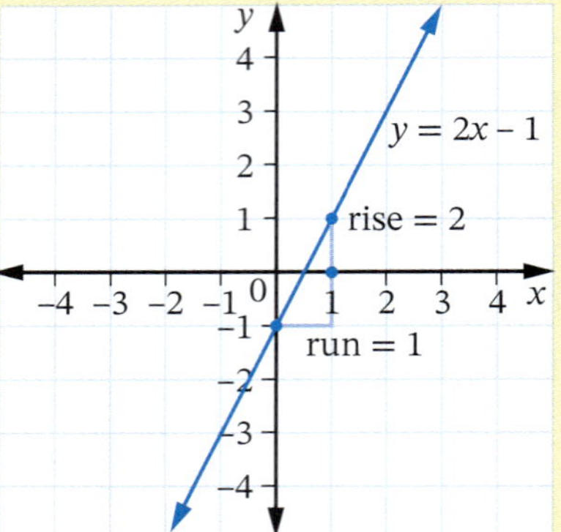

b $y = -2x + 5$ has a gradient of -2 and a y-intercept of 5.

- Plot the y-intercept 5 on the y-axis.
- Make a gradient of -2 by moving **across** 1 unit (**run**) and **down** 2 units (**'negative' rise**) and marking the point at (1, 3).
- Rule a line through this point and the y-intercept.

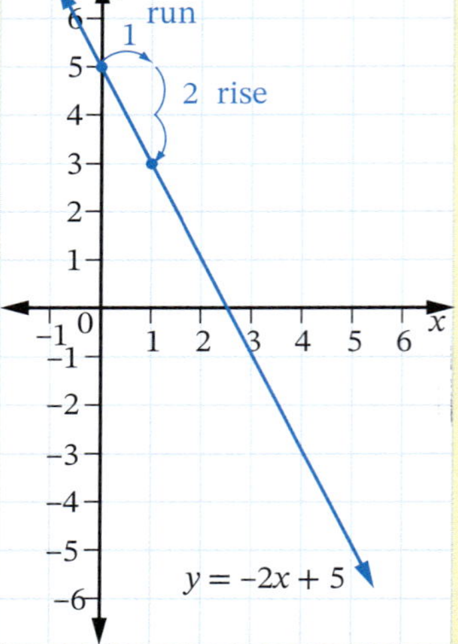

c $y = \frac{3}{4}x - 2$ has a gradient of $\frac{3}{4}$ and a y-intercept of -2.

- Plot the y-intercept -2 on the y-axis.
- Make a gradient of $\frac{3}{4}$ by moving **across** 4 units (**run**) and **up** 3 units (**rise**) and marking the point at (4, 1).
- Rule a line through this point and the y-intercept.

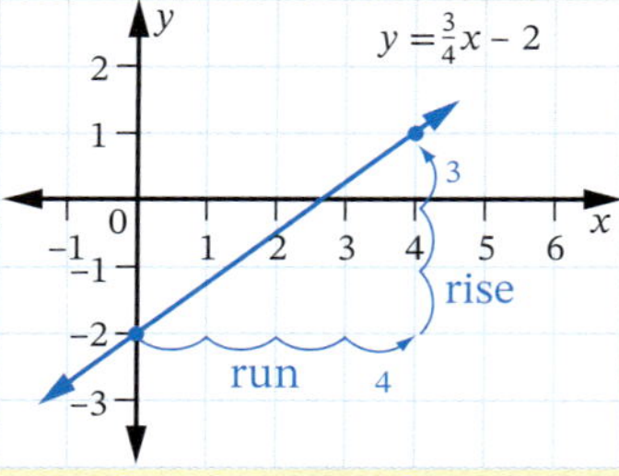

EXERCISE 11.05 ANSWERS ON P. 664

The gradient–intercept equation $y = mx + c$

U F

1 Find the gradient and y-intercept of the line with equation:

EXAMPLE 10

a $y = 3x - 2$ **b** $y = 2x + 7$ **c** $y = x + 4$ **d** $y = -x - 9$

e $y = \frac{3x}{4} + 6$ **f** $y = x$ **g** $y = \frac{x}{2} - 11$ **h** $y = \frac{2}{3}x + 6$

i $y = 8 - 2x$ **j** $y = 5 - x$ **k** $y = \frac{x}{4}$ **l** $y = -3x + 11$

m $y = 5$ **n** $y = \frac{2}{5}x + 1$ **o** $y = x - \frac{7}{2}$ **p** $y = \frac{3x}{4} + 1$

2 Which equation has a gradient of 5? Select the correct answer **A**, **B**, **C** or **D**.

A $y = x + 5$ **B** $y = 5 - x$ **C** $y = 5$ **D** $y = 5x$

3 Write the equation of the line with a gradient of:

EXAMPLE 11

a 2 and a y-intercept of -1 **b** $\frac{3}{4}$ and a y-intercept of 2

c -3 and a y-intercept of 5 **d** $-\frac{2}{5}$ and a y-intercept of 3

4 Which equation has a y-intercept of -1? Select the correct answer **A**, **B**, **C** or **D**.

A $y = 1 - x$ **B** $y = 1 + x$ **C** $y = x - 1$ **D** $y = -x$

5 Graph each linear function by finding the gradient and y-intercept first.

EXAMPLE 12

a $y = 2x + 1$ **b** $y = x - 4$ **c** $y = 3x + 2$ **d** $y = -4x$

e $y = -3x - 3$ **f** $y = \frac{x}{2} + 2$ **g** $y = 8 - \frac{3x}{4}$ **h** $y = \frac{5x}{2} - 1$

6 Which diagram shows the graph of $y = 2x - 1$? Select the correct answer **A**, **B**, **C** or **D**.

A

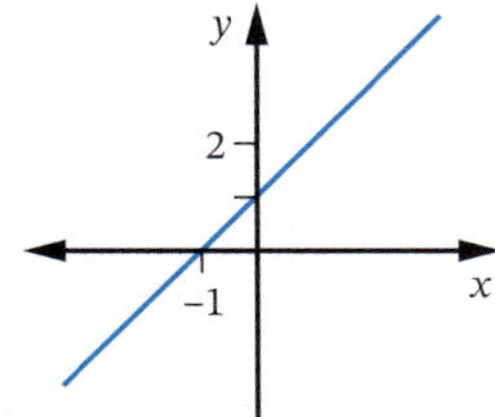

B

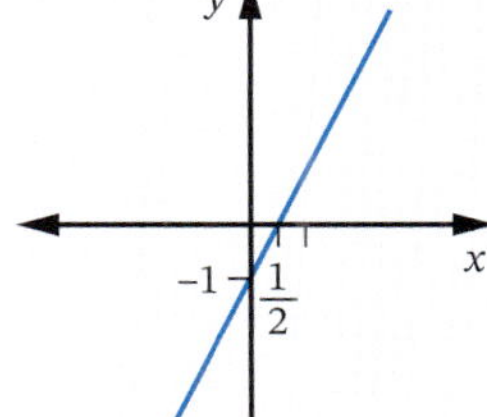

C

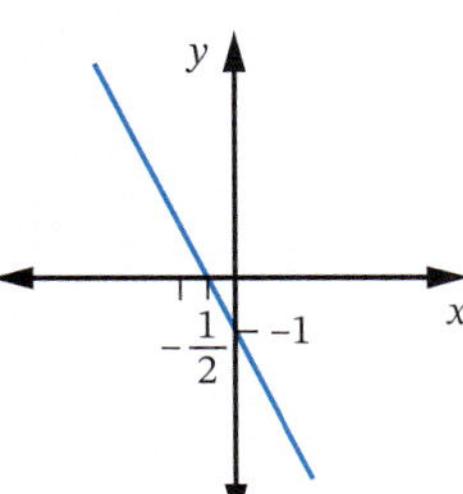

D

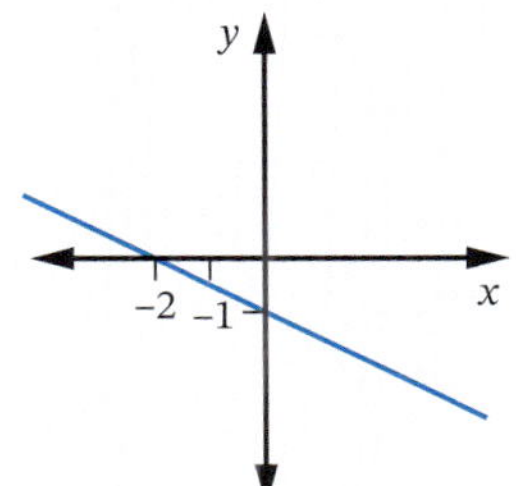

☐ Foundation ○ Standard ○ Complex

11.06 Finding the equation of a line

Video Finding the equation of a line

Worksheet A page of lines

Puzzle Straight-line equations

Example 13

Find the equation of each line.

a

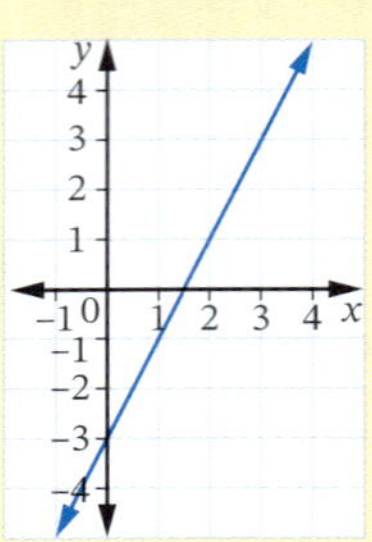

b

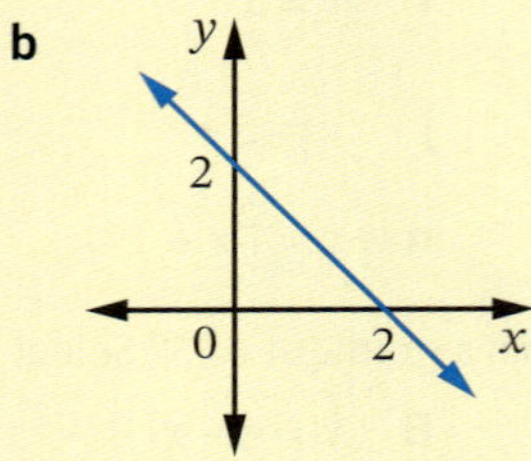

SOLUTION

a Select 2 points on the line to find the gradient, say (0, −3) and (2, 1).

Gradient $m = \frac{\text{rise}}{\text{run}} = \frac{4}{2} = 2$

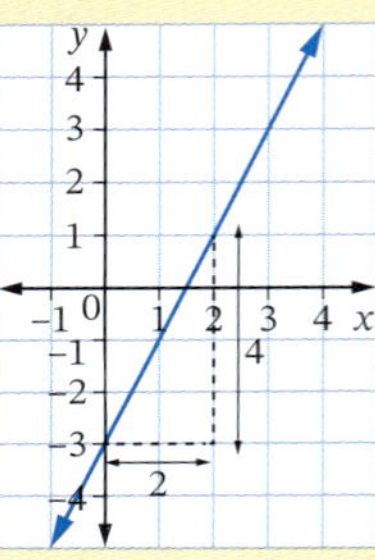

y-intercept $c = -3$.

From graph $y = mx + c$.

$\therefore$ The equation of the line is $y = 2x - 3$.

We can check that this equation is correct for another point, say (3, 3).

When $x = 3$, $y = 2 \times 3 - 3 = 3$.

b Find the gradient of the line passing through (0, 2) and (2, 0).

Gradient $m = \frac{\text{rise}}{\text{run}} = \frac{-2}{2} = -1$

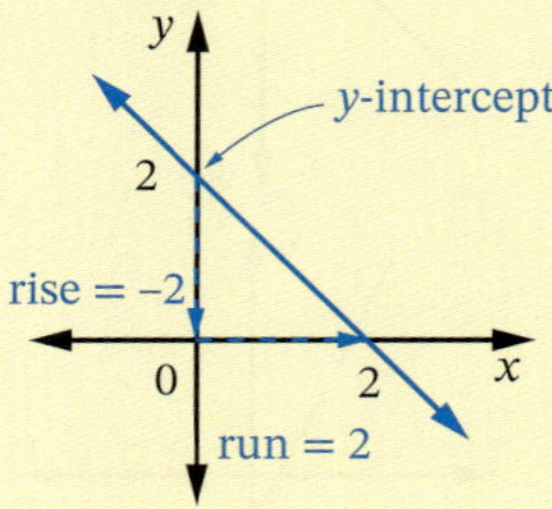

y-intercept $c = 2$.

From graph $y = mx + c$.

$\therefore$ The equation of the line is $y = -x + 2$.

Finding the equation of a line

U F

EXAMPLE 13

1 Find the equation of each line.

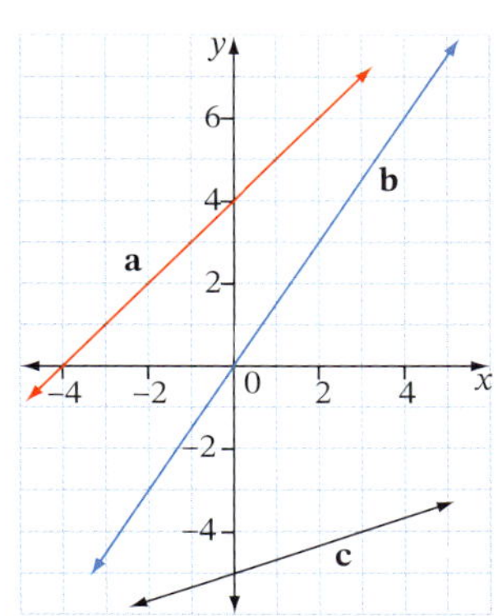

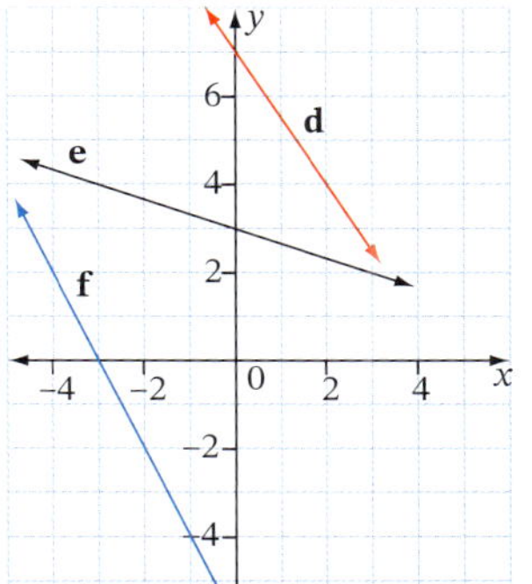

2 Find the equation of each line.

a

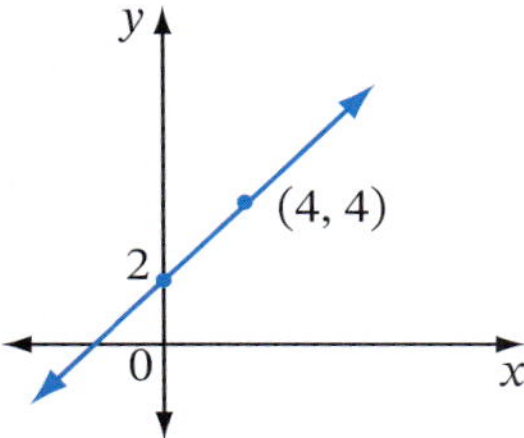

b

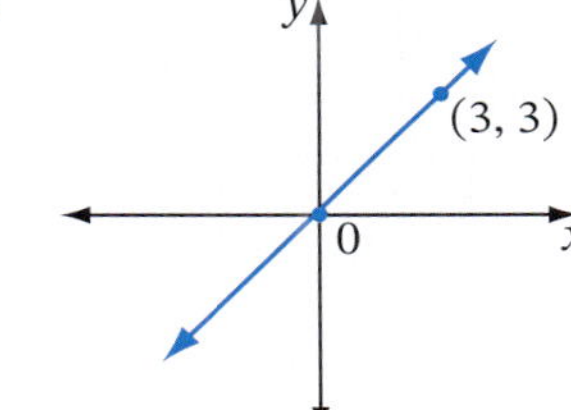

c

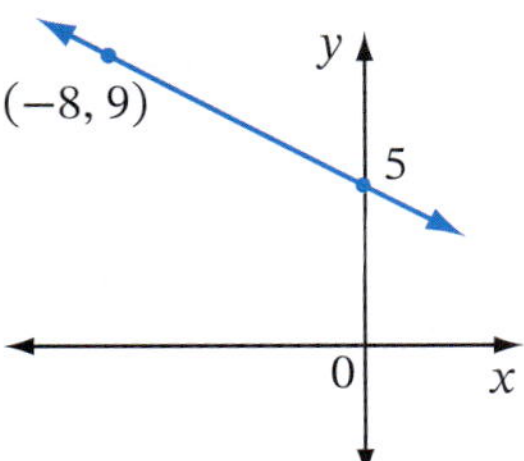

d

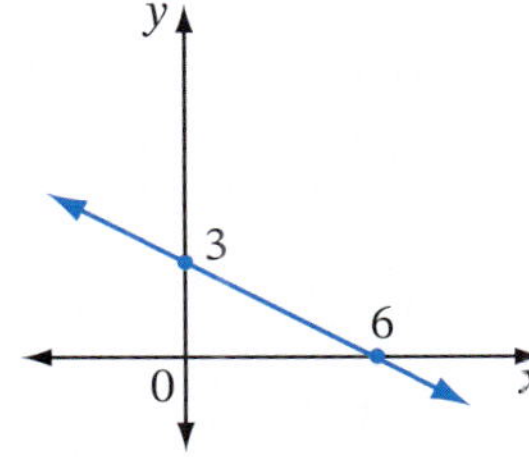

e

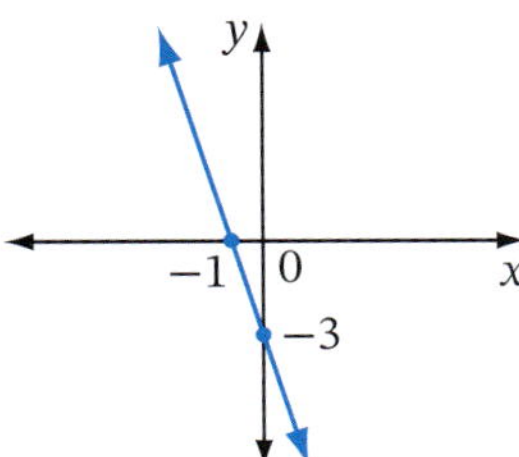

f

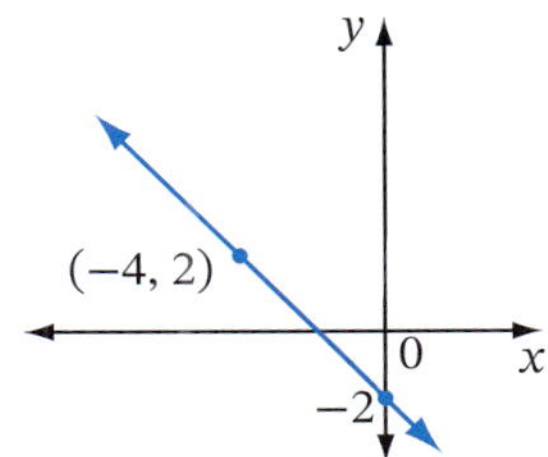

g

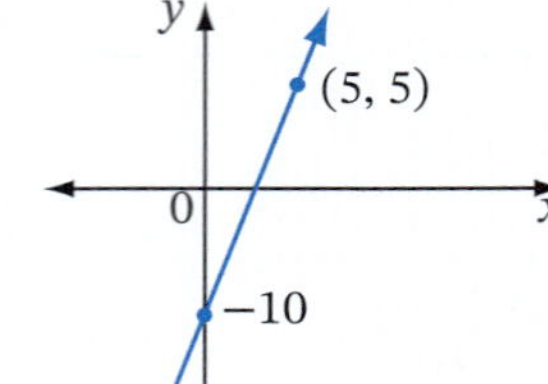

h

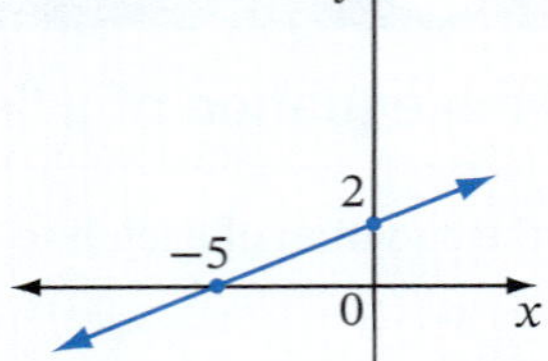

i

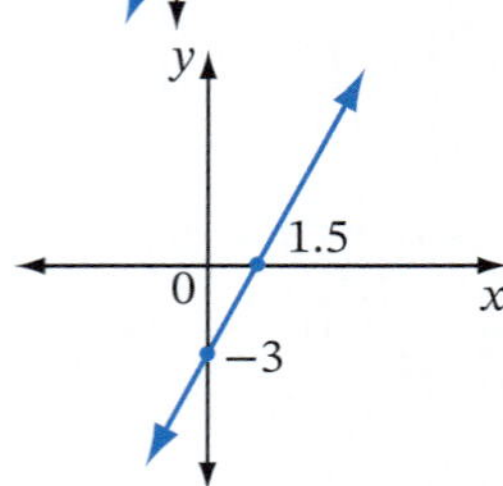

3 Match each linear function to its graph.

a $y = x - 1$ **b** $y = x + 1$ **c** $y = -x - 1$ **d** $y = -x + 1$

A

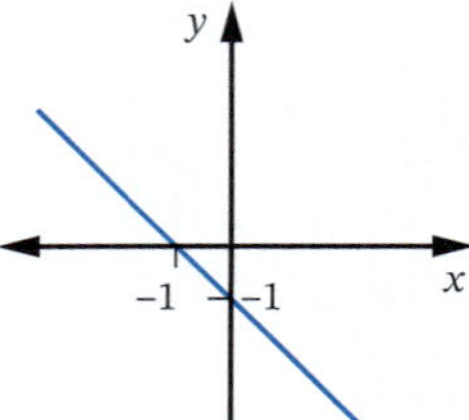

B

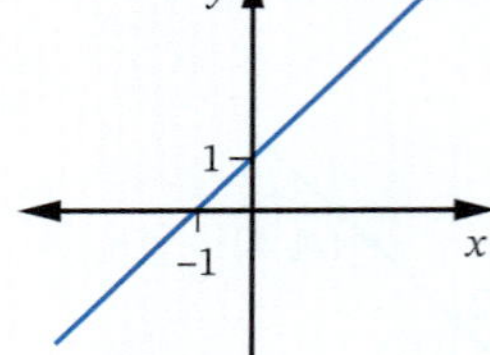

C

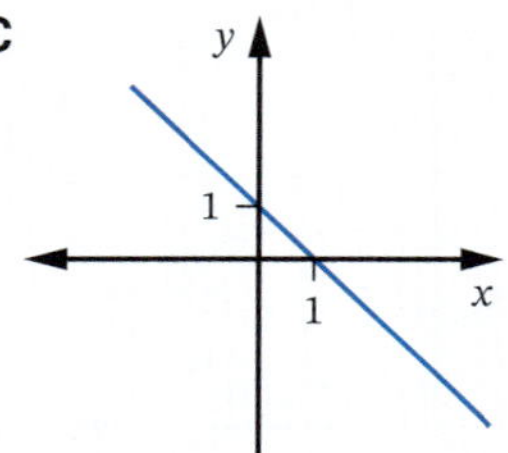

D

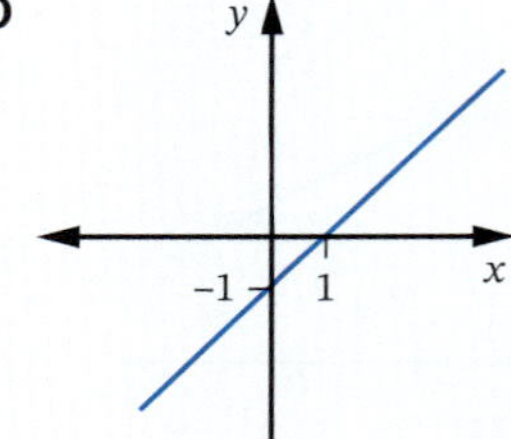

☐ Foundation ○ Standard ⬡ Complex

Solving linear equations graphically 11.07

In Chapter 7, *Equations and inequalities*, we solved equations **algebraically**. We can also solve an equation **graphically** by first graphing it on the number plane.

Example 14

Solve the equation $2x - 3 = 5$ graphically.

SOLUTION

Graph $y = 2x - 3$ (the LHS of the equation) on a number plane.

For every point on the line, the y-coordinate is the value of $2x - 3$. So to solve $2x - 3 = 5$, we need to find the point whose y-coordinate is 5.

Draw a dotted horizontal line at $y = 5$ and read off the coordinates of the point on $y = 2x - 3$ crossed by this line.

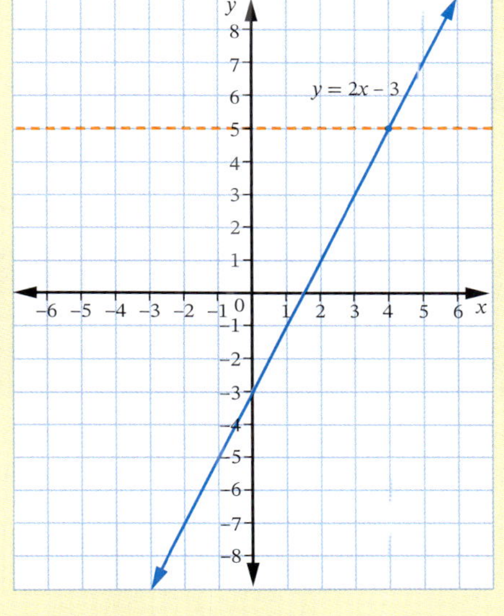

This point is (4, 5), which means the solution to $2x - 3 = 5$ is $x = 4$.

Check: $2 \times 4 - 3 = 5$

EXERCISE 11.07 ANSWERS ON P. 665

Solving linear equations graphically

U F

This exercise is best completed using graphing technology such as *dynamic geometry software or graphing websites.*

1 Use the graph in Example **14** above to solve each equation. Check your solutions. EXAMPLE 14

a $2x - 3 = 3$ **b** $2x - 3 = -7$ **c** $2x - 3 = 0$

2 Graph $y = 2x + 1$ and use it to solve each equation graphically.

a $2x + 1 = 7$ **b** $2x + 1 = 10$ **c** $2x + 1 = -5$

Foundation Standard Complex

3 Solve each equation in question **2** algebraically.

4 At which point do the graphs of $y = 2x + 3$ and $y = 1$ intersect? Select the correct answer **A**, **B**, **C** or **D**.

A (2, 1) **B** (−1, 1) **C** (4, −2) **D** (8, −4)

5 Graph $y = 3x - 2$ and use it to solve each equation graphically.

a $3x - 2 = 7$ **b** $3x - 2 = -11$ **c** $3x - 2 = -2$

6 Graph $y = -x + 1$ and use it to solve each equation graphically.

a $-x + 1 = -3$ **b** $-x + 1 = 4$ **c** $-x + 1 = -1$

7 Graph $y = -2x - 1$ and use it to solve each equation graphically.

a $-2x - 1 = -3$ **b** $-2x - 1 = 0$ **c** $-2x - 1 = 3$

8 Graph $y = \frac{1}{2}x + 3$ and use it to solve each equation graphically.

a $\frac{1}{2}x + 3 = 5$ **b** $\frac{1}{2}x + 3 = 1$ **c** $\frac{1}{2}x + 3 = 2$

11.08 Direct proportion

Ann-Marie noticed a relationship between the amount of petrol used by her car and the distance travelled by the car, as shown in the table of values.

Amount of petrol used (L)	10	14	22	32	36	42
Distance travelled (km)	135	189	297	432	486	567

When she graphed these values, she found a linear relationship.

Distance travelled and petrol used

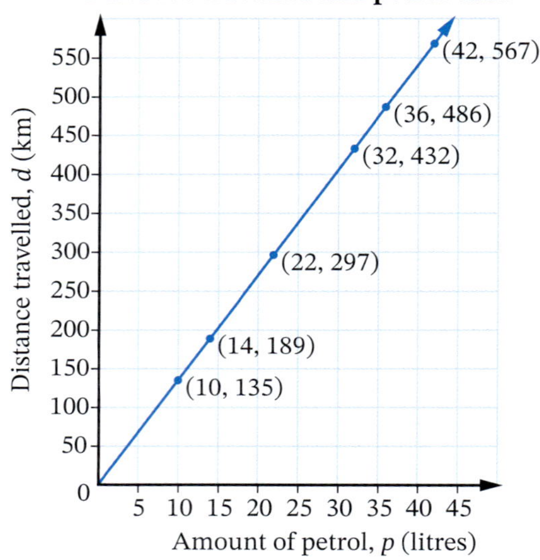

If p stands for the amount of petrol used and d stands for the distance travelled, then it can be shown that d is directly related to p by the formula $d = 13.5p$. This means that the distance travelled is a constant multiple of the amount of petrol used.

This is an example of **direct proportion** (or **direct variation**).

Two variables are **directly proportional** to each other if one variable is a constant multiple of the other, and when one variable changes, the other one changes by the same factor.

ⓘ Direct proportion

If y is **directly proportional** to x, then $y = kx$, where k is a constant (number) called the **constant of proportionality** or **constant of variation**.

- A direct linear relationship exists between x and y.
- If x increases (or decreases), y increases (or decreases).
- If x is doubled (or halved), y is doubled (or halved).
- Another way of saying 'y is directly proportional to x' is 'y **varies directly** with x'.
- A graph for direct proportion is a straight line going through (0, 0) with gradient k.

Example 15

Video
Direct proportion

The distance (d), in metres, travelled by a car is directly proportional to the number of rotations (r) of its tyres. After 540 rotations, a distance of 950 m is travelled.

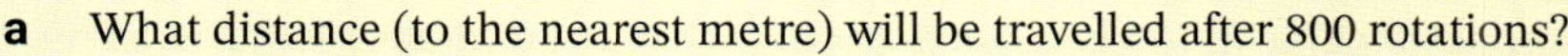

a What distance (to the nearest metre) will be travelled after 800 rotations?

b How many full rotations will be needed to cover 360 m?

SOLUTION

a d is directly proportional to r. — *Forming a proportion equation.*

$\therefore d = kr$

To find k, substitute the information given for r and d. — *Finding k.*

When $r = 540$ and $d = 950$,

$950 = k \times 540$

$k = \frac{950}{540}$

$= 1.759...$

$\therefore d = 1.759...\, r$ — *Rewriting the equation $d = kr$.*

When $r = 800$, — *Solving the problem.*

$d = 1.759... \times 800$

$= 1407.4074...$

≈ 1407 m

After 800 rotations, the distance travelled will be 1407 m.

b When $d = 360$, — *Solving the problem.*

$360 = 1.759...\, r$

$r = \frac{360}{1.759...}$

$= 204.661...$

≈ 205 rotations

For a distance of 360 m, there will be 205 rotations. — *Rounding up for full rotations.*

ⓘ Solving direct proportion problems

1 Identify the 2 variables (say x and y) and form a proportion equation, $y = kx$.
2 Substitute values for x and y to find k, the constant of proportionality.
3 Rewrite $y = kx$ using the value of k.
4 Substitute a value for x or y into $y = kx$ to solve the problem.

EXERCISE 11.08 ANSWERS ON P. 666

Direct proportion

EXAMPLE 15

1 The stretch, S cm, of a spring is directly proportional to the mass, M kg, pulling on it.

Mass, M (kg)	0	8	12	20	24	40
Stretch, S (cm)	0	5	7.5	12.5	15	25

a Graph the linear relationship between S and M.

b The equation of the line is $S = kM$. Use the table of values to find the value of k as a decimal, then rewrite the equation for S.

c Use the equation to find the stretch caused by a mass of 36 kg.

d What is the mass that will cause a stretch of 40 cm?

2 B is directly proportional to h. If when $h = 5$, $B = 16$, what is the value of k in $B = kh$? Select the correct answer **A**, **B**, **C** or **D**.

A 80 **B** 11 **C** 3.2 **D** 0.3125

3 Y is directly proportional to X. If when $X = 15$, $Y = 75$, find the constant of proportionality, then use the proportion equation to find Y when $X = 23$.

4 The mass, M, in grams of a chemical is directly proportional to its volume, V, in cm^3.

PS **a** Write the formula $M = kV$ where k is a decimal, given that $M = 140$ when $V = 80$.

R **b** Use the formula to calculate the mass of 406 cm^3 of the chemical.

c Calculate (correct to one decimal place) the volume of the chemical if the mass is:

i 28 g **ii** 0.36 kg

5 Find the linear formula for n in terms of m for this table of values.

m	3	6	9	12	15
n	7.5	15	22.5	30	37.5

6 p varies directly with q. If $q = 7$ when $p = 42$, find p when $q = 18$.

7 The distance, D, travelled by Pete, a marathon runner, varies directly with time, t. Pete runs 8.25 km in 44 minutes.

a How far will he run in 1 hour?

b How long will it take him to run 24 km?

▷

☐ Foundation ○ Standard ⬡ Complex

8 For an object that is cooling, the drop in temperature varies directly with time. If the temperature drops 8 °C in 5 minutes, how long would it take to drop 10 °C? Select the correct answer **A**, **B**, **C** or **D**.

A 6.25 min **B** 7 min **C** 12.8 min **D** 16 min

9 The amount of interest, I, earned on an investment account varies directly with the size of the deposit, D.

PS R

a If Hannah earns \$35 interest on an investment of \$1400, write an equation relating I to D.

b How much will she earn on an investment of \$2500?

c If Hannah doubles the size of her investment in **b**, how much will she earn in interest?

10 The volume V (in litres) of water in a tank being filled is proportional to the time taken t (in minutes), as shown on the graph.

a Find the formula for V.

b Calculate the amount of water in the tank after 8 minutes.

c Calculate how long it takes to pump 960 L of water.

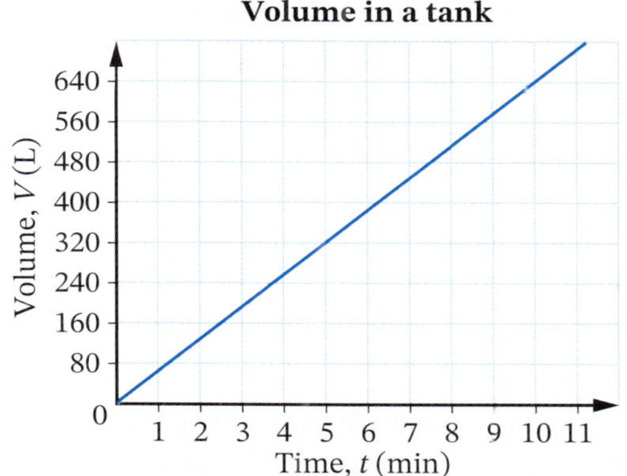

11 The weight of an astronaut on Mars is proportional to his weight on Earth. A 72 kg astronaut weighs 27.4 kg on Mars.

PS R

a Calculate how much a 60-kg astronaut weighs on Mars, correct to one decimal place.

b If an astronaut weighs 32 kg on Mars, calculate his weight on Earth, correct to one decimal place.

12 The pressure experienced by a scuba diver is directly proportional to her depth under the water. The pressure at 25 m is 147 kilopascals (kPa).

PS R

a What pressure is experienced at 40 m?

b At what depth is the pressure 833 kPa?

Foundation Standard Complex

11.09 Applying linear functions

We can model, interpret and analyse real world situations using linear functions.

ⓘ Solving problems using linear models

1. Form a linear function to represent the information in the problem.
   - The y-intercept represents the starting value in the problem.
   - The gradient represents the rate of increase or decrease in the problem. For an increase, the gradient is positive and for a decrease, the gradient is negative.
2. Graph the linear function on a number plane.
3. Use the graph to find the solution to the problem.
4. Check your solution by solving a linear equation algebraically.

Example 16

A taxi company charges a flagfall of \$2.90 and then charges \$2.17 per kilometre travelled.

a Write an equation to model the cost of a taxi trip for different distances.

b Construct a graph to model the costs for a trip from 0 to 50 km.

c Taksheel needs to use a taxi to get from his home to a friend's house 32 km away. Use your graph to determine the cost of this trip.

d Use the equation to determine the cost of this trip.

e Comment on any differences between your answers using the graph and equation.

SOLUTION

a Let $y = mx + c$ where y is the cost of the taxi trip in dollars.

You could also name the equation $C = md + c$ where C = cost and d = distance.

The flagfall is a fixed amount and will be the starting value, so $c = 2.90$, the y-intercept.

The charge per kilometre is a regular increase, so $m = 2.17$, which represents a positive gradient.

$y = 2.17x + 2.90$

b Construct a table of values that can be graphed. Distance goes up to 50 km so go up.

x	0	10	20	30	40	50
y	2.90	24.60	46.30	68.00	89.70	111.40

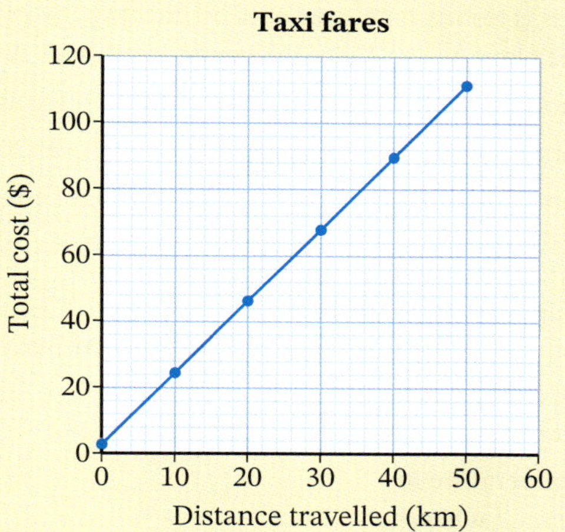

c Find 32 km on the x-axis, read up until you touch the graphed line and then read left to get the total cost.

Reading from the graph, the total cost for travelling 32 km is approximately \$72.00.

d Algebraically, substitute $x = 32$ into the equation:

$$y = 2.17x + 2.90$$
$$= 2.17 \times 32 + 2.9$$
$$= \$72.34$$

e The algebraic solution is more accurate and 34 cents higher than the graphical solution. This is due to the scale used when constructing the graph.

Applying linear functions

1 A school purchases a new photocopier for \$12 000. The value of this photocopier decreases by \$750 every year.

R C

- **a** Write an equation to model the value of the photocopier after different numbers of years.
- **b** Graph the equation on a number plane.
- **c** Use the graph to find the value of the photocopier after 5 years.
- **d** Use the equation to find the value of the photocopier after 5 years.
- **e** How long will it take for the value of this photocopier to become \$0?

2 A fertiliser business is testing a new product by tracking the growth of a plant. At the start of the test, the plant has a height of 6 cm and grows an average of 1.5 cm per day.

R C

a Write an equation to model the plant's growth over a number of days.

b Construct a graph to model the growth of the plant over 20 days.

c Use the graph to find the growth in height of the plant after Day 8.

d Use the equation to calculate the growth in height of the plant after Day 13.

3 This graph models the relationship between the conversion of degrees Celsius temperatures to degrees Fahrenheit temperatures.

R C

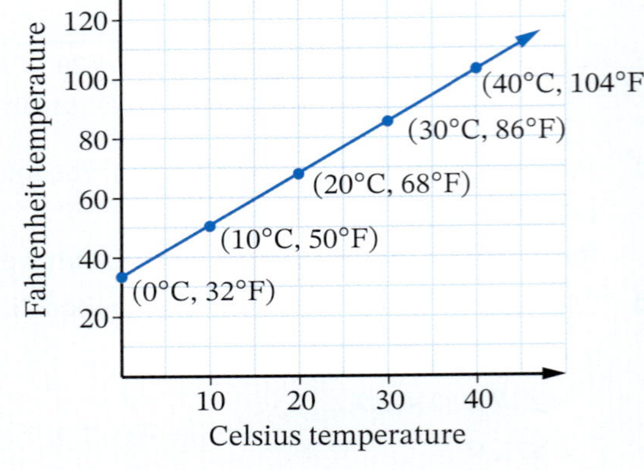

a What is the temperature in degrees Fahrenheit when it is 25 °C?

b Use the graph to develop a linear equation for converting temperature in degrees Celsius to degrees Fahrenheit.

c Use the equation to calculate the temperature in degrees Fahrenheit for 25 °C.

d Compare your answers to parts **a** and **c** above and comment on any differences.

4 A 5000-L water tank is full to the top. A small hole near the bottom of the tank causes a leak of 150 L per hour.

R C

a Write an equation to model the volume of water left in the tank over time.

b Construct a graph of this equation.

c Use your graph to determine the volume of water left in the tank after 15 hours.

d Use your equation to determine the volume of water left in the tank after 15 hours.

e Comment on any differences between your answers to parts **c** and **d** above.

5 Lucy opens up a bank account with the $150 she received for her birthday. Her mother has promised her that she will deposit $25 into the account every week.

R C

a Write an equation to model Lucy's account balance over time.

b Graph this equation for up to 10 weeks.

c Use your graph to determine how much money will be in Lucy's account after 6 weeks.

d How many weeks will it take for Lucy to have a balance of at least $500 in her account?

☐ Foundation ○ Standard ○ Complex

6 Dallas is tiling the floors of his home. The tiler charges an installation fee of $720 plus $12.50 per square metre of tile.

R C

a Write an equation to model the cost of re-tiling the floors.

b Graph this equation for up to 120 square metres.

c Use your graph to determine the cost of tiling 80 m^2 of floor space.

d If Dallas paid $2200 to the tiler, how much floor space was tiled?

7 When Yanina takes a cake out of the oven, its temperature is 98 °C. The temperature of the cake reduces by 1.6 °C every minute after being removed from the oven.

R C

a Write an equation to model the cake's temperature.

b Graph this equation for up to 60 minutes.

c Use your graph to determine the temperature of the cake 6 minutes after it has been removed from the oven.

d If the cake is completely cooled when it is at the room temperature of 26 °C, how long will it take this cake to completely cool?

8 An eScooter company charges an administrative fee of $50 for the hire of a scooter and then an extra $25 per hour for the use of the scooter. A second company does not charge administrative fees but the hourly cost of hiring their scooters is $37.50 per hour. Model both company's hiring charges on the same graph and then use the graph to determine the number of hours at which both companies charge the same cost.

PS R C

9 Tiarne opened up a savings account on the 1st of December. The graph models her progress with her savings. Tiarne wishes to buy a Bluetooth speaker which costs $160. Based on this graph, when will Tiarne be able to buy the speaker?

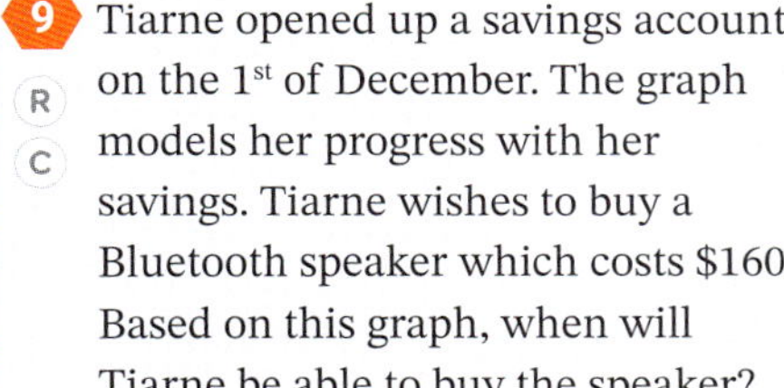

R C

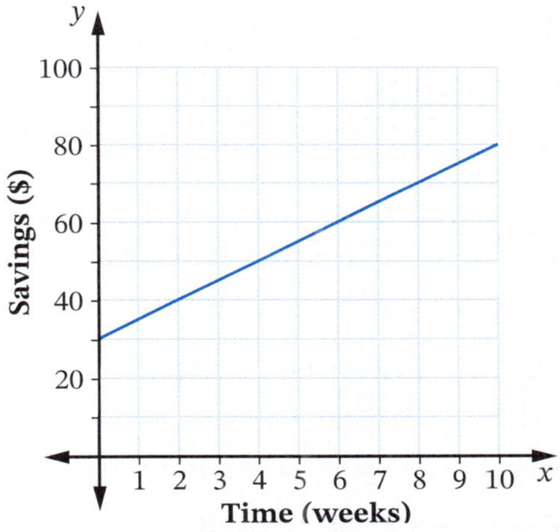

10 A plumber quotes $180 for a 2-hour job and $440 for a 6-hour job. Find the plumber's callout fee and hourly rate.

PS R C

☐ Foundation ○ Standard ⬡ Complex

9780170465618

INVESTIGATION

Graphing $y = x^2$ and $y = ax^2$

You will need: 1 cm grid paper. This activity can also be completed using graphing software.

1 Copy and complete this table of values for $y = x^2$.

x	−4	−3	−2	−1	−0.5	0	0.5	1	2	3	4
y											

2 Draw this number plane on grid paper.

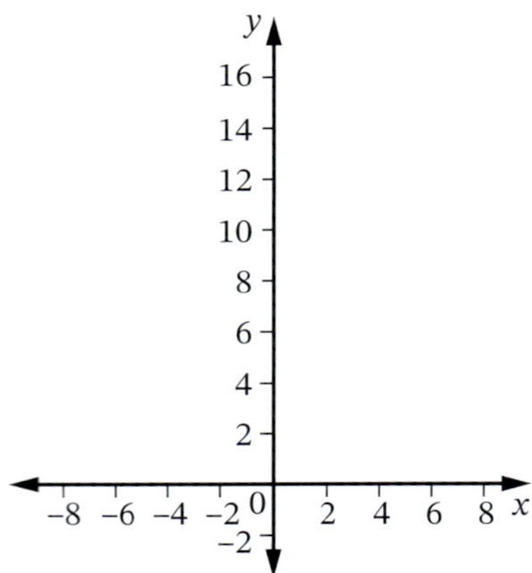

3 Plot the points from the table of values onto the number plane.

4 Draw a smooth curve through all the points and label the graph $y = x^2$.

5 Compare your graph with those of other students.

6 List as many features as you can about the graph of $y = x^2$ that you have drawn. Compare your list with those of other students.

7 Copy and complete this table of values for $y = 2x^2$.

x	−3	−2	−1	0	1	2	3
y							

8 Graph $y = 2x^2$ on the same number plane as $y = x^2$ and label it.

9 How is the graph of $y = 2x^2$ different from the graph of $y = x^2$? Give an explanation for the difference.

10 On the same set of axes, draw these graphs:

a $y = x^2$ b $y = \frac{1}{2}x^2$ c $y = \frac{1}{3}x^2$

11 What do you notice about the graphs you have drawn? Give an explanation.

12 Compare your observations and explanation with those of other students.

9780170465618

INVESTIGATION

Graphing $y = ax^2 + c$

You will need: 1 cm grid paper. This activity can also be completed using graphing software.

1 Graph $y = x^2$ on a number plane for values of x from –3 to 3.

2 Copy and complete this table of values for $y = x^2 + 2$.

x	−3	−2	−1	0	1	2	3
y							

3 Graph $y = x^2 + 2$ on the same number plane as $y = x^2$ and label it.

4 What do the graphs of $y = x^2$ and $y = x^2 + 2$ have in common?

5 How are the 2 graphs different?

6 Describe how the graph of $y = x^2 + 2$ could be drawn using the graph of $y = x^2$.

7 Graph each equation on the same number plane. Label your graphs clearly.

a $y = x^2 - 1$ b $y = x^2 + 4$ c $y = x^2 - 3$

8 Write a sentence to describe the effect of the constant term (which is added to, or subtracted from, x^2) on the shape and position of the parabola $y = x^2$.

9 Graph each equation on the same number plane. Label your graphs clearly.

a $y = -x^2$ b $y = -x^2 + 2$ c $y = -x^2 - 3$ d $y = -x^2 - 1$

10 What do you notice about the shape and position of each of the graphs in question **9**? Compare your results with those of other students.

Graphing quadratic functions $y = ax^2 + c$

11.10

A function in which the highest power of the variable is 2 is called a **quadratic function**, for example, $y = 2x^2 - 5$, $y = x^2 + 7x + 12$ and $y = -5x^2$.

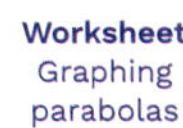

Worksheet Graphing parabolas

Technology Curve sketcher

Spreadsheet Curve sketcher

Interactive Graphing quadratics

The graph of $y = x^2$

The graph of $y = x^2$ is a smooth U-shaped curve called a **parabola**.

- The axis of symmetry, called the **axis of the parabola**, is the y-axis.
- The **vertex** or **turning point** is (0, 0).
- Because x^2 is always positive or zero, the minimum value of the parabola is 0.
- The parabola's shape is **concave up** (looks like a smile ☺).

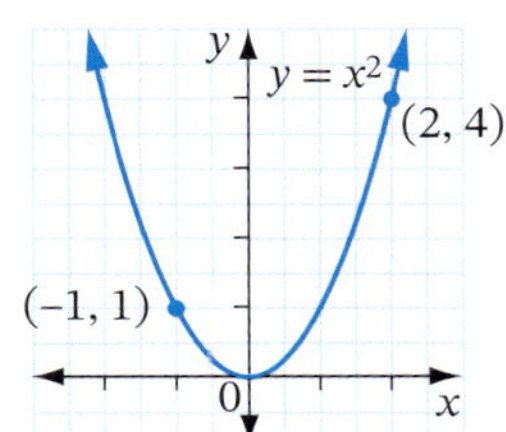

The graph of $y = -x^2$

The graph of $y = -x^2$ is the parabola $y = x^2$ reflected in the x-axis, because $-x^2$ is the 'negative' of x^2.

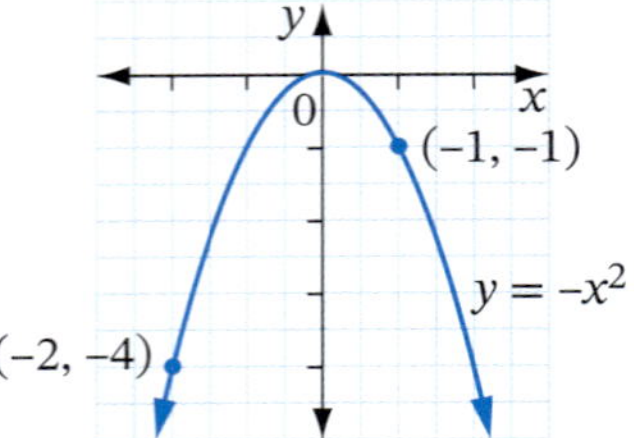

- The **axis of the parabola** is the y-axis.
- The **vertex** or **turning point** is (0, 0).
- Because $-x^2$ is always negative or zero, the maximum value of the parabola is 0.
- The parabola's shape is **concave down** (looks like a frown ☹).

The graph of $y = ax^2$

For the graph of a quadratic function in the form $y = ax^2$, where a is a constant (number), the size of a affects whether the parabola is 'wide' or 'narrow'.

As the size of a increases, the parabola becomes 'narrower' and as the size of a decreases, the parabola 'widens'. If a is negative, then the parabola is concave down.

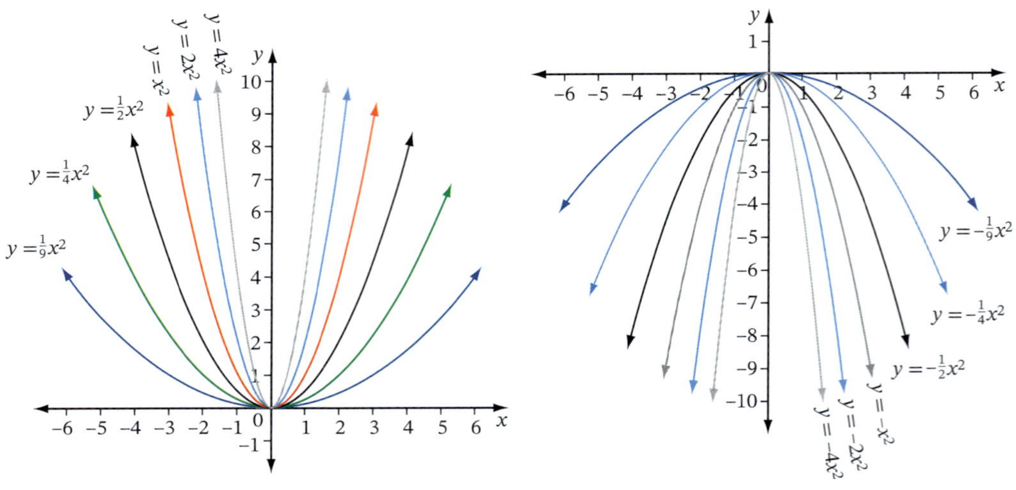

Example 17

a Graph $y = -x^2$ and $y = -2x^2$ on the same set of axes.

b What is the vertex of each parabola?

c Are the parabolas concave up or concave down?

d At what point do the 2 parabolas meet?

e Which parabola is 'wider'?

SOLUTION

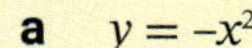

a $y = -x^2$

x	−3	−2	−1	0	1	2	3
y	−9	−4	−1	0	−1	−4	−9

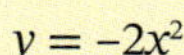

$y = -2x^2$

x	−3	−2	−1	0	1	2	3
y	−18	−8	−2	0	−2	−8	−18

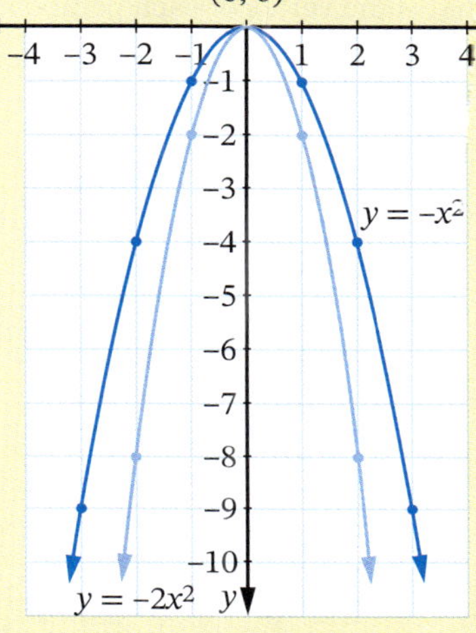

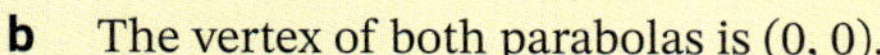

b The vertex of both parabolas is (0, 0).

c The parabolas are both concave down.

d The parabolas meet at (0, 0).

e The parabola $y = -x^2$ is wider.

The graph of $y = ax^2 + c$

For the graph of a quadratic function in the form $y = ax^2 + c$, where a and c are constants, the effect of c is to move the parabola up or down from the **origin**. Also, c is the y-intercept of the parabola.

Example 18

a Graph $y = x^2$, $y = x^2 + 2$, $y = -x^2 - 1$ and $y = -x^2 - 4$ on the same set of axes.

b Compare each parabola with the parabola $y = x^2$.

c What is the vertex of each parabola?

d Is each parabola concave up or concave down?

SOLUTION

a

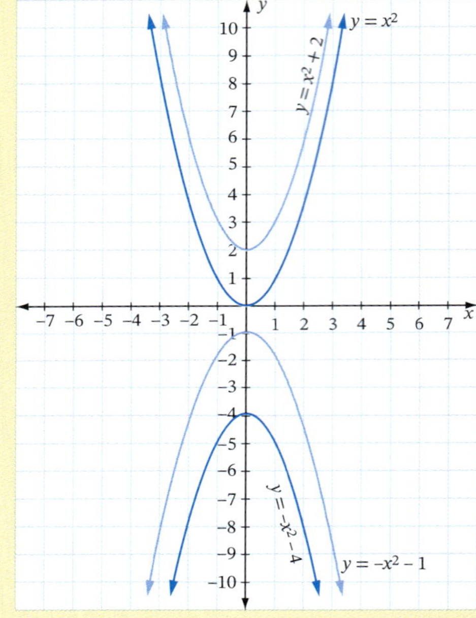

b $y = x^2 + 2$ is $y = x^2$ shifted up 2 units.

$y = -x^2 - 1$ is $y = x^2$ turned upside down and shifted down 1 unit.

$y = -x^2 - 4$ is $y = x^2$ turned upside down and shifted down 4 units.

c $y = x^2$ has vertex (0, 0). $y = -x^2 - 1$ has vertex (0, –1).

$y = x^2 + 2$ has vertex (0, 2). $y = -x^2 - 4$ has vertex (0, –4).

d $y = x^2$ and $y = x^2 + 2$ are concave up.

$y = -x^2 - 1$ and $y = -x^2 - 4$ are concave down.

EXERCISE 11.10 ANSWERS ON P. 667

Graphing quadratic functions $y = ax^2 + c$

U F R C

The questions in this exercise may also be completed using graphing software.

1 R C **a** Graph $y = 2x^2$ on a number plane after copying and completing this table of values.

EXAMPLE 17

x	−2	−1	0	1	2
y					

b Graph $y = \frac{1}{2}x^2$ on the same number plane after copying and completing this table of values.

x	−2	−1	0	1	2
y					

c Compare both parabolas. What features are the same? What are different?

2 Which statement is correct about the parabola $y = x^2$? Select **A**, **B**, **C** or **D**.

A It has a y-intercept of 1.

B It is symmetrical about the x-axis.

C It is concave down.

D Its vertex is (0, 0).

3 R C **a** Graph $y = -3x^2$ and $y = -0.1x^2$ on the same number plane after copying and completing this table of values for each equation.

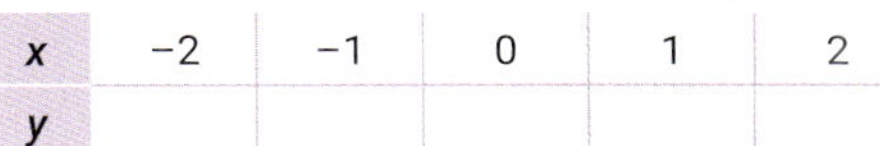

x	−2	−1	0	1	2
y					

b Compare both parabolas. What features are the same? What are different?

4 R C Graph each pair of parabolas on the same set of axes and state:

i the vertex of each parabola.

ii whether each parabola is concave up or concave down.

iii which parabola is 'wider'.

iv the point at which the 2 parabolas meet.

a $y = x^2$ and $y = \frac{1}{4}x^2$

b $y = 2x^2$ and $y = -2x^2$

c $y = -3x^2$ and $y = -\frac{1}{2}x^2$

d $y = 5x^2$ and $y = 0.2x^2$

5 Which statement is correct about the parabola $y = -6x^2$? Select **A**, **B**, **C** or **D**.

A It has a y-intercept of −6.

B It passes through (−1, 6).

C It is concave down.

D Its vertex is (0, −6).

☐ Foundation ○ Standard ⬡ Complex

EXAMPLE 18

6 Graph each pair of parabolas on the same set of axes and state:

R C

i the vertex of each parabola.

ii whether each parabola is concave up or concave down.

iii which parabola is 'wider'.

iv the y-intercept of each parabola.

a $y = x^2 + 1$ and $y = x^2 + 3$

b $y = -x^2 - 2$ and $y = 2x^2 + 4$

c $y = -5x^2 - 2$ and $y = -2x^2 - 1$

d $y = -3x^2 - 4$ and $y = 2x^2 + 2$

7 Which statement is correct about the parabola $y = x^2 + 2$?

A It has a maximum value of 2.

B It is symmetrical about the x-axis.

C It is concave up.

D Its vertex is (2, 0).

8 Match each quadratic function to its parabola.

a $y = x^2 - 1$

b $y = x^2 + 1$

c $y = -x^2 - 1$

d $y = -x^2 + 1$

A

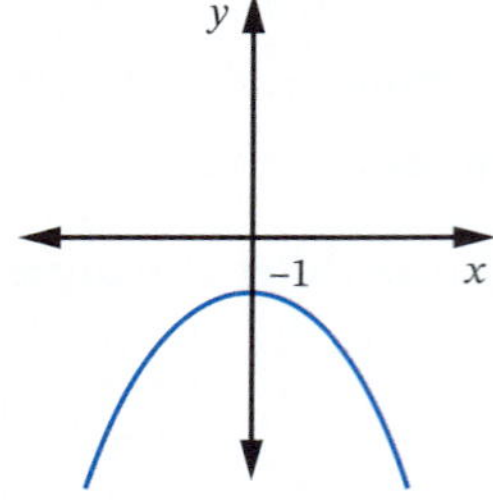

B

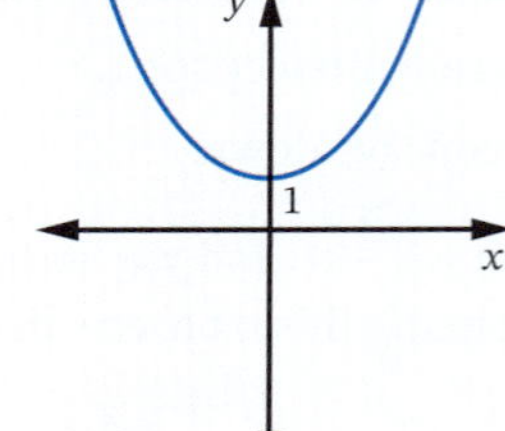

C

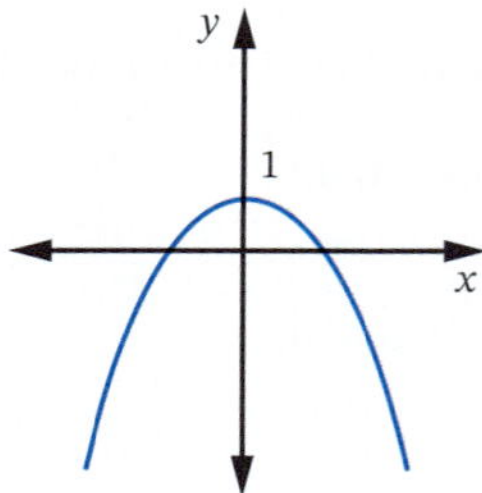

D

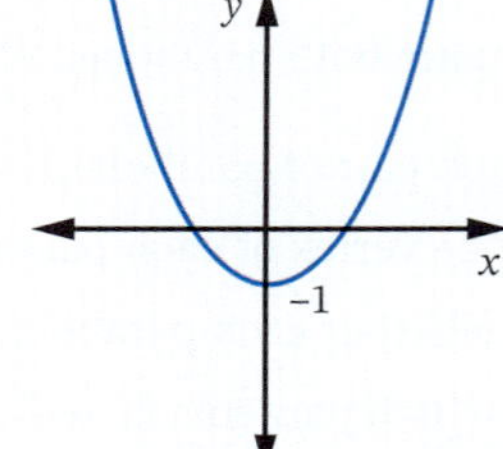

☐ Foundation ○ Standard ⬡ Complex

Graphing quadratic functions $y = ax^2 + bx + c$

11.11

Now we will graph parabolas of the general form $y = ax^2 + bx + c$.

Interactive Graphing quadratics

Video Quadratic functions

Worksheet Graphing quadratic functions

Graphing quadratics

Features of a parabola

Technology Curve sketcher

Spreadsheet Curve sketcher

Example 19

a Graph $y = x^2 - 4x + 1$.

b Write the coordinates of the vertex of the parabola.

SOLUTION

a $y = x^2 - 4x + 1$

x	−1	0	1	2	3	4	5
y	6	1	−2	−3	−2	1	6

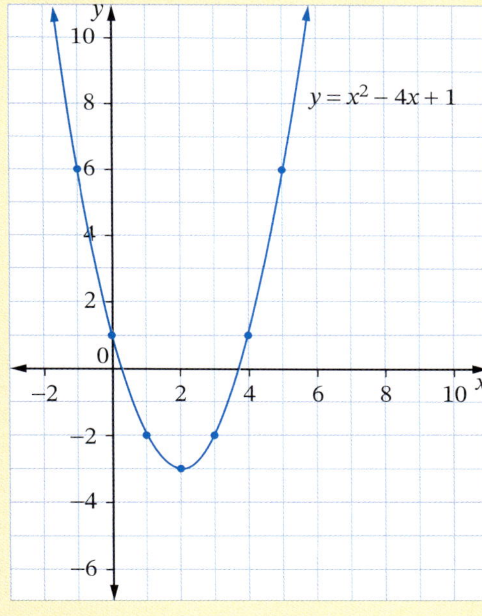

Note that this parabola is **concave up** because of the positive x^2.

b From the graph, the vertex is (2, −3).

Example 20

a Graph $y = -2x^2 + 6x - 4$.

b What are the x-intercepts of the parabola?

c What is the y-intercept?

d Find the coordinates of the vertex.

SOLUTION

a $y = -2x^2 + 6x - 4$.

x	−1	0	1	2	3	4
y	−12	−4	0	0	−4	−12

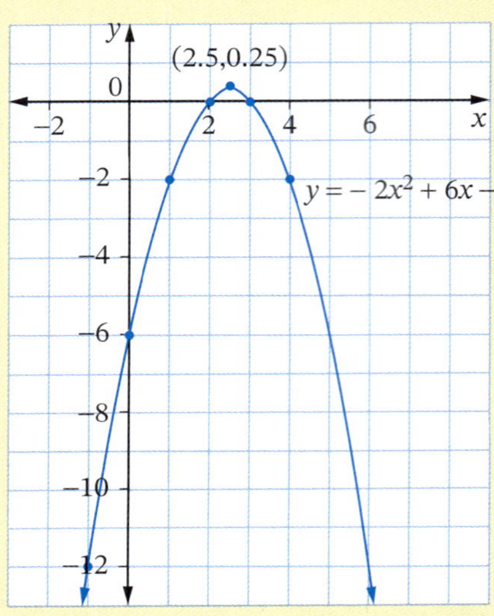

Note that this parabola is **concave down** because of the negative x^2.

b From the table or graph, the x-intercepts are 1 and 2. When $y = 0$.

c From the equation, table or graph, the y-intercept is −4. When $x = 0$, or the value of c in $y = -2x^2 + 6x - 4$.

d By symmetry, the x-coordinate of the vertex is halfway between the x-intercepts 1 and 2,

so $x = \frac{1+2}{2} = 1.5$.

When $x = 1.5$, $y = -2(1.5)^2 + 6(1.5) - 4 = 0.5$.

So the vertex is (1.5, 0.5).

ⓘ The graph of $y = ax^2 + bx + c$

- If a is positive, the parabola is **concave up**
- If a is negative, the parabola is **concave down**
- The ***y*-intercept** is where $x = 0$, so it is the value of c.
- The ***x*-intercepts** are where $y = 0$.

11.11

Example 21

a Find the x-intercepts of the graph of $y = x^2 + 6x + 5$.

b Find the y-intercept and vertex.

c Graph $y = x^2 + 6x + 5$.

SOLUTION

a Substitute $y = 0$ into the equation and solve:

$0 = x^2 + 6x + 5$

$x^2 + 6x + 5 = 0$

$(x + 5)(x + 1) = 0$ Factorising

$x + 5 = 0$ or $x + 1 = 0$

$x = -5$ or $x = -1$

The x-intercepts are -5 and -1.

b The y-intercept of $y = x^2 + 6x + 5$ is 5, the constant term, c, or the value of y when $x = 0$.

When $x = 0$, $y = 0^2 + 6(0) + 5 = 5$.

The x-value of the vertex is halfway between the x-intercepts -5 and 1.

$x = \frac{-5+(-1)}{2} = -3$.

When $x = -3$, $y = (-3)^2 + 6(-3) + 5 = -4$. The vertex is $(-3, -4)$.

c Graphing the vertex and the x- and y-intercepts:

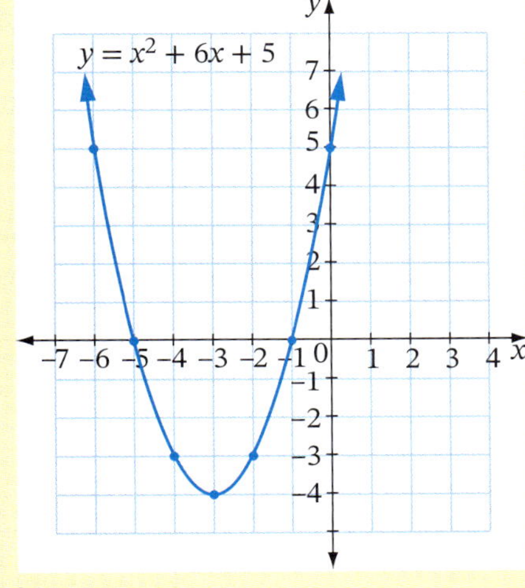

EXERCISE 11.11 ANSWERS ON P. 669

Graphing quadratic functions $y = ax^2 + bx + c$

U F R C

EXAMPLE 19

1 Graph each quadratic function, showing its vertex.

a $y = 2x^2 - 4x - 6$ **b** $y = \frac{1}{2}x^2 + x + \frac{1}{2}$ **c** $y = -3x^2 + 6x$

EXAMPLE 20

2 Graph each quadratic function, showing its vertex and x- and y-intercepts where possible.

R C

a $y = x^2 - 2x - 3$ **b** $y = -x^2 + 9x - 8$ **c** $y = 2x^2 - 4x + 5$

EXAMPLE 21

3 Find the x-intercepts of each quadratic function by substituting $y = 0$ and solving, then find its y-intercept and vertex, then sketch its graph.

R C

a $y = x^2 - 2x - 15$ **b** $y = -x^2 + 4x + 5$ **c** $y = x^2 - 4x$

11.12 Applying quadratic functions

Real-world situations can be modelled using quadratic functions and parabolas. Things like throwing a ball or shooting an arrow are referred to as 'projectile motion'. In these situations, the projectile goes up in the air, slows down as it travels to an almost stopping position, and then comes down again faster and faster. Quadratic functions are used to determine the position of these projectiles at all times.

Parabolas can be seen in our everyday lives. The surface of car headlights and satellite dishes are in the shape of a parabola, as are some bridges and structures. A parabola has the equation of a quadratic function $y = ax^2 + bx + c$, where a, b and c are constants.

□ Foundation ○ Standard ○ Complex

Example 22

An arrow is shot into the air from a height of 1.5 m and its height can be modelled by the function

$$y = -5x^2 + 29x + 1.5,$$

where y is the height above the ground in metres and x is the time in seconds after its release. Graph the function for values of x from 0 to 6 and use it to determine:

a the total time the arrow is in the air (correct to one decimal place).

b the maximum height that it reaches (correct to the nearest metre).

SOLUTION

Complete a table of values for values of x from 0 to 6 and increasing in steps of 0.5.

x	0	0.5	1	1.5	2	2.5	3	3.5	4	4.5	5	5.5	6
y	1.5	14.75	25.5	33.75	39.5	42.75	43.5	41.75	37.5	30.75	21.5	9.75	-4.5

Graph the table of values and draw a smooth curve to connect the points to draw the parabola. Graphing software can also be used to construct the graph.

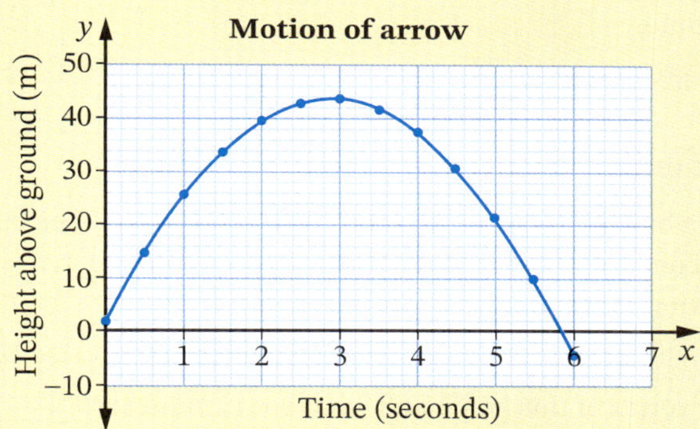

Note: This is not the graph of the path of the arrow. It is the graph of the height of the arrow at different times, x seconds.

a The arrow hits the ground when the height is 0, which is the x-intercept of the parabola. Looking at the graph, this occurs at just less than 6 seconds.

The total time the arrow is in the air is approximately 5.8 seconds.

b The maximum height that the arrow reaches will be at the vertex (or turning point) of the parabola. The maximum height is approximately 43 m (at time just under 3 seconds).

EXERCISE 11.12 ANSWERS ON P. 670

Applying quadratic functions

U F PS R C

The questions in this exercise may also be completed using graphing software.

EXAMPLE 22

1 The graph below shows the height of a golf ball, y metres, at time x seconds.

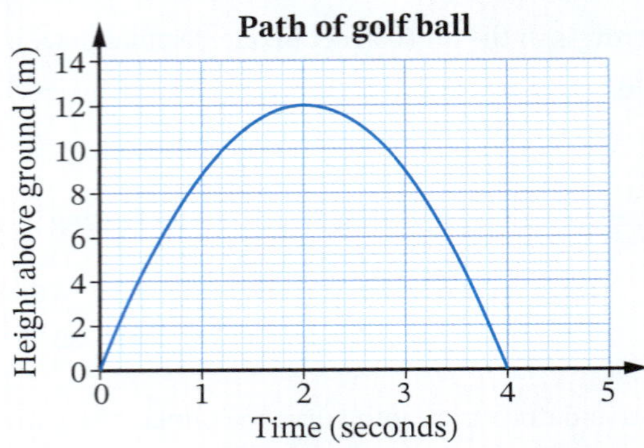

- **a** Identify the times at which the ball is at a height of 9 m and explain why there are 2 different times.
- **b** What is the greatest height the ball reaches and what is the time when it reaches this height?
- **c** How long is the ball in the air for?

2 Lawson throws a ball up in the air from his bedroom window. The ball's height, h metres, at t seconds after being thrown is given by the function $h = -4t^2 + 18t + 20$. Graph this function for the values of t from 0 to 6.

- **a** What is the y-intercept for this graph and what does this value represent?
- **b** What is the vertex of this graph and what does it represent?
- **c** When does the ball hit the ground?
- **d** How many metres does the ball travel upwards before coming back down?

3 A bridge crossing a river contains an arch in the shape of a parabola with a quadratic function of $y = 16 - 0.03x^2$, where x and y are in metres. Graph this function for x values between -30 and 30, going up in steps of 10 m.

- **a** Estimate the x-intercepts from your graph.
- **b** Determine the approximate width of the river if the x-intercepts sit on the edges of the river bank.
- **c** What is the highest point of the arch and how does it align with the position of the river?

Dreamstime.com/Rinus Baak

Foundation Standard Complex

4 This graph gives the height, h metres, at time t seconds, of a toy rocket that is fired into the air.

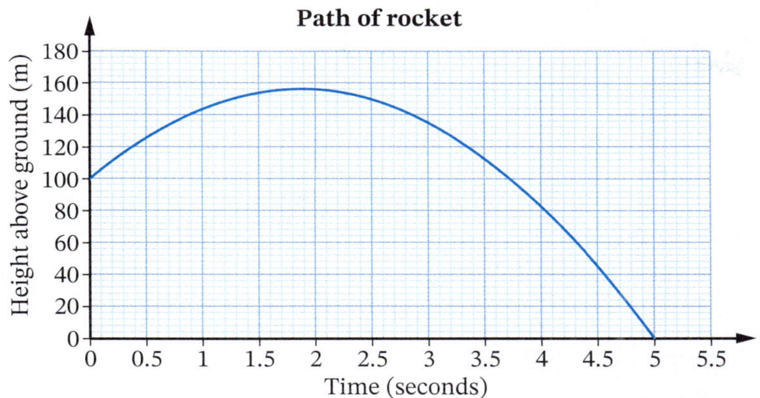

a Which part of the graph shows the height from which the rocket was launched? What was this height?

b How far up into the air did the rocket travel before coming back down to the ground? Explain how you obtained your answer.

c For how long did the rocket travel?

d How long did it take the rocket to reach the ground from its highest position?

5 A cricket ball is hit into the air and its height, h metres, above the ground after time, t seconds, is given by the function $h = -5t^2 + 24t$. Graph this function for t values from 0 to 6 seconds.

a How long does it take the ball to reach its maximum height?

b What is the maximum height of the ball?

c How long was the ball in the air for?

6 A rock is thrown from a cliff and its height, y metres, at time, x seconds, is $y = -2x^2 + 10x + 35$. Graph this function for x values between 0 and 10.

a From what height was the rock thrown?

b For how long did the rock stay in the air?

c How far up does the rock travel before starting to fall down?

☐ Foundation ○ Standard ⬡ Complex

7 PS R C The fuel consumption of a car travelling at a speed of s km/h is modelled by the function $C = 0.01s^2 - s + 33$, where fuel consumption C is measured in L/100 km. Graph this function to determine the travelling speed at which fuel consumption is most economical (meaning it is at its lowest), and state the fuel consumption at this speed.

8 PS R C The temperature, T °C, in a small rural town n hours after 6 a.m. is given by the function

$$T = -\frac{1}{6}n^2 + 2n + 14.$$

a Determine the approximate increase in temperature from 6 a.m. in the morning to the daily maximum temperature, and determine the approximate time at which the temperature reaches its maximum.

b Explain why this model is inappropriate for values of n greater than 13.

9 PS R C The profit from selling tickets for the school musical depends on the price of each ticket. Using data from previous years, the profit made can be modelled by the function $p = -2t^2 + 100t + 50$, where $\$p$ is the profit and t is the price of each ticket. Graph this function to determine the price that tickets should be sold at to obtain the maximum profit from the musical. Identify what profit can be made with this ticket price.

10 PS R C Each cable of the Golden Gate Bridge in San Francisco, California, is suspended in a parabolic shape between 2 towers that sit approximately 150 m above the roadway.

The cable length can be modelled using the function $y = \frac{19x^2}{52000}$, where y is the height of the cable above the ground in metres, and x is the horizontal distance in metres. Graph this function to determine the horizontal distance between the 2 towers that suspend the parabolic cable.

Shutterstock.com/Min C. Chiu

Graphing exponential functions $y = a^x$ 11.13

Videos
The exponential curve
The emperor's chess board

Worksheet
Graphing exponentials

A function of the form $y = a^x$, where a is a positive constant and the variable x is a power, is called an **exponential function**, for example, $y = 5^x$, $y = 2^x$ and $y = 3^x$. The graph of an exponential function is called an **exponential curve**.

The table of values and graph of $y = 4^x$ is shown.

x	−2	−1	0	1	2	3	4
y	$\frac{1}{16}$	$\frac{1}{4}$	1	4	16	64	256

- The y-intercept of $y = a^x$ is always 1 since $a^0 = 1$.
- As x increases (to the right, in the positive direction), a^x becomes very large. Graphically, this means that the graph of $y = a^x$ increases sharply with a steep positive gradient.
- As x decreases (to the left, in the negative direction), a^x approaches 0. This means that the graph of $y = a^x$ flattens out and approaches the x-axis as x becomes a large negative number. The x-axis is an **asymptote** because the curve approaches it but never touches it.

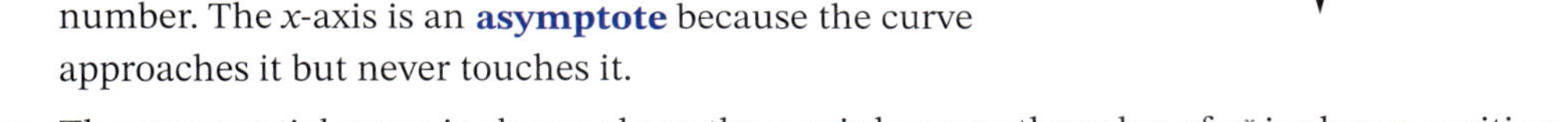

- The exponential curve is always above the x-axis because the value of a^x is always positive.

Example 23

Sketch each exponential function and mark the y-intercept on each curve.

a $y = 2^x$ **b** $y = 3^{-x}$

SOLUTION

a The y-intercept of $y = 2^x$ is 1.

When $x = 1$, $y = 2$, so $(1, 2)$ is a point on the curve.

As x increases (to the right in the positive direction), 2^x becomes very large (steep positive gradient).

As x decreases (to the left in the negative direction), 2^x approaches 0.

The x-axis is an asymptote.

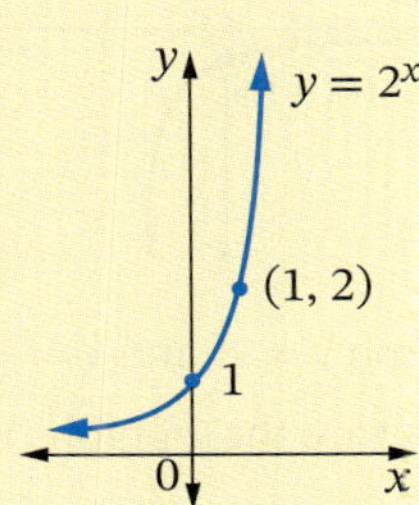

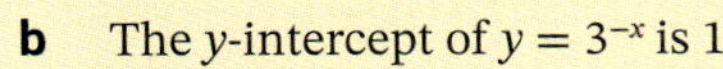

b The y-intercept of $y = 3^{-x}$ is 1

When $x = -1$, $y = 3$, so $(-1, 3)$ is a point on the curve.

As x decreases (to the left in the negative direction), 3^{-x} becomes very large (steep negative gradient).

As x increases (to the right in the positive direction), 3^{-x} approaches 0.

The x-axis is an asymptote.

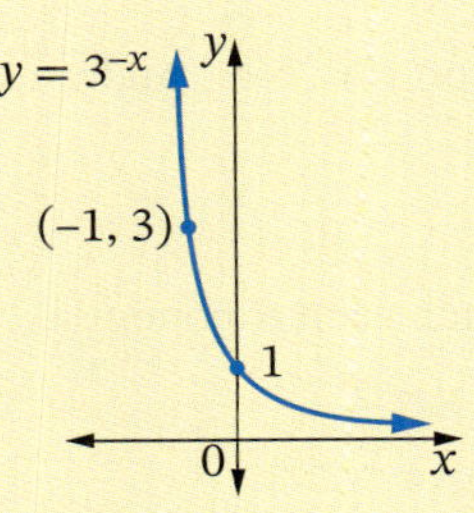

Note that the graph of $y = 3^{-x}$ (and of $y = a^{-x}$ in general) is decreasing, and is actually a reflection of $y = 3^x$ in the y-axis.

EXERCISE 11.13 ANSWERS ON P. 671

Graphing exponential functions $y = a^x$

U F R C

Some of this exercise may be completed using graphing technology.

EXAMPLE 23

1 **a** Graph each exponential function on the same axes.

R **i** $y = 2^x$ **ii** $y = 3^x$ **iii** $y = 5^x$

C **b** What is the y-intercept of each curve?

c Describe what happens to the graph of $y = a^x$ as a increases.

2 **a** Graph $y = 4^x$ and $y = 4^{-x}$ on the same axes.

b Copy and complete:

i The reflection of $y = 4^x$ in the y-axis is ...

ii The reflection of $y = a^x$ in the y-axis is ...

3 Which graph represents $y = 2^{-x}$? Select the correct answer **A**, **B**, **C** or **D**.

R C

A

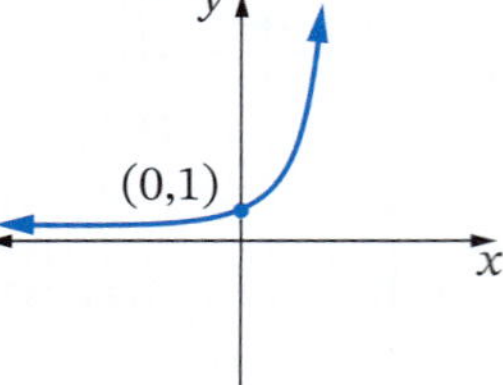

B

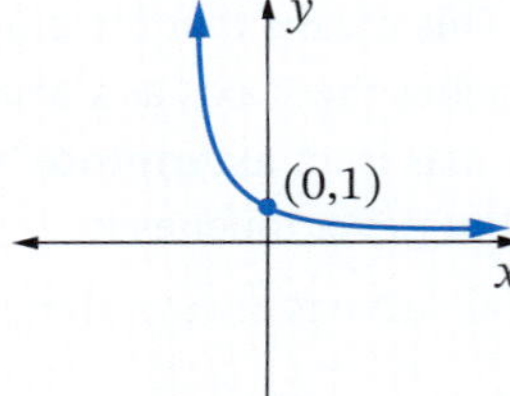

C

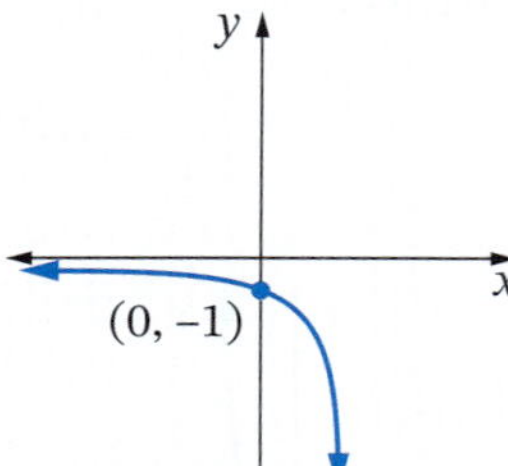

D

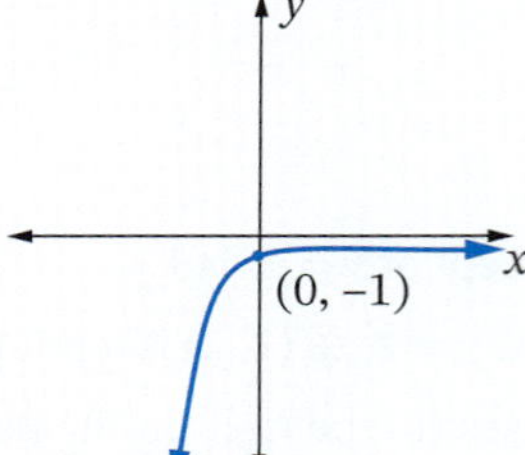

4 **a** Graph $y = 2^x$ and $y = -2^x$ on the same axes.

R **b** How are the 2 graphs related?

C **c** Copy and complete: The reflection of $y = a^x$ in the x-axis is ...

5 Graph $y = 3^x + 1$ and $y = 3^x - 1$ on the same axes and describe how they are related.

R C

6 Sketch each exponential curve, showing the y-intercept.

a $y = 2^x$ **b** $y = 3^{-x}$ **c** $y = -4^x$

d $y = -2^{-x}$ **e** $y = 4^x + 1$ **f** $y = 4^x - 1$

▷

☐ Foundation ○ Standard ⬡ Complex

9780170465618

7 Find an exponential function for this graph.

R C

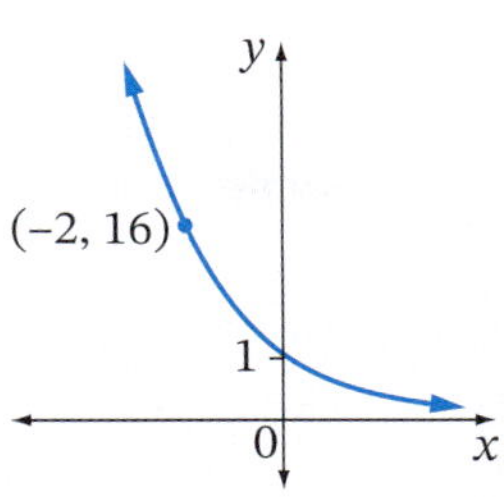

INVESTIGATION

Videos going viral

Use a spreadsheet to help you with this investigation.

With the invention of social media and social networks, we can now communicate with each other instantly. However, what often starts off as a simple message between friends or an online video post can quickly multiply at an alarming rate, be seen by the general public worldwide and 'go viral' (spread like a virus).

Suppose a video of an angry cat is shared with a friend who 1 minute later shares it with 2 other friends. This occurs every minute so that every minute the number of *new* views doubles. This is an example of something growing exponentially, according to the formula $y = 2^x$, where x represents number of minutes.

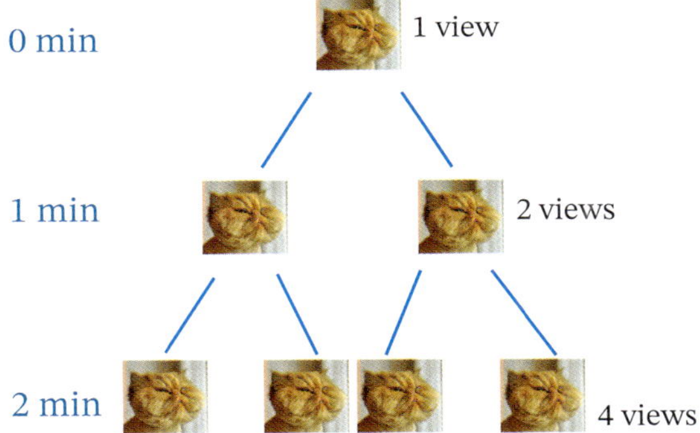

Shutterstock.com/Zanna Pesnina

1 Use a spreadsheet to calculate the number of new views after:

a	15 min	b	30 min	c	1 hour	d	2 hours
e	4 hour	f	10 hours	g	20 hours	h	30 hours

2 Find how long it will take until the total number of new views reaches:

a	64	b	256	c	over 500	d	over 1000
e	over 3000	f	over 10 000	g	over 1 million	h	over 4 million

3 Use the spreadsheet's **Graph Option** to graph the viral pattern. Describe the shape of the graph.

Foundation Standard Complex

DID YOU KNOW?

Exponential growth

When an increase can be described using an exponential function, it is called exponential growth. Examples include the growth of a virus, population (people or animals) and compound interest investments.

Population growth is monitored in different countries through the fertility (birth) and mortality (death) rates as well as migration. The data collected for these figures can often be modelled as an exponential function. By modelling, the changes in population, predictions of future changes in population can be simulated and towns and cities can prepare for possible expansion in the numbers of schools, hospitals, housing and other necessary infrastructure.

At what rate is the population of Australia growing? What about the world's population?

POWER PLUS ANSWERS ON P. 673

1 *ABCD* is a parallelogram. The coordinates of *A*, *B* and *C* are $(-1, 4)$, $(4, 6)$ and $(2, 7)$ respectively. Find the coordinates of *D*.

2 A line drawn through points $A(0, -2)$ and $B(3, 0)$ also goes through point *C* with x-coordinate 9. Find:

- a the equation of the line.
- b the y-coordinate of *C*.

3 $M(1, 1)$ is the midpoint of the interval *XY*. If the coordinates of *X* are $(-3, -2)$, what are the coordinates of *Y*?

4 A circle is drawn with its centre at $C(2, 3)$. The point $A(-1, -5)$ lies on the circumference.

- a Calculate the radius of the circle.
- b The point *P* is 10 units from *C*. Explain why *P* lies outside the circle.

5 The points of $R(2, 19)$, $S(-3, 7)$ and $T(10, 7)$ are joined to form a triangle. Show that the triangle is isosceles.

6 The point $(2, 1)$ is a vertex of a square. $M(5, 1)$ is the point of intersection between the diagonals of the square. Find the coordinates of the other 3 vertices of the square.

9780170465618

11 CHAPTER REVIEW

Quiz
Language of maths 11

Language of maths

asymptote	axis/axes	constant of proportionality	direct proportion
distance	exponential function	formula	gradient
horizontal	interval	length	linear function
midpoint	number plane	parabola	quadratic function
rise	run	surd	varies directly
vertex	vertical	x-intercept	y-intercept

1 What is the name given to the graph of:

a a quadratic function? **b** a linear function?

2 Is the line $y = 3$ horizontal or vertical?

3 What is another name for the steepness of a line?

4 What is the value at which a line crosses the y-axis called?

5 What are the coordinates of the **vertex** of the parabola with equation $y = -x^2$?

6 If $y = kx$, what is the constant of proportionality?

7 When the length of an interval needs to be given as an **exact** length, what does this mean?

8 What is the asymptote of the graph of the function $y = a^x$?

Topic summary

- In your own words, describe what a gradient is.
- What parts of this topic did you find difficult?

Worksheet
Mind map: Coordinate geometry and graphs

Print (or copy) and complete this mind map of the topic, adding detail to its branches and using pictures, symbols and colour where needed. Ask your teacher to check your work.

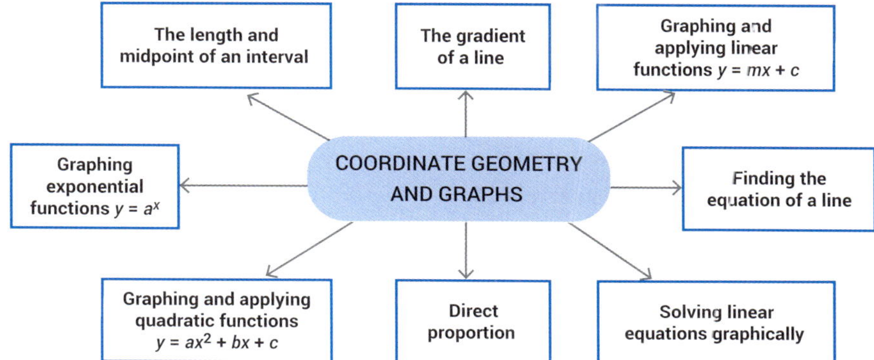

11 TEST YOURSELF ANSWERS ON P. 673

11.01

1 Find the length of the interval joining the points $H(7, 2)$ and $T(-2, 6)$. Express your answer:

a as a surd

b correct to one decimal place.

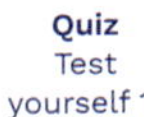
Quiz
Test yourself 11

11.02

2 Find the coordinates of the midpoint of the interval joining:

a $(8, 4)$ and $(-2, 10)$ **b** $(-1, -1)$ and $(-7, 9)$ **c** $(3, -6)$ and $(-5, 10)$

11.03

3 **a** Which intervals below have a gradient that is:

i positive? **ii** negative? **iii** 0?

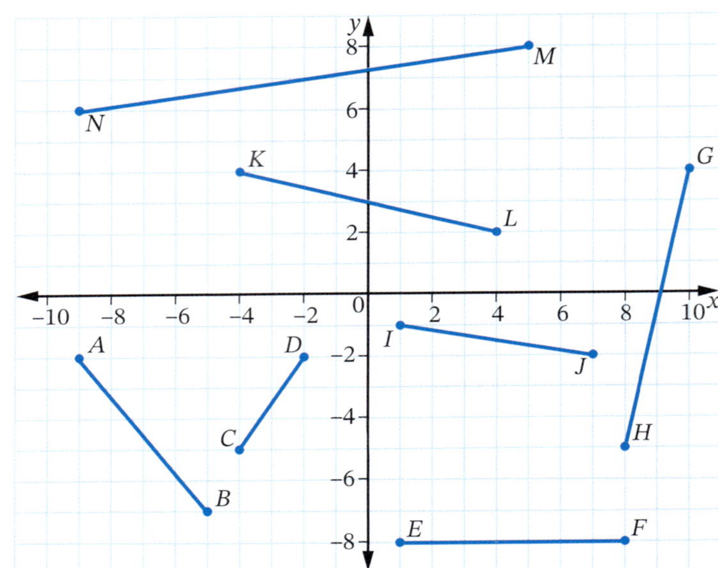

b Find the gradient of:

i interval NM **ii** interval AB **iii** interval KL.

c Find the midpoint of interval HG.

d Find the exact length of interval CD.

11.03

4 Find the gradient of the interval joining:

a $(1, 9)$ and $(3, 5)$ **b** $(6, 0)$ and $(8, -4)$ **c** $(-1, -1)$ and $(5, -3)$

11.04

5 Graph $y = -3x + 6$ on a number plane and write:

a its x-intercept **b** its y-intercept.

☐ Foundation ○ Standard ○ Complex

9780170465618

6 Test whether the point (4, –2) lies on the line with equation: 11.04

a $y = x - 6$ b $y = -2x + 6$ c $x - 3y = 10$

7 a Graph and label each line on the same number plane: 11.04

$x = -4$, $x = 3$, $y = -6$, $y = 2$

b What is the area of the rectangle enclosed by these lines?

8 Write the equation of a line that is: 11.04

a horizontal with a y-intercept of 4.

b parallel to the y-axis and passes through (–1, 7).

c the x-axis.

9 Find the gradient and y-intercept of the line with equation: 11.05

a $y = 3x + 1$ b $y = 8 - x$ c $y = \frac{x}{5} + 4$ d $y = -6x$

10 Graph each linear function by finding the gradient and y-intercept first. 11.05

a $y = -3x - 1$ b $y = \frac{x}{2} + 6$

11 Find the equation of each line. 11.06

a

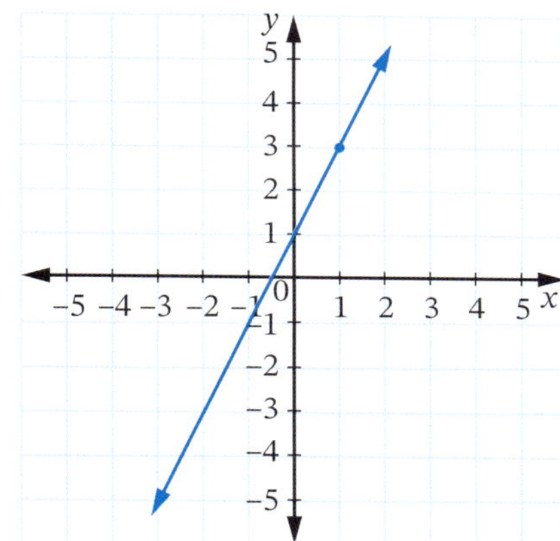

b

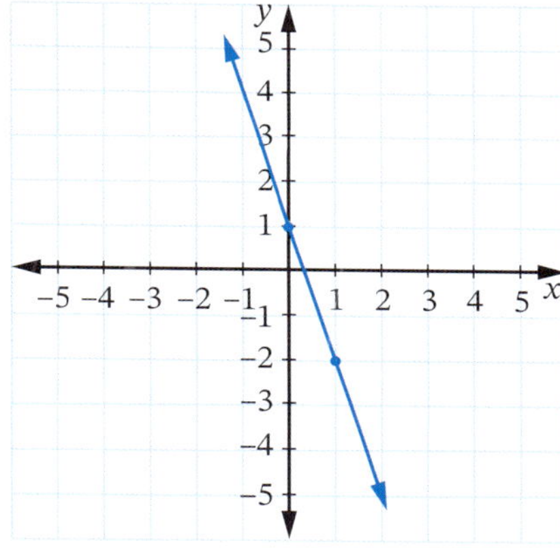

12 Solve the equation $3x - 4 = 5$ graphically.

13 During a thunderstorm , the time, T seconds, between seeing lightning and hearing thunder is directly proportional to the distance, d kilometres, of the observer from the centre of the storm. Thunder is heard 6 seconds after the lightning is seen when a storm is 35 km away.

a Find a formula for T in terms of d.

b Calculate, correct to the nearest kilometre, the distance of a storm when thunder is heard 10 seconds after the lightning.

14 An electrician charges a callout fee of \$75 and an hourly rate of \$55 for any work completed.

a Model this scenario using an equation and graph.

b Estimate how many hours the electrician worked for if he charged \$295 for a job.

Foundation Standard Complex

15 Graph $y = x^2 + 1$ and $y = -2x^2$ on the same set of axes and state:

a the vertex of each parabola.

b whether each parabola is concave up or concave down.

c which parabola is 'wider'.

16 Which equation below represents a concave up parabola with vertex (0, 0)? Select the correct answer **A**, **B**, **C** or **D**.

A $y = x^2 - 1$ **B** $y = 2x^2$ **C** $y = -x^2$ **D** $y = 3x^2 + 2$

17 Graph $y = -x^2 + 5x - 6$, showing its x-intercepts, y-intercept and vertex.

18 The running cost, $\$C$ per hour, of a ship travelling at an average speed of s knots, is given by the function $C = 2s^2 - 128s + 2400$. Graph this function to determine what the minimum running cost is and the average speed at which the ship travels to obtain this minimum running cost.

19 Sketch each exponential function.

a $y = 4^x$ **b** $y = 4^{-x}$ **c** $y = -4^x$ **d** $y = -4^{-x}$

9780170465618

12

PROBABILITY

Probability

Probability, the study of chance, was first used in France to solve problems in gambling in the 1600s. In 1654, the writer and gambler, the Chevalier de Méré, asked mathematician Blaise Pascal many questions involving gambling, including how many rolls of a pair of dice were required to have a better than 50% chance of rolling a double 6. Pascal wrote to another mathematician, Pierre de Fermat, about the problems, and the correspondence between them is regarded to be the beginning of probability theory.

Probability is now used widely in many areas, including the likelihood of a house fire or flood, and forecasting the weather for a week ahead.

iStock.com/itiero

Chapter outline

U = Understanding
F = Fluency
PS = Problem solving
R = Reasoning
C = Communication

Wordbank

Quiz Wordbank 12

complementary event The 'opposite' event, for example, the complementary event to selecting an Ace from a deck of cards is selecting a card that is not an Ace

relative frequency The frequency of an event over repeated trials as a fraction of the total number of trials

tree diagram A diagram that uses branches to list all the possible outcomes in a multi-step chance experiment

trial One go or run of a repeated probability experiment, for example, one roll of a die

two-step experiment A chance experiment with 2 steps or stages, such as rolling a pair of dice

two-way table A way of grouping items into 2 overlapping categories, such as gender and the ability to drive a car

Venn diagram A diagram of circles (usually overlapping) for grouping items into categories

Videos (9):

12.01 Probability • Complementary events
12.02 Experimental probability
12.03 Venn diagrams 2 • Venn diagrams 1
12.05 Tables and tree diagrams • Tree diagrams 3
12.06 Tree diagrams 1 • Tree diagrams 2

***Twig* videos (5):**

12.01 Why do share prices change? • The card counter • The prisoner's dilemma • The Monty Hall problem
12.03 Venn diagrams: Global habitats

***PhET* interactive (1):**

12.02 Plinko probability

Quizzes (5):

- Wordbank 12
- SkillCheck 12
- Mental skills 12
- Language of Maths 12
- Test Yourself 12

Worksheets (11):

12.01 Games of chance
12.02 Coins probability • Dice probability • Probability review • Relative frequencies • Final 6 investigation
12.03 Venn diagrams
12.04 Two-way tables • Two-way probability tables
12.05 Tree diagrams
Mind map: Probability

Puzzles (2):

12.03 Venn diagrams matching activity • Venn diagrams

Presentation (1):

12.05 Combined events

Nelson MindTap

To access resources above, visit **cengage.com.au/nelsonmindtap**

12 **In this chapter you will:**

- ✓ calculate the probabilities and relative frequencies of events, including complementary events.
- ✓ describe compound events using terminology such as 'and', exclusive 'or', inclusive 'or' and 'at least'.
- ✓ use Venn diagrams and two-way tables to represent sample spaces and compound events to solve probability problems.
- ✓ use tree diagrams and tables (arrays) to represent the sample space of two-step chance experiments, with and without replacement, to solve probability problems.
- ✓ conduct probability simulations to demonstrate the relationship between the probability of compound events and the individual probabilities.

SkillCheck

ANSWERS ON P. 675

Quiz SkillCheck 12

1 A bag contains 3 green lollies, 4 red lollies, 5 white and 8 yellow lollies.

a What fraction of the lollies are red?

b What percentage of the lollies are yellow?

c What percentage of the lollies are not yellow?

d What fraction of the lollies are green or yellow?

2 If a die is rolled, which one of the following events is most likely? Select the correct answer **A**, **B**, **C** or **D**.

A an even number **B** a number greater than 2

C a number greater than 3 **D** a number greater than 4

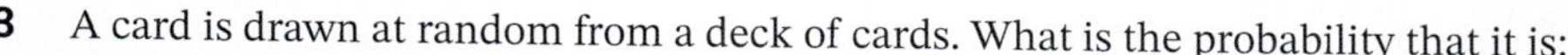

3 A card is drawn at random from a deck of cards. What is the probability that it is:

a a 7? **b** a heart card? **c** a black card?

4 Match each description (**a** to **f**) below with the correct set of all whole numbers (**A** to **F**) that meet that description.

a between 3 and 10 **b** at least 1 **c** more than 4

d at most 3 **e** 5 or less **f** not more than 4

A {1, 2, 3, 4, 5} **B** {4, 5, 6, 7, 8, 9} **C** {1, 2, 3}

D {1, 2, 3, 4} **E** {1, 2, 3, ...} **F** {5, 6, 7, ...}

5 A die is rolled. Find the probability that the number that comes up is:

a 1 **b** more than 2 **c** odd **d** composite

6 **a** Write the sample space for this spinner.

b Find, as a percentage, the probability that the number spun is:

i 5 **ii** at least 5

iii 5 or less **iv** more than 5

Probability 12.01

The language of probability

A **chance experiment** is an activity or process that involves chance, for example, rolling a die or tossing a coin.

An **outcome** is the result of a chance experiment. For example, when rolling a die, there are 6 possible outcomes – the numbers 1 to 6.

The **sample space** is the set of all possible outcomes of an experiment. When rolling a die, the sample space is {1, 2, 3, 4, 5 and 6}.

An **event** is one or more outcomes of an experiment. When rolling a die, one event may be rolling an odd number, which consists of the 3 outcomes 1, 3 and 5.

A **trial** is one go or run of an experiment, for example, one roll of the die.

In a **random** experiment, every possible outcome has an equal (fair) chance of occurring.

Videos Probability

Why do share prices change?

The card counter

The prisoner's dilemma

The Monty Hall problem

Worksheet Games of chance

Probability of an event

$P(E)$ means 'the **probability** of an event, E (occurring)'. If all possible outcomes are **equally likely**, then:

$$P(E) = \frac{\text{number of favourable outcomes}}{\text{total number of outcomes}}$$

$$\text{or } P(E) = \frac{\text{number of outcomes matching } E}{\text{number of outcomes in the sample space}}$$

A **favourable outcome** is one of the outcomes in the event that you want, whose probability you are calculating.

Example 1

A card is chosen at random from a set of cards numbered 0 to 9.

Find the probability of selecting:

a the card numbered 8.

b a number less than 4.

c a number 5 or more.

SOLUTION

There are 10 outcomes in the sample space.

a $P(8) = \frac{1}{10}$ — 1 chance in 10

b Out of the 10 numbers, there are 4 numbers less than 4: {0, 1, 2, 3}.

$P(\text{less than } 4) = \frac{4}{10} = \frac{2}{5}$ — 4 chances in 10

c Out of the 10 numbers, there are 5 numbers that are 5 or more: {5, 6, 7, 8, 9}.

$P(5 \text{ or more}) = \frac{5}{10} = \frac{1}{2}$ — 5 chances in 10

The range of probability

The probability of an event can be written as a fraction, decimal or percentage. Probability is measured on a scale from 0 to 1, or as a percentage from 0% to 100%. If an event has a probability of 0, it is **impossible** and it will not occur. If an event has a probability of 1, it is **certain** and it *must* occur.

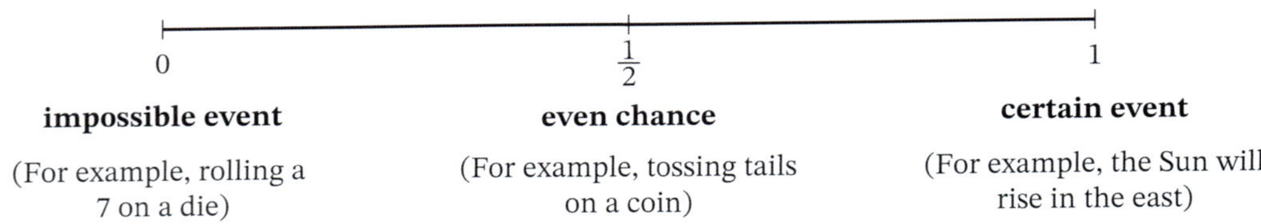

Complementary events

Video Complementary events

Complementary events are 2 events that together make up all the possible outcomes, such as a head and a tail when tossing a coin. The **complement** of an event E is made up of those outcomes in the sample space that are 'not E', usually written as $\overline{E}$. Because an event and its complement covers all possible outcomes in the sample space, the sum of their probabilities must equal 1.

ⓘ Complementary events

$P(\text{event}) + P(\text{complementary event}) = 1$

$P(E) + P(\overline{E}) = 1$

$P(\overline{E}) = 1 - P(E)$, where $\overline{E}$ is the complement of E.

Example 2

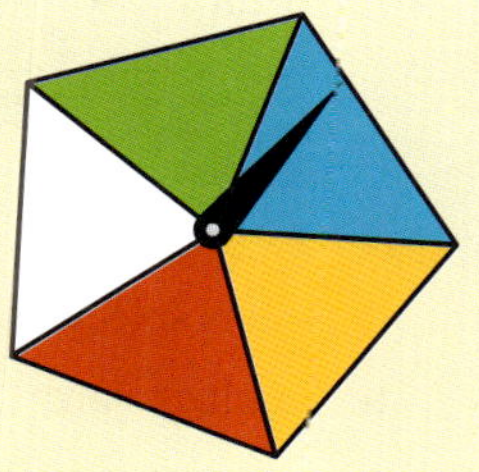

a List the sample space when this spinner is spun.

b Find $P(\text{red})$.

c What is the complementary event to the spinner landing on red?

d List the outcomes in the complementary event.

e Find $P(\text{not red})$.

SOLUTION

a Sample space = {red, yellow, blue, green, white}

b $P(\text{red}) = \frac{1}{5}$

c The complementary event to landing on red is landing on a colour other than red.

d The outcomes in the complementary event are {yellow, blue, green, white}.

e $P(\text{not red}) = 1 - P(\text{red})$

$= 1 - \frac{1}{5}$

$= \frac{4}{5}$

EXERCISE 12.01 ANSWERS ON P. 675

Probability

U F PS R C

1 A letter is drawn at random from the alphabet. EXAMPLE 1

PS **a** How many outcomes are there in the sample space?

R **b** Are all the outcomes equally likely?

c What is the probability of choosing a vowel?

d What is the probability of choosing a letter from the word PROBABILITY?

2 This spinner is spun. What is the decimal probability that the needle points to:

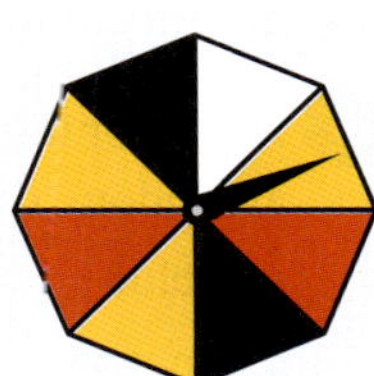

a yellow?

b black?

c red?

Foundation Standard Complex

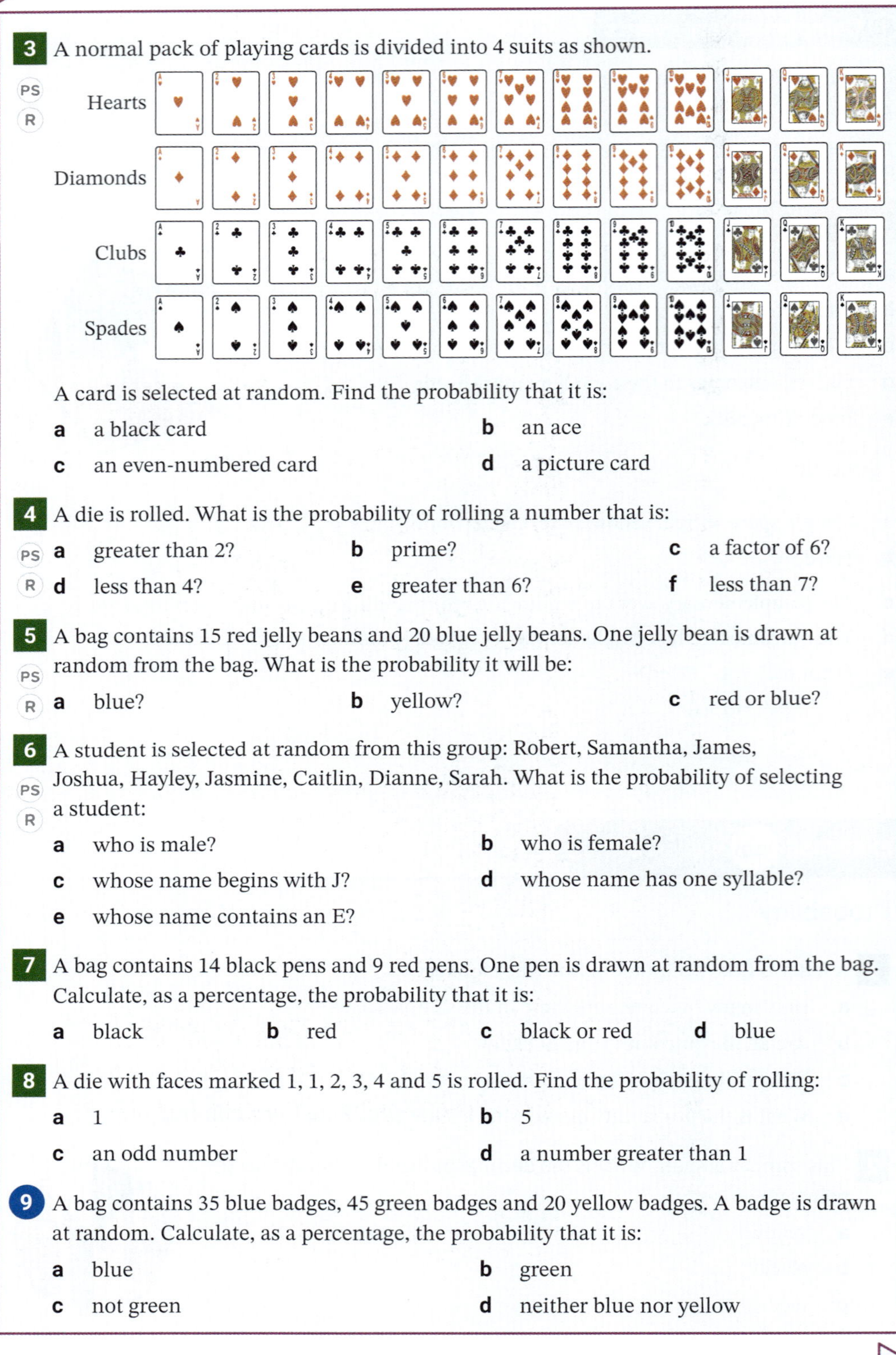

3 A normal pack of playing cards is divided into 4 suits as shown.

PS R

Hearts

Diamonds

Clubs

Spades

A card is selected at random. Find the probability that it is:

a a black card **b** an ace

c an even-numbered card **d** a picture card

4 A die is rolled. What is the probability of rolling a number that is:

PS R

a greater than 2? **b** prime? **c** a factor of 6?

d less than 4? **e** greater than 6? **f** less than 7?

5 A bag contains 15 red jelly beans and 20 blue jelly beans. One jelly bean is drawn at random from the bag. What is the probability it will be:

PS R

a blue? **b** yellow? **c** red or blue?

6 A student is selected at random from this group: Robert, Samantha, James, Joshua, Hayley, Jasmine, Caitlin, Dianne, Sarah. What is the probability of selecting a student:

PS R

a who is male? **b** who is female?

c whose name begins with J? **d** whose name has one syllable?

e whose name contains an E?

7 A bag contains 14 black pens and 9 red pens. One pen is drawn at random from the bag. Calculate, as a percentage, the probability that it is:

a black **b** red **c** black or red **d** blue

8 A die with faces marked 1, 1, 2, 3, 4 and 5 is rolled. Find the probability of rolling:

a 1 **b** 5

c an odd number **d** a number greater than 1

9 A bag contains 35 blue badges, 45 green badges and 20 yellow badges. A badge is drawn at random. Calculate, as a percentage, the probability that it is:

a blue **b** green

c not green **d** neither blue nor yellow

☐ Foundation ○ Standard ⬡ Complex

10 For each event, write the complementary event.

- **a** Selecting a consonant from the alphabet.
- **b** The weather being sunny tomorrow.
- **c** A traffic light showing red.
- **d** Selecting a spade from a normal deck of cards.
- **e** Hitting bullseye on a dartboard.
- **f** Selecting a blue ball from a bag containing blue, red and yellow balls.

EXAMPLE 2

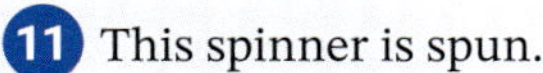

11 This spinner is spun.

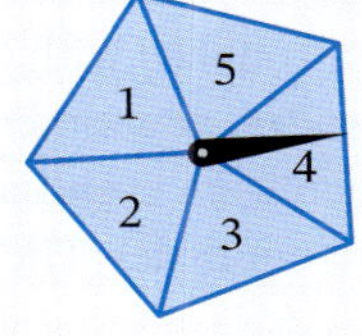

- **a** List the sample space.
- **b** Find $P(4)$.
- **c** What is the complementary event to spinning a 4?
- **d** List the outcomes in the complementary event.
- **e** If $P(\overline{4})$ means $P(\text{not } 4)$, find $P(\overline{4})$.

12 Dilone buys a ticket in a raffle in which 50 tickets are sold and there is only one prize.

- **a** What is Dilone's chance of winning the prize? Write your answer as a:
  i fraction **ii** decimal **iii** percentage.
- **b** Calculate the probability of Dilone not winning the prize. Write your answer as a:
  i fraction **ii** decimal **iii** percentage.

13 Priya buys tickets in a raffle in which 200 tickets are sold and there is only one prize. She has a 40% chance of winning the prize.

- **a** What is the probability that Priya will not win the prize?
- **b** How many tickets did Priya buy?

14 Tamin bought a ticket in a raffle in which 500 tickets were sold. If there are 5 prizes, what is the probability that Tamin does not win a prize? Select **A**, **B**, **C** or **D**.

A $\frac{1}{100}$ **B** 49% **C** $\frac{499}{500}$ **D** 0.99

15 The probability of rain on Tuesday is 0.35. What is the probability that it will not rain?

16 What is the probability that a student chosen at random from your school was not born in a month beginning with the letter M?

17 Stephanie has a 76% chance of hitting the bullseye when playing darts. What is the probability that she misses the bullseye? Select **A**, **B**, **C** or **D**.

A 24% **B** 26% **C** 34% **D** 76%

18 What is the probability that a mobile phone number selected at random does not begin with a 0?

DID YOU KNOW?

The history of playing cards

The earliest European cards were Tarot packs, which contained either 78 or 97 cards, and were used to play a game called Tarocco. The smaller 52-card pack may have become popular for practical reasons because cards were originally drawn and painted by hand.

Bridgeman Images/DeAgostini Picture Library/A. Dagli Orti

The names and symbols on cards have changed many times but in the 14th century a pack was introduced with the symbols for hearts, spades, diamonds and clubs, the symbols which were eventually adopted by most Western countries. The names for the symbols still differ from country to country.

The pictures used for the Kings, Queens and Jacks have undergone some changes. The 4 Kings originally represented Charlemagne (hearts), the emperor of most of Europe until 814 ce, David (spades), the Hebrew king mentioned in the Bible, Julius Caesar (diamonds) of ancient Rome and Alexander the Great (clubs) of ancient Greece.

Investigate who the original Queens and Jacks were.

12.02 Relative frequency

Video
Experimental probability

Worksheets
Coins probability
Dice probability
Probability review
Relative frequencies
Final 6 investigation

Interactive
Plinko probability

Probability that is calculated using the formula:

$$P(E) = \frac{\text{number of favourable outcomes}}{\text{total number of outcomes}}$$

is more specifically called the **theoretical probability**.

We can also determine probability based on the results of a trial that has been repeated many times, such as testing the effectiveness of 100 light globes, or rely on past statistics, such as weather patterns or the ages of drivers having car accidents. This type of probability is called **experimental probability** or **observed probability**, which is based on **relative frequency**, the number of times an event occurred as a fraction of the total frequency of outcomes.

ⓘ Experimental probability

$$P(E) = \frac{\text{number of times the event happened}}{\text{total number of trials}}$$

$$\text{or } P(E) = \frac{\text{frequency of } E}{\text{total frequency}}$$

Expected frequency is the expected number of times an event will occur over repeated trials.

ⓘ Expected frequency

Expected frequency = theoretical probability × number of trials.

Example 3

Connor tossed a coin 100 times and recorded the results in a table.

Outcome	Frequency
Head	46
Tail	54
	100

a Calculate, as a decimal,

i the experimental probability of tossing a head.

ii the theoretical probability of tossing a head.

b Are the experimental and theoretical probabilities similar?

c For 100 tosses, what is the expected frequency of heads based on the theoretical probability? How does this compare with the observed frequency?

SOLUTION

a **i** Experimental probability: $P(H) = \frac{46}{100} = 0.46$

ii Theoretical probability: $P(H) = \frac{1}{2} = 0.5$

b By comparing the decimals for the 2 answers, we see that the experimental and theoretical probabilities are similar.

c Expected frequency of heads $= 0.5 \times 100$ Probability × number of trials

$= 50$

From the table, the observed frequency of heads = 46, which is close to 50.

Note: When calculating experimental probability, the results will not always be consistent with what is expected, unless a very large number of trials are performed.

Example 4

Briana rolled a die 200 times and recorded her results.

Outcome	Frequency
1	38
2	40
3	29
4	24
5	35
6	34

a Find the experimental probability of rolling:

i an even number.

ii an even number or a number greater than 2.

iii an even number and a number greater than 2.

iv an even number not greater than 2.

b Calculate the probability of rolling a 5 or 6:

i as an experimental probability

ii as a theoretical probability.

c What is the expected frequency of rolling a 5 or 6 in 200 rolls of a die? How does this compare with Briana's observed frequency?

SOLUTION

a **i** Rolls of even numbers $= 40 + 24 + 34$ — Frequencies of 2, 4, 6.

$= 98$

Experimental $P(\text{even}) = \frac{98}{200} = \frac{49}{100}$

ii Rolls of even numbers or numbers greater than 2 — Frequencies of 2, 4, 6, 3, 5.

$= 40 + 24 + 34 + 29 + 35$

$= 162$

Experimental $P(\text{even or} > 2) = \frac{162}{200} = \frac{81}{100}$

iii Rolls of even numbers greater than $2 = 24 + 34$ — Frequencies of 4 and 6.

$= 58$

Experimental $P(\text{even and} > 2) = \frac{58}{200} = \frac{29}{100}$

iv Rolls of even numbers not greater than $2 = 40$ — Frequency of 2.

Experimental $P(\text{even and not} > 2) = \frac{40}{200} = \frac{1}{5}$

b **i** Rolls of 5 or $6 = 35 + 34 = 69$ — Frequencies of 5 and 6.

Experimental $P(5 \text{ or } 6) = \frac{69}{200}$

ii Theoretical $P(5 \text{ or } 6) = \frac{2}{6} = \frac{1}{3}$ — 2 chances out of 6.

c Expected frequency of 5 or $6 = \frac{1}{3} \times 200$ — Probability × number of trials

$= 66.666...$

≈ 67

From the table, the observed frequency $= 35 + 34 = 69$, which is close to 67.

EXERCISE 12.02 ANSWERS ON P. 675

Relative frequency

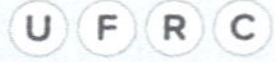

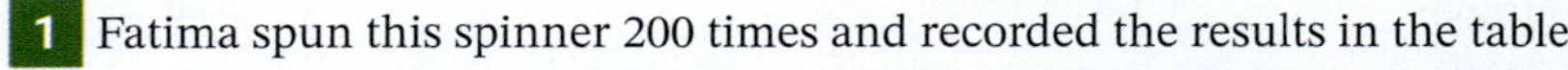

EXAMPLE 3

1 Fatima spun this spinner 200 times and recorded the results in the table.

R C

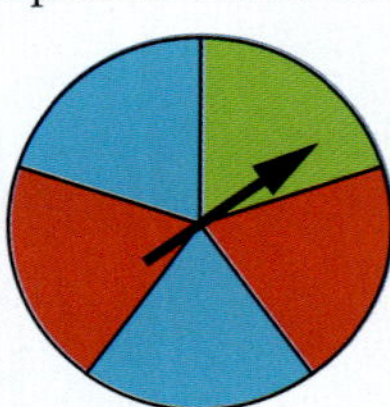

Event	Frequency
Red	90
Green	36
Blue	74
	200

a Calculate, as a decimal, the experimental probability that the needle points to:

i red **ii** green

b Calculate, as a decimal, the theoretical probability that the needle points to:

i red **ii** green

c Are the experimental and theoretical probabilities similar?

d For 200 spins, what is the expected frequency of red or blue based on the theoretical probability? How does this compare with the observed frequency?

2 R C Use cardboard and a toothpick to make a square spinner that has 4 colours as shown in the diagram below. Spin the spinner 50 times and record your results in a table similar to the one shown.

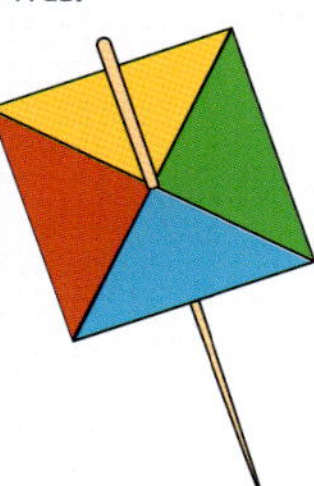

Event	Tally	Frequency
Red		
Blue		
Green		
Yellow		

a Calculate, as a percentage, the relative frequency of spinning:

i red **ii** blue

iii a colour that is not yellow **iv** blue or green

b Calculate, as a percentage, the theoretical probability of spinning blue. How does this compare with the relative frequency?

c For 50 spins, what is the expected frequency that blue should come up? How does this compare with the observed frequency?

3 R C A die is rolled.

a Predict the number of times a 2 would appear if the die was rolled 100 times.

b Roll a die 100 times and record your results in a table similar to the one shown.

c What is the relative frequency of rolling a 2?

d How does your predicted frequency compare with the observed frequency?

e What is the relative frequency of rolling an even number?

Event	Tally	Frequency
1		
2		
3		
4		
5		
6		
		100

4 R C Place 4 black counters, 5 red counters and 1 yellow counter in a bag. Select a counter at random, record the colour and return the counter to the bag. Run this experiment 50 times.

a What is the experimental probability of selecting:

i a black counter?

ii a red counter?

iii a yellow counter?

Event	Tally	Frequency
Black		
Red		
Yellow		

☐ Foundation ○ Standard ⬡ Complex

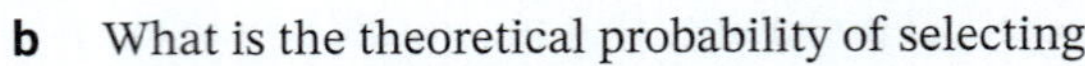

b What is the theoretical probability of selecting:

- **i** a black counter?
- **ii** a red counter?
- **iii** a yellow counter?

c Compare your experimental probabilities to the theoretical probabilities.

d Repeat the experiment another 50 times and find the relative frequency of each event. Are the results different to the results found in part **a**?

EXAMPLE 4

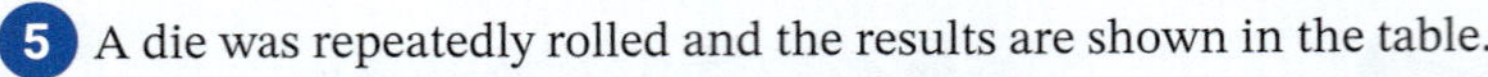

5 A die was repeatedly rolled and the results are shown in the table.

Outcome	Frequency
1	90
2	84
3	92
4	76
5	83
6	75

a What was the total number of trials?

b Find the experimental probability (as a decimal) of rolling:

- **i** an even number.
- **ii** a number greater than 3.
- **iii** a 3 or a 5.
- **iv** a number greater than 2 and even.
- **vii** a number greater than 3 or odd.

c Calculate the probability of rolling a number not less than 3:

- **i** as an experimental probability.
- **ii** as a theoretical probability.

d How do the 2 probabilities in part **c** compare?

6 Students at a high school were asked how they travelled to school. The results are presented in the table. Find the probability (correct to 2 decimal places) that a student chosen at random will:

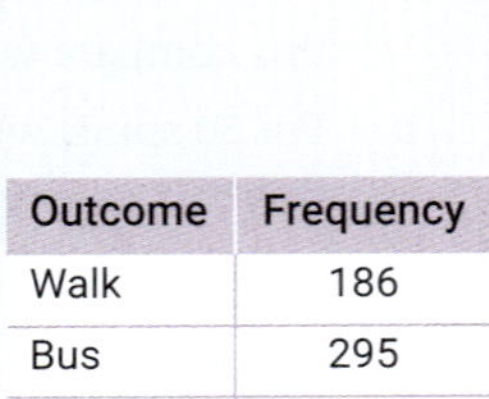

Outcome	Frequency
Walk	186
Bus	295
Car	103
Bicycle	74

a walk **b** be driven **c** catch a bus

d not ride a bicycle **e** be driven or catch a bus.

7 A die was rolled many times and the results are shown in the frequency histogram.

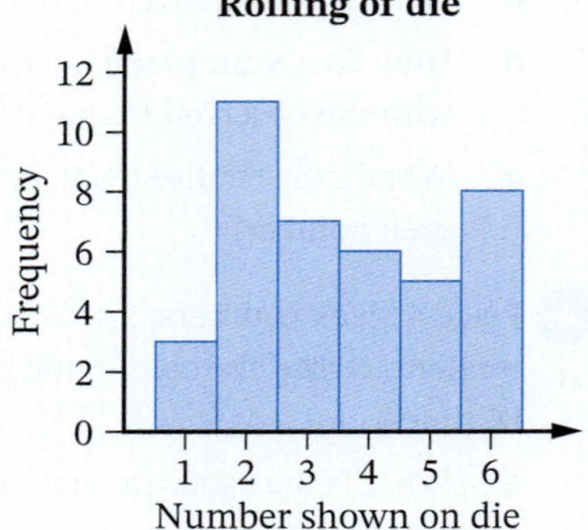

a What is the total number of trials?

b Find the experimental probability of rolling:

- **i** 2
- **ii** 5
- **iii** an even number or a number greater than 3.
- **iv** a number greater than 4 and odd.

12.02

8 The table shows the number of babies born per day at a hospital during the month of January. On any particular day, find the probability of the birth of exactly:

Number of babies per day	Frequency
0	1
1	4
2	7
3	10
4	5
5	4

a 1 baby **b** 3 babies
c 1 or 2 babies **d** **at least** 2 babies
e at most 1 baby **f** 4 or 5 babies.

9 This dot plot shows the number of goals per match scored by a soccer team during the season.

Goals	0	1	2	3	4	5	6	7	8	9	10
Dots	2	3	6	4	2	1	1	0	0	0	1

Number of goals per match

For a match selected from the season at random, find the probability of the team scoring:

a 2 goals **b** 4 goals per game **c** fewer than 2 goals
d 2 or 3 goals **e** at least 3 goals **f** not more than 4 goals

10 R C A card is drawn randomly from a normal pack of cards to see if it is an Ace. This trial is run 120 times, with each card being replaced in the deck before the next card is drawn. The results are shown in the table.

Outcome	Frequency
Ace	11
Not an Ace	109

a Calculate, correct to 3 decimal places:
  i the experimental probability of selecting an Ace.
  ii the theoretical probability of selecting and Ace.
b Are the experimental and theoretical probabilities similar?
c If a card is drawn 300 times, what is the expected frequency of an Ace being drawn? Select the correct answer **A**, **B**, **C** or **D**.

A 27 **B** 28 **C** 23 **D** 6

TECHNOLOGY

Tossing a coin

In this activity, a spreadsheet is used to simulate the tossing of a coin. The computer can quickly generate either a 1 (representing heads) or a 2 (tails).

1 In cell A1, enter the label 'Tossing a coin'.
2 In cell A2, enter the formula =INT(RAND()*2+1), press Enter and either 1 or 2 will appear in the cell.
3 Select cell A2 and Fill Down to cell A11 to simulate the tossing of a coin 10 times.
4 Copy the table and in the first blank row record your results for Trials 1–10 (the numbers of heads and tails in your first 10 'tosses').

Worksheet Coins probability

5 Simulate another 10 tosses of the coin by making the spreadsheet generate another set of random numbers. Do this by placing the cursor between the tops of columns A and B (so that it turns into a 'double arrow') and clicking.

6 Enter the results for Trials 11–20 in your table.

7 Repeat 8 more times so that you have 100 tosses of the coin recorded in the table.

8 a For 100 tosses of a coin, what is the expected frequency of heads based on the theoretical probability?

b Compare this with the actual frequency.

9 Calculate, as a decimal, the theoretical and experimental probabilities of tossing heads.

10 Compare your results with those of other students in your class. Briefly explain any differences and why they may have occurred.

11 Combine the results of students in your class to calculate the experimental probability of throwing heads, as a decimal. How does this compare with the theoretical probability?

Trial	Number of heads	Number of tails
1–10		
11–20		
21–30		
31–40		
41–50		
51–60		
61–70		
71–80		
81–90		
91–100		
Total		

TECHNOLOGY

Rolling dice

Worksheet Dice probability

In this activity, we will use a scientific calculator to simulate the rolling of a die.

1 Copy the table shown.

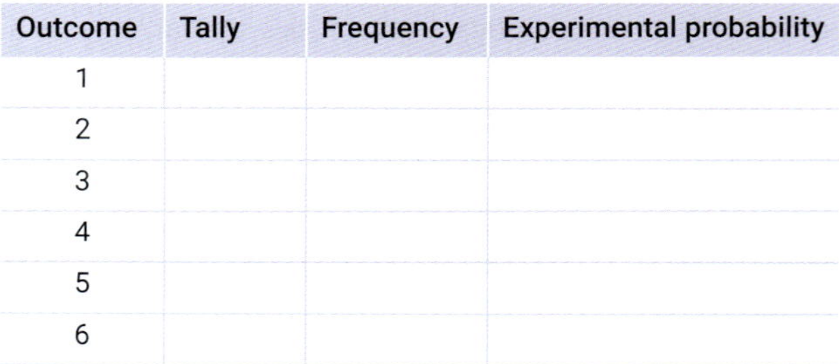

Outcome	Tally	Frequency	Experimental probability
1			
2			
3			
4			
5			
6			

2

Casio scientific calculator	**Sharp scientific calculator**
Enter the formula RanInt#(1,6) as shown: ALPHA . (1 SHIFT) 6) =	2nd F 7 (RANDOM), select 1 R-DICE, =

This generates a random number from 1 to 6. Press = more times to generate more random numbers. Alternatively, on a spreadsheet, type **=INT(RAND()*6+1)** into a cell and press the Enter key, then the F9 key repeatedly for more random numbers.

3 Run the simulation 50 times and record the results in the table.

4 Calculate the theoretical probability for each number being tossed on the die.

5 Compare the theoretical probability and experimental probability for each number. Are there similarities or differences? How could the experimental probabilities be more accurate? Explain.

6 Copy this table for rolling 2 dice and finding their sum.

Sum	Tally	Frequency	Experimental probability
2			
3			
4			
5			
6			
7			
8			
9			
10			
11			
12			

7 On a scientific calculator, rolling 2 dice is achieved by adding the result of rolling one die twice.

Casio scientific calculator	**Sharp scientific calculator**
Enter the formula RanInt#(1,6) + RanInt#(1,6) as shown below: ALPHA · (1 SHIFT) 6) + ALPHA · (1 SHIFT) 6) =	Enter the formula R.DICE + R.DICE as shown below: 2nd F 7 (RANDOM), select 1 R-DICE, + 2nd F 7 (RANDOM), select 1 R-DICE, =

This generates the sum of 2 random numbers from 1 to 6.

Press = more times to generate more sums.

8 Run the simulation 50 times and record the results in the table.

9 Which sum is the most likely to occur?

10 Which sum is the least likely to occur? Why?

11 Which sum is theoretically the:

a most likely?

b least likely?

Why?

12 Compare your answers to questions **9**, **10** and **11** and then compare them with the answers of other students in your class. Write a few sentences explaining your observations.

13 If you ran 50 more simulations, would you expect your results to be similar or different to your first 50 results? Explain.

12.03 Venn diagrams

Videos
Venn diagrams 2

Venn diagrams: Global habitats

Worksheet
Venn diagrams

Puzzles
Venn diagrams matching activity

Venn diagrams

A **Venn diagram** is a diagram of circles (usually overlapping) for grouping items into categories. A rectangle represents the whole group, while the circles represent categories. Items common to 2 or more categories are placed in the intersection (overlapping region) of the circles. The Venn diagram was invented in 1880 by English mathematician and priest, John Venn (1834–1923).

Example 5

This Venn diagram shows the results of a survey on how people travel to work (car (C), bus (B) or train (T)).

a How many people participated in the survey?

b How many people used all 3 methods: car, bus *and* train? Give an explanation on how this might happen.

c Calculate the probability of selecting a person from this survey who travels by:

- **i** car only or bus only
- **ii** car and train
- **iii** car and train only
- **iv** bus but not train.

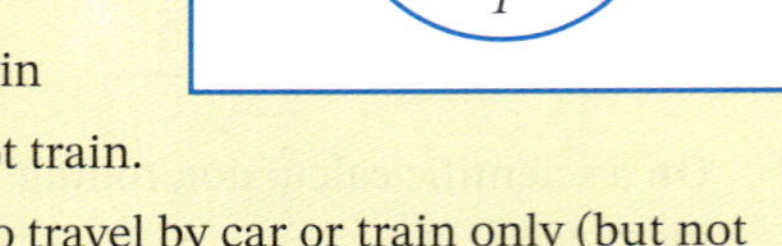

d A person is chosen at random from the people who travel by car or train only (but not bus). What is the probability that the person travels by car only?

e If 120 people actually participated in the survey, what extra information would be needed on the diagram?

SOLUTION

a Number of people $= 27 + 12 + 5 + 20 + 21 + 4 + 11$

$= 100$

b 5 people travelled by car, bus and train.

The region where the 3 circles intersect.

They may have driven to the station, caught a train and then a bus to the workplace, or used different methods on different days.

c **i** People travelling by car only or bus only

The regions in C and B that don't overlap.

$= 27 + 11$

$= 38$

$P(\text{car only or bus only}) = \frac{38}{100}$

$= \frac{19}{50}$

'Car only or bus only' is an example of a compound event.

ii People travelling by car and train

The region where C and T intersect.

$= 20 + 5$

$= 25$

$P(\text{car and train}) = \frac{25}{100}$

$= \frac{1}{4}$

'Car and train' is another example of a compound event.

iii People travelling by car and train only = 20

Where C and T intersect, but excluding B.

$P(\text{car and train only}) = \frac{20}{100}$

$= \frac{1}{5}$

iv People travelling by bus but not train

The circle B excluding T.

$= 11 + 12$

$= 23$

$P(\text{bus but not train}) = \frac{23}{100}$

d People travelling by car or train only = 27 + 20 + 21

$= 68$

People travelling by car only = 27.

$P(\text{car only}) = \frac{27}{68}$

e 120 – 100 = 20. This means that there were 20 people who did not use a car, bus or train. A '20' would need to be placed in the rectangle, but outside of the 3 circles.

When we combine 2 or more simple events, we get a **compound event**. In the above example, 'car and train' and 'bus but not train' are compound events.

'And' vs 'or'

For 2 categories or events A and B, the compound event **'A and B'** means to have both of them occurring together. For example, 'to drive a car' **and** 'to ride a bus' means to do both things.

If A and B are **overlapping**, the compound event **'A or B'** means to have A or B or both. For example, 'to drive a car' **or** 'to ride a bus' means to drive a car only, or to ride a bus only, or to do both. In this case, 'A or B' actually **includes** 'A and B' so this is an example of an **inclusive** 'or'.

If A and B are **mutually exclusive**, this means that they are **not overlapping** and on a Venn diagram they appear as 2 separate circles. For mutually exclusive categories or events, the phrase **'A or B'** means to have A only or B only (but not both). For example, 'male' **or** 'female' means to be male, or female, but not both. In this case, 'A or B' **excludes** 'A and B' so this is an example of an **exclusive** 'or'.

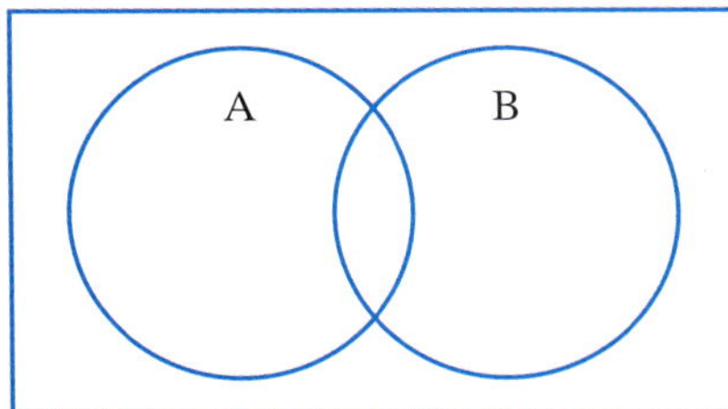

Overlapping events: 'A or B' means A or B or both

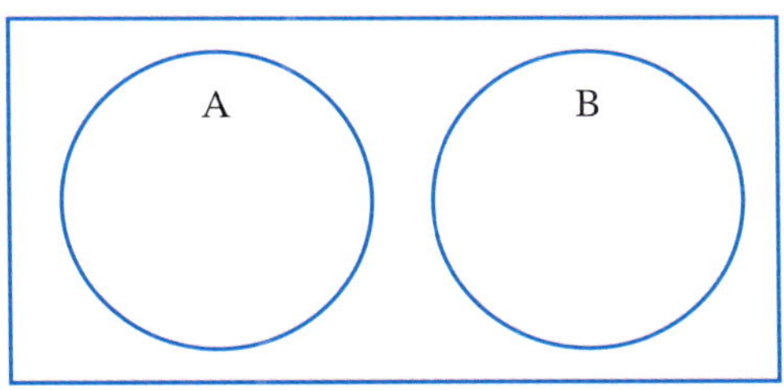

Mutually exclusive events: 'A or B' means A or B but not both

Video
Venn diagrams 1

Example 6

A survey of 110 students at Redford College showed that 61 students had dark hair, 42 students had brown eyes and 20 students had dark hair and brown eyes.

Left to right: Shutterstock.com/michaeljung, Alamy Stock Photo/blickwinkel, Shutterstock.com/Dorota Zietek

a Represent this information on a Venn diagram.

b How many students had brown eyes or dark hair but not both?

c What is the probability of randomly selecting from this group a student:

i with dark hair? **ii** with dark hair and brown eyes?

iii with dark hair or brown eyes? **iv** with neither dark hair nor brown eyes?

SOLUTION

a RC = students at Redford College

DH = students with dark hair

BE = students with brown eyes

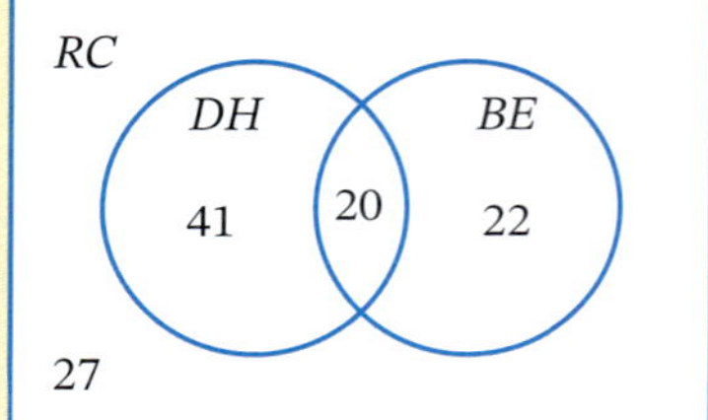

When completing a Venn diagram, always begin with the intersection.

There are 20 students with dark hair and brown eyes.

$\therefore$ Students with dark hair only $= 61 - 20$
$= 41$

$\therefore$ Students with brown eyes but not dark hair $= 42 - 20$
$= 22$

$\therefore$ (Remaining) Students with neither dark hair nor brown eyes $= 110 - 41 - 20 - 22$
$= 27$

b Number of students with brown eyes or dark hair only $= 22 + 41 = 63$

c **i** Students with dark hair $= 41 + 20$
$= 61$

$P(\text{dark hair}) = \frac{61}{110}$

ii $P(\text{dark hair and brown eyes}) = \frac{20}{110}$
$= \frac{2}{11}$

'Dark hair and brown eyes' is a compound event.

iii Number of students with dark hair or brown eyes $= 41 + 20 + 22$
$= 83$

$P(\text{dark hair or brown eyes}) = \frac{83}{110}$

iv $P(\text{neither dark hair nor brown eyes}) = \frac{27}{110}$

9780170465618

EXERCISE 12.03 ANSWERS ON P. 676

Venn diagrams

U F R C

EXAMPLE 5

1 R The Venn diagram shows the number of students who play either basketball (B) or volleyball (V).

B V
23 7 18

a Find the total number of students who play basketball or volleyball.

b Find the probability of randomly selecting a student who plays:

i basketball

ii volleyball.

c What is the probability of selecting a student who plays:

i basketball or volleyball?

ii basketball and volleyball?

iii basketball but not volleyball?

iv neither basketball nor volleyball?

d Of the students who play volleyball, find the probability of selecting a student who also plays basketball. Select the correct answer **A**, **B**, **C** or **D**.

A 0.28 **B** 0.375 **C** 0.72 **D** 0.92

2 R The Venn diagram shows the number of Year 9 students who achieved Credit grades in their Science (S) and History (H) exams.

H S 27
53 38 42

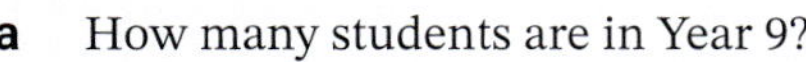

a How many students are in Year 9?

b Find the probability of selecting a student who achieved a Credit in:

i History

ii Science

iii History and Science

iv History or Science

v History or Science but not both

vi History only

vii Science only

c What is the percentage probability of selecting a student who did not achieve a Credit in both History and Science?

d Out of the students who achieved a Credit in History, what is the probability that the student also achieved a Credit in Science?

3 R C 40 people were asked whether they preferred watching movies at the cinema (C) or on a device (D) at home. The results are shown in the Venn diagram.

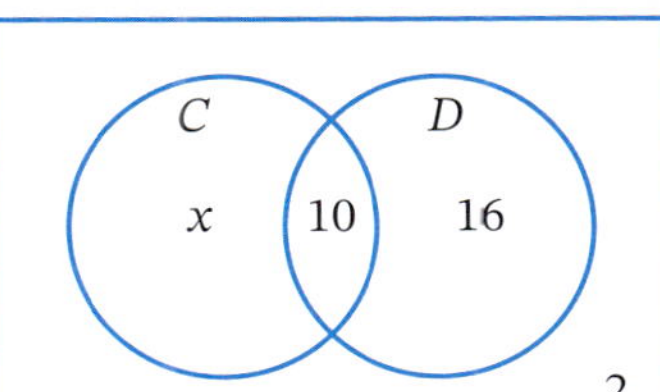

a Describe the category that contains the pronumeral x in the diagram.

b Find the value of x.

Foundation Standard Complex

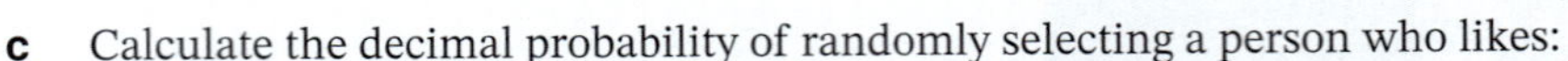

c Calculate the decimal probability of randomly selecting a person who likes:

- **i** the cinema and a device.
- **ii** a device only.
- **iii** the cinema or a device.
- **iv** neither the cinema nor a device.

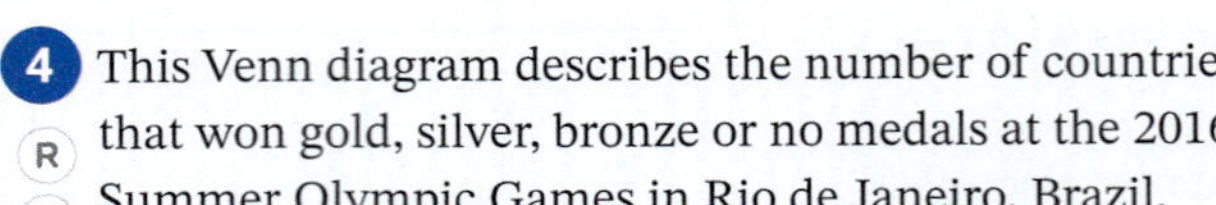

4 (R) (C) This Venn diagram describes the number of countries that won gold, silver, bronze or no medals at the 2016 Summer Olympic Games in Rio de Janeiro, Brazil.

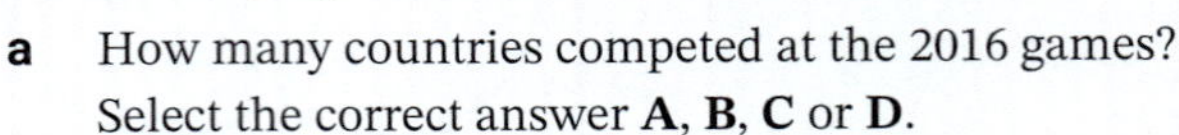

a How many countries competed at the 2016 games? Select the correct answer **A**, **B**, **C** or **D**.

A	86	**B**	183
C	189	**D**	196

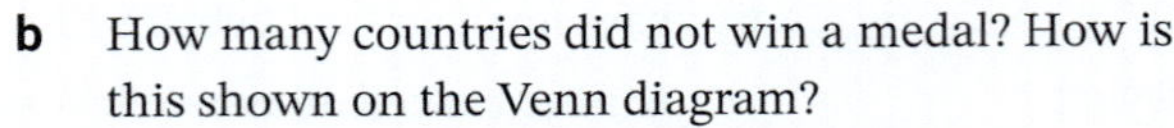

b How many countries did not win a medal? How is this shown on the Venn diagram?

c What is the probability of randomly selecting a country that:

- **i** won only gold medals?
- **ii** won gold or silver medals, but not both?
- **iii** won gold, silver and bronze medals?
- **iv** won gold and bronze only?
- **v** won gold only or silver only or bronze only?
- **vi** did not win a gold medal?
- **vii** did not win a gold or silver medal?

d Out of the countries that won medals, what is the probability of selecting a country that:

- **i** won gold medals?
- **ii** won silver but not bronze?

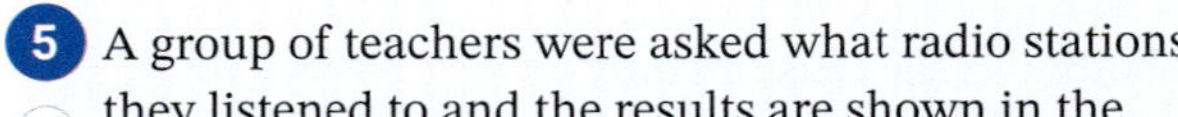

5 (R) A group of teachers were asked what radio stations they listened to and the results are shown in the Venn diagram.

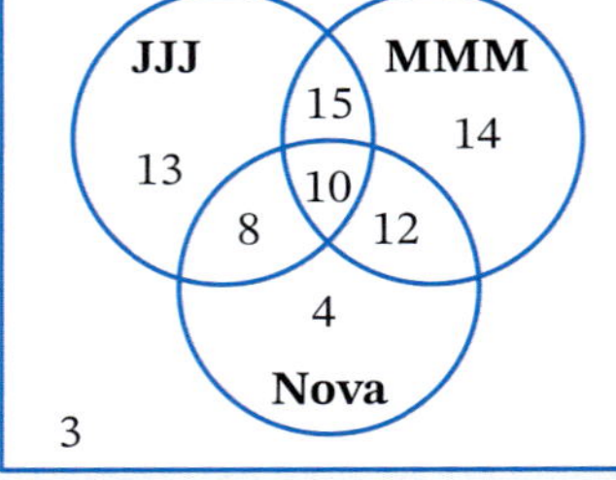

a How many teachers were surveyed?

b What is the probability of selecting a teacher who did not listen to the 3 stations?

c What is the probability of selecting a teacher who listens to:

- **i** JJJ?
- **ii** MMM?
- **iii** Nova?

d What is the probability of selecting a teacher who listens to at least 2 of the stations?

A $\frac{50}{79}$ **B** $\frac{45}{79}$ **C** $\frac{35}{79}$ **D** $\frac{30}{79}$

Foundation Standard Complex

12.03

e Find the probability of selecting a teacher who listens to:

i JJJ or MMM but not both.

ii all 3 stations.

iii MMM only.

EXAMPLE 6

6 In a drama group of 30 people, 18 people dance (D), 20 people sing (S) and 8 people dance and sing.

R C

a Show this information on a Venn diagram.

b Find the probability of selecting a person who:

i dances and sings.

ii sings but does not dance.

iii dances but does not sing.

iv dances or sings.

v dances or sings but not both.

7 A class of 30 students were given the option of playing an indoor sport (S) or watching a movie (M) when regular sport was cancelled due to bad weather. 15 students wanted to play an indoor sport and 24 students wanted to watch a movie.

R C

a How many students indicated that they would do either?

b Copy and complete a Venn diagram using the information given.

c For this Venn diagram, there are no values outside the circles. Explain why.

d What is the probability of selecting a student who prefers to:

i play an indoor sport or watch a movie, but not both?

ii watch a movie but not play an indoor sport?

iii play an indoor sport but not watch a movie?

8 80 families living in Mt Robinson were surveyed on whether they had a dog (D) or a cat (C) as a pet. 12 families had no pets, 35 had a dog and 47 had a cat.

R C

a Draw Venn diagram showing the information given.

b How many families had:

i a dog and a cat?

ii a dog but not a cat?

iii at least a dog or a cat?

iv a dog or a cat, but not both?

c Find the decimal probability of selecting a family with:

i a dog only

ii either a dog or a cat, but not both

iii at most one pet

iv at least one pet.

d Given that a family has a pet, what is the probability that it is a cat?

9 At Nelson River High School, the Student Representative Council (SRC) has 25 members, the Chess club has 20 members and the Environment club has 33 members. 5 students belong to all 3 groups, while 13 members belong to the SRC and no other club. 4 members of the Environment club are members of the SRC, but not the Chess club, while 10 members belong to both the Environment and Chess clubs. There are 8 members of the Chess club who are members of the SRC.

R C

Foundation Standard Complex

a Copy and complete the Venn diagram.

b How many students belong to the Environment club only?

c How many students belong to the Chess club and no others?

d Find the probability of selecting a student who belongs to exactly 2 clubs.

e What is the probability of a student belonging to at least 2 clubs?

f Find the probability of selecting a student who belongs to:

i the Environment club and the SRC, but not the Chess club.

ii only one club.

iii the SRC or the Chess club, but not the Environment club.

iv at most 2 clubs.

g Is it necessary to include the rectangle in this Venn diagram? Give reasons.

SRC C E

12.04 Two-way tables

Worksheet Two-way tables

Two-way probability tables

A **two-way table** is another way of grouping items into overlapping categories, especially when there are many overlaps that cannot easily be represented by Venn diagrams.

Example 7

This table categorises the heights and ages of 200 Year 9 students.

	Short	Tall
Aged 14	45	25
Aged 15	90	40

a How many students are:

i aged 15 and tall?

ii aged 14 or short?

b What fraction of students are neither tall nor aged 15?

c What is the probability of selecting a student at random who is:

i aged 14 and tall?

ii aged 15 but not tall?

iii tall or aged 14, but not both?

d If a short student is selected at random, what is the probability that he or she is aged 14?

SOLUTION

a i Students aged 15 and tall = 40

ii Students aged 14 or short $= 45 + 25 + 90$

$= 160$

Foundation Standard Complex

b Students who are neither tall nor 15 = 45 — The short 14-year-olds

∴ Fraction of students neither tall nor 15

$$= \frac{45}{200} = \frac{9}{40}$$

c **i** $P(\text{tall and }14) = \frac{25}{200}$

$= \frac{1}{8}$

ii $P(15\text{ but not tall}) = \frac{90}{200}$

$= \frac{9}{20}$

iii Students tall or aged 14, but not both = 40 + 45

= 85

$P(\text{tall or }14\text{, but not both}) = \frac{85}{200} = \frac{17}{40}$

d Short students = 90 + 45 = 135

Students who are short and 14 = 45.

$P(\text{short student who is }14) = \frac{45}{135} = \frac{1}{3}$

EXERCISE 12.04 ANSWERS ON P. 676

Two-way tables

U F R C

1 Households were surveyed at random to see whether or not they had dishwashers and clothes dryers. The results are in the table. EXAMPLE 7

R C

	Dishwasher	No dishwasher
Clothes dryer	75	108
No clothes dryer	49	68

a How many households were surveyed?

b How many households had a clothes dryer and a dishwasher?

c How many households had neither?

d What is the probability of selecting a household at random that has:

i a dishwasher?

ii a dishwasher and a dryer?

iii a dishwasher or a clothes dryer, but not both?

iv at least one of them?

2 People at a train station were asked whether they ate breakfast or lunch that day.

R C

	Ate lunch	Did not eat lunch
Ate breakfast	35	26
Did not eat breakfast	54	15

a How many people were surveyed?

b What percentage of people ate lunch?

Foundation Standard Complex

c Find the probability of picking a person at random who ate:

i breakfast

ii lunch

iii breakfast but not lunch

iv breakfast or lunch

v breakfast and lunch

vi breakfast or lunch, but not both

d What is the probability of picking a person who has eaten lunch given that the person ate breakfast? Select the correct answer **A**, **B**, **C** or **D**.

A $\frac{26}{61}$ **B** $\frac{35}{61}$ **C** $\frac{35}{89}$ **D** $\frac{89}{130}$

e What is the probability of picking a person who has not eaten lunch given that the person did not eat breakfast?

3 People at a shopping mall were asked if they watched football on TV.

R C

	Football	No football
Male	85	27
Female	36	92

a How many people were surveyed?

b How many people:

i watched football?

ii did not watch football?

c What percentage of people surveyed were females?

d What is the probability of randomly selecting a person who did not watch football and was male?

e Given that a person watched football, what is the probability that the person was female?

4 The results comparing gender and handedness of students at a primary school are shown in the table.

R C

	Male	Female
Left-handed	14	18
Right-handed	86	65

a How many students are at the school?

b What is the probability of selecting a student at random who is:

i male and left-handed?

ii male or left-handed?

c Find the probability, expressed as a percentage, of randomly selecting a right-handed female.

d Given that a student is a male, what is the probability that he is right-handed?

e What is the probability of selecting a female or a right-handed student?

5 After the 2019 Federal election, the members (politicians) of the House of Representatives in the Australian Parliament were as shown.

	Government	Opposition
Male	62	43
Female	15	31

☐ Foundation ◯ Standard ⬡ Complex

12.04

a How many members of parliament (MPs) were there in the House of Representatives?

b Find the percentage probability of randomly selecting an MP who is:

i male

ii female and in Opposition

iii in the Government or female

iv male and in Opposition.

c Find the probability of selecting an MP who is male and in Government.

d If a female MP is selected at random, what is the probability that she is in Government?

6 Students at Cengage Anglican College were surveyed on whether they played video games. The results are shown in the table.

R C

	Plays video games	Does not play video games
Female	178	267
Male	310	210

a How many students attended the college?

b Find the percentage probability that a student selected at random from the college:

i plays video games and is female.

ii doesn't play video games.

iii is male and doesn't play video games.

iv is female and doesn't play video games.

c If a female student is selected at random, what is the probability that she plays video games?

d If a male student is selected at random, what is the probability that he plays video games?

7 A group of retired people was surveyed on whether they exercised regularly.

R C

	Exercise	Don't exercise
Men	65	106
Women	74	75

a How many people were surveyed?

b What percentage of people exercised?

c What is the probability of selecting a person at random who doesn't exercise regularly?

d Of the women who were surveyed, what percentage exercised regularly?

e If a man was selected randomly from this survey, what is the probability that he exercises regularly? Select the correct answer **A**, **B**, **C** or **D**.

A 38.0% **B** 62.0% **C** 20.3% **D** 53.4%

☐ Foundation ○ Standard ⬡ Complex

INVESTIGATION

Using two-way tables

Work in groups of 3 or 4.

Investigate 2 characteristics, for example, gender and exercise, gender and eating habits, eating habits and exercise, playing sport and going to the movies, eye colour and hair colour.

Conduct a survey to obtain the information required, then summarise your results in a two-way table similar to this one:

	Eat regular meals	Don't eat regular meals
Play sport		
Don't play sport		

Note any patterns in your data. This may include answering questions such as 'What percentage of …' or 'What is the probability of picking a person who …'.

Quiz
Mental skills 12

☆ MENTAL SKILLS (12) ANSWERS ON P. 677 Maths without calculators

The unitary method with percentages

The unitary method is used when you are only given a *percentage* of an amount and you need to find the amount. It is called the unitary method because we find 1% of the amount first, then multiply that by 100 to find the whole (100%).

1 Study each example.

a If 8% of a number is 24, what is the number?

8% of the number = 24

$\therefore$ 1% of the number = $24 \div 8 = 3$

$\therefore$ 100% of the number = $3 \times 100 = 300$

The number is 300. *Check*: $8\% \times 300 = 24$

b If 15% of an amount is \$90, what is the whole amount?

15% of the amount = \$90

$\therefore$ 1% of the amount = \$90 ÷ 15 = \$6

$\therefore$ 100% of the amount = \$6 × 100 = \$600

The amount is \$600. *Check*: 15% × \$600 = \$90

2 Find the whole amount if:

a 5% of the amount is \$35

b 11% of the amount is \$88

c 20% of the amount is 80

d 6% of the amount is 42

e 90% of the amount is \$270

f 15% of the amount is \$60

g 40% of the amount is 100

h 120% of the amount of \$360

i 25% of the amount is \$75

j 8% of the amount is 40

b Calculate, as a decimal, the theoretical probability that the needle points to:

i red **ii** green

c Are the experimental and theoretical probabilities similar?

d For 200 spins, what is the expected frequency of red or blue based on the theoretical probability? How does this compare with the observed frequency?

2 R C Use cardboard and a toothpick to make a square spinner that has 4 colours as shown in the diagram below. Spin the spinner 50 times and record your results in a table similar to the one shown.

Event	Tally	Frequency
Red		
Blue		
Green		
Yellow		

a Calculate, as a percentage, the relative frequency of spinning:

i red **ii** blue

iii a colour that is not yellow **iv** blue or green

b Calculate, as a percentage, the theoretical probability of spinning blue. How does this compare with the relative frequency?

c For 50 spins, what is the expected frequency that blue should come up? How does this compare with the observed frequency?

3 R C A die is rolled.

a Predict the number of times a 2 would appear if the die was rolled 100 times.

b Roll a die 100 times and record your results in a table similar to the one shown.

c What is the relative frequency of rolling a 2?

d How does your predicted frequency compare with the observed frequency?

e What is the relative frequency of rolling an even number?

Event	Tally	Frequency
1		
2		
3		
4		
5		
6		
		100

4 R C Place 4 black counters, 5 red counters and 1 yellow counter in a bag. Select a counter at random, record the colour and return the counter to the bag. Run this experiment 50 times.

a What is the experimental probability of selecting:

i a black counter?

ii a red counter?

iii a yellow counter?

Event	Tally	Frequency
Black		
Red		
Yellow		

b What is the theoretical probability of selecting:

i a black counter?

ii a red counter?

iii a yellow counter?

c Compare your experimental probabilities to the theoretical probabilities.

d Repeat the experiment another 50 times and find the relative frequency of each event. Are the results different to the results found in part **a**?

5 A die was repeatedly rolled and the results are shown in the table.

Outcome	Frequency
1	90
2	84
3	92
4	76
5	83
6	75

a What was the total number of trials?

b Find the experimental probability (as a decimal) of rolling:

i an even number.

ii a number greater than 3.

iii a 3 or a 5.

iv a number greater than 2 and even.

vii a number greater than 3 or odd.

c Calculate the probability of rolling a number not less than 3:

i as an experimental probability.

ii as a theoretical probability.

d How do the 2 probabilities in part **c** compare?

6 Students at a high school were asked how they travelled to school. The results are presented in the table. Find the probability (correct to 2 decimal places) that a student chosen at random will:

Outcome	Frequency
Walk	186
Bus	295
Car	103
Bicycle	74

a walk **b** be driven **c** catch a bus

d not ride a bicycle **e** be driven or catch a bus.

7 A die was rolled many times and the results are shown in the frequency histogram.

a What is the total number of trials?

b Find the experimental probability of rolling:

i 2

ii 5

iii an even number or a number greater than 3.

iv a number greater than 4 and odd.

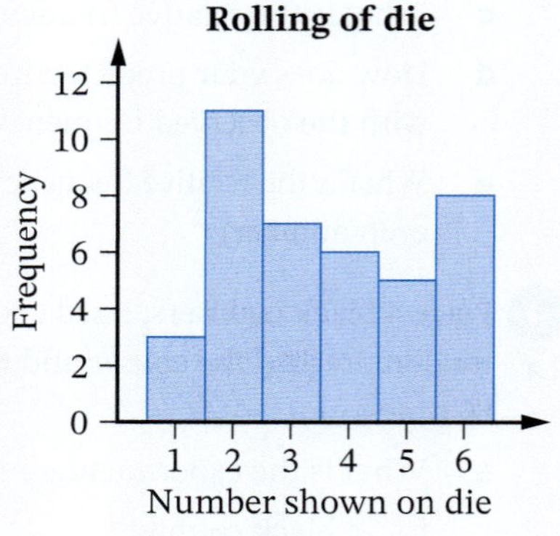

Foundation Standard Complex

8 The table shows the number of babies born per day at a hospital during the month of January. On any particular day, find the probability of the birth of exactly:

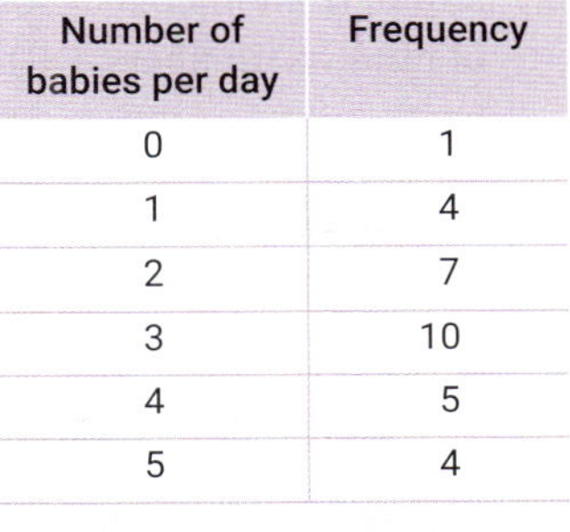

Number of babies per day	Frequency
0	1
1	4
2	7
3	10
4	5
5	4

a 1 baby **b** 3 babies
c 1 or 2 babies **d** **at least** 2 babies
e at most 1 baby **f** 4 or 5 babies.

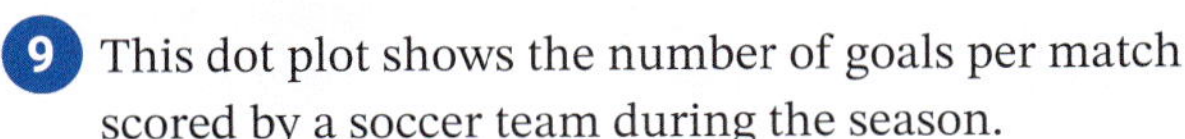

9 This dot plot shows the number of goals per match scored by a soccer team during the season.

Number of goals per match

For a match selected from the season at random, find the probability of the team scoring:

a 2 goals **b** 4 goals per game **c** fewer than 2 goals
d 2 or 3 goals **e** at least 3 goals **f** not more than 4 goals

10 R C A card is drawn randomly from a normal pack of cards to see if it is an Ace. This trial is run 120 times, with each card being replaced in the deck before the next card is drawn. The results are shown in the table.

Outcome	Frequency
Ace	11
Not an Ace	109

a Calculate, correct to 3 decimal places:

i the experimental probability of selecting an Ace.

ii the theoretical probability of selecting and Ace.

b Are the experimental and theoretical probabilities similar?

c If a card is drawn 300 times, what is the expected frequency of an Ace being drawn? Select the correct answer **A**, **B**, **C** or **D**.

A 27 **B** 28 **C** 23 **D** 6

TECHNOLOGY

Tossing a coin

In this activity, a spreadsheet is used to simulate the tossing of a coin. The computer can quickly generate either a 1 (representing heads) or a 2 (tails).

1 In cell A1, enter the label 'Tossing a coin'.

2 In cell A2, enter the formula =INT(RAND()*2+1), press Enter and either 1 or 2 will appear in the cell.

3 Select cell A2 and Fill Down to cell A11 to simulate the tossing of a coin 10 times.

4 Copy the table and in the first blank row record your results for Trials 1–10 (the numbers of heads and tails in your first 10 'tosses').

Worksheet
Coins probability

Foundation Standard Complex

5 Simulate another 10 tosses of the coin by making the spreadsheet generate another set of random numbers. Do this by placing the cursor between the tops of columns A and B (so that it turns into a 'double arrow') and clicking.

6 Enter the results for Trials 11–20 in your table.

7 Repeat 8 more times so that you have 100 tosses of the coin recorded in the table.

8 a For 100 tosses of a coin, what is the expected frequency of heads based on the theoretical probability?

b Compare this with the actual frequency.

9 Calculate, as a decimal, the theoretical and experimental probabilities of tossing heads.

10 Compare your results with those of other students in your class. Briefly explain any differences and why they may have occurred.

11 Combine the results of students in your class to calculate the experimental probability of throwing heads, as a decimal. How does this compare with the theoretical probability?

Trial	Number of heads	Number of tails
1–10		
11–20		
21–30		
31–40		
41–50		
51–60		
61–70		
71–80		
81–90		
91–100		
Total		

TECHNOLOGY

Rolling dice

Worksheet Dice probability

In this activity, we will use a scientific calculator to simulate the rolling of a die.

1 Copy the table shown.

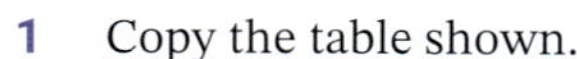

Outcome	Tally	Frequency	Experimental probability
1			
2			
3			
4			
5			
6			

2

Casio scientific calculator	Sharp scientific calculator
Enter the formula RanInt#(1,6) as shown: ALPHA · (1 SHIFT) 6) =	2nd F 7 (RANDOM), select 1 R-DICE, =

This generates a random number from 1 to 6. Press = more times to generate more random numbers. Alternatively, on a spreadsheet, type **=INT(RAND()*6+1)** into a cell and press the Enter key, then the F9 key repeatedly for more random numbers.

12.02

3 Run the simulation 50 times and record the results in the table.

4 Calculate the theoretical probability for each number being tossed on the die.

5 Compare the theoretical probability and experimental probability for each number. Are there similarities or differences? How could the experimental probabilities be more accurate? Explain.

6 Copy this table for rolling 2 dice and finding their sum.

Sum	Tally	Frequency	Experimental probability
2			
3			
4			
5			
6			
7			
8			
9			
10			
11			
12			

7 On a scientific calculator, rolling 2 dice is achieved by adding the result of rolling one die twice.

Casio scientific calculator	Sharp scientific calculator
Enter the formula RanInt#(1,6) + RanInt#(1,6) as shown below: ALPHA . (1 SHIFT) 6) + ALPHA . (1 SHIFT) 6) =	Enter the formula R.DICE + R.DICE as shown below: 2nd F 7 (RANDOM), select 1 R-DICE, + 2nd F 7 (RANDOM), select 1 R-DICE, =

This generates the sum of 2 random numbers from 1 to 6.

Press = more times to generate more sums.

8 Run the simulation 50 times and record the results in the table.

9 Which sum is the most likely to occur?

10 Which sum is the least likely to occur? Why?

11 Which sum is theoretically the:

a most likely?

b least likely?

Why?

12 Compare your answers to questions **9**, **10** and **11** and then compare them with the answers of other students in your class. Write a few sentences explaining your observations.

13 If you ran 50 more simulations, would you expect your results to be similar or different to your first 50 results? Explain.

12.03 Venn diagrams

Videos
Venn diagrams 2
Venn diagrams: Global habitats

Worksheet
Venn diagrams

Puzzles
Venn diagrams matching activity
Venn diagrams

A **Venn diagram** is a diagram of circles (usually overlapping) for grouping items into categories. A rectangle represents the whole group, while the circles represent categories. Items common to 2 or more categories are placed in the intersection (overlapping region) of the circles. The Venn diagram was invented in 1880 by English mathematician and priest, John Venn (1834–1923).

Example 5

This Venn diagram shows the results of a survey on how people travel to work (car (C), bus (B) or train (T)).

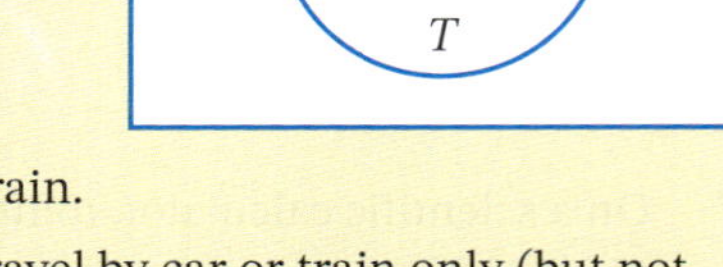

a How many people participated in the survey?

b How many people used all 3 methods: car, bus *and* train? Give an explanation on how this might happen.

c Calculate the probability of selecting a person from this survey who travels by:

i car only or bus only　　**ii** car and train

iii car and train only　　**iv** bus but not train.

d A person is chosen at random from the people who travel by car or train only (but not bus). What is the probability that the person travels by car only?

e If 120 people actually participated in the survey, what extra information would be needed on the diagram?

SOLUTION

a Number of people $= 27 + 12 + 5 + 20 + 21 + 4 + 11$

$= 100$

b 5 people travelled by car, bus and train.

The region where the 3 circles intersect.

They may have driven to the station, caught a train and then a bus to the workplace, or used different methods on different days.

c **i** People travelling by car only or bus only

The regions in C and B that don't overlap.

$= 27 + 11$

$= 38$

$P(\text{car only or bus only}) = \frac{38}{100}$

$= \frac{19}{50}$

'Car only or bus only' is an example of a compound event.

ii People travelling by car and train

The region where C and T intersect.

$= 20 + 5$

$= 25$

$P(\text{car and train}) = \frac{25}{100}$

$= \frac{1}{4}$

'Car and train' is another example of a compound event.

iii People travelling by car and train only = 20 — Where C and T intersect, but excluding B.

$P(\text{car and train only}) = \frac{20}{100}$

$= \frac{1}{5}$

iv People travelling by bus but not train — The circle B excluding T.

$= 11 + 12$

$= 23$

$P(\text{bus but not train}) = \frac{23}{100}$

d People travelling by car or train only = 27 + 20 + 21

= 68

People travelling by car only = 27.

$P(\text{car only}) = \frac{27}{68}$

e 120 – 100 = 20. This means that there were 20 people who did not use a car, bus or train. A '20' would need to be placed in the rectangle, but outside of the 3 circles.

When we combine 2 or more simple events, we get a **compound event**. In the above example, 'car and train' and 'bus but not train' are compound events.

'And' vs 'or'

For 2 categories or events A and B, the compound event **'A and B'** means to have both of them occurring together. For example, 'to drive a car' **and** 'to ride a bus' means to do both things.

If A and B are **overlapping**, the compound event **'A or B'** means to have A or B or both. For example, 'to drive a car' **or** 'to ride a bus' means to drive a car only, or to ride a bus only, or to do both. In this case, 'A or B' actually **includes** 'A and B' so this is an example of an **inclusive** 'or'.

If A and B are **mutually exclusive**, this means that they are **not overlapping** and on a Venn diagram they appear as 2 separate circles. For mutually exclusive categories or events, the phrase **'A or B'** means to have A only or B only (but not both). For example, 'male' **or** 'female' means to be male, or female, but not both. In this case, 'A or B' **excludes** 'A and B' so this is an example of an **exclusive** 'or'.

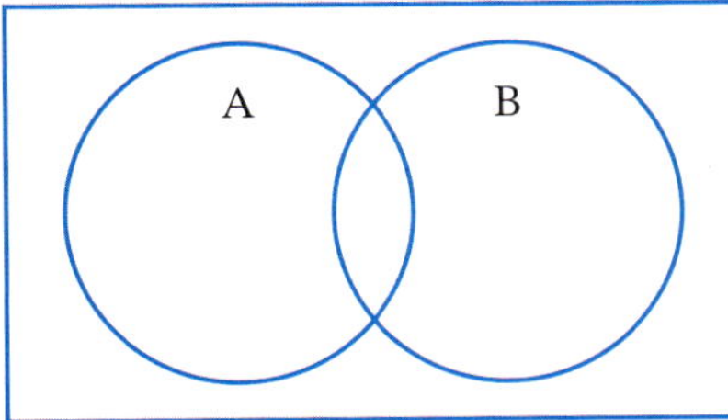

Overlapping events: 'A or B' means A or B or both

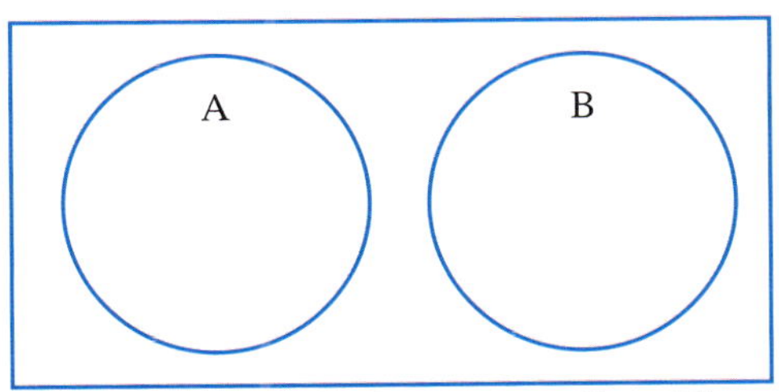

Mutually exclusive events: 'A or B' means A or B but not both

Video
Venn diagrams 1

Example 6

A survey of 110 students at Redford College showed that 61 students had dark hair, 42 students had brown eyes and 20 students had dark hair and brown eyes.

Left to right: Shutterstock.com/michaeljung, Alamy Stock Photo/blickwinkel, Shutterstock.com/Dorota Zietek

a Represent this information on a Venn diagram.

b How many students had brown eyes or dark hair but not both?

c What is the probability of randomly selecting from this group a student:

i with dark hair? **ii** with dark hair and brown eyes?

iii with dark hair or brown eyes? **iv** with neither dark hair nor brown eyes?

SOLUTION

a RC = students at Redford College

DH = students with dark hair

BE = students with brown eyes

When completing a Venn diagram, always begin with the intersection.

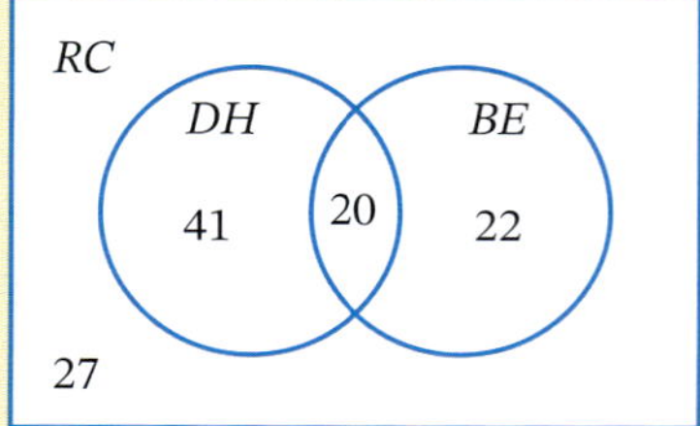

There are 20 students with dark hair and brown eyes.

$\therefore$ Students with dark hair only $= 61 - 20$
$= 41$

$\therefore$ Students with brown eyes but not dark hair $= 42 - 20$
$= 22$

$\therefore$ (Remaining) Students with neither dark hair nor brown eyes $= 110 - 41 - 20 - 22$
$= 27$

b Number of students with brown eyes or dark hair only $= 22 + 41 = 63$

c **i** Students with dark hair $= 41 + 20$
$= 61$

$P(\text{dark hair}) = \frac{61}{110}$

ii $P(\text{dark hair and brown eyes}) = \frac{20}{110}$
$= \frac{2}{11}$

'Dark hair and brown eyes' is a compound event.

iii Number of students with dark hair or brown eyes $= 41 + 20 + 22$
$= 83$

$P(\text{dark hair or brown eyes}) = \frac{83}{110}$

iv $P(\text{neither dark hair nor brown eyes}) = \frac{27}{110}$

9780170465618

EXERCISE 12.03 ANSWERS ON P. 676

Venn diagrams

U F R C

EXAMPLE 5

1 The Venn diagram shows the number of students who play either basketball (B) or volleyball (V).

R

a Find the total number of students who play basketball or volleyball.

b Find the probability of randomly selecting a student who plays:

i basketball

ii volleyball.

c What is the probability of selecting a student who plays:

i basketball or volleyball?

ii basketball and volleyball?

iii basketball but not volleyball?

iv neither basketball nor volleyball?

d Of the students who play volleyball, find the probability of selecting a student who also plays basketball. Select the correct answer **A**, **B**, **C** or **D**.

A 0.28 **B** 0.375 **C** 0.72 **D** 0.92

2 The Venn diagram shows the number of Year 9 students who achieved Credit grades in their Science (S) and History (H) exams.

R

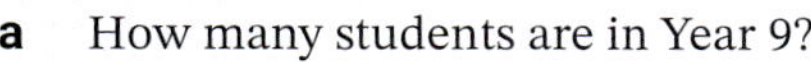

a How many students are in Year 9?

b Find the probability of selecting a student who achieved a Credit in:

i History

ii Science

iii History and Science

iv History or Science

v History or Science but not both

vi History only

vii Science only

c What is the percentage probability of selecting a student who did not achieve a Credit in both History and Science?

d Out of the students who achieved a Credit in History, what is the probability that the student also achieved a Credit in Science?

3 40 people were asked whether they preferred watching movies at the cinema (C) or on a device (D) at home. The results are shown in the Venn diagram.

R C

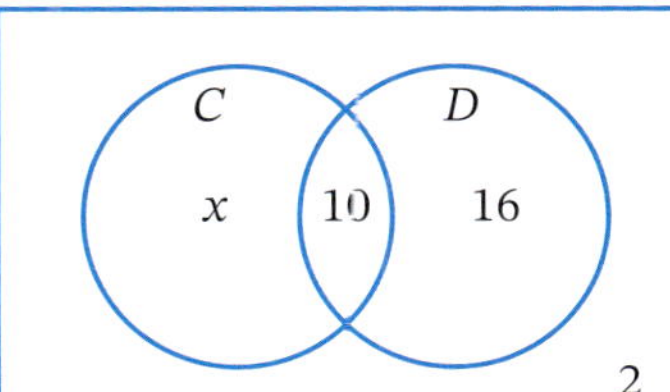

a Describe the category that contains the pronumeral x in the diagram.

b Find the value of x.

Foundation Standard Complex

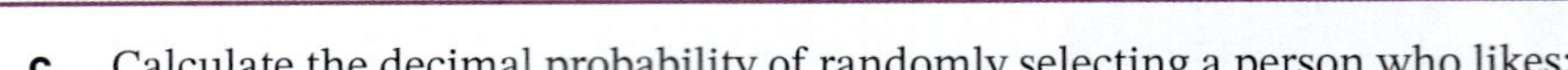

c Calculate the decimal probability of randomly selecting a person who likes:

- **i** the cinema and a device.
- **ii** a device only.
- **iii** the cinema or a device.
- **iv** neither the cinema nor a device.

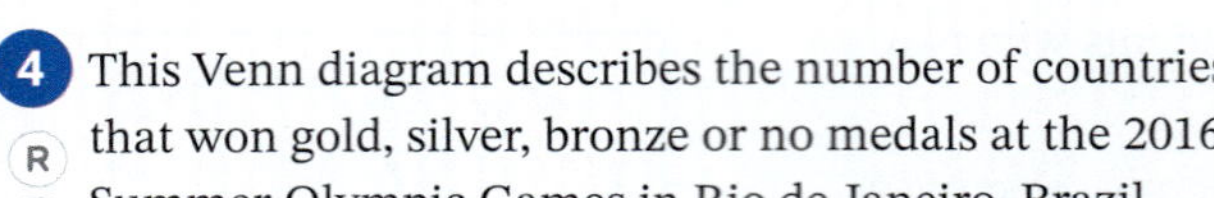

4 R C This Venn diagram describes the number of countries that won gold, silver, bronze or no medals at the 2016 Summer Olympic Games in Rio de Janeiro, Brazil.

G 6, *S* 7, *B* 13, G∩S only 6, G∩B only 4, S∩B only 7, G∩S∩B 43, outside 110

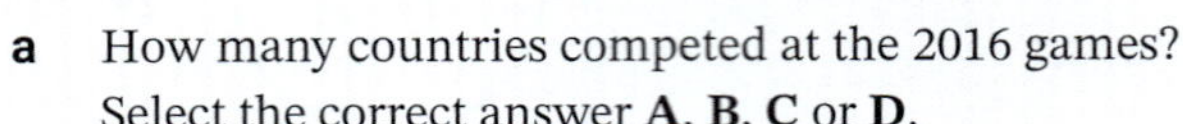

a How many countries competed at the 2016 games? Select the correct answer **A**, **B**, **C** or **D**.

A 86 **B** 183 **C** 189 **D** 196

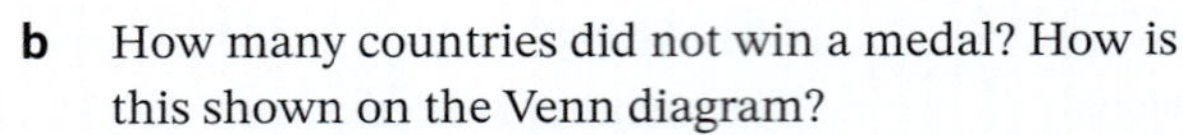

b How many countries did not win a medal? How is this shown on the Venn diagram?

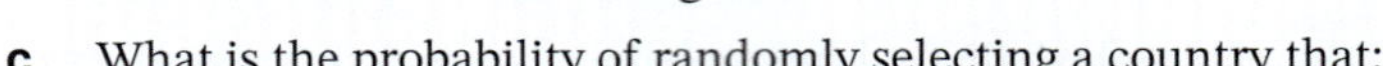

c What is the probability of randomly selecting a country that:

- **i** won only gold medals?
- **ii** won gold or silver medals, but not both?
- **iii** won gold, silver and bronze medals?
- **iv** won gold and bronze only?
- **v** won gold only or silver only or bronze only?
- **vi** did not win a gold medal?
- **vii** did not win a gold or silver medal?

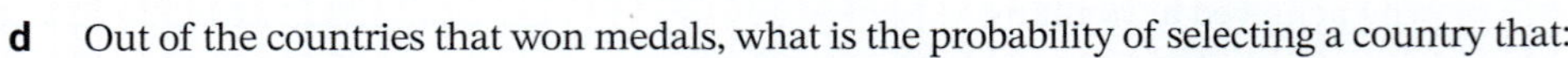

d Out of the countries that won medals, what is the probability of selecting a country that:

- **i** won gold medals? **ii** won silver but not bronze?

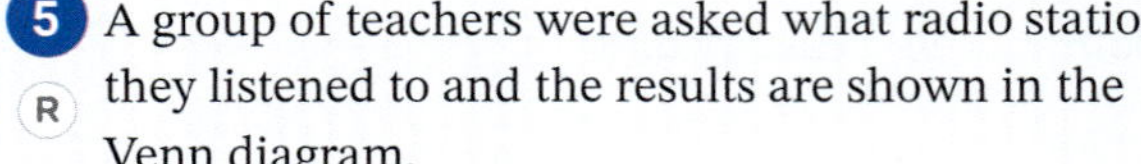

5 R A group of teachers were asked what radio stations they listened to and the results are shown in the Venn diagram.

JJJ 13, **MMM** 14, **Nova** 4, JJJ∩MMM only 15, JJJ∩Nova only 8, MMM∩Nova only 12, all three 10, outside 3

a How many teachers were surveyed?

b What is the probability of selecting a teacher who did not listen to the 3 stations?

c What is the probability of selecting a teacher who listens to:

- **i** JJJ? **ii** MMM? **iii** Nova?

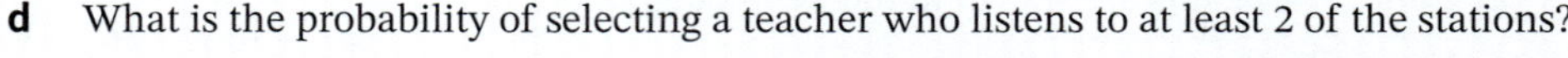

d What is the probability of selecting a teacher who listens to at least 2 of the stations?

A $\frac{50}{79}$ **B** $\frac{45}{79}$ **C** $\frac{35}{79}$ **D** $\frac{30}{79}$

Foundation Standard Complex

9780170465618

e Find the probability of selecting a teacher who listens to:

i JJJ or MMM but not both.

ii all 3 stations.

iii MMM only.

6 In a drama group of 30 people, 18 people dance (D), 20 people sing (S) and 8 people dance and sing.

R C

a Show this information on a Venn diagram.

b Find the probability of selecting a person who:

i dances and sings.

ii sings but does not dance.

iii dances but does not sing.

iv dances or sings.

v dances or sings but not both.

7 A class of 30 students were given the option of playing an indoor sport (S) or watching a movie (M) when regular sport was cancelled due to bad weather. 15 students wanted to play an indoor sport and 24 students wanted to watch a movie.

R C

a How many students indicated that they would do either?

b Copy and complete a Venn diagram using the information given.

c For this Venn diagram, there are no values outside the circles. Explain why.

d What is the probability of selecting a student who prefers to:

i play an indoor sport or watch a movie, but not both?

ii watch a movie but not play an indoor sport?

iii play an indoor sport but not watch a movie?

8 80 families living in Mt Robinson were surveyed on whether they had a dog (D) or a cat (C) as a pet. 12 families had no pets, 35 had a dog and 47 had a cat.

R C

a Draw Venn diagram showing the information given.

b How many families had:

i a dog and a cat?

ii a dog but not a cat?

iii at least a dog or a cat?

iv a dog or a cat, but not both?

c Find the decimal probability of selecting a family with:

i a dog only

ii either a dog or a cat, but not both

iii at most one pet

iv at least one pet.

d Given that a family has a pet, what is the probability that it is a cat?

9 At Nelson River High School, the Student Representative Council (SRC) has 25 members, the Chess club has 20 members and the Environment club has 33 members. 5 students belong to all 3 groups, while 13 members belong to the SRC and no other club. 4 members of the Environment club are members of the SRC, but not the Chess club, while 10 members belong to both the Environment and Chess clubs. There are 8 members of the Chess club who are members of the SRC.

R C

- **a** Copy and complete the Venn diagram.
- **b** How many students belong to the Environment club only?
- **c** How many students belong to the Chess club and no others?
- **d** Find the probability of selecting a student who belongs to exactly 2 clubs.

SRC C E

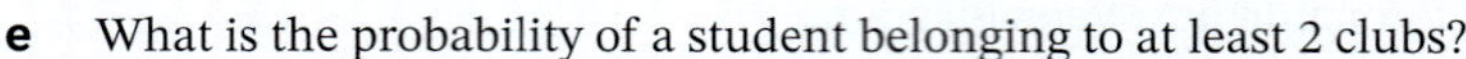

- **e** What is the probability of a student belonging to at least 2 clubs?
- **f** Find the probability of selecting a student who belongs to:
  - **i** the Environment club and the SRC, but not the Chess club.
  - **ii** only one club.
  - **iii** the SRC or the Chess club, but not the Environment club.
  - **iv** at most 2 clubs.
- **g** Is it necessary to include the rectangle in this Venn diagram? Give reasons.

12.04 Two-way tables

Worksheet
Two-way tables

Two-way probability tables

A **two-way table** is another way of grouping items into overlapping categories, especially when there are many overlaps that cannot easily be represented by Venn diagrams.

Example 7

This table categorises the heights and ages of 200 Year 9 students.

	Short	Tall
Aged 14	45	25
Aged 15	90	40

- **a** How many students are:
  - **i** aged 15 and tall?
  - **ii** aged 14 or short?
- **b** What fraction of students are neither tall nor aged 15?
- **c** What is the probability of selecting a student at random who is:
  - **i** aged 14 and tall?
  - **ii** aged 15 but not tall?
  - **iii** tall or aged 14, but not both?
- **d** If a short student is selected at random, what is the probability that he or she is aged 14?

SOLUTION

a **i** Students aged 15 and tall $= 40$

ii Students aged 14 or short $= 45 + 25 + 90$

$= 160$

☐ Foundation ○ Standard ⬡ Complex

b Students who are neither tall nor 15 = 45 — The short 14-year-olds

∴ Fraction of students neither tall nor 15

$$= \frac{45}{200} = \frac{9}{40}$$

c **i** $P(\text{tall and } 14) = \frac{25}{200} = \frac{1}{8}$

ii $P(\text{15 but not tall}) = \frac{90}{200} = \frac{9}{20}$

iii Students tall or aged 14, but not both = 40 + 45 = 85

$P(\text{tall or 14, but not both}) = \frac{85}{200} = \frac{17}{40}$

d Short students = 90 + 45 = 135

Students who are short and 14 = 45.

$P(\text{short student who is 14}) = \frac{45}{135} = \frac{1}{3}$

EXERCISE 12.04 ANSWERS ON P. 676

Two-way tables

U F R C

1 Households were surveyed at random to see whether or not they had dishwashers and clothes dryers. The results are in the table.

R C

EXAMPLE 7

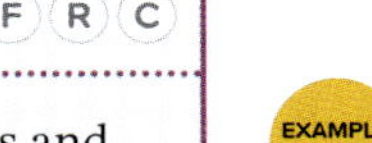

	Dishwasher	No dishwasher
Clothes dryer	75	108
No clothes dryer	49	68

a How many households were surveyed?

b How many households had a clothes dryer and a dishwasher?

c How many households had neither?

d What is the probability of selecting a household at random that has:

i a dishwasher?

ii a dishwasher and a dryer?

iii a dishwasher or a clothes dryer, but not both?

iv at least one of them?

2 People at a train station were asked whether they ate breakfast or lunch that day.

R C

	Ate lunch	Did not eat lunch
Ate breakfast	35	26
Did not eat breakfast	54	15

a How many people were surveyed?

b What percentage of people ate lunch?

c Find the probability of picking a person at random who ate:

i breakfast
ii lunch
iii breakfast but not lunch
iv breakfast or lunch
v breakfast and lunch
vi breakfast or lunch, but not both

d What is the probability of picking a person who has eaten lunch given that the person ate breakfast? Select the correct answer **A**, **B**, **C** or **D**.

A $\frac{26}{61}$ **B** $\frac{35}{61}$ **C** $\frac{35}{89}$ **D** $\frac{89}{130}$

e What is the probability of picking a person who has not eaten lunch given that the person did not eat breakfast?

3 People at a shopping mall were asked if they watched football on TV.

R C

	Football	No football
Male	85	27
Female	36	92

a How many people were surveyed?

b How many people:

i watched football?
ii did not watch football?

c What percentage of people surveyed were females?

d What is the probability of randomly selecting a person who did not watch football and was male?

e Given that a person watched football, what is the probability that the person was female?

4 The results comparing gender and handedness of students at a primary school are shown in the table.

R C

	Male	Female
Left-handed	14	18
Right-handed	86	65

a How many students are at the school?

b What is the probability of selecting a student at random who is:

i male and left-handed?
ii male or left-handed?

c Find the probability, expressed as a percentage, of randomly selecting a right-handed female.

d Given that a student is a male, what is the probability that he is right-handed?

e What is the probability of selecting a female or a right-handed student?

5 After the 2019 Federal election, the members (politicians) of the House of Representatives in the Australian Parliament were as shown.

	Government	Opposition
Male	62	43
Female	15	31

☐ Foundation ○ Standard ⬡ Complex

12.04

a How many members of parliament (MPs) were there in the House of Representatives?

b Find the percentage probability of randomly selecting an MP who is:

i male
ii female and in Opposition
iii in the Government or female
iv male and in Opposition.

c Find the probability of selecting an MP who is male and in Government.

d If a female MP is selected at random, what is the probability that she is in Government?

6 R C Students at Cengage Anglican College were surveyed on whether they played video games. The results are shown in the table.

	Plays video games	Does not play video games
Female	178	257
Male	310	210

a How many students attended the college?

b Find the percentage probability that a student selected at random from the college:

i plays video games and is female.
ii doesn't play video games.
iii is male and doesn't play video games.
iv is female and doesn't play video games.

c If a female student is selected at random, what is the probability that she plays video games?

d If a male student is selected at random, what is the probability that he plays video games?

7 R C A group of retired people was surveyed on whether they exercised regularly.

	Exercise	Don't exercise
Men	65	106
Women	74	75

a How many people were surveyed?

b What percentage of people exercised?

c What is the probability of selecting a person at random who doesn't exercise regularly?

d Of the women who were surveyed, what percentage exercised regularly?

e If a man was selected randomly from this survey, what is the probability that he exercises regularly? Select the correct answer **A**, **B**, **C** or **D**.

A 38.0%
B 62.0%
C 20.3%
D 53.4%

☐ Foundation ○ Standard ○ Complex

INVESTIGATION

Using two-way tables

Work in groups of 3 or 4.

Investigate 2 characteristics, for example, gender and exercise, gender and eating habits, eating habits and exercise, playing sport and going to the movies, eye colour and hair colour.

Conduct a survey to obtain the information required, then summarise your results in a two-way table similar to this one:

	Eat regular meals	Don't eat regular meals
Play sport		
Don't play sport		

Note any patterns in your data. This may include answering questions such as 'What percentage of …' or 'What is the probability of picking a person who …'.

Quiz
Mental skills
12

☆ MENTAL SKILLS 12 ANSWERS ON P. 677 Maths without calculators

The unitary method with percentages

The unitary method is used when you are only given a *percentage* of an amount and you need to find the amount. It is called the unitary method because we find 1% of the amount first, then multiply that by 100 to find the whole (100%).

1 Study each example.

a If 8% of a number is 24, what is the number?

8% of the number = 24

$\therefore$ 1% of the number = 24 ÷ 8 = 3

$\therefore$ 100% of the number = 3 × 100 = 300

The number is 300. *Check*: 8% × 300 = 24

b If 15% of an amount is \$90, what is the whole amount?

15% of the amount = \$90

$\therefore$ 1% of the amount = \$90 ÷ 15 = \$6

$\therefore$ 100% of the amount = \$6 × 100 = \$600

The amount is \$600. *Check*: 15% × \$600 = \$90

2 Find the whole amount if:

a 5% of the amount is \$35

b 11% of the amount is \$88

c 20% of the amount is 80

d 6% of the amount is 42

e 90% of the amount is \$270

f 15% of the amount is \$60

g 40% of the amount is 100

h 120% of the amount of \$360

i 25% of the amount is \$75

j 8% of the amount is 40

9780170465618

INVESTIGATION

Are the Lotto numbers equally likely?

In each Lotto game, 8 numbers are selected at random from the barrel: 6 main numbers, plus 2 supplementary numbers. Between January 2015 and October 2019, there were 2000 Lotto numbers drawn on the NSW Saturday game. The frequency of each number (from 1 to 45) was as follows:

1	53	10	50	19	36	28	36	37	38
2	37	11	53	20	45	29	49	38	47
3	39	12	41	21	43	30	47	39	47
4	41	13	44	22	53	31	42	40	35
5	53	14	41	23	43	32	48	41	39
6	44	15	47	24	47	33	45	42	45
7	46	16	51	25	39	34	45	43	35
8	41	17	41	26	52	35	39	44	45
9	40	18	57	27	52	36	46	45	43

a If 2000 numbers were drawn over the 5 years, and each number from 1 to 45 was equally likely, how often would you expect each number to come up?

b How many of the frequencies from the above table are equal to this expected number?

c Which Lotto number came up the:

i most often? **ii** least often?

d Do you think the Lotto numbers are equally likely, or is the Lotto draw biased? Give a reason for your answer.

Source: www.thelott.com

Tree diagrams 12.05

A **two-step experiment** or **two-stage experiment** is a chance experiment that has 2 parts or stages, for example:

- rolling 2 dice
- tossing a coin and die together
- drawing 2 prizes in a raffle
- observing the weather each day over a weekend.

The outcome of the second step or stage may or may not be affected by the outcome of the first step. The sample space for a two-step experiment can be displayed using lists, tables or tree diagrams.

Worksheet Tree diagrams

Presentation Combined events

Videos Tables and tree diagrams

Tree diagrams 3

Example 8

Two coins are tossed together.

a Find all the outcomes in the sample space:

  i by simply listing them **ii** using a table.

b What is the probability of tossing:

  i a head and a tail? **ii** 2 tails? **iii** at least one head?

SOLUTION

a Let H = heads and T = tails.

  i The possible outcomes are {HH, HT, TH, TT}. Finding the 4 outcomes by listing.

  ii

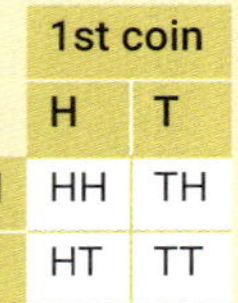

		1st coin	
		H	T
2nd coin	H	HH	TH
	T	HT	TT

Using a table ensures that all outcomes are counted.

b **i** $P(\text{a head and a tail}) = P(\text{HT or TH}) = \frac{2}{4} = \frac{1}{2}$ 2 chances out of 4.

  ii $P(\text{2 tails}) = P(TT) = \frac{1}{4}$ 1 chance out of 4.

  iii $P(\text{at least one head}) = P(\text{1 or 2 heads})$ 'At least one' means '1 or more'.

  $= P(\text{HT or TH or HH})$ 3 chances out of 4.

  $= \frac{3}{4}$

When there are a large number of outcomes in each step of a two-step experiment, it is best to use a table to list all the possible outcomes. The table is a systematic way of ensuring all possible outcomes have been included in the sample space.

Example 9

While staying at a hotel, John-Paul can choose one item from each course of a breakfast menu.

1st course	2nd course
Cereal (C)	Bacon and eggs (B)
Raisin toast (R)	Pancakes (P)
Watermelon (W)	Sausages and hash browns (S)
Yoghurt (Y)	

a Use a table to list all the different two-course breakfasts available.

b If one of these combinations is chosen at random, find the probability that it includes:

  i watermelon but not pancakes 'Watermelon, not pancakes' is a compound event.

  ii yoghurt or pancakes

  iii yoghurt and pancakes.

SOLUTION

a

		1st course			
		C	R	W	Y
2nd course	B	CB	RB	WB	YB
	P	CP	RP	WP	YP
	S	CS	RS	WS	YS

$4 \times 3 = 12$ different breakfasts possible.

b **i** $P(\text{watermelon, not pancakes}) = \frac{2}{12} = \frac{1}{6}$ WB or WS

ii $P(\text{yoghurt or pancakes}) = \frac{6}{12} = \frac{1}{2}$ Column Y + row P

iii $P(\text{yoghurt and pancakes}) = \frac{1}{12}$ YP

A **tree diagram** lists all the possible outcomes of each stage. Branches stretch out to show the possible pathways of outcomes at each step or stage. The outcomes column at the end of the diagram lists the sample space.

Example 10

From a set of cards numbered 1 to 5, two cards are drawn at random *with replacement* to make a 2-digit number.

'With replacement' means that the first card is returned to the set before the second draw, which means that the same card could be drawn twice.

a Show all outcomes in the sample space using a tree diagram.

b What is the probability of making a number:

i less than 32? **ii** divisible by 3?

SOLUTION

a This experiment has 2 steps: the 1st digit and the 2nd digit.

For the 1st digit, there are 5 possible digits: 1, 2, 3, 4 or 5.

For the 2nd digit, there are also 5 possible digits: 1, 2, 3, 4 or 5.

There are $5 \times 5 = 25$ possible 2 digit numbers in the sample space.

Using a tree diagram ensures that all outcomes are counted.

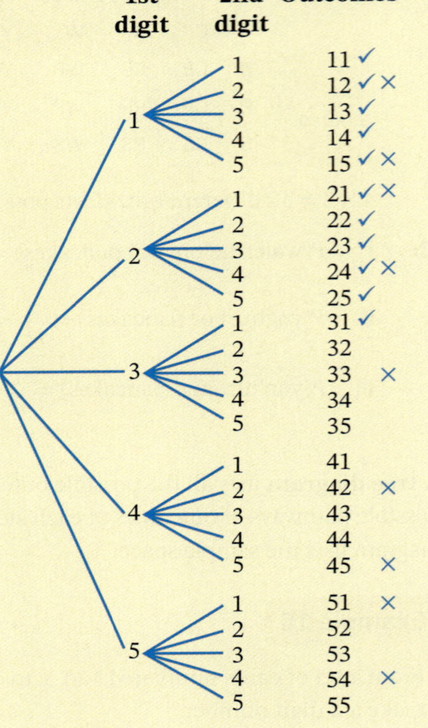

b **i** There are 11 numbers that are less than 32 (ticked in the tree diagram).

$\therefore P(\text{number} < 32) = \frac{11}{25}$.

ii There are 9 numbers that are divisible by 3 (crossed in the tree diagram).

$\therefore P(\text{number divisible by 3}) = \frac{9}{25}$.

Example 11 in the next section will revisit the above example but with the selection being 'without replacement', that is, the first card is *not* returned to the set.

EXERCISE 12.05 ANSWERS ON P. 677

Tree diagrams

U F PS R C

EXAMPLE 8

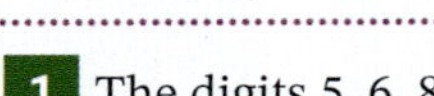

1 PS R C The digits 5, 6, 8 and 9 are written on cards and put in a bag. Two cards are drawn at random *with replacement* to make a 2-digit number.

a Make a list of all possible outcomes.

b How many 2-digit numbers are possible?

c Find as a decimal the probability that the number is:

i even **ii** greater than 60

iii between 60 and 90 **iv** divisible by 3

Foundation Standard Complex

EXAMPLE 9

12.05

2 A coin and a die are tossed together.

PS R C

		Die					
		1	2	3	4	5	6
Coin	H						
	T						

a Copy and complete the table to display the sample space.

b Find the probability of tossing:

i a head and 4

ii a tail and an even number

iii a tail and a number less than 5.

3 Two dice are rolled and the sum of the numbers is calculated.

PS R C

a Copy and complete this table to show all possible sums.

		1st die					
		1	2	3	4	5	6
2nd die	1						
	2					7	
	3						
	4		6				
	5						
	6						

b Find the probability of rolling a sum

i of 3

ii of 7

iii of at least 6

iv less than 10

v that is even

vi between 3 and 7.

EXAMPLE 10

4 A bag contains 2 blue balls, 1 yellow ball and 1 green ball. Two balls are drawn out of the bag at random, one at a time *with replacement*.

PS R C

a Copy and complete the tree diagram to list the sample space.

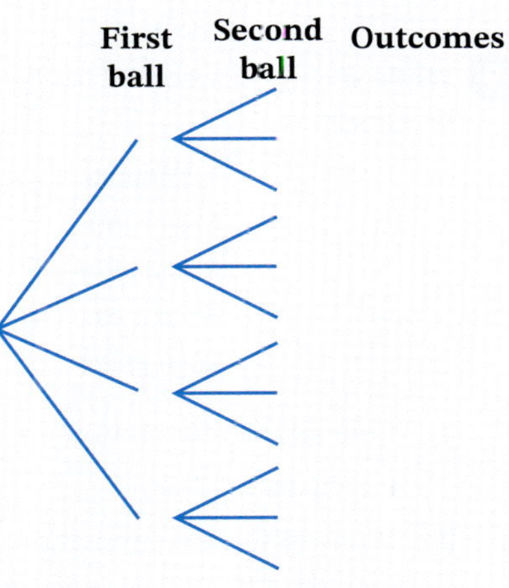

b Find the probability of drawing:

i a blue and green ball, in that order.

ii a yellow ball and a green ball, in any order.

iii a blue ball and a yellow ball, in any order.

iv at least one blue ball.

v 2 balls the same colour.

Foundation Standard Complex

5 The numbers 3, 4, 5 and 8 are written on separate cards and placed in a bag. Two cards are drawn at random with replacement to form a 2-digit number.

PS R C

a Use a tree diagram to show all possible outcomes if the cards are drawn *with replacement*, that is, the same card can be used twice.

b Find the probability that the two-digit number formed:

i ends in 4

ii is an odd number

iii is less than 45

iv begins with 3 and is even.

6 The weather this weekend will be either fine or rainy each day, with each outcome being equally likely.

PS R C

a Draw a tree diagram to show the possible outcomes for Saturday and Sunday.

b What is the probability that:

i it is fine on both days?

ii Saturday is fine and it rains on Sunday?

iii it rains on at least one day of the weekend?

7 Two cards are chosen at random with replacement from a set of 4 cards consisting of the 2 red Kings and 2 red Queens from a normal deck of cards. Calculate the probability that the cards are:

PS R C

a both Kings

b a King and a Queen

c a King and Queen of the same suit

d both Queens

e at least one King

f no Queens

8 Draw a table for the sample space when 2 dice are rolled. Find the probability of rolling:

PS R C

a 2 numbers the same

b a 3 and a 6 (in any order)

c at least one 2

d a 4 or a 5 on only one of the dice

9 While staying at a hotel, Phoebe can choose one item from each course of a breakfast menu.

PS R C

1st course	2nd course
Muesli (M)	Bacon and eggs (B)
Raisin toast (R)	Pancakes (P)
Watermelon (W)	Sausages and hash browns (S)
Yoghurt (Y)	

a Find all the different two-course breakfasts available:

i by listing them

ii using a tree diagram.

b If one of these combinations is chosen at random, find the probability that it includes raisin toast or bacon and eggs.

10 From a set of cards numbered 1 to 5, two cards are drawn at random with replacement to make a 2-digit number.

PS R C

a Show all outcomes in the sample space:

i by simply listing them **ii** using a table.

b What is the probability of making a number:

i that is odd? **ii** greater than 30?

INVESTIGATION

Simulating two-step experiments

For this investigation, you will need some different coloured counters (2 red, 3 blue and 1 green), a box or container to shake them in, and some paper.

Probability of symbols

1 Copy this table.

Result (any order)	Tally	Frequency	Probability (%)
☺ and ☺			
☺ and ★			
☺ and ▲			
★ and ★			
★ and ▲			
▲ and ▲			
Total		20	

2 Cut your piece of paper into 4 squares, draw a smiley face on 2 squares, a star on one square and a triangle on one square.

3 Fold each piece of paper up and place into the box or container.

4 Draw out at random one piece from the container, note the symbol, place the paper back in the container, give the container a little shake and then select a second piece and note the symbol.

5 Conduct this experiment 20 times, filling out the tally column in the table as you complete each trial.

6 Calculate the probability of each event in your simulation as a percentage and write the answer in the table.

7 Were there any results that were impossible to get? Explain your answer.

8 Copy the blank table again and repeat this experiment another 20 times BUT this time *do not replace* the paper after the first draw (without replacement). Record your results in the table.

9 Were there any results that were impossible to get in this second experiment? Explain your answer.

10 Compare the probabilities you calculated from both experiments (both with replacement and without replacement). What difference does not replacing the first choice make to the results? Would you expect there to be a difference? Why? Write a paragraph to explain your findings.

Probability of counters

1 Copy this table.

Result (any order)	Tally	Frequency	Probability (%)
R and R			
R and B			
R and G			
B and B			
B and G			
G and G			
Total		30	

2 Place the 2 red counters (R), 3 blue counters (B) and 1 green counter (G) in the box or container.

3 Select one counter at random from the container, note the colour, place the counter back in the container, give the container a little shake and then select a second counter and note the colour.

4 Conduct this experiment 30 times, filling out the tally column in the table as you complete each trial.

5 Calculate the probability of each event in your simulation as a percentage and write the answer in the table.

6 Were there any results that were impossible to get? Explain your answer.

7 Copy the blank table again and repeat this experiment another 30 times BUT this time do not replace the counter after the first draw. Record your results in the table.

8 Were there any results that were impossible to get in this second experiment? Explain your answer.

9 Compare the probabilities you calculated from both experiments (both with replacement and without replacement). What difference does not replacing the first choice make to the results? Would you expect there to be a difference? Why? Write a paragraph to explain your findings.

10 Look at your comparisons for both sections of this investigation and identify and similarities or differences between the results obtained with replacement compared to without replacement.

Selections without replacement

In the real world, selections with replacement are impractical. Once an item has been selected, it is usually kept and the total number of items left are reduced.

A tree diagram for a two-step experiment without replacement has fewer branches at the second step (usually one less for each branch than the first step), because the first item selected cannot be selected again.

The next example looks at Example **10** again with the 2-digit numbers but this time, the first card is not replaced.

Videos
Tree diagrams 1
Tree diagrams 2

Example 11

From a set of cards numbered 1 to 5, two cards are drawn at random (without replacement) to make a 2-digit number.

a Show all outcomes in the sample space using a tree diagram.

b What is the probability of making a number:

i less than 32? **ii** divisible by 3?

SOLUTION

a For the 1st digit, there are 5 possible digits: 1, 2, 3, 4 or 5.

For the 2nd digit, there are only 4 possible digits left because the 1st digit cannot be used again.

There are $5 \times 4 = 20$ possible 2-digit numbers in the sample space.

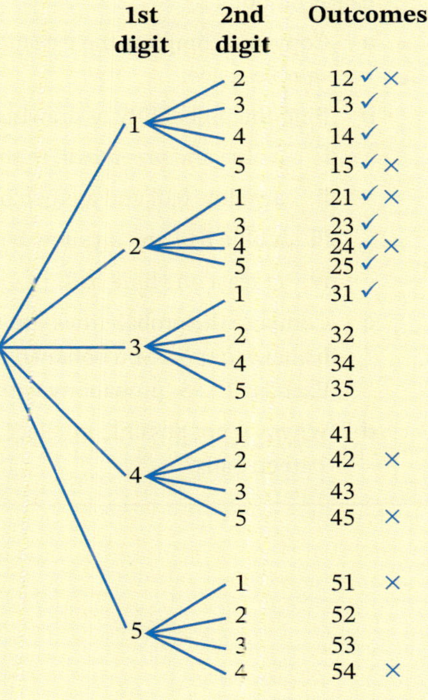

b **i** There are 9 numbers that are less than 32 (ticked in the tree diagram).

$\therefore P(\text{number} < 32) = \frac{9}{20}$.

ii There are 8 numbers that are divisible by 3 (crossed in the tree diagram).

$\therefore P(\text{number divisible by 3}) = \frac{8}{20}$

$= \frac{2}{5}$.

EXERCISE 12.06 ANSWERS ON P. 678

Selections without replacement

U F PS R C

1 Two people out of Ben, Anita, William, Lachlan and Noreen are to be selected for the positions of president and secretary in a tennis club.

PS R C

a List the possible pairings for the positions of president and secretary.

b What is the probability of the committee consisting of:

i Ben and Anita? **ii** 2 males?

iii a female president? **iv** at least one female?

2 The digits 5, 6, 8 and 9 are written on cards and put in a bag. Two cards are drawn at random (without replacement) to make a 2-digit number.

PS R C

a Make a list of all possible outcomes.

b How many 2-digit numbers are possible?

c Find as a decimal (**iv** to 2 decimal places) the probability that the number is:

i even **ii** greater than 60

iii between 60 and 90 **iv** divisible by 3

3 A bag contains 2 blue balls, 1 yellow ball and 1 green ball. Two balls are drawn out of the bag at random.

PS R C

a Copy and complete the tree diagram to list the sample space.

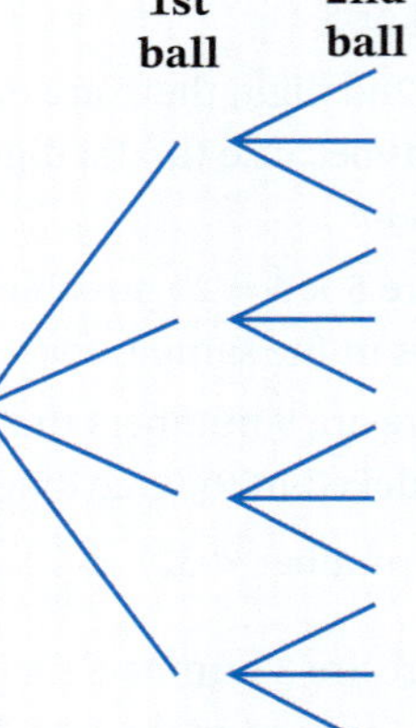

b Find the probability of drawing:

i a blue and green ball, in that order.

ii a yellow ball and a green ball, in any order.

iii a blue ball and a yellow ball.

iv at least one blue ball.

c Compare the probabilities you obtained in part **b** above to those you obtained in question **4b** of Exercise **12.05** (probability with replacement).

d Why is it not possible to select 2 yellow balls or 2 green balls?

4 Two cards are chosen at random from the 5 cards shown.
PS R C
Use a tree diagram to determine the probability of selecting:

a a 5 and then an 8

b a 5 and an 8 in any order

c 2 kings

Dreamstime.com/Noahgolan

5 Two class representatives need to be selected to participate in a school survey. There are 5 possible candidates from a particular class: Nicole, Marty, Sarah, Patrick and Colette. What is the probability of both a boy and a girl being selected to participate in the survey?
PS R C

6 A prize box contains 2 toys, 1 highlighter and 3 erasers. Jorja picks 2 prizes at random from the box. Determine the probability of Jorja selecting 2 different prizes.
PS R C

7 A lucky dip contains 3 gold gift vouchers of $200 and 4 silver gift vouchers of $50. Two winners select a prize each from the lucky dip. Determine the probability of both participants selecting a gold gift voucher.
PS R C

Foundation Standard Complex

POWER PLUS ANSWERS ON P. 678

1 Jenny needs to roll a 6 with a die to start moving on a board game. What is the probability that Jenny will start the game:

a on the first roll of the die?

b on the second roll?

c on the third roll?

2 Consider a family with 4 children. Use a tree diagram to list the 16 possible orders of boys and girls in the family. Find the probability of the family having:

a 3 boys and 1 girl

b no boys

c 2 boys and 2 girls

d at least 1 girl

e at most one girl

f more than 2 boys

3 A bag contains 5 yellow marbles and 4 red marbles. Two draws are made *without replacement*. Calculate the probability of drawing:

a 2 yellow marbles.

b a yellow marble first and a red marble second.

c a red marble and a yellow marble in any order.

d at most one yellow marble.

4 Set theory can be used to describe different groups and represent calculations using special notation.

If we consider the set $S = \{2, 4, 6, 8, 10, 12, 14, 16, 18, 20\}$ that represents all the even numbers from 1 to 20, then $n(S) = 10$, where $n(S)$ represents the *number* of elements in set S.

a Let $E = \{3, 6, 9, 12, 15, 18\}$, which represents the multiples of 3 from 1 to 20.

Find $n(E)$.

b i $S \cup E$ represents the **union** of the 2 sets. It consists of all of the elements that are in S or in E or in both S and E.

Copy and complete: $S \cup E = \{ ... \}$

ii $S \cap E$ represents the **intersection** of the 2 sets. It consists of the elements that are common to both S and E.

Copy and complete: $S \cap E = \{ ... \}$

c Given F represents the set of multiples of 5 from 1 to 20, copy and complete:

i $F = \{ ... \}$

ii $n(F) =$ ____

iii $S \cap F = \{ ... \}$

12 CHAPTER REVIEW

Language of maths

Quiz
Language of maths 12

at least
at most
chance experiment
complementary event
compound event
die/dice
even chance
event
expected frequency
experimental probability
favourable
frequency
mutually exclusive events
observed frequency
random
relative frequency
replacement
sample space
theoretical probability
tree diagram
trial
two-step experiment
two-way table
Venn diagram

1 Which term from the above list is used to calculate **experimental probability**?

2 What is the meaning of **relative frequency**?

3 What happens to the experimental probability of an event as the number of trials becomes very large?

4 Which term describes the set of all possible outcomes of an experiment?

5 Name the event that is **complementary** to winning first prize in Lotto.

6 What is drawn using 2 or more circles, often overlapping?

Topic summary

Worksheet
Mind map: Probability

Print (or copy) and complete this mind map of the topic, adding detail to its branches and using pictures, symbols and colour where needed. Ask your teacher to check your work.

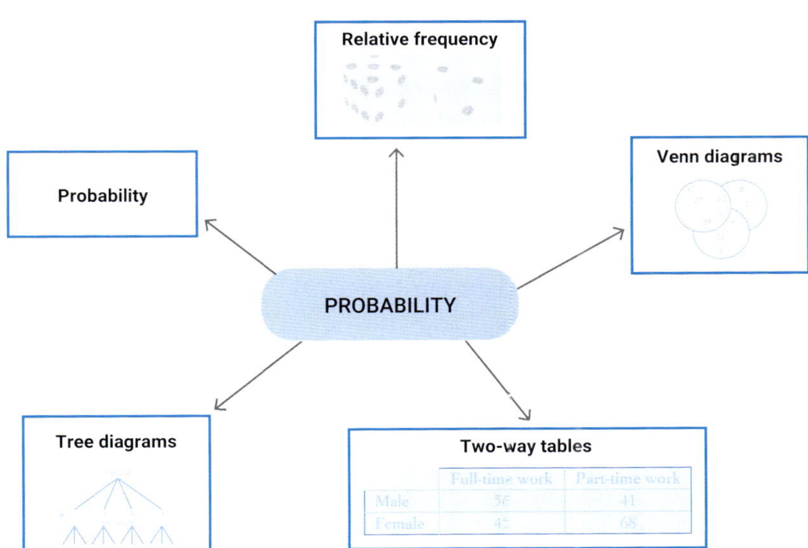

12 TEST YOURSELF

ANSWERS ON P. 679

Quiz
Test yourself 12

1 A jar contains only red and yellow marbles. The probability of selecting a red marble from the jar is 40%.

a What is the probability of selecting a yellow marble?

b If there are 40 marbles in the bag, how many red marbles are there?

2 A card is drawn randomly from a normal deck of cards. Find the probability that it is:

a a red card.

b a King or a Queen card.

c a card with a number greater than 6.

3 For the spinner shown, the red sector is twice as large as each of the other sectors. The spinner was spun 200 times and the results are shown in the table.

Outcome	Frequency
Blue	29
Green	35
Yellow	30
Purple	34
Red	72

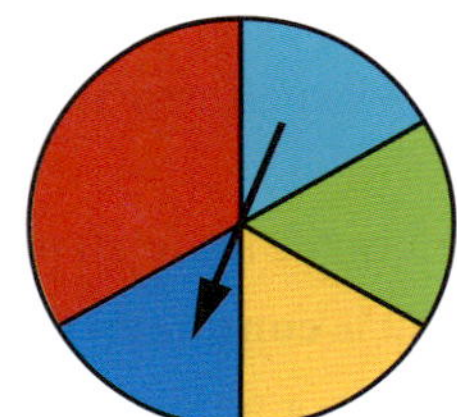

a Find the experimental probability (as a decimal) that the arrow lands on:

i red **ii** blue or green **iii** not purple **iv** green.

b Find the theoretical probability (as a decimal) that the arrow lands on:

i red **ii** blue or green **iii** not purple **iv** green.

c Are the experimental probabilities from part **a** similar to the theoretical probabilities from part **b**?

d If the spinner is spun 500 times, calculate the expected frequency of spinning red using the theoretical probability. Select the correct answer **A**, **B**, **C** or **D**.

A 100 **B** 85 **C** 167 **D** 200

4 30 teenagers were surveyed on whether they preferred to listen to music (M) or watch TV (T). 21 students preferred music and 18 liked TV, while 4 students did not like either.

a How many students preferred both music and TV?

b Copy and complete the Venn diagram.

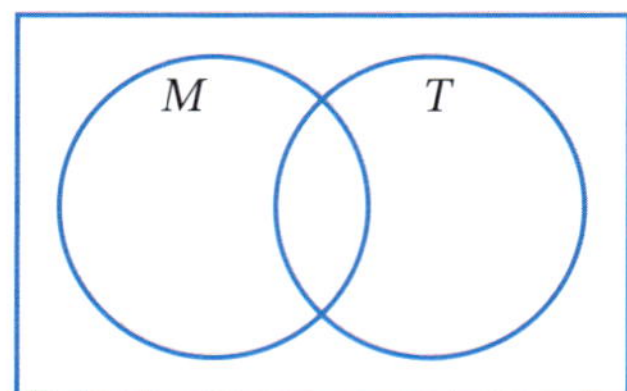

c Find the probability of randomly selecting a teenager who likes:

i music but not TV shows

ii TV shows

iii music or TV shows, but not both

iv both or neither music nor TV shows.

☐ Foundation ○ Standard ⬡ Complex

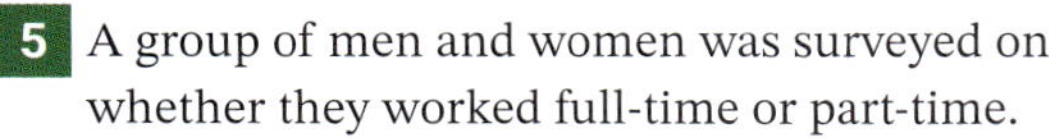

5 A group of men and women was surveyed on whether they worked full-time or part-time. 

	Full-time work	Part-time work
Male	56	41
Female	45	68

a How many people were surveyed?

b Find the probability (as a percentage, correct to one decimal place) of selecting a person who:

i works full-time

ii is female and works part-time

iii does not work full-time

iv works part-time or is female

c If a part-time worker is selected at random, what is the probability that the person is female?

6 Two dice are rolled and the difference between the numbers is calculated. 

a Copy and complete the table to show the sample space.

b Find the probability of rolling a difference of:

i 0

ii at least 2

ii 4 or 5.

		Die 1					
		1	2	3	4	5	6
Die 2	1						
	2						
	3						3
	4						
	5				1		
	6		4				

6 − 3 = 3

5 − 4 = 1

6 − 2 = 4

7 Four cards labelled with the letters A, B, C and D are placed in a bag. 

Two cards are drawn from the bag at random without replacement.

a List all of the possible outcomes.

b Copy and complete the tree diagram.

c Use the tree diagram to find the probability of drawing 2 cards with:

i the same letter.

ii the letters A and B.

iii the letters A or B or both.

iv no letter D.

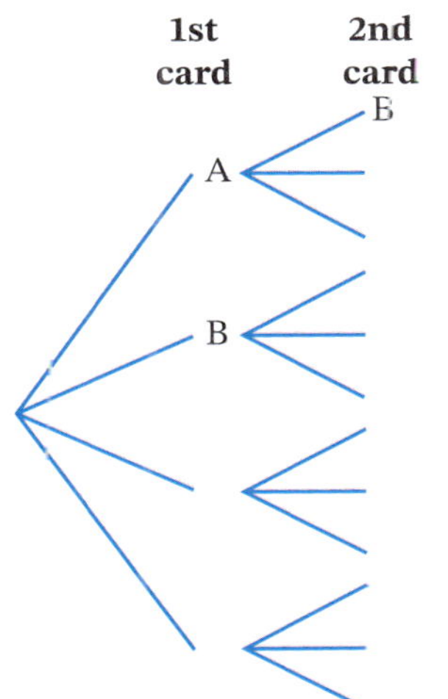

13

SPACE AND GEOMETRY

Congruent and similar figures

MC (Maurits Cornelis) Escher (1898–1972) was a Dutch graphic artist who used congruent figures to create tessellations (repeated 'tile' patterns). Although he had no mathematical training, Escher was talented in visualising and creating complex geometrical patterns, especially those involving 'impossible' shapes and structures.

Chapter outline

Chapter outline	Proficiencies				
13.01 Congruent figures	U	F		R	C
13.02 Tests for congruent triangles	U	F		R	C
13.03 Using congruence to prove geometrical properties	U	F		R	C
13.04 Similar figures	U	F		R	C
13.05 Properties of similar figures	U	F	PS	R	
13.06 Scale diagrams	U	F	PS	R	
13.07 Areas of similar figures	U	F		R	
13.08 Tests for similar triangles	U	F		R	C

U = Understanding
F = Fluency
PS = Problem solving
R = Reasoning
C = Communication

Wordbank

congruence test One of 4 tests for proving that triangles are congruent: SSS, SAS, AAS and RHS

congruent Identical; exactly the same (symbol: ≡)

enlargement An increase in the size of a shape

image A transformed shape after it has been enlarged or reduced

included angle The angle between 2 given sides of a shape, usually a triangle

scale factor The amount by which a shape has been enlarged or reduced, equal to $\frac{\text{image length}}{\text{original length}}$

similar To have the same shape but not necessarily the same size, an enlargement or reduction (symbol: |||)

superimpose To place one figure on top of another figure so that sides and angles match

Quiz
Wordbank 13

Videos (5):

13.03 Congruent triangles proofs
13.05 Finding an unknown side in similar figures
13.06 Scale drawings
13.07 Areas of similar figures
13.08 Tests for similar triangles

Twig videos (5):

13.06 Jai Singh (sundial) • Modelling the Spitfire • Queen Hatshepsut's ship • The history of the golden ratio • Fractals: The Koch snowflake

Quizzes (5):

- Wordbank 13
- SkillCheck 13
- Mental skills 13
- Language of maths 13
- Test yourself 13

Skillsheet (1):

13.05 Finding sides in similar triangles

Worksheets (9):

13.01 Congruent or different triangles?
13.03 Proving properties of quadrilaterals
13.04 A page of similar figures • Enlargements and reductions • Enlarging a logo
13.06 Problems involving scale drawings
13.07 Investigating paper sizes
13.08 Congruence and similarity review • Investigating paper sizes

Mind map: Congruent and similar figures

Puzzles (3):

13.04 Cartoon enlargement
13.05 Similar triangles • Nested similar triangles

Technology (2):

13.06 Converting map scales to ratios • Area of similar shapes

Spreadsheets (2):

13.06 Converting map scales to ratios • Area of similar shapes

Nelson MindTap

To access resources above, visit **cengage.com.au/nelsonmindtap**

13 **In this chapter, you will:**

- ✓ identify and use congruent and similar figures.
- ✓ investigate the properties of congruent and similar figures.
- ✓ investigate the properties of quadrilaterals using congruent triangles and solve related geometry problems.
- ✓ formulate proofs involving congruent triangles and angle properties.
- ✓ use scale factors to find unknown sides in similar figures.
- ✓ measure and calculate using scale diagrams.
- ✓ investigate ratio of areas of similar figures.
- ✓ identify and use the 4 tests for similar triangles.

SkillCheck

ANSWERS ON P. 679

Quiz SkillCheck 13

1 Match each pair of congruent (identical) figures below.

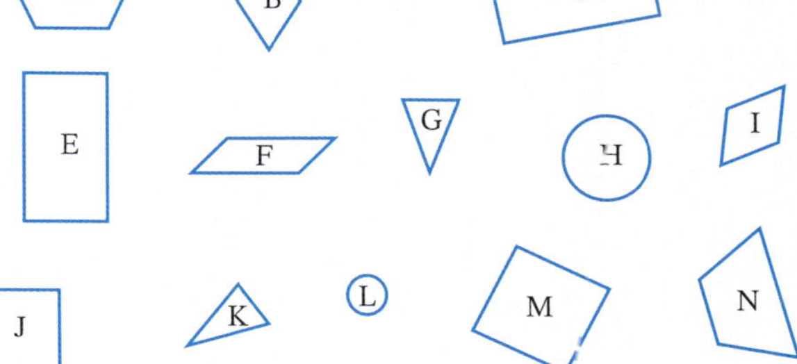

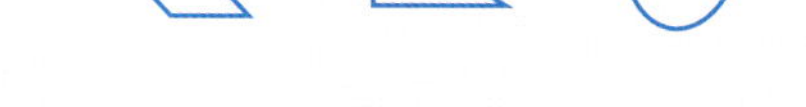

2 For each pair of congruent figures, list all pairs of:

i equal angles

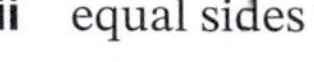

ii equal sides

a

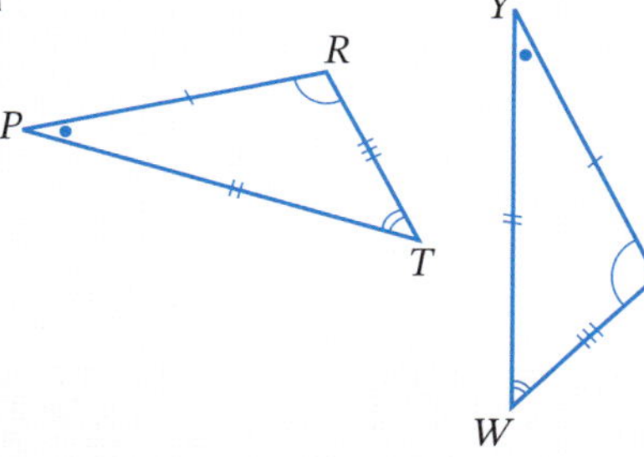

b

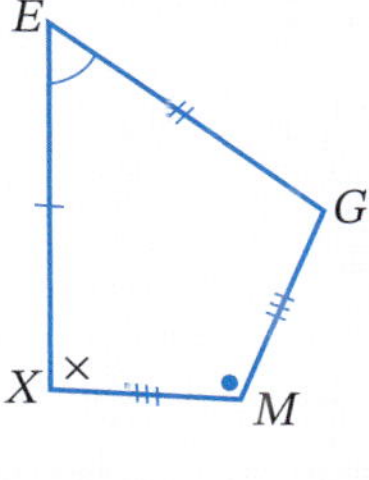

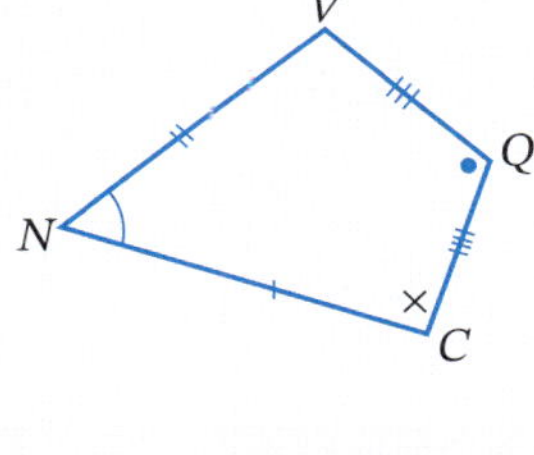

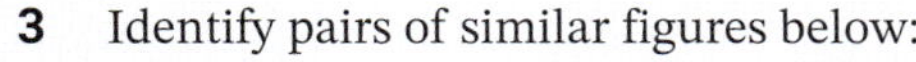

3 Identify pairs of similar figures below:

a

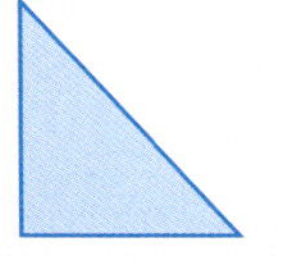

b

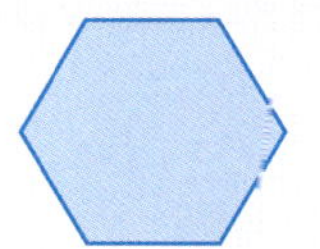

c

d

e

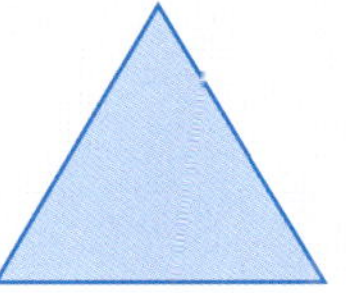

f

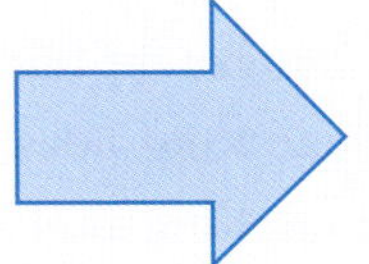

g

h

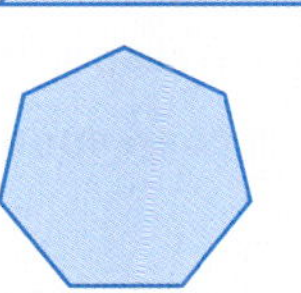

i

j

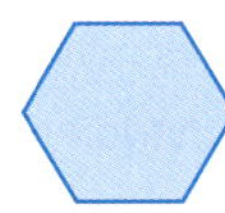

k

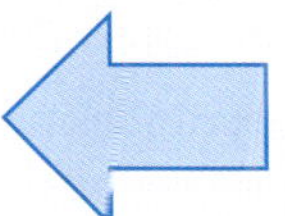

l

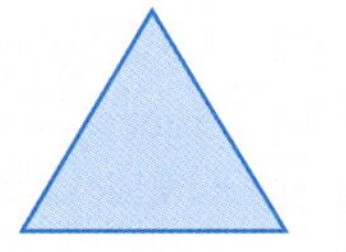

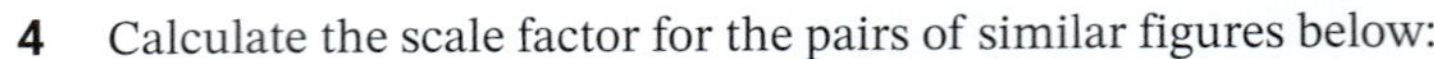

4 Calculate the scale factor for the pairs of similar figures below:

a

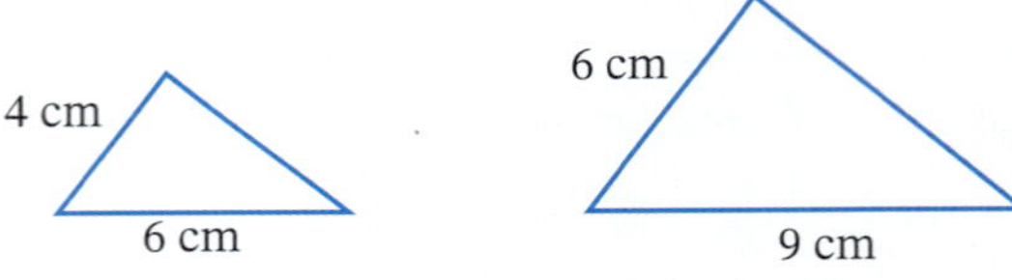

b

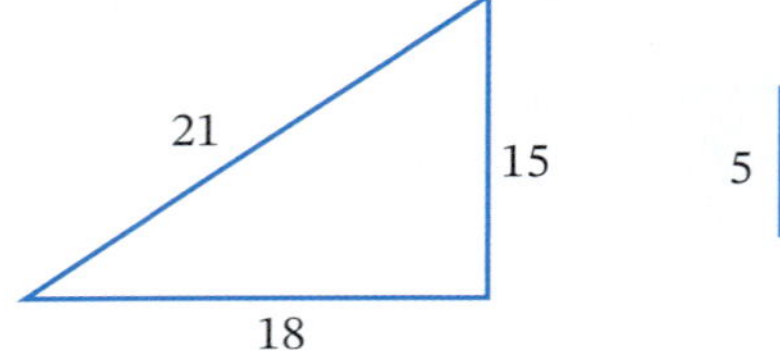

c

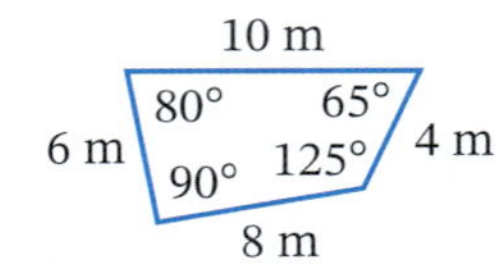

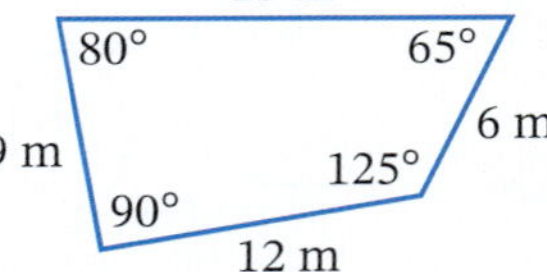

d

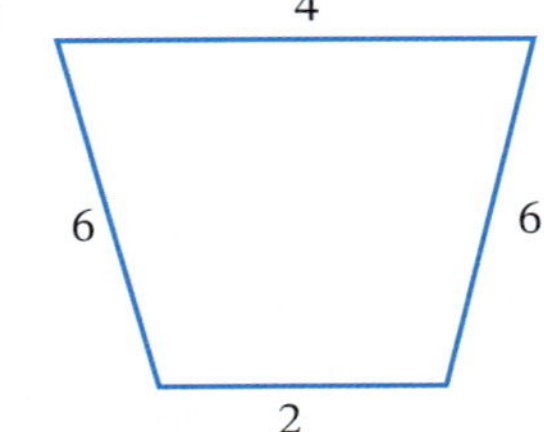

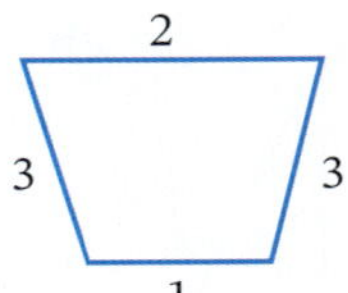

13.01 Congruent figures

ⓘ Congruent figures

Two figures are **congruent** if they are identical in shape and size.

For **congruent figures**:

- **matching sides** are equal
- **matching angles** are equal

Matching sides of congruent figures are **corresponding sides** that are in the same position.

Matching angles of congruent figures are **corresponding angles** that are in the same position.

One way of testing whether 2 figures are congruent is to **superimpose** one shape onto the other, that is, to move it to a position on top of the other so that sides and angles match.

Congruent figures may be identified by superimposing them through a combination of translations (slides), rotations (turns) and reflections (flips).

Alamy Stock Photo/View Stock

Shutterstock.com/Sketchart

Congruent triangles are commonly seen in the construction of bridges and buildings because they are considered more stable and strong to use.

Example 1

The 2 quadrilaterals *ABCD* and *EFGH* are congruent. List all pairs of matching sides and matching angles.

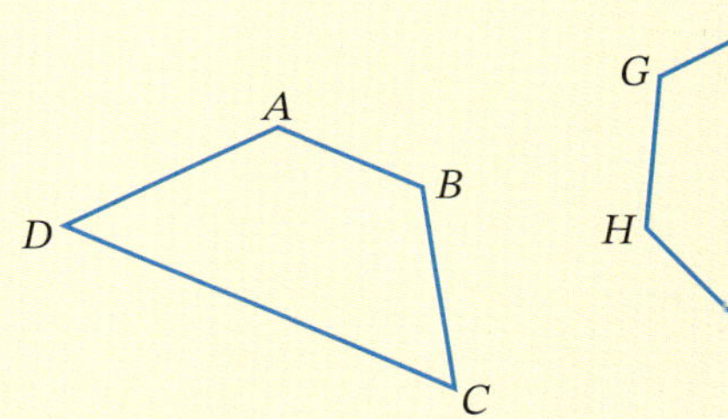

SOLUTION

By rotating the figure *ABCD* anticlockwise, its shape can be superimposed exactly on *EFGH*.

The pairs of matching sides are:

AB and *HG*

BC and *GF*

CD and *FE*

AD and *HE*

The pairs of matching angles are:

$\angle A$ and $\angle H$

$\angle B$ and $\angle G$

$\angle C$ and $\angle F$

$\angle D$ and $\angle E$

EXERCISE 13.01 ANSWERS ON P. 679

Congruent figures

U F R C

1 Is each pair of figures congruent? Why?

PS R

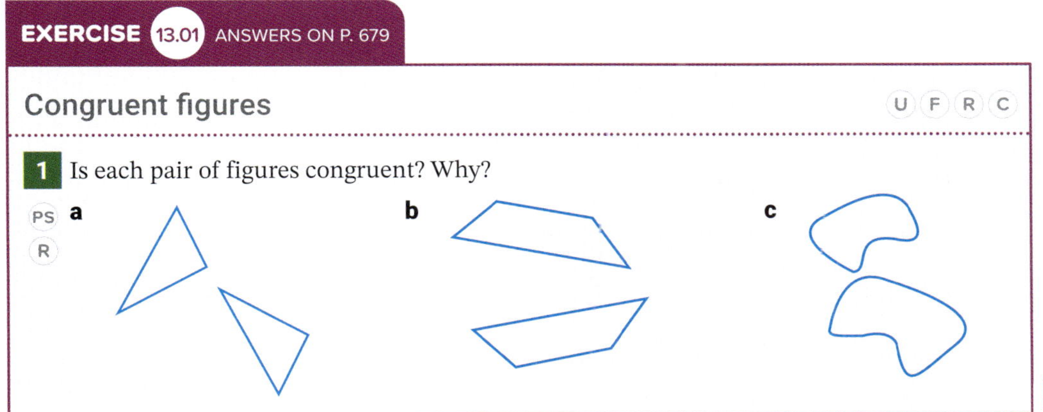

Foundation Standard Complex

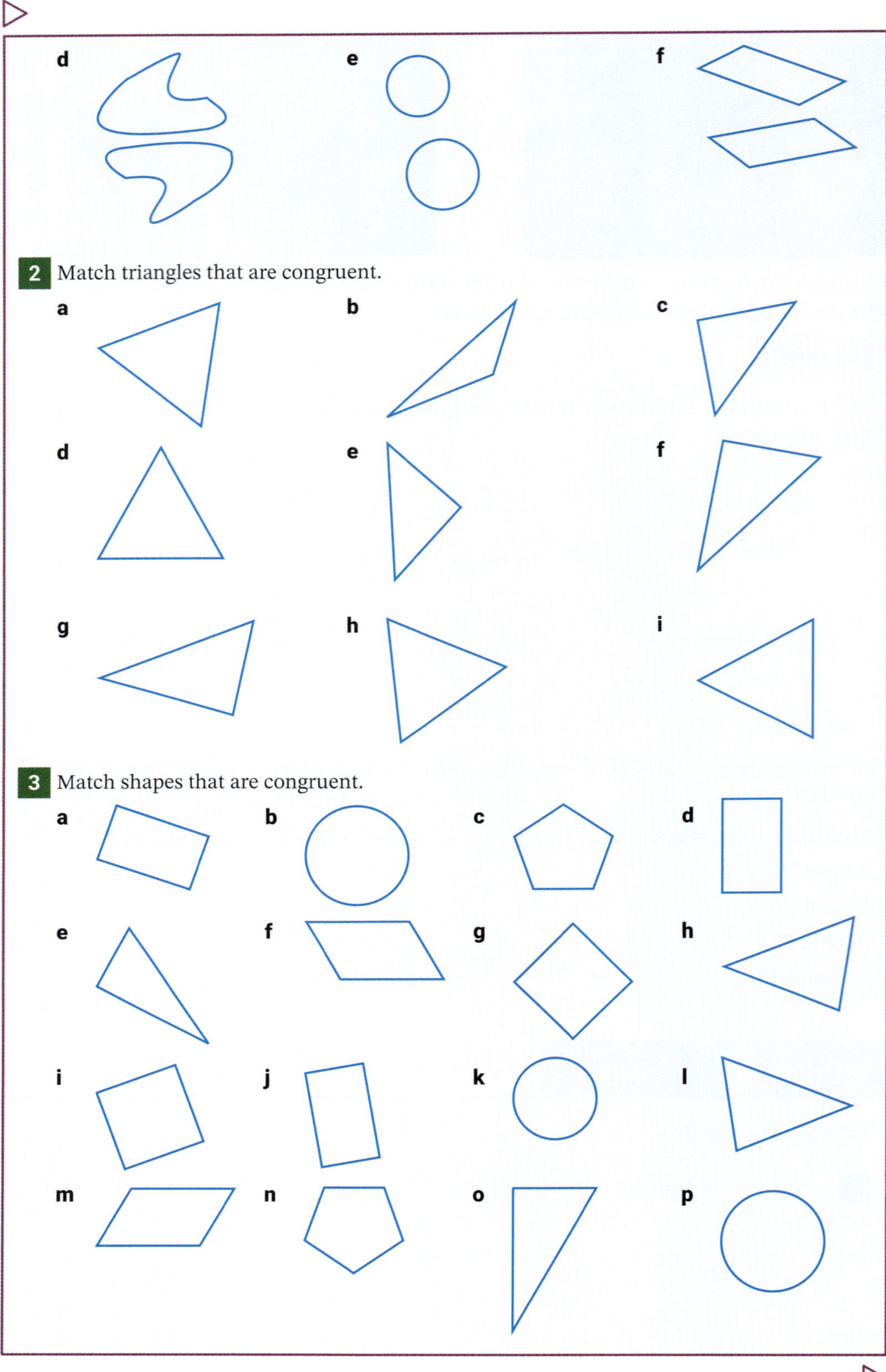
d
e
f
2 Match triangles that are congruent.
a
b
c
d
e
f
g
h
i
3 Match shapes that are congruent.
a
b
c
d
e
f
g
h
i
j
k
l
m
n
o
p

EXAMPLE 1

13.01

4 For each pair of congruent figures, list all pairs of:

i matching sides **ii** matching angles

a

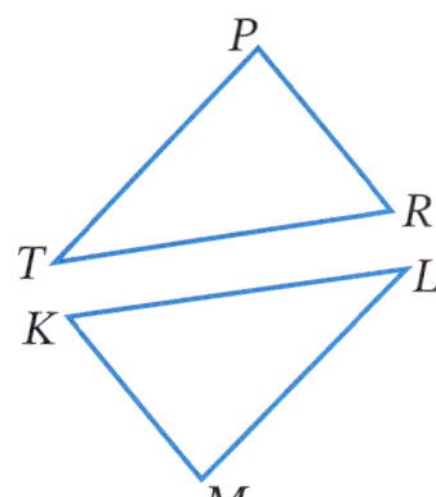

b

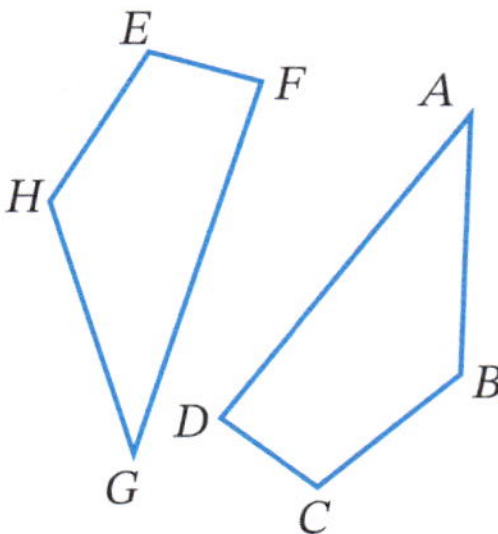

c

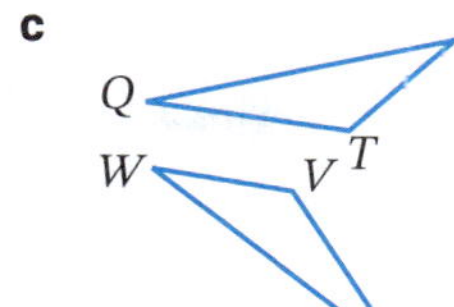

5 **a** For each rectangle, measure:

i its perimeter **ii** its area

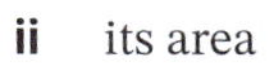

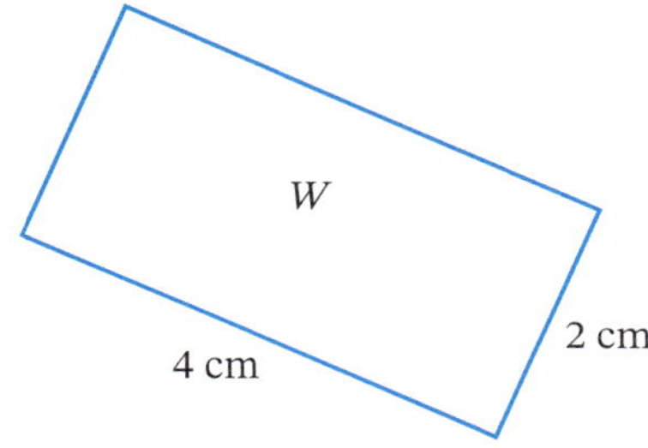

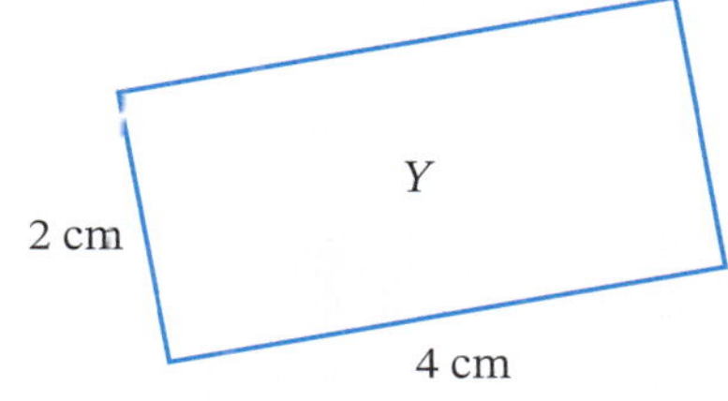

b Are the 2 rectangles congruent?

6 **a** For each triangle (drawn on 5 mm grid), measure:

i its perimeter **ii** its area.

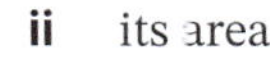

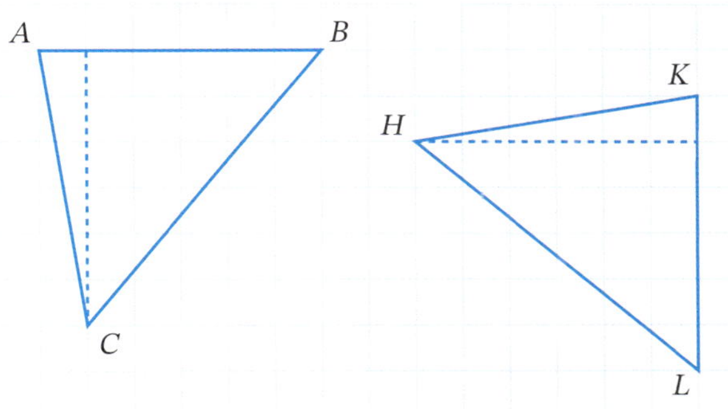

b Are the triangles congruent?

7 Copy and complete:

The p__________ and a_____ of congruent figures are e__________.

Foundation Standard Complex

8 Draw 2 triangles that have the same perimeter but are not congruent.

R C

9 Draw 2 triangles that have the same area but are not congruent.

R C

10 Write True or False for each statement.

R C

a If the perimeters of 2 rectangles are equal, then the 2 rectangles must be congruent.

b If the areas of 2 rectangles are equal, then the 2 rectangles must be congruent.

c If the perimeters and areas of 2 rectangles are equal, then the 2 rectangles must be congruent.

d Congruent figures must have the same perimeters and areas.

TECHNOLOGY

Tests for congruent triangles

Use dynamic geometry software to test whether 2 triangles are congruent. It is best to do this activity in small groups.

Given 3 sides: SSS

1 Construct a triangle with side lengths 2 cm, 6 cm and 7 cm.

2 Can you construct a **different** triangle with the same 3 side lengths? What do you notice about the second triangle? Is it congruent (exactly the same shape and size) to the first triangle?

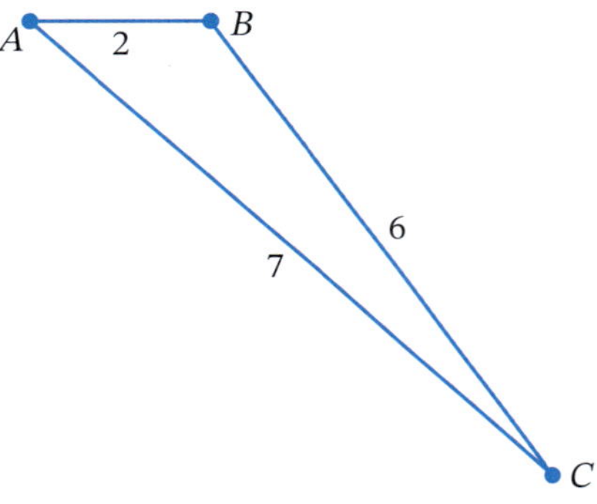

Given 2 sides and the included angle: SAS

1 Construct a triangle with sides of length 3 cm and 4 cm, and the angle between them to be exactly 80°.

2 Can you construct a different triangle with the same 2 sides and included angle of 80°? Or is it congruent to the first triangle?

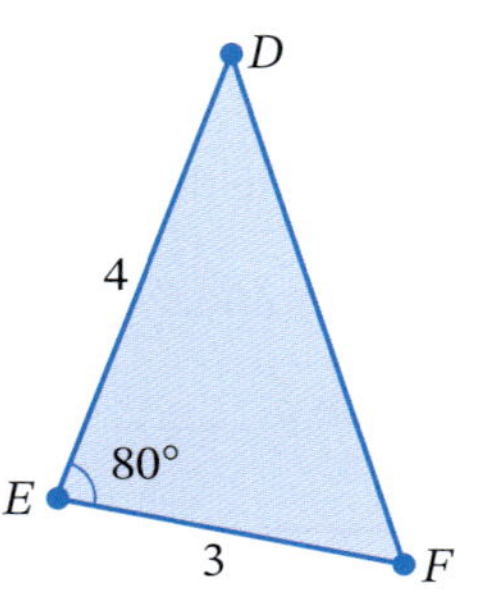

Given 2 angles and a side: AAS

1 Construct a triangle with a side length of 5 units between 2 angles of size 30° and 70°.

2 Can you construct a different triangle with the same 2 angles and a matching side of 5 cm? Or is it congruent to the first triangle?

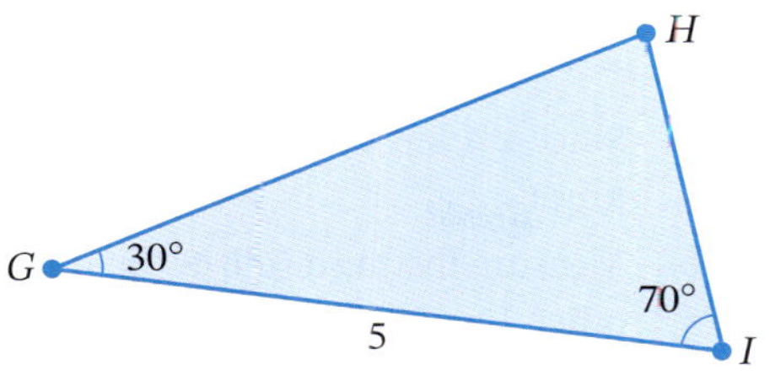

Given a right angle, hypotenuse and side: RHS

1 Construct a right-angled triangle with one of the shorter sides being 3 cm and the hypotenuse 5 cm.

2 Can you construct a different triangle with the same two sides and right angle? Or is it congruent to the first triangle?

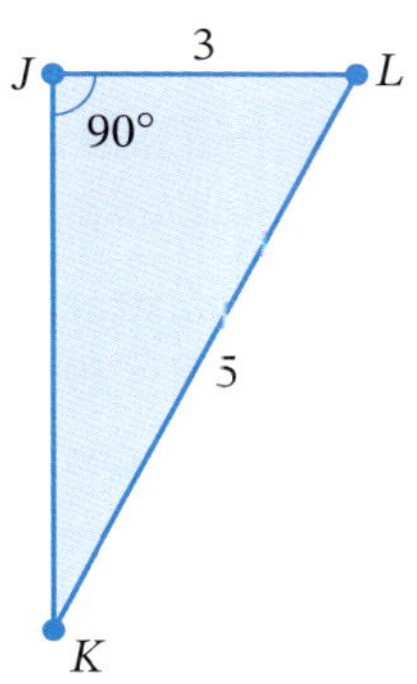

Given 3 angles: AAA

1 Construct a triangle with angles 77°, 62° and 41°.

2 Can you construct a different triangle with the 3 angles? Or is it congruent to the first triangle?

3 Are all triangles with angles 77°, 62° and 41° congruent?

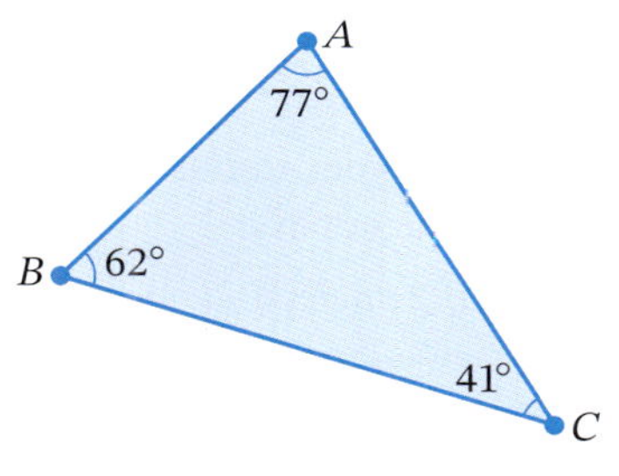

INVESTIGATION

Congruent triangles

Work in groups of 3 or 4 to complete this investigation.

You will need: paper, a ruler, a pencil, a protractor, compasses and scissors.

1 a Individually, construct a triangle with side lengths 5 cm, 7 cm and 9 cm.

b Use your protractor to measure the angles of the triangle.

c Compare your results with those of the students in your group. Are all your group's triangles congruent?

d Use your results to decide whether each statement is true or false.

i *If 3 sides of one triangle are respectively equal to the 3 sides of another triangle, then the matching angles of the triangles are equal.*

ii *If all sides of one triangle are equal to all sides of another triangle, then the 2 triangles are congruent.*

2 **a** Construct $\angle ABC$ so that $AB = 5$ cm, $AC = 7$ cm and $\angle A = 65°$.

Since $\angle A$ lies between AB and AC, $\angle A$ is called the **included angle**.

b Measure the third side of the triangle, and the sizes of the other 2 angles.

c Compare your results with those of others in your group.

d Use your results to decide whether this statement is true or false:

If 2 sides and the included angle of one triangle are respectively equal to 2 sides and the included angle of another triangle, then the 2 triangles are congruent.

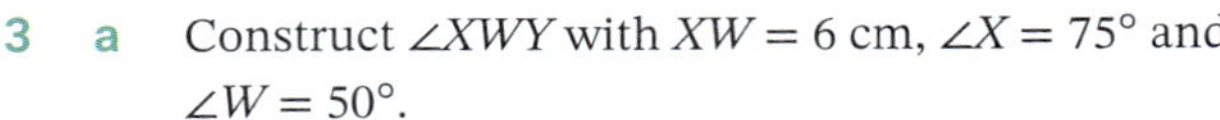

3 **a** Construct $\angle XWY$ with $XW = 6$ cm, $\angle X = 75°$ and $\angle W = 50°$.

b Measure $\angle Y$ and the other 2 sides of the triangle.

c Compare your results with those of your group members.

d Now construct $\angle DEF$ with $DE = 6$ cm, $\angle D = 50°$ and $\angle F = 75°$.

e Are $\angle XWY$ and $\angle DEF$ congruent? Give reasons for your answer.

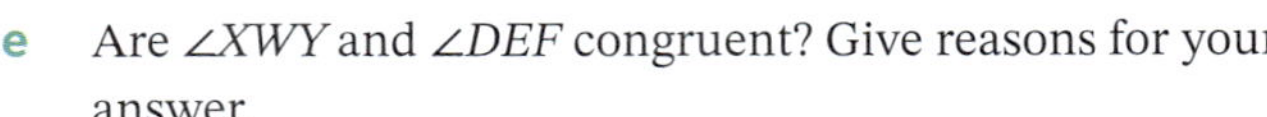

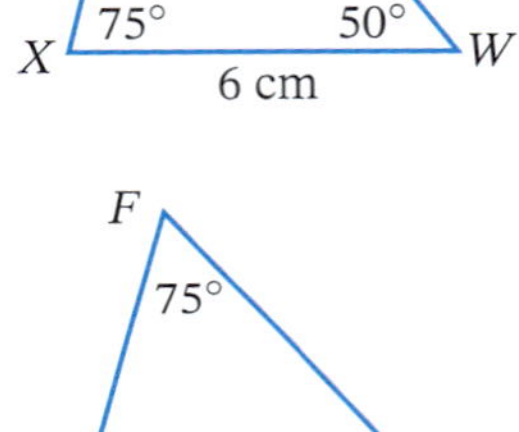

f Use your results to decide whether this statement is true or false:

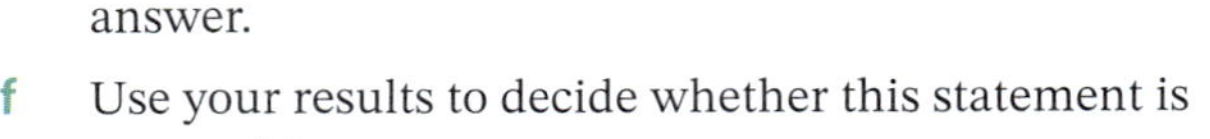

If 2 angles and one side of one triangle are respectively equal to 2 angles and the matching side of another triangle, then the 2 triangles are congruent.

4 **a** Construct $\angle GHK$ with $\angle H = 90°$, $GH = 8$ cm, $GK = 5$ cm.

b What is another name for the side GH?

c Measure the other side and angles of the triangle and compare your results with those of other members in your group. Are their triangles congruent to your triangle?

d Use your results to decide whether this statement is true or false:

If the hypotenuse and a second side of one right-angled triangle are respectively equal to the hypotenuse and a second side of another right-angled triangle, then the triangles are congruent.

5 **a** Draw a triangle with angles 40°, 63° and 77°.

b Compare your triangle with those of other members of your group. Are the triangles congruent? Give reasons.

6 Use your results to determine what information is needed to prove that 2 triangles are congruent.

Tests for congruent triangles

13.02

There are 4 sets of conditions that can be used to determine if 2 triangles are congruent. These are called the **tests for congruent triangles** or **congruence tests**.

Congruence tests

There are 4 tests for congruent triangles: **SSS**, **SAS**, **AAS** or **RHS**.

Two triangles are congruent if:

- the 3 sides of one triangle are respectively equal to the 3 sides of the other triangle **(SSS rule)**
- 2 sides and the **included angle** of one triangle are respectively equal to 2 sides and the **included** angle of the other triangle **(SAS rule)**
- 2 angles and one side of one triangle are respectively equal to 2 angles and the matching side of the other triangle **(AAS rule)**
- They are right-angled and the hypotenuse and another side of one triangle are respectively equal to the hypotenuse and another side of the other triangle **(RHS rule).**

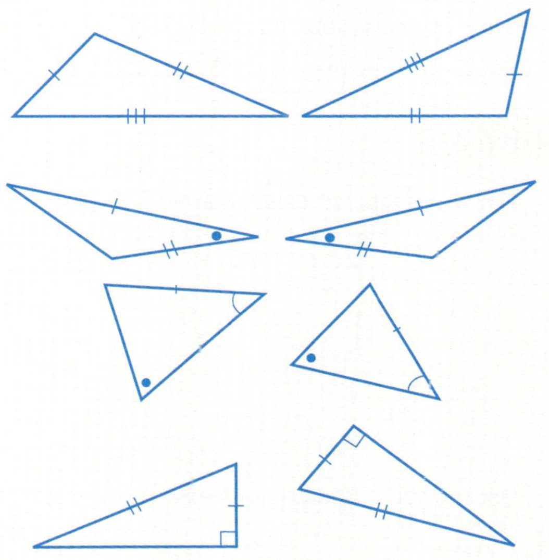

Example 2

Which congruence test (SSS, SAS, AAS or RHS) can be used to prove that each pair of triangles are congruent?

a

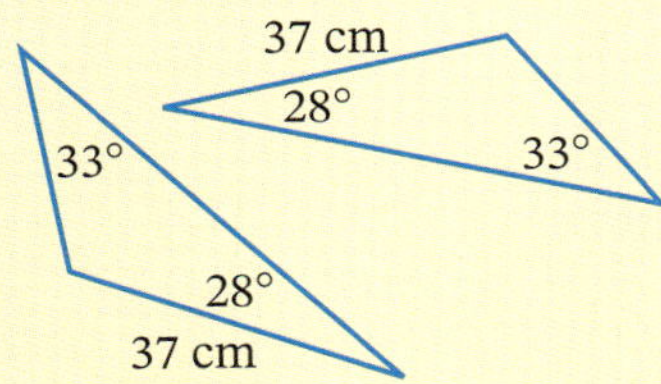

b

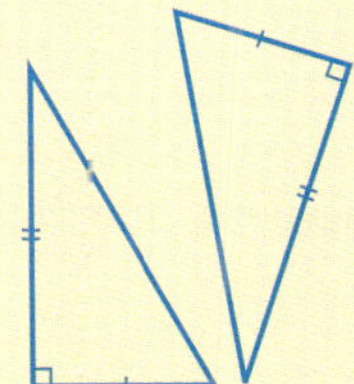

SOLUTION

a Two angles and one side of one triangle are equal to 2 matching angles and one matching side of the other triangle.

∴ The congruence test is AAS.

b Two sides and the included angle of one triangle are equal to 2 matching sides and the included angle of the other triangle.

∴ The congruence test is SAS.

The congruence symbol ≡

The symbol for 'is congruent to' is a special equals sign, written as '≡' (which also means 'is identical to'). The 2 triangles below are congruent, so we can write $\angle \boldsymbol{ABC} \equiv \angle \boldsymbol{XYZ}$, which is read: 'triangle *ABC* is congruent to triangle *XYZ*'.

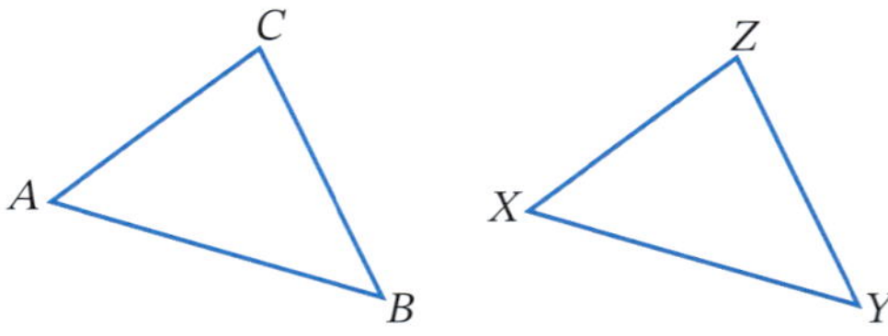

When using this notation, we must make sure that the vertices (and angles) of the congruent figures are written in matching order:

$\angle ABC \equiv \angle XYZ$ means $\angle A = \angle X$, $\angle B = \angle Y$, $\angle C = \angle Z$

Example 3

a Decide which 2 of these triangles are congruent and state the congruence test used.

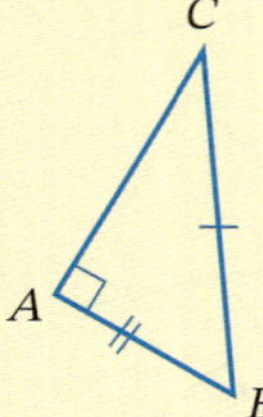

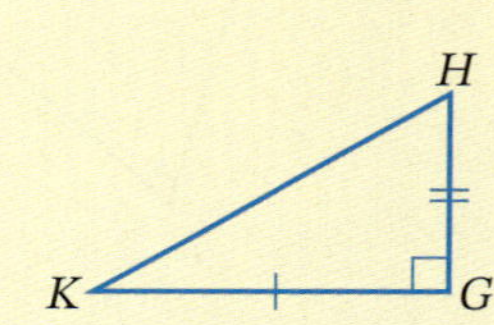

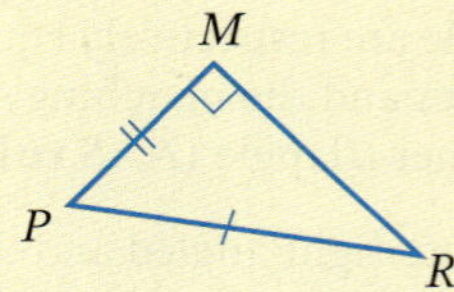

b Use the correct notation to write a congruency statement relating the two triangles.

SOLUTION

a The 3 right-angled triangles all have 2 equal sides, but *BC* and *PR* are both hypotenuses while *KG* is not.

$\therefore \angle CAB$ and $\angle MPR$ are congruent by the RHS rule.

b $\angle A$ matches with $\angle M$, $\angle B$ matches with $\angle P$, $\angle C$ matches with $\angle R$.

$\therefore \angle ABC \equiv \angle MPR$ Matching order of vertices

Example 4

If the 2 triangles are congruent, find the value of each variable.

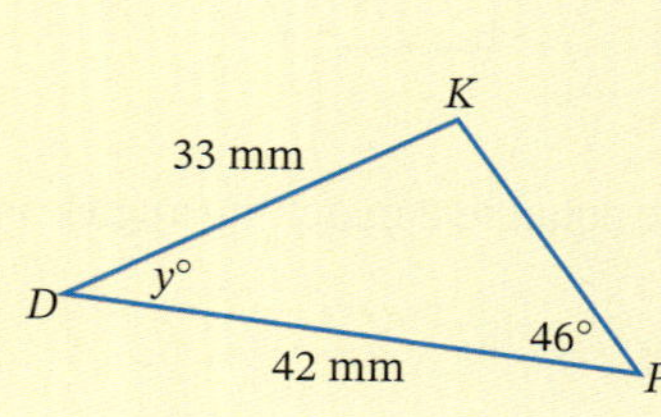

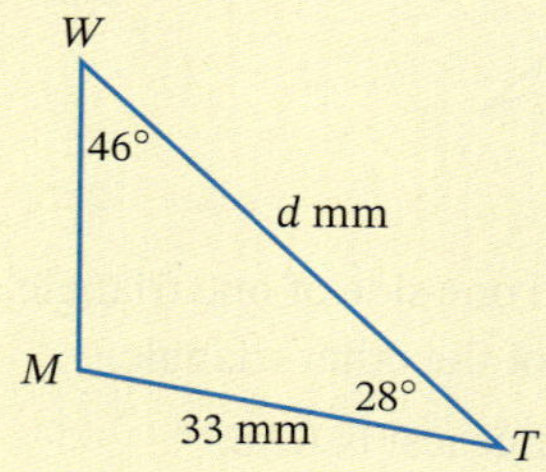

SOLUTION

Since $\angle DPK \equiv \angle TWM$,

$DP = TW$ Matching sides of congruent triangles.

$\therefore d = 42$

$\angle T = \angle D$ Matching angles of congruent triangles.

$\therefore y = 28$

EXERCISE 13.02 ANSWERS ON P. 679

Tests for congruent triangles

U F R C

1 Which congruence test (SSS, SAS, AAS or RHS) can be used to prove that each pair of triangles are congruent?

R C

EXAMPLE 2

a

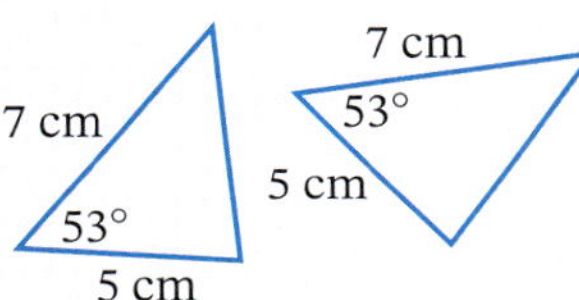

b

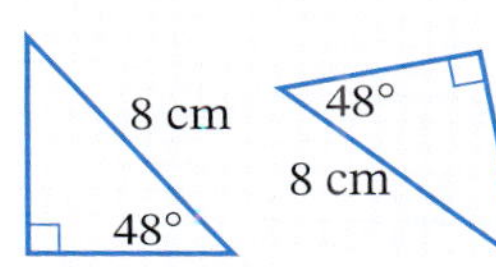

c

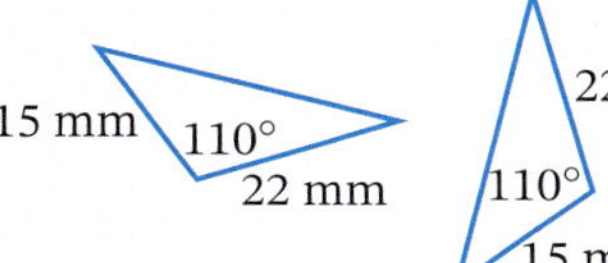

d

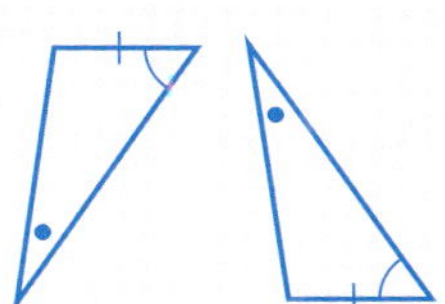

e

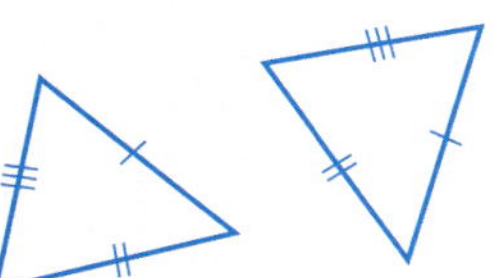

f

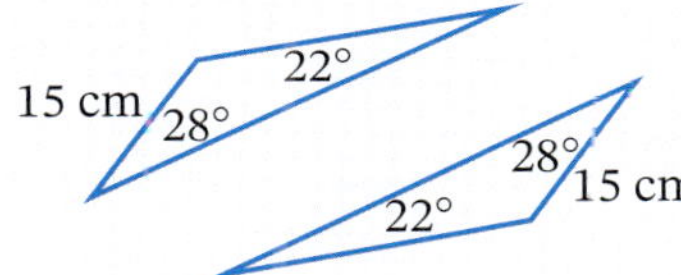

g

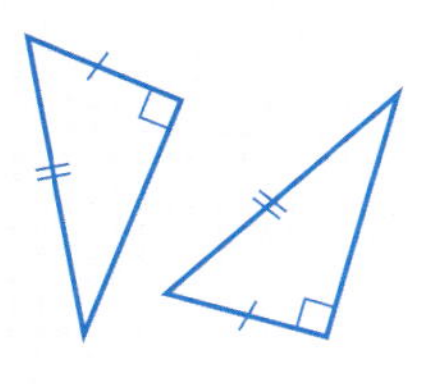

h

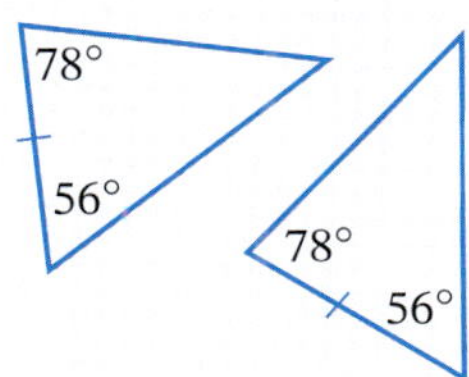

i

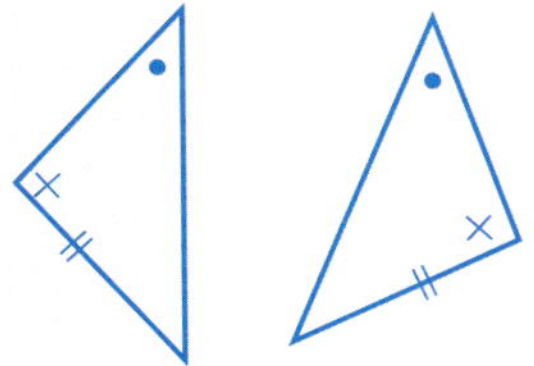

j

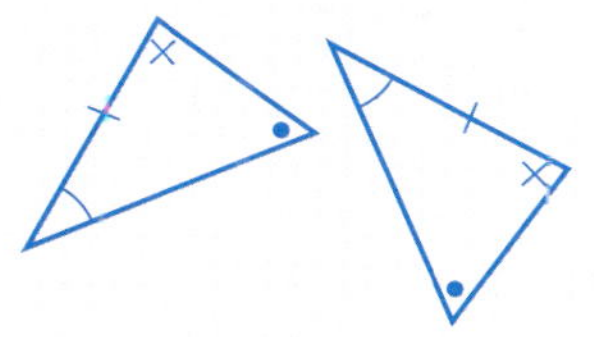

 Foundation  Standard Complex

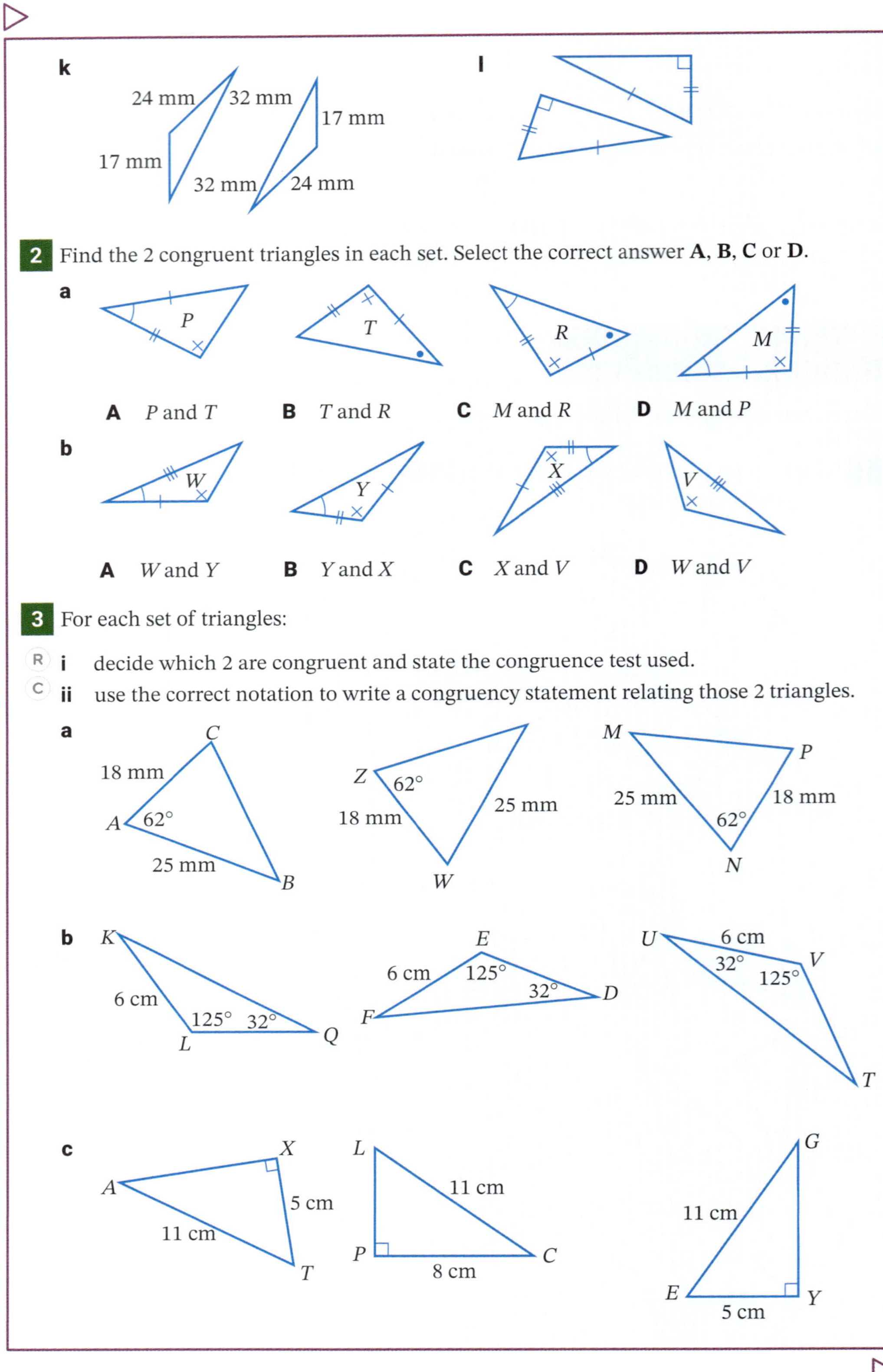

Foundation Standard Complex

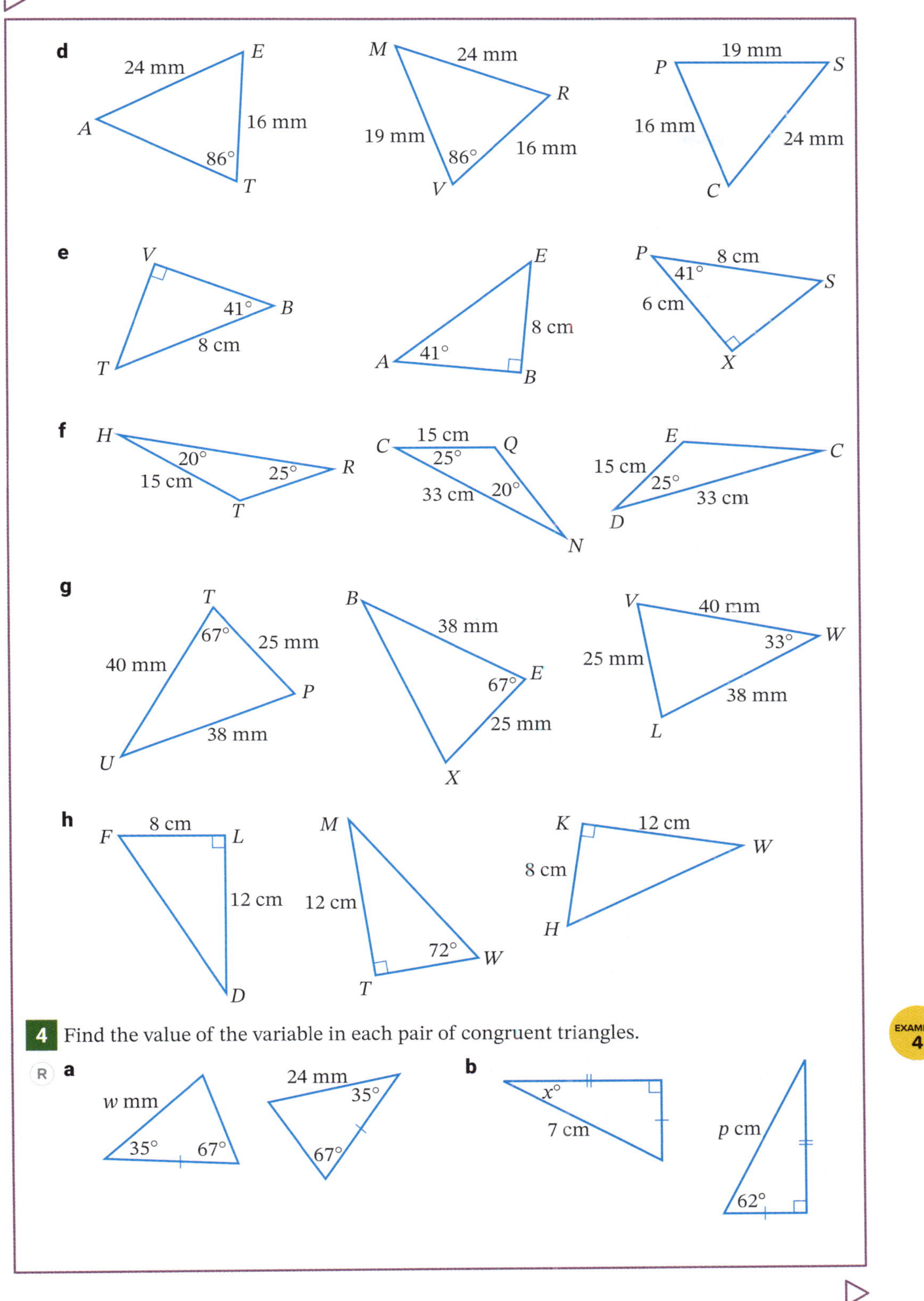

4 Find the value of the variable in each pair of congruent triangles.

EXAMPLE 4

R **a**

b

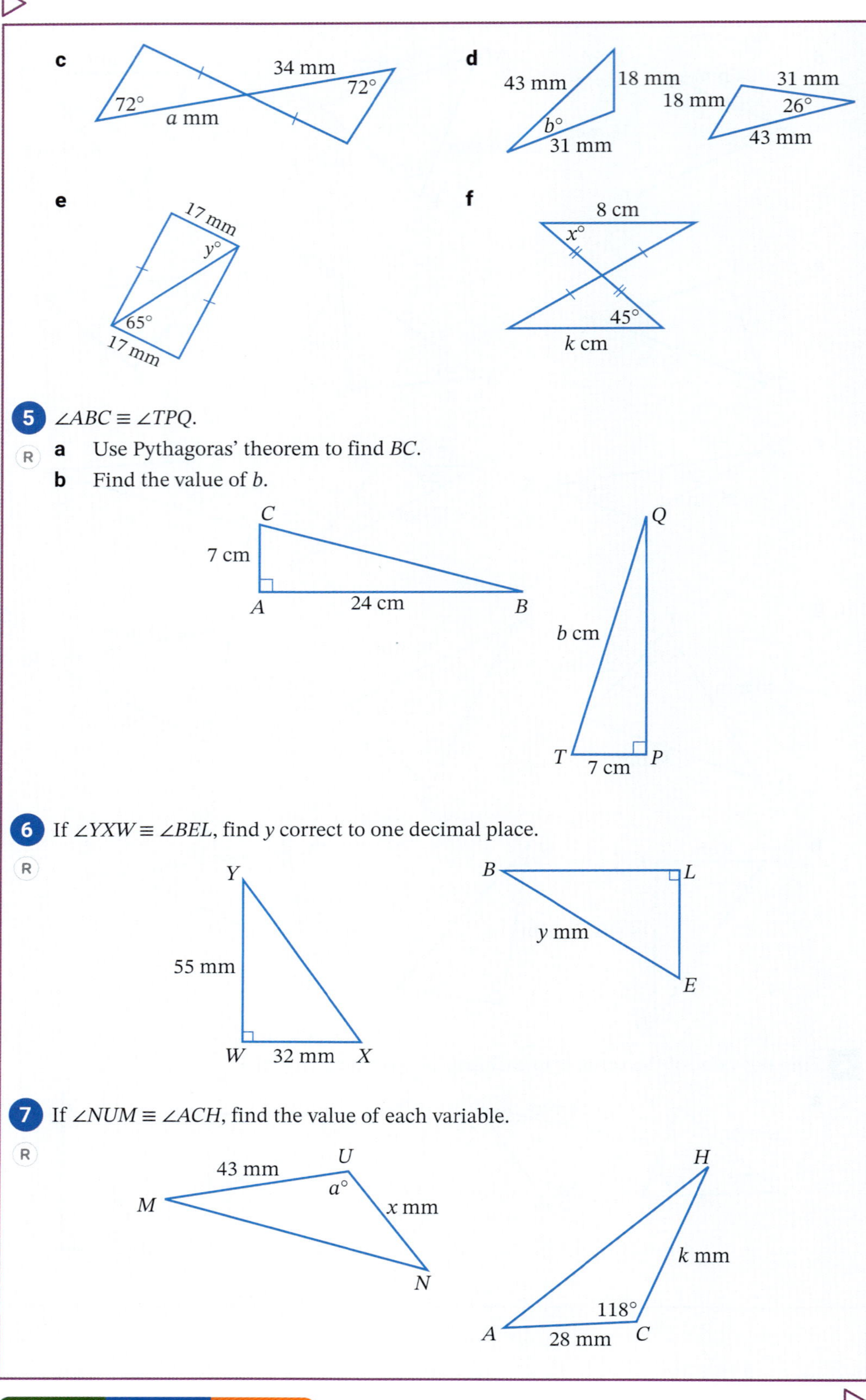

5 $\angle ABC \equiv \angle TPQ$.

R

a Use Pythagoras' theorem to find BC.

b Find the value of b.

6 If $\angle YXW \equiv \angle BEL$, find y correct to one decimal place.

R

7 If $\angle NUM \equiv \angle ACH$, find the value of each variable.

R

Foundation Standard Complex

8 The diagram shows 4 right-angled triangles joined together. State whether each pair of triangles are congruent, and if so, state the test used.

R C

a W and X **b** W and Y

c W and Z **d** X and Y

e X and Z **f** Y and Z

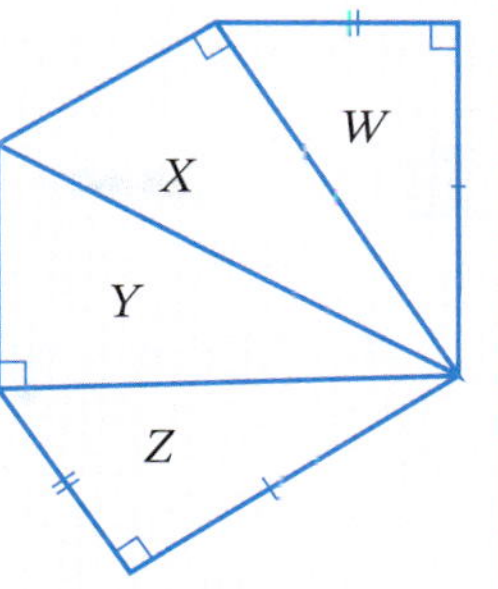

Using congruence to prove geometrical properties

13.03

Properties of triangles and quadrilaterals can be proved using the congruence tests.

Video
Congruent triangles proofs

Worksheet
Proving properties of quadrilaterals

Example 5

$KLMN$ is a rectangle, so all angles are right angles and opposite sides are equal.

a Which congruence test can be used to prove that $\angle KLN \equiv \angle LKM$?

b Explain why $KM = LN$.

c What geometrical result about rectangles does this prove?

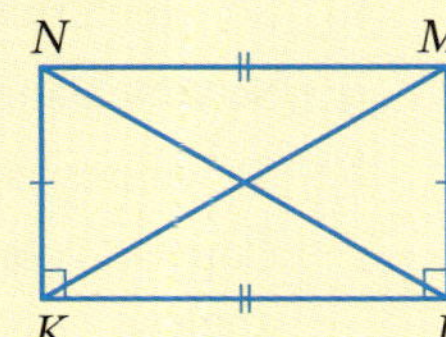

SOLUTION

a For $\angle KLN$ and $\angle LKM$:

KL is common to both triangles.

$KN = LN$

$\angle NKL = \angle MLK = 90°$.

$\therefore$ The congruence test is SAS.

b $KM = LN$ because they are matching sides of congruent triangles.

c The diagonals of a rectangle are equal.

EXERCISE 13.03 ANSWERS ON P. 680

Using congruence to prove geometrical properties

U F R C

1 *ABCD* is a kite, so adjacent sides are equal.

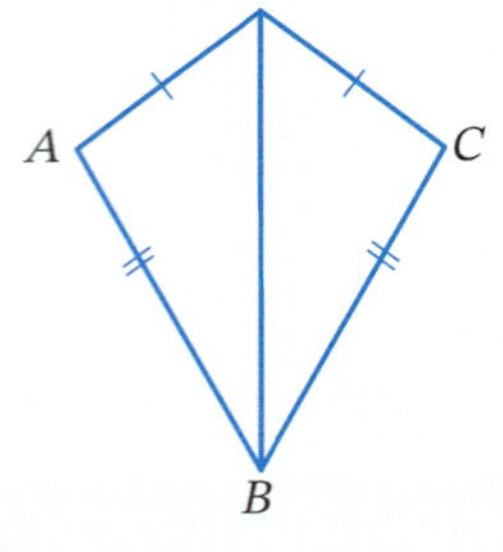

R **a** Which congruence test proves that $\angle ABD \equiv \angle CBD$?

C **b** Copy and complete:

$\angle DAB =$ ____________

$\angle ADB =$ ____________

$\angle ABD =$ ____________

c Copy and complete:

i *AD* and ____________ are adjacent equal sides.

ii *AB* and ____________ are adjacent equal sides.

iii The angles between the unequal adjacent sides, $\angle A$ and $\angle C$, are e____________.

iv The angles between the equal adjacent sides, $\angle ADC$ and $\angle ABC$ are b____________ by the diagonal ________.

d Draw the other diagonal *AC*, intersecting *DB* at point *X*.

e This creates 4 triangles. Looking at your diagram, which congruence test proves that $\angle DAX \equiv \angle DCX$?

f So which side is equal to *AX*? Mark both sides with 3 dashes.

g So which angle is equal to $\angle DXA$?

h So what is the size of $\angle DXA$? Mark this on your diagram.

i Copy and complete: The diagonal *BD* bisects the ____________ *AC* at ____________ angles.

2 *WXYZ* is a parallelogram with parallel opposite sides.

R C **a** Copy the diagram into your book.

b On your diagram, show 2 pairs of equal alternate angles.

c Which congruence test proves that $\angle WXZ \equiv \angle YZX$?

d What angle is equal to $\angle W$?

e Draw the other diagonal *WY* and mark 2 pairs of alternate angles.

f Why is $\angle WXY \equiv \angle YZW$?

g Why is $\angle WXY = \angle YZW$?

h Copy and complete: The opposite a__________ of a parallelogram are e_________.

9780170465618

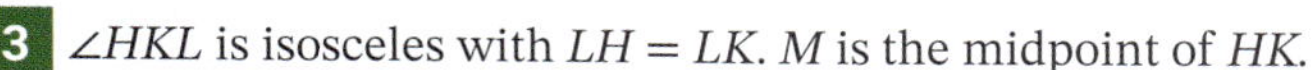

3 $\angle HKL$ is isosceles with $LH = LK$. M is the midpoint of HK.

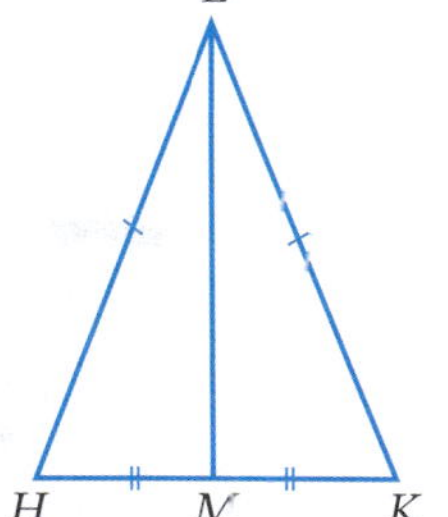

R C **a** Which congruence test can be used to prove that $\angle HML \equiv \angle KML$?

b Explain why $\angle H = \angle K$.

c Copy and complete: The angles opposite the e________ sides in an isosceles triangle are e_________.

d What angle is equal in size to $\angle HML$?

e Hence, what is the size of $\angle HML$?

4 $CDEF$ is a rectangle whose diagonals CE and DF intersect at T.

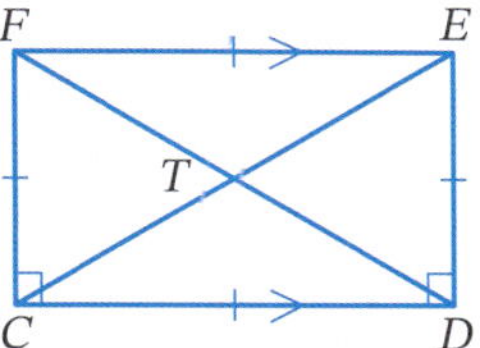

R C **a** Copy the diagram and mark 2 pairs of equal alternate angles.

b Why is $CD = FE$?

c Which congruence test proves that $\angle CDT \equiv \angle EFT$?

d Explain why $CT = TE$ and $FT = TD$.

e Copy and complete: The d__________ of a rectangle b_________ each other.

5 $STUV$ is a rhombus so all sides are equal.

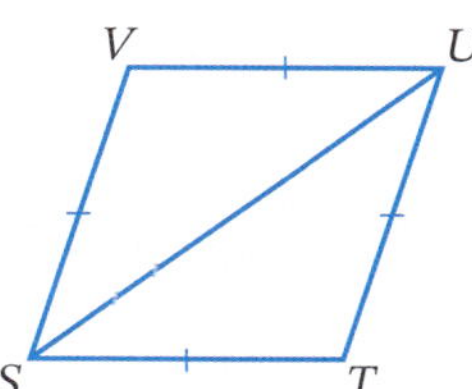

R C **a** Which congruence test can be used to prove that $\angle VUS \equiv \angle TUS$?

b Copy and complete:

i $\angle VUS = \angle TUS$ and $\angle VSU = \angle$__________.

ii $\angle VUT$ is b__________ by the d__________ SU.

iii The angles of a rhombus are b__________ by the d__________.

6 $ABCD$ is a parallelogram with opposite sides parallel.

R C **a** Copy the diagram and mark 2 pairs of equal alternate angles.

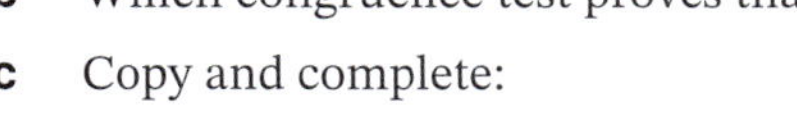

b Which congruence test proves that $\angle AED \equiv \angle CDB$?

c Copy and complete:

i $AB = CD$ and $AD = CB$

(m__________ s__________ of c__________ triangles).

ii O__________ sides of a parallelogram are e__________.

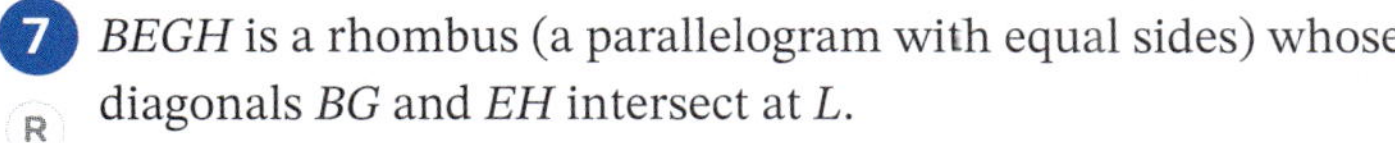

7 $BEGH$ is a rhombus (a parallelogram with equal sides) whose diagonals BG and EH intersect at L.

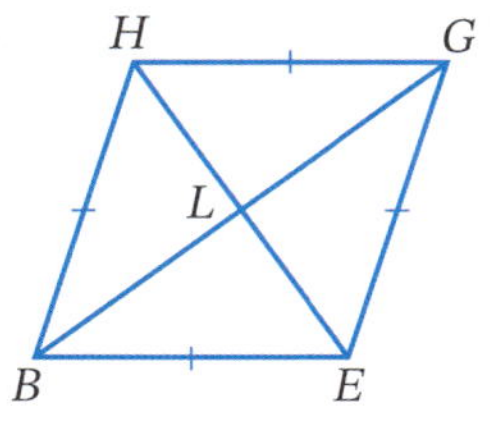

R C **a** Explain why $\angle BEL = \angle GHL$ and $\angle EBL = \angle HGL$.

b Why is $BE = GH$?

c Which congruence test can be used to prove that $\angle BEL = \angle GHL$?

Foundation | Standard | Complex

d Copy and complete:

i $BL =$ ____ and $EL =$ _____ (matching sides of congruent triangles)

ii The d__________ of a rhombus b__________ each other.

8 ΔABC is an equilateral triangle ($AB = BC = AC$). X is the midpoint of BC.

R C **a** Which congruence test can be used to prove that $\angle ABX \equiv \angle ACX$?

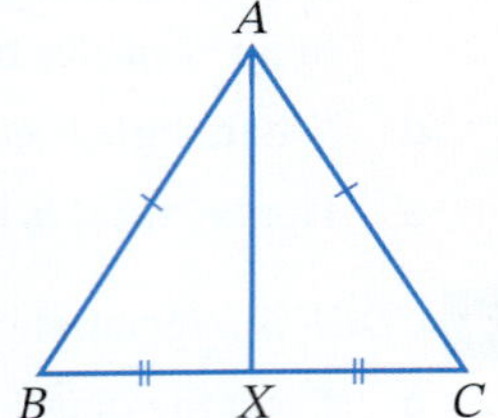

b Explain why $\angle B = \angle C$.

c In this diagram, $\angle ABC$ is redrawn so that Y is the midpoint of AC. Which congruence test can be used to prove $\angle BAY \equiv \angle BCY$?

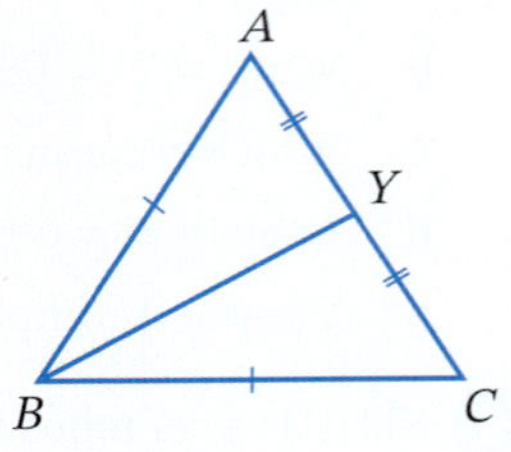

d Is $\angle A = \angle C$? Why?

e Calculate the sizes of the 3 angles of $\angle ABC$.

f Copy and complete: In an e__________ triangle, each angle is __________.

INVESTIGATION

Same shape, different size

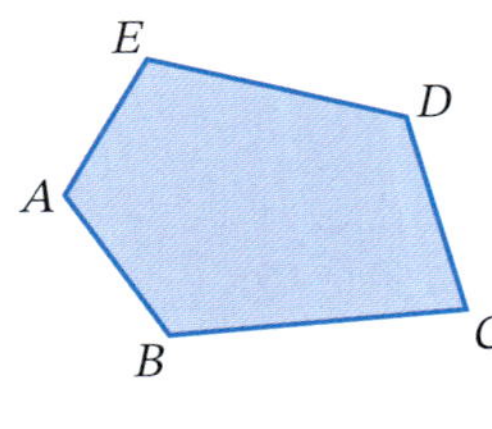

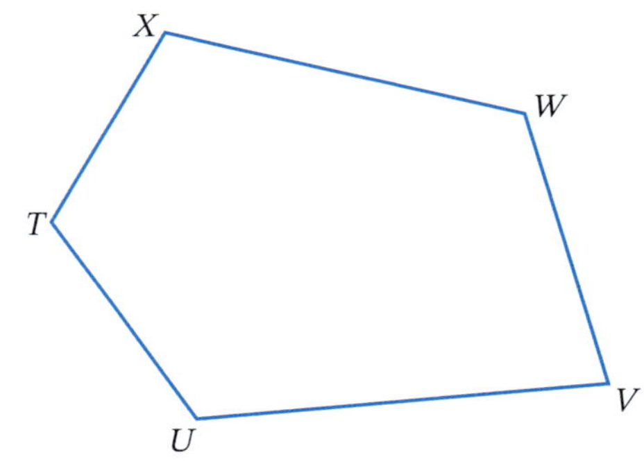

1 **a** Consider the pentagons $ABCDE$ and $TUVWX$. Do they look similar?

b What do you think is meant by the word 'similar'?

c Copy and complete this table by measuring each angle and side in the 2 'similar' pentagons.

Angles		Sides (in mm)	
$A =$	$T =$	$AB =$	$TU =$
$B =$	$U =$	$BC =$	$UV =$
$C =$	$V =$	$CD =$	$VW =$
$D =$	$W =$	$DE =$	$WX =$
$E =$	$X =$	$AE =$	$TX =$

d What do you notice about the sizes of the matching angles?

e What do you notice about the sizes of the matching sides?

☐ Foundation ○ Standard ○ Complex

2 $\angle PTR$ has been reduced to produce the similar triangle $\angle DEF$.

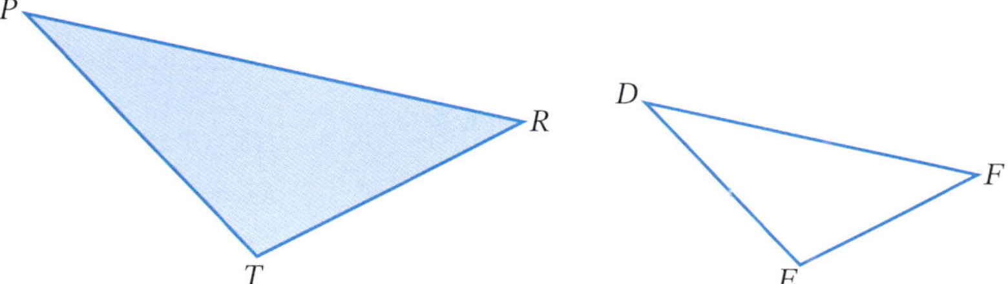

a List the 3 pairs of matching angles and measure the size of each one.

b List the 3 pairs of matching sides and measure the length of each one.

3 Copy and complete:

a For similar figures, m__________ angles are e________.

b Similar figures have the s________ shape, but not the same s________.

Similar figures 13.04

Similar figures have the same shape but are not necessarily the same size.

When a photograph is enlarged or reduced, all parts of the picture remain the same shape, but all lengths and distances will be respectively larger or smaller than the corresponding lengths and distances on the original.

Worksheets
A page of similar figures
Enlargements and reductions
Enlarging a logo

Puzzle
Cartoon enlargement

iStock.com/aluxum

When a figure is enlarged or reduced, a **similar figure** is created. The original figure is called the **original**, while the enlarged or reduced figure is called the **image**.

The scale factor

The **scale factor** describes by how much a figure has been enlarged or reduced. For example, if the quadrilateral *DEFG* is enlarged by doubling each side, the scale factor is 2.

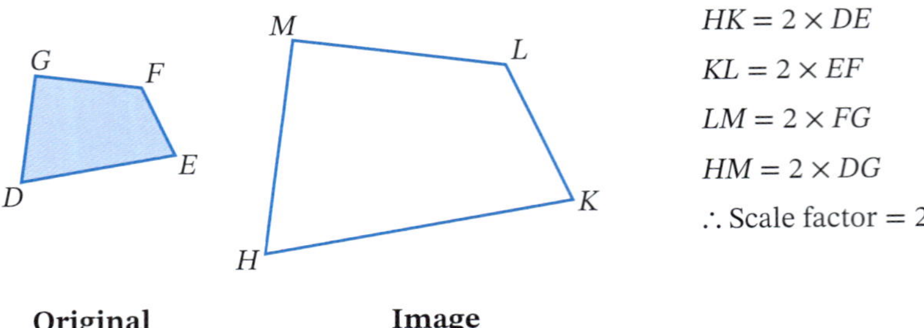

$HK = 2 \times DE$

$KL = 2 \times EF$

$LM = 2 \times FG$

$HM = 2 \times DG$

$\therefore$ Scale factor $= 2$

In a similar manner, if the arrow figure is reduced by halving each side, the scale factor is $\frac{1}{2}$.

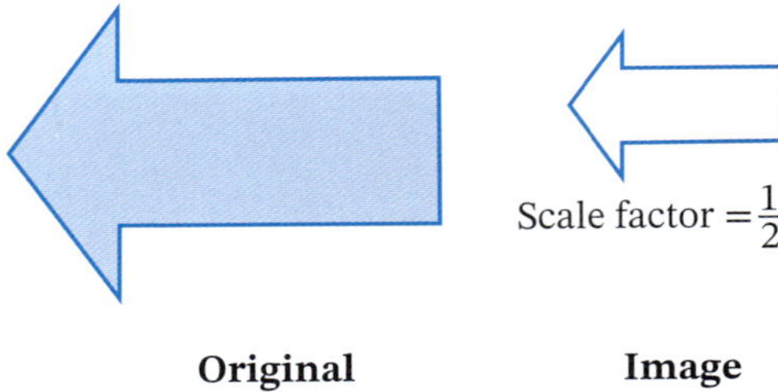

ⓘ Scale factor

$$\text{Scale factor} = \frac{\text{image length}}{\text{original length}}$$

- If the scale factor is greater than 1, then the image is an enlargement.
- If the scale factor is between 0 and 1, then the image is a reduction.

Example 6

Find the scale factor for each pair of similar figures.

In all questions, assume the left figure is the original and the right figure the image.

a

12 mm, 20 mm; 18 mm, 30 mm

b

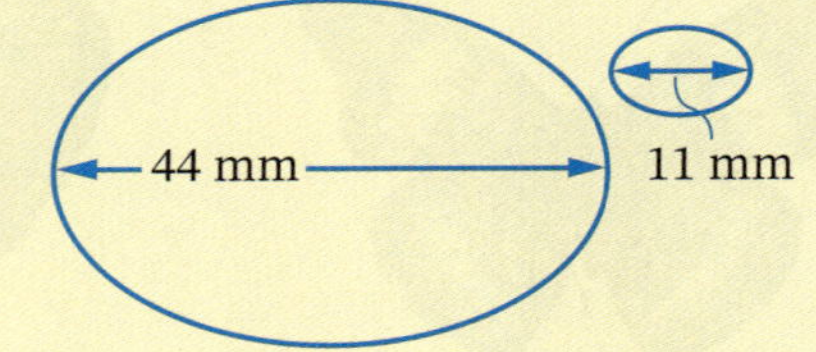

SOLUTION

a Scale factor $= \frac{30}{20}$ (or $\frac{18}{12}$) $\qquad \frac{\text{image length}}{\text{original length}}$

$= 1\frac{1}{2}$

b Scale factor $= \frac{11}{44}$

$= \frac{1}{4}$

Example 7

Construct the image of the figure shown if the scale factor is:

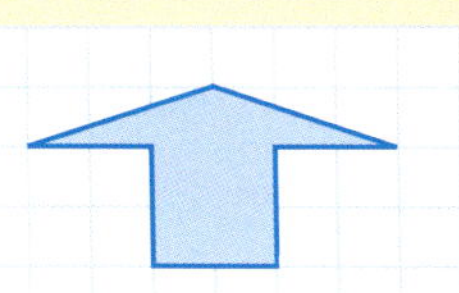

a 3 **b** $\frac{1}{2}$

SOLUTION

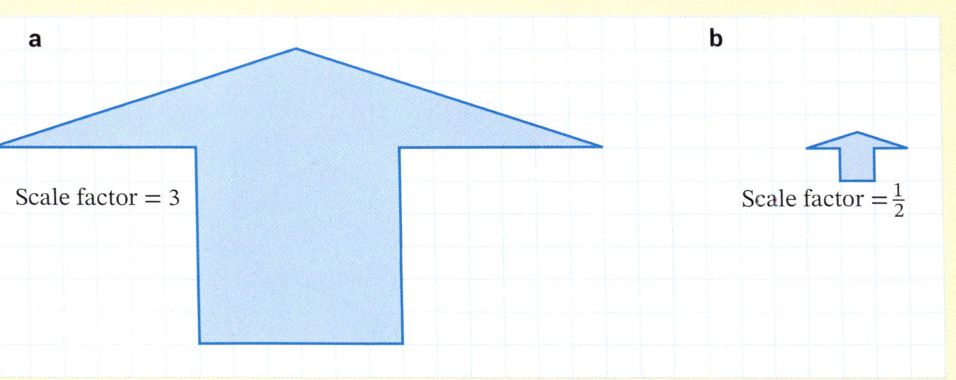

The similarity symbol |||

The symbol for 'is similar to' is '|||'. As with congruence notation, we must make sure that the vertices (or angles) of the similar figures are written in matching order.

Example 8

Test whether each pair of figures are similar by measuring them.

a

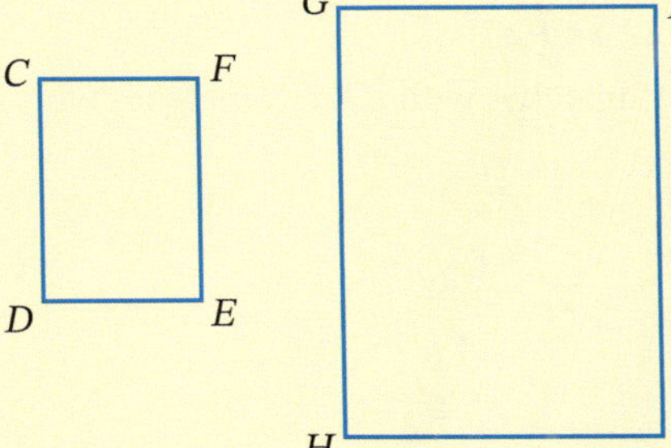

b

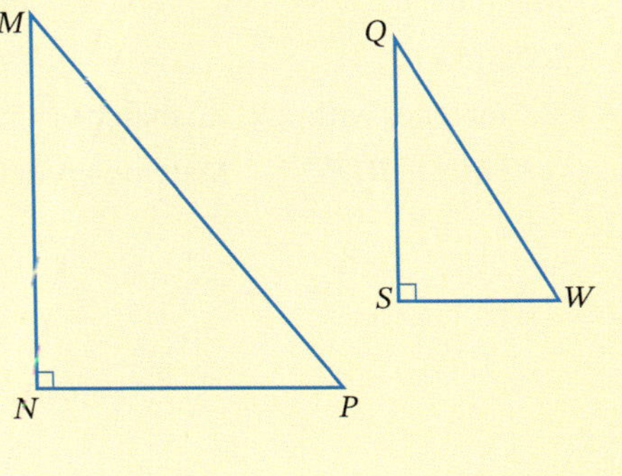

SOLUTION

a The rectangle $GHKL$ is an enlargement of rectangle $CDEF$.

$$\frac{GH}{CD}=\frac{36}{18}=2$$

$$\text{Scale factor} = \frac{HK}{DE}=\frac{26}{13}=2$$

$$\therefore \frac{GH}{CD}=\frac{HK}{DE}$$

$\therefore CDEF \mid\mid\mid GHKL$ Matching order of vertices.

b ΔQWS is not a reduction of ΔMPN.

$$\frac{QS}{MN}=\frac{28}{40}=\frac{7}{10}$$

$$\frac{SW}{NP}=\frac{16}{32}=\frac{1}{2}$$

$$\frac{QS}{MN}\neq\frac{SW}{NP}$$

$\therefore \Delta MNP$ is not similar to ΔQSW.

Example 9

The 2 quadrilaterals $KLMN$ and $PQRT$ are similar.

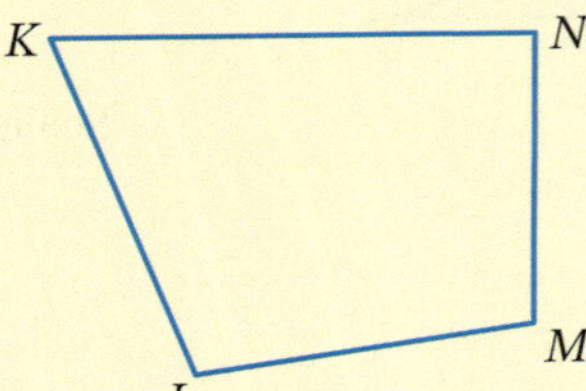

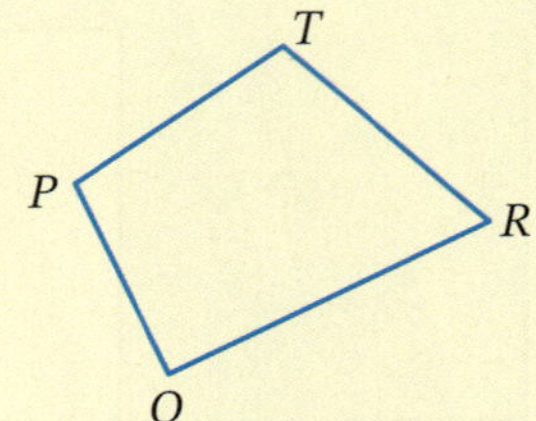

a List all pairs of matching sides and matching angles.

b Use the correct notation to write a similarity statement relating these 2 quadrilaterals.

SOLUTION

a By rotating the figure $KLMN$, its shape can be matched with $PQRT$.

The pairs of matching sides are:	The pairs of matching angles are:
KN and QR	$\angle K$ and $\angle R$
MN and PQ	$\angle N$ and $\angle Q$
ML and PT	$\angle M$ and $\angle P$
LK and TR.	$\angle L$ and $\angle T$.

b $\angle K$ matches with $\angle R$, $\angle L$ matches with $\angle T$, $\angle M$ matches with $\angle P$, $\angle N$ matches with $\angle Q$.

$\therefore KLMN \mid\mid\mid RTPQ$ Matching order of vertices

EXERCISE 13.04 ANSWERS ON P. 680

Similar figures

1 By measurement, find the scale factor for each pair of similar figures.

EXAMPLE 6

a

b

c

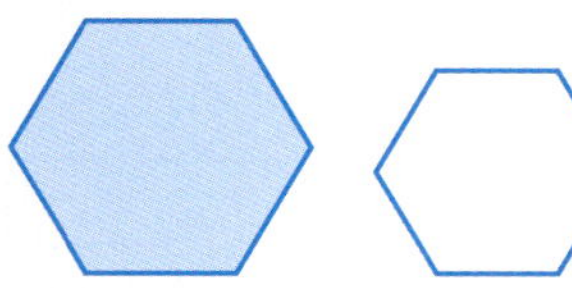

d

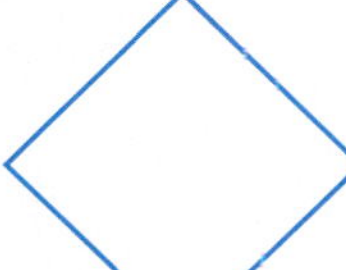

e

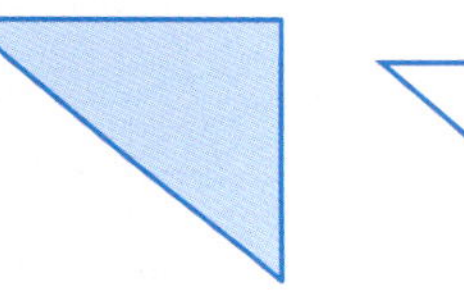

f

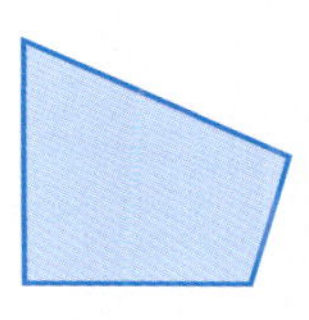

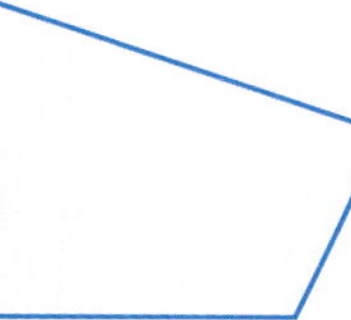

g

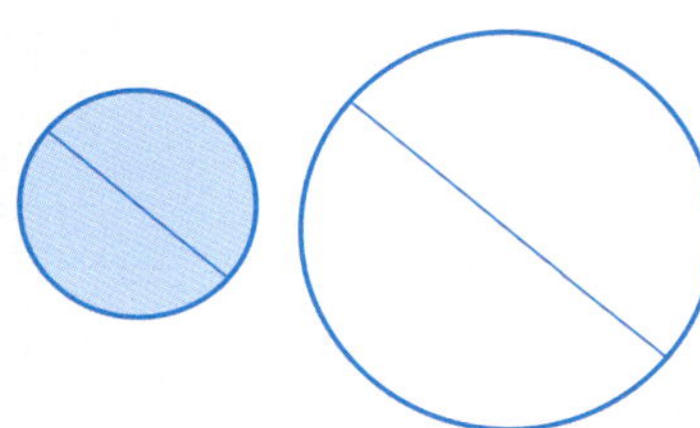

h

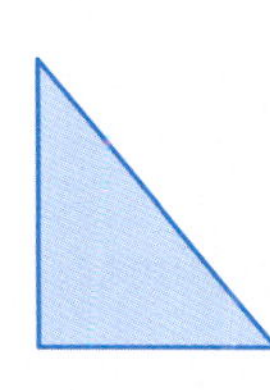

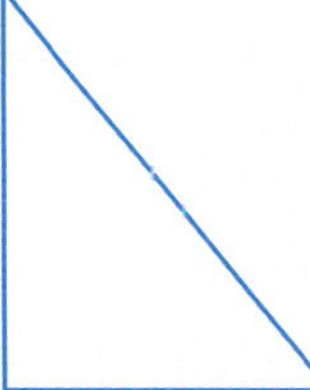

2 Copy each figure onto graph paper and draw its image using the given scale factor.

EXAMPLE 7

a Scale factor $= 2$

b Scale factor $= \frac{1}{3}$

c Scale factor $= 2$

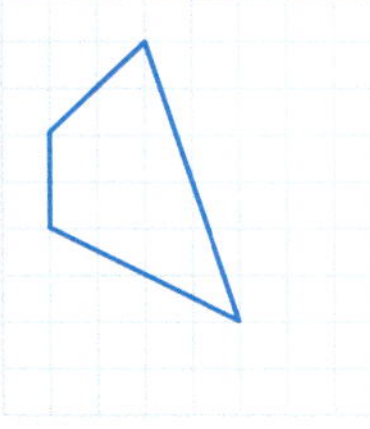

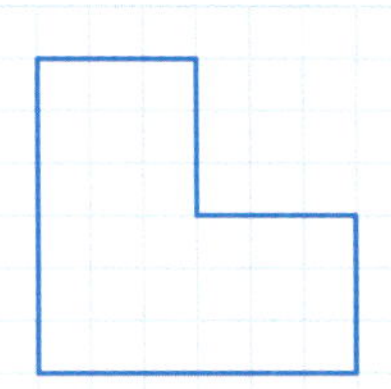

d Scale factor $= 3$

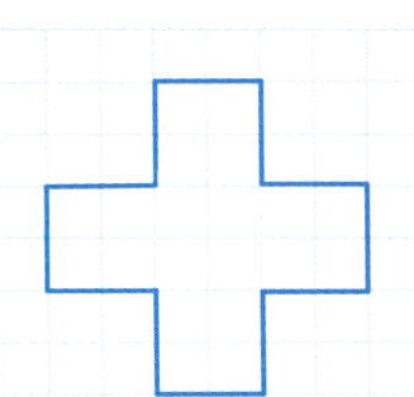

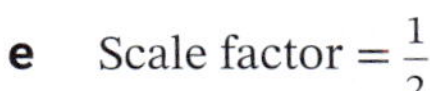

e Scale factor $= \frac{1}{2}$

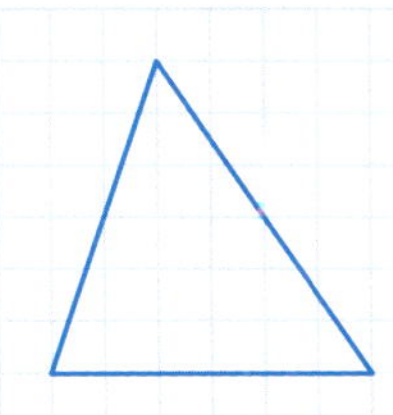

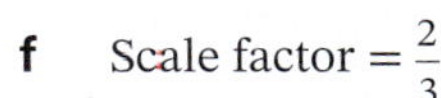

f Scale factor $= \frac{2}{3}$

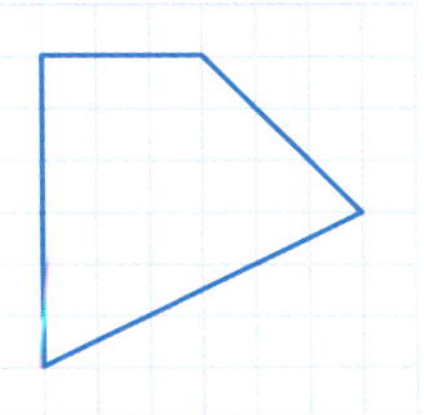

Complex

EXAMPLE 8

3 Are each pair of figures similar?

a

b

c

d

e

f

g

h

EXAMPLE 9

4 For each pair of similar figures:

R **i** list all pairs of matching angles.

C **ii** list all pairs of matching sides.

iii use the correct notation to write a similarity statement relating them.

a

b

c

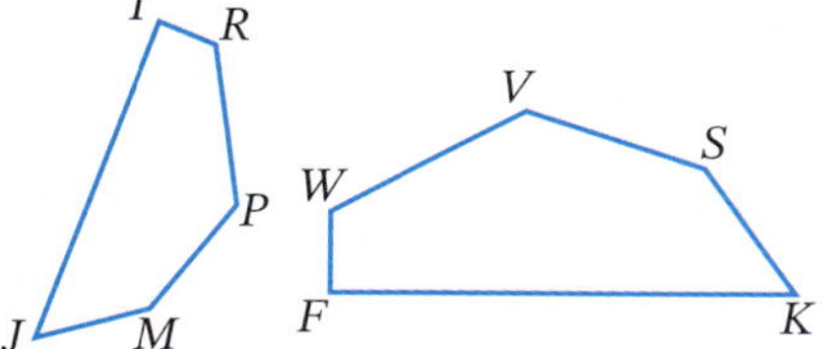

d

Q K G C B N

e

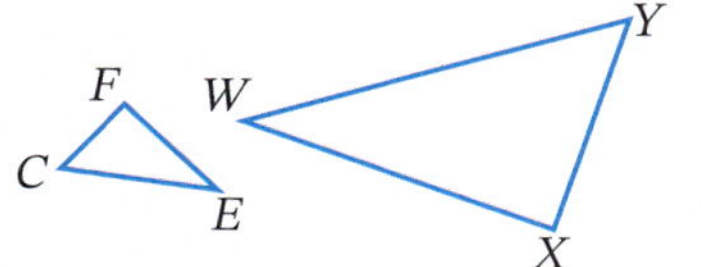

f

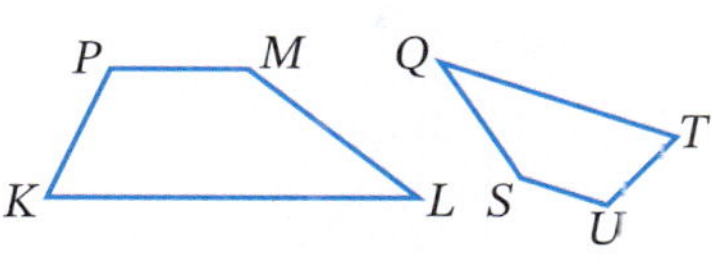

13.04

☆ MENTAL SKILLS (13) ANSWERS ON P. 681 Maths without calculators

Quiz Mental skills 13

Simplifying fractions and ratios

When simplifying a fraction or a ratio, look for a common factor to divide into both the numerator and the denominator, preferably the highest common factor (HCF). Remember the divisibility tests:

A number is divisible by:	if:
2	it is even (its last digit is 2, 4, 6, 8 or 0)
3	the sum of its digits is divisible by 3
4	its last 2 digits form a number divisible by 4
5	its last digit is 0 or 5
6	it is even and the sum of its digits is divisible by 3
9	the sum of its digits is divisible by 9
10	its last digit is 0

1 Study each example.

a Simplify $\frac{27}{45}$.

$\frac{27^9}{45_{15}} = \frac{9}{15}$ dividing numerator and denominator by 3

$\frac{9^3}{15_5} = \frac{3}{5}$ dividing numerator and denominator by 3 again

Note: This fraction could be simplified in one step if you divided by 9, the highest common factor (HCF) of 27 and 45.

b Simplify $\frac{160}{400}$.

$\frac{160}{400} = \frac{16}{40}$ dividing numerator and denominator by 10

$\frac{\cancel{16}^{2}}{\cancel{40}_{5}} = \frac{2}{5}$ dividing numerator and denominator by 8

Note: This fraction could be simplified in one step if you divided by 80, the HCF of 160 and 400.

c Simplify 24 : 36.

$24:36 = \cancel{24}^{4} : \cancel{36}^{6} = 4:6$ dividing both terms by 6

$\cancel{4}^{2} : \cancel{6}^{3} = 2:3$ dividing both terms by 2

Note: This fraction could be simplified in one step if you divided by 12, the HCF of 160 and 400.

d Simplify 135 : 90.

$135:90 = \cancel{135}^{27} : \cancel{90}^{18} = 27:18$ dividing both terms by 5

$\cancel{27}^{3} : \cancel{18}^{2} = 3:2$ dividing both terms by 9

e Calculate $\frac{3}{8} \times \frac{2}{15}$ in simplest form.

$\frac{3}{8} \times \frac{2}{15} = \frac{3}{\cancel{8}^{4}} \times \frac{\cancel{2}^{1}}{15}$ dividing 2 and 8 by 2

$= \frac{\cancel{3}^{1}}{\cancel{8}^{4}} \times \frac{\cancel{2}^{1}}{\cancel{15}^{5}}$

$= \frac{1}{20}$

f What fraction is 36 minutes of 1 hour?

$\frac{36}{1\text{ h}} = \frac{36\text{ min}}{60\text{ min}} = \frac{3}{5}$

2 Now simplify each fraction or ratio.

a	$\frac{10}{15}$	**b**	$\frac{16}{20}$	**c**	$\frac{30}{42}$	**d**	$\frac{8}{16}$
e	$\frac{20}{80}$	**f**	$\frac{6}{36}$	**g**	$\frac{20}{24}$	**h**	$\frac{12}{30}$
i	20 : 36	**j**	25 : 45	**k**	18 : 40	**l**	28 : 35
m	27 : 21	**n**	16 : 12	**o**	$\frac{5}{6} \times \frac{18}{25}$	**p**	$\frac{12}{50} \times \frac{10}{21}$

3 Express each as a simplified fraction.

a	425 g of 1 kg	**b**	8 months of 1 year	**c**	64 cm of 1 m
d	750 mL of 3 L	**e**	10 hours of 2 days	**f**	80c of $10

INVESTIGATION

13.04

Properties of similar figures

1 The quadrilaterals $ABCD$ and $HKLM$ are similar.

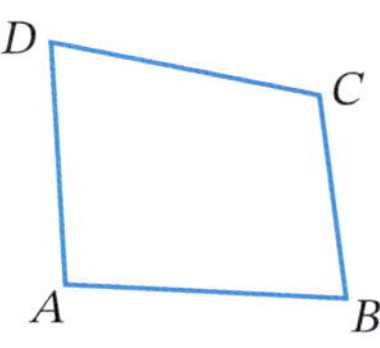

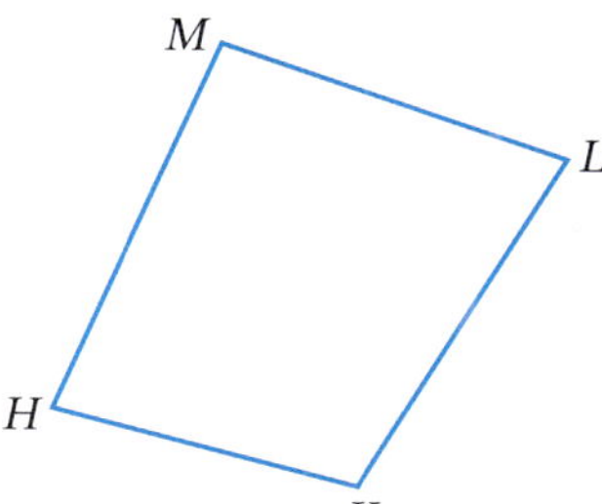

a Measure the sizes of each pair of matching angles. What do you notice?

b Measure the lengths of each pair of matching sides.

c Calculate each ratio of matching sides, as a decimal correct to one decimal place.

i $\frac{MH}{AB}$ ii $\frac{HK}{BC}$ iii $\frac{KL}{CD}$ iv $\frac{ML}{AD}$

What do you notice?

d Copy and complete.

i The matching a________ of similar figures are e________.

ii Matching s________ of similar figures are in the same r________.

2 Consider the parallelograms $DEKL$ and $ANWX$.

a Measure the sizes of each pair of matching angles. Are they equal?

b Measure the lengths of each pair of matching sides.

c Calculate each ratio of matching sides, as a decimal correct to one decimal place:

i $\frac{AN}{DE}$ ii $\frac{AX}{DL}$

Are they equal?

d $DEKL$ is not similar to $ANWX$. Explain why.

13.05 Properties of similar figures

Skillsheet Finding sides in similar triangles

Puzzle Similar triangles

ⓘ Properties of similar figures

- **Matching angles** are equal
- **Matching sides** are in the same ratio

Example 10

Test whether each pair of figures are similar.

a

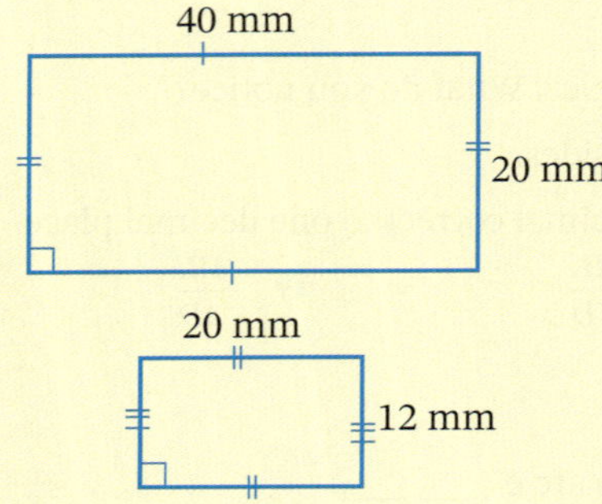

b

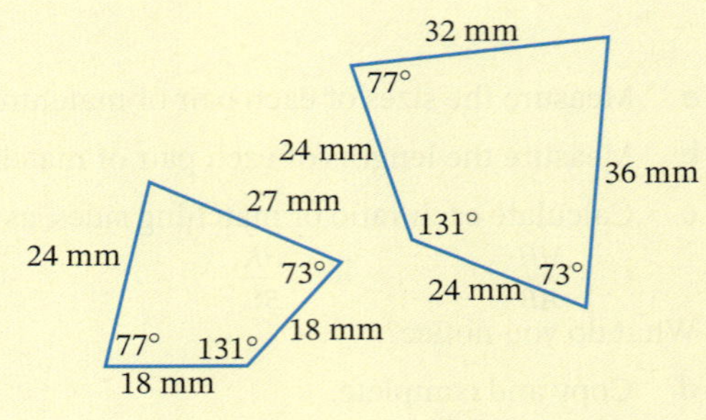

SOLUTION

a For the 2 rectangles, matching angles are equal (90°), but the ratios of the matching sides are not equal ($\frac{20}{40} = \frac{1}{2}$ but $\frac{12}{20} = \frac{3}{5}$).

$\therefore$ The rectangles are not similar.

b For the 2 quadrilaterals, matching angles are equal and the ratios of matching sides are equal $\left(\frac{32}{24} = \frac{36}{27} = \frac{24}{18} = \frac{4}{3}\right)$.

$\therefore$ The quadrilaterals are similar.

Video Finding an unknown side in similar figures

Example 11

These 2 triangles are similar. Find the value of x and y.

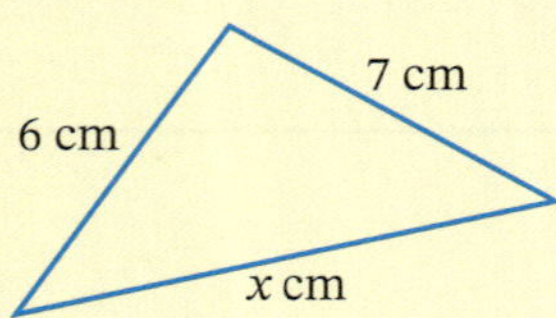

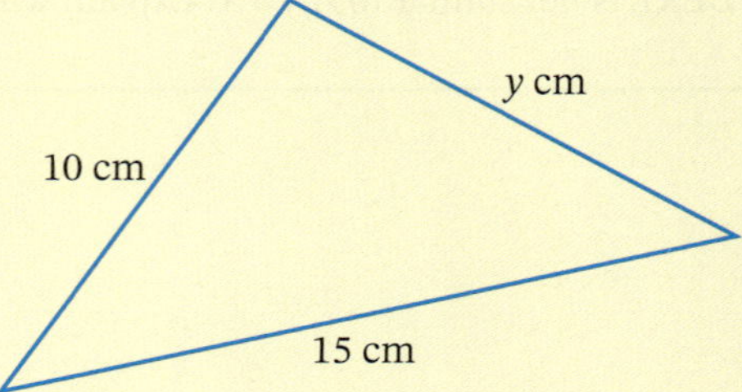

13.05

SOLUTION

Since the triangles are similar, the ratios of matching sides are equal.

$\frac{x}{15}=\frac{6}{10}$

$x=\frac{6}{10}\times 15$

$=9$

$\frac{y}{7}=\frac{10}{6}$

$y=\frac{10}{6}\times 7$

$=11\frac{2}{3}$

Alternative method:

Scale factor $=\frac{10}{6}=\frac{5}{3}$

$x=15\div\frac{5}{3}$

$=9$

$y=7\times\frac{5}{3}$

$=11\frac{2}{3}$

Example 12

$\angle ABC\ |||\ \angle ADE$. Find the value of k.

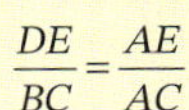

SOLUTION

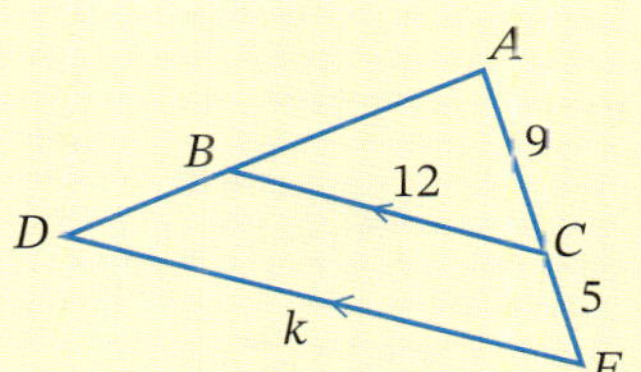

$\frac{DE}{BC}=\frac{AE}{AC}$ Ratios of matching sides are equal.

$\frac{k}{12}=\frac{14}{9}$ $AE=9+5=14$

$k=\frac{14}{9}\times 12$

$=18\frac{2}{3}$

Puzzle Nested similar triangles

EXERCISE 13.05 ANSWERS ON P. 681

Properties of similar figures

U F PS R

EXAMPLE 10

1 Test whether each pair of figures are similar.

R **a**

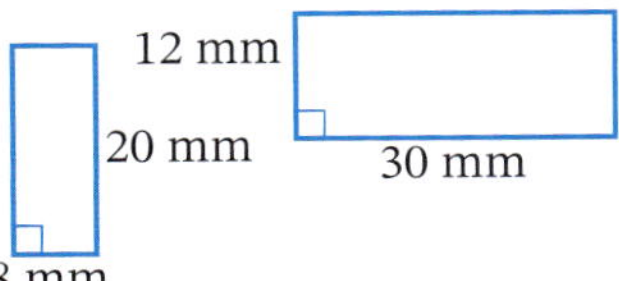

b

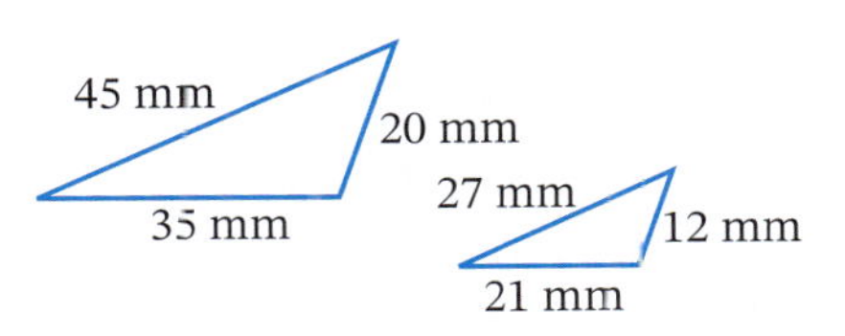

Foundation Standard Complex

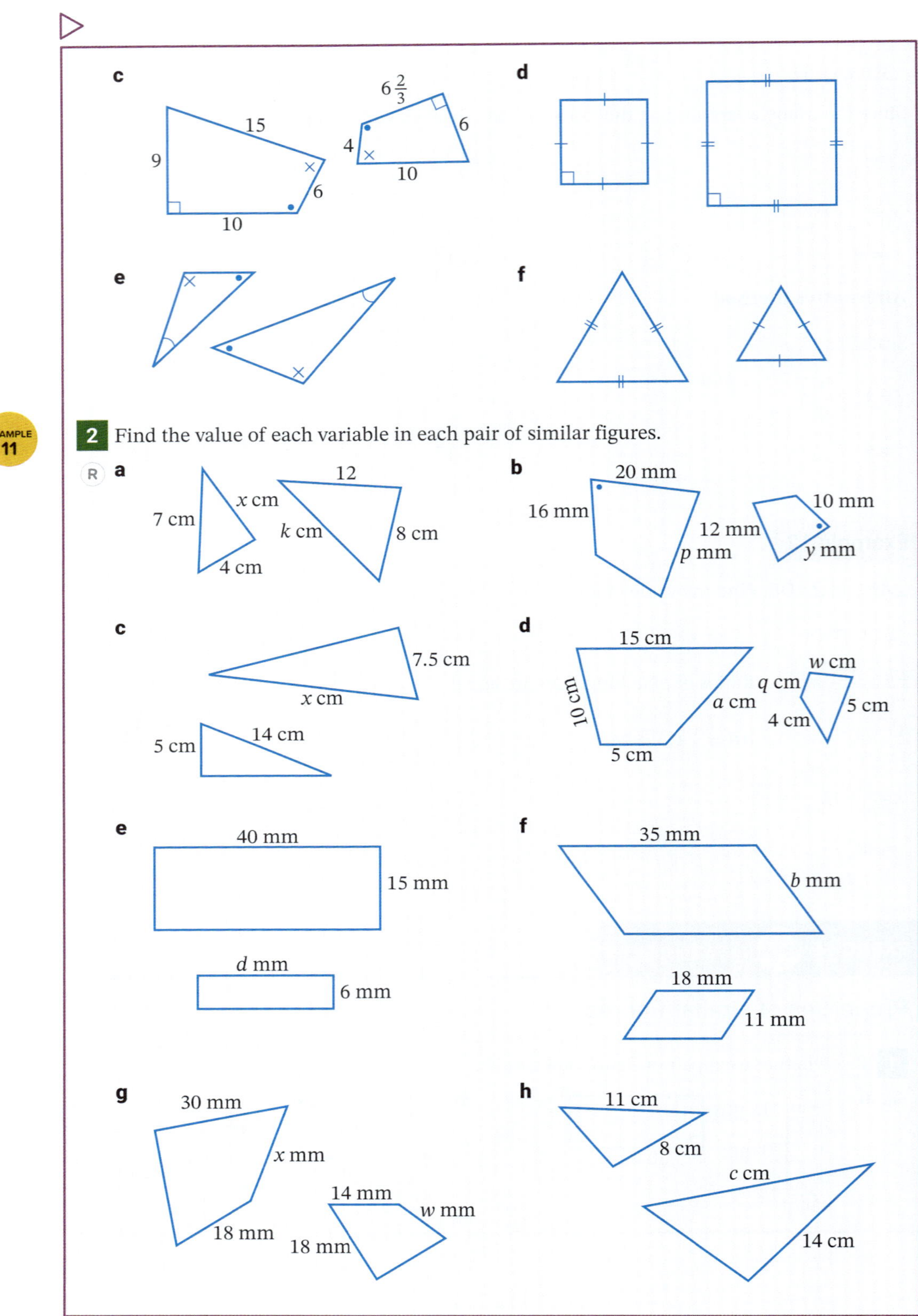

Foundation Standard Complex

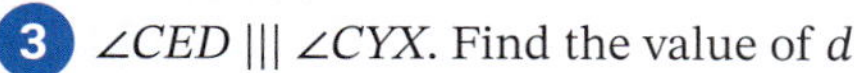

3 R $\angle CED \mid\mid\mid \angle CYX$. Find the value of d.

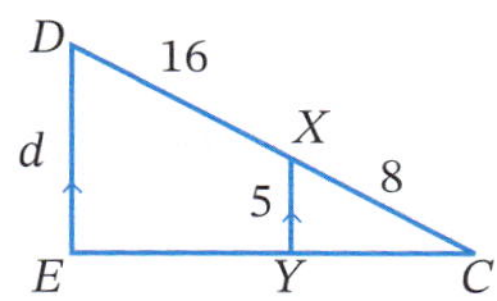

4 R $\angle LMN \mid\mid\mid \angle LPR$. Find the value of a.

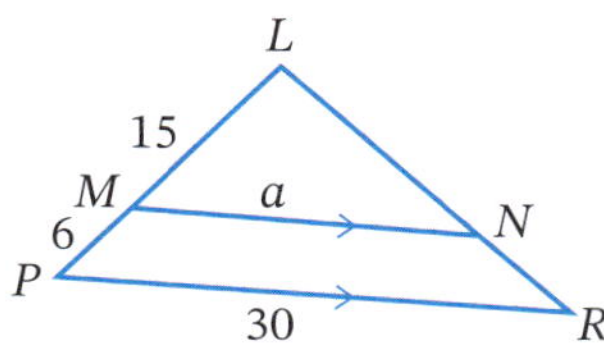

5 R A metre stick casts a shadow of length 0.6 m when a tree casts a shadow of 8 m. Find the height h of the tree.

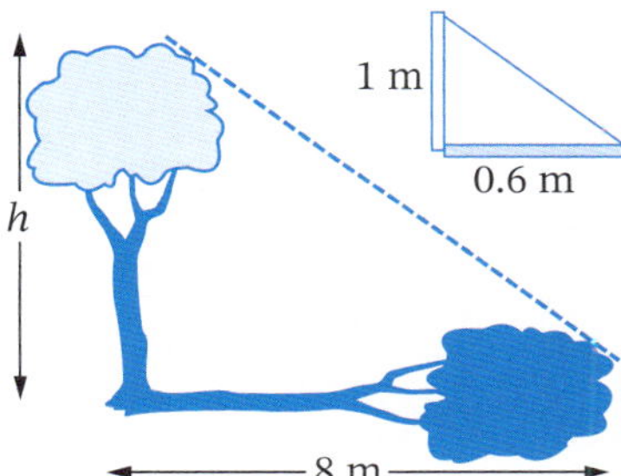

6 R Ayon, who is 1.9 m tall, casts a 1.5 m long shadow. At the same time, a light pole casts a shadow 7 m long. What is the height of the light pole, correct to one decimal place?

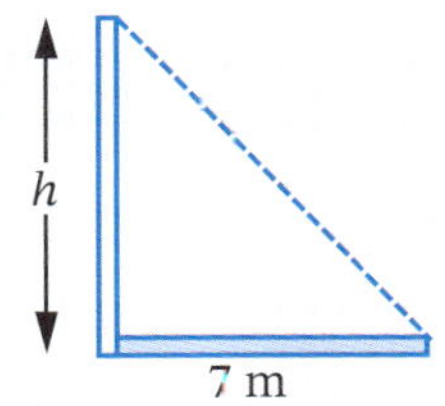

7 PS R Emily holds up a small 40 cm long stick from her eye so that it appears to be the same height as the flagpole. She is standing 28 m from the flagpole and the stick is 30 cm long. Find the height of the flagpole.

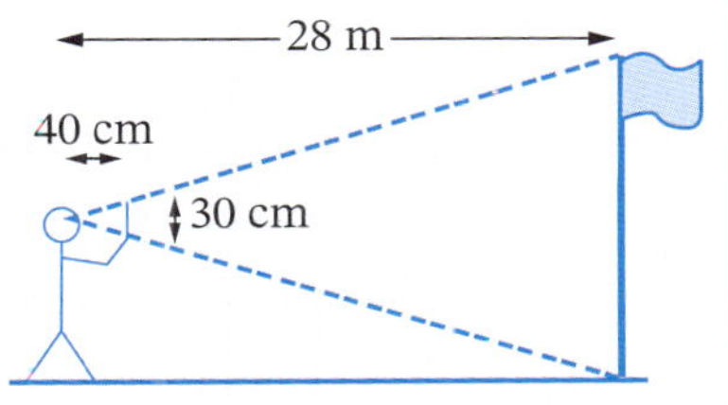

8 For each set of figures, find the pair of similar figures and the scale factor between them.

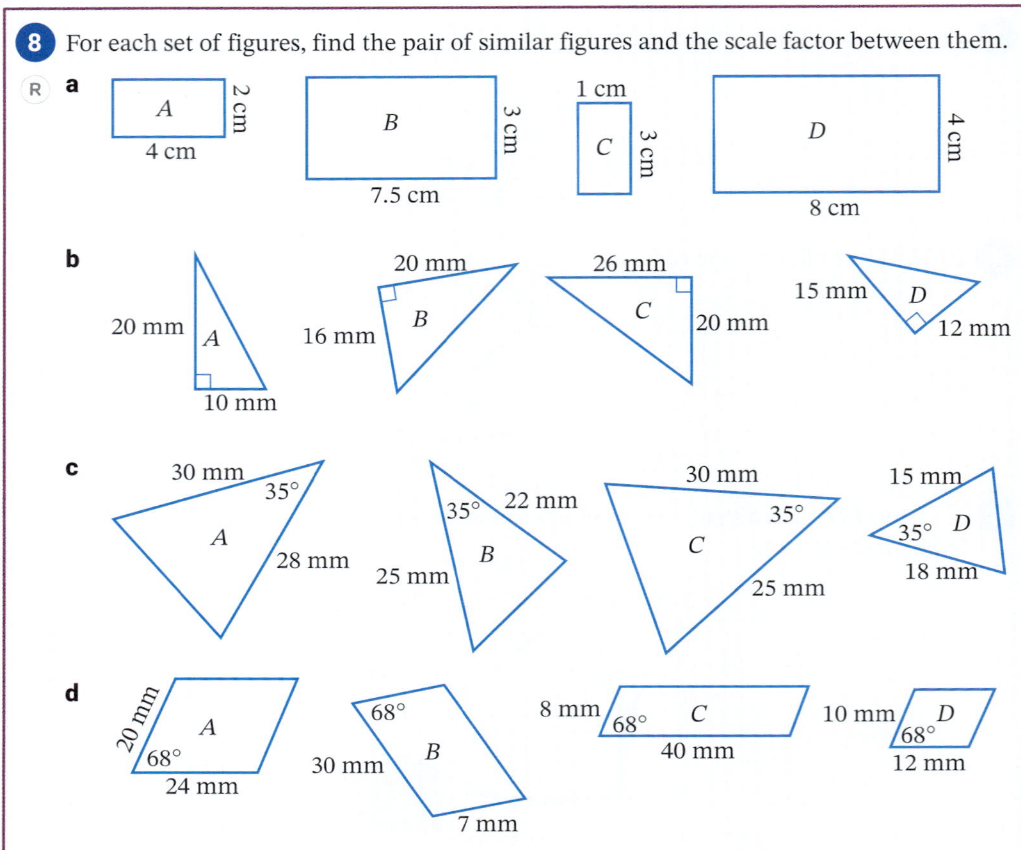

INVESTIGATION

Are all shapes similar?

Work in groups of 3 or 4 to complete this activity.

You will need: a ruler, a pencil, scissors.

1 a Draw 2 equilateral triangles, using a different side length for each triangle.

b Are the 2 equilateral triangles similar? Give reasons.

c Compare your triangles with those of other members of your group.
Are all equilateral triangles similar? Give reasons.

2 a Draw a pair of similar isosceles triangles.

b Are all isosceles triangles similar? Give reasons.

3 a Draw a pair of rectangles that are similar.

b Draw a pair of rectangles that are *not* similar.

c When are rectangles similar? Discuss your answer with other students.

Foundation Standard Complex

4 State whether the following statements are true or false. Give reasons for your answers. (Making a few sketches will help.)

a All parallelograms are always similar.

b All circles are always similar.

c All squares are always similar.

d All rhombuses are always similar.

Scale diagrams 13.06

A **scale diagram** accurately represents a larger or smaller object such that the lengths and distances on it are in the same ratio as the real lengths and distances. This means that the scale diagram is either an enlargement or reduction of the real object. The scale factor used is called the **scale** of the diagram.

A scale may be represented in different ways.

- 1 cm = 10 m or 1 cm to 10 m (a pair of corresponding measurements)
- 1 : 50 (a ratio)
- (a line drawn to scale)

0 1 2 3 4 5 km

Worksheet Problems involving scale drawings

Technology Converting map scales to ratios

Spreadsheet Converting map scales to ratios

Videos Scale drawings

Jai Singh (sundial)

Modelling the Spitfire

Queen Hatshepsut's ship

The history of the golden ratio

Fractals: The Koch snowflake

ⓘ Scaled length

The scale on a scale diagram is written as a ratio of **scaled length : real length**, where **scaled length** is the length on the diagram.

For example, a scale of 1 : 100 means that the real lengths are 100 times larger than the lengths on the diagram.

Example 13

A scale diagram of a rectangular block of land has been drawn. What scale has been used?

20 mm

35 mm

SOLUTION

By measurement, the length and width of the scale drawing are 35 mm and 20 mm.

Scale = 35 mm : 35 m (or 20 mm : 20 m) — Scaled length : real length

= 35 mm : 35 000 mm (or 20 mm : 20 000 m) — 1 m = 1000 mm

= 1 : 1000

Example 14

A SIM card for a mobile phone has a length of 25 mm and a width of 15 mm.
A scale diagram of the SIM card is shown below.

By measurement, determine what scale has been used.

SOLUTION

By measurement, the length and width of the scale diagram are 75 mm and 45 mm respectively.

Scale = 75 mm : 25 mm (or 45 mm : 15 mm)   *scaled length : real length*

= 3 : 1

Example 15

The scale on a map is 1 : 500 000. If the distance from Caboolture to Brisbane on the map is 9 cm, calculate the actual distance from Caboolture to Brisbane.

SOLUTION

Scaled distance = 9 cm

Actual distance = 9 × 500 000 cm

= 4 500 000 cm

= 45 000 m   *1 m = 100 cm*

= 45 km   *1 km = 1000 m*

EXERCISE 13.06 ANSWERS ON P. 681

Scale diagrams

U F R

1 If the length of a netball court is 30.5 m, what scale has been used on this diagram?

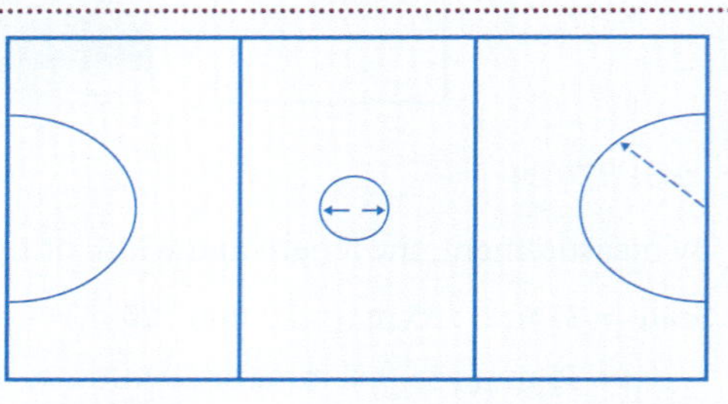

2 What scale has been used on this diagram of a car if its actual length is 3.92 m?

3 The picture of the Giant Brown Bull Ant has been enlarged. If its actual length is 25 mm, what scale has been used?

EXAMPLE 14

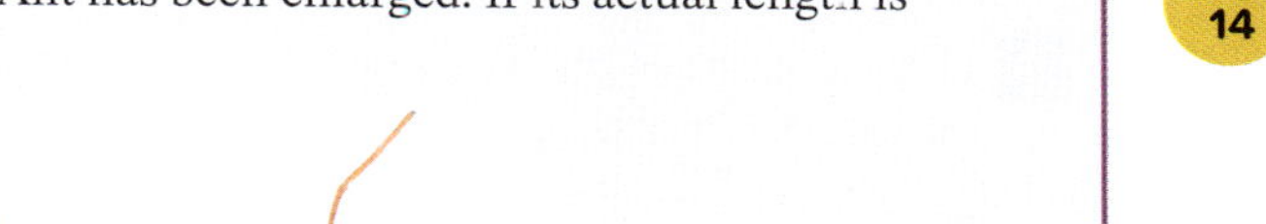

Nature Picture Library/Kim Taylor

4 A scale image of a microchip is shown. Determine the scale that has been used if its real length is 8 mm.

iStock.com/Fotokot197

5 The Q1 Tower on the Gold Coast, Queensland, is 322 m high and is Australia's tallest building. What scale has been used in this photo if the scaled height is 40 mm?

Shutterstock.com/Steven Bostock

6 This photo of the Sydney Harbour Bridge has a scale of 1 : 9000.

R By measurement and calculation, find:

a the height of the top of the bridge above the water.

Shutterstock.com/EerikSandstrom

b the length of the arch span of the bridge.

7 This is a photograph of a mosquito, at a scale of 15 : 1. Calculate, to the nearest 0.1 mm, the length of the wing.

R

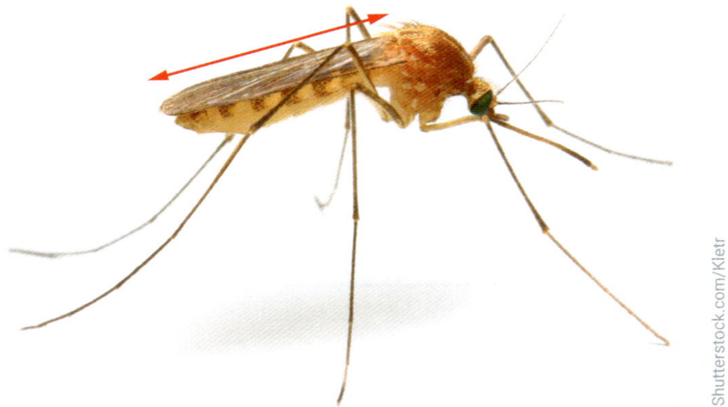
Shutterstock.com/Kletr

8 This box of cereal is 16 cm long and 22 cm tall. Construct a scale drawing of the front of the box, using a scale of 1 : 5.

Shutterstock.com/Nor Gal

9 **a** Construct a scale drawing of this flagpole using a scale of 1 : 300.

R **b** Use the scale to calculate the actual height of the flagpole.

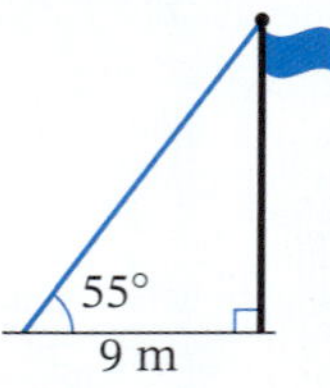

Foundation Standard Complex

EXAMPLE 15

13.06

10 The scale on a map of Australia is 1 : 10 000 000. If the distance from Perth to Adelaide on the map is 21.2 cm, calculate the actual distance in kilometres.

PS R

11 Use the scale on this map to find, to the nearest kilometre, the distance from:

R

- **a** Nowra to Batemans Bay
- **b** Bega to Moruya
- **c** Mollymook/Ulladulla to Mogo
- **d** Nowra to Eden.

INVESTIGATION

Floor plans

Construct a scale diagram of your bedroom or classroom.

Measure the dimensions of the room and each piece of furniture in it. Remember to measure the location of immovable objects such as windows, doors, wardrobes and whiteboards. Include a list of your measurements and the scale used. You may also do an Internet search of architectural symbols (bed, lounge, sink etc.) and include these in your room, where applicable.

Foundation Standard Complex

INVESTIGATION

Areas of similar figures

Technology
Area of similar shapes

Spreadsheet
Area of similar shapes

1 a The 2 rectangles A and B are similar. Explain why.
  b What is the ratio of their matching sides (A to B)?
  c What is the ratio of their areas (A to B)?

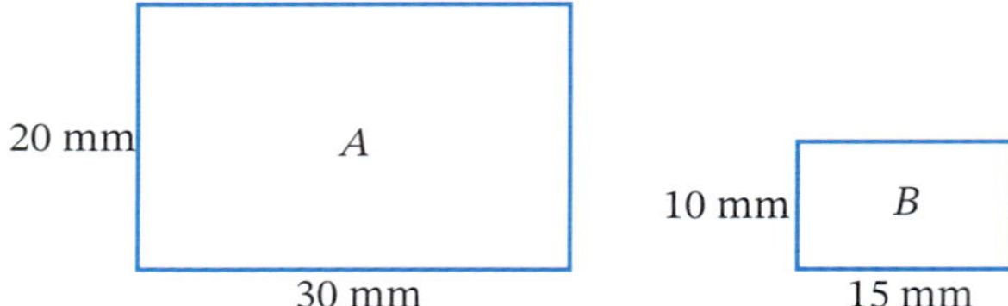

2 Triangles K and L are similar.

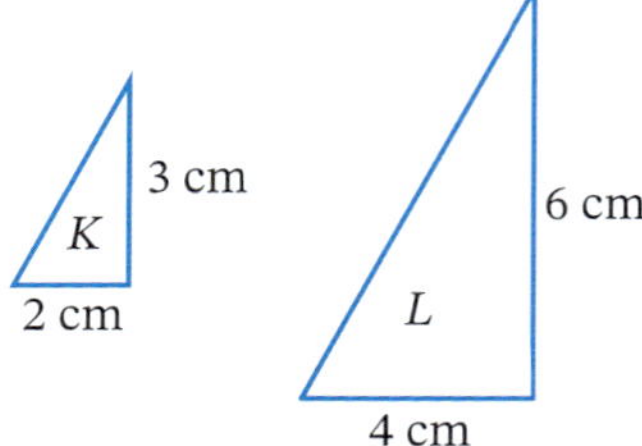

  a What is the ratio of their matching sides (K to L)?
  b Find the areas of both triangles.
  c What is the ratio of their areas (K to L)?
  d Copy and complete: When you double the lengths of a figure, the area is increased _______ times.

3 a What is the area of a rectangle with length 5 cm and width 4 cm?
  b The length and width of the rectangle are both multiplied by 3. What is the area of the enlarged rectangle?
  c What is the ratio of the matching sides of the 2 rectangles from parts **a** and **b**?
  d What is the ratio of the areas of the 2 rectangles?
  e Copy and complete: When you triple the lengths of a figure, the area is increased _______ times.

4 a What is the area of a circle with radius 2 cm?
  b The radius of this circle is multiplied by 4.
  c What is the area of the enlarged circle?
  d What is the ratio of the areas of the 2 circles?
  e Copy and complete: When you multiply the lengths of a figure by 4, the area is increased _______ times.

5 If the sides of a square are multiplied by 5, by how many times is the area increased?

6 If the sides of a figure are multiplied by k, by how many times is the area increased?

TECHNOLOGY

Areas of similar figures

For this activity, use dynamic geometry software to draw similar figures and compare their areas.

1 Construct a square. An example is shown.

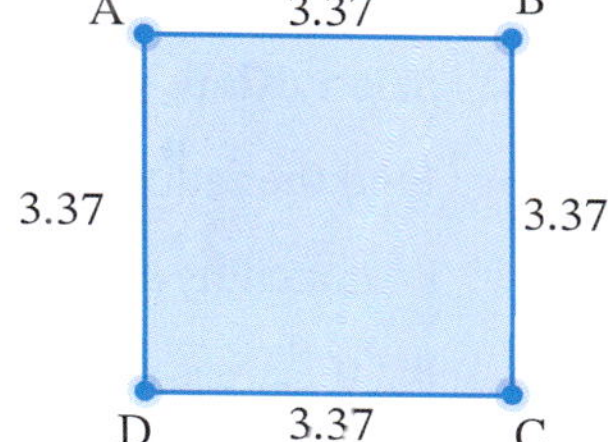

2 Calculate the area of the square you have constructed in question 1.

3 Dilate (enlarge) the square by a scale factor of 2.

4 Use the tools to:

a measure the length of each side.

b calculate the area of the enlarged square.

c What is the ratio of the sides $DC : GF$?

d What is the ratio of the areas of the squares $ABCD : AEFG$?

e Compare the 2 ratios, **c** and **d**. How are they related?

A 3.37 B E
3.37 3.37
D 3.37 C
G F

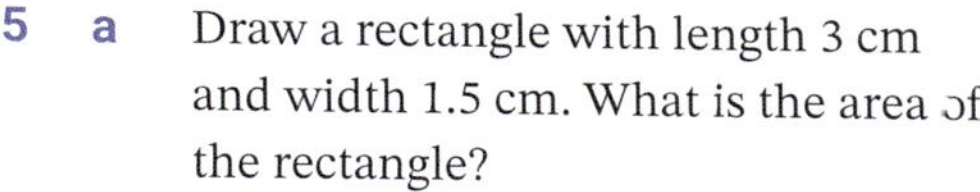

5 a Draw a rectangle with length 3 cm and width 1.5 cm. What is the area of the rectangle?

b Dilate the rectangle by a scale factor of 2 to enlarge the rectangle.

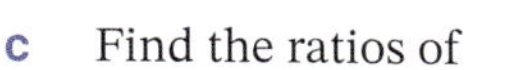

c Find the ratios of

i sides $DC : GF$

ii sides $BC : EF$

iii areas of rectangles $ABCD : AEFG$.

d Compare the ratio of the sides to the ratio of the areas. How are they related?

6 a Draw a triangle with $AB = 8$ cm. It should look like triangle ABC shown.

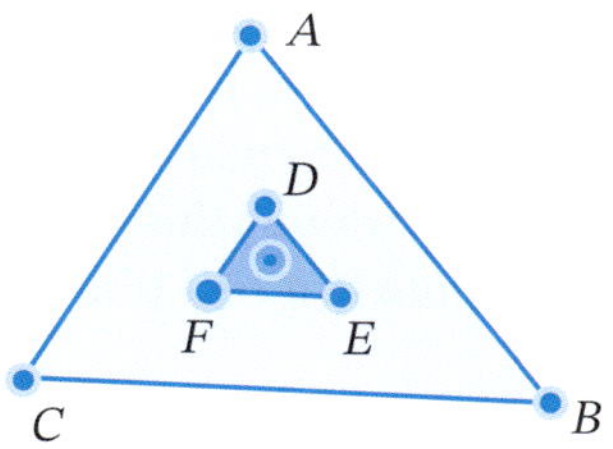

b Dilate (reduce) the triangle by a scale factor of $\frac{1}{4}$ to reduce the triangle. Dilation can also be done by creating a point inside triangle ABC (it doesn't have to be one of the vertices, as shown in the diagram above).

c Find the ratios of:

i sides $AB : DE$ ii areas of triangles $ABC : DEF$

d Compare the ratio of the sides to the ratio of the areas. How are they related?

7 a Draw any polygon.

b Select **dilate** and either enlarge or reduce the polygon.

c Find the ratios of:

i matching sides of the similar shapes.

ii the areas of the similar shapes.

d Compare the ratio of the sides to the ratio of the areas. How are they related?

DID YOU KNOW?

Australia's Big Things

Unique to Australia are a number of tourist attractions known as the 'Bigs' or 'Biggies':

- the Big Merino (Goulburn, NSW)
- the Big Banana (Coffs Harbour, NSW)
- the Big Golden Guitar (Tamworth, NSW)
- the Big Pineapple (Woombye, Qld)
- the Big Orange (Berri, SA)
- the Big Crocodile (Wyndham, WA)
- the Big Captain Cook (Cairns, Qld)
- the Big Mango (Bowen, Qld)
- the Big Rocking Horse (Gumeracha, SA)
- the Giant Koala (Dadsells Bridge, Vic).

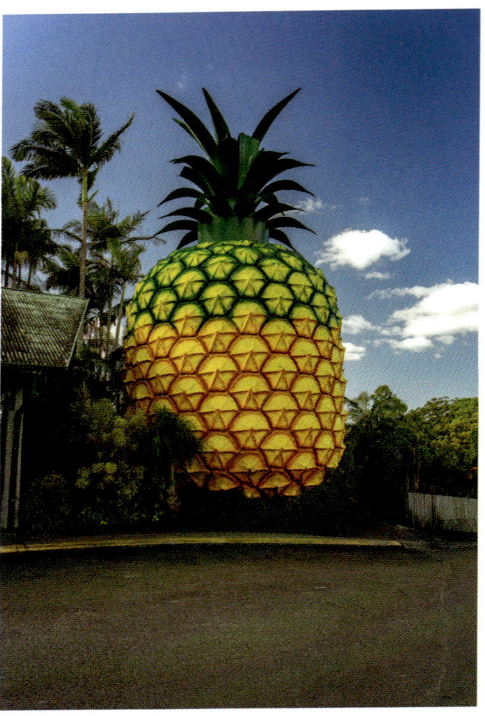

Shutterstock.com/Alex Cimbal

The Big Pineapple on Queensland's Sunshine Coast is 16 metres high. If the height of an average pineapple is 30 cm, work out the scale of the Big Pineapple to a real pineapple.

Areas of similar figures

ⓘ Areas of similar figures

If the matching sides of 2 similar figures are in the ratio $1 : k$, then their areas are in the ratio $1 : k^2$.

If the matching sides are in the ratio $m : n$, then their areas are in the ratio $m^2 : n^2$.

Worksheet
Investigating paper sizes

Example 16

a Two similar triangles have sides in the ratio 1 : 5. Find the ratio of their areas.

b Two similar parallelograms have sides in the ratio 3 : 1. Find the ratio of their areas.

SOLUTION

a Ratio of sides $= 1 : 5$ $1 : k^2$

Ratio of areas $= 1 : 5^2$

$= 1 : 25$

b Ratio of sides $= 3 : 1$

Ratio of areas $= 3^2 : 1$

$= 9 : 1$

Example 17

What is the ratio of the areas of the similar quadrilaterals shown?

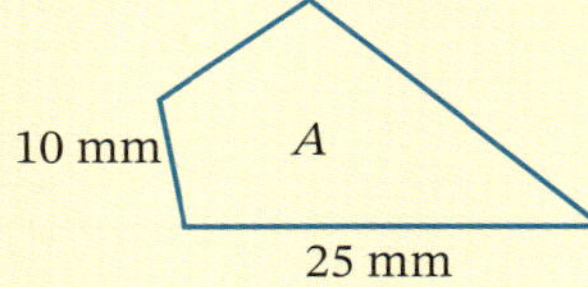

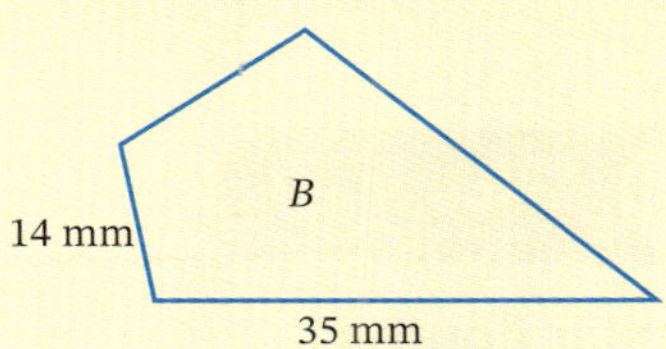

Video
Areas of similar figures

SOLUTION

Ratio of matching sides $(A \text{ to } B) = 25 : 35$ or 10 : 14

$= 5 : 7$

Ratio of areas $(A \text{ to } B) = 5^2 : 7^2$

$= 25 : 49$

Example 18

Two similar trapeziums have their areas in the ratio 25 : 16. Find the ratio of the lengths of the matching sides.

SOLUTION

Ratios of areas $= m^2 : n^2 = 25 : 16$

$\therefore$ Ratio of sides $= m : n = \sqrt{25} : \sqrt{16} = 5 : 4$

Example 19

Two similar triangles P and Q have matching sides in the ratio 3 : 4. If the area of the larger triangle is 288 cm², find the area of the smaller triangle.

SOLUTION

Drawing a diagram of the problem is useful.

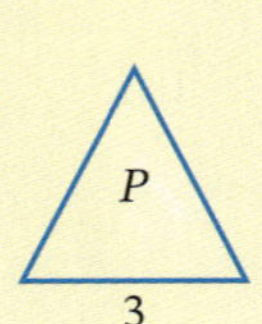

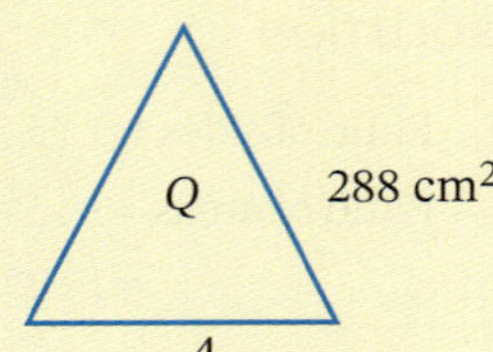

Ratio of matching sides = 3 : 4

Ratio of areas $= 3^2 : 4^2 = 9 : 16$

$\therefore \frac{\text{Area of } P}{288} = \frac{9}{16}$

$\text{Area of } P = \frac{9}{16} \times 288$

$= 162 \text{ cm}^2$

The area of the smaller triangle is 162 cm².

EXERCISE 13.07 ANSWERS ON P. 681

Areas of similar figures

U F R

1 Two similar rectangles have sides in the ratio 1 : 4. Find the ratio of their areas.

2 Two similar circles have radii in the ratio 1 : 2.5. Find the ratio of their areas.

3 The ratio of the matching sides for pairs of similar figures are as shown. Find the ratio of the areas for each pair.

a	10 : 1	**b**	5.5 : 1	**c**	3 : 4	**d**	5 : 2
e	4 : 7	**f**	1 : 10	**g**	8 : 5	**h**	2 : 9

EXAMPLE 17

4 Find the ratio of the areas for each pair of similar figures.

a

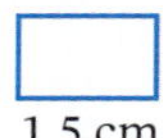

b

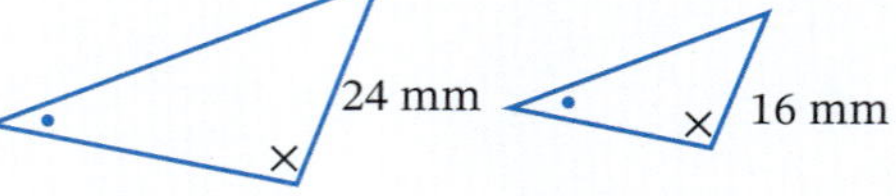

c

28 mm

16 mm

d

2 cm

4.4 cm

e

5 cm

3 cm

f

28 mm

20 mm

35 mm

25 mm

5 For each ratio between the areas of a pair of similar figures, find the ratio of the lengths of the matching sides.

a 64 : 49 **b** 36 : 1 **c** 9 : 400 **d** 1 : 20.25

6 Find the ratio of the length of matching sides for each pair of similar figures.

a

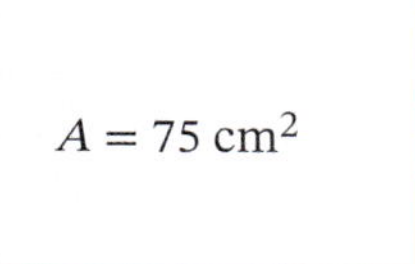

$A = 27$ cm^2

b

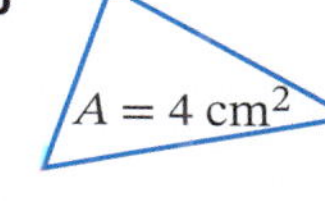

$A = 12.25$ cm^2

EXAMPLE 19

7 Find the value of x if these 2 triangles are similar.

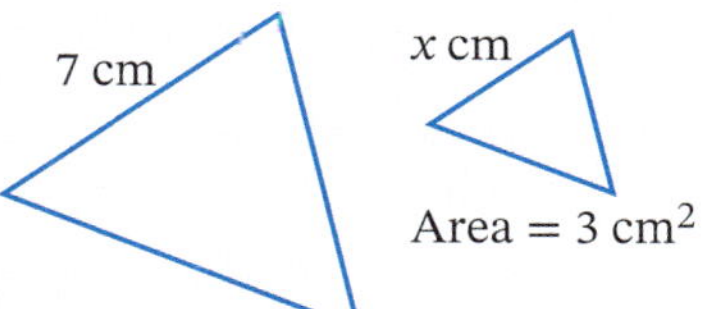

8 (R) Two similar triangles have matching sides in the ratio 4 : 1. If the area of the smaller triangle is 192 cm^2, find the area of the larger triangle.

9 (R) Two squares have sides in the ratio 3 : 4. If the area of the larger square is 288 cm^2, find the area of the smaller square.

10 (R) Two similar quadrilaterals have matching sides in the ratio 8 : 5. If the area of the smaller quadrilateral is 400 mm^2, find the area of the larger quadrilateral.

11 (R) Two circles have radii in the ratio 3 : 7. Find the area of the smaller circle if the area of the larger circle is 154 m^2.

☐ Foundation ○ Standard ⬡ Complex

12 Two similar rectangles have areas in the ratio 36 : 121. If the width of the smaller rectangle is 84 mm, find the width of the larger rectangle.
R

13 Two similar triangles have areas in the ratio 4 : 9. If the length of the base of the smaller triangle is 15 mm, find the length of the base of the larger triangle.
R

14 Two squares have areas in the ratio 49 : 25. The length of the larger square is 17.5 cm. Find the length of the smaller square.
R

15 If the radius of a circle is tripled, how has its area changed?
R

16 If the area of a square is reduced by a factor of 4, how have the sides changed?
R

17 The area of an equilateral triangle is decreased to $\frac{9}{16}$ of its original amount. How does its new area compare to its original area?
R

18 A rectangle is reduced using a scale factor of $\frac{1}{2}$. If the area of the rectangle was 200 cm^2, what is the area of the reduced rectangle?
R

INVESTIGATION

Similar triangles

Work in groups of 3 or 4 to complete this investigation.

You will need: paper, a ruler, a pencil, a protractor, compasses and scissors.

1 a Draw $\angle ABC$ with $AB = 25$ mm, $AC = 35$ mm and $BC = 20$ mm.

b How many different triangles can be drawn with sides of 25 mm, 35 mm and 20 mm? Is this triangle unique?

c Draw $\angle DEF$ with $DE = 75$ mm, $DF = 105$ mm and $EF = 60$ mm.

d Are the matching sides of the triangles in the same ratio?

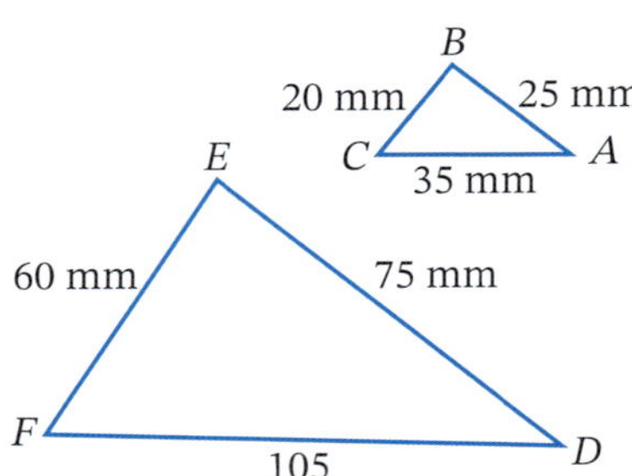

e Use your protractor to measure the angles of both triangles. Are matching angles equal?

f Is $\angle ABC$ similar to $\angle DEF$?

2 **a** Draw $\angle GHK$ with $\angle G = 72°$ and $\angle H = 45°$.

b Draw $\angle TPR$ with $\angle T = 72°$ and $\angle P = 45°$. Is $\angle GHK \mid\mid\mid \angle TPR$?

c Compare your triangles with those of other students. Are they all similar?

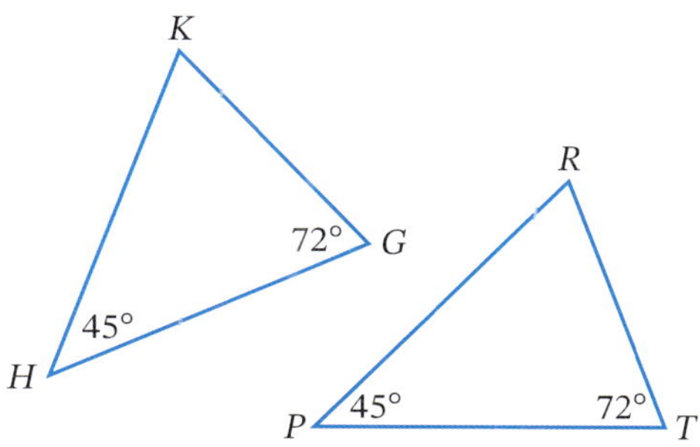

3 **a** Draw $\angle LMP$ with $LM = 30$ mm, $LP = 40$ mm and $\angle L = 55°$ (the included angle).

b Draw $\angle XWY$ with $XW = 45$ mm, $XY = 60$ mm and $\angle X = 55°$.

c Find the ratio of 2 pairs of matching sides $\frac{LM}{XW}$ and $\frac{LP}{XY}$. Are they equal?

d Are the 2 triangles similar? Explain.

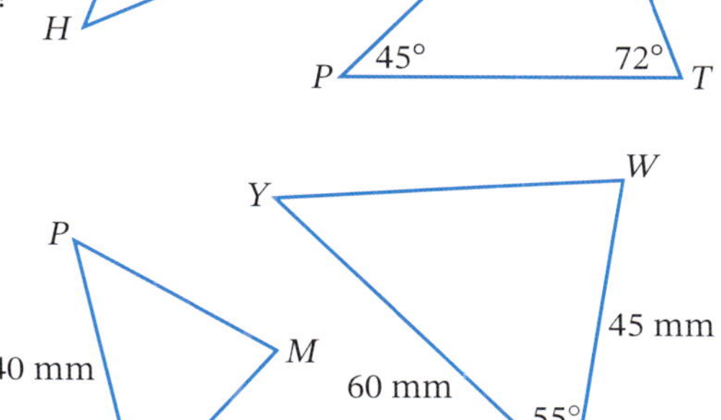

4 **a** Construct $\angle PQS$ with $PQ = 60$ mm, $\angle P = 90°$ and $QS = 80$ mm (the hypotenuse).

b Construct $\angle TXW$ with $TX = 45$ mm, $\angle T = 90°$ and $XW = 60$ mm.

c Calculate the ratios of matching sides $\frac{PQ}{TX}$ and $\frac{QS}{XW}$.

d Are the triangles similar? Explain.

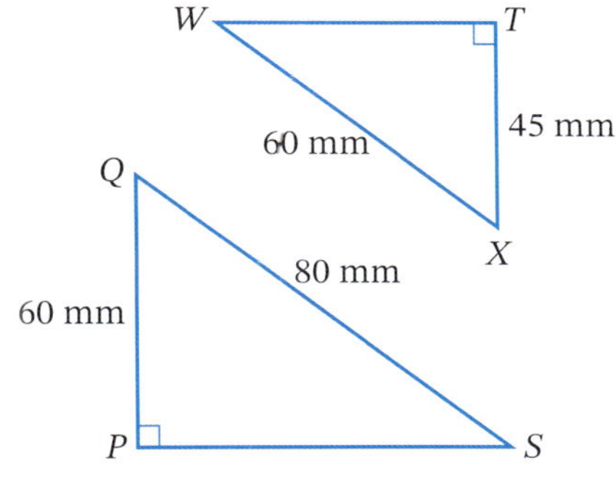

5 **a** Compare and discuss your results with other students.

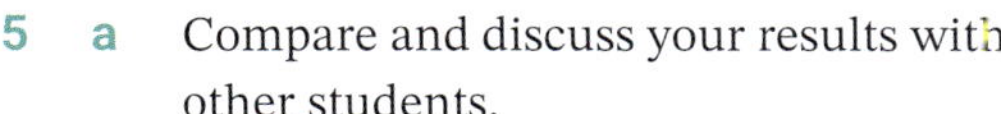

b What tests are necessary for 2 triangles to be similar?

c How do these conditions compare to the conditions for congruent triangles?

13.08 Tests for similar triangles

There are 4 sets of conditions that can be used to determine if 2 triangles are similar. These are called the **tests for similar triangles** or **similarity tests**.

Worksheet
Congruence and similarity review

ⓘ Similarity tests

There are 4 tests for similar triangles.

Two triangles are similar if:

- the 3 sides of one triangle are proportional to the 3 sides of the other triangle **('SSS')**

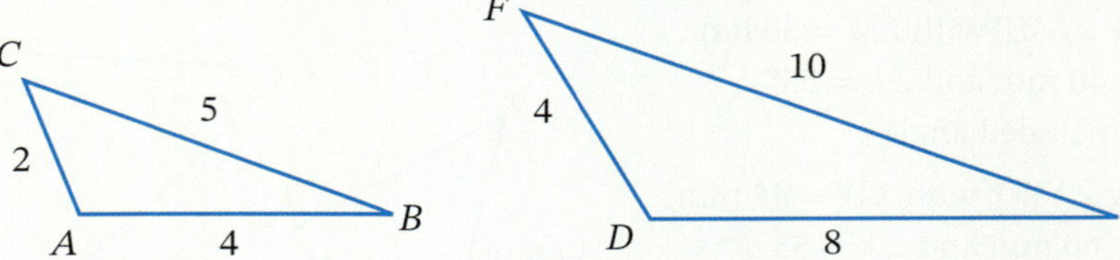

- 2 sides of one triangle are proportional to 2 sides of the other triangle, and the **included** angles are equal **('SAS')**

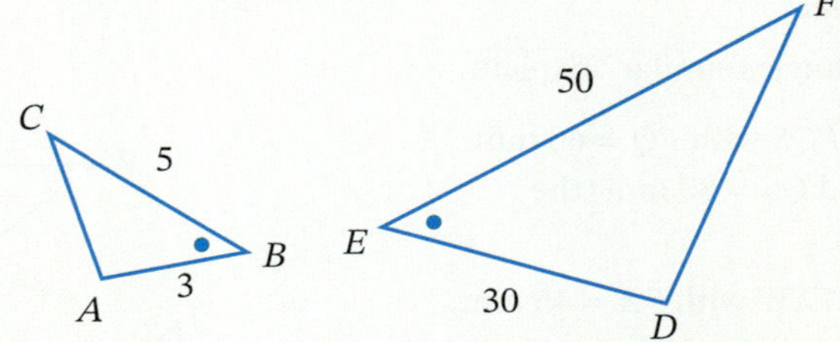

- 2 angles of one triangle are equal to 2 angles of the other triangle **('AA' or 'equiangular')**

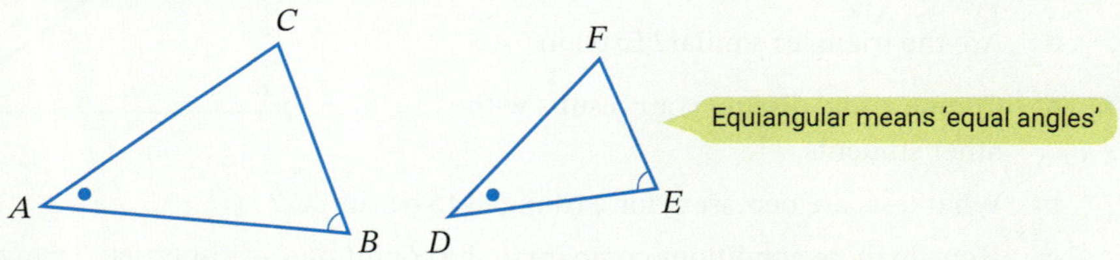

- they are right-angled and the hypotenuse and a second side of one triangle are proportional to the hypotenuse and a second side of the other triangle **('RHS').**

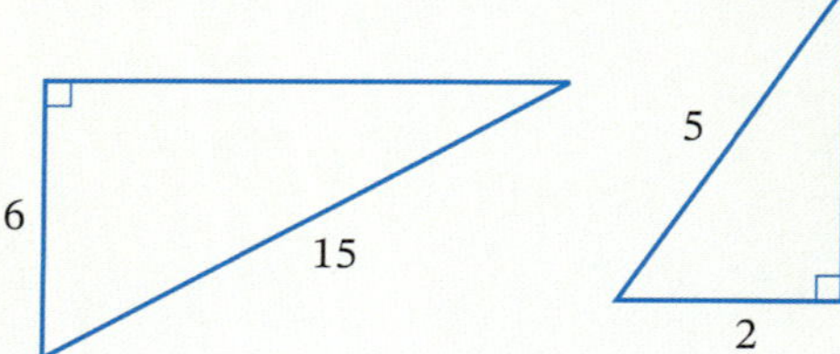

Example 20

Video Tests for similar triangles

Which test can be used to prove that each pair of triangles are similar?

a

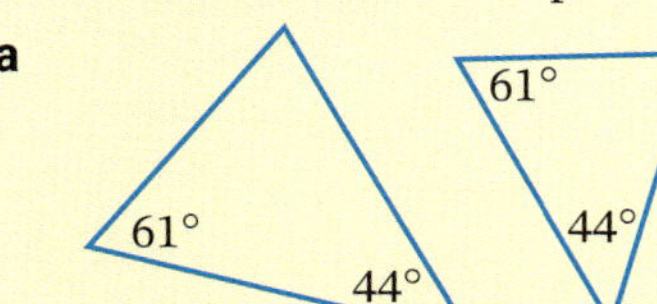
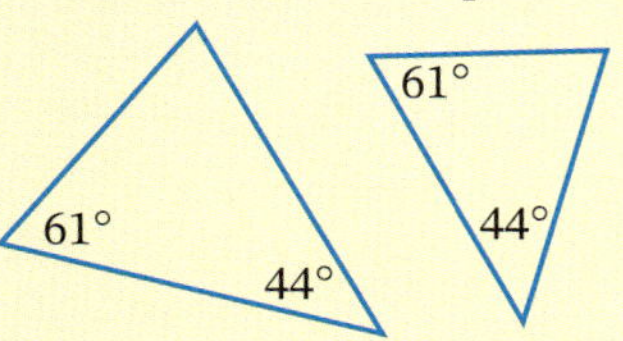

b

c

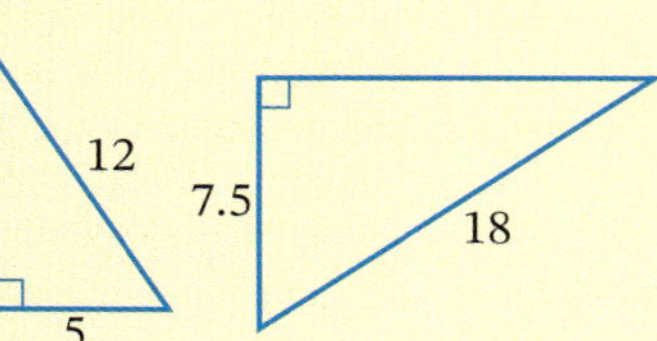

d

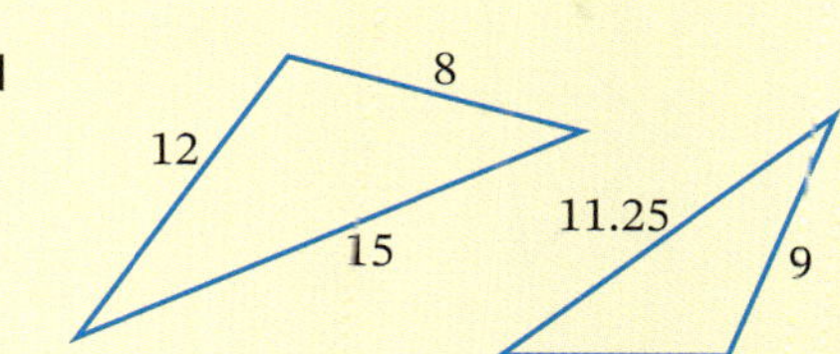

SOLUTION

a Two pairs of angles are equal, or equiangular ('AA').

b Two pairs of matching sides are in the same ratio $\frac{15}{5} = \frac{24}{8} = 3$ and the included angles in both triangles are 105° ('SAS').

c In both right-angled triangles, the pairs of hypotenuses and second sides are in the same ratio $\frac{7.5}{5} = \frac{18}{12} = \frac{3}{2}$ ('RHS').

d All 3 pairs of matching sides are in the same ratio $\frac{11.25}{15} = \frac{9}{12} = \frac{6}{8} = \frac{3}{4}$ ('SSS').

EXERCISE 13.08 ANSWERS ON P. 682

Tests for similar triangles

U F R C

EXAMPLE 20

1 Which test can be used to prove that each pair of triangles are similar?

R C

a

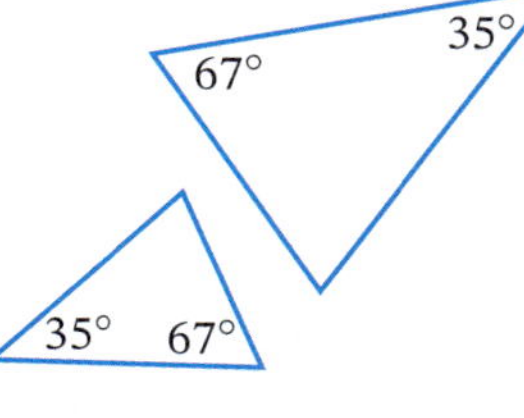

b

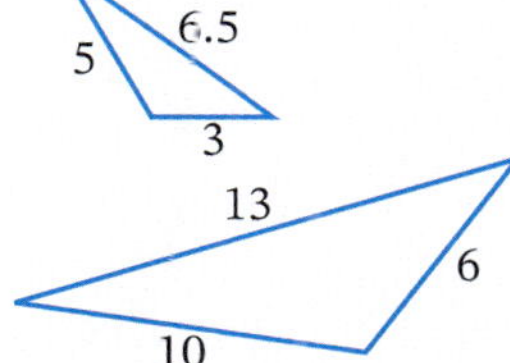

c

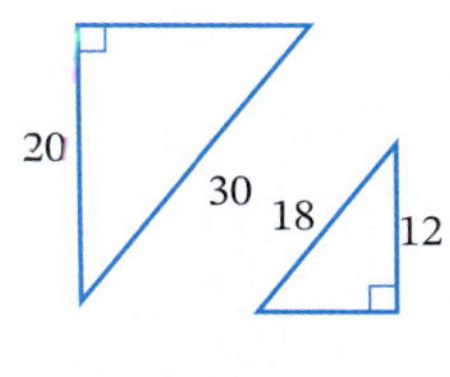

d

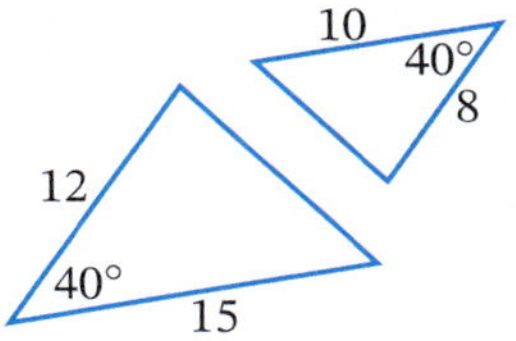

e

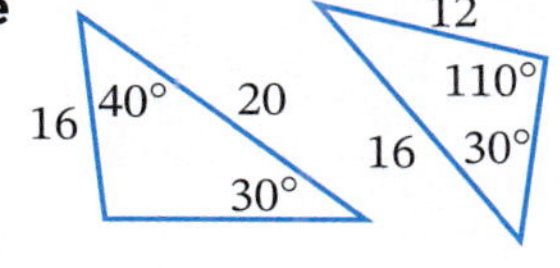

f

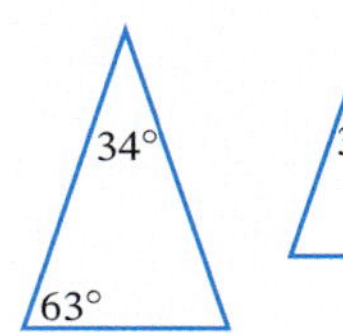

Foundation Standard Complex

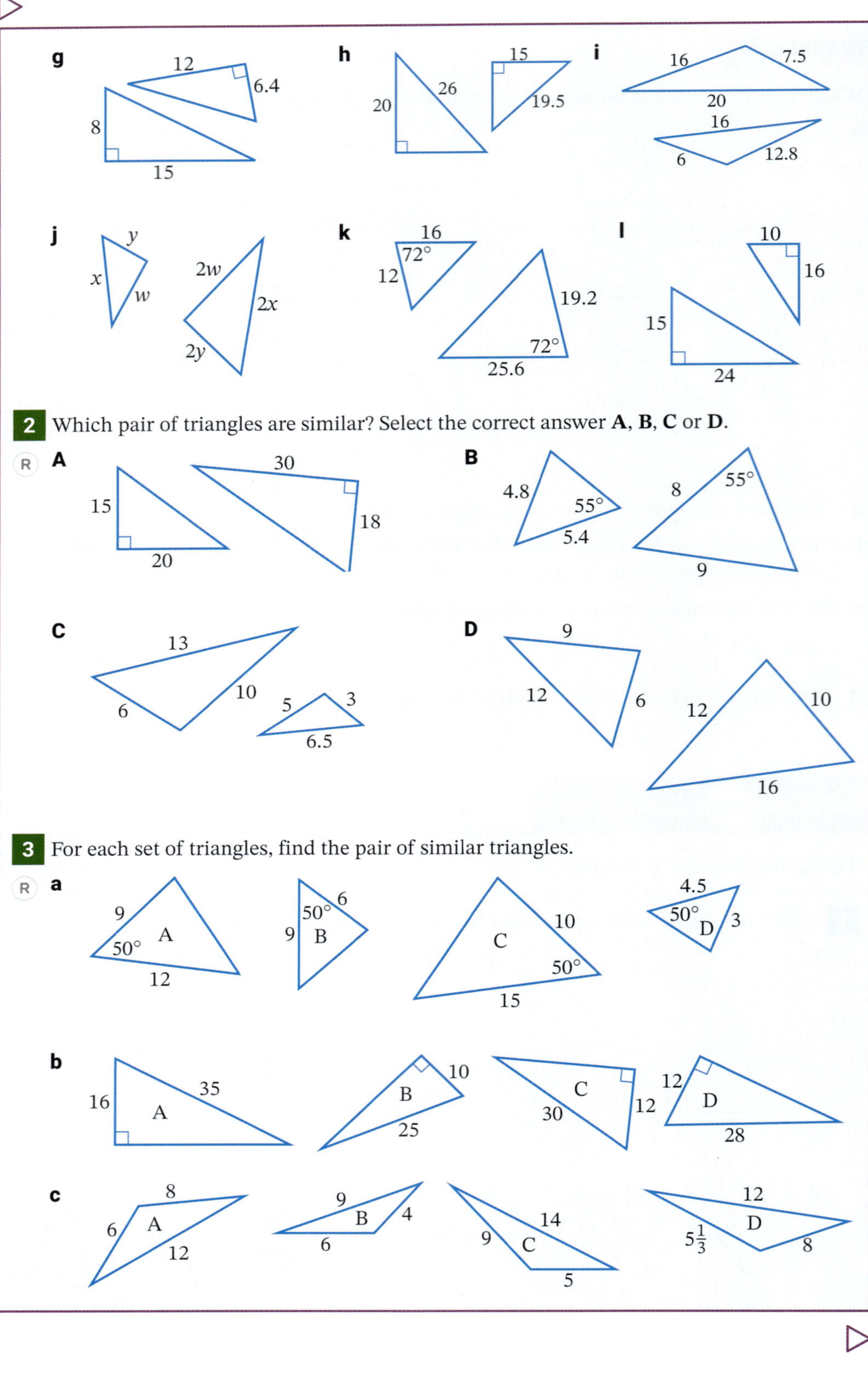
g
12
6.4
8
15
h
15
20
26
19.5
i
16
7.5
20
16
6
12.8
j
y
x
w
2w
2x
2y
k
16
72°
12
19.2
72°
25.6
l
10
16
15
24
2 Which pair of triangles are similar? Select the correct answer A, B, C or D.
A
30
15
18
20
B
4.8
55°
5.4
8
55°
9
C
13
10
6
5
3
6.5
D
9
12
6
12
10
16
3 For each set of triangles, find the pair of similar triangles.
a
9
A
50°
12
50°
6
9
B
C
10
50°
15
4.5
50°
D
3
b
16
A
35
B
10
25
C
30
12
12
D
28
c
8
6
A
12
9
B
4
6
14
9
C
5
12
D
$5\frac{1}{3}$
8

d

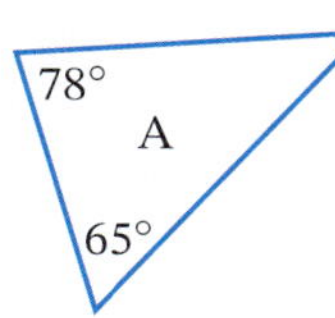

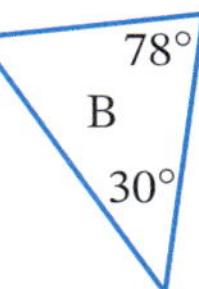

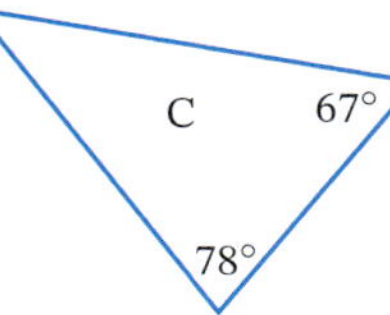

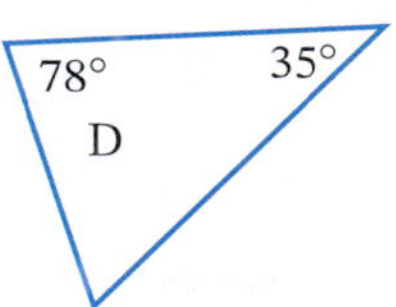

4 Use the correct notation to write a similarity statement relating each pair of similar triangles.

R C **a**

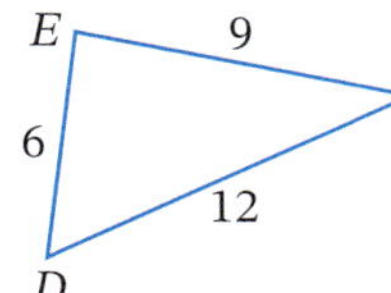

b

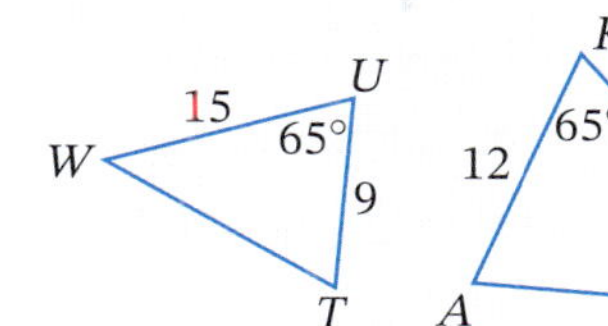

c

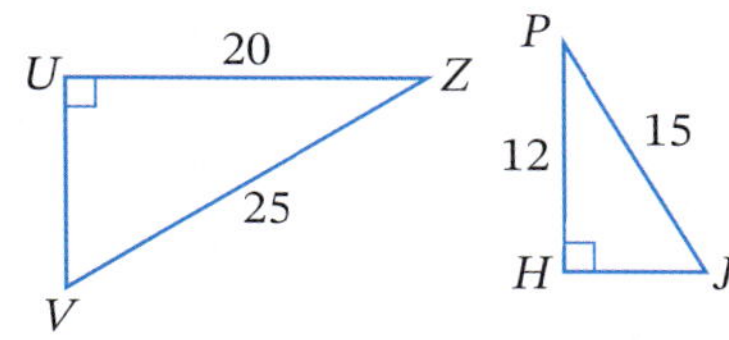

d

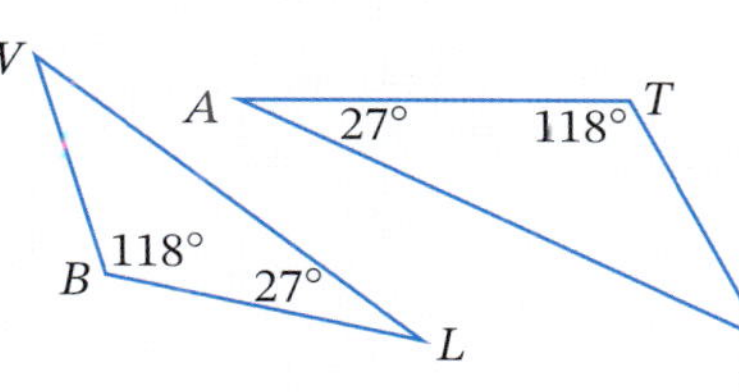

e

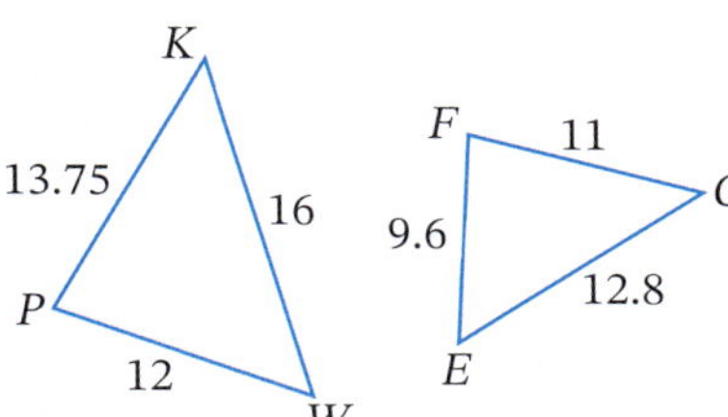

f

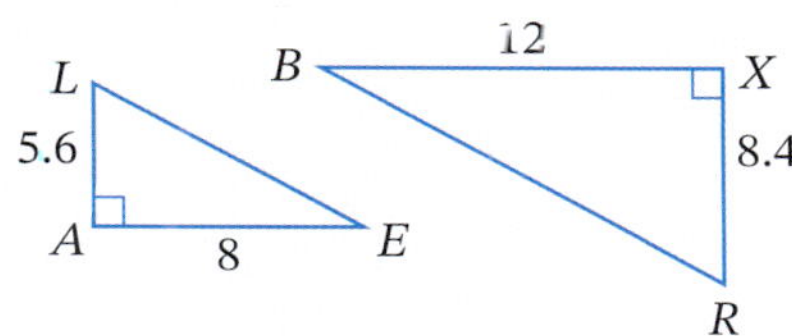

5 Similar triangles have 3 pairs of matching angles. So why is the similarity test AA and not AAA?

R C

INVESTIGATION

Congruence and similarity

1 a What is the first test for congruent triangles?

b What is the first test for similar triangles?

c Use the tests to determine whether each pair of triangles below are congruent or similar.

i

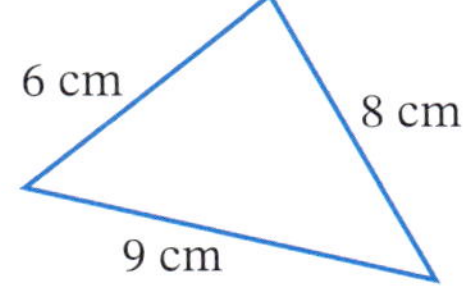

9 cm
8 cm
6 cm

ii

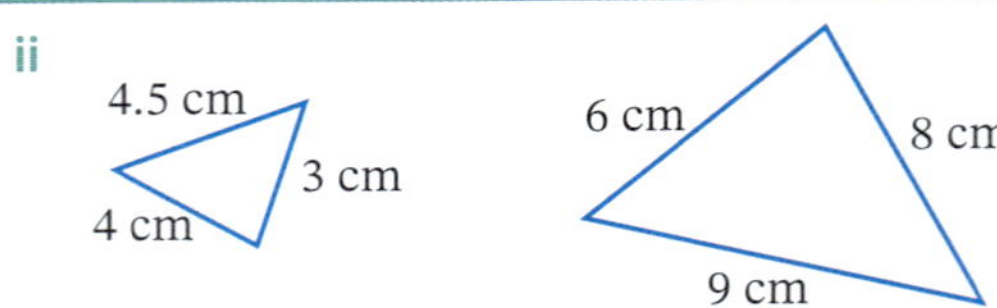

d How is the first test for congruent triangles and the first test for similar triangles:

i the same? ii different?

2 a What is the second test for congruent triangles?

b What is the second test for similar triangles?

c Use the tests to determine whether each pair of triangles are congruent or similar.

i

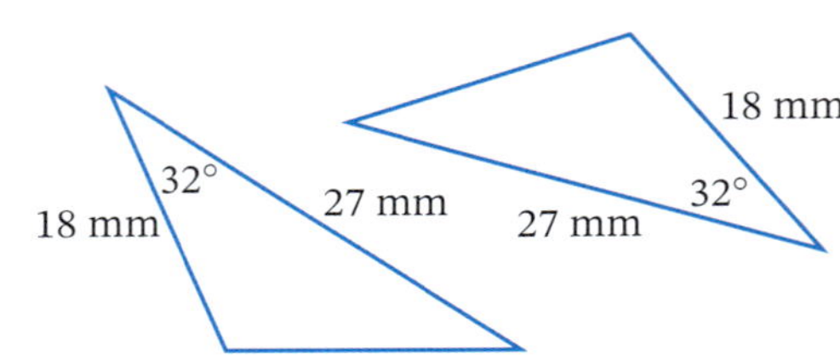

ii

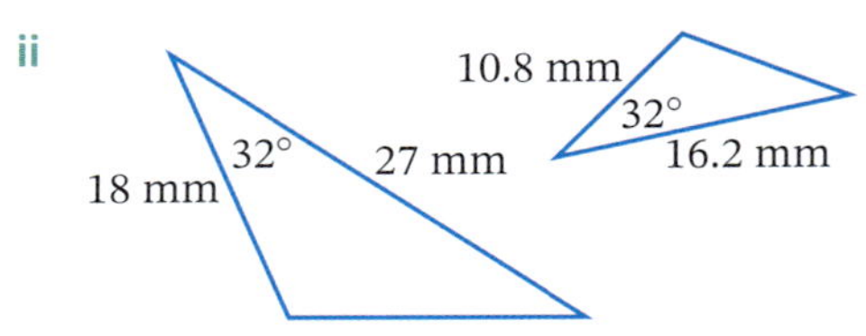

d How is the second test for congruent triangles and the second test for similar triangles:

i the same? ii different?

3 a The abbreviation of the third test for congruent triangles is AAS, and for similar triangles it is AA. What do they mean?

b Use the tests to determine whether each pair of triangles are congruent or similar.

i

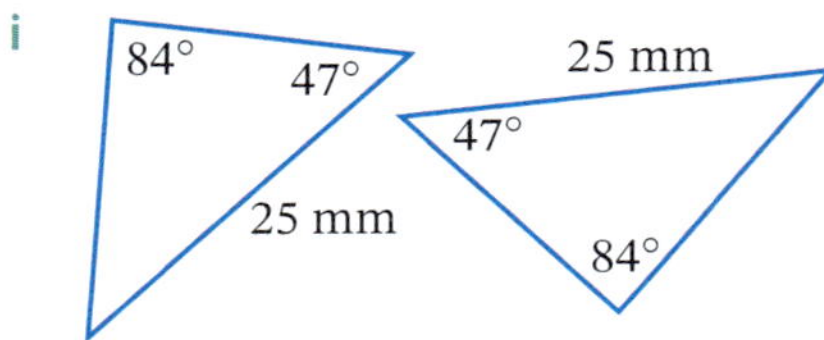

ii

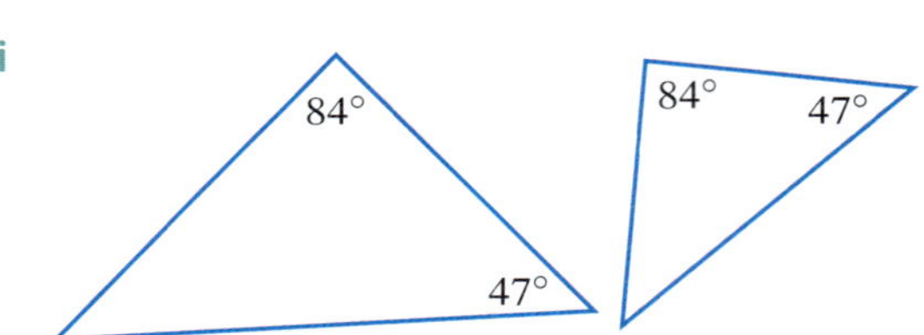

c How is the third test for congruent triangles and the third test for similar triangles:

i the same? ii different?

4 a The fourth tests for congruent and similar triangles can be written as RHS. What is the fourth test for congruent triangles and for similar triangles?

b Use the tests to determine whether each pair of triangles are congruent or similar.

i

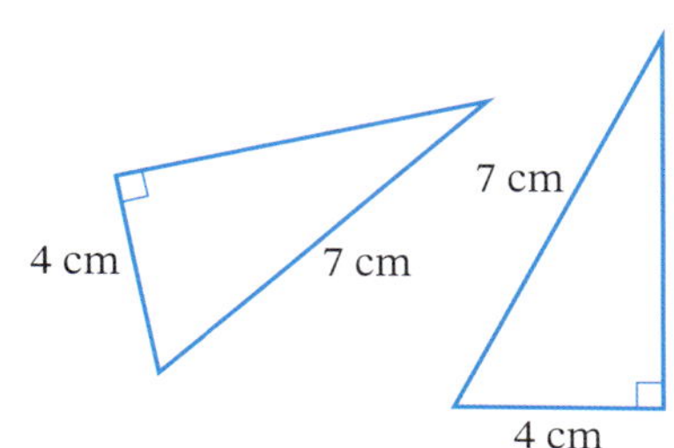

ii

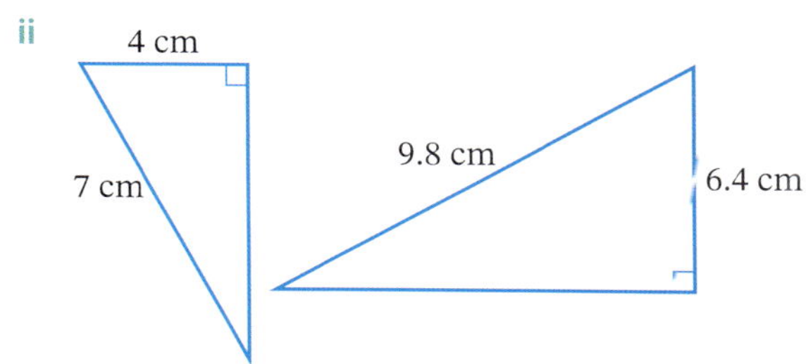

c How is the fourth test for congruent triangles and the fourth test for similar triangles:

i the same? ii different?

5 Use your answers to questions **1** to **4** to summarise what is the same and what is different about the tests for congruent and similar triangles.

POWER PLUS ANSWERS ON P. 682

Worksheet
Investigating paper sizes

1 Construct $\angle JPT$ by:

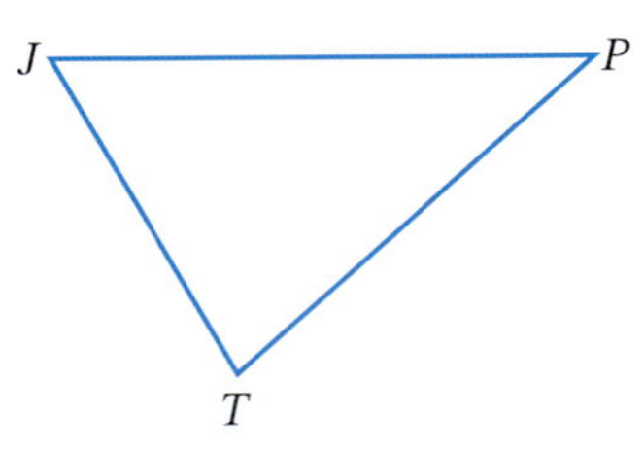

a using compasses and the lengths of the 3 sides.

b using the length of one side and the size of the 2 angles.

c using the length of 2 sides and the size of one angle.

2 a List the information given in this diagram.

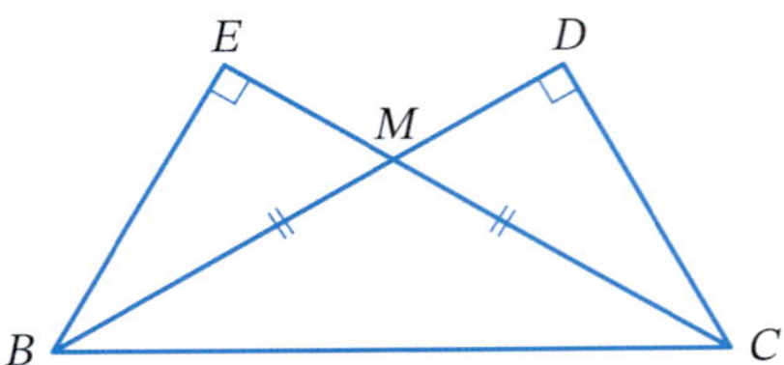

b Why is $\angle BME = \angle CMD$?

c Show that $\angle BME \equiv \angle CMD$.

d Why is $EM = DM$?

e Why is $\angle EBM = \angle DCM$?

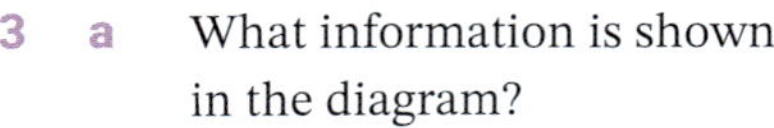

3 a What information is shown in the diagram?

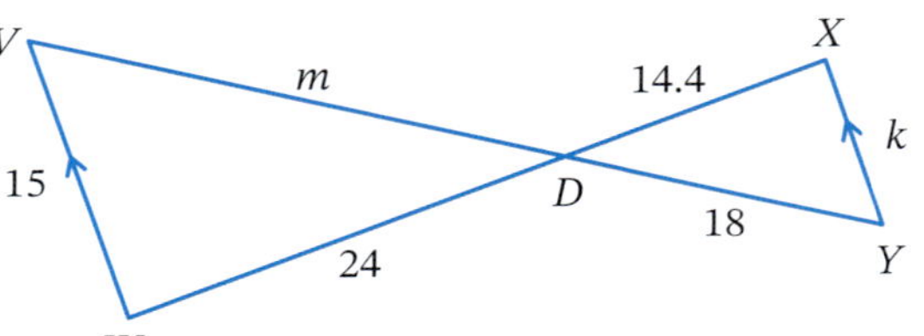

b Why is $\angle V = \angle Y$ and $\angle W = \angle X$?

c Show that $\angle VWD \mid\mid\mid \angle XYD$.

d Find the values of k and m.

4 Alexa places a mirror on the ground. She stands 1 m from the mirror and then bobs down until she can see the top of a building in the mirror. Her eyes are 95 cm above ground level when she can see the building. If the mirror is 35 m from the base of the building, what is the height of the building?

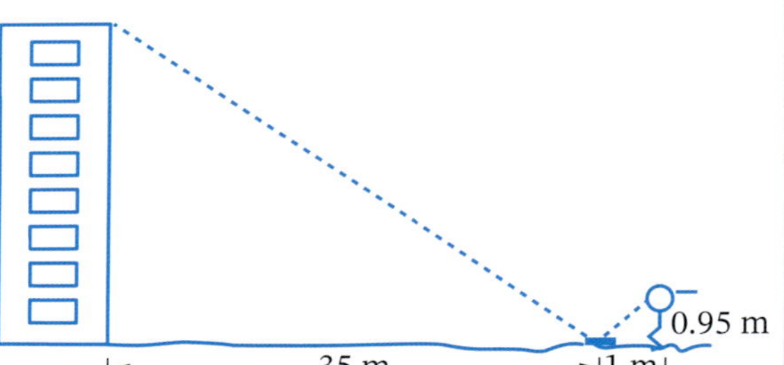

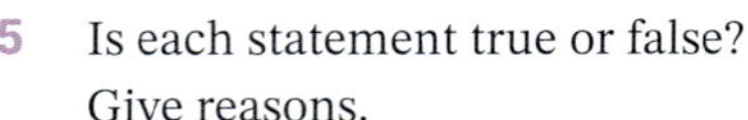

5 Is each statement true or false? Give reasons.

a All equilateral triangles are similar.

b All isosceles triangles are similar.

c All circles are similar.

d All rectangles are similar.

e All squares are similar.

f All regular pentagons are similar.

13 CHAPTER REVIEW

Language of maths

AAS	bisect	congruence test	congruent (≡)
diagonal	enlargement	hypotenuse	image
included angle	matching	original	parallel
proportional	ratio	reduction	RHS
SAS	scale	scale diagram	scale factor
similar (|||)	similarity test	SSS	vertex/vertices

Quiz
Language of maths 13

1 When an original figure is enlarged or reduced, what is the new figure called?

2 What is the angle between 2 given sides of a triangle called?

3 What is the symbol and meaning of 'is congruent to'?

4 What is the symbol and meaning of 'is similar to'?

5 Which word from the above list is another name for 'corresponding'?

6 What happens to a shape that is changed by a scale factor of 3?

7 What are the 4 tests for congruent triangles?

8 What does RHS stand for:

a in congruent triangles?

b in solving equations?

Topic summary

Worksheet
Mind map: Congruent and similar figures

Print (or copy) and complete this mind map of the topic, adding detail to its branches and using pictures, symbols and colour where needed. Ask your teacher to check your work.

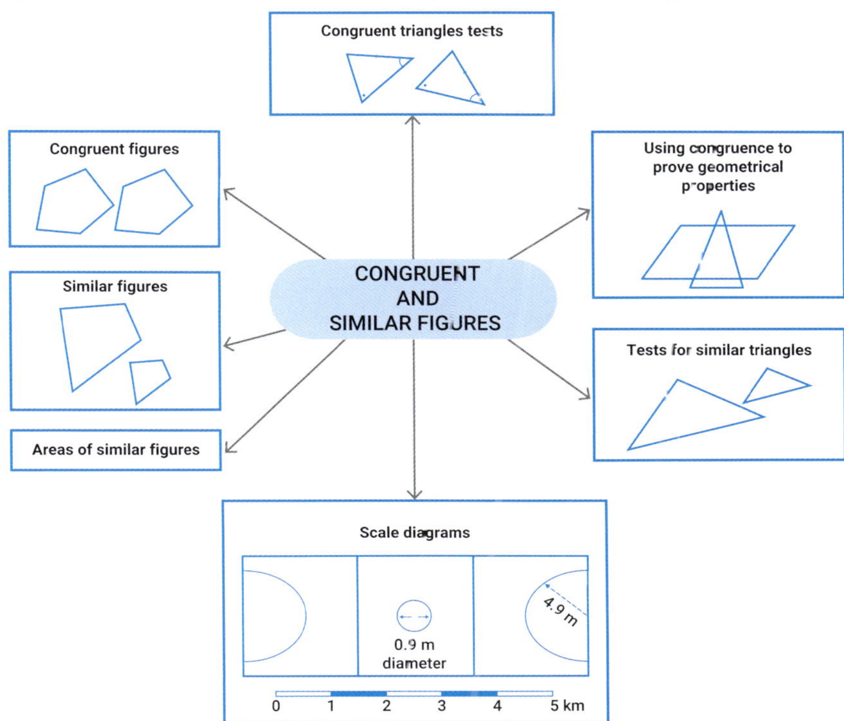

CHAPTER 13 REVIEW

13 TEST YOURSELF

ANSWERS ON P. 682

Quiz
Test yourself 13

1 Match pairs of shapes that are congruent.

A 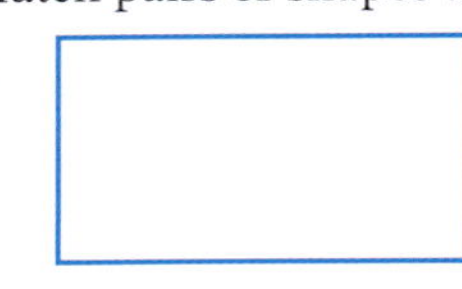

B 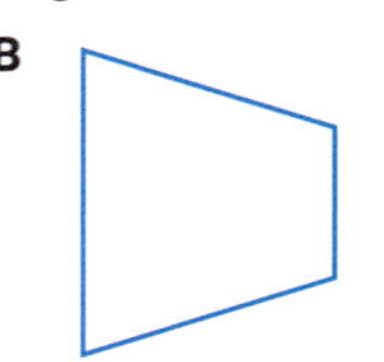

C 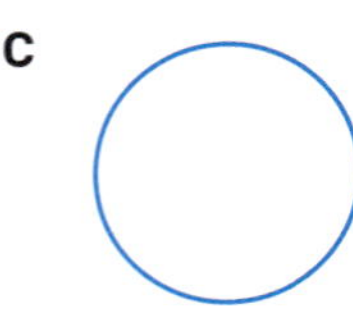

D 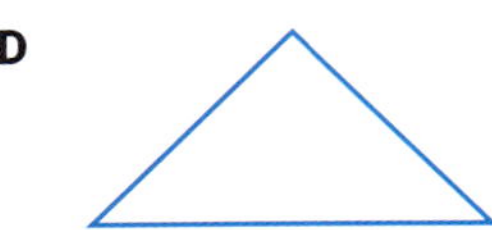

E 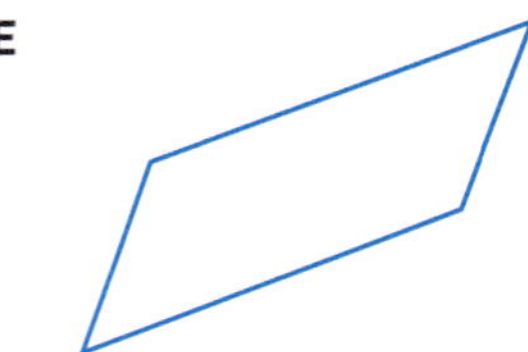

F 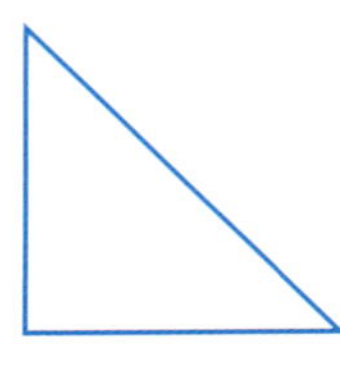

G 

H 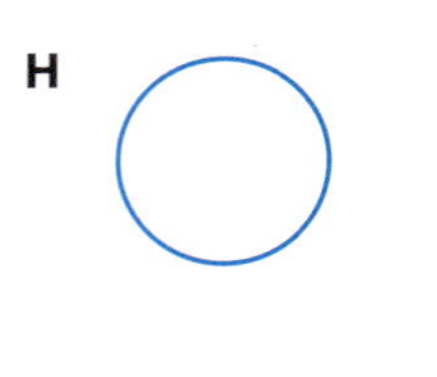

I 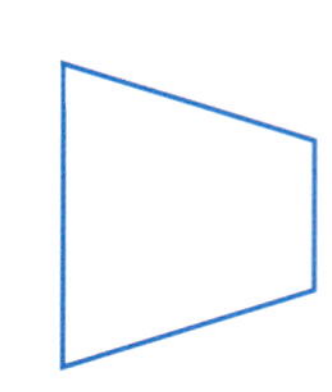

J 

K 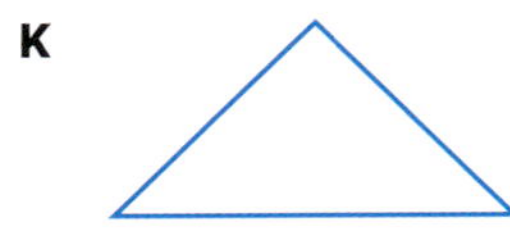

L 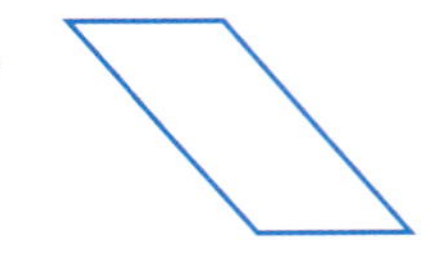

2 The quadrilaterals *FECD* and *TRMP* are congruent.

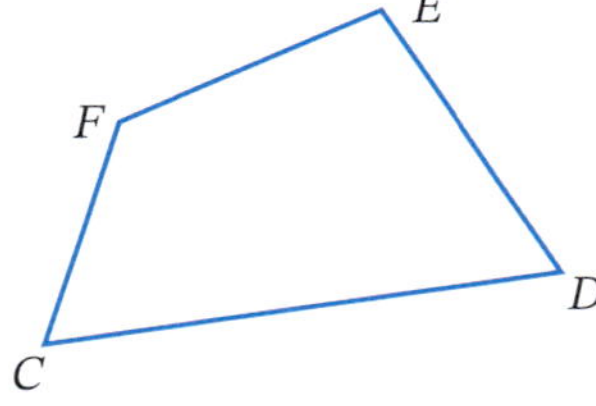

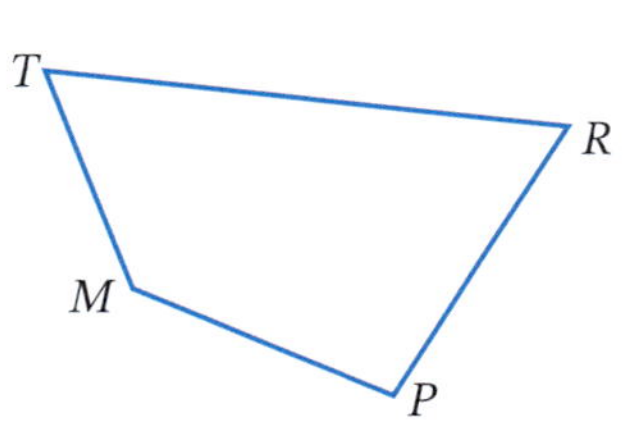

a List all pairs of matching sides.

b List all pairs of matching angles.

c Use the correct notation to write a congruency statement relating the 2 quadrilaterals.

◻ Foundation ◯ Standard ⬡ Complex

9780170465618

3 Which congruency test (SSS, SAS, AAS or RHS) can be used to prove that each pair of triangles are congruent? (13.02)

a

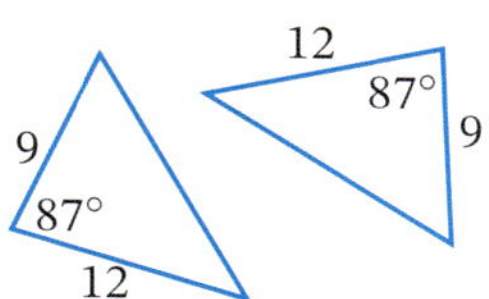

b

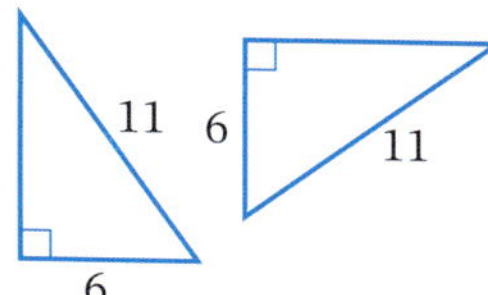

c

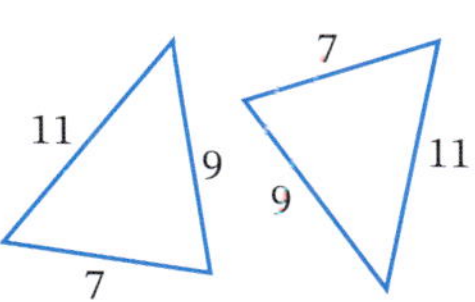

4 Which pair of triangles are congruent? Select the correct answer **A**, **B**, **C** or **D**. (13.02)

A

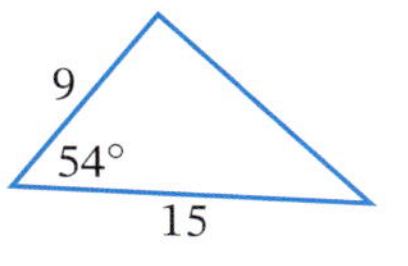

B

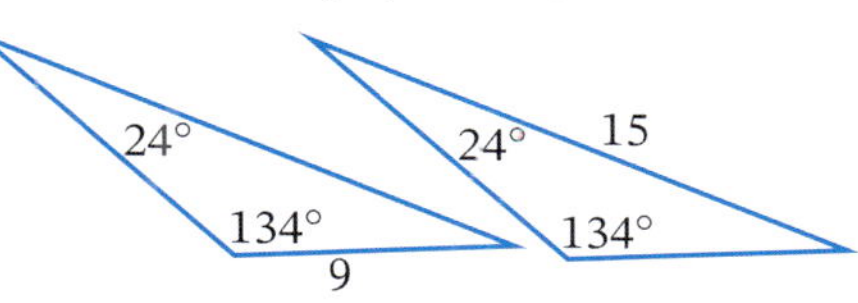

C

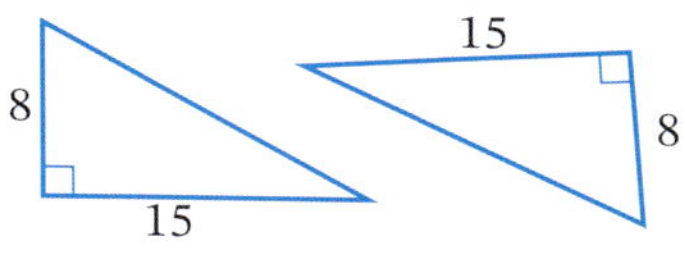

D

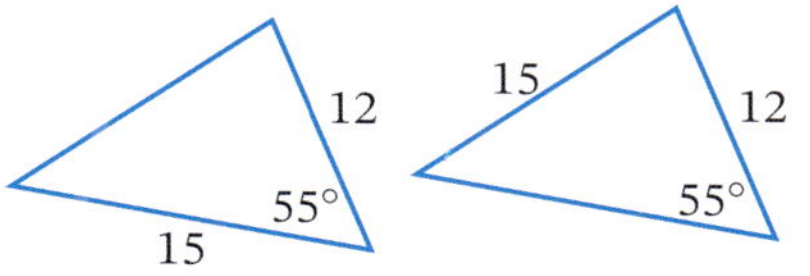

5 $ABCD$ is a rectangle. We want to prove that $\angle DAB \equiv \angle CBA$. (13.03)

a Why is $DA = CB$?

b Which side is common to both triangles?

c Why is $\angle DAB = \angle CBA$?

d Hence, which congruency test proves that $\angle DAB \equiv \angle CBA$?

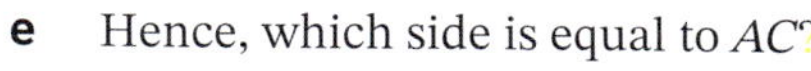

e Hence, which side is equal to AC?

f What does this prove about the diagonals of a rectangle?

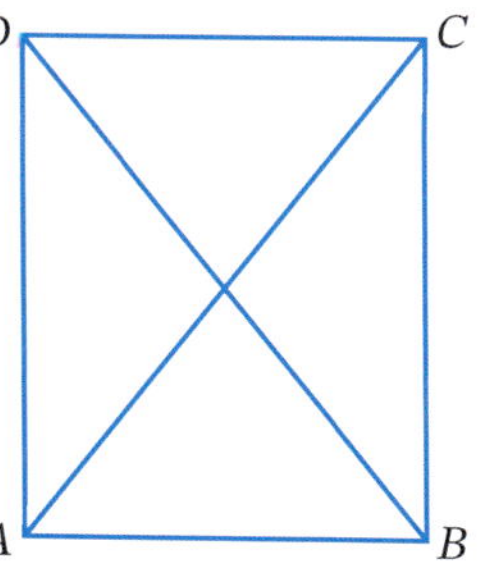

6 By measurement, find the scale factor for each pair of similar figures. (13.04)

a

b

c

7 Copy each figure onto graph paper and draw its image if: (13.04)

a scale factor = 3

b scale factor = $\frac{1}{2}$

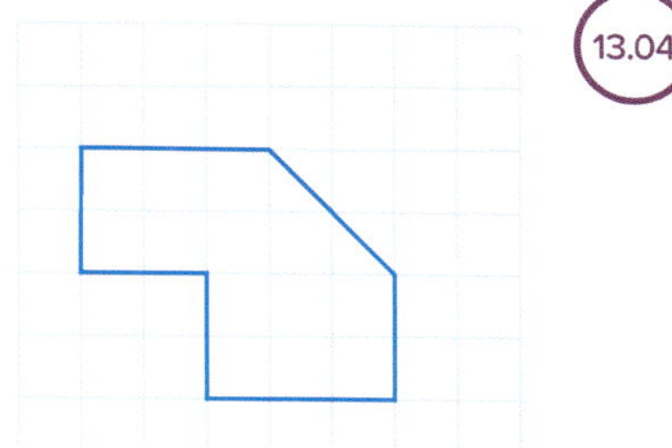

CHAPTER 13 TEST YOURSELF

Foundation Standard Complex

8 Test whether each pair of figures are similar.

a

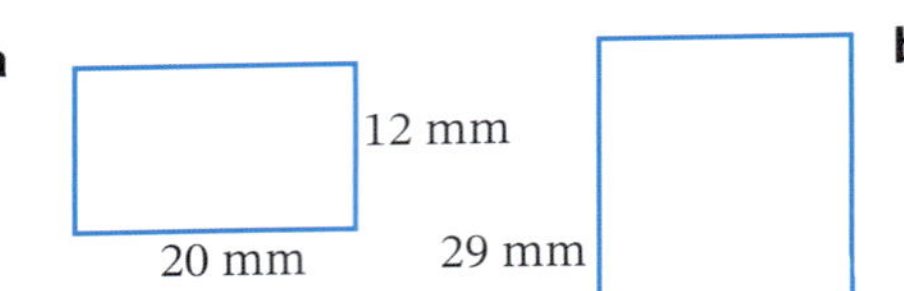

b

8 cm, 7 cm, 10 cm; 6 cm, 7.5 cm, 5.25 cm

13.05 **9** Find the value of each variable in each pair of similar figures.

a **b**

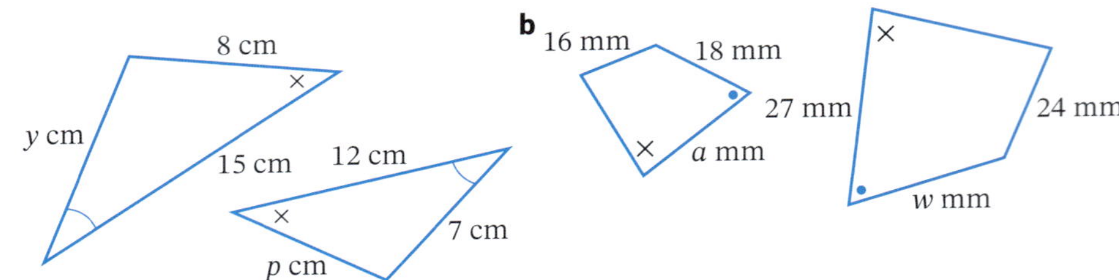

13.05 **10** A flagpole's shadow is 20 m long. At the same time, a stick 60 cm tall casts a shadow 150 cm long.

What is the height, h, of the flagpole? Select the correct answer **A**, **B**, **C** or **D**.

A 12 m **B** 50 m **C** 8 m **D** 9 m

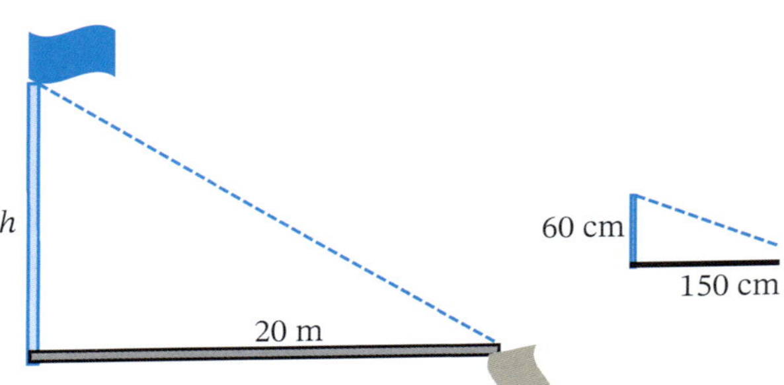

11 This floor plan of a holiday house is drawn to a scale of 1 : 200. What is the actual (internal) length of:

a AB? **b** CD?

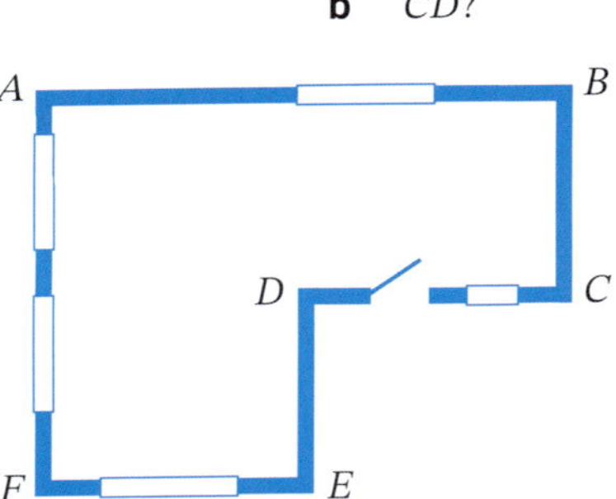

12 Find the ratio of the areas for each pair of similar figures.

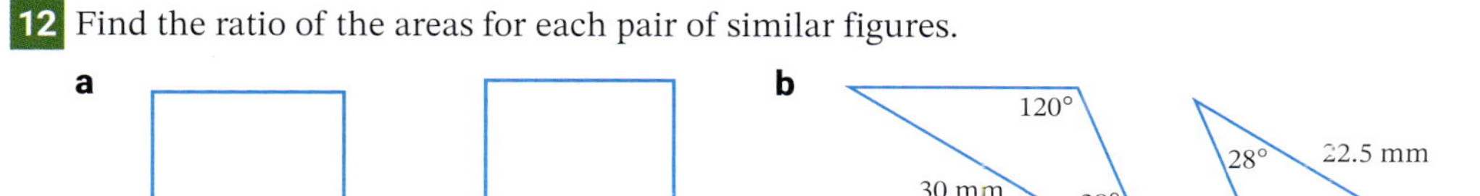

13 Two similar trapeziums have areas in the ratio 25 : 4. If the perpendicular height of the smaller trapezium is 40 cm, find the perpendicular height of the larger trapezium.

14 Which test can be used to prove that each pair of triangles is similar? 13.08

a

27, 15, 10, 18

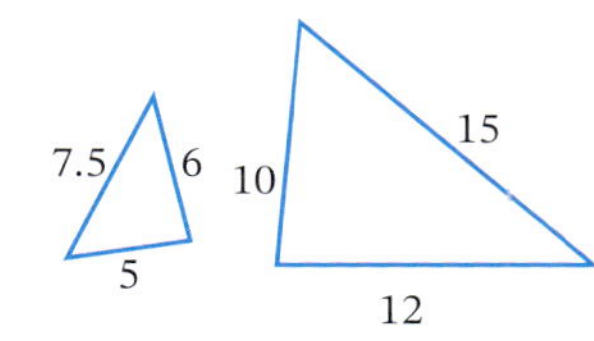

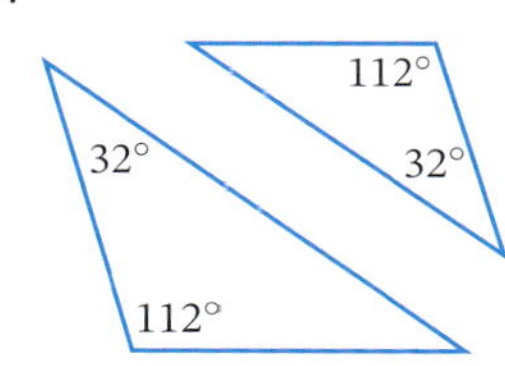

Practice set 4

ANSWERS ON P. 682

1 Test whether the point (2, 3) lies on each line.

a $2x + y = 4$ **b** $y = 2x - 1$ **c** $x + y = 5$

2 The rectangle on the left has been enlarged to make the rectangle on the right.

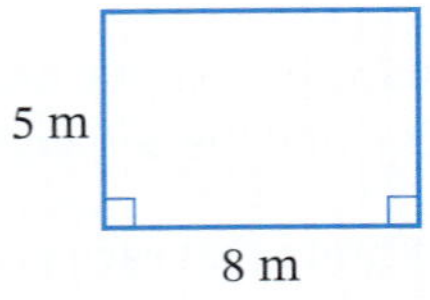

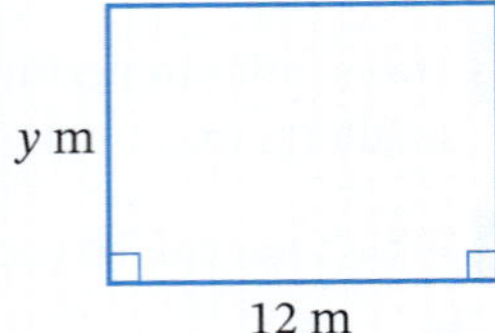

a What is the scale factor?

b Find the value of y.

3 The probability of Eva hitting a bullseye on a target is 0.2. Find the probability of Eva missing the bullseye.

4 A class was surveyed on whether its students could swim. The results are shown.

	Can swim	Cannot swim
Boys	13	2
Girls	9	3

a Find the probability of randomly selecting a student who is:

i a boy **ii** a boy or cannot swim

iii a girl who cannot swim **iv** a girl or can swim, but not both.

b If a swimmer is selected at random, what is the probability that it is a boy?

13.06 **5** The scale on a tourist map of Sydney is 2 cm = 500 m. If the scaled distance from Circular Quay station to the Opera House is 2.5 cm, what is the actual distance?

Questions **6** to **9** refer to this triangle with vertices $P(-1, 2)$, $Q(4, 3)$ and $R(3, -2)$.

6 Find the length of:

a side PQ as a surd

b side PR, correct to one decimal place.

11.01 **7** Show that ΔPQR is isosceles.

11.02 **8** Find the coordinates of the midpoint of side PR.

11.03 **9** Find the gradient of:

a PQ **b** PR

11.04 **10** Graph $y = 4 - 2x$ on a number plane and write:

a its x-intercept **b** its y-intercept

☐ Foundation ○ Standard ⬡ Complex

11 Daniel rolled a biased die 90 times and the number 6 came up 24 times. 

a What is the experimental probability of rolling a 6?

b If the same die was rolled 400 times, what is the expected frequency of rolling a 6?

12 **a** Sketch the graph of $y = x^2 - 2$. 

b Write the coordinates of the vertex.

c Is the graph concave up or concave down?

13 Which congruence test (SSS, SAS, AAS or RHS) can be used to prove that each pair of triangles are congruent? 

a

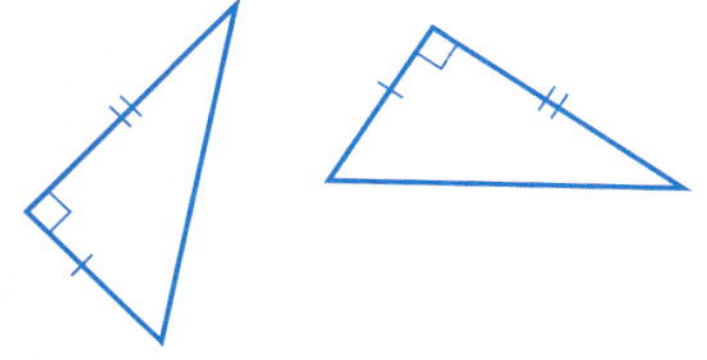

b

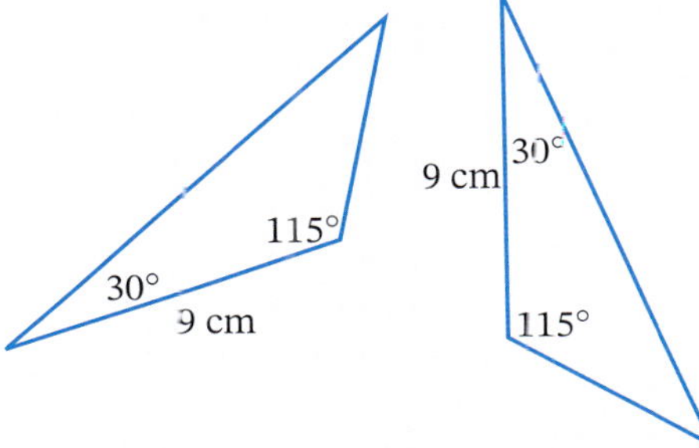

c

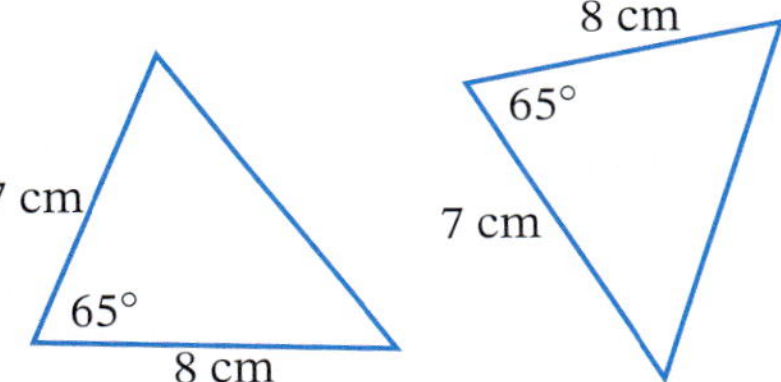

d

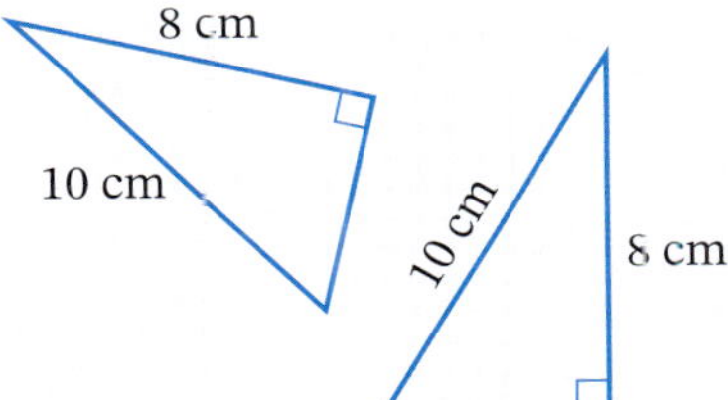

14 What is the equation of the horizontal line passing through (−2, 4)? 

15 The numbers 1 to 30 are written on cards and the cards are placed in a bag. If one card is drawn at random from the bag, what is the probability that it is: 

a a multiple of 5?

b greater than 10?

16 **a** Find the gradient and y-intercept of the line with equation $y = \frac{1}{2}x + 3$. 

b Graph a line with gradient 2 and y-intercept −5 and write its equation.

17 The increase in pressure experienced by a scuba diver is directly proportional to their depth under the water. The increase in pressure at depth 40 m is 235 kPa (kilopascals).  

a What increase in pressure is experienced at a depth of 32 m?

b At what depth, to the nearest metre, is the increase in pressure 735 kPa?

☐ Foundation ○ Standard ⬡ Complex

9780170465618

12.05

18 Two dice are rolled together.
Copy and complete the table to show the sample space.

Find the probability of rolling:

a a 3 and a 6 (in any order)

b at least one 2

c 2 numbers the same

d a 4 or a 5, but not both.

		1st die					
		1	2	3	4	5	6
2nd die	1						
	2					5, 2	
	3						
	4		2, 4				
	5						
	6						

19 What is the value of k if these triangles are similar? Select **A**, **B**, **C** or **D**.

A 8 **B** 9 **C** 10 **D** 12

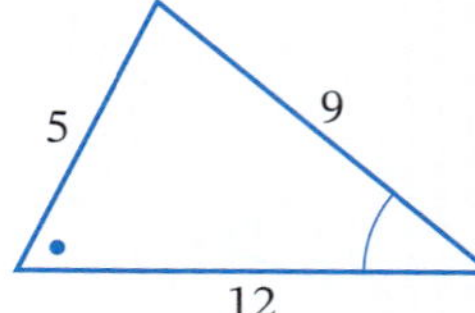

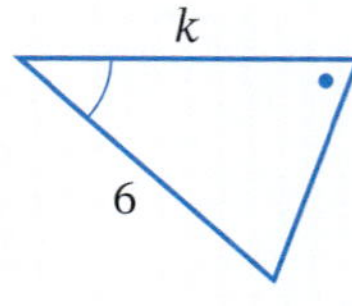

20 **a** Find the gradient of this line.

b Find the equation of this line.

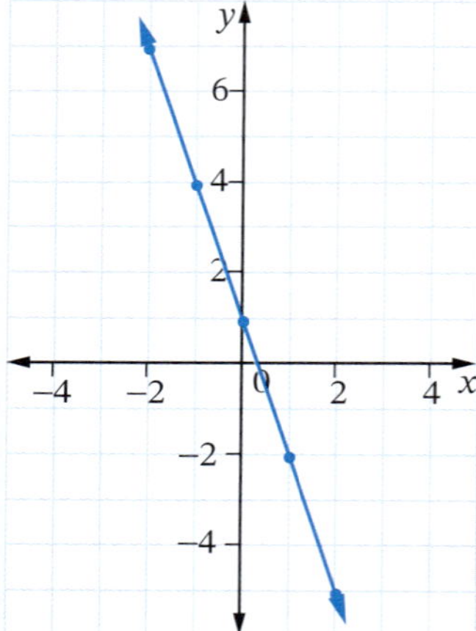

21 Tanya has $25 in her piggy bank. Her father has promised to give her $5 every week if she empties the dishwasher. Construct a graph to represent Tanya's piggy bank balance and the use the graph to determine how much Tanya will have in her piggy bank after 7 weeks.

22 40 students at a school camp could select kayaking or bushwalking as an activity, or both. A total of 35 students chose kayaking, 20 chose bushwalking, and some chose both. There weren't any students who chose neither.

a Copy and complete the Venn diagram representing these students.

b What is the probability that a student selected at random:

i chose kayaking and bushwalking?

ii chose kayaking or bushwalking, but not both?

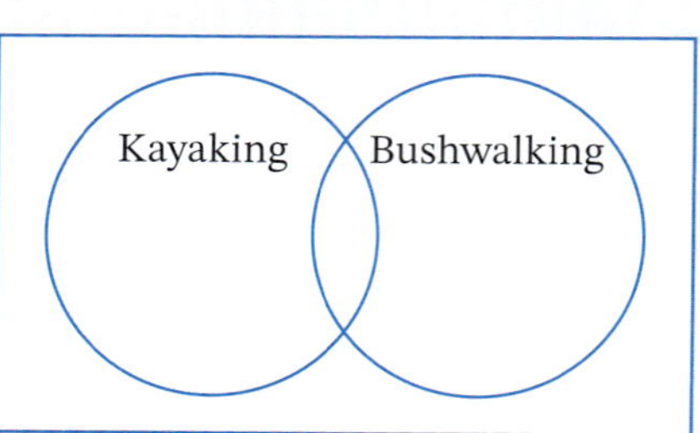

☐ Foundation ○ Standard ⬡ Complex

9780170465618

23 The diagonals of this rhombus divide the rhombus into 4 congruent triangles. 13.03

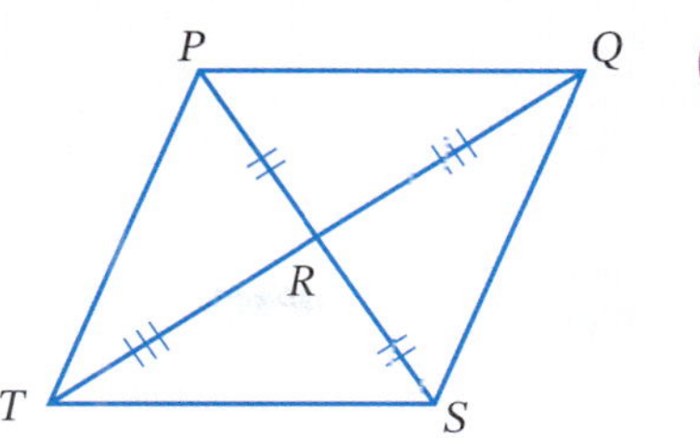

a Which test proves that the 4 triangles are congruent?

b Hence find the size of $\angle PRQ$.

c Copy and complete: The diagonals of a rhombus __________ each other at __________ __________.

24 Which test can be used to prove that this pair of triangles are similar? 13.08

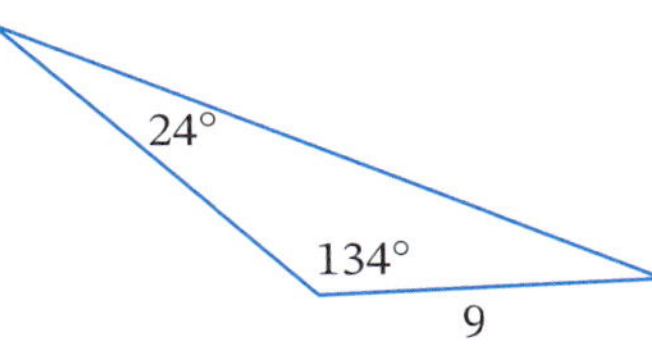

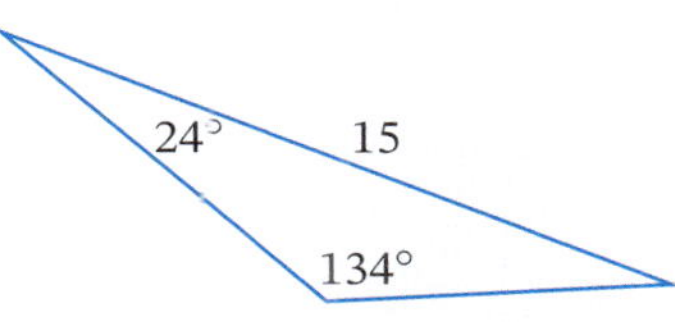

25 Match each equation to the correct description of its graph. 11.05 11.10 11.13

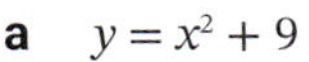

a	$y = x^2 + 9$	**A**	straight line
b	$y = x + 9$	**B**	exponential curve
c	$y = 9^x$	**C**	parabola

26 A pebble is thrown from a hill and its height, y metres, above the ground at time, x seconds, is given by the function $y = -x^2 + 6x + 16$. Graph this function for x values between 0 and 10. 11.12

a At what height is the pebble thrown from?

b For how long does the pebble stay in the air?

c How far up does the pebble travel before starting to fall back to the ground?

27 Find an equation for this exponential graph. 11.13

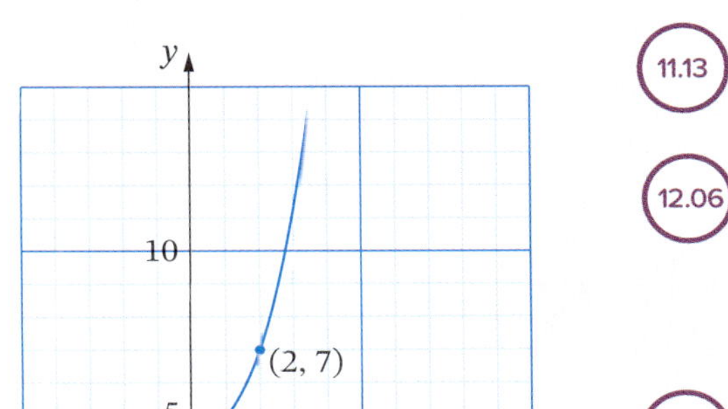

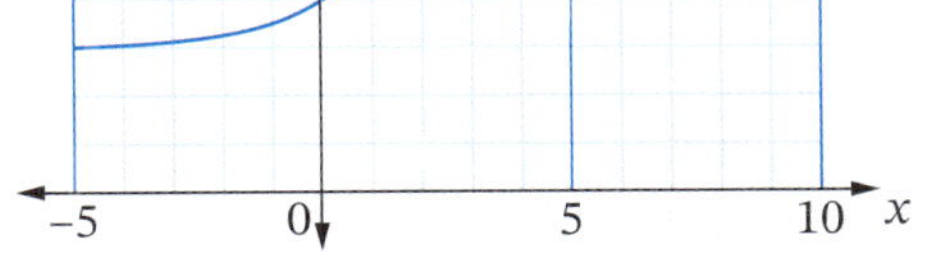

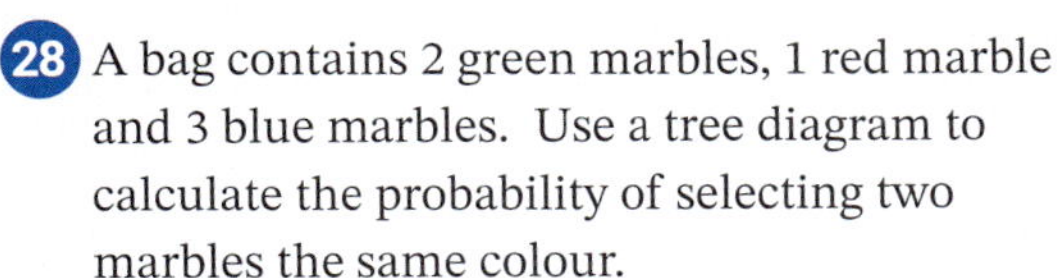

28 A bag contains 2 green marbles, 1 red marble and 3 blue marbles. Use a tree diagram to calculate the probability of selecting two marbles the same colour. 12.06

29 Two similar rectangles has sides in the ratio 1 : 7. Find the ratio of their areas. 13.07

☐ Foundation ○ Standard ○ Complex

PRACTICE SET 4

General practice ANSWERS ON P. 687

1 Find the volume of each solid, correct to one decimal place where necessary.

a

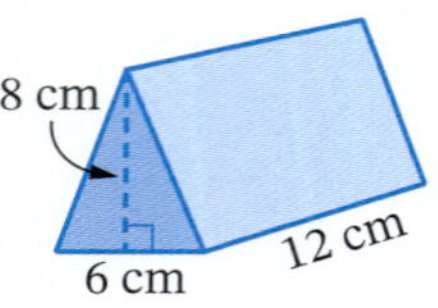

b

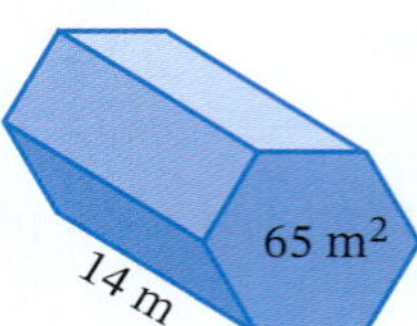

c

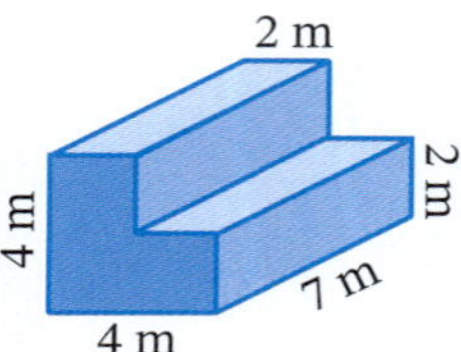

d

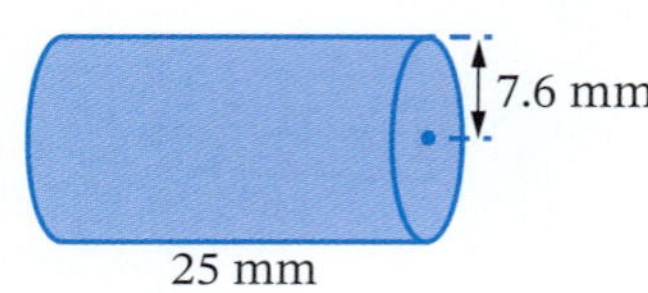

2 Simplify $(3m^4n)^2$. Select the correct answer **A**, **B**, **C** or **D**.

A $9m^8n^2$ **B** $6m^6n^2$ **C** $9m^6n^2$ **D** $6m^8n^2$

3 Simplify each expression.

a $7d \times 3d$ **b** $\frac{20rs^2}{25r^2s}$ **c** $2y - 5e + 6y - e$

4 This dot plot shows the project results (out of 10) of a Year 9 class.

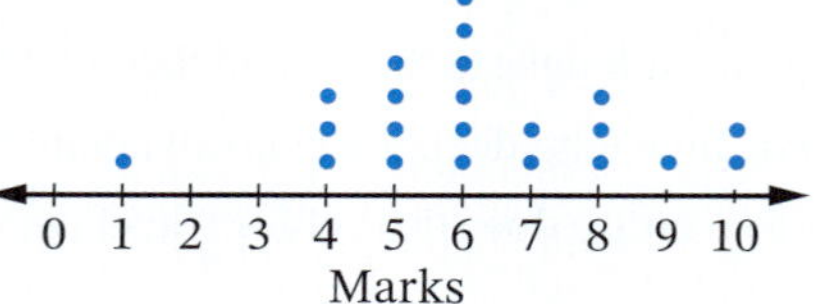

a How many students were in the class?

b Find:

i the range **ii** the mode

iii the mean, correct to one decimal place **iv** the median

c Which of the scores can be considered as an outlier?

d Are there any clusters? Identify them.

5 Classify each type of data as categorical (C), numerical discrete (ND) or numerical continuous (NC).

a number of taxis on the road **b** rating of an app (1 star to 5 stars)

c masses of babies in a maternity ward **d** brands of mobile phones

☐ Foundation ○ Standard ⬡ Complex

9780170465618

6 Which congruence test (SSS, SAS, AAS or RHS) can be used to prove that each pair of triangles are congruent? CHAPTER 13

a

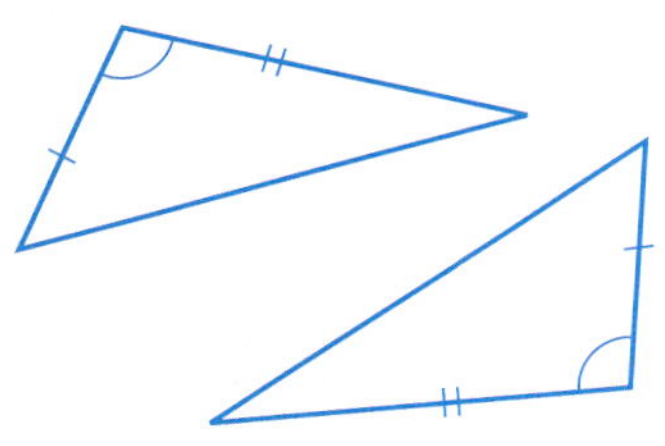

b

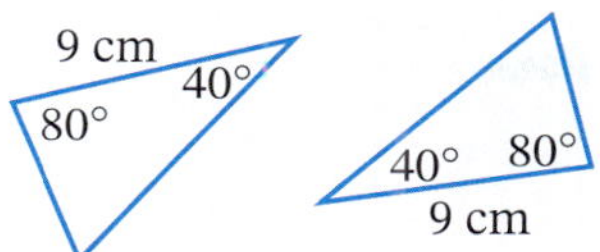

c

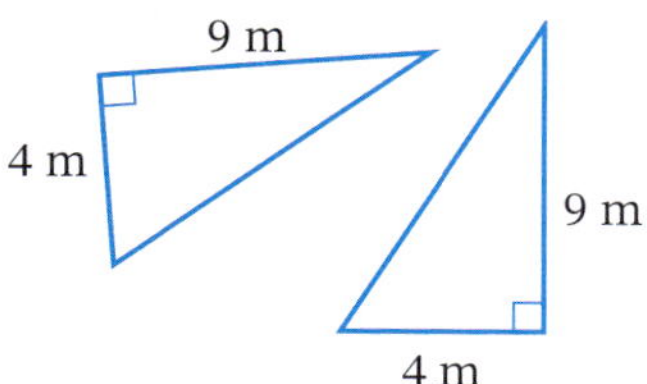

d

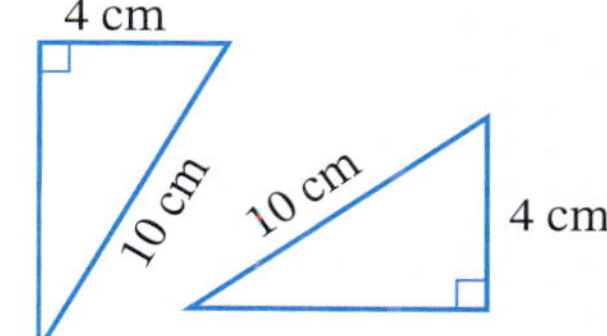

7 A survey was conducted among a sample of 30 Australians to find out whether they had travelled to Japan or Vietnam. The results are shown below: CHAPTER 12

Japan 9 Vietnam 18 Japan and Vietnam 5

a Copy and complete the Venn diagram to show this information.

b What is the probability that a person chosen at random from this sample:

i has not been to either country?

ii has been to Japan or Vietnam, but not both?

iii has been to Vietnam but not Japan?

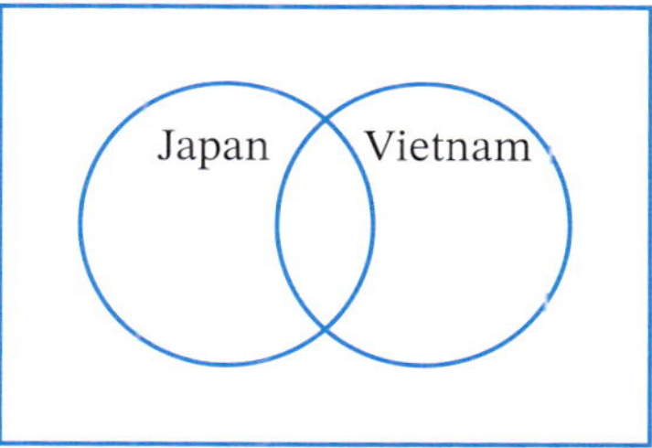

8 Convert: CHAPTER 3

a 40 m/s to km/h **b** 10 g/s to kg/min

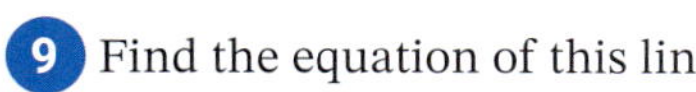

9 Find the equation of this line. CHAPTER 11

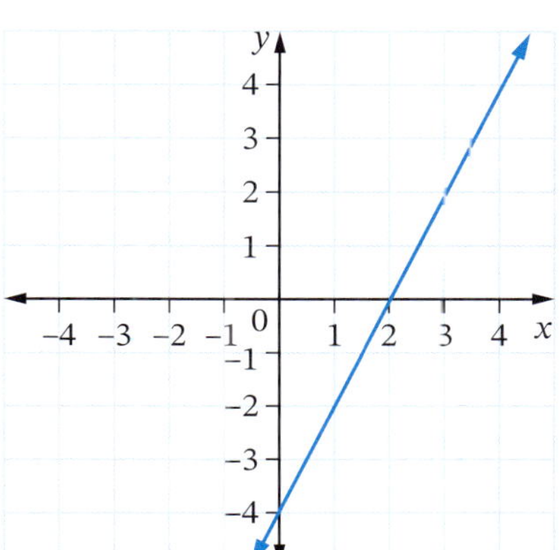

10 Jotika earns a salary of $82 000 p.a. How much (correct to the nearest cent) did she earn: CHAPTER 8

a per week? **b** per month?

☐ Foundation ○ Standard ⬡ Complex

11 Find each shaded area, correct to one decimal place.

a

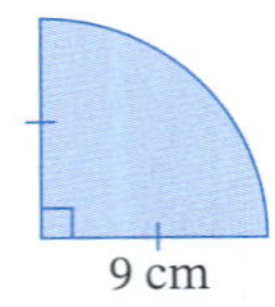

b

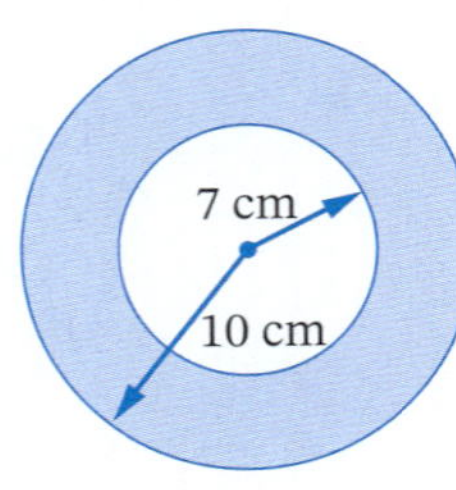

c

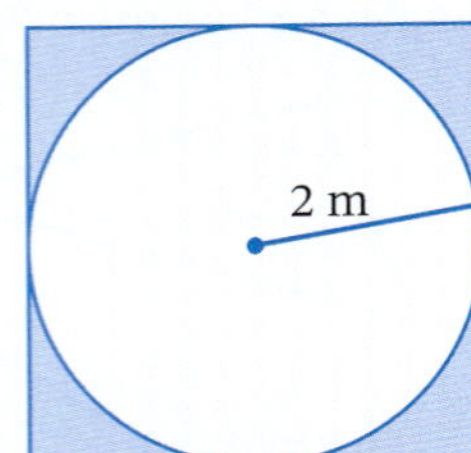

12 Find the value of x in each diagram.

a

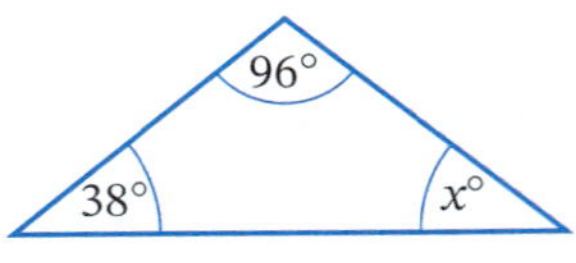

b

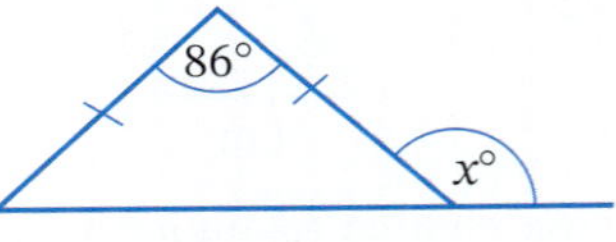

c

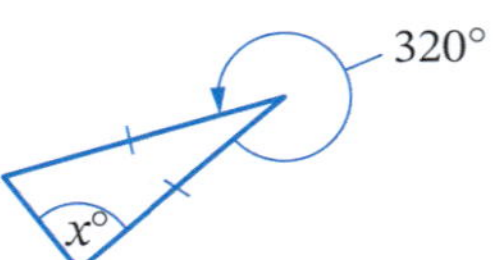

d

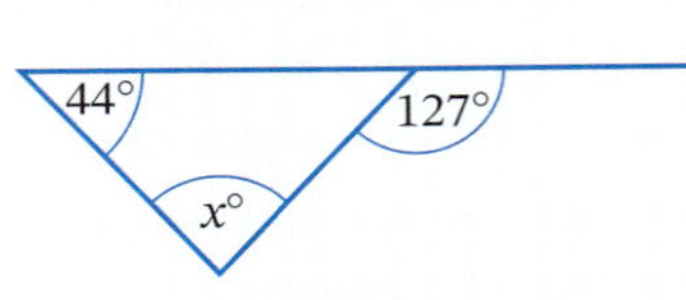

13 Find the area of each shape.

a

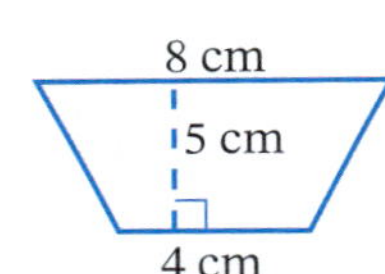

b

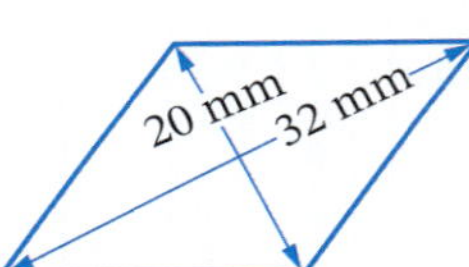

c

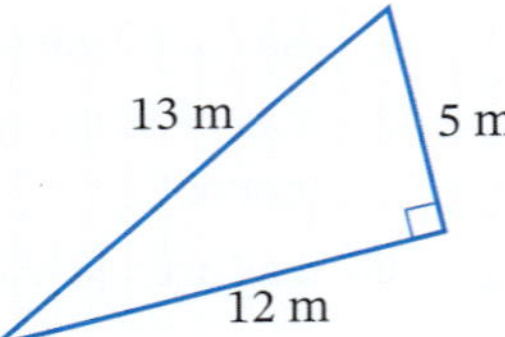

14 George earns $21.45 per hour. How much does he earn for working 38 hours at normal rate and 6 hours at time-and-a-half?

15 These 2 triangles are congruent.

a Which side in $\triangle QRS$ matches side NM?

b Which angle in $\triangle LMN$ matches $\angle S$?

c Copy and complete: $\triangle LMN \equiv$ ______

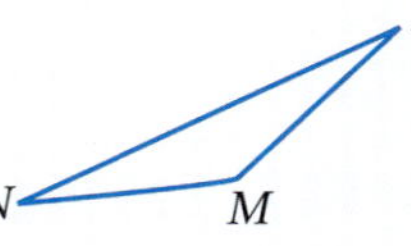

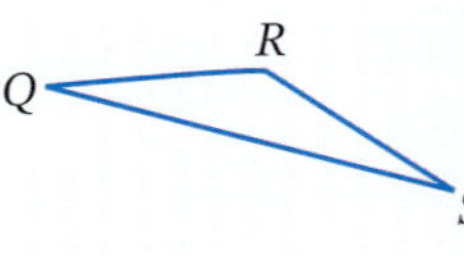

16 Simplify each expression.

a $k^7 \times k^5$ **b** $d^{10} \div d^5$ **c** $3p^2q \times 5pq^3$

d $20hk^5 \div 5hk$ **e** $\left(\frac{2r}{5}\right)^3$ **f** $5c^0$

☐ Foundation ○ Standard ○ Complex

17 **a** Complete a frequency table for this data, including an 'fx' column.

53	50	50	52	47	49	49	50	48	50
52	49	47	51	51	52	50	53	50	52
50	48	52	49	50	49	51	45	52	50
51	52	51	50	52	50	51	52	49	51

b Find the mean of this data.

c Show this data in a frequency histogram and polygon.

d Find the mode.

e Find the median.

18 Expand each expression. CHAPTER 1

a $3(1-2y)$ **b** $x(2x-7)$ **c** $-4m(1-m)$ **d** $(5x+7)(x-4)$

19 Expand and simplify $3(2d-5)-2(4+d)$. CHAPTER 1

20 What are the values of m and p in this diagram? CHAPTER 6

Select the correct answer **A**, **B**, **C** or **D**.

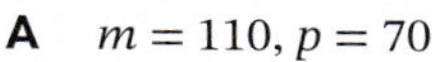

A $m = 110, p = 70$

B $m = 125, p = 70$

C $m = 110, p = 55$

D $m = 125, p = 55$

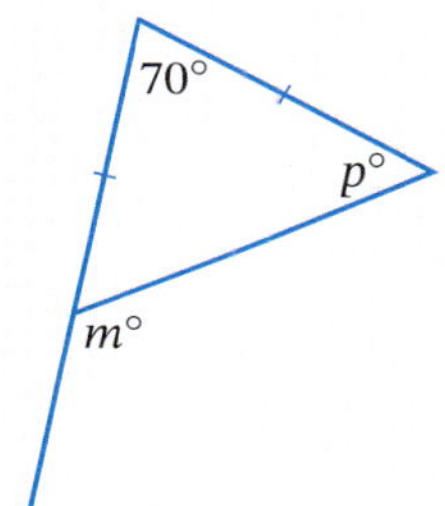

21 For the quadrilateral $LMNP$, calculate its: CHAPTER 2

a perimeter, correct to 2 decimal places.

b area.

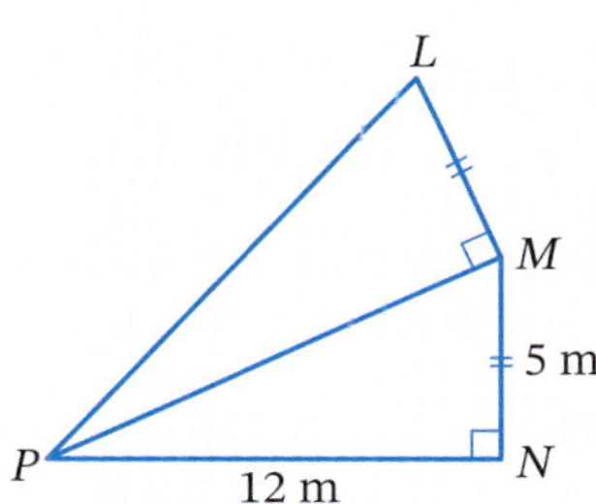

22 The number of goals scored by Damien's hockey team in 18 games were: CHAPTER 9

1	3	1	3	4	1	3	3	2
1	1	1	2	2	3	3	3	4

If 5 goals were scored in the 19th game, which statistical measure will not change?

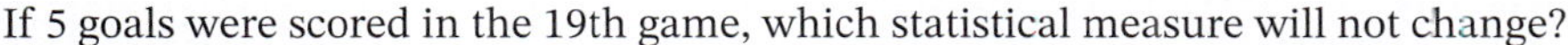

A The range **B** The mean **C** The median **D** The mode

23 Tahlia is paid a weekly commission of 4.3% on the value of her perfume sales over \$1500. Calculate her commission if her sales during one week totalled \$19 200.

24 Factorise each expression. CHAPTER 1

a $2y+8$ **b** $5a-15ab$ **c** $-8ab+2b$

25 Find, correct to 2 decimal places, the value of each variable.

a

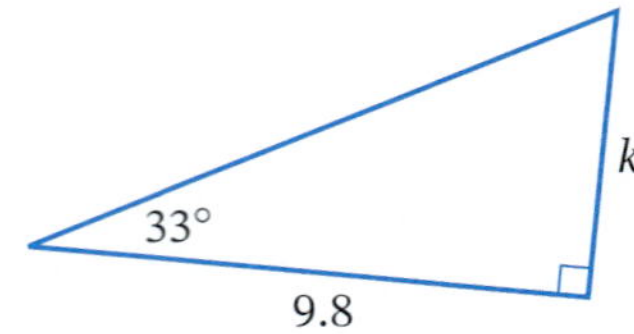

b

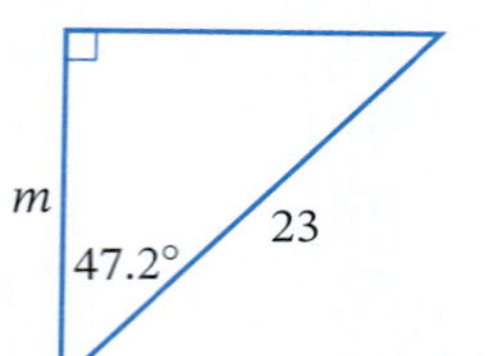

c

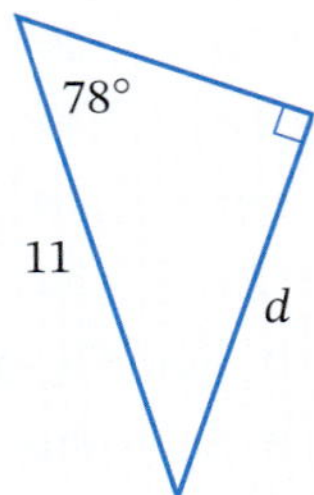

26 Romesh buys a car for \$11 000 and sells it for \$5200 2 years later. Calculate the percentage loss, correct to one decimal place.

27 Find the value of each variable, giving reasons.

a

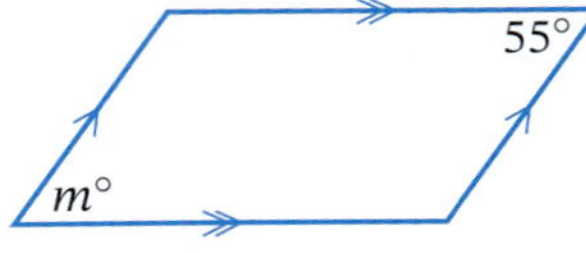

b

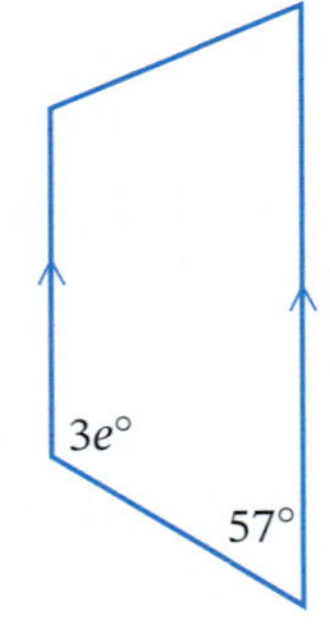

c

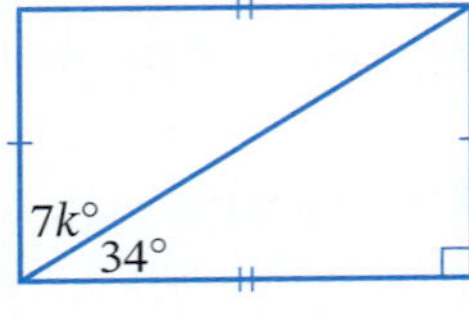

28 What is the equation of the horizontal line that passes through the point $(-2, 5)$?

29 For the interval joining the points $(5, 3)$ and $(-4, 2)$, find:

a its midpoint **b** its length in exact form **c** its gradient

30 The scale of a plan is 1 : 2000. What distance in centimetres on the plan would represent a real distance of 50 m?

31 Name the most general shape that matches each description.

a A quadrilateral with equal diagonals.

b A triangle with 3 equal sides.

c A quadrilateral with opposite sides equal and 2 axes of symmetry.

d A triangle with 2 equal angles.

32 The median of the data shown in this stem-and-leaf plot is 154.

a What is the missing value shown by □?

b Find the range.

c Describe the shape of the distribution.

Stem	Leaf
13	2 3
14	0 5 6 7
15	1 □ 6 8 8 9
16	3 4 4 5

33 Which triangle is right-angled given the side lengths described?

A

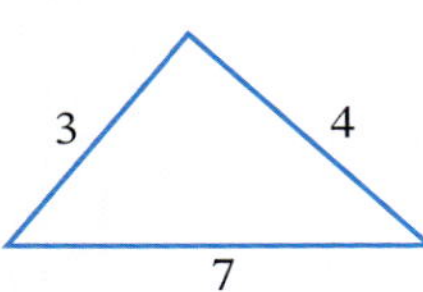

B

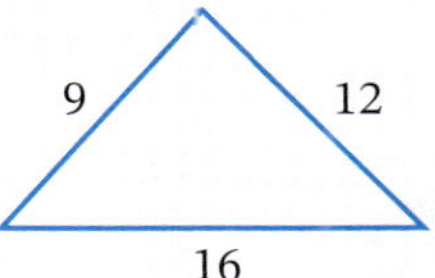

C

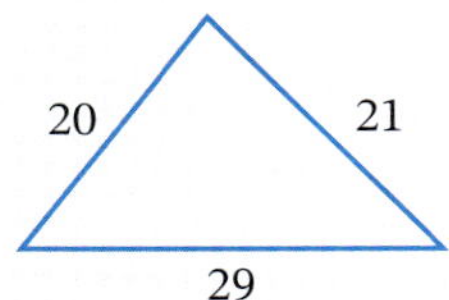

D

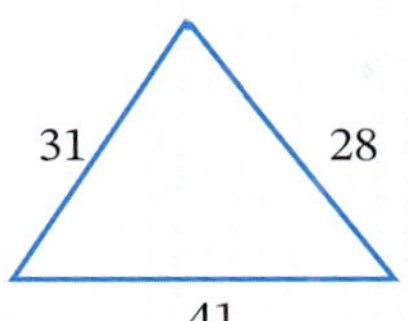

34 **a** Graph each equation on a separate number plane.

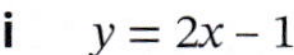

i $y = 2x - 1$ **ii** $y = -x^2 + 3$ **iii** $y = 4^x + 1$

b What is the gradient and y-intercept of the graph of $y = 2x - 1$?

c What are the coordinates of the vertex of the graph of $y = -x^2 + 3$?

d What is the asymptote of the graph of $y = 4^x + 1$?

35 Calculate the simple interest on \$8750 invested at 5.5% for 4 years.

36 The cost of a computer, including 10% GST, is \$968. What is the amount of the GST?

37 Which point lies on the line $y = 2x - 5$? Select **A**, **B**, **C** or **D**.

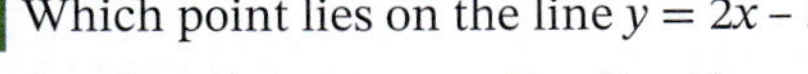

A $(0, -3)$ **B** $(1, -3)$ **C** $(-3, 0)$ **D** $(-3, 1)$

38 Find the surface area for each solid, correct to the nearest whole number where necessary. All measurements are in centimetres.

a

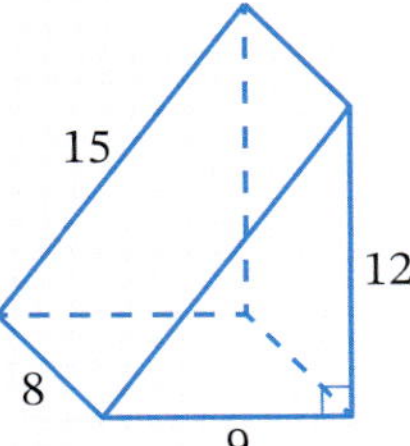

b

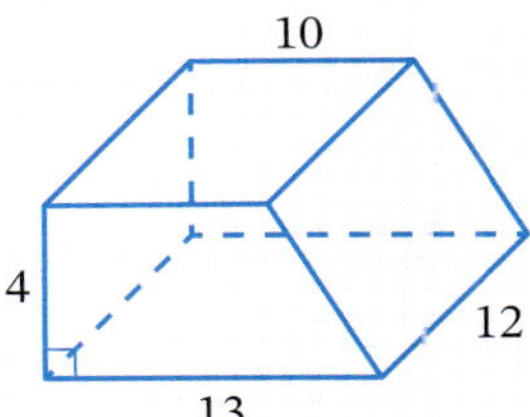

c

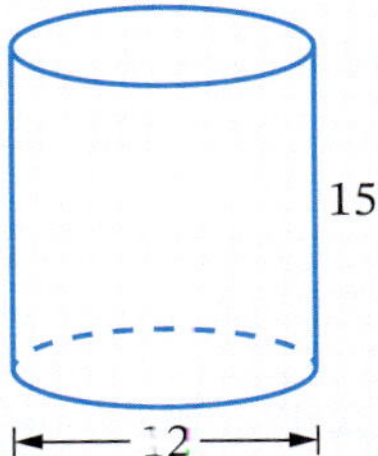

39 Solve each equation.

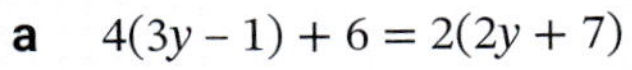

a $4(3y - 1) + 6 = 2(2y + 7)$ **b** $3x^2 = 27$

c $-6(12 - x) = 48$ **d** $\frac{3a - 8}{2} = 7$

40 Express each number in scientific notation.

a 714 000 **b** 0.000 35 **c** 15 763 000

☐ Foundation ○ Standard ⬡ Complex

41 Evaluate each expression in scientific notation, correct to 2 significant figures.

a $(4.89 \times 10^7)^2$ **b** $\frac{9.3\times10^{-3}}{2.2\times10^{8}}$

CHAPTER 4

42 Find the value of θ, correct to the nearest degree.

a

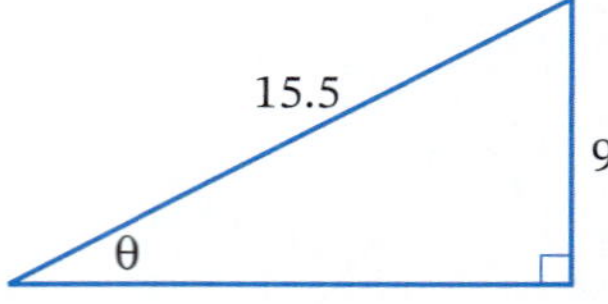

b

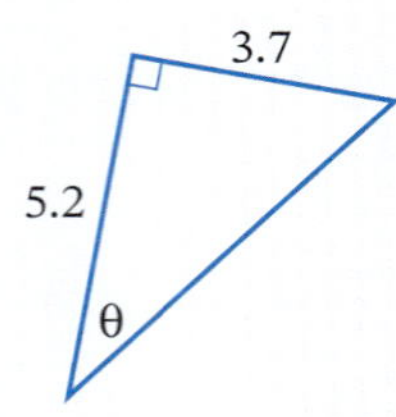

c

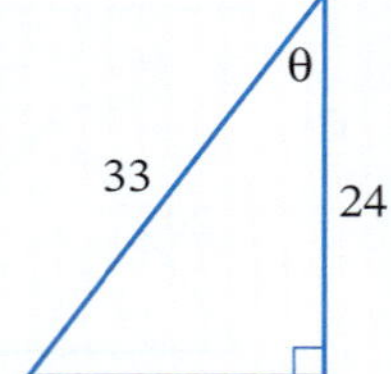

43 A die was repeatedly rolled and the results are shown in this table.

Outcome	Frequency
1	40
2	35
3	53
4	45
5	50
6	47

a Find, correct to 2 decimal places, the experimental probability of rolling:

i a 1 **ii** a 4

iii an even number **iv** a number less than 5

v neither a 2 nor a 6

b Based on the theoretical probability, what is the expected frequency of rolling a number less than 5? How does the observed frequency compare with the expected frequency?

CHAPTER 6

44 **a** What is the angle sum of a regular octagon?

b What is the size of one angle in a regular octagon?

CHAPTER 8

45 Luan has a gross income of $1225.63 per week. His deductions are $307.21.48 tax, $36.35 for private health insurance and $42.07 for superannuation. Calculate Luan's:

a net income.

b total deductions as a percentage of his gross income (correct to one decimal place).

46 What type of statistical graph could you use to represent the number of siblings of students in a particular class at school?

CHAPTER 9

47 How would you use systematic sampling to choose 7 employees from a workforce of 350 employees?

CHAPTER 10

48 Joel estimates there are 52 000 people at a concert. If there were actually 49 725 at the concert, what is the percentage error in Joel's prediction, correct to one decimal place?

☐ Foundation ○ Standard ⬡ Complex

49 A contractor charges a callout fee of \$80 and an hourly rate of \$45 for any work completed. Model this charge using a linear function and then calculate how many hours the contractor worked for a job if he charged a client \$350.

50 Sketch the graph of $y = 2^x + 2$.

51 Two similar triangles have matching sides in the ratio 3 : 5. If the area of the smaller triangle is 225 cm², determine the area of the larger triangle.

52 Write a simplified algebraic expression for the area of this rectangle.

53 Identify the surds in the list below:

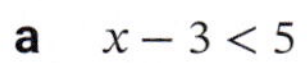

$\sqrt{225}$ $\sqrt{150}$ $\sqrt{36}$ $\sqrt{70}$

54 Solve each inequality:

CHAPTER 7

a $x - 3 < 5$ **b** $3x \geq 36$ **c** $2x + 5 \leq 17$ **d** $\frac{x}{2} - 4 > 8$

GENERAL PRACTICE

☐ Foundation ○ Standard ⬡ Complex

Answers

CHAPTER 1

SkillCheck

1 **a** $m + n$ **b** mn **c** $n + 5$ **d** $5 - n$

2 **a** 10 **b** 12 **c** 16 **d** 160

3 **a** like **b** like **c** like
d unlike **e** unlike **f** unlike

4 **a** $11n$ **b** $6h$ **c** $9pq$ **d** $5x^2$
e $4w$ **f** $40x$ **g** $6gh$ **h** $18y^2$
i $5x$ **j** 6 **k** $2y$ **l** $\frac{3n}{5}$

5 **a** $\frac{11}{15}$ **b** $\frac{7}{20}$ **c** $\frac{5}{16}$ **d** $\frac{14}{5}$ or $2\frac{4}{5}$
e $\frac{5}{22}$ **f** $\frac{29}{24}$ or $1\frac{5}{24}$ **g** $\frac{40}{27}$ or $1\frac{13}{27}$ **h** $\frac{1}{12}$

Exercise 1.01

1 **a** L **b** M **c** B **d** P
e H **f** I **g** D **h** F
i N **j** G **k** E **l** Q
m O **n** A **o** J **p** K
q C

2 **a** $x + 5$ **b** $m - 3$ **c** $4k$
d $\frac{20}{w}$ **e** $7 + d$ **f** hk
g $9g - 4y$ **h** $\frac{5a}{4}$ **i** $3d - 2w$
j k^2 **k** $\frac{t}{r}$ **l** $2q - 7p$
m $2a - 2$ **n** $\frac{d}{2} + 5$ **o** $\frac{gh}{3}$
p $3n - 7$ **q** $2n - 4$ **r** $10 - b$

3 **a** $3t + 5$ **b** $3(t + 5)$ **c** $7w - 4$
d $7(w - 4)$ **e** $\frac{k+8}{5}$ **f** $\frac{k}{5}+8$
g $\frac{d}{6}-2$ **h** $\frac{d-2}{6}$

4 Teacher to check.

5 **a** $16 + y$ **b** $16 - k$

6 D

7 **a** \$31.40 **b** \$(3.80 +2.30D)

8 B

9 **a** \$$pQ$ **b** \$$(40 + t)$
c $1000h$ m **d** $\frac{m+p+v+w}{4}$
e \$$\frac{y}{100}$ **f** \$$(950 - x)$
g \$$(100 - 12w)$ **h** \$$\frac{450}{n}$
i $(2a + 2b)$ cm **j** s^2 mm^2

10 **a** 25° **b** $180° - a° - b°$

11 **a** 55° **b** $\frac{(180-x)°}{2}$

12 **a** $d + 2$ **b** $d + 1$

13 **a** 12 **b** −8
c $x + 1$ **d** $n + 5$

14 **a** 8, 9, 10, 11
b −3, −2, −1, 0
c $y + 4, y + 5, y + 6, y + 7$
d $n - 4, n - 3, n - 2, n - 1$

15 **a** 5, 7, 9, 11 **b** 29, 31, 33, 35
c $t, t + 2, t + 4, t + 6$ **d** $k - 1, k + 1, k + 3, k + 5$

Exercise 1.02

1 **a** 52 **b** 108 **c** −18 **d** −137

2 **a** 36 **b** 225 **c** 64 **d** 225

3 **a** −28 **b** −16 **c** 0.5 **d** 38

4 **a** 8 **b** $10\frac{1}{2}$ **c** 0 **d** −8

5 **a** 7.5 **b** 26 **c** 24
d 2 **e** 9 **f** 54

6 **a** 4 **b** 5 **c** −15 **d** −5
e −17 **f** −1 **g** 0 **h** 2
i 1 **j** 20 **k** 21 **l** −1

7 14.4 cm 8 100 m^2 9 280 cm^3

10 90 cm^2 11 B 12 4.1

13 31.25 14 C

Exercise 1.03

1 **a** $11y$ **b** $5m$ **c** $11t$
d $14q$ **e** $-3g$ **f** $11r^2$
g 0 **h** $-5xy$ **i** $5n^2 + 3n$
j $21gh$ **k** $5my^2 + 2m^2y$ **l** $-12d$

2 C

3 **a** fg **b** $9uvw$ **c** $8t + 3$
d $3x + 10y$ **e** $2g + h$ **f** $3f - 2fg$
g $5p^2 - 3$ **h** $4p - 4q$ **i** $3 - 2r$
j $-10a + 4b$ **k** $6v + 10$ **l** $3x + 3y$
m $6p - 9q$ **n** $15 - 5w$ **o** $2 - 10n$
p $6e - 7f$ **q** $5l^2 - 4l$ **r** $-6b^2 - 14b$
s $2p^3 - 4p^2$ **t** $-2 - 17k$ **u** $6 - 13d$

4 C

5 **a** correct **b** incorrect **c** incorrect
d correct **e** correct **f** incorrect

6 **a** $4m + 6p$ **b** $16d$ **c** $4x + 5y$
d $6p + 3$ **e** $90h$ **f** $12a + 18c$

7, 8 Teacher to check

Exercise 1.04

1 **a** $40x$ **b** $15n$ **c** $5ab$ **d** $42mt$
e $-12w$ **f** $-6kp$ **g** $12de$ **h** $24vw$
i $88q^2$ **j** $20a^2$ **k** $24by$ **l** $-36w^2$
m $72chk$ **n** $8abc$ **o** $-30dew$ **p** $25n^2$
q $9r^2$ **r** a^2t^2 **s** $-10ad^2$ **t** ek^2
u $4x^2y^2$

2 **a** $4x$ **b** 15 **c** 9 **d** $3n$
e $\frac{1}{6}$ **f** $4ac$ **g** -6 **h** $-5z$
i $-\frac{1}{5q}$ **j** $3yz$ **k** $\frac{r}{4}$ **l** $-6k$
m $\frac{3k^2}{a}$ **n** $\frac{4w}{3}$ **o** $\frac{h}{3}$

3 C

4 **a** $30abc$ **b** $24pq$ **c** $8n$ **d** $\frac{1}{5}$
e $-36hk$ **f** $40k^2$ **g** $\frac{d}{2}$ **h** $\frac{1}{7}$
i $-11m$ **j** $\frac{1}{5}$ **k** $\frac{4a}{c}$ **l** $-6e^2f^2$
m $\frac{2c}{3}$ **n** $-24pq^2$ **o** $\frac{4vw}{3}$

5 D

6 **a** correct **b** incorrect **c** correct
d incorrect **e** correct **f** correct
g correct **h** incorrect **i** correct
j correct

7 **a** $6ad$ **b** $25m^2$ **c** $5hk$
d $12py$ **e** $3a^2$ **f** $10ck$

8 **a** $8m^3$ **b** $60dty$ **c** $12kw^2$

Mental skills 1A

2 **a** 160 **b** 70 **c** 240 **d** 900 **e** 2600
f 900 **g** 140 **h** 300 **i** 180 **j** 770
k 18 **l** 34 **m** 46 **n** 26 **o** 18
p 12 **q** 40 **r** 8 **s** 14 **t** 24

Exercise 1.05

1 **a** $3x + 15$ **b** $8e + 72$ **c** $2n^2 + 6n$

2 **a** $4h + 24$ **b** $15 + 5t$ **c** $6b + 21$
d $40 + 15x$ **e** $x^2 + 9x$ **f** $5p + p^2$
g $6r^2 + 3r$ **h** $4y + 12y^2$ **i** $20e^2 + 40ef$

3 **a** $10b - 20$ **b** $12y - 18$ **c** $20t^2 - 16t$

4 **a** $3t - 6$ **b** $35 - 7d$ **c** $64g - 24$
d $w^2 - 7w$ **e** $a^2 - a$ **f** $18h^2 - 6h$
g $12x^2 - 4x$ **h** $10x^2 - 20xy$ **i** $12ab - 21a^2$

5 C

6 **a** incorrect **b** correct

7 **a** $-a + 5$ **b** $-a - 5$ **c** $-2x - 12$
d $-2x + 12$ **e** $-11 - w$ **f** $-11 + w$
g $-y^2 - 9y$ **h** $-8x + xz$ **i** $-4t^2 + 32t$

8 B

9 D

10 **a** $19m + 10$ **b** $3 - 9e$
c -10 **d** $43 - 10x$
e $14a + 43$ **f** $d - 11$
g $29g - 19$ **h** $22 - 18w$
i $39c - 27$ **j** $t^2 + 7t + 12$
k $12 + 11h - 2h^2$ **l** $15e - 11$
m $6x^2 + 23x + 20$ **n** $2v^2 - 3v - 6$
o $w^2 - 8w + 3$ **p** $6y^2 - 29y + 35$

11 C

Exercise 1.06

1 **a** $x^2 + 18x + 80$ **b** $12h^2 + 31h + 7$

2 **a** $x^2 + 3x - 54$ **b** $n^2 - 12n + 35$

3 **a** $2y^2 + 11y + 15$ **b** $w^2 - 2w - 15$
c $p^2 - 8p - 15$

4 **a** $n^2 + 9n + 13$ **b** $t^2 + 11t + 30$ **c** $p^2 - 100$
d $x^2 - 5x - 24$ **e** $b^2 + 7b - 18$ **f** $u^2 - 15u + 56$
g $15 + 14r - r^2$ **h** $a^2 - 19a + 90$ **i** $c^2 - 8c + 15$
j $t^2 + t - 2$ **k** $y^2 + 6y - 40$ **l** $n^2 + 2n - 99$
m $e^2 - 4e + 4$ **n** $25 + 10w + w^2$ **o** $g^2 + 22g + 121$

5 D

6 A

7 **a** $2x^2 + 11x + 15$ **b** $12e^2 + 23e + 10$
c $3p^2 + 7p - 10$ **d** $49d^2 - 28d + 4$
e $6f^2 + 4f - 10$ **f** $12m^2 + 5m - 25$
g $6 + 7h - 20h^2$ **h** $8p^2 - 38p + 35$
i $-10m^2 + 23m - 12$ **j** $6t^2 + 7t - 5$
k $25y^2 - 25$ **l** $-4a^2 + 24a - 36$
m $9k^2 + 48k - 64$ **n** $16c^2 - 40c + 25$
o $6dx - 2d + 27x - 9$ **p** $5h + 15 - 2gh - 6g$
q $49 - 56w + 16w^2$ **r** $12n^2 + 7n - 45$

8 D

9 **a** $x^2 - x - 6$ **b** $9y^2 - 42y + 49$
c $4k^2 + k - \frac{3}{2}$ **d** $15p^2 - 22p + 8$
e $7m^2 - \frac{29}{2}m - \frac{15}{2}$ **f** $2d^2 + 12d + 10$

10 **a** $100 + x$ **b** $75 + x$
c $A = x^2 + 175x + 7500$ **d** $x^2 + 175x$
e 17.51 cm^2

11 **a** $l = 4 + x, w = 3 + y$
b $A = (4 + x)(3 + y)$
c $A = 12 + 4y + 3x + xy$
d $4y + 3x + xy$

12 Proofs: see worked solutions

Exercise 1.07

1 **a** 6 **b** 8 **c** 5 **d** 3 **e** m
f c **g** $6p$ **h** $8w$ **i** $(x - 5)$

2 **a** $4(a + 3)$ **b** $a(a + 10)$ **c** $10(b - 4)$
d $6(3p^2 + 4)$ **e** $3m(3m + 1)$ **f** $10d(2 - 3d)$
g $2xy(3x - 4)$ **h** $(x - 2)(x + 5)$ **i** $(3 + n)(n - 3)$
j $(v + 1)(1 - w)$

3 **a** $3(f + 2)$ **b** $4(m - 7)$ **c** $9(n + 3)$
d $8(2 + 3t)$ **e** $6(2q - 3)$ **f** $6(4x + 5)$
g $4(5g^2 - 16)$ **h** $y(x + 1)$ **i** $n(mn - 3)$
j $y(2 - y)$ **k** $4r(4r - 3)$ **l** $3t(2t + 9)$

4 **a** $4x(3xy - 4)$ **b** $2p(9p + 8r)$ **c** $4mn(m - n)$
d $7bc(2a + 3)$ **e** $7vw(4 - 3vw)$ **f** $9rt(5 + 6r)$

5 A

6 a $(a-3)(a+6)$ b $(8+t)(t-3)$
c $(b+5)(b-2)$ d $(2-y)(x-6)$
e $(a-7)(5+b)$ f $(a+3)(s+4)$
g $(g+5)(3p-2)$ h $(2-3m)(5+4n)$
i $(8+r)(r+1)$ j $(y-6)(1-y)$

7 a $-4(q+2)$ b $-9(u-2)$ c $-3(g-2)$
d $-6(3a-2)$ e $-(n+1)$ f $-(n-1)$
g $-y(y+9)$ h $-2t(3-5t)$ i $-3a(a+2b)$
j $-a(p-q)$ k $-2e(10e+11)$ l $-3m(3-m)$

8 B

Mental skills 1B

2 a 176 b 363 c 261 d 405
e 682 f 707 g 1818 h 3564
i 152 j 540 k 2142 l 588
m 288 n 693 o 3939 p 852

Exercise 1.08

1 a $-1, -6$ b $-3, 4$ c $-5, 3$
d $3, 4$ e $-5, -4$ f $-2, 7$
g $-2, 5$ h $-5, 5$ i $-2, 1$
j $-9, 2$

2 a $(m+3)(m+4)$ b $(x+7)(x+5)$
c $(k+4)(k+1)$ d $(p+5)(p+2)$
e $(w+5)(w+4)$ f $(b+1)(b+5)$
g $(e+2)(e+3)$ h $(r+2)^2$
i $(n+10)(n+1)$ j $(a+6)(a+5)$
k $(d+4)(d+6)$ l $(y+4)(y+11)$

3 a $(p-4)(p-1)$ b $(v-5)(v-6)$
c $(y-4)(y-3)$ d $(x-10)(x-2)$
e $(a-3)(a-6)$ f $(e-6)(e-2)$
g $(c-8)(c-6)$ h $(w-2)(w-3)$
i $(d-11)(d-7)$ j $(u-2)(u-1)$
k $(n-4)(n-4)$ l $(b-7)(b-8)$

4 a $(x+4)(x-1)$ b $(g+8)(g-3)$
c $(y+1)(y-3)$ d $(r-7)(r+2)$
e $(m+5)(m-3)$ f $(a+2)(a-1)$
g $(m-4)(m-5)$ h $(h-4)(h+1)$
i $(d-3)^2$ j $(w+6)(w-2)$
k $(a-6)(a+2)$ l $(k-2)(k+7)$
m $(c-9)(c+2)$ n $(e-9)(e+3)$
o $(u-5)(u+11)$ p $(p+12)(p-3)$
q $(x-4)(x+13)$ r $(q-14)(q+5)$

5 a $(m+2)^2$ b $(p+10)^2$ c $(a-5)^2$
d $(y-7)^2$ e $(w+12)^2$ f $\left(a-1\frac{1}{2}\right)^2$

6 a $3(m+1)(m+2)$ b $2(y+2)(y-1)$
c $5(t-10)(t+8)$ d $5e^2(e+8)(e-3)$
e $x(x-11)(x+10)$ f $4(b-7)(b+6)$
g $4(w+4)(w-3)$ h $3a(a-4)(a+1)$
i $2(e+5)(e+4)$ j $-(t+8)(t-3)$
k $-(u-7)(u+6)$ l $-(x-7)(x+4)$
m $-(b+4)(b-3)$ n $-(k-3)(k-4)$
o $-(x-5)(x-7)$

7 a $(h-6)(h+3)$ b $-(t+9)(t-2)$
c $(w+1)(w+7)$ d $(k+3)(k-15)$
e $(v+10)(v-2)$ f $3(c+5)^2$
g $(q-1)(q-5)$ h $(a-3)(a-1)$
i $6(x+8)(x-2)$ j $(x-8)^2$
k $8(u-4)(u+1)$ l $(b+5)(b+6)$
m $(y-6)^2$ n $5(r-2)(r+1)$
o $4(l-4)(l+2)$ p $(g-20)(g-4)$
q $-2(d-6)(d+9)$ r $-(n+2)(n-13)$

Power plus

1 a $p+1, p+2, p+3$
b $w+2, w+3, w+4$
c $2n+2, 2n+3, 2n+4$
d $h-2, h-1, h$

2 a $-\frac{1}{3}$ b $-\frac{80}{3}$ c 15
d 2 e -20 f $-\frac{6}{5}$
g 29 h 395 i -55

3 a $u+4t$ b $2u+3t$
c $10u+14t$ d $4u^2+8ut-2t^2$

4 a $10a$ b 18 c $16m$ d $20p^2$
e $12h$ f $13t$ g $-12x$ h $4c$
i $8r-1$ j $16r$

5 a $2-2m$ b $3x^2-x$
c $9q-1$ d $5-5m$
e $4a^2+20a+25$ f y^2-2y+2

6 $\frac{5r+6}{2}$

7 a $A=y^2-5y$ b $P=4y-10$

8 $5t$ hours

9 a $4(2a+3b-1)$ b $5(4xy+y-2x^2)$
c $mt(m+1+t)$ d $4(8-6y+y^2)$
e $4(3v^2+vw-4w^2)$ f $-3(4+3f+2f^2)$
g $\frac{1}{2}(p+q)$ h $\frac{k}{4}(3-t)$

TEST YOURSELF 1

1 a $\$nP$ b $\frac{m}{3}$ c $\$(5A+8B)$
d $\$(100-y)$ e $60H$ min f $\frac{m}{1000}$ L
g $4P+5$ h $4(T+5)$ i $\frac{Q}{4}-5$
j $\frac{Q-5}{4}$

2 a 5 b 6 c 19 d 11
e 12 f 2.5 g 5 h 18

3 a $16y$ b $-3ab$ c $8x^2$
d $2p+2q$ e $6w-5$ f $3x^2-13x$
g $7m+10n$ h $3u+4$ i $12d^2-2d+4$

4 a $18n$ b $3gh$ c ad
d $-6r$ e $4y$ f $2x$
g 4 h 3 i $-4p$

5 **a** $5r - 10$ **b** $a^2 - 10a$ **c** $6xy - 15y^2$
d $12p^2 - 21pq$ **e** $-6n^2 + 54n$ **f** $-6t^2 - 42t$
g $b^2 + 6b + 5$ **h** $6x^2 - 31x + 35$ **i** $30 - 12x$

6 $3d^2 + 17d + 20$

7 **a** $a^2 + 5a + 6$ **b** $y^2 - 4y - 21$
c $t^2 + 5t - 24$ **d** $h^2 - 9h + 20$
e $12 + g - g^2$ **f** $6p^2 + 11p + 3$
g $20x^2 - 11x - 3$ **h** $12r^2 + r - 20$
i $-6q^2 + 13q - 5$ **j** $9m^2 + 30m + 25$
k $49 - 28y + 4y^2$ **l** $-3w^2 + 11w - 10$

8 **a** $4(m - 3)$ **b** $x(y + z)$
c $-3(g + 10)$ **d** $8w(2w + 3)$
e $-8(q - 2)$ **f** $5k(3 - 2hk)$
g $(y - 6)(3 + y)$ **h** $(2x - 1)(3x - 2)$
i $(p + 6)(p - 2)$

9 **a** $(m - 12)(m + 1)$ **b** $(y - 3)(y - 1)$
c $(h - 13)(h + 5)$ **d** $(p + 9)(p - 3)$
e $3(n + 1)(n + 2)$ **f** $4(y + 2)(y - 1)$

CHAPTER 2

SkillCheck

1 **a** 500 **b** 2 **c** 1400
d 6.2 **e** 8.5 **f** 0.012

2 **a** 16 **b** 106.09 **c** 34
d 28 **e** 7 **f** 11

3 **a** 2.5 **b** 831.9 **c** 372.7

4 **a** 28.78 **b** 7.95 **c** 536.86

5 **a** 25 cm **b** 28 cm **c** 76 mm

6 **a** 40 cm^2 **b** 49 cm^2 **c** 100 mm^2

Exercise 2.01

1 **a** 16 **b** 256 **c** 39.69 **d** 0.49
e 163.84 **f** 0.0004 **g** 1 **h** 81
i 169 **j** 1.44 **k** 53 **l** 449
m 165 **n** 44 **o** 0.25 **p** 7.99

2 **a** 14 **b** 30 **c** 25 **d** 19
e 1.3 **f** 1.7 **g** 3.2 **h** 12.5

3 **a** 3.46 **b** 21.21 **c** 31.64 **d** 18.03
e 12.37 **f** 83.68 **g** 30.63 **h** 6.48

4 C

5 **a** 8.60 **b** 12.53 **c** 12.04 **d** 7.42

6 **a** 4.5 **b** 10.3 **c** 31.5

7 C

8 B

9 $\sqrt{32}, \sqrt{33}, \sqrt{4.9}, \sqrt{52}, \sqrt{200}$

10 C

11 No, only if $x \geq 0$

12 Yes, as $\sqrt{x}$ is defined for $x \geq 0$

Exercise 2.02

1 D

2 **a** p **b** z **c** c
d PR **e** ST **f** TU

3 **a** 22 **b** 36 **c** 33 **d** 35
e 36 **f** 41 **g** 39

4 **a** $k^2 = m^2 + n^2$ **b** $d^2 = p^2 + c^2$
c $x^2 = w^2 + y^2$ **d** $k^2 = h^2 + t^2$
e $z^2 = c^2 + r^2$ **f** $q^2 = f^2 + d^2$

5 **a** A **b** C **c** C **d** C

6 **a** $5^2 = 4^2 + 3^2$ **b** $h^2 = 12^2 + 5^2$
c $12.5^2 = 10^2 + 7.5^2$ **d** $17^2 = 15^2 + x^2$
e $6.5^2 = 2.5^2 + 6^2$ **f** $10^2 = 8^2 + b^2$

7 **e** Yes
f The area of the square on the hypotenuse is equal to the sum of the areas of the squares on the other two sides of the triangle. The area of any square is its length squared. So this proves Pythagoras' theorem: the square of the hypotenuse in a right-angled triangle is equal to the sum of the squares of the other two sides.

Exercise 2.03

1 **a** 10 cm **b** 13 m **c** 30 cm **d** 37 cm
e 26 mm **f** 34 cm **g** 20 cm **h** 5.8 m
i 25 cm **j** 20.5 cm **k** 61 mm **l** 19.5 m

2 **a** $\sqrt{325}$ cm **b** $\sqrt{377}$ cm **c** $\sqrt{3541}$ mm
d $\sqrt{4405}$ mm **e** $\sqrt{117}$ cm **f** $\sqrt{4122}$ mm

3 **a** 82.9 mm **b** 2.6 m **c** 107.4 mm
d 30.6 cm **e** 6.5 m **f** 85.4 mm

4 B

5 4.8 m

6 3.4 m

7 **a** 69.1 cm **b** $\sqrt{19.45}$ m **c** 134 mm
d 179.6 mm **e** 12.8 m **f** 6.26 m

Exercise 2.04

1 **a** 12 mm **b** 16 cm **c** 8 m
d 15 mm **e** 18 cm **f** 12 mm

2 **a** 18.1 cm **b** 14.4 cm **c** 93.7 cm
d 24.1 m **e** 35.0 cm **f** 48.4 m

3 **a** $\sqrt{5031}$ cm **b** $\sqrt{5520}$ m or $4\sqrt{345}$ m
c $\sqrt{1344}$ m or $8\sqrt{21}$ m **d** $\sqrt{10\,32}$ cm
e $\sqrt{14.03}$ m **f** $\sqrt{37\,127}$ m

4 B **5** B

6 C **7** 10 39 cm

8 212 mm

Mental skills 2A

2 **a** 895.4 **b** 37 **c** 83.1 **d** 4200
e 5271.6 **f** 1561 **g** 31 840 **h** 6430
i 22.4 **j** 48.94 **k** 7389 **l** 1142

4 **a** 73.34 **b** 0.94 **c** 6.52 **d** 0.104
e 0.704 **f** 1.985 **g** 0.02 **h** 4.159
i 12.3 **j** 0.007 58 **k** 0.0849 **l** 0.0251

Exercise 2.05

1 D 2 B

3 a 21.0 cm b 4.7 m c 120.0 mm d 11.9 cm
e 27.1 mm f 28.0 m g 8.5 cm h 1.9 m
i 40.1 m j 4.0 m k 11.7 cm l 13.9 m

4 4.58 m

5 153 cm

6 92.2 M

7 7 m

8 a 5 b $\sqrt{58}$

9 a 15.81 cm b 11.79 m c 40 mm

10 a 38.3 cm b 12.21 c $\sqrt{31}$

11 110 mm

12 22.8 m

Exercise 2.06

1 a Yes b Yes c No d Yes
e No f Yes g No h Yes
i No j Yes k No l Yes

2 $\sqrt{17^2+21^2} = \sqrt{730} = 27.01... \neq 30$
$\therefore$ not a right-angled triangle

3 C

4 B

5 The pole is vertically upright because {2.9, 2.1, 2} follows Pythagoras' theorem.

Exercise 2.07

1 a Yes b Yes c Yes
d No e Yes f No
g Yes h Yes i No

2 D

3, 4 Teacher to check.

5 a $b = 12, c = 13$
b $\sqrt{12^2+5^2} = \sqrt{169} = 13$ $\therefore$ {5, 12, 13} is a triad.

6 a {7, 24, 25} b {11, 60, 61}
c {15, 112, 113} d {4, 7.5, 8.5}
e {9, 40, 41} f {19, 180, 181}
g {10, 49.5, 50.5} h {51, 1300, 1301}

7 Proof: see worked solutions

Exercise 2.08

1 4.7 m

2 a 257 m b 182 m c 75 m

3 637.1 m

4 752 m

5 a 21.8 m b 23 m

6 10.5 m

7 4.43 m

8 a 4.387 m b 6.652 m

9 25.5 m

10 a 48 cm b 31.4 cm c 170 cm
d 6.5 m e 35.9 cm f 127.9 mm

11 a 192 cm^2 b 95.0 cm^2 c 84 mm^2
d 1.3 m^2 e 108 m^2 f 22 m^2

12 23.09 cm 13 43.3 cm 14 12.72 m

Mental skills 2B

2 a 16 000 b 210 c 9900 d 600
e 16 000 f 400 g 490 h 52
i 1500 j 1160 k 180 l 13 200

4 a 8 b 5 c 8 d 22
e 4 f 50 g 9 h 9.3
i 8.5 j 4.23 k 9.6 l 24

Power plus

1 BD, BC, CE, DE, CD 2 $\sqrt{128}$ m or $8\sqrt{2}$ m

3 a 22.6 b 10.6

4 a 16.2 m b 16.9 m

5 5.8

6 a 21.21 cm b 25.98 cm

7 $\sqrt{340}$ cm or $2\sqrt{85}$ cm

8 Proof: see worked solutions.

TEST YOURSELF 2

1 a 72.25 b 36 c 202

2 a 12.08 b 13.42 c 21.75

3 a BC b q c EG

4 a $(AB)^2 + (AC)^2 = (BC)^2$ b $n^2 + p^2 = q^2$
c $(EF)^2 + (FG)^2 = (EG)^2$

5 B

6 a 175 mm b 113 mm c $\sqrt{98.72}$ m

7 B

8 a 34.9 m b 12.2 cm c 824.9 m

9 a $\sqrt{161}$ b $\sqrt{244}$ or $2\sqrt{61}$ c $\sqrt{649}$

10 168 cm

11 a

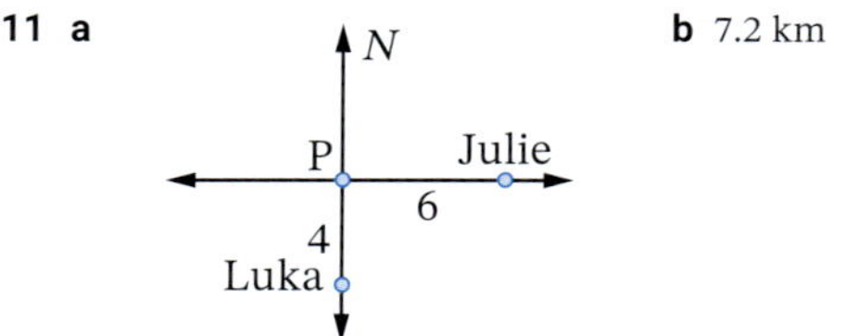

b 7.2 km

12 a No b Yes c Yes

13 a Yes b No c Yes d Yes

14 a 432 m b 1.296 km

15 B

16 48.5 cm

CHAPTER 3

SkillCheck

1 a 56 b 7 c 7 d 7
e 54 f 1 g 54 h 28
i 48 j 9 k 36 l 0
m 4 n 7 o 35 p 9

2 a $\frac{1}{2}$ b $\frac{9}{10}$ c $\frac{9}{20}$ d $1\frac{2}{5}$

9780170465618

3 **a** 16 **b** 31 **c** 0
d 4 **e** 20 **f** 10

4 **a** 25% **b** 30% **c** 7% **d** 60%

Exercise 3.01

1 **a** < **b** < **c** > **d** >

2 −13, −5, −1, 0, 2, 8

3 6, 2, −2, −7, −10, −15

4 **a** 3 **b** 2 **c** −8 **d** −4
e −4 **f** 5 **g** −15 **h** −4
i −20 **j** −9 **k** −18 **l** 0
m 8 **n** 7 **o** −8 **p** −18

5 −10 °C

6 **a** 74°C **b** 59°C

7 **a** Teacher to check, for example, −5 and 3, 4 and −6.
b Teacher to check, for example, 7 and 10, −5 and −2.

8 Teacher to check, for example, 5 and −10, 9 and −6. More than one answer.

9 **a** −15 **b** −8 **c** −80 **d** 60
e 81 **f** −240 **g** −64 **h** −450
i −5 **j** 16 **k** −8 **l** −6
m 4 **n** −20 **o** 6 **p** −13

10 **a** −56 **b** 20 **c** 22 **d** 100
e −9 **f** −20 **g** 13 **h** −12
i 20 **j** 80 **k** −3 **l** −64
m 11 **n** −32 **o** −3 **p** −19
q −20 **r** 4

11 **a** Teacher to check, for example, 4 and −5, −10 and 2.
b Teacher to check, for example, −20 and 5, 8 and −2.

Exercise 3.02

1 **a** 0.8 **b** 0.31 **c** 0.456 **d** 0.57
e 0.03 **f** 0.0092 **g** 0.06 **h** 0.0124

2 **a** $\frac{7}{20}$ **b** $\frac{1}{125}$ **c** $\frac{2}{5}$ **d** $\frac{17}{10\,000}$

3 **a** 0.7, 0.7007, 0.707, 7.007
b 0.002, 0.022 02, 0.202, 0.22

4 **a** 6.33, 6.3033, 6.1, 6.03, 0.6003
b 3.66, 3.606, 3.6, 3.0667

5 Teacher to check, for example, 5.32, 5.35, 5.37.

6 **a** 7.96 **b** 17.801 **c** 1.221 **d** 28.14
e 16.74 **f** 36.982 **g** 2.718 **h** 18.544
i 249.967 **j** 18.56 **k** 15.521 **l** 45.593

7 Teacher to check, for example, 8.15 and 3.6.

8 **a** 8.3 and 7.1 **b** 38.6

9 **a** 2.15 m **b** 58.59 s **c** 3.94 s

10 **a** 72.3 **b** 1501.2 **c** 53.2
d 8920 **e** 0.3567 **f** 0.97
g 0.0987 **h** 0.2345 **i** 0.000 54

11 **a** 9.2 **b** 0.72 **c** 2.092
d 0.024 **e** 0.0333 **f** 0.25

12 **a** 6.9 **b** 6.04 **c** 0.52
d 1.5 **e** 16 **f** 200.01

13 **a** 0.009 **b** 15.4 **c** 0.246

14 C

15 **a** 55.35 **b** 325.8 **c** 5.0
d 0.070 **e** 201 **f** 23.855
g 20 **h** 46.1 **i** 7.60
j 245.7 **k** 36.00 **l** 301.984

16 A

17 **a** $0.1\dot{6}$ **b** $0.\dot{2}7\dot{3}$ **c** $0.\dot{8}$ **d** $0.20\dot{9}$

18 **a** 0.109 810 98... **b** 0.171 717 17...
c 0.2555... **d** 8.318318...

19 **a** 0.875, terminating **b** $0.\dot{2}\dot{7}$, recurring
c $4.\dot{8}$, recurring **d** 2.6, terminating
e 0.76, terminating **f** $0.\dot{6}$, recurring
g 0.65, terminating **h** $1.58\dot{3}$, recurring
i 8.325, terminating **j** $0.708\dot{3}$, recurring

20 **a** $0.\dot{3}$ **b** **i** $0.\dot{6} = \frac{2}{3}$ **ii** $0.\dot{9} = 1$

21 **a** 8.525 **b** 8.534

22 James' answer is more accurate as he multiplied the numbers and then rounded the answer while Aditi rounded the numbers before multiplying and also rounded her answer after multiplying.

23 **a** 0.000 001 **b** 0.008 **c** 0.0081

Exercise 3.03

1 **a** $\frac{5}{3}$ **b** $\frac{9}{2}$ **c** $\frac{13}{5}$ **d** $\frac{111}{20}$

2 **a** $2\frac{1}{3}$ **b** $1\frac{4}{7}$ **c** $3\frac{1}{5}$ **d** $3\frac{7}{10}$

3 **a** $\frac{7}{8}$ **b** $\frac{3}{4}$ **c** $\frac{7}{8}$

4 **a** $\frac{6}{7}, \frac{3}{4}, \frac{3}{5}$ **b** $\frac{2}{3}, \frac{1}{2}, \frac{5}{11}$ **c** $\frac{7}{8}, \frac{17}{20}, \frac{4}{5}$

5
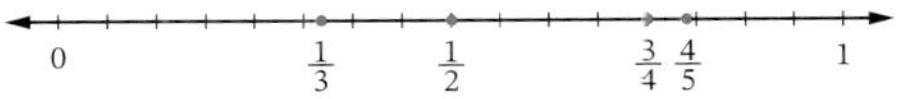

6 **a** $\frac{2}{3}$ **b** $\frac{3}{4}$ **c** $\frac{5}{11}$ **d** $\frac{4}{5}$

7 **a** $1\frac{5}{21}$ **b** $1\frac{1}{10}$ **c** $\frac{11}{36}$ **d** $\frac{7}{30}$
e $1\frac{3}{40}$ **f** $\frac{1}{12}$ **g** $\frac{4}{33}$ **h** $\frac{43}{100}$
i $4\frac{3}{20}$ **j** $6\frac{1}{30}$ **k** $1\frac{2}{15}$ **l** $2\frac{22}{35}$

8 **a** $\frac{15}{28}$ **b** $\frac{3}{10}$ **c** $4\frac{1}{2}$ **d** $4\frac{2}{3}$
e $\frac{7}{12}$ **f** 21 **g** $\frac{7}{36}$ **h** $6\frac{2}{5}$
i 8 m **j** 15 L **k** $350 **l** 57 kg
m $5\frac{3}{4}$ **n** $14\frac{11}{15}$ **o** 1 **p** $6\frac{1}{4}$

9 **a** $1\frac{1}{2}$ **b** $\frac{5}{6}$ **c** $\frac{3}{35}$ **d** $1\frac{2}{7}$
e $\frac{5}{8}$ **f** $\frac{45}{64}$ **g** $3\frac{1}{8}$ **h** $2\frac{4}{9}$

10 Teacher to check, for example, $\frac{19}{30}$.

11 a B　b C　c C

12 $2\frac{9}{10}$　13 $10\frac{1}{2}$　14 $2\frac{3}{4}$ hours

15 14 lessons　16 $\frac{1}{8}$

17 a Teacher to check, for example, $4\frac{2}{3}$ and $5\frac{5}{6}$.

b Teacher to check, for example, $4\frac{1}{2}$ and $2\frac{1}{3}$.

18 924 students

19 4 min 27 s

Mental skills 3A

2 a 625　b 3025　c 2025　d 4225
e 5625　f 6.25　g 20.25　h 42 025
i 42.25　j 56.25　k 75 625　l 119 025

4 a 441　b 10 201　c 961　d 8281
e 26.01　f 6561　g 3721　h 40 401
i 4.41　j 82.81

6 a 3481　b 4761　c 7921　d 361
e 11 881　f 34.81　g 47.61　h 14 161
i 1521　j 79.21

Exercise 3.04

1 a $\frac{3}{50}$　b $\frac{9}{20}$　c $\frac{21}{25}$　d $\frac{1}{50}$
e $\frac{1}{200}$　f $1\frac{1}{2}$　g $\frac{51}{100}$　h $\frac{9}{200}$
i $\frac{7}{80}$　j $\frac{1}{3}$　k $\frac{1}{8}$　l $\frac{19}{400}$
m $\frac{1}{16}$　n $\frac{11}{200}$　o $\frac{3}{10}$　p $\frac{1}{6}$

2 a 0.85　b 0.051　c 0.1　d 0.0845
e 0.065　f 0.1225　g 0.004　h 4.5

3 a 7%　b 237.5%　c 48%　d 5.5%
e 72%　f 1.6%　g $66\frac{2}{3}$%　h 75%
i 44%　j 73%　k 20%　l $63\frac{7}{11}$%
m 101.5%　n $16\frac{2}{3}$%　o 90%　p 90%

4 a F　b T　c T　d T

5 a $\frac{2}{5}$, 0.427, $\frac{3}{7}$, 46%　b 70%, $\frac{29}{40}$, $\frac{8}{11}$, 0.73

6 a $\frac{4}{11}$, $\frac{1}{3}$, 33.33%, 0.333　b 0.58, $\frac{4}{7}$, 55%, $\frac{1}{2}$

7 a $12.75　b 72 kg　c 7.2 m　d $8.40
e $4.20　f 11.4 kg　g $21.75　h 24 L
i $106.88　j $2180　k 1.5625 t　l $14.88

8 a 29%　b 24 534

9 186　10 140 g　11 106.25 kg

Exercise 3.05

1 a $500.50　b $92.40　c $2040
d 82 kg　e 45.5 m　f 288 L

2 a $539.50　b 114 min　c 79.2 L
d 5.1 m　e 106.6 kg　f $86.85

3 a $43　b $903

4 $20.90

5 $6.69

6 $944.23

7 $14 520

8 a 68.8%　b 66.7%　c 64.3%　d 83.3%
e 66.7%　f 15%　g 14.6%　h 12.5%
i 17%　j 12.5%　k 17%　l 14%

9 English 80%, Maths 78%. English is the better mark.

10 33.3%　11 19.8%

12 a i 5%　ii 62%　b 64%

13 51 goals　14 $8400　15 750 g

16 $83 654　17 1% decrease　18 1% decrease

Exercise 3.06

1 a $600　b 92.3%

2 a $1500　b 9.4%

3 a $4500　b $1700　c 37.8%

4 a $130 000　b 15.3%

5 40%　6 $33\frac{1}{3}$%　7 $91\frac{2}{3}$%

8 a $1.56　b $1.95　c $7.15　d $10.39

9 C

10 C

11 $16.58　12 23.5%　13 21.8%

14 a 31.3%　b 48.7%

15 $140　16 $3295　17 B

18 $289

19 a $2090, $22 990　b $181, $1991
c $14.50, $159.50　d $18, $198

20 a $4.20, $42　b $67.20, $672　c $2.85, $28.50
d $5.99, $59.90　e $84, $840　f $18, $180

21 a $1329.30　b $1262.85
c $636.15　d 33.5%

22 a $599　b $509.15
c $483.70　d $479.20
e No, the 15% discount followed by the 5% discount is the same as a 19.25% discount.

Exercise 3.07

1 a $6300　b $300.96　c $660
d $708.75　e $2529.60　f $1825

2 $21 386

3 a $94.50　b $46.72　c $254.79
d $305.67　e $13 407.69　f $1185.67

4 a 6.3% p.a.　b 7.7% p.a.　c 4.7% p.a.

5 6 years　6 100 months　7 144 days

8 a $1400　b 10% p.a.

9 a $6144　b $2144　c 13.4% p.a.

Mental skills 3B

2 a 2.8　b 4.2　c 5.3　d 10.5
e 8.9　f 5.6　g 3.5　h 8.1
i 8.7　j 5.4　k 6.3　l 11.2

Exercise 3.08

1 a 12 : 7 b 5 : 12 c 25 : 12 d 3 : 2
e 2 : 1 f 12 : 5 g 17 : 6 h 5 : 18
i 25 : 4 j 15 : 16 k 15 : 2 l 5 : 6 : 2
m 3 : 4 : 1 n 6 : 1 : 4 o 5 : 18 p 1 : 2

2 a 3 : 16 b 25 : 4 c 8 : 3
d 3 : 10 e 1 : 3 f 1 : 5
g 3 : 20 h 14 : 1 i 8 : 1

3 9

4 $312 000

5 16

6 20 m^3 gravel, 15 m^3 sand, 5 m^3 cement

7 1125

8 Shahid $27 600, Bridie $32 400

9 D

10 a 67.5 km/h b $8.40/kg c 56 words/min
d 12.1 km/L e 7.7 m/s f 41.7 g/m^2
g 22.5 km/day h $71.67/h

11 a 162.5 km b 552.5 km

12 a $27.50/h b $1100 c 8 days

13 a $15.99/kg b $47.97 c 2.5 kg

14 a 4680 beats/hour b 12 min 49 seconds

15 a i 3 : 5 ii 3 : 1 b 29.2 g

16 a 74.24 L b 6.728 L c 16.24 L

17 3.4 L/100 km 18 32 500 19 A

20 a 144.8 persons/km^2
b 3.4 people/km^2
c 1 113 801 600

Exercise 3.09

1 a 18 000 m/h b 108 000 m/h c $0.01/g
d 1200 mL/h e 2400 mL/h f 18 000 mL/h
g 50 000 kg/ha h 2500 kg/day i $39 000/year
j 30 km/h k 360 sheep/day l $420/week

2 a 18 km/h b 1260 m/h c $100/kg
d 2.4 L/h e 666.7 mL/s f 432 L/day
g 250 kg/ha h 60 000 kg/day i 30.6 m/s
j 50 g/cm k 8 m/mL l $42/h

3 a 97.2 km/h b 349.2 km/h c 11.88 km/h
d 108 km/h e 82.8 km/h

4 a 15 m b 20 m c 27.5 m

Exercise 3.10

1 a 7 h b 7 h c 3 h 30 min
d 4 h 40 min e 8 h 43 min f 6 h 20 min
g 18 h 30 min h 4 h 47 min i 10 h 56 min

2 a 03:40 b 17:00 c 00:35
d 07:18 e 21:51 f 12:24

3 a 10:15 p.m. b 8:30 a.m. c 1:44 p.m.
d 6:52 p.m. e 2:05 a.m. f 7:27 p.m.

4 a 8 h b 6 h 30 min c 6 h 15 min
d 11 h 22 min e 14 h 8 min f 1 h 15 min
g 3 h 6 min h 4 h 24 min i 5 h 37 min

5 a 1:10 p.m. b 9:50 a.m. c 3:33 a.m.
d 15:50 e 22:35 f 22:15

6 11:26 a.m.

7 C

8 a 12 h b 10 h 30 min
c 37 h d 14 h 30 min

9 a 10:35 p.m. b 9:45 a.m.
c 9:10 a.m. d 1:40 p.m.

10 a 1 h 52 min b 10:22 p.m. c 11:23 a.m.

Power plus

1 $A = \frac{17}{80}, B = \frac{9}{40}, C = \frac{19}{80}$

2 $21 000

3 $44.\dot{4}\%$

4 $11.68

5 a 0.667 L
b 0.001 L = 1 mL, and measuring cups do not measure to the nearest mL

6 $108

7 11.1 s

8 25 : 28

9 $350

10 Jordan arrives first. Jordan takes $\frac{D}{50}$ h and James takes $\frac{D}{48}$ h, where D is the distance travelled.

11 a 20 years b 20 years

TEST YOURSELF 3

1 a 3 b −5 c −21
d −25 e 4 f 7

2 4.001, 0.401, 0.056 35, 0.0483

3 a 12.56 b 7.718 c 21 d 0.1035
e 1.2565 f 1609 g 7432 h 3.457
i 0.0605 j 341 000
k 856 030 l 0.000 034 6

4 D 5 4.995 6 $0.\dot{3}\dot{6}$ 7 $\frac{2}{5}$

8 a $1\frac{7}{40}$ b $-\frac{7}{33}$ c $1\frac{6}{35}$
d $\frac{5}{16}$ e $130.40 f $\frac{4}{3} = 1\frac{1}{3}$
g $6\frac{2}{3}$ h $2\frac{6}{25}$ i $\frac{17}{30}$

9 a $\frac{37}{50}$ b $\frac{1}{200}$ c $\frac{2}{3}$ d $\frac{37}{250}$

10 a 0.214 b 0.45 c $0.08\dot{3}$ d 0.002

11 a 24.5% b 87.5% c 127.5% d 76%

12 a $243.60 b 7.425 kg c $8.77

13 a 162 L b $21.88

14 a 8.3% b 16.7%

15 773 runs

16 D

17 a $380 000 b 152%

18 $272.73

19 a $1932 b $54.46

20 13%

21 **a** 4 : 9 **b** 7 : 400 **c** 20 : 3
22 Bill: $476 190.48, Ben: $323 809.52
23 **a** 83 km/h **b** 332 km **c** 12 h
24 **a** 2880 mL/h **b** 0.14 kg/m^2 **c** $8\frac{1}{3}$ m/s **d** 24 cm/min
25 **a** 10 h 10 min **b** 11 h 47 min

PRACTICE SET 1

1 **a** 0.625, terminating **b** $0.\dot{5}3846\dot{1}$, recurring **c** $0.\dot{4}$, recurring **d** 4.75, terminating
2 **a** $5m - 2$ **b** $20p + m$ **c** $3n^2 - 10n$
3 **a** 2 **b** 5 **c** 13 **d** −15 **e** 60 **f** −5 **g** 25 **h** 33 **i** −29 **j** 12
4 $\sqrt{98}$, $\sqrt{160}$, $\sqrt{52}$
5 8.30
6 **a** $g = 24.2$ **b** $e = 4.2$ **c** $k = 21.2$
7 $2\frac{11}{12}$
8 B
9 **a** 934.5 **b** 0.178 **c** 9300 **d** 0.2347 **e** 23.56 **f** 19.86 **g** 27.04 **h** 3.6 **i** 36.8 **j** 4.02 **k** 113 **l** 2.25
10 5.85, 5.095, 0.59, 0.5867
11 **a** $y = 6$ **b** $p = 8$ **c** $k = 12.5$
12 **a** $r + 7$ **b** $T - 15$ **c** $4k$ **d** $\frac{10}{z}$
13 **a** $\frac{17}{20}$ **b** $\frac{1}{30}$ **c** $4\frac{13}{15}$ **d** $1\frac{7}{10}$ **e** $\frac{12}{35}$ **f** $\frac{15}{16}$ **g** $\frac{4}{15}$ **h** $3\frac{3}{14}$
14 $\frac{3}{5}, \frac{5}{8}, \frac{2}{3}$
15 D
16 **a** 67.5% **b** 27% **c** $27\frac{3}{11}$% **d** 8.5%
17 **a** $102 **b** $35 **c** $23.45
18 **a** No, $50^2 \neq 40^2 + 14^2$ **b** Yes, $25^2 = 7^2 + 24^2$ **c** Yes, $6.1^2 = 1.1^2 + 6^2$
19 **a** $104.45 **b** 31.6 L
20 $485
21 **a** 10 **b** −50 **c** 1 **d** 24 **e** 1 **f** 7
22 **a** $8a^3$ **b** $40p^2d$ **c** $27b^3d^2$
23 **a** $3m$ **b** $\frac{3x}{y}$ **c** $\frac{wv^2}{2}$
24 **a** $\frac{7}{20}$ **b** 0.002 14
25 33.3%
26 **a** 5 : 3 **b** 4 : 1 **c** 85 words/min
27 2.3 m
28 **a** 110.5 kg **b** $5.25
29 **a** $40 - 8x$ **b** $6h - 21h^2$ **c** $18cd - 27c^2$
30 **a** $22w - 2$ **b** $1 - 16k$ **c** $26t - 23$
31 Klaas $90, David $150, Sarah $210
32 40 320 km/h
33 4 h 33 min
34 $\sqrt{74}$ cm
35 $1875
36 **a** $p^2 - 4p - 32$ **b** $6n^2 + 13n + 5$ **c** $49 - y^2$ **d** $16g^2 - 24g + 9$ **e** $4e^2 + 20e + 25$ **f** $15 - 7x - 2x^2$
37 B
38 **a** $3(p - 5)$ **b** $(m + 4)(5 - t)$ **c** $-4xy(2x + 3)$
39 **a** $\frac{1}{3}$ **b** 8 hours
40 D
41 13.8 m

CHAPTER 4

SkillCheck

1 **a** $\frac{3}{5}$ **b** $\frac{3}{4}$ **c** $\frac{2}{5}$
2 **a** 0.375 **b** 0.714 **c** 0.222
3 **a** AC **b** w **c** PQ or 10
4 **a** 35 **b** 33.2 **c** 5
5 **a** 85 **b** 16 **c** 3.9
6 **a** 8 h **b** 4 h **c** 2 h
7 **a** 4 h 42 min **b** 2 h 15 min **c** 6 h 51 min

Exercise 4.01

1 **a** hyp = 13, opp = 12, adj = 5
b hyp = BC, opp = AC, adj = AB
c hyp = v, opp = u, adj = w
d hyp = SR, opp = TR, adj = ST
e hyp = x, opp = y, adj = z
f hyp = 41, opp = 9, adj = 40
2 **a** $\angle K$ **b** $\angle L$ **c** $\angle M$ **d** $\angle M$ **e** $\angle L$
3 C
4 **a** $\angle\theta$ EF, $\angle\phi$ DE **b** $\angle\theta$ 12, $\angle\phi$ 9 **c** $\angle\theta$ f, $\angle\phi$ e
5 **a** $\angle\theta$ IJ, $\angle\phi$ HI **b** $\angle\theta$ 27, $\angle\phi$ 36 **c** $\angle\theta$ b, $\angle\phi$ c
6 B
7 **a**
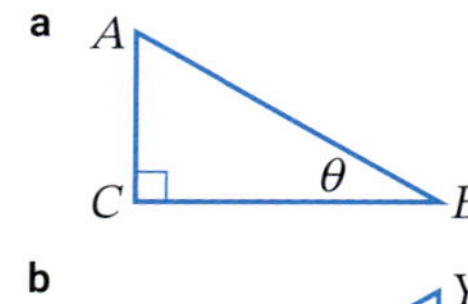

b
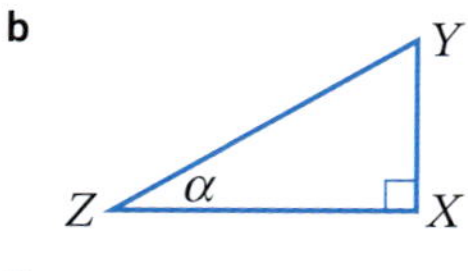

c
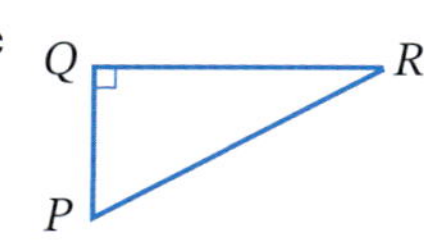

d
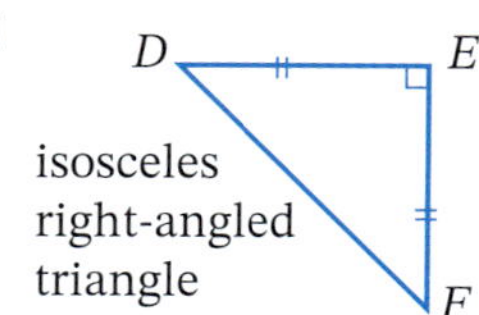

Exercise 4.02

1 **a** $\sin = \frac{55}{73}$ $\cos = \frac{48}{73}$ $\tan = \frac{55}{48}$

b $\sin = \frac{e}{f}$ $\cos = \frac{g}{f}$ $\tan = \frac{e}{g}$

c $\sin R = \frac{ST}{RT}$ $\cos R = \frac{RS}{RT}$ $\tan R = \frac{ST}{RS}$

d $\sin = \frac{n}{k}$ $\cos = \frac{m}{k}$ $\tan = \frac{n}{m}$

e $\sin = \frac{3.6}{8.5}$ $\cos = \frac{7.7}{8.5}$ $\tan = \frac{3.6}{7.7}$

f $\sin W = \frac{XY}{WY}$ $\cos W = \frac{WX}{WY}$ $\tan W = \frac{XY}{WX}$

2 **a** $\frac{4}{5}$ **b** $\frac{4}{3}$ **c** $\frac{3}{5}$ **d** $\frac{4}{5}$

3 **a** α **b** β **c** β

d α **e** α **f** β

4 **a** **i** $\frac{24}{7}$ **ii** $\frac{7}{25}$ **iii** $\frac{24}{25}$ **iv** $\frac{7}{24}$

b **i** $\frac{u}{v}$ **ii** $\frac{v}{w}$ **iii** $\frac{u}{w}$ **iv** $\frac{v}{u}$

c **i** $\frac{FG}{FH}$ **ii** $\frac{FH}{HG}$ **iii** $\frac{FG}{HG}$ **iv** $\frac{FH}{FG}$

d **i** $\frac{84}{13}$ **ii** $\frac{13}{85}$ **iii** $\frac{84}{85}$ **iv** $\frac{13}{84}$

e **i** $\frac{b}{c}$ **ii** $\frac{c}{a}$ **iii** $\frac{b}{a}$ **iv** $\frac{c}{b}$

f **i** $\frac{QS}{RS}$ **ii** $\frac{RS}{QR}$ **iii** $\frac{QS}{QR}$ **iv** $\frac{RS}{QS}$

5 **a** $\tan Y = \frac{60}{11}$

b $\tan X = \frac{11}{60}$

c $\sin X = \frac{11}{61}$ or $\cos Y = \frac{11}{61}$

d $\cos X = \frac{60}{61}$ or $\sin Y = \frac{60}{61}$

6 C

7 A

8 **a**

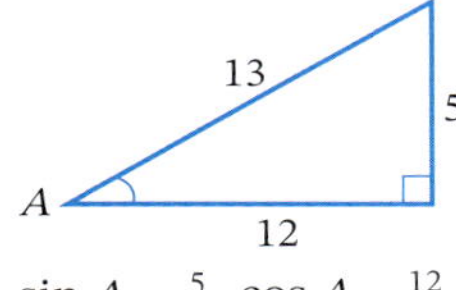

$\sin A = \frac{5}{13}, \cos A = \frac{12}{13}$

b

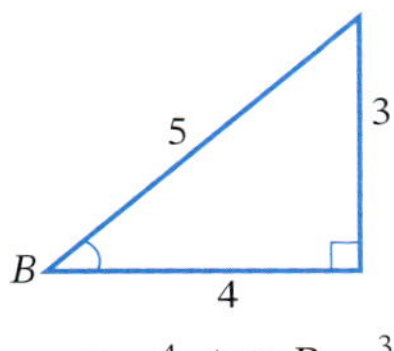

$\cos B = \frac{4}{5}, \tan B = \frac{3}{4}$

c

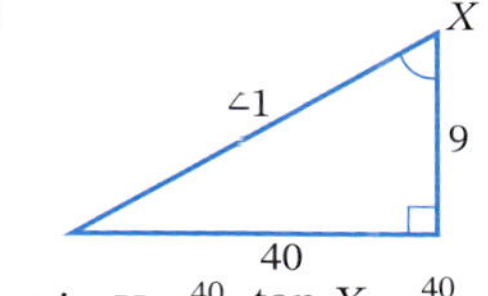

$\sin X = \frac{40}{41}, \tan X = \frac{40}{9}$

d

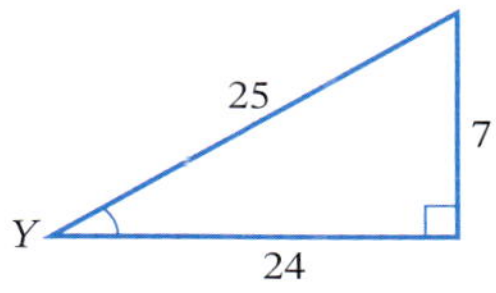

$\cos Y = \frac{24}{25}, \tan Y = \frac{7}{24}$

Exercise 4.03

1 **a**

	Side length (mm)			Trigonometric ratio		
	XZ (opp)	ZY (adj)	XY (hyp)	$\sin Y = \frac{XZ}{XY}$	$\cos Y = \frac{ZY}{XY}$	$\tan Y = \frac{XZ}{ZY}$
1	45	26	52	0.87	0.50	1.73
2	24	14	28	0.86	0.50	1.71
3	51	30	59	0.86	0.51	1.70
4	62	36	72	0.86	0.50	1.72

b All are approximately the same (0.9).

c The table answer is an approximation of the calculator answer.

d All are approximately the same (0.5).

e The table answer is an approximation of the calculator answer.

f All are approximately the same (1.7).

g The table answer is an approximation of the calculator answer.

2 **a** Teacher to check.

b, c, d The table answer is an approximation of the calculator answer.

3 Teacher to check.

Exercise 4.04

1 **a** 9.51 **b** 0.97 **c** 1.07 **d** 0.54

e 0.97 **f** 0.50 **g** 0.83 **h** 0.95

i 29.86 **j** 11.61 **k** 32.97 **l** 10.26

m 47.15 **n** 9.56 **o** 3.58 **p** 33.11

q 19.34 **r** 244.80 **s** 12.26 **t** 16.83

2 **a** 28° **b** 41° **c** 19° **d** 33°

e 34° **f** 56° **g** 30° **h** 44°

3 **a** 68°40′ **b** 54°22′ **c** 24°47′ **d** 18°30′

e 9°11′ **f** 47°59′ **g** 3°46′ **h** 57°10′

4 **a** 55°30′ **b** 14°9′ **c** 72°23′ **d** 33°46′

e 66°25′ **f** 7°53′ **g** 23°7′ **h** 31°3′

i 34°27′ **j** 71°5′ **k** 5°29′ **l** 69°27′

5 **a** 52° **b** 14° **c** 52° **d** 60°

e 33° **f** 28° **g** 36° **h** 89°

Mental skills 4

2 **a** 330 **b** 160 **c** 1580 **d** 260
e 420 **f** 200 **g** 410 **h** 740
i 6600 **j** 16 000 **k** 280 **l** 70

4 **a** 29 **b** 15 **c** 180 **d** 400
e 3 **f** 4 **g** 20 **h** 5
i 6 **j** 25

Exercise 4.05

1 **a** tan θ **b** cos θ **c** tan θ
d sin θ **e** tan θ **f** cos θ

2 C

3 **a** 6.55 cm **b** 4.33 m **c** 27.63 m
d 44.99 m **e** 13.90 cm **f** 16.97 cm
g 16.17 cm **h** 26.68 m **i** 16.07 m

4 B

5 A

6 **a** 38.71 **b** 7.38 **c** 17.21
d 51.59 **e** 24.49 **f** 132.57
g 33.59 **h** 12.06 **i** 36.06

7 D

8 B

9 **a** 4.3 m **b** 141.4 m **c** 671.4 m
d 1053.1 cm **e i** 10 km **ii** 17.3 km

10 D **11** 12.3 m **12** 10 cm

13 5.76 m **14** 3.4 m **15** 933 cm

16 5738 mm **17** 48 m

18 364 m by 265 m

19 0.47 m **20** 5.1 m **21** C

22 B **23** 319 cm

Exercise 4.06

1 **a** 102.3 **b** 62.9 **c** 95.6
d 42.6 **e** 79.3 **f** 24.8

2 **a** 21.5 **b** 378.3 **c** 245.7

3 325 mm

4 12.4 m

5 **a** 88.97 **b** 1188.24 **c** 45.70 **d** 187.70

6 6.88 m

7 B

8 A

9 54.4 cm

10 **a** 27.5 **b** 109.4 **c** 299.5
d 5.7 **e** 88.2 **f** 20.8

11 D

12 **a** 1203.3 cm **b** 414.2 mm **c** 6.7 m
d i 3.8 km **ii** 8.4 km **e** 1044.3 cm

13 402.8 m **14** 782 cm **15** 12.1 m

16 4.9 km **17** 6.88 m **18** B

19 77.3 m

20 **a** 18.0 cm **b** 8.4 cm

Exercise 4.07

1 **a** 41° **b** 64° **c** 60° **d** 81°
e 61° **f** 30° **g** 6° **h** 45°
i 60° **j** 82° **k** 84° **l** 14°

2 **a** 64°59′ **b** 54°35′ **c** 36°52′ **d** 20°10′
e 43°32′ **f** 72°33′ **g** 27°2′ **h** 61°39′
i 86°12′ **j** 81°47′ **k** 35°16′ **l** 51°3′

3 **a** 34° **b** 51° **c** 60°
d 29° **e** 67° **f** 54°

4 C

5 **a** 25°12′ **b** 36°52′ **c** 31°30′
d 26°47′ **e** 48°11′ **f** 66°33′

6 **a** 12° **b** 68° **c** 27°
d 27° **e** 30° **f** 88°

7 39° **8** 38.9° **9** 10°56′

10 52°8′ **11** 11° **12** 37°

13 B **14** 71° **15** 11°

16 50° **17** 38° **18** 29°48′

Power plus

1 **a i** 0.342, 0.342 **ii** 0.731, 0.731
iii 0.819, 0.819 **iv** 0.996, 0.996
b Each pair of trigonometric ratios has the same value.
c The pairs of angles are complementary (add to 90°).
d 60°
e i 15° **ii** sin **iii** 18°
iv cos **v** 25° **vi** 32°
f $\sin x = \cos (90° - x)$ **g** Teacher to check.

2 **a** 1932 m **b** 31°

3 **a** $\frac{\sqrt{3}}{2}$ **b** $\frac{1}{\sqrt{3}}$

4 **a** 24°12′26″ **b** 64°9′41″

5, 6 Teacher to check.

TEST YOURSELF 4

1 opp = 15, adj = 20, hyp = 25

2 opp = v, adj = u, hyp = w

3 **a** $\frac{33}{65}$ **b** $\frac{33}{56}$ **c** $\frac{56}{65}$ **d** $\frac{33}{65}$

4 $\cos \alpha = \frac{77}{85}$ $\tan \alpha = \frac{36}{77}$

5 **a** $\tan 42° \approx 0.900$ **b** $\cos 42° \approx 0.743$
c $\sin 42° \approx 0.669$

6 **a** 64° **b** 26° **c** 12°

7 **a** 50°19′ **b** 31°56′ **c** 64°19′

8 **a** 0.8480 **b** 0.7677 **c** 0.1539
d 64.9839 **e** 13.9884 **f** 13.7044

9 **a** 45°48′ **b** 33°11′ **c** 5°21′

10 **a** 14.49 **b** 72.50 **c** 8.71
d 25.13 **e** 54.82 **f** 106.86

11 **a** 98.9 cm **b** 21.1 m **c** 232.0 mm

12 **a** 69° **b** 55° **c** 70°

13 **a** 33° **b** 28° **c** 38°

14 68°13′

9780170465618

CHAPTER 5

SkillCheck

1 **a** **i** 3 **ii** 5 **iii** three to the power of five
b **i** 4 **ii** 8 **iii** four to the power of eight
c **i** h **ii** 5 **iii** h to the power of five
d **i** 7 **ii** y **iii** seven to the power of y

2 **a** 25 **b** 343 **c** 10 000 **d** 64
e $\frac{8}{27}$ **f** 81 **g** 81 **h** 216
i 256 **j** $\frac{1}{8}$ **k** $\frac{9}{16}$ **l** 125

3 **a** 2^5 **b** $3^4 \times 7^3$ **c** $5^6 \times 8^2$
d $\frac{3^4}{7^4}$ or $\left(\frac{3}{7}\right)^4$ **e** $10^1 \times x^5$ **f** $6^3 \times k^2$
g $x^3 \times y^2$ **h** $a^2 \times b^3$ **i** $5^2 \times n^3$
j $p^2 \times q^4$

4 **a** $9 \times 9 \times 9$ **b** 7×7
c $d \times d \times d \times d \times d$ **d** $k \times k$
e $\frac{4}{5} \times \frac{4}{5} \times \frac{4}{5}$

5 **a** 1024 **b** 10 000 **c** 729 **d** 1
e 9 **f** 15 625 **g** 8 **h** 64

6 **a** 3 **b** 4 **c** 3 **d** 2
e 6 **f** 4 **g** 6 **h** 4

Exercise 5.01

1 B

2 **a** 10^5 **b** 2^5 **c** 3^7 **d** 7^5
e 8^8 **f** 5^9 **g** 6^{10} **h** 4^{12}
i 3^{11} **j** x^5 **k** g^8 **l** w^8
m b^{13} **n** p^{20} **o** r^2 **p** y^{12}
q m^{19} **r** n^{10}

3 D

4 B

5 **a** 10^2 **b** 8^4 **c** 20^{10} **d** 5^6
e 9^9 **f** 2^{24} **g** 7^1 **h** 2^{19}
i 11^0 **j** p^5 **k** n^6 **l** w^{18}
m h^{16} **n** y^6 **o** a^8 **p** b^1
q w^{24} **r** m^0

6 D

7 C

8 A

9 **a** $6p^7$ **b** $12y^{12}$ **c** $18m^9$
d $5h^{11}$ **e** $24q^9$ **f** $10a^7$
g $30n^{16}t^5$ **h** $30a^2b^4$ **i** $3e^{10}g^5$
j $32p^7m^{10}$ **k** $48q^8r^8$ **l** $54u^4v^3w^8$

10 **a** $2y^{12}$ **b** $5w^6$ **c** $8r^7$
d $30x$ **e** $5m^9$ **f** $\frac{g^6}{2}$
g $2d^2h^9$ **h** x^5y^4 **i** $\frac{e^{20}d^{36}}{3}$
j $\frac{3qt}{4}$ **k** $9a^5b^8$ **l** $\frac{3pq^2r^4}{2}$

11 B

12 **a** 3^{10} **b** 10^6 **c** 7^4 **d** 8^5 **e** a^{14}
f m **g** w^{13} **h** x^4 **i** p^5 **j** $30c^{14}$
k $3y^8$ **l** $120d^{11}$ **m** $2q^2$ **n** $3g^2$ **o** $30h^{12}$
p $\frac{3v^3}{2}$ **q** $\frac{8b^2}{5}$ **r** $2k$

13 D

14 **a** a^0 **b** y^{-1} **c** n^1 or n
d h^{-5} **e** b^0 **f** r^{-2}

Exercise 5.02

1 C

2 **a** 4^6 **b** 5^{16} **c** 3^{12} **d** 2^{28}
e 2^2 **f** 9^3 **g** 10^0 **h** 6^{20}
i 5^{15} **j** e^8 **k** t^{25} **l** y^{21}
m c^5 **n** m^{35} **o** y^{16} **p** h^0
q q^{18} **r** w^4 **s** $1024x^{10}$ **t** $390\,625n^{24}$
u $64d^9$ **v** $-k^{45}$ **w** d^{12} **x** $256a^{64}$

3 B

4 **a** $16d^{12}$ **b** $25m^6$ **c** $16y^{10}$ **d** $81x^8$
e $3125u^{30}$ **f** $8w^{15}$ **g** $10\,000d^{20}$ **h** $27e^3$
i $2b^4$ **j** $36d^{12}$ **k** $243f^{20}$ **l** $1024c^{30}$
m $16r^4$ **n** $-125t^3$ **o** $9m^6$ **p** y^{36}
q $-x^3$ **r** m^{30} **s** $256w^{20}$ **t** $-243f^5$
u $-27p^6$ **v** $81h^{20}$ **w** $100k^2$ **x** $-8y^3$

5 A

6 **a** d^{-3} **b** x^{-4} **c** $w^{\frac{1}{4}}$ **d** z^2
e u^1 or u **f** p^1 or p

Exercise 5.03

1 A

2 **a** a^3b^3 **b** $x^{10}y^5$ **c** $l^{18}m^{30}$
d $1296d^4p^8$ **e** $64k^8y^{10}$ **f** $243m^{10}n^5$
g e^3k^9 **h** $-w^{21}x^{28}$ **i** $64d^6y^{10}$
j $256b^8c^{12}$ **k** $-27a^9d^3$ **l** $16p^8q^{12}$

3 **a** $x^4y^2w^6$ **b** $h^8y^{12}m^{20}$ **c** $9d^{14}k^4p^6$
d $125a^9q^9r^{12}$ **e** $32d^{20}e^5w^{15}$ **f** $k^4m^{12}p^8x^{24}$

4 D

5 **a** $\frac{36}{49}$ **b** $\frac{m^3}{8}$ **c** $\frac{625}{x^4}$
d $\frac{256n^{40}}{p^8}$ **e** $\frac{w^{10}}{t^{15}}$ **f** $\frac{m^8}{256n^4}$
g $\frac{16}{81}$ **h** $-\frac{125h^3}{216}$ **i** $\frac{49k^8}{100}$
j $\frac{9r^2}{t^4}$ **k** $\frac{a^8b^4}{d^{20}}$ **l** $-\frac{32}{243c^{10}}$

6 **a** $40x^{32}y^{48}$ **b** $1000x^{36}y^{54}$ **c** $2q^3r^4$
d $36q^8r^{12}$ **e** $27a^{12}x^{15}$ **f** $3ax^2$
g $32p^8h^{29}$ **h** $8p^4h^{11}$ **i** $64p^{10}h^{38}$
j $8a^3d^9$ **k** $49e^4y^2$ **l** $5xy$
m $\frac{3h^3k^3}{4}$ **n** $\frac{27m^{11}w^7}{2}$ **o** $81c^{12}d^4$

Exercise 5.04

1 **a** 1 **b** 1 **c** −1 **d** 1
e −1 **f** 1 **g** 1 **h** 7
i −1 **j** 1 **k** 1 **l** −2
m 1 **n** 1 **o** 1 **p** 1
q 1 **r** −1 **s** −6 **t** 6
u 1 **v** −1 **w** −3 **x** 1

2 **a** 2 **b** 0 **c** 3 **d** 4
e 7 **f** −5 **g** −3 **h** 0
i 1 **j** 9 **k** $1\frac{3}{4}$ **l** 2
m 6 **n** 4 **o** 125 **p** −1
q 12 **r** 3 **s** $\frac{1}{3}$ **t** 27
u −3 **v** 11 **w** 11 **x** 1

Mental skills 5

2 **a** 117 **b** 740 **c** 138 **d** 140 **e** 174
f 100 **g** 436 **h** 109 **i** 200 **j** 312

3 **a** 160 **b** 700 **c** 90
d 2200 **e** 72 **f** 1600
g 2100 **h** 480 **i** 132

Exercise 5.05

1 **a** $\frac{1}{6^2}$ **b** $\frac{1}{5^7}$ **c** $\frac{1}{3^1}$ **d** $\frac{1}{10^2}$
e $\frac{1}{g^5}$ **f** $\frac{1}{z^1}$ **g** $\frac{1}{n^3}$ **h** $\frac{1}{t^2}$
i $\frac{1}{a^4}$ **j** $\frac{1}{5^3}$ **k** $\frac{1}{y^d}$ **l** $\frac{1}{r^m}$

2 **a** $\frac{1}{9}$ **b** $\frac{1}{625}$ **c** $\frac{1}{6}$ **d** $\frac{1}{49}$
e $\frac{1}{25}$ **f** $\frac{1}{128}$ **g** $\frac{1}{64}$ **h** $\frac{1}{1000\,000}$
i $\frac{1}{1024}$ **j** $\frac{1}{27}$ **k** $\frac{1}{36}$ **l** $\frac{1}{6561}$

3 **a** n^{-2} **b** n^{-1} **c** 8^{-3} **d** 8^{-1}
e 10^{-5} **f** $2a^{-4}$ **g** 3^{-1} **h** $-b^{-1}$
i $6a^{-1}$ **j** $4t^{-2}$ **k** $-2w^{-5}$ **l** $5d^{-3}$

4 **a** $\frac{5}{h}$ **b** $\frac{2}{b^5}$ **c** $\frac{3}{e^3}$ **d** $\frac{4}{n^2}$
e $\frac{p}{b^2}$ **f** $\frac{r^2}{s^4}$ **g** $\frac{y}{w^2}$ **h** $\frac{y^3}{d^3}$
i $\frac{1}{2m}$ **j** $\frac{1}{xy}$ **k** $\frac{1}{16h^2}$ **l** $\frac{1}{125k^3}$
m $\frac{3m^3}{p^2}$ **n** $\frac{15}{kw^4}$ **o** $\frac{12}{x^2y^3}$ **p** $\frac{12y^3}{x^2}$
q $\frac{1}{9h^2}$ **r** $\frac{1}{64k^3}$ **s** $\frac{1}{16c^4}$ **t** $\frac{1}{8y}$
u $\frac{4p}{q^3}$ **v** $\frac{4}{pq^3}$ **w** $\frac{v}{m^2}$ **x** $\frac{1}{vm^2}$

5 **a** $\frac{7}{2}$ **b** $\frac{5}{8}$ **c** $\frac{10}{9}$ **d** $\frac{2}{3}$
e $-\frac{4}{3}$ **f** $\frac{2}{5}$ **g** $\frac{3}{x}$ **h** $\frac{a}{5}$
i $-\frac{2}{m}$ **j** $\frac{4}{5r}$ **k** $\frac{3z}{2}$ **l** v

Exercise 5.06

1 **a** 3^{11} **b** 2^3 **c** 4^8 **d** 9^7
e 8^{12} **f** 5^9 **g** 6^{12} **h** 3^4
i 2^{19} **j** 4^{10} **k** 8^5 **l** 10^1

2 **a** a^7 **b** t^9 **c** n^6 **d** p^2
e w^8 **f** g^{18} **g** $6b^7$ **h** $20d^{13}$
i $6c^4$ **j** $625b^{16}$ **k** $3m^2$ **l** $9a^2$

3 **a** 1 **b** 1 **c** 7 **d** 1
e −8 **f** 9 **g** 625 **h** 128
i 1 **j** 64 **k** $\frac{1}{64}$ **l** 1
m 25 **n** $\frac{1}{10\,000}$ **o** 1 **p** 1

4 **a** $\frac{1}{25}$ **b** $\frac{1}{32}$ **c** $\frac{1}{20}$ **d** $\frac{1}{1000}$

5 **a** $9m^2n^6$ **b** $40a^5w^9$ **c** $256a^8b^{20}$ **d** $4pq^2$
e $\frac{64}{125}$ **f** $6c^2$ **g** $3u^4v^3$ **h** $\frac{9x^2}{100}$
i $-64n^6t^3$ **j** $\frac{16}{81}$ **k** $\frac{y^3}{5x^3}$ **l** $\frac{p^{10}}{81y^2}$
m $32p^{15}q^{10}$ **n** $\frac{1}{343n^3}$ **o** −2 **p** $\frac{a^{11}}{b}$

6 **a** $\frac{1}{8^7}$ **b** $\frac{1}{3^5}$ **c** $\frac{1}{y}$ **d** $\frac{1}{x^3}$
e $\frac{1}{25b^2}$ **f** $\frac{5}{b^2}$ **g** $\frac{1}{ab}$ **h** $\frac{a}{b}$
i $\frac{4}{t^8}$ **j** $\frac{1}{1331t^3}$ **k** $\frac{p^3}{q^5}$ **l** $\frac{m}{w^3}$
m $\frac{8}{u^3v^4}$ **n** $\frac{-2r^6}{y^5}$ **o** $\frac{10f^3}{e}$ **p** $\frac{n^7}{2k^4}$

7 **a** $\frac{4}{7}$ **b** $\frac{2}{5}$ **c** $\frac{3}{2}$ **d** 7
e $\frac{8}{r}$ **f** $10p$ **g** $\frac{z}{6y}$ **h** $\frac{5a}{2}$

8 **a** 4^{-3} **b** 2^{-1} **c** 10^{-4} **d** 9^{-2}
e k^{-1} **f** $9k^{-4}$ **g** $-x^{-7}$ **h** $5p^{-3}$

9 **a** q^3 **b** d^4 **c** $\frac{1}{m^{11}}$ or m^{-11}
d t^2 **e** $30g^2$ **f** $24a$
g $\frac{28}{x}$ or $28x^{-1}$ **h** $\frac{4}{p^3}$ or $4p^{-3}$ **i** $16q^3$
j $\frac{t^4}{2}$ **k** $\frac{2}{b^4}$ or $2b^{-4}$ **l** $\frac{1}{9h^2}$

Exercise 5.07

1 **a** 3 **b** 2 **c** 2 **d** 5
e 1 **f** 1 **g** 3 **h** 5
i 2 **j** 2 **k** 3 **l** 6

2 **a** 37.6 **b** 9440 **c** 168
d 2.81 **e** 16.0 **f** 60 500
g 1 770 000 **h** 386 000 **i** 10.3

3 **a** 0.064 **b** 0.90 **c** 0.085
d 0.000 16 **e** 0.0076 **f** 0.039
g 0.28 **h** 0.019 **i** 0.31

4 B

5 C

6 **a** 9 **b** 60 **c** 0.04
d 0.008 **e** 0.5 **f** 10 000
g 2000 **h** 0.0003 **i** 60 000 000

7 **a** \$36 000 000, 2 **b** \$40 000 000, 1

8 5

9 21 600

10 C

11 **a** 180 **b** –5 **c** 6.16 **d** –4.176
e 0.06 **f** 0.67 **g** 33 300 **h** 0.6565

Exercise 5.08

1 **a** 2.4×10^3 **b** 7.86×10^5 **c** 5.5×10^7
d 9.5×10^1 **e** 7.8×10^0 **f** 3.48×10^8
g 5.967×10^4 **h** 1.5×10^1 **i** 3×10^9
j 8×10^1 **k** 7.63×10^2 **l** 1×10^1
m 3.5×10^{-2} **n** 7.6×10^{-5} **o** 8×10^{-1}
p 7.13×10^{-2} **q** 3×10^{-6} **r** 9.13×10^{-1}
s 7.146×10^{-6} **t** 9×10^{-3} **u** 1×10^{-7}
t 8.9×10^{-4} **w** 7.8×10^{-8} **x** 1×10^{-1}

2 **a** 1.3×10^5 **b** 2.9×10^{-7} **c** 8×10^{-5}
d 3×10^8 **e** 4×10^{13} **f** 1.2×10^{-4}
g 6.1×10^2 **h** 1.06×10^{-10} **i** 7.7×10^{20}
j 1×10^{-6} **k** 2.2×10^6

3 **a** 600 000 **b** 7100
c 302 000 000 **d** 3.14
e 0.000 06 **f** 0.0071
g 0.000 000 030 2 **h** 0.000 000 000 59
i 1 100 000 000 000 **j** 0.0004
k 5000 **l** 0.000 476
m 0.803 **n** 63 200
o 0.016 **p** 0.000 000 22
q 9 000 000 **r** 0.111

4 **a** 4×10^5 **b** 2.7×10^6 **c** 1.3×10^7
d 6.3×10^{-7} **e** 9.3×10^9 **f** 9.3×10^2
g 3.04×10^0 **h** 4.5×10^{-5} **i** 2×10^{-15}
j 9.7×10^{-5}

5 **a** $3.8 \times 10^9, 5.5 \times 10^9, 7.3 \times 10^9$
b $5.8 \times 10^{-6}, 2.2 \times 10^{-4}, 7 \times 10^{-4}$
c $3.5 \times 10^0, 4.9 \times 10^2, 5.3 \times 10^2$

6 **a** $6 \times 10^5, 8 \times 10^2, 2.9 \times 10^2$
b $6.3 \times 10^2, 8.1 \times 10^{-4}, 1.2 \times 10^{-9}$
c $9.5 \times 10^{-1}, 4.1 \times 10^{-1}, 6.4 \times 10^{-3}$

Exercise 5.09

1 **a** 6×10^8 **b** 2×10^5 **c** 8×10^{15}
d 3×10^6 **e** 2.4×10^{16} **f** 5×10^4
g 1.024×10^{18} **h** 3.01×10^{-6} **i** 6.3×10^9
j 2.5×10^{-11} **k** 2×10^{-3} **l** 3.81×10^{13}

2 **a** 1.31×10^9 **b** 2.96×10^{13} **c** 2.84×10^{15}
d 1.44×10^{16} **e** 6.99×10^{19} **f** 9.05×10^3
g 2.00×10^{14} **h** 2.14×10^9

3 3.78×10^{19} atoms

4 2.25×10^{-2} mm

5 **a** 7.0×10^{30} **b** 1.3×10^{17} **c** 8.1×10^{13}
d 2.3×10^{36} **e** 5.5×10^{50} **f** 5.2×10^{22}
g 1.8×10^{-4} **h** 9.5×10^2 **i** 3.1×10^6
j 9.5×10^{13} **k** 1.2×10^{-3} **l** 1.7×10^{-5}
m 3.9×10^8 **n** 5.9×10^0

6 **a** 500 s **b** 8 min 20 s

7 1.9×10^{14} t

8 1.7×10^0 km/s

9 9.5×10^{12} km

10 **a** 0.000 1 s **b** 91 s

11 **a** $9.999\,999\,999 \times 10^{99}$
b $1.000\,000\,000 \times 10^{-99}$

Power plus

1 **a** $\frac{1}{5}$ **b** $-\frac{1}{8}$ **c** 32 **d** 1000
e $\frac{25}{8}$ or $3\frac{1}{8}$ **f** $\frac{4}{25}$ **g** $\frac{1}{400}$ **h** $\frac{1}{1600}$

2 **a** 5^{2a} **b** 1 **c** 2^{2y} **d** 10^3
e 3^{3p} **f** 3^{3p} **g** 8^{3n} **h** 6^{2t}
i $2p^2$ **j** $3n^2$ **k** $\frac{q}{2}$

3 **a** 4.382×10^{11} **b** 5.2×10^{-8} **c** 4×10^8
d 2.013×10^0 **e** 5.78×10^4 **f** 5.78×10^{-2}
g 6.7×10^{-6} **h** 3.2×10^9 **i** 3.2×10^{-9}

4 If $a = b$ there are an infinite number of solutions, otherwise $a = 2$, $b = 4$

5 **a** $2^n + 1$ where $n = 1, 2, 4, 8, \ldots$ (doubling the Power each time), or $2^{2^n} + 1$ for $n = 1, 2, 3, \ldots$
b 4 294 967 297

TEST YOURSELF 5

1 **a** 10^{10} **b** 4^{16} **c** a^{10} **d** h^{10}
e $12n^4$ **f** $2d^{12}$ **g** $5m^8$ **h** $6v^7w^7$
i $15x^6y^3$ **j** $8t^4h^6$ **k** p^4q^8 **l** $20ab^2$

2 **a** 2^6 **b** k^{25} **c** x^4
d $1024y^{30}$ **e** $25t^4$ **f** $1000g^3$
g -32 **h** $-32k^5$ **i** $25m^6$

3 **a** a^4b^8 **b** $25x^6y^4$ **c** $64t^6$
d $-64h^6g^3$ **e** $\frac{a^4}{2401}$ Z $\frac{a^4}{2401}$ **f** $32p^5q^5r^5$
g $\frac{243m^5}{32}$ **h** $81n^4p^8$ **i** $\frac{16a^{28}}{b^4}$
j $512t^{14}u^{16}$ **k** $\frac{b^{18}y^{15}}{512}$ **l** $5c^4d^4$

4 **a** 1 **b** 1 **c** 1 **d** 1 **e** –1
f h **g** 1 **h** 1 **i** $\frac{2}{3}$

5 **a** $\frac{1}{8^3}$ **b** $\frac{1}{19^2}$ **c** $\frac{1}{x}$ **d** $\frac{1}{p^5}$

6 **a** $\frac{1}{4m}$ **b** $\frac{1}{16m^2}$ **c** $\frac{1}{5b}$ **d** $\frac{5}{b}$
e $\frac{-2}{x^4}$ **f** $\frac{5a}{3}$ **g** $\frac{d^2}{c^4}$ **h** $\frac{9}{100}$

7 **a** 10^{-3} **b** r^{-5} **c** r^{-1} **d** $3b^{-1}$

8 **a** $4u^4t^4$ **b** $25a^2b^2$ **c** $20h^2d^9$
d e^{12} **e** r^{10} **f** p
g $\frac{a^4}{c}$ **h** $\frac{5m^2}{n^4}$ **i** $\frac{9t^4}{u^6v^3}$

9 **a** 8.6 **b** 15 700 **c** 500
d 0.007 127 **e** 0.904 **f** 300 000
g 4810 **h** 0.0009 **i** 70 000 000

10 **a** 3.7×10^4 **b** 6.1×10^{-1} **c** 2.5×10^5
d 4.9×10^{-4} **e** 1.3×10^1 **f** 8×10^{-11}

11 **a** 8100 **b** 60 000 000 **c** 3.075
d 0.0081 **e** 0.000 000 6 **f** 0.030 75

12 9.1×10^{-8}, 2.4×10^{3}, 3×10^{3}

13 **a** 2.701×10^{31} **b** 4×10^{5}
c 1.25×10^{17} **d** 2.5×10^{-4}

14 **a** 9.8×10^{16} **b** 3.7×10^{18} **c** 9.6×10^{8}

CHAPTER 6

SkillCheck

1 **a** acute **b** right **c** reflex
d revolution **e** reflex **f** obtuse
g reflex **h** straight

2 D **3** B **4** D

Exercise 6.01

1 **a** $m = 70$ **b** $w = 33$ **c** $h = 50$
d $k = 100$ **e** $x = 46$ **f** $a = 65$
g $d = 70$ **h** $k = 70$ **i** $y = 34$
j $m = 60$ **k** $h = 150$ **l** $r = 45$

Reasons: **a–f**, **i** (angle sum of a triangle), **g**, **h**, **k**, **l** (isosceles triangle, angle sum of a triangle), **j** (equilateral triangle)

2 C **3** A, D

4 **a** 115 **b** 112 **c** 85
d 68 **e** 67.5 **f** 130

5 B **6** B

7 **a** 125° **b** 110° **c** 67°
d 50° **e** 60° **f** 56°

8 Teacher to check: 4 possible triangles, depending on whether 6 cm is one of the equal sides, and 40° is one of the equal angles.

9 $a = 65$, $c = 50$, $b = 65$

10

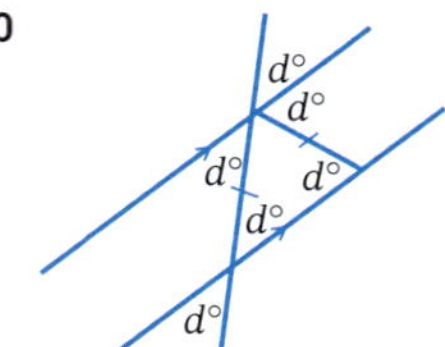

11 **a** $t = 35$ **b** $m = 26$ **c** $h = 30$
d $y = 110$ **e** $g = 40$ **f** $x = 11$

Exercise 6.02

1 **a** $m = 74$ **b** $m = 96$ **c** $m = 30$
d $m = 30$ **e** $m = 221$ **f** $m = 75$

2 See table below

Properties	Trapezium	Kite	Parallelogram	Rhombus	Rectangle	Square
One pair of opposite sides parallel	✓					
Opposite sides parallel			✓	✓	✓	✓
Opposite sides equal			✓	✓	✓	✓
All sides equal				✓		✓
Two pairs of adjacent sides equal		✓				
Diagonals equal					✓	✓
Diagonals bisect each other			✓	✓	✓	✓
Diagonals meet at right angles		✓		✓		✓
Diagonals bisect the angles of the shape				✓		✓
Opposite angles equal			✓	✓	✓	✓
One pair of opposite angles equal		✓				
All angles 90°					✓	✓
Axes of symmetry	0	1	0	2	2	4

3 a trapezium, parallelogram
b trapezium
c square, rhombus
d rectangle, square
e parallelogram, rectangle, square, rhombus
f square
g rhombus, square
h kite
i parallelogram, rectangle, square, rhombus
j rectangle, square
k rectangle, rhombus
l parallelogram, rectangle, square, rhombus
m parallelogram, rectangle, square, rhombus
n rhombus, kite, square

4 a rhombus, parallelogram, trapezium
b 9 rhombuses, 6 parallelograms, 18 trapeziums
c 13

5 A, C, D, E, G, I

6 a parallelogram b rhombus
c trapezium d square
e rhombus f kite
g rectangle h square
i rhombus

7 a $k = 70$ (opposite angles of parallelogram are equal)
b $a = 45$ (angles of square bisected by diagonals)
c $p = 55$ (co-interior angles, opposite sides of rhombus parallel)
d $w = 112$ (opposite angles of parallelogram are equal, then angles on a straight line)
e $t = 55$ (angles of a rectangle equal 90°)
f $n = 67$ (corresponding angles, opposite sides of a rhombus are parallel)
g $a = 55$ (co-interior angles)
h $c = 9$ (diagonals bisect angles of a square)
i $r = 110$ (angles between unequal sides of a kite are equal)

8 a $\angle PQR = 72°$ (alternate angles, $RW \parallel PV$)
b $\angle TQU = \angle TUQ = 53°$ (angles in an isosceles triangle)
$\angle QUP = 53°$ (alternate angles, $QT \parallel PU$)
$\angle UPQ = \angle PQU = 63.5°$ (angles in an isosceles triangle)
$\therefore \angle PQR = 63.5°$ (alternate angles, $RQ \parallel PU$)
c $(30° + \angle QRT) + 110° = 180°$ (co-interior angles, $QT \parallel PR$)
$\angle QRT = 40°$
$\therefore \angle PQR = 40°$ (alternate angles, $PQ \parallel RT$)
d $\angle PRQ = 35°$ (vertically opposite angles)
$\therefore \angle PQR = 72.5°$ (angles in isosceles $\triangle PQR$, diagonals of rectangle are equal and bisect each other)
e $\angle QRP = 75°$ (angle sum of straight line)
$\angle QPR = 75°$ (angles in isosceles triangle)
$\therefore \angle PQR = 30°$ (angle sum of triangle)
f $\angle TPQ = 60°$ (alternate angles, $RT \parallel ?Q$)
$\therefore \angle PQR = 30°$ (angle sum of triangle, diagonals of rhombus meet at 90°)

9 sample answers
a
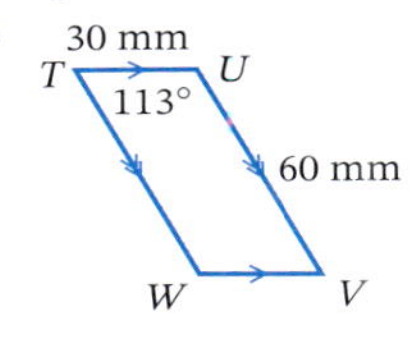

b
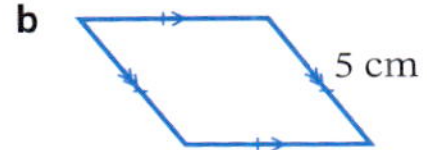

c
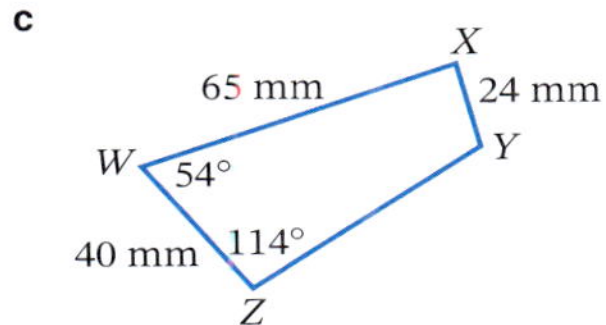

d
200°
4.5 cm

Exercise 6.03

1 a 6 b 4 c 9
d 10 e 7 f 5
g 12 h 8 i 11

2 a hexagon b quadrilateral
c pentagon d triangle
e regular decagon f heptagon
g regular octagon h nonagon

3 a a, b, d, e, f, g b e, g c c, h

4 D

5 a equilateral triangle
b square

6 a 540° b 1080° c 2340° d 1800°
e 900° f 1440° g 3240° h 1620°

7 a $a = 95$ b $b = 110$
c $d = 108$ d $c = 160$
e $g = 140$ f $x = 90$
g $e = 135$ h $m = 120$, $n = 60$
i $y = 222$

8 a 14 b 11 c 24 d 17

9 B

10 a 135° b 144° c 150° d 120°

11 a 40 b 171°

12 a 24 b 36 c 10

Mental skills 6

2 **a** 0.5 **b** 90 **c** 8 **d** 0.9
e 90 **f** 0.7 **g** 8 **h** 30
i 0.08 **j** 3.5 **k** 4 **l** 0.3

4 **a** 16 **b** 160 **c** 1.6 **d** 1.6
e 1.6 **f** 16 **g** 160 **h** 1.6
i 0.016 **j** 160 **k** 0.16 **l** 0.0016

Exercise 6.04

1

a

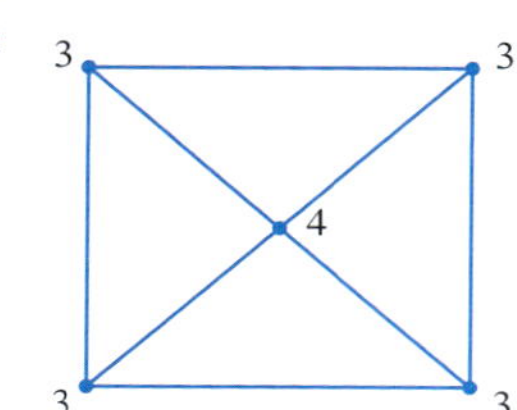

b

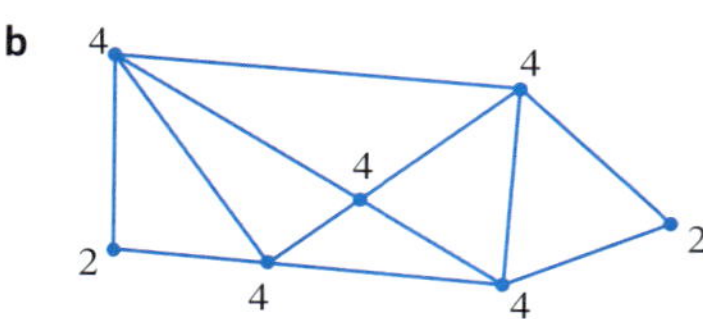

2 **a**

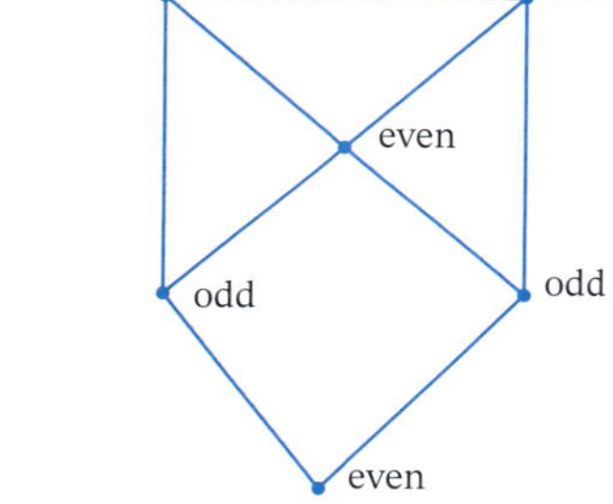

b

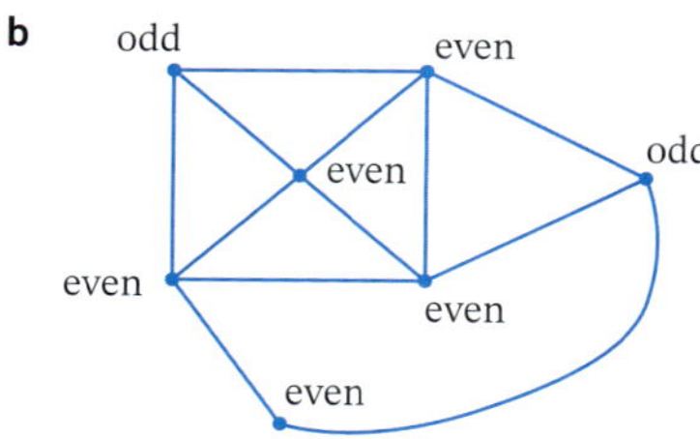

3 **a** 7 faces **b** 5 faces

4 **a**

Network	Number of faces (F)	Number of vertices (V)	$F + V$	Number of edges (E)
1 a	5	5	10	8
1 b	7	7	14	12
2 a	5	6	11	9
2 b	7	7	14	12
3 a	7	7	14	12
3 b	5	6	11	9

b $F + V - E = 2$ or $F + V = E + 2$

5 **a**

Network	Traversable?	Number of odd vertices
1 a	No	4
1 b	Yes	0
2 a	No	4
2 b	Yes	2
3 a	No	4
3 b	Yes	2

6 No, because there are 4 odd vertices. Can only traverse if there are 0 or 2 odd vertices.

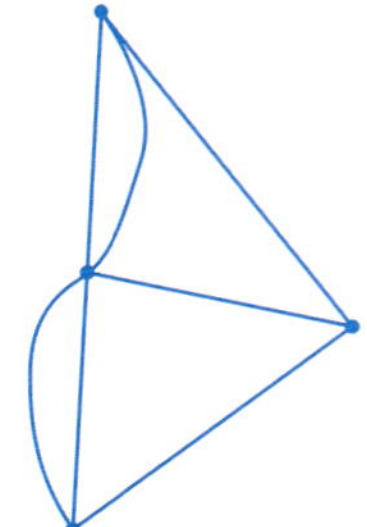

7 7 faces, teacher to check network.

8 No, teacher to check network.

Exercise 6.05

2 **a** dodecahedron
b tetrahedron
c cube or octahedron
d icosahedron
e cube
f dodecahedron
g dodecahedron
h octahedron

3 **a**

Platonic solid	Number of faces (F)	Number of verti-ces (V)	$F + V$	Number of edges (E)
Tetrahedron	4	4	8	6
Cube	6	8	14	12
Octahedron	8	6	14	12
Dodecahedron	12	20	32	30
Icosahedron	20	12	32	30

b $F + V - E = 2$ or $F + V = E + 2$

9780170465618

4

	Polyhedron	Number of faces (F)	Number of vertices (V)	$F + V$	Number of edges (E)	2	$E + 2$
a	Triangular prism	5	6	11	9	2	11
b	Hexagonal prism	8	12	20	18	2	20
c	Square pyramid	5	5	10	8	2	10
d	Pentagonal prism	7	10	17	15	2	17
e	Trapezoidal prism	6	8	14	12	2	14

5

Polyhedra	Number of faces (F)	Number of vertices (V)	$F + V$	Number of edges (E)	Euler's formula true?
a	5	6	11	9	Yes
b	4	4	8	6	Yes
c	10	16	26	24	Yes
d	6	8	14	12	Yes
e	7	7	14	12	Yes

6

Polyhedra	Number of Faces (F)	Number of vertices (V)	$F + V$	Number of edges (E)
a	6	6	12	10
b	4	4	8	6
c	6	8	14	12
d	7	10	17	15
e	8	6	14	12

7 Teacher to check.

Power plus

1 **a** 2 **b** 20 **c** 54

2 **a** square, rhombus
b parallelogram, rhombus, rectangle, square
c kite, rhombus, square
d rectangle, square
e rhombus, square
f square, rectangle

3 Some answers are abbreviated to fully show the reasons required to reach the final answer. Other reasons are also possible.
a $h = 113$ (vertically opposite angles)
$m = 67$ (co-interior angles on parallel lines)
$p = 67$ (vertically opposite angles)
$k = 67$ (corresponding angles)
b $w = 73$ (exterior angle of triangle is equal to the sum of the two opposite interior angles)
c $e = 97$ (co-interior angles on parallel lines, then alternate angles on parallel lines)
d $y = 61$ (angle sum of a triangle, then corresponding angles on parallel lines)
e $f = 56$ (alternate angles on parallel lines, then angle sum of isosceles triangle)
f $x = 78$ (alternate angles, then exterior angle of triangle result)
g $t = 34$ (alternate angles on paravllel lines)
$n = 117$ (opposite angles of parallelogram)
$v = 29$ (angle sum of a triangle)
h $a = 90$ (diagonals of rhombus meet at right angles)
$d = 65$ (alternate angles on parallel lines, then angle sum of a triangle)
i $c = 153$ (angle sum of quadrilateral, then angles at a point)
j $x = 75$ (construct a line parallel to the other lines, then sum of 2 alternate angles)
$y = 285$ (angles at a point)
k $x = 75$ (alternate angles)
$y = 50$ (corresponding angles)
$z = 55$ (either angle sum of a triangle or angles on a straight line)
l $x = 68$ (angles on a straight line, isosceles triangle, then alternate angles)
$y = 248$ (alternate angles, angles at a point)

TEST YOURSELF 6

1 **a** $k = 47$ (angle sum of triangle)
b $e = 75$ (exterior angle of a triangle)
c $x = 24$ (equal angles in an isosceles triangle)
d $m = 18$ (angle sum of triangle)
e $t = 107.5$ (equal angles in an isosceles triangle, exterior angle of a triangle)
f $b = 60$ (equilateral triangle , alternate angles on parallel lines)
g $y = 140$ (exterior angle of a triangle)
h $y = 48$ (angles on a straight line, isosceles triangle)
i $y = 78$ (co-interior angles on parallel lines)

2 **a** all 60° **b** 90°, 45°, 45°

3 **a** $x = 110$ (angle sum of quadrilateral)
b $y = 112$ (corresponding angles on parallel lines), $m = 68$ (co-interior angles on parallel lines)

c $g = 60$ (angles on a straight line, diagonals of a rectangle bisect each other, isosceles triangle)
d $x = 35$ (alternate angles on parallel lines) $y = 55$ (diagonals of a rhombus bisect at right angles, angle sum of a triangle)
e $m = 97$ (one pair of opposite angles equal in a kite, angle sum of a quadrilateral)
f $e = 36$ (co-interior angles on parallel lines)

4 a parallelogram, rectangle, square, rhombus
b square, rectangle
c square, rhombus
d square, rectangle, rhombus, parallelogram

5 a

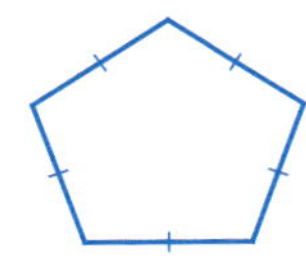

b

c

6 a $180(10 - 2) = 1440°$ b $140°$

CHAPTER 7

SkillCheck

1 a −20 b 6 c −5 d 16
e 19 f −15 g 7 h 1

2 a subtracting b multiplying
c dividing d adding

3 a $4n$ b $11x$ c $-5r$ d $-6t$

4 a $8k + 28$ b $15d - 5$ c $-3 - 6y$
d $-4b + 32$ e $15d + 18$ f $7a + 20$
g $19d + 1$

5 a $m = 6$ b $a = 9$ c $y = 8$
d $w = 50$ e $y = -9$ f $m = 23$
g $a = -7$ h $a = 12$ i $w = -30$

6 a $7n$ b n^2 c $5(n + 8)$
d $n - 20$ e $6n - 9$ f $n + 2$

7 a $(x + 3)(x + 5)$ b $(m + 4)(m - 1)$
c $(b + 4)(b + 7)$ d $(t - 6)(t + 2)$
e $(u - 2)(u - 5)$ f $(h + 8)(h + 9)$

Exercise 7.01

1 a $x = 2$ b $p = 4$ c $w = 4\frac{1}{2}$ d $k = 1$
e $x = -\frac{3}{7}$ f $m = \frac{2}{3}$ g $y = -4$ h $x = -1$
i $h = 2$ j $q = -1\frac{1}{5}$ k $w = -2$ l $d = 3\frac{1}{2}$

2 B

3 a $n = 6$ b $x = 2\frac{4}{5}$ c $a = 9\frac{1}{3}$ d $y = 36$
e $n = 5.4$ f $p = 7\frac{1}{3}$ g $x = 22$ h $n = 10$
i $d = 1$ j $a = -9$ k $n = 2$ l $x = 11$

4 Clockwise: $3k$, 20, ÷ 3, 5

5 Clockwise:
a $3c$, $3c + 6$, − 6, −30, ÷ 3, −10
b $5n$, $\frac{5n}{3}$, × 3, 30, ÷ 5, 6
c $\frac{m}{5}$, $\frac{m}{5}$−12, +12, 15, × 5, 75

6 a $w = 2$ b $a = 48$ c $m = 60$
d $a = -3$ e $x = 16$ f $a = 14$

7 D

8 a $x = 10$ b $w = 0$ c $y = -33$
d $k = -12$ e $x = -45$ f $d = 1$

9 A

Exercise 7.02

1 a $w = 6$ b $q = 6$ c $x = 5$ d $n = -2$
e $y = 10$ f $m = 2$ g $a = 6$ h $x = -1$
i $y = 0$ j $u = -1$ k $x = -1\frac{3}{4}$ l $x = -5$

2 a B b C

3 a $w = -2$ b $t = -1\frac{1}{2}$ c $a = 1$
d $y = -3$ e $y = -1\frac{1}{2}$ f $y = \frac{3}{5}$
g $t = 1\frac{3}{8}$ h $y = 2\frac{1}{2}$ i $k = 1\frac{2}{3}$

4 C

5 a i $3x = 8 + 2x$, $x = 8$
b i $2x + 3 = x + 5$, $x = 2$
c i $3 + 6x = 3x + 6$, $x = 1$
d $2x + 1 = 4x$, $x = \frac{1}{2}$

Exercise 7.03

1 a $m = 1$ b $x = 2$ c $y = -1$ d $k = 4$
e $a = -1$ f $p = \frac{1}{9}$ g $h = -24$ h $m = 9$
i $u = 11$ j $y = 1\frac{3}{7}$ k $p = -1\frac{1}{5}$ l $x = 1\frac{13}{14}$

2 A

3 LHS = RHS = 48

4 LHS = RHS = 16

5 a $m = 3$ b $y = 7$ c $x = 2$
d $p = 4$ e $n = -4$ f $x = 3\frac{1}{3}$
g $w = -8$ h $x = 6$ i $y = \frac{5}{9}$

6 LHS = RHS = 30

7 a $m = -7$ b $y = 7$ c $y = -10$ d $x = 12$
e $y = -26$ f $y = -1\frac{2}{3}$ g $m = -7$ h $y = -\frac{71}{89}$

Exercise 7.04

1 11 2 12 3 24
4 30 5 40 cm × 10 cm
6 a $4w + 14 = 94$ b 27 cm × 20 cm
7 $x = 33$
8 a $y = 13$ b 110°
9 82°, 98° 10 43, 44 11 28, 29, 30
12 54, 56, 58 13 23, 25, 27
14 Dean is 10, Franco is 50, Helen is 2.
15 Rebekah is 14.

16 78 **17** \$4.30

18 Ashvin: 140 cm, Sureti: 70 cm, father: 170 cm.

19 26.5 cm × 17 cm

20 Brett is 17, Vila is 51, Amanda is 12, Janine is 45.

21 $d = 5.7$ m **22** 240

Mental skills 7A

2 **a** 14 **b** 9 **c** 7 **d** 8 **e** 3
f 8 **g** 10 **h** 5 **i** 33 **j** 5
k 15 **l** 18 **m** 26 **n** 6 **o** 36
p 6 **q** 14 **r** 160 **s** 14 **t** 2

Exercise 7.05

1 **a** $m = \pm 12$ **b** $x = \pm 20$ **c** $y = \pm 15$
d $k = \pm\sqrt{59}$ **e** $y = \pm\sqrt{10}$ **f** $w = \pm 6$
g $x = \pm 5$ **h** $t = \pm 3$ **i** $a = \pm 4$
j $k = \pm\sqrt{8}$ **k** $w = \pm\sqrt{10}$ **l** $d = \pm 12$
m $k = \pm 4$ **n** $w = \pm\sqrt{70}$ **o** $x = \pm\frac{1}{2}$
p $m = \pm 9$ **q** $y = \pm\sqrt{5}$ **r** $p = \pm 3$
s $k = \pm 4$ **t** $y = \pm\sqrt{55}$ **u** $x = \pm\sqrt{7}$

2 **a** $m = \pm 4.5$ **b** $b = \pm 4.1$ **c** $v = 2.4$
d $p = \pm 4.2$ **e** $k = \pm 2.6$ **f** $x = \pm 6.3$
g $k = \pm 9.8$ **h** $u = \pm 1.7$ **i** $y = \pm 3.7$
j $w = \pm 5.2$ **k** $a = \pm 4.1$ **l** $y = \pm 6.2$

3 $\sqrt{-25}$ doesn't exist for real numbers; there isn't a real number whose square is –25.

4 **a**, **c** and **f**. They all involve square roots of negative numbers.

Exercise 7.06

1 **a** $x = -2, -1$ **b** $y = -4, -1$ **c** $y = -4, -12$
d $x = -4, 3$ **e** $x = -3, 1$ **f** $x = -8, 5$

2 **a** $x = 6, -5$ **b** $x = 4$ **c** $x = 11, -6$
d $d = 0, 2$ **e** $x = 5, -2$ **f** $n = 0, -4$
g $k = 0, 7$ **h** $y = 0, 5$ **i** $v = 0, 12$

3 **a** A **b** C

Exercise 7.07

1 **a** $y = 28$ **b** $y = 11$ **c** $b = 25$
d $b = -1$ **e** $x = 3$ **f** $x = 16$

2 **a** 95°F **b** 14°F **c** 60.8°F

3 **a** 37.8°C **b** –42.8°C **c** 25.6°C

4 **a** $A = 72$ cm^2 **b** $h = 20$ cm **c** $p = 32.5$ m

5 C

6 **a** \$3100 **b** \$9700
c 10 hours **d** 18 hours

7 **a** \$7300 **b** 536

8 **a** \$2255 **b** \$6030 **c** 112 people

9 D

10 **a** 65 °C **b** 91.25 °C **c** 4 h

11 **a** 182 cm **b** 39 cm

Exercise 7.08

1

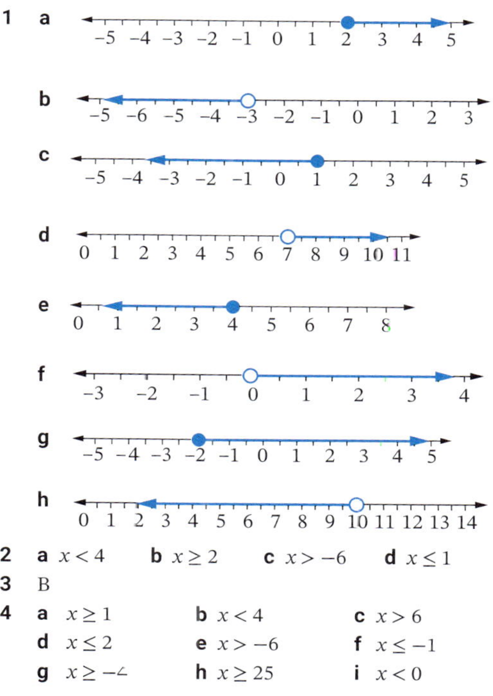

2 **a** $x < 4$ **b** $x \geq 2$ **c** $x > -6$ **d** $x \leq 1$

3 B

4 **a** $x \geq 1$ **b** $x < 4$ **c** $x > 6$
d $x \leq 2$ **e** $x > -6$ **f** $x \leq -1$
g $x \geq -4$ **h** $x \geq 25$ **i** $x < 0$

Exercise 7.09

1 **a** $x > 7$

b $y \geq 4$

c $m \leq -2$

d $x \geq -100$

e $x < 5$

f $y > -4$

g $a \geq \frac{1}{2}$

h $w \le -10$

i $a \ge 5$

j $a \le 3$

k $a \ge -1$

l $w < -3$

m $a \le -6$

n $y \le 3$

o $a < 2\frac{1}{2}$

2 A

3 **a** $x \ge 1$ **b** $m \le 6$ **c** $y \le -8$

d $w > 0$ **e** $w \le 0$ **f** $m \ge 3\frac{1}{2}$

g $m \ge -2$ **h** $x \le 5$ **i** $w > -3$

j $a < 4$ **k** $a \ge 6$ **l** $m \le 3\frac{1}{2}$

m $m > 3\frac{2}{5}$ **n** $m \ge -2\frac{1}{2}$ **o** $x < 35$

4 D

5 **a** $x \ge 3$

b $y > -8$

c $k > -11$

d $m \le 0$

e $p < -6$

f $t \le -4$

6 **a** $x > -3$ **b** $k \le -12$ **c** $t < -2\frac{2}{5}$

d $x \ge 12$ **e** $w < -1$ **f** $y \ge -2$

g $x \le 4$ **h** $a > 1$ **i** $d < -5\frac{1}{2}$

j $w < -11$ **k** $x \le -8$ **l** $p < -4$

Mental skills 7B

2 **a** 11 **b** 40 **c** 7 **d** 24

e 23 **f** 6 **g** 43 **h** 80

i 18 **j** 15 **k** 40 **l** 65

m 11 **n** 14 **o** 12 **p** 135

Power plus

1 **a** $y = 14$ **b** $s = 3\frac{1}{6}$ **c** $t = 4\frac{1}{5}$

d $x = -4$ **e** $m = -6\frac{1}{2}$ **f** $d = 17\frac{2}{5}$

g $x = 5$ **h** $y = 35\frac{1}{2}$ **i** $y = \frac{5}{28}$

j $r = 3\frac{1}{2}$ **k** $y = \pm\sqrt{48}$ or $\pm 4\sqrt{3}$

l $m = \pm 5$

2 **a** $W = \frac{1}{5}$ **b** $W = -3$

c $X = -6\frac{1}{2}$ **d** $Y = -\frac{2}{3}$

3 26 **4** Wendi is 20

5 $B = 30$

6 **a**

b

c

TEST YOURSELF 7

1 **a** $x = 5$ **b** $y = 8.5$ **c** $y = 1.5$

d $m = 32$ **e** $x = 33$ **f** $x = 6$

g $x = -7.5$ **h** $a = -3$ **i** $d = -6.5$

2 **a** $x = 1$ **b** $x = -5$ **c** $w = 1\frac{1}{3}$

d $x = 3$ **e** $n = -4$ **f** $d = -9$

g $t = 1\frac{1}{2}$ **h** $j = -8.5$ **i** $q = -1$

3 Teacher to check.

4 **a** $w = 7$ **b** $n = 1$ **c** $p = -1$

d $x = \frac{1}{3}$ **e** $y = 5$ **f** $x = 1.5$

5 **a** $a = 1$ **b** $g = 14\frac{2}{3}$

6 22 cm × 16 cm

7 **a** $x = 15.5$ **b** $a = 46$

8 1

9 **a** $d = \pm 8$ **b** $p = \pm 6$ **c** $z = \pm\sqrt{35}$
d $x = -6, 3$ **e** $m = 2, 8$ **f** $b = -4, 9$
g $t = 7$ **h** $u = -8, -3$ **i** $h = -1, 8$

10 **a** BMI = 21 **b** 105.84 kg

11 **a** \$41 **b** $d = 30.5$ km

12 **a** $x \geq 0$

−4 −3 −2 −1 0 1 2 3 4

b $x < 3$

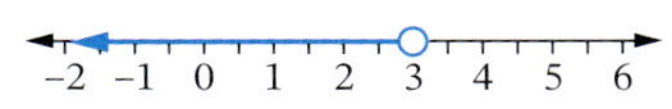

c $x \leq -2$

−5 −4 −3 −2 −1 0 1 2 3

d $x > -5$

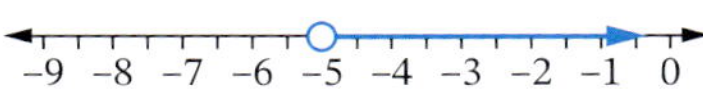

13 **a** $y \geq 16$ **b** $y \leq -7\frac{1}{2}$ **c** $a > -5$
d $x > -3$ **e** $a < -16$ **f** $x \leq -3$

PRACTICE SET 2

1 **a** y^{10} **b** $12g^{15}$ **c** $18p^5q^3$ **d** k^{12} **e** x^4
f h^9 **g** $9k^4$ **h** h^2y^3 **i** $\frac{2m^4n}{3}$

2 **a** 6 **b** 1 **c** 1 **d** x **e** 1

3 **a** 9.2 **b** 2.0 **c** 40.7 **d** 168.9

4 **a** 120° **b** decagon

5 **a** 6 faces, 4 vertices, 8 edges

b

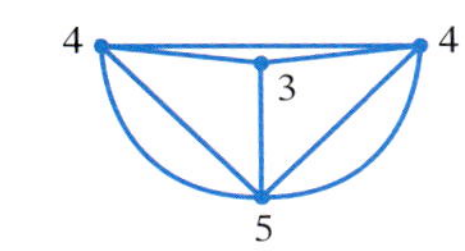

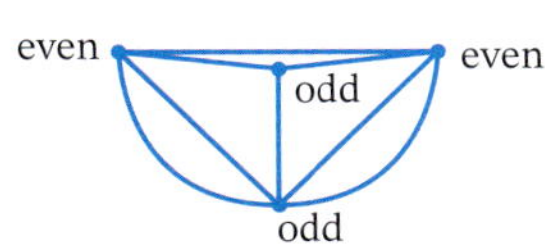

c Traversable as it has 2 odd vertices.

6 **a** $\frac{1}{2}$ **b** $\frac{1}{125}$ **c** $\frac{1}{100}$

7 **a** 2 **b** $\frac{3}{8}$ **c** $\frac{5}{4d}$

8 **a** 20.50 **b** 33.34 **c** 3.31

9 **a** 2^{15} **b** 4^{12} **c** 3^2p^{14} **d** 2^3w^{15}

10 **a** 15 400 000 **b** 0.0032
c 0.9837 **d** 24 000

11 **a** 53° **b** 78° **c** 59°

12 **a** 77°53′ **b** 64°9′ **c** 56°19′

13 **a** n^{-1} **b** 2^{-3} **c** k^{-5} **d** $7y^{-2}$

14 7.28×10^6

15 **a** 2.7×10^{16} **b** 1.2×10^{-10}

16 **a** 39° **b** 52° **c** 44°

17 **a** $x = 4$ **b** $x = -2\frac{1}{4}$ **c** $x = -1\frac{1}{4}$
d $y = 13$ **e** $a = 18$ **f** $x = 28\frac{1}{2}$

18 **a** $m = 103$ (angle sum of a triangle)
b $f = 54$ (equal angles of an isosceles triangle, angle sum of a triangle)
c $d = 132$ (exterior angle of a triangle)

19 A

20 **a** 5.3×10^9 **b** 3.73×10^5

21 **a** $m = 17.9$ **b** $d = 11.7$ **c** $g = 160.2$

22 **a** $w = 103$ **b** $w = 32$ **c** $w = 35$

23 C **24** 2.38 m or 238 cm

25 12 edges

26 **a** $16m^{12}n^8$ **b** $\frac{4h^2}{25}$ **c** $343r^3t^6$

27 **a** $x < 42$ **b** $x \geq 2$

28 **a** 40° **b** 37° **c** 39°

29 **a** 225 min or 3 h 45 min **b** 3.125 kg

30 **a**

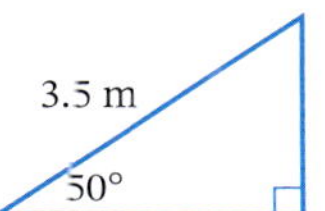

b **i** 2.7 m **ii** 2.2 m

31 square

32 **a** $p = 11$ **b** $y = 1$

33 **a** 19, 20, 21, 22 **b** 127 \$2 coins

34 50 m **35** A, C, D

36 **a** $m = \pm 4$ **b** $x = \pm 6$

37 40 cm **38** D

CHAPTER 8

SkillCheck

1 0.082 **2** \$2.25

3 **a** 365 **b** 52 **c** 28 **d** 36
e 14 **f** 13 **g** 60 **h** 26

4 **a** \$7.68 **b** \$21.42 **c** \$524.74
d \$92.36 **e** \$3085.64 **f** \$106.34

5 **a** \$733.59 **b** \$6285.83 **c** \$3150
d \$1198.21 **e** \$694.67 **f** \$14 911.22
g \$1339.02 **h** \$2343.50 **i** \$2293.20

6 \$5590.25

Exercise 8.01

1 **a** \$1049.60 **b** \$1332.10
c \$895.20 **d** \$321.30

2 \$103 316.40 **3** D **4** \$79 000.22

5 a $3034.50 b $6068.99 c $13 195
6 a $117.80 b $139.50 c $130.20
7 Danica, by $31.26.
8 Fortnightly income pays more
9 a Marisol by $1.23/h
b Paula by $3.26/h
c Fabian by $2.50/h
d Ranjan by $2.46/h
10 a $544 b $2720 c $11 786.67
11 a $127 680 b $2446.91
12 a $7393.33 b $1700.27 c $40.48
13 $1252.80 14 $1054.13
15 $1026
16 $1114.12 17 $28.95/h
18 $3495.59

Exercise 8.02

1 a $802.20 b $959.78 c $1217.63
d $1239.11 e $1547.10 f $1590.08
2 C 3 $1141.98 4 $2033.94
5 $1118.15 6 $1341.31 7 B
8 a $782.10 b $865.05 c $1007.25
d $1220.55 e $953.93 f $1102.05
9 a $1136.80 b $1365 c $1401.40
d $1300 e $1976 f $1445.40
10 a $28.69/h b $29.69/h c $24.29/h
d $21.98/h e $25.92/h f $13.68/h

Exercise 8.03

1 $736.80 2 $1783.49 3 $27 506
4 Perry: $2600, Amy: $4050
5 C
6 a $25 000 b $31 250 c $38 750
7 $4520
8 a $337.50 b $230 c $555.75
d $316 e $405 f $480
g $45 h $738
9 a i $124.10 ii $172.55
b 540
c i 236 ii 730
10 $120
11 a $4389 b $7866 c $4433.65
d $2455.18 e $4987.50 f $5132.66
12 a $1096 b $767.20 c $5151.20
13 a $9658.87 b $3380.60 c $22 698.34
14 a $397.90 b $584.90 c $2069.90
15 a $1142.50 b $190.42 c $22.40/h
16 $38 800
17 a $1032 b $4850.40

Exercise 8.04

1 $90 634
2 a $74 673 b $14 735.60
3 a $127 525 b $32 251.25 4 C
5 a $65 262 b $11 677.15
6 a $100 270 b $23 054.75
7 a $86 285.58 b $86 096 c $18 448.20
8 D 9 $101 450.50 10 $13 594.36

Mental skills 8

2 a $408 b 99 c 200
d $404 e 672 f $81
g $517 h 225 i 560
j $84 k $350 l 84
4 a 330 b $240 c 1600 d $425
e $225 f $60 g $63 h 76
i $68 j $3762 k $374 l $100

Exercise 8.05

1 a $3200 b $704
2 a $8064 b $1764
3 $240
4 a $43 b 22.1% c $152
5 Phoebe, by $24.
6 Electronics Store offers higher net pay by $24 over the 6 weeks.
7 a $158 b $435.90 c 37.6%
8 C
9 a $812.03 b 27.9%
10 a $670.98 b 16.4%
11 a gross weekly income = $1094.40, total deductions = $339.60, net weekly income = $754.80
b gross weekly income = $1527.50, total deductions = $504.10, net weekly income = $1023.40
12 a $251.70 b 24.5%
13 a $2943.66 b $956.69
c $1208.31 d $1735.35
14 a $519.12 b $4092.60
c $87 856.28 d 28.3%

Power plus

1 $260.08/min 2 $55.56 3 $288/h
4 Felicity: $514; Rose: $454.80; Difference: $59.20
5 $1368 6 $21.50/h 7 $855
8 $841.50 9 $127 200 10 18 h 40 min

TEST YOURSELF 8

1 a $1675.16 b $3350.33 c $7284.17
2 $875.05 3 $1700.79
4 a $940.50 b $1200.38
5 $4200 6 $1950
7 a $750 b $1237.50 c $1242.50
8 $11 400 9 $8394.20
10 a $185 454 b $54 121.30
11 a $182 b $1208.96 c 12.2%
12 a $1071.60 b $359.60 c $712

CHAPTER 9

SkillCheck

1 **a** 39 **b** 6 **c** 43.6%

2 **a** 38 **b** 6 **c** 36
d 14 **e** 39.5%

3 **a**

Length of queue	Frequency
1	5
2	4
3	7
4	9
5	7
6	3
7	1

b

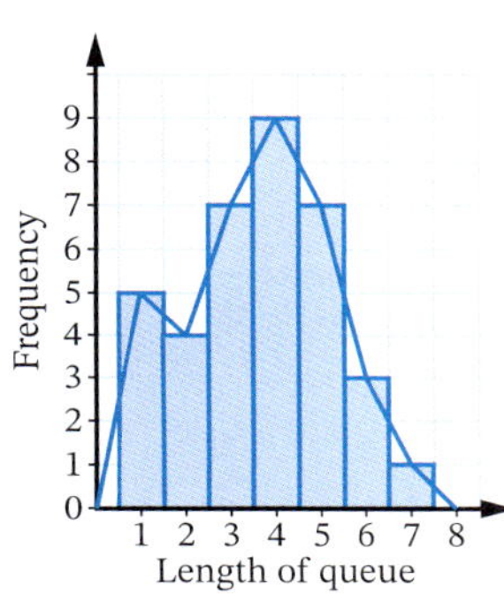

c 4 **d** 20

Exercise 9.01

1 **a** **i** 18.2 **ii** 18 **iii** 15 **iv** 16
b **i** 6.9 **ii** 7 **iii** 7 **iv** 14
c **i** 34 **ii** 35 **iii** 32, 45 **iv** 33
d **i** 8.7 **ii** 8.8 **iii** 8.8 **iv** 1
e **i** 0.3 **ii** −0.5 **iii** −2 **iv** 11

2 **a** max $\bar{x}$ = 22.9 °C, min $\bar{x}$ = 15.6 °C
b 7.3°
c max mode = 23 °C, min mode = 15 °C
d 8°

3 **a** 415 mm **b** 168 mm **c** 92 mm

4 **a** 6.3 **b** 7 **c** 7 **d** 8

5 **a** 1 is an outlier as it is away from the main body of data.
b 8, 9 **c** 7.3 **d** 8
e **i** 56.7% **ii** 33.3%

6 **a** Yes, 7
b **i** $\bar{x}$ = 2, median = 2
ii $\bar{x}$ = 2.3, median = 2
c The mean.

7 **a** $180 000
b **i** $\bar{x}$ = $86 944, median = $90 000
ii $\bar{x}$ = $96 250, median = $91 500
c The median, as it is only $1500 higher with the outlier included, while the mean is $9306 higher.
d **i** mode = $93 000, range = $35 000
ii mode = $93 000, range = $110 000
e Range

8 **a** 6.4 **b** 14.8 **c** 66.8

9 76

10 **a** 370 **b** 462 **c** 92

11 No, as she needs to score 103%.

12 **a** 10, 9, 28, 30, 24
$\Sigma f = 25$, $\Sigma fx = 101$, $\bar{x} = 4.04$
b 66, 184, 168, 225, 130, 108
$\Sigma f = 36$, $\Sigma fx = 881$, $\bar{x} = 24.47$

13 **a** 34.4 **b** 12.9

14 **a** 2, 7, 18, 25, 35, 44, 50
i 5.5 **ii** 4 **iii** 6
b 3, 11, 31, 46, 57, 64
i 20 **ii** 19 **iii** 5

15 **a**

Number of seedlings per punnet	Frequency
4	4
5	2
6	4
7	8
8	11
9	8
10	4
11	3
12	1

Median = 8

b 8 **c** 8 **d** 7.8

Exercise 9.02

1 **a**

Results (Score)	Frequency
2	1
3	2
4	2
5	7
6	5
7	5
8	5
9	2
10	1

b

c 6.1 **d** 6 **e** 5 **f** 8

g No outliers but clustering at 5–8.

2 a

Children per family	Frequency
1	5
2	8
3	9
4	5
5	4
6	3
7	4
8	1
9	1

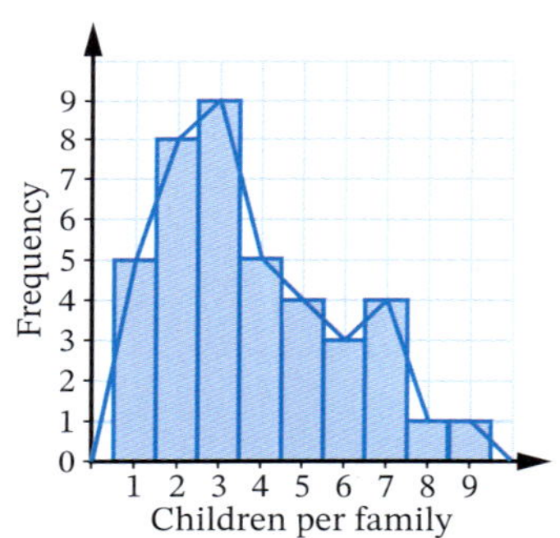

b mode = 3, range = 8

c no outliers, clustered near 2–3 children

d 3 **e** 3.8

3 a 6 It's the highest column.

b 5.96 **c** 6 **d** 5

4 a 73 **b** 14

c 2 It's the number of calls with the highest frequency, shown by the peak on the frequency polygon.

d 5 **e** 2 **f** 2.43

g clustering at 2–3 calls per minute

5 a 18.5 **b** 25.5 **c** 47.5

6 a

Class interval	Class centre	Tally	Frequency
0.00 – 0.04	0.02	III	3
0.05 – 0.09	0.07	𝍸 II	7
0.10 – 0.14	0.12	𝍸 IIII	9
0.15 – 0.19	0.17	𝍸 II	7
0.20 – 0.24	0.22	II	2
0.25 – 0.29	0.27	II	2

b

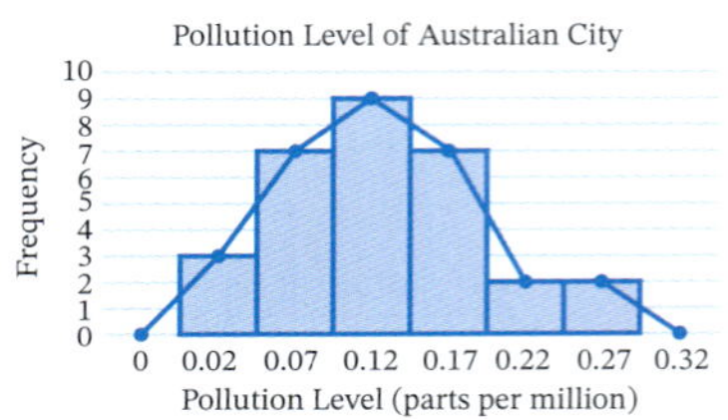

c 0.10 – 0.14

7 a 62

b

Class interval	Class centre	Tally	Frequency
30 – 39	34.5	I	1
40 – 49	44.5	IIII	4
50 – 59	54.5	𝍸	5
60 – 69	64.5	𝍸 III	8
70 – 79	74.5	𝍸 IIII	9
80 – 89	84.5	II	2
90 – 99	94.5	I	1

c

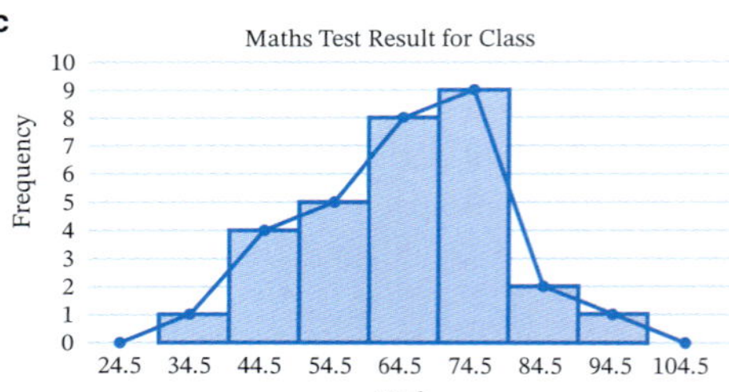

8 a 45

b 139.5 – 146.5 ; 146.5 – 153.5 ; 153.5 – 160.5 ; 160.5 – 167.5 ; 167.5 – 174.5 ; 174.5 – 181.5 ; 181.5 – 188.5

c 160.5 – 167.5

9 a

Class interval	Class centre	Tally	Frequency
60 – 64	62	I	1
65 – 69	67	𝍸 𝍸 II	12
70 – 74	72	𝍸 𝍸	10
75 – 79	77	𝍸 𝍸	8
80 – 84	82	II	2

b

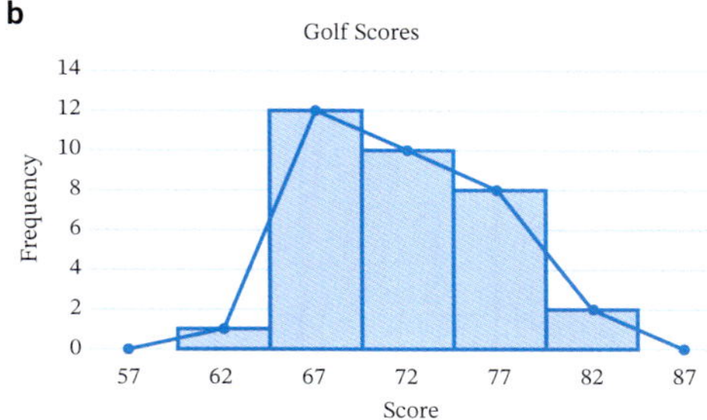

c 65 – 69

d 70 – 74

10 a

Class interval	Class centre	Tally	Frequency
131 – 140	135.5	III	2
141 – 150	145.5	卌 II	7
151 – 160	155.5	卌 IIII	10
161 – 170	165.5	卌 II	17
171 – 180	175.5	II	11
181 – 190	185.5	II	3

b

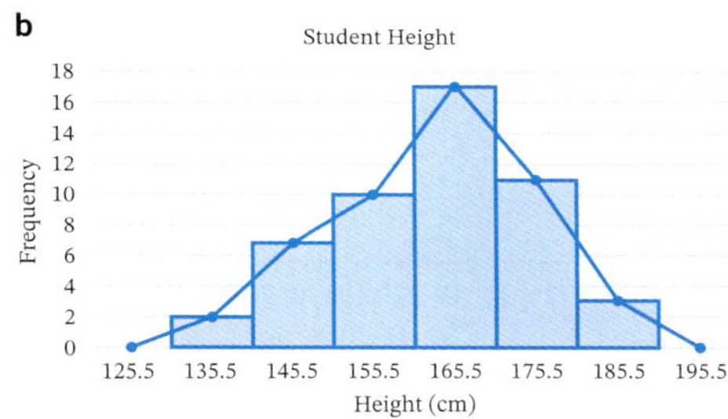

c 161 – 170

d 28%

Exercise 9.03

1 a

Stem	Leaf
4	0 1 5
5	1 3 5 6 8
6	3 4 5 6 7 7 7 8
7	1 2 4 7 9
8	2 7 8
9	0 2 4

b 54 **c** Clustering occurs in the 60s.

d 27 **e** 67.85 **f** 67

2 a 52 **b** 68 **c** 60.0 **d** 63

3 a

Stem	Leaf
11	8 9
12	1 5 6
13	1 5 5 7
14	3 4 5 6 9 9
15	0 0 0 1 4 4 7 7
16	1 1 3 3
17	6
18	3 7

b i no outliers

ii clustering occurs near 14–15 seconds

c 8 **d** 70%

e i 43.3% **ii** 50%

4 a 54, 55 or 56 **b** 22 **c** 59.5

5 a Rockets: 3, Blues: 13

b Rockets: 52, Blues: 57

c The Blues, 40.

d The Blues performed better as they had more games in which they scored more than 30 points.

e The Blues' median was higher – 37.5 compared to Rockets' median of 27.5.

6 a

Robyn	Stem	Anthony
7	1	
7 3	2	8 9
7 7 7 6 5 3	3	4 5 5 6 6 7 9
5 2 2 1	4	2 3 7
4	5	1 5

b Anthony (547) sold more than Robyn (506).

c 7 **d** Robyn, 54

e Clustering for both is in the 30's for both.

f Anthony

7 a

Stem	Leaf
1	1 5 9
2	3 5 7
3	1 2 3 4 5 6 8
4	0 2 5 5 7 8
5	2 5
6	5
12	5

b Yes, 12.5 – this value is away from the main body of data.

c Clustering occurs in the 3–4 km distances.

d 4.01 **e** 3.6

f The median is the better as it is not affected by the outlier of 12.5 km.

8 a

Girls	Stem	Boys
8	1	0 6
	2	6 8 9
9 6	3	2
8 5	4	0
9 9 6 1	5	0 0 5 7
7 0	6	3 7
8 6 6 4 3	7	2 4 5 6
5 2 0	8	2 8
4	9	5

b 95, boy

c i 63.5 **ii** 56

d 45% **e** 50%

f i 62.8 **ii** 54.25

g i no **ii** no

Exercise 9.04

1 a N **b** N **c** N **d** C **e** N

f N **g** N **h** C **i** C **j** N

2 C

3 a D **b** C **c** C **d** C

e D **f** D **g** C **h** D

4 **a** column graph, sector graph or picture graph
b working with categorical data
c teacher to check **d** see part **a**

5 **a** line graph
b will allow trend/changes to be easily observed
c teacher to check

6 **a** dot plot, column graph, sector graph
b working with discrete data
c teacher to check

7 **a** sector graph, picture graph or column graph
b working with categorical data

8 No, because this is grouped continuous data. A frequency histogram or polygon should have been constructed.

9 **a** picture graph, column graph or sector graph
b working with categorical data

10 **a** Clustered column graph
b Representing two sets of data

Exercise 9.05

1 **a** **i** symmetrical and bimodal
ii no outliers, some clustering near 5, 11
b **i** symmetrical and bimodal
ii 10 is an outlier, clustering near 4, 6
c **i** negatively skewed
ii no outliers, clustering in the 70s and 80s
d **i** positively skewed
ii no outlier, clustering near 21, 22
e **i** not symmetrical, bimodal
ii no outliers, some clustering 12–13, 16
f **i** positively skewed
ii no outliers, clustering near 1, 2
g **i** symmetrical
ii no outliers, some clustering near 51
h **i** negatively skewed
ii 10 is an outlier, clustering at 17–19

2 **a** 30 **b** in the 30s and 40s
c positively skewed
d Test was too hard or students did not prepare for the test.
e $\bar{x} = 48.6$, median = 45, mode is 28, 38, 45, 47, 64

3 **a**

Children per family	Frequency
1	4
2	7
3	9
4	6
5	4
6	3
7	3
8	1
9	1

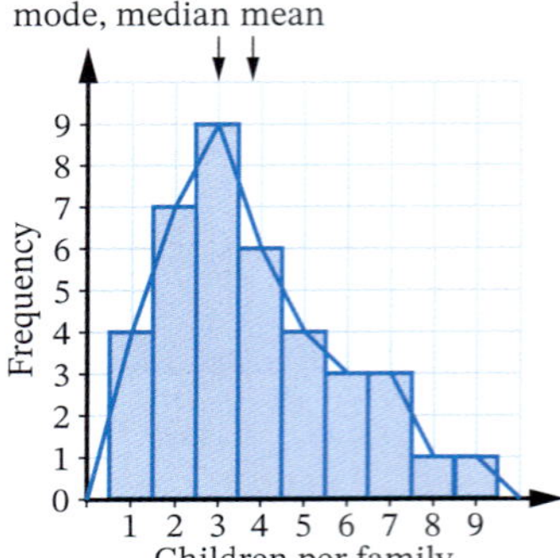

b No **c** positively skewed
d 2 to 4
e mode = 3, median = 3, $\bar{x} = 3.8$

4 C

5 **a** symmetrical **b** 56, 64, 83
c 76 **d** 72.8 **e** 72
f No, because the range has been affected by the outlier of 26. Without the outlier, the range is 45.

6 **a** **i**

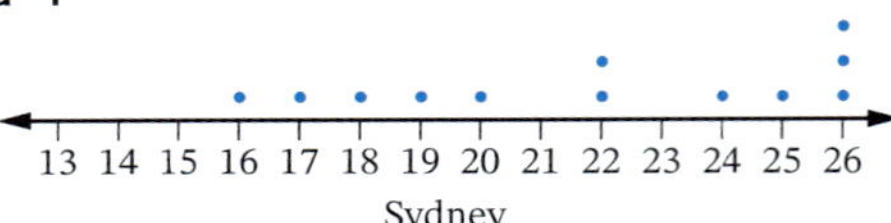

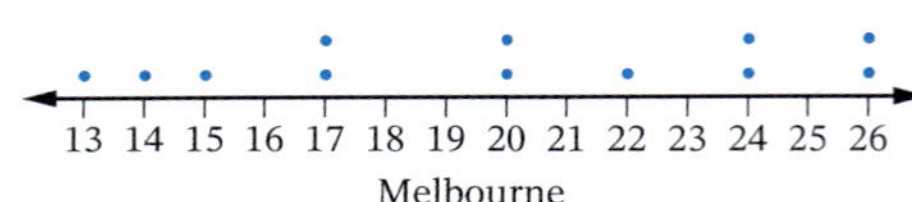

ii Both are slightly negatively-skewed.
b Sydney: $\bar{x} = 21.75$, median = 22, mode = 26, range = 10
Melbourne: $\bar{x} = 19.83$, median = 20, modes = 17, 20, 24, 26, range = 13
c Sydney is generally warmer – higher mean, median and lower range.

Exercise 9.06

1 **a** Brooke: 59.58 s Ryan: 57.54 s
b Brooke: 59.60 s Ryan: 57.33 s
c Brooke: 1.71 s Ryan: 4.55 s
d Ryan, his mean time is lower.
e Brooke, as her range is 1.71 s compared to Ryan's range of 4.55 s.

2 **a** Sydney: 13.1 °C Adelaide: 13.3 °C
b Sydney: 27.9 °C Adelaide: 26.6 °C
c Sydney: 29.2 °C Adelaide: 26.2 °C
d 8 days
e Sydney. Its mean and median were higher.

3 **a** Test 1: 53 Test 2: 64
b Test 1: 58.3 Test 2: 68.0
c Test 1: 59 Test 2: 54

d Test 1 is positively skewed. Test 2 is slight positively skewed.

e Yes, the mean and median have increased significantly.

4 a Logan City

b Logan City: 27.25 °C Wollongong: 26.13 °C

c Logan City: 11 °C Wollongong: 12 °C

d Logan City, slightly lower range and temperatures clustered from 26 °C to 28 °C.

e Logan City: 27 °C Wollongong: 26 °C

f Wollongong, its mean and median are slightly lower and it had a temperature of 26 °C or more on only 8 days compared to Logan City's 14 days.

5 a

Men	Stem	Women
9 9 9 8 6	2	
8 7 3 2 2 1 1 0	3	4 5 5 6 6 7 7 9 9
9 8 7 4 3	4	0 0 2 3 5 5 8
5 3	5	1 1 4 7

b No, the range is too high.

c i Different ($\bar{x} = 37.2$, median $= 32.5$).

ii Different ($\bar{x} = 42.2$, median $= 40$), but the difference is less.

d Men: 29 Women: 35, 36, 37, 39, 40, 45, 51

e Both are positively skewed. Clustering occurs in 30s for both.

f Men, as their mean and median are less than that for women.

g Men. Their range is 29 compared to a range of 23 for women.

6 Teacher to check.

Mental skills 9

2	a 8	b 0.3	c $12	d 0.8
	e $48	f 1.2	g 5.6	h $18
	i $3	j $7.20	k 77	l 26
4	a 12.5	b $4.50	c $10.80	d 9.9
	e $15	f 4.8	g $30	h $90
	i $14	j $7.20	k 5.7	l $10.50

Exercise 9.07

1 a census b sample c sample
d census e census

2 a every 5 years

b 2026 or 2031 depending on the current year

3 a 872

b i 17.3% ii 15.1%

c i 17 ii 15

4 No, as these students may live close to school or they may be early because they come by private transport.

5 No, as they only come from one shopping centre.

6, 7 Teacher to check

8 6 Australians, 4 Indians, 3 Europeans and 2 Asians

9 2 Gingernut Snaps, 1 Scotch Finger, 3 Milk Arrowroot, 2 Monte Carlo and 2 Chocolate Wafer packets

10 7 year 7s, 6 year 8s, 5 year 9s, 5 year 10s, 4 year 11s and 3 year 12s

Exercise 9.08

1 Teacher to check.

2 a People at a football match would not be thinking and/or interested in the amount of time spent on homework.

b Many sports are/have body contact to some degree. Spectators may be less likely to complain about violence.

c Most of the cars at the car show will use petrol.

d People at a rock concert will usually favour rock music.

3 Advantages – high accessibility as most people have a telephone

– good quality control as questions are asked the same way

– anonymity of people being surveyed

– data can be processed quickly

Disadvantages – time taken for survey is limited so questions are restricted to short answer type ones

– people who screen calls may not answer when called

– if the survey is about a product, participants cannot see, taste or feel the product

4 a Depends on the number of students in the school. For a large school, more students need to be surveyed.

b Can be given out at random to students in a school, may only take a short time to compete.

c May take too much time to complete, some of the questions may require a lengthy response.

d Personal interview – students are asked at random at recess or lunch.

5 C as all students in the school are asked. Answer A means only Year 9 students are asked and Answer B would only provide an estimate.

6 a The question is not precise enough. A better question might be:

'Circle the amount of time you spend studying per week:

0 – less than 2 hours 2 – less than 4 hours 4 hours or more '

b The question is too general. A better question could be:

'Select your favourite sport from the list below' and include 'Other' in the list.

c The question is too general. A better question could be:

'What is your income?

A $15 000 – <$25 000

B $25 000 – < $40 000

C $40 000 – < $60 000 ......'

d, e The question is too general. A better question could be:

'Select your favourite food/activity from the list below' and include 'Other' in the list.

f, g The question is not precise enough. A better question might be:

'Circle the amount of time you spend exercising/ on social media per day or week:

0 – less than 1 hours 1 hour – less than 2 hours
2 hours or less than 4 hours more than 4 hours '

h The question is too general. A better one might be:

'How many text messages do you send in one day?

☐0 – 10 ☐11 – 20
☐21 – 40 ☐more than 40

7 Teacher to check.

8 **a** Students from each year group are chosen at random.

b Choose students at random as they leave the canteen.

c, d Choose students from each year group.

e Randomly select some classes from your school or from different schools.

f Telephone survey or personal interview in a shopping centre.

Power plus

1 **a**

Class interval	Frequency	Cumulative frequency
16–21	3	3
22–27	8	11
28–33	7	18
34–39	7	25
40–45	7	32
46–51	2	34

b

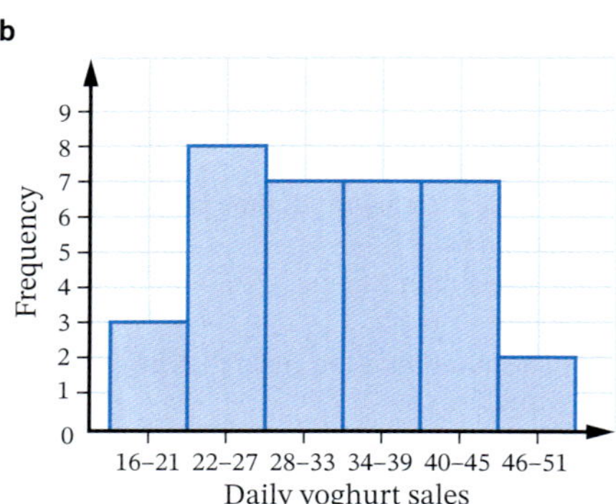

Shape is nearly symmetrical. No outliers or clusters.

c 22–27 **d** 28–33

2 If the range is too large.

3 **a** **i**

Stem	Seedlings
0	3 4 4 5 7 7 8
1	1 1 2 3 4 4 6 6 7 7 7 8 8 8 8 9 9 9
2	0 0 3 3 4 4 5 5 6 7 8 8
3	0 1 3

ii

Stem	Seedlings
0^0	3 4 4
0^5	5 7 7 8
1^0	1 1 2 3 4 4
1^5	6 6 7 7 7 8 8 8 8 9 9 9
2^0	0 0 3 3 4 4
2^5	5 5 6 7 8 8
3^0	0 1 3

b **i** Positively skewed, no outliers, clustered in the 10s stem.

ii Symmetrical, no outliers, clustered in 15–19 stem.

c Plot **ii** gives the better indication. It gives a better breakdown of how many seedlings grow per box (modal group is 15–19).

TEST YOURSELF 9

1 $\bar{x} = 7$, median = 8, mode = 8, range = 6

2 **a** $525 100 **b** $393 500

c The median is the better measure as it is not affected by the high price of $1 750 000.
If $1 750 000 is excluded, the mean is $389 000.

d Because the mean is affected by either very high or very low prices (outliers).

3 **a**

Stem	Leaf
2	2
3	8
4	2 3 5 7
5	4 6 8 9
6	2 3 4 4 7
7	0 0 1 2 3 5 7 8 8
8	3 4 5 6 9
9	0

b $\bar{x} = 65.5$, median = 68.5, modes = 64, 70 and 78, range = 68

c Yes, 22

d The median, as the mean is affected by the outlier 22.

4 **a** 6 **b** 1 **c** 1.93 **d** 2

e The mean as there are no outliers.

5 **a**

Northbridge	Stem	Westvale
9 9 7	0	5 7 7 8 9
8 6 6 5 1	1	3 5 5 8 8
8 7 7 3 2	2	2 4 4 4 4 5 6
8 7 7 5 2 0 0	3	3 5 6 7 7
7 5 2	4	3 4
3	5	

b Northbridge: 27.25 min, 27.5 min, 46 min
Westvale: 22.875 min, 24 min, 39 min

c No outliers, some clustering for Northbridge in the 30s and for Westvale in the 20s.

d The waiting times for Westvale are less than for Northbridge and more consistent.

6 **a** numerical, continuous

b numerical, discrete

c categorical

d categorical
e numerical, continuous

7 a Clustered column graph
b Working with two data sets
c Teacher to check

8 a positively skewed, bimodal, no outliers, clustering at 12, 14.
b symmetrical, no outliers, clustering near 100 – 109.
c negatively skewed, no outliers, clustering near 15 – 17.

9 a

Girls	Stem	Boys
8 1	13	4
8 7 5 2	14	8
6 5 3 0 0	15	0 4
3 2 0	16	4 4 5
	17	0 1 5 6 6 8
	18	0

b i Nearly symmetrical
ii negatively skewed
c No outliers or clusters for girls, 134 is an outlier for the boys, clustering in the 170s for boys.
d Girls: 150, 150;
Boys: 164.64, 167.5
e For the boys, the median is the better measure. For the girls, both measures are good.
f Girls
g Boys. Their range is 46 cm compared to the range for girls, which is 32 cm.

10 a sample b sample
c census d sample

11 5 doctors, 16 nurses, 11 administrative staff and 13 kitchen staff should be included in the survey.

12 a Not all teens necessarily play computer games.
b Only those customers interested in completing the survey are sampled.
c Need to sample workers as well, not only management who may have a biased view.
d People at a premiere are not the usual moviegoers.
e It is restricted to one radio station playing a specific type of music.

13 a The question is too general and many responses could be given. Instead, give choices such as always, most of the time, sometimes, half the time, not often, rarely, never.
b The question is too general. Give choices such as excellent, good, satisfactory, poor.
c The question encourages bias towards saying yes because people have fought for the flag. A better question might be 'Do you think Australia's flag should be changed?' with a Yes/No response and then a choice of reasons such as:
❑ Australia is independent of the United Kingdom
❑ Australia is a multicultural nation
❑ Australia is 'part' of Asia
❑ Other

CHAPTER 10

SkillCheck

1 a i 88 cm ii 384 cm^2
b i 112 cm ii 336 cm^2
c i 37.7 cm ii 113.1 cm^2
d i 70 cm ii 200 cm^2
e i 84 cm ii 336 cm^2
f i 94.2 cm ii 706.9 cm^2

2 a 34 000 mm b 9.25 m
c 1.5 km d 3.75 L
e 380 mm f 7500 mL

3 a 50% b 15.5% c 8.3%
d 2.3% e 38.9% f 15%

4 a $\frac{1}{2}$ b $\frac{1}{4}$ c $\frac{1}{3}$ d $\frac{23}{36}$

5 a 180 cm^3 b 2197 cm^3
c 1440 cm^3 d 4500 cm^3

6 a i 5
ii 3 rectangles and 2 triangles
iii Triangular prism
b i 8
ii 6 rectangles and 2 hexagons
iii Hexagonal prism
c i 6
ii 4 rectangles and 2 trapeziums
iii Trapezoidal prism

7 a 41 b 18 c 12.6

Exercise 10.01

1 a cm b t c km
d L e kL f hours

2 a 750 000 mg b 36 000 L
c 67 000 g d 1500 mL
e 67 mm f 3500 m
g 9400 kg h 3750 mg
i 425 cm j 200 g
k 26 200 mL l 300 m
m 3 000 000 mL n 5760 min
o 54 000 s p 270 000 cm
q 5 000 000 mg r 700 mm
s 29 375 000 g t 18 140 000 mm
u 3270 mg

3 B

4 a B b B c A
d C e B

5 a 8 kL b 9 cm c 3 min
d 4 h e 0.375 L f 14.7 t
g 8.6 g h 7.5 m i 22.3 cm
j 0.8 kg k 0.125 kL l 0.0035 t
m 17 kg n 5 days o 45 t
p 40 h q 1 day r 0.025 km
s 0.075 kL t 34.7 m u 0.000 628 ML

6 B

7 503 250 000 000 L or 5.0325×10^{11} L

ANSWERS

8 **a** 4800 kB **b** 2100 GB **c** 8500 MB
d 10 800 000 B **e** 2.91 kB **f** 0.74 GB
g 1.05 MB **h** 5.9 TB **i** 0.000 94 MB
j 57 000 MB **k** 430 000 kB **l** 21 000 GB
m 7800 B **n** 2.18 GB **o** 0.042 MB
p 940 000 kB **q** 640 000 000 kB **r** 0.278 GB

9 9900 kB

10 225 MB

11 12 800 files

12 **a** 520 000 000 = 5.2×10^8 km
b 6 s **c** 750 y
d 0.012 m **e** 6.4×10^8 m
f 2.0×10^{13} km **g** 70 ms
h 250 000 years **i** 4.8 s
j 5290 Mm **k** 6.2×10^{12} km
l 6000 μm **m** 940 millennia
n 8.5 centuries **o** 85 ly
p 80 nm **q** 6.1 pc
r 0.16 Ma

Exercise 10.02

1 **a** 5 cm **b** 0.1 kg **c** 7 mL **d** 9 minutes

2 **a** 24 mL **b** 4 cm **c** 4 kg **d** 50°
e 600 g **f** 18 mL **g** 3 cm **h** 127°

3 **a** 0.4 mL **b** 0.08 cm **c** 0.09 kg **d** 0.5°
e 8 g **f** 0.8 mL **g** 0.02 cm **h** 0.5°

4 **a** 0.017 **b** 0.020 **c** 0.022 **d** 0.010
e 0.014 **f** 0.047 mL **g** 0.007 **h** 0.004

5 **a** Angle on protractor (h) **b** 0.8 mL liquid (f)

6 6.54%

7 4.90%

8 Percentage error of Lisa's estimate = 2.67%
Percentage error of Aryan's estimate = 4.67%
Lisa's estimate is more accurate.

9 Percentage error of Melissa's estimate = 3.38%
Percentage error of Sandeep's estimate = 2.66%
Sandeep's estimate is more accurate.

10 **a** **i** 1 mm **ii** 0.5 mm
b **i** 1 °C **ii** 0.5 °C
c **i** 1 min **ii** 0.5 min = 30 s
d **i** 100 g **ii** 50 g
e **i** 1 cm **ii** 0.5 cm = 5 mm
f **i** 1° **ii** 0.5°
g **i** 5 km/h **ii** 2.5 km/h
h **i** 250 g **ii** 125 g
i **i** 10 mL **ii** 5 mL
j **i** 1 kL **ii** 0.5 kL = 500mL
k **i** $\frac{1}{4}$ tank **ii** $\frac{1}{8}$ tank

11 D

12 **a** 1 cm **b** 0.5cm
c 0.5 mm **d** 26.75 to 26.85 mm

13 1%

14 1.25%

15 No, actual value = \$19 450.15 so the painting cannot be valued greater than \$20 000

16 range is $0.5625 \leq x \leq 0.6375$

Exercise 10.03

1 **a** **i** 122 cm **ii** 690 cm²
b **i** 76 m **ii** 211 m²
c **i** 324 mm **ii** 3761 mm²
d **i** 52 cm **ii** 126 cm²
e **i** 92 cm **ii** 516 cm²
f **i** 136 cm **ii** 864 cm²

2 C

3 **a** 255 cm² **b** 217.5 cm² **c** 297 cm²
d 410 cm² **e** 204 cm² **f** 134 cm²
g 36 cm² **h** 73.5 cm² **i** 536 cm²
j 1640 cm² **k** 340.24 cm² **l** 744 cm²

4 B

5 **a** 38.72 cm² **b** 27.2 cm²

6 **a** 525 000 m²
b **i** 0.525 km² **ii** 52.5 ha

7 B

Exercise 10.04

1 **a** 48 cm² **b** 54.4 m² **c** 340 mm²
d 3 cm² **e** 84 cm² **f** 950 mm²
g 1048 mm² **h** 91.52 cm² **i** 52 m²

2 **a** 102 cm² **b** 1519 mm² **c** 210 cm²
d 837 mm² **e** 1000 mm² **f** 356.5 cm²
g 406 cm² **h** 63 m² **i** 288 m²

3 **a** 84 m² **b** 52.25 m² **c** 800 mm²
d 190 m² **e** 54 cm² **f** 3.15 m²
g 84 cm² **h** 144.6 m² **i** 840 mm²

4 B

5 **a** 46 m² **b** 140 cm² **c** 24 m²
d 168 cm² **e** 100.8 cm² **f** 29.04 m²

6 **a** 160 m² **b** 141 m² **c** 7.8 m²

Mental skills 10A

2 **a** 19 **b** \$7.50 **c** 87.5
d \$20.20 **e** \$3.76 **f** 40
g \$0.925 **h** 89.6 **i** \$270
j \$0.38 **k** \$152.76 **l** \$7.25
m 315.4 **n** \$1.07 **o** 42.6
p \$2431.76

4 **a** 10 **b** 124 **c** \$490
d \$1.72 **e** 14.4 **f** \$2540
g 78 **h** \$1.16 **i** \$9
j \$16.80 **k** \$920 **l** 64

6 **a** 100 **b** \$0.60 **c** 2.5
d \$1.35 **e** \$1.84 **f** \$4.20
g 40 **h** 6.5 **i** \$0.48
j \$6.90 **k** \$3.60 **l** 42

Exercise 10.05

1 a i 50.27 cm ii 201.06 cm^2
b i 32.67 cm ii 84.95 cm^2
c i 94.25 cm ii 706.86 cm^2
d i 88.59 cm ii 624.58 cm^2

2 a i 92.5 m ii 508.9 m^2
b i 35.7 m ii 78.5 m^2
c i 4.1 m ii 1.0 m^2
d i 136.0 m ii 1055.6 m^2
e i 221.5 m ii 2565.9 m^2
f i 15.1 m ii 9.4 m^2
g i 17.4 m ii 5.6 m^2
h i 303.9 m ii 4618.1 m^2

3 D
4 D

5 a 182.83 cm b 72.85 cm c 121.12 mm
d 129.12 m e 554.20 cm f 1416.67 m
g 103.98 m h 46.57 cm i 56.55 cm

6 B

7 a 2228 m^2 b 293 m^2 c 952 m^2
d 442 m^2 e 63 m^2 f 71 m^2
g 503 m^2 h 151 m^2 i 193 m^2
j 743 m^2 k 4664 m^2 l 471 m^2

8 a 20 056 m^2 b $4001

9 a 40 074 km b 42 587 km

10 1.2 m^2

11 a 34.56 m^2 b $5702.40

12 4.9 m

13 a 15.708 cm b 157 m

14 a 39 m^2 b 157% increase

15 a Izabella 399.3 m b Noah 405.6 m
c 6.3 m

16 a 207.35 cm b 2411

Exercise 10.06

1 C

2 Other answers possible.

a

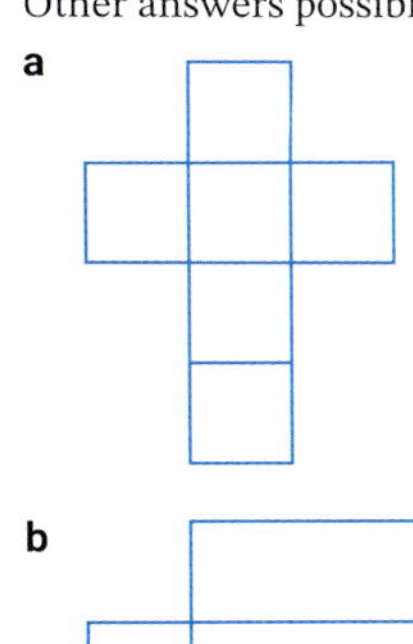

b

c

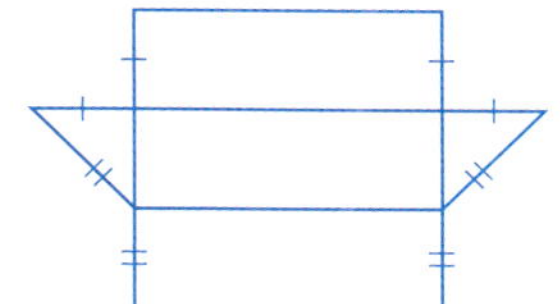

d

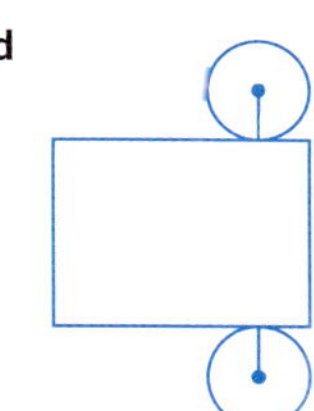

3 a rectangular prism b trapezoidal prism
c cube d triangular prism
e square pyramid f cylinder
g triangular prism h rectangular prism

4 a cube, $A = 2646$ m^2
b rectangular prism, $A = 684$ m^2
c right triangular prism, $A = 9720$ m^2
d trapezoidal prism, $A = 7236$ m^2

5 a 253.5 m^2 b 4140 cm^2 c 103 680 cm^2
d 125.36 cm^2 e 9600 mm^2 f 9680 cm^2

6 a 408 m^2 b 348 m^2 c 2310 cm^2
d 71.4 cm^2 e 360 m^2 f 648 m^2

7 a 145 140 cm^2 b $174.17

8 D

9 a $h = 8.9$ m b 71.2 m^2 c 862.4 m^2

10 a 2804 cm^2 b 19 440 cm^2 c 3740 cm^2

11 a 151.33 m^2 b 251.33 m^2

Exercise 10.07

1 a 747.70 m^2 b 2550.97 cm^2
c 206.47 m^2 d 15.83 m^2

2 a 904.8 cm^2 b 3545.1 mm^2

3 a 2159 .8 m^2 b 3.9 m^2 c 1894.5 m^2
d 6.5 m^2 e 14.4 m^2 f 1131.0 m^2

4 a $A = 56.5$ m^2 b 7 L of paint

5 The triangular prism tent (a) has 7.6 m^2 more surface area.

6 2073.5 cm^2

Exercise 10.08

1 a 50 m^3 b 13 200 cm^3 c 60 000 mm^3
d 32 768 mm^3 e 1.296 m^3 f 5880 cm^3

2 D
3 A

4 a 288 000 cm^3 = 288 000 ÷ 100^3 m^3 = 0.288 m^3
b $l = 1.2$ m, $b = 0.6$ m, $h = 0.4$ m, $V = 1.2 \times 0.6 \times 0.4 = 0.288$ m^3
It is easier to convert each measurement to the required units before doing the calculation.

5 a 504.8 m^3 b 1045 cm^3 c 19 968 cm^3

6 a 3016 cm^3 b 3958 mm^3 c 3 m^3
d 57 m^3 e 63 m^3 f $48\,384\text{ cm}^3$
7 a 9.7 m^3 b $94\,247.8\text{ m}^3$ c 7.4 m^3
d 135.7 m^3 e 5026.5 m^3 f 216.9 m^3
g 42.3 m^3 h 107.5 m^3 i 146.3 m^3
8 B 9 10.5 m 10 23.76 L
11 $13\,666\text{ cm}^3$ 12 6 m^3 13 300 kL
14 B 15 678.58 cm^3 16 13 mm
17 a 6.7 cm
b radius = 3.43 cm, height = 20.58 cm
c i If the radius and height were rounded down to 3.4 cm and 20.5 cm instead, the balls would not fit in the container. Radius = 3.5 cm, height = 20.6 cm.
ii 792.8 cm^3
d i length = 28 cm, width = 21 cm, height = 20.6 cm (after rounding up)
ii $V = 12\,112.8\text{ cm}^3$
e 2599.2 cm^3 f 21.5%

Mental skills 10B

2 a 18 b \$126 c 39 d \$30.30
e \$7.50 f 10.8 g \$27 h 60
i \$240 j \$3.30 k 900 l \$52.50
4 a 10 b 166 c \$50 d \$22
e 37.5 f \$5.80 g 135 h \$22.60
6 a 500 b \$20 c 4.5 d \$6.25
e 81 f \$35 g 16.5 h 74.5
i \$195 j \$425 k \$31.50 l 290
8 a 160 b \$1.50 c 7.5 d \$32.50
e \$67.50 f \$31.25 g 38 h 170

Power plus

1 a 157.08 cm b 50π cm
2 a 70.83 m b 125.66 m^2
3 60 cm 4 628 m^2
5 a i $A = 52.3\text{ cm}^2$ ii $P = 33.4$ cm
b i $A = 121.1\text{ cm}^2$ ii $P = 46.6$ cm
c i $A = 185.1\text{ cm}^2$ ii $P = 50.7$ cm
6 160 cm^3 7 3375 cm^3 8 40 mm
9 1629.2 cm^3

TEST YOURSELF 10

1 a 8000 m b 900 s c 700 mm
d 250 kg e 8400 MB f 9500 kg
g 5 h h 34 kB i 125 000 L
j 0.025 km k 17 000 t l 8.9 TB
m 10.5 s n 9100 years o 2 600 000 m
p 2.838×10^{13} km q 0.4 µs
r 75 m
2 65 jellybeans
3 6.89%
4 a i 1 mm ii 0.5 mm
b i 1° ii 0.5°
c i 1 min ii 0.5 min
d i 10 km/h ii 5 km/h
5 a 52 m b 74 cm c 84 mm d 114 cm
6 a 432 mm^2 b 64 m^2 c 16 cm^2 d 2340 mm^2
7 a 52 cm^2 b 468 mm^2 c 72 m^2 d 246 mm^2
e 120 m^2 f 260 cm^2
8 a i $P = 34.6$ cm ii $A = 95.0\text{ cm}^2$
b i $P = 171.4$ mm ii $A = 2056.8\text{ mm}^2$
c i $P = 59.4$ m ii $A = 253.1\text{ m}^2$
d i $P = 57.3$ m ii $A = 205.3\text{ m}^2$
e i $P = 7.7$ m ii $A = 3.5\text{ m}^2$
f i 54.8 cm ii 56.5 cm^2
9 a 14.14 m^2 b 162.62 mm^2 c 326.73 cm^2
10 a 61.44 m^2 b 152 m^2 c 760 m^2 d 2256 m^2
11 a 32.768 m^3 b 176 m^3 c 960 m^3 d 5040 m^3
12 a 7389.03 m^2 b 1437.28 mm^2 c 104.33 cm^2
13 a $48\,490.48\text{ m}^3$ b 4064.44 mm^3 c 109.93 cm^3
14 123.75 L 15 $11\,830\text{ cm}^3$

PRACTICE SET 3

1 a 12.5 h b 1.35 L c 0.512 GB
d 3005 g e 2700 m f 250 kB
2 a \$551.60 b 320
3 a \$87 420 b \$18 878.50
4 a

Shoe size	Frequency
5	5
6	8
7	6
8	5
9	3
10	2
11	1

b

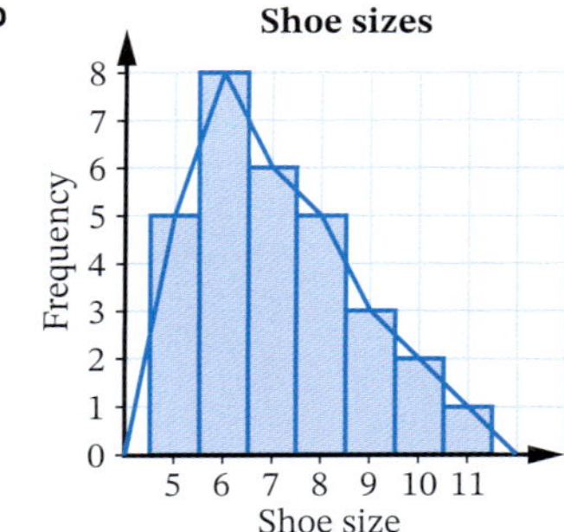

c 63.3% d mode = 6, range = 6
e 7 f 7.1
5 1.5%
6 a \$1720.20
b \$3440.40
c \$7480
7 a \$1653 b \$1157.10 c \$3884.55
8 0.08kg (or 80g)

8 **a** **i** 46, 48 **ii** 48.6 **iii** 48 **iv** 6
b **i** 70 **ii** 70.7 **iii** 70 **iv** 10

10 **a** numerical, discrete
b categorical
c numerical, continuous
d numerical, continuous
e numerical, discrete
f categorical

11 **a** Perimeter = 72 cm, Area = 212 cm^2
b Perimeter = 66 cm, Area = 231 cm^2

12 **a** $2028.50 **b** 66.8%

13 C **14** $1589.20

15 125 g

16 **a** column graph, pie chart, picture graph
b working with categorical data
c teacher to check

17 **a** 200 mm^2 **b** 240 mm^2 **c** 30 cm^2

18 **a** census **b** sample **c** sample

19 $17 468

20 **a** **i** 20.6 m **ii** 25.1 m^2
b **i** 20.5 m **ii** 26.2 m^2
c **i** 28.6 m **ii** 44.6 m^2

21 **a** Positively skewed **b** At 2, 3
c Mode = 3, range = 8

22 **a** 158 m^2 **b** 360 cm^2 **c** 4086 cm^2

23 **a** 120 m^3 **b** 300 cm^3 **c** 15 390 cm^3

24 **a** 'People at a park' does not represent everybody, as employed people are less likely to be at the park.
b Not very specific as they are different meanings and categories of being employed, information obtained as responses will just be Yes or No.
c Which of the following applies to you?
- ❏ going to school/TAFE/University
- ❏ working part-time
- ❏ working full-time
- ❏ retired
- ❏ carer for children and/or other person
- ❏ on unemployment benefits

(Other answers possible)

25 24 740 mm^3 **26** 4010 mm^2

27 **a** Mon: positively skewed Wed: negatively skewed
b Mon: 8.79 Wed: 9.67
c Mon: 8.5 Wed: 10
d Monday
e Wednesday

28 10.1 cm

29 1 manager, 7 sales assistants and 2 cleaning staff

30 $2181.20

31 **a** $\bar{x}$ = $32.95, median = $32.50
b $48.90
c $\bar{x}$ = $31.50, median = $32.50
d If the outlier is included, the mean increases and the median is the same.

CHAPTER 11

SkillCheck

1 $Z(-2, 1)$ $D(3, -2)$ $Q(0, -5)$ $W(5, 0)$ $P(-4, -3)$

2 **a**

x	−2	−1	0	1	2
y	5	3	1	−1	−3

b

x	−2	−1	0	1	2
y	0	$\frac{1}{2}$	1	$\frac{1}{2}$	2

3 **a** 7.8 m **b** 10.8 mm **c** 12.1 m

4 **a** $\frac{6}{5}$ or $1\frac{1}{5}$ **b** $\frac{2}{5}$ **c** $\frac{11}{5}$ or $2\frac{1}{5}$

5 **a** 2 **b** 2 **c** 8.5

Exercise 11.01

1 **a** 7.3 **b** 7.2 **c** 8.2
d 14.3 **e** 7.1 **f** 14.3

2 **a** **i** $\sqrt{45}$ or $3\sqrt{5}$ **ii** 6.7
b **i** $\sqrt{117}$ or $3\sqrt{13}$ **ii** 10.8
c **i** $\sqrt{128}$ or $8\sqrt{2}$ **ii** 11.3
d **i** $\sqrt{68}$ or $2\sqrt{17}$ **ii** 8.2

3 **a** C **b** B

4 **a** 7.071 **b** 5 **c** 6.403 **d** 18.474

5 A

6 **a** $RS = \sqrt{20}$, $ST = \sqrt{8}$, $RT = \sqrt{20}$
b isosceles triangle

7 **a** trapezium **b** $P = 28$ units **c** $A = 38$ units^2

8 **a** square **b** $\sqrt{52}$ or $2\sqrt{13}$ units
c 28.8 units **d** 52 units^2

Exercise 11.02

1 A

2 **a** (4, 4) **b** (5, 3) **c** (1, 1)

3 **a** (8, 7) **b** (7, 8) **c** $\left(4\frac{1}{2}, 4\right)$
d (2, 3) **e** $\left(4, 4\frac{1}{2}\right)$ **f** $\left(7, \frac{1}{2}\right)$
g $\left(\frac{1}{2}, 3\frac{1}{2}\right)$ **h** $\left(-3, -\frac{1}{2}\right)$ **i** $\left(-1, -6\frac{1}{2}\right)$
j (−1, −1) **k** (7, −8) **l** $\left(-5\frac{1}{2}, -6\right)$

4 **a** $P(2, 3)$ **b** $Q(-1, 1)$ **c** $R(5, 5)$

5 D

6 **a** **i** $\left(-\frac{1}{2}, -1\right)$ **ii** $\left(-\frac{1}{2}, -1\right)$
b They are the same point. The diagonals bisect each other.

7 **a** $M\left(\frac{1}{2}, 4\frac{1}{2}\right)$
b **i** $\sqrt{50}$ or $5\sqrt{2}$ **ii** $\sqrt{4.5}$ **iii** $\sqrt{17}$
c **i** 7.5 units^2 **ii** $P = 15.3$ units

Exercise 11.03

1 **a** + **b** − **c** −
d − **e** + **f** +
g + **h** − **i** +

2 **a** 2 **b** $-\frac{1}{4}$ **c** −5
d −1 **e** $\frac{3}{7}$ **f** $\frac{7}{3}$
g $\frac{2}{3}$ **h** $-\frac{8}{7}$ **i** $\frac{2}{5}$

3 **a** $-\frac{1}{2}$ **b** 5 **c** $\frac{2}{3}$ **d** 7
e $-\frac{1}{6}$ **f** $\frac{5}{2}$ **g** $-\frac{7}{2}$

4 C

5 **a** 1 **b** $\frac{1}{3}$ **c** −5
d $-\frac{1}{2}$ **e** 0 **f** $-\frac{1}{2}$
g $\frac{7}{4}$ **h** $\frac{1}{2}$ **i** 0

6 A **7** $k: \frac{1}{5}, l: -\frac{1}{2}$ **8** B

9 **a** $m_{PQ} = -\frac{1}{2}, m_{QR} = \frac{5}{2}, m_{RS} = -\frac{1}{2}, m_{PS} = \frac{5}{2}$
b QR and PS, PQ and RS. They are parallel.
c parallelogram

Exercise 11.04

1 **a**

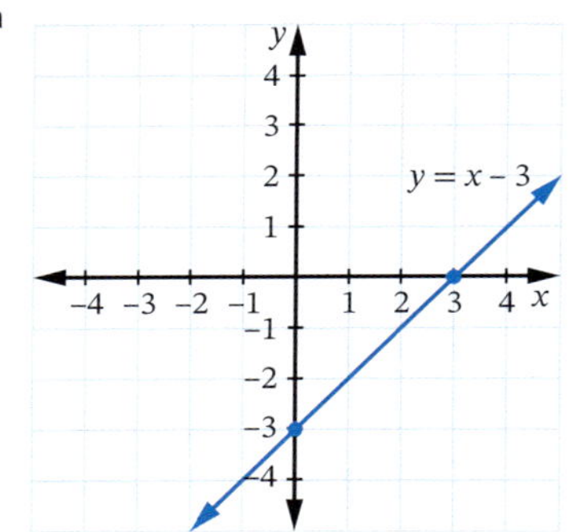

 **i** 3  **ii** −3

b

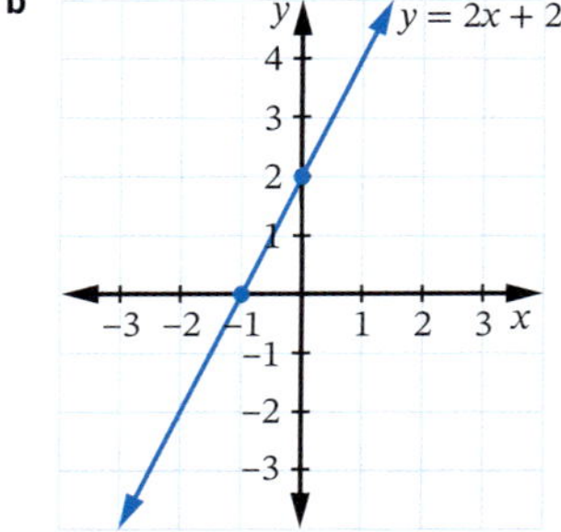

i −1 **ii** 2

c

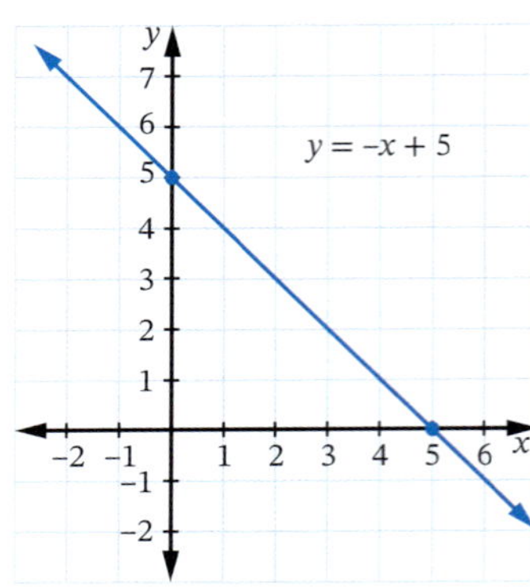

i 5 **ii** 5

d

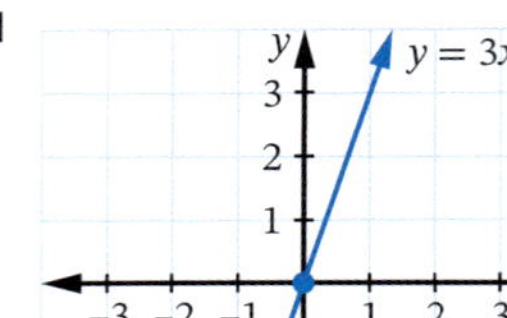

i 0 **ii** 0

e

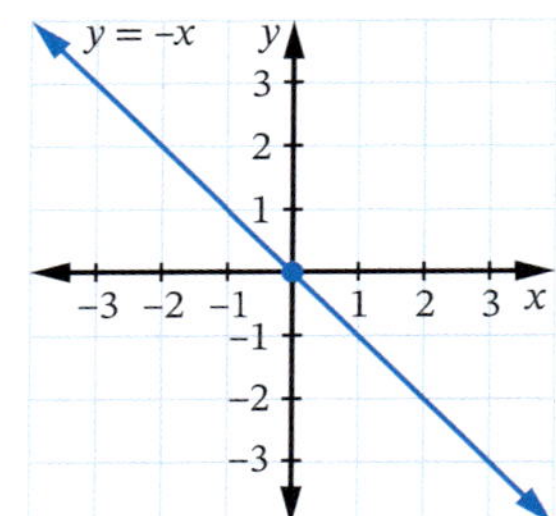

i 0 **ii** 0

f

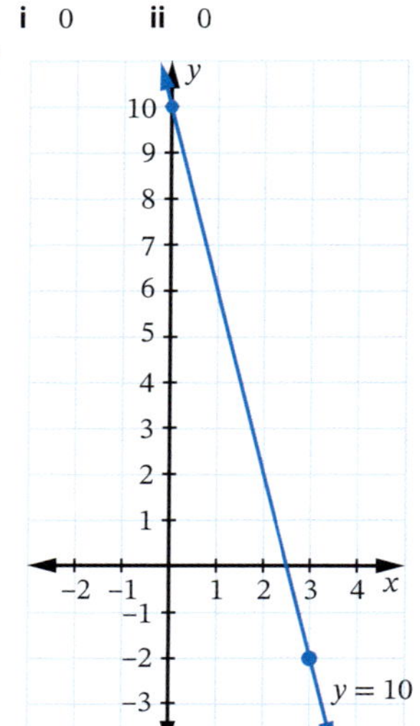

i $2\frac{1}{2}$ **ii** 10

g

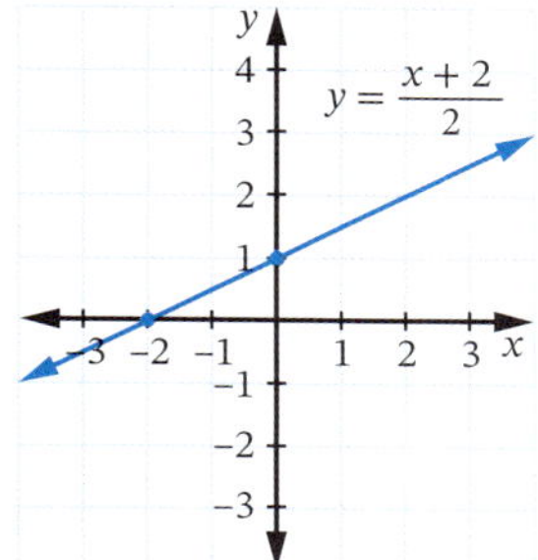

i −2 **ii** 1

h 

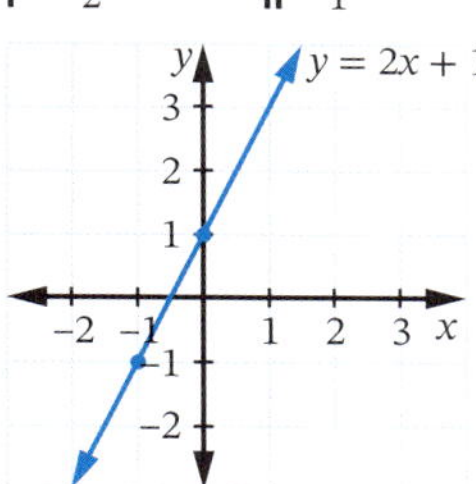

i $-\frac{1}{2}$ **ii** 1

i

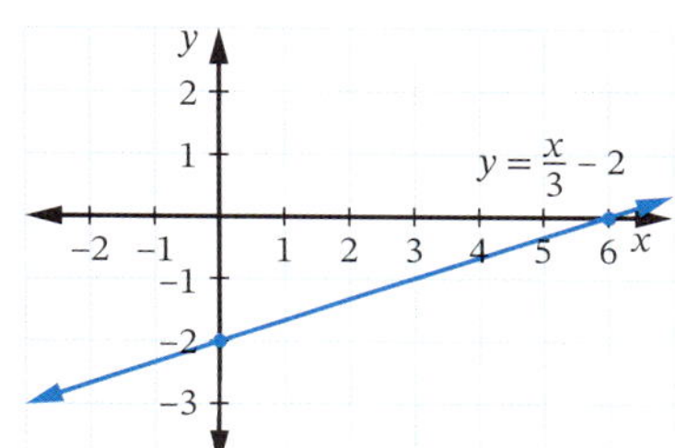

i 6 **ii** −2

j

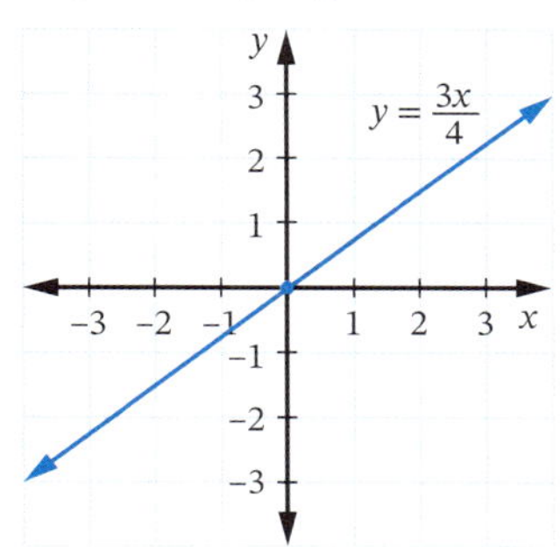

i 0 **ii** 0

k

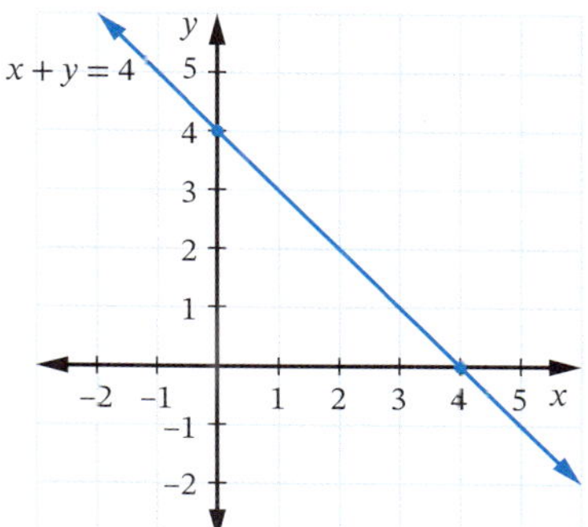

i 4 **ii** 4

l

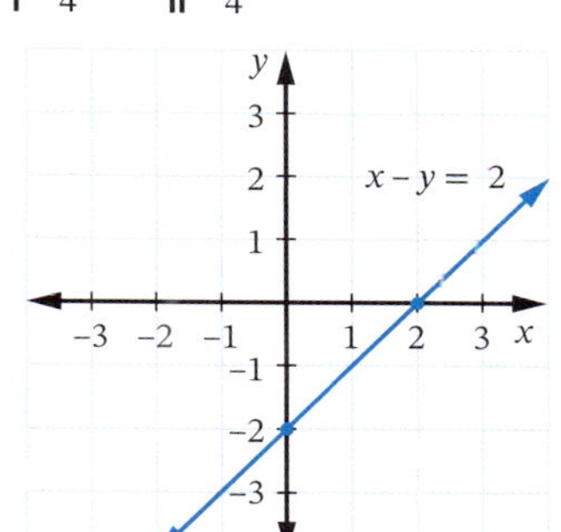

i 2 **ii** −2

2 B

3 **a** yes **b** no **c** yes
d yes **e** yes **f** no

4 C

5 **a**

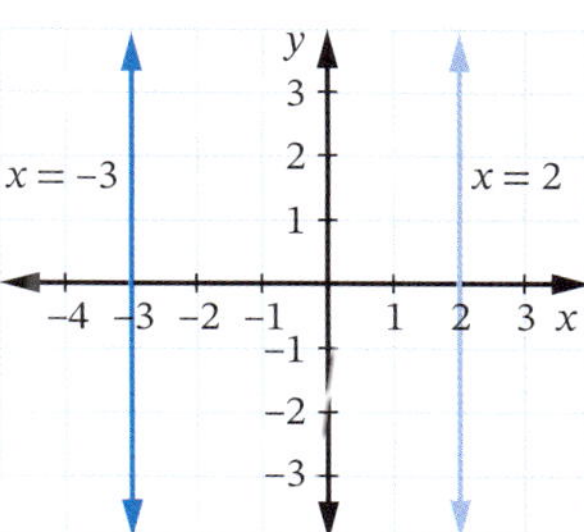

b

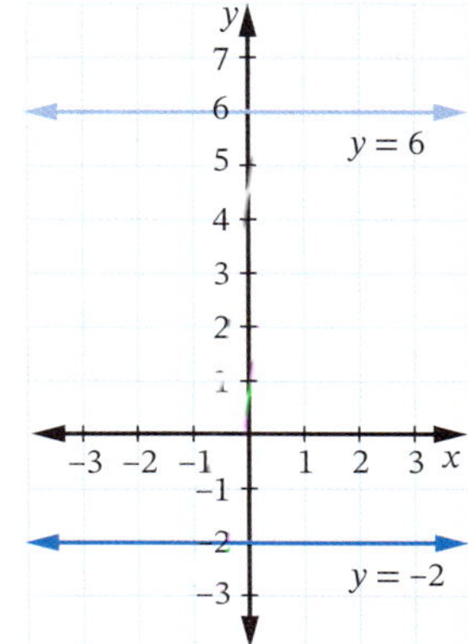

c

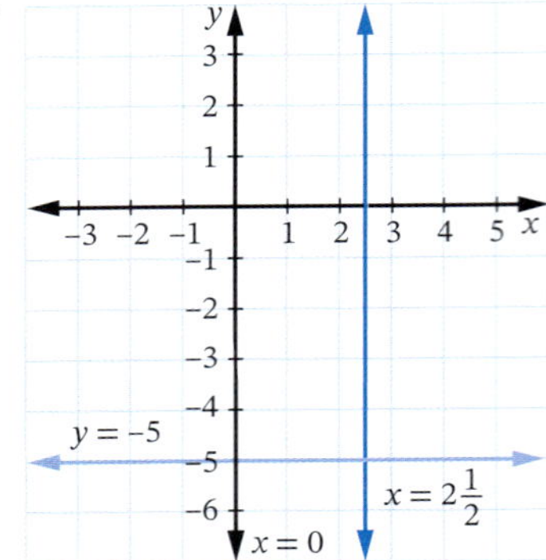

d

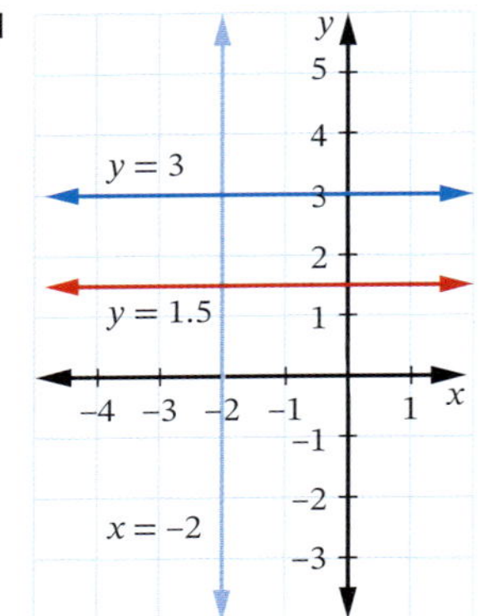

6 **a** $y = 6$ **b** $x = -4$ **c** $y = 3$ **d** $x = 5$
e $y = -5$ **f** $x = 3$ **g** $y = -3$ **h** $x = -1$

7 **a** LHS = RHS = 3 for $x = 2y - 1$; LHS = RHS = 5 for $x + y = 5$
b (3, 2)

8 LHS = RHS = −12 for $y = -3x + 6$;
LHS = RHS = −12 for $y = -2x$

9 **a** $y = 6$ **b** $x = -5$ **c** $y = 2$ **d** $x = -5$
e $y = -2$ **f** $y = -1$ **g** $x = -1$

10 **a** $x = 0$ **b** $y = 0$

Mental skills 11

2 **a** 2, 5 **b** 3 **c** 2, 3, 6 **d** 3, 5
e 2 **f** 2, 3, 5, 6 **g** 5 **h** 3

3 **a** 4, 9 **b** 4, 10 **c** 9 **d** 10
e 4, 9 **f** None **g** 9 **h** 4, 9, 10

Exercise 11.05

1 **a** $m = 3, c = -2$ **b** $m = 2, c = 7$
c $m = 1, c = 4$ **d** $m = -1, c = -9$
e $m = \frac{3}{4}, c = 6$ **f** $m = 1, c = 0$
g $m = \frac{1}{2}, c = -11$ **h** $m = \frac{2}{3}, c = 6$
i $m = -2, c = 8$ **j** $m = -1, c = 5$
k $m = \frac{1}{4}, c = 0$ **l** $m = -3, c = 11$
m $m = 0, c = 5$ **n** $m = \frac{2}{5}, c = 1$
o $m = 1, c = -\frac{7}{2}$ **p** $m = \frac{3}{4}, c = 1$

2 D

3 **a** $y = 2x - 1$ **b** $y = \frac{3}{4}x + 2$
c $y = -3x + 5$ **d** $y = -\frac{2}{5}x + 3$

4 C

5 **a**

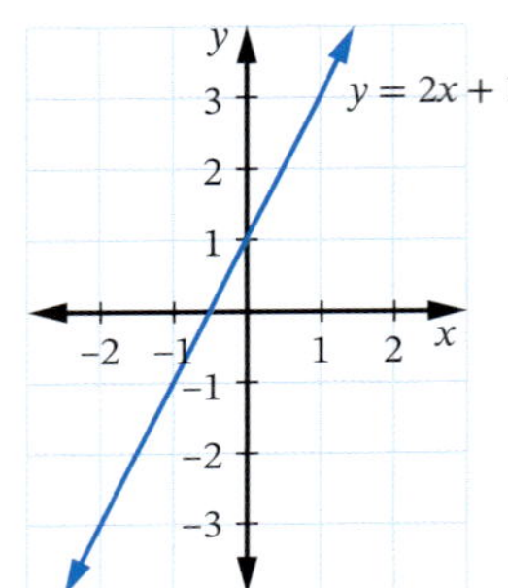

b

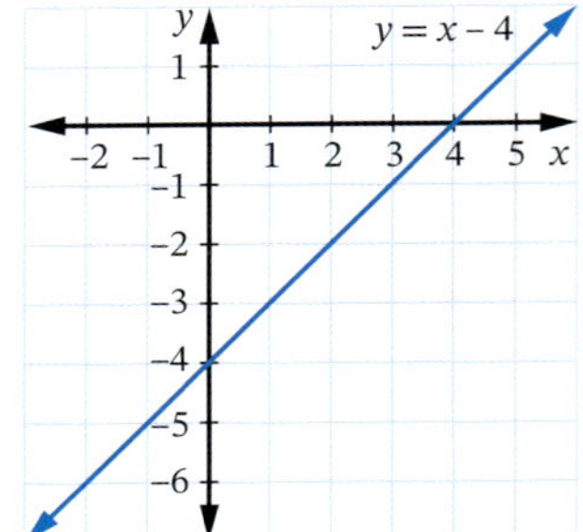

c

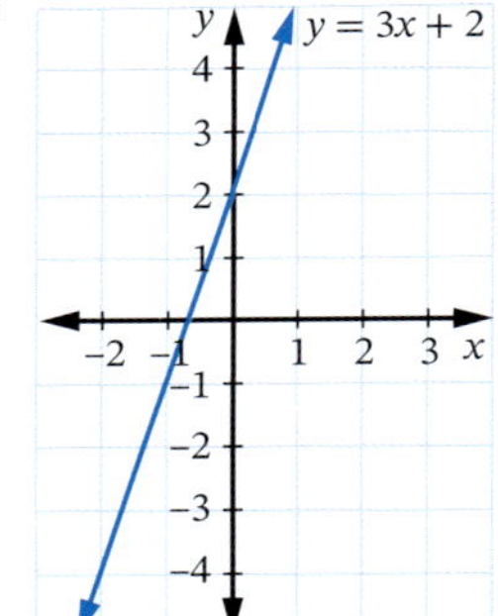

d

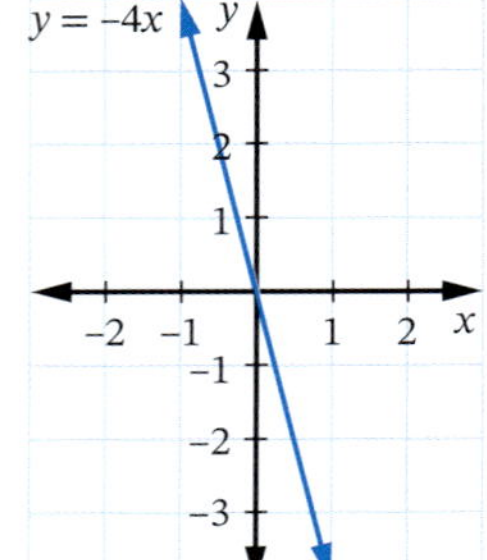

e

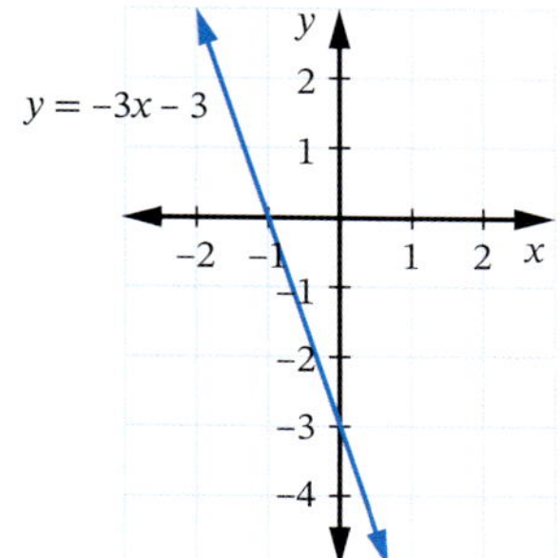

f

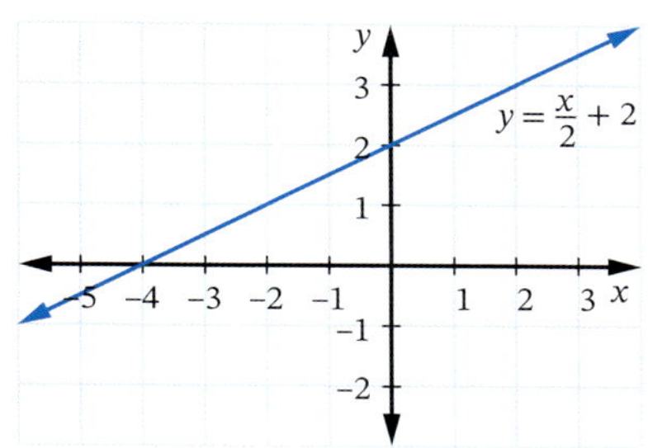

g

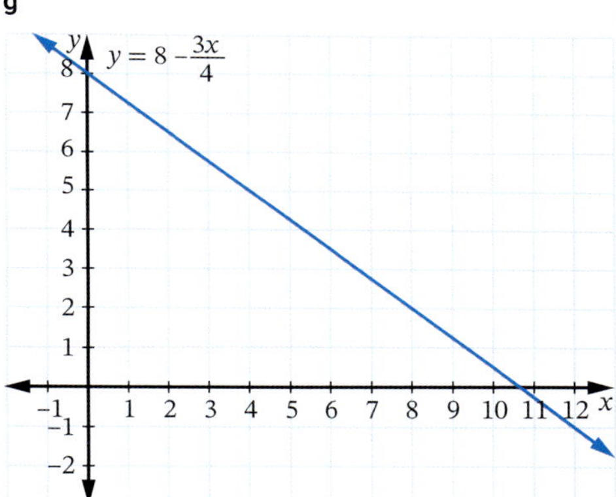

h

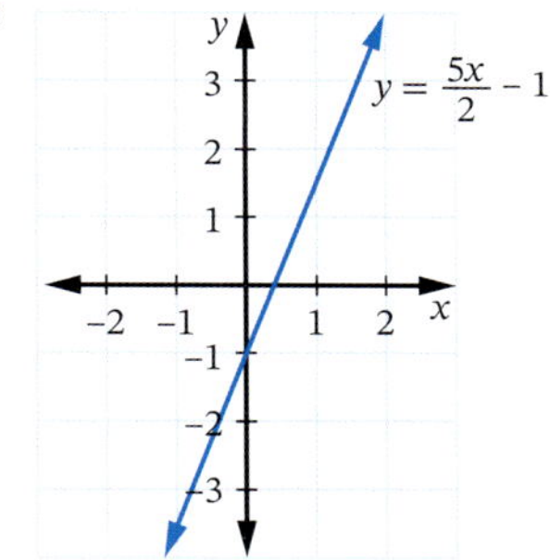

6 B

Exercise 11.06

1 **a** $y = x + 4$ **b** $y = \frac{3x}{2}$ **c** $y = \frac{x}{3} - 5$

d $y = \frac{-3x}{2} + 7$ **e** $y = -\frac{x}{3} + 3$ **f** $y = -2x - 6$

2 **a** $y = \frac{x}{2} + 2$ **b** $y = x$ **c** $y = -\frac{x}{2} + 5$

d $y = -\frac{x}{2} + 3$ **e** $y = -3x - 3$ **f** $y = -x - 2$

g $y = 3x - 10$ **h** $y = \frac{2x}{5} + 2$ **i** $y = 2x - 3$

3 **a** D **b** B **c** A **d** C

Exercise 11.07

1 **a** $x = 3$ **b** $x = -2$ **c** $x = 1\frac{1}{2}$

2 **a**

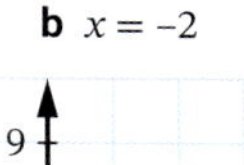

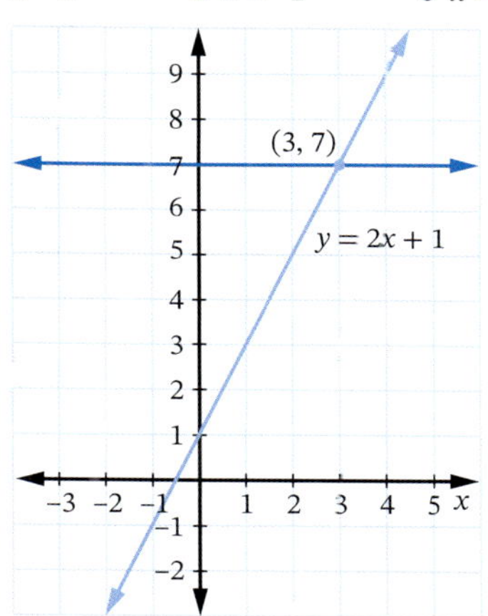

$x = 3$

b $x = 4.5$ **c** $x = -3$

3 **a** $x = 3$ **b** $x = 4\frac{1}{2}$ **c** $x = -3$

4 B

5 **a**

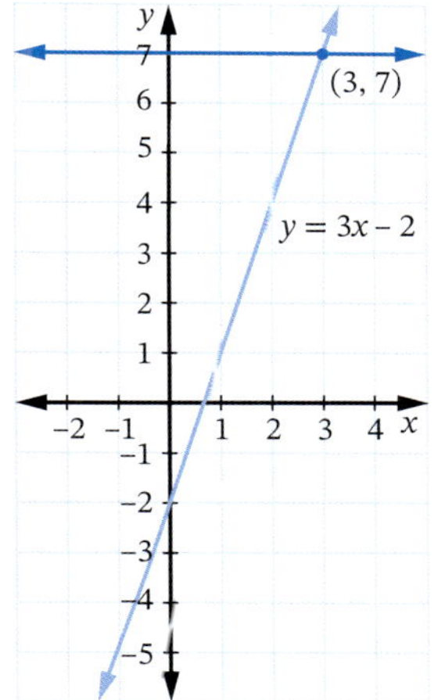

$x = 3$

b $x = -3$ **c** $x = 0$

6 a

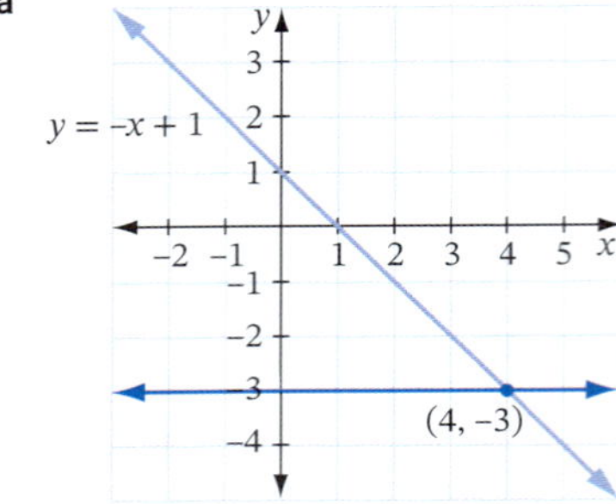

$x = 4$

b $x = -3$ c $x = 2$

7 a

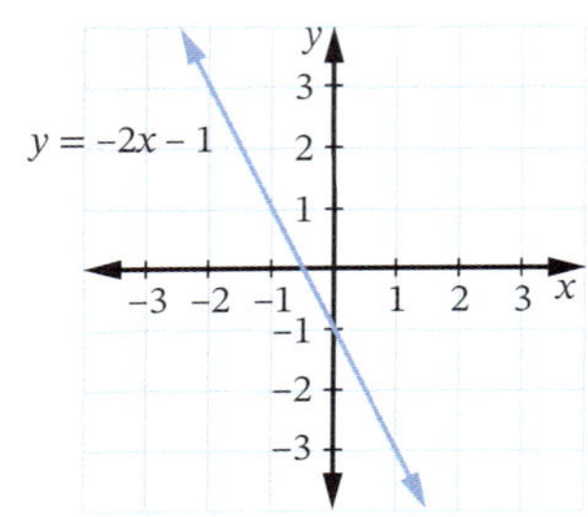

$x = 1$

b $x = -0.5$ c $x = -2$

8 a

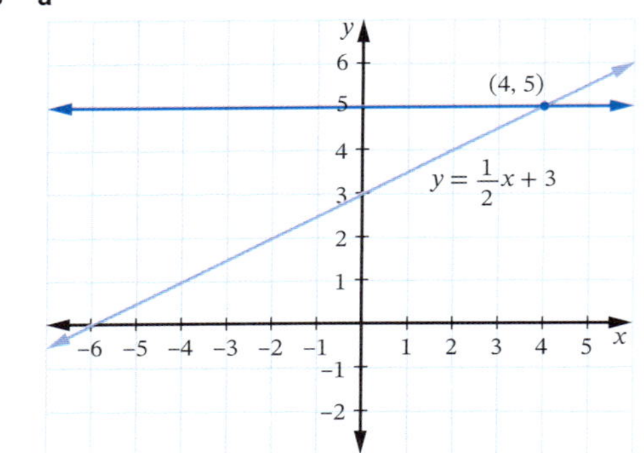

$x = 4$

b $x = -4$ c $x = -2$

Exercise 11.08

1 a

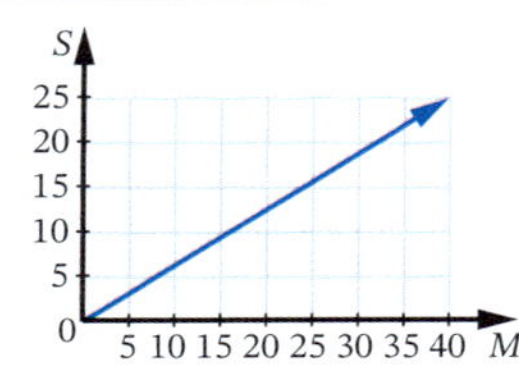

b $k = 0.625$, $S = 0.625\,M$ c 22.5 cm d 64 kg

2 C 3 $k = 5$, $Y = 115$

4 a $M = 1.75V$ b 710.5 g

c i 16 cm^3 ii 205.7 cm^3

5 $n = 2.5m$ 6 $p = 108$

7 a 11.25 km b 2 h 8 min

8 A

9 a $I = 0.025D$ b $62.50 c $125

10 a $V = 64\,t$ b 512 L c 15 min

11 a 22.8 kg b 84.1 kg

12 a 235.2 kPa b 141.7 m

Exercise 11.09

1 a $y = -750x + 12\,000$

b

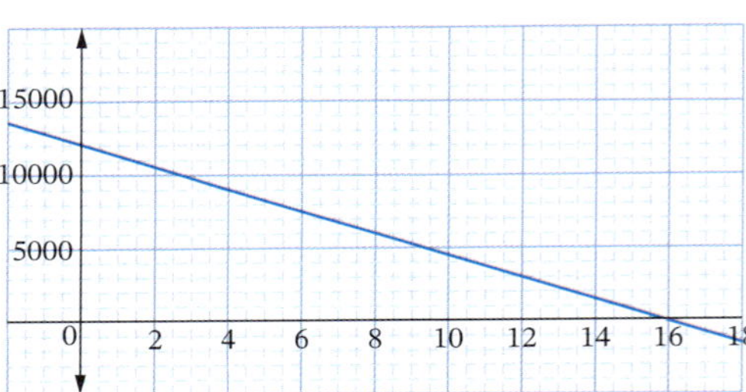

c $8250

d $y = -750 \times 5 + 12\,000$; $y = \$8250$

e 16 years

2 a $y = 1.5x + 6$

b

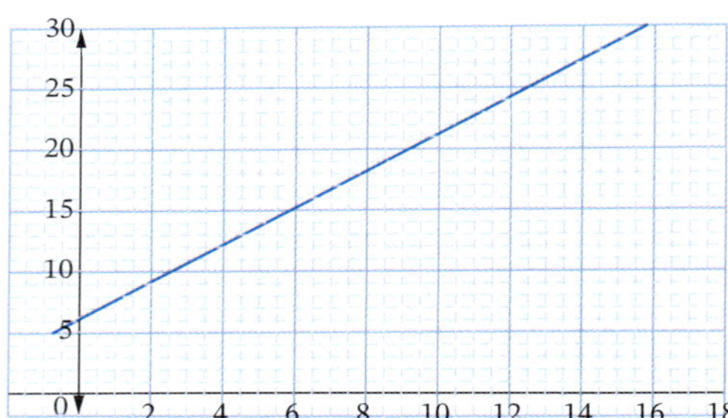

c increase in height = 18 – 6 = 12cm

d $y = 1.5 \times 13 + 6$; $y = 25.5$cm

3 a approximately 77 °F

b $y = 1.8x + 32$

c 77 °F

d Teacher to check.

4 **a** $y = -150x + 5000$

b

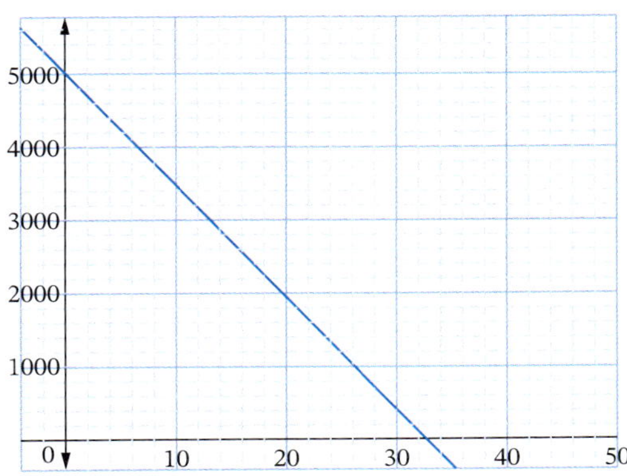

c 2700 L

d $y = -150 \times 15 + 5000$; $y = 2750$ L

e Teacher to check.

5 **a** $y = 25x + 150$

b

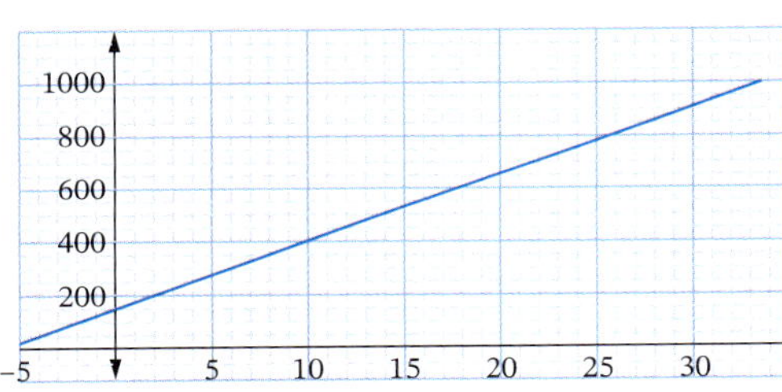

c \$300 **d** 14 weeks

6 **a** $y = 12.5x + 750$

b

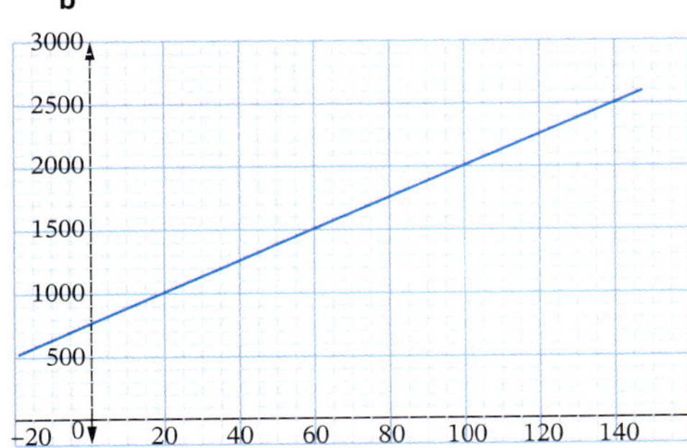

c \$1750 **d** 116 m^2 of flooring was tiled

7 **a** $y = -1.6x + 98$

b

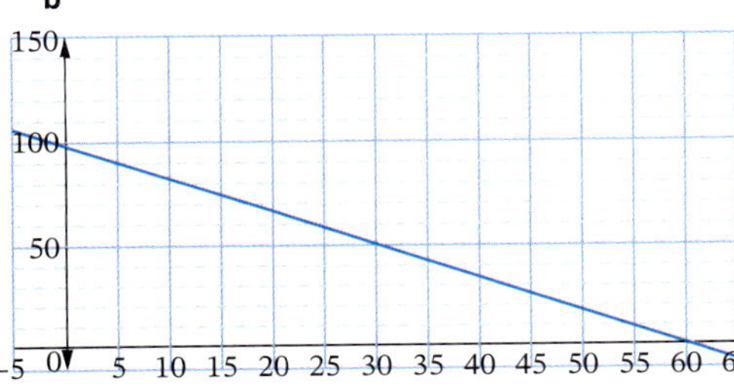

c approximately 87 °C

d approximately 45 minutes

8

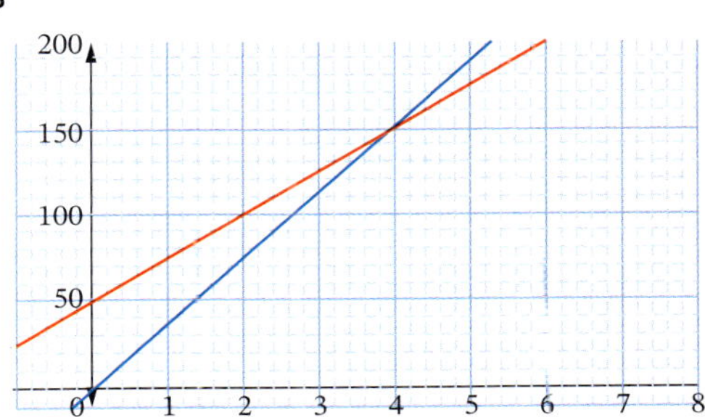

Both companies charge the same amount of \$150 for 4 hours of hire.

9 Equation of function based on graph is $y = 5x + 30$. If $y = \$160$ then $x = 26$ weeks. She can buy the speaker 27 weeks after the 1st of December. On approximately the 8th of June 2022.

10 Call out fee is y-intercept and hourly rate is gradient. Graph the function using the two co-ordinates provided:

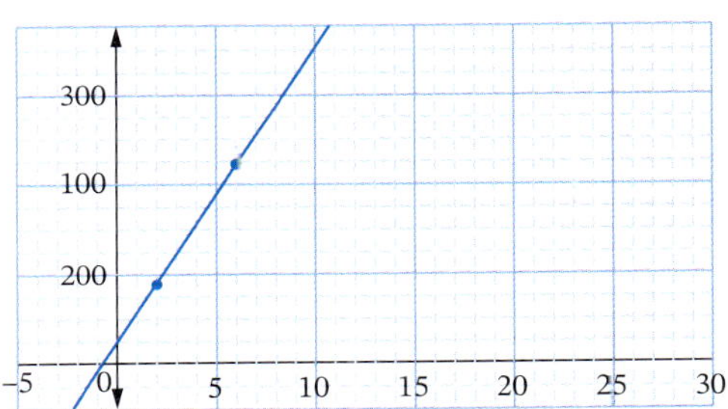

y-intercept = (0,50) and gradient = 65, therefore the call out fee is \$50 and the hourly rate is \$65.

Exercise 11.10

1 **a, b**

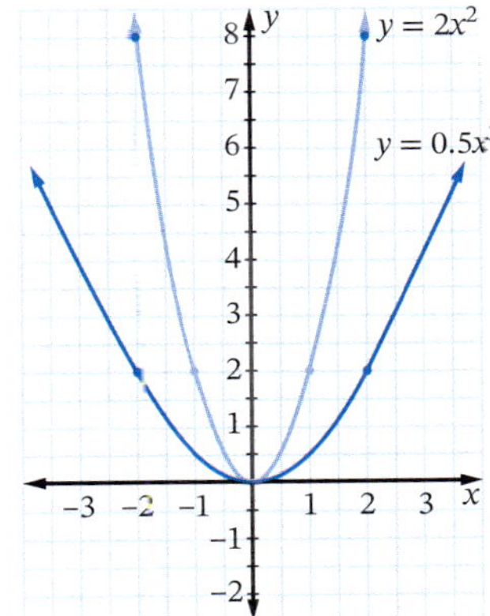

c Same: both concave up, vertex (0, 0)
Different: $y = 2x^2$ is narrower than $y = \frac{1}{2}x^2$

2 D

3 **a**

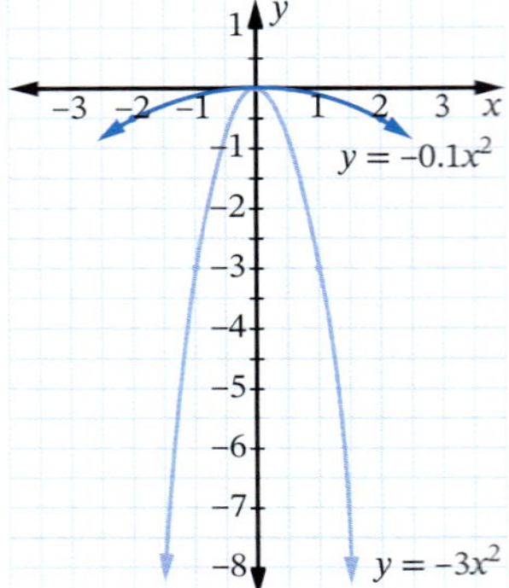

b Same: concave down, vertex (0, 0)

Different: $y = -3x^2$ narrower than $y = -0.1x^2$

4 **a** **i** vertex (0, 0)

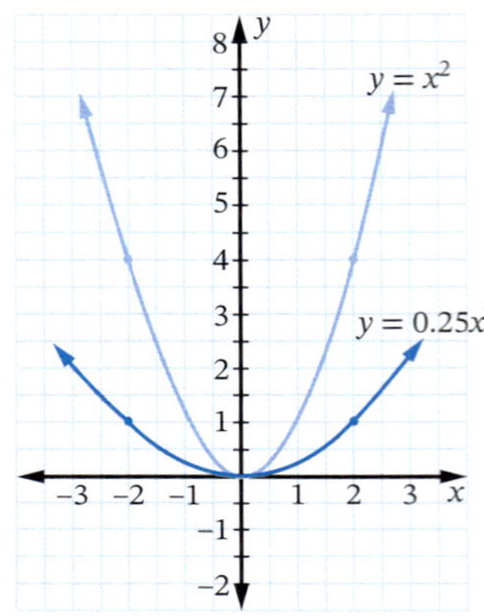

ii Both concave up. **iii** $y = \frac{1}{4}x^2$ **iv** (0, 0)

b **i** vertex (0, 0)

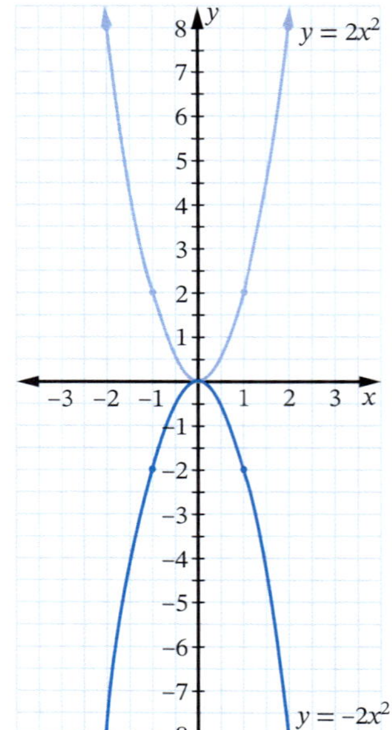

ii $y = 2x^2$ concave up, $y = -2x^2$ concave down

iii same width

iv (0, 0)

c **i** vertex (0, 0)

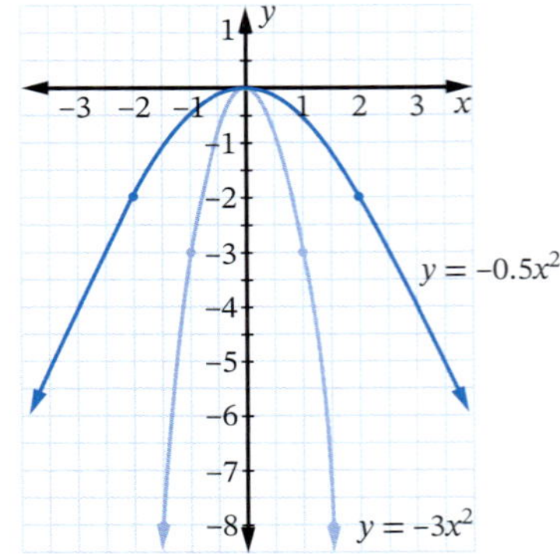

ii both concave down

iii $y = -\frac{1}{2}x^2$ wider

iv (0, 0)

d **i** vertex (0, 0)

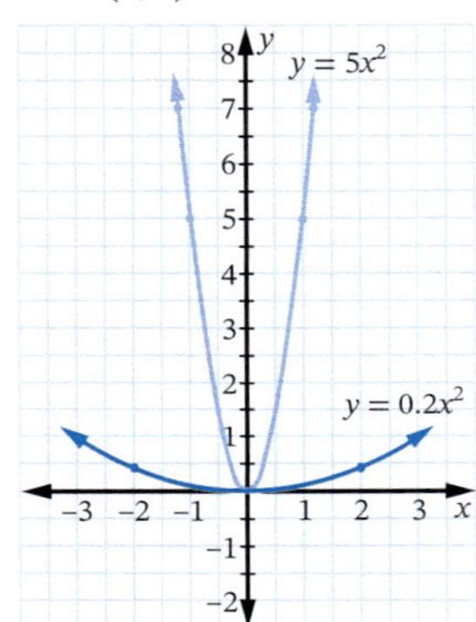

ii both concave up

iii $y = 0.2x^2$ is wider

iv (0, 0)

5 C

6 **a** **i** $y = x^2 + 1$, vertex (0, 1); $y = x^2 + 3$, vertex (0, 3)

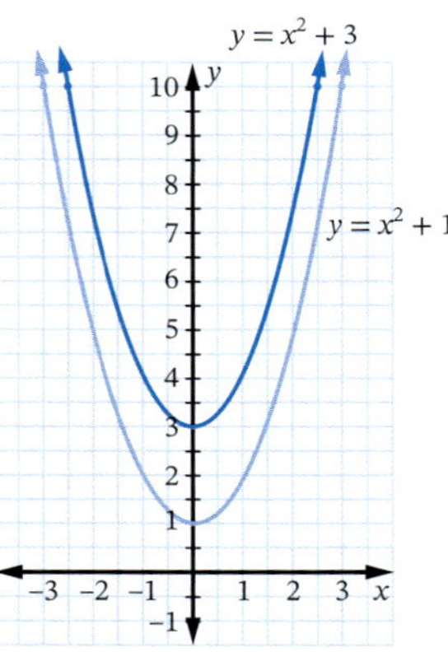

ii both concave up **iii** both same width

iv $y = x^2 + 1$, y-int: 1; $y = x^2 + 3$, y-int: 3

b **i** $y = -x^2 - 2$, vertex $(0, -2)$; $y = 2x^2 + 4$, vertex $(0, 4)$

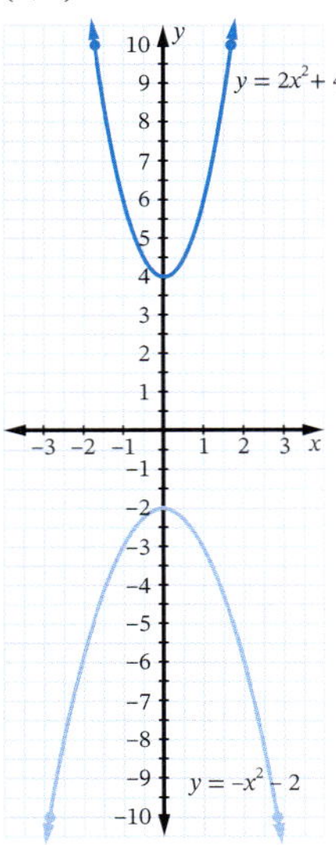

ii $y = 2x^2 + 4$ concave up; $y = -x^2 - 2$ concave down

iii $y = -x^2 - 2$

iv $y = -x^2 - 2$, y-int: -2; $y = 2x^2 + 4$, y-int: 4

c **i**

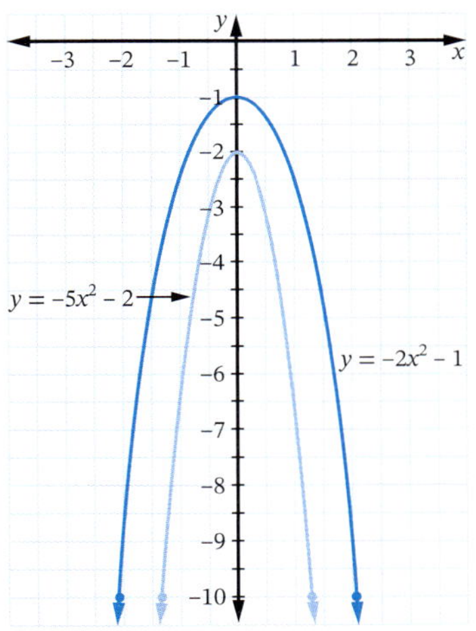

$y = -5x^2 - 2$, vertex $(0, -2)$; $y = -2x^2 - 1$, vertex $(0, -1)$

ii Both concave down.

iii $y = -2x^2 - 1$

iv $y = -5x^2 - 2$, y-int: -2; $y = -2x^2 - 1$, y-int: -1

d **i** $y = -3x^2 - 4$, vertex $(0, -4)$; $y = 2x^2 + 2$, vertex $(0, 2)$

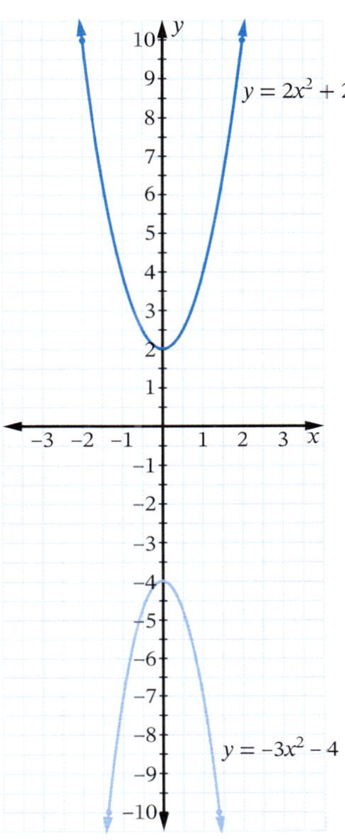

ii $y = 2x^2 + 2$ concave up; $y = -3x^2 - 4$ concave down

iii $y = 2x^2 + 2$

iv $y = 2x^2 + 2$, y-int: 2; $y = -3x^2 - 4$, y-int: -4

7 C

8 **a** D **b** B **c** A **d** C

Exercise 11.11

1 **a** Vertex $(-1, -4)$

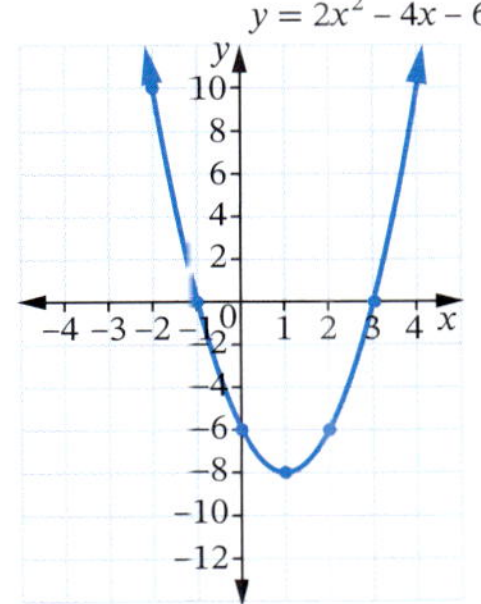

b Vertex (–1, 0)

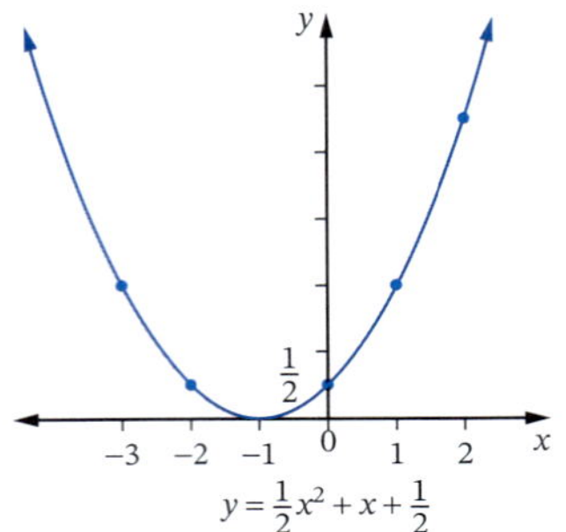

c Vertex (1, 3)

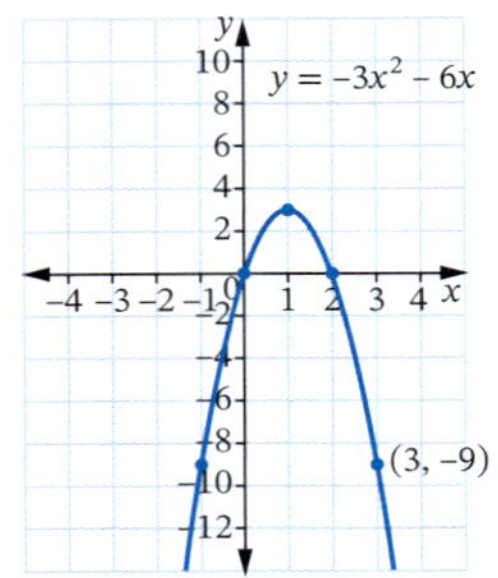

2 **a** Vertex (1, –4), x-intercepts –1 and 3, y-intercept 3.

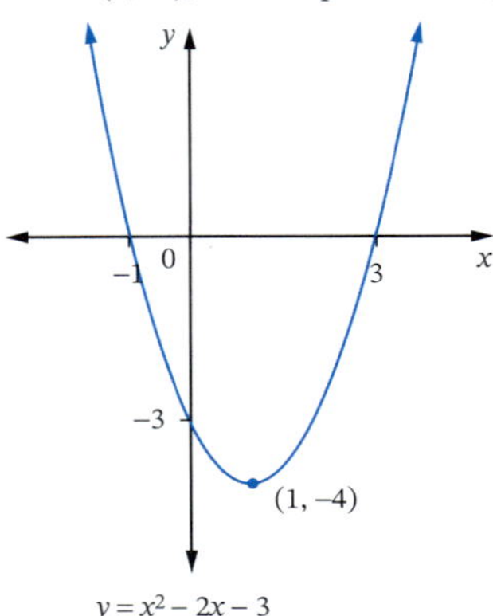

b Vertex $\left(4\frac{1}{2}, 12\frac{1}{4}\right)$, x-intercepts 1 and 8, y-intercept -8.

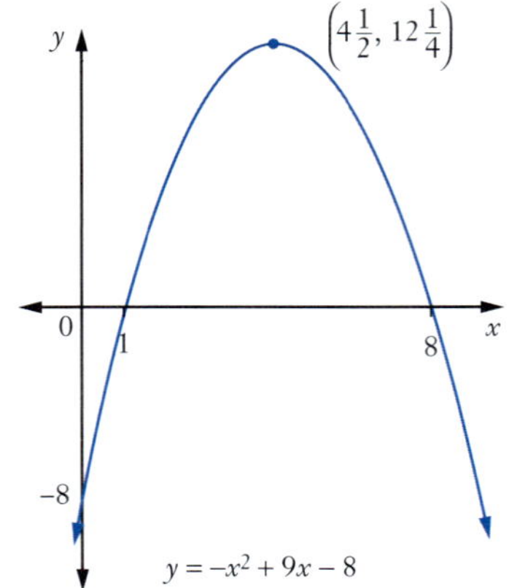

c Vertex **(1, 3)**, no x-intercepts, y-intercept 5.

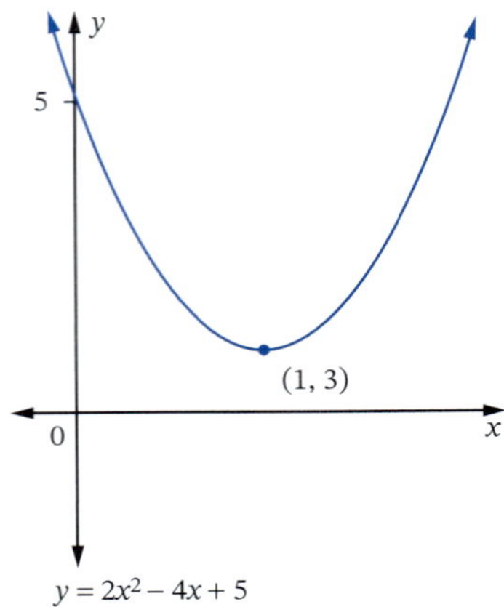

3 **a** x-intercepts –3 and 5, y-intercept –15, vertex (1, –16).

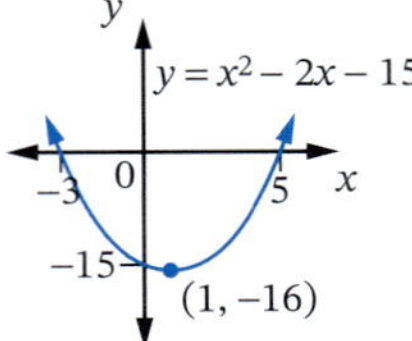

b x-intercepts –1 and 5, y-intercept 5, vertex (2, 9).

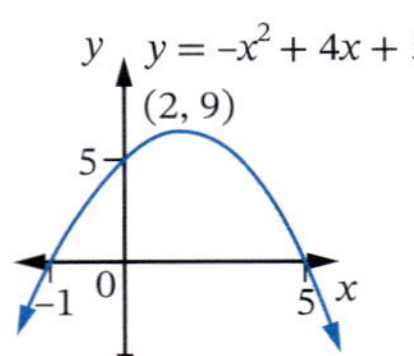

c x-intercepts 0 and 4, y-intercept 0, vertex (2, –4).

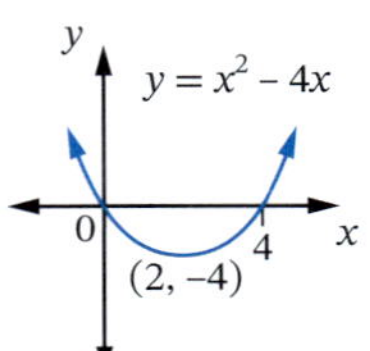

Exercise 11.12

1 **a** The ball is at a height of 9 m at times 1 second and 3 seconds. There are two different times as it reaches this height while going up and is also at this height while coming down.

b Greatest height is 12 m at a time of 2 seconds.

c Ball is in the air for approximately 4 seconds.

2 **a** 20 m is the height of bedroom window above the ground.

b Vertex = (2.25, 40) ball reaches a maximum height of 40 m at a time of 2.25 seconds after throwing.

c Ball hits the ground at approximately 5.4 seconds.

d Travels 20 m upwards

3 **a** (−23, 0) and (23, 0)

b River is approximately 46 m wide.

c The highest point of the arc is 16 m and it aligns with the centre of the river as it is the position of the y-intercept.

4 **a** The y-intercept of the graph. Rocket was launched from a height of approximately 100 m.

b Approximately 55 m. Vertex height is approximately 155 m and rocket was launched from 100 m therefore difference of 55 m is the gain in vertical height.

c Approximately 2 seconds

d Approximately 3 seconds

5 **a** About 2.4 seconds

b Approximately 29 m

c Approximately 4.8 seconds

6 **a**

a The rock is thrown from a height of 35 m

b The rock is in the air for approximately 7.3 to 7.4 seconds

c The rock travels approximately 12.5 m upwards before starting to fall to the ground

7 Most economical speed is 50 km/h with a fuel consumption of 8 L/100 km.

8 **a** 6 a.m. = 14 °C, daily maximum is at vertex = 20 °C, therefore increase in temperature = 6 °C. The maximum temperature is reached at 12:00 midday (6 hours after 6 a.m.).

b This model is inappropriate for 'n' values greater than 13 as it will be 7 p.m. in the afternoon and the temperature will not continue to decrease as the model shows. Once 24 hours pass, the temperature will start to rise again.

9 A ticket price of $25 will get a maximum profit of $1300.

10 The distance between the towers is approximately 1280 m

Exercise 11.13

1 **a**

$y = 5^x$ $y = 3^x$ $y = 2^x$

b 1

c becomes steeper, in 1st quadrant

2 **a**

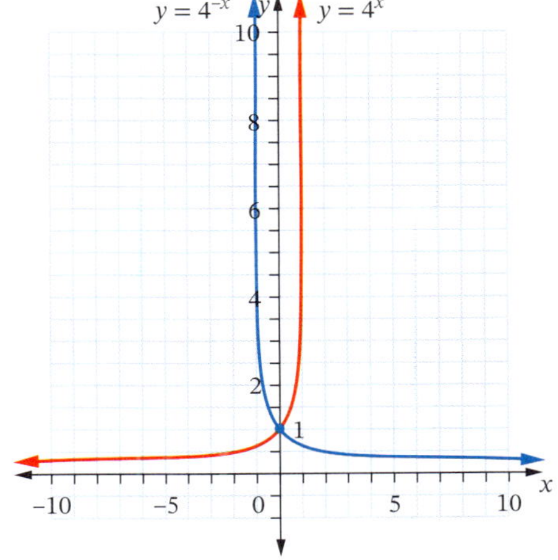

b **i** $y = 4^{-x}$ **ii** $y = a^{-x}$

3 B

4 **a**

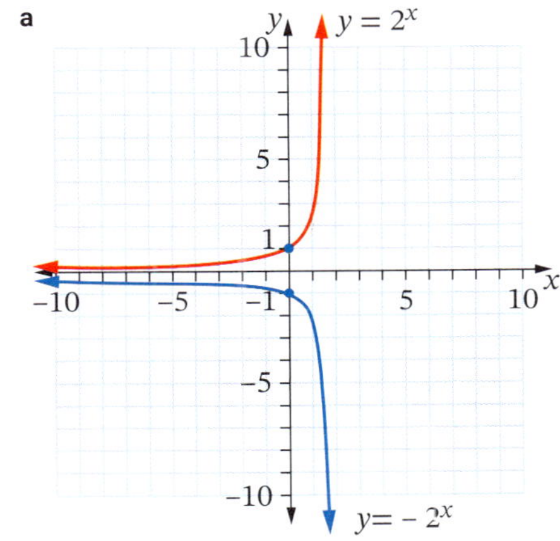

b one is the reflection of the other in the x-axis

c $y = -a^x$

5 Same shape, shifted down 2 units.

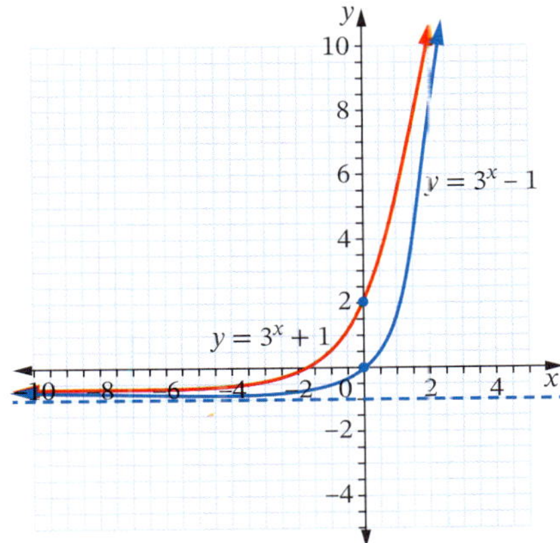

6 **a**

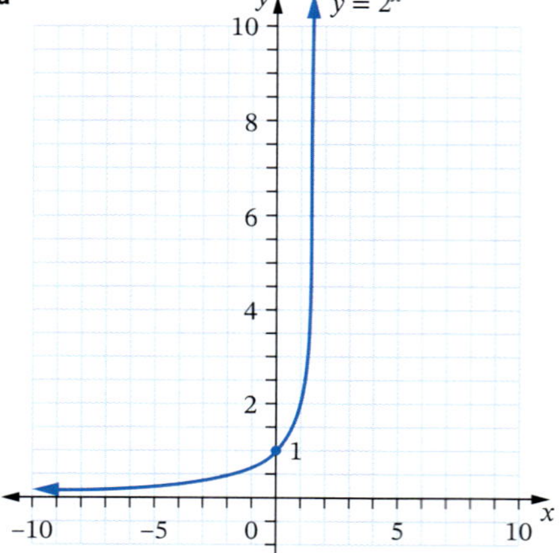

b

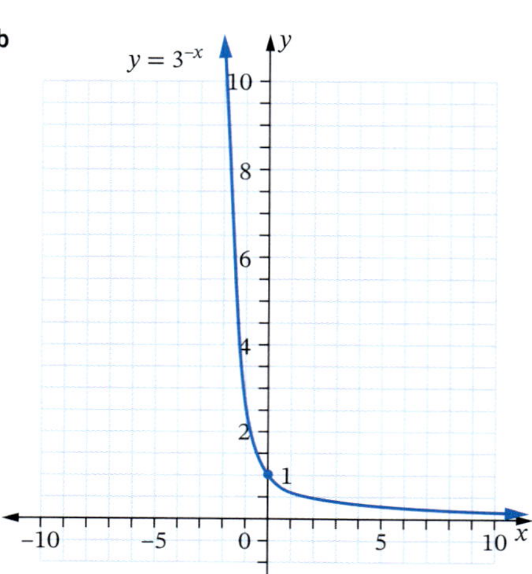

c

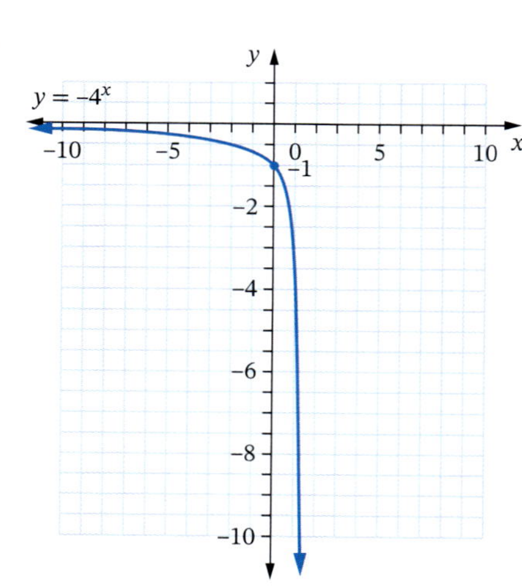

d

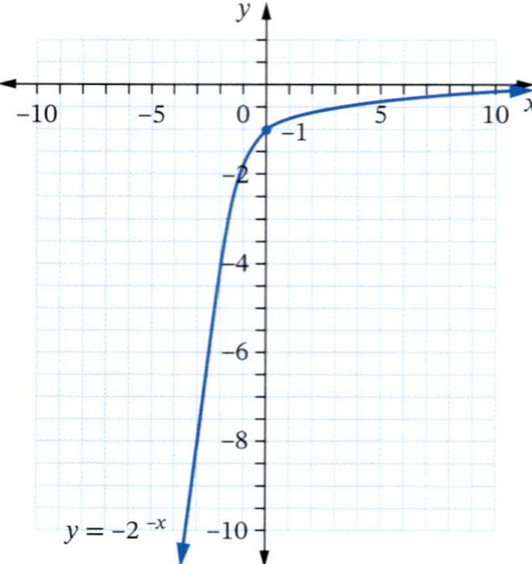

e

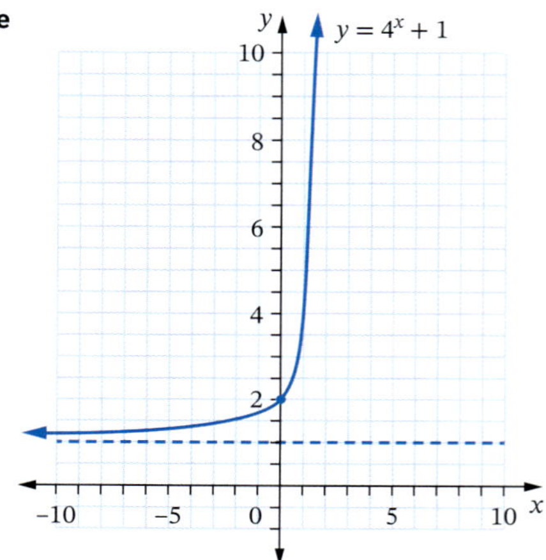

f

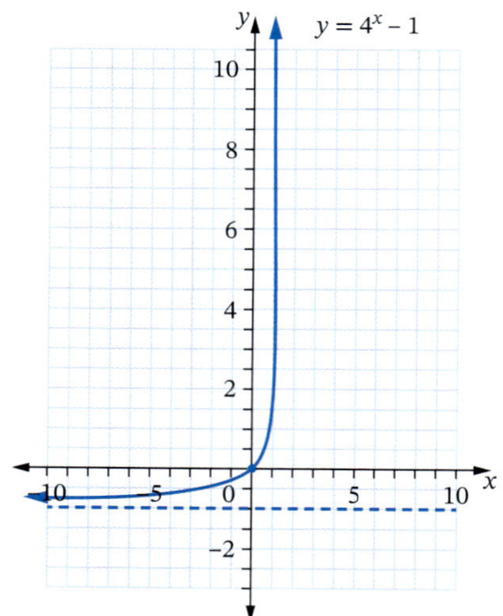

7 $y = 4^{-x}$

Power plus

1 $D(-3, 5)$

2 **a** $y = \frac{2}{3}x - 2$ **b** 4

3 $Y(5, 4)$

4 **a** $r = \sqrt{73}$

b $r = \sqrt{73} \approx 8.54$; $10 > 8.54$, $\therefore P$ lies outside the circle as it is further from origin

5 $RS = 13$, $ST = 13$, $RT = \sqrt{208}$ $\therefore RS = ST = 13$ units $\therefore$ isosceles triangle

6 $(5, 4), (8, 1), (5, -2)$

TEST YOURSELF 11

1 **a** $\sqrt{97}$ **b** 9.8

2 **a** $(3, 7)$ **b** $(-4, 4)$ **c** $(-1, 2)$

3 **a** **i** MN, CD, GH **ii** KL, IJ, AB **iii** EF

b **i** $\frac{1}{7}$ **ii** $-\frac{5}{4}$ **iii** $-\frac{1}{4}$

c $\left(9, -\frac{1}{2}\right)$ **d** $\sqrt{13}$

4 **a** $m = -2$ **b** $m = -2$ **c** $m = -\frac{1}{3}$

5

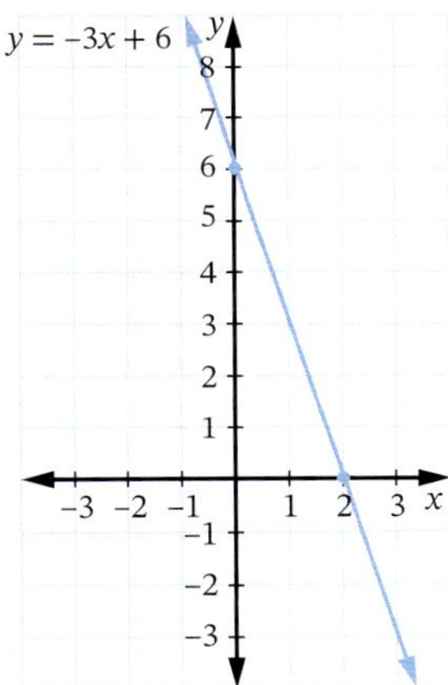

a 2 **b** 6

6 **a** yes **b** yes **c** yes

7 **a**

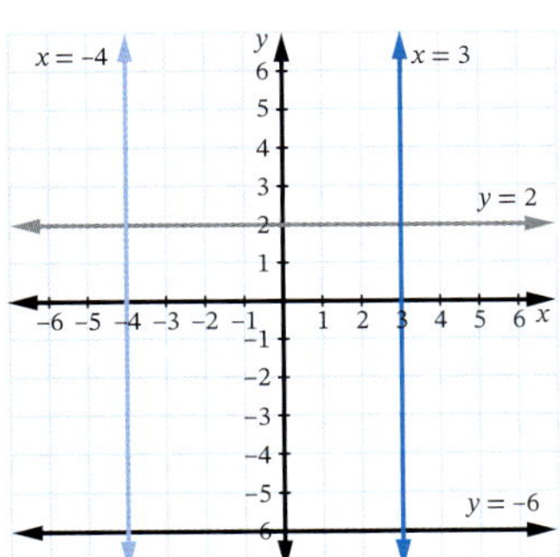

b 56 units^2

8 **a** $y = 4$ **b** $x = -1$ **c** $y = 0$

9 **a** $m = 3, c = 1$ **b** $m = -1, c = 8$

c $m = \frac{1}{5}, c = 4$ **d** $m = -6, c = 0$

10 **a**

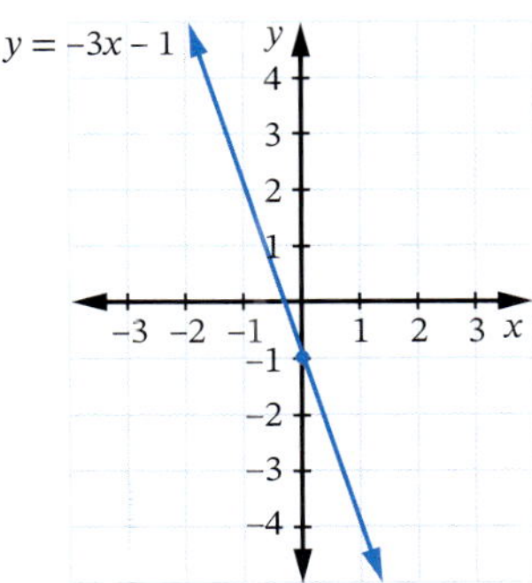

$m = -3, c = -1$

b

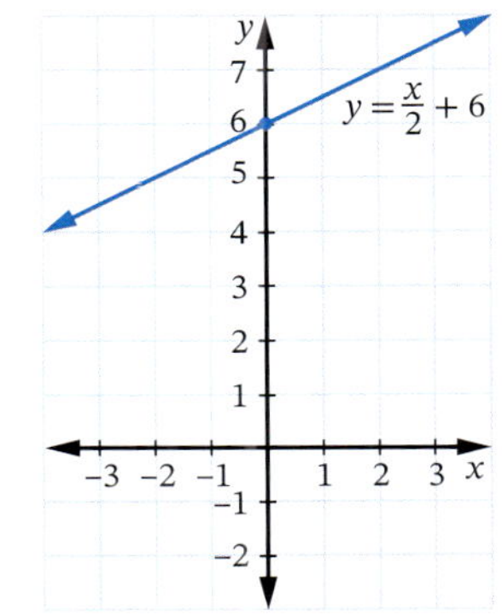

$m = \frac{1}{2}, c = 6$

11 **a** $y = 2x + 1$ **b** $y = -3x + 1$

12 $x = 3$

13 **a** $T = \frac{6}{35}a$ **b** $d = 58$ km

14 **a**

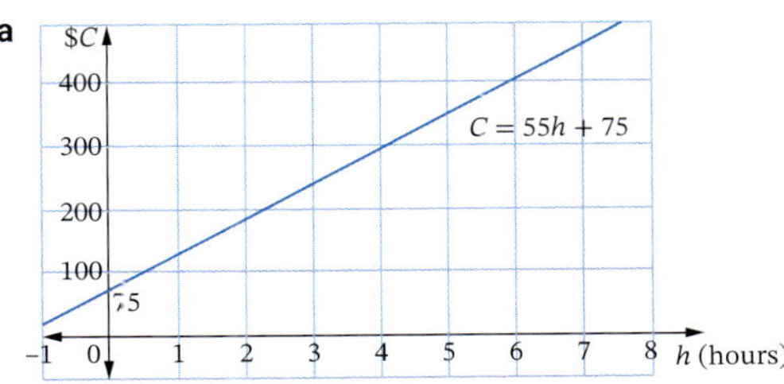

b The electrician worked for approximately 4 hours.

15

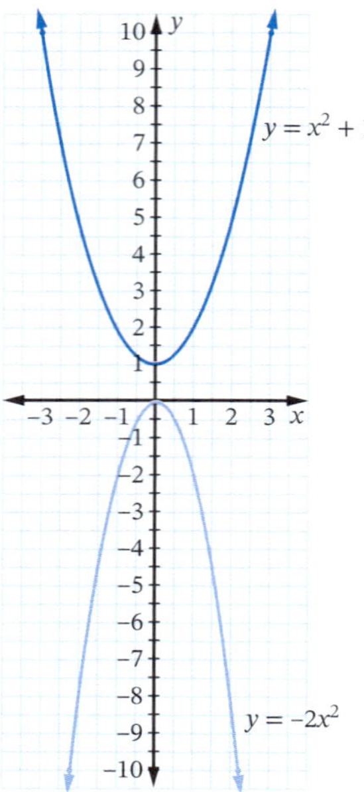

a $y = x^2 + 1$, vertex (0, 1): $y = -2x^2$, vertex (0, 0)
b $y = x^2 + 1$ concave up; $y = -2x^2$ concave down
c $y = x^2 + 1$

16 B

17 x-intercepts 2 and 3, y-intercept −6, vertex (2.5, 0.25),

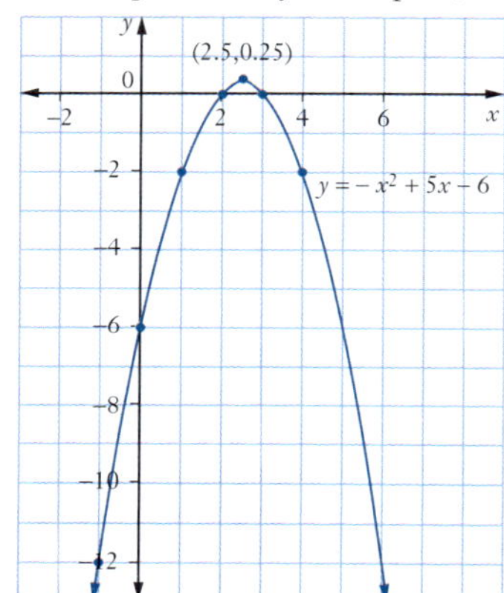

18

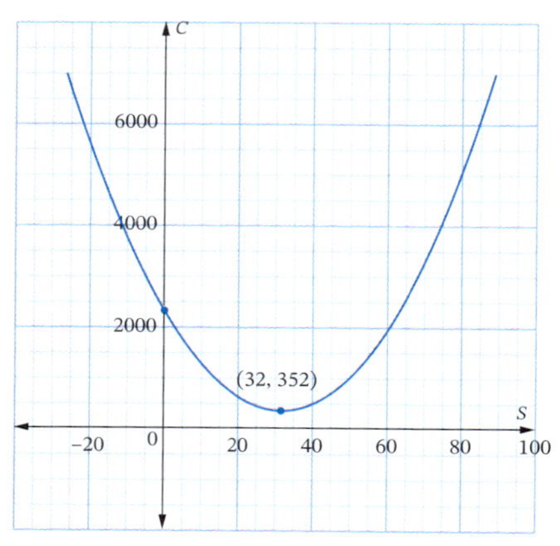

Minimum running cost = \$352 per hour ;
Average speed = 32 knots

19 **a**

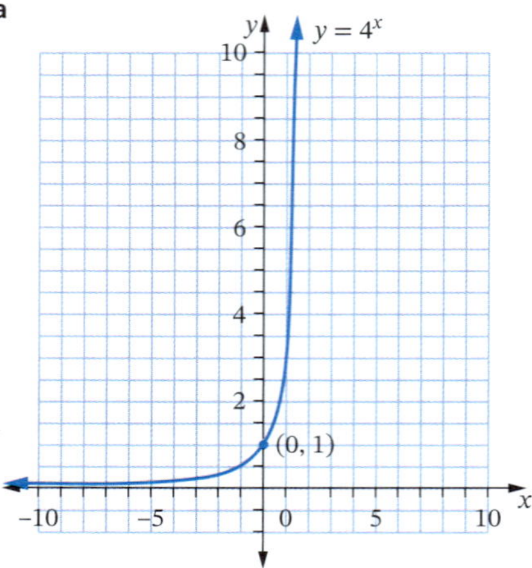

b

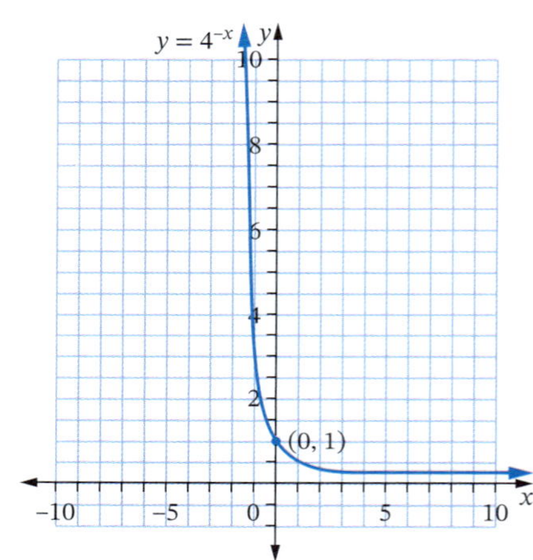

c

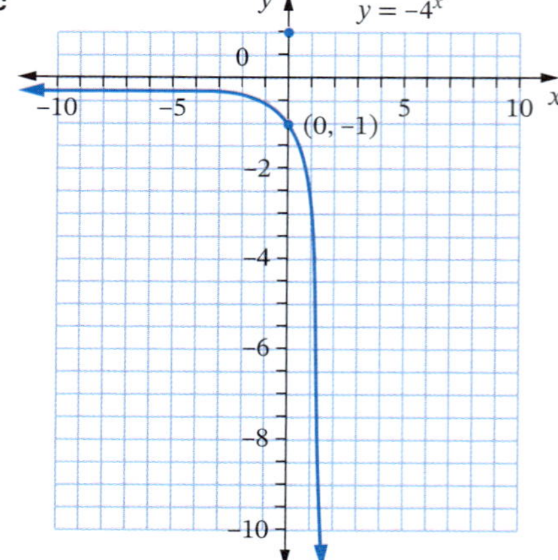

d

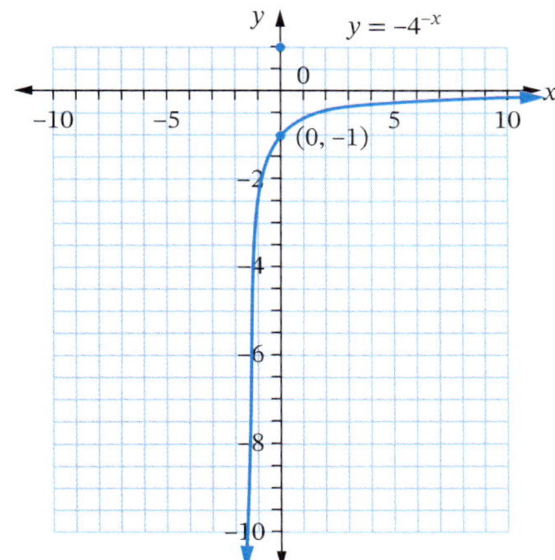

CHAPTER 12

SkillCheck

1 a $\frac{1}{5}$ b 40% c 60% d $\frac{11}{20}$
2 B
3 a $\frac{1}{13}$ b $\frac{1}{4}$ c $\frac{1}{2}$
4 a B b E c F
d C e A f D
5 a $\frac{1}{6}$ b $\frac{2}{3}$ c $\frac{1}{2}$ d $\frac{1}{3}$
6 a {5, 6, 7, 8}
b i 25% ii 100% iii 25% iv 75%

Exercise 12.01

1 a 26 b yes c $\frac{5}{26}$ d $\frac{9}{26}$
2 a 0.375 b 0.25 c 0.25
3 a $\frac{1}{2}$ b $\frac{1}{13}$ c $\frac{5}{13}$ d $\frac{3}{13}$
4 a $\frac{2}{3}$ b $\frac{1}{2}$ c $\frac{2}{3}$
d $\frac{1}{2}$ e 0 f 1
5 a $\frac{4}{7}$ b 0 c 1
6 a $\frac{1}{3}$ b $\frac{2}{3}$ c $\frac{1}{3}$
d $\frac{1}{9}$ e $\frac{5}{9}$
7 a 60.9% b 39.1% c 100% d 0 %
8 a $\frac{1}{3}$ b $\frac{1}{6}$ c $\frac{2}{3}$ d $\frac{2}{3}$
9 a 35% b 45% c 55% d 45%
10 a selecting a vowel
b weather not being sunny or weather being overcast
c traffic light showing amber or green
d selecting a heart, diamond or club
e not hitting the bullseye
f selecting a red or yellow ball
11 a {1, 2, 3, 4, 5} b $\frac{1}{5}$
c Spinning a number other than 4 (spinning a 1, 2, 3 or 5).
d {1, 2, 3, 5} e $\frac{4}{5}$
12 a i $\frac{1}{50}$ ii 0.02 iii 2%
b i $\frac{49}{50}$ ii 0.98 iii 98%
13 a 60% b 80
14 D
15 0.65
16 $\frac{5}{6}$
17 A
18 0 as all mobile numbers begin with 0.

Exercise 12.02

1 a i 0.45 ii 0.18
b i 0.4 ii 0.2
c They are similar but not equal.
d 160, which is close to the observed frequency of 164.
2 a Teacher to check b 25%, teacher to check.
c 12 or 13, teacher to check.
3 a 16 or 17 b–e Teacher to check.
4 a Teacher to check.
b i 0.4 ii 0.5 iii 0.1
c, d Teacher to check.
5 a 500
b i 0.47 ii 0.468 iii 0.35
iv 0.302 v 0.832
c i 0.652 ii 0.667
d The probabilities are close.
6 a 0.28 b 0.16 c 0.45 d 0.89 e 0.60
7 a 40
b i $\frac{11}{40}$ ii $\frac{1}{8}$ iii $\frac{3}{4}$ iv $\frac{1}{8}$
8 a $\frac{4}{31}$ b $\frac{10}{31}$ c $\frac{11}{31}$
d $\frac{26}{31}$ e $\frac{5}{31}$ f $\frac{9}{31}$
9 a $\frac{3}{10}$ b $\frac{1}{10}$ c $\frac{1}{4}$
d $\frac{1}{2}$ e $\frac{9}{20}$ f $\frac{17}{20}$
10 a i 0.092 ii 0.077
b They are similar. The experimental probability is just higher.
c C

Exercise 12.03

1 **a** 48

b **i** $\frac{5}{8}$ **ii** $\frac{25}{48}$

c **i** 1 **ii** $\frac{7}{48}$ **iii** $\frac{23}{48}$ **iv** 0

d A

2 **a** 160

b **i** $\frac{91}{160}$ **ii** $\frac{1}{2}$ **iii** $\frac{19}{80}$ **iv** $\frac{133}{160}$

v $\frac{19}{32}$ **vi** $\frac{53}{160}$ **vii** $\frac{21}{80}$

c 76.25% **d** $\frac{38}{91}$

3 **a** People who preferred watching movies only at the cinema.

b 12

c **i** 0.25 **ii** 0.4 **iii** 0.95 **iv** 0.05

4 **a** D

b 110 – inside the rectangle but outside of the circles.

c **i** $\frac{3}{98}$ **ii** $\frac{6}{49}$ **iii** $\frac{43}{196}$ **iv** $\frac{1}{49}$

v $\frac{13}{98}$ **vi** $\frac{137}{196}$ **vii** $\frac{123}{196}$

d **i** $\frac{59}{86}$ **ii** $\frac{13}{86}$

5 **a** 79 **b** $\frac{3}{79}$

c **i** $\frac{46}{79}$ **ii** $\frac{51}{79}$ **iii** $\frac{34}{79}$

d B

e **i** $\frac{47}{79}$ **ii** $\frac{10}{79}$ **iii** $\frac{14}{79}$

6 **a**

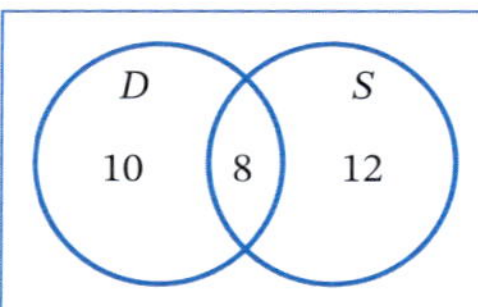

b **i** $\frac{4}{15}$ **ii** $\frac{2}{5}$ **iii** $\frac{1}{3}$ **iv** 1 **v** $\frac{11}{15}$

7 **a** 9

b

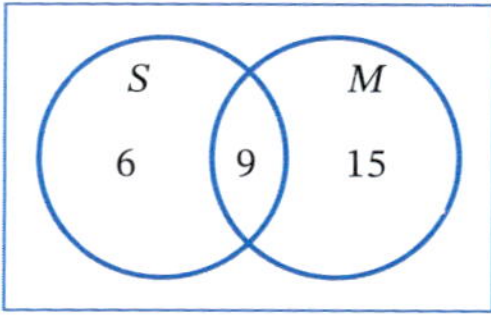

c Because neither playing sport nor watching a movie was not an option.

d **i** $\frac{7}{10}$ **ii** $\frac{1}{2}$ **iii** $\frac{1}{5}$

8 **a**

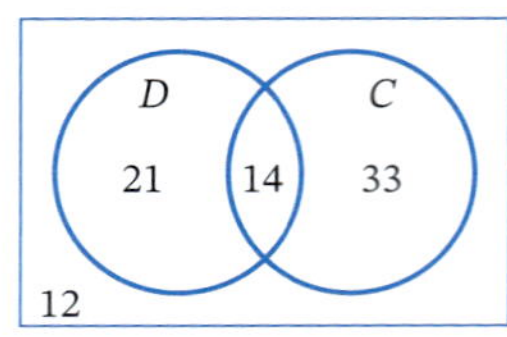

b **i** 14 **ii** 21 **iii** 68 **iv** 54

c **i** 0.2625 **ii** 0.675 **iii** 0.825 **iv** 0.85

d $\frac{47}{68}$

9 **a**

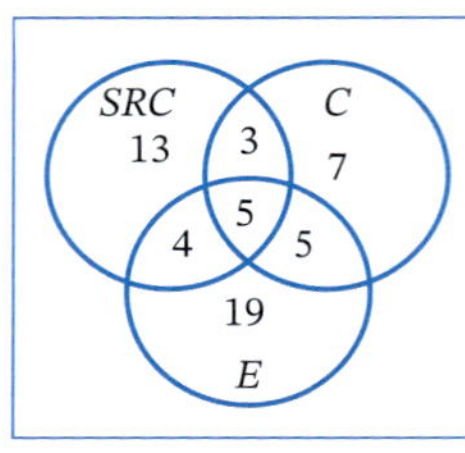

b 19 **c** 7 **d** $\frac{3}{14}$ **e** $\frac{17}{56}$

f **i** $\frac{1}{14}$ **ii** $\frac{39}{56}$ **iii** $\frac{23}{56}$ **iv** $\frac{51}{56}$

g No, because information is only given about numbers belonging to the clubs and not about the numbers attending the school.

Exercise 12.04

1 **a** 300 **b** 75 **c** 68

d **i** $\frac{31}{75}$ **ii** $\frac{1}{4}$ **iii** $\frac{157}{300}$ **iv** $\frac{58}{75}$

2 **a** 130 **b** 68.5%

c **i** $\frac{61}{130}$ **ii** $\frac{89}{130}$ **iii** $\frac{1}{5}$

iv $\frac{23}{26}$ **v** $\frac{7}{26}$ **vi** $\frac{8}{13}$

d B **e** $\frac{5}{23}$

3 **a** 240 **b** **i** 121 **ii** 119

c $53.\dot{3}\%$ **d** $\frac{9}{80}$ **e** $\frac{36}{121}$

4 **a** 183 **b** **i** $\frac{14}{183}$ **ii** $\frac{118}{183}$

c 35.5% **d** $\frac{43}{50}$ **e** $\frac{169}{183}$

5 **a** 151

b **i** 69.5% **ii** 20.5% **iii** 71.5% **iv** 28.5%

c $\frac{62}{151} = 41.1\%$ **d** $\frac{15}{46} = 32.6\%$

6 **a** 965

b **i** 18.4% **ii** 49.4% **iii** 21.8% **iv** 27.7%

c $\frac{2}{5}$ **d** $\frac{31}{52}$

7 **a** 320 **b** 43.4% **c** $56.6\% = \frac{181}{320}$

d 49.7% **e** A

Mental skills 12

2 **a** \$700 **b** \$800 **c** 400 **d** 700
e \$300 **f** \$400 **g** 250 **h** \$300
i \$300 **j** 500

Exercise 12.05

1 **a** 55, 56, 58, 59, 65, 66, 68, 69, 85, 86, 88, 89, 95, 96, 98, 99

b 16

c **i** $\frac{8}{16}=0.5$ **ii** $\frac{12}{16}=0.75$ **iii** $\frac{8}{16}=0.5$ **iv** $\frac{4}{16}=0.25$

2 **a**

		Die					
		1	2	3	4	5	6
Coin	H	H1	H2	H3	H4	H5	H6
	T	T1	T2	T3	T4	T5	T6

b **i** $\frac{1}{12}$ **ii** $\frac{3}{12}=\frac{1}{4}$ **iii** $\frac{4}{12}=\frac{1}{3}$

3 **a**

		1st die					
	+	1	2	3	4	5	6
2nd die	1	2	3	4	5	6	7
	2	3	4	5	6	7	8
	3	4	5	6	7	8	9
	4	5	6	7	8	9	10
	5	6	7	8	9	10	11
	6	7	8	9	10	11	12

b **i** $\frac{2}{36}=\frac{1}{18}$ **ii** $\frac{6}{36}=\frac{1}{6}$ **iii** $\frac{26}{36}=\frac{13}{18}$

iv $\frac{30}{36}=\frac{5}{6}$ **v** $\frac{18}{36}=\frac{1}{2}$ **vi** $\frac{12}{36}=\frac{1}{3}$

4 **a**

1st ball 2nd ball Outcomes

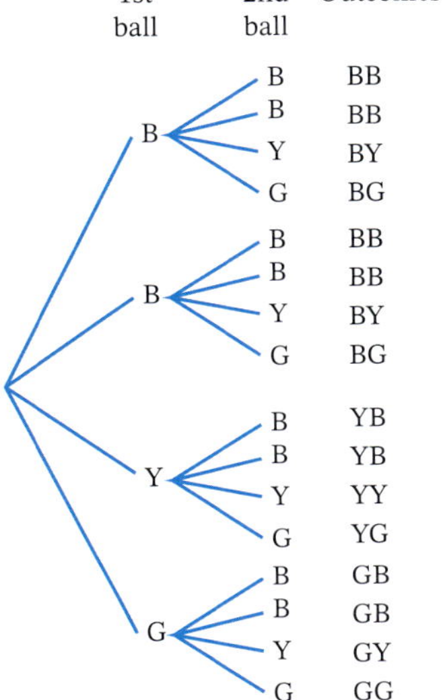

b **i** $\frac{2}{16}=\frac{1}{8}$ **ii** $\frac{2}{16}=\frac{1}{8}$ **iii** $\frac{4}{16}=\frac{1}{4}$

iv $\frac{12}{16}=\frac{3}{4}$ **v** $\frac{6}{16}=\frac{3}{8}$

5 **a**

1st draw 2nd draw Outcomes

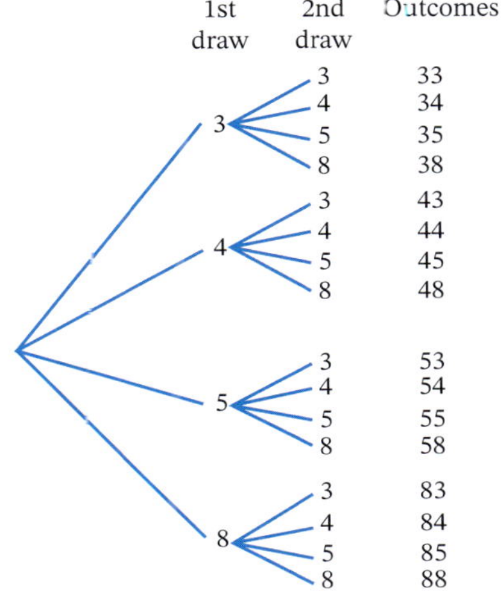

b **i** $\frac{4}{16}=\frac{1}{4}$ **ii** $\frac{8}{16}=\frac{1}{2}$ **iii** $\frac{6}{16}=\frac{3}{8}$ **iv** $\frac{2}{16}=\frac{1}{8}$

6 **a**

Saturday Sunday Outcomes

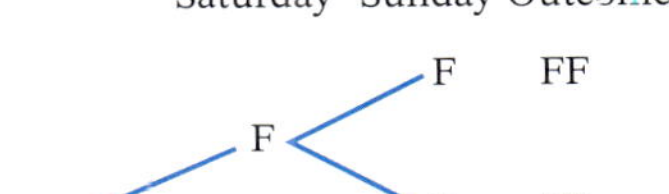

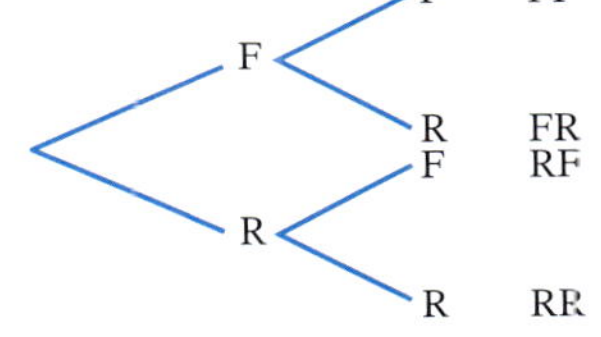

b **i** $\frac{1}{4}$ **ii** $\frac{1}{4}$ **iii** $\frac{3}{4}$

7 **a** $\frac{1}{4}$ **b** $\frac{1}{2}$ **c** $\frac{1}{4}$

d $\frac{1}{4}$ **e** $\frac{3}{4}$ **f** $\frac{1}{4}$

		1st card			
		K♥	K♦	Q♥	Q♦
2nd card	K♥	K♥ K♥	K♦ K♥	Q♥ K♥	Q♦ K♥
	K♦	K♥ K♦	K♦ K♦	Q♥ K♦	Q♦ K♦
	Q♥	K♥ Q♥	K♦ Q♥	Q♥ Q♥	Q♦ Q♥
	Q♦	K♥ Q♦	K♦ Q♦	Q♥ Q♦	Q♦ Q♦

8

		1st die					
		1	2	3	4	5	6
2nd die	1	1,1	2,1	3,1	4,1	5,1	6,1
	2	1,2	2,2	3,2	4,2	5,2	6,2
	3	1,3	2,3	3,3	4,3	5,3	6,3
	4	1,4	2,4	3,4	4,4	5,4	6,4
	5	1,5	2,5	3,5	4,5	5,5	6,5
	6	1,6	2,6	3,6	4,6	5,6	6,6

a $\frac{1}{6}$ **b** $\frac{1}{18}$ **c** $\frac{11}{36}$ **d** $\frac{4}{9}$

9 a i MB, MP, MS, RB, RP, RS, WB, WP, WS, YB, YP, YS

ii

1st course	2nd course	Outcomes
M	B	MB
	P	MP
	S	MS
R	B	RB
	P	RP
	S	RS
W	B	WB
	P	WP
	S	WS
Y	B	YB
	P	YP
	S	YS

b $\frac{1}{2}$

10 a i 11, 12, 13, 14, 15, 21, 22, 23, 24, 25, 31, 32, 33, 34, 35, 41, 42, 43, 44, 45, 51, 52, 53, 54, 55

ii

		1st digit				
		1	2	3	4	5
2nd digit	1	11	21	31	41	51
	2	12	22	32	42	52
	3	13	23	33	43	53
	4	14	24	34	44	54
	5	15	25	35	45	55

b i $\frac{3}{5}$ **ii** $\frac{3}{5}$

Exercise 12.06

1 a BA, BW, BL, BN, AB, AW, AL, AN, WB, WA, WL, WN, LB, LA, LW, LN, NB, NA, NW, NL

b i $\frac{2}{20}=\frac{1}{10}$ **ii** $\frac{6}{20}=\frac{3}{10}$ **iii** $\frac{8}{20}=\frac{2}{5}$ **iv** $\frac{14}{20}=\frac{7}{10}$

2 a 56, 58, 59, 65, 68, 69, 85, 86, 89, 95, 96, 98

b 12

c i 0.5 **ii** 0.75 **iii** 0.5 **iv** 0.17

3 a

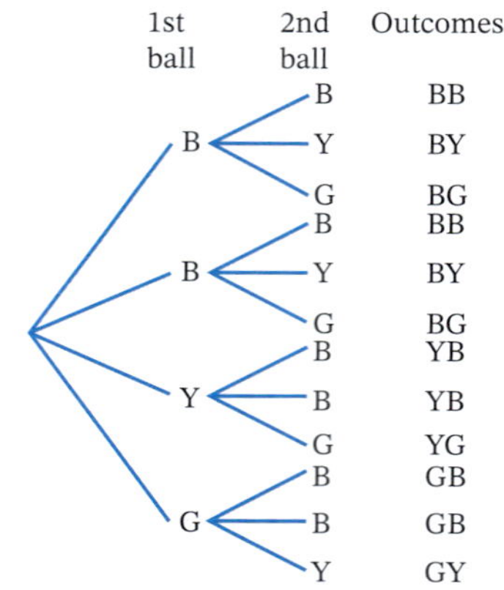

b i $\frac{2}{12}=\frac{1}{6}$ **ii** $\frac{2}{12}=\frac{1}{6}$ **iii** $\frac{4}{12}=\frac{1}{3}$ **iv** $\frac{10}{12}=\frac{5}{6}$

c The probabilities obtained without replacement are slightly larger than the probabilities with replacement. This is because the total number of elements reduces at each step.

d Because there is only 1 yellow ball and 1 green ball so if either of these are selected at the first step, that colour will not be available to select at the second step.

4 a $\frac{1}{20}$ **b** $\frac{2}{20}=\frac{1}{10}$ **c** $\frac{2}{20}=\frac{1}{10}$

5 $\frac{12}{20}=\frac{6}{10}$

6 $\frac{25}{30}=\frac{5}{6}$

7 $\frac{6}{42}=\frac{1}{7}$

Power plus

1 a $\frac{1}{6}$ **b** $\frac{5}{36}$ **c** $\frac{25}{216}$

2 Teacher to check the tree diagram.

a $\frac{1}{4}$ **b** $\frac{1}{16}$ **c** $\frac{3}{8}$

d $\frac{15}{16}$ **e** $\frac{5}{16}$ **f** $\frac{5}{16}$

3 a $\frac{5}{18}$ **b** $\frac{5}{18}$ **c** $\frac{5}{9}$ **d** $\frac{13}{18}$

4 a 6

b i $S \cup E = \{2, 3, 4, 6, 8, 9, 10, 12, 14, 15, 16, 18, 20\}$

ii $S \cap E = \{6, 12, 18\}$

c i $F = \{5, 10, 15, 20\}$

ii $n(F) = 4$ **ii** $S \cap F = \{10, 20\}$

TEST YOURSELF 12

1 **a** 60% **b** 16

2 **a** $\frac{1}{2}$ **b** $\frac{2}{13}$ **c** $\frac{4}{13}$

3 **a i** 0.36 **ii** 0.32 **iii** 0.83 **iv** 0.175
b i $0.\dot{3}$ **ii** $0.\dot{3}$ **iii** $0.8\dot{3}$ **iv** $0.1\dot{6}$
c They are similar but not equal. **d** C

4 **a** 13
b

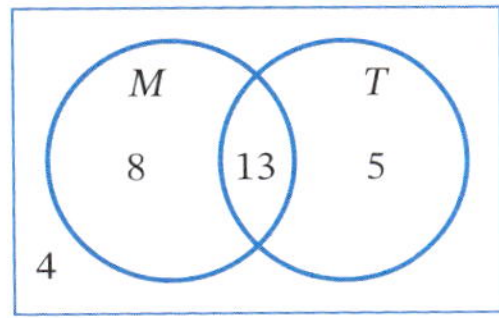

c i $\frac{4}{15}$ **ii** $\frac{3}{5}$ **iii** $\frac{13}{30}$ **iv** $\frac{17}{30}$

5 **a** 210
b i 48.1% **ii** 32.4% **iii** 51.9% **iv** 73.3%
c i $\frac{68}{109} = 62.4\%$

6 **a**

Die 2 \ Die 1	1	2	3	4	5	6
1	0	1	2	3	4	5
2	1	0	1	2	3	4
3	2	1	0	1	2	3
4	3	2	1	0	1	2
5	4	3	2	1	0	1
6	5	4	3	2	1	0

b i $\frac{1}{6}$ **ii** $\frac{5}{9}$ **iii** $\frac{1}{6}$

7 **a** AB, AC, AD, BA, BC, BD, CA, CB, CD, DA, DB, DC
b

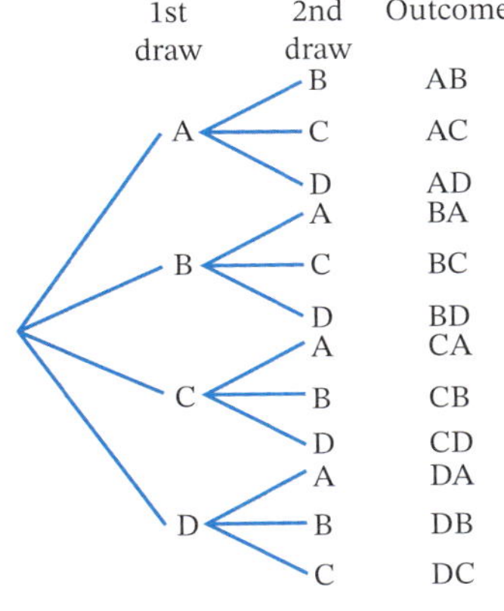

c i 0 **ii** $\frac{1}{6}$ **iii** $\frac{5}{6}$ **iv** $\frac{1}{2}$

CHAPTER 13

SkillCheck

1 B and I, C and E, D and L, F and P, G and K, J and T, Q and S

2 **a i** $\angle P$ and $\angle Y$, $\angle R$ and $\angle B$, $\angle T$ and $\angle W$
ii *PT* and *YW*, *PR* and *YB*, *RT* and *BW*
b i $\angle E$ and $\angle N$, $\angle G$ and $\angle V$, $\angle M$ and $\angle Q$, $\angle X$ and $\angle C$
ii *EX* and *NC*, *XM* and *CQ*, *MG* and *QV*, *GE* and *VN*

3 b and j ; c and g ; d and i ; e and l ; f and k

4 **a** SF = 1.5 **b** SF = $\frac{1}{3}$
c SF = 1.5 **d** $\frac{1}{2}$

Exercise 13.01

1 **a** Yes, same size and shape.
b Yes, same size and shape.
c No, different size. **d** Yes, same size and shape.
e No, different size. **f** Yes, same size and shape.

2 **A** and **H**, **C** and **E**, **D** and **I**, **F** and **G**

3 **a**, **d** and **j**, **b** and **p**, **c** and **n**, **f** and **m**, **g** and **i**, **h** and **l**

4 **a i** *PR* and *MK*, *RT* and *KL*, *LM* and *TP*
ii $\angle R$ and $\angle K$, $\angle P$ and $\angle M$, $\angle T$ and $\angle L$
b i *EF* and *CD*, *FG* and *DA*, *GH* and *AB*, *HE* and *BC*
ii $\angle E$ and $\angle C$, $\angle F$ and $\angle D$, $\angle G$ and $\angle A$, $\angle H$ and $\angle B$
c i *ST* and *WV*, *SQ* and *WC*, *QT* and *CV*
ii $\angle T$ and $\angle V$, $\angle S$ and $\angle W$, $\angle Q$ and $\angle C$

5 **a i** *W*: 12 cm, *Y*: 12 cm
ii *W*: 8 cm^2, *Y*: 8 cm^2
b Yes

6 **a i** $\triangle ABC$: 9.9 cm, $\triangle KLH$: 9.9 cm
ii $\triangle ABC$: 4.5 cm^2, $\triangle KLH$: 4.5 cm^2
b Yes

7 perimeter, area, equal

8, 9 Teacher to check.

10 **a** F **b** F **c** T **d** T

Exercise 13.02

1 **a** SAS **b** AAS **c** SAS **d** AAS
e SSS **f** AAS **g** RHS **h** AAS
i AAS **j** AAS **k** SSS **l** RHS

2 **a** B **b** B

3 **a** $\triangle ABC \equiv \triangle NMP$ (SAS)
b $\triangle KLQ \equiv \triangle FED$ (AAS)
c $\triangle AXT \equiv \triangle GYE$ (RHS)
d $\triangle MVR \equiv \triangle SPC$ (SSS)
e $\triangle BTV \equiv \triangle PSX$ (AAS)
f $\triangle CNQ \equiv \triangle DCE$ (SAS)
g $\triangle PTU \equiv \triangle LVW$ (SSS)
h $\triangle FLD \equiv \triangle HKW$ (SAS)

4 **a** $w = 24$ **b** $x = 28, p = 7$ **c** $a = 34$
d $b = 26$ **e** $y = 65$
f $x = 45, k = 8$

5 **a** 25 cm **b** $b = 25$

6 63.6

7 $a = 118, k = 43, x = 28$

8 **a** No **b** No **c** Yes, SAS
d Yes, RHS **e** No **f** No

Exercise 13.03

1 **a** SSS
b $\angle DCB, \angle CDB, \angle CBD$
c **i** DC **ii** CB
iii equal **iv** bisected, BD
e SAS **f** CX **g** $\angle DXC$ **h** 90°
i diagonal, right

2 **b**

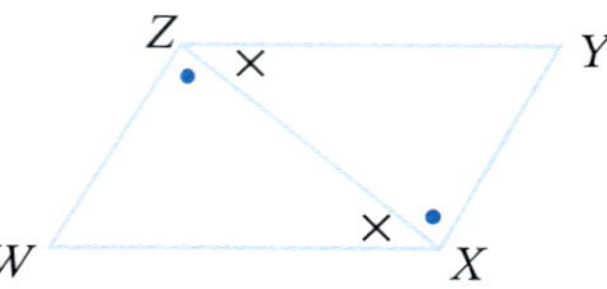

c AAS **d** $\angle Y$ **f** AAS
g matching angles of congruent triangles
h angles, equal

3 **a** SSS
b Matching angles of congruent triangles.
c equal, equal **d** $\angle KML$
e 90°

4 **a** $\angle CDF = \angle EFD, \angle DCE = \angle FEC$
b Opposite sides of a rectangle.
c AAS
d Matching sides of congruent triangles.
e diagonals, bisect

5 **a** SSS
b **i** TSU **ii** bisected, diagonal
iii bisected, diagonals

6 **a**

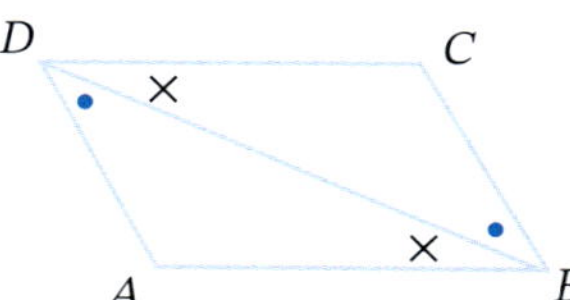

b AAS
c **i** **matching** sides, congruent
ii Opposite, equal

7 **a** $\angle BEL = \angle GHL$ (alternate angles, $BE \parallel HG$),
$\angle EBL = \angle HGL$ (alternate angles, $BE \parallel HG$)
b Equal sides of a rhombus **c** AAS
d **i** GL, HL **ii** diagonals, bisect

8 **a** *SSS*
b Matching angles of congruent triangles.
c SSS
d Yes, matching angles of congruent triangles.
e 60° **f** equilateral, 60°

Exercise 13.04

1 **a** 2 **b** $\frac{2}{3}$ **c** $\frac{4}{5}$ **d** $\frac{8}{3}$
e $\frac{2}{3}$ **f** $\frac{4}{3}$ **g** $\frac{9}{5}$ **h** $\frac{3}{2}$

2 **a**

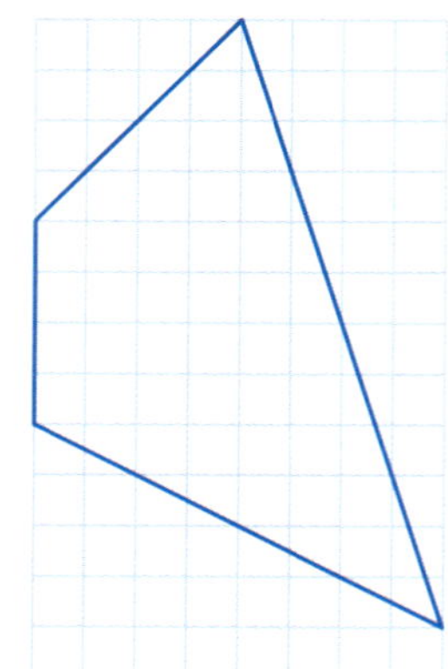

b

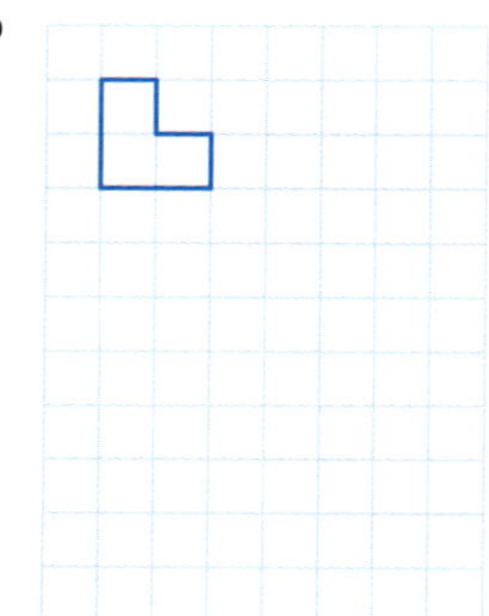

c

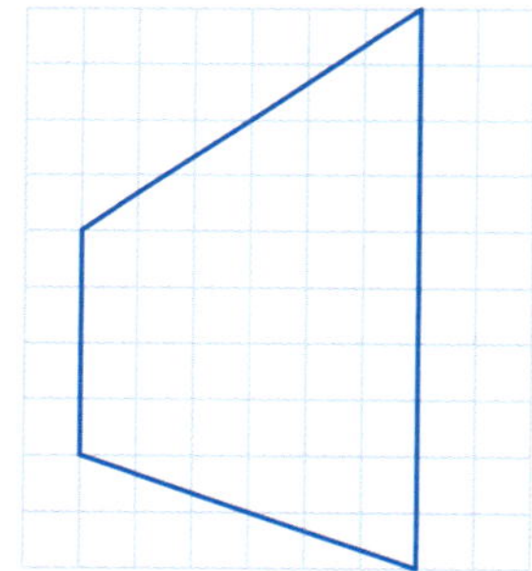

d

e

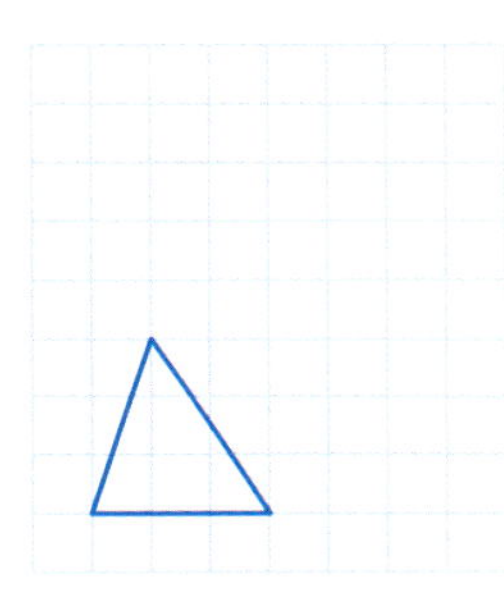

f

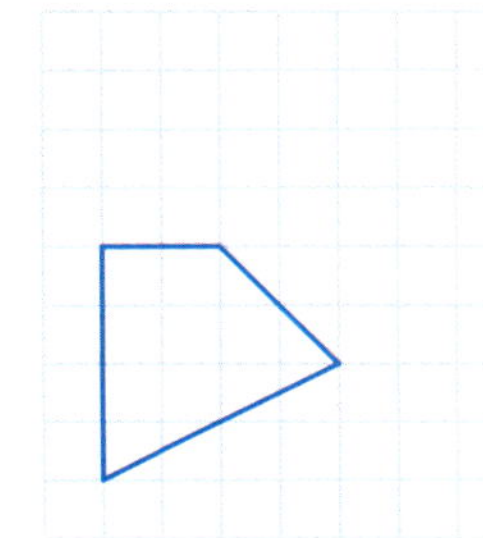

3 **a** No **b** Yes **c** No **d** Yes
e No **f** Yes **g** No **h** Yes

4 **a** **i** $\angle A$ and $\angle L$, $\angle B$ and $\angle M$, $\angle C$ and $\angle K$
ii AC and LK, AB and LM, BC and MK
iii $\triangle ACB \mid\mid\mid \triangle LKM$
b **i** $\angle W$ and $\angle D$, $\angle X$ and $\angle E$, $\angle Y$ and $\angle G$, $\angle Z$ and $\angle H$
ii WX and DE, XY and EG, YZ and GH, WZ and DH
iii $ZYXW \mid\mid\mid HGED$
c **i** $\angle T$ and $\angle F$, $\angle R$ and $\angle W$, $\angle P$ and $\angle V$, $\angle M$ and $\angle S$, $\angle J$ and $\angle K$
ii TR and FW, RP and WV, PM and VS, MJ and SK, TJ and FK
iii $TRPMJ \mid\mid\mid FWVSK$
d **i** $\angle Q$ and $\angle N$, $\angle K$ and $\angle C$, $\angle B$ and $\angle G$
ii QB and NG, BK and GC, KQ and CN
iii $\triangle QKB \mid\mid\mid \triangle NCG$
e **i** $\angle F$ and $\angle X$, $\angle E$ and $\angle W$, $\angle C$ and $\angle Y$
ii FE and XW, CE and YW, CF and YX
iii $\triangle FEC \mid\mid\mid \triangle XWY$
f **i** $\angle K$ and $\angle T$, $\angle L$ and $\angle Q$, $\angle M$ and $\angle S$, $\angle P$ and $\angle U$
ii KL and TQ, KP and TU, PM and US, LM and QS
iii $PMLK \mid\mid\mid USQT$

Mental skills 13

2 **a** $\frac{2}{3}$ **b** $\frac{4}{5}$ **c** $\frac{5}{7}$ **d** $\frac{1}{2}$
e $\frac{1}{4}$ **f** $\frac{1}{6}$ **g** $\frac{5}{6}$ **h** $\frac{2}{5}$
i 5 : 9 **j** 5 : 9 **k** 9 : 20 **l** 4 : 5
m 9 : 7 **n** 4 : 3 **o** $\frac{3}{5}$ **p** $\frac{4}{35}$

3 **a** $\frac{17}{40}$ **b** $\frac{2}{3}$ **c** $\frac{16}{25}$
d $\frac{1}{4}$ **e** $\frac{5}{24}$ **f** $\frac{2}{25}$

Exercise 13.05

1 **a** Yes $(\frac{8}{12} = \frac{20}{30} = \frac{2}{3})$
b Yes $(\frac{12}{20} = \frac{27}{45} = \frac{21}{35} = \frac{3}{5})$
c Yes $(\frac{4}{6} = \frac{5}{9} = \frac{6\frac{2}{3}}{10} = \frac{10}{15} = \frac{2}{3})$
d Yes, all squares are similar.
e Yes, matching angles are equal.
f Yes, all equilateral triangle are similar.

2 **a** $k = 14, x = 6$
b $y = 12\frac{1}{2}$ $p = 19\frac{1}{5}$ **c** $x = 21$
d $a = 12, q = 1\frac{2}{3}, w = 3\frac{1}{3}$ **e** $d = 16$
f $b = 21\frac{7}{18}$ **g** $w = 10\frac{4}{5}, x = 23\frac{1}{3}$
h $c = 19\frac{1}{4}$

3 $d = 15$ **4** $a = 21\frac{3}{7}$

5 $13\frac{1}{3}$ m **6** 8.9 m **7** 21 m

8 **a** A and D, scale factor = 2
b B and D, scale factor = $\frac{3}{4}$
c C and D, scale factor = $\frac{3}{5}$
d A and D, scale factor = $\frac{1}{2}$

Exercise 13.06

1 1 : 500 **2** 1 : 35 **3** 2 : 1
4 3 : 1 **5** 1 : 8050
6 **a** 135 m **b** 378 m
7 2.1 mm
8 Scaled dimensions: 3.2 cm long by 4.4 cm tall
9 **a** Teacher to check: base 3 cm
b approx. 13 m
10 2120 km **11** **a** 6.2 m **b** 4 m
12 **a** 100 km **b** 80 km
c 54 km **d** 224 km

Exercise 13.07

1 1 : 16 **2** 1 : 6.25 or 4 : 25
3 **a** 100 : 1 **b** 30.25 : 1 or 121 : 4
c 9 : 16 **d** 25 : 4 **e** 16 : 49
f 1 : 100 **g** 64 : 25 **h** 4 : 81
4 **a** 1 : 9 **b** 9 : 4 **c** 49 : 16
d 25 : 121 **e** 25 : 9 **f** 16 : 25
5 **a** 8 : 7 **b** 6 : 1 **c** 3 : 20 **d** 1 : 4.5
6 **a** 5 : 3 **b** 4 : 7
7 3.5 **8** 3072 cm^2 **9** 162 cm^2
10 1024 mm^2 **11** $28\frac{2}{7}$ m^2 **12** 154 mm

ANSWERS

13 22.5 mm

14 12.5 cm

15 Increased 9 times

16 reduced by a factor of 2

17 9 : 16

18 50 cm^2

Exercise 13.08

1 **a** AA **b** SSS **c** RHS **d** SAS
e SAS **f** AA **g** SAS **h** RHS
i SSS **j** SSS **k** SAS **l** SAS

2 C

3 **a** B and C (SAS) **b** B and C (RHS)
c B and D (SSS) **d** C and D (AA)

4 **a** $\triangle EFD \,|||\, \triangle PMR$ (SSS)
b $\triangle UTW \,|||\, \triangle KAM$ (SAS)
c $\triangle UZV \,|||\, \triangle HPJ$ (RHS)
d $\triangle WLB \,|||\, \triangle XAT$ (AA)
e $\triangle KWP \,|||\, \triangle GEF$ (SSS)
f $\triangle LEA \,|||\, \triangle RBX$ (SAS)

5 In two triangles, if two pairs of matching angles are equal, then the remaining pair of matching angles must also be equal, due to the angle sum of all triangles being 180°.

Power plus

1 Teacher to check.

2 **a** $\angle E = \angle D = 90°$, $BM = CM$
b Vertically opposite angles are equal.
c $\angle E = \angle D = 90°$
$\angle BME = \angle CMD$ (see part **b**)
$BM = CM$
$\therefore \triangle BME \equiv \triangle CMD$ (AAS)
d Matching sides of congruent triangles.
e Matching angles of congruent triangles.

3 **a** $VW \,||\, XY$
b Alternate angles are equal, $VW \,||\, XY$
c $\angle V = \angle Y$ (proved in **b**)
$\angle W = \angle X$ (proved in **b**)
$\therefore \triangle VWD \,|||\, \triangle XYD$ (AA)
d $k = 9$, $m = 30$

4 33.25 m

5 **a** true **b** false **c** true
d false **e** true **f** true

TEST YOURSELF 13

1 B and I, C and J, D and K.

2 **a** CF and TM, CD and TR, DE and RP, FE and MP.
b $\angle F$ and $\angle M$, $\angle E$ and $\angle P$, $\angle D$ and $\angle R$, $\angle C$ and $\angle T$.
c $FEDC \equiv MPRT$

3 **a** SAS **b** RHS **c** SSS

4 C

5 **a** opposite sides of a rectangle
b AB
c All angles in a rectangle are 90°
d SAS
e BD **f** They are equal.

6 **a** 2 **b** $\frac{3}{5}$ **c** 1.5

7 **a**

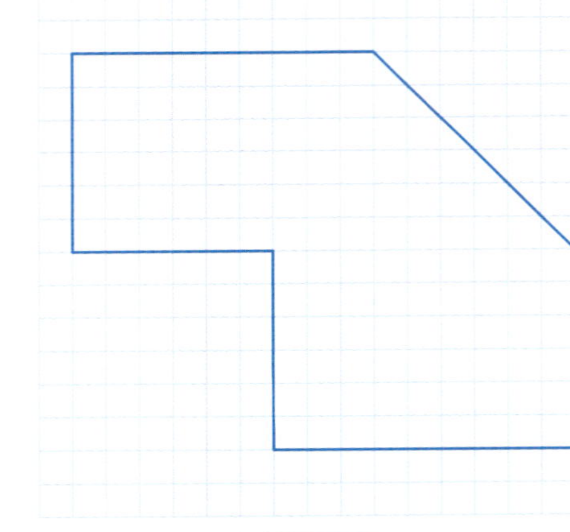

b

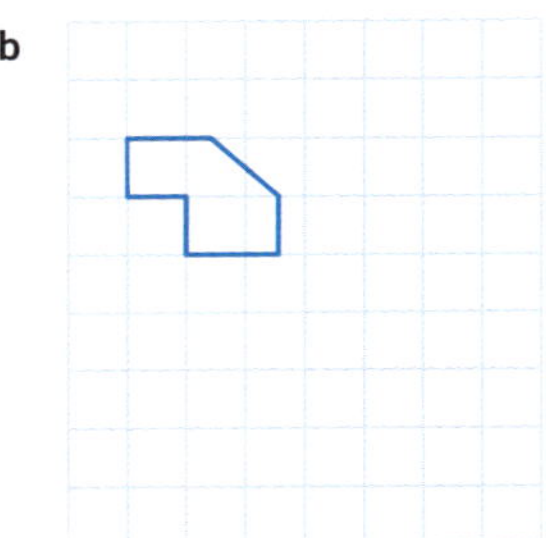

8 **a** No, $\frac{12}{18} \neq \frac{20}{29}$ **b** Yes, $\frac{5.25}{7} = \frac{6}{8} = \frac{7.5}{10}$

9 **a** $y = 8.75$, $p = 6.4$ **b** $a = 18$, $w = 27$

10 C

11 **a** 8 m **b** 4 m

12 **a** 25 : 81 **b** 16 : 9

13 100 cm

14 **a** SAS **b** SSS **c** AA

PRACTICE SET 4

1 **a** no **b** yes **c** yes

2 **a** $\frac{3}{2}$ **b** 7.5 m

3 0.8

4 **a i** $\frac{5}{9}$ **ii** $\frac{2}{3}$ **iii** $\frac{1}{9}$ **iv** $\frac{16}{27}$
b $\frac{13}{22}$

5 625 m

6 **a** $\sqrt{26}$ units **b** 5.7 units

7 $PQ = QR = \sqrt{26}$; Since two sides are equal, it is an isosceles triangle.

8 (1, 0)

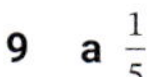

9 **a** $\frac{1}{5}$ **b** −1

10

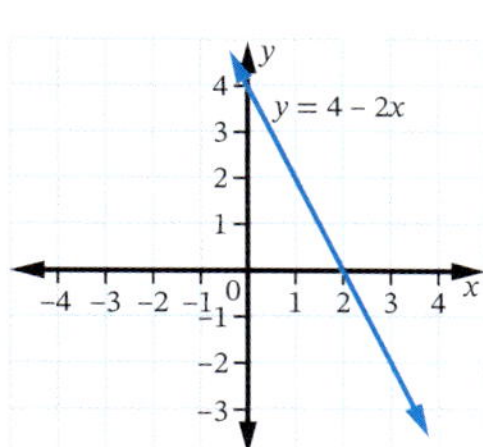

a 2 **b** 4

11 **a** $\frac{4}{15}$ **b** 107

12 **a**

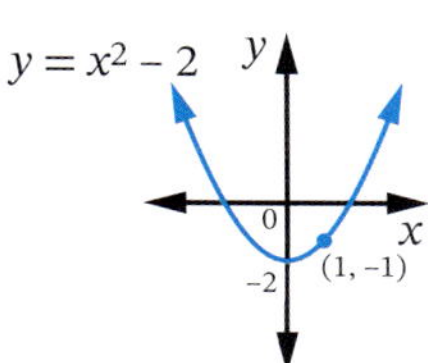

b (0, −2) **c** concave up

13 **a** SAS **b** AAS **c** SAS **d** RHS

14 $y = 4$

15 **a** $\frac{1}{5}$ **b** $\frac{2}{3}$

16 **a** gradient $\frac{1}{2}$, y-intercept 3

b $y = 2x - 5$

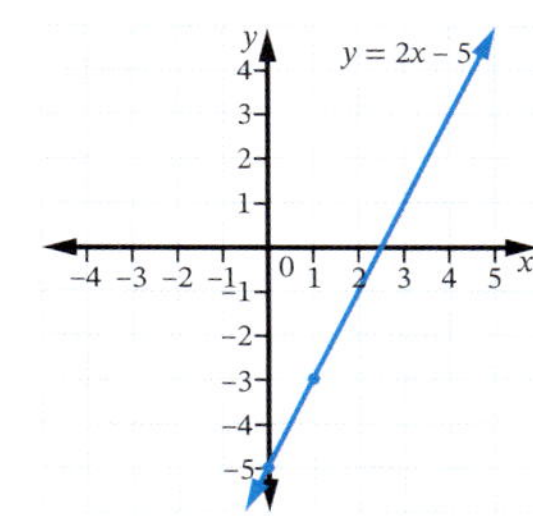

17 **a** 188 kPa **b** 125 m

18

		1st die					
		1	2	3	4	5	6
2nd die	1	1, 1	2, 1	3, 1	4, 1	5, 1	6, 1
	2	1, 2	2, 2	3, 2	4, 2	5, 2	6, 2
	3	1, 3	2, 3	3, 3	4, 3	5, 3	6, 3
	4	1, 4	2, 4	3, 4	4, 4	5, 4	6, 4
	5	1, 5	2, 5	3, 5	4, 5	5, 5	6, 5
	6	1, 6	2, 6	3, 6	4, 6	5, 6	6, 6

a $\frac{1}{18}$ **b** $\frac{11}{36}$ **c** $\frac{1}{6}$ **d** $\frac{1}{2}$

19 A

20 **a** −3 **b** $y = -3x + 1$

21 Tanya will have \$60 in her piggy bank after 7 weeks.

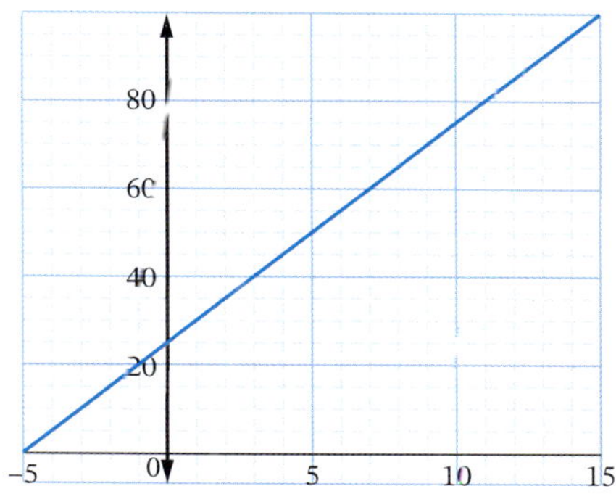

22 **a**

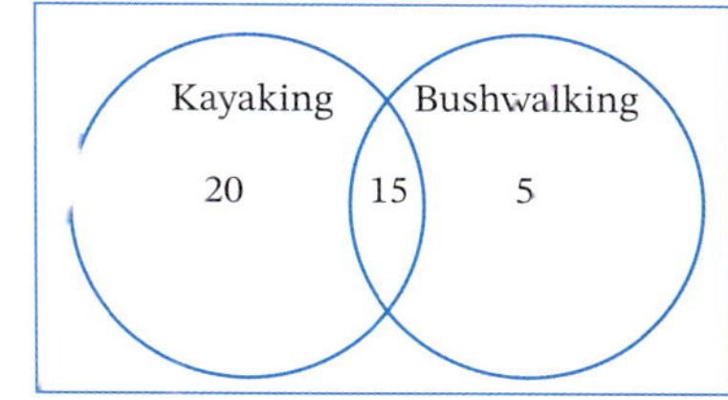

b **i** $\frac{3}{8}$ **ii** $\frac{5}{8}$

23 **a** SSS **b** 90°

c bisect, right angles

24 equiangular or AA

25 **a** C **b** A **c** B

25

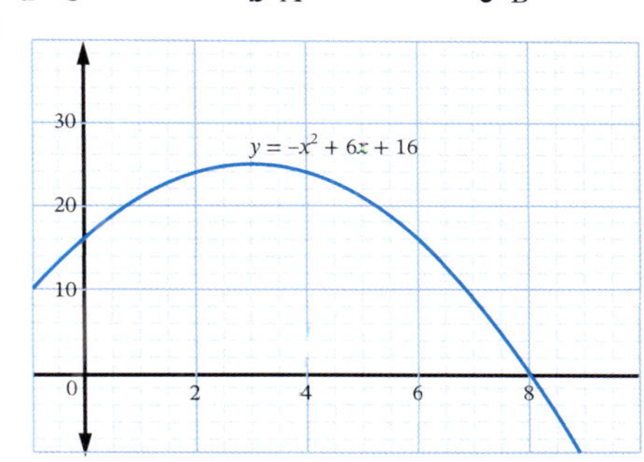

a 16 m **b** 8 seconds **c** 9 m

27 $y = 2^x + 3$

28 $\frac{8}{36} = \frac{2}{9}$

29 1:49

GENERAL PRACTICE

1 **a** 288 cm³ **b** 910 m³

c 84 m³ **d** 4536.5 mm³

2 A

3 **a** $21d^2$ **b** $\frac{4s}{5r}$ **c** $8y - 6e$

4 **a** 23
b **i** 9 **ii** 6 **iii** 6.2 **iv** 6
c 1 **d** Yes, around 5 and 6.

5 **a** ND **b** ND **c** NC **d** C

6 **a** SAS **b** AAS **c** SAS **d** RHS

7 **a**

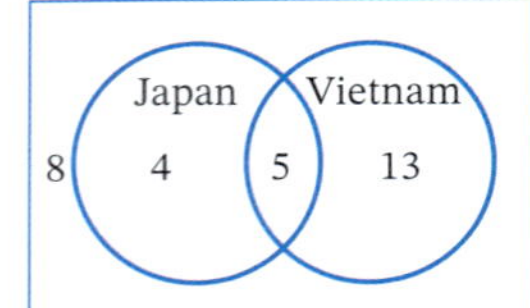

b **i** $\frac{4}{15}$ **ii** $\frac{17}{30}$ **iii** $\frac{13}{30}$

8 **a** 144 km/h **b** 0.6 kg/min

9 $y = 2x - 4$

10 **a** \$1571.48 **b** \$6833.33

11 **a** 30 cm^2 **b** 320 mm^2 **c** 30 m^2

12 **a** 63.6 cm^2 **b** 160.2 cm^2 **c** 3.4 m^2

13 **a** 46° **b** 133° **c** 70° **d** 83°

14 \$1008.15

15 **a** QR **b** $\angle L$ **c** $\triangle SRQ$

16 **a** k^{12} **b** d^5 **c** $15p^3q^4$
d $4k^4$ **e** $\frac{8r^3}{125}$ **f** 5

17 **a**

Score	Tally	f	fx
45	|	1	45
47	||	2	94
48	||	2	96
49	~~||||~~ |	6	294
50	~~||||~~ ~~||||~~ |	11	550
51	~~||||~~ ||	7	357
52	~~||||~~ ||||	9	468
53	||	2	106
	Total	40	2010

b 50.25

c

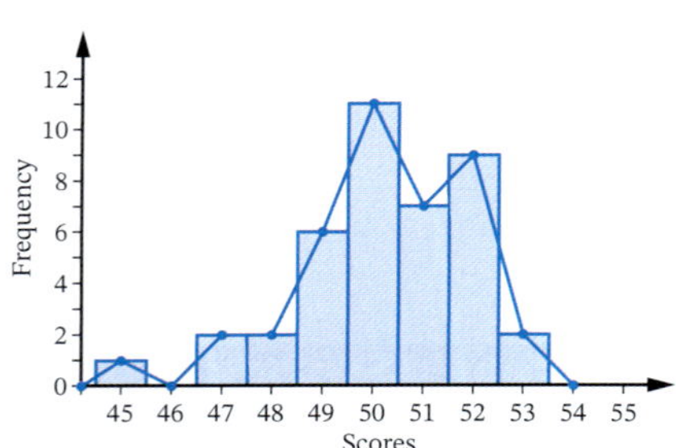

d 50 **e** 50

18 **a** $3 - 6y$ **b** $2x^2 - 7x$
c $-4m + 4m^2$ **d** $5x^2 - 13x - 28$

19 $4d - 23$

20 D

21 **a** 35.93 m **b** 62.5 m^2

22 D

23 \$761.10

24 **a** $2(y + 4)$ **b** $5a(1 - 3b)$ **c** $-2b(4a - 1)$

25 **a** 6.36 **b** 15.63 **c** 10.76

26 52.7%

27 **a** $m = 55$ (opposite angles of a parallelogram)
b $e = 41$ (co-interior angles between parallel lines)
c $k = 8$ (right angle in a rectangle, adjacent angles)

28 $y = 5$

29 **a** $\left(\frac{1}{2}, 2\frac{1}{2}\right)$ **b** $\sqrt{82}$ **c** $\frac{1}{9}$

30 2.5 cm

31 **a** rectangle **b** equilateral triangle
c rectangle **d** isosceles triangle

32 **a** 2 **b** 33
c negatively skewed

33 C

34 **a** **i**

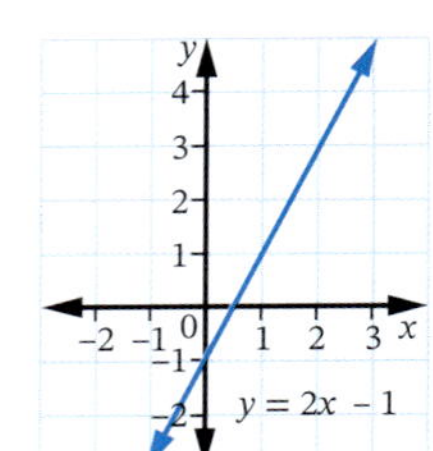

ii

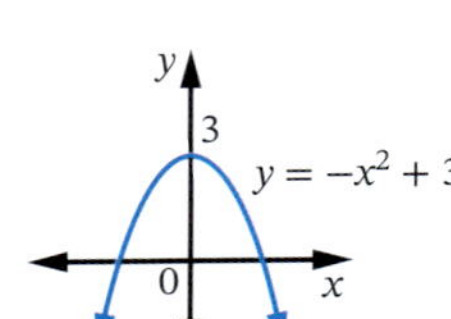

iii

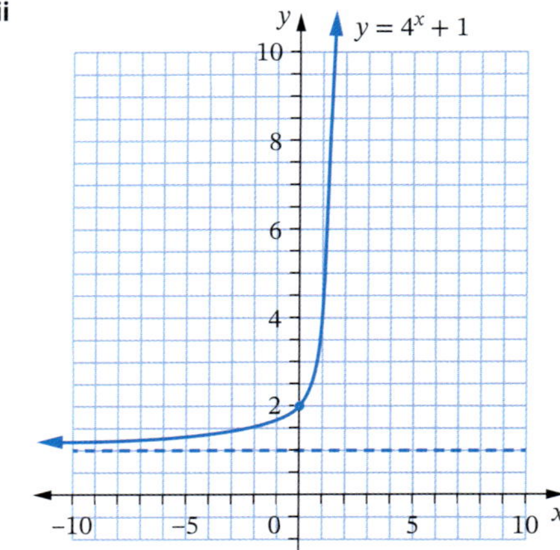

b gradient 2, y-intercept -1
c (0, 3) **d** $y = 1$

35 \$1925

36 \$88

37 B

38 **a** 396 cm^2 **b** 476 cm^2 **c** 792 cm^2

39 **a** $y = 1\frac{1}{2}$ **b** $x = \pm 3$

c $x = 20$ **d** $a = 7\frac{1}{3}$

40 **a** 7.14×10^5 **b** 3.5×10^{-4} **c** 1.5763×10^7

41 **a** 2.4×10^{15} **b** 4.2×10^{-11}

42 **a** 35° **b** 35° **c** 43°

43 **a** **i** 0.15 **ii** 0.17 **iii** 0.47 **iv** 0.64 **v** 0.70

b 180; the observed frequency of 173 is close to this.

44 **a** 1080° **b** 135°

45 **a** $840 **b** 31.5%

46 Number of siblings is numerical discrete data and this would be plotted against frequency. Acceptable graphical displays can include column graphs and dot plots.

47 Choose every 50th employee from a full list of employees.

48 4.6%

49 $C = 45h + 80$ where C is the cost and h is the number of hours worked.

Worked for 6 hours.

50

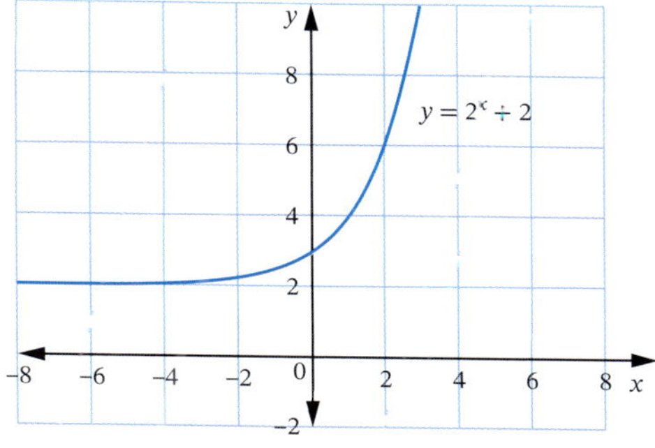

51 625 cm^2

52 $10x^2 + 11x + 3$

53 $\sqrt{150}, \sqrt{70}$

54 **a** $x < 8$ **b** $x \geq 12$

c $x \leq 6$ **d** $x > 24$

ANSWERS

Glossary and index

absolute error The difference between the actual value and the measured value of a quantity, as a positive value. Also a measure of the precision of an instrument, calculated as 0.5 of the smallest unit on the instrument. (p. 395)

acute-angled triangle A triangle with all 3 angles acute. (p. 222)

adjacent side In a right-angled triangle, the side 'next to' the given angle, leading to the right angle. (p. 141)

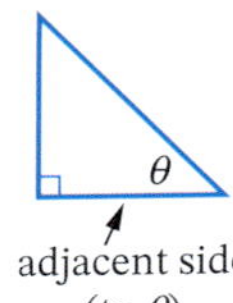

algebraic expression A number written in algebraic form using variables, for example, $2xy + 4y - 5$. (p. 5)

algebraic term See term (**of an expression**). (p. 12)

allowable (**tax**) **deduction** A part of a person's annual income that is not taxed, such as work-related expenses or donations to charities. All deductions are subtracted from annual income to determine **taxable income.** (p. 316)

angle sum The total of the sizes of the angles in a shape. The angle sum of a triangle is 180°. (p. 222)

annual leave loading (or **holiday loading**) Extra payment to a worker during annual leave that is 17.5% of 4 weeks' pay. (p. 313)

annulus A ring shape between 2 different-sized circles with the same centre. (p. 415)

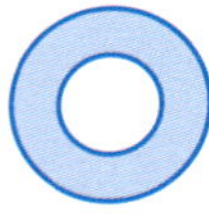

arc Part of the circumference of a circle. (p. 414)

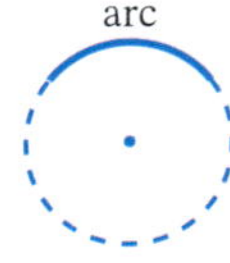

area The amount of surface covered by a shape, measured in square units. (p. 402)

ascending order Going up, increasing, from smallest to largest (1-2-3). The opposite of **descending order**. (p. 88)

asymptote A line that a curve gets very close to but never touches, for example, the x-axis is an asymptote of the exponential curve. (p. 509)

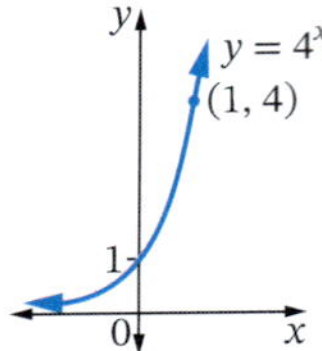

at least Referring to the smallest number, for example, 'at least 2' means 2, 3, 4, ..., or '2 or more'. (p. 531)

average (p. 331) See **mean**.

backtracking method A method of solving equations by 'undoing', or performing inverse (opposite) operations in reverse order. (p. 257)

balancing method A method of solving equations by performing inverse (opposite) operations on both sides. (p. 257)

base (**in index notation**) When a number is raised to a power, the number raised is the base. In the expression 3^5, the 3 is called the base. (pp. 184, 432)

base (**of a flat shape**) The bottom side of a shape.

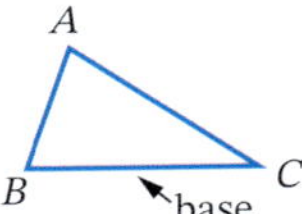

base (**of a solid**) The bottom or end face of a solid shape. (p. 432)

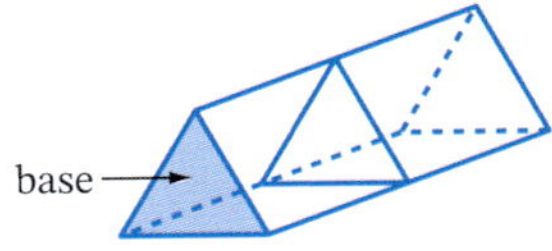

bias In statistics, something that causes a sample to not truly represent the population. (p. 374)

bimodal distribution A statistical distribution that has 2 peaks. (p. 359)

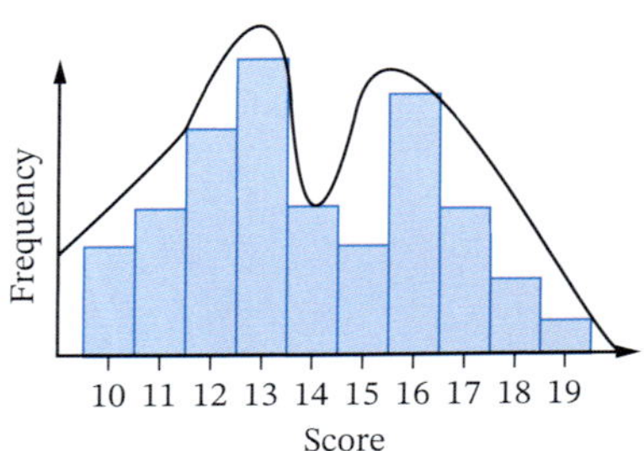

binomial expression An algebraic expression with 2 terms, for example, $x + 9$, $2y - 12$. (p. 23)

bisect To cut in half. (p. 219)

bonus Extra pay for achieving a high quality or volume of work, such as meeting an important quota, goal or deadline. (p. 312)

byte (B) A unit of digital (computer) memory required to store one character of data, such as the letter 'a'. (p. 390)

C

capacity The amount of material (usually liquid) that a container can hold, measured in millilitres (mL), litres (L), kilolitres (kL) and megalitres (ML). (p. 431) See also **volume**.

Cartesian plane Another name for **number plane**.

categorical data Data that can be classified into categories, such as hair colour, favourite radio station or postcode. Data that is not **numerical**. (p. 352)

census A survey of the entire **population** of people or items, not just a survey of a **sample**. (p. 368)

chance experiment An activity or process that involves chance, for example, rolling a die or tossing a coin. (p. 521)

circumference The perimeter of a circle. $C = \pi d$ or $C = 2\pi r$, where C is the circumference, π is pi, d is the diameter and r is the radius. (p. 412)

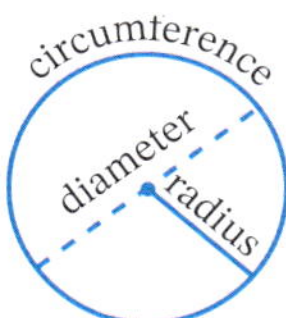

cluster A group of data values that are bunched or close together. (p. 347)

commission Pay earned by salespeople and agents, calculated as a percentage of the value of items sold or income made. (p. 311)

complementary event All the outcomes that are *not* the event; the 'opposite' event. For example, the complementary event to rolling 1 on a die is rolling a number that is not 1. (p. 522)

composite shape A shape made up of 2 or more basic shapes. (p. 402)

compound event A chance event that is a combination of 2 or more simple events, for example, 'female or left-handed'. (p. 535)

congruent Identical, exactly the same. The symbol '≡' means 'is congruent to' or 'is identical to'. (p. 564)

congruent figures Identical figures, having the same shape and size. (p. 564)

congruence test One of 4 tests for proving that 2 triangles are congruent: SSS, SAS, AAS and RHS. (p. 571)

consecutive numbers Any series of integers that follow each other in order, for example, 8, 9 and 10. (p. 8)

constant of proportionality See **direct proportion**. (p. 487)

continuous data Numerical data that can be measured on a smooth scale without any gaps, and can take on a full range of values, such as the height of people. Continuous data is measured on a scale without 'gaps', unlike **discrete data**. (p. 352)

converse The 'reverse' of a rule, the rule written 'back-to-front' or 'turned around'. The converse of Pythagoras' theorem is that if the square of the longest side of a triangle is equal to the sum of the squares of the other 2 sides, then the triangle is right-angled. (p. 67)

convex polygon A polygon whose vertices all point outwards. All diagonals lie within the shape, and all angles are less than 180°. (p. 237)

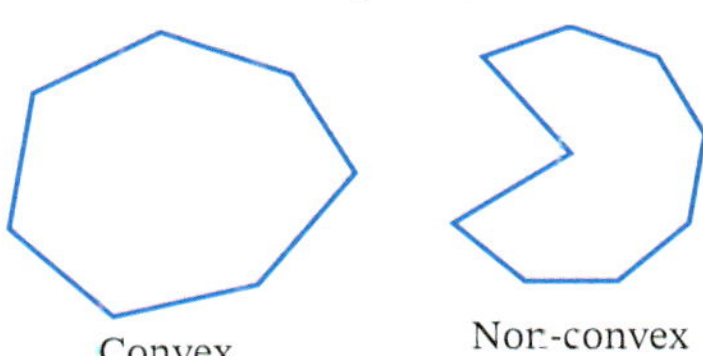

cosine A ratio in a right-angled triangle:

$$\cos\theta = \frac{\text{side adjacent to } \theta}{\text{hypotenuse}}$$

(p. 144) See also **sine** and **tangent**.

cost price The price it costs a retailer (shop) to buy an item (for reselling). The cost price is usually less than the **selling price**. (p. 113)

cross-section A 'slice' of a solid cut across it rather than along it, as shaded in the diagram. (pp. 246, 431)

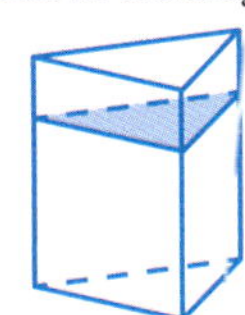

cumulative frequency A progressive or running total of frequencies, the sum of frequencies of a particular data value and all values below it. (p. 335)

cylinder A can-shaped solid with ends that are circles. (p. 428)

D

data Information, a collection of facts. (p. 331)

degree (of a vertex) The number of edges coming out of the vertex in a network. (p. 243)

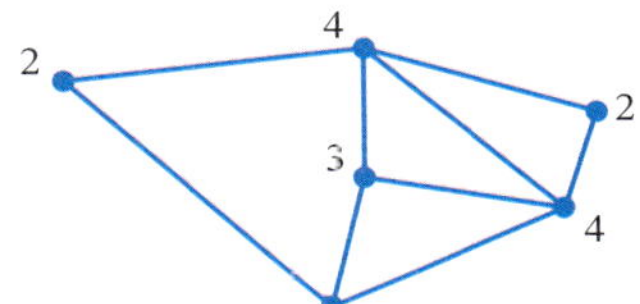

denominator The number below the line in a fraction. The denominator of $\frac{2}{3}$ is 3. (pp. 15, 93)

descending order Going down, decreasing, from largest to smallest (3-2-1). The opposite of **ascending order**. (p. 88)

diagonal An interval joining 2 non-adjacent vertices of a shape. (p. 227)

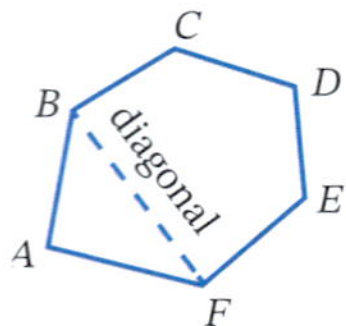

diameter An interval joining 2 points on the circumference and passing through the centre of a circle, and the length of that interval. The diameter is double the **radius**. (p. 412)

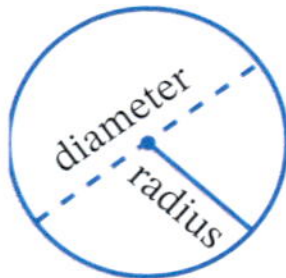

discount The saving made that is the difference between the original price of an item and the reduced price. (p. 111)

discrete data Numerical data that are counted or measured and only take on distinct, separate values, such as the number of children in a family (0, 1, 2, ...). Discrete data has a scale with 'gaps' or jumps, unlike **continuous data**. (p. 352)

direct proportion (or **direct variation**) A relationship between 2 variables, where one variable is a constant multiple of the other, that is, $y = kx$, where k is a constant called the **constant of proportionality**. For example, if $y = 8.5x$, then y is directly proportional to x. (p. 487)

distributive law A law of arithmetic that says you can multiply by a number by splitting it into the sum or difference of 2 other numbers. For example, $27 \times 12 = 27 \times (10 + 2) = 27 \times 10 + 27 \times 2$. Algebraically, $a(b + c) = ab + ac$ for any 3 numbers a, b and c. (p. 20)

dot plot A graph that uses dots above a number line to represent the frequencies of data values. (p. 333)

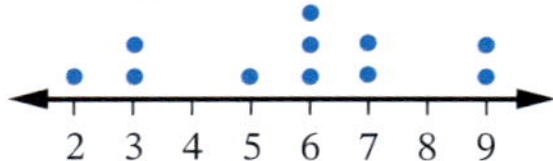

double time Overtime pay that is calculated at 2 times the normal pay rate. (p. 306)

edge (of a network) The connecting lines that join vertices together. (p. 242)

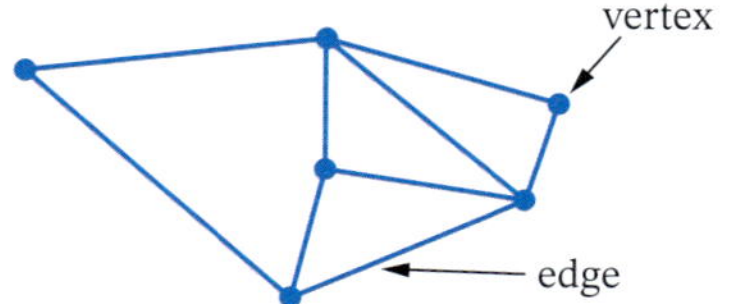

edge (of a solid or polyhedron) The straight lines or borders of the faces of the solid. (p. 246)

equation A mathematical statement that 2 quantities are equal. For example, $8 + 2 = 10$ or $3b - 7 = 5$. (p. 257)

equilateral triangle A triangle with all 3 sides equal. (p. 221)

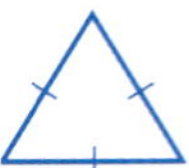

Euler's formula (for networks and polyhedra) $F + V = E + 2$, where F is the number of faces, V is the number of vertices, E is the number of edges. (p. 251)

expected frequency The expected number of times an event will occur over repeated trials, calculated by multiplying the probability of the event by the number of trials. (p. 526)

experimental probability An estimate of theoretical probability; the **relative frequency** of an event in repeated trials of an experiment, found using the formula

$P(E) = \frac{\text{frequency of } E}{\text{total frequency}}$ (p. 526)

exponential curve The graph of an exponential equation $y = a^x$. (p. 509) See **asymptote** for a diagram.

exponential function An equation of the form $y = a^x$, where a is a positive constant and the variable x is a power, for example, $y = 4^x$. (p. 509)

exterior angle An 'outside' angle of a shape created by extending one of the sides of the shape. (p. 223)

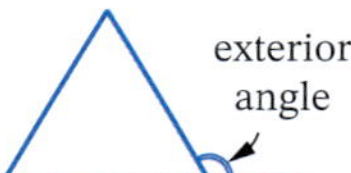

event In probability, a result involving one or more outcomes. For example, when rolling a die, the event 'rolling an even number' contains the 3 outcomes {2, 4, 6}. (p. 521)

face (of a network) A region bounded by the edges of the network, and the region outside the network. (p. 243)

face (of a solid or polyhedron) A surface of the solid. (p. 246)

factor or **divisor** (**of a number**) A value that divides evenly into a given number. For example, the factors of 15 are 1, 3, 5 and 15. (p. 28)

formula (plural: **formulas** or **formulae**) A rule written as an algebraic equation, using variables. The formula for the area of a triangle is $A = \frac{1}{2}bh$. (pp. 9, 276)

fraction A number written in the form $\frac{a}{b}$, where a and b are integers and $b \neq 0$. (pp. 15, 97)

frequency The number of times a value appears in a set of data, or the number of times an event occurs in repeated trials of a probability experiment. (p. 334)

frequency histogram A column graph that shows the frequencies of numerical data. There are no spaces between the columns. (p. 341)

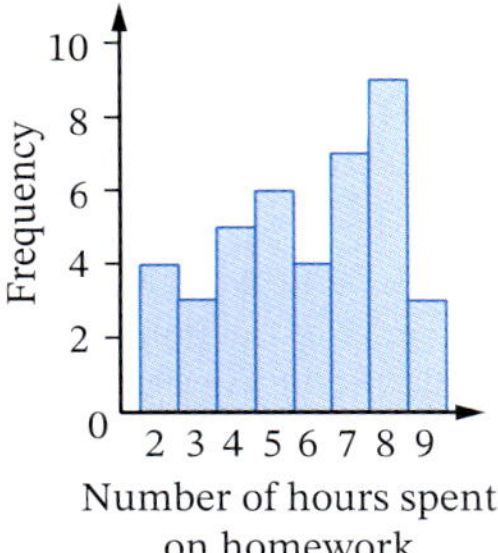

frequency polygon A line graph that shows the frequencies of numerical data. It can be made by joining the midpoints of the tops of the columns of a histogram. (p. 341)

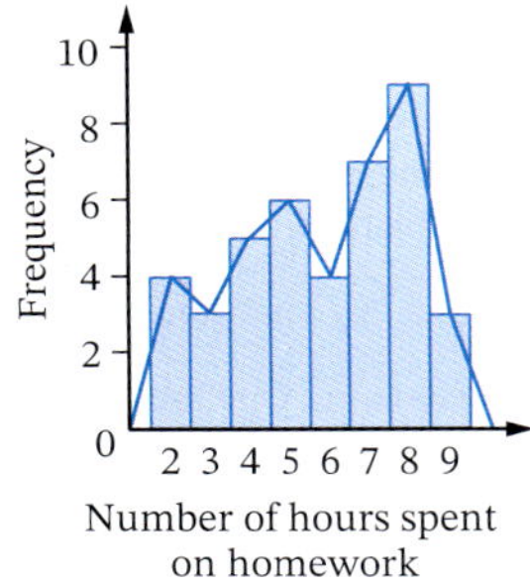

frequency (distribution) table A table listing the frequency of each value in a set of data, with columns for Score (x), Frequency (f) and sometimes Tally and fx. (p. 334)

giga- A prefix meaning one billion (1 000 000 000), represented by the letter G. (p. 390)

gigabyte (GB) A unit of digital memory equal to one billion bytes. (p. 390)

goods and services tax (GST) A 10% tax added to the original price of an item or service. (p. 114)

gradient The steepness of a line or interval, measured by the fraction $\frac{\text{rise}}{\text{run}}$. (p. 465)

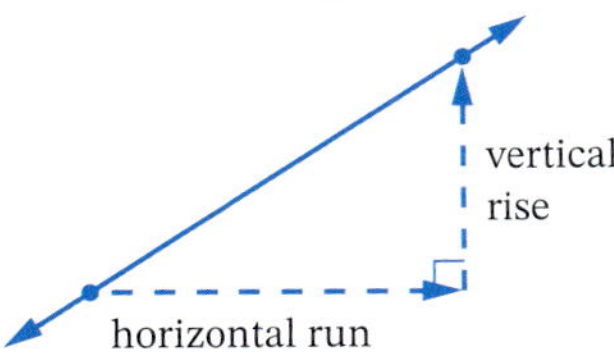

gradient–intercept form (of a linear equation) The equation of a straight line $y = mx + c$, where m is the **gradient** and c is the **y-intercept**. (p. 479)

greatest common divisor (GCD) See **highest common factor**. (p. 28)

gross pay Pay received before tax and other deductions are taken out. (p. 320)

GST See **goods and services tax**. (p. 114)

hectare A large metric unit of area, equivalent to 10 000 m^2 or to the area of a square 100 m × 100 m. (p. 403)

highest common factor (HCF) or **greatest common divisor (GCD)**. The largest factor shared by 2 or more numbers or algebraic terms. For example, the HCF of 36 and 8 is 4 and the HCF of $6xy$ and $12y^2$ is $6y$. (p. 28)

horizontal Going across, sideways, flat. (p. 472)

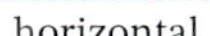

hypotenuse The longest side of a right-angled triangle, opposite the right angle. (pp. 46, 141)

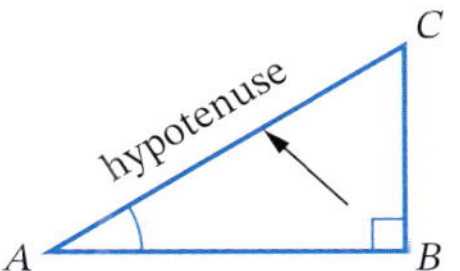

image A transformed shape after it has been enlarged or reduced. (p. 581)

improper fraction A fraction whose numerator is greater than or equal to its denominator, such as $\frac{7}{4}$. (p. 97)

included angle The angle between 2 given sides of a shape, usually a triangle. For example, the included angle for sides AC and CB in this triangle is the marked angle $\angle C$. (p. 570)

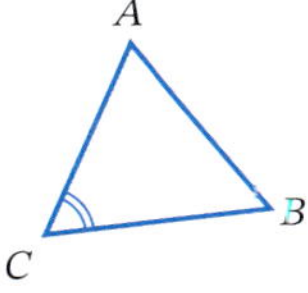

income tax A tax paid to the government based on the size of a person's income. (p. 316)

index (plural: **indices**) See **power**. (p. 184)

index notation A way of writing repeated multiplication using indices (powers), in the form a^n. Index notation for $2 \times 2 \times 2 \times 2$ is 2^4. (p. 184)

index law An algebraic rule for simplifying expressions involving powers of the same base, for example, $a^m \times a^n = a^{m+n}$. (p. 184)

inequality A mathematical statement that 2 quantities are not equal, involving algebraic expressions and an inequality sign ($>$, $\geq$, $<$, or $\leq$), for example, $-3 > -10$ or $2x - 7 \leq 15$. (p. 279)

integer A positive or negative whole number or zero. The numbers {..., −3, −2, −1, 0, 1, 2, 3, ...} are integers. (p. 87)

interval A line segment (section of a line) with a definite length, such as *AB* shown. (p. 455)

inverse operation The opposite or reverse operation, used when solving equations. For example, the inverse operation to adding is subtracting, the inverse operation to dividing is multiplying. (p. 170)

irrational number A number such as π or $\sqrt{2}$ that cannot be expressed as a fraction (rational number). In decimal form, its digits run endlessly without repeating. (p. 43, 93, 413) See also **rational number**.

isosceles triangle A triangle with 2 equal sides. (p. 221)

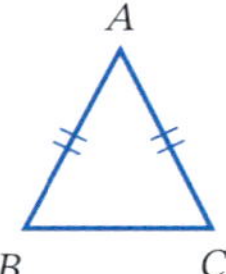

kilo- A prefix meaning one thousand (1000), represented by the symbol 'k'. For example, one kilogram is 1000 grams. (p. 389)

kilobyte (kB) A unit of digital memory equal to 1000 bytes. (p. 390)

kite A quadrilateral with 2 pairs of equal adjacent sides. (p. 228)

LHS The left-hand side (of an equation). (p. 262)

like terms Algebraic terms that have exactly the same variables. For example, $5xy$ and $2xy$ are like terms, $3xy$ and $4x^2$ are not like terms. (p. 13)

linear equation A formula or function whose graph is a straight line, or an equation involving a variable that is not raised to a power, such as $2x + 9 = 17$. (p. 257)

linear function A function with equation $y = mx + c$, whose graph is a straight line. (p. 471)

loss The money lost when selling an item at a lower price, when the selling price is less than the cost price. The opposite of **profit**. (p. 113)

mean The average of a set of data, represented by $\overline{x}$, calculated by dividing the sum of the data values by the number of values. (p. 331)

measure of central tendency An average, middle or typical value of a set of data, such as the **mean**, **median** and **mode.** (p. 331)

median The middle value when the values of a data set are arranged in order. If the number of values is even, then the median is the average of the 2 middle values. (p. 331)

mega- A prefix meaning one million (1 000 000), represented by the letter 'M'. A megalitre is one million litres. (p. 389)

megabyte (MB) A unit of digital memory equal to 1 000 000 bytes. (p. 390)

micro- A prefix meaning one-millionth, represented by the Greek letter μ. One microsecond is one-millionth of a second. (p. 390)

midpoint The point in the middle of an interval or halfway between 2 given points. (p. 462)

milli- A prefix meaning one-thousandth $\left(\frac{1}{1000}\right)$, represented by the letter m. One millimetre is one-thousandth of a metre. (p. 390)

minute (symbol ′) A measure of angle size. $\frac{1}{60}$ of a degree. $1° = 60'$. (p. 151)

mode The most common or frequent value(s) in a set of data. (p. 331)

mutually exclusive events Events or categories that have no items in common. (p. 535)

nano- A prefix meaning one-billionth, represented by the letter 'n'. One nanosecond is one-billionth of a second. (p. 390)

net pay Pay received after deductions from gross pay; 'take-home' pay. (p. 320)

network A collection of points called vertices connected together by lines called edges. (p. 242)

number plane or **Cartesian plane** A coordinate grid system based on 2 number lines that cross at right angles: a horizontal line called the *x*-axis and a vertical line called the *y*-axis. (p. 455)

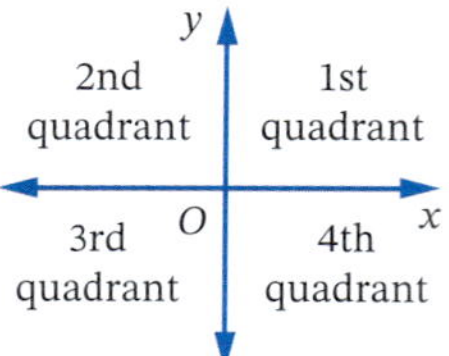

numerator The number above the line in a fraction. The numerator of $\frac{2}{3}$ is 2. (pp. 15, 93)

numerical data Data that can be measured or counted, such as a person's height or number of goals scored. Data that is not **categorical**. (p. 352)

obtuse-angled triangle A triangle with one obtuse angle (between 90° and 180°). (p. 222)

opposite side In a right-angled triangle, the side directly facing the given angle. (p. 141)

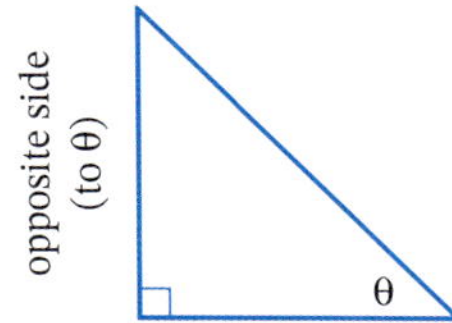

origin The point $O(0, 0)$ at the centre of the number plane, where the x-axis and y-axis cross. (p. 497)

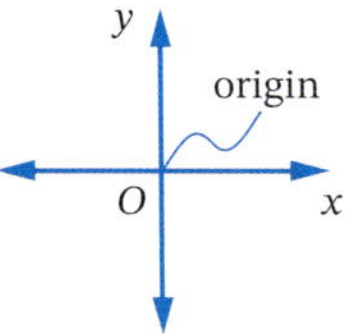

outcome In probability, a result of a situation or experiment. For example, when rolling a die, one possible outcome is rolling a 4. (p. 521)

outlier An extreme data value that is much different from the other values in a set. (p. 332)

overtime Time worked beyond normal working hours, such as at night or on weekends, at a higher rate of pay. (p. 306)

parabola A U-shaped curve that is the graph of a quadratic function, such as $y = x^2$. (p. 495)

parallelogram A quadrilateral in which the opposite sides are parallel. (p. 228)

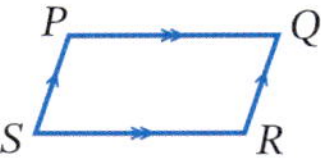

PAYG (Pay As You Go) tax Income tax deducted from your pay each payday by your employer. (p. 320)

percentage error The absolute error of a measurement as a percentage of the actual value, the relative error expressed as a percentage. (p. 395)

perimeter The distance around the outside of a shape; the sum of the lengths of its sides. (p. 402)

per annum (p.a.) Per year. (pp. 117, 301)

perpendicular height The height of a shape, measured at right angles to its base. (p. 432)

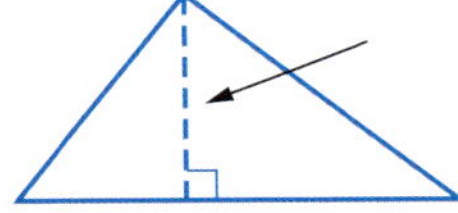

pi (π) A special number, approximately 3.1416, used in calculating circle measurements. The circumference of any circle is found by multiplying π by the diameter of the circle. (p. 412)

piecework Earnings based on the number of items processed, made or delivered, paid at a rate per item rather than on the number of hours worked. (p. 312)

Platonic solid (or regular polyhedron) A polyhedron whose faces are identical regular polygons (e.g. a cube). There are only 5 Platonic solids. (p. 246)

polygon Any flat shape made up of straight sides. (p. 237)

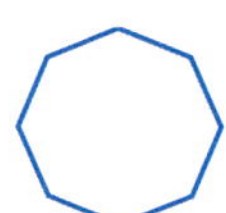

polyhedron (plural. polyhedra or polyhedrons) A solid shape with flat faces made up of polygons (no curved surfaces). (p. 246)

population In statistics, all of the items being studied the entire group. (p. 368)

power (or index or exponent) The number of times a base is multiplied by itself. In 2^5, the power is 5. (p. 184)

principal An amount of money invested or borrowed, on which interest is calculated. (p. 117)

prism A solid shape with identical cross-sections with straight sides, such as the hexagonal prism below. (p. 431)

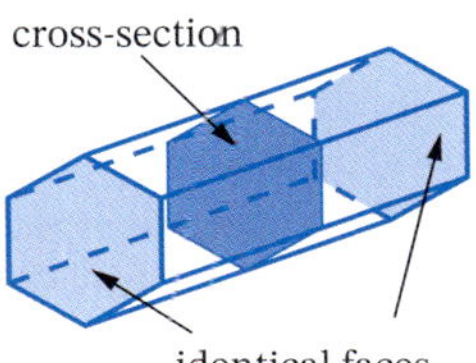

probability The chance of an event occurring, measured as a fraction, decimal or percentage between 0 and 1. (p. 521)

product The result of multiplication. The product of 7 and 3 is 21. (p. 5)

profit The amount made when selling an item at a higher price, when the selling price is more than the cost price. The opposite of **loss**. (p. 113)

pronumeral Another name for **variable.** (p. 5)

proportion See **direct proportion**. (p. 487)

Pythagoras' theorem The formula $c^2 = a^2 + b^2$ for a right-angled triangle, where c is the length of the hypotenuse and a and b are the lengths of the other 2 shorter sides. (p. 47)

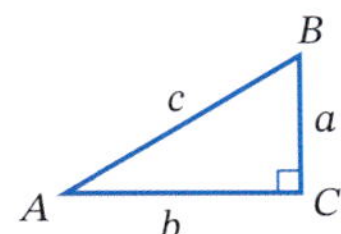

GLOSSARY AND INDEX

Pythagorean triad A set of 3 numbers that follow Pythagoras' theorem, such as {3, 4, 5}. (p. 47)

quadrant (of a circle) A sector that is a quarter of a circle, containing a right angle. (p. 413)

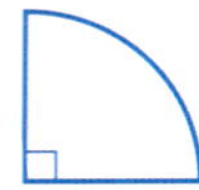

quadrant (of a number plane) See **number plane**.

quadratic equation An equation in which the highest power of the variable is 2, that is, a variable squared, for example, $3x^2 - 6 = 69$, or a quadratic function such as $y = 3x^2 - 6$. (p. 272)

quadratic function A function in which the highest power of the variable is 2, such as $y = 3x^2 - 6$, whose graph is a **parabola**. (p. 495)

quadrilateral Any polygon with 4 sides. (p. 227)

radius (plural: **radii**) An interval joining the centre of a circle to the circumference, and the length of that interval. The radius is half of the **diameter**. (p. 412)

random In probability, describing a situation where every possible outcome has an equal chance, or is equally likely. (p. 521)

random sampling In statistics, selecting a sample in which every person or item in the population has an equal chance of being selected. A sample should be random to be truly representative of the population. (p. 368)

range In a set of data, the difference between the highest and lowest values. (p. 331)

rate A relationship between 2 quantities measured in different units. For example, a speed of 107 km/h compares distance travelled (in kilometres) with time (in hours). (p. 123)

ratio A relationship between quantities measured in the same units. For example, the ratio of 3 teachers to 40 students is 3 : 40 (read '3 to 40'). (p. 121)

rational number A number that can be written as a fraction in the form $\frac{a}{b}$, where a and b are integers and $b \neq 0$. (pp. 43, 87, 97) See also **irrational number**.

reciprocal The product of any number and its reciprocal is 1. The reciprocal of any number is found by first writing the number as a fraction and then swapping the numerator with the denominator. The reciprocal of 5 is $\frac{1}{5}$ and the reciprocal of $\frac{2}{3}$ is $\frac{3}{2}$. (pp. 100, 200)

rectangle A quadrilateral with 4 right angles. (p. 228)

recurring (or repeating) decimal A decimal with one or more digits that repeat endlessly. For example 0.1666..., abbreviated as $0.1\dot{6}$. (p. 92)

regular polygon A polygon that has all sides equal and all angles equal. For example, this regular pentagon has 5 equal sides and 5 equal angles. (p. 238)

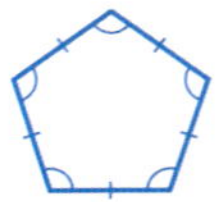

relative error The absolute error of a measurement as a fraction of the actual value. (p. 395)

relative frequency The number of times an event or data value occurred, written as a fraction of the total number of events or values. (p. 526) See also **experimental probability**.

retainer A fixed amount paid to a salesperson before **commission** is added. (p. 311)

rhombus A quadrilateral with 4 equal sides. (p. 228)

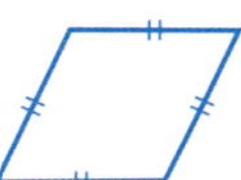

RHS The right-hand side (of an equation). (p. 262)

right-angled triangle A triangle with one 90° angle. (pp. 46, 222)

rise Short for 'vertical rise', the change in vertical position between 2 points on a line or interval, the number of units 'going up', used with the **run** to calculate the **gradient** of a line or interval. See **gradient** for diagram. (p. 465)

run Short for 'horizontal run', the change in horizontal position between 2 points on a line or interval, the number of units 'going right', used with the **rise** to calculate the **gradient** of a line or interval. See **gradient** for diagram. (p. 465)

salary A fixed annual amount of money that is paid weekly, fortnightly or monthly, not dependent on the number of hours worked. (p. 301)

sample In statistics, a group of people or items selected from a population for study. (p. 368)

sample space In a probability situation, the set of all possible outcomes. (p. 521)

scale diagram A diagram that is in proportion to the object it represents, such as a house plan or map, that is drawn according to a scale (ratio). (p. 595)

scale factor The amount by which a shape has been enlarged or reduced, equal to $\frac{\text{image length}}{\text{original length}}$ (p. 582)

scaled length The length on a scale drawing, map or plan, not the actual length. (p. 595)

scalene triangle A triangle with no equal sides. (p. 221)

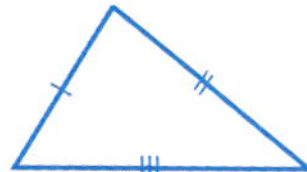

scientific notation A shorter way of writing very large or very small numbers using powers of 10, for example, 9 460 000 000 000 in scientific notation is 9.46×10^{12}. (p. 207)

second (″) A measure of angle size. $\frac{1}{60}$ of a minute. $1' = 60''$. (p. 151)

sector A region of a circle cut off by 2 radii. (p. 413)

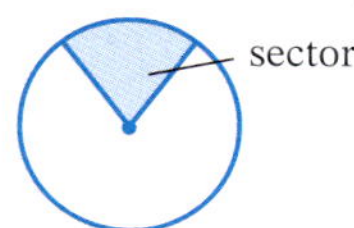

selling price The price at which an item is sold by the retailer (shop). The shop buys the item for the **cost price**. (p. 114)

shape of a distribution The way the data in a frequency distribution is spread, can be **symmetrical**, **positively skewed** or **negatively skewed**. (p. 359)

significant figures The meaningful digits in a number that show its level of accuracy, the first non-zero digits, for example, 31 487 000 has 5 significant figures. (p. 204)

similar To have the same shape but not necessarily the same size, an enlargement or reduction. The symbol '|||' means 'is similar to'. (p. 581)

similarity test One of 4 tests for proving that 2 triangles are similar. (p. 608)

simple interest Interest that is calculated as a percentage of the original principal. (p. 117)

sine A ratio in a right-angled triangle:
$\sin \theta = \frac{\text{side opposite to } \theta}{\text{hypotenuse}}$ (p. 144) See also **cosine** and **tangent**.

skewed distribution A statistical distribution in which most of the data values are clustered at one end, creating a 'tail' at the other end. The tail determines whether the skew is positive or negative. (p. 359) See also **symmetrical distribution.**

Negatively skewed

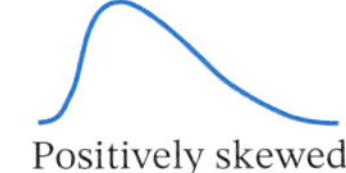

Positively skewed

solution In algebra, the value that makes an equation or inequality true. (p. 257)

square A quadrilateral with 4 equal sides and 4 right angles. (p. 228)

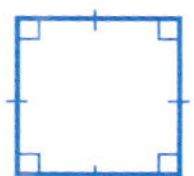

stem-and-leaf plot A 'number graph' that lists all the data values in groups. Each value is split into a 'stem' and a 'leaf'. This stem-and-leaf plot shows 12 test marks, from 42 to 82. (p. 347)

Stem	Leaf
4	2 5
5	0 2 8
6	6 7
7	3 5 7 7
8	2

Key: 5|8 stands for 58

subject of a formula The variable for which a formula is written, the variable on the left-hand-side of a formula. For example, the subject of the formula $A = \frac{1}{2}bh$ is A. (p. 276)

superimpose To place one figure on top of another figure so that sides and angles match. (p. 564)

surd A square root (or other root) whose exact value cannot be found because it is **irrational**, such as $\sqrt{10}$ or $\sqrt[3]{7}$. (pp. 43, 456)

surface area The total area of all the faces of a solid shape. (p. 421)

symmetrical distribution A statistical distribution in which all data values are distributed equally on both sides of the centre, its shape having line symmetry. See also **skewed distribution**. (p. 359)

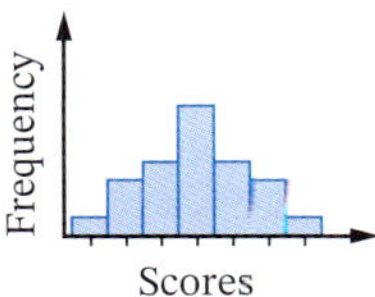

tangent A ratio in a right-angled triangle:
$\tan \theta = \frac{\text{side opposite to } \theta}{\text{side adjacent to } \theta}$ (p. 144) See also **sine** and **cosine**.

tax deduction See **allowable deduction**. (p. 316)

taxable income The part of a person's income that is taxed, equal to annual income minus allowable deductions. (p. 316)

terabyte (TB) A unit of digital memory equal to one trillion bytes. (p. 390)

terminating decimal A decimal whose digits do not repeat endlessly, for example, 0.125. (p. 92)

term (of an expression) A part of an algebraic expression. For example, $b^2 + 6b - 9$ has 3 terms: b^2, $6b$ and -9. (p. 12)

theoretical probability Probability calculated using the formula

$$P(E) = \frac{\text{number of favourable outcomes}}{\text{total number of outcomes}}$$ (p. 526)

time-and-a-half Overtime pay that is calculated at 1.5 times the normal pay rate. (p. 306)

trapezium A quadrilateral with one pair of opposite sides parallel. (p. 228)

tree diagram A diagram of branches for listing all of the possible outcomes in a multi-step chance experiment. (p. 547)

trial One go or run of a repeated probability experiment, for example, one roll of a die. (p. 521)

two-step experiment (or **two-stage experiment**) A chance experiment with 2 steps or stages, such as rolling a pair of dice. (p. 545)

two-way table A table that shows the number of items belonging to overlapping categories. (p. 540)

	Can swim	Cannot swim
Boys	13	2
Girls	9	3

unitary method A method for finding a quantity by finding the size of one part or 1% first. (pp. 110, 121)

variable A symbol, usually a letter of the alphabet, that stands for a number. Also called a **pronumeral** or **unknown.** (p. 5)

variation See **direct proportion**. (p. 487)

Venn diagram A diagram of circles (usually overlapping) for grouping items into categories. (p. 534)

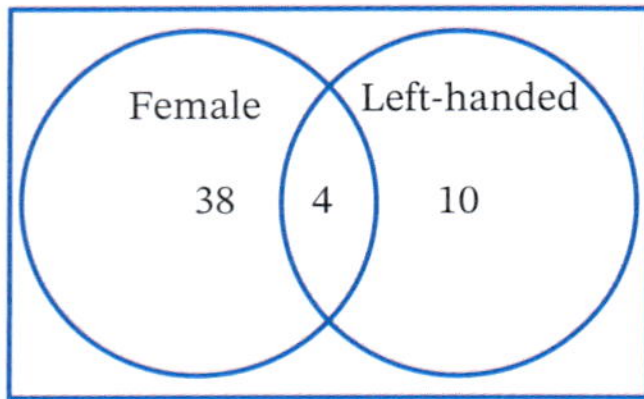

vertex (plural: **vertices**) A corner of a shape, polyhedron or curve. (pp. 242, 246)

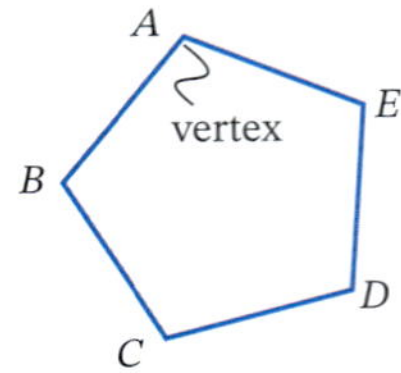

vertex (of a network) Points representing items that are connected by lines called edges. See edge for diagram. (p. 246)

vertical Going up and down, at a right angle to the **horizontal**. (p. 472)

volume The amount of space taken up by a solid object, measured in cubic units. (p. 431)

wage An amount of money paid to people for work, calculated on the number of hours worked. (p. 301)

***x*-axis** The horizontal axis of a number plane (running across, see diagram below). (p. 471)

***x*-intercept** The x-value at which a graph cuts the x-axis. (p. 471)

***y*-axis** The vertical axis of a number plane (running up and down). (p. 471)

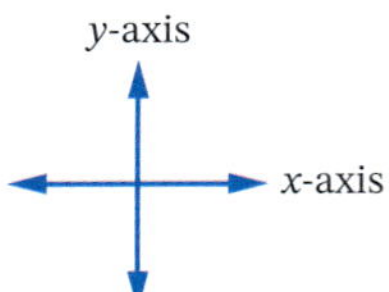

***y*-intercept** The y-value at which a line cuts the y-axis. (p. 471)